# RUSSIAN-ENGLISH DICTIONARY OF ELECTROTECHNOLOGY AND ALLIED SCIENCES

**PAUL MACURA**

Department of Foreign Languages
and Literature
University of Nevada, Reno

WILEY INTERSCIENCE
a Division of John Wiley & Sons, Inc.
New York · London · Sydney · Toronto

# PREFACE

Because of Soviet scientific, industrial, and technical advances, institutions of higher education and research outside the U.S.S.R. are becoming more and more interested in scientific publications in the Russian language.

This interest has created a need for various comprehensive, up-to-date Russian-English dictionaries, such as this dictionary of radar, radio, television, electronics, and electrotechnology, to assist scientists, translators, teachers, students, engineers, and others who deal with Russian-language publications in these fields.

In addition to the technical vocabulary, this dictionary contains the following features:

1. A large number of Russian abbreviations found in scientific literature dealing with subjects represented in this dictionary.

2. Russian transliteration of the names of foreign scientists.

3. English equivalents of many Russian compound words not appearing in any other Russian-English dictionary.

4. Almost all of the participles found in scientific literature dealing with subjects covered here.

5. Many synonyms and synonymous expressions.

The dictionary has approximately 60,000 entries, thus constituting the most extensive Russian-English vocabulary listing in the subjects enumerated below. It also includes almost all nontechnical terms occurring in scientific publications, so that the need for a general Russian-English dictionary is minimal.

The following are subjects represented in this dictionary: Accumulators, acoustics, aeronautical applications, aeronavigation, analog computers, antennas and waveguides, architectural acoustics, astronautics, astronomy, atmospheric acoustics, atmospheric physics, atomics, automatic control and regulation, automatic control systems, automatic guiding and control, automation and instruments, batteries, beacons, cable, communications, computers, computing technique, cybernetics, dial telephone systems, direction finding, electroacoustics, electrochemistry, electrometallurgy, electromedical applications, electron devices, electronics, electrotechnology, facsimile, geophysics, high-frequency techniques, holography, hydroacoustics, industrial electronics, interlocking and automatic control, lighting, machines and transformers, marine applications, mathematical terminology (fringe terms), measurements and measuring instruments, modulation systems, musical acoustics, navigation (radio aids to), optics, oscillographs, photo applications, physics (fringe terms), physiological acoustics, programing techniques, pulse work, radar, radio, radiolocation, radiology, remote control, railroad signals, scientific and industrial measuring instruments, semiconductors, servosystems, signaling, sound recording, static convertors, switchboards and apparatus for connection and regulation, switches, telegraphy, telemechanics, telemetry, telephony, television, transistors, tubes, and waveguides.

All Russian entries are arranged alphabetically according to the Russian alphabet. Entries are alphabetized letter by letter to the end of the word. In many cases the key entry is followed by one or more subentries, which are given in the following sequence: 1. those in parentheses, if any; 2. plurals, if any; 3. those involving any gender, number, or case variation of the key word; 4. all those involving the key word in unchanged form, abbreviated to its initial letter. Every word is taken into account including names, abbreviations, prepositions, and conjunction. All definitions preceding and following a key entry are arranged in strictly alphabetical order, and then word by word to the last word. In the various subitems following the key entry, the repeated part of the main item, occurring initially, in the middle, or at the end of such subitem, is abbreviated to the initial letter:

**антенна** *f.* antenna; **а. вытекающей волны** leaky wave antenna; **линзовая а. из проволочной сетки** wire-grid lens antenna; **пароболо-цилиндрическая а.** cylindrical parabola antenna

A dictionary of this nature cannot include all nontechnical vocabulary. I assume that the user knows it and that he will consult a general dictionary as well.

The user is also assumed to have a basic knowledge of the Russian grammar. I do not list separately irregular or inflected forms of nouns, adjectives, participles, and verbs. For the latter, the reader should consult Powers' *Dictionary of Irregular Russian Verb Forms.* Additional scientific vocabulary can be found in Callahan's *Russian-English Technical and Chemical Dictionary* and Emin's *Russian-English Physics Dictionary.*

All nouns are listed in the nominative singular, accompanied where necessary by instances of other cases in the singular or plural.

Adjectives and participles are listed in the masculine nominative singular.

All participles are designated as adjectives in this dictionary.

The imperfective and perfective verbs are listed together with the corresponding English translation. The perfective verb is marked as *pf. of . . . See* refers to a synonym or a synonymous expression; such cross references are used sparingly, however.

In parentheses are included explanations, the full text of Russian abbreviations, and words that may be omitted without changing the meaning.

I will be very grateful for any suggestions regarding omissions and errors.

In conclusion, I express my sincere gratitude to everyone who aided me in any way in the preparation of this dictionary. I am especially indebted to the linguistic editor who reviewed the manuscript and contributed valuable counsel, leading to several improvements.

I express my feeling of indebtedness also to Professor Eugene V. Kosso, Department of Electrical Engineering, University of Nevada, for his valuable suggestions and criticism, and for his reading of the English equivalents of the Russian entries of the manuscript. I thank the Desert Research Institute of University of Nevada and to the University of Nevada Graduate School for their financial contributions which helped make the compilation of this dictionary possible.

To all my friends and colleagues who have taken an interest in this work, I offer my sincere gratitude.

Last but not least, I thank my wife and daughter for their infinite understanding and moral support.

PAUL MACURA

*University of Nevada*
*Reno, Nevada*
*June 1970*

# Abbreviations

| | |
|---|---|
| *abbr.* | abbreviation |
| *acc.* | accusative |
| *acoust.* | acoustics |
| *adj.* | adjective |
| *adv.* | adverb |
| *aero.* | aeronautics |
| *anat.* | anatomy |
| *ant.* | antenna |
| *astron.* | astronomy |
| *aut.* | automation, automatic |
| *av.* | aviation |
| *batt.* | battery |
| *biol.* | biology |
| *chem.* | chemistry |
| *comm.* | communication |
| *comp.* | computer, computer technology |
| *compar.* | comparative |
| *conj.* | conjunction |
| *dat.* | dative |
| *decl.* | declension |
| *dts.* | dial telephone system(s) |
| *elec.* | electricity, electrical engineering |
| *elec.-chem.* | electrochemistry |
| *elec. comm.* | electrical communication |
| *f.* | feminine |
| *genit.* | genitive |
| *impf.* | imperfective |
| *indecl.* | indeclinable |
| *instr.* | instrumental |
| *interr.* | interrogative |
| *m.* | masculine |
| *mach.* | machinery |
| *math.* | mathematics |
| *meas.* | measurements and instruments |
| *mech.* | mechanics |
| *med.* | medicine, electromedical |

| *met.* | metals, metallurgy |
| *meteor.* | meteorology |
| *mil.* | military |
| *n.* | neuter |
| *nav.* | navigation |
| *num.* | number, numeral |
| *osc.* | oscillography |
| *pf.* | perfective |
| *phot.* | photography |
| *phys.* | physics |
| *pl.* | plural |
| *prep.* | preposition |
| *prepos.* | prepositional |
| *pron.* | pronoun |
| *poss.* | possessive |
| *rad.* | radio |
| *rdr.* | radar |
| *rel.* | relative |
| *RR, rr.* | railroad |
| *sing.* | singular |
| *superl.* | superlative |
| *tbs.* | electron tubes |
| *tgphy.* | telegraphy |
| *tmech.* | telemechanics |
| *tphny.* | telephony |
| *TV, tv* | television |
| *v.* | verb |

# Russian-English Dictionary
# of Electrotechnology
# and Allied Sciences

# A

Å *abbr.* (ангстрем, ангстром) ångstrom
а *abbr.* (ампер) ampere
а *conj.* but, and; **а так как** and since, now as; **а то** or else
**АА** *abbr.* (армейская авиация) army aviation
**ААЕ** *abbr.* (Авиационный астрономический ежегодник) Air almanac
**ААИ** *abbr.* (амплитудный анализатор импульсов) pulse height analyzer
**АБ** *abbr.* 1 (авиационная база, авиабаза) air base; 2 (атомная бомба) atom bomb
**аб** *abbr.* (авиационная бригада, авиабригада) air brigade
**абажур** *m.* lampshade; screen
**абажуродержатель** *m.* gallery, ornamental ring to support a lampshade
**абака** *f.* 1. abacus; 2. abaca (Manila hemp)
**абаксиальный** *adj.* abaxial
**абампер** *m.* abampere
**абасси** *n. indecl.* abassi (white Egyptian cotton)
**Аббе** Abbe
**абвольт** *m.* abvolt
**абгенри** *m. indecl.* abhenry
**абелев** *poss. adj.*, **абелевский** *adj.* Abelian
**аберрационный** *adj.* aberration(al)
**аберрация** *f.* aberration, deflection, deviation; **а. высших порядков** higher aberration; **а. света от тяготения** gravitational aberration; **скоростная а.** chromatic aberration
**абиетин** *m.* abietene
**абиссальный** *adj.* abyssal
**абкулон** *m.* abcoulomb
**аблиц** *m.*, **аблицовка** *f.* flashing
**аблятограф** *m.* ablatograph
**абляционный** *adj.* ablating, ablation, ablative
**абляция** *f.* ablation
**абмо** *n.* abmho
**Абней** Abney
**абом** *m.* abohm
**абонемент** *m.* subscription; **временный а.** temporary service contract; **а. с**

поврёменной оплатой flat rate subscription; **а. с поразговорной оплатой** message rate subscription
**абонент** *m.* subscriber; customer, consumer, user, party; **далеко расположенный а. учрежденского коммутатора** outside PBX extension; **дальный а.** suburban teletype subscriber; **а. извещен о предстоящем переговоре** PN (party notified); **а. местной станции, местный а.** local subscriber; **многоговорящий а.** high calling rate subscriber; **а. обобществленного сектора** business subscriber; **отдельный а.** calling unit; **отдельный телефонный а.** single line subscriber; **прямой междугородный а.** LD terminal (long-distance terminal); **а. с добавочным аппаратом** extension subscriber; **а. с основным аппаратом** direct line subscriber, individual line subscriber; **а. с повремённой оплатой** flat rate subscriber; **а. с поразговорной оплатой** message rate subscriber; **телеграфный а.** teletype subscriber; **а. электросети** user (of electric power)
**абонентский** *adj.* subscriber; customer
**АБР** *abbr.* (авиационная баллистическая ракета) airborne ballistic rocket
**абразив** *m.* abrasive, abradant
**абразивный** *adj.* abrasive, abradant
**абразионный** *adj.* abrasion, abrasive
**абразия** *f.* abrasion
**абрис** *m.* contour, outline
**абс.** *abbr.* (абсолютный) absolute
**абс. выс.** *abbr.* (абсолютная высота) absolute elevation
**абс. ед.** *abbr.* (абсолютная единица) absolute unit
**абсолютно** *adv.* absolutely
**абсолютно-разрешительный** *adj.* absolute-permissive
**абсолютно-сходящийся** *adj.* absolute convergent
**абсолютный** *adj.* absolute
**абсорбент** *m.* absorbent

**абсорбер** *m.* absorber; absorbent
**абсорбированный** *adj.* absorbed
**абсорбировать** *v.* to absorb; to occlude
**абсорбируемый** *adj.* absorbable
**абсорбирующий** *adj.* absorbing, absorptive, absorbent
**абсорбциальный** *adj.* absorptial
**абсорбциометр** *m.* absorptiometer
**абсорбциометрический** *adj.* absorptiometric
**абсорбционный** *adj.* absorption, absorptive
**абсорбция** *f.* absorption; occlusion
**абстатампер** *m.* abstatampere
**абстрагировать** *v.* to abstract
**абстрактный** *adj.* abstract
**абстракция** *f.* abstraction
**абсцисса** *f.* abscissa; **а. центра тяжести** longitudinal center of gravity
**абс. эл. ст. ед.** *abbr.* (**абсолютная электростатическая единица**) absolute electrostatic unit
**абфарада** *f.* abfarad
**АВ** *abbr.* (**арктический воздух**) arctic air
**ав** *abbr.* (**авиационный**) aviation, aircraft
**ав, а-в** *abbr.* (**ампервиток**) ampere-turn
**аванкамера** *f.* forebay; **а. у напорного трубопровода** penstock forebay
**аварийно-защитный** *adj.* emergence safety
**аварийность** *f.* breakdown, susceptibility
**аварийный** *adj.* emergency, accident, breakdown, damage, fault, standby; **для аварийных случаев** emergency, crash (*adj.*)
**авария** *f.* emergency; failure, fault, breakdown, accident, trouble; **общая а.** general average (*elec.*); **а. производства** breakdown of service, operating trouble; **частная а.** particular average (*elec.*)
**авгитовый** *adj.* augite, augitic
**авентуризировать** *v.* to punctiform flash
**авиа-** *prefix* air, aerial, aviation, airplane
**авиабаза** *f.* air base
**авиабомба** *f.* aerial bomb; **а. управляемая на расстоянии** television bomb
**АВИАВНИТО** *abbr.* (**Авиационное всесоюзное научно инженерно-техническое общество**) All-Union Aeronautical Engineering and Technical Society
**авиагоризонт** *m.* gyrohorizon, flight indicator, artificial horizon
**авиакосмический** *adj.* aerospace
**авиалиния** *f.* airline
**авиамагистраль** *f.* major airline
**авиаматка** *f.* aircraft carrier
**авиамаяк** *m.* aeronautical beacon, air beacon, **световой а.** aeronautical beacon
**авиамотор** *m.* aircraft engine
**авианосец** *m.* aircraft carrier
**авиапеленгатор** *m.* azimuth finder
**авиапром** *abbr.* (**авиационная промышленность**) aviation industry
**авиарадиостанция** *f.* aeronautical station
**авиарадист** *m.* aircraft radio operator
**авиаразведка** *f.* aircraft search
**авиационно-космический** *adj.* aerospace
**авиационный** *adj.* aeronautical, aviation, avio, air(borne)
**авиация** *f.* aviation, aeronautics
**авионика** *f.* avionics
**авиопатрульный** *adj.* air-patrol
**авиоперехват** *m.* air interception
**Авогадро** Avogadro
**авометр** *m.* avometer; **электронный а.** electronic volt-ohm-mA-meter
**авт.** *abbr.* 1 (**автоматический**) automatic; 2 (**автономный**) autonomous
**автентический** *adj.* authentic
**авто-** *prefix* auto-, automatic; automobile
**автоаларм** *m.* automatic-alarm receiver
**автоблокировать** *v.* to block automatically
**автоблокировка** *f.* automatic blocking; automatic block system; self-locking; automatic shut-off; **двусторонняя а.** single-track block system; **комбинированная а.** either direction traffic automatic block system; **односторонняя а.** double-track automatic block system; **точечная а.** intermittent automatic block system
**автобрадер** *m.* autobrader
**автобус** *m.* bus
**автовозбуждение** *n.* self-excitation
**автовыключатель** *m.* automatic switch, twilight switch
**автогенератор** *m.* self-excited oscillator; **а. Колпитса с последовательной настройкой** Clapp oscillator; **а.**

разрывных колебаний relaxation oscillator; **а. с индуктивной** (*or* **трансформаторной) обратной связью** tickler-coil oscillator; **а., стабилизированный линией** resonant-line oscillator; **а. электрических колебаний** self-maintained circuit

**автогенерирующий** *adj.* self-oscillating

**автогенный** *adj.* autogenous

**автогетеродин** *m.* autoheterodyne

**автогетеродинный** *adj.* autodyne, autoheterodyne

**автография** *f.* autography; **контактная а.** contact radioautography

**автодин** *m.* autodyne, autoheterodyne, self-heterodyne

**автодинный** *adj.* autodyne, autoheterodyne, self-excited

**автодиск** *m.* dial

**автожир** *m.* autogyro, gyroplane

**автозал** *m.* auto-room, automatic switch room

**автоимпульсный** *adj.* self-pulse

**автоиндуктивный** *adj.* autoinductive

**автоионизация** *f.* autoionization

**автокартограф** *m.* autocartograph

**автокатализ** *m.* autocatalysis

**автокаталитический** *adj.* autocatalytic

**автоклав** *m.* autoclave; **а. для гидрирования жиров** hydrogenisator

**автоклавировать** *v.* to autoclave

**автокластический** *adj.* autoclastic

**автокоалесценция** *f.* autocoalescence

**автокод** *m.* autocode

**автоколебание** *n.* self-excited vibration, natural vibration, self-oscillation; *pl.* hunting

**автоколебательный** *adj.* self-oscillatory self-oscillating, self-excited, astable

**автоколлиматор** *m.* autocollimator

**автоколлимационный** *adj.* autocollimating, autocollimation

**автоколлимация** *f.* autocollimation

**автокомпаундированный** *adj.* autocompounded

**автоконвертор** *m.* autoconvertor

**автоконденсация** *f.* autocondensation

**автокондукция** *f.* autoconduction

**автокоррелограмма** *f.* autocorrelogram

**автокоррелятор** *m.* autocorrelator

**автокорреляционный** *adj.* autocorrelation

**автокорреляция** *f.* autocorrelation

**автокран** *m.* truck crane, lorry-mounted crane

**автоманипулятор** *m.* automatic manipulator, automatic keying device

**автомат** *m.* automat, robot, automaton, automatic machine; automatic breaker, circuit breaker; **бесклапанный откачной а.** valveless exhaust machine; **временной а.** automatic time switch; **грозозащитный а.** lightning arrester, lightning protector; **а. давления** pressure switch; **денежный а.** prepayment coin box; **а. для загрузки катодов в рамку** cathode spray bar loading machine; **а. для навивки сеток** grid machine, grid lathe; **а. для опаливания монтированных ножек** bulb blowing machine; **а. для опускания монет** coin collector; **а. для пересчёта координат на требуемый курс** offset-course computer; **а. для продевания выводов и пайки штырьков** base threading and soldering machine; **а. для размена денег** coin changer; **а. для решения пари** coin matching machine; **а. для скоростной проверки ракеты перед запуском** rapid automatic check-out equipment; **а. для смены пластинок** automatic record changer; **а. для сортировки** sorter; **а. для штамповки пуговичной ножки** button stem machine; **заварочно-откачной а., запаечно-откачной а.** sealing-exhaust machine, sealex (machine); **защитный а.** automatic circuit-breaker; **конечный а.** finite automaton; **контрольно-сортировочный а.** inspection machine; **а.-лунник** moon robot; **максимальный а.** maximum cutout, overload circuit-breaker; **максимальный а. с кнопочным возвратом** automatic breaker; **максимальный токовый а.** current limiter; **малый а.** automatic breaker; **местный монетный а.** public telephone-booth for local service; **а. минимального тока** *see* **а. нулевого тока; минимальный а.** minimum cutout, zero current cutout; **монетный (телефонный) а.** prepayment coin box, coin-box (*or* public) telephone, monetary telephone, coin collector; **а. нулевого напряжения** no-volt release relay; **а. нулевого тока** automatic minimum circuit-breaker; **нулевой а.** zero cutout;

no-load switch; **а., отключающий при обрыве ленты** tape-breakage cutoff; **а. питания** automatic feeder; **пламечувствительный а.** burner-flame controller; **а. повторного включения (цепи)** (circuit) recloser; **предохранительный а.** automatic breaker; **радиотелемеханический а.** radio robot; **расцепляющий а.** release; **а. сбрасывания дипольных отражателей** chaff dispenser, chaff scattering device, dispersing machine; **сварочный а.** welding machine, automatic (arc) welder; **сетевой а.** network protector; **тепловой а.** thermal breaker, bimetallic-strip-type switch; **торпедный а. стрельбы** torpedo data computer; **а., управляемый по радио** radio robot; **а. усилий** artificial feel system, feel system; **фотоэлектронный а. для изготовления клише** scan-a-graver; **центробежный а.** centrifugal switch; **электрический разливочный а.** drink dispenser; **электронный сортировочный а.** electronic sorter

**автоматизация** *f.* automation, automatization

**автоматизированный** *adj.* automated, mechanized; **полностью а.** full-automatic

**автоматизировать** *v.* to automate, to automatize

**автоматизм** *m.* automatism, automaticity, automatic performance

**автоматика** *f.* automation, automatics; automatic equipment; **а. автоматического повторного включения, а. АПВ** automation of automatic reclosing; **а. сопровождения или наведения** automatic (radar) guidance system

**автоматически** *adv.* automatically, self-

**автоматический** *adj.* automatic, auto-, automatically adjusting; self-acting, power-operated, power-driven, power; **полностью а.** full-automatic

**автоматичность** *f.* automaticity

**автоматный** *adj.* automatic, auto-

**автомашина** *f.* car, automobile, vehicle; truck; **грузовая а.** truck; **звукофицированная а.** public address car, sound truck; **а. с громкоговорящей установкой** sound truck; **а. со звукоусилительной установкой** public-address car; sound truck

**автомобиль** *m.* automobile, car, motor vehicle; **а., оборудованный телевизионной установкой** television car; **а.-радиостанция** radio car; **а. с пусковой установкой** launcher truck

**автомобильный** *adj.* automobile, car carborne

**автомодельный** *adj.* self-simulating, self-similar, automodel, self-progressive; progressing, progressive (motion)

**автомодулировать** *v.* to self-modulate

**автомодулируемый** *adj.* self-pulsed, self-modulated

**автомодуляция** *f.* self-modulation; self-pulsed modulation

**автоморфизм** *m.* automorphism

**автоморфический, автоморфный** *adj.* automorphic

**автомотор-генератор** *m.* automotor-generator

**автомотриса** *f.* railway motor car, railcar; **аккумуляторная а.** battery powered railcar

**автонастраивать** *v.* to tune automatically

**автонастройка** *f.* autotune, autotuning

**автономно** *adv.* autonomously, independently

**автономность** *f.* noninteraction; **а. плавания** self-sustained period

**автономный** *adj.* autonomous, independent, noninteracting, self-contained, self-consistent; self-powered; self-reacting; off-line

**автоответчик** *m.* automatic identification device (teletype systems)

**автопеленгация** *f.* self-bearing

**автопилот** *m.* autopilot, automatic pilot, mechanical pilot, radio pilot; **а.-автоштурман** autopilot-navigator; **а. с датчиками ускорений** inertial autopilot; **а. с демпфирующими гироскопами** rate-sensitive autopilot; **а. с дистанционным управлением** electronic pilot, remote-control autopilot

**автопилотный** *adj.* autopilot

**автоплекс** *m.* autoplex

**автоподатчик** *m.* power feed

**автоподстройка** *f.* automatic control, automatic tuning

**автопроводимость** *f.* autoconduction

**автопрограммирование** *n.* self-programming

**автопрокладчик** *m.* dead reckoning tracer

**автор** *m.* author; **а. сценария** continuity writer

**авторадиограмма** *f.* autoradiograph

**авторадиограф** *m.* autoradiograph

**авторадиография** *f.* autoradiography

**авторадиоснимок** *m.* autoradiograph

**авторегулирование** *n.* automatic control, self-regulation

**авторегулировка** *f.* automatic control (-system), automatic regulation, self-regulation; **фазовая а. в цветном телевизоре** quadrature information correlator

**авторегулируемый** *adj.* automatically controlled, auto-controllable

**авторегулятор** *m.* automatic regulator, automatic controller, automatic control; **вращающийся а.** Rototrol; **а. усиления (с контрольным каналом)** pilot regulator

**авторентгеноснимок** *m.* autoradiograph

**авторизация** *f.* authorization, privilege

**авторизованный** *adj.* authorized

**авторитетный** *adj.* authoritative

**авторотация** *f.* autorotation

**авторулевой** *m. adj. decl.* automatic pilot, autopilot

**автосин** *m.* autosyn, selsyn, synchro-

**автосинный** *adj.* autosyn, selsyn, synchro-

**автосинхронизация** *f.* self-synchronization, autosynchronization

**автосмещение** *n.* automatic bias, self-bias

**автосообщение** *n.* motor communication

**автосопровождение** *n.* automatic following

**автостабилизатор** *m.* autostabilizer

**автостартёр** *m.* autostarter

**автостоп** *m.* automatic stop device, automatic lock, self-locking device

**автосцепка** *f.* automatic coupling, automatic coupler; **жёсткая а.** tight-lock coupler

**автотаймер** *m.* autotimer

**автотипия** *f.* autotype

**автоторможение** *n.* automatic train control application

**автотормоз** *m.* automatic brake

**автотрансдуктор** *m.* autotransductor

**автотракторный** *adj.* automotive

**автотрансформатор** *m.* autotrans-former, autoformer; one-coil transformer, volt box; **вч а. связи** auto-jigger; **а. для сдвига фаз** phase shifter; **а. колебаний** auto-jigger; **плавнорегулируемый а.** variac; **а. по схеме «вилки»** forked autotransformer; **поворотный а.** induction (voltage) regulator, static induction motor; **пусковой а.** starting compensator, induction starter, auto-starter transformer; **а. с регулируемым передачным числом** adjustable radio autotransformer; **а. связи** auto-jigger

**автотрансформаторный** *adj.* auto-transformer

**автоуправление** *n.* automatic control

**автоустановка** *f.* auto-plant, automatic installation, automatic outfit

**автоустройство** *n.* automatic equipment, automatic device (or apparatus)

**автофазировка** *f.* automatic phase stabilization; phase stability

**автофлаг** *m.* autoflag, wigwag signal

**автофургон** *m.* van, truck

**автоштурман** *m.* avigraph, flight-log, flight path recorder; **а. с автоматической коррекцией** self-correcting autonavigator; **а. с наглядной индикацией** airborne pictorial plotter

**автошина** *f.* tire

**автоэлектронный** *adj.* autoelectronic

**автоэмиссия** *f.* autoelectronic emission, field emission

**АВФ** *abbr.* (Академия воздушного флота) Air Force Academy

**агальматолит** *m.* agalmatolite; pagodite

**агар-агар** *m.* agar-agar

**агармоничность** *f.* inharmonicity

**агаровый** *adj.* agar

**агатовый** *adj.* agate

**агломерат** *see* агломерат

**агломерационный** *adj.* agglomeration

**агломерация** *f.*, **агломерирование** *n.* agglomeration, sintering

**агглютинативность** *f.* agglutinability

**агглютинативный** *adj.* agglutinative

**агглютинация** *f.* agglutination

**агглютинирование** *n.* agglutination

**агглютинировать** *v.* to agglutinate

**агглютинирующий** *adj.* agglutinating

**аггрегат** *see* агрегат

**агент** *m.* agent; factor; **агенты, образующие кислородные вакансии**

reducing agents; **охлаждающий а.** coolant; **тормозящий а.** inhibitor; **а., управляющий централизационным аппаратом** leverman

**агентство** *n.* agency

**агеострофический** *adj.* ageostrophic

**агитатор** *m.* agitator, stirrer

**агитационный** *adj.* agitation

**агитация** *f.* agitation, stirring

**агитировать** *v.* to agitate, to stir

**агломерат** *m.* agglomerate, sinter; **цилиндрический а.** bobbin

**агломерационный** *adj.* agglomeration

**агломерация** *f.* agglomeration, sintering

**агломерировать(ся)** *v.* to agglomerate

**аглютинировать** *v.* to agglutinate

**агогический** *adj.* agogic

**агометр** *m.* agometer

**агона** *f.* agonic line, agonic curve

**агональный** *adj.* agonic

**агонистический, агонический** *adj.* agonic

**агрегат** *m.* set, unit, outfit; assembly; apparatus, machine; plant; aggregate; *pl. oft.* accessories; **а. азимутальной наводки** deflection assembly; **а. аэродромного электропитания** mobile generator unit; **бортовой а. реактивного снаряда** missile-borne unit; **ветроэлектрический зарядный а.** wind charger; **вольтодобавочный а.** booster, auxiliary set, additional aggregate (*or* set); **двухполюсный громкоговорящий а.** woofer-tweeter; **а. для собственных нужд** house set; **а. из камеры** camera chain; **а. Ильгнера** flywheel set; **а. конденсаторов** gang capacitor; **ламповый а.** lighting set; **машинный а.** generator set, generating set; **наборный а.** stacking unit; **а. наддува** supercharger; **а. настройки** tuner; **одноместный сборочный а.** single station assembly machine; **а. питания** power plant, power generator, excitation set; **преобразовательный а.** motor generator set, conversion unit; **а. резервного питания** emergency power plant, emergency power set; **тепловой а.** heat engine set; **умформерный а.** motor generator set; **а. управления автоматом сбрасывания дипольных отражателей** chaff dispenser control subsystem

**агрегативный** *adj.* aggregative

**агрегатированный** *adj.* ganged

**агрегатировать** *v.* to aggregate

**агрегатная** *f. adj. decl.* motor room, generator room; instrument room, equipment room

**агрегатный** *adj.* aggregate, aggregation; ganged

**агрегация** *f.* aggregation

**агроклиматический** *adj.* agroclimatic

**«агфаколор»** *m.* Agfacolor

**адалин** *m.* adaptive linear neuron

**Адамар** Hadamar

**адаптация** *f.* adaptation

**адаптер** *m.* adapter; (sound) pickup, phonograph pickup, sound box, scanner; **а. для воспроизведения звуков** sound box

**адаптивный** *adj.* adaptive

**адаптированный** *adj.* adapted

**адаптировать** *v.* to adapt

**адаптометр** *m.* adaptometer

**адванс** *m.* advance (copper-nickel alloy)

**адвективный** *adj.* advective, advection

**адвекция** *f.* advection

**адгезивный, адгезионный** *adj.* adhesive, adhesion

**адгезия** *f.* adhesion, adhesiveness, adhesive power

**АДД** *abbr.* (авиация дальнего действия) long-range aviation, long-range air force

**аддавертор** *m.* addavertor

**аддитивность** *f.* additivity

**аддитивный** *adj.* additive

**аддитрон** *m.* additron

**адекватно** *adv.* adequately, sufficiently; equally

**адекватность** *f.* adequacy

**адекватный** *adj.* adequate, sufficient

**аджойнт** *m.* adjoint

**адиабата** *f.* adiabat, adiabatic curve; **а. конденсации** saturation adiabat

**адиабатический, адиабатичный, адиабатный** *adj.* adiabatic

**адиактинический** *adj.* adiactinic

**адиатермический** *adj.* adiathermal

**Адкок** Adcock

**административный** *adj.* administrative

**администратор** *m.* administrator

**администрация** *f.* administration

**адмиттанс, адмиттанц** *m.* admittance

**адмиттанционный** *adj.* admittance

**адрес** *m.* address; **а. для телеграмм** registered address for telegrams; **а. запоминающего устройства** memory address, storage address; **а. запоминающей ячейки** storage position;

**исполнительный а.** effective address; **истинный а.** absolute address; **а. команды** instruction address, location of instruction; **а. компоненты операции** operand address; **а.-операнд** immediate address; **а. переключения** reference address; **подвижный а.** floating address; **синтезированный а.** generated address, synthetic address; **условный а.** symbolic address; **а. ячейки** storage position; **а. ячейки запоминающего устройства, а. ячейки памяти** storage location, memory location

**адресат** *m.* addressee

**адресация** *f.* addressing

**адресный** *adj.* address

**адресование** *n.* addressing

**адресовать** *v.* to address, to direct

**АДС** *abbr.* (**адрес**) address

**адсорбат** *m.* adsorbate

**адсорбент** *m.* adsorbent

**адсорбер** *m.* adsorber; adsorbent

**адсорбирование** *n.* adsorption

**адсорбированный** *adj.* adsorbed

**адсорбировать** *v.* to adsorb

**адсорбирующий** *adj.* adsorbent, adsorptive

**адсорбционный** *adj.* adsorption

**адсорбция** *f.* adsorption; **а. в повреждённых местах** partial adsorption, adsorption of inhomogeneous spots

**АДУ** *abbr.* (**аппаратура дистанционного управления**) remote control equipment

**адуроль** *m.* adurol

**адхезия** *see* **адгезия**

**адьюнкт** *m.*, **адьюнкта** *f.* adjunct, cofactor

**адьюнкт-детерминант** *m.* adjunctive (or adjoint) determinants

**адьюнктный** *adj.* adjunct

**адэкватный** *adj.* adequate

**ажурный** *adj.* openwork; skeleton; jig(saw)

**азартный** *adj.* audacious; hazardous

**азбука** *f.* alphabet; code; **нотная а.** scale, gamut; **слоговая а.** syllabary

**«Аздик»** *m.* asdic

**азеотропический** *adj.* azeotropic

**азимут** *m.* azimuth, azimuth angle, bearing, bearing bar; radio bearing, horizontal direction; **а. в делениях угломера** mil azimuth; **выработанный магнитный а. цели** generated true target bearing; **геодезический а.**

**пройденного направления** back azimuth, forward azimuth; **исправленный а.** exact azimuth; **исходный а.** reference azimuth; **а. контрольной точки наводки** azimuth to reference point; **а. курса на перехват** intercept bearing; **прямой а.** forward azimuth; **а. сон** radar azimuth; **средний а.** general bearing, midazimuth; **а. точки наводки** azimuth datum point; **а. упреждённой точки** collision heading; **упреждённый а.** azimuth at future position, future azimuth; **упреждённый а. гироскопа** gyro of the predicted intercept point; **а. цели по источнику активных (радио) помех** target bearing by the jamming source

**азимутально-отклоняющий** *adj.* azimuth deflection

**азимутальный** *adj.* azimuth, azimuthal; bearing

**азот** *m.* nitrogen, azote

**азотирование** *n.* nitration, nitriding, nitrogen case hardening

**азотированный** *adj.* nitrated; nitrided

**азотировать** *v.* to nitrify, to nitrate; to nitride

**азотистый** *adj.* nitrogenous, nitrous, nitrate (of), azotic

**азотический, азотный** *adj.* nitric, nitrogen, nitrogenous, azotic

**АЗС** *abbr.* (**автоматическая защита сети**) automatic circuit protection

**АИМ** *abbr.* (**амплитудно-импульсная модуляция**) PAM, pulse-amplitude modulation

**Айри** Airy

**айри** *m. indecl.* airy

**Айртон** Ayrton

**айсберг** *m.* iceberg

**айсолантайт** *m.* isolantite

**АК** *abbr.* (**авиационная коротковолновая станция**) aeronautical shortwave radio station

**академик** *m.* academician

**академист** *m.* academist

**академический** *adj.* academic

**академия** *f.* academy; **Военноэлектротехническая а.** Military Electrotechnical Academy; **Инженерно-Техническая а. связи им. Подбельского** the Podbelsky Engineering and Technical Academy of Communication

**акажу** *n. indecl.* acajou

**акароид** *m.* acaroid

**акароидный** *adj.* acaroid

**акваграф** *m.* aquagraph

**аквадаг** *m.* Aquadag

**аквадаговый** *adj.* aquadag

**акванол** *m.* acquanol

**Аквариды** *pl.* Aquarids

**аквасил** *m.* aquaseal

**акватория** *f.* aquatory

**акклиматизация** *f.*, **акклиматизирование** *n.* acclimatization

**акклиматизованный** *adj.* acclimatized

**аккомодация** *f.* accomodation, adjustment

**аккомодировать** *v.* to accomodate, to adjust

**аккомпанемент** *m.* accompaniment

**аккомпанировать** *v.* to accompany

**аккорд** *m.* chord; accord, concord; **диссонансный а.** discord; **консонансный а.** concord

**аккордеон** *m.* accordion

**аккреция** *f.* accretion

**аккумулирование** *n.* accumulation, storage, cumulation, congestion

**аккумулированный** *adj.* accumulated, stored

**аккумулировать** *v.* to accumulate, to store, to cumulate, to collect; to retain, to register

**аккумулятивный** *adj.* accumulative, accumulation

**аккумулятор** *m.* accumulator, accumulator cell, storage battery, storage cell; storage system, storage device; **аэродромный а.** ground battery; **буферный а.** by-pass accumulator, line accumulator; **а. в зарядно-разрядном режиме** floating battery; **а. в кислотонепроницаемом сосуде** nonspillable accumulator; **водолазный а.** weighted accumulator; **а. для запала, а. для системы зажигания** ignition accumulator; **кислотный а.** lead battery; **кислотный свинцовый а.** lead accumulator, lead-acid cell; **контрольный а.** pilot cell; **пневмогидравлический а.** air lead accumulator; **пороховой а. давления** solid fuel pressure accumulator, solid-reactant gas generator, cartridge gas generator; **а. с виндзейлем** bladder accumulator; **а. с эластичным мембраном** flexible separator accumulator; **серебряно-цинковый а.** silver-cell; **а. солнечной энергии** solar boiler; **тепловой а.** storage-type A-battery, storage heater; **электрический а.** accumulator; **электрический а. горячей воды** electrical hot-water boiler; **а. ядерной энергии** nuclear energy storage battery

**аккумуляторная** *f. adj. decl.* battery room; storage-battery plant

**аккумуляторный** *adj.* accumulator, battery, storage battery, storage cell; storage system; battery testing

**аккумуляция** *f.* storage, accumulation, cumulation

**аккуратно** *adv.* accurately, exactly

**аккуратность** *f.* precision, accuracy, exactness

**аккуратный** *adj.* precise, accurate, exact

**аклина, аклиналь** *f.* aclinic line

**аклинальный, аклинический** *adj.* aclinic

**А-коммутатор** *m.* A-switch board, A-position, outgoing position

**акрелеиновый** *adj.* acrolein

**акрибометр** *m.* acribometer

**акрил** *m.* acryl

**акрилатовый** *adj.* acrylate

**акрилил** *m.* acrylyl

**акрилнитрил** *m.* acrylic nitrile

**акриловый** *adj.* acryl, acrylic

**акровакс** *m.* acrowax

**акролеин** *m.* acrolein

**акролит** *m.* acrolite

**акронал** *m.* acronal

**акселератор** *m.* accelerator

**акселерограф** *m.* accelerograph

**акселерометр** *see* **акцелерометр**

**аксессуары** *pl.* adjuncts, accessories

**аксиальносимметричный** *adj.* axially symmetric(al), axisymmetric(al)

**аксиальный** *adj.* axial

**аксиома** *f.* axiom; principle, postulate

**аксиоматика** *f.* axiomatics; **содержательная а.** informal axiomatics

**аксиоматический** *adj.* axiomatic, postulational

**аксиометр** *m.* axiometer

**аксиотрон** *m.* axiotron

**аксис** *m.* axis

**аксометр** *m.* axometer

**аксон** *m.* axon

**аксонометрический** *adj.* axonometric

**аксонометрия** *f.* axonometry

**акт** *m.* act, event

**акт.** *abbr.* (**активный**) active

**актив** *m.* assets; **а. и пасив** assets and liabilities

**активатор** *m.* activator, promoter; sensitizer; **основной а.** dominant activator; **а., способствующий повышению люминесценции** luminescent activator

**активаторный** *adj.* activator

**активационный** *adj.* activation

**активация** *f.* activation; sensitization, sensitizing; **повторная а.** reactivation, reconditioning; **а. прожектора** gun activation

**активирование** *n.* activation; reconditioning; sensitization; **а. эмиттера** coating activation

**активированный** *adj.* activated; sensitized

**активировать** *v.* to activate, to promote; to accelerate; to sensitize; to start

**активирующий** *adj.* activating

**активнодействующий** *adj.* active

**активно-реактивный** *adj.* impulse-reaction

**активность** *f.* activity; radioactivity

**активный** *adj.* active; operating; radioactive (sample); wattful

**актинизм** *m.* actinism

**актиний** *m.* actinium, Ac

**актинический** *adj.* actinic

**актиничность** *f.* actinism; **а. экрана** screen actinic efficiency

**актиничный** *adj.* actinic

**актино-** *prefix* actino-

**актинограмма** *f.* actinogram

**актинограф** *m.* actinograph

**актинодиэлектрический** *adj.* actinodielectric

**актинолайт** *m.* actinolite

**актинометр** *m.* actinometer

**актинометрический** *adj.* actinometric

**актинометрия** *f.* actinometry

**актиноскоп** *m.* actinoscope

**актиноскопия** *f.* actinoscopy

**актинотерапия** *f.* actinotherapy

**актиноуран** *m.* actinouranium, AcU

**актиноурановый** *adj.* actinouranium

**актино-электрический** *adj.* actinoelectric

**актиноэлектричество** *n.* actinoelectricity

**актуальность** *f.* actuality

**актуальный** *adj.* actual; present

**актуатор** *m.* actuator

**акуметр** *m.* acoumeter

**акустик** *m.* soundman

**акустика** *f.* acoustics, phonics; acoustic engineering, audio-engineering; **а. помещений** room acoustics; **а. ультразвуковых частот** supersonics

**акустико-механический** *adj.* acoustic-mechanical

**акустико-электрический** *adj.* acousto-electric, acoustic-electric

**акустиметр** *m.* acoustimeter

**акустически** *adv.* acoustically

**акустический** *adj.* acoustic(al), sonic, sound(ing); phonic; aural

**акустическо-механический** *adj.* acoustic-mechanical

**акустогравитационный** *adj.* acoustic-gravity

**акустодинамический** *adj.* acousto-dynamic

**«акустолит»** *m.* Acoustolith (tile)

**акцелератор** *m.* accelerator

**акцелерация** *f.* acceleration

**акцелерограмма** *f.* accelerogram

**акцелерограф** *m.* accelerograph; **а. поперечных ускорений** lateral accelerograph; **а. тензометрического типа** strain-gage-type accelerograph

**акцелерометр** *m.* accelerometer, acceleration pickup, acceleration detector; **вертикальный а.-датчик импульсов для счётчика** vertical counting accelerometer; **вибротронный а.** vacuum-tube accelerometer; **комбинированный а.** V.G. recorder; **посадочный а.** touchdown impact accelerometer; **а. с пьезоэлементом, работающий на изгиб** bender-type accelerometer; **тензометрический а.** strain-gage accelerometer; **а. фирмы «Галтон Индастрис»** Glennite accelerometer

**акцент** *m.* accent, accentuation

**акцентировать** *v.* to accent, to accentuate

**акцентный** *adj.* accent

**акцентуационный** *adj.* accentuation

**акцентуация** *f.* accentuation; pre-emphasis

**акцепт** *m.* acceptance

**акцептировать** *v.* to accept

**акцептование** *n.* acceptance

**акцептованный** *adj.* accepted

**акцептовать** *v.* to accept

**акцептор** *m.* acceptor

**акцепторный** *adj.* acceptor, acceptor-type

акцепция *f.* acceptance
акцессорный *adj.* accessory
акцидентный *adj.* accidental
алгебра *f.* algebra; **а. высказаний** propositional algebra; **а. релейных схем** Boolean algebra
алгебраический *adj.* algebraic
**АЛГОЛ** ALGOL
алгоритм *m.* algorithm, algorism; **а. извлечения квадратного корня** square rooting algorithm; **а. распределения** scheduling algorithm; **а. составления блок-схем** (*or* **программ**) flow-synthesis algorithm
алгоритмический *adj.* algorithmic
алгорифм *m.* algorithm, algorism
алдрей *m.* aldrey
алейрометр *m.* aleurometer
**Александерсон** Alexanderson
алетевтика *f.* aletheutics
алидада *f.* alidade, vernier plate
ализонит *m.* alisonite
алифатический *adj.* aliphatic
алициклический *adj.* alicyclic
алкали *n. indecl.* alkali
алкализация *f.* alkalization
алкализировать *v.* to alkalize
алкалиметр *m.* alkalimeter
алкалиметрический *adj.* alkalimetric
алкалиметрия *f.* alkalimetry
алкатрон *m.* alcatron
алкидный *adj.* alcyde
алкил *m.* alkyl; **галоидный а.** alkyl halide
алкоголь *m.* alcohol
алкогольный *adj.* alcoholic, alcohol
алкомакс *m.* alcomax
аллигаторный *adj.* alligator
аллобар *m.* allobar
алломорф *m.* allomorph
алломорфизм *m.* allomorphism
алломорфный *adj.* allomorphic
аллосема *f.* alloseme
аллотриоморфный *adj.* allotriomorphic
аллотропический *adj.* allotropic
аллотропия *f.* allotropy
аллохроматический *adj.* allochromatic
алмаз *m.* diamond
алмазный, *adj.* diamond
алмелек *m.* almelec
алнико *n.* alnico
алсимаг *m.* alsimag
алудур *m.* aludur
алумель *m.* alumel
алунд *m.* alundum

алфавит *m.* alphabet; code; **а. с контролем по чётности** parity-check alphabet
алфавитно-цифровой *adj.* alphanumeric
алфавитный *adj.* alphabetic(al)
алфакс *m.* alphax
алфенол *m.* alfenol
алфер *m.* alfer
альбедо *n.* albedo; **а. земной поверхности** earth's reflectivity
альбедометр *m.* albedometer
альберит *m.* alberit
альбертат *m.* albertat
альбертит *m.* albertite
альбертол *m.* albertol
альболит *m.* albolit
альбуминометр *m.* albuminometer
альвеола *f.* alveola
альвеолярный *adj.* alveolar
альдегид *m.* aldehyde
альдегидный *adj.* aldehyde
альдрей *m.* aldrey
альдрейный *adj.* aldrey
альклад *m.* alclad
алькомакс *m.* alcomax
альмукантарат *m.* almucantarat, altitude circle
альнико *n.* alnico
альсимаг *m.* alsimag
альсинт *m.* alsint
альсифер *m.* alsifer
**Альстрем** Allström
альт *m.* alto, counter-tenor; viola, tenor-viola
альтазимут *m.* altazimuth
альтазимутальный *adj.* altazimuth(al)
альтерация *f.* alteration
альтерированный *adj.* altered
альтернатива *f.* alternative, choice; **альтернативы дилеммы** horns of dilemma
альтернативный *adj.* alternative, alternate
альтернатор *m.* alternator; **двухфазный а.** diphaser; **однофазный а.** uniphaser
альтернаторный *adj.* alternator
альтернация *f.* alteration; alternation
альтернирующий *adj.* alternating
альтиграф *m.* recording altimeter
альтиметр *m.* altimeter, altitude finder; **абсолютный а.** radio altimeter, terrain-clearance indicator; **а. с непрерывным отсчётом** terrain-clearance indicator

**альтовый** *adj.* alto

**альфа** *f.* alpha; **а. Кассиопеи** Cassiopeia A; **а. Лебедя** Cygnus A

**альфаграничный** *adj.* alpha-cutoff

**альфакс** *m.* alphax

**альфаметр** *m.* conductivity-type gas analyzer

**альфатрон** *m.* alphatron; **а. точки полного торможения** impact alpha-tron

**альфачастичный** *adj.* alpha-particle

**Альфвен** Alfven

**Альфен** Alphen

**альфина-частицы** *pl.* alphina particles

**альфолевый** *adj.* alfol, aluminum-foil

**альфоль** *m.* alfol, aluminum foil

**Альфорд** Alford

**альформатор** *m.* allformator

**альфталат** *m.* alftalate

**алюзиль** *m.* alusil

**алюминат** *m.* aluminate

**алюминиево-медный** *adj.* aluminum-copper

**алюминиево-никелево-кобальтовый** *adj.* alnico, Al-Ni-Co

**алюминиево-никелевый** *adj.* alni, Al-Ni

**алюминиевый** *adj.* aluminum

**алюминизированный** *adj.* aluminized

**алюминий** *m.* Aluminum, Al

**алюминийтриметил** *m.* trimethylaluminum

**алюминирование** *n.* aluminizing

**алюминировать** *v.* to aluminize

**АМ** *abbr.* (**амплитудная модуляция**) AM, amplitude modulation

**амальгама** *f.* amalgam

**амальгаматор** *m.* amalgamator

**амальгамационный** *adj.* amalgamating

**амальгамация** *f.*, **амальгамирование** *n.* amalgamation

**амальгамированный** *adj.* amalgamated

**амальгамировать** *v.* to amalgamate

**амальгамирующий** *adj.* amalgamating

**аматёр** *m.* amateur

**амбиполярный** *adj.* ambipolar

**амблиакозия** *f.* amblyacousia

**амброин** *m.* ambroin

**амбушюр** *m.* mouthpiece (of telephone) lip, embouchure, opening; **микрофонный а.** mouthpiece (of a telephone transmitter)

**амер.** *abbr.* (**американский**) American

**американский** *adj.* American

**америций** *m.* americium, Am

**А-место** *n.* A-position, outgoing position

**аметропия** *f.* ametropia

**амианайт** *m.* amianite

**амиант** *m.* amianthus, asbestos

**амиантовый** *adj.* amianthus, asbestos

**амидоль** *m.* amidol

**амикрон** *m.* amicron

**амилацетатный** *adj.* amyl-acetate

**амиловый** *adj.* amyl

**амилоидный** *adj.* amyloid

**амин** *m.* amine

**амино-** *prefix* amino-

**аминовый** *adj.* amine

**аминопласты** *pl.* plastic amines, amino-plastics

**Амичи** Amici

**амметр** *m.* ammeter

**аммиак** *m.* ammonia

**аммиачный** *adj.* ammonium, ammonia, ammoniacal

**аммоний** *m.* ammonium

**амортизатор** *m.* shock absorber, bumper, damping spring, damper, amortisseur, vibrashock, vibration absorber; baffler, deadener, silencer, mute, deafener; **а. колебаний** vibration damper; **раздвижной а.** telescopic shock absorber; **резиновый а.** rubber cushion assembly; **а. рулевого механизма** steering damper; **а. шума** noise killer

**амортизаторный** *adj.* shock absorber

**амортизационный** *adj.* shock-absorbing, shock absorber; amortization; damper

**амортизация** *f.* amortization, depreciation, absorption (of shock), shock mount, cushioning, insulation; damping, vibration control, buffer action, resilience

**амортизированный** *adj.* shockproof; damped; cushioned

**амортизировать** *v.* to damp; to absorb (shock); to cushion; to amortize

**амортизование** *n.* amortization, depreciation; absorption (of shock), cushioning; damping; buffer action

**амортизованный** *adj.* shockproof; damped; cushioned

**амортизующий** *adj.* (shock) absorptive; resilient

**аморфизм** *m.* amorphism

**аморфический** *adj.* amorphous

**аморфность** *f.* amorphism

**аморфный** *adj.* amorphous

**амп** *abbr.* (ампер) ampere, amp.

**ампер** *m.* ampere

**ампераж** *m.* amperage, current

**ампервесы** *pl.* ampere balance

**ампер-виток** *m.* ampere-turn; *pl.* **ампер-витки** ampere-turns, number of ampere turns, ampere-wires; **а.-в. возбуждения** field ampere-turns; **встречные а.-в.** back ampere-turns; **намагничивающие а.-в.** magnetic flux; flow (of radio frequency); circulation; **противодействующие а.-в.** back ampere-turns; **реактивные а.-в.** choking turns

**ампер-вольт-оммер** *m.* avometer, multimeter

**амперит** *m.* amperite, ballast resistor tube

**амперметр** *m.* ampere meter, ammeter, current meter; **антенный а.** radiator ammeter; **втычной а.** clip-on ammeter, tong-test ammeter; **двусторонний а.** polarized ammeter, central zero ammeter; **дистанционный а.** teleammeter; **а. для надевания на провод** hook-on ammeter; **а. закрытый в плотный кожух** ironclad ammeter; **образцовый а.** standard ammeter; **а. по принципу ампервесов** steel-yard ammeter; **а. с безнулевой шкалой** depressed-zero ammeter; **а. с квадратичной шкалой** current square meter; **а. с переключателем шкал** multirange ammeter; **а. с противовесом** gravity ammeter; **а. с растянутой шкалой** platform scale ammeter

**ампер-минута** *f.* ampere-minute

**амперный** *adj.* ampere

**амперометрический** *adj.* amperometric

**ампер-проводник** *m.* ampere-conductor, ampere-wire

**ампер-секунда** *f.* ampere-second

**ампер-фут** *m.* ampere-foot

**ампер-час** *m.* ampere-hour

**амперчасовый** *adj.* ampere-hour

**амплидин** *m.* amplidyne

**амплидинный** *adj.* amplidyne

**амплистат** *m.* amplistat

**амплитрон** *m.* amplitron

**амплитронный** *adj.* amplitron

**амплитуда** *f.* amplitude, range; crest, peak, crest value; pulse (height); maximum; **а. вертикальной качки** height of heave; **а. всплеска на импульсе, а. выброса на фронте**

импульса pulse-spike amplitude; **гиперболическая а.** Guderman function; **двойная а.** peak-to-peak value; double amplitude, total amplitude, peak-to-peak separation; **двойная а. напряжения пульсации** peak-to-peak ripple voltage; **а. деформации** strain amplitude; **а. звуковых колебаний** sound energy density; **а. изменений регулируемой величины** range-ability; **а. кривой намагничения** magnetic amplitude; **максимальная а. дальности** peak pulse amplitude; **максимальная а. тропического прилива** great tropic range; **а. отклонения (за предписанные пределы)** maximum variation of the controlling quantity; **перигейная а.** perigean range; **а. пики на импульсе** pulse-spike amplitude; **а. рыскания** hunting zone; **а. сигнала накачки** pump amplitude; **а. сизигийного прилива** spring amplitude; **а. тропического прилива** small tropic range; **а. цикла напряжений** amplitude of stress

**амплитудноимпульсный** *adj.* pulse-amplitude

**амплитудномодулированный** *adj.* amplitude-modulated

**амплитудноограниченный** *adj.* amplitude-limited

**амплитудночастотный** *adj.* amplitude-frequency, gain-frequency

**амплитудный** *adj.* amplitude, crest, peak (value), maximum, volume

**амплитудо-частотный** *see* **амплитудночастотный**

**амплификатор** *m.* amplifier

**амплификация** *f.* amplification, amplifying

**ампула** *f.* ampoule, ampulla, bulb

**ампульный** *adj.* ampoule, ampulla, bulb

**АМС** *abbr.* (автоматическая межпланетная станция) automatic interplanetary station

**амфенол** *m.* amphenol

**амфидромический** *adj.* amphidromic

**амфойл** *m.* amfoil

**амфолит** *m.* ampholyte

**амфотерность** *f.* amphoteric character

**амфотерный** *adj.* amphoteric

**АН** *abbr.* (Академия наук) Academy of Sciences

**анабатический** *adj.* anabatic

анагалактический *adj.* anagalactic, extra-galactic

анаглиф *m.* anaglyph

анаглифный *adj.* anaglyph

анализ *m.* analysis, analyzing; **анодно-лучевой а.** positive-ray analysis; **а. бесконечно малых** infinitesimal calculus; **а. «в потоке»** running analyzer; **а. двухмерного потока** two-dimensional analysis; **дискретный а.** sampling analysis; **а. звукового спектра путём среза частот** reduction of harmonic content in sound analysis; **а. импульсов** pip interpretation; **а. импульсов на экране индикатора** scope reading; **а. методом узловых точек** nodal analysis; **а. отражённого сигнала по частоте и фазе** frequency and phase analysis of a radar; **а. по непосредственно составляющим** immediate continuents analysis; **а. размещения узлов напряжения (в волноводах)** nodal analysis (of waveguides, etc.); **рентгеноструктурный а.** X-ray crystal analysis; **а. с заменой характеристик отрезками прямых линий** broken-line analysis; **а. смешанных вторых моментов** analysis of covariance; **технический а.** proximate analysis; **фазовый а. данных рассеивания** phase shift analysis of the scattering; **а. частотно-фазовой характеристики** frequency response analysis

анализатор *m.* analyzer, analyst; comparator; **векторный а.** complex plane analyzer; **гетеродинный а. со скользящей зондирующей частотой** sweep frequency heterodyne-type analyzer; **дифференциальный а. дискретного типа** digital differential analyzer; **а. задач, звязанных с комплексной плоскостью** complex-plane analyzer; **а. импульсов с преобразованием: амплитуда-время** pulse-height analyzer; **а., использующий повторяющуюся звукозапись** loop recording analyzer; **а. комплексной проводимости** admittance analyzer; **а. координат положения** dead reckoning analyzer; **а. кристаллических триодов** transistor analyzer; **механический язычковый а.** vibrating reed analyzer; **а. непрерывного действия, непрерывный а.** continuous ana-lyzer; **а. относительного содержания** ratio analyzer; **а. прохождения сигналов** signal tracer; **реактивный а.** jet analyzer; **а. с отрицательной обратной связью** degenerative analyzer; **а. с периодизацией решения** repetitive analyzer; **а. с постоянной относительной шириной полосы частот** constant-percentage bandwidth analyzer; **а. со смежными полосами частот** contiguous-band analyzer; **узкополосный а. с постоянной относительной шириной полосы частот** narrow-band proportional bandwidth analyzer; **цифровой дифференциальный а. на транзисторах** transistorized real-time incremental computer

анализирование *n.* analysis, analyzing, analyzation

анализированный *adj.* analyzed

анализировать *v.* to analyze

аналитически *adv.* analytically

аналитический *adj.* analytical

аналитичность *f.* analiticity

аналматик *m.* Analmatic

аналог *m.* analogue; **магнитный а. диэлектрического преобразования** dynamometric transducer

аналогический, аналогичный *adj.* analogous, similar, corresponding; analog (computer)

аналогия *f.* analogy, similarity, resemblance; **а. для течений** analogies to mechanism of flow; **обобщённая а.** extended analogy

аналоговый *adj.* analogous, similar

аналого-дискретный *adj.* semidigital

аналого-цифровой *adj.* analog-digital

анаморфизм *m.* anamorphism

анаморфный *adj.* anamorphous, anamorphic

анаморфоз *m.* anamorphosis

анастигмат *m.* anastigmatic lens, anastigmatic objective

анастигматический *adj.* anastigmatic

анафорез *m.* anaphoresis

анафронт *m.* anafront

ангармонический *adj.* anharmonic

ангармоничность *f.* anharmonicity

англ. *abbr.* (английский) English, British

английский *adj.* English, British

ангстрем *m.* angström

Андерсон Anderson

анемо- *prefix* anemo-

**анемобиаграф** *m.* anemobiagraph
**анемограмма** *f.* anemogram
**анемограф** *m.* anemograph, self-registering anemometer
**анемоклинограф** *m.* anemoclinograph
**анемоклинометр** *m.* anemoclinometer
**анемометр** *m.* anemometer, wind gage; **манометрический а.** pressure-tube anemometer; **а. с вертушками** propeller anemometer; **тепловой а.** hot-wire anemometer
**анемометрический** *adj.* anemometric
**анемометрия** *f.* anemometry
**анеморумбограф** *m.* anemorumbograph
**анемоскоп** *m.* anemoscope
**анемотахометр** *m.* air-speed meter; **самопишущий а.** air-speed recorder
**анероид** *m.* aneroid, aneroid barometer
**анероидный** *adj.* aneroid
**анероидограф** *m.* aneroidograph
**анестезия** *f.* anesthesia
**анизо-** *prefix* aniso- (disimilar, unequal; not)
**анизодемический** *adj.* anisodemic
**анизометрический** *adj.* anisometric
**анизопия** *f.* anisopia
**анизотропический** *adj.* anisotropic
**анизотропия, анизотропность** *f.* anisotropy
**анизотропный** *adj.* anisotropic, aeolotropic
**анизо-упругость** *f.* anisoelasticity
**анилиновый** *adj.* aniline
**анион** *m.* anion
**анионный** *adj.* anion, anionic
**анионообменник** *m.* anion exchanger
**анионообменный** *adj.* anion-exchange
**анисейкон** *m.* aniseikon
**анкер** *m.* anchor; rod, stay, tie, guy; **а. держателя** lead anchor; **а. оттяжки** guy anchor; **а. с ершом** expanding anchor
**анкераж** *see* **анкеровка**
**анкерный** *adj.* anchor(-type); tension chain; anchoring, stay
**анкерование** *n.* anchoring; **а. контактной линии** anchoring of trolley line
**анкеровать** *v.* to anchor, to guy, to stay, to tie, to grapple
**анкеровка** *f.* anchoring; bracing, stay(ing), guy(ing), grappling; **а. проводов** termination of wires on terminal poles; **проволочная а.** wirenail stay
**анкеровочный** *adj.* anchor
**аннекс** *m.* annex
**аннексировать** *v.* to annex

**аннигилировать** *v.* to annihilate
**аннигиляционный** *adj.* annihilation
**аннигиляция** *f.* annihilation; nucleonic transformation
**аннулирование** *n.* annulment, cancellation, nullification; extinguishment; **а. заказа** cancellation of a call; **а. отсчёта** metering-pulse suppression, cancellation of metering
**аннулированный** *adj.* annulled, canceled, nullified
**аннулировать** *v.* to cancel, to annul, to nullify, to abrogate, to destroy, to extinguish; to defer
**аногенный** *adj.* anogene
**анод** *m.* anode, plate; wing; **а. вторичной цепи катушки Румкорфа** prepondering anode; **газовый а.** wiremesh type plate (*or* anode); **дежурный а.** excitation anode, keep-alive electrode; **действующий а.** virtual anode; **добавочный а.** intensifying ring; **жидкий а.** anode layer; **а. записи** writing plate; **защищённый а.** hooded anode; **кольцевой а.** orificed anode; **а.-кратер** plate-target; **неразрезной а.** solid anode, heavy anode; **нечернёный а.** clear plate; **а. последующего ускорения, послеускоряющий а.** post accelerating anode, post-deflection anode; **а.-регулятор сходимости** convergence anode; **а. с отдельным выводом** separated anode; **а. с ушками** lug-ear anode; **а., создающий послеускорение** post-accelerating anode; **ускорительный а.** accelerator; **а. ускорителя** second anode, second accelerator; **цилиндрический а.** can anode; **шунтирующий а.** by-pass anode, relieving anode, transition anode
**анодизация** *f.* anodizing, anodic oxidation
**анодик** *m.* anodic
**анодирование** *n.* anodizing, anodic oxidation
**анодировать** *v.* to anodize
**анодно-лучевой** *adj.* anode-ray
**анодномодулированный** *adj.* anode-modulated, plate-modulated
**анодносеточный** *adj.* grid-anode, grid-plate
**анодный** *adj.* anode, anodic, anodal, plate; inverse (voltage of tube); **с анодной связью** plate-coupled

анодосеточный *adj.* grid-plate, grid-anode

анолит *m.* anolyte

аномалистический *adj.* anomalistic

аномалия *f.* anomaly, abnormality; **остаточная а. сил тяжести** residual gravity anomaly; **средняя а. вэпоху** mean anomaly at time of epoch

аномальный *adj.* anomalous, abnormal, irregular, unusual

анормальность *f.* abnormality; **а. осей** abnormality of axes

анормальный *adj.* abnormal, anomalous

анортит *m.* anorthite

анортитовый *adj.* anorthite

анотрон *m.* anotron

АНС *abbr.* (аэронавигационная служба) air-navigation service

ансамбль *m.* ensemble, aggregate, assembly

«анскоколор» *m.* Anscocolor

анталголь *f.* antalgol, antalgol star

антарктический *adj.* Antarctic

антенна *f.* antenna, aerial, antenna wire, transmitting wire; **автомобильная а. с нижним расположением** under-car antenna; **а. в виде усечённого параболоида** Cindy antenna, halfcheese antenna; **веерная шлейф-а.** folded-fan antenna; **вертикальная а. с параллельным питанием** shunt-fed vertical antenna; **в. «вертикальное полотно»** vertical curtain; **вибраторная а.** dipole antenna; **висячая а.** trailing antenna; **внешняя стержневая а.** whip-type external antenna; **а., возбуждаемая излучением** parasitic antenna; **а., возбуждаемая щелью** slot-fed termination; **воздушная а.** elevated antenna, high antenna, outdoor antenna; **волноводная а., волноводная** (*or* **волноводно**) **щелевая а.** leaky-pipe antenna, radiating waveguide; **выбрасываемая а.** trailing antenna; **выдвижная а.** vertical antenna; **выдвижная штыревая а.** collapsible whip antenna; **выпускная (самолётная) а.** drag antenna, trailing-wire antenna; **высотная а.** height-finding antenna; **гармоническая направленная а.** harmonic wire projector; **гибкая штыревая а.** whip antenna; **главная а. для коллективного приёма телевидения** master com-munity antenna; **главная радиолокационная а.** main scanner; **Г-образная а.** gamma-type antenna, inverted L-type antenna; **а. головки самонаведения** homing scanner; **горизонтальная синфазная а.** broadside array; **гофрированная а.** corrugated-surface antenna; **групповая а.** master antenna system; **двускатная а.** inverted-V antenna; **двухдипольная Н-а.** "Lazy H" antenna, **диапазонная а.** multi-tuned antenna, broadband antenna; **дипольная а., запитываемая коаксиальным кабелем** sleeve-dipole antenna; **дипольная а. из двух коллинеарных диполей** "Lazy H" antenna; **дипольная а. с применением согласующего четверть волнового шлейфа** Q-a antenna; **дипольная а. с симметричным питанием** Delta-matched impedance antenna; **диэлектрическая а. с фронтальным излучением** broadside dielectric antenna; **диэлектрическая стержневая а.** dielectric rod-radiator; **а. для дальнего приёма** fringe antenna; **а. для наблюдения за спутниками Земли** satellite tracking antenna; **а. для надводных и надземных целей** sea/land antenna (for radar); **а. для определения истинного направления** sense antenna; **а. для определения угла места** elevation position-finding antenna; **а. для разнесённого приёма** fading-reducing antenna, diversity antenna, antifading antenna; **а. для связи на рассеянном излучении** scatter antenna; **а. для телевидения** television antenna, clover-leaf antenna; **а. доплеровского радиолокатора** Doppler antenna; **а. доплеровской радиолокационной станции** fence aerial; **дуплексная а.** duplexer; **ёлочная а.** christmastree antenna, fishbone antenna; **а. запросчика-ответчика** IFF antenna; **а.-звёздочка** star antenna; **а.-змей** kite(-hoisted) antenna; **зонтичная вертикальная а.** umbrella-loaded vertical antenna; **а. из горизонтальных параллельных проводов** wave antenna; **а. из двух параллельных диполей** "Lazy H" antenna; **а. из наклонных проводов** lattice-wire broadside antenna; **а. из скрещён-**

ных рамок crossed-coil antenna, Bellini-Tosi antenna; **а. изготовленная из сплошного металлического прутка** flagpole antenna; **а., излучающая вдоль оси** end-on directional antenna; **а., излучающая равномерно во всех направлениях** unipole antenna; **а., имеющая форму перевёрнутой буквы V** Bruse antenna, Grecian type antenna; **а.-имитатор цели** phantom-target antenna; **индуктивная а.** loop antenna, ferrite (loop) antenna; **каркасная щелевая укв-а.** skeleton slot antenna; **качающаяся а.** nutating antenna, rockinghorse antenna; **качающаяся поисковая а.** rockinghorse scanner; **квадрупольная шлейф-а.** folded quadrupole antenna; **колбасная а., колбасообразная а.** cage antenna; **коллективная а.** block antenna, common antenna, community aerial, master antenna system, combined antenna; **коллинеарная а.** collinear antenna, Franklin antenna; **а., комбинированная из рамки и вертикального штыря** combined V and H antenna; **корабельная а. станции орудийной наводки** shipborne fire control antenna; **кормовая а.** aft(er) antenna; **коробочная а., коробчатая а.** box antenna; **кратная горизонтальная а.** multiple range antenna, multiband antenna; **крестообразная а.** crossed-coil antenna, crossed-loop antenna; **а. кругового обзора** (*or* **поиска**) all-round looking antenna; **кругонаправленная а., кругообразная а.** omnidirectional antenna; **крыльевая а.** skid-fin antenna, wing-cap antenna; **крыловидная турникетная а.** superturnstile antenna; **крышевая сетка-а.** mesh-roof antenna (automobile type); **крышная а.** overhouse antenna; **куполообразная а.** barrel antenna; **курсовая а.** homing antenna; **а. курсового маяка** localizer antenna; **«лестничная» а.** fishbone antenna; **а. летательного аппарата** airborne (*or* airplane) antenna; **линзовая а. из плексигласа** perspex lens; **линзовая а. из полиэтилена** polythene lens; **лопастная а.** turnstile antenna; **манипуляторная а.** probe antenna; **а. маяка-ответчика**

beacon antenna; **маячная а. сантиметровых волн** "bups" antenna; **миниатюрная рамочная а.** loopstick antenna; **мнимая а.** image antenna; **многовибраторная а.** echelon antenna, multiunit antenna, multidipole antenna, antenna array; **многодипольная многоярусная а.** christmastree antenna, pinetree array; **многократно настроенная а.** multiple-tuned antenna; **многолучевая а.** multi-wire antenna, multiple-wire antenna; **многорядная а.** stacked antenna, mattress antenna; **многосекционная турникетная а.** superturnstile antenna; **многоугольковая а., многоугольная а.** multi-V antenna; **многоэтажная турникетная а.** superturnstile antenna; **многоярусная а.** stacked antenna, stacked dipole array, musa antenna; **а. моноимпульсной РЛС** monopulse antenna; **а., монтированная заподлицо** flush-mounted antenna; **а. на краткие волны** multiband antenna, two-wave horizontal broadside array; **а. на подвижном объекте** vehicular antenna; **а. навигационной системы «Такан»** Tacan antenna; **а. наземной установки** ground antenna, earth antenna; **напольная а.** outdoor antenna; **наружная стержневая а.** car-top rod antenna; **настроенная а. (стоячей волны)** modulated antenna, periodic antenna; **настроенная рамочная а.** spiral loop; **а. невыступающего типа** surface-type antenna; **ненаправленная по азимуту дипольная решётка а.** "bups" antenna; **непосредственно связанная а.** plain antenna; **неудлинённая (самоиндукцией) а.** unloaded antenna; **нижняя автомобильная а.** running-board antenna; **ножевая а.** blade antenna; **носовая а.** forward antenna; **общая а. для передачи и приёма** duplexer; **однополюсная шлейф-а.** folded unipole antenna; **а. опознавательного устройства (типа «свой-чужой»)** IFF antenna; **а. определения знака направления, а., определяющая однозначность пеленга** sense antenna, sensing antenna; **отражённая а.** image antenna; **пеленгаторная а. с разнесёнными рамками** spaced-frame loop antenna;

перегнутая а. folded dipole antenna; а. передатчика радиолокационных помех ECM antenna; а., переключаемая искровым разрядником spark-gap switched antenna; перестраиваемая а. multiple-tuned antenna; а. переходного типа от волноводного к типу «сыр» hoghorn (antenna); а. Пистолькорса V antenna; питаемая а. active antenna, driven antenna; а., питаемая в узле vortex-fed antenna; питаемая в центре полуволновая а. doublet (antenna); а., питаемая в фокусе focal-line fed antenna; плашмя-направленная а. broadside-directional antenna, broadside array; плоско-пластинчатая а. flat-plate antenna; поверхностная невыступающая а. flush antenna; подкузовная а. see нижняя автомобильная а.; а. поиска, поисковая а. scanning antenna, search antenna, radar scanner, probe-type antenna; полуволновая а. с симметричным питанием Delta-matched impedance antenna, Y-antenna; полувыступающая а. quasi-flush-mounted antenna; поперечно-направленная а. broadside antenna; а. при отсутствии модуляции quiescent antenna; а. приёма сигналов управления radio control antenna; а. приёмника системы посадки по приборам ILS antenna; а. приёмного устройства missile receiver antenna; приёмно-передающая а. break-in antenna; продольно-направленная а. end-fire antenna; прожекторная а. beam antenna; простейшая а. plain antenna; а. радиодальномера DME antenna; а. радиокомпаса ADF antenna; радиолокационная а. обзора radar scanner; радиолокационная а. с ножевым лучом beavertail antenna; радиолокационная а. с узкой диаграммой направленности pencil-beam radar scanning antenna; а. радиолокационного координатора цели системы самонаведения antenna of the radar homing unit; а. радиолокационной головки самонаведения homing dish; а. радиопеленгатора ADF antenna; а. радиостанции, регистрирующей промахи и попадания scoring antenna; раз-

несённая а. (против замираний) diversity antenna; а., расположенная в двух плоскостях two-plane (or stacked) antenna; а., расположенная по потоку rearward-facing antenna; ребром-направленная а. end-on directional antenna; а. регистрации результатов стрельбы scoring antenna; рупорно-зеркальная а. horn reflector antenna; а. с витковым облучателем loop-fed antenna; а. с вращающейся диаграммой направленности rotary beam antenna; а. с горизонтальной поляризацией ground-plane antenna; а. с двойной характеристикой dual antenna, two-pattern antenna; а. с ёмкостной нагрузкой top-capacitor antenna, top-loaded antenna; а. с игольчатой диаграммой направленности pencil-beam antenna; а. с излучением вдоль оси end-on fire antenna; а. с качанием луча scanning antenna; а. с качающимся лучом "Eagle" antenna; а. с квадратичной косекансной характеристикой saxophone (antenna); а. с короткой щелью keyhole-slot antenna; а. с малым излучением на боковых лепестках диаграммы low sidelobe antenna; а. с механическим развёртыванием rockinghorse antenna; а. с минимальным горизонтальным излучением folded top antenna; а. с настроечным шлейфом stub antenna; а. с неизменной реактивностью unloaded antenna; а. с параллельным питанием shunt-excited antenna; а. с переключаемой настройкой switch tuned antenna; а. с переключением диаграммы lobe-switching antenna; а. с полем круговой поляризации rotating field antenna; а. с прижатым (к земле) излучателем low-altitude dish, low-angle dish; а. с прямой связью plain antenna; а. с равномерным излучением во всех направлениях unipole antenna, spherical antenna; а. с управляемой диаграммой направленности steerable antenna; а. с устройством против замирания diversity antenna; а. с экранированным снижением screened antenna; самолётная крылевая а. skid-fin antenna; самолётная невыступающая щелевая **а.** pocket antenna; **сдвоенная крестообразная рамочная а.** double (or spaced) crossed-loop antenna; **сегментно-параболическая а.** parallel-plate antenna; **сетевая а.** lightline antenna, mains antenna; **а. системы опознавания в гребне на фюзеляже** IFF aerial in dorsal spine; **скрытая самолётная а.** skin antenna; **слабонаправленная а.** low-gain antenna; **а. со «змеем»** kite antenna, kite-hoisted antenna; **а. со скачкообразной настройкой** switch tuned antenna; **согласованная У-образная а.** Y-antenna, delta-matched impedance antenna; **спирально-сферическая а.** helisphere antenna; **суррогатная а.** random antenna, auxiliary antenna; **а. Татаринова** (horizontal) collinear array; **а. типа «бабочка»** turnstile antenna, doughnut antenna; **а. типа «волновой канал»** Yagi antenna, director antenna; **а. типа «муза»** multiple-unit steerable antenna; **а. точной посадочной радиолокационной системы** PAR (precision approach radar) antenna; **трёхсекционная турникетная а.** superturnstile antenna; **угломестная а.** elevation antenna, position finding antenna; **уголковая а. Пистолькорса** V-antenna; **уголковая петлевая а., уголковая шлейф-а.** folded-dipole V antenna; **удлинённая а.** loaded antenna, extended antenna; **а., укороченная ёмкостью** top-capacitor antenna; **утопленная а.** suppressing antenna, suppressive antenna; **фидерная а.** TL antenna; **цельноволновая а.** dipole antenna; **цилиндрически-параболическая а.** pill-box antenna; **четырёхвибраторная а.** square-loop antenna; **широкоугольная а.** all-band antenna; **шлейф-а.** folded dipole antenna; **«штопорная» а.** helical antenna; **штыревая а.** flagpole antenna, spike antenna, pivoting antenna, (collapsible-)whip antenna; **щелевая трубчатая а.** leaky-pipe antenna; **эквивалентная а.** artificial antenna, dummy aerial, mute antenna, phantom antenna; **а. электронного музыкального инструмента** etherophone antenna; **а. «энд-он»** end-on directional array;

**эхорезонаторная а.** echo box antenna

**антеннаплекс** *m.* antennaplex
**антенный** *adj.* antenna, aerial
**анти-** *prefix* anti-; counter: non-
**антианод** *m.* antianode
**антиапекс** *m.* antiapex
**антиатом** *m.* antiatom
**антибатный** *adj.* antibate
**антивещество** *n.* antimatter
**антивибратор** *m.* shock absorber, vibration mount, vibration assembly, antivibrator
**антивибрационный** *adj.* antivibration; antimicrophonic
**антиволноводный** *adj.* antiwaveguide
**антигомоклиналь** *m. and f.* antihomocline
**антигравитационный** *adj.* antigravity
**антигравитация** *f.* antigravity
**антидетонатор** *m.* knock-reducer, antiknock
**антидетонационный** *adj.* antiknock
**антидин** *m.* antidyne
**антидинатронный** *adj.* suppressor (grid)
**антидюна** *f.* antidune
**антизапорный** *adj.* antibarrier
**антизатухание** *n.* antidamping
**антиизоморфизм** *m.* anti-isomorphism
**антиинтерференционный** *adj.* beat-interference, anti-interference
**антикатод** *m.* anticathode, target cathode
**антиклиналь** *m. and f.* anticline; **опрокинутая а.** recumbent anticline
**антиклинальный** *adj.* anticlinal
**антиклинорий** *m.* anticlinorium, composite anticline
**антикогерер** *m.* anticoherer
**антикоммутативный** *adj.* anticommutative
**антикоммутатор** *m.* anticommutator
**антикорона** *f.* anticorona
**антикоррозионный** *adj.* anticorrosive, corrosion-resisting, oxidation-proof
**антикруговой** *adj.* anticircular
**антилогарифм** *m.* antilogarithm
**антилокационный** *adj.* antiradar; anti-detection
**антимагнитный** *adj.* antimagnetic, nonmagnetic
**антиматерия** *f.* antimatter, contraterrene matter
**антиметрический, антиметричный** *adj.* antimetric

**антимикрофонный** *adj.* antimicrophonic, jar-proof, nonmicrophonic
**антимода** *f.* antimode
**антимонат** *m.* antimonate
**антимонид** *m.* antimonide
**антинейтрон** *m.* antineutron
**антинейтрино** *n.* antineutrino
**антиномия** *f.* antinomy
**антинуклон** *m.* antinucleon
**антиобледенитель** *m.* defroster, de-icer; **электротермический a.** resistive de-icer
**антипаразитный** *adj.* antiparasitic
**антипараллельный** *adj.* antiparallel, inversely parallel
**антипассат** *m.* anti-trade wind, return trade wind
**антипеленгационный** *adj.* anti-DF, antidirection-finder
**антипеленгация** *f.* antidirection finding
**антипереполнение** *n.* underflow
**антипод** *m.* antipode; **a. эпицентра** anticenter
**антиподальный** *adj.* antipodal
**антипроизводная** *f. adj. decl.* anti-derivative
**антипротон** *m.* antiproton
**антирадиолокационный** *adj.* antiradar, anti-radiolocation
**антиракета** *f.* antirocket, antimissile, contrarocket; **самонаводящаяся a.** missile seeker
**антирезонанс** *m.* antiresonance
**антирезонансный** *adj.* antiresonance
**антисегнетоэлектрик** *m.* antiferroelectric
**антисегнетоэлектрический** *adj.* antiferroelectric
**антисейсмический** *adj.* antiseismic
**антисептик** *m.* antiseptic material; wood preservative
**антисимметризация** *f.* antisymmetrization
**антисимметрический, антисимметричный** *adj.* antisymmetric
**антисимметрия** *f.* antisymmetry
**антисовпадение** *n.* anticoincidence
**антисолнечный** *adj.* antisolar
**антистоксов** *poss. adj.* antistokes, contradictory to Stokes rule
**антисульфатный** *adj.* antisulfating
**антитело** *n.* antibody
**антитетический** *adj.* antithetic(al)
**антиузел** *m.* antinode
**антифединговый** *adj.* antifading

**антиферромагнетизм** *m.,* antiferromagnetism
**антиферромагнетик** *m.* antiferromagnet
**антиферромагнитный** *adj.* antiferromagnetic
**антиферроэлектрический** *adj.* antiferroelectric
**антиферроэлектричество** *n.* antiferroelectricity
**антифонный, антифоновый** *adj.* anti-hum
**антифрикционный** *adj.* antifriction, antifrictional
**антифриттер** *m.* anticoherer
**антицентр** *m.* anticenter
**антициклогенез** *m.* anticyclogenesis
**антициклолиз** *m.* anticyclolysis
**антициклон** *m.* anticyclone
**антициклональный, антициклонический, антициклонный** *adj.* anticyclonic, anticyclone
**антициклотрон** *m.* anticyclotron
**античастица** *f.* antiparticle
**антишумовой** *adj.* antinoise
**антиэлектрон** *m.* anti-electron
**АНТО** *abbr.* (**Авиационное научно-техническое общество**) Aviation Scientific and Technical Society
**антрацен** *m.* anthracene
**антраценовый** *adj.* anthracene
**антропометр** *m.* anthropometer
**антропометрический** *adj.* anthropometric
**аншпуг** *m.* crowbar
**Аншютц** Anschütz
**анэлектрический** *adj.* anelectric
**анэлектротонус** *m.* anelectronus
**анэстез** *m.* anastase
**АОО** *abbr.* (**аэродромное осветительное оборудование**) airfield lighting equipment
**АП** *abbr.* 1 (**автоматическая подстройка, автоподстройка**) automatic tuning; 2 (**автопилот**) automatic pilot
**апастрон** *m.* apastron
**АПВ** *abbr.* (**автоматическое повторное включение**) automatic reclosing
**апекс** *m.* apex
**апериодический** *adj.* aperiodic, non-periodic, dead-beat; nonrecurrent, nonoscillatory; nontunable
**апериодичность** *f.* aperiodicity
**апериодичный** *see* **апериодический**
**апертометр** *m.* apertometer, aperture

meter

**апертура** *f.* aperture, opening, orifice; **а. диафрагмы** masking aperture; **а. луча передающей трубки** quantizing aperture; **а. луча приёмной трубки** synthesizing aperture

**апертурметр** *m.* aperture meter

**апертурный** *adj.* aperture, opening, orifice

**апикальный** *adj.* apical

**АПЛ** *abbr.* (**авиоподогревательная лампа**) aircraft heater lamp

**апланат** *m.* aplanat, aplanatic lens

**апланатический, апланатный** *adj.* aplanatic

**апликатор** *m.* applicator

**апликатура** *see* **аппликатура**

**апо-** *prefix* apo-

**апоастр** *m.* apastron

**апогей** *m.* apogee; peak, culmination; **в апогее** at apogee

**аполлониев** *poss. adj.* Apollonic

**аполярный** *adj.* apolar, nonpolar

**апостериорный** *adj.* a posteriori, posterior

**апостильб** *m.* apostilb

**апофема** *f.* apothem

**апохромат** *m.* apochromat

**апохроматический** *adj.* apochromatic

**апоцентр** *m.* apocenter

**апп.** *abbr.* (**аппарат**) apparatus

**аппарат** *m.* apparatus, device, instrument, set, appliance, unit, equipment; aid, mechanism, vehicle; camera; often not translated or replaced by -er or -or ending in name of device, e.g. **буквопечатающий а.** printer (*lit.* printing apparatus); **летательный а.** aircraft (*lit.* flying device); **в корпусе аппарата** built-in; **абсорбционный а.** absorber; **автоматический а. службы времени** automatic time announcer; **аритмичный а.** start-stop apparatus; **а. АТС** dial telephone set; **а. бдительности** acknowledger; **а. без противоместной схемы** sidetone telephone set; **биржевой телеграфный а.** stock exchange telegraph, stock ticker; **блокировочный а.** block-post equipment; **буквопечатающий а. для биржи** stock ticker; **буквопишущий а.** teleprinter, tele(type)printer; **а. в учреждении** business telephone; business station; **взрывобезопасный а.** flameproof apparatus; **воздушно-**

**космический летательный а.** aerospace vehicle; **а. грамзаписи** transcription turntable; **далеко расположенный добавочный а.** outside extension; **дальнопишущий а.** telewriter, teleprinter, ticker; **двухкратный а.** double multiplex insulator; double apparatus; **а. для воспроизведения накопленной информации** reproducer; **а. для двусторонней связи** intercommunicator; **а. для дистанционного управления** positioner; **а. для дозирования** dispensing apparatus; **а. для записи речи** dictaphone; **а. для записи речи по телефону** telescribe; **а. для звукозаписи** transcriber, sound recorder; **а. для звукозаписи на ленту** tape recorder; **а. для звукозаписи на пластинку** phonograph recorder; **а. для измерения искажений** distortion set; **а. для испытания на продавливание** (*or* **на прорыв**) Mullen tester; **а. для обнаружения подводных лодок ультразвуком** ASDIC; **а. для обучения коду Морзе** omnigraph; **а. для определения неоднородности линии** singing-point tester; **а. для параллельного включения с другим** bridging set; **а. для передачи кодовых слов радиомаяков** "baby blabbermouth"; **а. для перемешивания речи** scrambler; **а. для подслушивания** detectophone; **а. для разложения амальгамы** denuder; **а. для регистрации данных** logger machine; **а. для точечной сварки** spot welder; **а. для тугоухих** deaf-aid; **записывающий а. с обратной связью** feedback cutter; **звукозаписывающий по телефону а.** telescribe; **кассовый а.** cash register; **кинокопировальный а.** film printer; **киносъёмочный а. многократного использования** recoverable film camera; **коммутационный а.** switch apparatus, switchgear; **коммутационный а. для угольного забоя** gate-end switchgear; **копировальный а.** duplicator, (film) printer; **копировальный а. прерывистого действия** step printer; **космический а.** spacecraft; **летательный а. с электрической силовой установкой** electrical propulsion apparatus; **а. максимальной защиты** overload protector; **маслонаполнен-**

ный **a.** oil-immersed apparatus; **a. МБ** local battery telephone set; **a. местного коммутатора** residence station; **микротелефонный слуховой a.** Acousticon; **монтажный a. для магнитофонных лент** editing (tape) recorder; **направляющий a.** stator, guide wheel, inlet guide vane; **настольный a.** desk set; portable apparatus; **опросный a.** answering equipment, operator's apparatus; **основной a.** main station, PBX operator's position; intercom master set; **передающий фототелеграфный a.** facsimile transmitter, picture transmitter; **пишущий a.** inker, telegraphic register; **полевой буквопечатающий a.** portable teletype; **последовательно включённый (телефонный) a.** series-connected station, telephone instrument for an intercommunication system; **построчный печатающий a.** line printer; **почтовый испытательный (телефонный) a.** central-office PBX test-set; **предвключённый a.** preset; adapting apparatus, adapter; **прерывистый копировальный a.** step printer; **приёмный a.** recorder; receiving apparatus; telegraph printer; **a., присоединяемый к сети** power unit, power pack; **a. противоточного действия** counter stream cooler; **прямой телеграфный a.** deciphering printer, secondary telegraph; **пусковой a.** starter; **радиобуквопечатающий a.** radio-teletypewriter; **радиоуправляемый летательный a.** radio-guided vehicle; **рентгеновский a. высокого напряжения** hard X-ray machine; **рентгеновский a. низкого напряжения** soft X-ray machine; **рентгеноскопический a.** radioscope, fluoroscope; **роликовый a. для нанесения припоя** roller-coater; **рулонный телеграфный a.** roll telegraph; **рулонный телеграфный пишущий a.** page-printing telegraph; **a. с втычными контактами** plug-in unit; **a. с выдержкой времени** time-lag apparatus; **a. с двумя линиями для повторного опроса** PBX subscriber station equipment with direct central office trunk, extension line and holding key; **a. с зависимой выдержкой времени** inverse time-lag apparatus;

**a. с кнопкой для наведения справок** call-back telephone set; **a. с независимой выдержкой времени, a. с независимым замедлением** definite time-lag apparatus; **a. с оконечным усилителем** loudspeaker telephone set; **a. с отдельным микрофоном** solid-back transmitter; **a. с противоместной схемой** antisidetone telephone set; **сборный a.** bastard set; **a. световой телефонии** photophone; **светокопировальный a.** copying (or prototype) apparatus; **светосигнальный a.** blinker; **a. связи** communication apparatus; transmitting apparatus; **синтетический голосовой a.** voder, voice operation demonstrator; **a. службы времени** speaking clock; **слуховой a.** hearing aid, Acousticon; **слуховой телеграфный a.** sounder; **списывающий a.** transcriber; **страничный a.** page printer; **съёмочный a.** camera; **телеграфный a.** telegraph set, telegraphic apparatus; **телеграфный a. для автографической передачи и приёма** telescriptor; **телеграфный a. по схеме мостового дуплекса** duplex bridge telegraph; **телепишущий a.** telescriptor; **телефонный a. без устранения слышимости собственного микрофона** sidetone telephone set; **телефонный a. системы центральной батареи** common-battery telephone set; **телефонный a. с кнопкой** ring-back telephone set; **телефонный a. с линиевыбирателем** inter-communication telephone; **телефонный звукозаписывающий a.** telescribe; **a. теплового действия** thermally-operated device; **фотографический a.** camera; **фотозвукозаписывающий a.** photographic sound recorder; **фотопечатающий a.** flash photographic printer; **фототелеграфный a.** telectrograph, facsimile machine, radio facsimile set, copying telegraph; **франкирующий a.** postage permit printer; **a. ЦБ** common battery set, CB set; **централизационный a.** interlocker, interlocking machine; **циферблатный телеграфный a.** ABC telegraph; **чернопишущий a.** ink writer, Morse inker; **электрический a. для сжигания металлов** deflagrator; **электрический**

слуховой а. для тугоухих earphone; электросварочный а. welder; а. Эпштейна Epstein hysteresis tester

**аппаратная** *f. adj. decl.* control room, instrument room, equipment room, control booth; **кинопроекционная а.** telefilm room, movie projection booth; **а. монтажа записей** recording room; **техническая а.** maintenance apparatus room

**аппаратный** *adj.* apparatus, instrument, device, machine

**аппаратура** *f.* equipment, outfit, apparatus, gear; facilities, control device, instrumentation, package; **абонентская а.** station apparatus; **а. абонентского ввода** service equipment; **автоматическая а. радиоразведки и помех** automatic electronic countermeasures equipment; **а. автоматического управления огнём** automatic weapon launching gear; **а. АТС** dial system equipment; **аэрофотосъёмная а.** aerial equipment; **а. бдительности** acknowledger; forestaller; **а. ближней радионавигации** short-range aids; **бортовая а.** internal instrumentation; missile-borne system, missile-contained system; **бортовая а. наведения** airborne guidance package, guidance component, receiver and guidance package; robot pilot; **бортовая а. обнаружения воздушных целей** air-to-air detector; **бортовая а. самонаведения** brain unit; **а. дискретного анализа** digital evaluation equipment; **а. для автоматической обработки и отображения данных** automatic data processing and display equipment; **а. для включения машин на параллельную работу** paralleling apparatus; **а. для непрерывного измерения величины pH** flow assembly; **а. для определения отклонения ракеты от оси луча** means of seeking the center of the radar beam; **а. для отображения воздушной обстановки** plotting and event recording equipment, display equipment; **а. для предпусковой регламентной проверки** missile checkout set, missile exerciser; **а. для предпусковых регламентных работ** missile ground handling equipment,

missile handling equipment; **а. для противодействия средствам обнаружения** antidetection equipment, detection-foiling equipment; **а. для радиосвязи в сети «борт-борт»** interplane radio; **а. для радиосвязи при помощи рассеянного распространения** transhorizon equipment; **а. для создания помех радиолокационным станциям** antiradar equipment, detection-foiling device; **а. для создания радиопомех** jamming equipment; **а. для фотографирования по методу полос** schlieren apparatus; **а. искусственной речи** voder; **киносъёмочная а. для последующей телевизионной передачи** intermediate film equipment; **комплексная полигонная контрольно-измерительная а.** chain instrumentation; **комплексная радиоэлектронная а.** integrated avionics package; **контрольная а.** test gear; supervisory equipment, monitoring apparatus; **контрольно-измерительная а.** instrumentation; **контрольно-испытательная а., контрольно-проверочная а.** checkout equipment; **а. наведения на передатчики (радио)помех** jamming homing set; **а. наведения оружия на истребителе-перехватчике** armament-laying gear; **а. наведения ракет** rocket-laying equipment; **наземная а. радиоразведки и радиопротиводействия** ground electronics countermeasures equipment; **а. обнаружения и сопровождения** acquisition element; **а. обнаружения подводных лодок по выхлопным газам** sniffer gear; **оконечная вч а. (в системе с несущей)** carrier terminal; **подсобная а.** auxiliary equipment; **а. помех самолётным радиолокационным бомбардировочным прицелам** bombsight jamming equipment; **производственная контрольно-измерительная а.** process instrumentation; **радиолокационная а. дальнего обнаружения и наведения** early-warning radar control and guidance equipment; **радиолокационная а. обнаружения и прицеливания** blind-flying fire-control system; **а. радиопротиводействия** electronic warfare equipment, electronic countermeasures device; **a. радиоразведки и помех радиолокационным станциям** radar detection and jamming equipment, radar countermeasures equipment; **радиоэлектронная a. огневой батареи** weapons battery terminal equipment; **a. разведки и помех радиовзрывателям** variable time fuse countermeasures equipment; **a. размечения (информации) на лентах** tape-editing machinery; **a. распределительного устройства** switchgear; **a. регистрации результатов стрельбы** scoring equipment; **самолётная a. радиоразведки радиолокационных средств ПВО** electronic countermeasures airborne signal analyzing equipment; **сигнально-опросная a.** answering equipment; **a. телекино** telecine equipment; **телекинопроекционная a.** film-scanning equipment; **a. уплотнения** multiplex apparatus; **a. управления взводом пусковых установок** section control equipment; **a. управления воздушной мишенью** drone guidance equipment; **a. управления планшетом** plotting board controls; **a. управления полётом** control-electronics package; **управляющая a.** motion-sensitive element; **фототелеграфная штриховая a.** facsimile; **шумопеленгаторная a.** passive sonar apparatus; **электронная a. телеуправления** control electronics

**аппаратурный** *adj.* apparatus, outfit, equipment
**аппарель** *m.* ramp, approach
**аппендикс** *m.* appendix
**Аплетон** Appleton
**аппликата** *f.* z-coordinate
**аппликатор** *m.* applicator
**аппликатура** *f.* application; finger guide (of musical instrument)
**аппликация** *f.* application
**аппрет** *m.* filling
**аппретирование** *n.* filling; sizing; dressing, finishing
**аппретирован(н)ый** *adj.* finished; sized
**аппретировать** *v.* to dress, to finish
**аппретура** *f.* filler; size; finish(ing), dressing
**аппретурный** *adj.* finishing
**апретура** *see* **аппретура**
**априори** *adv.* a priori

**априорный** *adj.* a priori, prior, antecedent
**апробация** *f.*, **апробирование** *n.* approbation; test, testing
**апробировать** *or* **апробовать** *v.* to approbate, to approve; to test
**апроксимативный** *adj.* approximate
**апроксимация** *f.* approximation; **a. соприкасающимися конусами** tangent-cone approximation
**апроксимирование** *n.* approximation
**апроксимированный** *adj.* approximated, approximate
**апроксимировать** *v.* to approximate
**апроксимирующий** *adj.* approximating, approximate
**АПС** *abbr.* (**аппаратура преобразования сигнала**) signal-conversion equipment
**апсида** *f.* apsis, apse
**апсидальный, апсидный** *adj.* apsidal
**аптекарский** *adj.* apothecary
**АПФ** *abbr.* (**автоматическая подстройка [частоты и] фазы**) automatic (frequency and) phase control
**АПЧ** *abbr.* (**автоматическая подстройка частоты**) AFC, automatic frequency control
**аралдит** *m.* araldite
**арбитр** *m.* arbitrator, referee
**арбитражный** *adj.* arbitrary, referee
**АРГ** *abbr.* (**автоматическая регулировка громкости**) AVC, automatic volume control
**Арганд** Argand
**аргентан** *m.* argentan, Ar
**аргентометр** *m.* argentimeter
**аргиро-** *prefix* argyro-
**аргиродит** *m.* argyrodite
**аргирометрический** *adj.* argyrometric
**аргирометрия** *f.* argyrometry
**аргон** *m.* argon, Ar
**аргонно-дуговый** *adj.* argon-arc
**аргонно-ртутный** *adj.* argon-mercury
**аргонный, аргоновый** *adj.* argon, argonal
**аргоно-ртутный** *adj.* argon-mercury
**аргумент** *m.* argument, amplitude; reasoning, argument; **a. импеданса** impedance angle; **a. функции** independent variable
**ардометр** *m.* ardometer
**ардорит** *m.* ardorit
**ардостан** *m.* ardostan
**аренда** *f.* lease; **повременная a.** time-hire

**арендованный** *adj.* leased, rented
**арендовать** *v.* to lease, to rent
**ареографический** *adj.* areographic
**ареография** *f.* areography
**ареометр** *m.* areometer, densimeter, hydrometer; **а. для (определения жёсткости) кислоты** acidometer; **а. для определения концентрации соли в растворе** halometer
**ареометрический** *adj.* areometric
**ареометрия** *f.* areometry, hydrometry
**ареопикнометр** *m.* areopycnometer
**ареоцентрический** *adj.* areocentric
**аридность** *f.* aridity
**аридный** *adj.* arid
**аристотипный** *adj.* aristo (paper)
**аритмический, аритмичный** *adj.* arrhythmic, not rhythmic
**арифметика** *f.* arithmetic; **а. порядков** exponent arithmetic
**арифметический** *adj.* arithmetic(al)
**арифмограф** *m.* arithmograph
**арифмометр** *m.* adding machine, adder, calculating machine, arithmometer, calculator; **а. для извлечения корней** mechanism for conversion of quadratic quantities into linear quantities
**АРК** *abbr.* (**автоматический радиокомпас**) automatic radio compass, automatic direction finder
**арка** *f.* arch, arc
**аркан** *m.* check cable, safety stop cable loop; rope
**аркатом** *m.* atomic-hydrogen arc welding
**арккосеканс** *m.* arc cosecant
**арккосинус** *m.* arc cosine
**арккотангенс** *m.* arc cotangent
**аркообразный** *adj.* arched
**аркотрон** *m.* arcotron
**арксеканс** *m.* arc secant
**арксинус** *m.* arc sine, antisine
**арктангенс** *m.* arc tangent, antitangent
**Арктика** *f.* Arctic
**арктический** *adj.* arctic, polar
**арматура** *f.* fittings, fixtures, hardware; accessories; armature; mounting; outfit; **а. для ввода** connection material; **железная а.** iron-sheathing; tape armoring; **корпусная а. гидроакустической станции** asdic hull outfit; **промежуточная защитная а.** partition insulator-chain; **рудничная а.** bulkhead fittings; **а.**

**слюдяных изоляторов** mica clip; **стоячная осветительная а.** standard fittings
**арматурный** *adj.* fitting, fixture; armature
**армейский** *adj.* army
**армиллярный** *adj.* armillary
**армирование** *n.* armoring, armature, sheathing, reinforcement
**армированный** *adj.* armored, reinforced, sheathed
**армировать** *v.* to armor, to reinforce, to sheathe
**армировка** *f.* armoring, armature
**армия** *f.* army
**армко-железо** *n.* armco iron
**АРМС** *abbr.* (**автоматическая радиометеорологическая станция**) automatic radio meteorological station
**Армстронг** Armstrong
**ароматический, ароматичный, ароматный** *adj.* aromatic
**Арон** Aron
**арочный** *adj.* arched, arch
**Арп** *abbr.* (**автоматический радиопеленгатор**) automatic radio direction finder
**арпеджио** *n. indecl.* arpeggio
**арретир** *m.* arrester, arrest(ment), lock, stop, stopping device, checking device; locking device, catch
**арретирный** *adj.* locking, arresting, stop, stopping
**арретирование** *n.* locking, stopping, arresting, caging
**арретированный** *adj.* arrested, locked, stopped
**арретировать** *v.* to arrest, to stop, to check; to secure, to lock; to cage
**арретировка** *f.* arresting, locking, stopping; arresting device, locking device
**арретировочный** *adj.* arresting, stopping, stop
**арретирующий** *adj.* arresting, stopping
**арсенид** *m.* arsenide
**артезианский** *adj.* artesian
**артериальный** *adj.* arterial
**артерия** *f.* artery
**артикль** *m.* article
**артикуляционный** *adj.* articulation
**артикуляция** *f.* articulation, intelligibility; **а. в полосе разговорных частот** band articulation; **логатомная а.** syllabic articulation, CCIT syllable articulation
**артиллерийский** *adj.* artillery

**артиллерия** *f.* artillery
**АРУ** *abbr.* (**автоматическая регулировка частоты**) AGC, automatic gain control
**АРФ** *abbr.* (**автоматическая регулировка фазы**) automatic phase control
**арфа** *f.* harp
**арфообразный** *adj.* harp(-shaped)
**архив** *m.* archives, records; **магнитофонный а. лент** magnetic tape record
**архивольт** *m.* archivolt
**Архимед** Archimedes
**архимедов** *poss. adj.* Archimedean, Archimedes
**архитектоника** *f.* architectonics
**архитектура** *f.* architecture
**архитектурный** *adj.* architecture, architectural
**АРЧ** *abbr.* (**автоматическая регулировка чувствительности**) automatic sensitivity control
**АС** *abbr.* (**авиационная радиостанция, авиарадиостанция**) aeronautical station
**а-с** *abbr.* (**ампер-секунда**) ampere-second
**асб** *abbr.* (**апостильб**) apostilb
**асбест** *m.* asbestos
**асбестин** *m.* asbestine
**асбестит** *m.* asbestos insulation
**асбестовидный** *adj.* asbestiform, fibrous
**асбестоволокно** *n.* asbestos fiber
**асбестовый** *adj.* asbestos
**асбестон** *m.* asbestone
**асбестоцементный** *adj.* asbestos-cement
**асбокартон** *m.* asbestos board
**асбоцемент** *m.* eternit, asbestos-cement
**асбоцементный** *adj.* eternit, asbestos-cement
**асбошифер** *m.* artificial slate
**асейсмический** *adj.* aseismic
**АСИ** *abbr.* (**амплитудный селектор импульсов**) pulse height selector
**асидометр** *m.* acidometer; **сифонный а.** acid syphon
**асимметр** *m.* asymmeter
**асимметрический** *adj.* asymmetric(al)
**асимметрично** *adv.* asymmetrically
**асимметричность** *f.* asymmetry
**асимметричный** *adj.* asymetric(al)
**асимметрия** *f.* asymmetry, dissymmetry, unbalance, asymmetrical effcet
**асмптота** *f.* asymptote

**асимптотически** *adv.* asymptotically
**асимптотический** *adj.* asymptotic
**асинхронический, асинхроничный, асинхронный** *adj.* asynchronous, non-synchronous; start-stop
**аскарел** *m.* ascarel
**асколой** *m.* Ascoloy
**АСП** *abbr.* (**авиационный сигнальный пост**) air signal station
**аспазил** *m.* aspasil
**аспект** *m.* aspect, appearance
**аспирационный** *adj.* aspiration, suction
**аспирация** *f.* aspiration
**ассигнование** *n.* assigning, assignment, allotment, appropriation
**ассигновать** *v.* to assign, to allot, to appropriate
**ассимилированный** *adj.* assimilated
**ассимилировать** *v.* to assimilate
**ассимилируемость** *f.* assimilability
**ассимилируемый** *adj.* assimilable
**ассимилятор** *m.* assimilator
**ассимиляторный** *adj.* assimilatory
**ассимиляция** *f.* assimilation
**ассистор** *m.* assistor
**ассонанс** *m.* assonance
**ассортимент** *m.* assortment, selection; set, outfit
**ассоциативность** *f.* associativity
**ассоциативный** *adj.* associative
**ассоциационный** *adj.* association
**ассоциация** *f.* association; combination; molecular association; **а. молодых звёзд** young association
**ассоциированный** *adj.* associated
**ассоциировать(ся)** *v.* to associate, to join, to unite
**аст** *abbr.* (**астрономический; астрономия**) astronomical; astronomy
**астазирование** *n.* astatization, astatizing
**астазированный** *adj.* unstable, astatized
**астазировать** *v.* to astatize
**астазирующий** *adj.* astatizing
**астатизированный** *adj.* astatized
**астатин** *m.* astatine, At, alabamine, helvetium
**астатический** *adj.* astatic, unstable, floating
**астатичность** *f.* astaticism
**астеносфера** *f.* astenosphere
**астеризм** *m.* asterism
**астероид** *m.* asteroid
**астигмат** *m.* astigmatic lens

**астигматизатор** *m.* astigmatizer
**астигматизм** *m.* astigmatism
**астигматический** *adj.* astigmatic
**Астон** Aston
**астонов** *poss. adj.* Aston's, (of) Aston
**Астр. обс** *abbr.* (**астрономическая обсерватория**) astronomic observatory
**астралон** *m.* astralon
**астральный** *adj.* astral
**астрионика** *f.* astrionics
**астробаллистический** *adj.* astroballistic
**астробиология** *f.* astrobiology
**астроботаника** *f.* astrobotany
**астрогравиметрический** *adj.* astrogravimetrical
**астрограмма** *f.* astrogram
**астрограф** *m.* astrograph, astrographic camera
**астрографический** *adj.* astrographic
**астрография** *f.* astrography
**астроинерционный** *adj.* inertial-celestial, astro-inertial, celestial-inertial
**астрокамера** *f.* astrocamera
**астроклимат** *m.* astroclimate
**астрокомпас** *m.* celestial compass, astrocompass
**астролит** *m.* astrolite
**астролог** *m.* astrologer
**астрология** *f.* astrology
**астролябический** *adj.* astrolabe
**астролябия** *f.* astrolabe, equiangular
**астрометеорология** *f.* astrometeorology
**астрометрический** *adj.* astrometric
**астрометрия** *f.* astrometry
**астрон** *m.* astron
**астронаведение** *n.* stellar guidance, celestial guidance, star-tracking
**астронавигация** *f.* astronavigation, celestial navigation
**астронавт** *m.* astronaut
**астронавтика** *f.* astronautics, astronomical navigation
**астронавтический** *adj.* astronautical
**астроном** *m.* astronomer
**астрономический** *adj.* astronomical, celestial
**астрономия** *f.* astronomy
**астрономо-геодезический** *adj.* astrogeodetic
**астроориентатор** *m.* astrotracker
**астроориентировка** *f.* celestial orientation
**астропеленгатор** *m.* celestial direction finder, astrotracker
**астрорадионавигация** *f.* astro radio-navigation
**астрорадиопеленгация** *f.* celestial radio tracking
**астроскоп** *m.* astroscope
**астроспектроскопия** *f.* astrospectroscopy
**астротрекер** *m.* star tracker
**астрофизик** *m.* astrophysicist
**астрофизика** *f.* astrophysics
**астрофотография** *f.* astrophotography
**астрофотокамера** *f.* astrophotocamera
**астрофотометр** *m.* astrophotometer
**астрофотометрия** *f.* astrophotometry
**астроэкология** *f.* astroecology
**асфальт** *m.* asphalt; **кабельный а.** cable compound
**асфальтирование** *n.* asphalting
**асфальтированный** *adj.* asphalted, bituminized
**асфальтировать** *v.* to asphalt
**асфальтовый** *adj.* asphalt(ic)
**асферический** *adj.* aspherical
**А-телефонистка** *f.* A-operator, subscriber's operator, outward operator
**ат.** *abbr.* 1 (**атмосфера техническая**) technical atmosphere; 2 (**атомный**) atomic
**ата.** *abbr.* (**атмосфера абсолютная**) absolute atmosphere
**атака** *f.* attack; **ложная а.** deceptive attack; **а. на кривой погони** pass pursuit attack; **а. на пересекающихся курсах** straight pass attack, beam attack
**атаковать** *v.* to attack, to assault
**атаксит** *m.* ataxite
**Атвуд** Atwood
**атекс-плита** *f.* Atex wood-wool slab
**ати.** *abbr.* (**атмосфера избыточная**) atmospheric excess pressure
**атипический** *adj.* atypical, irregular
**Аткинсон** Atkinson
**атлантический** *adj.* atlantic
**атлас** *m.* atlas
**атм.** *abbr.* (**атмосфера**) atmosphere
**атмалой** *m.* atmalloy
**атмидометр** *m.* atmidometer
**атмометр** *m.* atmometer, atmidometer
**атмометрия** *f.* atmometry
**атмораdiograf** *m.* atmoradiograph
**атмостойкий** *adj.* weatherproof
**атмосфера** *f.* atmosphere; air; **а. земного шара** total atmosphere; **избыточная а.** atmospheric excess pressure; **ощутимая а.** sensible atmosphere; **реальная а.** meteorological

atmosphere; **а. с барометрическим давлением в один бар** C.G.S. atmosphere; **а. с изменяющейся по высоте температурой** temperature-stratified atmosphere; **а. с конечной толщиной** finite atmosphere

**атмосферики** *pl.* atmospheric disturbances; strays; atmospherics

**атмосферический, атмосферичный, атмосферный** *adj.* atmospheric; air

**атмосферостойкий** *adj.* weatherproof

**атолл** *m.* atoll

**атом** *m.* atom; **а., лишённый внешних электронов** stripped atom; **а., меняющий направление движения** recoil atom

**атомарный** *adj.* atomic; nascent

**атомизатор** *m.* atomizer

**атомизация** *f.* atomization

**атомизм** *m.* atomism

**атомикрон** *m.* Atomichron

**атомистика** *f.* atomics, atomic theory

**атомистический** *adj.* atomic; atomistic

**атомихрон** *m.* Atomichron

**атомический** *adj.* atomic

**атомно-водородный** *adj.* atomic-hydrogen

**атомно-лучевой** *adj.* atomic-ray, atomic-beam

**атомность** *f.* atomicity, valence, valency

**атомный** *adj.* atomic; nuclear

**атональный** *adj.* atonal

**атрибут** *m.* attribute

**атрибутивный** *adj.* attributive

**атрогенный** *adj.* athrogene

**АТС** *abbr.* (**автоматическая телефонная станция**) Automatic Telephone Station, automatic exchange, automatic office; **блочная АТС** unit dial office, unit automatic exchange; **АТС внутренней связи, внутридомовая АТС** dial intercommunication system; **внутрирайонная АТС** community dial office, rural automatic exchange, САХ; **вспомогательная АТС с районным выключателем** discriminating satellite office; **домовая АТС** dial intercommunication system; **малая АТС, АТС малой ёмкости** unattended automatic office; **АТС с районным выключателем** switching-selector-repeater office; **частная АТС соединённая с городской** dial PBX

**аттенюатор** *m.* attenuator, attenuator pad; attenuation; expander; resistance network; **а. волн длиннее критической** beyond-cutoff attenuator; **волноводный а.** cutoff attenuator; **а. на активных сопротивлениях, а. на рассеивающих элементах** dissipative attenuator; **нерегулируемый а.** pad; **а. ножевого типа** flap attenuator; **ножевой а.** guillotine attenuator; **а. поршневого типа, постоянный а.** piston attenuator; **предельный а.** cutoff attenuator; **предельный а. с диафрагмовой связью** iris-coupled cutoff attenuator; **проволочный регулируемый а.** slidewire attenuator; **продольный а.** series equalizer; **а. тонмейстра** fade attenuator; **флажковый а.** vane attenuator

**аттенюаторный** *adj.* attenuator

**аттенюация** *f.* attenuation

**аттенюометр** *m.* attenuation meter, transmission measuring set

**атто** *n. indecl.* atto

**АУ** *abbr.* (**автономное управление**) independent control

**аугетрон** *m.* augetron

**аугзиометр** *m.* auxiometer

**аугментация** *f.* augmentation

**аудиограмма** *f.* audiogram

**аудиология** *f.* audiology

**аудиометр** *m.* audiometer, acuity meter; **а. по методу маскировки** screening audiometer; **а. с плавным изменением частоты** sweep-frequency audiometer

**аудиометрический** *adj.* audiometric

**аудиометрия** *f.* audiometry; **а. на речевом сигнале** speech audiometry; **а. по естественному уху** simple air-conduction audiometry; **а. с помощью записи звука речи** recorded voice audiometry

**аудион** *m.* audion, triode, electric tube

**аудитория** *f.* auditorium

**аудифон** *m.* audiphone

**ауксанометр** *m.* auxanometer

**ауксограф** *m.* auxograph

**ауксометр** *m.* auxiometer, auxometer

**аурипигмент** *m.* orpiment

**АУРС** *abbr.* (**авиационный управляемый реактивный снаряд**) guided aircraft missile

**аускультоскоп** *m.* auscultoscope

**аустенит** *m.* austenite

**аустенитный, аустенитовый** *adj.* austenite, austenitic
**ауто-** *prefix* auto-; automobile
**аутокатализ** *m.* autocatalysis
**аутооксидация** *f.* autooxidation
**Ауэр** Auer
**афазия** *f.* aphasia
**афелий** *m.* aphelion (point)
**афокальный** *adj.* afocal
**афферентный** *adj.* afferent
**аффикс** *m.* affix
**аффинный** *adj.* affine, related
**аффинор** *m.* affinor
**аффинорный** *adj.* affinor, dyadic
**аффинность** *f.* affinity
**АФХ** *abbr.* (**амплитудно-фазовая характеристика**) amplitude-phase characteristic
**ахромат** *m.* achromat, achromatic lens
**ахроматизация** *f.* achromatization
**ахроматизировать** *v.* to achromatize
**ахроматизм** *m.* achromaticity, degree of achromatic correction
**ахроматический** *adj.* achromatic, colorless, monochrome
**АЦВМ** *abbr.* (**автоматическая цифровая вычислительная машина**) automatic digital computer
**ацеотропический** *adj.* azeotropic
**ацетат** *m.* acetate
**ацетатный** *adj.* acetate
**ацетил** *m.* acetyl
**ацетилен** *m.* acetylene
**ацетиленовый** *adj.* acetylene
**ацетилцеллулоид** *m.* acetylcelluloid
**ацетилцеллюлоза** *f.* cellulose acetate, acetylcellulose
**ацетилцеллюлозный** *adj.* cellulose acetate, acetate
**ацетометр** *m.* acetometer
**ацетон** *m.* acetone
**ацидометр** *m.* acidometer; **сифонный а.** hydrometer syringe, acidometer
**ациклический** *adj.* acyclic, nonperiodic
**ач., а-ч., а/ч** *abbr.* (**ампер-час**) amp-hr, ampere-hour
**Ачесон** Acheson
**АЧР** *abbr.* (**автоматическая частотная разгрузка**) automatic frequency relief
**Ашкинасс** Aschkinass
**АЭР** *abbr.* 1 (**аэродром**) airport, airfield; 2 (**аэростат**) aerostat, balloon
**аэратор** *m.* aerator
**аэрация** *f.* aeration
**аэро-** *prefix* air-, aero-; airborne
**аэроакустика** *f.* aeroacoustics

**аэробаллистика** *f.* aeroballistics
**аэробиология** *f.* aerobiology
**аэробный** *adj.* aerobic
**аэрограмма** *f.* aerogram
**аэрограф** *m.* aerograph
**аэрография** *f.* aerography
**аэродинамика** *f.* aerodynamics; **а. дозвуковых скоростей** subsonic aerodynamics; **а. околозвуковых скоростей** transonic aerodynamics; **а. сверхзвуковых потоков** supersonics
**аэродинамически** *adv.* aerodynamically
**аэродинамический** *adj.* aerodynamic
**аэродром** *m.* airport, airfield, airdrome
**аэродромный** *adj.* airport, airfield, airdrome
**аэрозолоскоп** *m.* aerosoloscope
**аэрозоль** *m.* aerosol
**аэрозондирование** *n.* aerial sounding
**аэроизыскание** *n.* aerial survey
**аэрокамера** *f.* aero chamber
**аэроклиматология** *f.* aeroclimatology
**аэрокосмический** *adj.* aerospace
**аэролиния** *f.* airway
**аэролит** *m.* aerolite
**аэролитный** *adj.* aerolite, aerolyth
**«Аэролог»** *m.* Aerolog
**аэрологический** *adj.* aerological
**аэрология** *f.* aerology
**аэромагнитный** *adj.* aeromagnetic
**аэромагнитометр** *m.* aeromagnetometer
**аэромаяк** *m.* aeronautical beacon
**аэрометеорограф** *m.* aerometeorograph
**аэрометр** *m.* aerometer; **самопишущий а.** recording densimeter, aerometeorograph
**аэрометрический** *adj.* aerometric
**аэрометрия** *f.* aerometry
**аэронавигационный** *adj.* aeronautical, aeronavigation, air navigation
**аэронавигация** *f.* aerial navigation, aeronavigation, air navigation; **импульсная а.** aircraft pulse navigation
**аэронавт** *m.* aeronaut
**аэронавтика** *f.* aeronautics
**аэронавтический** *adj.* aeronautical
**аэрономический** *adj.* aeronomic
**аэрономия** *f.* aeronomy
**аэропауза** *f.* aeropause
**аэропорт** *m.* airport
**аэропортовый** *adj.* airport
**аэропрофилограф** *m.*, **радиоэлектронный а.** airborne profile recorder
**аэропсихограф** *m.* aeropsychograph
**аэрорадиометрический** *adj.* aeroradiometric

**аэроразведка** *f.* airborne prospecting
**аэросолоскоп** *m.* aerosoloscope
**аэростат** *m.* aerostat, balloon, airship; **привязной а.** captive balloon, tethered balloon
**аэростатика** *f.* aerostatics
**аэростатный** *adj.* aerostat, balloon, airship
**аэросфера** *f.* aerosphere
**аэросъёмка** *f.* aerial surveying
**аэротермодинамика** *f.* aerothermodynamics
**аэротермодинамический** *adj.* aerothermodynamic
**аэротермоупругость** *f.* aerothermoelasticity
**аэротранспортабельный** *adj.* air-transportable
**аэротрон** *m.* aerotron
**аэроупругий** *adj.* aeroelastic
**аэроупругость** *f.* aeroelasticity

**аэроф.** *abbr.* (**аэрофотометрия**) aerial photometry
**аэрофара** *f.* aerophare
**аэрофон** *m.* aerophone
**аэрофотоаппарат** *m.* aerial camera
**аэрофотограмметрический** *adj.* aerophotogrammetric
**аэрофотограмметрия** *f.* aerophotogrammetry
**аэрофотография** *f.* aerial photography
**аэрофотокамера** *f.* aerial camera
**аэрофотоснимок** *m.* aerial photograph
**аэрофотосъёмка** *f.* aerial photographic survey(ing)
**аэрофототриангуляция** *f.* aerial triangulation
**аэроэластичность** *f.* aeroelasticity
**аэрп.** *abbr.* (**аэропорт**) airport
**АЭС** *abbr.* (**атомная электростанция**) nuclear power station

# Б

**б** *abbr.* 1 (**бар**) bar; 2 (**бел**) bel; 3 (**бывший**) former; 4 *see* **бы**
**+б** (**плюс батареи**) B+
**—б** (**минус батареи**) B—
**Бабинэ** Babinet
**бабка** *f.* head, headstock (of lathe)
**бабочка** *f.* butterfly; cloverleaf; **кривая б.** butterfly curve
**багор** *m.* cant hook, pole hook; pike pole
**Бадер** Bader
**бадья** *f.* bucket, pail, tub; kettle
**база** *f.* base, basis, foundation, pedestal; baseline; array; **б. акустического измерения расстояния** sound-ranging base; **б. арктических радиолокационных станций обнаружения и наведения истребителей** aircraft control and warning type arctic base; **вторая б.** drain; **групповая б. гидроакустических преобразователей** transducer system, array; **б. импульса** frame; **б. кода** code length; **круговая компенсированная б.** delayed circular array; **многоэлементная б.** multispot array (of hydrophones); **направленная групповая б. гидроакустических преобразователей** directional transducer system; **б. приёмников(-гидрофонов) шумопеленгатора** hydrophone receiving array; **б. СОН** radar-directing point parallax; **электрически компенсированная групповая б. гидроакустических преобразователей** electrically trained transducer system
**базальный** *adj.* basal
**базальт** *m.* basalt
**базальтовый** *adj.* basalt, basaltic
**базилярный** *adj.* basilar
**базировать** *v.* to base (on)
**базис** *m.* basis, base, foundation, baseline
**базисный** *adj.* base, basis, foundation, basal; reference
**базовый** *adj.* basic, base, basal, reference

**байпас** *m.* by-pass, secondary line
**байпасный** *adj.* bypass-type
**байт** *m.* byte
**бак** *m.* tank, vat, tub, container, vessel; reservoir; cell; **мягкий б.** flexible cell; **б. питания (для маслонаполненных кабелей)** breather; **б. предварительного разрежения** interstage reservoir; **расширительный б. для масла** header tank, oil conservator; **б. с отдельными отсеками** cell-type tank; **топливный б. с наддувом** gas-pressurized tank; **уравнительный б.** surge tank; **форвакуумный б.** backing-up tank
**бакелизация** *f.* bakelization
**бакелизированный, бакелизованный** *adj.* bakelized, treated with bakelite
**бакелит** *m.* bakelite
**бакелитированный** *adj.* bakelized, treated with bakelite
**бакелитобумага** *f.* turbonit
**бакелитовый** *adj.* bakelite
**бакелитокартон** *m.* pertinax
**бакен** *m.* beacon, buoy; **маркировочный б.** channel marker
**баковый** *adj.* tank, vat, tub; pool
**бактерицид** *m.* bactericide
**бактерицидный** *adj.* bactericidal
**бактерия** *f.* bacterium
**балалайка** *f.* balalaika
**баланс** *m.* balance, equilibrium; balancing network, compensating network; **акустический б.** aural null; **б. для настройки** balancing network; **дуплексный б.** duplex artificial line; **«парящий» б.** floating balance; **б. по белому** white balance; **слуховой б.** aural null
**балансер** *m.* balancer
**балансир** *m.* bob, balance bob; beam, balance beam, equalizer, balancer (in a direction finder); cradle; balance-wheel; **поперечный б.** transverse equalizer
**балансирный** *adj.* bob, balance bob; beam, balance beam, equalizer
**балансирование** *n.* balancing

**балансированный** *adj.* balanced, in equilibrium

**балансировать, сбалансировать** *v.* to balance, to compensate; to imitate

**балансировка** *f.* balancing, balance, trim, balancing adjustment; register, in-register; **электронная динамическая б.** dynetric balancing

**балансировочный** *adj.* balancing

**балансирующий** *adj.* balancing

**балансный** *adj.* balancing, balance, compensation; balanced

**балка** *f.* beam, bar, girder, rafter; **двустенная б.** box girder; **двутавровая б.** I-beam; **двухопорная б.** simple beam; **б., зажатая на обоих концах** fixed-fixed beam; **консольная б.** cantilever bar; **консольная б., зажатая на одном и свободная на другом конце** fixed-free band; **неразрезная б.** continuous girder; **несущая б.** main beam; **однопролётная б.** simple beam; **б., опёртая на одном и зажатая на другом конце** fixed-hinged beam; **б., опёртая на одном и свободная на другом конце** hinged-free beam; **б., опёртая обоими концами** hinged-hinged beam; **б., работающая на изгиб** transverse beam; **решётчатая б.** frame girder, lattice girder; **составная б.** split beam; **сплошная б.** web girder; **тавровая б.** T-beam, T-girder

**балкон** *m.* balcony; **б. кросса** distributing frame balcony

**балл** *m.* mark; point, number; class

**Баллантин** Ballantine

**балласт** *m.* ballast, stabilizer; inert material

**балластёр** *m.* ballasting equipment

**балластирование** *n.* ballasting

**балластированный** *adj.* ballasted

**балластировать** *v.* to ballast

**балластировка** *f.* ballasting

**балластировочный** *adj.* ballasting

**балластный** *adj.* ballast

**балластовыгружатель** *m.* ballast unloader

**балластрон** *m.* ballastron

**баллистика** *f.* ballistics

**баллистический** *adj.* ballistic

**баллистокардиограмма** *f.* ballistocardiogram

**баллистокардиография** *f.* ballistocardiography

**баллон** *m.* bulb, shell, envelope; (gas) cylinder, carboy, demijohn, flask; **б. для кислоты** carboy; **б. для подъёма антенны** skyhook; **б. лампы** tube housing; **б. предварительного вакуума** backing-up tank; **б. системы воздухоснабжения** pressure tank of the air regeneration system; **стеклянный б.** glass envelope, glass bulb; carboy, demijohn; **сферический б.** storage sphere; **трубчатый б.** bulb

**баллонный** *adj.* bulb, shell, envelope

**баллоэлектричество** *n.* balloelectricity

**балочка** *f.* arm, small beam

**балсовый** *adj.* balsa

**бальзам** *m.* balsam, balm

**Бальмер** Balmer

**бальмеровский** *adj.* Balmer's, (of) Balmer

**бальный** *adj.* mark, point

**банановый** *adj.* banana

**бандаж** *m.* bandage, band, binding, binding band; belt; **б. ротора** retaining band

**бандажирование** *n.*, **бандажировка** *f.* bandaging, banding

**бандажный** *adj.* bandage, band, binding

**банка** *f.* jar, can, tin; cell bucket

**банковский** *adj.* bank

**баня** *f.* bath

**бар** *m.* bar (unit of pressure); bar, ledge

**барабан** *m.* drum; roll, cylinder; barrel; reel; kettledrum; **б. внешнего запоминающего устройства** file drum; **б. внешнего накопления с высокой плотностью записи** HD file drum; **б. вч блока настройки** turret; **б. для наматывания (проволоки)** coil form, wire drum; **б. для прокладки (кабеля)** paying-out drum; **желобчатый б. для передачи каната** winch head; **заводной б.** spring drum; spring barrel; **запоминающий б.** drum memory; **зубчатый б.** sprocket; **канатный б.** winding drum; **катушечный б.** turret-coil-set; **контактный б.** controller cylinder; **малый б.** side drum; **б. полётного времени** time-of-flight cylinder; **развёртывающий б.** scanning drum, drum scanner; **разматывающий б.** take-off drum; **самоустанавливающийся б.** self-indexing drum; **турецкий б.** bass drum; **числовой б.** figure pointing drum

**барабанный** *adj.* drum, barrel(-type); (ear)drum, tympanum, tympanic

**барабанчик** *m.* small drum; **зубчатый б.** sprocket

**баранка** *f.* doughnut

**баранок** *m.* jumble

**барашек** *m.* wing nut; kink (of wire)

**барашковый** *adj.* wing-nut; thumb (screw)

**барботаж** *m.* bubbling, splashing

**барботажный** *adj.* bubbling, bubble, splash

**барботёр** *m.* bubbler; diffuser

**барботировать** *v.* to bubble

**барботирующий** *adj.* bubbling

**бареттер** *m.* barretter, ballast resistor, ballast tube, resistance tube, current regulator tube; **железоводородный б.** (iron filament) ballast tube

**бареттерный** *adj.* barretter

**бариево-стронциево-оксидный** *adj.* barium-strontium-oxide

**бариевый** *adj.* barium

**барий** *m.* barium, Ba

**барико-циркуляционный** *adj.* barometric-circulation

**барированный** *adj.* barium-coated, bariated

**баритон** *m.* baritone (voice)

**барицентр** *m.* barycenter

**барицентрический** *adj.* barycentric

**барический** *adj.* baric, barometric, pressure

**бария** *f.* barye (unit of pressure)

**баркгаузеновский** *adj.* Barkhausen

**Баркхаузен** Barkhausen

**баркхаузеновский** *adj.* Barkhausen

**Барлоу** Barlow

**барн** *m.* barn (unit of nuclear cross section)

**Барнет** Barnett

**баро-** *prefix* baro-

**барограмма** *f.* barogram

**барограф** *m.* barograph, recording barometer; **малогабаритный б.** pocket barograph, small-clearance barograph

**барокамера** *f.* altitude chamber, pressure chamber

**бароклинность** *f.* baroclinity

**бароклинный** *adj.* baroclinic

**бароключатель** *m.* baroswitch (in a radio sounding balloon)

**барометр** *m.* barometer, barometer gage; **чашечный б.** cap barometer, cistern barometer

**барометрический** *adj.* barometric

**баромиль** *m.* baromil

**барорезистор** *m.* baroresistor

**барореле** *n. indecl.* pressure relay, air switch, baroswitch

**бароскоп** *m.* baroscope

**баростат** *m.* barostat

**баротермограф** *m.* barothermograph

**баротропия** *f.* barotropy

**баротропный** *adj.* barotropic

**бароциклонометр** *m.* barocyclometer

**барратрон** *m.* barratron

**баррел** *m.* barrel

**барретирование** *n.* current regulation

**барреттер** *see* **бареттер**

**Бартельс** Bartels

**Бартлетт** Bartlett

**бартлеттов** *poss. adj.* Bartlett, Bartlett's

**архан** *m.* barchan

**барьер** *m.* barrier, barrier layer; enclosure, bar, division; hill; **б. из радиогидроакустических буёв** sonobuoy barrier; **фоторелейный б., фотоэлектрический б.** light barrier (photocell installation)

**барьерный** *adj.* barrier, bar

**бас** *m.* bass voice, bass; **баритональный б.** bass-baritone

**басистый** *adj.* bass, deep

**басовый** *adj.* bass

**басок** *m.* bass-string; **б. скрипки** g string

**бассейн** *m.* basin, tank, reservoir, chamber, well; pool, pond; **головной б.** forebay; **напорный б. на канале** conduit forebay; **питательный б.** forebay; **портовый б.** dock basin

**бассейновый** *adj.* basin, tank, reservoir, chamber, well; pool, pond

**бассетгорн** *m.* basset horn, alto clarinet

**батал** *m.* batalum, barium-tantalum

**баталовый** *adj.* batalum

**батальон** *m.* battalion; **б. связи** signal battalion

**батарейка** *f.* flashlight battery, baby battery; **сеточная б., б. сеточного смешения** C-battery

**батарейная** *f. adj. decl.* battery room

**батарейный** *adj.* battery

**батарея** *f.* battery; set, bank, range; **сухие батареи** assembled batteries; **аккумуляторная б.** storage battery, secondary battery, accumulator (battery), wet battery, accumulator plant; **аккумуляторная б. для нужд**

тяги traction battery; **аккумуляторная б. для освещения** light battery; **аккумуляторная б. для подвижного состава** traction battery; **аккумуляторная б. с вольтодобавочной машиной** battery-booster set; **бортовая аккумуляторная б.** missile battery; **водоналивная б.** wet battery; **б. для кратковременной работы** open-circuit battery; **б. для питания несколько цепей** banked battery; **б. для сигнальных фонарей** railroad-lantern battery; **железо-никелевая аккумуляторная б.** Ni-Fe battery; **б. задаваемого напряжения** bias battery; **б. из наливных элементов** bottle battery; **б. из серебряно-кадмиевых элементов** silver-oxide-cadmium alkaline battery; **б. из элементов** trough battery; **карманная б.** pocket lamp battery, flash lamp battery; **монтированная б.** assembled battery; **нормальная б.** calibration battery; **б. оперативного тока** control battery; **б. покоя, пробельная б.** spacing battery; **б. продолжительного действия** closed circuit battery; **противодействующая б.** bucking supply; **б. работы** marking battery; **б. с параллельным соединением элементов** banked battery; **б. с подъёмными электродами** plunge battery; **б. сеточного смещения** C-battery, grid-bias battery, control-grid battery; **станционная б.** exchange battery; **трубчатая б.** flashlight battery; **уравновешенная б.** floating-trickle; **центральная б. ЦБ** exchange battery; **цилиндрическая б.** flashlight battery; **электрохимическая б.** secondary battery; thermal battery
**бати-** *prefix* bathy-, deep
**батиальный** *adj.* bathyal
**батикондактограф** *m.* bathyconductograph
**батиметр** *m.* attenuation meter, transmission measuring set
**батиметрический** *adj.* bathymetric
**батиметрия** *f.* bathymetry
**батипитометр** *m.* bathypitometer
**батискаф** *m.* bathyscaphe
**батист** *m.* batiste, cambric; **пропитанный б.** Kabak cloth
**батистовый** *adj.* batiste, cambric
**батисфера** *f.* bathysphere

**батитермограмма** *f.* bathythermogram
**батитермограф** *m.* bathythermograph
**бато-** *prefix* batho-, depth, height
**батолитовый** *adj.* batholitic
**батометр** *m.* water sampler, bathometer
**Баттерворт** Butterworth
**батчобработка** *f.* batch processing
**Баушингер** Bauschinger
**Бауэршмидт** Bauerschmidt
**бафтинг** *m.* buffeting
**бахрома** *f.* fringe, edging
**бахромочный** *adj.* fringe
**бахромчатость** *f.* fringing
**бахромчатый** *adj.* fringed
**бачок** *m.* can, small tank; **охлаждающий б.** cooling jacket
**башенный** *adj.* tower
**башмак** *m.* shoe; socket; clip; **полюсный б. с закруглёнными краями** beveled pole shoe, chamfered pole shoe; **рельсовый б. прижимного рельса** rail clip; **скользящий б.** slipper; **токоприёмный б.** trolley head; **токоснимательный б., б. токосъёмника** collector shoe; trolley shoe; **тормозной б.** brake head; **тормозной б. вагона** car stop; **тормозной б. замедлителя** retarder brake shoe
**башня** *f.* tower, turret; **аэропортовая диспетчерская б.** airport traffic control tower; **напорная б.** standpipe; **напорная б. с нижним присоединением** floating standpipe
**бдительность** *f.* vigilance, watchfulness, alert(ness)
**бдительный** *adj.* vigilant, watchful, alert
**беватрон** *m.* bevatron
**Беверидж** Beveridge
**бег** *m.* run, running, course
**бегать** *v.* to run
**бегающий** *adj.* running, moving; scanning
**Беггеров** Beggerow
**беглый** *adj.* running; rapid; brief; superficial
**беговой** *adj.* running; working
**бегом** *m.* begohm
**бегунок** *m.* race; jockey pulley; roller, (idler) runner; cursor
**бегучесть** *f.* running, traveling
**бегучий, бегущий** *adj.* running; moving; traveling, transient; scanning
**Бёдекер** Boedeker

**бедственный** *adj.* distress; disastrous
**бедствие** *n.* distress; emergency
**без** *prep. genit.* without, minus, less, free (of), non-
**без-** *prefix* in-, ir-, un-; non-; -less, -free; *see also* **бес-**
**безаберрационный** *adj.* free of aberration(s), aberration-free, non-aberrational, without aberration
**безаварийность** *f.* without failure, fail-safety
**безаварийный** *adj.* trouble-free, faultless, free of damage, trouble-proof, fail-safe
**безадресный** *adj.* no-address, zero-address
**безалаберность** *f.* inconsistency, lack of order
**безалаберный** *adj.* inconsistent, without system
**безантенный** *adj.* wireless, direct
**безарретирный** *adj.* non-stopping, non-arresting, unchecked
**безбатарейный** *adj.* non-battery, self-powered
**безбрежный** *adj.* unlimited, vast
**безбурный** *adj.* calm, stormless
**безвариантный** *adj.* invariant
**безваттный** *adj.* wattless, idle, reactive (component)
**безвернерьный** *adj.* vernierless
**безвестность** *f.* obscurity
**безвестный** *adj.* obscure
**безветренный** *adj.* windless, calm
**безветрие** *n.* calm(ness)
**безвибрационный** *adj.* vibrationless, vibration-free
**безвихревой** *adj.* irrotational; vortex-free, eddy-free; noncircuital
**безвлажный** *adj.* arid, dry
**безводность** *f.* aridity, dryness
**безводный** *adj.* arid, dry; anhydrous
**безводородный** *adj.* hydrogen-free
**безвозвратно** *adv.* irrevocably
**безвозвратный** *adj.* irrevocable, irreversible
**безвоздушный** *adj.* vacuous, vacuum, void, airless, evacuated
**безвредно** *adv.* harmlessly
**безвредность** *f.* harmlessness
**безвредный** *adj.* harmless, noninjurious
**безвременно** *adv.* prematurely
**безвременность** *f.* prematureness, untimeliness
**безвременный** *adj.* premature, untimely

**безвыгодность** *f.* disadvantage
**безвыгодный** *adj.* disadvantageous
**безвыходный** *adj.* hopeless, helpless
**безвязкостный** *adj.* nonviscous
**безгласный** *adj.* voiceless, mute
**безгранично** *adv.* infinitely
**безграничность** *f.* infinity, limitlessness
**безграничный** *adj.* infinite, limitless, unbounded (function)
**бездейственность** *f.* inactivity; ineffectiveness
**бездейственный** *adj.* inactive, inert, passive; idle, inoperative
**бездействие** *n.* inaction, inertia; idle, idling, nonoperation, inaction; standstill; failure; **в бездействии** out of commission
**бездействующий** *adj.* inactive; idle, nonoperative, inoperative, out-of-commission; quiescent
**бездеструктивный** *adj.* nondestructive
**бездефектный** *adj.* faultless, correct
**бездеятельно** *adv.* passively
**бездеятельность** *f.* inactivity, inertia
**бездеятельный** *adj.* inactive, passive, inoperative
**бездивергентный** *adj.* non-divergent
**бездиодный** *adj.* diodeless
**бездиректорный** *adj.* nondirector (-type)
**бездоказательный** *adj.* unproved, unsubstantiated
**бездонный** *adj.* fathomless, bottomless
**бездуговой** *adj.* non-arcing, arcless, arcfree
**бездымный** *adj.* smokeless
**безжелезный** *adj.* iron-free
**беззаветный** *adj.* unlimited, unrestrained, supreme; allowed
**беззаконный** *adj.* illegal
**беззарядный** *adj.* uncharged, neutral
**беззатейливый, беззатейный** *adj.* simple
**беззащитный** *adj.* unprotected
**беззвучный** *adj.* soundless, noiseless, quiet, silent, tuneless
**безизлучательный** *adj.* radiationless
**безимпендансный** *adj.* impedanceless, zero-impedance
**безиндуктивный, безиндукционный** *adj.* noninductive
**безинерционный** *adj.* inertialess; electronic
**безискристый, безискровой** *adj.* sparkless

**безламповый** *adj.* tubeless; solid-state
**безлично** *adv.* impersonally
**безличный** *adj.* impersonal
**безлопастный** *adj.* bladeless
**безлучевой** *adj.* unpronged; rayless
**безлюфтовой** *adj.* free from play, free from backlash
**безмасляный** *adj.* dry-type; oilless
**безмембранный** *adj.* no-diaphragm, diaphragmless
**безмен** *m.* spring balance, Roman balance, steelyard balance
**безмерно** *adv.* immeasurably
**безмерность** *f.* immensity, excess, immeasurableness
**безмерный** *adj.* immeasurable, immense, excessive; illimitable, infinite, boundless
**безмоментный** *adj.* free from momentum; zero-torque
**безмоторный** *adj.* motorless
**безнагрузочный** *adj.* standby, idle
**безнасосный** *adj.* pumpless
**безномерный** *adj.* numberless; innumerable; non-numerical
**безнулевый** *adj.* suppressed-zero
**безо** *see* **без**
**безобидный** *adj.* inoffensive, harmless; fair
**безоблачность** *f.* cloudlessness
**безоблачный** *adj.* cloudless
**безобмоточный** *adj.* inert (inductor)
**безобрывность** *f.* continuity
**безобрывный** *adj.* continuous
**безопасно** *adv.* safely, without danger
**безопасность** *f.* safety, security, reliability of operation; **б. при продольном изгибе** breaking safety; **б. эксплуатации** reliability of operation
**безопасный** *adj.* safe, secure, reliable; foolproof; safety; **б. от прикосновения** shockproofed
**безопорный** *adj.* cantilever
**безосколочный** *adj.* shatterproof, splinterproof
**безосновательно** *adv.* without basis, groundless
**безосновательный** *adj.* groundless, unfounded
**безостановочно** *adv.* continuously
**безостановочность** *f.* continuousness
**безостановочный** *adj.* continuous, ceaseless, uninterrupted, nonstop
**безотбойность** *f.* permanent (*tphny.*)
**безотбойный** *adj.* pressing, urgent

**безотказность** *f.* dependability, infallibility, no-failure operation
**безотказный** *adj.* dependable, infallible, reliable, trouble-free, trouble-proof, no-failure
**безоткатный** *adj.* recoilless
**безотлагательно** *adv.* immediately, urgently, without delay
**безотлагательность** *f.* urgency
**безотлагательный, безотложный** *adj.* urgent, pressing
**безотменный** *adj.* irrevocable
**безотносительно** *adv.* irrespectively (of), without reference (to)
**безотносительность** *f.* irrespectivity
**безотносительный** *adj.* irrespective (of), irrelative; absolute
**безотраженный** *adj.* reflectionless
**безотрывный** *adj.* without separation, non-detachable
**безошибочно** *adv.* correctly, without error
**безошибочность** *f.* infallibility; accuracy
**безошибочный** *adj.* infallible, correct, faultless; accurate, error-free
**безподогревный** *adj.* heaterless
**безрадиоактивный** *adj.* non-radioactive
**безраздельный** *adj.* inseparable
**безразлично** *adv.* irrespectively, with indifference
**безразличный** *adj.* irrespective, indifferent, neutral
**безразмерность** *f.* zero dimension
**безразмерный** *adj.* dimensionless, non-dimensional
**безразрядный** *adj.* nondischarge
**безрамный** *adj.* frameless
**безреактивный** *adj.* nonreactive
**безрегистровый** *adj.* non-register, non-director; direct
**безрезультатно** *adv.* without result, in vain
**безрезультатность** *f.* ineffectiveness, inefficiency, futility
**безрезультатный** *adj.* ineffective, inefficient, futile; inoperative
**безрельсовый** *adj.* railless, trackless
**безроторный** *adj.* irrotational
**безрупорный** *adj.* hornless
**безсердечниковый** *adj.* coreless
**безубыточный** *adj.* without loss
**безударный** *adj.* smooth, steady, shockless; hammerless
**безудержный** *adj.* unchecked, unrestrained

безуклонный *adj.* straight, undeviating
безукоризненный *adj.* irreproachable, faultless
безупречный *adj.* faultless, irreproachable; unobjectionable; flawless, sound; acceptable, tolerable
безусадочный *adj.* nonshrinking
безусловно *adv.* unconditionally, absolutely, positively
безусловность *f.* certainty
безусловный *adj.* unconditional, absolute, positive
безуспешно *adv.* unsuccessfully
безуспешность *f.* failure, lack of success
безуспешный *adj.* unsuccessful
безустанный *adj.* incessant, unremitting
безфлюсовый *adj.* fluxless
безшовный *adj.* seamless, jointless
безщёточный *adj.* brushless
безъёмкостный *adj.* non-capacitive
безъядерный *adj.* anuclear
безызвестность *f.* uncertainty, obscurity
безызлучательный *adj.* radiationless, non-radiative
безымпедансный *adj.* impedanceless
безындикаторный *adj.* scopeless
безындуктивный, безындукционный *adj.* non-inductive
безынерционный *adj.* inertialess, unslugged
безыскровой *adj.* sparkless, nonsparking
безысходный *adj.* perpetual, endless, continual; irreparable
безэлектродный *adj.* electrodeless
безэлектронный *adj.* electronless, non-electronic
безэховый *adj.* anechoic
Бейес Bayes
бейесовский *adj.* Bayes', (of) Bayes
Бек Beck
бекар *m.* natural (note)
бекацит *m.* beckacite
Бекеши Bekesy
Беккерель Becquerel
Бекман Beckmann
бекмановский *adj.* Beckmann's, (of) Beckmann
бел *m.* bel
«белее белого» whiter than white
Белен Belin
беление *n.* bleaching; электролитическое б. electrolitic bleaching

белёный *adj.* bleached
беленье *n.* bleaching
белеть, побелеть *v.* to become white, to whiten, to turn white
белизна *f.* whiteness, white
белила *pl.* whiting, white mineral pigment; баритовые б. barium sulphate; титановые б. titanium white
белильный *adj.* bleaching
беличий *adj.* squirrel
белковый *adj.* albuminous, albumin
Белл Bell
Беллини Bellini
Белов Below
беловатость *f.* whitishness
беловатый *adj.* whitish, albescent
белое *n. adj. decl.* white (color); б. в изображении picture white; равно-энергетическое б. E equal-energy white; *see also* белый
белокалильность *f.* incandescence
белокалильный *adj.* incandescent, glowing
белокурый *adj.* blizzard-head (*tv*); blond, fair
белонит *m.* acicular bismuth, needle ore, belonite
белый *adj.* white; pure; идеально б. equal-energy white; основной б. characteristic white; б.-чёрный white-to-black
белящий *adj.* bleaching
Бем Behm
бемоль *m.* flat (*mus.*)
Бемпорад Bemporad
Бен Ben
Бенардос Benardos
бензилцеллюлоза *f.* benzylcellulose
бензин *m.* gasoline; benzine
бензиновый *adj.* gasoline; benzine
бензиномер *m.* gasoline gauge, petrol (*or* gas) level gauge
бензино-электрический *adj.* gas-electric
бензол *m.* benzole, benzene
бензостойкий *adj.* gasoline-proof
бензоэлектрический *adj.* gas-electric
Бенинг Böning
бентонит *m.* bentonite
бентонитовый *adj.* bentonite, bentonitic
бентонический, бентонный *adj.* benthonic
бентос *m.* benthos
Бергман Bergman

**бергманский** *adj.* Bergman's, (of) Bergman
**Берд** Baird
**берег** *m.* shore, coast, seacoast, seashore
**береговой** *adj.* shore, coastal; harbor
**берегущий** *adj.* keeping, preserving
**бериллиевый** *adj.* beryllium
**бериллий** *m.* beryllium, Be
**беркелий, берклий** *m.* berkelium, Bk
**берковец** *m.* berkowetz (163.80 kg.)
**бернотар** *m.* bernotar polarizer
**бернский** *adj.* (of) Berne
**Бернулли** Bernoulli
**бернуллиев** *poss. adj.* Bernoulli's, (of) Bernoulli
**«беролит»** *m.* beroliet
**беррит** *m.* berrite (*insulator*)
**Бертело, Бертло** Berthelot
**Бертон** Berthon
**Бертран** Bertrand
**берущий** *adj.* taking; sharing
**бес-** *prefix*; *see* **без-**
**бескатушечный** *adj.* coilless, non-coil
**бескварцевый** *adj.* quartz-free, non-crystal
**бескислородный** *adj.* oxygen-free
**бескислотный** *adj.* free from acid, nonacid
**бесклапанный** *adj.* valveless
**бесключный** *adj.* keyless
**бесколебательный** *adj.* dead-beat
**бескоммутаторный** *adj.* without switch-board; commutatorless
**бескомпрессорный** *adj.* compressorless, non-compression
**бесконечно** *adv.* endlessly, infinitely, forever
**бесконечность** *f.* infinity, endlessness; infinity point
**бесконечный** *adj.* infinite, endless, everlasting, perpetual
**бесконтактный** *adj.* noncontact(ing), contactless
**бесконтекстный** *adj.* context-free
**бесконтрольно** *adv.* without control
**бесконтрольный** *adj.* uncontrolled
**бескрылый** *adj.* wingless
**беспаечный, беспайный** *adj.* solderless
**бесперебойность** *f.* steadiness, continuity (of service), serviceability
**бесперебойный** *adj.* steady, continuous, non-stop, uninterrupted, trouble-free
**беспеременно** *adv.* without change
**беспеременный** *adj.* changeless, unchangeable

**беспилотный** *adj.* pilotless, unmanned, robot
**беспламенный** *adj.* flameless
**бесплатно** *adv.* free of charge
**бесплатный** *adj.* free, free of charge
**бесповоротность** *f.* irreversibility
**бесповоротный** *adj.* irreversible
**бесподогревный** *adj.* heaterless
**бесподшипниковый** *adj.* without a bearing, without a bushing
**беспоисковый** *adj.* non-searching, non-scanning, non-rotating
**беспокойный** *adj.* disturbed, turbulent
**бесполезно** *adv.* in vain, uselessly
**бесполезность** *f.* uselessness; ineffectiveness
**бесполезный** *adj.* useless, inutile; ineffective, inefficient
**бесполюсный** *adj.* poleless
**беспористость** *f.* density
**беспористый** *adj.* dense, nonporous
**беспорочный** *adj.* faultless
**беспорядок** *m.* disorder, confusion; random, scrambling, inversion
**беспорядочно** *adv.* in disorder, in confusion; at random, randomly
**беспорядочность** *f.* disorderliness, randomness
**беспорядочный** *adj.* disorderly, disordered, haphazard, incoherent; random, irregular
**беспосадочный** *adj.* non-stop (flight)
**беспотерный** *adj.* loss-free, lossless, dispersionless, no-leakage, nondissipative
**беспошлинно** *adv.* duty-free
**беспредельно** *adv.* infinitely
**беспредельность** *f.* infinity
**беспредельный** *adj.* infinite, illimitable, limitless; unbounded
**беспрепятственно** *adv.* without hindrance, free (to)
**беспрепятственный** *adj.* unobstructed, free, unimpeded, unchecked
**беспрерывно** *adv.* continuously, uninterruptedly
**беспрерывность** *f.* continuance, continuity, continuousness
**беспрерывный** *adj.* continuous, uninterrupted, ceaseless, incessant
**беспрестанный** *adj.* continual, incessant
**беспримерно** *adv.* without example
**беспримерный** *adj.* unexampled, unparalleled, unmatched

**беспримесный** *adj.* pure, unalloyed, uncontaminated, i-type; intrinsic(al)

**беспристрастно** *adv.* without bias, impartially

**беспристрастный** *adj.* unbiased, impartial

**бесприцельный** *adj.* aimless, random

**беспроводный, беспроволочный** *adj.* wireless

**беспыльный** *adj.* dustless, dust-free

**бессальниковый** *adj.* packless

**бессвязность** *f.* incoherence, inconsistency

**бессвязный** *adj.* incoherent, inconsistent, disconnected

**бесселев** *poss. adj.* Besselian, (of) Bessel

**Бессель** Bessel

**Бессемер** Bessemer

**бессемеров** *poss. adj.* Bessemer('s)

**бессемерование** *n.* Bessemer process

**бессемеровать** *v.* to bessemerize

**бессемеровский** *adj.* (of) Bessemer

**бессердечниковый** *adj.* coreless

**бессеточный** *adj.* gridless

**бессистемно** *adv.* at random, without system, unsystematically

**бессистемность** *f.* unsystematic character, absence of pattern; chance

**бессистемный** *adj.* unsystematic, unmethodical, haphazard

**бесскачковый** *adj.* shock-free

**бесскобочный** *adj.* parenthesis-free

**бесследный** *adj.* traceless, without a trace

**бессменно** *adv.* at a stretch, continuously, without change

**бессменность** *f.* permanency, changelessness

**бессменный** *adj.* permanent, changeless, set, fixed; continuous

**бессмысленность** *f.* senselessness, nonsense, absurdity; vacancy, inanity

**бессмысленный** *adj.* senseless, nonsensical, absurd; meaningless

**бесснежный** *adj.* snowless; bare

**бессодержательный** *adj.* empty

**бесстебельный** *adj.* stemless

**бесстолкновительный** *adj.* collisionless

**бесструктурный** *adj.* formless, structureless

**бесступенчато-регулируемый** *adj.* continuously variable

**бесступенчатый** *adj.* stepless, without stages

**бессчётный** *adj.* innumerable, countless

**бестеневой** *adj.* shadowless

**бестигельный** *adj.* crucibleless

**бестолчковый** *adj.* jerkless

**бестоновый** *adj.* unpitched (sound)

**бестрансформаторный** *adj.* transformerless

**бесфланцевый** *adj.* bald, flush; flangeless

**бесфоновый** *adj.* humless, hum-free

**бесформенно** *adv.* without form, without shape

**бесформенность** *f.* formlessness, shapelessness

**бесформенный** *adj.* formless, shapeless

**бесхвостный** *adj.* tailless; ecaudate

**бесхлопотный** *adj.* trouble-free

**бесцветность** *f.* colorlessness, achromatism

**бесцветный** *adj.* colorless, achromatic; blank; no top (*acoust.*)

**бесцельно** *adv.* at random, aimlessly

**бесцельность** *f.* aimlessness; idleness

**бесцельный** *adj.* aimless, haphazard, purposeless; idle

**бесцентровый** *adj.* centerless

**бесциркуляционный** *adj.* without circulation

**бесцокольный** *adj.* baseless

**бесчисленность** *f.* innumerability, numberlessness

**бесчисленный** *adj.* countless, innumerable, numberless; infinite

**бесшнурный, бесшнуровый** *adj.* cordless

**бесшовный** *adj.* seamless, jointless

**бесшумность** *f.* quietness

**бесшумный** *adj.* noiseless, no-noise, silent, quiet, mute, muting, dumb; inaudible

**бесщелевой** *adj.* slitless

**бесщёточный** *adj.* brushless

**бета** *f.* beta; **б. созвездия Лиры** Beta Lyrae stars

**бетаактивный** *adj.* beta-active

**беталучевой** *adj.* beta-ray

**бетаметр** *m.* betameter

**бетарадиоактивный** *adj.* beta-radioactive

**бетатопический** *adj.* betatopic

**бетатрон** *m.* betatron, induction accelerator, rheotron

**бетатронный** *adj.* betatron

**бетаустойчивый** *adj.* beta-stable

**Бете** Bethe

**бетон** *m.* concrete; **предварительно напряжённый б.** prestressed concrete
**бетонирование** *n.* concreting
**бетонированный** *adj.* concrete, concreted, treated with concrete
**бетонировать** *v.* to concrete, to treat with concrete
**бетонный** *adj.* concrete; mortar
**Бехштейн** Bechstein
**беч** *m.* batch, charge
**бечёвка** *f.* string, cord, binder, twine
**би-** *prefix* bi-, di-
**библиотека** *f.* library; **б. лент** library of tapes; **б. программ** program library, routine library
**библиотечный** *adj.* library
**бивариант** *m.* bivariant
**бивариантный** *adj.* bivariant
**бивибратор** *m.* double resonator, double dipole, folded dipole, double doublet
**биграмма** *f.* digram
**бидон** *m.* can, container, vessel
**биение** *n.* beating, beat, pulse, pulsation, throb, throbbing; surge, heterodyne; wobble; **нулевые биения** dead beats; **б. высокочастотных сигналов** continuous wave interference
**биквадрат** *m.* biquadrate
**биквадратный** *adj.* biquadratic
**биркварц** *m.* biquartz
**биконический** *adj.* biconical, double-cone
**биконусный** *adj.* double-cone
**бикристалл** *m.* bicrystal
**бикрон** *m.* bicron
**билатеральный** *adj.* bilateral
**билет** *m.* ticket, pass
**билинейный** *adj.* bilinear
**биллион** *m.* milliard; billion; **б. электронвольт** billion electron volt
**билогический** *adj.* bilogical
**бильдаппарат** *m.* picture transmitter, facsimile transmitter, telewriter, teleautograph
**бильдпередача** *f.* transmission of photographs
**бильдсвязь** *f.* phototelegraphy
**бильдтелеграмма** *f.* wirephoto, phototelegram, facsimile
**бильдтелеграф** *m.* facsimile telegraph, wirephoto
**бильдтелеграфия** *f.* phototelegraphy, facsimile telegraphy, facsimile transmission, picture transmission
**бимагматический** *adj.* bimagmatic

**биметалл** *m.* bimetal, compound metal
**биметаллизм** *m.* bimetallism
**биметаллический** *adj.* bimetallic
**бимодальный** *adj.* bimodal
**бимолекулярный** *adj.* bimolecular
**биморфный** *adj.* bimorph
**бинант** *m.* semicircular segment of Dolezalek electrometer
**бинарно-кодированный** *adj.* binary-coded
**бинарный** *adj.* binary
**бинауральность** *f.* binaural effect
**бинауральный** *adj.* binaural
**биндер** *m.* binder
**бинистор** *m.* binistor
**бинод** *m.* binode
**бинодаль** *f.* binodal
**бинокль** *m.* binocular(s); spectacles
**бинокулярный** *adj.* binocular
**бином** *m.* binomial
**биномиальный, биноминальный, биномный** *adj.* binomial, binominal
**бинормаль** *f.* binormal
**биоакустика** *f.* bioacoustics
**биоастронавтика** *f.* bioastronautics
**биогенный** *adj.* biogenic
**биогеограф** *m.* biogeographer
**биогеографический** *adj.* biogeographic
**биогеография** *f.* biogeography
**биокибернетика** *f.* biocybernetics
**биоклиматология** *f.* bioclimatology
**биокосмонавтика** *f.* bioastronautics
**биолог** *m.* biologist
**биологически** *adv.* biologically
**биологический** *adj.* biologic(al)
**биология** *f.* biology; **космическая б.** bioastronautics, space biology; **математическая б.** biomathematics
**биолюминесценция** *f.* bioluminescence
**биомагнетизм** *m.* biomagnetism
**биоматематика** *f.* biomathematics
**биомедицина** *f.* biomedicine
**биомедицинский** *adj.* biomedical
**биометрика** *f.* biometrics
**биометрический** *adj.* biometric
**биометрия** *f.* biometry
**биомеханика** *f.* biomechanics
**бионика** *f.* bionics
**бионический** *adj.* bionic
**биономия** *f.* bionomy, bionomics
**биоразъём** *m.* biopatch
**Био-Савар** Biot-Savart
**биосинтез** *m.* biosynthesis
**биосистема** *f.* biosystem
**биостатика** *f.* biostatics
**биостатистика** *f.* biostatistics

**биосфера** *f.* biosphere
**биотелеметрия** *f.* biotelemetry
**биотехника** *f.* bioengineering, biotechnology
**биотит** *m.* biotite, black mica
**биотический** *adj.* biotic; biological
**биоток** *m.* nerve-action current, biocurrent
**биотопливо** *n.* biofuel
**биотрон** *m.* biotron
**биофизика** *f.* biophysics
**биофотический** *adj.* biophotic
**биохимик** *m.* biochemist
**биохимический** *adj.* biochemical
**биохимия** *f.* biochemistry
**биоэлектрический** *adj.* bio-electric
**биоэлектроника** *f.* bionics, bioelectronics
**«бипак»** *m.* bi-pack
**бипирамида** *f.* bipyramid
**бипирамидальный** *adj.* bipyramidal
**бипланарный** *adj.* biplanar
**биполярность** *f.* bipolarity
**биполярный** *adj.* bipolar; two-way
**бипризма** *f.* biprism, optical cube
**бипятерный** *adj.* biquinary
**Бир** Beer
**Бирг** Birge
**биржа** *f.* exchange, stock exchange
**биржевой** *adj.* exchange, stock exchange
**бирка** *f.* tag, ticket, nameplate; score, rating plate; **контрольная б.** test card
**бирмингамский** *adj.* Birmingham
**биротативный** *adj.* birotary
**бирюза** *f.* turquoise
**бирюзовый** *adj.* turquoise
**бисектриса** *see* **биссектриса**
**бисекция** *f.* bisection
**бисер** *m.* (glass) beads
**бисериальный** *adj.* biserial
**бисерный** *adj.* bead, beaded
**бисквит** *m.* biscuit; **б. квадратного электрометра** vane
**бисквитный** *adj.* biscuit
**бисманол** *m.* bismanol (*alloy*)
**биспираль** *f.* double spiral, coiled coil, coiled-coil filament, double helix
**биспиральный** *adj.* coiled-coil, double-spiral
**биссектриса** *f.* bisector, bisecting line; bisectrix
**бистабильный** *adj.* bistable
**бистатический** *adj.* bistatic
**бисульфат** *m.* bisulfate

**бисфеноид** *m.* bisphenoid
**бисфеноидальный** *adj.* bisphenoidal
**бит** *m.* bit, information bit; **последовательный б. за битом** bit by bit; **потерянный б.** dropped bit
**битовый** *adj.* bit
**Биттер** Bitter
**биттит** *m.* bittite
**битум** *m.* bitumen, asphalt
**битумизация** *f.* asphalt grouting, hot asphalt grouting
**битумизированный** *adj.* bituminized
**битуминизация** *f.* bituminization
**битуминозный, битумный** *adj.* bituminous, asphalt(ic)
**бить, побить** *v.* to strike, to beat; to wobble, to play; to hammer; to break
**биться** *v.* to beat, to pulsate, to pulse
**бифилярный** *adj.* bifilar, double-wound
**бифокальный** *adj.* bifocal
**бифуркальный** *adj.* bifurcated
**бифуркация** *f.* bifurcation, branching
**бихромат** *m.* dichromate
**бихроматический** *adj.* dichromate, dichromatic
**бициклический** *adj.* bicyclic
**бициркулярный** *adj.* bicircular
**биэллиптический** *adj.* bi-elliptical
**БКГ** *abbr.* (**баллистокардиограмма**) ballistocardiogram
**б-коммутатор** *m.* B-switchboard, B-position, incoming (trunk) position
**БКШ-теория** *f.* Bardeen-Cooper-Schrieffer theory
**Блавье** Blavier
**благодаря** *prep. dat.* owing (to), thanks (to)
**благонадежный** *adj.* reliable, dependable
**благополучный** *adj.* safe; satisfactory
**благоприятно** *adv.* favorably
**благоприятный** *adj.* favorable, opportune; contributory; **наиболее б.** optimum
**благоприятствовать** *v.* to favor, to foster
**благородный** *adj.* noble; precious
**благоуспешный** *adj.* successful
**бланк** *m.* blank, form; recording blank; recording disk; **буквенный б.** letter blank; **запросный б.** request-sheet, Rq-sheet
**бланкирование** *n.* blanking; beam suppression; **б. луча** beam suppression; **б. полукадров** frame suppression
**бланкирующий** *adj.* blanking

**бланковый** *adj.* blank, form

**бланфикс** *m.* blanc fixe, permanent white

**Блатон** Blaton

**Блаттгаллер** Blatthaller

**«блаттгаллер»** *m.* Blatthaller loudspeaker

**бледнеть, побледнеть** *v.* to grow pale, to turn pale

**бледный** *adj.* pale; faint, weak

**бленда** *f.* lens hood, lens screen, flag, gobo, blind; diaphragm; blende, asphalerite; **б. на объекте** lens hood; **экранирующая б.** light shield, viewing hood

**бленкер** *m.* indicator, drop, automatic drop, target, indicating disk, annunciator; **вызывной б.** drop-indicator shutter, calling indicator

**блеск** *m.* brilliance, brilliancy, brightness, lustre, glitter, flash, flashing; glance; gloss; **б. в видимых лучах** optical brightness; **железный б.** hemetite; **б. звукозаписи** bounce brilliancy; **ослепительный б.** glare; **свинцовый б.** galena, lead glance, lead sulphide

**блескомер** *m.* glossmeter

**блескообразователь** *m.* brightener

**блёскость** *f.* glare, brilliance, lustre

**блестеть, блеснуть** *v.* to flash, to shine, to glitter, to glint, to glisten

**блестящий** *adj.* bright, brilliant, shiny, luminous, sparkling, glossy

**блеф** *m.* bluff

**ближайший** *superl. of* **близкий** nearest, near, next, closest, immediate

**ближний** *adj.* near, close, neighboring; short-range, short-distance

**близ** *prep. genit.* near, close (to), at hand, in the vicinity (of)

**близкий** *adj.* near, close, close by; imminent

**близко** *adv.* near, close, closely, near by

**близкодействие** *n.* proximity effect, close-range action

**близкодействующий** *adj.* short-range

**близлежащий** *adj.* nearby, adjacent, near, neighboring

**близмеридиальный** *adj.* ex-meridial

**близорукость** *f.* myopia

**близостный** *adj.* proximity, proximate, near, close

**близость** *f.* nearness, closeness, proximity, vicinity

**блик** *m.* high-light, spot of light, patch of light; optical pattern; flare; **б. освещаемой пластинки** christmas-tree pattern; *pl.* **блики** flare light, highlights; **блуждающие блики** ghosts, ghost images; **блики в объективе** lens flare

**бликовать** *v.* to flash, to lighten

**блинкер** *m.* blinker, visual indicator (signal), annunciator

**блинкерный** *adj.* blinker

**блинк-компаратор** *m.* blink-comparator

**блинк-микроскоп** *m.* blink-microscope

**блинчатый** *adj.* cake, cake-like

**блистание** *n.* shining, glittering

**блок** *m.* block, pulley; unit, assembly, package, set, pack; turret; organ, element, component; box tackle; slug; *see also* **тали; блоки вычислительной машины** computer components; **блоки для натяжки антенны** antenna snatch block, halyard, tackle rope; **блоки для натяжения проводов** regulation table; **натяжные блоки** block and tackle; **основные блоки вычислительной машины** central computer; **сложные блоки** block and tackle; **блоки с лягушкой** draw-tongs; **б. автоматического контроля** automatic check; **б. автоматической настройки нуля** automatic tare unit; **б. агрегата** machine bay; **б. аккумуляторных пластин** sub-assembly of positive and negative accumulator plates; **б. анероидных коробок** sylphon; **б. аппаратуры самонаведения** homing guidance package; **арифметический б. последовательного действия** serial arithmetic unit; **барабанный вч б. настройки** turret tuner; **бортовой автономный б.** built-in self-reacting device; **вводящий б.** assertion box; **вентильный б.** gate unit; **взаимозаменяемый стандартизованный б.** modulator unit; **б. вибропреобразователя** vibrator power-pack; **б. вращающегося коммутатора** rotary sampling switch; **вспомогательный б.** slave unit; **б. выборки данных** reader, data selection unit; **б. выдачи данных** data unit; **вынесенный б. видеоусилителя** remote video unit; **выносной б. передачи данных** remote data box; **б., вырабатывающий гасящие** (*or* **запирающие**) импульсы black-out unit; **б. выработки упреждённого азимута** future azimuth mechanism; **выходной б. накопителя** output block; **герметизированный б.** unit under pressure, package; **герметизованный б.** potted unit, sealed unit; **б. головок** head assembly; **готовый б.** building block; **б. дальномера** radar range unit; **б.-датчик высоты** altitude data unit; **б. для кабельной магистрали** trunk line conduit; **б. для натягивания проводов** snatch block, block and tackle, pulley block; **б. для передачи дальности с индикатора** range transmission unit; **задающий б. осуществления ведомого режима** master slaving unit; **б. запирающих сигналов** black-out unit; **б. записания (информации) с (магнитным) барабаном** drum storage unit: **б. из трёх конденсаторов переменной ёмкости** three-gang (tuning) condenser; **измерительный б. гидролокатора** sonar test set; **б. имитации гистерезиса** hysteresis unit; **б. имитации давления в функции высоты** height-pressure unit; **б. имитации зоны нечувствительности** inert zone unit, backlash unit; **б. импульсного генератора** pulser box; **б. индекс-регистров** B-box, B-line; **б. информации** message; **б. кадровой радиотелеметрической записи** histogram; **канатный б.** pulley, rope wheel; **б. коммутации выходов синхрогенераторов** sync generator changeover switch; **конденсаторный б.** gang capacitor; **б. конденсаторов переменной ёмкости** gang(ed) condenser; gang tuning capacitor; **б. конденсаторов постоянной ёмкости в корпусе** encased fixed condenser; **контрольно-планшетный б.** monitor charting unit; **б. контроля отработки координат цели** target control assembly; **б. манипулятора** keying unit; **б.-механизм** block and lock; **многоотверстный б. канализации** multiple duct; **б. моделирования (имитации) параметров радиолокатора** radar parameter unit; **б. моделирующей установки** computing element; **б., моделирующий аппаратуру управления** guidance and control equip-

ment simulator; **б. модификаторов** B-box, B-line; **модульный б. оптической системы управления** optical guidance module; **б. модулятора** keying unit; **б. на штепселях** pluggable unit; **неразборный б.** solid block, "throw away" unit; **одноканальный б. из однородного материала** monolithic conduit; **б. определения стороны** sign-determining unit, sign-sensing unit; **б. отклоняющей системы** yoke assembly; **б. отработки упреждённых координат** prediction unit; **б. отсчёта** read-out unit; **первый задающий б.** first driver; **б.-перегон** block; **б. передачи данных (на транзисторах)** data transmission package; **б. перемены знака** phase inverter, phase-inverting amplifier; **пересчётный б.** matrix unit; **б. питания** power pack; power unit, supply unit, AB power pack, power supply unit, power strip, power stage; **б. питания взвода пусковых установок** section power cabinet; **б. питания накала, анода и сетки** ABC power unit; **поворотный б.** curve sheave; **б. подъёма антенны** aerial elevation pawl; **подъёмный б.** davit; **полный б. питания** AB power pack; **предварительно собранный б.** subassembly; **б. предупреждения об опасности столкновения** proximity warner; **б. преобразования видеосигналов** colorplexer; **б. приёма и переработки входной информации** input store; **б. проверки исполнения команд управления** reference guidance section; **б. промежуточной частоты** I.F. section (or amplifier); **б. развёртки по звуковой частоте** audio scanner; **б. разделения ступеней** separation gear; **ранцевый б.** pack unit; **б. регистров переадресации** B-box, B-line; **б. регулировки нелинейности** gamma control unit; **б.-резина** cushion sheet; **б. реле системы управления** recovery relay box; **б. с приводом от сельсина** synchro-driven mount; **б. светоделения** dichroic block; **секционированный б.** bin unit; **селекторный б.** strobe unit; **б. следящих систем** servomechanism unit; **б. слежения по дальности** range follow-up unit; **сложный б.**

pulley block, tackle; **сменный б.** plug-in unit, plug-in package, pluggable package; **б. сменных катушек** plug-in coil(s); **соединительный б.** conduit coupling (in a duct system); **б. сопровождения по азимуту и углу места** azimuth and elevation tracking unit; **б. сопряжения** automatic fire-control coupler; **б. сопряжения радиолокационного прицела с автопилотом** fire-control coupler; **б. сохранения** recovery package; **б. сочетания координат** coordinate multiplexing circuit; **б. сочетания и накопления данных** coordinate multiplexing storage circuits; **б.-станция** isolated generating plant; **стоечный б.** rack-mounted unit; **б. телеуправления** beep-box; **б. тормозных двигателей** retropackage; **б. указателя кругового обзора** plan position indicator unit; **б. универсальных функциональных преобразователей** universal function unit; **б. управления** function unit, control unit, control block, control mechanism; **б. управления взводом пусковых установок** section control cabinet, section control assembly; **б. управления элеронами** aileron servo; **б. усиления промежуточной частоты с расстроенными контурами** staggertuned intermediate frequency strip; **б. усилителя канала крена** roll amplifier unit; **б. ускорителей** cluster of boosts; **б.-установка** isolated plant; **б. формирования сигнала цветного изображения** color coder, colorplexer; **фотоэлектрический считывающий б.** photoelectric reader, photoelectric tape reader; **б. цепочки сквозного переноса** ripple through carry unit; **цифровой б. связи с автопилотом** digital auto-pilot coupler; **цифровой (стандартный) б.** digital module; **б. чисел в запоминающем устройстве** storage block; **широкополосный вч б. настройки** broadband tuner

**блокгауз** *m.* blockhouse

**блокинг** *m.* blocking, squegging

**блокинг-генератор** *m.* blocking oscillator, squegging oscillator, self-pushing oscillator; **ждущий б.-г., задержанный б.-г.** biased blocking oscillator; **запертый б.-г., заторможённый б.-г.** single-shot blocking oscillator; **б.-г. на плоскостных триодах** junction transistor blocking oscillator; **одноразовый б.-г.** single-swing blocking oscillator (*or* generator); **б.-г. с самовозбуждением** free-running blocking generator; **синхронизируемый б.-г.** triggered blocking generator; **синхронизованный б.-г.** monostable blocking oscillator; **управляемый б.-г.** triggered blocking generator

**блокинг-процесс** *m.* squegging

**блокинг-трансформатор** *m.* blocking-oscillator transformer

**блокирование** *n.* blocking; paralysis; **б. обратного движения** reversal prevention

**блокированный** *adj.* blocked, obstructed; locked, jammed, interlocked

**блокировать** *v.* to block, to obstruct; to lock, to interlock; to guard; to stop, to shut; to put out of service; **б. в разомкнутом состоянии** to lock open

**блокировка** *f.* blocking, obstruction; locking, interlocking, holding; plugging up, barring; holding-on, seizing; holding device, locking device; eliminator, suppressor; suspension; **с блокировкой фазы** in locked phase, phase-locked; **б. вч** carrier (channel) blocking; **зависимая б.** block proving; **б. луча** ray-locking device; **неавтоматическая б.** manual block system, lock and block; **обратная б. вызова, б. от встречного вызова** bar for originating call; **б. передатчиков** alarm-box rewinding interlock; **полуавтоматическая б.** manually controlled block system; **б. против случайного стирания** erase interlock; **путевая б.** block signaling; **б. цепи вызвавшего абонента** number checking arrangement; **б. шарнира** gimbal lock

**блокировочный** *adj.* block(ing); holding, locking, interlocking; lock-out

**блокируемый** *adj.* blocking; interlocking

**блокирующий** *adj.* block, blocking, stop; holding, locking, interlocking; disabling; suppressing; plugging-up

**блокконтакт** *m.* auxiliary contact

**блоковый** *adj.* block, pulley

**блокстанция** *f.* block system

**блок-схема** *f*. block diagram, logic(al) diagram, functional diagram; **б.-с. программы** flow chart, flow diagram

**блок-участок** *m*. block section; **абсолютно-разрешительный б.-у.** absolute-permissive block

**Блондель** Blondel

**Блох** Bloch

**блочный** *adj*. block, unit-type, unitized; lumped; plug-in

**блуждание** *n*. wandering, straying, walk; **неограниченное случайное б.** unrestricted random walk

**блуждать, проблуждать** *v*. to wander, to stray

**блуждающий** *adj*. wandering, stray (-ing), erratic; transient

**блюм** *m*. bloom

**блюминг** *m*. blooming, blooming mill

**бляха** *f*. metal plate; badge

**б-место** *n*. B-position, incoming (trunk) position

**бобина** *f*. reel, spool, bobbin; (ignition) coil; **б. для наматывания киноплёнки** film spool; **подающая б.** feed reel, feed spool

**бобиновый** *adj*. reel, spool, bobbin

**бобровый** *adj*. beaver

**бобышка** *f*. spool; boss, lug

**богатый** *adj*. rich; abundant

**бод** *m*. baud

**Боде** Bode

**Бодо** Baudot

**боевой** *adj*. combat, fighter; war

**боеголовка** *f*. warhead; **б. с атомным зарядом** atomic warhead

**боезапас** *m*., **боезапасы** *pl*. ammunition

**Бозе** Bose

**бозон** *m*. boson

**бой** *m*. cullet, broken glass; combat, battle; **стеклянный б.** cullet; **электродный б.** broken electrode

**Бойз** Boys

**Бойль** Boyle

**бок** *m*. side, flank; wall; **бок о бок** abreast, alongside, side by side

**Бокли** Buckley

**боковой** *adj*. side, lateral; sideward, sidelong; marginal, secondary, accessory; adjacent

**боком** *adv*. sideways, sidewise

**бокорезцы-плоскогубцы** *pl*. diagonal cutting pliers

**бокс** *m*. box; blimp; dividing box; **испытательный б.** test cell, test pit;

**кабельный б. с маслом** oil-filled pothead

**болванка** *f*. block, mold, slug, pig; rod, bar, ingot

**Болдуин** Baldwin

**болевой** *adj*. pain, painful

**боле(е)** *adv*. more; -er (as compar. ending) e.g. **б. высокий** higher; **б. высоких типов** higher mode

**болезненный** *adj*. painful

**болезнь** *f*. sickness, illness, disease; **воздушная б.** aero-embolism; **кессонная б.** dysbarism, aeremia

**болид** *m*. fire-ball, bolide

**бологрaмма** *f*. bologram

**болограф** *m*. bolograph

**болометр** *m*. bolometer; **б. типа предохранителя** little-fuse bolometer

**болометрический** *adj*. bolometric, bolometer-type

**болотистый** *adj*. bog, boggy, marshy, swampy

**болото** *n*. bog, marsh, swamp

**болт** *m*. bolt, screw; pin; **анкер(овоч)-ный б.** anchor bolt, stay bolt, tie bolt, wall screw; **контактный б.** stud; **крепежный б.** carriage bolt; **полюсный б.** terminal pillar; **б. с гайкой** bolt; **б. с головкой** set bolt; **б. с ушком** eye bolt; **б. с цилиндрической головкой** round-head(ed) screw; **б. с чекой** cotter bolt; **стяжной б.** maneton bolt; **токоведущий б.** contact stud; **фундаментальный б.** stone stud, wall screw

**болтание** *n*. shaking, stirring

**болтанка** *f*. jigging, jiggling; blimp; bumpiness; **б. изображения** jigging, jiggling (*tv*)

**болтать, взболтать** *v*. to shake, to stir

**болтаться, взболтаться** *v*. to swing, to fluctuate

**болтающийся** *adj*. swinging, fluctuating

**болтик** *see* **болт**

**болтовня** *f*. chatter; **«детская» б.** "baby blabbermouth"; **«обезьянья» б.** monkey chatter, adjacent channel interference

**болтовой** *adj*. screw, bolt; pin

**больничный** *adj*. hospital

**Больцманн** Boltzmann

**больцманновский** *adj*. Boltzmann

**больше** *compar. of* **большой** more; larger, bigger; **б.-сигнальный** large-signal

**больший** *adj., compar. of* **большой** greater, larger, major

**большинство** *n.* majority

**большой** *adj.* big, large, great, high; heavy; major; strong; **с большим запасом прочности** conservatively loaded; **с большим затуханием** heavily damped

**бомба** *f.* bomb; **реактивная б.** rocket bomb; **телеуправляемая б.** tele-controlled bomb; **б., управляемая с самолёта только по азимуту** Azon bomb

**бомбардировать** *v.* to bombard; to bomb

**бомбардирование** *n.*, **бомбардировка** *f.* bombardment, bombarding; bombing; **б. катода** filament bombardment

**бомбардировочный** *adj.* bombing

**бомбардировщик** *m.* bomber

**бомбардируемый** *adj.* bombed, struck, bombarded; target

**бомбардирующий** *adj.* bombarding

**бомбёжка** *f.* bombing, bombardment

**бомбовый** *adj.* bomb

**бомбометание** *n.* bombing, bomb dropping; **б. по расчётной дальности** range bombing; **б. с кабрирования** loft bombing

**бомбоприцел** *m.* bombsight

**бомбоупорный** *adj.* bombproof

**Боме, Бомэ** Baumé

**Бонд** Bond

**бор** *m.* boron, B; drill

**Бор** Bohr; Baur

**боран** *m.* borane, boron hydrite

**борат** *m.* borate

**боридный** *adj.* boride

**бормотание** *n.* babble, babbling, mumbling; **несвязное б.** (incoherent) babble

**борнит** *m.* bornite

**борный** *adj.* boric, boracic

**боров** *m.* horizontal flue; breaching

**боровковый** *adj.* flue

**бороводород** *m.* boron hydride

**боровок** *m.* baffle, baffle plate, baffler

**боровский** *adj.* Bohr's, (of) Bohr

**бородок** *m.* hollow punch, puncher; broach

**борозда, бороздка** *f.* groove, slot, flute; trail; furrow; cut, incision; ditch

**бороздчатый** *adj.* grooved, slotted, channeled

**бороскоп** *m.* boroscope

**бороться, побороться** *v.* to fight, to combat

**бороуглеродистый, бороуглеродный** *adj.* boron-carbon

**борт** *m.* ledge, edge, rim, border, hem; skirt; board, side; flange, bead, crimp; cornice

**бортик** *see* **борт**; **завальцованный б.** rolled lock

**бортинженер** *m.* flight engineer

**бортовой** *adj.* edge, rim, border, hem; board, aboard, airborne, on-board, shipboard; flange, bead, crimp

**бортрадист** *m.* aircraft radio operator

**борьба** *f.* struggle, fight, combat; control; abatement (of noise); **б. за живучесть электрооборудования** electrical casualty control; **б. с активными (радио)помехами** anti-jamming; **б. с мерами противодействия противника** counter-countermeasures; **б. с передатчиками (радио)помех** jammer hunting; **б. с помехами** interference suppression, radio shielding

**Босанке** Bosanquet

**босон** *m.* boson

**бот** *m.* boat, skiff

**Бофорт** Beaufort

**боченкообразный** *adj.* beer-barrel-shaped

**боченок** *m.* barrel, keg

**бочка** *f.* barrel, drum, cask, keg; tubby, bottom, bass; tub

**бочкообразный** *adj.* barrel-shaped, barrel

**бочонок** *m.* barrel, keg

**бочоночный** *adj.* (beer-)barrel-shaped

**Бош** Bosch; Bausch

**бра** *n. indecl.* bracket, wall bracket, scone, lamp bracket; **б. вектора** bra vector

**Браве, Бравэ** Bravais

**Брагг** Bragg

**брагговский** *adj.* Bragg's, (of) Bragg

**брадисейсмы** *pl.* bradyseisms

**браз.** *abbr.* (**бразильский**) Brazilian

**бразильский** *adj.* Brazil, Brazilian

**брак** *m.* refuse, scrap; rejected material, rejects, discard, spoilage; shrinkage; flaw

**бракет** *m.* bracket; **поворотный б.** swinging bracket

**бракованный** *adj.* rejected, condemned; defective, faulty; waste, refuse

**браковать, забраковать** v. to reject, to refuse, to condemn, to discard

**браковка** f. rejection, rejecting, condemning; sorting; inspection

**браковочный** adj. rejection

**браковщик** m. inspector, quality checker

**Брандт** Brandt

**Бранли** Branley

**брать, взять** v. to take; to remove, to withdraw; **б. в вилку** to bracket; bracketing; **б. пеленг** to take a bearing, to bear; **б. производную** to derive

**браться, взяться** v. to undertake, to tackle; **б. за решение задачи** to tackle the problem

**Браун** Braun; Brown

**брауновский** adj. Brownian (movement); Braun's, (of) Braun (or Brown)

**брахистохрона** f. brachistochrone

**Брегг** Bragg

**Бреге** Breguet

**брезент** m. canvas, tarpaulin, (car) tilt

**брезентовой** adj. canvas, tarpaulin

**Брейс** Brace

**Бремер** Bremer

**бренный** adj. perishable, fragile

**брешь** f. flaw, gap, breach, break, notch

**бригада** f. crew, gang, squad, brigade; personnel, team; **артикуляционная б.** testing personnel, testing crew; **б. для исправления повреждений** trouble crew; **комплексная б.** multiple-skill crew

**бригадир** m. brigade foreman, headman

**бригадный** adj. brigade, crew, team

**Бригг** Brigg

**бриггов** poss. adj., **бригговский** adj. Brigg's, (of) Brigg

**бриз** m. breeze

**бризантность** f. brisance

**бризантный** adj. high explosive; shattering, disruptive, brisant

**брикет** m. briquette; cake, preform, compact; biscuit

**брикетированный** adj. briquetted

**брикетировочный, брикетный** adj. briquette

**Бриллуэн** Brillouin

**бриллуэновский** adj. Brillouin's, (of) Brillouin

**Бриллюэн** Brillouin

**бриллюэновский** see **бриллуэновский**

**Бринель** Brinell

**британка** f. Britannia joint, Britannia splice

**британский** adj. British

**бровка** f. lip, brow, crest, peak

**Бродхун** Brodhun

**Бройль(е), Брольи, де** de Broglie

**бром** m. bromine, Br

**бромид** m. bromide

**бромирование** n. bromination, bromating

**бромированный** adj. brominated

**бромировать** v. to bromate

**бромистый, бромный** adj. bromine; bromide

**бромосеребряный** adj. silver bromide

**броневой** adj. armor, armoring, armored, sheathing

**бронекабель** m. armored cable

**бронелента** f. armoring tape

**бронепроволока** f. armoring wire

**бронза** f. bronze; **алюминиевая б.** albronze; **бериллиевая б.** berillium copper; **твёрдая б.** hard bronze, durobronze; **фосфористая б.** phosphor bronze

**бронзирование** n. bronzing, brass plating

**бронзированный** adj. bronzed

**бронзировать** v. to bronze

**бронзировка** f. bronzing, brass plating

**бронзовый** adj. bronze, bronzy, bronzed

**бронирование** n. armoring, armor-plating, sheathing; inhibiting; **ленточное б.** hoop-iron sheathing

**бронированный** adj. armored, armor-plated, sheathed, steel-clad, iron-clad; jacketed

**бронировать** v. to armor, to armor plate, to sheathe

**бронирующий** adj. armoring, sheathing; inhibitory

**бронтограф** m. brontograph

**бронтометр** m. brontometer

**броня** f. armor, armoring, armature, casing, jacket, shell; **фасонно-проволочная б.** interlocking-wire armoring

**бросание** n. throwing, casting; projection; abandonment

**бросать, бросить** v. to throw, to cast; to project; to abandon, to drop; **б. в лом** to scrap

**бросаться, броситься** v. to plunge, to throw oneself; to rush

**бросовый** adj. worthless

# Russian-English Dictionary
## of Electrotechnology
## and Allied Sciences

**бросок** *m.* kickback, kick; throw, hurl; rush, surge; bump; **б. нагрузки** load swing; **б. обратной мощности** kickback; **б. стрелки** kick of a pointer, throw, needle throw, elongation, deflection of the needle; **б. тока** bump of current, rush of current, surge, inrush

**Броун** Brown

**Броунинг** Browning

**броунов** *poss. adj.*; **броуновский** *adj.* Brownian, Brown's

**брошенный** *adj.* abandoned; thrown

**Брукс** Brooks

**брукхейвенский** *adj.* Brookhaven

**брус** *m.* bar, block; beam, girder, squared beam; **б. автосцепки** drawbar; **ведущий б.** guide rail; **отбойный б.** buffer block; **поперечный б.** cross bolster, cross-sill; **стрелочный б.** head block

**брусковый** *adj.* bar, rod, block

**брусовой** *adj.* girder, beam, joist

**брусок** *m.* bar, pig; block, slug; filler piece; rod, rail; **бруски между концевыми витками** edge blocks; **отбойные бруски** tree guard; **АТ-б. из Х-сечения кварца** AT-bar; **ВТ-б. из Х-сечения кварца** BT-bar; **выводной б.** terminal bar, terminal yoke; **корректирующий железный б.** Flinders bar; **правильный б.** regular bar

**брусочек** *m.* bar; *see* **брусок**; **изоляционный б.** cross-over block; **магнитный пробный б.** magnetic proof piece

**брутто** *adv.* gross; **б.-момент** gross torque; **б.-формула** empirical formula

**брызгальный** *adj.* splashing, sprinkling; sprinkler

**брызгание** *n.* splashing, spatter; sprinkling; spraying

**брызгать, брызнуть** *v.* to splash, to spatter; to spray

**брызги** *pl.* splash, spatter; spray

**брызгозащитный** *adj.* splash-proof; antispray

**брызгозащищённый, брызгонепроницаемый, брызгостойкий** *adj.* splash-proof; dirt-proof

**брызгоуловитель** *m.* spray trap

**брызнуть** *pf. of* **брызгать**

**Брэгг** Bragg

**брэгговский** *adj.* Bragg's, (of) Bragg

**Брэдли** Bradley

**Брэкетт** Brackett

**Брюстер** Brewster

**б-телефонистка** *f.* B-operator, incoming (trunk) operator

**Буге** Bouguer

**бугель** *m.* bow, hoop, loop; bow-collector, bow pantograph, bow trolley; clip; strap, clamp; brace; **закорачивающий б.** shorting bar, jumper wire, U-link; **б. наушников** head band, head-phone bow; **угольный скользящий б.** carbon runner

**бугорчатый** *adj.* nodular

**будильник** *m.* alarm clock; ringer; **мембранный б.** diaphragm-type bell (*or* ringer)

**будить, разбудить** *v.* to wake, to rouse, to call; to ring

**будка** *f.* booth, stall, cabin, penthouse, hut; cab; **звуконепроницаемая переговорная б.** silence cabin; **переговорная б.** telephone booth, telephone cabin; **переговорная телефонная б.** telephone station

**будничный** *adj.* weekday

**буёк** *m.* buoy

**Буже** Bouger; Bouguer

**буй** *m.* buoy, float, balize; beacon; **гидроакустический б.** sonobuoy; **б. для передачи по радио подводных шумов** sonoradiobuoy; **звуковой б.** sonobuoy; **поддерживающий б.** antenna float; **радиогидроакустический б.** radiosonic buoy, radio sonobuoy, sonoradiobuoy; **радиогидроакустический б. ненаправленного действия** nondirectional sonobuoy; **б.-радиолокатор** sonobuoy, radar buoy; **б. с уголковым отражателем** radar marker float; **б., указывающий место** station buoy; **якорный гидроакустический б.** anchored radio sonobuoy

**буйно** *adv.* violently; vigorously

**буйность** *f.* violence, turbulence

**буйный** *adj.* violent, turbulent; vigorous

**буква** *f.* letter, character; **буквы, обозначающие порядок кода операций, буквы порядка кода операции** function letters; **пропускная б.** capital letter; **строчная б.** lower case letter

**буквально** *adv.* word for word, literally, verbatim

**буквальный** *adj.* word-for-word, literal

**букварный** *adj.* alphabetic

**буквенно-цифровой** *adj.* alpha-(nu)meric

**буквенный** *adj.* alphabetical, by letter, letter, literal

**буквопечатающий, буквопишущий** *adj.* printing, type-printing; printer

**букса** *f.* box; bushing, bush, brush, sleeve, liner; **б. колеса** bushing; **неразъёмная б.** journal

**буксир** *m.* tow, towline; tugboat

**буксирный** *adj.* tow, towing, hauling

**буксирование** *n.* towing; hauling, haul (-age)

**буксировать** *v.* to tow, to tug, to haul

**буксировочный** *see* **буксирный**

**буксируемый** *adj.* towed

**буксование** *n.* slipping, slippage, skidding

**буксовать** *v.* to slip, to skid, to spin

**булавка** *f.* pin

**булавовидный** *adj.* mace-shaped

**булавочный** *adj.* pin

**булев** *poss. adj.*, **булевский** *adj.* Boolean **Буль** Boole

**булькание** *n.* thump

**бумага** *f.* paper; **бакелизированная прессованная б.** geax; **битая б.** papier-mache; **б. для определения знака полюсов** pole-finding paper; **б. для построения графика** coordinate paper; **б. для регистрирующего прибора** recording chart; **б. из волокон манильской пеньки** Manilla paper; **б. из древесной целлюлозы с содержанием серы** Kraft paper; **изоляционная б.** Empire paper, insulating paper; **кабельная б.** cable binding paper; **кабельная б. с предварительной пропиткой** preimpregnated paper; **конденсаторная б. из льняного тряпья** condenser tissue paper; **миллиметровая б.** plotting paper, cross-section paper, ruled paper; **наждачная б.** emery paper; **натронная б.** soda pulp paper; **неразмеченная б.** virgin paper; **обёрточная б.** wrapping paper; **оклеечная б.** core disc paper; **б.-основа** body paper (for condensers); **полюсная б.** pole-finding paper, polarity tester, pole test paper; **полюсная реагентная б.** polarity test-paper; **проводящая б.** resistance paper; **прозрачная б.** tracing paper; **реактивная б.** test paper; **светокопировальная б.** tracing paper; **слюдяная б.** mica paper, paper and mica; **токопроводящая б.** resistance paper; **тонкая б.** tissue paper; **тонкая прочная глянцевитая б.** tissue paper; glazed paper; **хлопчатая б.** cotton, cotton fabric; **чистая б.** virgin paper; **шелковистая б.** tissue paper, glazed paper

**бумагообмоточный** *adj.* paper lapping

**бумажный** *adj.* paper; cotton; **с бумажной изоляцией** cotton-covered

**буна** *f.* buna, synthetic rubber

**Бунзен** Bunsen

**бунзеновский** *adj.* Bunsen's, (of) Bunsen

**бункер** *m.* bunker, bin; silo

**бункерный** *adj.* bunker, bin; silo

**бунт** *m.* bunch, bundle; pack; coil (of wire)

**бунчирование** *n.* bunching

**бур** *m.* bit, boring bit; drill, borer, perforator; **вращательный б.** attack drill, rotary drill; **б. для линейных опор** post-hole digger; **закрытый клапанный б.** auger; **канатный б.** churn drill; **колонковый б.** columnal drill; **коронковый б.** cross-mouthed chisel; **ложечный б.** gimlet for boring; **остроконечный б.** V-drill; **б. с концом в виде ласточкина хвоста** dowel borer; **б. с коронкой** cross-mouthed chisel; **ударный б.** piercer, percussion borer, sinker drill; **штанговый б.** pole drill

**бура** *f.* borax

**бурав** *m.* borer, drill, perforator; **испытательной б. для столба** pole tester; **кольцевой б.** tubular drill; **плотничий б.** wood boring drill, bit

**буравить, пробуравить** *v.* to bore, to drill, to perforate, to pierce

**буравление** *n.* boring, drilling

**буравчик** *m.* gimlet, borer; corkscrew; **б. для пробы дерева** increment borer, pole tester, timber tester; **б. для пробы столбов** pole auger

**буран** *m.* blizzard; buran

**Бурдон** Bourdon

**бурение** *n.* boring, drilling

**бурильный** *adj.* boring

**бурить, пробурить** *v.* to bore, to drill

**бурление** *n.* swirling

**бурливость** *f.* turbulence

**бурливый** *adj.* turbulent, wild

**бурлить, взбурлить** *v.* to swirl, to bubble

**бурлящий** *adj.* turbulent, wild
**бурность** *f.* violence, intensity
**бурный** *adj.* violent, vigorous, turbulent, brisk; squally
**буровой** *adj.* drilling, boring
**буровый** *adj.* borax, boracic
**бурт, буртик** *m.* bead, crimp, collar, rib, shoulder
**бурый** *adj.* brown
**буря** *f.* storm; **снежная б.** blizzard
**бус** *m.* bead; spacer
**бусина** *f.* bead
**бусинка** *f.* (small) bead; *pl.* **бусинки** fillets; **наплавленная б.** stem bead
**бусинковый** *adj.* bead
**буссоль** *f.* compass, surveying compass, box compass; **б. магнитного наклонения** dip circle; **б. наклонения** inclinometer; **б. склонения** declinometer; **б.-угломер** aiming circle
**бустер** *m.* booster, servomotor-actuator; compensator; **б. для (подъёма напряжения) конца линии** tail-end booster; **добавляющий б.** *see* **повысительный бустер; повысительно-понизительный б.** buck-boost generator; **повысительный б.** positive booster, pressure controller; **понизительный б.** negative booster; **поперечный б.** quadrature booster; **фидерный б.** tail-end booster
**бустерный** *adj.* booster, power-assisted, servocontrolled, auxiliary
**бутадиен** *m.* butadiene
**бутадиеновый** *adj.* butadiene
**бутанол** *m.* butanol, *n*-butyl alcohol
**бутафория** *f.* dummy, dummies; properties; props
**бутербродный** *adj.* sandwich, sandwich-type
**бутирометр** *m.* butyrometer
**бутстрэп** *m.* bootstrap circuit
**бутстрэпный** *adj.* bootstrap
**бутылка** *f.* bottle; **б. для взятия проб** sampling bottle
**бутылочный** *adj.* bottle
**бутыль** *f.* large bottle; carboy, demijohn; vessel; (gas) cylinder
**буфер** *m.* buffer, float(ing) bumper, shock absorber, cushion, dashpot; **б. с матрицей** die cushion
**буферизация** *f.* buffering
**буферность** *f.* buffering, buffer action
**буферный** *adj.* buffer, bumper, floating
**бухгалтер** *m.* book-keeper, accountant

**бухгалтерия** *f.* book-keeping
**бухгалтерский** *adj.* book-keeping, account(ing)
**Бухгольц** Buchholz
**бухолит** *m.* bucholith
**бухта** *f.* coil, bundle, flake (of wire), hank; bay, inlet; **б. во льду** bight; **положительная б.** positive bay
**бухтообразный** *adj.* bay(-shaped)
**бушинг** *m.* bushing; **б. укороченного типа, укороченный б.** re-entrant bushing
**бы** *conditional particle;* **было бы** would be
**бывший** *adj.* former, late, ex-
**быстрина** *f.* rapid, race, swift course
**быстро** *adv.* rapidly, quickly, swiftly, with speed
**быстровращающийся** *adj.* highly rotating, fast spinning; high-speed
**быстровяжущий** *adj.* quick-setting
**быстрогорящий** *adj.* fast-burning
**быстродвижущийся** *adj.* fast, high-speed, fast-moving, quick
**быстродействие** *n.* quick operation, speed, speed of operation; **б. арифметического устройства** arithmetic speed; **б. запоминающего устройства** memory speed, storage speed; **б. регулятора** control rate; **среднее б.** average speed
**быстродействующий** *adj.* high-speed, rapid-action, fast, quick-acting, quick-operating, immediate-action, ultraspeed
**быстрозакрывающийся** *adj.* quick-closing
**быстрозамыкающий(ся)** *adj.* quick-make
**быстролетящий** *adj.* fast-moving, transient
**быстроменяющийся** *adj.* rapidly changing, rapidly alternating
**быстронадеваемый** *adj.* quick-done
**быстронакальный** *adj.* quick-heating
**быстрооткрывающийся** *adj.* quick-opening; quick-acting
**быстроотпускающий** *adj.* quick-release
**быстропротекающий** *adj.* high-speed
**быстропроходящий** *adj.* transitory, transitional
**быстроразгоняемый** *adj.* quick-to-accelerate
**быстроразмыкающий(ся)** *adj.* quick-break, quick-make
**быстроразъёмный** *adj.* quick-disconnecting

**быстрораспадающийся** *adj.* short-lived
**быстросгорающий** *adj.* free-burning, fast-burning
**быстросканирующий** *adj.* rapid-scanning
**быстросменный** *adj.* quick-change
**быстросохнущий** *adj.* quick-drying
**быстросрабатывающий** *adj.* quick-operating; high-speed
**быстросхватывающийся** *adj.* quick-hardening, rapid-setting
**быстросходящийся** *adj.* rapidly converging
**быстросъёмный** *adj.* quick-release
**быстрота** *f.* rate; speed, velocity, rapidity, quickness; frequency; **б. ответа** answering interval; **б. реагирования оператора** operator attention factor; **б. реагирования прибора** responsiveness of an instrument; **б. установления стрелки** transient time of an instrument pointer, responsiveness of an instrument
**быстротвердеющий** *adj.* quick-hardening, fast-hardening
**быстротечный** *adj.* transient; rapid
**быстроток** *m.* chute
**быстроуплотняющийся** *adj.* quick-seal(ing)
**быстроустанавливающийся** *adj.* quick-adjusting
**быстроходность** *f.* speed
**быстроходный** *adj.* high-speed, fast, quick-running
**быстрочередующийся** *adj.* rapidly alternating
**быстрый** *adj.* quick, fast, rapid
**бытовой** *adj.* domestic ; everyday

**быть** *v.* to be, to exist; **б. в резонансе** to resonate; **б. раскалённым** to glow
**Бьенэмэ** Bienaymé
**бьеф** *m.* reach of a canal (between two locks); mill-race; **верхний б.** head race (*or* water), upper water, upper bay; **нижний б.** tail-water, aft(er)-bay, under water
**бьющийся** *adj.* beating, pulsating, pulsatory
**Бэв** *abbr.* (**биллион электрон-вольт**) Bev, billion electron volts
**бэватрон** *m.* bevatron
**Бэйли** Baily
**бэр** *abbr.* (**биологический эквивалент рентгена**) roentgen equivalent, man; rem
**Бэрд** Baird
**Бэрнс** Barnes
**БЭСМ** *abbr.* (**быстродействующая электронная счётная машина**) high-speed electronic computer
**БЭТ** Brunauer, Emmett, and Teller, BET
**бюгель** *see* **бугель**
**Бюи-Балло** Buys-Ballot
**бюллетень** *m.* bulletin; **зенитный метеорологический б. с радиозондированием** radiosonde message
**бюретка** *f.* burette, dropping glass
**бюро** *n. indecl.* bureau, office, department; desk; **авиаметеорологическое б.** Airway Weather Central; **контрольное б.** filter center; **б. повреждений** repair service, fault clerk's desk, fault complaint service, service observation; **чертёжное б.** drawing room

# В

в *prep. acc.* in, into; *prep. prepos. case* in; on; at; per; **в действии** in service, in work, in operation

в. *abbr.* (вольт) volt

ва *abbr.* (вольтампер) va, volt-ampere

**Вагнер** Wagner

вагон *m.* wagon, truck; (railroad) car, coach; **аккумуляторный в.** battery powered motor car, accumulator car; **диэлектрический моторный в.** Diesel-electric rail motorcar; **в. закрытого типа** box car; **моторный в.** self-propelled (rail) car; **моторный в. по системе единиц** multiple-unit car; **прицепной в.** trailer (*elec. rr.*); **прицепной в. с кабиной управления** control trailer; **самозагружающийся в.** dumping car, self-dumping car

вагон-вышка *f.* tower car; **монтажный в.-в.** electric tower car

вагонетка *f.* truck, lorry, wagon, car; buggy

вагонный *adj.* car, wagon

вагон-цистерна *f.* tank car

важнейший *superl. adj.* major, paramount

важность *f.* importance, significance; gravity, consequence

важный *adj.* important, significant

вазелин *m.* vaseline, petroleum jelly

вазелиновый *adj.* vaseline

вайербарс, вайрбарс *m.* wire bars

вайретрон *m.* wiretron

вайтаскэн *m.* vitascan

вакансия *f.* vacancy, vacant site; vacancy defect

вакантный *adj.* vacant, empty

вактроллер *m.* vactroller

вакуметр *m.* vacuometer, vacuum gauge

вакуоля *f.* vacuole

вакуум *m.* vacuum, emptiness; **предельно высокий в.** highest attained vacuum; **предельный в.** ultimate vacuum

вакуумбак *m.* vacuum tank

вакууметр *m.* vacuometer, vacuum gauge; **вязкостный в.** viscosity manometer

вакуумирование *n.* evacuation

вакуумировать *v.* to evacuate

вакуум-наполнительный *adj.* vacuum-filling

вакуумнасос *m.* suction pump; vacuum pump

вакуум-непроницаемый *adj.* vacuum-tight

вакуумноплотный *adj.* airtight, vacuum-tight

вакуумный *adj.* vacuum, void, vacuous, evacuated

вакуумоплотный, вакуумплотный *adj.* vacuum-tight, airtight, leak-proof

вакуум-пропитка *f.* vacuum impregnation

вакуум-спектограф *m.* vacuum spectograph

вакуум-фактор *m.* vacuum factor, gas ratio

вал *m.* shaft, spindle, axle, arbor; roll, roller; drum; **включающий в.** interrupter shaft; selector shaft; **в., вращаемый часовым механизмом** time shaft; **коленчатый в.** crankshaft; **в. контроллёра** interrupter shaft; **кулачковый в.** camshaft; **в. обратной связи** resetting shaft; **отрабатывающий в. (следящей системы)** resolver shaft; **приводной в.** power shaft, drive shaft, gear shaft; **приводной (линейный) в.** squaring shaft; **промежуточный в.** jack shaft; **в. с кулачками** camshaft; **трансмиссионный в.** line shaft; **упрощённый электрический в.** semi synchro-tie; **упрощённый электрический компенсирующий в.** simplified electrical differential driving gear; **шлицевой в.** spline shaft; **электрический в.** synchro, selsyn, autosyn, electric shaft, electric driving gear

**Валенси** Valensi

валентность *f.* valency, valence; performance of filter networks

валентный *adj.* valence

валик *m.* roll, roller, cylinder, reel; shaft, spindle; ridge; **в. выверки по**

**высоте** correction wedge shaft; **в. измерения дальности** range knob; **измерительный в. дальности** observer's measuring knob; **колеблющийся в.** oscillating rotor; **контактный в.** controller cylinder; **мускульный в.** muscular ledge; **наплавленный в.** bead, welding seam; **в. с чекой** cotter bolt

**валить, повалить** *v.* to bring down, to throw down, to overthrow

**валопровод** *m.* shafting

**вал-цифра** *f.* shaft-to-digital

**вальцевать** *v.* to roll

**«вамистор»** *m.* vamistor

**вамоскоп** *m.* Wamoscope tube

**ВАН** *abbr.* (**Вестник Академии Наук**) Bulletin of the Academy of Sciences

**ванадиевый** *adj.* vanadium, vanadic

**ванадий** *m.* vanadium, V

**ванадистый** *adj.* vanadium, vanadous

**ванадовый** *adj.* vanadium, vanadic

**Ван-де-Грааф** van de Graaf

**Ван-дер-Ваальс** van der Waals

**Ван дер Монд** van der Mond

**Ван дер Поль** van der Pol

**ванна** *f.* tank, tub, vat; bath; **в. для контактного осаждения** strike bath; **в. для приготовления катодных основ** stripper tank; **паяльная в.** cable jointing box; **в. регенерации электролита** liberator tank; **в. с расплавленным припоем** (soldering) dipper; **сильно кипящая в.** wild bath; **электролечебная в.** hydroelectric bath (*medicine*)

**вапориметр** *m.* vaporimeter

**вар** *m.* var (volt-ampere-reactive); var (visual-aural-range)

**варактор** *m.* varactor, varactor diode, variable reactor

**вариак** *m.* variac, transtat

**вариант** *m.* variant, variation, version, modification, alternate, alternative; **в. многократного использования** reusable version; **операционный в.** interpretative version; **в. перестановок** ordering (*math.*)

**вариантность** *f.* variance

**вариантный** *adj.* variant, varying, alternate, alternative

**вариатор** *m.* variator; buncher; input resonator; **механический в.** mechanical variable speed drive

**вариационный** *adj.* variational

**вариация** *f.* variation, change, alternation; **в. по скорости** velocity modulation; **в. показания измерительного прибора** variation of instrument reading

**вариирование** *n.* variation; bunching

**вариировать** *see* **варьировать**

**варикап** *m.* variable-capacitance diode, voltage-variable capacitor

**вариконд** *m.* varicap, voltage-variable capacitor

**варимю** varimu, variable mu

**вариокуплер** *m.* varicoupler, variometer-type coupler

**вариометр** *m.* variometer, adjustable inductor, variable inductor; **в. высоты** rate-of-change indicator; **переменный в.** rotating-coil variometer; **в. связи** coupler (variometer)

**вариообъектив** *m.* Zoomar

**варистор** *m.* varistor, voltage-dependent resistance, variable resistor

**варитрол** *m.* varitrol

**варитрон** *m.* varitron

**варметр** *m.* varmeter, idle-current wattmeter

**варочный** *adj.* cooking, boiling, digestion

**ВАРУ** *abbr.* (**временная автоматическая регулировка усиления**) periodic automatic gain control

**вар-час** *m.* var-hour

**варьирование** *n.* variation; bunching

**варьированный** *adj.* changed, modified, diversified; bunched

**варьировать** *v.* to variate, to vary, to change, to diversify, to modify

**вата** *f.* cotton, cotton wool, wadding; (glass) wool; **минеральная в.** mineral wool, stonefelt

**ватерпас** *m.* level, leveling instrument

**ватерпасный** *adj.* level, horizontal

**ватность** *f.* dullness (of sound)

**ватный** *adj.* cotton, wadded

**ватт** *m.* watt; **отдаваемые ватты** watts-out; **потребляемые ватты** watts-in; **реактивный в.** reactive volt-amperes

**ватт-варметр** *m.* watt- and varmeter

**ваттметр** *m.* wattmeter; **в. без железа с вращающейся катушкой** coreless wattmeter, dynamometer-type wattmeter; **в. для измерения проходящей мощности** transmission dynamometer; **замедленный максимальный в.** lagged-demand meter; **тепловой**

**в.** hot-wire wattmeter; **флажковый в.** vane wattmeter; **цанговый в.** prong-type wattmeter

**ваттметровый** *adj.* wattmeter

**ваттность** *f.* wattage

**ваттсекунда** *f.* watt-second, joule

**ватт-час** *m.* watt-hour

**ваттчасовой** *adj.* watt-hour

**вафельный** *adj.* wafer, waffle

**вафля** *f.* wafer, waffle

**вахтенный** *adj.* lookout

**ваш** *poss. pron. adj.* your

**вб** *abbr.* (вебер) weber

**вбегать, вбежать** *v.* to run in, to run into, to rush in, to flow in, to flow into

**вбивать, вбить** *v.* to drive in, to drive, to hammer; to ram

**вбитый** *adj.* driven in

**вбить** *pf. of* вбивать

**вблизи** *adv.* close by, near, at hand, in the vicinity, in the neighborhood (of)

**вбрызгивание** *n.* injection

**вбрызгивать, вбрызнуть** *v.* to inject; to spray in

**вбрызнутый** *adj.* injected

**вбрызнуть** *pf. of* вбрызгивать

**вв** *abbr.* (взрывчатые вещества) explosives

**ВВ** *abbr.* (выдержка времени) time lag

**ВВА** *abbr.* (военно-воздушная академия) Air Force Academy

**введение** *n.* leading in, insertion, putting in, breaking in, introduction; injection; establishment,; **в. в заблуждение** deception; **в. в телевизионную программу заставок** (*or* **титров**) cueing; **в. поправок в направление линии прицеливания** in-flight bore sighting; **в. примеси, в. присадки** doping; **в. узких импульсов** peaking; **в. усилителя** repeater insertion; **в. числа в машину** entry insertion

**введённый** *adj.* lead in, introduced, fed (into), injected; inserted, driven in

**ввернуть** *pf. of* ввёртывать

**ввёртный** *adj.* screw-in, screw

**ввёртывать, ввернуть** *v.* to screw in, to turn in

**вверх** *adv.* up, upward(s)

**ввести** *pf. of* вводить

**ввиду** *prep. genit.* in view (of)

**ввинтить** *pf. of* ввинчивать

**ввинченный** *adj.* screwed in

**ввинчивание** *n.* screwing in

**ввинчивать, ввинтить** *v.* to screw in

**в-во** *abbr.* (вещество) substance, matter

**ввод** *m.* inlet, lead, lead-in, entry, entrance, intake, input, insertion; fair-lead; seal (in); lead-in wire, terminal(s); leg (of a tube); bushing; **абонентский в.** drop wire, telephone drop, service entrance, substation; **в. в стекло** glass bead; **в.-вывод** in-out; **в. данных о высоте цели** height input; **двойной абонентский в.** subscriber's shared service; **кабельный в. с эластичным уплотнением** compression gland; **в. линии** leading-in (point); **в. потребителя** service entrance; **в. программы** program display; **сквозной в. в стекле** glass-bead seal; **в. со штыковым контактом** plug bushing; **станционный в.** leading-in cable; **управляющий в.** drive wire; **штепсельный в.** inlet-plug and socket, connector

**вводимый** *adj.* introducible; input; injecting

**вводить, ввести** *v.* to introduce, to lead in, to enter, to feed in; to insert, to drive in; to inject; to gather; to mount (information); to gate in; **в. в заблуждение** to confuse; **в. в синхронизм** to synchronise; **в. в эксплуатацию** to put in operation, to put in service; **в. гамма-корекцию** to gammate; **в. задачу в вычислительную машину** to lead the computer; **в. подпрограмму** to assemble, to compile

**вводный** *adj.* incoming, inlet, lead-in, leading, leading in; input; inserted; parenthetic

**вводящий** *adj.* leading-in; insertion

**ввосьмеро** *adv.* eight times

**ввосьмером** *adv.* eight (together)

**ВВР** *abbr.* (водо-водяной реактор) water-moderated water-cooled reactor

**ВВС** *abbr.* 1 (военная воздухоплавательная станция) military airship station; 2 (военно-воздушные силы) Air Force; 3 (высоковольтная сеть) high-voltage network

**ввязываться, ввязаться** *v.* to interfere

**ВГ** *abbr.* 1 (вибратор, горизонтальный) horizontal dipole; 2 (вибрационный гальванометр) vibration galvanometer

**ВГД** *abbr.* (вибратор, диапазонный горизонтальный) (wide-)band horizontal dipole

**вгон** *m.* driving in

**вгонять, вгонить, вогнать** *v.* to force, to force in, to drive, to drive in

**вдавить** *pf. of* вдавливать

**вдавление** *see* вдавливание

**вдавленный** *adj.* pressed in; sunken, depressed

**вдавливание** *n.* pressing in, caving in; depression, impression

**вдавливать, вдавить** *v.* to press in; to force, to force in, to imbed, to imprint

**вдвижной** *adj.* movable, sliding

**вдвое** *adv.* double, twice; twofold; **в. больше** twice as much; **в. мечьше** half, half as much

**вдвойне** *adv.* double, doubly, twice, twofold

**вдевятеро** *adv.* ninefold

**вделанный** *adj.* fitted in, set in, built-in; inlaid

**вделывать, вделать** *v.* to fit in, to set in, to embed, to incase, to seal in; to install, to build in, to mount

**вдесятеро** *adv.* tenfold

**вдоль** 1 *adv.* lengthwise, longitudinally; down; **в. по** along; 2 *prep. genit.* along, by; **в. раскрыва** across the aperture

**вдуваемый** *adj.* blown in

**вдувание** *n.* injection, blast, blowing in

**вдувать, вдуть** *v.* to blow in, to force in, to inject

**вдутый** *adj.* blown in, injected

**вдуть** *pf. of* вдувать

**вебер** *m.* weber

**веберметр** *m.* flux meter

**ведение** *n.* guiding, leading; working; practice; management, supervision; knowledge; **в. программы** programming; **в. программы передач** presentation; **в. чёрно-белой программы** monochrome programming

**ведённый** *adj.* led, conducted

**ведомость** *f.* list, record, register, roll; journal, report; **инвентаризационная в.** inventory; **платёжная в.** payroll, paysheet; **поразговорная в.** call-order log, toll log; **расчётная в.** payroll, paysheet

**ведомственный** *adj.* departmental

**ведомство** *n.* office, department, authorities

**ведомый** *adj.* guided, controlled; driven; dependent (on); slave

**ведущий** *adj.* leading, guiding; driving; steering, pilot; master, control

**веер** *m.* fan

**веерный** *adj.* fan, fanned

**веерообразный** *adj.* fan-shaped, fan-type, radiating

**Вейерштрасс** Weierstrass

**Вейлер** Weiler

**Вейль** Weyl

**Веймар** Weimar

**Вейсс** Weiss

**Вейтч** Veitch

**век** *m.* age; century; lifetime; **в. автоматики** automation age

**вековечный** *adj.* everlasting

**вековой** *adj.* secular

**вектограф** *m.* vectograph

**вектор** *m.* vector; phasor; **составляющие векторы** vectorial combination; **в. абсолютной начальной скорости** initial resultant vector; **в. абсолютной скорости снаряда** resultant projectile velocity; **в. вертикальной скорости цели в направлении наклонной дальности** range-rate diving speed vector; **волновой в.** propagation vector; **главный в.** resultant vector; **доминирующий собственный в.** dominant eigenvector; **в. излучения** Poyntings vector; **в. на плоскости** two-dimensional vector; **в. пути** vector distance; **в.-радиус текущего положения точки** position vector; **в. разностного сигнала цветности** color-difference phasor; **в. скорости поражающих элементов боевой части ракеты** vector warhead emission velocity; **собственный в.** eigenvector, characteristic vector; **строчный в.** row vector; **в. тяги** thrust vector, thrust line; **в. цветовой информации** phasor

**вектографф** *m.* vectorgraph

**векториальный** *adj.* vectorial

**векторлайзер** *m.* vector analyzer, vectorlyzer

**векторный** *adj.* vector, vectorial

**вектороскоп** *m.* vectorscope

**вектор-потенциал** *m.* vector potential

**векторскоп** *m.* vectorscope

**велер** *m.* selector

**Вёлер** Wöhler

**величина** *f.* value, magnitude, quantity, amount; size, dimension, scale, measure; volume, bulk; degree, extent, scope; intensity; bigness; rate; figure; level; often omitted in transl., e.g. **переменная в.** variable (=variable quantity); **гиперболические величины согласования** hyperbolic matching equivalents; **некоррелированные величины** uncorrelated variables; **подынтегральные величины интегралов** integrands of integrals; **попарно-сопряжённые величины** conjugate pairs; **величины, являющиеся функциейв ремени** time variable data; **в. атмосферных помех** static level; **базисная в.** base value; datum quantity; **в. бокового перемещения** distance off track; **влияющая в.** limiting quantity; **вносимая в.** reflected value; **входная в. в моделирующем устройстве** analog input; **выгоднейшая в.** optimal value; **выходная переменная в.** output variable; **данная в.** datum, given value, given quantity; **в. действия** controlled quantity; **допустимая в. мёртвого хода** tolerable backlash; **допустимая в. поля рассеяния** stray field; **заданная в.** desired value, index value; **заранее заданная в.** predetermined value; **в. затягивания частоты** pulling figure; **в. изменения** increment; **в. изменения курсового угла цели** increment of change of target; **в. изменения угла места** complimentary error (angle); **в., измеренная балансным методом** equality value; **измеряемая в.** measurand, measured value; **в. искрового промаха** air-gap clearance; **в. количества движения в выходном сечении сопла** nozzle exit momentum; **контрольная входная в.** reference input; **линейная в. изменения угла места** linear rate of change of elevation; **линейная в. ошибки наведения** guidance miss-distance; **максимальная в.** peak, peak value, peak amplitude, crest factor; **в. междужелезного пространства** air-gap clearance; **в. мёртвой зоны** skip distance; **мешающая в.** annoyance value; **минимальная в.** valley value; **минимальная в. срабатывания** minimum pickup value, pick-

up (of a relay); **многоволновая переменная в.** periodically-variable quantity; **направленная в.** dressed quantity; **направляющая в.** determining value (of control systems); **независимо регулируемая взаимносвязанная переменная в.** independent controlled variable; **некоторая заданная в.** specified value; **непрерывная в.** anolog quantity; **несоизмеримая в.** incommensurate value; **в. нестабильности фазы** phase pushing figure; **номинальная допустимая в. тока** rated current-carrying capacity; **обратная в.** multiplicative inverse, reciprocal; **в., обратная времени перемагничивания** inverse reversal time, inverse switching time; **в., обратная диэлектрической постоянной** electivity; **в., обратная ёмкости** elastance; **в., обратная крутизне характеристики (лампы)** mutual resistance, transimpedance; **в., обратная «нормализованному полному сопротивлению»** normalized admittance, reduced admittance; **в., обратная сопротивлению** inverse; **в., обратная удельной ёмкости** elasticity; **в., обратная ускорению** deceleration; **обратная в. времени перемагничивания** reciprocal of switching time; **обратная в. коэффициенту возврата (реле)** pick-up-to-drop-off ratio; **в., обратно пропорциональная расстоянию** inverse distance; **ориентированная в.** directional quantity; **в. отклонения от заданной траектории** trajectory error; **в. отклонения снаряда от цели** standard miss distance; **относительная в. выброса** overshoot ratio; **в., относящаяся к состоянию покоя** quiescent value; **в. отсчётов по счётчику (Гейгера)** counting rate; **в. переходных искажений** crosstalk volume; **пиковая в. на выходе** peak output; **в. подъёма** rate of upward gradient; **положительная выходная в.** positive output; **в. порога переключения** switching value; **приведённая в. выхода** effective output; **в. проводимости** conductance; **в. проводимости при слабом сигнале** small-signal forward transadmittance; **программная в.** desired value; **в.**

**промаха** miss distance; aim bias; **в. просвета** air-gap; **в. пульсации** per-cent ripple, ripple factor; **пульсирующая в.** pulsating quantity, oscillating quantity, undulating quantity; **в., равная (1-кпд)** deficiency; **регулируемая переменная в.** process variable; **в. регулирующего воздействия** control value; **синусоидальная в. с затуханием по экспоненциальному закону** exponentially damped sinusoidal quantity; **слабо пульсирующая в.** ripple quantity; **случайная в.** variate; stray parameter; **собственная в. энергии** eigenvalues of energy, inherent values of energy; **в. «сползания» нуля** drift rate; **средневзвешенная в.** weighted mean; **средняя в.** mean, average quantity; **средняя квадратная в.** r.m.s. value; **текущая в. массы** instantaneous mass; **требуемая в.** nominal value; **в. трогания** pickup value; **в. угла** wave-length constant; phase constant; **в. угла места** elevation rate; **угловая в. изменения угла места** angular of elevation; **в. упреждения** lead distance, predicted interval; **установившаяся входная в.** rate response; **установленная в.** fixture; **в. ухода частоты при изменении режима (магнетрона)** pushing figure (of a magnetron); **физическая в. моделируемого процесса** analog quantity; **численная в. удельного веса** relative weight; **чисто мнимая в.** pure imaginary

**велодайн, «велодин»** *m.* velodyne

**велосиметр** *m.* velocity pickup, velocimeter

**велосипедный** *adj.* bicycle

**велоситрон** *m.* velocitron

**Вельте** Welte

**Венельт** Wehnelt

**венец** *m.* crown, rim, corona; **компенсирующий магнитный в.** magnetic field equalizer; **контактный в.** contact group, contact ring; **линзовый в.** lens drum

**венецианский** *adj.* Venetian

**венечный** *adj.* crown, corona, rim

**вентилирование** *n.* ventilation

**вентилированный** *adj.* ventilated

**вентилировать, провентилировать** *v.* to air, to ventilate

**вентилируемый** *adj.* ventilated

**вентиль** *m.* valve; rectifier; gate, stopper, check; vent; isolator; **в. ввода** in-gate; **в. ввода второго слагаемого** addend-in gate; **в. ввода первого слагаемого** augend-in gate; **входной в.** in-gate; **в. выдачи, выходной в.** read-out gate, out-gate; **германиевый в.** germanium diode, germanium diode for high inverse voltage; **запорновыпускной в.** shut-off valve; **ионный в.** ionic valve, cold cathode rectifier tube, gas-filled rectifier; **ламповый в.** valve tube; **многоанодный в.** multi-anode rectifier; **в. множимого** multiplicand gate; **в. «не и»** NAND gate; **в. «не или»** NOR gate; **в. несовпадения** except gate; **в. органной трубы** pallet (*mus.*); **потенциальный в.** voltage sensitive gate; **проходной в.** globe valve; **ртутный в.** mercury-arc rectifier; **в. с приводом от гидравлического поршня** thrustor valve; **в. синхронизирующих импульсов** clock pulse gate; **в. со смещением поля** field displacement isolator; **в. считывания** read-out gate; **управляемый в.** geared valve, mechanically operated valve; **в., управляемый чётными членами** even-controlled gate; **в. установки на нуль** zero gate; **ферритовый в.** ferrite isolator

**вентильный** *adj.* valve; rectifier; gate, gated

**вентилятор** *m.* ventilator, blower, fan; blast; **вдувной в.** induction fan, supply fan; **вытяжной батарейный в.** battery exhaust blower; **в., забирающий (свежий) воздух** make-up fan; **нагнетательный в.** blast fan, forced-draught fan; **струйный в.** air jet

**вентиляторный** *adj.* ventilator, fan, blower

**вентиляционный** *adj.* ventilation, ventilating; vent (hole)

**вентиляция** *f.* ventilation, aeration, airing; **вытяжная в.** exhaust ventilation; **нагнетательная в.** supply ventilation; **в. обдувом** shell ventilation

**Вентури** Venturi

**Вентцель** Wentzel

**ВЕП** *abbr.* (высота единицы передачи) height of a transfer unit

**Вердэ** Verdet

**верёвка** *f.* cord, rope, line, string; **в. для спуска (антенны)** downhaul

**верёвочный** *adj.* cord, rope, line, string; funicular

**вереница** *f.* row, line; **в. импульсов** pulse train; **в. колебаний** wave train

**веретенный** *adj.* spindle

**веретено** *n.* axis, axle, pivot, spindle

**веретенообразный** *adj.* spindle-shaped; tapered, taper; fusiform; ogive

**верещание** *n.* squawk, squawking

**верзор** *m.* versor (*math.*)

**верзус** *see* **косинус-верзус**; **синус-верзус**

**верикон** *m.* image vericon

**вернистат** *m.* Vernistat

**верность** *f.* correctness, accuracy, precision; fidelity (of reproduction); **в. воспроизведения** high fidelity; **в. воспроизведения звука** acoustic fidelity; **в. воспроизведения цвета** color fidelity; **в. воспроизведения электрических сигналов** electric fidelity; **высокая в. воспроизведения** high fidelity; **в. звуковоспроизведения** acoustic fidelity

**вернуть(ся)** *pf. of* **вертеть(ся)**, **возвращать(ся)**

**верный** *adj.* correct, accurate; reliable; true; significant

**верньер** *m.* vernier, micrometer screw; **механический в.** vernier arrangement; **электрический в.** vernier condenser

**верньерный** *adj.* vernier

**вероятнейший** *superl. adj.* most probable

**вероятностный** *adj.* probable, likely

**вероятность** *f.* probability, likelihood, expectancy; potential; theoretical frequency; prospect; **в. безопастности полёта ракеты** missile survival probability; **в. безотказной работы** probability of survival; **в. воспроизведения** fidelity (*see also* **верность**); **в. востановления** repairability probability; **в. встречи** probability of capture; **в. выполнения самонаведения** homing probability; **высокая в. воспроизведения** high fidelity; **в. выхода истребителя в область возможных атак ракетами** probability of achieving a launching position; **доверительная в.** fiducial probability; **заданная в. поражения** desired destruction probability; **в. захвата лучом** probability of capture; **в. захвата цели** lock-on probability, probability of acquisition; **в. захвата цели, равная единице** probability of acquisition of unity; **интегральная в.** cumulative probability; **в. на единичный объём пространства колебаний** probability per unit volume of waveform space; **в. нахождения устойчивого режима** probability of stability; **в. неправильного срабатывания**, **в. несрабатывания** malfunction probability; **в. неуязвимости** probability of survival; **общая доверительная в.** overall confidence; **в. одного попадания** probability for one kill; **полная в.** composite probability; **в. 100% попаданий** probability of zero misses; **в. (появления) диграмм** digram probability; **в., равная единице** probability of unity; **в. сближения ракеты с целью до радиуса поражения** probability of propelling the missile to the range of the target; **суммарная в. непоражения** cumulative survival probability; **в. туннельного перехода** tunneling probability; **удовлетворительная в.** pertinent probability; **в. эффекта Оже** Auger yield

**вероятный** *adj.* probable, likely; stochastic

**версайн** *m.* versine

**верситрон** *m.* versitron

**версия** *f.* version

**верстак** *m.* file bench, (work)bench

**Вертгеймер** Wertheimer

**вертекс** *m.* vertex

**вертеть, повертеть, вернуть** *v.* to turn; to spin; to reverse

**вертеться, повертеться, вернуться** *v.* to turn around, to revolve, to rotate, to spin; to roll

**вертикал** *m.* vertical; vertical flue

**вертикаль** *f.* vertical, vertical line; **экваториальная в. в плоскости полёта снаряда** vertical in the track plane at the equator

**вертикально** *adv.* vertically

**вертикальноотклоняющий** *adj.* vertically deflecting

**вертикальнополяризованный** *adj.* vertically polarized

**вертикально-сверлильный** *adj.* vertical boring

**вертикальность** *f.* verticality
**вертикальный** *adj.* vertical, perpendicular, upright, erect
**вертолёт** *m.* helicopter, copter; **в. с опускаемым гидролокатором** sonar-dunking helicopter
**вертушка** *f.* dial; turntable; impeller, rotor, rotator; current meter; ventilator; vane; **цветная в.** color wheel
**вертячий** *adj.* rotary, whirling
**вертящийся** *adj.* rotating, revolving
**верх** *m.* summit, top, upper part; *pl.* **верхи** highs, treble, tops
**верхнепропускающий** *adj.* high-pass
**верхний** *adj.* upper, top, overhead; superior
**верховой** *adj.* top, upper
**верховый** *adj.* upper
**верхушка** *f.* top, summit; head; **плоская в. импульса** flatness of wave
**вершина** *f.* apex; point, cutting point (of a tool); peak (of a curve, wave); top, head; vertex; **в. гасящего импульса** sync pedestal; **в. импульса** pulse tilt, horizontal part of the pulse; **в. кривой** peak response
**вершинный** *adj.* apical
**вес.** *abbr.* (весовой) weight, by weight
**вес** *m.* weight, gravity; *pl.* 1 **веса** weights (*abstract*); 2 **весы** scales, balance; **в. без топлива** zero-fuel weight; **в. в данный момент** instantaneous weight; **нетабличный в.** nonstandard weight; **погонный в.** linear loading density; **в. полезной нагрузки** payload weight; **суммарный в. двигателя и топлива** filled propulsion system weight; **суммарный в. силовой установки** wet weight; **в. тары** empty weight; **удельный в. топлива** propellant bulk density; **в. упаковки** dead weight; **в. элементов конструкции** inert component weight
**весенний** *adj.* spring, vernal
**веслообразный** *adj.* oar-shaped
**весовой** *adj.* weight, gravimetric
**весовщик** *m.* weigher; balancer
**весомость** *f.* weight, weightiness; ponderability
**весомый** *adj.* weight; ponderable
**вести, повести** *v.* to conduct, to direct; to lead (to); **в. радиотехническую разведку** to see electronically; **в. разведку мёртвых зон** to reconnoiter radar blind area; **в. счёт** to score

**вестибулярный** *adj.* vestibular
**Вестон** Weston
**весы** *pl.* scales, balance; *see also* **вес**; **аэродинамические в.** wind tunnel balance; **в. для определения плотности твёрдых тел** Westphal balance, Mohr balance; **в. Кельвина** Thompson balance; **в. Кельвина от 0,01 до 1 А** centi-ampere balance; **мостовые в.** platform balance; **поршневые в.** piston-cylinder balance; **радиационные в.** radio-balance; **в. с проектируемой шкалой** piston-cylinder balance, shadowgraph scales; **токовые в.** electrodynamic balance; **токоые в. от 0,1 до 10 А** deci-ampere balance; **токовые в. от 0,1 до 100 А** deca-ampere balance; **электростатические в.** attracted disc electrometer, Kelvin electrometer; **электротензометрические крановые в.** electric crane weighing machine
**весь (вся, всё, все)** *pron.* all, the whole, total, complete
**весьма** *adv.* extremely, very, much; **в. жидкий** semi-fluid
**ветвление** *n.* branching
**ветвь** *f.* branch, branching, arm, tapping, junction; side; path; **сопряжённые ветви четырёхполюсника** conjugate branches of a network; **в. (в радиолокации)** loop (radiodirection finding); **в. камертона** prong of a tuning fork; **в. обмотки** phase belt; **в. обмотки якоря** path of winding; **параллельная в. обмотки** winding path; **положительная в.** R-branch; **предполагаемая нисходящая в. траектории** extended trajectory of the incoming missile; **продольная в. (скрещённой рамки)** longitudinal loop (of a cross-frame loop antenna); **в. ремня** ply; **в. якорной обмотки** winding path
**ветвящийся** *adj.* branching
**ветер** *m.* wind; **по ветру** downwind; **акустический в.** quartz wind; **в. в верхних слоях атмосферы** wind aloft; **встречный в.** head wind; **попутный в.** following wind; **порывистый в.** gusty wind; **продольный в.** range wind; **в., учтённый в ПУАЗО** wind applied in the director; **фиктивный в.** apparent wind; **электрический в.** aura, breeze static
**ветровой** *adj.* wind

**ветродвигатель** *m.* windmill, wind motor, wind-driven generator; fan

**ветроуказатель** *m.* wind indicator, wind cone, wind tetrahedron; **световой в.** wind indicator

**ветроуловитель** *m.* air scoop

**ветрочёт** *m.* course and distance calculator, vector computer

**ветро-электрический** *adj.* anemo-electric

**ветряк** *m.* windmill, wind motor, wind turbine

**ветряница** *f.* vent fingers

**ветрянка** *f.* air vane; fan governor

**ветряной** *adj.* wind

**ветрячный** *adj.* wind-driven, fan-driven

**веха** *f.* surveying rod; beacon, pole, rod, stake, staff; landmark

**вечный** *adj.* endless, everlasting, perpetual

**вешать, повесить** *v.* to hang, to hang up, to suspend

**вешка f.** marking post

**вещание** *n.* broadcasting; **всеобщее европейское телевизионное в.** Eurovision; **одновременное в. через несколько станций** chain broadcasting, simultaneous broadcasting; **отечественное в.** home broadcasting; **в. по проводам** rediffusion on wire, rediffusing broadcasting; **проволочное в.** radio relay, wire broadcasting, broadcast relay; **программное в.** programming; **радио-проводное в.** (a-f) rediffusion; **фототелеграфное в.** facsimile service

**вещательный** *adj.* broadcast(ing)

**вещать** *v.* to broadcast

**вещественный** *adj.* real; substantial, material

**вещество** *n.* substance, material, matter, stuff; agent; composition; **атомные взрывчатые вещества** atomic explosives; **белковые вещества** albuminoids; **буферные вещества** buffers, buffer salts (*el. chem.*); **волокнистые вещества** fibres, fibers; **индикаторные вещества** tracer elements; **эпоксидные клейкие вещества** epoxy adhesives; **бризантное взрывчатое в.** high explosive; **взрывчатое в.** explosive; **в., гасящее люминесценцию** quencher; **долгоживущее в.** long-half-life material; **заполняющее в.** filling medium; **звукопогло-**

**щающее в.** sound absorber, vibration absorber; **в., имеющее малую работу выхода** low-work-function material; **испаряемое в.** evaporant; **клеивающее в.** adhesive, bond; **космическое в.** interstellar dust; **легковоспламеняющееся в.** heat sensitive material; **липкое в.** adhesive; **осаждающее в.** precipitant; **охлаждающее в.** coolant, refrigerating agent; **пироскопическое в.** temperature markers; **полупроводящее в.** semiconductor; poor conductor; partial conductor; **постороннее в.** foreign matter; **пропитывающее в.** impregnant; **простое в.** element; **распыляемое в.** evaporant; **в. с большим периодом полураспада** long-half-life material; **в. с электрическими потерями** lossy material; **в., содержащее радиоактивную примесь** tracer agent

**вживление** *n.* implantation

**вживляемый** *adj.* implantable

**вжигать, вжечь** *v.* to bake

**взад** *adv.* back, backwards

**взаимно** *adv.* mutually, reciprocally; inter- (*e.g.* **в. заменимый** interchangeable); **в. перпендикулярный** orthogonal; **в. простой** coprime; **в. соединять** to interlock

**взаимнодополнительный** *adj.* complementary

**взаимноиндуктивность** *f.* mutual inductance

**взаимно-исключающий** *adj.* mutually-exclusive

**взаимнокорреляционный** *adj.* cross-correlation

**взаимномодуляция** *f.* intermodulation

**взаимнообменный** *adj.* interchangeable

**взаимнообратно** *adv.* mutually opposing

**взаимнооднозначный** *adj.* one-to-one, one-one

**взаимноортогональный** *adj.* mutually orthogonal

**взаимнопроникающий** *adj.* interpenetrating

**взаимносвязанный** *adj.* linked, coupled, interrelated

**взаимность** *f.* reciprocity; duality

**взаимный** *adj.* mutual, reciprocal, inter-, cross(-)

**взаимо-** *prefix* inter-; *see also* **взаимно**

**взаимоаннулирование** *n.* cancellation

**взаимовидность** *f*. mutual visibility

**взаимовлияние** *n*. interference

**взаимодействие** *n*. interaction, reciprocity, self-interaction, reciprocal action, mutual effect; cooperation, liaison; reaction; **в. между полями** coupling between fields; **в. между электронами** beam coupling; **в. различных видов колебаний** coupling of modes; **спинорбитальное в.** spin-orbit relation

**взаимодействовать** *v*. to interact; to react; to interfere with

**взаимодействующий** *adj*. interacting; reacting

**взаимозависимость** *f*. interdependence

**взаимозависимый** *adj*. interdependent

**взаимозаменимость** *f*. reciprocity

**взаимозаменяемость** *f*. interchange, interchangeability

**взаимозаменяемый** *adj*. interchangeable, replaceable; duplicate

**взаимозаместимость** *f*. reciprocity

**взаимозамкнутость** *f*. interlock

**взаимозамыкание** *n*. interlocking

**взаимозамыкающий** *adj*. interlocking

**взаимоиндуктивность** *f*. mutual inductance

**взаимоиндукция** *f*. mutual induction, mutual inductance

**взаимоисключающий** *adj*. incompatible

**взаимомодуляционный** *adj*. intermodulation

**взаимонезаменяемый** *adj*. non-interchangeable

**взаимообмен** *m*. interchange, exchange, replacing

**взаимообменный** *adj*. interchangeable

**взаимоотношение** *n*. interrelation, correlation, relation(ship)

**взаимоотталкивание** *n*. mutual repulsion

**взаимоположение** *n*. relative position

**взаимопонимание** *n*. "terms of reference"; understanding, quality of reception

**взаимопревращаемый** *adj*. interconvertible

**взаимопроникающий** *adj*. interpenetrating

**взаимосвязанность** *f*. coherence

**взаимосвязанный** *adj*. interrelated, interdependent; interconnected, coupled; interacting; coherent

**взаимосвязывать, взаимосвязать** *v*. to cohere

**взаимосвязь** *f*. interaction, interrelation(ship), correlation, interdependence, intercoupling; intercoordination

**взамен** *adv*. instead, in exchange, in return

**взбалтывание** *n*. agitation, shaking (up)

**взбалтывать, взболтать** *v*. to agitate, to shake, to shake up

**взболтать(ся)** *pf*. *of* **взбалтывать, болтать(ся)**

**взбурлить** *pf*. *of* **бурлить**

**взведение** *n*. reset, resetting; leading (up, to)

**взвесить** *pf*. *of* **взвешивать**

**взвесь** *f*. suspension

**взвешенный** *adj*. suspended, in suspension; weighed; weighted

**взвешивание** *n*. weighing; weighting; suspension

**взвешивать, взвесить** *v*. to weigh; to suspend

**взвешивающий** *adj*. weighing; weighting; suspending

**взвизгивание** *n*. squeal, squawk

**взволновать** *pf*. *of* **волновать**

**взвыть** *pf*. *of* **выть**

**ВЗД** *abbr*. (**взрыватель замедленного действия**) delay fuse, time fuse

**вздваивать, вздвоить** *v*. to double up

**вздутие** *n*. bulging, bulge, swelling, inflation

**взлёт** *m*. take-off, upward flight; **в. с ракетным ускорителем** rocket-assisted take-off

**взлетать, взлететь** *v*. to fly up, to take off

**взлетающий** *adj*. taking off, flying up, rising

**взлетевший** *adj*. airborne

**взлетеь** *pf*. *of* **взлетать**

**взлетно-посадочный** *adj*. landing and take-off

**взлётный** *adj*. take-off

**взлом** *m*. intrusion, breaking in, break

**взмах** *m*. stroke, sweep, swing

**взойти** *pf*. *of* **восходить**

**взорвать** *pf*. *of* **взрывать**

**взрастать, взрасти** *v*. to grow, to increase

**взрыв** *m*. explosion, detonation; blast, burst, blow-up, rupture; failure; implosion; **мгновенный в.** instantaneous action; **солнечный в.** solar outburst; **в. ядра** nuclear explosion

**взрываемость** *f*. explosiveness

**взрывание** *n.* explosion; blasting, blowing up; **электрическое в.** electric blast, electric firing

**взрыватель** *m.* detonator, fuse; destructor; igniter; **всеубойный в.** always fuse; **в., выраженный целым числом** integral fuse length, integral fuse range; **головной в.** nose fuse, point detonating fuse; **дистанционный в. комбинированного действия** time and percussion fuse; **дистанционный в. с самоликвидатором** self-destruction time fuse; **дистанционный регулируемый в.** variable-time fuse; **в. зенитного снаряда** anti-aircraft fuse; **в.-ловушка** antidisturbance fuse; **в. малопрочной конструкции** delicate fuse; **неконтактный в.** influence fuse, proximity fuse, ambient fuse; **в. предохранительного типа** bore-safe fuze; **радиолокационный в.** radio-proximity fuse; **радиолокационный в. в зенитном снаряде** proximity fuse; **в. с предварительной установкой** precut fuse; **в. сближения** proximity fuse; **в. ударного действия** percussion fuse; **штатный в.** design fuse

**взрывать, взорвать** *v.* to explode, to detonate, to blast; to demolish

**взрывающий** *adj.* exploding, explosive

**взрывной** *adj.* explosive; mute (*ling.*)

**взрывобезопасный** *adj.* explosion-proof, safe

**взрывоопасность** *f.* explosion-hazard, explosive risk

**взрывоопасный** *adj.* dangerously explosive

**взрывостойкий** *adj.* explosion-proof

**взрывочный** *adj.* explosive

**взрывчатость** *f.* explosiveness

**взрывчатый** *adj.* explosive

**взятие** *n.* taking

**взятый** *adj.* taken; given (value); **в. с запасом** conservative

**взять(ся)** *pf. of* **брать(ся)**

**вибрант** *m.* vibrant (human voice)

**вибрато** *n.* vibrato (*mus.*); **в. амплитуд** amplitude vitrato, tremolo

**вибратор** *m.* vibrator, dipole, doublet; oscillator; transducer; resonator; vibration exciter; **вибраторы, расположенные вдоль одной прямой линии, соосные вибраторы** collinear dipoles; **вибраторы, расположенные рядом на нескольких параллельных линиях** collateral dipoles; **активный в.** directly fed antenna; **в. в коаксиальной оболочке** sleeve-dipole antenna; **в. в параболоиде** dipole-fed paraboloid; **в. гидролокатора** hydroacoustic vibrator, sonar projector; **глубинный в.** internal vibrator; **в. из полиэтилена** rod of polystyrene; **в. из цилиндрической связки проводов** cage dipole; **в., изготовленный из смеси нескольких диэлектриков** polycrystalline transducer; **коаксиальный в.** sleeve dipole, sleeve stub (half of a sleeve-dipole antenna); **пассивный в.** parasitic antenna, parasitic element; **петлевой в.** bent dipole; **продольноколеблющийся в.** longitudinal vibrator; **резонансный в.** electrodynamic shaker; **в. с несимметричным питанием** end-fed antenna; **в. с симметрирующим устройством** skirt dipole; **в. с фокусированием излучения** focusing transducer; **сантиметровый в., в. сантиметровых волн** microwave dipole; **смещённый в.** offset dipole; **стержневой в.** needle vibrator; **шлейф-в., шлейфовый в.** folded dipole, bent dipole; **электродинамический измерительный в.** electrodynamic calibrator

**вибраторный** *adj.* vibrator, dipole; oscillator; resonator

**вибратрон** *m.* vibratron

**вибрафон** *m.* vibraphone, vibes

**вибрационный** *adj.* vibration, vibrating, shaking

**вибрация** *f.* vibration, oscillation, vacillation, swing; rumble, jarring, shaking; bounce, bouncing; flutter effect; chatter(ring); time jitter; **вибрации дозвуковой частоты** infrasonic vibrational waves; **автоколебательная в.** self-excited vibration; **(синфазная) в. стенки в целом** drumskin action; **в. станка** при резании chatter; **в. щётки** wiper chatter

**вибрирование** *n. see* **вибрация**

**вибрировать** *v.* to vibrate, to shake, to swing, to jar, to jostle, to jolt; to oscillate; to chatter

**вибрирующий** *adj.* vibrating, vibratory, vibration, shaking; jostling; oscillating; chattering

**виброанализатор** *m.* vibration analyzer

**вибровыпрямитель** *m.* vibrating-reed rectifier

**виброгаситель** *m.* vibration damper, damper; **в. с вспомогательной массой** auxiliary mass damper

**виброграмма** *f.* vibro-record

**виброграф** *m.* vibrograph; **в. ударных смещений** shock-displacement gauge

**вибродатчик** *m.* vibration pickup, vibration transducer; vibration detector

**вибродвижущий** *adj.* vibromotive

**виброизмерительный** *adj.* vibration-measuring

**виброизолирующий** *adj.* vibration isolation

**виброизоляция** *f.* vibration insulation

**вибромер, виброметр** *m.* vibration-measuring device, vibration meter, vibrometer

**вибропередача** *f.* (vibration) transmissibility

**виброплекс** *m.* vibroplex

**вибропоглотитель** *m.* vibration absorber

**вибропреобразователь** *m.* vibropack, vibrating rectifier, vibrator power pack, vibrating interrupter and rectifier, chopper, chopping inverter, (electro-)magnetic rectifier, vibratory converter, oscillating contact-rectifier

**вибропреобразовывать** *v.* to chop

**вибропрочность** *f.* vibration strength, dynamic strength

**вибропрочный** *adj.* shock-resistant, shockproof, vibration resistant

**виброскоп** *m.* vibroscope

**вибростенд** *m.* vibration table, shaker

**вибростойкий** *adj.* vibration-proof, shockproof, shake-proof, antivibration, jar-proof

**вибростойкость** *f.* resistance to vibration, shockproofability; vibration resistance, dynamic strength

**вибростол** *m.* vibration table

**вибротрон** *m.* vibrotron, mechanically controlled tube, vacuum tube

**вибротронный** *adj.* vibrotron

**виброустойчивость** *f.* vibration strength

**виброустойчивый** *adj.* vibration-proof, shatter-proof, shakeproof

**ВИВ** *abbr.* (**величина изменения высоты**) magnitude of height variation

**Виганд** Wiegand

**вигнеровский** *adj.* Wigner's, (of) Wigner

**вид** *m.* appearance, look, aspect, form, figure, shape, pattern; condition, state; type; mode (of oscillation); sight, view; make; **запертые виды колебаний** trapped modes; **виды применения аккумуляторов** methods of storage-battery operation; **в виде** (+ *gen. case*) in the form (of); as, by way (of); **в виду** (+ *gen. case*) in view (of); **внешний в. кристаллов** habits of crystals; **вырождающийся в. колебаний** evanescent mode; **иметь в виду** to keep in mind; **в. индикатора** scope pattern; **в. колебаний падающей волны** incident mode; **в. обработанной поверхности** surface finish; **особый в. телевизионных помех** chirps; **перспективный в.** expanded view; **побочный в. импульсов** spurious pulse mode; **преобладающий в. распространения** dominant mode of propagation; **в. работы счётно-решающего устройства** computer mode; **в. с торца** front elevation; end view, end-on view; **в. сигнала** signal aspect; **сквозной в.** phantom view; **скрученный в.** torsional mode; **сосредоточенный в. колебаний** trapped mode; **в. феррита** ferrospinel, ferramic, ferroxcube

**Видеман** Wiedemann

**видение** *n.* vision, sight; viewing; **в. в инфракрасных лучах** infrared television; **в. в темноте** noctovision

**видео-** *prefix* video-, television

**видеоаттенюатор** *m.* video attenuator

**видеоблок** *m.* video unit

**видеовставка** *f.* video insert, video inset; **электронная в.** chroma-key electronic video insert

**видеогеничный** *adj.* telegenic

**видеоголовка** *f.* video head

**видеодемодулятор** *m.* video modulator, picture modulator

**видеоденситометр** *m.* video densitometer

**видеодетектирование** *n.* video detection, baseband detection

**видеодетектор** *m.* video detector, video crystal, baseband detector; signal detector; **кристаллический в.** crystal video rectifier

**видеозапись** *f.* video recording

**видеоимпульс** *m.* video pulse, baseband pulse, picture synchronizing signal, frame-synchronizing (im)pulse

**видеоиндикатор** *m.* video indicator

**видеоинженер** *m.* technical director, video engineer, video operator, vision control supervisor

**видеоинтегрирование** *n.* video integration

**видеоискатель** *m.* view finder

**видеокабель** *m.* video cable

**видеоканал** *m.* television channel, video channel

**видеолампа** *f.* video tube

**видеолиния** *f.* video line

**видеомагнитофон** *m.* videotape recorder

**видеомикшер** *m.* video mixer, vision mixer; **в. центральной аппаратной** central vision mixer

**видеомодуляция** *f.* video modulation

**видеомонитор** *m.* picture monitor, television monitor

**видеонапряжение** *n.* video voltage

**видеонесущая** *f. adj. decl.* video carrier

**видеооборудование** *n.* video equipment

**видеоограничитель** *m.* video limiter

**видеооператор** *m.* video operator

**видеопередатчик** *m.* videotransmitter, visual transmitter

**видеопомехи** *pl.* video crosstalk, picture interference, picture noise, picture contamination

**видеопреобразователь** *m.* video converter

**видеораспределительный** *adj.* video distribution

**видеосигнал** *m.* video signal; **входной в.** video input; **выходной в.** video output; **нестробированный в.** raw video; **в., подаваемый на приёмную трубку** video drive; **в. с постоянной составляющей** unidirectional video signal

**видеосистема** *f.* base band system

**видеосмесительный** *adj.* video monitoring and mixing

**видеотелефон** *m.* phonovision system, videotelephone, picture-phone, television-telephone system, audovisual system

**видеотелефония** *f.* phonovision, visual communication

**видеотехника** *f.* video techniques

**видеоток** *m.* video current, light current

**видеотрак** *m.* television (*or* video) channel

**видеотрансформатор** *m.* video transformer

**видеотрон** *m.* videotron

**видеотэйп** *m.* videotape

**видеоусилитель** *m.* video amplifier; **в., вносящий заданные искажения** video distorter; **добавочный (линейный) в.** video-line booster; **многорезонансный в.** stagger-peaked video amplifier; **предварительный в.** video head, video power; **в. с корекцией индуктивности** shunt-compensated video amplifier; **в. со взаимно сдвинутыми максимумами частотных характеристик** (*or* **сдвинутыми настроенными контурами**) stagger-peaked video amplifier

**видеоусилительный** *adj.* video amplifier

**видеофильтр** *m.* video filter

**видеоцепь** *f.* video circuit

**видеочастота** *f.* video frequency, vision frequency, picture frequency

**видеоэкран** *m.* picture screen, fluorescent screen

**видиак** *m.* vidiac

**видикон** *m.* vidicon, vidicon pickup tube; **в. с развёрткой пучком медленных электронов** slow-scan vidicon

**видиконный** *adj.* vidicon

**видимость** *f.* visibility, visual range, vision, sight; **без видимости цели** blind, instrument; **в. в шумах** subchatter visibility; **предельная в.** marginal visibility conditions; **прямая в.** line-of-sight coverage; **в. строчной структуры** scanning line structure visibility

**видимый** *adj.* visible, apparent

**видность** *f.* luminosity factor, visibility factor, luminance, luminosity; **в. излучения** luminous efficiency; **в. монохроматического света** visibility factor

**видный** *adj.* visible, observable, prominent

**видно** 1. *pred. adj.* visible, obvious, clear; it can be seen; 2. *adv.* visibly

**видоизменение** *n.* modification, alteration, change, modulation, conversion

**видоизмененный** *adj.* modified, changed, altered

**видоизменить(ся)** *pf. of* **видоизменять (ся)**

**видоизменяемость** *f.* changeability, variability

**видоизменяемый** *adj.* changeable, modifiable

**видоизменять, видоизменить** *v.* to modify, to change, to alter

**видоизменяться, видоизмениться** *v.* to undergo a change, to alter

**видоискатель** *m.* viewfinder, visor, viewer, finder

**визг** *m.* squeal(ing), screech; scream, shriek

**визгнуть** *pf. of* визжать

**визжание** *n.* squeal(ing), squeak; spluttering (of arc)

**визжать, визгнуть** *v.* to squeal, to squeak, to screech, to shriek

**визибилиметр** *m.* visibility meter

**визикордер** *m.* visicorder

**визиовещание** *n.* telecasting

**визир** *m.* cross wire, hairline (pointer), visor, sight; telescope; finder; **в. грубой наводки** open sight; **в. для контроля наводки** check sight; **механический в.** mechanical index; **в. наводки по азимуту** azimuth sighting telescope; **в. наводки по углу места** elevation sighting telescope, vertical tracking telescope, elevation telescope; **в. ПУАЗО АА** director telescope, director sighting telescope

**визирный** *adj.* sighting

**визирование** *n.* sighting, sight; **в. через канал ствола** bore sighting

**визировать, завизировать** *v.* to sight; to backsight, to collimate; to level

**визитрон** *m.* visitron

**визкозиметр** *see* вискозиметр

**визкозиметрия** *see* вискозиметрия

**визуализатор** *m.* visualizer; **в. речи** visible speech analyzer

**визуализация** *f.* visualization

**визуально-двойной** *adj.* visual-binary

**визуально-звуковой** *adj.* visual-aural

**визуально-спектральный** *adj.* visual-spectral

**визуальный** *adj.* visual; direct (reading)

**«викаллой»** *m.* cobalt-vanadium steel alloy

**виктрон** *m.* victron

**вилка** *f.* fork; bracket; two-pin plug, electric connector; jaw; cradle, receiver rest; **в. (в шарнирном соединении)** jaw (of a hinge point); *pl.*

вилки V-end connections; **в. для микротелефона** switch-hook; **в. для переводки ремня** belt fork, belt guider; **контактная в.** contact clip; нерегулируемая **в.** solid jaw; **в. ножевого контакта** female contact; однополюсная **в.** single-pin plug, wander plug; **ответвительная в.** distribution plug, socket-outlet adapter; **переходная в.** plug adapter; регулируемая **в.** screw jaw; **соединительная штепсельная в.** interconnecting plug; **штепсельная (двухконтактная) в.** two-pin plug, wall plug, plug

**вилковый** *adj.* fork, forked

**вилкообразный** *adj.* forked, fork-shaped, Y-shaped

**Виллард** Villard

**Виллари** Villary

**виллемит** *m.* willemite; **в. с зелёным свечением** green willemite phosphor

**виллис** *m.* jeep

**вилообразный** *adj.* forked, bifurcate

**вилочка** *f.* Y-connector

**вилочный** *adj.* fork, forked

**Вильсон** Wilson

**вильчатый** *adj.* fork, forked, Y-shaped

**Вильямс** Williams

**ВИМ** *abbr.* (**временная импульсная модуляция**) pulse-time modulation

**ВИН** *abbr.* (**величина изменения направления**) azimuth lead

**Вин** Wien

**винил** *m.* vinyl

**винилбутиловый** *adj.* vinyl butyl

**винилит** *m.* vinylit

**винилитовый, виниловый** *adj.* vinyl

**виннокислый** *adj.* tartaric acid, tartrate

**винный** *adj.* tartaric

**винт** *m.* screw; (propeller) screw; helix; **барашковый в.** wing nut; **бесконечный в.** spiral, worm; **боковой в. цоколя** side-terminal; **воздушный в.** air screw, propeller, **гребной в.** propeller screw, screw propeller; **в. дифференциальной установки** differential screw; **в. для точной регулировки** micrometer screw, vernier; **запорный в.** screw plug; **затяжной в.** clamping screw, turnbuckle; **в. коррекции** correcting pin; **мерительный в.** micrometer gauge; **многоходный в.** multiworm screw; **нажимной в.** thrust bolt, pressing bolt, clamping screw; **в. настройки** regulating screw; **нетеряемый зажимной в.** non-losable clamp screw; **подающий в.** lead screw, feed screw; **потайной в.** set screw, grub screw; **в. регулирования положения сердечника** core screw; **в. с изолирующим ушком** insulated screw eye; **в. с камнем** jewel bearing; **в. с крыльчатой головкой** winged screw, winged bolt, butterfly screw; **в. с накатной головкой** knurled screw; **в. с подвижным стержнем в головке** capstan-headed (thumb) screw; **в. с потайной головкой** countersunk screw; **в. с шестигранным шлицем** Allen screw; **стопорный в.** stop screw, locking screw; **стяжной в.** clamping screw; **упорный в.** stop screw; **в. установки в горизонтальное положение** leveling screw; **установочный в.** set screw, adjusting screw, mounting screw, tuner screw; fixing screw, fastening screw; recalibration screw; setting screw; **ходовой в.** lead screw

**винтовка** *f.* rifle

**винтовой** *adj.* screw, spiral, helical

**винтом** *adv.* spirally

**винтообразный** *adj.* helical, spiral, screw-shaped

**виньетирование** *n.* vignetting

**виньетка** *f.* vignette

**виола** *f.* viol, viola

**Виолль** Violle

**виолончель** *f.* violoncello, bass viol

**ВИП** *abbr.* (**величина изменения пеленга**) bearing variation

**ВИР** *abbr.* (**величина изменения расстояния**) magnitude of distance variation

**вириал** *m.* virial

**вириальный** *adj.* virial

**виртуальный** *adj.* virtual

**вискозиметр** *m.* viscosimeter, viscometer, fluidimeter; **в. Лепенау** leptometer; **стержневой в.** bar viscometer

**вискозиметрия** *f.* viscometry

**вискоэлектрический** *adj.* electroviscous

**висмут** *m.* bismuth, Bi

**висмутовый** *adj.* bismuth, bismuthic

**висячий** *adj.* hanging, suspended; pendulous, pendant; trailing

**висящий** *adj.* hanging, suspended

**вита-лучи** *pl.* vita rays

**витамин** *m.* vitamin; **в. Ку** vitamin Q

**витаскан, витаскэн** *m.* Vitascan

**витафон** *m.* vitaphone

**витковый** *adj.* turn; loop

**витой** *adj.* twisted, spiral

**виток** *m.* turn, convolution; loop, coil; winding; wrap; *pl.* **витки** *oft.* band; **витки в последовательной обмотке** series turns; **обесточенные витки катушки** flared ends; **бифилярный в.** double-back loop; **в. катушки** coil segment; **короткозамкнутый в. на полюсе** pole shader; **в. обратной связи** regenerative loop, feedback loop; **центральный в. связи** center-coupled loop

**витрамон** *m.* vitramon

**витрит** *m.* vitrite

**витритовый** *adj.* vitrite

**«витром»** *m.* vitrohm

**Витстон** Wheatstone

**витый** *adj.* twisted, spun

**вить, свить** *v.* to twist, to spin, to wind

**витье** *n.* twining, twisting; torsion

**Вихерт** Wiechert

**вихревой** *adj.* vortical, vortex; turbulent, eddy, whirling; cyclical, circuital, rotational

**вихреобразование** *n.* vortex formation

**вихр** *m.* vortex; whirl, whirlwind, eddy, eddying, curl, rotation

**вкапывание** *n.* digging in

**вкапывать, вкопать** *v.* to dig in, to drive in

**ВКБ** *abbr.* (Wenzel, Kramers, Brillouin) WKB

**вкл** *abbr.* 1 (**включено**) on, switched on, cut in; 2 (**включительно**) inclusively

**вклад** *m.* deposit; contribution; ante

**вкладка** *f.* inset, insert

**вкладывание** *n.* insertion, laying in, enclosing

**вкладыш** *m.* bearing, bearing axle; bush; brass; insert, insertion piece; **в. подшипника** bearing bush, bearing liner, bearing sleeve; **в. слухового аппарата** earpiece; **сменный в.** launcher adapter, rocket adapter; **угольный в.** carbon button; **штампованный в.** wafter base; **в. штепсельного разъёма** male contact

**вклейка** *f.* inset, insert, glued-in piece; timing tape; spacer

**включатель** *m.* switch, circuit closer; **пробующий в.** feeler switch; **в. цепи** circuit closer

**включать, включить** *v.* to switch on, to cut in, to turn on, to plug in, to connect, to energize, to throw in; to include, to insert, to enclose, to embrace; to join up, to close, to engage, to put in gear; to activate; to fade-in, to blend in, to mix in; **в. в работу** to put into operation; **в. в соответсвии с правилами** to set at normal; **в. линию в цепь наблюдения** to remove the circuit for observation; **в. многократно** to multiple(*tphny.*); **в. по мостовой схеме** to bridge across; **в. по ступеням** to stagger; to connect in multiples; **в. сквозно** to jumper; **в. совместно** to interconnect; **в. ступенями** to step forward; **в. ток питания бортовой аппаратуры** to energize the round; **в. цепь управления запуском ракеты** to connect a round for firing; **в. штепсель** to plug in

**включаться, включиться** *v.* to engage, to interlock; to cut in, to enter, to listen in, to intervene; to offer; **в. в цепь** to cut in a circuit; **в. на линию** to enter a circuit

**включающий** *adj.* enclosing, including; cut-in; engaging; actuating; selector

**включение** *n.* connection, switching on, turning on; cutting in, interconnection; engaging; inclusion, insertion; lineage; actuation, activation; pull-in; implication; **автоматическое повторное в.** automatic reclosing; **аддитивное в. обмоток трансформатора** additive polarity of a transformer; **бисквитное в.** symmetrical heterostatic circuit; **в. бортовой аппаратуры наведения** guidance initiation; **в. в ответвление** shunting; **в. в состав** integration; **воздушное в.** air slug; **встречное в.** connection in opposition; **встречное последовательное в.** inverse series connection, opposed series connection; **двойное в.** idiostatic circuit, idiostatic method; **дистанционное в. приёмника** receiver distant-control, remote control of a receiver; **в. для предварительного просмотра** preview switching; **в. для транзитного разговора** through position; **в. для циркулярной передачи** conference circuit; **в. звезда-треугольник** wye-delta; **в. звеньями** recurrence network; chain connection (of electrical equipment); consecutive

connection of outside calls to several extensions; **в. и выключение** turn-on, start-stop; **квадратное в.** asymmetrical heterostatic circuit; **левое в.** link circuit; inductive neutralization; **неверное в., неправильное в.** misconnection; **ножевое в.** knife switch; **парное в.** dual-subscriber connection; **перехваченное в.** commoning; **периодическое в. и выключение** intermittent operation; **в. по местной схеме** side-tone wiring; **в. по противоместной схеме** side-tone reduction wiring; **повторное в.** reclosure, reclosing; reset; **последовательное в. через контакты реле** relay chain circuit; **последовательное встречное в. катушек** series opposing; **прямое в.** straight multiple; **прямое в. на сеть** direct on line switching; **сдвинутое многократное в.** slip multiple; **секретное в.** secrecy, privacy; **спаренное в.** bussing arrangement; **транзитное в.** built-up connection; **угловое в. синхронного двигателя** angle switching; **цепное в.** relay chain circuit; **в. четвёркой** phantom connection, superposed circuit

**включено** *adv.* on, switched on, cut in; **в. в эфир** on-the-air; **в. обходом** in bypass, in bridge; **в. параллельно** across

**включено-выключено** *adv.* on-off

**включённый** *adj.* switched on, cut in, on, connected; engaged; included, incorporated, enclosed; **в. в обход** by-passing; **в. в цепь** on, switched on, connected; **в. поперёк** cross-connected; **последовательно в.** in tandem

**включительно** *adv.* inclusively

**включить(ся)** *pf. of* **включать(ся)**

**вкопать** *pf. of* **вкапывать**

**вкрапление** *n.* impregnation, dissemination

**вкрапленный** *adj.* impregnated, disseminated, ingrained

**ВКС** *abbr.* (Всесоюзный комитет стандартов) All-Union Standard Committee

**ВКУ** *abbr.* (видеоконтрольное устройство) video monitor

**вл.** *abbr.* (влажность) humidity

**влага** *f.* moisture, humidity, dampness

**влаго-** *prefix* hygro-, moisture; water

**влагоизоляция** *f.* waterproofing

**влагомер** *m.* hygrometer; **электрический в.** electrical moisture meter

**влагонепроницаемый** *adj.* moisture-resistant, moistureproof, moisture-repellent, moisture-impermeable

**влагоотталкивающий** *adj.* moisture-repellent

**влагопоглощение** *n.* moisture absorption

**влагосодержание** *n.* moisture content

**влагостойкий** *adj.* moisture-resistant, moistureproof

**влагостойкость** *f.* moisture resistance

**влагоупорный** *adj.* moisture-impermeable, moisture-resistant, moisture-proof, dampproof

**влагоустойчивый** *adj.* moisture-resistant

**влагочувствительный** *adj.* moisture-sensitive, humidity-sensitive

**владелец** *m.* owner, possessor

**влажность** *f.* humidity, moisture, dampness, damp; moisture content

**влажный** *adj.* moist, damp, humid, wet

**влезать, влезть** *v.* to climb (in), to get in; to intrude

**влетать, влететь** *v.* to fly in, to fly into

**влетающий** *adj.* entering, incoming, oncoming

**влететь** *pf. of* влетать

**влеченный** *adj.* attracted, drawn

**влечь, повлечь, увлечь** *v.* to draw, to attract; to imply

**вливание** *n.* injection, infusion, pouring in

**вливать, влить** *v.* to pour (in), to inject, to infuse, to run in; to blend, to merge

**влияние** *n.* influence, effect, action; electrostatic induction; **в. анодного детектирования при сеточном детектировании** influence of anode rectification in a grid detector; **взаимное в.** interference; **в. вредной ёмкости** stray-capacity effect; **в. ёмкости руки, ёмкостное в. руки** body effect, hand capacitance; **в. качания антенны на контрастность изображения** flutter effect; **в. лобового сопротивления** drag effect; **в. мёртвых витков** (*or* концов) dead-end effect; **мешающее в.** interference; **в. на измерительный прибор, в. на показания прибора** influence upon an instrument; **в. наполнения чернильницы** inkwell influence; **в. неуравновешенности** position influence; **обратное в.** reaction; retroaction; reactive effect; feedback; **в. отражателя** reflector savings; **в. полётного времени на различные факторы стрельбы** time-of-flight problem; **в. помех** interference effect; **в. представления земной поверхности плоской** plane-earth factor; **в. различного наполнения чернильницы** inkwell influence; **в. ракурса(цели)** aspect effect; **в. руки** hand capacity effect, body effect; **в. слоя скачка** layer effect; **в. соседней станции** kindred effect; **в. спутной струи** wake effect; **в. упреждения на дальность** range effect of deflection; **в. холостых витков (индуктивной катушки)** dead-end effect; **в. числа M** Mach effect

**влиятельный** *adj.* influential

**влиять, повлиять** *v.* to influence, to affect, to control, to induce; to act, to operate, to function

**вложение** *n.* embedding, putting in; **инвариантное в.** invariant embedding

**ВЛТШ** *abbr.* 1 (**военно-лётная техническая школа**) technical school for aerial warfare; 2 (**высшая лётная техническая школа**) higher technical school for aviators

**вм.** *abbr.* 1 (**вместе**) together; 2 (**вместо**) instead

**ВМ** *abbr.* (**вычислительная машина**) computer

**вместе** *adv.* together, jointly; **в. с тем** in addition to that, moreover; at the same time

**вместилище** *n.* container, vessel; tank, storage tank

**вместимость, вместительность** *f.* capacity, holding capacity; room; space, volume; content; containment

**вместительный** *adj.* large, roomy

**вместить** *pf. of* вмещать

**вместо** *prep. genit.* instead (of), in place (of), for

**вмешательство** *n.* interference, intervention; meddling, break-in; **в. справочной телефонистки** operator intercepting

**вмешивать, вмешать** *v.* to add, to mix in

вмешиваться, вмешаться *v.* to interfere, to intervene (in), to cut in, to break (in, into)

вмещать, вместить *v.* to hold, to contain, to comprise; to put in, to insert

вмещение *n.* insertion, putting in

вмещённый *adj.* inserted; contained, held, housed

вмонтированный *adj.* built-in, fixed, stationary; self-contained

вмонтировать *v.* to build-in, to fix

вмороженный *adj.* trapped

ВМС *abbr.* (военно-морские силы) Navy, Naval Forces

ВМТ *abbr.* (верхняя мёртвая точка) upper dead center

вмятина *f.* dent, hollow; scab

ВН *abbr.* 1 (вакуум-насос) vacuum-pump; 2 (высокое напряжение) high voltage, high tension

ВНАБ *abbr.* (воздушное наблюдение) air observation, aerial observation

внакладку *see* внахлёстку

внаклон *adv.* slant-wise, at an angle

внакрой, внапуск *see* внахлёстку

ВНАР *abbr.* (воздушное наблюдение и разведка) air observation and reconnaissance

внахлёстку *adv.* overlap, overlapping, lap, lapped

вне 1. *prep. genit.* outside, out of, beyond, without; regardless of; 2. *prefix* ex-, extra-

внеатмосферный *adj.* extraterrestrial, extra-atmospheric

вневихревой *adj.* nonvortical, irrotational

внегалактический *adj.* extragalactic

внегородский *adj.* out-of-town

внедрение *n.* adoption (of a system), introduction, injection; implantation, instillation; в. АТС adoption of dial system

внедрять, внедрить *v.* to adopt, to introduce, to inject; to implant, to instill

внезапно *adv.* suddenly, unexpectedly

внезапный *adj.* sudden, unexpected, abrupt

внеземной *adj.* extraterrestrial

внемеридиональный *adj.* ex-meridian

внеосевой *adj.* extra-axial, off-axis

внеочередной *adj.* extra, special, top-priority

внеочередность *f.* top priority

внепиковый *adj.* off-peak

внесение *n.* introduction, insertion, bringing in, entry, entering; в. дырок в полупроводник hole injection into a semiconductor

внесённый *adj.* introduced, brought in, inserted, entered; added

внести *pf. of* вносить

внестудийный *adj.* outside, out-of-studio; remote, outdoor

внефокусный *adj.* defocused, out of focus

внецентренно *adv.* eccentrically

внецентренный *adj.* eccentric(al), non-central

внешкальный *adj.* off-scale

внешнеобразованный *adj.* externally generated

внешний *adj.* outer, external, exterior, peripherical, outward, outside; foreign, extraneous, extrinsic, alien; superficial; male (screw thread)

внешность *f.* surface, exterior, appearance, outside form, outward form

внеядерный *adj.* extranuclear

вниз *adv.* down, downwards, underneath

внизу *adv.* below, beneath, under, underneath

внимание *n.* attention, care, regard

ВНИТО *abbr.* (Всесоюзное научное инженерно-техническое общество) All-Union Scientific, Engineering and Technical Society

вновь *adv.* again, anew, one more, afresh; re-

ВНОР, ВНОРиЭ *abbr.* (Всесоюзное научно-техническое общество радиотехники и электросвязи) The All-Union Engineering Society of Radio and Electrocommunication

ВНОС *abbr.* (воздушное наблюдение оповещение и связь) civil defense warning network, intelligence net, reporting chain

внос *see* внесение

вносимый *adj.* introduced, added; inserted; insertion

вносить, внести *v.* to bring in, to carry in; to introduce, to add; to insert, to inject, to enter; в. затухание to attenuate

вносящий *adj.* bringing in; introducing, adding; inserting, injecting, entering; contributing

внутр. *abbr.* (внутренний) internal

**внутренне-** *prefix* inside, internal

**внутренний** *adj.* internal, inner, inside, interior, endo-; indoor; intrinsic, inherent; home, domestic; self-residual (stress); cut-away, cut-open

**внутренность** *f.* core; inside, interior

**внутреродность** *f.* endomorphism

**внутри** 1. *adv.; prep. genit.* inside, in, within, inward, into; **в. лунной орбиты** cislunar; 2. *prefix* endo-, intra-

**внутриатомный** *adj.* intra-atomic, subatomic

**внутригрупповой** *adj.* inter-group

**внутридолинный** *adj.* intravalley

**внутридомовый** *adj.* inside the house

**внутризаводский** *adj.* interplant, intrafactory

**внутризонный** *adj.* intraband

**внутриклеточный** *adj.* intracellular

**внутрикристаллический** *adj.* intracrystalline

**внутримолекулярный** *adj.* intramolecular

**внутриобластный** *adj.* intraregional, "oblast" (-wide)

**внутриохлаждаемый** *adj.* internally cooled

**внутрипазовый** *adj.* inner

**внутрипанельный** *adj.* back-panel

**внутрирайонный** *adj.* intraregional, community (-wide), "raion" (-wide)

**внутрирезонаторный** intracavity, incavity

**внутрисистемный** *adj.* intrasystem

**внутристанционный** *adj.* inside a station, intrastation

**внутританковый** *adj.* inside a tank, within a tank

**внутрь** *adv.; prep. genit.* in, inside, inward(s), into

**внятно** *adv.* distinctly; audibly

**внятность** *f.* intelligibility, distinctness, audibility

**внятный** *adj.* distinct, intelligible, audible

**во** *see* **в**

**вобулированный** *adj.* wobbulated

**вобулировать** *v.* to wobble

**вобулируемый** *adj.* wobbled

**вобулятор** *m.* wobbulator, wobbler; wobbler telegraph-key

**вобулять** *see* **вобулировать**

**вобуляция** *f.* wobbulation, wobble; **заполняющая в. электронного луча** fill-in-spot wobble; **синхронная в.**

электронного луча с подсветкой sampled synchronous spot wobble; **в. электронного луча** spot wobble

**вовлекать, вовлечь** *v.* to draw in, to implicate

**вовлечение** *n.* implication, drawing in

**вовлечённый** *adj.* implicated, drawn in

**вовлечь** *pf. of* **вовлекать**

**вогнать** *pf. of* **вгонять**

**вогнуто-** *prefix* concavo-, concave

**вогнуто-выпуклый** *adj.* concavo-convex

**вогнутость** *f.* concavity

**вогнутый** *adj.* concave, bent in

**вода** *f.* water

**водитель** *m.* driver, leader; **в.-автомат** automatic driver

**водить** *v.* to lead, to conduct; to carry on; to take (somewhere)

**водка** *f.* vodka; **царская в.** aquaregia

**водн.** *abbr.* (**водный**) water, aqueous

**водно-** *prefix* water, aqueous

**водный** *adj.* water, aqueous, hydrous

**водо-** *prefix* water, aqua-, hydraulic

**водовод** *m.* water pipe, water conduit, canal, flume

**водоводяной** *adj.* water-to-water; water-moderated water-cooled

**водоворот** *m.* eddy, vortex, whirl, whirlpool

**вододействующий** *adj.* hydraulic

**вододисперсионный** *adj.* water-dispersible

**водоём** *m.* reservoir, cistern, tank, water tank, well, basin; pond; water; **открытый в.** open water

**водозащищённый** *adj.* watertight; hose-proof

**водокачка** *f.* pumping station, water supply station; water tower

**водолазный** *adj.* diving; diver

**водомер** *m.* water flowmeter, water gauge, hydrometer

**водомерный** *adj.* water meter, water gauge

**водомёт** *m.* water-jet, fountain

**водомётный** *adj.* water-jet, fountain

**водоналивной** *adj.* water-filling

**водонапорный** *adj.* water-pressure

**водонепромокаемость, водонепроницаемость** *f.* waterproofness

**водонепроницаемый** *adj.* watertight, waterproof

**водоотталкивающий** *adj.* water-repellent

**водоохладительный** *adj.* water-cooling
**водоохлаждаемый** *adj.* water-cooled
**водопоглотитель** *m.* water absorber
**водопоглощающий** *adj.* water-absorbing
**водоподготовка** *f.* water treatment
**водоподогреватель** *m.* water heater
**водопровод** *m.* water conduit, water pipe, aqueduct
**водопроводный** *adj.* water-conducting; tap (water)
**водопроницаемость** *f.* permeability of water
**водопроницаемый** *adj.* permeable (to water)
**водород** *m.* hydrogen, H
**водородистый** *adj.* hydrogen; hydride (of)
**водородный** *adj.* hydrogen, hydrogenous
**водородоподобный** *adj.* hydrogen-like
**водородосодержащий** *adj.* hydrogenous
**водосброс, водослив** *m.* spillway, spillover, overfall, overflow
**водоспуск** *m.* water drain; floodgate
**водоспускной** *adj.* water drain; floodgate
**водостойкий** *adj.* water-resistant, waterproof
**водостойкость** *f.* water resistance, water-resisting property
**водоструйный** *adj.* water-jet
**водоток** *m.* flow, current of water
**водоточный** *adj.* flowing, running
**водотрубный** *adj.* water-tube
**водоуказатель** *m.* water gauge, water-level indicator
**водоуказательный** *adj.* water-level (indicator)
**водоупорность** *f.* water resistance
**водоупорный** *adj.* water-resistant, waterproof, watertight, impermeable
**водоэмульгирующий** *adj.* water-emulsifying
**водянистый** *adj.* watery, aqueous, hydrous
**водяной** *adj.* water, aqueous; aquatic; hydraulic
**водящий** *adj.* leading, guiding
**воедино** *adv.* together, jointly
**воен.-мор** *abbr.* (**военноморский**) naval, navy
**военноморский** *adj.* naval, navy
**военный** *adj.* military
**вожатый** *m. adj. decl.* driver

**вождение** *n.* leading, driving; **в. самолёта по заданной траектории** trajectory control
**возбудимость** *f.* excitability, irritability
**возбудитель** *m.* driver, exciter, oscillator, actuator, activator; energizer; erector; stimulus, stimulator; **в. возбудителя** pilot exciter; **высокочастотный в. радиомаяка** beacon beating oscillator; **отдельно стоящий в., расположенный вне плиты в.** outboard exciter; **в. с искуственной линией** artificial line bootstrap driver; **в. с несколькими обмотками возбуждения** multiple exciter; **в. с противокомпаудной обмоткой** differential compound exciter; **в. с слабым насыщением магнитной системы** undersaturated exciter; **стабилизованный линией в.** resonant-line oscillator
**возбудительный** *adj.* exciter; exciting, stimulating
**возбудить** *pf. of* возбуждать
**возбуждаемый** *adj.* driven, excitable, energized, actuated, fed
**возбуждать, возбудить** *v.* to excite, to rouse, to arouse, to stir, to provoke, to stimulate, to call forth; to give rise (to), to create; to activate, to actuate, to energize, to establish; to induce; to launch
**возбуждающий** *adj.* exciting, excitory, actuating, stimulating, driving, energizing
**возбуждение** *n.* excitation, exciting, drive, driving; activation; pump; excitement, agitation, stimulation; energizing; launching; swing; firing; magnetizing; disturbance; **встречносмешанное в.** differential excitation; **в. действия пограничного слоя** continuous excitation of boundary layer; **в. колебаний основного вида** principal-mode excitation; **в. местным гетеродином** local-oscillator drive; **в. микрофона звуком громкоговорителя** acoustical feedback; **в. напряжением на сетке** grid excitation; **независимое в.** separate excitation, external driving source; **низкочастотное в.** audio howl; **номинальное в.** rated excitation; **в. от задающего генератора** master(-oscillator) drive; **в. потоком** arma-

ture excitation; **в., регулируемое местным гетеродином** adjustable local-oscillator drive; **смешанное в. генератора с пологой характеристикой** level-compound excitation (of a generator); **согласное встречное смешанное в.** cumulative (differential) compound excitation; **ударное в. колебаний** impulsing; **ударное в. люминесценции** impact fluorescence

**возбуждённый** *adj.* excited, stimulated, perturbed

**возведение** *n.* raising, erection; **в. в квадрат** squaring; **в. в степень** involution, exponentiation

**возводить, возвести** *v.* to raise, to erect; to elevate; to derive (from) to trace to; **в. в квадрат** to square; **в. в куб** to cube; **в. в степень** to raise to power (*math.*); **в. в третью степень** to cube

**возврастающий** *adj.* ascending; recovering; returning

**возврат** *m.* return, returning; reset, resetting; link, linkage; **без возврата** non-homing; **в. в линейное положение** homing action; **в. луча** flyback (of a beam), retrace; **в. показателя циклов в исходное положение** cycle reset; **в. тока по рельсам** track return; **электрический в. энергии** electric pumpback

**возвратимый** *adj.* recoverable, retrievable; revertible

**возвратить(ся)** *pf. of* **возвращать(ся)**

**возвратнопетлевой** *adj.* reversed-loop

**возвратно-поступательно** *adv.* back and forth

**возвратно-поступательный** *adj.* reciprocating

**возвратный** *adj.* return, returning, recurring, retrogressive, reciprocal

**возвращаемый** *adj.* recoverable

**возвращать, вернуть, возвратить** *v.* to return; to recover; to restore; to reset; **в. в исходное положение** to reset

**возвращаться, вернуться, возвратиться** *v.* to return, to come back, to revert

**возвращающий** *adj.* returning, back-moving; recovering

**возвращающийся** *adj.* return, returning; resurgent

**возвращение** *n.* return, returning; restoration, restoring; reversion, re-

currence; resetting; **апериодическое в. (к заданной величине)** aperiodic recovery; **в. в исходное положение** return to normal, homing reset; regression; unset; **в. стрелки пружиной** spring control; **в. энергии в сеть** recuperation

**возвышать, возвысить** *v.* to elevate, to raise, to lift up

**возвышаться** *v.* to be raised; to rise

**возвышение** *n.* altitude, elevation, height, raising; rise, increase

**возвышенность** *f.* elevation, height, eminence; altitude

**возвышенный** *adj.* elevated, raised, increased

**возглавлять, возглавить** *v.* to head

**возглавляющий** *adj.* head, superior

**возгонка** *f.* volatilization; sublimation

**возгораемый** *adj.* inflammable, combustible

**возгорание** *n.* ignition, inflammation; **повторное в. (дуги)** restrike

**возгорающийся** *adj.* combustible, ignitable, inflammable

**возд.** *abbr.* (**воздушный**) aerial, air

**воздвигание** *n.* erection, raising

**воздвигать, воздвинуть** *v.* to erect, to raise, to set up

**воздействие** *n.* effect, reaction, action, influence; stimuli, actuation; **перекрёстные воздействия** cross-talk; **астатическое в.** floating response; **внешнее в.** external interference; **возмущающее в.** disturbing quantity; forcing function; **импульсное в.** sampling action; **мгновенное в.** quick-action of contacts; **отдалённое в.** ultimate effect; **периодическое в.** sampling action; **в. по отклонению** proportional action; **в. по ускорению** second derivative action; **в. солнечных лучей** insolation; **стабилизирующее в.** antihunt action; **суммарное управляющее в.** summation action; **управляющее в. по скорости** rate response

**воздействовать** *v.* to influence, to act (on, upon), to activate, to work (upon); to affect; to actuate

**воздействующий** *adj.* actuating; influencing

**воздух** *m.* air; atmosphere; **в.-воздух** air-to-air; **входящий в.** air-in; **выходящий в.** air-out; **в.-земля** air-to-ground; **первичный в.** combustion air; **приточный в.** additional air; **в., сжатый за счёт скоростного напора** ram air

**воздухо-** *prefix* air-

**воздуходувка** *f.* blower; blast engine; ventilator, ventilating fan; air compressor; bellows; **центробежная в.** turbo-blower

**воздуходувный** *adj.* air-blowing, blast

**воздухозаборник** *m.* air duct, air inlet, air scoop, air intake

**воздухозаборный** *adj.* air-inlet, air-intake

**воздухозаправщик** *m.* air servicer

**воздухомер** *m.* aerometer

**воздухонепроницаемый** *adj.* airtight, airproof

**воздухоотсасывающий** *adj.* exhaust, air-ejector

**воздухоохладитель** *m.* air cooler

**воздухоохладительный** *adj.* air-cooling

**воздухоохлаждаемый** *adj.* air-cooled

**воздухоочиститель** *m.* air strainer

**воздухоочистительный** *adj.* air-filtration

**воздухоприёмный** *adj.* air induction

**воздухопровод** *m.* air duct, air conduit, air pipe, air passage

**воздухопроводящий** *adj.* air-conducting

**воздухоразделительный** *adj.* air-fractioning; air-deflecting

**воздухоснабжение** *n.* air regeneration

**воздухосос** *m.* air sucker

**воздухостойкий** *adj.* air-proof

**воздухоэквивалентный** *adj.* air-equivalent; air-wall

**воздушно-** *prefix* air

**воздушно-водяной** *adj.* air-to-water

**воздушно-космический** *adj.* aerospace air-space, space-air

**воздушноохлаждаемый** *adj.* air-cooled

**воздушно-подземный** *adj.* overhead-underground

**воздушно-реактивный** *adj.* air-breathing jet (engine)

**воздушносухой** *adj.* air-dry, air-dried

**воздушноэквивалентный** *adj.* air-equivalent; air-wall

**воздушный** *adj.* air; aerial; pneumatic; overhead; air-insulated

**возмещать, возместить** *v.* to compensate, to make up, to supply; to replace, to substitute

**возможность** *f.* possibility, opportunity, feasibility; admissibility; facility; capacity, ability, capability; -ability, -ibility; **возможности по преодолению ПВО на больших высотах** high-altitude penetration capability; **предельные возможности** saturation level; **в. дальнего радиолокационного обнаружения** early-warning electronics capability; **в. дистанционного применения оружия** stand-off capability; **в. наводки** trainability; **в. повторного применения** recoverability; **в. применения (радио)помех** jamming threat; **в. продолжения работы на резервных машинах** fall-back possibility; **в. функционирования** sentencehood

**возможный** *adj.* possible, feasible, potential, practicable; probable; available; optional

**возмущать, возмутить** *v.* to agitate, to rouse, to stir up, to disturb

**возмущающий** *adj.* disturbing; driving

**возмущение** *n.* disturbance, trouble, disorder, perturbance, perturbation; **в. угла тангажа** pitch angle deviation

**возмущённый** *adj.* perturbed, disturbed

**вознести** *pf. of* **возносить**

**возникание** *see* **возникновение**

**возникать, возникнуть** *v.* to arise, to rise, to spring up, to crop up, to emerge, to develop, to originate

**возникающий** *adj.* originating, rising; nascent

**возникновение** *n.* origin, beginning, rise, emergence, formation; onset; **в. дуги** arcing; **в. электрического разряда** striking (of an arc, of a spark)

**возникнуть** *pf. of* **возникать**

**возносить, вознести** *v.* to raise

**возобновить** *pf. of* **возобновлять**

**возобновление** *n.* renewal, restoration; resumption

**возобновленный** *adj.* renewed, restored

**возобновляемый** *adj.* renewable, restorable

**возобновлять, возобновить** *v.* to renew, to restore; to resume

**возобновляющий** *adj.* regenerative, renewing, restoring

**возраст** *m.* age

**возрастание** *n.* increase, growth, rise, rising; **в. лобового сопротивления воздуха** drag-rise

**возрастать, возрасти** *v.* to grow, to increase, to rise, to accelerate; to augment; **скачкообразно в.** to jump

**возрастающий** *adj.* increasing, rising, ascending; progressive

**возрасти** *pf. of* **возрастать**

**возрастной, возрастный** *adj.* age; chronological

**возрождать, возродить** *v.* to regenerate, to renew; to reactivate

**возрождающий** *adj.* regenerative

**воинский** *adj.* military, army

**вой** *m.* howl, howling, singing, wail; warble

**войлок** *m.* felt

**война** *f.* war, warfare

**войти** *pf. of* **входить**

**вокал** *m.* vocal

**вокальный** *adj.* vocal

**вокодер** *m.* vocoder

**вокруг** *adv. and prep. genit.* round, around, circum; **в. Земли** circumtellurian; around the earth; **в. Луны** circumlunar; around the moon; **в. Солнца** circumsolar; around the sun

**воллостоновский** *adj.* Wollaston's, (of) Wollaston

**волна** *f,* wave; surge; **аксиально-симметричные волны вдоль проводов** axial line waves; **блуждающие электромагнитные волны естественного происхождения** atmospherics; **головные волны** Mintrop waves; **короткие ультрафиолетовые волны** far ultraviolet waves; **магнитные волны** TE waves; **волны растяжения в пластине** (*or* **стержне**) dila(ta)tional waves; **волны сверхвысокой частоты** microwaves; **электрические волны атмосферного происхождения** natural waves; **электромагнитные волны** TEM waves; **электромагнитные волны менее 1 см** dwarf waves; **бегущая в.** moving wave, traveling wave, progressive wave; **бегущая в. с крутым фронтом** surge; **блуждающая в.** stray wave, freak wave; transient wave; **в. возбуждения** field wave; **в. возмущений** Mach wave; **головная в.** bow wave, bow shock; **головная ударная в.** Mach wave; **в. для закрепления** fixing wave; **в. для передачи ответа** answering wave; **в. $E_{01}$, в. $E_{11}$** TM-mode; **колебательная преходящая в.** oscillatory

surge; **контрольная в.** pilot wave; **паразитная в.** disturbing wave; **в. паузы, в. покоя** spacing wave; **поперечная электрическая в. в круглом (в прямоугольном) волноводе** $TE_{m,n}$ wave in a circular (in a rectangular) waveguide; **предельная в.** shortest usable wavelength, maximum usable wavelength, maximum usable frequency; electromagnetic wave in the range from 100-200 mtrs.; **преломлённая в.** refracted wave; **преходящая в.** surge; **пробельная в.** spacing wave; **в. разрежения** negative pressure wave; Mach fan; **сантиметровая в.** microwave; **связанная в.** associated wave, bound wave, guided wave; **связанная в. пространственного заряда** coupled space charge wave; **срезанная в.** chopped wave; **в. TE с осевой симметрией** circular electric wave; **в. типа «Е»** E wave, TM wave, TM mode of transmission; **в. типа Е и Н** E-H wave, TEM wave; **в. ТМ с осевой симметрией** circular magnetic wave; **шаровая в. Герца** Hertzian derivation of the spherical wave; **эквиволюминальная в.** isometric wave

**волнение** *n.* disturbance, agitation

**волнистость** *f.* waviness, undulation; ripple factor, percent ripple; cabling; corrugation; **двойная в.** double-humped response (of a band-pass amplifier)

**волнистый** *adj.* wavy, undulating, rolling; corrugated; warped, buckled; fluted, sinous

**волно-** *prefix* ondo-, wave

**волновать, взволновать** *v.* to disturb, to agitate, to stir; to upset

**волновод** *m.* waveguide, guide wave duct; plumbing; **в.-антенна с продольной щелью** leaky waveguide; **в.-вибратор** full-wave dipole; **гребневой в.** ridge waveguide; **изогнутый в.** E-bend; H-bend; **изогнутый прямоугольный в.** edgewise bend; flatwise bend; **«мостиковый» в.** ridge waveguide; **в. Н-образного сечения** double-ridge waveguide; **однопроводной в.** guide wire; **панцирный в.** vertebrate waveguide, vertebrae; **в. переменного сечения, в. плавно меняющегося сечения** tapered waveguide; **в. П-образного сечения** bridge waveguide; **полосковый в.** microstrip; **в., работающий на частоте ниже критической** cut-off waveguide; **в. с гантелеобразным поперечным сечением** dumbbell waveguide; **в. с направляющими листами** flared radiating guide; **в. с переменным сечением** tapered waveguide; **в. с разомкнутым концом** open-ended waveguide; **«хребтовый» в.** ridge waveguide; **шланговый в.** metal-hose waveguide

**волноводно-коаксиальный** *adj.* waveguide-to-coaxial, coaxial-to-waveguide

**волноводный** *adj.* waveguide; tubular

**волновой** *adj.* wave, undulatory

**волноискатель** *m.* wave detector

**«волнолом»** *m.* wave trap

**волномер** *m.* wavemeter, ondometer; frequency meter, cymoscope, cymometer; **интерференциальный в.** heterodyne wavemeter, heterodyne frequency meter; **в. поглощающего типа на полупроводниковых приборах** absorption-type frequency meter with transistor; **проходной в.** two-entry wavemeter, transmission frequency meter; **в. с индикацией по (минимуму)тока сетки** grid-dip wavemeter; **в. с ламповым детектором** rectifier-type wavemeter; **в. с прямым отсчётом** point frequency meter; **в. с цилиндрическим резонатором** cylindrical-cavity wavemeter; **в. со связью через диафрагму** iris-coupled wavemeter

**волномер-контролёр** *m.* meter-monitor; **комбинированный в.-к. частоты** frequency meter-monitor

**волнометр** *m.* wavemeter

**волнообразный** *adj.* wave, wave-like, wavy, undulating, ripple; pulsating

**волнообразователь** *m.* oscillator (*rdr.*)

**волноповышающий** *adj.* booster

**волнопреобразователь** *m.* wave converter; **в. с изменяющейся формой** sheath-reshaping converter

**волнопробойный** *adj.* wave-cut

**волностойкий** *adj.* surge-proof

**волноуказатель** *m.* wave detector; **электрический в. Шлемильха** electrolytic detector

**волноуловитель** *m.* wave detector

**волноформирующий** *adj.* wave-shaping

**волокнистый** *adj.* fibrous, fiber, stringy, filamentous

**волокно** *n.* fiber, filament, thread; **стеклянное в.** fiberglass (insulation)

**волоконный** *adj.* fiber, fibrous, filamentous

**волокончатый** *adj.* filamentary

**волос** *m.* hair; **измерительный в.** reading hair; **в. смычка** bow hair

**волосистый** *adj.* fibrous

**волосковый** *adj.* hair; filament

**волосной** *adj.* capillary; hair; hairline

**волосовина** *f.* spill; fine crack, seam

**волосок** *m.* catwhisker, hair; filament, fiber; hair spring; **контактный в.** whisker; **в. предохранителя** fuse wire

**волосяной** *adj.* hair; capillary

**волочащийся** *adj.* dragging, trailing

**волочение** *n.* dragging, drag, haul; drawing

**волоченный** *adj.* drawn; dragged

**волочильный** *adj.* (wire-)drawing, pulling

**волочить, поволочить** *v.* to drag, to draw, to pull; to trail, to lug; to prolong

**«Волскэн»** *m.* Volscan

**волчий** *adj.* wolf

**волчок** *m.* top, tip-top; gyroscope, gyrostat

**волынка** *f.* bagpipes; droning (*rad.*)

**вольт** *m.* volt; **10-8 в.** abvolt

**Вольта** Volta

**вольтаж** *m.* voltage, tension

**вольтаический** *adj.* voltaic

**вольталак** *m.* voltalac

**вольтаметр** *m.* voltmeter, coulometer

**вольтаметрический** *adj.* voltametric

**вольтамметрия** *f.* voltammetry

**вольтампер** *m.* volt-ampere; **намагничивающие вольтамперы** quadrature exciting watts

**вольтамперметр** *m.* voltammeter, voltameter, avometer, volometer

**вольтамперный** *adj.* volt-ampere, current-voltage

**вольтамперометрический** *adj.* voltammetric

**вольтамперометрия** *f.* voltammetry

**вольтаскоп** *m.* voltascope

**вольтастат** *m.* voltastat

**вольтлиния** *f.* volt line

**вольтметр** *m.* voltmeter; **амплитудный в.** peak-to-peak indicator, peak-to-peak voltmeter; **двойной амплитуд-**

ный в. peak-to-peak voltmeter; в. для грубых измерений volt gage; карманный в. battery gage; катодный в. *see* ламповый в.; компенсационный в. slide-back voltmeter; ламповый в. electronic voltmeter, (electron-)tube voltmeter, VTVM, thermionic voltmeter; ламповый в. с анодным детектированием plate-detection voltmeter, anode-bend voltmeter; ламповый в. с установкой на нуль slide-back vacuum-tube voltmeter; ламповый пиковый в. electronic peak-reading voltmeter; в. магнитного потока flux voltmeter; в. манипулятора diode-probe-type voltmeter; многокамерный (электростатический) в. multicellular voltmeter; многошкальный в. multivoltmeter, multirange voltmeter; в. на полупроводниковых триодах transistor voltmeter; низковольтный в. cell tester; нулевой в. synchronizing voltmeter; в., основанный на явлении ионного ветра ionic wind voltmeter; в. по двойной амплитуде peak-to-peak voltmeter; проходной в. interconnectable output indicator; в. с двумя диапазонами измерения double-range voltmeter; в. с квадратичной характеристикой square-law voltmeter; в. с квадратичным детектированием square-law rectifier voltmeter; сетевой в. pilot voltmeter, line voltmeter; статический в. electrostatic voltmeter, statometer; узкополосный избирательный в. narrow-band-selective voltmeter; усредняющий в. average voltmeter; электромагнитный в. moving-iron voltmeter; электроннолучевой в. cathode-ray voltmeter; электронный в. для пиковых значений напряжения переменного тока slide-back voltmeter

**вольтметровый** *adj.* voltmeter

**вольтмиллиамперметр** *m.* volt-milammeter, avommeter, multimeter

**вольтов** *poss. adj.* voltaic

**вольтовый** *adj.* volt, voltage; voltage-type

**вольтодобавочный** *adj.* voltage-booster, booster

**вольтомметр** *m.* volt-ohmmeter, volt-ohmyst; **ампер-в.** avometer

**вольтоскоп** *m.* spark plug tester

**вольтотрансформатор** *m.* voltage transformer, transformer

**вольтсекунда** *f.* volt second

**Вольф** Wolf

**вольфрам** *m.* tungsten, W

**вольфрамат** *m.* tungstate

**вольфрамистый** *adj.* tungsten

**вольфрамит** *m.* wolframite

**вольфрамо-** *prefix* tungsten

**вольфрамовый** *adj.* tungsten, tungstic

**вольфрамо-кремниевый** *adj.* tungsten-silicon

**волюменометр** *m.* volumenometer

**волюметр** *m.* volumeter

**волюминометр, волюмометр** *m.* volumenometer, volumeter

**«вомакс»** *m.* vomax

**вомоскоп** *m.* wamoscope

**воображаемый** *adj.* imaginary, fictitious, imagined, hypothetical, unreal

**воображать, вообразить** *v.* to imagine, to figure

**воображение** *n.* imagination, idea

**воображённый** *adj.* imagined

**вообразить** *pf. of* **воображать**

**вообще** *adv.* in general, generally

**вооружать, вооружить** *v.* to arm; to outfit

**вооружение** *n.* armament, armature, arming; fitting out; **реактивное в.** missile armament

**вооружённый** *adj.* armed; fitted out

**вооружить** *pf. of* **вооружать**

**вопреки** *prep. dat.* despite, in spite (of), regardless (of), contrary

**вопрос** *m.* question, interrogation; inquiry; examination, scrutiny; **альтернативный в.** yes-no question

**воронка** *f.* funnel; eddy; cone (of depression); hopper; **вводные воронки** leading-in cups; **кабельная в.** cable terminal; **мёртвая в.** dead arc; dead area, cone of silence

**воронкообразный** *adj.* funnel, funnel-shaped, funneled; cone-shaped

**ворот** *m.* reel, winch, capstan; shaft, drum; collar; pull, hoist; **подъёмный в.** reel, windlass

**воротить** *pf. of* **ворочать**

**воротник** *m.* collar, lip, flange

**воротный** *adj.* winch, windlass

**вороток** *m.* tap wrench, twisting pliers; tuning key

**ворочать, воротить** *v.* to reverse; to turn, to roll

**ворсистость** *f.* fluffiness
**ворсистый** *adj.* fluffy, fluffed
**вортекс** *m.* vortex
**восемнадцатеричный** *adj.* octodenary
**воск** *m.* wax; **горный в.** ozokerite, native paraffin; **в. ископаемого происхождения** mineral wax
**восковка** *f.* tracing paper; wax paper
**восковой** *adj.* wax, waxy
**воспламенение** *n.* combustion, ignition, inflammation, firing; burning, kindling; fusing; initiating combustion; **в. с помощью электрозапала** glow-plug ignition
**воспламенитель** *m.* igniter, lighting device; detonator, blasting cap; ignition charge; **пиротехнический в.** pinwheel igniter, powder igniter; **пиротехнический пороховой в.** powder-charge igniter; **форкамерный в.** igniter-chamber assembly, pilot-flame igniter
**воспламенить** *pf. of* **воспламенять**
**воспламеняемость** *f.* combustibility, flammability, inflammability
**воспламеняемый** *adj.* inflammable, combustible, ignitable
**воспламенять, воспламенить** *v.* to ignite, to flame, to inflame
**воспламеняющийся** *adj.* combustible, inflammable, ignitable
**восполнять, восполнить** *v.* to fill in, to complete, to make up; to supply
**воспользоваться** *pf. of* **пользоваться**
**воспрепятствовать** *pf. of* **препятствовать**
**воспретительный** *adj.* prohibitive
**воспрещать, воспретить** *v.* to deter; to prohibit
**воспрещение** *n.* denial, prohibition
**восприимчивость** *f.* susceptibility, receptivity, susceptiveness; **удельная магнитная в. на единицу массы** coefficient of magnetization
**восприимчивый** *adj.* receptive, susceptible, sensitive
**воспринимаемость** *f.* perceptibility
**воспринимаемый** *adj.* perceptible, discernible; sensed
**воспринимание** *see* **восприятие**
**воспринимать, воспринять** *v.* to take, to assume, to receive; to absorb; to perceive; to sense
**воспринимающий** *adj.* receiving, pickup; sensing, detecting

**восприятие** *n.* perception, sensing, sensation, perceptibility; pickup; **в. слов** word association
**воспроизведение** *n.* reproduction, reproducing, regeneration; playback; rendering, rendition; display, representation; recovery; *see also* **верность; в. изображения поражённого шумами** noisy reproduction; **в. низких частот** bass response; **прямое в.** instantaneous playback, play-over; **в. с выдержкой времени** playback delay, delayed relay action; **в. сигнала** playback
**воспроизвести** *pf. of* **воспроизводить**
**воспроизводимость** *f.* reproducibility, repeatability
**воспроизводимый** *adj.* reproducible
**воспроизводительный** *adj.* reproductive, reproducing
**воспроизводить, воспроизвести** *v.* to reproduce; to produce; to render; to display; to transcribe
**воспроизводство** *n.* reproduction, regeneration
**воспроизводящий** *adj.* reproducing
**воспротивление** *n.* resistance
**воссоединение** *n.* recombination
**воссоединять, воссоединить** *v.* to reunite, to recombine
**воссоединяться, воссоединиться** *v.* to reunite (with), to recombine
**восстанавливаемость** *f.* restorability, restorability ratio; reducibility
**восстанавливаемый** *adj.* restorable; reducible; remade
**восстанавливать, восстановить** *v.* to restore, to reestablish, to recreate; to recondition, to reinstate; to regenerate, to recover, to reclaim; to reduce; *oft.* re- + verb, e.g. **в. нейтральное положение** to recentralize; **в. нулевое положение** to rezero
**восстанавливающий** *adj.* restoring, regenerative, regenerating; reducing; righting; reconstruction
**восстанавливающийся** *adj.* regenerative
**восстановимый** *adj.* restorable, reducible
**восстановитель** *m.* restorer, regenerator, renovator, restituent; reinserter; reducing agent, reducer; **в. импульсов** regenerator; **в. постоянной составляющей** d-c restorer; **в. постоянной составляющей сигнала**

цветности chrominance direct-current restorer

**восстановительный** *adj.* regenerating, restoration; reducing, reduction

**восстановить** *pf. of* **восстанавливать**

**восстановление** *n.* restoration, re-establishment, renewal; reactivation, recovery; resetting; recombination; restitution; reduction; reinsertion; resumption; **в. дискретной информации** staticising; **в. окисления** redox; **в. прежнего соединения** reconnection; **в. сеточного управления (тиратрона)** grid recovery time; **в. спектра** de-emphasis; **в. чувствительности** suppression recovery; **в. энергетической сети** power restoration

**восстановленный** *adj.* restored; recovered; reinserted; reconstructed

**восстановляемость** *f.* reducibility

**восстановляемый** *adj.* reducible

**восстановлять** = **восстанавливать**

**восстановляющий** *adj.* reducing

**восстановляющийся** *adj.* reducible

**вост.** *abbr.* 1 (**восток**) east; 2 (**восточный**) east, eastern

**восток** *m.* east

**восточный** *adj.* east, eastern

**восход** *m.* rise, rising, ascent

**восходить, взойти** *v.* to rise, to ascend

**восходящий** *adj.* rising, ascending, anabatic; upward

**восхождение** *n.* ascension, ascent; **прямое в.** right ascension

**восьмая** *f. adj. decl.* eighth; quaver (*mus.*)

**восьмеричный** *adj.* octuple, octal, octonary

**восьмёрка** *f.* eight; figure eight, figure-of-eight; double-phantom circuit

**восьмёрочный** *adj.* figure-of-eight; octuple; eight-

**восьми-** *prefix* oct-, octa-, octo-; eight

**восьмивалентный** *adj.* octavalent

**восьмигранный** *adj.* octahedral

**восьмизначный** *adj.* eight-digit, eight-place

**восьмикратный** *adj.* octuple; eightfold

**восьмиотрезный** *adj.* eight-bar

**восьмиполюсник** *m.* eight-terminal network, eight-terminal circuit

**восьмиполюсный** *adj.* eight-terminal

**восьмиразрядный** *adj.* eight-digit

**восьмиричный** *adj.* octal, octonary

**восьмисегментный** *adj.* eight-segment

**восьмисложный** *adj.* octosyllabic

**восьмиугольный** *adj.* octagonal

**восьмиштырьковый** *adj.* octal (base)

**воткнутый** *adj.* driven (in, into); mounted (into)

**воткнуть** *pf. of* **втыкать**

**вощение** *n.* waxing

**вощёный** *adj.* waxed

**воющий** *adj.* howling; warble

**впадать, впасть** *v.* to fall in; to discharge (into), to flow

**впадающий** *adj.* falling into; discharging (into); flowing into

**впадение** *n.* inflow; **в. в синхронизм** lock-in

**впадина** *f.* trough (of wave); valley (of curve); hollow, cavity, depression; gap, notch, space; sag; crevasse; concavity, dent; **в. на горизонтальной площадке импульса** pulse valley

**впаиваемый** *adj.* sealed in, soldered in

**впаивание** *n.* sealing-in, soldering in

**впаивать, впаять** *v.* to seal in, to solder in

**впай** *m.* soldering (in), sealing (in), seal, seal-in; press lead; lead; **зажимный в.** hook lead; **простой в.** tip lead; **в. штенгеля в колбу** stem-to-bulb seal, sealing-in joint

**впайка** *f.* sealing in, soldering in; **дисковая в.** disk seal

**впалость** *f.* hollowness, concavity

**впалый** *adj.* hollow, sunken, concave

**впараллель** *adv.* in parallel; (connected) across; **в. линии** in parallel with the line; across the line

**впасть** *pf. of* **впадать**

**впаянный** *adj.* sealed in, soldered

**впаять** *pf. of* **впаивать**

**вперёд** *adv.* forward, on, ahead, forth, onward; first

**впереди** *adv.* in front, before, ahead (of)

**вперекидку** *adj.* pile (winding)

**вперекрой, вперекрышку** *adv.* overlapping

**впечатление** *n.* impression, sensation

**впечатлительность** *f.* susceptibility; impressibility, sensitiveness

**впечатлительный** *adj.* susceptible; impressionable, sensitive

**вписанный** *adj.* inscribed, written in, entered

**впитанный** *adj.* absorbed

**впитать(ся)** *pf. of* **впитывать(ся)**

**впитывание** *n.* absorption, soaking in

**впитывать, впитать** *v.* to absorb, to take up

**впитываться, впитаться** *v.* to be absorbed

**впитывающий** *adj.* absorbing, absorbent

**вплавление** *n.* fusing, fusion

**вплавленный** *adj.* fused

**вплеск** *m.* splash

**вплескивать, вплеснуть** *v.* to splash in, to dump in

**вплотную** *adv.* close, close by, closely, up to

**вплоть** *adv.* up to, till; close; **в. до** down to, up to

**вполне** *adv.* completely, fully, entirely, totally, wholly, quite, perfectly

**вполунакрой, вполунахлест, вполунахлёстку** *adv.* half-lap

**впотай** *adv.* flush, even

**ВПП** *abbr.* (**взлётно-посадочная полоса**) runway, landing and take-off strip

**впрессование** *n.* stamping

**впрессованный** *adj.* stamped in, pressed in, built in, set in, embedded; pressfitted

**впритык** *adv.* butt, butt-joint; against, flush, end to end

**впрочем** *adv.* however, though; on the other hand; nevertheless

**впрыск** *m.* injection, injecting; **в. (в камеру) против потока** backshot injection

**впрыскать** *pf. of* **впрыскивать**

**впрыскивание** *n.* injection, injecting

**впрыскивательный** *adj.* injection

**впрыскивать, впрыскать, впрыснуть** *v.* to inject, to spray in

**впрыскивающий** *adj.* injecting, spray, spraying in

**впрыснутый** *adj.* injected

**впрыснуть** *pf. of* **впрыскивать**

**ВПУ** *abbr.* (**внутреннее переговорное устройство**) intercommunication system

**впуск** *m.* inlet, intake, admission, admittance, letting in, introduction, induction, inflow, afflux

**впускание** *n.* admission, admittance, letting in, introduction, inlet, intake; **в. носителей** carrier injection

**впускать, впустить** *v.* to let in, to introduce, to inject, to admit

**впускной** *adj.* admission, intake, inlet

**впустить** *pf. of* **впускать**

**впущенный** *adj.* admitted, injected, let in

**ВР** *abbr.* (**воздушная разведка**) air reconnaissance

**враждебный** *adj.* hostile, antagonistic

**вражеский** *adj.* hostile, enemy's

**вразбежку** *adv.* alternate, alternately

**вразброс** *adv.* scattered, haphazard

**вразрез** *adv.* contrary; in series

**вразумительный** *adj.* clear, articulated, intelligible

**врассечку** *adv.* in series; contrary

**врассыпную** *adv.* in all directions, in confusion, in disorder

**врастяжку** *adv.* flat, at fulllength, prone

**вращеп** *adv.* split, forked

**вращаемый** *adj.* rotatable

**вращатель** *m.* rotator, spinner, (aircraft) scanner; **в. грамофонных пластинок** phonograph mechanism

**вращательный** *adj.* rotary, rotatory, rotatable; rotation(al); vortical

**вращать** *v.* to turn, to rotate, to revolve; to drive; to wind; **в. рукоятку** to crank

**вращаться** *v.* to whirl, to spin, to revolve, to rotate, to turn, to circle, to gyrate; to roll; to run, to move

**вращающий** *adj.* rotating, turning; **в. момент** torque

**вращающийся** *adj.* rotating, rotary, rotable, rotative, revolving, turning, spinning, gyrating; running, moving

**вращение** *n.* rotation, revolution, rotary motion, revolving, turn, turning, gyration, spin; running; **в. вправо** clockwise rotation; **в. источника излучения вокруг пациента** rotational therapy; **круговое в.** revolution; **в. (машины) в перегретом состоянии** running hot; **в. относительно оси X**-rotation; **в. при помощи мотора** power slewing; **в. фазы** phase shift

**вращполе** *n.* rotating field

**ВРД** *abbr.* (**воздушно-реактивный двигатель**) thermal jet engine, thermojet, air breather

**вред** *m.* damage, injury

**вредительский** *adj.* harmful

**вредить, повредить** *v.* to harm, to damage, to hurt, to injure

**вредность** *f.* harm, damage, injury

**вредный, вредоносный** *adj.* harmful, bad, injurious, deleterious, detrimental; ill(effect)

**врезание** *n.* incision, notch, gash
**врезанный** *adj.* notched, incised
**врезать** *pf. of* **врезывать**
**врезной** *adj.* set in, fit in; cut in
**врезывать, врезать** *v.* to cut, to cut in, to engrave; to fit in, to fit into; to notch; to embed
**временник** *m.* timer; annals
**временно** *adv.* temporarily
**временноимпульсный** *adj.* pulse-time
**временной** *adj.* temporal, provisional; interim; time
**временный** *adj.* temporal, temporary, provisional; transitory, transitional; auxiliary
**время** *n.* time; period; **в. блокировки (эхозаградителя)** hangover time (of an echo-suppressor); **введённое в. задержки** preset delay time; **в. включения** turn-on time, time-on, make-time; **в. возбуждения люминофора** rise time; **в. возврата в устойчивое положение** recovery time; **в. возможного срабатывания радиовзрывателя** armed period; **в. восстановления управления по сетке** grid-recovery time; **в. входа ПУАЗО в режим** director settling time; **в. выборки данных, в. выдачи информации** access time; **в. выдачи кода** word time; **в. выделения сигнала вспышки** burst sampling time; **в. выпадения (радиоактивных осадков)** fallout arrival time; **в. высвечивания** fluorescent lifetime; **в. гашения** blanking interval; **в. гашения (газового разряда)** deionization time (of gas discharges); **в. гашения луча** blanking time; **в. горения дуги** arcing time; **в. готовности** readiness time; reaction time; **в. движения кадра** moving period; **в. действия** response time, working time, time of operation; **в. действия дуги** arc time; **декретное в.** daylight saving time; **в. до и после запуска снаряда** X-time; **в. заглушения приёмника** blocking time; **заданное в. опережения** procurement lead time; **в. заданной последовательности** numerical time; **в. задержки на устранение неисправности** repair delay time; **в. заказа** *see* **в. приёма заказа; в. замыкания** closing time, make time; **в. занятия** holding period, holding time; **в. записи с**

предварительной очисткой (ячейки) clear-write time; **в. заправки жидким кислородом** loxing time; **в. изодрома** integral action time, reset time; **инерционное в.** coast time; **инерционное в. (тиратрона)** pulse-firing time; **в. интенсивного обмена** busy hours; **в. кадрового обратного хода** frame flyback time, image flyback time; **лишнее в.** overtime period; **в. ложного отсутствия целей, в. ложного спокойствия** false quietness time; **максимальное в. ожидания** storage cycle period; **максимальное в. хранения** maximum retention time, storage time; **в. манипуляции** keying cycle; **минимальное в. ожидания** mimimum latency; **в. на набор высоты для перехвата** climb-to-intercept altitude time; **в. на перехват цели** detection-to-interception time lapse; **в. на приведение в готовность средств ПВО** reaction time; **в. на радиолокационное обнаружение и опознавание** radar log time; **в. на развёртывание** setting-up time, set up time; **в. набора высоты** (*or* **подъёма**) time of climb; **назначенное в. разговора** booking time (*tphny.*); **в. наибольшей нагрузки** busy period, busy hour; **в. наработки на отказ** error-free running time; **в. нарастания (уровня) сигнала** attack time, build-up time; **в., необходимое для передачи цифры** one-digit time; **в. неправильной работы** down-time; **непродуктивное в. при исправной машине** no charge non-machine-fault time; **номинальное в.** time rating; **нормированное производственное в.** standard duty periods, standard running periods; **в. обратного хода** return interval, scan flyback, flyback period, return time, retrace time, beam dead time; **в. обращения информации** access time; **общепринятое в.** standard time; **объёмное в. жизни** volume lifetime, bulk life; **в. определения местонахождения** fix time; **в. остановки** down time; **в. отзвука** duration of audibility; **в. отладки** debug time; **относительное в. включения** relative working time, relative duty cycle; **относительное в. обратного** хода развёртки retrace ratio; в. отработки дальности position learning time; в. отставания dead time; в. пассивного полёта coast time; в. передачи (информации в накопитель) access time; в. передней площадки front-porch interval; в. переключения в закрытое состояние turn-off time; в. переключения в открытое состояние turn-on time; в. перемещения time of travel; в. перехода из нормального состояния в сверхпроводящее normal-superconducting transition time; в. перехода процесса transient period; в. перехода триггера binary transition time; плавное в. работы scheduled operating time; в. подготовки к работе setup time; в. подготовки системы управления к действию readiness time; в. подрегулирования adjusting time; в. подрыва боевой части burst time; в., покрываемое счётом billing period; в. полёта по инерции coasting time; в. полёта по орбите за один оборот orbital period; полное в. обработки floor-to-floor period, floor-to-floor time; полное в. отключения total break-time, interrupting time (of a switching device); в. полной загрузки линии critical moment of traffic flow (of telephone exchange systems); в. последействия hangover time (of an echo-suppressor); в. послесвечения decay time; в. послесвечения экрана after-glow time, screen persistence time, screen storage time; в. посылки импульсов keying cycle; в. появления сигнала epoch; предельное полётное в. (для ПУАЗО) director time-of-flight limit; в. предупреждения при ракетном ударе МБР warning time on an ICBM attack; в. приёма заказа (на переговор) booking time (of a call), filing time; в. пробега transit time, travel time; transit time of an electron; sound transit-time; running time; в. продёргивания плёнки (в кинопроекторе) pull-down time; в. проекции opening time (of movie projectors); playing time; в. произвольной выборки random access time; в. пропуска поля field-suppression

period, frame suppression period; в. простоя standby unattended time, dead (or idle) time; в. простоя радиостанции off-air time; в. прохождения переднего фронта leading-edge pulse time; в. прямого хода развёртки trace interval, video interval; в. работы двигателя burning duration (of an engine), combustion duration; в. работы (двигателя) на полной мощности full flow endurance; в. работы (двигателя) с максимальной тягой full thrust duration; в. работы машины без ошибок good time; в. работы на номинальном режиме rated duration; в. развёртывания (станции) build-up time (of a station), setting-up time; в. разговора holding time, ticket time; в. разгона start time, rise time, response time; settling time; в. разложения (изображения) по вертикали vertical scanning interval; разрешающее в., в. разрешения resolving time, time resolution; в. раскачивания контуров, в. раскачки контура build-up time (of a tuned circuit); в. решения computing time; в. сбегающего края импульса trailing-edge pulse time; сверхурочное в. overtime; в. свободного искания hunting time; в. свободного полёта coasting period; в. слабого обмена slack hours; в. смены одного другим change-over period; собственное в. отключения opening time; собственное в. (устройства) intrinsic time; в., соответствующее задней площадке гасящего импульса back-porch interval; в., соответствующее передней площадке гасящего импульса front-porch interval; в. спада, в. спадания release time, releasing time; decay time; fall time; в. спадания интенсивности вдвое half-intensity period; в. срабатывания operating time (of a relay); transit time (of armature); actuation time, response time; clearing time, opening time; pickup time; в. срабатывания (пружины) (relay-spring) rebounding time; в. срабатывания (счётчика частиц) resolving time (of a radiation counter); среднее в. безошибочной работы mean free time between errors; в.

существования спутника orbit life-time; **точное в. запуска (снаряда)** T time; **в. транспорта киноплёнки** pull-down time; **в. тренировки лампы** (*or* **трубки**) burning-in period; **в. трогания (реле)** time for motion to start; **в. удержания абонента** period of number reservation (in long-distance service); **в. упреждения** rate time, derivative action time; lead time; **упреждённое полётное в.** future time of flight; **в. успокоения измерительного прибора** response (of a measuring instrument); **в. успокоения (стрелки)** decay time (of an instrument pointer); **в. установления** setup time, settling time, transition time, transient time, time of rise; **в. установления регулируемого параметра** correction time; **в. фиксации уровня** clamping interval; **в. формирования катодов** heating-up period of cathodes; **элементарное в. срабатывания** elementary reaction time

**«времянка»** *f.* temporary connection, temporary bridge, jumper, haywire; bug trap

**ВРМ** *abbr.* (**всенаправленный радиомаяк**) omnidirectional beacon, omnidirectional radio range

**вровень** *adv.* level, flush (with)

**вроде** *prep. genit.* like, such as

**вронскиан** *m.* Wronskian

**Вронский** Wronski

**врс.** *abbr.* (**верста**) verst (3,500.00 feet)

**ВРС** *abbr.* (**внутрирайонная связь**) "raion"(-wide) communication service

**ВРТС** *abbr.* (**внутрирайонная телефонная станция**) "raion"(-type) telephone office, community office, community exchange

**вруб** *m.* cut, notch, channel

**врубать, врубить** *v.* to cut in; to switch on, to throw in; to groove, to notch

**врубка** *f.* groove, slot, notch

**врубленный** *adj.* cut in, grooved, notched; switched on

**вручную** *adv.* manual, manually, hand, by hand

**ВС** *abbr.* (**Вооружённые Силы**) Armed Forces

**всасываемость** *f.* absorbability

**всасываемый** *adj.* suction; absorbable

**всасывание** *n.* absorption, soaking-in, suction; intake, induction

**всасывать, всосать** *v.* to absorb, to suck in; to draw in; to surge

**всасывающий** *adj.* sucking, suction; absorbing; intake; absorptive

**все** *pl. see* **весь**; all, everyone, everybody

**всё** *pron. see* **весь**; 2. *adv.* always, all the time; only; still

**все-** *prefix* all-, omni-, pan-

**всеволновый** *adj.* all-wave, multipleband, all-band

**всегда** *adv.* always, constantly, ever

**вседиапазонный** *adj.* all-band

**ВСЕК** *abbr.* (**вольтсекунда**) volt second

**Вселенная** *fl. adj. decl.* universe, world

**вселенский** *adj.* universal

**всемирный** *adj.* universal, global; world, world-wide

**всенаправленный** *adj.* omnidirectional all-directional; omnibearing; omni-range

**всеобщий** *adj.* common, universal, general

**всеохватывающий** *adj.* all-embracing, comprehensive

**всепогодный** *adj.* all-weather

**всес.** *abbr.* (**всесоюзный**) All-Union

**всесветный** *adj.* universal, common

**всесоюзный** *adj.* All-Union, Union

**всесторонне** *adv.* comprehensively, in detail, thoroughly; closely

**всесторонний** *adj.* comprehensive, detailed, thorough; close

**всецело** *adv.* completely, wholly, entirely

**всецелый** *adj.* complete, whole, entire, full

**всечасный** *adj.* hourly

**вскружить** *pf. of* **кружить**

**вскрывать, вскрыть** *v.* to open, to reveal

**вскрытие** *n.* opening; stripping; discovery

**вскрыть** *pf. of* **вскрывать**

**вслед** *adv. and prep. dat.* after, following; **в. за** behind

**вследствие** *prep. genit.* in consequence of, on account of, owing to

**вслепую** *adv.* blind(ly)

**всосанный** *adj.* absorbed; drawn in; suction

**всосать** *pf. of* **всасывать**

**вспенивание** *n.* foaming, frothing

**вспенивать, вспенить** *v.* to foam, to froth

**всплеск** *m.* surge; splash, splashing, dash; bump; kick; *pl.* coronals; **в. напряжения** surge, tension impulse, voltage pulse; **в. от разрыва снаряда** shell splash

**всплескивание** *n.* splashing

**всплёскивать, всплеснуть** *v.* to splash

**всплывать, всплыть** *v.* to ascend, to rise to the surface, to emerge, to float (up)

**всплытие** *n.* ascent, emersion, floating (up)

**всплыть** *pf. of* **всплывать**

**всползание** *n.* creeping

**вспомогательный** *adj.* auxiliary, accessory, subsidiary; spare, emergency; booster; standby; adjunct; ancillary; slave

**вспучивание** *n.* buckling, bulging; swelling; mushrooming

**вспыхивание** *n.* flash, flashing, flare up, flame up, flickering, blinking

**вспыхивать, вспыхнуть** *v.* to flash, to flare up, to flicker, to blink; to scintillate

**вспыхивающий** *adj.* flashing, flaring

**вспыхнуть** *pf. of* **вспыхивать**

**вспышка** *f.* flash, flare, burst, blink; splash; scintillation; outburst, burst pulse, subcarrier pulse, color burst sync signal; lightning (flash); **в. кругового огня** buckover; **в. сигнала на экране радоилокатора** radar signal glint; **хромосферная в.** chromospheric eruption

**вспышковый** *adj.* flashing

**вспышкообразный** *adj.* flash, flash-like

**вставить** *pf. of* **вставлять**

**вставка** *f.* insert, insertion piece, inset, embedding; cue, inlay; obstacle; spacer; mounting installation, assembly; **дисковая в. на электроде** contact point insert; **изоляционная в. с плавкой проволокой** fuse carrier, fuse holder; **коаксиальная аттенюаторная в.** coaxial pad insert; **легкоплавкая в.** heating fuse; **плавкая в.** fuse link, fuse cartridge, fuse; **разделительная в.** sectionalizer; **разъединительная в.** isolating (or disconnecting) link; **шлейфовая в. (осциллографа)** wattmeter loop (of loop oscillographs)

**вставкодержатель** *m.* fuse holder

**вставление** *n.* insertion, introduction, installation

**вставленный** *adj.* inserted, set in; inlaid; mounted, installed

**вставлять, вставить** *v.* to insert, to plug in, to introduce, to put in, to set, to set in, to fit, to fit in, to fix in, to embed; to secure; **в. втулку** to bush; **в. штепсель** to plug, to plug in

**вставной** *adj.* inserted, insertion; plug-in, push-in; slide-in; detachable

**вставной-вытяжной** *adj.* push-pull

**вставочный** *adj.* insertion (piece)

**встраивать, встроить** *v.* to build in, to install, to mount

**встретить(ся)** *pf. of* **встречать(ся)**

**встреча** *f.* meeting, encounter; incidence; rendezvous

**встречать, встретить** *v.* to meet, to encounter

**встречаться, встретиться** *v.* to meet, to face; to be found, to occur

**встречно-параллельный** *adj.* anti-parallel

**встречноштыревой** *adj.* interdigital, interdigitated

**встречный** *adj.* counter; contrary, opposite; opposing, opposition; head-on; back

**встроенный** *adj.* built-in, fixed; housed; embedded, integrated

**встроить** *pf. of* **встраивать**

**встройка** *f.* mounting; **в. в корпус** cavity mounting

**встряхивание** *n.* conditioning (of microphone); shaking, jar, jarring; stirring, jogging; **в. когерера** tapping back

**встряхиватель** *m.* shaker; scrambler

**встряхивать, встряхнуть** *v.* to shake (up), to agitate; to jolt

**встряхивающий** *adj.* shaking, jolting

**встряхнуть** *pf. of* **встряхивать**

**вступать, вступить** *v.* to initiate (conversation), to originate; to enter, to step in

**вступающий** *adj.* entering, incoming

**вступительный** *adj.* introductory, ingoing, incoming

**вступить** *pf. of* **вступать**

**вступление** *n.* arrival (of a wave); entry; introduction; opening, inlet; **в. в связь** netting, establishing the communication

**встык** *adv.* butt; **приваренный в.** buttwelded

**всунутый** *adj.* inserted, put in

**всхлипывание** *n.* motor-boating

**всход** *m.* ascent, rise

**всыпной** *adj.* random, haphazard; pouring

**вся** *see* **весь**

**ВТ** *abbr.* 1 (**ватт**) watt; 2 (**воздушная тревога**) air raid warning

**втекание** *n.* inflow, inflowing, influx, introduction, afflux, induction

**втор-** *abbr.* (**вторичный**) secondary

**вторгаться, вторгнуться** *v.* to break in, to intrude; to invade; to intervene

**вторжение** *n.* infringement, intrusion, invasion, incursion; **в. космических лучей** burst of cosmic rays

**вторично** *adv.* second time, again; secondarily; re-

**вторично-электронный** *adj.* secondary-electron; secondary-emission

**вторично-эмиттирующий** *adj.* secondary-emission

**вторичный** *adj.* secondary; reiterative; subsidiary; derivative; branch-, side-

**второй** *adj.* second; alternate

**второразрядный** *adj.* secondary, inferior

**второстепенный** *adj.* secondary, minor, accessory

**ВТ-С** *abbr.* (**ватт-секунда**) watt-second

**ВТС** *abbr.* (**военно-техническое снабжение**) military technical supply

**ВТСЕК** *abbr.* (**ватт-секунда**) watt-second

**ВТУ** *abbr.* (**временные технические условия**) provisional technical specifications

**ВТУЗ** *abbr.* (**высшее техническое учебное заведение**) technical college, school of higher technical education

**втулка** *f.* bush, bushing, shell, sleeve; plug, insert, hub, bolster; **в. анода** anode lug; **в. для укрепления кабельных консолей в колодце** cable supporting sleeve; **замедляющая в.** copper slug (of a relay), copper head; **в. колеса** wheel hub; **кондукторная в.** guide socket, guide box; **в. масляника** breaker shell; **навивная водонепроницаемая в.** watertight gland; **направляющая в.** journal bearing, collar bearing, guide socket; **опорная в.** bush, hub; **переменяющаяся капсюльная в. (взрывателя)** slider; **в. под контактные кольца** slip-ring bush; **в. подпятника**

thrust collar; **проходная в.** wall bush, grommet; **разъёмная в.** plug hub; **в. ролика токосъёмника** trolley bush; **в. сальника** gland; **штепсельная контактная в.** plug socket

**втулочка** *f.* bush, bushing, (insulating) collet (in a relay)

**втулочный** *adj.* bush, bushing, socket, sleeve, collar, boss; insert, plug

**втч, вт-ч** *abbr.* (**ватт-час**) watt-hour

**в т. ч.** *abbr.* (**в том числе**) among them

**втыкать, воткнуть** *v.* to plug in, to drive in, to thrust into

**втычной** *adj.* plug-in; clip-on, tong-test

**втягивание** *n.* pull-in, pulling in, drawing in; pickup; **в. в синхронизм** picking-up (synchronous speed, synchronous motors)

**втягивать, втянуть** *v.* to pull in, to pull into, to draw in; to absorb

**втягиваться, втянуться** *v.* to be drawn in

**втягивающий** *adj.* pull-in; retract, retractable

**втягивающийся** *adj.* retracting, drawing in; plunging

**втяжной** *adj.* suction, sucking; plunger; push (*see also* **вытяжной**)

**втянуть(ся)** *pf. of* **втягивать(ся)**

**ву** *abbr.* (**вертикальный угол**) vertical angle

**вуалирование** *n.* fogging, veiling effect; masking effect; hazing, blooming

**вуалирующий** *adj.* masking; fogging

**вуаль** *f.* fog (area) (*tv*); haze; film, veil; turbidity; fuzzy (sound); **«серая» в.** grayout; **фотографическая в.** haze (*photography*)

**Вуд** Wood

**ВУЗ** *abbr.* 1 (**военно-учебное заведение**) military school; 2 (**высшее учебное заведение**) college

**вулканизация** *f.* vulcanization, cure, curing

**вулканизированный** *adj.* vulcanized, rubberized

**вулканизировать** *v.* to vulcanize

**вулканизование** *n.* vulcanization

**вулканизованный** *adj.* vulcanized, rubberized

**вулканизовать** *v.* to vulcanize

**вулканит** *m.* vulcanite

**Вульф** Wulf

**вурцит** *m.* wurzite

**вход** *m.* input, "in"; intake, inlet; entry, entrance, access; port; point of entry (of a circuit); **в. в луч** initial entry; **в. восстановления в приёмнике поднесущей** local sub-carrier input; **в. на запись, в. на чтение** entry; **невозбуждённый в.** inhibitory input; **номинальный в.** rated input; **постоянно возбуждённый в.** live input; **в. ПУАЗО в режим** director settling; **в. пусковых импульсов** trigger input; **в. сигналов от наземных наблюдателей** manual input; **в. сигналов синхронизации, в. синхронизации** lock input, sync input; **счётный в. (триггера)** complementing input, inverting input; **в. эталона** reference input

**входить, войти** *v.* to enter, to go in, to come in, to get in; to pull in; to penetrate; to fall in; **в. в зону радиолокационного обнаружения** to hit the radar net; **в. в луч радиолокационной станции** to enter the radar; **в. в синхронизм** to pull in step, to lock in synchronism

**входной** *adj.* input; entry, inlet, entrance, entering; approach; accepted; read-in

**входной-выходной** *adj.* input-output

**входящий** *adj.* incoming, ingoing, entering, entrance

**вхождение** *n.* entering, entry; occurrence; **в. в связь** netting, establishing the communication; **в. в синхронизм** coming into step; **вырожденное в.** vacuous occurrence; **невырожденное в.** nonvacuous occurrence

**вхолостую** *adv.* no-load, idle, empty

**ВЧ** *abbr.* 1 (**высокая частот**) high-frequency, radio frequency; 2 (**высокочастотный**) high-frequency, h-f

**ВЧС** *abbr.* (**высокочастотная связь**) high-frequency communication

**ВЧТ** *abbr.* (**вычислительная техника**) computer technology

**выбалансировать** *v.* to balance

**выбег** *m.* running out, running-down; overshoot, overswing; drift

**выбегать, выбежать** *v.* to run out

**выбивание** *n.* knocking out, dislodging

**выбивать, выбить** *v.* to knock out, to dislodge, to drive out; to punch

**выбивка** *f.* knockout, knocking out, dislodging

**выбирание** *n.* selection

**выбиратель** *m.* selector; selector switch

**выбирать, выбрать** *v.* to select, to choose, to pick, to pick out; **в. из** to choose from; to recall from; **в. команду** to take an instruction; **в. параметры с запасом** to overdimension

**выбирающий** *adj.* selecting, choosing; sampling

**выбитый** *adj.* knocked out; knocked on

**выбить** *pf. of* **выбивать**

**выбоина** *f.* hollow, dent, indentation, spot; rut; pit, pitting

**выбор** *m.* selection, choice; sample; drawing; option; dialing; **в. информации без её разрушения** nondestructive readout; **в. кратного** multiple selection; **в. наивыгоднейшего решения** optimization; **в. образца** sampling; **пристрастный в.** biased sampling; **в. системы наведения** guidance selection; **управляемый в. дискретных данных** controlled sampling; **установочный в.** locked selection

**выборка** *f.* sampling, sample; selection; recovery; access; readout; **в. группы информации** block transfer; **в. данных** access; **жёсткая в.** nonrandom access; **в. из выборки** subsample; **в. информации** access, readout; **необработанная в.** crude sampling; **непрерывная в. дискретных данных** continuous sampling; **в. по адресу** addressing; **последовательная в. информации** serial access; **последовательная в. команд** control sequence; **почти произвольная в. данных** quasi-random access; **предварительная в.** pilot survey; **в. с большим временем** non-random access

**выборочный** *adj.* selective; sample, sampling; random, spot (check)

**выбранный** *adj.* selected, chosen

**выбрасываемый** *adj.* thrown out, ejected; ejective

**выбрасывание** *n.* throwing out, dropping, discarding; projection; ejection; knocking out; screening out; **в. пламени** emitting of flame

**выбрасыватель** *m.* ejector, extractor, shedder, pick-up fingers, lifting-out device, knock-out rod

**выбрасывать, выбрасать, выбросить**

*v.* to throw out, to discard, to reject; to eject, to knock out; to project

**выбрасывающий** *adj.* throwing out, rejecting; ejecting; projecting

**выбрать** *pf. of* **выбирать**

**выброс** *m.* ejection; overshooting, overshoot, preshoot, miss; overswing, blip, pip, peaked trace, spike; surge; overrunning; **выбросы во время обратного хода кадровой развёртки** vertical peaking; **выбросы обратного хода строк** line fly-back pulses; **отрицательные выбросы переходной характеристики** underthrow distortion; **хромосферные выбросы** spicules; **в. амплитуды** amplitude excursion; **белый в., предшествующий сигналу** leading white; **в. за шкалу** overshooting; **импульсный в.** peaking pulse; **индуктивный в.** inductive kick; **конечный в.** undershot; **в. кривой** overshoot; **в. на прямоугольном импульсе** horn on a square wave; **одиночный в.** (single) count; **основной в.** main bang, pip; **острый в.** peaked trace; **предварительный в.** preshoot; **пусковой в.** trigger pip; **расщеплённый в.** split blip; **согласующий в., хронирующий в.** timing pip; **чёрный в., предшествующий сигналу** leading black

**выбросать, выбросить** *pf. of* **выбрасывать**

**выброшенный** *adj.* thrown out, ejected, rejected

**выведение** *n.* pullout, drawing out, removal; cutting-out; **в. на орбиту** orbital injection, orbital insertion

**выведенный** *adj.* brought out, exposed, uncovered

**выверенный** *adj.* adjusted, aligned, calibrated

**выверить** *pf. of* **выверять**

**выверка** *f.* adjustment, alignment, lining up, centering; gaging, calibration; **в. (дальномера) по бесконечно удалённой точке** adjustment upon an artificial infinity

**вывернутый** *adj.* reversed, turned out; inverted

**выверочный** *adj.* aligning, straightening, adjustment

**выверять, выверить** *v.* to adjust, to align, to straighten; to test, to check; to calibrate, to gage

**выверяющийся** *adj.* adjustable

**вывести** *pf. of* **выводить**

**выветривание** *n.* airing, ventilation; seasoning; weathering, disintegration

**выветривать, выветрить** *v.* to air, to ventilate; to season

**выветрившийся** *adj.* weathered

**выветрить** *pf. of* **выветривать**

**вывод** *m.* outlet, escape; leading out, lead(-out), tail; bushing; leg (of tube); inference, deduction, corollary; tapping; pin: output; readout; terminal, cap; **антенный в.** sensing aperture; **базовый кольцевой в.** base ring; **в. в заданный район по неизменному азимуту** directional homing; **волноводный в.** waveguide adapter; **в. второй сетки** second-grid cup; **высокопотенциальный в. отклоняющей системы** hot side of yoke **в. данных на экран электроннолучевой трубки** cathode-ray tube display; **в. двигателя на режим** thrust buildup; **в. для сложения** add output; **в. из обтекателя** dome output (of a waveguide); **в. из работы** disabling, putting out of service; **в. из строя** disablement, incapacitation; **в из строя радиотехнических средств ПВО** electronic defense knockout; **коаксиальный в.** coaxial fitting; **в. на орбиту** entering an orbit; **в. на порт** harbor approach; **в. наружу** bringing out; **в. от средней точки** center tap; **в. памяти на печать** memory dump; **пистонный в.** stack-mounting terminal; **в. при вычитании** subtract output; **в. при сложении** add output; **в. (самолёта) на аэродром** approach navigation; **в. (самолёта) по приборам** instrument approach; **ступенчатый в. двигателя на режим** sequence starting; **трансформаторный в.** transformer bushing

**выводимость** *f.* deducibility

**выводимый** *adj.* deducible; taken out, brought out

**выводить, вывести** *v.* to lead out, to take out, to withdraw, to remove, to bring out; to conclude, to deduct, to deduce, to infer; to derive; to develop; to dump; **в. из строя** to disable; **в. из строя радиолокационную систему** to blanket the radar system; **в. из фазы** to dephase; **в. на экран** to display

**выводной** *adj.* lead-out, leading out, lead; throw-out; terminal

**выводящий** *adj.* outgoing, exit

**выворачивание** *n.* reversing

**выворачивать, выворотить** *v.* to reverse, to turn inside out

**выворотный** *adj.* inverted, reversed

**вывороченный** *adj.* inverted, reversed

**выгиб** *m.* bend, flexure, outward curve

**выгибание** *n.* arching

**выгибатель** *m.* adjuster

**выгибать, выгнуть** *v.* to bend, to curve, to buckle

**выгнутость** *f.* convexity

**выгнутый** *adj.* curved, convex

**выгнуть** *pf. of* **выгибать**

**выговаривание** *n.* pronunciation, enunciation

**выговаривать, выговорить** *v.* to pronounce, to utter

**выговор** *m.* pronunciation

**выговорить** *pf. of* **выговаривать**

**выгодность** *f.* efficiency, utility

**выгодный** *adj.* profitable, advantageous, favorable, gainful; efficient

**выгорание** *n.* burning, burning out, dying out; burn-out, burn-in; burn-up; pitting; **местное в. части меди** "picking-up" of copper; **в. мишени** target burn; **ступенчатое в. топлива** graded irradiation

**выгорать, выгореть** *v.* to burn out, to burn down; to die out, to fade

**выгорелый** *adj.* burned out; faded

**выгореть** *pf. of* **выгорать**

**выгородка** *f.* compartment; **в. гидролокатора** asdic compartment

**выгружать, выгрузить** *v.* to dump, to drop, to discharge, to unload

**выгрузка** *f.* discharge, discharging, unloading

**выгрузочный** *adj.* discharge, discharging, unloading

**выдаваемый** *adj.* issued, presented, delivered, distributed, given out

**выдавать, выдать** *v.* to issue, to deliver, to distribute, to give out

**выдаваться, выдаться** *v.* to protrude, to jut out, to project; to occur

**выдавить** *pf. of* **выдавливать**

**выдавление** *n.* stamping, pressing; embossing; squeezing out

**выдавленный** *adj.* stamped, pressed out; embossed; squeezed out

**выдавливать, выдавить** *v.* to press, to press out, to squeeze out, to crush out

**выдать(ся)** *pf. of* **выдавать(ся)**

**выдача** *f.* distribution, delivery; output, production; conveying; presentation, read-out; **автоматическая в.** automatic discharge; **почти произвольная в.** quasi-random access; **принудительная в. ленты** ejection of tape; **в. результатов** readout

**выдающийся** *adj.* outstanding, prominent; protruding, salient, projecting

**выдвиг** *m.* advance

**выдвигающий** *adj.* outstanding, prominent

**выдвижение** *n.* advance, advancement, promotion

**выдвижной** *adj.* sliding; pull-out, draw-out, extension-type, telescopic, extendable; retractable

**выделение** *n.* isolation, separation; precipitation; formation; settling, deposit; evolution, generation, liberation; emission; loss, escape; recovery; discrimination; elimination, extraction; sampling; take-off; detection; **в. амплитуд** amplitude selection; **в. атомных средств** nuclear allocation; **в. газа (в лампах)** gassing, liberation of gas; **в. дискретных значений сигнала при помощи узких импульсов** narrow-pulse sampling; **в. звука, в. звукового сигнала** (*or* **сопровождения**) sound take-off; **в. импульса от ракеты** missile identification; **в. импульсов синхронизации из полного телевизионного сигнала** synchronizing-pulse separation; **в. красного цвета** red gamut; **в. сигнала звукового сопровождения** sound take-off; **в. синхроимпульсов из полного телевизионного сигнала** synchronizing-pulse separation; **хаотическое в. (во времени) значений сигнала** random sampling; **в. целей** tracing (of targets); **в. цели на общем фоне** discrimination

**выделенный** *adj.* separated, isolated; formed; evolved; given off, liberated; preferred; precipitated

**выделившийся** *adj.* separated; evolved; precipitated

**выделимое** *n. adj. decl.* extractée

**выделимый** *adj.* extracted

**выделитель** *m.* discriminator, separator; eliminant; **в. вызова** ringing-current selector

**выделить(ся)** *pf. of* **выделять(ся)**

**выделяемость** *f.* separability; precipitability

**выделяемый** *adj.* separable; precipitable

**выделять, выделить** *v.* to separate, to isolate, to educe; to deposit, to precipitate; to liberate, to give off; to yield; to distinguish, to discriminate; to extract; to gate out

**выделяться, выделиться** *v.* to separate out; to be given off, to be liberated; to escape, to emanate, to segregate, to stand out; to precipitate

**выделяющий** *adj.* giving off; yielding; sorting

**выделяющийся** *adj.* separating out, sorting; outstanding, prominent, selective

**выдержанный** *adj.* sustained, kept, maintained; postponed

**выдержать** *pf. of* **выдерживать**

**выдерживание** *n.* seasoning; holding, keeping; withstanding; **в. повышенной мощности** power-handling ability; **в. положения в пространстве** attitude hold

**выдерживать, выдержать** *v.* to bear, to endure, to stand, to withstand; to hold, to keep, to maintain, to sustain

**выдерживающий** *adj.* bearing, enduring, withstanding; keeping, maintaining, sustaining; lasting

**выдержка** *f.* firmness, endurance; aging; time of exposure, duration (of exposure); delay, (time) lag; **в. времени** delay, time lag, time element; pause, retardation; **независимая в. времени** definite time (element), definite time lag; **обратно-зависимая в. времени** inverse-time element, inverse-time delay; **последовательная в.** sequential time delay

**выдувание** *n.* blowing out, blow-off, blowing, discharge; deflation

**выдувать, выдуть** *v.* to exhaust, to blow out; to deflate

**выемка** *f.* hollow, recess, dent, depression; groove, channel, furrow, chamfer; notch, gap; housing; sample; **с выемкой** grooved, fluted, slotted; **цокольная в.** basing dimple

**выжелобить** *pf. of* **желобить**

**выжечь** *pf. of* **выжигать**

**выжженный** *adj.* burned out, burnt, seared, cauterized; branded

**выживание** *n.* survival; driving out

**выживать, выжить** *v.* to survive; to drive out

**выжигание** *n.* branding, burning out; bake-out; scorching; burn; **в. мишени** target burn; **в. экрана** image burn, screen burning; **в. (экрана** *or* **мишени) потоком ионов** ion burn

**выжигать, выжечь** *v.* to burn, to burn out, to burn up; to brand

**выжить** *pf. of* **выживать**

**вызванный** *adj.* induced; produced, caused (by); called; actuated

**вызвать** *pf. of* **вызывать**

**вызов** *m.* call, calling, call-in, ring, ringing; buzz; **акустический в.** audible call, phonic call; **в. в системе с автоматическим печатанием ярлыков, в. в системе с поразговорной оплатой** registered call, automatic ticketing call; **взаимный в.** reverting call; **групповой в.** conference call; **диспетчерский в.** individual signal; **единичный в.** call unit, selective ringing; **индуктивный в., индукторный в.** magneto call, magneto ringing, generator call; **в. кабины** landing call; **контрольный в.** revertive ringing, revertive signaling; **в. лампой** lamp signal; **ложный в.** permanent signal, permanent call (*or* loop); false start call, false ring; rubout signal; **машинный в.** power ringing **в. между местными районными станциями** junction call; **в. на очереди** call on hand, call on waiting list; **наложенный в.** superposed ringing; **в., не закончившийся разговором** lost call; **непосредственный в.** ringdown, generator signaling; **неучитываемый в.** no-charge call; **обратный в.** revertive ringing, recall; **в. определённого лица** person-to-person call; **в. определённого номера** station-to-station call; **в. переменным током** generator signaling; **в. переменным током от генератора** power ringing; **платный в.** charged call, metered call, registered call; **в. по коду Морзе** keyed call; **в. по радио** aircall; **в. по цепи** ringdown, generator signaling; **в. посредством индуктора** extension bell call; **предварительный в.** advance calling; **предварительный трансокеанский в.** preliminary transoceanic call; **прямой в.** ringdown, generator signaling; **пятисекундный в.** five-second machine ringing; **в. с помощью соединения шлейфом** loop ringing; **светосигнальный в.** lamp signal; **в. станции ЦБ** common battery signaling; **в. станции через дополнительный звонок** extension bell call; **тональный в.** voice-frequency ringing; **тональный избирательный в.** harmonic selective signaling; **в. ЦБ** common battery signaling; **чрезвычайный в., экстренный в.** emergency call

**вызывной** *adj.* call

**вызываемый** *adj*, called, call; effected, caused

**вызыватель** *m.* ringer; **звонковый в.** ringer

**вызывать, вызвать** *v.* to call (up), to phone up, to ring up; to stir, to provoke; to give rise (to), to cause; to elicit, to call forth, to initiate, to produce, to induce; to generate, to exite

**вызывающий** *adj.* causing, bringing about; calling, ringing; call (number)

**вызывной** *adj.* call, calling; ringing

**выигрывать, выиграть** *v.* to win, to gain, to profit

**выигрыш** *m.* gain, profit, win, payoff; **в. антенны оптического телефона** reflector gain of an optophone; **в. в мощности при полном согласовании** completely matched power gain; **в. при передаче** modulating gain

**выигрышный** *adj.* winning; advantageous

**выиск** *m.* search

**выискивать, выискать** *v.* to hunt (for); to discover

**выйти** *pf. of* **выходить**

**выкачанный** *adj.* pumped out

**выкачать** *pf. of* **выкачивать**

**выкачивание** *n.* evacuation, pumping, pumping out, exhaustion; **в. воздуха** exhaustion, evacuation, degassing, de-aeration

**выкачивать, выкачать** *v.* to pump, to pump out, to evacuate, to exhaust, to empty

**выкачка** *see* **выкачивание**

**выкл.** *abbr.* (**выключено**) off, cut off, out, turned off, switched off

**выкладка** *f.* computation, calculation; unpacking

**выключаемый** *adj.* capable of being switched off, turn-off

**выключатель** *m.* switch, (contact) breaker, circuit breaker, disconnecting switch, automatic time switch, automatic interrupter, cutout; release; actuator; **выключатели, размещённые в последовательном порядке** cascade breakers; **автогазовый в.** hard-gas circuit-breaker; **автоматический в.** automatic circuitbreaker, breaker, automatic cutout, automatic cutoff, automatic interrupter; **автоматический в. перегрузки** maximum cutout; **автоматический в. с жёстким соединением ручного привода** fixed-handle breaker; **автоматический в. с механизмом свободного расцепления** free-handle breaker; **баковый масляный в.** dead-tank oil circuit-breaker, bulk-oil circuit-breaker; **безопасный в.** dead-front switch; **в., безопасный для прикосновения** shock-proof switch; **безынерционный в.** high-speed circuit breaker; **блокированный в.** mechanically interlocked switch; (door) lock alarm contact; **в. блокировки** interlocking switch; **в. возбуждения** field switch; **воздушный в. с дугогасящими рогами** air-break switch; **газогенерирующий в.** hard-gas breaker; **гидравлический в.** pressure-operated switch; **главный защитный в.** master contactor; **двухступенчатый в. с сопротивлением** resistance switch; **в., действующий при приближении** proximity switch; **в. дефектного напряжения** leakage protective switch; **дистанционный в. установки** remotely controlled lock-in contactor; **в. для бесшумной настройки** muting switch; **в. для включения (гетера) на параллельную работу** paralleling switch; **в. для защиты сетей** circuit breaker; **в. для осветительного прибора** fixture switch; **в. для отключения приёмника и включения передатчика** break-in key; **в. для секционирования обмотки возбуждения** field break-up switch, field-dividing switch; **в. для скрытой проводки** flush switch; **закрытый в.** box switch, ironclad switch; **в., запираемый на замок с кожухом** locked cover switch, locking switch; **защитный в.** circuit breaker; **измерительный в.** instrument rectifier; **в. канала цветности** color killer; **качающийся в.** bat-handle switch; **квитирующий в.** position control switch; **комнатный в.** snap switch; **конечный в. главного тока** limit-master switch; **в. контрольного устройства** monitor switch; **в. контрольной лампы рабочего места** pilot lamp switching key; **концевой в.** limit switch; elevation down relay; **в. магистрали** main switch; **малоёмкостный в.** anti-capacity switch; **междушинный в.** bar coupler, bus-tie switch; **мембранный в.** pressure switch; **миниатюрный (цилиндрический) в.** nut switch, microswitch; **минимальный в.** underload switch, no-load switch; **многобаковый в.** multiple single-pole breaker; **в. многопетлевой сети** back-current breaker for network protection; **моментный в.** snap switch, quick maker; **в. на ножках** feet switch; **надстроенный в.** base-mounted switch, built-on switch; **нажимаемый пальцем в.** thumb switch; **в. непринятия сигнала костановке** non-stop switch; **ножевой в.** chopper switch; **ножной в.** foot switch, foot-operated switch; **ночной в.** day-and-night transfer key; **в. ночной сигнализации** night alarm switch; **общий в.** master switch; **однобаковый в.** single-tank breaker, single-tank switch; **оттяжной в. со шнурком** burglar-alarm pull switch; **пакетный в.** multisection-type rotary switch, packet-type switch, rotary switch; **пальцевый в.** thumb switch; **в. перекидного типа, перекидной в.** flip-flop switch, toggle switch, tumbler switch, trigger switch, lever switch; **перекрёстивый в.** intermediate switch; **в. переподъёма** overwinding breaker; **пластинчато-роликовый в.** roller-leaf actuator; **пластинчатый в.** leaf actuator; **в. поверх штукатурки** switch for surface mounting; **поворотный в. со съёмным ключом** rotary lock switch; **в. под действием протекающего масла** oil blast circuit-breaker; **полу-утопленный (в штукатурке) в.** semi-recessed switch; **поплавковый в.** float

switch; tank switch; **последовательный в., постепенный в.** electrolier switch; **предельный в.** limit switch, terminal switch; **предельный в. лифтового типа** track-type limit switch; **программный в.** timer; **простой в.** single-throw switch; **путевой в.** position switch; **путевой конечный в.** track-limit switch; **рабочий в.** service switch; **районный в.** switch selector repeater, discriminator; **в. расцепляющей катушки** trip switch; **расширительный в.** expansion circuit-breaker, air-blast circuit-breaker; **рубящий в.** knife switch, blade switch; **рычажный в.** lever switch, knife switch; **рычажный в. с перекидной головкой** toggle switch; **в. с гашением дуги в водяных парах** circuit-breaker employing water-vapor for deionizing; **в. с закрытыми токоведущими частями** dead-front switch; **в. с запирающимся на замок кожухом** asylum switch; **в. с защитой против токов повреждения** fault-current protective switchgear; **в. с контактной щёткой** laminated-brush switch; **в. с непосредственным разрывом дуги** plain-break breaker; **в. с плоским контактным ходом** wafer switch; **в. с подводом проводов сверху** top-entry switch; **сверхбыстродействующий в.** ballistic breaker; **«сдвижной» рычажный в.** finger switch; **секционный в.** sectionalizing switch, coupled switch, linked switch, coupling breaker; **сквозной подвесной в.** feed through switch; **в. со смотровым окошком в крышке** indicating switch; **в. со штангой** push-pull control (switch); **ступенчатый автоматический в. с реактором** reactance breaker; **тормозной в.** plugging switch; **в., управляемый резонансной настройкой** electrosyntonic switch; **флажковый в.** flag switch; **в. холостого хода** motor disconnect-switch; **холостой в.** disconnect switch; **в. цепей возбуждения** field break switch; **в. цепи** circuit breaker; **цепной в.** pull switch; **шахтный предельный в.** hatching limit switch; **шпиндельный концевой в.** screw limit switch; **щелчковый в.** snap switch; **экстрен-**

ный в. emergency switch; emergency release; этажный в. floor switch

выключать, выключить v. to turn off, to switch off, to cut off, to cut out, to shut off; to break (contact), to de-energize; to disengage, to disconnect, to uncouple, to release; to exclude; to throw out (of gear); в. ток питания бортовой аппаратуры to de-energize the round

выключающий adj. cut-off; disconnecting, releasing, disengaging, tripping

выключение n. switching (off), cutting out, cutting, cut-off; turning off, shutting off, shut-down; disconnecting, disengaging, switching out, tripping out, drop-out; (color) killing; в. в две степени resistance switching; в. воспроизведения цвета killing of color reproduction; в. двигателя thrust cutoff; зависимое в. intertripping; в. измерительного трансформатора от тока transformer-type tripping; в. местного разговора, производимое междугородной станцией breaking of local calls for trunk calls, through clearing; нулевое в. no-voltage release; в. питания dump; в. поиска search disconnect; в. прямоугольного импульса gate turnoff; в. прямоугольного импульса высоты height-gate turnoff; в. развёртки дальности range turnoff; в. реле relay reset; в. электропитания power dump

выключенный adj. out, switched off, cut off, turned off, disconnected; в. из схемы inoperative

выключено adv. off, switched off

выключить pf. of выключать; в. линию to clear the circuit

выкопать pf. of копать

выкружка f. fillet, groove, recess

вылеживание n. aging

вылет m. departure, escape, flying out; в. на калибровочный облёт радиолокационных станций radar-calibration sortie; в. на подавление радиолокационных станций radar neutralization flight; в. на радиоразведку и постановку (радио)помех electronic countermeasures flight; в. по индикатору кругового обзора PPI-departure

вылетать, вылететь v. to fly out, to escape

вылетающий adj. outgoing, outcoming, escaping, emerging, emitted

вылететь pf. of вылетать

вымененный adj. interchanged, exchanged

выменивать, выменить v. to interchange, to exchange

выменять = выменивать

вымеривание n. measuring

вымеривать, вымерить v. to measure, to gage

вымерять = вымеривать

вымораживание n. freezeout

вынесенный adj. remote, extension (-type), outlying, distance-type; taken-out; external; advanced

вынести pf. of выносить

вынимаемый adj. removable; draw-out

вынимание n. withdrawal, removal, take-off

вынимать, вынуть v. to take out, to remove, to carry out, to withdraw, to draw out; to cannibalize

вынос m. carrying out, bearing out; branch office, out-office, out-station; outlying point; outside PBX extension

выносить, вынести v. to carry out, to take out; to stand, to endure, to sustain, to hold out, to tolerate

выносливость f. endurance, resistance, strength, hardiness; tolerance

выносливый adj. hardy, enduring, rugged, sturdy

выносной adj. remote

вынуждать, вынудить v. to force, to compel

вынуждающий adj. forcing, compelling, driving

вынужденный adj. forced, constrained, driven

вынутый adj. removed, withdrawn

вынуть pf. of вынимать

вып. abbr. (выпуск) issue

выпадание n. drop-out; falling, falling out; fallout

выпадать, выпасть v. to drop out, to come out, to fall out, to precipitate

выпадающий adj. falling out, dropping out; fallout

выпадение n. fall, falling out, falling, fallout, drop-out; precipitation; outage, loss; tripping; в. сети mains failure

выпадка see выпадение

выпаренный *adj.* evaporated

выпаривание *n.* evaporation, volatilization

выпаривательный *adj.* evaporating

выпаривать, выпарить *v.* to evaporate; to steam

выпарной, выпарный *adj.* evaporating

выпасть *pf. of* выпадать

выпилить *pf. of* пилить

выписанный *adj.* written-out, copied

выписать *pf. of* выписывать

выписка *f.* extract, copy; summary; **в. счётов** billing

выписывать, выписать *v.* to write out, to copy; to order (by mail)

выплавить *pf. of* выплавлять

выплавка *f.*, выплавление *n.* melting, extraction; smelting

выплавленный *adj.* smelted, extracted, melted out

выплавлять, выплавить *v.* to smelt (down); to melt, to extract

выплавной *adj.* smelted; melted out, extracted

выползание *n.* creeping out; **в. листков из миканта** oozing of mica

выползать, выползти *v.* to creep out, to get out

выполнение *n.* fulfillment, performance, execution, completion, achievement; **в. нескольких команд одновременно** multiplexing; **повторное в.** rerun procedure

выполненный *adj.* executed, achieved, performed, fulfilled, made, carried out

выполнимость *f.* satisfiability; feasibility, practicability; executability

выполнимый *adj.* executable, workable; feasible, practicable

выполнять, выполнить *v.* to execute, to fulfil, to effect, to carry out, to perform; to obey

выполняющий *adj.* performing, executing

выпотевание *n.* sweating, exudation

выправить *pf. of* выправлять

выправка *f.* correction, straightening

выправление *n.* correcting, correction, straightening; **в. расфокусирования** debunching correction

выправлять, выправить *v.* to correct, to straighten, to set right, to true up; to direct; to prepare, to arrange, to dress

выпрямитель *m.* rectifier, detector;

**в. амплитудного значения** peak rectifier; **анодный в.** HT supply; **в. без сеточного управления** straight rectifier; **в. в форме синхронного вращающегося коммутатора** Ferranti's rectifier; **в. высокой частоты** detector; **газовый в.** discharge-tube rectifier, gas-filled tube rectifier; **газоразрядный в. с сеткой** grid-pool tube, grid-pool tank; **газосветноламповый в.** glow-rectifier, glow-discharge rectifier; **газотронный в.** rare-gas rectifier; tungar rectifier, tube rectifier; **грецовский в.** bridge-circuit rectifier; **двухполупериодный в.** full-wave rectifier, biphase rectifier, diametric rectifier; **двухтактный в.** *see* **двухполупериодный в.; в. для защиты слуха** acoustic-shock absorbing rectifier; **в. для непрерывной зарядки** trickle charger; **в. для питания анода переменным током** B-eliminator; **кенотронный в.** vacuum-tube rectifier; electronic type rectifier, kenotron; **коллекторный в.** rectifying commutator; **линейный в.** line equalizer; **в. на кристаллических диодах** crystal rectifier; **в. накала** "A"-rectifier; **однополупериодный в.** half-wave rectifier; **в. питания** mains unit; **в. питания накала** A-eliminator; **в. питания от сети** mains rectifier, full wave rectifier; **в., питающий цепи смещения** bias rectifier; **пластинный в.** plate rectifier; **в. по схеме Латура** voltage-doubler rectifier; **ртутный в. с управляющей сеткой** plomatron; **в. с жидким катодом** pool-type tube; **в. с запирающим слоем** barrier-layer rectifier, barrier-film rectifier; **в. с комбинированной схемой коммутации** multiple rectifier; **в. с лампой тлеющего разряда** neon tube rectifier, cold-cathode rectifier; **в. с простой схемой коммутации** simple rectifier; **в. с регулируемым сопротивлением** varistor rectifier; **сверхмощный в.** pumped ignitron; **сетевой в.** full wave rectifier; anode voltage rectifier, power rectifier; **силовой в.** mains rectifier, full wave rectifier; **в. смещения** bias rectifier; **в., собранный по схеме двойной звезды** double-V connected rectifier; **стаби-**

лизированный в. controlled rectifier, regulated rectifier; стеклянный ртутный в. с сетками grid-pool tube; твёрдый в. dry(-contact) rectifier, dry-disk rectifier; трёхфазный полупериодный в. zigzag rectifier; универсальный в. universal mains unit; в., управляемый сельсином direction rectifier; фланговый в. fm slope detection; электролитический свинцово-алюминиевый в. Nodon tube

**выпрямительный** *adj.* rectifier, rectifying

**выпрямить** *pf. of* **выпрямлять**

**выпрямление** *n.* rectifying, rectification, straightening; redressing; demodulation, detection; двухполупериодное в., двухтактное в. full-wave rectification; в. провода stretching of wire

**выпрямленный** *adj.* rectified, straightened; redressed; clipped

**выпрямлять, выпрямить** *v.* to rectify, to rectificate, to straighten; to redress; to level off (a curve)

**выпрямляющий** *adj.* rectifying, straightening; leveling off

**выпукло** *adv.* convexly

**выпукло-вогнутый** *adj.* convexo-concave

**выпуклость** *f.* bulge, bulging; convexity, curvature; embossing; prominence

**выпуклый** *adj.* arched, convex; bulging; prominent; embossed; dished, gibbous

**выпуск** *m.* outlet, escape; release; emission, discharge, discharging; launching; production, yield, output, delivery; холостой в. relief

**выпускание** *n.* discharge, discharging; emission

**выпускать, выпустить** *v.* to discharge, to empty, to let out, to run out; to deliver, to put out, to manufacture; to eject

**выпускной** *adj.* outlet, discharge, run out, let out; release; drag; evacuation, exhaust; eduction

**выпустить** *pf. of* **выпускать**

**выпученный** *adj.* bulging, protruding

**выпучивание** *n.* bulging, bulge, swelling; buckling

**выпучивать, выпучить** *v.* to bulge, to protrude

**выпущение** *n.* discharge, discharging; emission

**выпущенный** *adj.* let out, discharged

**выпячивание** *n.* protruding, protrusion; buckling

**вырабатывать, выработать** *v.* to produce, to manufacture; to develop, to work; to generate

**вырабатывающий** *adj.* producing, generating

**выработанный** *adj.* produced, manufactured; worked; worked out; generated; spent

**выработать** *pf. of* **вырабатывать**

**выработка** *f.* yield, output, (productive) capacity, generation; preparation, elaboration; production; exhaustion; wear; в. графика scheduling; децентрализованная в. энергии load-area generation; в. упреждённого угла возвышения (*or* места) elevation prediction; в. энергии асинхронным генератором induction generation

**выравненность** *f.* uniformity

**выравненный** *adj.* matched

**выравнивание** *n.* line-up, lining-up, alignment, equalization, equalizing, leveling, balancing, trimming, smoothing, padding; matching, fiting; в. волнового сопротивления planing of characteristic impedance; в. данных data smoothing, rate smoothing; ёмкостное в. характеристики линии capacitance loading (of a line); в. колебаний нагрузки compensation of load variations; в. коммутирующего сигнала gate balancing; в. контуров по диапазону band compensation, padding; в. координат и параметров движения цели rate control; в. напряжений compensation of voltage; параллельное частотное в. shunt loading; последовательное частотное в. series loading; в. потенциального (*or* зарядного) рельефа за счёт утечки charge pattern leakage; в. стопки перфокарт joggling; в. фазового фронта линзовой антенны zoning of a lens antenna; в. фона изображения shading correction, shading compensation

**выравниватель** *m.* equalizer, compensator, balancer; flattener; phase compensator; в. затухания с параллельным включением parallel equalizer, shunt equalizer

**выравнивать, выравнить, выравнять** *v.* to level, to align, to line (up); to trim, to adjust, to equalize, to balance, to dress, to flatten; to match

**выравнивающий** *adj.* compensating, equalizing, balancing, aligning; leveling, straightening; padding

**выравнить, выравнять** *pf. of* **выравнивать**

**выражаемый** *adj.* expressed, indicated, conveyed, conveyable

**выражать, выразить** *v.* to express, to convey

**выражение** *n.* expression, term, equation; representation, formula; connective; **аналитическое в.** formula; **в. во времени** time domain; **девятикомпонентное в. тензора** nine-term form (matrix algebra); **иррациональное в.** surd, irrational term; **в. мощности (в децибелах) по отношению к мощности 1** bm volume unit; **определяемое в.** definiendum; **определяющее в.** definiens; **параметрическое в. функции** uniformization; **первичное в.** primary; **подкоренное в.** radicand; **подставляемое в.** substituend; **подынтегральное в.** integrand, element of integration; **приближённое в.** approximation; **числовое в. помех** disturbing value

**выражённый** *adj.* expressed; pronounced, marked

**выразительный** *adj.* indicative, expressive, significant

**выразить** *pf. of* **выражать**

**вырастание** *n.* growing

**вырастать, вырастить** *v.* to grow; to increase; to arise, to appear

**вырасти** *pf. of* **расти**

**вырастить** *pf. of* **выращивать, выращать**

**выращать** = **выращивать**

**выращенный** *adj.* grown; rate-grown; pulled

**выращивание** *n.* growing, raising, cultivation

**выращивать, вырастить** *v.* to grow, to raise

**вырвать** *pf. of* **вырывать**

**вырез** *m.* cut, groove, notch, opening, slot

**вырезание** *n.* cutting, cutting out, grooving, engraving

**вырезанный** *adj.* cut out, engraved

**вырезать** *pf. of* **вырезывать**

**вырезающий** *adj.* eliminating, elimination

**вырезывание** *n.* cutting out; engraving, carving; elimination; excision, ablation; **в. пучка** beam masking

**вырезывать, вырезать** *v.* to cut, to cut out; to eliminate; to engrave

**выровненный** *adj.* leveled, straightened; aligned; trimmed

**вырождаться, выродиться** *v.* to degenerate

**вырождающийся** *adj.* degenerating; degenerative

**вырождение** *n.* degeneration, degeneracy

**вырожденный** *adj.* degenerated, degenerate

**вырожденческий** *adj.* degenerative

**выруб** *m.* cut, notch

**вырубание** *n.* cutting, cutting out; punching

**вырубать, вырубить** *v.* to cut out; to cut off; to cut down

**вырубка** *f.* cutting out; notching, indentation; punching

**вырубной** *adj.* cutting

**вырывание** *n.* extraction, drawing; **в. электронов полем** field-induced emission, strong-field emission

**вырывать, вырвать** *v.* to extract, to draw; to tear (out), to pull out

**выс.** *abbr.* (**высота**) height, altitude

**высадить** *pf. of* **высаживать**

**высаживание** *n.* precipitation

**высаживать, высадить** *v.* to precipitate

**высаживающийся** *adj.* precipitating

**высасывание** *n.* exhaustion, drawing out

**высасывать, высосать** *v.* to exhaust, to evacuate, to draw off (air), to draw out

**высверленный** *adj.* drilled, bored

**высверливание** *n.* drilling, boring

**высверливать, высверлить** *v.* to drill out, to bore out

**высветить** *pf. of* **высвечивать**

**высветка** *f.* light; **в. основного объекта передачи** hot light, key light

**высвечивание** *n.* scintillation; de-excitation

**высвечивать, высветить** *v.* to flash on

**высвобождать, высвободить** *v.* to set free, to liberate, to release

**высвобождение** *n.* liberation, release

**высвобожденный** *adj.* liberated, set free, released

**высекать, высечь** *v.* to cut, to cut out; to punch, to stamp

**высказанный** *adj.* expressed

**высказать** *pf. of* **высказывать**

**высказывание** *n.* expression; statement; proposition

**высказывать, высказать** *v.* to express; to state

**выслушать** *pf. of* **выслушивать**

**выслушивание** *n.* listening, hearing; examination, auscultation

**выслушивать, выслушать** *v.* to listen, to hear; to sound, to auscultate

**высовывать, высунуть** *v.* to put out, to thrust out, to protrude

**высокий** *adj.* high, high-pitched, sharp, acute; high, toll

**высоко-** *prefix* high, highly

**высокоактивный** *adj.* highly active; hot, high-activity, highly radioactive

**высоковакуумный** *adj.* high-vacuum

**высоковольтный** *adj.* high-voltage, high-tension, high-potential

**высоковязкий** *adj.* high-viscosity

**высокоградусный** *adj.* high-grade; highly concentrated

**высокодисперсный** *adj.* highly dispersed

**высокодобротный** *adj.* high-Q, high-quality

**высокоинтенсивный** *adj.* high-intensity, high (-flux)

**высокоинформативный** *adj.* high-information

**высокоионизированный** *adj.* highly ionized

**высококалорийный** *adj.* high-calorie; high-energy

**высококачественный** *adj.* high-fidelity; high-quality, fine, high-grade; high-strength; high-definition; high-performance

**высококипящий** *adj.* high-boiling

**высококонтрастный** *adj.* high-contrast

**высококонцентрированный** *adj.* highly concentrated

**высококоэрцитивный** *adj.* high-coercivity

**высококремнеземистый** *adj.* high-silica, high-silicon

**высококремнистый** *adj.* high-silicon

**высоколетящий** *adj.* high-flying

**высоколинейный** *adj.* ultra-linear

**высокомолекулярный** *adj.* highly molecular

**высокомощный** *adj.* high-power, high-capacity; heavy duty

**высоконапорный** *adj.* high-pressure

**высоконасыщенный** *adj.* highly saturated

**высокоогнеупорный** *adj.* highly refractory

**высокоомный** *adj.* high-resistance, highly resistive, high-impedance

**высокоплавкий** *adj.* high-melting

**высокополимерный** *adj.* highly polymeric

**высокопотенциальный** *adj.* high-potential, "hot"

**высокопробный** *adj.* high-grade; fine

**высокопродуктивный** *adj.* highly efficient, highly productive

**высокопроизводительный** *adj.* highly efficient, highly productive, high-performance

**высокопроницаемый** *adj.* high-permeability

**высокопроходный** *adj.* high-pass

**высокопроцентный** *adj.* high-percentage, of high percentage

**высокопрочный** *adj.* high-strength

**высокорадиоактивный** *adj.* highly radioactive, hot

**высокоразвитый** *adj.* highly developed

**высокорезистивный** *adj.* high-resistance

**высокоскоростный** *adj.* high-speed, high-velocity

**высокосортный** *adj.* high-quality, high-grade, fine

**высокостабильный** *adj.* highly stable, very stable

**высокотемпературный** *adj.* high-temperature

**высокоточный** *adj.* high-precision, precision

**высокоуглеродистый** *adj.* high-carbon

**высокочастотный** *adj.* high-frequency, high-cycle; high-speed, fast

**высокочувствительный** *adj.* highly sensitive, supersensitive, high-performance

**высокоэвакуированный** *adj.* highly exhausted

**высокоэластичность** *f.* Mackian elasticity

высокоэнергетический *adj.* high-energy

высокоэффективный *adj.* highly effective

высосать *pf. of* высасывать

высота *f.* height, elevation, altitude, level; pitch (of sound); depth; degree (of temperature); **в. амплитуды** vertical height; **барометрическая в.** constant-pressure altitude, pressure altitude; **в., близкая к потолку** extreme altitude; **в. боевого применения** operating altitude, operational altitude; **в. выключения двигателя** cutoff altitude; **в. забора** intake head, suction lift; **в. запайки** sealing length; **заранее определённая в.** estimated altitude; **заранее рассчитанная в.** precomputed altitude; **заранее установленная в.** estimated altitude; **в. засоса** intake head, suction lift; **в., измеренная с помощью радиолокатора** radar altitude; **в. к моменту полного выгорания топлива** all-burnt altitude; **классическая в. строя** classic(al) pitch; **контрольная в. изображения** reference image height; **в. линзы** button height; **максимально возможная в. перехвата** higher interception limit; **в. местности** altitude; **минимальная в. полёта по приборам** IFR minimum height, IFR minimum flight-level; **в. монтированной ножки** mount height; **наиболее вероятная в. перехвата** (*or* **стрельбы**) critical height, critical altitude; **в. напора** fall, drop, descent; fall of potential, potential gradient; **в. непросматриваемого пространства** gap height; **оркестровая в. строя** philharmonic pitch; **в. отделения (снаряда от самолёта-носителя)** release altitude; **в. подачи** delivery head; **в. подхода к аэродрому** initial approach altitude; **в. подъёма** delivery head; **в. подъёма клапана** valve lift; **в. подъёма плунжера** valve stroke; **в. полёта при преодолении системы ПВО** penetration altitude; **в. положения** altitude, level, height, elevation; **программированная в. полёта** preset altitude; **в. (радиолокационной станции) с антенной в походном положении** height (scanner towed); **в. (радиолокационной станции) с установленной антенной** height

(scanner erect); **в. ракеты в конце активного участка полёта** end-of-boost phase altitude; **в. расчёта** rated altitude; design altitude; **в. ртутного столба** barometric pressure, barometer reading; **в. светила** celestial altitude; **в. срабатывания взрывателя** arming altitude; **в. столба (жидкости)** pressure head; **стратосферная в.** ultrahigh altitude; **текущая в.** present height, altitude at present position; **теодолитная в.** camera altitude; **в. тона** pitch (of note), pitch of beat, pitch of tone; **в. управляемого полёта** controllable altitude; **упреждённая в.** altitude at future position, altitude at future height

высотность *f.* altitude, height, altitude capability, altitude range; **в., определяемая тягово-оруженностью** ceiling parameters

высотный *adj.* height, altitude; high-altitude, high-level; tall; high-flying; upper-air

высотомер *m.* altimeter, height finder; **в. для измерения наддува двигателя** engine altimeter; **прецизионный в.** statoscope, sensible altimeter; **тарировочный в.** altimeter calibrator

высотомерный *adj.* altimeter, height-finder

высотомерщик *m.* height-finder observer

высохнуть *pf. of* высыхать

выставить *pf. of* выставлять

выставка *f.* show, display

выставленный *adj.* exposed, set out, put out

выставлять, выставить *v.* to show, to display; to expose, to put out

выставочный *adj.* exhibition, display

выстраивать, выстроить *v.* to line, to line up, to align; to build, to erect

выстрогать *pf. of* строгать

выстроить *pf. of* выстраивать

выступ *m.* ledge, shelf, ridge; lug, boss, cam, horn, cog; flange, rib; projection, protrusion, jut, overhang; notch, pedestal; baffle; **выступы на поверхности гладкого якоря** driving horns; **полюсные выступы** pole horns; **в. аккумуляторной пластины** plate lug; **контактный в.** contact lug; **набегающий в. полюса** leading pole horn; **опорный в.** seating boss;

**сбегающий в. полюса** following horn; trailing pole horn; **спиральный костный в.** spiral ledge; **установочный в.** locating boss; **цокольный в.** base boss; **чашечный в.** ferrule boss; **штенгельный в.** exhaust boss

**выступать, выступить** *v.* to protrude, to jut, to jut out, to project; to come forward

**выступающий** *adj.* exposed, projecting, protruding, outstanding; salient

**выступить** *pf. of* **выступать**

**высунутый** *adj.* protruding, thrust out

**высунуть** *pf. of* **высовывать**

**высушенный** *adj.* dried, desiccated

**высушивание** *n.* drying, desiccation; baking

**высушивать, высушить** *v.* to dry, to desiccate

**высушивающий** *adj.* drying, desiccating

**высушить** *pf. of* **высушивать**

**высчитывать, высчитать** *v.* to compute, to calculate

**высший** *superl. adj.* higher, superior; highest, supreme

**высыхание** *n.* drying, desiccation

**высыхать, высохнуть** *v.* to run dry, to dry, to dry up; to fade

**высыхающий** *adj.* drying

**выталкивание** *n.* ejection, expulsion extrusion, pushing out

**вытаскивание** *n.* drawing out, pulling out, extraction, withdrawal (of a plug)

**вытаскивать, вытащить** *v.* to pull out, to pull, to draw, to draw out

**вытачивать, выточить** *v.* to bore out, to cut (grooves)

**вытащить** *pf. of* **вытаскивать**

**вытекание** *n.* efflux, effluence, running out; outlet, discharge; **в. жидкости** leaking, outflow; **в. несжимаемого газа** exhaustion of an incompressible gas

**вытекать, вытечь** *v.* to flow out, to run out, to leak out, to issue, to discharge, to escape; to flow, to ensue

**вытекающий** *adj.* flowing out, running out, discharging; resultant; sequent

**вытеснение** *n.* displacement, substitution, dislodgment

**вытесненный** *adj.* displaced, dislodged, supplanted; expelled

**вытеснитель** *m.* displacer

**вытеснить** *pf. of* **вытеснять**

**вытесняемый** *adj.* displaced

**вытеснять, вытеснить** *v.* to force out, to eject, to expel; to displace, to dislodge; to replace; to liberate, to crowd out, to exclude

**вытечь** *pf. of* **вытекать**

**выточить** *pf. of* **вытачивать**

**выточка** *f.* groove, recess; bore

**вытравить** *pf. of* **вытравлять, травить, вытравливать**

**вытравка** *f.*, **вытравление** *n.* etching, erosion

**вытравленный** *adj.* etched, eroded

**вытравливание** *n.* etching, erosion

**вытравливать, вытравлять** *pf.* **вытравить** *v.* to corrode, to pickle, to etch, to erode; to remove

**вытравной** *adj.* corrosive, corrodent

**вытрамбовать** *pf. of* **трамбовать**

**вытрясти** *pf. of* **трясти**

**выть, взвыть** *v.* to howl; to wail

**вытягиваемый** *adj.* removable

**вытягивание** *n.* extraction, drawing out; pulling (out), extension, drawing; exhaustion; **в. провода** wire drawing; straightening of wire

**вытягивать, вытянуть** *v.* to strain, to stretch, to elongate; to extract, to remove, to draw out

**вытягивающий** *adj.* extraction

**вытяжка** *f.* extension, expansion; extract; **в. с несколькими переходами** deep-drawing in steps

**вытяжной** *adj.* drawing; extraction, pull; **в. и втяжной** push-pull

**вытянутый** *adj.* elongated, stretched out, prolonged; drawn out, extracted

**вытянуть** *pf. of* **вытягивать**

**выхлоп** *m.* exhaust, discharge, expulsion; outlet; **в. реактивного двигателя** jet exhaust; **световой в.** light yield

**выхлопной** *adj.* exhaust, escape; waste

**выход** *m.* output, "out"; terminator, outlet, exit, egress, outflow, escape, discharge; departure (of wave); approach; emergence, yield, product, production, output; **анодный в. по току** anode efficiency; **в. в город** PBX trunk; **в. в начальной стадии работы** initial failure; **в. в эфир** on-air shot; **занятый в.** busy trunk; **в. из диапазона** derating; **в. из калибра** overage; **в. из синхронизма** holdoff; **в. из строя** failure, collapse, mortality; **в. из строя во время приработки** infant mortality (*comp.*); **в. из строя системы наведения** guidance failure; **в. из-под контроля** runaway; **в. информации** output; **катодный в. по току** cathode efficiency; **квантовый в. (фотолюминесценции)** quantum yield; **в. коммутирующего сигнала** gate output; **в. на аэродром (по приборам)** final approach, approach navigation; **в. на ГТС** connection to a central office; **в. на цель по приборам** blind approach; **номинальный в.** rated output, rated production; **нормальный в. двигателя на режим** smooth start; **в.-перенос (в следующий разряд)** carry output; **в. по напряжению** voltage efficiency; **в. по току** current efficiency; **в. по энергии** energy efficiency; **в. ракеты из сферы притяжения Земли** rocket escaping; **в. самолёта вслепую** blind approach; **слепой в. на маяк** blind approach beacon system; **в. строчной развёртки** line output; **в. суммы за пределы счётчика** overflow (*comp.*); **шнуровой в. от линейного комплекта** lineswitch trunk output; **в. ядерных реакций** efficacy of bombardment in nuclear reactions

**выходить, выйти** *v.* to go out, to come out, to come forth, to emerge, to issue; to appear; to escape; to yield; to face; **в. из синхронизма** to fall out of step; **в. из строя** to fail

**выходной** *adj.* output; discharge; outgoing; exit; read-out

**выходящий** *adj.* outgoing, coming out, emerging; escaping

**выхождение** *n.* coming out, going out

**выцветание** *n.* fading, discoloration, bleaching; **в. стекла** solarization

**выцветать, выцвести** *v.* to fade, to bleach

**выцветший** *adj.* faded, discolored

**вычёркивание** *n.* cancellation; omission, deleting, obliteration

**вычёркивать, вычеркнуть** *v.* to cross out, to strike out, to cancel, to delete

**вычеркнутый** *adj.* canceled; crossed out, deleted

**вычеркнуть** *pf. of* **вычёркивать**

**вычертить** *pf. of* **вычерчивать**

**вычерчивание** *n.* plotting, tracing, laying out, mapping

**вычерчиватель** *m.* plotter, tracer, plotting device; **дистанционный автоматический в.** teleplotter

**вычерчивать, вычертить** *v.* to trace, to trace out, to draw, to map out, to lay out

**вычерчивающий** *adj.* tracing, mapping, plotting

**вычесть** *pf. of* **вычитать**

**вычет** *m.* residue, deduction; **за вычетом** less, minus

**вычисление** *n.* calculation, computation, computing, figuring, reckoning, evaluation, rating; calculus; determination; measurement; **приближённые вычисления** calculus of approximation; **автоматическое в. данных для занятия исходного положения и атаки цели** automatic approach and attack computing; **быстрое в.** real-time computation; **в. курса** track analysis; **в. методом неявных функций** implicit computation; **в. на моделирующем устройстве, непрерывное в.** analog computation; **основное в.** main (frame) computing; **в. последовательными приближениями** trial and error calculation, trial calculations, calculation by successive approximations; **в. траектории баллистической ракеты** targeting; **в. траектории полёта** trajectory computation; **в. упреждений** lead computing; **в. упреждения** deflection prediction; **в. упреждённой точки** lead prediction; **в. упреждённых баллистических координат** ballistic prediction; **в. факториала** factorization

**вычисленный** *adj.* computed, calculated, estimated

**вычислимый** *adj.* computable

**вычислитель** *m.* computer; calculator; **в. азимут-дальность** R-Theta computer; **в. данных наведения** vectoring computer; **механический в. треугольника скоростей** avigraph; **в. наводки и прицеливания** elevation computer; **в. поправки на параллакс между станцией обнаружения и СОН** parallax computer between search and fire-control radar; **в. поправок азимута на параллакс** train parallax computer; **в. прицеливания** elevation computer; **в. самолётного радиолокационного при-**

**цела** fire-control computer; **в. угла крена** rocket roll calculator

**вычислительный** *adj.* computing, computational; calculating

**вычислить** *pf. of* **вычислять**

**вычисляемый** *adj.* countable, enumerable

**вычислять, вычислить** *v.* to compute, to calculate, to figure out, to estimate, to determine; to account for; to enumerate

**вычитаемое** *n. adj. decl.* subtrahend

**вычитаемый** *adj.* deductible

**вычитание** *n.* subtraction; deduction

**вычитатель** *m.* subtractor

**вычитательный** *adj.* subtractive

**вычитать, вычесть** *v.* to subtract, to deduct, to take away, to take from, to take off

**вычитающий** *adj.* subtractive

**вышедший** *adj.* having gone out; **в. из строя** inoperative, disabled; **в. из употребления** obsolete

**вышеизложенный** *adj.* foregoing, above-stated

**вышеисчисленный** *adj.* above-numbered, above-calculated

**вышеназванный** *adj.* aforesaid, above-named

**вышеприведённый** *adj.* afore-cited, given above, shown above

**вышесказанный** *see* **вышеуказанный**

**вышестоящий** *adj.* higher

**вышеуказанный, вышеупомянутый** *adj.* foregoing, above, above-mentioned, aforesaid

**вышибание** *n.* breaking out, knocking out

**вышибать, вышибить** *v.* to eject, to dislodge, to knock out, to drive out, to break out

**вышибка** breaking out, knocking out

**вышина** *f.* height, altitude, elevation, level

**вышка** *f.* tower; pulpit; **в. для запуска реактивных снарядов** guiding tower; **пусковая в.** firing tower

**вышлифовать** *pf. of* **шлифовать**

**выштампованный** *adj.* stamped out, punched

**выштамповать** *v.* to stamp out, to punch; to eliminate

**выщелачивание** *n.* leaching, depletion, lixiviation

**выщелачивать, выщелочить** *v.* to leach, to leach out, to lixiviate, to extract

**выщелоченный** *adj.* leached out, lixiviated

**выщелочить** *pf. of* **выщелачивать**

**выявить** *pf. of* **выявлять**

**выявление** *n.* exposure; development; **аналитическое в. характеристик кривых** curve discussion (*math.*); **в. неисправностей** trouble shooting

**выявлять, выявить** *v.* to develop; to expose, to show

**выяснение** *n.* explanation, clarification

**выясненный** *adj.* explained, cleared up

**выяснять, выяснить** *v.* to clear up, to explain, to elucidate; to inquire, to examine, to investigate; to prove, to ascertain; to develop

**вьючный** *adj.* pack

**вьюшка** *f.* damper; reel

**ВЭИ** *abbr.* (**Всесоюзный электротехнический институт**) All-Union Electrotechnical Engineering Institute

**ВЭТА** *abbr.* (**Военно-электротехническая академия**) Military Electrotechnical Academy

**вяжущий** *adj.* binding, cementing, astringent

**вязальный** *adj.* binding

**вязаный** *adj.* bound

**вязательный** *adj.* binding

**вязать, связать** *v.* to bind, to tie

**вязаться, связаться** *v.* to be tied; to be connected, to be joined

**вязка** *f.* binding, tying; binder twine, tie wire

**вязкий** *adj.* viscous, viscid, glutinous; sticky; tough, tenacious, ductile

**вязкость** *f.* viscosity, viscidity; adhesiveness; strength, ductility, toughness, tenacity; **в. в запиле** (notched bar) impact resistance

**вязкотекучесть** *f.* viscosity

**вязкоупругий** *adj.* viscoelastic

**вязкоэластичный** *adj.* viscoelastic

**вязнуть, завязнуть** *v.* to stick

**вязочный** *adj.* binding, tying

**вялость** *f.* flabbiness, limpness; inertia, sluggishness

**вялый** *adj.* flabby, flaccid, limp; sluggish, inert, dull, slack

# Г

In foreign words taken over into Russian, the letter h is generally represented by Russian г: *see* **Гамильтон, гексод, гидрат** *etc.*

**Г** *abbr.* 1 (**гига**) giga; 2 [**грамм (сила)**] gram (force)

**г** *abbr.* 1 (**грамм**) gram; 2 (**год**) year

**Гааз, де** de Haas

**габарит** *m.* dimension, size, overall size; clearance; profile; loading gauge; **габариты труб для электропроводки** conduit diameters, conduit sizes; **габариты электродов** dimensions of electrodes; **г. свободного прохода** equipment clearance line

**габаритный** *adj.* dimension, size, overall size; clearance; profile

**Габер** Haber

**«габриты»** *pl.* Gabrite (trade name)

**гавайский** *adj.* Hawaiian

**гаванный** *adj.* harbor, port

**гавань** *f.* harbor, port

**ГАВБ** *abbr.* (**главная авиабаза**) main air base

**Гаген** Hagen

**Гагенбах** Hagenbach

**гадолиний** *m.* gadolinium

**гадолиновый** *adj.* gadolinium, gadolinic

**гаечный** *adj.* screw-nut; inside (thread), female

**газ** *m.* gas; gauze; **газы возникающие при воспламенении** ignition gases; **благородный г.** noble gas; **водной г.** water gas; **выделенный г.** swept-off gas; **выхлопный г.** burned gas, combustion gas; exhaust gas; **(генераторный) г. с избытком окислителя** oxidizer-rich gas; **гремучий г.** oxyhydrogen gas; **диэлектрический г.** octafluorocyclobutane; **защитный г.** gas shielding, protective gas; **«кислый» г.** oxidizer-rich gas; **г. Монда** Mond gas; **г., поглощённый стенками вакуумного прибора** occluded gas; **радиоактивный г.** radon; **реальный г.** imperfect gas; **светильный г.** coal gas; **г., создающий давление**

pressurizing gas; **углекислый г.** carbonic acid; **г. Ферми** Fermi gas

**газирование** *n.* gassing

**газированный** *adj.* gassed

**газировать** *v.* to gas, to aerate

**газифицирующийся** *adj.* gasifiable

**газный** *adj.* gassy

**газо-** *prefix* gas

**газоанализатор** *m.* gas analyzer; **акустический г. на метан** acoustic metanometer; **г. Орса** Orsat gas analysis apparatus

**газобаллон** *m.* gas cylinder

**газобаллонный** *adj.* gas cylinder; gas pressurized

**газовать** *v.* to gas

**газовидный** *adj.* gaseous

**газовыделение** *n.* gassing, boiling, ebullition

**газовый** *adj.* gas, gaseous

**газогенератор** *m.* gas producer, gas generator, producer, decomposer, decomposition chamber; **г., работающий на перекиси водорода** peroxide gas generator; **г., работающий на химическом топливе** chemical gas generator; **г. с газобаллонной системой подачи топлива** gas-pressurized generator; **г. с избытком топлива** fuel-rich gas generator

**газогенераторный** *adj.* gas-generator; gas-producing

**газогенерирующий** *adj.* gas-producing, gas-generating

**газодинамика** *f.* gasdynamics

**газодиффузионный** *adj.* gas-diffusion

**газойль** *m.* gas oil

**газокалильный** *adj.* incandescent

**газомер** *m.* gas meter; **пузырьковый г.** bubble gauge; **самопишущий г.** gas flow recorder

**газометр** *see* **газомер**

**газометрический** *adj.* gasometric

**газонаполненный** *adj.* gas-filled

**газонепроницаемый** *adj.* gastight, gas-proof, vapor-proof

**газообменник** *m.* gas exchanger

**газообразность** *f.* gaseousness

**газообразный** *adj.* gaseous, gasiform
**газообразование** *n.* gassing, gas formation, boiling, ebullition, volatilization; **бурное г.** gassing, boiling
**газообразователь** *m.* gas generator, gas producer
**газообразующий** *adj.* gas-forming
**газоотвод** *m.* gas issue, gas vent; gas bleeder
**газоотводный** *adj.* bleeding, (of a) vent
**газоотделение** *n.* liberation of gases, gassing
**газоотделитель** *m.* gas separator
**газоотсасывающий** *adj.* exhaust, gas-suction
**газоохладитель** *m.* gas condenser
**газоочиститель** *m.* gas purifier
**газоочистительный** *adj.* gas-purifying
**газоочистка** *f.* gas purification, gas cleaning
**газоплотный** *adj.* gas-tight, gasproof
**газопоглотитель** *m.* getter, getter pellet, gas absorber; **г., наносимый слоем** coating getter; **объёмный г.** bulk getter; **поверхностный г.** flash getter; **таблеточный г.** pellet getter
**газопоглощение** *n.* gas occlusion, gas clean-up, getter action
**газоподводящий** *adj.* gas-feed, gas-intake
**газополный** *adj.* gas-filled
**газопривод** *m.* gas feed
**газопроизводитель** *m.* gas generator
**газопроницаемость** *f.* gas permeability
**газопроницаемый** *adj.* gas-permeable
**газоразрядный** *adj.* gas-discharge
**газосборник** *m.* gas collector
**газосварка** *f.* gas welding
**газосветный** *adj.* gas-discharge, luminous discharge
**газостойкий** *adj.* gas-proof
**газоструйный** *adj.* gas-jet
**газотрон** *m.* gas-filled tube rectifier, hot-cathode gas-discharge rectifier, phanotron, gas diode; **двухтактный г.** full-wave rectifier; **плоский г.** flat gas-filled diode; **ртутный г.** (hot-cathode) mercury-vapor rectifier, mercury vapor tube; **г. с ртутными парами** hot-cathode mercury-vapor rectifier; **г. с холодным катодом** cold-cathode diode, tactron
**газотурбина** *f.* gas-driven turbine
**газотурбинный** *adj.* gas-turbine
**газоудерживающий** *adj.* gas-retaining

**газоуловитель** *m.* gas trap, gas collector, gas take
**газоупорный** *adj.* gasproof
**газофокусированный** *adj.* gas-focused
**газоэлектрический** *adj.* gas-electric
**газящий** *adj.* gassy; gassing
**гайдроп** *m.* hauling rope, guide rope
**гайка** *f.* (screw) nut, female screw; **барашковая г., г.-барашек** flynut, wing nut; **зажимная г., натяжная г.** coupling nut, tension female-screw; **г. с кольцом** ring locket; **г. с накаткой** knurled nut; **г. с трубной резьбой** tube nut; **стяжная г.** turn buckle
**гайперник** *m.* Hypernic
**гакабортный** *adj.* taffrail, poop
**Гакеталь** Hackethal
**гал** *m.* gal (unit of acceleration)
**галактика** *f.* Galaxy
**галактический** *adj.* galactic
**галакто-** *prefix* galacto-
**галактометр** *m.* galactometer, lactodensitometer
**галалит** *m.* Galalith (trade name)
**гален** *m.* galena, lead glance
**галенит** *m.* galena, galenite
**галеновый** *adj.* galena
**галерея** *f.* gallery; tunnel; **подземная г.** adit
**Галеркин** Galerkin; Halerkin
**галета** *f.* disc coil, pancake coil
**галентный** *adj.* wafer(-type); disc (-type)
**Галилей** Galilei; Galilean
**Галицин** Galitzin
**галлалит** *m.* gallalith
**Галлей** Halley
**галий** *m.* gallium, Ga
**галло-иттриевый** *adj.* yttrium-gallium
**галлон** *m.* gallon
**галлонный** *adj.* gallon
**галлюцинация** *f.* hallucination, delusion
**гало** *n. indecl.* halo
**галовакс** *m.* halowax, chloro-naphthalene wax
**галоид** *m.* halogen; halide
**галоидный** *adj.* halogen; halide
**галтовка** *m.* ball burnishing
**галь** *see* **гал**
**гальванизатор** *m.* galvanizer
**гальванизация** *f.* galvanism, galvanization; **г. рисунка** plating
**гальванизирование** *n.* plating, electroplating

гальванизированный *adj.* plated, galvanized

гальванизировать *v.* to plate, to galvanize, to electroplate

гальванически *adv.* by galvanization

гальванический *adj.* galvanic, voltaic; electro-

гальвано *n. indecl.* electrotype

гальванограф *m.* galvanograph

гальванография *f.* galvanography

гальванокаустика *f.* galvanocautery

гальванолюминесценция *f.* galvanoluminescence

гальванометр *m.* galvanometer; астатизированный г. Broca galvanometer; баллистический г. quantometer, ballistic galvanometer; батарейный г. battery gauge (tester); г. Дарсонваля d'Arsonval galvanometer; дифференциальный г. для сравнения напряжений comparative indicator; г. для баллистических измерений quantometer; г. для измерения количества электричества quantity galvanometer; г. для приёма сигналов по подводному кабелю speaking galvanometer; г. для применения на линии lineman's detector; г. для распознавания концов многожильного кабеля wire finder; карманный г. battery gauge; магнитоэлектрический г. Deprez-d'Arsonval galvanometer, moving coil galvanometer; магнитоэлектрический г. без железа внутри подвижной катушки Ayrton-Mather galvanometer; г. с успокоением масляным катарактом oil vessel galvanometer; г. со световым «зайчиком» light-spot galvanometer, light-band galvanometer; г. со стрелкой в вертикальной плоскости vertical galvanometer, upright galvanometer; струнный г. Einthoven galvanometer, string galvanometer

гальванометрический *adj.* galvanometric, galvanometer-type

гальванометрия *f.* galvanometry

гальванопласт *m.* galvanizer

гальванопластика *f.* electro-forming, galvanoplastics, electrotyping

гальванопластический, гальванопластичный *adj.* galvanoplastic, electroformed

гальванопокрытие *n.* electroplating, electrodeposition; г. в колокольном аппарате barrel plating

гальваноскоп *m.* galvanoscope, current detector

гальваноскопический *adj.* galvanoscopic

гальваностегировать *v.* to electroplate

гальваностегический *adj.* electroplating, electrolytic plating

гальваностегия *f.* electroplating, galvanostegy

гальванотаксис *m.* galvanotaxis

гальванотермомагнитный *adj.* galvanothermomagnetic

гальванотехника *f.* electroplating, electrolytic metallurgy

гальванотипия *f.* electrotype

гальванотропизм *m.* galvanotropism

гальватрон *m.* galvatron

галька *f.* gravel, pebble, grit; nodule

Гальс Hals

Гальтон Galton

гамба *f.* viola da gambe

Гамильтон Hamilton

гамильтониан *m.* Hamiltonian, Hamiltonian operator

гамильтонов *poss. adj.* Hamiltonian, (of) Hamilton

гамильтоновский *adj.* Hamiltonian

гамма *f.* gamma, scale; gamut, range; г. воспроизводящего устройства display gamma; китайская г. pentatonic scale; натуральная г. just scale; общая г. системы overall gamma; г. приёмной трубки display gamma; г. с выравненными интервалами tempered scale; г. с правильными интервалами just scale; г. цветов, цветовая г. color gamut

гаммаграф *m.* gammagraph

гаммаграфия *f.* gammagraphy

гамма-дефектоскопия *f.* gamma-ray materiology

гамма-корректор *m.* gamma corrector; г.-к. в канале зелёного (красного, синего) цветоделённого изображения green (red, blue) gamma operator

гамма-спектр *m.* gamma-ray spectrum

«гамматрон» *m.* gammatron

гамма-фактор *m.* gamma

Гаммонд Hammond

гамовский *adj.* Gamov's, (of) Gamov

ГАМС *abbr.* 1 (Главная авиаметеорологическая станция) main air weather station; 2 (гражданская авиационная метеорологическая станция) civil air weather station

**ганновский** adj. (of) Gunn, Gunn's

**гантелеобразный** adj. dumbbell-like

**гантели** pl. dumbbell

**гантель-модель** f. model of molecular structure

**гарантийный** adj. guarantee

**гарантирование** n. guaranteeing

**гарантированный** adj. guaranteed

**гарантировать** v. to secure; to guarantee

**гарантия** f. warranty, guarantee, security

**Гарвей** Harvey

**гаргревский** adj. Hargreaves

**Гаргривс** Hargreaves

**гармодотрон** m. harmodotron

**гармонизация** f. harmonization

**гармоника** f. harmonic; harmonic curve; harmonic wave; higher harmonics, harmonic overtone; fundamental component, partial; musical glasses; accordion; 1 sing.: **вторая г.** quadratic component; **губная г.** mouth organ, French harp; **г. (звука вращения) лопасти** blade harmonic; **г. колебания** harmonic components, harmonics; **обратная г.** fractional frequency; **растяжная г.** accordion; **г. синусоидального колебания** multifrequency sinusoid. 2 pl.: **гармоники** harmonics, harmonic oscillations, upper harmonics; **г. высшего порядка, высшие г.** upper harmonics, ultraharmonics, harmonic oscillations; **зубцовые г.** tooth ripples; **г. кварца** crystal harmonics; **г. кривой напряжения во времени** harmonics; **г. пьезокварцевого резонатора** see **гармоники кварца**; **субъективные г.** aural harmonics; **г. третьего порядка, трёхкратные г.** triple-frequency harmonics

**гармониковый** adj. harmonic

**гармоническая** f. adj. declen. harmonic (phys.); **высшая г.** harmonic wave

**гармонически** adv. harmonically, harmonicly

**гармонический** adj. harmonic, harmonious

**гармоничность** f. harmonicity; **г. звукосочетаний** sound relation

**гармоничный** adj. harmonious, harmonic

**гармония** f. harmony, concord

**гармошка** f. bellows; corrugated waveguide

**гарнитур** m. fittings, mountings, trimmings; set; **головной г.** headset; **телефонный г.** telephone headset assembly

**гарнитура** f. fittings, mountings, outfit, set; breast-plate (of a telephone operator); **многократная г. телефонистки** (operator's) multiple answering equipment

**гарнитурный** adj. breast-plate

**гарпиус** m. rosin, white resin

**гарпиусный** adj. rosin

**гартблей** m. hard lead, type metal

**Гартман** Hartman

**гартованный** adj. hard-drawn

**гасило** n. extinguisher; damper

**гасильный** adj. quenching, damping, extinguishing

**гасимый** adj. extinguishable; quenched

**гаситель** m. extinguisher, quencher; damper; **г. заземляющей дуги** arcing ground suppressor; **механический г. колебаний** mechanical damper; **г. пламени** flame arrester; **г. фосфоресценции** scotophor; **г. энергии** energy absorber

**гасительный** adj. quenching, damping, extinguishing

**гасить, загасить** or **погасить** v. to extinguish, to quench, to suppress; to damp; to kill; to put out, to cut off; to clear; to blank off, to blank out; **г. дугу** to quench an arc

**гаснуть, загаснуть** or **погаснуть** v. to go out, to die out; to be extinguished

**гаснущий** adj. going out, dying out, sinking; quenched

**гасящий** adj. damping; quenching, extinguishing; blanking

**г-ат, г-атом** abbr. (грамм-атом) gram-atom

**Гаугвиц** Haugwitz

**Гаунт** Gaunt

**гаус** see **гаусс**

**гаусзистор** m. gaussistor

**Гаусс** Gauss

**гаусс** m. gauss (unit of magnetic induction)

**гауссметр** m. gaussmeter

**гауссов** poss. adj. gaussian, Gauss

**гауссовый** adj. Gaussian, Gauss

**гафний** m. hafnium, Hf

**гашение** n. quenching, extinguishing, putting out, cancellation; extinction;

clearance, clearing; suppression; blanking; **г. луча** beam suppression, black-out (of a beam); **г. луча по строке** horizontal blanking; **нормальное г.** normal clearance; **г. обратного хода** return-trace (or retrace) blanking; **г.о.х. кадровой развёртки, г.о.х. луча по кадру** (or **по кадрам**) field retrace blanking, frame blanking, frame fly-back suppression; vertical blanking; **г.о.х. луча по полю** field (or frame) blanking; **г.о.х. (луча) по строкам, г.о.х. строчной развёртки** horizontal (or line) blanking; **г.о.х. по полю развёртки** frame blanking; **г. обратных ходов развёртки камеры** camera blanking; **плавное г. изображения** dissolve; **полное г. колебаний** absolute damping; **г. поля** field discharge; **постепенное г. одного изображения с последующим постепенным усилением другого** vision fading; **г. пучка** beam suppression; **раздельное г.** split clearance; **г. разряда** quenching of the discharge, inhibition of reignition; **г. счётчика** clearance; **г. фосфоресценции** tenebrescence

**гашеный** *adj.* extinguished, quenched; damped

**гашетка** *f.* firing button

**гб** *abbr.* (**гильберт**) gilbert

**гвоздевой** *adj.* nail

**гвоздь** *m.* nail; pin; peg; tack; **деревянный г.** peg; **желобчатый г.** grooved rivet, notched rivet; **г. с буквой (для столбов)** letter nail (for poles); **г. с цифрой (для столбов)** number nail (for poles); **фасонный г.** shaping nail

**гвоздяной** *adj.* nail

**гвт** *abbr.* (**гектоватт**) hectowatt

**гвт-ч** *abbr.* (**гектоватт-час**) hectowatt-hour

**ГВФ** *abbr.* (**Гражданский воздушный флот**) Civil Aviation

**ГГ** *abbr.* (**газогенератор**) gas generator, gas producer

**гг.** *abbr.* 1 (**годы**) years; 2 (**города**) cities

**ГГУ** *abbr.* (**Горьковский государственный университет**) Gorki State University

**ГД** *abbr.* (**горизонтальная дальность**) horizontal range

**где** *adv.* where, wherever

**«геваколор»** *m.* Gevacolor (tradename)

**гевея** *f.* hevea

**Гей** Gay; Hay

**Гейгер** Geiger

**гейгеровский** *adj.* Geiger

**Гейзенберг** Heisenberg

**гейзенберговский** *adj.* Heisenberg

**Гейланд** Heyland

**Гейслер** Geissler

**гейслеров** *adj.* Geissler

**Гейтлер** Heitler

**гекаллой** *m.* gecalloy

**гекса-** *prefix* hexa-

**гексагональный** *adj.* hexagonal

**гексадекаэдроид** *m.* hexadecahedroid

**гексакизоктаэдр** *m.* hexakisteoctohedron

**гексакисоктаэдр** *m.* hexakisoctahedron

**гексакистетраэдр** *m.* hexakistetrahedron

**гексакосиэдроид** *m.* hexacosihedroid

**гексаметилентетрамин** *m.* hexamethylene-tetramine

**гексациклический** *adj.* hexacyclic

**гексаэдр** *m.* hexahedron

**гексаэдрический** *adj.* hexahedral, cubic

**гексо-** *prefix* hexo-

**гексод** *m.* hexode, six-electrode tube; **г. регулируемого усиления** fading hexode; **г. с переменной (or удлинённой) крутизной (or характеристикой)** variable-mu hexode

**гексод-смеситель** *m.* mixing hexode

**гекто-** *prefix* hecto-

**гектоватт** *m.* hectowatt

**гектоватт-час** *m.* hectowatt-hour

**гектограмм** *m.* hectogram

**гектограф** *m.* hectograph, copying apparatus

**гектографический** *adj.* hectographic

**гектолитр** *m.* hectoliter

**гектометр** *m.* hectometer

**гектометрический, гектометровый** *adj.* hectometric

**гектопьеза** *f.* hectopiezoelectric unit

**гелиево-водяной** *adj.* helium-water

**гелиевый** *adj.* helium

**гелий** *m.* helium, He

**геликальный** *adj.* helical

**геликоид** *m.* helicoid

**геликоидальный** *adj.* helical, spiral, helicoid

**геликометр** *m.* helicometer

**геликон** *m.* helicon; helicon wave

**геликотерма** *f.* helicoterma

**геликс** *m.* helix
**гелио-** *prefix* helio-
**гелиобатарея** *f.* solar battery
**гелиограф** *m.* heliograph
**гелиографический** *adj.* heliographic
**гелиография** *f.* heliography
**гелиометр** *m.* heliometer
**гелиоскоп** *m.* helioscope
**гелиостат** *m.* heliostat
**гелиосфера** *f.* heliosphere
**гелиотермометр** *m.* heliothermometer
**гелиотехника** *f.* heliotechnics, solar energy technology
**гелиотроп** *m.* heliotrope
**гелиотропизм** *m.* heliotropism
**гелиотропин** *m.* heliotropine
**гелиофизика** *f.* heliophysics
**гелиоцентрический** *adj.* heliocentric
**гелиоцинкография** *f.* heliozincography
**гелиоэлектростанция** *f.* solar generator
**гелиоэлемент** *m.* solar cell
**гелитрон** *m.* Helitron tube
**Гелл** Hell
**гель** *m.* gel
**Гельмгольц** Helmholtz
**гельмгольцевый** *adj.* Helmholtzian, Helmholtz
**гемералопия** *f.* hemeralopia, night-blindness
**геми-** *prefix* hemi-, semi-
**гемиморфный** *adj.* hemimorphic
**гемит** *m.* hemit (insulating material)
**гемитропический** *adj.* hemitropic, twinned, half-coiled
**гемиэдр** *m.* hemihedron
**гемиэдрический** *adj.* hemihedral
**гемиэдрия** *f.* hemihedrism
**гемоглобинометр** *m.* hemoglobinometer
**гемодинамика** *f.* hemodynamics
**гемодинамометр** *m.* hemodynamometer
**гемотахометр** *m.* hemotachometer
**гемоцитометр** *m.* hemocytometer
**ген** *m.* gene
**ген.** *abbr.* (**генеральный**) gen., general
**генеральный** *adj.* general
**генератор** *m.* generator, dynamo; oscillator; (gas) producer; **автомодулируемый г. импульсов** self-pulsed oscillator; **аммиачный молекулярный г.** NH₃ maser; **апериодический г. пилообразных напряжений** aperiodic sawtooth generator; **асинхронный г.** induction generator, asynchronous alternator; **атомный г.** atomic oscillator; **г., балансируемый с помощью моста** bridge-balancing

oscillator; **балансный г.** balanced oscillator; **баркгаузеновский г., г. Баркхаузена** Barkhausen oscillator; **батарейный вызывной г.** battery ringer; **г. биений** beat-frequency oscillator; **г. Блонделя** Blondel oscillator; **г. Брэдли** Bradley oscillator; **бустерный г. переменного тока** booster alternator; **бутстрэпный г. развёртки** bootstrap sawtooth generator, bootstrap(-sweep) generator; **г. в приёмнике** receiver oscillator; **г. Ван де Граафа** belt generator; **г., варьированный по скорости** velocity-modulated oscillator; **ведомый г.** slave oscillator; **г. вертикального отклонения** vertical-deflection oscillator; **г. вертикальной развёртки** vertical scanning generator, vertical oscillator; **ветровой зарядный г.** wind charger; **г. видеочастоты** photo-audio generator; **г. вобуляции пятна** spot wobble generator; **г. возмущающей электродвижущей силы** function generator; **вольтодобавочный г.** booster (generator); **вольтодобавочный г. двустороннего действия** buck-boost generator; **вольтодобавочный г. переменного тока** synchronous booster; **г. воющего тона** warble tone oscillator, audio-howler; **воющий г.** warbler; **г. временных меток** pulse-timing marker oscillator; **г. временных отметок** time mark generator; **всеволновой г.** all-wave oscillator; **г. вспомогательной частоты** quenching oscillator; **(вспомогательный) г. с пропорциональным ускорению напряжением** acceleration generator; **г. вспышки** burst generator, color burst generator; **г., выдающий треугольные сигналы** triangular-wave oscillator; **г. вызывного тока** ringing generator, ringing and signal machine; **вызывной г. звуковой (*or* низкой, *or* тональной) частоты** voice-frequency ringing generator, low-frequency signaling set, low-frequency ringer; **г., выполненный по мостовой схеме** bridge-balancing oscillator; **высокостабильный г. с четырёхплечевым мостом в цепи обратной связи** Meacham bridge oscillator; **высокочастотный г.** radio-frequency generator, high-frequency oscillator; **высокочастотный г. Александерсона** Alexanderson alternator; **высокочастотный г. для телевизионного приёма** R-F high-voltage generator; **газовый г., работающий на окислителе и горючем** (*or* **на унитарном топливе**) liquid-reactant (monoreactant) gas generator; **газоструйный г.** fluid generator; **г. гармоник** harmonic oscillator, harmonic generator; **гармонический г.** harmonic oscillator; **г. главных синхронизирующих импульсов** clock, master clock; **г. горизонтального отклонения** horizontal-deflection oscillator; **граммофонный (вч) г.** phonograph oscillator; **г. двойной строчной частоты** twice-horizontal frequency oscillator; **двухобмоточный синхронный г.** double-winding alternator; **двухполюсный г.** two-terminal oscillator; **двухполюсный г. с якорем ниже обмотки возбуждения** inverted dynamo, overtype dynamo; **двухтактный г.** push-pull oscillator; **двухфазный г.** two phase alternator; **динатронный г.** dynatron oscillator; **г. для визуального регулирования приёмников** visual alignment generator, visalgen; **г. для непосредственного соединения с паровой турбиной** direct-connected turbogenerator; **г. для питания приводных двигателей судовых винтов** propulsion generator; **г. для питания собственных нужд электростанции** house generator; **добавочный г.** booster; **дозаряжающий г.** milking booster, milker; **дуговой г.** arc converter; **г. единиц, г. единичных импульсов** one-generator; **ёмкостно-резистивный** (*or* **-реостатный**) **г.** capacitance-resistance oscillator, R-C oscillator; **ждущий г.** triggered oscillator; **задающий г.** master oscillator, driving oscillator, self-oscillator, exciter; master clock; **задающий многочастотный г.** injection oscillator; **г. задержанных импульсов** delayed-pulse oscillator, delayed-gate generator; **г. задержанных пусковых импульсов** delayed-trigger generator; **запирающий г.** blanking oscillator; **г. запускаемый импульсами** pulsed oscillator; **запускающий г. модели-**

**рующего устройства** analog generator, analog stimulus generator; **звуковой г. типа сирены** siren transducer; **измерительный г. высокой частоты** R-F signal generator; **г. имитации цели** target generator; **г., имитирующий движение цели** target-motion generator; **г. имитирующий флуктуационные колебания отражённого от цели сигнала** scintillation noise generator; **импульсный г.** pulse generator, pulser; pulsed oscillator, discharge impulse oscillator; lightning generator; **г. импульсов, возвращающих схему в исходное положение** reset pulse generator; **г. импульсов высокого напряжения** pulse extra-high tension generator; **г. импульсов тока** key sender; **г. испытательной таблицы** TV pattern generator; **испытательный г. с индикацией резонанса по спадению тока в цепи сетки** grid-dip meter; **г. калибровочных меток** fixed range mark generator; **карманный испытательный г.** pocket tracer; **г. «качания» луча** "spot wobbler" generator; **г. качающейся частоты для регулирования приёмников** alignment oscillator; **г. качающейся частоты для съёмки характеристики задержки огибающей** envelope delay sweep generator; **г. качающейся частоты с линейной зависимостью частоты от координаты** linear sweep generator; **квантомеханический г. инфракрасного диапазона волн** iraser; **квантомеханический г. оптического диапазона волн** laser; **г. колебаний** oscillator; **кольцевой г.** ring oscillator; **г. коммутирующих импульсов, предназначенный для электронной вставки в цветном телевидении** chromakeyer; **г. коммутирующих сигналов** off-on wave generator; **г. компенсации чёрного пятна** shading generator; **ламповый г. с одновременным возбуждением колебаний двух частот** diplex generator; **магнитоэлектронный г.** magneto alternator; **маломощный импульсный г.** low-level pulser; **г. маркёрных импульсов** marker oscillator; **г. масштабных импульсов** range-marker oscillator, marker oscillator, time mark generator; **г. масштабных импульсов дальности** range-marker generator; **г. многофазных сигналов** multiphase clock; **г. моделирования движения цели** target-motion generator; **г. моделированный импульсами** pulse-modulated oscillator; **молекулярный г.** maser oscillator; **мощный г. диапазонов L, S, C, и X** stabilotron; **г. на биениях** beat-frequency oscillator, heterodyne oscillator, beat buzzer; **г. на боковой полосе** side-band generator; **г. на кристаллических триодах** transistor inverter; **г. на триод-триоде** tri-tet oscillator; **г. незатухающих колебаний** (*or* **сигналов**) CW oscillator, continuous wave signal generator; **г. нелинейных функций** Vernistat; **низкочастотный г. биений** L.F. beat oscillator, low-frequency generator; **нормальный г.** standard-level generator; **«обращённый» синхронный г.** inverted alternator; **г. одиночных импульсов** single-pulse device; **одноламповый г.** single-tube oscillator, squegger; **г. опорного напряжения** reference generator; **г. опорной цветной поднесущей** color reference generator; **г. опорных напряжений** pedestal generator; **оптический квантовый г.** laser; **г. отметки** notch generator; **г. «пакета» колебаний поднесущей частоты** burst generator, color burst generator; **г. пакета кратковременных импульсов** burst generator; **параметрический г.** parametric oscillator, parametron; **г. парогаза, работающий на двухкомпонентном топливе** bireactant gas generator; **г. переменного тока** alternator, a-c generator; **г. переменной частоты** variable-frequency oscillator, sweep generator, sweep oscillator, cycle generator, VFO; **переносный г.** service oscillator; **г. пилообразных колебаний развёртки точного отсчёта** fine triangular wave-form generator; **г. П-импульсов** rectangular pulse generator, square-wave generator; **г., питаемый от батареи** battery generator; **г. помех** jamming transmitter, jammer, interference unit; **г. постоянного момента** torque generator; **г. постоянного тока** d-c generator, constant-

current generator; **г. постоянно-переменного тока** double-current generator; **г. прерывистого действия** chopping oscillator; **прерывистый вч г.** blocking oscillator, squegging oscillator, self-pushing oscillator; **прерывно действующий г.** chopping oscillator; **пусковой г.** trigger generator; **«радужный» г.** rainbow generator; **г. развёртки с катодным повторителем и линеаризирующей цепью обратной связи** bootstrap sweep circuit generator; **г. размывания пятна** "spot wobbler" generator; **г. разрывных колебаний** relaxation oscillator; **г. реактивной мощности** reactive oscillator; **реактивный г.** reluctance generator; **г. регулируемой частоты** variable oscillator; **регулируемый г. развёртки** controllable sweep generator; **г. релаксационного напряжения** blocking oscillator; **ремённый (электростатический) г.** belt generator; **реостатно-ёмкостный г. трапецеидального напряжения** R-C trapezoidal voltage generator; **г. с большим пролётным временем** transit-time oscillator; **г. с добавочными полюсами** interpole generator; **г. с заземлённым анодом** plate-return oscillator; **г. с импульсной автомодуляцией** self-pulsed oscillator; **г. с индуктивной обратной связью** tickler-coil oscillator, Meissner oscillator; **г. с индуктором без обмотки** inductor generator; **г. с искусственной линией** "bootstrap" generator; **г. с кольцевой схемой (включения ламп)** ring oscillator; **г. с модуляцией частоты** f-m oscillator; **г. с «ответвлёнными полюсами», соединёнными с плавными полюсами магнитными перемычками** diverter pole generator; **г. с отводами на катушке колебательного контура** tapped-down generator; **г. с паровой машиной** steam-electric generating set; **г. с постоянными магнитами** magneto, magneto generator, hand generator; **г. с самовозбуждением** master oscillator, self-oscillator, self-excited oscillator, self-excited alternator, free-running generator; **г. с электрической настройкой** strophotron, electronically-tuned oscillator; **г. с элек-**

тронной перестройкой voltage-tuning oscillator; **г. сантиметровых волн** microwave oscillator; **сварочный г.** arc-welding generator; **сверхгенеративный г.** quench generator; **г. селекторных импульсов** gating pulse generator; **г. серии импульсов** pulse train generator; **г. серий импульсов** pulse-series generator; **г. сигнала (для испытания приёмников), г. сигналов** signal generator; **г. сигналов звуковой частоты** note oscillator; **г. сигналов цветности** chroma oscillator; **сигнальный г.** signal generator; **г. символов, работающий по принципу использования фигур Лиссажу** Lissajous symbol generator; **синхронизированный г.** locked oscillator; **синхронизированный г. пусковых импульсов** synchronized trigger generator; **синхронизирующий г.** synchronizing generator, timing-wave generator, timing(-pulse) generator; **синхронный г.** synchronous generator, alternator; **синхронный г. переменного тока** synchronous alternating-current generator; **синхронный г. с вращающимся искровым промежутком** alternator disc set; **г. синхросигналов** synchronizing generator, sync-signal generator; **г. смещения** displacement generator, shifting oscillator; **стабилизированный г. меток** fixed range mark generator; **г. стандартного сигнала** signal generator; **статический г.** statitron; **стендовый г.** bench oscillator; **г. тока** current generator, dynamo; **трёхточечный г. (по схеме Гартлея)** Hartley oscillator; **управляемый извне г. развёртки** sweep amplifier with external excitation; **г. фиксированных масштабных импульсов** fixed range mark generator; **шумовой г. искусственных помех** interference unit; **эквивалентный г.** Thevenin's generator; **электронный г. сигналов времени** time signal set; **эталонный г. высокочастотных колебаний основной частоты и её гармоник** multivibrator

**генераторно-усилительный** *adj.* oscillator-amplifier

**генераторный** *adj.* generator; generating

**генератриса** *f.* generatrix

**генерация** *f.* generation; oscillation; reaction; lasing; **акустическая паразитная г.** acoustic howling; **паразитная г.** spurious oscillation; **прерывистая г.** squitter, squegging oscillation; **ритмичная г. приёмника** poping; **г. сеточного смещения** manner of obtaining grid bias voltage; **слышимая г.** howl

**генерирование** *n.* generation, generating; oscillation, oscillating; **г. мощных звуков полей для цели маскировки собственного звукового излучения цели** acoustic blanket

**генерированный** *adj.* generated

**генерировать** *v.* to generate, to produce; to oscillate; to lase

**генерируемый** *adj.* generated; lasing

**генерирующий** *adj.* generating, generative; oscillating, self-oscillating

**генетика** *f.* genetics

**генетический** *adj.* genetic

**«генлок»** *m.* Genlock

**генри** *m. indecl.* henry (unit of induction)

**генриметр** *m.* henrymeter, inductance meter

**Генричи** Henrici

**гео-** *prefix* geo-

**геоакустика** *f.* geoacoustics

**геогнозия** *f.* geognosy

**географический** *adj.* geographical

**география** *f.* geography

**геодезический** *adj.* geodetic, geodesic

**геодезия** *f.* geodesy

**геодиметр** *m.* geodimeter

**геоид** *m.* geoid

**геоидальный** *adj.* geoidal

**геоизотермы** *pl.* geo-isotherms

**геокосмический** *adj.* geocosmic

**геокосмология** *f.* geocosmology

**геологический** *adj.* geological

**геология** *f.* geology

**геомагнетизм** *m.* terrestrial magnetism

**геомагнитный** *adj.* geomagnetic, terrestrial-magnetic

**геометрический** *adj.* geometric, geometrical

**геометрия** *f.* geometry; configuration; **г. выходного сечения** exit geometry; **г. камеры сгорания** chamber geometry; **г. корпуса** casing configuration

**геометрооптический** *adj.* geometrical-optics

геопотенциал *m.* geopotential
геопотенциальный *adj.* geopotential
геоскоп *m.* geoscope
геострофический *adj.* geostrophic
геотермический *adj.* geothermal
геотермия *f.* geothermy
геотермометр *m.* geothermometer
геофизика *f.* geophysics
геофизический *adj.* geophysical
геофон *m.* geophone
геоцентр *m.* geocenter
геоцентрический *adj.* geocentric
геоэлектрический *adj.* geoelectric
гептан *m.* heptane
гептод *m.* heptode, pentagrid
Гербер Gerber
Гёргес Goerges
Герке Gehrcke
германиевый *adj.* germanium, germanic
германий *m.* germanium, Ge; **г. с собственной проводимостью** intrinsic germanium
Гермес (астероид) Hermes
герметизация *f.* hermetic sealing, sealing; capsulation, encapsulation; pressurization; "potting"; **г. пластмасой** plastic capsulation; **погодостойкая г.** weather seal
герметизированный *adj.* sealed, hermetically sealed; encapsulated; potted; pressure-proof, pressurized
герметизировать *v.* to seal off; to pressurize
герметизируемый *adj.* sealing
герметизованный *see* **герметизированный**
герметически *adv.* hermetically (sealed)
герметический *adj.* hermetically sealed
герметичность *f.* hermetic state, tightness; containment
герметичный *adj.* hermetically sealed, airtight; sealed in; pressurized; leakproof, leak-tight
гермошлем *m.* pressurized helmet, space helmet; **г. скафандра** full pressure helmet
Герни Garney
Герц Hertz
герц *m.* hertz, cycle per second
герцметр *m.* frequency indicator
герцовый *adj.* Hertzian, Hertz
Герцшпрунг Hertzsprung
Геспер Hesperus
Гесс Hess
Гессе Hesse
гессит *m.* hessite, Ag$_2$Te

гетеро- *prefix* hetero-
гетерогальванометр *m.* heterogalvanometer
гетерогенизация *f.* heterogenization
гетерогенность *f.* heterogeneity
гетерогенный *adj.* heterogeneous
гетеродин *m.* heterodyne; oscillator, local oscillator, receiver oscillator, frequency-controlled oscillator; **г. дающий колебания по амплитуде равные приходящим** equal heterodyne; **измерительный г.** test oscillator, signal generator; **навигационный г.** beacon oscillator; **г. стабилизированный кварцем** quartz (-crystal) oscillator
гетеродинирование *n.* heterodyning, beat, superposition, superheterodyne; **двойное г.** double super-effect
гетеродинировать *v.* to heterodyne, to superpose
гетеродинный, гетеродиновый *adj.* heterodyne(-type)
гетеродирование *n.* heterodyning
гетероколлектор *m.* hetero-collector junction
гетеропереход *m.* heterojunction, heterogenous junction
гетерополярный *adj.* heteropolar
гетеростатический *adj.* heterostatic
гетеросфера *f.* heterosphere
гетерохроматный, гетерохромный *adj.* heterochromatic
гетероциклический *adj.* heterocyclic
гетинакс *m.* pertinax, laminated insulation, isolit, turbonit, micarta
геттер *m.* getter; **г. наносимый слоем** coating getter
геттерирование *n.* gettering (effect)
геттерный *adj.* getter
Гефнер Hefner
ГИ *abbr.* 1 (**генератор импульсов**) pulse generator; 2 (**групповой искатель**) group selector; **первый ГИ** district selector
Гиады Hyades
Гиббс Gibbs
гибель *f.* ruin, loss, destruction; **г. (лампы) от импульса сверхтока** surge-current breakdown (of a tube)
гибельный *adj.* destructive, catastrophic
гибернация *f.* hibernation
гибкий *adj.* flexible, pliable, pliant, bendable, ductile; elastic, springy

**гибкость** *f.* flexibility, flexure, pliability, pliancy, ductibility; elastance, elasticity; plasticity; compliance; **акустическая г.** acoustic compliance; **безразмерная г.** specific elastance; **г. вычисления** computing flexibility; **г. конца иглы** needlepoint compliance; **г. края диафрагмы** diaphragm edge compliance

**гибочный** *adj.* bending

**гибрид** *m.* hybrid

**гибридный** *adj.* hybrid

**гига-** *prefix* giga-

**гигагерц** *m.* gigahertz, gigacycle, kilomegacycle

**гигагерцовый** *adj.* gigahertz

**гигантский** *adj.* giant; gigantic, huge

**гигро-** *prefix* hygro-

**гигрограф** *m.* hygrograph, recording hygrometer

**гигрометр** *m.* hygrometer, air humidity indicator; **ёмкостный г.** humidity-sensitive capacitor; **конденсационный г.** hygrodeik; **г. основанный на определении точки росы, г. по точке росы** dewpoint hygrometer

**гигрометрический** *adj.* hygrometric

**гигрометрия** *f.* hygrometry

**гигроном** *m.* hygronom

**гигроскоп** *m.* hygroscope

**гигроскопический** *adj.* hygroscopic, moisture-absorbing

**гигроскопичность** *f.* hygroscopicity

**гигроскопичный** *see* **гигроскопический**

**гигростат** *m.* hydrostat, humidistat

**гигротермограф** *m.* hygrothermograph

**гигроэлектрометр** *m.* hygroelectrometer

**гидр-** *prefix* hydr-

**гидравлика** *f.* hydraulics

**гидравлический** *adj.* hydraulic, liquid-operated

**гидравличность** *f.* hydraulicity

**гидразин** *m.* hydrazine, diamine

**гидразингидрат** *m.* hydrazine hydrate

**гидразиниевый, гидразиновый** *adj.* hydrazine

**гидразон** *m.* hydrazone

**гидрат** *m.* hydrate; **г. окиси** hydroxide; **г. окиси калия** potassium hydroxide

**гидратация** *f.* hydration

**гидратцеллюлоза** *f.* hydrated cellulose

**гидрид** *m.* hydride

**гидро-** *prefix* hydro-, hydr-, water, hydraulic

**гидроаккумулятор** *m.* hydraulic accumulator

**гидроакустик** *m.* sound-gear man, sonarman; **авиационный г.** airborne sonarman; **базовый г., береговой г.** harbor defense sonarman; **г. надводного корабля** surface sonarman

**гидроакустика** *f.* hydroacoustics, underwater acoustics

**гидроакустический** *adj.* hydroacoustic

**гидрогенератор** *m.* hydraulic generator, water-wheel type alternator

**гидрогенный** *adj.* hydrogenous, hydrogen

**гидрографический** *adj.* hydrographic

**гидрография** *f.* hydrography

**гидродинамика** *f.* hydrodynamics; **магнитная г.** magnetohydrodynamics

**гидродинамический** *adj.* hydrodynamic

**гидродинамометр** *m.* hydrodynamometer

**гидродром** *m.* seadrome

**гидродромный** *adj.* seadrome

**гидрокаучук** *m.* hydrogenated rubber, hydrogenated caoutchouc

**гидроклапан** *m.* hydrovalve, pressure-operated valve

**гидрокран** *m.* hydrovalve

**гидроксид** *m.* hydroxide

**гидролиз** *m.* hydrolysis

**гидролитический** *adj.* hydrolytic

**гидролокатор** *m.* echo-ranging sonar, sonar, active sonar, sonar detection gear, hydrolocator, hydro-acoustic range finder; **глубинный г.** depth-determining sonar; **г. для поиска подводных лодок** subsearch sonar; **низкочастотный г.** subsonic sonar; **опускаемый г.** dipping sonar, dipping asdic; **г. работающий на принципе получения эха от взрыва** depth-charge sonar

**гидролокационный** *adj.* hydrolocation, echo-ranging, sonar

**гидролокация** *f.* asdic, hydrolocation, subaqueous sound ranging, sonar localizing; **звуковая г.** sound fixing and ranging

**гидромагнетизм** *m.* hydromagnetism

**гидромагнитный** *adj.* hydromagnetic

**гидрометаллургия** *f.* hydro-metallurgy

**гидрометеоры** *pl.* hydrometeors, precipitation elements

**гидрометр** *m.* hydrometer

**гидрометрический** *adj.* hydrometric

**гидромеханический** *adj.* hydromechanical

**гидромотор** *m.* hydraulic actuator

**гидромуфта** *f.* fluid clutch, fluid coupling

**гидроокись** *f.* hydroxide

**гидропневматический** *adj.* hydropneumatic

**гидропровод** *m.* hydraulic drive

**гидростанция** *f.* hydro-power station

**гидростатика** *f.* hydrostatics

**гидростатический** *adj.* hydrostatic

**гидросфера** *f.* hydrosphere

**гидротрансформатор** *m.* torque converter

**гидротурбина** *f.* hydraulic turbine, water turbine

**гидротурбоагрегат** *m.* hydroelectric generating set

**гидротурбогенератор** *m.* water-wheel type generator, hydroelectric generating set

**гидроусилитель** *m.* booster

**гидроустановка** *f.* hydroelectric plant

**гидрофон** *m.* hydrophone, sonic detector, submarine detector; **г. гидроакустической станции подводной лодки** submarine hydrophone; **г. установленный на корпусе корабля** hull-mounted hydrophone

**гидрофонный** *adj.* hydrophone

**гидрохинон** *m.* hydroquinone

**гидроэлектрический** *adj.* hydroelectric

**гидро(электро)привод** *m.* fluid drive

**гидроэлектростанция** *f.* hydroelectric power station

**гил** *m.* gill

**Гилл** Gill

**Гиллемин** Guillemine

**Гилль** Gill

**гильберт** *m.* gilbert (unit of magnetomotive force); **г. на сантиметр** gilbert per centimeter

**гильбертов** *poss. adj.* Hilbert's, (of) Hilbert

**гильза** *f.* sleeve, bush, bushing, slug (of a relay), sheath, case, shell; cartridge; socket; muff; **г. (цоколя лампы)** shell; **г. (штепсельного гнезда)** bushing; **бумажная г.** paper sleeve; **внутренняя г.** inner shell; **г. гнезда** sleeve of a jack, jack socket, bushing; **г. для сращивания (проводов)** jointing sleeve, splicing sleeve (for wires); **замедляющая г.** copper slug (of a relay); **изолирующая г.**

**(между обмотками)** barrier; **г. с фланцем** pipe adapter

**гильзовый** *adj.* sleeve, bush, socket

**гильотинный** *adj.* guillotine-shaped

**г-ион** *abbr.* (грамм-ион) gram-ion

**гипер-** *prefix* hyper-, super-

**гипербаризм** *m.* hyperbarism

**гипербола** *f.* hyperbola

**гиперболический** *adj.* hyperbolic

**гиперболоид** *m.* hyperboloid

**гипервентиляция** *f.* hyperventilation, overventilation

**гипергеометрический** *adj.* hypergeometric

**гиперголь** *m.* hypergol, hypergolic fuel

**«гипергон»** *m.* trade-name for an outdated wide-angle objective

**гипердин** *m.* hyperdyne

**гиперзвук** *m.* hypersonics

**гиперзвуковой** *adj.* hypersonic, hyperacoustic, superaerodynamic

**гиперкапния** *f.* hypercapnia

**гиперкоммутация** *f.* accelerated commutation, overcommutation

**гиперкомпаунд** *m.* overcompound generator

**гиперкомпаундирование** *n.* overcompounding

**гиперкомпаундированный** *adj.* overcompounded

**гиперкомпаундный** *adj.* overcompound

**гиперкон** *m.* hypercon

**«гиперме»** *n. indecl.* hyperme (tradename for ferromagnetic alloys)

**гиперник** *m.* hypernik

**«гиперокс»** *m.* hyperox (trade-name for ferrite materials)

**гипероксемия** *f.* hyperoxemia

**гипероксипатия** *f.* hyperoxypathy

**гиперон** *m.* hyperon

**гипероскуляция** *f.* extended osculation

**гиперосмия** *f.* hyperosmia

**гиперповерхность** *f.* hypersurface

**гиперпространство** *n.* hyperspace

**гиперсинхронный** *adj.* hypersynchronous

**гиперсфера** *f.* hypersphere

**гипертензия** *f.* hypertension

**гипертермия** *f.* hyperthermia

**гипертермометр** *m.* hyperthermometer

**гипертонический** *adj.* hypertonic

**гипертония** *f.* hypertonia

**гиперфокальный** *adj.* hyperfocal

**гиперхромный** *adj.* hyperchromic

**гиперэкспоненциальный** *adj.* hyperexponential

**гиперэллиптический** *adj.* hyperelliptic
**гипо-** *prefix* hypo-, sub-, under
**гипобаризм** *m.* hypobarism
**гипобаропатия** *f.* hypobaropathy
**гиповентиляция** *f.* hypoventilation
**гиподинамия** *f.* hypodynamia
**гипокапния** *f.* hypocapnia
**гипоксемия** *f.* hypoxemia
**гипоксипатия** *f.* hypoxipathy
**гипоксия** *f.* hypoxia; **г. лёгких** pneumo-hypoxia
**гипометаболизм** *m.* hypometabolism
**гипомнезия** *f.* hypomnesia
**гипопепсия** *f.* hypopepsia
**гипопное** *n. indecl.* hypopnoe
**гипорексия** *f.* hyporexia
**гипосинергия** *f.* hyposynergia
**гипосинхронный** *adj.* hyposynchronous
**гипосмия** *f.* hyposmia
**гипотаксия** *f.* hypotaxia
**гипотеза** *f.* hypothesis; **(катастрофическая) г. образования солнечной системы** collision hypothesis, hypothesis of dynamic encounter; **г. образования солнечной системы в результате очень тесного сближения с другой звездой** tidal theory of the solar system; **г. образования солнечной системы в результате столкновения светила с одним из компонентов двойной звезды** double star collision hypothesis
**гипотензия** *f.* hypotension
**гипотенуза** *f.* hypotenuse
**гипотермальный** *adj.* hypothermal
**гипотермия** *f.* hypothermia
**гипотетический** *adj.* hypothetic(al)
**гипотонический** *adj.* hypotonic
**гипотония** *f.* hypotonia, hypopiesia
**гипохлоремия** *f.* hypochloremia
**гипохромный** *adj.* hypochromic
**гипоциклоида** *f.* hypocycloid, internal epicycloid
**гипс** *m.* gypsum
**гипсовый** *adj.* gypsum, gypseous; plaster
**гипсометр** *m.* hypsometer
**гипсометрический** *adj.* hypsometric
**гипсометрия** *f.* hypsometry
**гипсотермометр** *m.* hypsometer, hypsometric altimeter
**гипсохром** *m.* hypsochrome
**гипсохромный, гипсохромовый** *adj.* hypsochromic
**гираллой** *m.* gyralloy

**гиратор** *m.* gyrator; microwave gyrator
**гирлянда** *f.* chain, garland; **г. для оттягивания провода от опорной конструкции** tie-down string; **изоляторная г.** chain insulator, insulator string
**гирляндный** *adj.* chain, garland
**гиро-** *prefix* gyro-; spiral; ring
**гировертикаль** *f.* vertical gyroscope
**гиродатчик** *m.* gyroscope pickup, gyroscope; **г. положения** altitude gyroscope; **г. тангажа** pitch gyroscope; **г. угловой скорости** rate gyroscope
**гиродинамика** *f.* gyrodynamics
**гироида** *f.* gyroid
**гирокомпас** *m.* gyrocompass, gyrostabilized compass, earth rate directional reference
**гиромагнетический** *adj.* gyromagnetic
**гиромагнитный** *adj.* gyromagnetic
**гиромагнитостробный** *adj.* gyro flux-gate
**гирометр** *m.* gyrometer
**гиропилот** *m.* gyropilot
**гироплатформа** *f.* gyroscope-stabilized platform
**гирополукомпас** *m.* azimuth gyro, directional gyroscope
**гиропреобразователь** *m.* gyro converter
**гирорама** *f.* gimbal system
**гироскоп** *m.* gyroscope; **г. автопилота** altitude gyroscope; **г. изменений по крену** roll rate gyroscope; **г. с маятниковой корекцией** pendulum gyroscope; **трёхстепенный г.** altitude gyroscope
**гироскопический** *adj.* gyroscopic, gyro
**гиростабилизатор** *m.* gyrostabilizer
**гиростабилизация** *f.* gyrostabilization
**гиростабилизированный** *adj.* gyroscope-stabilized
**гиростабилизируемый, гиростабилизованный** *adj.* gyro-stabilized
**гиростат** *m.* gyrostat
**гиротрон** *m.* gyrotron
**гиротропный** *adj.* gyrotropic(-medium)
**гироустройство** *n.* gyrosystem
**гироцентраль** *f.* master gyroscope
**гирочастота** *f.* gyro frequency, Larmor frequency
**гироэдр** *m.* gyrohedron, icositetrahedron, leucitohedron
**гироэлектромагнитный** *adj.* gyroelectric-magnetic

**гиря** *f.* weight (of balance); **заводная г.** driving weight

**гистерезиграф** *m.* hysteresigraph

**гистерезиметр** *m.* hysteresimeter; **самопишущий г.** hysteresis curve recorder

**гистерезис** *m.* hysteresis, (magnetic) lag, lagging; **г. альтиметра** altimeter fatigue; **вязкий г.** hysteresis lag, magnetic creeping, viscous hysteresis; **магнитный г. при переменном поле** linear hysteresis; **ползучий г.** *see* **вязкий г.; тепловой г., термический г.** thermal hysteresis

**гистерезисный** *adj.* hysteretic, hysteresis, lag

**гистерезисограф** *m.* hysteresisograph

**гистерезометр** *m.* hysteresimeter

**гистограмма** *f.* bar chart, bar graph, histogram; **г. с площадью столбцов, пропорциональной значениям функций** area histogram

**гитара** *f.* guitar

**Гитторф** Hittorf

**ГК** *abbr.* (**герметическая кабина**) pressurized cabin

**гл.** *abbr.* (**глава**) chapter; head

**глав.** *abbr.* (**главный**) main, chief, principal

**глава** *f.* chapter; head, chief

**«главарь»** *m.* Häuptling (code-name for type of radar IFF)

**главнейший** *adj.* predominant

**главный** *adj.* principal, chief, main, primary, leading, predominant; major; master; **главным образом** chiefly, mainly, principally, for the most part

**главщит** *m.* (= **главный щит переключений**) master control board

**глагол** *m.* verb

**глагольный** *adj.* verb

**гладилка** *f.* steel burnisher, polisher; sleeker, trowel

**гладило** *n.* burnisher, polisher

**гладить, сгладить** *v.* to smooth, to plane, to polish

**гладкий** *adj.* flat, even, plane, smooth

**гладкость** *f.* flatness; smoothness

**глаз** *m.* eye; **«кошачий» г.** "Cat's Eye"

**глазирование** *n.* glazing, glaze; varnish; frosting

**глазированный** *adj.* glazed; varnished; frosted

**глазировать** *v.* to glaze; to varnish; to frost

**глазной** *adj.* eye, optic, ocular

**глазовидный** *adj.* ocellated, ocellate

**глазок** *m.* eye, eyelet; slot, aperture; lug, ear; iris

**глазомер** *m.* measuring by sight, estimation by sight

**глазомерный** *adj.* by eye, approximate

**глазообразный** *adj.* eye-shaped

**глазуренный, глазурованный** *adj.* glazed; enameled

**глазурь** *f.* glaze, glazing

**гласный** *m. adj. decl.*; *adj.* vowel (sound), vowel; vowel

**глаукизировать** *v.* to gleam in a greenish color

**Глауэрт** Glaubert

**Гледстон** Gladstone

**глёт** *m.* litharge, massicot

**глетовый** *adj.* litharge

**глина** *f.* clay; **огнеупорная г.** fire clay

**глиняный** *adj.* clay, clayey

**глиптал** *m.* glyptal resins

**глипталевый** *adj.* glyptal

**глиссада** *f.* glide path, glide slope; landing beam; **г. планирования** glide path

**глиссадный** *adj.* glide; landing

**глифталь** *m.* glyphtal

**глицерин** *m.* glycerin

**глицериновый** *adj.* glycerin

**глицин** *m.* glycine developer

**гл. о.** *abbr.* (**главным образом**) chiefly, mainly, principally

**глобальный** *adj.* global; entire

**«глобар»** *m.* globar (trade-name)

**гл. обр.** *abbr.* (**главным образом**) chiefly, mainly, principally

**глобула** *f.* globule

**глобулит** *m.* globulite

**глобус** *m.* globe, sphere

**глобусный** *adj.* globe, sphere

**глория** *f.* glory

**глоссарий** *m.* glossary

**глотка** *f.* pharynx; throat, gullet

**глоточный** *adj.* pharyngeal; throat

**глоттохронологический** *adj.* glottochronological

**глоттохронология** *f.* glottochronology

**глубже** *compar. of* **глубокий, глубоко** deeper

**глубина** *f.* depth; intensity; profundity; degree (of conversion); interior; **г. вкапывания столба** burying depth; **допустимая г. модуляции** modulation capability; **г. залегания скачка** (*or* **слоя**) layer depth; **максимальная**

г. модуляции без недопустимых помех modulation capability; г. обратной связи amount of feedback; г. проникновения электронов в виллемит (в сульфид кадмия, в сульфид цинка) through willemite (cadmium sulphide, zinc sulphide) penetration; г. расположения снимаемых объектов, г. резко изображаемого пространства depth of field; г. частотной модуляции warble rate; г. ясного различения деталей depth of field

глубинный adj. depth; deep, deepwater; remote; abyssal; internal

глубиномер m. depthometer, fathometer, depth sounder; metron; звуковой г. sonic depth-finding instrument

глубокий adj. deep, penetrating, thorough-going; profound; depth

глибоко adv. deep, deeply; profoundly

глубоко- prefix deep-; low, intense

глубоководный adj. deep-water, deep-sea, deep-depth

глубокоизлучатель m. narrow-angle lighting fitting, narrow-angle fluorescent lamp, focussing reflector

глубокополимеризованный adj. highly polymerized

глубокость f. depth, profundity

глубомер m. depthometer, fathometer

глубь f. depth, deep, profound, profundity, bottom

глухо adv. dully

глухой 1 (noun) m. adj. decl. deaf (person); 2 adj. deaf; dull, toneless; dead-end, blind (passage); blank; dead, deadened; anechoic

глухонемой adj.; m. adj. declen. deaf-mute

глухость f. dullness

глухота f. deafness; г. вследствие повреждения звуковоспринимающего нервного аппарата cochlear deafness; г. вследствие поражения звукопроводящего аппарата conductive deafness; г. типа нервита nerve deafness

глушение n. amortization, damping; jamming, blackout, barrage; choking; extinguishing; г. звука attenuation of noise, silencing of noise; г. звучания sound insulation; г. качаний anti-hunting action; г. обратного хода кадровой развёртки field

retrace blanking; прицельное г. spot jamming, selective jamming

глушёный adj. damped; jammed; opalescent

глушилка f. blank plug

глушитель m. silencer, muffler, damper; baffler, buffer, attenuator; jammer, interference generator, deadener, killer; amortisseur; antihum device; detonating chamber; автоматический г. поисковых станций automatic search jammer; г. звука sound damper, sourdine; г. поиска search jammer; г. с перегородками baffle-plate silencer; г. сетевого шума hum suppressor; г. шума silencer

глушительный adj. jamming; deadening; suppressing; deafening

глушить, поглушить, заглушить v. to attenuate; to deafen; to deaden; to suppress; to quench; to jam; to plug; to choke

глыба f. lump, chunk, block, clump; mother crystal

глыбистый, глыбоватый adj. lumpy

гляделка f. peephole

глянцевание n. glazing, polishing

глянцеватый adj. shiny

глянцевать v. to polish, to gloss

глянцевитый adj. glossy, shiny, lustrous

глянцевый adj. glazed; glossy

глянцемер m. glossmeter

глянцованный, глянцовый adj. polished, glossed

Гмелин Gmelin

ГМК abbr. (гиромагнитный компас) gyromagnetic compass

г-моль abbr. (грамм-моль) gram molecule

гн abbr. (генри) henry

гнать, погнать v. to drive; to race (engine)

гнездный adj. jack; (plug-in) socket

гнездо n. socket; couple, jack; housing (of machine); mesh (of screen); receptacle, plug-in socket; bunch; hub; depression, pit; cradle, nest; гнёзда в рассечке проходной линии looping bridge; штепсельные гнёзда switch springs; вводное г. introducing box; introducing socket; г. выхода программ program-exit hub, program-output hub; выходное г. видеосигнала video output socket; г. гарнитуры телефонистки operator's telephone jack; **двадцати-штырьковое г.** duodecal socket; **г. для головки звукоснимателя в тонарме** pickup bush; **г. для контрольного телефона** phonotest jack; **г. для присоединения абонентов** operator's (service) jack; **г. для присоединения шнура** cord jack; **г. для проигрывателя граммофонных пластинок** gramophone socket; **г. для сигналов времени** time jack; **г. добавочного аппарата** extension line jack; **капсюльное г. патрона** primer holder; **клапанное г.** valve cone; **коаксиальное г.** coaxial socket; **г. местного поля, местное г.** answering jack; calling jack; extension jack; **г. настроечного прибора** tuning-meter jack; **одиночное г., однополюсное г.** tip jack; **опросное г.** answering jack; calling jack; listening jack; **г. подачи feed pawl; последовательное г.** break jack, cutoff jack; **промежуточное г.** trunk junction jack; **г. синхронизации** sync jack, lockout jack; **г. сквозной связи** through jack; **служебное г., г. служебной линии** order-wire jack, ancillary jack

**гнездовой** adj. nest, nested, nesting; jack; hollow, pit; socket

**гнести, нагнести** v. to press

**гнёт** m. pressure, weight

**гниение** n. rotting, putrefaction, decay

**гнилой** adj. rotten, putrid, decayed

**гниль** f. rot, decay; **красная г.** dry rot

**гнить, погнить** or **сгнить** v. to rot, to putrefy, to decay

**гниющий** adj. rotting, decaying

**гномон** m. gnomon

**гномонический** adj. gnomonic

**ГНТК** abbr. (Государственный научно-технический комитет) State Scientific and Technical Institute

**гнусавый** adj. nasal

**гнутый** adj. bent, curved

**гнуть** pfs. **загнуть, погнуть, согнуть** v. to bend, to curve, to flex; to deflect

**гнущийся** adj. flexible, elastic

**ГО** abbr. (грубый отсчёт) coarse reading

**гобой** m. oboe, hautboy; British code-name for a bomber guidance-system

**говоритель** m. speaker, talker; **настольный г.** table talker (rad.)

**говорить, поговорить, сказать** v. to speak, to talk; to say; **г. переходно** to crosstalk

**говорной** adj. speaking, speech; talk

**говорящий** adj. speaking, talking

**Гоген** Gaugian

**год** m. year

**годиться, пригодиться** v. to fit, to suit, to be useful

**годичный** adj. annual, yearly

**годность** f. fitness, suitability

**годный** adj. able, suitable, fit, useful, proper; adaptable, applicable; effective

**годовой** adj. annual, yearly

**годограф** m. hodograph; root locus, polar plot; **корневой г.** root locus; **г. сопротивлений** transmission-line calculator; **г. сопротивлений Смита** Smith chart; **г. функции передачи** transfer locus; **г. функции передачи разомкнутой системы** feedback transfer locus

**годоскоп** m. hodoscope

**ГОИ** abbr. 1 (Государственный океанографический институт) State Institute of Oceanography; 2 (Государственный оптический институт) State Optical Institute

**Голей** Golay

**Голиаф** Goliath

**голова** f. head; chief; **«индийская» г.** Indian head

**головка** f. head; cap; knob; end, tip attachment; (drill) bit; block; header; **автоматическая астронавигационная г.** automatic star-tracking device; **агрегатная г.** power pack; **боевая г.** warhead; **болометрическая г.** bolometer mount, thermistor mount; **винторезная г.** screw plate; **волноводная болометрическая г.** waveguide bolometer; **воспроизводящая г. звукозаписи** pick-up head; **воспроизводящая магнитная г.** play-back head, reproducing head; **высокочастотная г. перестраиваемого приёмника** variable-frequency tuner; **г. жидкостного ракетного двигателя** injection assembly; **звукозаписывающая г.** cutter, recording head; **г. ключа, г. кнопки** key button; **г. лампы** header; **магнитная читающая г.** video head; **г. магнитной записи** magnetic tape recorder head; **массивная металлическая г.** heavy metal cap; **г. мачты** pole top, pinnacle; **многосопловая г.** multiple-hole injector; **многосопловая г. со**

сталкивающимися струями multi-hole impinging stream; **г. предварительного считывания** preread head; **г. пробки** screw-plug cartridge fuse carrier; **проблесковая г.** flashing light unit; **радиолокационная г. самонаведения** radar-homer, radar-homing device; **г. ракетного двигателя** rocket motor injector; **распылительная г.** injector head, burner cup; **револьверная г.** multilens head; **револьверная г. (со сменными) объективами** lens turret; **г. регулятора** control knob; **резьбонарезная г.** tapping unit; **г. с волнистым ободком** fluted knob; **сверлильная г.** drilling head; **г. свинцового пресса** extrusion block, lead press box; **силовая г.** power pack; **г. считывания** playback head; reading head; sensing head; **считывающая г. звукозаписи** pickup head; **считывающая магнитная г.** reading head; **тепловая г. самонаведения** heat seeker, infrared homing head; **г. управления** control knob; **г. штепселя** tip of a plug

**головной** *adj.* head; leading; **г. телефон** headset

**голово-** *prefix* head

**головокружение** *n.* vertigo, dizziness

**головообразный** *adj.* head-shaped

**голограммный** *adj.* holographic

**голодание** *n.* deficiency; starvation

**гололёд** *m.* rime, glaze, glazed frost, sleet, ice deposit (on lines); **г. на проводах** sleet on wires, ice deposit on wires

**голоморфный** *adj.* holomorphic

**голос** *m.* voice; **высокий г.** high-pitched voice; **низкий г.** chest voice; **синтетический г.** voder, voice operation demonstrator; **хриплый г.** beery voice

**голосовой** *adj.* vocal, voice

**голоэдр** *m.* holohedron

**голоэдрия** *f.* holohedrism

**голубоватый** *adj.* bluish

**голубой** *adj.* blue, azure

**голый** *adj.* bare, uncovered, naked; uninsulated

**Гольборн** Hohlborn

**гольмий** *m.* holmium

**гомак** *m.* gohmak

**гомальный** *adj.* homal

**гомео-** *prefix* homeo-, homo

**гомеоморфизм** *m.* homeomorphism

**гомеоморфный** *adj.* homeomorphous

**гомеополярный** *adj.* homeopolar, homopolar

**гомеостазис** *m.* homeostasis

**гомеостат** *m.* homeostat

**гомеостатический** *adj.* homeostatic

**гомеостатичность** *f.* ultrastability

**гомо-** *prefix* homo-

**гомогенность** *f.* homogeneity

**гомогенный** *adj.* homogeneous

**гомодинамический** *adj.* homodynamic

**гомодинный** *adj.* homodyne

**гомологический** *adj.* homologous

**гомология** *f.* homology

**гомометрический** *adj.* homometric

**гомоморфизм** *m.* homomorphism

**гомоморфный** *adj.* homomorphous, homomorphic

**гомопауза** *f.* homopause

**гомополярный** *adj.* homopolar

**гомосфера** *f.* homosphere

**гомотетичный** *adj.* homothetic

**гомоциклический** *adj.* homocyclic

**гонг** *m.* gong, (large) circular bell

**гондола** *f.* nacelle, car, gondola; **г. гидролокатора** sonacelle

**гондольный** *adj.* nacelle, basket, car, gondola

**гониасмометр** *m.* goniasmometer

**«гонио»** *n. indecl.* "Gonio" for maritime radio directionfinding stations

**гониометр** *m.* goniometer, Helmholtz coil; **азимутальный г.** azimuth indicating goniometer; **прикладный г.** protractor

**гониометрический** *adj.* goniometric

**гониометрия** *f.* goniometry, direction finding

**ГОНТИ** *abbr.* (Государственное объединенное научнотехническое издательство) State United Technical Publishing House

**гончарный** *adj.* ceramic, earthenware; clay

**гонять, погонять** *v.* to drive; to race (a motor)

**Гопкинсон** Hopkinson

**гор.** *abbr.* (городской) city, municipal

**горб** *m.* hump, hunch, camber; **кривой г.** hump

**горбатый** *adj.* humpbacked, hunchbacked

**горбина** *f.* hump

**горбылёвый** *adj.* slab

горбыль *m.* slab

Гордон Gordon

горелка *f.* burner; (welding) torch; **г. для отпайки** sealing-off burner, falling-off burner, tipping torch; **однопламенная г.** single flame orifice burner; **отпаечная г.** tipping torch; **г. с калильной сеткой** Auer burner

горелый *adj.* burnt, scorched

горение *n.* burning, combustion, blazing; **г. (не)бронированного порохового заряда** (un)restricted burning; **неустойчивое г. с низкой частотой вибрации** rumble; **г. по всей поверхности** unrestricted burning; **г. по наружной поверхности** external burning; **г. по части поверхности** restricted burning; **г. при неизменной тяге** neutral burning, neutral combustion; **г. с торца** cigarette burning; **г. топливой смеси с избытком кислорода** oxidizer-rich combustion; **г. шашки твёрдого топлива по (внешней и) внутренней поверхности** internal(-external) burning grain

гореть, сгореть *v.* to burn; to shine

горизонт *m.* horizon; level, floor; layer; **верхний г.** upper level; **математический г.** celestial horizon, rational horizon

горизонтали 1. *pl. of* **горизонталь**; 2. *pl.* contour lines

горизонталь *f.* horizontal, horizontal line; contour line

горизонтально *adv.* horizontally

горизонтально-поляризованный *adj.* horizontally-polarized

горизонтальность *f.* horizontal position

горизонтальный *adj.* horizontal, level, flat

горло *n.* throat; neck (of vessel)

горловина *f.* throat; neck (of vessel); entrance, orifice, mouth, vent; manhole; **г. для заливки жидкого кислорода** oxygen filling point; **г. для одноточечной заправки топливом** single-point fueling adapter; **заливная г. бака** tank filler; **заправочная г.** filler cap; **заправочная г. для топлива** fuel filling point; **г. колодца** manhole chimney; **г. рупора** horn, throat; **удлиненная г. сопла ракетного двигателя** rocket motor tube

горловой *adj.* throat; neck

горлышко *n.* neck, spout; throat

горн *m.* furnace, hearth; kiln

Горнбостель Hornbostel

Горнер Horner

горновой, горновый *adj.* hearth, furnace

горный *adj.* mining; mountain, mountainous

Горовиц Horowitz

город *m.* town, city

городской *adj.* city, civil, municipal

гороптер *m.* horopter

горочный *adj.* hump

гортанный *adj.* throat; laryngeal; guttural

гортань *f.* larynx, throat

горшковый *adj.* pot

горшкообразный *adj.* pot-shaped, pot; cylindrical

горшок *m.* pot, vessel; **сборный г.** storage receiver; **термостатический конденсационный г.** thermostatic steam trap

горьковский *adj.* Gorki

горючее *n. adj. declen.* fuel; **высококалорийное г. на основе бороводородов** zip fuel; **г. на основе гидридов** hydrid fuel; **суспензированное г.** slurry fuel

горючесть *f.* inflammability, combustibility

горючий *adj.* inflammable, combustible

горячий *adj.* hot; hot, highly radioactive; combustible, inflammable

горящий *adj.* burning

ГОС *abbr.* (**гибкая обратная связь**) flexible feedback

гос. *abbr.* (**государственный**) state

Госиздат *m. abbr.* (**Государственное книгоиздательство**) State Publishing House

ГосНИИ *abbr.* (**Государственный научно-иследовательный институт** State Scientific Research Institute

господство *n.* supremacy, domination, superiority; **г. в воздухе и космосе** aerospace superiority; **г. в космосе** cosmic space supremacy; **г. на море** maritime supremacy

господствование *see* **господство**

господствовать *v.* to dominate, to predominate; to govern

господствующий *adj.* dominant, predominant, prevalent

ГОСТ *abbr.* (**Государственный общесоюзный стандарт**) All-Union State Standard

Госторг *abbr.* (Государственная экспортная-импортная контора) State Export and Import Company

государственный *adj.* state; public, government; national

Госхимтехиздат *m. abbr.* (Государственное химико-техническое издательство) State Chemical-Technical Publishers

Гото Goto

готовность *f.* stand-by, readiness, preparedness; availability; alert; двухминутная г. к вылету stand-by alert; г. к вылету в течение трёх часов reserve alert; г. к вылету в течение часа backup alert; г. на взлетной полосе runway alert; г. на запасной дорожке аэродрома strip alert; пятиминутная г. readiness alert

готовый *adj.* ready, prepared, finished, fabricated

Гофман Hoffmann

гофрировальный *adj.* crimping

гофрирование *n.* corrugation; crimping

гофрированный *adj.* corrugated, undulated; crimped; fluted, grooved

гофрировать *v.* to corrugate; to crimp

Гофти *abbr.* (Государственный физико-технический институт) State Physical-Technical Institute

ГПК *abbr.* (гирополукомпас) directional gyroscope

г-р *abbr.* (грамм-рентген) gram-roentgen

гр. *abbr.* (группа) group; radical

гравий *m.* gravel, grit, pebble, pyrite

гравийный *adj.* gravel, grit

гравиметр *m.* gravimeter

гравиметрический *adj.* gravimetric

гравиметрия *f.* gravimetry

гравирецептор *m.* gravireceptor

гравировальный *adj.* engraving

гравирование *n.* etching, engraving

гравировать *v.* to etch, to engrave

гравировка *f.* engraving; электрическая г. electrography

гравитационный *adj.* gravitation(al), gravity

гравитация *f.* gravitation; gravity concentration

гравитировать *v.* to gravitate

гравитирующий *adj.* gravitating

гравитометр *m.* gravity meter

гравитон *m.* graviton; *pl. also* gravitational quanta

град. *abbr.* (градус) degree

град *m.* grad, grade (hundreth of a right angle); hail; г. в радарном экранном изображении masking effect of hail in radar display

градация *f.* gradation, graduation, grading, scale; shading, shade; градации в самых светлых местах изображения highlight tones; градации полутонов tonal range; средние градации mid-tone; г. светлого highlight tone; г. яркости tonal gradation, tonal value

градиент *m.* gradient, grade, slope; lapse rate; адиабатический вертикальный г. температуры adiabatic lapse rate; г. концентрации компонентов composition gradient; г. модуля упругости bulk elasticity gradient; г. напряжения voltage stress; г. противодействующего момента restoring torque gradient; г. характеристики передачи transfer gradient; г. числа M Mach number gradient

градиентный *adj.* gradient

градиометр *m.* gradiometer

градуатор *m.* induction coil; graduator

градуирование *n.* graduation, calibration, division; rating

градуированный *adj.* graduated, calibrated, divided; graded, marked

градуировать *v.* to graduate, to calibrate, to divide; to gage, to standardize

градуировка *f.* graduation, calibration, division; г. с (измерительной) камерой, г. с переходной камерой coupler calibration; г. шкалы graduation

градуировочный *adj.* graduation, calibration

градус *m.* degree, grade; новый г. centesimal graduation

градусник *m.* thermometer

градусный *adj.* degree

гражданский *adj.* civil, civic

гразнить, загрязнить *v.* to contaminate, to pollute, to poison; to choke

грамзапись *see* граммзапись

грамм *m.* gram

Грамм Gramme

грамма *f.* gram; diffraction pattern

грамматика *f.* grammar; grammatics

грамматически *adv.* grammatically

грамматический *adj.* grammatical

грамм-атом *m.* gram-atom

**граммзапись** *f.* disc recording; transcription; **г. переменной плотности** variable density (sound recording) process

**грамм-ион** *m.* gram-ion

**грамм-молекула** *f.* gram-molecule

**грамм-молекулярный** *adj.* gram-molecular

**граммовый** *adj.* gram

**граммофон** *m.* record player, phonograph

**граммофонный** *adj.* record-player, phonograph

**грамм-рентген** *m.* gram-roentgen

**грамм-эквивалент** *m.* gram-equivalent, val

**граммпластинка** *f.* phonograph record, disc, disk record, platter (16″); **г. для непосредственного воспроизведения** instantaneous disk

**граммприставка** *f.* phonograph attachment

**гран** *m.* grain (weight)

**гранат** *m.* garnet

**гранатоэдр** *m.* rhombododecahedron

**гранецентрированный** *adj.* face-centered

**граница** *f.* boundary, border; line, limit, limitation, termination, end, threshold; bound, wall; **на границе затухания** (*or* **устойчивости**) critically damped; **границы изменения потенциала** potential boundaries; **г. Блоха** Bloch wall; **верхняя г. полосы пропускания** (*or* **г. спектра частот**) upper cut-off frequency; **внешняя г. района цели** outer limit of target area; **водная г.** water termination; **воздушная г.** air termination; **г. двух сред** interface; **длинноволновая г. фотоэффекта** photoelectric red threshold; **г. домена** domain wall, magnetic domain wall; **г. (домена) с поперечными связями** cross-tie wall; **г. достоверности** confidence level; **г. зуммирования** singing point; **г. между зёрнами** grain-to-grain boundary; **г. Нееля** Neel wall; **номинальная г. верхней видеочастоты** nominal cut-off; **г. области управления самолётами** clearance limit; **плоская г. сред** plane interface; **г. поглощения (света в полупроводниках), г. полосы поглощения** absorption edge; **г. пропускания** transmission cutoff, threshold; **урав-**

**нительная г.** neutral point, trough

**граничащий** *adj.* adjacent, adjoining

**граничить, сграничить** *v.* to border (on), to be contiguous (to)

**граничный** *adj.* boundary, border, limiting; cut-off; marginal, barrier; threshold

**гранула** *f.* granule

**гранулирование** *n.* granulation, granulating

**гранулированный** *adj.* granulated, granular

**гранулировать** *v.* to granulate

**гранулометр** *m.* granulometer

**гранулометрический** *adj.* granulometric

**гранулопения** *f.* granulopenia (*med.*)

**гранулоцит** *m.* granulocyte (*med.*)

**гранулярный** *adj.* granular

**грануляционный** *adj.* granulation, granulating

**грануляция** *see* **гранулирование**

**грань** *f.* face, facet, edge, heel, side; margin; bound; **большая г. ромбоэдра** A-face; **вершинная г., головная г.** cap face (of crystal)

**Грасгоф** Grashof

**Грассо** Grassot

**граф** *m.* graph

**графа** *f.* column; range

**графекон** *m.* graphecon

**графема** *f.* grapheme

**графемика** *f.* , **графемология** *f.* graphemics

**графехон** *m.* graphechon

**график** *m.* graph, plot, curve; diagram; chart, table; schedule; **графики для определения сдвига фазы** phase-shift curves; **графики Ферми-Кюри** K-plots; **г. в полулогарифмическом масштабе** semilog plot; **вставной г.** insert curve; **монтажный г. для натяжки проводов** erection chart; **г. последовательности операций** countdown profile; **пространственный г. нагрузки** load diagram in three-dimensional representation; **г. профилактических мероприятий** servicing schedule; **равноконтрастный цветовой г.** uniform chromacity scale color triangle; **стандартный цветовой г. МКО** standard chromacity diagram; **г. стрел провеса и натяжения проводов** sag-tension chart, stress deflection chart; **технологический г. подготовки ракеты к**

пуску countdown profile; **г. функции** plotted function; **цветовой г. МКО** CIE color chart, ICI chromacity diagram
**графика** *f.* graphing, plotting; curve, diagram
**графит** *m.* graphite; **г.-аквадаг** Aquadag graphite coating
**графитизация** *f.*, **графитирование** *n.* coating with graphite; graphitization
**графитированный** *adj.* graphitized, coated with graphite
**графитировать** *v.* to graphite
**графитный** *adj.* graphite
**графитование** *see* **графитирование**
**графитовый** *adj.* graphite; **с графитовым замедлителем (отражателем)** graphite-moderated (-reflected)
**графитообразный** *adj.* graphitic
**графически** *adv.* graphically
**графический** *adj.* graphic, schematic, diagrammatic
**графленый** *adj.* ruled; divided into columns
**графология** *f.* graphology
**графометр** *m.* graphometer
**графопостроитель** *m.* dataplotter, plotting device, plotter
**графостатика** *f.* graphostatics
**гребёнка** *f.* terminal block, distributing block, connecting strip, comb; **контактные гребёнки для пересчёта** translation field; **волноводная г. с прорезами** slotted ridge guide; **г. для расшивки кабеля** cable fan; **г. ножки** stem press; **г. электростатической машины** comb collector
**гребёночный** *adj.* comb
**гребенчатый** *adj.* comb, comb-like, comb-shaped; pinched; interdigital, interdigitated; edge
**гребень** *m.* comb; peak, crown, ridge; crest; ledge; hump; **г. волны** wave crest, peak of wave, ridge of wave; **г. гарнитура телефонистки** plug for operator's headset
**гребешок** *m.* crest; comb; **коллекторный г.** commutator lug, commutator riser; **г. ножки** stem press
**гребневидный** *see* **гребенчатый**
**гребневой** *adj.* ridge; comb
**гребной** *adj.* paddle; propeller, propelling
**Грей** Gray; Grey
**Грейнахер** Greinacher

**грейфер** *m.* bucket, grab bucket; gripper
**грелка** *f.* heater; **электрическая г.** electric pad, electric heater
**греметь, загреметь** *v.* to rattle, to rumble; to detonate
**гремучий** *adj.* rattling; detonating, fulminating
**греть, нагреть** *or* **согреть** *v.* to heat, to warm
**Греффе** Graeffe
**Грехем** Graham
**Грец** Grätz, Graetz
**грецовский** *adj.* Grätz, Graetz
**греющий** *adj.* heating
**ГРИ** *abbr.* (**Государственный радиевый институт**) State Radium Institute
**гриб** *m.* mushroom; fungus
**грибковый** *adj.* mushroom, mushroom-like, fungous
**грибовидный** *adj.* mushroom, mushroom-shaped, fungoid
**грибок** *m.* fungus, molds
**грибообразный** *adj.* mushroom-shaped, fungiform
**грибостойкий** *adj.* fungusproof
**грибостойкость** *f.* fungus-resistance, fungusproof
**гридистор** *m.* gridistor
**гридлик** *m.* grid leak
**Гриммингер** Grimminger
**Грин** Green
**гринвичский** *adj.* Greenwich
**гриф** *m.* neck, handle, touch; fingerboard
**Гриффитс** Griffiths
**ГРМ** *abbr.* (**глиссадный радиомаяк**) glide-path beacon
**Грове** Grove
**гроза** *f.* thunderstorm, storm
**гроздевой, гроздовый** *adj.* cluster, clustered, bunch, bunched
**грозовой** *adj.* thunderstorm; lightning
**грозозащита** *f.* lightning protection, lightning protective conductor; **г. линии** protection against line lightning
**грозозащитный** *adj.* lightning-protective
**грозоотметчик** *m.* storm-indicator
**грозописец** *m.* brontograph
**грозоразрядник** *m.* lightning arrester; **искровой г.** air gap protector; **роговой г.** horn lightning arrester
**грозостойкий, грозоупорный** *adj.* lightning-proof, surgeproof

**гром** *m.* thunder
**громкий** *adj.* loud, noisy
**громкоговоритель** *m.* loudspeaker, speaker, speaker unit, sound reproducer; **громкоговорители высшего качества** monitoring loudspeaker(s); **басовый г.** woofer; **г. большой мощности с плоской поршневой мембраной** Blatthaller loudspeaker; **внешний г.** extension loudspeaker; **г. воспроизводящий нижние частоты** woofer; **вспомогательный г. звуковых эффектов** effect loudspeaker; **выносной г.** extension loudspeaker; **г. высоких частот** treble loudspeaker, tweeter; **добавочный г. для улучшения воспроизведения звука** effect loudspeaker; **г. для (воспроизведения) верхних частот** tweeter; **г. для звукового сопровождения телевизионной передачи** background loudspeaker; **ионный г.** ionophone; **г. микшера** monitoring loudspeaker; **г. низких частот** boomer, bass loudspeaker; **отдельный г.** cabinet loudspeaker; **(рупорный) г. уличной системы озвучения** morning-glory horn; **г. с плоской поршневой мембраной** Blatthaller speaker; **сдвоенный г.** woofer-tweeter, twin loudspeaker, duplex loudspeaker; **студийный г. (для воспроизведения звукового сопровождения)** playback loudspeaker, background loudspeaker; **уличный г.** outdoor-type horn; **шунтирующий г.** loudspeaker capacitor
**громкоговорящий** *adj.* loudspeaker; loud-speaking
**громкость** *f.* loudness, volume of sound; signal strength; **г. звука в фонах** decibel
**громовой** *adj.* thunder
**громоздкий** *adj.* cumbersome, awkward, unwieldy, massive, bulky
**громоотвод** *m.* lightning conductor, discharging rod; **воздушный г.** air-gap lightning arrester; **конденсаторный г.** carbon lightning rod protector; **линейный г.** leakage conductor; **г. с предохранителем** fuse and protector block
**громоотводный** *adj.* lightning conductor
**громыхание, громыханье** *n.* rumble, rumbling, rumblers

**громыхать, громыхнуть** *v.* to rumble, to rattle
**Гросер** "Groser" (British jamming transmitter)
**грохот** *m.* thunder, roar; (turntable) rumble, rattle, grating; rumblers; **электродинамический вибрационный г.** electrodynamic shaker
**грохотанье** *n.* rolling; rumble, rumbling
**грохотать, загрохотать** *v.* to roll; to rumble; to rattle
**грубеть, огрубеть** *v.* to grow rough; to roughen, to harden
**грубо** *adv.* roughly, coarsely
**грубый** *adj.* rough, coarse; crude; mass, gross
**грудной** *adj.* chest, breast; thoracic
**груз** *m.* weight, load; (pendulum) bob; **грузы, подвешиваемые к изоляторным гирляндам** suspension insulator weights; **подъёмный г.** carrying capacity; **полезный г.** pay load; carrying capacity
**грузик** *m.* weight
**грузить, нагрузить** *v.* to load
**грузовик** *m.* truck; **г. с прицепом для строительных бригад** construction gang truck
**грузовой** *adj.* freight; load
**грузоподъёмник** *m.* freight elevator, goods lift
**грузоподъёмность** *f.* lifting capacity, carrying capacity; load capacity
**грузоподъёмный** *adj.* load-lifting, hoisting
**грузоспособность** *f.* capacity
**грунт** *m.* bottom, ground, base
**грунтовка** *f.* undercoat, ground coat, prime coat, first coat of paint
**грунтовой** *adj.* ground, soil
**грунтовочный** *adj.* prime, priming; ground, soil
**группа** *f.* group, cluster, bunch; bank; set, assembly; band; block; gang, crew, team; **вертикальные и горизонтальные группы расходящихся штрихов для определения чёткости** vertical and horizontal definition wedges; **группы движения** groups formed by screw motion; **г. блоков памяти** memory bank; **дипольная г.** stacked dipole array; **г. дорожек** band; **г. дорожек (с записью)** traffic group; **г. единиц** unit digit; **завершающая г. информации** trailer block; **звездообразная г. (радио-**

маяков) star chain; **звездообразная г. станции** star chain (of offices); **земная г. планет** inner planets, terrestrial planets; **г. из десяти знаков** decade; **г. из трёх зёрен люминофоров с красками, синим и зелёным свечениями** trio dots, phosphor trio; **г. из 4 радиомаяков в виде V с ведущей станцией в центре** star chain; **г. излучателей** array, beam antenna; speaker group; **импульсная контактная г. на размыкание** pulse springs make; **г. кабелей** bunched cables; **г. кодов** call number; **кодовая г.** word (group); **г. колонек на перфокарте** field, card field; **контактная г.** spring pile-up, spring assembly, spring set, contact set; **контактная замыкающая (размыкающая) г.** single make (break) contacts; **г. контактных колец** collector; **линейная г. микрофонов** line microphones; **нормальная г. дорожек** normal band; **г. одинаковых станций** chain of stations; **г. рабочих мест** supervisor's section; **г. символов команды** instruction word; **г. согласования (технических требований к новым видам оружия)** phasing group; **г. стативов** bay of racks; **г. студийных светильников** board, studio light boards; **г. судов, несущая службу дальнего радиолокационного обнаружения** fishing fleet (*sl.*); **г. токособирательных колец** collector; **г. цифр команды** instruction word; **чередующаяся г.** alternate block (of words or numbers); **г. электрических ванн** section multiple system; **г. элементов информации** block of information; **г. элементов информации, передаваемых как одно целое** message

**группирование** *n.* grouping, bunching, batching; concentration; **г. (постов на сборных агрегатах)** adding; **г. в электронном потоке, испытывающем отражение** reflex bunching; **г. четвёрток (кабеля)** squaring

**группированный** *adj.* grouped, bunched; classified

**группирователь** *m.* buncher, input resonator; **мощный г. электронных пучков** relativistic electron bunching accelerator

**группировать, сгруппировать** *v.* to group, to bunch, to bank; to classify

**группировка** *f.* grouping, bunching; concentration; **г. в параллелограмм** parallel wiring; **г. в прямоугольник** rectangular wiring

**группирующий** *adj.* grouping, bunching

**групповой** *adj.* group, gang

**групповыбиратель** *m.* selector, group selector; **г. для ускоренного сообщения** group selector for no-delay service, toll group selector; **междугородный г.** intercity group selector for no-delay service

**группообразование** *n.* transposing, twisting grouping, layout, trunking

**груша** *f.* bulb; pear push

**грушевидный, грушеобразный** *adj.* pear-shaped

**Грэй** Grey

**Грюнейзен** Grüneisen

**грязевой** *adj.* mud, muddy

**грязный** *adj.* dirty, soiled; impure

**грязь** *f.* dirt; impurity, contamination; mush, random noise; **анодная г.** anode mud

**гс.** *abbr.* (**гаусс**) gauss

**ГСВ** *abbr.* (**Гринвичское среднее время**) Greenwich mean time

**ГСС** *abbr.* (**генератор стандартных сигналов**) standard signal generator

**ГТ** *abbr.* 1 (**газовая турбина**) gas turbine; 2 (**телефонный, голый**) telephone-type, bare (lead-covered)

**ГТВД** *abbr.* (**газотурбинный винтовой двигатель**) turbopropeller engine, turbojet engine

**ГТД** *abbr.* (**газотурбинный двигатель**) gas turbine engine

**ГТДД** *abbr.* (**газотурбинный реактивный двигатель двойного действия**) double-action turbojet engine

**ГТРД** *abbr.* (**газотурбинный реактивный двигатель**) gas-turbine jet engine, turbojet engine

**ГТС** *abbr.* 1 (**городская телефонная сеть**) city telephone exchange; 2 (**городская телефонная станция**) city central office, local central office; **координатная ГТС** crossbar central office, crossbar local office

**ГТУ** *abbr.* (**газотурбинная установка**) gas-turbine unit

**ГУ** *abbr.* 1 (**главный узел**) regional center (*tphny.*); 2 (**государственное учреждение**) state office; 3 (**группо-**

вая установка) branch house exchange, branch group installation exchange (*tphny.*)

**гуанидин** *m.* guanidine

**гуанидин-ванадный** *adj.* guanidine vanadium

**губа** *f.* lip; jaw; bay, gulf

**губка** *f.* sponge; jaw; bit; **г. тисков** clamping jaw

**губковатый** *see* **губчатый**

**губно-зубной** *adj.* labiodental

**губной** *adj.* labial, lip

**Губо** Goubou

**губовидный** *adj.* lip-shaped

**губчатость** *f.* sponginess

**губчатый** *adj.* sponge, spongy; fungous; porous

**гудение** *n.* hum, humming, buzzing, droning, wobbled audio frequency; **заунывное г.** droning

**гудеть, загудеть** *v.* to hum, to buzz, to drone

**гудок** *m.* horn, howler; whistle; hooter; hooting

**гудрон** *m.* tar, asphalt

**гудронирование** *n.* tarring, asphalting

**гудронированный** *adj.* tarred, asphalted

**гудронировать** *v.* to tar, to asphalt

**Гудсмит** Goudsmit

**Гук** Hook

**гул** *m.* boom, rumble; buzz, hum, humming; din

**гулкий** *adj.* hollow, resonant; resounding, booming

**гулкость** *f.* boominess, reverberation

**Гулстад** Gulstad

**гумми** *n. indecl.* gum

**гуммиарабик** *m.* gum arabic, mucilage

**гуммон** *m.* gummon (insulating material)

**Гунд** Hund

**ГУП** *abbr.* (**гамма-установка, промышленная**) industrial gamma-unit

**Гурвиц** Hurwitz

**ГУС** *abbr.* (**Государственный учёный совет**) The State Scientific Council

**гусёк** *m.* bucket (of dam)

**гусеничный** *adj.* caterpillar-track, track; interlocked-type

**густой** *adj.* dense, thick; deep; fine

**густота** *f.* density, thickness, viscosity; depth; consistency

**гуськом** *adv.* tandem

**ГУТ** *abbr.* (**гамма-установка, терапевтическая**) therapeutic gamma-unit

**гуттаперча** *f.* guttapercha

**ГФО** *abbr.* (**Главная физическая обсерватория**) The Main Physical Observatory

**гц** *abbr.* (**герц**) hertz, cycle per second

**г-частица** *f.* gram molecule

**Гэв** *abbr.* (**гигаэлектрон-вольт**) gigaelectron-volt, billion electron-volt

**г-экв** *abbr.* (**грамм-эквивалент**) gram-equivalent

**ГЭС** *abbr.* 1 (**гидроэлектростанция**) hydroelectric power plant; 2 (**государственная электрическая станция**) State power plant

**ГЭТ** *abbr.* (**Государственный Электротехнический Трест**) The State Electrotechnical Trust

**гэтинакс** *see* **гетинакс**

**ГЭЦ** *abbr.* (**гидроэлектроцентраль**) central hydroelectric power plant

**Гюйгенс** Huygens

**Гюккель** Hückel

**Гюльберг** Guldberg

**Гюльден** Guldin

# Д

д *abbr.* 1 (дальность) range, distance; 2 (деци-) *prefix* deci-

давать, дать *v.* to give, to supply, to hand in; to yield, to produce; д. в обход to divert the traffic, to reroute (*tphny.*); д. отбой to ring off; д. связь по обходу *see* д. в обход; д. усадку to shrink

Дависсон Davisson

давить, задавить, раздавить *v.* to press, to squeeze

давление *n.* pressure; compression; push, thrust; stress; д. воздуха на входе в губные инструменты blow pressure of flute pipes; д. на мембрану, создаваемое отражённым сигналом echo push; номинальное звуковое д. (громко)говорителя speaker pressure rating; д. пара из мелких капель vapor pressure of small droplets; д. перед клапаном upstream pressure; д. после клапана downstream pressure; д. резца stylus pressure; световое д. radiation pressure; д., соответствующее нулю децибел reference pressure; эквивалентное д. шумов преобразователя transducer equivalent noise pressure

давленый *adj.* pressed; crumpled

дагерротип *m.* daguerreotype

дагерротипия *f.* daguerreotype

дагерротипный *adj.* daguerreotype

дазиметр *m.* dasymeter

дазиметрический *adj.* dasymetric, density-measuring

дайатрайн *m.* diatrine

дайнаквод *m.* dynaquad

Дайнс Dines

Дайсон Dyson

Д' Аламберт, Даламберт d'Alembert

даламбер(т)ов *poss. adj.* d'Alembert's

Даландер Dahlander

далёкий *adj.* distant, far, far away, remote

далеко *adv.* far, far off; by far, much

даль *f.* distance, remoteness

дальневидение *n.* television

дальнейший *adj.* furthest, furthermost; ulterior

дальний *adj.* distant, remote, far off, long; long-distance; дальнего действия long-range; дальнее обнаружение early warning

дально- *prefix* tele-, distance

дальнобойность *f.* long range

дальнобойный *adj.* long-range

дальнодействие *n.* remote control

дальнодействующий *adj.* remote-control; remote-range

дальноизмерение *n.* telemetering, telemetry, remote metering

дальномер *m.* range, range finder, range meter, telemeter, range measuring system, range unit, distance gage, diastimeter; зенитный д. antiaircraft range finder; радиолокационный д. range-only radar; д. с длинным базисом и двумя точками наблюдения long-base range finder, two-station range finder; д. с малым базисом и одной точкой наблюдения monostatic range finder, short-base range finder; д. с непрерывным излучением continuous-wave distance finder; д. с совпадением изображений coincidence range finder; д. со смещёнными полями изображения split-field range finder

дальномерный *adj.* range-measuring, distance-measuring

дальнопишущий *adj.* teletype, teleprinter

дальностный *adj.* distance, range

дальность *f.* distance, remoteness; length; clearance; range, compass, radius; д. активного (*or* активной системы) самонаведения active homing guidance range; д. действия operating range, range; coverage; air-line distance; (maximum) range of radio waves; (maximum) transmission range (of a telephone system); д. действия полуактивной системы самонаведения semiactive homing guidance range; **д. действия по маякам** beacon range; **д. действия (станции) по морю** range over sea; **д. меньше стандартной** substandard range; **наклонная д.** slant range, slant distance, air-to-ground distance, true distance; **нормальная д. распространения сантиметровых волн** microwave horizon; **д. обнаружения цели гидроакустической станцией** sonar range, asdic range; **д. обнаружения** (*or* **сопровождения**) **целей** radar range; **д., превышающая стандартную** superstandard range; **д. прямой видимости** visual range, optical range, visibility range, geometrical horizon; **д. распространения, меньшая стандартной** substandard (propagation) range; **д. распространения, превышающая стандартную** superstandard (propagation) range; **уменьшенная д. действия** lower limit of radio propagation range; **эффективная д. обнаружения цели гидроакустической станцией** effective asdic range

**дальноуправляемый** *adj.* remote-controlled

**дальтонизм** *m.* daltonism, color blindness

**дамба** *f.* dam, dike

**даммар** *m.* dammar (gums)

**даммаровый** *adj.* dammar

**Даниэль** Daniell

**данные** *pl.* data, facts, performance data; **д. абонента (в списке)** listing, address name (*tphny.*); **д. в двоичной системе** (*or* **форме**) binary data; **д. в десятичной системе** (*or* **форме**) decimal data; **д. в функции времени** time variable data; **входные д. о высоте цели** height input data; **выдаваемые д.** data out, data presented; **выходные дискретные д.** digital output; **дискретно-непрерывные д.** sampled analog data; **д. измерения уровня звукового давления** sound-level meter data; **исходные д.** raw data; **исходящие д.** data out; **коммутированные д.** gated information; **д. коэффициентов звукопоглощения** sound-absorption coefficient data; **непрерывные д.** analog data; **нерегулярно поступающие д.** fluctuating data; **номинальные д.** rating; **д. о полёте, полученные**

**телеметрическим путём** telemetered flight-test data; **д., определяющие местоположения** position finding results; **основные д. снаряда** missile parameters; **д. от сельсинов** synchro data; **паспортные д.** rating, rated values, nominal data; **д. по отысканию неисправностей** trouble-shooting data; **последующие д.** future data; **д. прессованного порошка (для сердечников)** powdered core-material data; **пробные непрерывные д.** sampled analog data; **протокольные д.** performance data; **расчётные д.** design values, design data; **д. с беспорядочными колебаниями** randomly fluctuating data; **справочные д.** reference data; **текущие д. радиолокатора** live radar information; **упорядоченные д.** ranked data; **упреждённые д.** predicted data; **уточнённые д.** specified data; **эксплоатационные д.** service data

**данный** *adj.* given, present, in question

**дараф** *m.* daraf

**даркфлекс** *m.* darkflex

**д'Арсонвализация, дарсонвализация** *f.* d'Arsonvalism, high-frequency electrical treatment

**Дарсонваль** D'Arsonval

**дата** *f.* date

**дататрон** *m.* datatron

**датировать** *v.* to date

**датрак** *m.* datrac

**датчик** *m.* pickup, pickup unit, sensing device (*or* unit), sensor, feeler; detecting element; data unit; transducer; transmitter, sender; monitor; controller; generator; actuator; "pickoff"; **д. без усилителя-преобразователя в линии передачи** direct transmitter; **д. видеосигналов с бегущим лучом** flying-spot video generator; **д. времени** control timer, timer; **д. времени интервалов** time interval generator; **д. времени с самовозвратом** self-resetting timer; **гироскопический д. угловой скорости** rate-of-turn gyroscope; **д. давления** pressure transmitter, (chamber-)pressure pickup; **динамометрический д.** load cell; **д. дискретной информации** true digital transducer; **дифференциально-индуктивный д.** ampere balance, current balance, Kelvin balance; **д. для изме-**

рительных приборов instrument transducer; д. заданного параметра desired value transducer; заполняющий д. storage transmitter; д.-измеритель pickup, sensory element; д. импульсов metering-pulse sender, impulser; д. испытательной таблицы с бегущим лучом flying-spot pattern generator; кнопочный д. (абонентских) номеров subscriber dial-pulse repeater; д. кодированных сигналов code oscillator; д. команд data computer; д. криволинейной траектории maneuverable-path generator; д. курса course setting device, track selector; д. линейного перемещения и углов поворота linear-and-angular-movement pickup; д. линейных ускорений linear accelerometer; д. ломано-линейной траектории dog-leg path generator; магнитный д. импульсов magnetic pulse generating device; манипуляторный д. keyboard transmitter; д. (моделирующей) функции simulation generator; д., моделирующий диаграмму направленности beam pattern generator; д. модульной конструкции modular actuator; д. наименования automatic identification device (in teletype systems); д. направления airstream direction transducer; д. непрерывной информации true analog transducer; д. номеров call sender; параметрический д. с реактивным сопротивлением variable-reactance transducer; параметрический индуктивный д. variable-inductance transducer; параметрический электронновакуумный д. vacuum-tube transducer; д. перемещения рулевых органов control surface pickup; д. перепада давления differential pressure transmitter, differential pressure pickup; д. периодических импульсов cycle-repeat timer; д. постоянного напряжения auto-repeater; потенциометрический д. variable-potentiometer transducer; потенциометрический д. давления resistance pressure transmitter; потенциометрический д. рассогласования potentiometer-type error "pickoff"; д.-преобразователь transducer; д. пробного сигнала test-signal gener-

ator (tv); проволочный д. electric strain gage; программированный д. program transmitter; д. прямого действия direct-acting transducer; д. прямолинейной траектории straight-line path generator; д. разности давлений pressure-difference transducer, pressure-difference transmitter; д. рассогласования error "pickoff"; резистивный д. по мостовой схеме resistance bridge-type pickup; д. с воздушным демпфированием air-damped pickup; д. с динамическим конденсатором vibration (capacitor) pickup; д. с жидкостным демпфированием fluid damped pickup; д. с переменной индуктивной связью mutual inductance pickup; сельсин-грубый д. азимута coarse azimuth transmitting selsyn; д. сигналов цветного телевидения с бегущим лучом flying-spot color signal generator; д. системы цветного телевидения с последовательным чередованием цветов по полям field-sequential color transmitter; тактовый д. cadence tapper; д. тактовых импульсов clock multivibrator; телевизионный д. с бегущим пятном flying-spot scanner; телевизионный д. с бегущим пятном для цветного телевидения color flying-spot scanner; термисторный д. thermistor heat detector cell; д. тока current-sensing device; д. угла angle-data transmitter; д. угловой скорости rate gyroscope; д. угловых вибрации angular vibration pickup; д. уровня жидкости liquid level transmitter; д. ускорений (or ускорения) accelerometer, acceleration transducer, acceleration pickup; фотоэлектрический д. photocell pickup; д. числа оборотов rate-of-turn transducer; электронный д. длительности импульса electronic timer

**дать** pf. of давать

**Дау** Dow

**Даунс** Downs

**дача** f. giving; portion, ration; д. квитанции receipting

**дающий** adj. giving; data

**дб** abbr. (децибел) decibel, db

**ДБС** abbr. (дальнобойный баллистический снаряд) long-range ballistic missile

**ДВ** *abbr.* 1 (длинноволновый) long-wave; 2 (длинные волны) long waves

**два** *num.* two

**двадцатигранник** *m.* icosahedron

**двадцатиразрядный** *adj.* 20-digit

**двадцатиштырьковый** *adj.* duodecal

**дважды** *adv.* twice, twofold, double

**двенадцатеричный** *adj.* duodecimal

**двенадцатигранник** *m.* dodecahedron

**двенадцатиканальный** *adj.* twelve-channel, duodenary

**двенадцатиричный** *adj.* duodecimal

**двенадцатифазный** *adj.* twelve-phase

**дверной** *adj.* door, gate

**дверца** *f.* door, gate; **откидная д.** trap, trap door

**дверь** *f.* door

**двигатель** *m.* motor, engine; driver, propeller; motive power; power plant; **асинхронный д. с контактными кольцами** slip-ring induction motor; **асинхронный д. с повышенным скольжением** high-slip induction motor; **асинхронный д. с фазным ротором** wound rotor induction motor; **асинхронный д. с чашеобразным ротором** drag cup induction motor; **д. без принадлежностей** bare motor; **д. без редуктора** gearless motor; **д. в режиме противовключения** (*or* **противотока**) stalled torque motor; **вентильный д.** thyratron motor; **д. вертикального типа** vertical shaft motor; **видимый насквоз д.** skeleton(-type) motor; **д. вращения катушки** coil-drive motor; **врубовый д.** coal-cutting motor; **д.-генератор** motor-generator set, motor-alternator, dynamotor; **д.-датчик** drive motor; **д. длительного режима** long-hour motor; **д. для вентиляторной нагрузки** fan-duty motor; **д. для кантовки** turning motor; **допускающий пуск под нагрузкой д.** load-start motor; **исполнительный д.** servo-motor, servomechanism; **кантовальный д.** barring motor; **коллекторный д. с изменяющимся полем** varying-field commutator motor; **коллекторный д. с регулировкой скорости сдвигом щёток** Deri brush-shifting motor; **коллекторный сериесный д. двойного питания** doubly fed series motor; **компаундный д. с встречно включёнными обмотками возбужде-** ния differential compound motor; **компаундный д. с согласно включёнными обмотками возбуждения** cumulative compound motor; **короткозамкнутый д. нормальной конструкции** plain squirrel-cage motor; **короткозамкнутый д. с гидромуфтой** fluid drive motor; **д. кратковременного режима** short-hour motor; **д. малой мощности, маломощный д.** subminiature motor; **моментный д.** torque motor; **д. мощностью меньше 1 л. с.** fractional (h.p.) motor; **д. на лапах с жёстким креплением** rigid-foot motor; **д. нажимного механизма** screw-down motor; **наклонно установленный д.** tilted motor; **д. напорного механизма** crowding motor; **«обращённый» синхронный д.** duosynchronous motor; **однофазный д. с включёнными конденсаторами на время пуска и работы** capacitor-start-and-run motor; **однофазный конденсаторный д. с постоянно расщеплённой фазой** permanent-split capacity motor; **д. параллельного возбуждения** shunt-conduction motor, shunt-wound motor; **питаемый непосредственно от контактной сети д.** line-fed motor; **д. по обратной связи** reaction motor; **поворотный д.** torque motor; **подсобный д.** servomotor; **д. постоянного тока с питанием от управляемого выпрямителя** d-c electronic motor; **д.-приёмник** follower-motor; **рассчитанный для работы с обдувом д.** airstream-rated motor; **реверсивный д. для пуска в обоих направлениях** externally reversible motor; **регулируемый д.** varispeed motor; varying-speed motor; **регулируемый д. с падающей характеристикой** adjustable varying-speed motor; **редукторный д.** motorized reducer; **редукторный д. с планетарной передачей** concentric-drive gearmotor; **редукторный д. с повысительным редуктором** step-up gearmotor; **репульсионный д. с питанием ротора от сети через щётки и коллектор** inverted repulsion motor; **д. с водонепроницаемой изоляцией статора** submersible motor; **д. с глубоким пазом** deep slot motor; **д. с двумя обмотками возбуждения**

dual-capacitor motor; **д. с диффе-ренциальным смешанным возбуждением** differential motor; **д. с добавочными полюсами** interpole motor; **д. с изменяемым напряжением на якоре** armature-controlled motor; **д. с мягкой характеристикой** drooping speed motor; **д. с неизменяющимся числом об/мин** constant-speed motor; **д. с обмоткой на роторе** wound-rotor motor; **д. с падающей характеристикой** varying speed motor; **д. с питанием от выпрямителя** rectifier-driven motor; **д. с подвеской на выступе** nose-suspension motor; **д. с последовательным возбуждением** series-wound motor; **д. с постоянно налегающими щётками** brush-riding motor; **д. с постоянным моментом** torque motor; **д. с поступательно-возвратным движением** reciprocating motor; **д. с прямым пуском от сети** across-the-line motor; **д. с пуском от полного напряжения** line-start motor; **д. с регулируемым числом оборотов** variable speed motor; **д. с секционированной обмоткой возбуждения** tap-field motor; **д. с сериесной характеристикой** inverse-speed motor; **д. с управлением по схеме «генератор-двигатель»** armature-controlled motor; **д. с чередующимися полюсами** consequent-pole(s) motor; **сетевой д.** all-mains motor; **синхронизированный асинхронный д.** synchronous induction motor; **д. со встроенным понизительным редуктором** back-geared motor, gearmotor; **д. со смешанным возбуждением** dual-field motor; **д. совершенно закрытого исполнения** fully enclosed motor; **тепловой д.** piston-type prime movers, prime movers; **цельнокорпусный д.** box-frame motor; **шаговый д.** pecking motor, stepping motor, quantized motor

**двигатель-генераторный** *adj.* motor-generator

**двигательный** *adj.* engine, motor; motive; power

**двигать, двинуть** *v.* to move, to push, to slide, to drive; **д. по кругу** to rotate, to circulate, to revolve

**двигаться, двинуться** *v.* to move, to run, to operate, to work

**движение** *n.* movement, motion, running; traffic; action; travel; **автомодельное д.** self-simulating motion; **беспорядочное д.** agitation; **д. в большом** motion, in the large; **д. в пространстве** three-dimensional motion; **вихревое д.** eddy, whirl; **возвратно-поступательное д.** back and forth motion, oscillating motion; **волнообразное д.** undulation; **вынужденное д. (искателя)** impulse action (of a selector); **завихренное д.** eddy-current flow; **д. изображения вверх и вниз** bounce, bouncing (*tv*); **д. изображения по вертикали** vertical hunting (*tv*); **интерференционное д. отметки (на трубке индикатора)** bobbing (*rdr.*); **качательное д.** nodding action; **маятникообразное д.** pendular motion; **обратное д.** return movement, back travel; retrace of sawtooth pulse; rewind; retrace (*tv*); **отбойное д.** return travel, return movement; **перекрещивающееся д.** crisscross motion; **переменно-ускоренное д.** variable increasing motion; **д. по инерции** running down, deceleration, gradual stopping; **д. подачи** feeding movement; **подъёмное д. (щёток искателя)** lifting movement (of wipers), vertical motion; **прерывистое д.** Geneva motion; **пространственное д.** three-dimensional motion; **равномерно-ускоренное д.** uniform acceleration

**движимый** *adj.* movable, mobile; actuated, moved

**движок** *m.* cursor, slide, slider, runner; contact blade; arm (of apparatus); wiper; **д. дальности** range cursor; **д. потенциометра** wiper, potentiometer movable arm; **д. программного регулятора** timing index

**движущий** *adj.* moving, motive, driving, propulsive, propelling; operating

**движущийся** *adj.* moving, running, operating, working

**двинутый** *adj.* moved, set in motion

**двинуть(ся)** *pf. of* **двигать(ся)**

**двое** *num.* two

**двоекратный** *adj.* twofold

**двоение** *n.* doubling; dividing

**двоечный** *adj.* binary

**двоично-десятичный** *adj.* binary-decimal, bini-ten

**двоично-кодированный** *adj.* binary-coded

**двоично-пятеричный** *adj.* biquinary

**двоичный** *adj.* binary, dyadic, dual, double

**двойка** *f.* two, pair; two-digit group

**двойник** *m.* double, twin, duplicate

**двойники** *pl.* twin crystals

**двойникование** *n.* twinning

**двойниковый, двойничный** *adj.* twin, twinning, twinned, duplicate

**двойной** *adj.* double, twofold, binary, dual, duplex, twin, di-, duo, pair

**двойственность** *f.* duality, duplicity; ambiguity

**двойственный** *adj.* dual, double; ambiguous

**Дворжак** Dvořak

**двоякий** *adj.* double, twofold, duplex

**двояковогнутый** *adj.* concavo-concave, double concave, biconcave

**двояковыпуклый** *adj.* convexo-convex, biconvex, lenticular

**двоякопреломляющий** *adj.* birefringent, double-refracting

**ДВС** *abbr.* (**двигатель внутреннего сгорания**) internal combustion engine

**дву-** *prefix* di-, bi-, duo-, two-, double

**двубазовый** *adj.* double-base

**двувалентность** *f.* bivalence

**двувалентный** *adj.* bivalent

**двувариантный** *adj.* bivariant

**двувидный** *adj.* dimorphous

**двугорбый** *adj.* double-humped, double-peaked

**двугранный** *adj.* dihedral

**двужидкостный** *adj.* two-fluid

**двужильный** *adj.* twin-core

**двузвездно-скрученный** *adj.* spiral-eight, double-star quad

**двузвёздный** *adj.* double-star

**двузвенный** *adj.* two-section, two-mesh

**двузначность** *f.* ambiguity

**двузначный** *adj.* double-valued; two-digit; ambiguous

**двузонтичный** *adj.* double-umbrella

**двуконусный** *adj.* biconical

**двуконусный-зонтичный** *adj.* double-cone-umbrella-type

**двукратно** *adv.* doubly

**двукратный** *adj.* twofold, double, two-stage

**двукристальный** *adj.* two-crystal

**двумерный** *adj.* two-dimensional

**двунаправленный** *adj.* bidirectional; bilateral

**двунитный, двуниточный** *adj.* bifilar, double-wound

**двуобмоточный** *adj.* two-winding

**двуокись** *f.* dioxide

**двуосновный** *adj.* dibasic, diatomic, dihydric

**двуосный** *adj.* biaxial

**двупитаемый** *adj.* double-feed

**двупламенный** *adj.* double-flame

**двуплечий** *adj.* double-arm

**двуплоскостной** *adj.* diplane, biplanar

**двуполярный** *adj.* bidirectional, bipolar, ambipolar

**двупреломление** *n.* birefringence

**двупреломляющий** *adj.* double-refracting

**двусеточный** *adj.* double-grid, bigrid, two-grid, duo-grid

**двусигнальный** *adj.* double-signal

**двускатный** *adj.* with two sloping surfaces, inverted-V

**двускоростный** *adj.* dual-speed

**двуслойный** *adj.* two-layer, two-play, double-braid

**двусмысленность** *f.* ambiguity

**двусмысленный** *adj.* ambiguous, obscure, doubtful

**двусопряжённый** *adj.* biconjugate

**двустадийный** *adj.* two-stage, two-step

**двусторонне** *adv.* bilaterally

**двусторонне-согласованный** *adj.* bilaterally matched

**двусторонний** *adj.* bilateral, bidirectional, two-way, bipolar, double-sided, both-way; double, double-ended, double-pointed; duplex; two-faced, two-surface; dual(-control); single-track (*rr.*)

**двустрелочный** *adj.* cross-pointer, double-needle

**двуступенный, двуступенчатый** *adj.* two-step, two-stage, double-stage

**двуустойчивый** *adj.* bistable

**двуухий** *adj.* binaural, double-ear

**двуфантомный** *adj.* double-phantom

**двух-** *prefix* bi-, di-, two, double

**двухадресный** *adj.* two-address

**двуханодный** *adj.* double-anode

**двухатомный** *adj.* diatomic, bivalent

**двухбазовый** *adj.* double-base

**двухбарабанный** *adj.* double-drum

**двухбороздчатый** *adj.* double-groove, double-furrow

**двухбуквенный** *adj.* digram

**двухвалентный** *adj.* divalent

**двухвариантный** *adj.* bivariant

двухвершинный *adj.* bimodal
двухвидность *f.* double mode
двухвидовой *adj.* two-mode
двухвитковый *adj.* two-loop, double-turn, double-coil
двухволновой *adj.* dual-frequency
двухвходный *adj.* two-input
двухгрупповой *adj.* two-group
двухдвигательный *adj.* dual-motor
двухдетекторный *adj.* dual-detector
двухдиапазонный *adj.* dual-range, two-range, two-band
двухдисковый *adj.* two-disk
двухдиффузионный *adj.* double-diffused
двухдиффузорный *adj.* duocone
двухдолинный *adj.* two-valley
двухдорожечный *adj.* dual-track, double-track
двухжелобчатый *adj.* double-groove, double-grooved
двухжидкостный *adj.* two-fluid, two-liquid
двухжильный *adj.* two-core, twin-core, twin-cored, twin cord, twin (cable), two-wire, two-cable; bifilar
двухзаконченный *adj.* doubly terminated
двухзвенный *adj.* double-section
двухзначный *adj.* bivalent
двухзональный *adj.* two-region, two-zone
двухимпульсный *adj.* two-pulse
двухинтеграторный *adj.* two-integrator
двухкабельный *adj.* two-cable
двухкамерный *adj.* two-chamber
двухканальный *adj.* two-channel, twin-channel; double-track, two-path, two-way
двухкапсюльный *adj.* double-button
двухкаркасный *adj.* double-skeleton
двухкаскадный *adj.* two-stage
двухкатушечный *adj.* two-coil
двухквадрантный *adj.* two-quadrant
двухквантовый *adj.* double-quantum
двухколейный *adj.* double-track
двухколенчатый *adj.* double-knee, double-throw
двухколлекторный *adj.* double commutator
двухкольчатый *adj.* dicyclic; binuclear
двухкомпонентный *adj.* two-component; two-variable
двухконечный *adj.* double-end(ed), double-pointed

двухконтактный *adj.* double-contact, two-pin, double-prong
двухконтурный *adj.* two-circuit, double-circuit, two-loop
двухконусный *adj.* biconical, double-cone
двухкратно *adv.* doubly
двухкратнозамкнутый *adj.* doubly re-entrant
двухкратный *adj.* double, twin, twofold
двухкристальный *adj.* two-crystal
двухкулачковый *adj.* two-lobe; double-jaw(ed)
двухкурсовой *adj.* two-course
двухламповый *adj.* two-tube
двухленточный *adj.* double-strand
двухлинейный *adj.* bi-linear
двухлинзовый *adj.* two-lens
двухлистный *adj.* two-sheeted
двухлопастный *adj.* double-vane
двухлучевой *adj.* double-beam, two-beam, double-trace
двухмагнитный *adj.* two-magnet
двухмерный *adj.* two-dimensional
двухниточный *adj.* bifilar, double-thread, double-wound
двухножевой *adj.* double-knife, double-blade
двухобмоточный *adj.* two-coil(ed), double-coiled, double-wound
двухопорный *adj.* double-seat, double-beat
двухосновный *adj.* dibasic
двухосный *adj.* biaxial
двухотверстный *adj.* dual-aperture
двухпентодный *adj.* two-pentodes
двухперекидный *adj.* double flip-flop
двухпериодический *adj.* biperiodic
двухперьевый *adj.* two-pen
двухпластинчатый *adj.* double-slab, double-vane
двухплечий *adj.* double-arm(ed), two-arm(ed)
двухплоскостный *adj.* two-plane
двухповерхностный *adj.* two-surface
двухпозиционный *adj.* two-position, two-step; on-off; bistatic
двухположенный *adj.* two-position, two-step; on-off
двухполосный *adj.* two-way, two-band, double-band, bi-band, bipolar; two-wire transmission; double-sideband; double-channel
двухполостный *adj.* two-cavity
двухполупериодный *adj.* full-wave (rectifier)

**двухполюсник** *m.* one-terminal pair network, single-terminal pair, two-terminal network; impedor, twopole; dipole; **реактивный д.** two-terminal (*or* two-pole) network made up by reactive elements

**двухполюсный, двухполярный** *adj.* double-pole, two-pole, bipolar, two-terminal

**двухпоршневой** *adj.* double-piston

**двухпотоковый** *adj.* two-stream

**двухпоточный** *adj.* double-flow

**двухпредельный** *adj.* double-range

**двухприборный** *adj.* two-element

**двухпроводной** *adj.* double-wire, two-wire, double-line, two-line, two-conductor, twin-lead

**двухпрожекторный** *adj.* two-gun, two-projector

**двухпролётный** *adj.* double-transit

**двухпроходный** *adj.* two-way

**двухпутевой, двухпутный** *adj.* double-track, two-way

**двухраздельный** *adj.* two-part

**двухразмерный** *adj.* bidimensional

**двухразрывный** *adj.* double-break

**двухрежимный** *adj.* double-mode

**двухрезонансный** *adj.* double-resonance

**двухрезонаторный** *adj.* double-resonator, two-cavity

**двухрельсовый** *adj.* double-track, double-rail

**двухромовокислый** *adj.* bichromate

**двухрядный** *adj.* double, double-row, two-row, two-series, double-line, two-range

**двухседельный** *adj.* double-seat

**двухсекторный** *adj.* two-sector; two-section

**двухсекционный** *adj.* two-section

**двухсеточный** *adj.* bigrid, twin-grid

**двухсигнальный** *adj.* bi-signal

**двухсистемный** *adj.* dual-system

**двухскоростной** *adj.* dual-speed, double-speed

**двухслойный** *adj.* double-braid, two-ply, two-layer

**двухсрезный** *adj.* double-shear

**двухставочный** *adj.* two-rate, two-part

**двухстанинный** *adj.* double-sided, double-standard

**двухстаторный** *adj.* double-stator

**двухстержневой** *adj.* two-stub, two-legged

**двухстенный** *adj.* double-walled

**двухсторонний** *adj.* two-way, bilateral, double-surface, on both sides; double-ended, bipolar

**двухстрелочный** *adj.* cross-pointer

**двухстрендовый** *adj.* double-strand

**двухступенный, двухступенчатый** *adj.* two-stage, two-step, double-stage, two-level

**двухтактный** *adj.* duple, in double measure; push-pull; two-cycle, two-stroke, dual-action

**двухтарифный** *adj.* two-rate, double-tariff, dual-tariff

**двухтоковый** *adj.* double-current

**двухтональный** *adj.* two-tone

**двухточечный** *adj.* double-dot

**двухударный** *adj.* double-hit

**двухуровневый** *adj.* two-level

**двухустойчивый** *adj.* bistable

**двухфазный, двухфазовый** *adj.* bi-phase, diphase, two-phase

**двухфокусный** *adj.* bifocal, double-focus

**двухходовой** *adj.* two-way, double-throw; double-thread

**двухцветность** *f.* dichroism, dichromatism

**двухцветный** *adj.* two-color(ed), dichromatic, dichroic

**двухцелевой** *adj.* dual-purpose

**двухцепной** *adj.* double-circuit; double-chain

**двухцилиндровый** *adj.* two-cylinder

**двухцокольный** *adj.* double-ended

**двухчастичный** *adj.* two-body

**двухчастный** *adj.* two-part

**двухчастотный** *adj.* two-frequency, dual-frequency, two-band, two-circuit

**двухчленный** *adj.* two-term

**двухшейковый** *adj.* double-neck; double-groove

**двухшкальный** *adj.* double-dial, double-scale

**двухшлейфовый** *adj.* double-stub

**двухшнуровой** *adj.* double-cord, two-cord

**двухшпиндельный** *adj.* duplex; double-spindle

**двухштырьковый** *adj.* two-pin

**двухщелевой** *adj.* two-hole

**двухъядерный** *adj.* binuclear

**двухъякорный** *adj.* double-armature

**двухъярусный** *adj.* double-tier, double-level, two-story

**двухэлектродный** *adj.* two-electrode, diode

двухэлементный *adj.* two-element
двухэмиттерный *adj.* two-emitter
двухэтажный *adj.* two-storeyed
двуцветный *adj.* two-color(ed), dichroic, dichromatic
двучлен *m.* binomial
двучленный *adj.* binomial, binary
двушкальный *adj.* double-scale, double-dial, double-range
двущелевой *adj.* double-slot
двуюбочный *adj.* double-shed, double-shell
двуякорный *adj.* double-armature
дг *abbr.* (дециграмм) decigram
ДГМК *abbr.* (дистанционный гидромагнитный компас) distant-reading gyromagnetic compass
ДГТС *abbr.* (двусторонняя групповая телефонная связь) duplex conference call
ДД *abbr.* (дальнего действия) long-range
де- *prefix* de-, des-
де Соти de Sauty
деактивация *f.* deactivation
деаэратор *m.* deaerator
деаэрация *f.* deaerating
деаэризационный *adj.* deaerating, deaeration
деаэрированный *adj.* deaerated
деаэрировать *v.* to deaerate
дебаевский *adj.* Debye's
Дебай Debye
дебит *m.* yield, output; discharge, flow; д. источника source productiveness; д. источника звука productiveness of sound generator
деблокирование *n.* unblocking, clearing, releasing
деблокированный *adj.* clear, cleared, released
деблокировать *v.* to clear, to release, to unlock, to disconnect
деблокировка *see* деблокирование
деблокировочный *adj.* line-freeing, clearing, cutoff
деблокирующий *adj.* releasing, unblocking, clearing
Дебройль De Broglie
дебустер *m.* debooster
Деварда Devarda
девиационный *adj.* deviation
девиация *f.* deviation, swing, deflection, shift; д. пеленгатора, д. радиокомпаса, д. радиопеленгатора direction-finder deviation

девиометр *m.* deviometer
девственный *adj.* native, virgin
девятеричный *adj.* nonary, novenary
девятизначный *adj.* nine-character (-type)
девятикомпонентный *adj.* nine-term; nine-component
девятикратный *adj.* ninefold
девятиштырьковый *adj.* noval; nine-prong(ed)
девятнадцатеричный *adj.* novendenary
де Гааз de Haas
дегазационный *adj.* decontaminating
дегазация *f.* outgassing, degassing, degasification, decontaminating; gettering; д. от положительных ионов degassing through positive-ion bombardment, sputtering; д. с помощью высокой частоты R-F degassing
дегазирование *n.* degassing, degasification
дегазированный *adj.* degassed, degasified
дегазировать *v.* to degasify, to degas, to out-gas
дегауссизация *f.* degaussing
дегенеративность *f.* degeneracy
дегенеративный *adj.* degenerate
дегенерация *f.* degeneration; negative feedback, negative reaction
дегенерированный *adj.* degenerated, degenerate
дегенерировать *v.* to degenerate
дегидрация *f.* dehydration
дёготь *m.* tar, pitch
дегтярный *adj.* tar, tarred
« дегуссит » *m.* Degussit
дедуктивный *adj.* deductive
дедукция *f.* deduction
деж. *abbr.* (дежурный) man on duty
дежурить *v.* to be on duty; д. по частоте to watch the frequency, to stand by
дежурный *adj. and m. adj. decl.* on duty; guard, man on duty, attendant, operator, supervisor; keep-alive (anode); д. междугородной станции call office attendant; д. переговорного пункта call office attendant
дежурство *n.* attendance, duty, being on duty
дез- *prefix* des-, de-, dis-
дезактивационный *adj.* deactivation
дезактивация *f.*, дезактивирование *n.* decontamination; deactivation; de-excitation

дезактивированный *adj.* decontaminated; deactivated

дезактивировать *v.* to decontaminate; to deactivate

дезактивирующий *adj.* decontaminating; deactivating

дезакцентировка *f.* de-emphasis

дезинтеграция *f.* disintegration

дезориентация *f.* disorientation

дезориентирующий *adj.* confusing, disorienting

деионизационный *adj.* deionization

деионизация *f.* deionization, scavenging

деионизировать *v.* to deionize

деионизирующий *adj.* deionizing

**Дейль** Dale

действенность *f.* activity, efficiency, effectiveness

действенный *adj.* active, effective, efficient, operative

действие *n.* action, effect, operation, work, performance, running (of machine), functioning; **действием** by means (of), by the action (of); **вакуумного действия** vacuum operated; **дальнего действия** long-range; **двойного действия** double-acting; **двустороннего действия** bilateral; **замедленного действия** sluggish, slugged; **непрерывного действия** analog; **пневматического действия** air-operated, pneumatically operated; **простого действия** single-acting; **с ёмкостным действием** capacitance operated; **с плавным действием** smooth-acting; **автодинное д.** self-acting converter; **арифметическое д. с учётом порядков** floating point arithmetic; **астатическое д. с несколькими скоростями** multispeed floating mode; **астатическое усреднённое д.** floating average position action; **д. без выдержки времени** nondelayed action; **д. без подвижных частей** static operation; **вентильное д.** gating; **возвращающее д.** homing action; **групповое направляющее д.** directivity due to grouping of nondirectional elements; **дефокусирующее д. отклонения** deflection defocussing; **д., комбинирующее две величины** binary operation; **д. на близкое расстояние** current displacement-effect (in conductors); **д. на расстоянии** distant

effect; **направленное д.** directional operation, directional effect; directive force, versorial force; **непрерывное д.** analog; **обратное д.** reaction, retroaction; **д. обратной связи** retroactive effect; **отсроченное д.** delayed action; **д. по двум производным** double derivative action; **д. по отклонению и по производной** proportional plus derivative action; **д. по скорости** rate response; **д. поездной авторегулировки** operation of an automatic train control; **д. поперечного магнитного поля** cross induction; **прерывистое регулирующее д.** discontinuous controller action; **преувеличенно-заграждающее д. поездной авторегулировки** false-restrictive operation of an automatic train control; **преувеличенно-разрешительное д. поездной авторегулировки** false-proceed operation of an automatic train control; **противоколебательное д., противопомпажное д.** antihunt(ing) action; **регулирующее д. по закону обратной производной** inverse derivative control action; **д. с выдержкой времени** delayed action; **д. с двумя величинами** binary operation; **д. с рабочим током** open-circuit working; **следящее д.** automatic control; **тепловое д. тока** heating effect of current; **термическое д. ультразвука** ultrasonic thermal action; **д. фриттера** coherer action; **шунтирующее д. ёмкости** capacitive-shunting effect

действительный *adj.* real, actual, true; effective, efficient; valid; net; intrinsic

действовать, подействовать *v.* to act, to operate, to function, to work, to run, to proceed; to effect; to react, to attack

действующий *adj.* effective, efficient; active, acting, working, functioning, operating, operational, actuating; virtual; **д. вручную** manually operated; **замедленно д.** sluggish; **д. навстречу** antagonistic; **д. от батареи** battery-driven; **д. от рассогласования** error-actuated; **плавно д.** smooth-acting

дейтерий *m.* deuterium, D

**Дейтерс** Deiters

дейтон, дейтрон *m.* deuteron, deutron

дека- *prefix* deca-, ten

**дека** *f.* sounding board, sound board

**декаграмм** *m.* decagram

**декада** *f.* level, decade; ten-day period; ten, decade multiple, decade step; digit, step, stage; **д. многократного поля** level multiple; **незадействованная д., неиспользуемая д.** unfitted level, spare level, vacant level, dead level

**декадно-счётный** *adj.* decimal-counting

**декадно-шаговый** *adj.* ten-step

**декадный** *adj.* decade; decimal

**декалесценция** *f.* decalescence

**декалитр** *m.* decaliter

**декаметр** *m.* decameter

**декаметровый** *adj.* decametric

**декапирование** *n.* pickling, pickle, dip

**декапировать** *v.* to pickle, to dip, to scour

**Декарт** Descartes

**декартов** *poss. adj.*; **декартовый** *adj.* Cartesian

**декатрон** *m.* dekatron

**декатронный** *adj.* dekatron

**декаэдр** *m.* decahedron

**декаэдрический** *adj.* decahedral

**Декка** Decca

**деклинатор** *m.* declinometer, declination compass

**деклинация** *f.* declination; magnetic declination

**деклинометр** *see* **деклинатор**

**декогерер** *m.* decoherer; tapper

**декогерирование** *n.* decoherence

**декогерировать** *v.* to decohere

**декодер** *m.* decoder

**декодирование** *n.* decoding

**декодировать** *v.* to decode, to decipher

**декодирующий** *adj.* decoding

**декометр** *m.* decometer, Decca indicator

**декомпенсированный** *adj.* unbalanced

**декомпозиция** *f.* decomposition

**декомпрессионный** *adj.* decompression, compression-release; relief

**декомпрессия** *f.* decompression

**декомпрессор** *m.* decompressor

**декоративный, декорационный** *adj.* decorative, ornamental

**декорация** *f.* **декорирование** *n.* decoration, scenery

**декорировать** *v.* to decorate

**декохерер** *m.* decoherer; tapper

**декремент** *m.* decrement, decrease; **д. затухания** decrement of damping, decay time-constant, attenuation ratio, damping factor; **д. затухания, обусловленный теплопроводностью** heat conduction decrement; **д. затухания пузырька** decrement of bubble pulsation; **линейный д.** subsidence ratio

**декрементный** *adj.* decrement

**декреметр** *m.* decremeter

**дектра** *f.* Dectra

**дел.** *abbr.* (**деление**) division, fission

**Деландр** Deslandres

**делать, сделать** *v.* to make, to do, to produce, to perform; **д. выкладки** to compute; **д. гальваническое покрытие** to electroplate, to galvanize; to treat by anodic etching; **д. ответвление** to branch, to tap off; **д. отвод** to tap; **д. перемычку** to tie, to connect, to bridge

**деление** *n.* division, dividing; scaling, scaling-down process; parting, splitting; indexing; graduation; point; unit, interval; dial; **верхние деления шкалы** high end of scale; **нижние деления шкалы** lower end of scale; **д. без восстановления** nonrestoring division; **д. в двоичной системе** binary division; **д. интерференционной частоты** frequency division by interference; **д. коллектора** unit interval at the commutator; **д. на градусы** graduation, division in degrees; **д. на участки** sectionalization; **д. нацело** exact division; **нулевое д.** zero mark; **д. обмотки якоря** unit interval (in a winding); **полюсное д.** pole pitch; **д. пополам** halving, bisection; **д. при помощи умножения на обратную величину** division by reciprocal multiplication; **д. с использованием метода неявных функций** division by implicit computation; **сотенное д.** centesimal graduation; **д. угломера** mil/fission; **д. фазы** phase splitting; **д. частот** (*or* **частоты**) frequency demultiplication, scaling-down process; **д. частоты импульсов** count-down, counting down of pulses, scaling down, repetition-rate scaling; **д. частоты повторения импульсов** pulse-rate division; **д. частоты пополам** frequency halving; **д. частоты следования импульсов** skip keying; **д. ядра** nuclear fission; **д. якоря** unit interval in winding

**делённый** *adj.* divided, split; indexed
**делимое** *n. adj. decl.* dividend
**делимость** *f.* divisibility
**делимый** *adj.* divisible; fissionable
**делитель** *m.* divisor, denominator; separator, divider; measure (*math.*); attenuator; demultiplier; scaler; **д. анодного напряжения** voltage-divider, bleeder; **анодный д. тока** current divider; **двоичный д. частоты с переменным коэффициентом деления** variable binary scaler; **ёмкостный д. напряжения** capacitor voltage divider; **импульсный д. напряжения** pulsed attenuator; **ламповый д. частоты** tube scaler, locked oscillator, transrectifier; **д. мощности на полосковых линиях передачи** strip transmission line power divider; **д. мощности с регулируемым коэффициентом деления** variable-ratio power divider; **д. на 16** scale-of-sixteen; **д. на два** halver; **д. на омических сопротивлениях** resistance divider; **д. напряжения** voltage divider, potential divider, bleeder resistor, static balancer, balancing coil; **д. напряжения для трёхпроводной системы** three-wire compensator; **д. напряжения на выходе** output voltage-divider, output attenuator; **д. напряжения на сопротивлениях и ёмкостях** resistance-capacitance divider; **омический д. напряжения** potentiometer-type resistor; **д. пополам** halver; **сдвинутый д.** shifted divisor; **секционный д. напряжения** adjustable voltage divider; **д. со скользящим контактом** slide divider; **ступенчатый д.** step attenuator; **д. фаз** phase-splitting circuit; **д. частоты импульсов** counter-down; **д. частоты на два** two-to-one frequency divider, frequency halver; **д. частоты на триггерной схеме** flip-flop frequency divider; **д. частоты повторения** repetition-rate divider
**делительный** *adj.* dividing, separating, division; indexing, index; scaling; fission
**делить, поделить, разделить** *v.* to divide, to part, to apportion, to share, to graduate, to split; **д. пополам** to halve, to bisect
**делиться, поделиться, разделиться** *v.*

to be divided; to split; to share; to contain (*math.*)
**Деллингер, Деллинджер** Dellinger
**дело** *n.* matter, business, thing, point; work, act; **приёмочное д.** acceptance practice
**Делон** Delon
**Делопиталь** de l'Hospital
**дельта** *f.* delta
**дельтамакс** *m.* deltamax
**дельтапоток** *m.* delta-flux
**дельтоиддодекаэдр** *m.* deltoid dodecahedron
**делящий** *adj.* dividing
**делящийся** *adj.* fissionable
**демагнетизатор** *m.* demagnetizer
**дематериализация** *f.* annihilation radiation
**Дембер** Dember
**демберовский** *adj.* (of) Dember, Dember's
**демография** *f.* demography
**демодулированный** *adj.* demodulated
**демодулировать** *v.* to demodulate
**демодулирующий** *adj.* demodulating
**демодулятор** *m.* demodulator, detector; staticiser; **д. звукового сопровождения** sound-track demodulator; **д. на полупроводниковых триодах** transistor demodulator; **д. по произведению** product demodulator; **д. устройства многократной связи** multiplex demodulator; **цветовой д. телевизора** receiver sampler (*tv*); **д. частоты** FM detector
**демодуляционный** *adj.* demodulation
**демодуляция** *f.* demodulation, detection, rectification; debunching; staticising; **д. пучка за счёт кулоновых сил** space charge debunching
**демонстрационный** *adj.* demonstrating, demonstration
**демонстрация** *f.* demonstration
**демонстрировать** *v.* to demonstrate
**демонтаж** *m.*, **демонтирование** *n.* dismantling, dismounting, stripping
**демонтировать** *v.* to dismantle, to dismount, to strip, to take down, to take apart, to remove, to disassemble
**демпфер** *m.* damper, damper winding; silencer, damper, shock absorber, deadener, amortisseur; dashpot; mute (*mus.*); **азимутальный д.** azimuth antihunt; **воздушный д. с вращающейся крыльчаткой** air vane retarder; **д. на принципе вихревых**

**токов** eddy current retarder; **д. с вязким трением** viscous damper
**демпферный** *adj.* damper, damping
**демпфирование** *n.* damping, shock absorption, attenuation, deadening, buffer action; antihunt; quenching; **воздушное д.** air friction damping; **д. короткозамкнутым медным проводником** copper damping; **д. по углу места** elevation damping; **д. с помощью вязкого трения** viscous damping; **угломестное д.** elevation antihunt (of an antenna); **д. федингов в радиорелейной связи** signal-attenuation in microwave links caused by fading effect; **чрезмерное д.** overdamping
**демпфированный** *adj.* damped, deadbeat; oscillatory damped; decadent; **д. с сохранением колебаний** oscillatory damped; **сильно д.** deadbeat
**демпфировать** *v.* to damp, to mute, to attenuate
**демпфирующий** *adj.* damping, shock-absorbing; antihunt
**демультипликатор** *m.* demultiplier, reducing gear
**денатурированный** *adj.* denatured
**денатурировать** *v.* to denature
**дендрит** *m.* dendrite; **дендриты и наросты** trees and nodules
**дендрометр** *m.* dendrometer
**денежный** *adj.* money
**дензитометр** *see* **денсиметр**
**денотат** *m.* denotation
**денсиметр, денситометр** *m.* densimeter, opacity meter, densitometer
**денситометрия** *f.* densitometry
**день** *m.* day; **д. осеннего равноденствия** autumnal equinox
**депеша** *f.* message, dispatch, wire, telegram; **д. для вручения** message for delivery; **циркулярная д.** multiple-call message
**деплистор** *m.* deplistor
**депозит** *m.* deposit
**деполимеризация** *f.* depolymerization
**деполимеризованный** *adj.* depolymerized
**деполимеризовать** *v.* to depolymerize
**деполяризатор** *m.* depolarizer
**деполяризационный** *adj.* depolarizing
**деполяризация** *f.* depolarization, depolarizing
**деполяризировать, деполяризовать** *v.* to depolarize

**депрессиометр** *m.* depression meter
**депупинизация** *f.* uncoiling, unwinding, unreeling
**дёрганный** *adj.* twitched, jerked, pulled
**дёрганье** *n.* pulling, pull, jerking, twitching, jitter, jogging; **д. частоты** transient frequency jitter, scintillation
**дёргать, дёргнуть** *v.* to twitch, to jerk, to pull
**дерево** *n.* wood, timber, lumber; tree; tree circuit
**деревовидный** *adj.* tree, tree-like
**деревянный** *adj.* wood, wooden
**державка** *f.* holder, hold, support, stand, bracket, mounting
**держание** *n.* hold, holding, keeping, maintaining
**держанный** *adj.* held, supported, kept, maintained
**держатель** *m.* holder, base, mount, fastener, support, bracket, jaw, cartridge, grip, adapter, arm; ear; **д. арматуры** mount support; **д. колбы** bulb jaw; **д. колпачка** cap support; **д. контактного провода** contact wire hanger; **д. ножки** mount support; **д. плавкой вставки** fuse-carrier; **потолочный д.** car-shed hanger; **тарелкообразный д.** cup-shaped base; **д. экрана** disk support; shield-base
**держать, подержать** *v.* to hold, to keep; **д. занятой соединительную линию, д. соединение с АТС** holding of an outside call for local inquiry
**держаться, подержаться** *v.* to hold, to hold on, to adhere, to cling, to be held up (by), to be supported (by); **д. на канифоли** "rosin joint"
**держащий** *adj.* holding
**дериват** *m.* derivative
**деривометр** *m.* derivometer
**дерматит** *m.* dermatitis; **лучевой д.** radio-dermatitis, X-ray dermatitis
**деррик** *m.* derrick
**десенсибилизация** *f.* desensitization
**десинхронизация** *f.* desynchronizing
**дескриптор** *m.* descriptor
**дескрипция** *f.* description
**« десмодур »** *m.* Desmodur
**десмотропия** *f.* desmotropism, tautomerism
**десмотропный** *adj.* desmotropic
**« десмофен »** *m.* Desmophen
**Дессауер** Dessauer

**дестиляция** *f.* distillation
**дестрибютор** *m.* distributor
**десятеричный** *adj.* denary, tenfold
**десяти-** *prefix* deca-, ten
**десятизначный** *adj.* ten-character
**десятиканальный** *adj.* ten-channel
**десятиклавишный** *adj.* ten-key
**десятикратный** *adj.* tenfold
**десятисантиметровый** *adj.* ten-centi-meter
**десятисекундник** *m.* ten-second in-terrupter, rheotone
**десятисекционный** *adj.* ten-bin
**десятично-двоичный** *adj.* decimal-binary
**десятичный** *adj.* decimal; decade
**десяток** *m.* ten, decade
**деталь** *f.* detail, part, component, piece, limb, element (of a tube), member, feature; **детали волноводов** components for waveguides; **детали (изображения) в затенённых** (*or* **в тёмных**) **участках** shadow details; **детали полосковых передающих линий** strip-line components; **сменные детали** renewal parts; **точно обработанные детали** precision parts; **загромождённый** (*or* **насыщенный,** *or* **перегружённый**) **деталями** busy (of a picture); **д., выдерживающая высокую температуру** red hot component; **запрессованная д.** insert; **изолирующая д.** rod insulator, pin insulator; **исполнительная д.** executive component; **д. красного каления** red hot component; **крепёжная д.** bracket, attachment; **обрабатываемая д.** workpiece; **ответственная д.** critical piece; **соединяющая д.** coupler; **строительная д.** component (part), construction unit, subassembly, construction material
**детально** *adv.* in detail
**детальность** *f.* detail; **д. в углах изображения** corner detail
**детальный** *adj.* detail, detailed, fine, minute
**детектирование** *n.* detection, detecting; rectification, rectifying, demodulation; **анодное д.** anode rectification, grid-bias detection, trans-rectification; **двухполупериодное д.** double-wave detection; **квадратичное д.** square-law detection; **д. с восстановленной несущей** exalted-carrier detection; **сеточное д.** cumu-

lative rectification, grid-leak rectification, grid-resistor-and-capacitor rectification, grid-current detection; **д. сигнала ошибки** (*or* **рассогласования**) error-signal detection; **д. сигналов изображения** video signal detection, video detection
**детектированный** *adj.* rectified; detected
**детектировать** *v.* to detect, to find, to catch; to rectify
**детектирующий** *adj.* detecting; rectifying
**детектор** *m.* detector, pickup; rectifier; demodulator; **до детектора** pre-detector; **анодный д. в режиме класса B** class-B anode detector; **анодный д. с отрицательной обратной связью** reflex detector; **балансный д.** ring-type modulator; **д. в канале сигналов звука и цветности** sound-and-chroma detector; **д. для контроля передачи с эфира** air check detector; **дробный д.** ratio detector; **д. из свинцового блеска** galena detector; **избирательный д.** switch detector; **д. измерительной линии** slotted-line standing-wave detector; **д. канала яркостного сигнала** brightness-signal detector; **коммутируемый д., реагирующий на фазу** phase-sensitive keyed detector; **кусочно-аппроксимированный квадратурный д.** piecewise quadratic detector; **ламповый д.** electron-tube detector, thermionic detector, trans-rectifier; **« д. лжи »** lie detector, psycho-integroammeter; **д. напряжения в линиях** live line tester; **д. огибающей** envelope demodulator; **первый д.** heterodyne detector; **д. по произведению** product demodulator; **д. промежуточной частоты сигнала звукового сопровождения** sound intermediate-frequency detector; **пропорциональный д.** ratio detector; **д., работающий на наклонном участке резонансной кривой** slope detector; **д. радиоволн** cymoscope; **разностный д., д. рассогласования** difference detector; **д. с квадратичной характеристикой** square-law detector; **д. с линейной характеристикой** linear rectifier; **д. с несколькими точками соприкосновения** multiple-contact detector;

**д. с одной контактной точкой** single-contact detector; **д. с перемещением среднего** moving average detector; **д. с постоянной точкой** average-reading detector, fixed-crystal detector; **сверхрегенеративный д.** self-quenching detector; **сеточный д. (с гридликом)** grid-leak capacitor detector, grid-current rectifier, cumulative grid detector; **сеточный д. с обратной связью** regenerative detector; **д. сигнала звукового сопровождения** sound detector; **д. сигналов изображений** video detector, picture video detector; **д.-смеситель** detector-converter; **суммирующий д.** count detector; **тормозящий д.** electron-oscillating detector, retarding field detector; **д. углового рассогласования** angle error demodulator; **д.-умножитель** product demodulator; **частотный д.** FM discriminator; **д. яркостного сигнала** luminance detector

**детекторно-усилительный** *adj.* detector-amplifier

**детекторный** *adj.* detecting, detector; rectifying, rectifier

**детектофон** *m.* detectophone

**детерминант** *m.* determinant (*math.*); **д. общих четырёхполюсников** determinant of a universal four-terminal network

**детерминанта** *f.* determinant (*math.*)

**детерминизм** *m.* determinism

**детерминированность** *f.* determinancy

**детерминированный** *adj.* determinate

**детерминистический** *adj.* deterministic

**детонатор** *m.* detonator, primer, percussion cup

**детонаторный** *adj.* detonator, shot-firing

**детонационный** *adj.* detonation, explosion

**детонация** *f.* detonation, explosion; clap, crack; **высокочастотная д.** flutter; **медленная д.** drift; **низкочастотная д.** wow

**детонировать** *v.* to detonate, to explode; to be out of tune

**детонирующий** *adj.* detonating, exploding; knocking

**дефазирование** *n.* out-phasing

**дефазированный** *adj.* dephased, out of phase

**дефазировать** *v.* to dephase

**дефект** *m.* flaw, defect, blemish, failure, fault, drawback, damage; error; disturbance, interference; breakdown; **дефекты изображения в виде белых секторов** sectoring; **д. в месте склейки фонограмы** blooping notch

**дефективный, дефектный** *adj.* defective, faulty, imperfect, damaged

**дефектоскоп** *m.* flaw detector, defectoscope, crack detector; **д. для обнаружения трещин** crack detector; **звуковой д.** reflectoscope; **д. с двумя преобразователями** two-transducer flaw detector

**дефектоскопический** *adj.* flaw-detector

**дефектоскопия** *f.* detection of defects, flaw detection, nondestructive testing, materiology, defectoscopy; **магнитная д. ферромагнетиков** magnaflux method

**дефинитный** *adj.* definite

**дефицит** *m.* shortcoming

**дефицитный** *adj.* deficient, deficit; critical, bad, poor

**дефлегматор** *m.* dephlegmator

**дефлегмирование** *n.* dephlegmation

**дефлегмировать** *v.* to dephlegmate

**дефлектометр** *m.* deflectometer

**дефлектор** *m.* deflector, deflecting electrode

**дефлекторный** *adj.* deflector; deflection

**дефлектрон** *m.* deflectron

**дефлекционный** *adj.* deflection

**дефокусирование** *n.* defocusing

**дефокусированный** *adj.* defocused, out of focus

**дефокусировать** *v.* to defocus

**дефокусировка** *f.* defocus(ing), out of focus, haziness; telegraph-pulse distortion; **д. вследствие отклонения, д. при развёртке** compensatory focusing of cathode-ray beam by means of deflection

**дефокусирующий** *adj.* defocusing

**дефокусно-пунктирный** *adj.* defocus-dot

**дефокусно-фокусный** *adj.* defocus-focus

**деформационный** *adj.* deformation, strain, distortion, warping; elasticity; deflection

**деформация** *f.* deformation, strain, straining, distortion, warping; elasticity; deflection; **исчезающая д.** elastic deformation; **д. при растя-**

жении deformation by tensile-stress; д. **среза** shear(ing)

**деформирование** *see* **деформация**

**деформированный** *adj.* deformed, strained, distorted, warped

**деформировать** *v.* to deform, to strain, to distort

**деформироваться** *v.* to become deformed, to be distorted, to become strained, to warp, to buckle

**деформирующий** *adj.* deforming, distorting, straining

**дефосфорация** *f.* dephosphorization

**децелерация** *f.* deceleration

**децелерометр** *m.* decelerometer

**децелит** *m.* Decelith

**децентрализация** *f.* decentralization, scattering

**децентрализовать** *v.* to decentralize, to scatter

**децентрирование** *n.* off-centering

**децентрированный** *adj.* off-center

**децентрирующий** *adj.* off-centering

**деци-** *prefix* deci-

**децибел** *m.* decibel; **децибелы на основе 1 миллиметра** decibels referred to one milliwatt

**децибелметр** *m.* decibel meter

**децибельный** *adj.* decibel

**дециграмм** *m.* decigram

**децил** *m.* decyl

**дециметр** *m.* decimeter

**дециметричный, дециметровый** *adj.* decimetric, decimeter

**децимиллиметровый** *adj.* submillimetric

**децинепер** *m.* decineper

**децинормальный** *adj.* decinormal

**дешифратор** *m.* decoder, decipherer, decoder switch; discriminator, converter unit; matrix gate; **д. на выходе телетайпа** paper tape decoder; **д. переноса** transfer interpreter; **д. с линией задержки** delay line decoder; **д. с накопителем** storage decoder; **д. сигналов от перфоленты** paper tape decoder; **цифровой д. последовательного действия** serial digital decoder

**дешифраторный** *adj.* decoder

**дешифрирование** *n.* decoding, deciphering, decipherment

**дешифрировать** *v.* to decipher

**дешифрирующий** *adj.* decoding, deciphering

**дешифрование** *n.,* **дешифровка** *f. see*

дешифрирование

**Дешман** Dushmann

**деэлектризация** *f.* deionization

**« деэмфазис »** *m.* de-emphasis, post-emphasis

**деятельность** *f.* activity, work

**дж** *abbr.* (джоуль) joule

**Джайльс** Giles

**джар** *m.* jar (unit of capacitance)

**Джедд** Judd

**« джек »** *m.* jack; (pushbutton type) key

**джемпер** *m.* jumper, non-tension bow

**Дженкин** Jenkin

**Джерелл** Jerrell

**джи** Gee (medium range hyperbolic radio aid to navigation)

**Джиль** Gill

**Джин** Jean

**Джинс** Jeans

**джозефсоновский** *adj.* (of) Josephson, Josephson's

**Джолли** Jolly

**« Джон Гильпин »** John Gilpin radio beacon

**Джонсон** Johnson

**Джорджи** Giorgi

**джоулев** *poss. adj.* Joule's, (of) Joule

**джоулометр** *m.* joule meter

**джоул(ь)** *m.* joule

**джоульметр** *m.* joule meter

**Джоши** Joshi

**джут** *m.* jute

**джутовый** *adj.* jute

**дзета-функция** *f.* zeta function

**ди-** *prefix* di-, bi-

**диабетометр** *m.* diabetometer

**диагноз** *m.* diagnosis

**диагностический** *adj.* diagnostic

**диагометр** *m.* diagometer

**диагональ** *f.* diagonal, diagonal line; **вспомогательная д.** secondary diagonal

**диагональный** *adj.* diagonal, oblique

**диаграмма** *f.* diagram, pattern, plan, drawing, figure; graph; chart, plot, characteristic curve; **азимутальная д. направленности** azimuth pattern; **д. амплитуды и угла** log magnitude angle diagram; **д. антенны без побочных лепестков** clean pattern; **д. антенны « на приём »** receiving pattern; **д. антенны, смещённая относительно оси отражателя** off-center lobe; **асимметричная треугольная д. направленности** asymmetrically flared beam; **д. в четырёх**

квадрантах quadrantal diagram; **векторная д. в полярных координатах** polar-phase diagram; **векторная д. в прямоугольных координатах** clock-phase diagram; **(векторная) д.** импеданса impedance locus; **векторная д. мощности** geometric power diagram; **векторная д. сигналов яркости** luminance vector diagram; **векторная д. цветовой поднесущей** color phase chart; **вертикальная д. излучения** vertical plane directional pattern; **д. видимости** coverage diagram; **восьмёрочная д. направленности** double circle diagram, figure-eight diagram, figure-eight pattern; **д. временных соотношений** time diagram; **д. движения электронов в лампе с модуляцией скорости** Applegate diagram; **д. для расчёта линий передач** transmission-line chart; **золотниковая д.** slide valve diagram; **игольчатая д. направленности** pencil-beam pattern (of an antenna); **д. излучения в вертикальной плоскости** vertical radiation pattern, vertical plane directional pattern; **д. излучения поверхностной волны** ground-wave pattern; **д. импеданса в виде петли** cusp-shaped impedance locus; **д. импульса** sphygmogram; **исходная д.** load diagram; **качающаяся д. (излучения)** scanning pattern; **д. комплексной проводимости** admittance diagram; **координированная с картой векторная д.** map coordinated phasor diagram; **круговая д. внесённого сопротивления** motional impedance circle; **круговая д. магнетрона** Rieke diagram; **круговая д. полной проводимости** admittance circle; **круговая векторная д.** clock diagram; **д. механических напряжений в функции действующих сил** stress-strain diagram; **д. мощности в функции времени** power-time diagram; **д. на половинной мощности** half-amplitude beam; **нагрузочная д.** load diagram, load curve, load impedance diagram, rating chart; **д. направленности** directional diagram, directivity pattern, antenna radiation pattern, directional (response) pattern, free space pattern, lobe pattern, reception diagram; **д. на-**

**правленности (антенны) в вертикальной плоскости** vertical-coverage diagram, vertical radiation characteristic, elevation radiation pattern; vertical diagram; **д. направленности без побочных лепестков** clean pattern; **д. направленности в свободном пространстве** free-space diagram (of an antenna); **д. направленности диполя в форме восьмёрки** figure-of-eight pattern of a dipole; **д. направленности излучения шумов** noise radiation pattern; **д. направленности на приём** reception diagram; **д. направленности облучателя антенны** feed pattern, primary pattern; **д. направленности по мощности** power pattern, power directivity pattern; **д. направленности по углу места** elevation radiation pattern; **д. направленности приёмной антенны** receiving pattern of an antenna; **д. направленности чувствительности** directional response pattern; **д. неопределённости** ambiguity diagram; **ножевая д. антенны** beavertail beam; **обычная векторная д.** conventional phasor diagram; **д. перекрывания** coverage diagram; **д. по времени** time base diagram; **д. полей трансформатора** field distribution diagram of a transformer; **д. полной проводимости** admittance diagram; **д. при неизменном токе** constant-current diagram; **д. производственного тока** flow chart; **д. пульса** sphygmogram; **рабочая д.** indicator diagram; **д. растяжения** stress-strain diagram; **д. самописца** chart of a recording instrument; **д. семейства кривых** nomograph chart; **д. сообщений** traffic chart (*tphny.*); **д. сопротивлений и проводимости** immittance chart; **д. сопротивления разрыву** tenacity diagram; **треугольная д. направленности** flat-top flared beam; **д. упругости пара** vapor diagram, steam diagram; **д. уровня на протяжении системы передачи** transmission level diagram; **д. фазовых состояний** phase diagram; **д. энергетических уровней** energy-level diagram; **яркостная цветовая д.** color brightness diagram, Russell diagram

диаграммный *adj.* chart; diagrammatic; coordinate (paper)
диаграф *m.* diagraph
диада *f.* dyad, two-digit group
диадический *adj.* dyadic
диазо- *prefix* diazo-
диак *m.* diac
диакисдодекаэдр *m.* diplododecahedron, dyakisdodecahedron
« диакон » *m.* Diacon
диакритика *f.* diacritic
диакритический *adj.* diacritic(al)
диактинический *adj.* diactinic
диакустика *f.* diacoustics
диакустический *adj.* diacoustic
диализ *m.* dialysis, ultrafiltration
диализатор *m.* dialyzer
диализированный, диализованный *adj.* dialyzed
диализирующий *adj.* dialyzing
диалит *m.* dialite
диалитический *adj.* dialytic
диам. *abbr.* (диаметр) diameter
диамагнетизм *m.* diamagnetism
диамагнетики *pl.* diamagnetic substances
диамагнитность *f.* diamagnetism
диамагнитный *adj.* diamagnetic
диаметр *m.* diameter; absolute value of a quantity in the complex plane; the arithmetic average of the sound velocities in a liquid and its saturated vapor; bore, caliber; д. в свету overall diameter; внутренний д. (цилиндра) drill hole, borehole, bore, boring; гидравлический д. hydraulic equivalent diameter; критический д. (волновода) cut-off diameter (of a waveguide); наружный д. каркаса (*or* катушки) bobbin outer diameter; д. окружности воздушного зазора gap diameter, air-gap diameter; опорный д. seating diameter; предельный д. cut-off diameter; средний д. pitch diameter; д. якоря с зубцами gap diameter, air-gap diameter
диаметрально *adv.* diametrically
диаметральный *adj.* diametral, diametric(al), diameter
диамидиорезорцин *m.* diamide resorcin(ol)
диапазон *m.* (sound) diapason, range; compass, scope; pitch; band (frequency); span; д. амплитуд изображения amplitude range of video signal; всеблокирующий д. common

suppressed frequency band; д. высот range of pitch; д. гармоник frequency band of formants; д. голоса voice-frequency range; д. громкости volume range; д. громкости звука volume of sound; д. громкости звуков речи speech volume; д. действия радиолокатора range of indication in radar apparatus; десятисантиметровый д. S-band; д. естественных звуковых явлений intensity range of natural sound, volume range of natural sound; д. запирания suppressed frequency band, attenuation band; д. захватывания частот frequency lock-in range; д. изменения stagger ratio; д. изменения несущей частоты carrier frequency range; д. использования usable range; д. колебаний frequency range of an oscillator; д. контрастности передаваемой сцены scene contrast range; д. модуляции workable control range, range of uniform control, drive range; д. настройки по частоте frequency-tuning range; д. несущих (частот) carrier-frequency range; одно-сантиметровый д. K-band; д. отклонений variation range; д. переменной частоты swept band; д. поддерживания hold-in range; полночастотный д. записи high-fidelity recording band; д. порядков order range; д. приёма frequency-range of receiver, interception range; пропускаемый д. частот, д. пропускания pass-band width, band-pass width; рабочий д. счётчика по силе тока effective current range of a meter, accurate current range of a meter; разбросанный д. speckled band; д. регулировки tapping range; рекламный д. commercial broadcast band; д. с участками scattered band, speckled band; д. синхронизации lock-in range, collecting zone; д. считывания full-scale value; трёхсантиметровый д. X-band; д. удерживания hold-in range; д. уровня ограничения limiting level range; частичный д. sub-band; д. частот pass-band, bandwidth, frequency band, frequency range; д. частот канала связи channel bandwidth; д. частот передачи transmission band; д.

частот, пропускаемых системой передачи frequency range of transmission system; **д. яркостного контраста** brightness contrast range; *see also* **частота**

**диапазонный** *adj.* diapason, band, range; scope, compass; broadband, multituned

**диапозитив** *m.* diapositive, slide, transparency; **д. с изображением испытательной таблицы** test pattern slide

**диапозитивный** *adj.* diapositive, slide

**диапроектор** *m.* slide projector (*or* scanner)

**диаскоп** *m.* slide projector

**диаскопический** *adj.* slide-projector

**диастатический** *adj.* diastatic

**диатермический** *adj.* diathermic, diathermal, diathermy

**диатермичность** *f.* diathermy, diathermance

**диатермичный** *see* **диатермический**

**диатермия** *f.* diathermy

**диатермометр** *m.* diathermometer

**диатонический** *adj.* diatonic

**диатрон** *m.* diatron

**диафанометр** *m.* diaphanometer

**диафон** *m.* audio-visual recorder

**диафонический** *adj.* cross-talk

**диафония** *f.* crosstalk

**диафрагма** *f.* diaphragm, membrane, iris, blind, lens hood, aperture, stop, dimmer, septum; brake sack; light barrier; window; **ирисовые диафрагмы для круглых волноводов** stops for circular waveguides; **поперечные диафрагмы** transverse septa; **быстросменная д. с отверстием** quick-change orifice plate; **измерительная д.** orifice; **ирисовая д. с дистанционным управлением** remote-control iris; **камерная д.** carrier-ring orifice; **острая д.** thin-plate orifice; **д. поля зрения** picture aperture; **поперечная д.** transverse septum; **расходомерная д.** plate orifice; **д. с эксцентрично расположенным круглым проходным отверстием** eccentric orifice plate

**диафрагменный** *adj.* diaphragm, membrane, stop

**диафрагмирование** *n.* iris action, irising, beam masking; equipping with septa; **д. пучка** beam masking

**диафрагмированный** *adj.* diaphragm-type, septate, appertured

**диафрагмировать** *v.* to screen off, to dim, to stop down, to diaphragm

**диафрагмовый** *adj.* diaphragm, membrane

**дивариантный** *adj.* divariant

**дивергентный** *adj.* divergent

**дивергенция** *f.* divergence, divergency; **д. плоскости** areal divergence

**дивергировать** *v.* to diverge

**дивергирующий** *adj.* divergent

**дивертер** *m.* diverter

**дивизионный** *adj.* division

**дивизия** *f.* division

**дивизор** *m.* divisor

**дигексагональный** *adj.* dihexagonal

**дигептальный** *adj.* diheptal

**дигитрон, диджитрон** *m.* digitron

**диез** *m.* sharp (*mus.*)

**дизель** *m.* diesel-engine

**дизель-генераторный** *adj.* diesel-generating

**дизельный** *adj.* diesel

**Дизельхорст** Dieselhorst

**дизельэлектрический** *adj.* diesel-electric

**дизъюнктивный** *adj.* disjunctive

**дизъюнкция** *f.* disjunction; **неразделительная д.** inclusive disjunction; **разделительная д.** exclusive disjunction

**ДИК** *abbr.* (**дистанционный индукционный компас**) induction telecompass

**дикий** *adj.* odd, extravagant; random, wild

**Дикке** Dicke

**диксонак** *m.* dixonac

**диктафон** *m.* dictaphone; **д. для предупреждений о разговоре** apparatus for announcements

**диктовальный** *adj.* dictating

**диктование** *n.* dictation

**диктованный** *adj.* dictated

**диктограф** *m.* dictograph

**диктор** *m.* announcer; talker, speaker

**дикторский** *adj.* announcer; speaker

**диктофон** *see* **диктафон**

**диктующий** *adj.* dictating

**дикция** *f.* diction, enunciation

**дилатометр** *m.* dilatometer

**дилатометрический** *adj.* dilatometer

**дилатометрия** *f.* dilatometry

**дилемма** *f.* dilemma, perplexity

**димер** *m.* dimer

**димеризация** *f.* dimerization

**димерный** *adj.* dimeric

**диметилкремний** *m.* dimethylsilicone

**дин** *abbr.* (**дина**) dyne

**дина** *f.* dyne

**« динаген »** *m.* Dynagen

**динаквод** *m.* dynaquad

**динамакс** *m.* dynamax

**диаметр** *m.* dynameter

**динамизм** *m.* dynamism

**динамик** *m.* dynamic loudspeaker; **контрольный д.** monitoring (dynamic) loudspeaker

**динамика** *f.* dynamism, dynamic range, volume range; dynamics

**динамически** *adv.* dynamically

**динамический** *adj.* dynamic

**динамный** *adj.* dynamo

**динамо** *n. indecl.* dynamo

**динамограф** *m.* dynamograph

**динамодвигатель** *m.* motor-generator set

**динамомашина** *f.* dynamo, dynamo-generator, dynamo-electric machine

**динамометр** *m.* dynamometer; **балансный д.** cradle dynamometer; **крутильный д.** tension dynamometer, torquemeter, torsionmeter, torsion indicator; **тяговый д.** drawbar dynamometer

**динамометрический** *adj.* dynamometric

**динаморегулятор** *m.* dynamo governor

**динамотор** *m.* dynamotor, rotary transformer

**динамофон** *m.* dynamophone

**динамоэлектрический** *adj.* dynamo-electric, electrodynamic

**« динатон »** *m.* dynatone

**динатрон** *m.* dynatron

**динатронный** *adj.* dynatron

**Дингслей** Dingslay

**динистор** *m.* dynistor, dinistor, diode thyristor, reverse blocking diode thyristor

**динод** *m.* dynode; **улучшенный д.** superdynode

**диод** *m.* diode, diode tube; **противовключённые диоды** back-to-back diodes; **диоды с противоположным включением** opposed diodes; **д. в цепи заряда** charging diode; **вольтодобавочный д.** booster diode, series-efficiency diode; **д.-восстановитель** restorer-diode; **д. восстановления постоянной составляющей** D.C. clamp diode, D.C. restorer diode; **д.-гептод** diode heptode; **д.-детектор**

Fleming tube; **запирающий д.** hold-off diode; **коммутирующий д.** gating diode; **кремниевый д. с вплавленным контактом** silicon alloy diode; **кремниевый полупроводниковый д.** silicon transistor; **кремниевый сплавной д.** silicon junction diode; **д., образованный сеткой-катодом лампы** grid-cathode diode; **обращённый д.** backward diode, tunnel rectifier; **д. ограничения, ограничивающий д.** limiter diode, bootstrap diode, clamping diode; **д.-ограничитель** diode clipper; **опорный д.** voltage-reference diode, Zener diode; **отключающий д.** isolating diode; **отсекающий д.** pick-off diode; **д. параметрического усилителя** parametric diode; **плоскостной д.** junction-type diode, drawn junction diode, p-n junction diode; **д., « поддерживающий » напряжение модулятора** hold-off diode; **д. подстройки** diode for frequency-control; **последовательный д.** series diode; **д. « привязки », привязывающий д.** D.C. clamp diode, D.C. restorer diode, clamping diode, catching diode; **рентгеновский д.** hot-cathode X-ray diode; **д. с золотой связкой, д. с золотым контактом** gold-bonded diode; **д. с малым временем восстановления** fast recovery diode; **д. с переменным сопротивлением** leak diode; **д. с плоскостным переходом** junction diode, planar diode; **д. с реактивным сопротивлением** reactance diode; **связующий д.** diode coupler; **д. со сверхмалым временем восстановления** ultrafast recovery diode; **д. со скачкообразным восстановлением проводимости** step recovery diode; **д. схемы восстановления постоянной составляющей, д. схемы фиксации уровня сигнала** D.C. restorer diode, D.C. clamp diode; **д. схемы центрирования** centering diode; **счётный д.** computer diode; **точечный д.** point-contact diode; **д. фиксирующей схемы, фиксирующий д.** clamping diode, catching diode, bootstrap diode, limiter diode, D.C. restorer diode; **экономический д.** booster diode

**диодно-ёмкостный** *adj.* diode-capacitor

**диодно-конденсаторный** *adj.* condenser-diode, diode-capacitor
**диодно-резисторно-конденсаторный** *adj.* capacitor-resistor-diode
**диодно-транзисторный** *adj.* diode-transistor
**диодный** *adj.* diode
**диод-пентод** *m.* diode-pentode; **двойной д.-п. с переменной крутизной характеристики** double diode variable-mu pentode
**диоптр** *m.* diopter, sight vane, alidade
**диоптрика** *f.* dioptrics
**диоптрический** *adj.* dioptric
**диоптрия** *f.* diopter
**диоптрометр** *m.* dioptrometer
**диорит** *m.* diorite
**диотрон** *m.* diotron, dyotron
**диофантов** *poss. adj.* diophantine
**диплекс** *m.* diplex
**диплексер** *m.* diplexer
**диплексный** *adj.* diplex, diplexer
**дипло-** *prefix* diplo-
**диплоэдр** *m.* diploid, disdodecahedron, dyakisdodecahedron
**диполь** *m.* dipole, dipole antenna, doublet; **курковые диполи** snapping dipoles; **соосные диполи** collinear dipoles; **горизонтальный д., питаемый однопроводным фидером** Windom antenna; **д. из двух полуволновых вибраторов** full-wave dipole; **д. из двух цилиндрических вибраторов, колбасообразный д.** cage dipole; **крестообразный д.** turnstile dipole; **д. Надененко** (wide-) band horizontal dipole, cage dipole; **направляющий д.** director (dipole); **распорный д.** V-dipole; **треугольный д. с сопротивлением, включённым посредине** delta-type antenna; **эксцентрично расположённый д., эксцентричный д.** off-center dipole
**дипольный** *adj.* dipole, dipolar
**диполюс** *m.* dipole
**диполярный** *adj.* dipolar
**Дирак** Dirac
**директива** *f.* instruction, directive
**директивный** *adj.* directive, instruction, directional
**директор** *m.* director, sender (*ant.*); computer, alotter; director, manager; **«д.-система»** director system, translator system
**директорный** *adj.* director
**директорский** *adj.* master, chief

**директриса** *f.* directrix
**Дирихле** Dirichlet
**дисбаланс** *m.* unbalance
**дисдодекаэдр** *m.* disdodecahedron
**диск** *m.* disk, plate; turntable; slice, cake, washer; dial; dot; **диски у антенного изолятора, диски у первого изолятора антенны** (*or* **ввода**) antispark disks; **вращающийся д. с прорезами** episcotister; **двухкулачковый д.** two-lobe cam; **д. для модуляции света** interrupter disk; **д. для перекрытия** cover-disk; **д. для получения цифрового кода** digital code wheel; **д. для развёртывания изображения** exploring disk (*tv.*); **дроссельный д.** disk of throttle valve; **заведённый наборный д.** finger-plate off normal; **заводный д.** dial plate; **криволинейный д.** cam plate, cup shaped cam disk; **наборный д.** dial, finger plate; **номерной д.** dial, telephone dial; number plate; **номерной падающий д.** drop indicator disk; **д. проигрывателя** turntable; **распорный д.** spacer, separator piece, spacing piece; **д. с многократной спиралью** multi-spiral scanning disk; **д. с цветными (свето-) фильтрами** color disk, color filter disk; **д. ступицы** nave plate; **д. сухого выпрямителя** rectifier disk; **счётный д.** circular slide rule; **тормозной д. для испытания торможения** pressure plate for breaking tests; **торцевой д.** end disk; **успокоительный д.** drag disk, retarding disk; **установочный д. десятков** feeder tens dial; **установочный д. единиц** feeder units dial; **установочный д. сотен** feeder hundreds dial; **цветовой секторный д.** Abney color sensitometer; **д. шкалы** dial, graduated dial
**дискант** *m.* treble, descant voice
**дискантовый** *adj.* treble
**дискование** *n.* dialing
**дисковый** *adj.* disk, plate, pan-shaped
**дисконический, дисконусный** *adj.* discone
**дискообразный** *adj.* disk-shaped, disk-like, discoidal
**дискрета** *f.* sample
**дискретизация** *f.* digitization
**дискретизированный** *adj.* sampled
**дискретно-временной** *adj.* time-digital

**дискретно-непрерывный** *adj.* sampled-analog

**дискретный** *adj.* discrete, distinct; digital

**дискриминант** *m.* discriminant

**дискриминантный** *adj.* discriminant

**дискриминатор** *m.* discriminator, discriminator tube; **амплитудный д.** amplitude-discriminator circuit; **двухтактный д. с расстроенным контуром** off-tune type discriminator; **д. импульсов по длительности** pulse width discriminator; **односкатный д., д. работающий на одном спаде резонансной кривой** single-ended discriminator; **д. с малой постоянной времени** short-time constant discriminator

**дискриминация** *f.* discrimination

**дислокация** *f.* dislocation, disturbance

**диспаратный** *adj.* disparate

**диспенсерный** *adj.* dispenser

**диспергирование** *n.* dispersion, scattering; standard deviation (in statistics); **д. вещества ультразвуком** ultrasonic material dispersion

**диспергированный** *adj.* dispersed

**диспергировать** *v.* to disperse, to diffuse, to scatter

**диспергирующий** *adj.* dispersive, dispersing

**дисперсионный** *adj.* dispersion, dispersing

**дисперсия** *f.* dispersion, scattering; standard deviation (in statistics); variance (*math.*); **д. ошибки** error variance; **д. пластмасс** dispersion mediums of synthetic materials; **д. режимов** dispersion of behavior

**дисперсность** *f.* dispersion, degree of dispersion, division

**дисперсный** *adj.* dispersed, disperse, dispersive

**дисперсоид** *m.* dispersoid

**дисперсоидный** *adj.* dispersoid

**диспетчер** *m.* dispatcher; phaser; **д. по управлению воздушным движением** air traffic operator

**диспетчеризация** *f.* traffic control; dispatching

**диспетчерская** *f. adj. decl.* control room

**диспетчерский** *adj.* dispatcher, dispatching, central switching

**диспетчирование** *n.* dispatching, dispatcher's control

**диспечер** *see* **диспетчер**

**диспрозий** *m.* dysprosium, Dy

**диссектор** *m.* dissector tube, image dissector tube; **д. с полупрозрачным фотокатодом** image dissector; **д. с полупрозрачным фотокатодом и вторично-электронным умножителем** multiplier-type image dissector

**диссекторный** *adj.* dissector

**диссиметричный** *adj.* asymmetric, unsymmetrical

**диссипативный** *adj.* dissipative

**диссипация** *f.* dissipation, dispersion, diffusion

**диссонанс** *m.* dissonance, discord

**диссонансный** *adj.* dissonance, discord

**диссоциативный** *adj.* dissociative

**диссоциация** *f.* dissociation

**дистанциометрирование** *n.* ranging

**дистанциометрический** *adj.* distance-measuring

**дистанционирующий** *adj.* spacing, spacer

**дистанционно** *adv.* remotely

**дистанционно-управляемый** *adj.* remote-controlled

**дистанционный** *adj.* distant, remote, tele-; long-distance, distance; remote-control; range; space, spacer; time(-delayed)

**дистанция** *f.* distance, range; interval; clearance; **текущая д.** sonar range

**дисторсия** *f.* distortion

**дисторсионный** *adj.* distortion

**дистрибутивный** *adj.* distributive, distributional

**дисфеноид** *m.* bisphenoid

**дитетрагональный** *adj.* ditetragonal

**дитетрод** *m.* duo-tetrode, double tetrode

**дитригон** *m.* ditrigon

**дитригональный** *adj.* ditrigonal

**дитрон** *m.* dytron

**диф-** *prefix* differential

**дифгенератор** *m.* differential generator; **д. сельсина** synchro differential generator

**диференциальный** *adj.* differential

**дифманометр** *m.* differential manometer

**дифманометрический** *adj.* differential-manometer

**дифмотор** *m.* differential motor

**дифонический** *adj.* diphonic

**дифония** *f.* diphonia

**дифрагированный** *adj.* diffracted

**дифрактометр** *m.* difractometer

дифракционный *adj.* diffraction

дифракция *f.* diffraction; **д. света в ультразвуковом волновом поле, д. света на ультразвуковых волнах в среде** ultrasonic light diffraction

дифреле *n. indecl.* differential relay, balanced relay

дифсельсин *m.* differential selsyn system

дифсхема *f.* differentiator

дифтонг *m.* diphthong

дифферентометр *m.* draught difference indicator

дифференциал *m.* differential; differential coefficient, derivative; differential gear assembly; **частные дифференциалы** partial differential-quotients (*math.*); **д. с коническими шестернями** bevel gear differential; **д. с червячной передачей** screw differential

дифференциально-компаундированный *adj.* differentially compound

дифференциально-разностный *adj.* differential-difference

дифференциальный *adj.* differential, variational

дифференциатор *m.* differentiator; **д. с обратной связью** feedback differentiator

дифференциация *f.* differentiation

дифференцирование *n.* differentiation; peaking; derivation; **неявное д., д. неявной функции** implicit differentiation; **д. по частям** partial differentiation

дифференцированный *adj.* differentiated

дифференцировать *v.* to differentiate, to distinguish

дифференцироваться *v.* to differentiate

дифференцируемость *f.* differentiability

дифференцируемый *adj.* differentiable

дифференцирующий *adj.* differentiating

дифформный *adj.* difform

дифрактометр *m.* diffractometer

дифракционный *adj.* diffraction

диффузанс *m.* diffusance

диффузиометр *m.* diffusiometer

диффузионно-базовый *adj.* diffused-base

диффузионно-сплавной *adj.* diffusion-alloy

диффузионно-эмиттерный *adj.* diffused-emitter

диффузионный *adj.* diffused, diffusion

диффузия *f.* diffusion; **д. в твёрдом теле** solid-state diffusion; **д. вихревых потоков** eddy diffusion; **д. несущей волны** carrier diffusion, carrier wave diffusion; **д. носителей тока** charge carrier diffusion

диффузно-отражающий *adj.* diffusely reflecting

диффузно-сплавной *adj.* alloy-diffuse

диффузность *f.* diffusivity

диффузный *adj.* diffuse, scattered; random

диффузометр *m.* diffusometer

диффузор *m.* diffuser, cone; **д. с криволинейной образующей** curvilinear cone; **свободно подвешенный д.** floating cone

диффузорный *adj.* diffuser, cone; diffusion

диффузодержатель *m.* diffuser holder, diffuser bracket

диффундировать *v.* to diffuse, to spread, to scatter

диффундируемый *adj.* diffusible

дихотомия *f.* dichotomy

дихроизм *m.* dichroism

дихроический *adj.* dichroic

дихроичность *f.* dichroism

дихроичный *adj.* dichroic

дихромазия *f.* dichromasy

дихроскоп *m.* dichroscope

дициандиамидовый *adj.* dicyandiamide

дициклический *adj.* dicyclic

диэлектрик *m.* dielectric, nonconductor; insulator, insulating material; **д. из твёрдой резины** solid-rubber dielectric; **многослойный д.** structural dielectric; **д. с (большими) потерями** lossy dielectric

диэлектриковый *adj.* dielectric

« диэлектрин » *m.* dielectrine

диэлектрит *m.* dielectrite

диэлектрический *adj.* dielectric, nonconducting

дк *abbr.* (дека-) deca-

дкг *abbr.* (декаграмм) decagram

ДКИМ *abbr.* (дифференциальная кодово-импульсная модуляция) differential pulse code modulation

дкл *abbr.* (декалитр) decaliter

дкм *abbr.* (декаметр) decameter

дл *abbr.* 1 (децилитр) deciliter; 2 (длина) length

длина *f.* length; stretch; width; longitude; run (of wire); **во всю длину** at full length, lengthwise, longitudinally; **двойной длины** double-length; **на единицу длины** linear; **д. вихревого тока** eddy length; **д. волны в волноводе** guide wavelength, wavelength in pipe; **д. волны дополнительного цвета** complementary wavelength; **д. волны напряжения селекции цвета** sampling wavelength (*tv*); **граничная д. волны** cutoff wavelength, minimum wavelength, quantum limit; **д. линейного участка шкалы** lineal scale length; **д. линии развёртки** sweep length (*tv*); **неуравновешенная д.** incomplete transposition section; **д. пологого участка характеристики** plateau length; **предельная д. волны** shortest usable wavelength, greatest wavelength, which is radiated in waveguides; **д. преобладающей волны** hue wavelength, dominant wavelength; **приведённая д.** corrected length; **д. пробега** range; **д. разрыва** gap between the contact members, length of break; **д. свободного пробега** (*or* **пути**) free path; **собственная д. волны** natural wavelength, unloaded wavelength; **средняя д. свободного пути** mean free path; **строительная д. (кабеля)** factory length (of a cable), shipping length, completed length; **д. усилительного участка** repeater section; **д. хода поршня** piston stroke; **эффективная д. линии** electrical length of a transmission line

**длиннобазисный** *adj.* long-base
**длинноволновый** *adj.* long-wave
**длинноклювый** *adj.* long-nosed
**длиннопробежный** *adj.* long-range
**длиннопроводный** *adj.* long-wire
**длиннотянутый** *adj.* long-drawn
**длиннофокусный** *adj.* long-focus, long-focal
**длинный** *adj.* long, lengthy
**длительно** *adv.* long, long time
**длительность** *f.* duration, length, time, period; width (of pulse); **д. возбуждённых колебаний** ring(ing) time; **д. времени эксплуатационной готовности** usable time; **д. гасящего импульса** blanking interval, blanking time; **д. действия** reacting duration; **д. заднего фронта импульса** pulse trailing edge, pulse-decay time;

**д. замкнутого состояния** make time; **д. зрительного восприятия** duration of vision; **д. кадрового гасящего импульса** frame blanking period, vertical blanking pulse period, vertical black-out period; **д. кадрового синхронизирующего импульса** vertical synchronizing pulse period; **д. квантовой посылки** quantization time; **д. неустановившегося режима** transitory period; **д. области перехода** width of transition steepness; **д. обратного хода** retrace interval, return interval, fly-back period; **д. обратного хода по кадрам** frame-return period, vertical fly-back period; **д. обратного хода по строкам** horizontal fly-back period, line-return period; **д. обратного хода развёртки** scan fly-back interval; **д. отбоя** clearing time; **д. отклонения** time delay, time lag; **относительная д. обратного хода** retrace ratio; **относительная д. обратного хода кадровой развёртки** vertical retrace ratio; **относительная д. обратного хода строчной развёртки** horizontal retrace ratio; **д. отпирающего импульса** gate length; **д. отсчёта** sampling period; **д. переднего фронта импульса** leading-edge pulse time; **д. переключения** switching time of a relay, operating time of a relay; picture feed interval; **д. площадки гасящего импульса перед синхроимпульсом** front-porch interval; **д. площадки гасящего импульса позади синхроимпульса** back-porch interval; **д. прямого хода** trace interval, video interval; **д. сигнала вспышки** color burst period, chrominance burst period; **д. срабатывания** reacting duration; **д. строб-импульса** gate width; **д. строчного гасящего импульса** horizontal black-out period, horizontal blanking pulse period; **усреднённая д.** integrated duration

**длительный** *adj.* prolonged, long, lasting, continuous, steady, sustained, long-term, long-range
**длиться, продлиться** *v.* to range; to last
**для** *prep. genit.* for, to; in phrases oft. transl. by English adj.: **генератор для нагрева** heating generator (*lit.* generator for heating)

**ДМ** *abbr.* 1 (**действительное место**) true position; 2 (**Дизельхорст-Мартин**) Dieselhorst-Martin, multiple-twin quad; 3 (**дельта-модуляция**) delta modulation

**дм** *abbr.* (**дециметр**) decimeter

**ДМВ** *abbr.* (**дециметровые волны**) decimetric waves

**ДМР** *abbr.* (**дифференциальное минимальное реле**) reverse-power relay

**ДМУ** *abbr.* (**дирекционный магнитный угол**) grid magnetic azimuth

**дн** *abbr.* (**дина**) dyne

**дневной** *adj.* diurnal, daytime, day

**днище** *n.* base, bottom; face-plate; **д. колбы** bulb face, face-plate (of an electron-beam tube)

**дно** *n.* bottom, ground; face, blank (of a cathode-ray tube); **штампованное плоское д.** pressed base seal

**до** *prep. genit.* to, up to, as far as; before; until, till; about

**добавить** *pf. of* **добавлять**

**добавка** *f.* addition, introduction; additional part, item; gadget, accessory; applique; instrument multiplier; additional agent; dopant; **смачивающая д.** wetting agent

**добавление** *n.* addition, adding; supplement; extension; insertion; **компенсационное д. примеси** doping compensation; **д. сигналов синхронизации и бланкирования к видеосигналу** signal insertion (*tv*)

**добавленный** *adj.* added, mixed (with)

**добавляемое** *n. decl.* addend (*math.*)

**добавлять, добавить** *v.* to add, to supplement, to admix, to fill up; to boost; to enclose, to attribute

**добавляющийся** *adj.* addend

**добавочный** *adj.* additional, supplementary, accessory, auxiliary, extra; booster, boosting; after(effect); extension; admixed, added; filling, filler

**добиваться, добиться** *v.* to aim (for), to strive for, to tackle

**доброкачественность** *f.* factor of merit, figure of merit; soundness, good quality

**добропорядочный** *adj.* well-behaved

**добротность** *f.* figure of merit, factor of merit, quality factor, Q-factor, energy factor; "goodness" (of a receiver); quality, grade, class; gain-bandwidth product; legibility (of

telegraph signals); **д. антенного контура** quality rating (of antenna); **д. нагруженного контура** Q loaded, loaded Q; **д. ненагруженного контура** Q unloaded, unloaded Q

**добропорядочный** *adj.* well-behaved

**добрый** *adj.* good

**добыча** *f.* yield, output; production; gain

**« доваль »** *m.* Dowal

**доведение** *n.* bringing (to); finishing (up); **д. до наименьшего значения** minimization (*math.*)

**доверие** *n.* confidence; reliability

**доверительный** *adj.* confidential, confidence, fiducial

**доводить, довести** *v.* to reach, to attain; to bring (to), to lead, to reduce; to finish

**доводка** *f.* lapping; grinding; finish, finishing, sizing; final adjustment, refinement, debugging; **д. усиления** spotting gain control

**доводочный** *adj.* development(al)

**довольно** *adv.* rather, fairly; enough, sufficiently

**Довэ** Dove

**догнать** *pf. of* **догонять**

**договор** *m.* agreement, contract; **д. на абонемент** subscriber's agreement (*tphny.*); **д. на аренду линии** circuit lease contract (*tphny.*)

**догонять, догнать** *v.* to gain (upon, on), to overtake, to catch up

**додар** *m.* dodar

**додекафония** *f.* dodecaphony, dodecaphonism

**додекаэдр** *m.* dodecahedron

**доделать** *pf. of* **доделывать**

**доделка** *f.* debugging

**доделывание** *n.* completion

**доделывать, доделать** *v.* to finish, to retouch, to complete

**додетекторный** *adj.* pre-detection

**дождемер** *m.* rain gage, udometer, ombrometer, hyetometer; **самопишущий д.** hyetograph, ombrograph

**дожденепроницаемый** *adj.* rain-tight

**дождь** *m.* rain; rain clutter, cascade, shower

**дожигание** *n.* afterburner

**доза** *f.* dose, portion, dosage; **биологическая д. излучения** RBE dose; **индивидуальная д.** personnel dose; **д. на выходе** exit dose; **д. половин-**

**ной выживаемости** median lethal dose

**дозаряд** *m.* milking (*elec.*)

**дозаряжающий** *adj.* milking (generator or buster)

**дозатор** *m.* dosimeter, dosing apparatus; batcher; timer; **д. времени** timing unit; **д. импульсов** pulsation timer; **часовой д.** switch timer

**дозваниваться, дозвониться** *v.* to reach by phone, to put through a call

**дозвуковой** *adj.* subsonic, presonic

**дозиметр** *m.* dosimeter, dose meter, dosage meter, radiacmeter; **карманный д. в форме автоматического пера** radiometer fountain pen dosimeter

**дозиметрист** *m.* radiological safety officer

**дозиметрический** *adj.* dosimetric

**дозиметрия** *f.* dosimetry, radiation control

**дозирование** *n.* dispensing, proportioning, dosing, dosage, dosage measurement; batching; monitoring

**дозированный** *adj.* measured out, proportioned

**дозировать** *v.* to dose, to measure out, to proportion

**дозировка** *see* **дозирование**

**дозировочный** *adj.* dosage, dosing; metering, proportioning

**дозирующий** *adj.* dispensing; proportioning, metering

**дозор** *m.* watch; patrol

**дозорный** *adj.* watch; patrol

**ДОИ** *abbr.* (**детектор огибающей импульсов**) pulse-envelope detector

**доказанность** *f.* validity

**доказательство** *n.* proof, evidence, demonstration, proving; **ветвящееся д., д. в форме дерева** tree form proof

**доказать** *pf. of* **доказывать**

**доказуемый** *adj.* demonstrable

**доказывать, доказать** *v.* to prove, to demonstrate, to show

**доквантовый** *adj.* pre-quantum

**документ** *m.* document; **д. с рукописными пометками** hand-marked document

**документалист** *m.* documentalist

**документалистика** *f.* documentation

**документальный** *adj.* documentary

**документация** *f.* documentation

**долбежный** *adj.* grooving, slotting

**долбить, долбнуть, продолбить** *v.* to chisel, to cut, to pick, to hollow

**долбяк** *m.* pestle; gear-wheel cutter

**долг.** *abbr.* (**долгота**) longitude

**долгий** *adj.* long, protracted

**долговечность** *f.* durability, lifetime, life (of machine), operating life, longevity; **номинальная д.** rated life; **д. при хранении** shelf life

**долговечный** *adj.* long-lasting, long-lived, durable, stable, solid, time-proof

**долговременный** *adj.* long-term, of long duration; durable, lasting; nonvolatile

**долгоиграющий** *adj.* long-playing

**долгопериодический** *adj.* long-period

**долгостойкий** *adj.* passivated

**долгострочный** *adj.* long-termed, long-range, lasting

**долгота** *f.* longitude; length, duration

**долготный** *adj.* longitude, longitudinal

**долевой** *adj.* per unit

**Долежалек** Dolezalek

**должностной** *adj.* official, functional

**доливать, долить** *v.* to fill up, to refill, to add (by pouring)

**долина** *f.* valley, trough; length (of wave)

**долиноорбитальный** *adj.* valley-orbit

**долить** *pf. of* **доливать**

**«долли»** *n. indecl.* dolly (*tv*)

**дологический** *adj.* prelogical

**долото** *n.* chisel

**доля** *f.* part, portion, share, allotment; segment; lot; particle, fraction; **выборочная д.** sampling fraction; **наибольшая допустимая д. дефектности в партии** lot tolerance fraction defective

**домашний** *adj.* domestic, home-made, home

**домен** *m.* domain

**доменный** *adj.* domain

**доменообразование** *n.* nucleation

**доминанта** *f.* dominant

**доминантный** *adj.* dominant

**доминирование** *n.* dominance

**доминировать** *v.* to dominate

**доминируемый** *adj.* dominated

**доминирующий** *adj.* dominating, dominant

**домкрат** *m.* jack, winch, hoist

**домовый** *adj.* house

**донатор** *m.* donor, donator

**донаторный** *adj.* donor, donator

**Дондерс** Donders

**донесение** *n.* message, dispatch, report;

циркулярное д. multiple-call message

**донный** *adj.* bottom, ground; base (-mounted)

**донор** *m.* donor, donator

**донорный** *adj.* donor, donator, donor-type

**донышко** *n.* head (of tube or lamp); bottom

**дополнение** *n.* complement, supplement, addition, supplementation; additive inverse (*math.*); **дополнения до девяти** nine's complements (*math.*); **д. до двух** two's complements; **поразрядное д.** base minus one complement, diminished radix complement; **точное д.** nought's complement, radix complement, ten complement

**дополненный** *adj.* supplemented, augmented, added, complemented

**дополнитель** *m.* complementer; **д. в счётных машинах** complementer in counting devices

**дополнительно** *adv.* in addition

**дополнительность** *f.* complementarity

**дополнительный** *adj.* additive, additional, supplementary, extra, spare, subsidiary, alternate, auxiliary; admixed, added; complementary; side (effect)

**дополнять, дополнить** *v.* to supplement, to add, to complement, to fill up, to complete, to make up, to bring up to date

**дополняющий** *adj.* complementing, complementary, supplementing

**Допплер** Doppler

**допплеровский** *adj.* Doppler

**допробойный** *adj.* preconduction

**допроизводственный** *adj.* pre-production

**допуск** *m.* tolerance, admittance, admission; clearance limit; permissible variation; allowance; **допуски экспозиции** range of exposure; **д. в верхней части** ceiling margin; **д. в нижней части** floor margin; **жёсткий д.** close tolerance, severe tolerance; **д. на впадину** valley tolerance; **д. на нелинейное искажение** nonlinear distortion tolerance, harmonic tolerance; **д. на перекос** tilt tolerance; **д. на провал** valley tolerance; **д. по частоте** frequency tolerance; **строгий д.** close tolerance; **д. усиления с**

точки зрения зуммирования singing margin, margin of stability

**допускаемость** *f.* admissibility

**допускаемый** *adj.* admissible, permissible

**допускать, допустить** *v.* to allow, to permit, to tolerate; to admit, to receive, to accept

**допускающий** *adj.* permitting; accessible

**допустимость** *f.* admissibility

**допустимый** *adj.* admissible, permissible, admissive, acceptable, allowable, safe; tolerable

**допустить** *pf. of* **допускать**

**допущение** *n.* tolerance, allowance, permission; admission; assumption

**допущенный** *adj.* tolerated, allowed, permitted; admitted

**Доргело** Dorgelo

**дорелятивистский** *adj.* pre-relativity

**дорн** *m.* pin, mandrel, spindle

**дорога** *f.* road, way, passage, path

**дорожка** *f.* track, path, trail; groove; **д. автоматического управления** servo-control track; **д. ведущей информации** feed track; **д. для посадки по приборам** instrument landing runway; **д. записи изображения** video track; **д. записи положения** sprocket channel; **односторонняя звуковая д.** unilateral-area sound track

**Досааф** *abbr.* (**Добровольное общество содействия армии, авиации и флоту**) the Voluntary Society for Aiding the Army, the Air Force, and the Navy

**доска** *f.* plank, board; panel; slab, plate; **волочильная д.** drawing die, wire plate, draw knife; **д. для временных схем на штепселях** patch board; **д. для измерительных приборов** gage board; **калиберная д.** wire gage; **ключевая д.** keyboard (*tphny.*); **коммутационная д.** patch board, peg board, plug board, problem-board; switchboard, switchboard panel; **коммутационная д. с каркасом** frame-type switchboard; **д. лабораторной схемы** breadboard; **операционная д. с каркасом** skeleton-type switchboard; **д. основных данных** status board; **отражательная д. (динамика)** baffle (of a dynamic speaker); **планшетная д.**

plot table, plotting table; **приборная д.** cockpit panel, instrument panel, panel board, gage board, dashboard; **прогарная д.** baffle, baffle plate; **д. с планом путей** track indicator chart; **д. состояния** status board; **схемная д.** patch board

**досрочно** *adv.* ahead of schedule

**доставить** *pf. of* **доставлять**

**доставка** *f.* supply, procuring; transportation, conveyance; delivery

**доставлять, доставить** *v.* to supply; to deliver, to forward, to transmit, to relay over, to procure; to bring, to give, to yield

**доставщик** *m.* messenger, delivery man, supplier

**достаточно** *adv.* sufficiently, satisfactorily, adequately

**достаточность** *f.* adequacy, sufficiency

**достаточный** *adj.* adequate, sufficient, satisfactory

**достигаемый** *adj.* accessible, attainable, approachable, practicable

**достижение** *n.* achievement, attainment, advance, improvement; **д. максимума** maximizing

**достижимый** *see* **достигаемый**

**достоверность** *f.* authenticity, validity, truth; confidence level

**достоверный** *adj.* authentic, reliable, certain, proved, trustworthy; significant

**достоинство** *n.* merit, quality, virtue

**доступ** *m.* access, entrance, inlet, admission, approach, passage

**доступно** *adv.* accessibly, easily, simply

**доступность** *f.* accessibility, access

**доступный** *adj.* accessible, within reach, practicable, approachable; available; addressable

**досягаемость** *f.* accessibility, attainability, reach, range

**досягаемый** *adj.* accessible, attainable, within reach

**дотягивание** *n.* inching, jogging

**Доу** Dow

**Дохерти** Doherty

**дощечка** *f.* small plank, small board; plate, slab; **д. с присоединительными зажимами** connecting block

**ДПВРД** *abbr.* (**дозвуковой прямоточный воздушно-реактивный двигатель**) subsonic ramjet engine

**ДПК** *abbr.* (**дистанционный пневматический компас**) pneumatic tele-compass

**ДПРМ** *abbr.* (**дальний приводной радиомаяк**) outer marker beacon

**ДПС** *abbr.* (**двухперекидная схема**) double flip-flop circuit

**драгоценный** *adj.* noble, precious, valuable

**« драйвер »** *m.* driver, driver stage

**драпировка** *f.* drapery

**дребезжание** *n.* jarring, rattling, racking, chatter, trembling, flutter, blasting; distortion, speaker distortion; **д. звука** flutter effect

**дребезжащий** *adj.* trembling, vibrating

**древесина** *f.* wood, timber, lumber; **слоистая прессованная д.** pressed plywood

**древесноволокнистый** *adj.* cellulose-fiber

**древесный** *adj.* woody, wood; wood-pulp; fibrous

**древко** *n.* shaft, pikestaff **д. смычка** bow, fiddlestick

**древовидный** *adj.* tree, tree-shaped, tree-like; woody

**Дрейпер** Draper

**дрейф** *m.* drift, drifting; **д. дырок** hole drift; **д. нуля** zero drift; **д. нуля за счёт накала** heater drift; **поперечный д.** cross drift

**дрейфовый** *adj.* drift

**дрейфующий** *adj.* drifting

**дрель** *f.* drill, drill bore

**дренаж** *m.* drain, drainage

**дренажировать** *v.* to drain, to drain off

**дренажный** *adj.* drainage, drain, bleeder

**дренирование** *n.* draining, drainage

**дренированный** *adj.* drained

**Дрепер** Draper

**дрифтанс** *m.* driftance

**дробить, раздробить** *v.* to split, to divide; to crush, to grind, to granulate

**дробление** *n.* granulation, breaking, crushing, grinding; splitting

**дробноатомный** *adj.* subatomic

**дробность** *f.* divisibility

**дробношаговый** *adj.* fractional-pitch, short-pitch, short-chord

**дробный** *adj.* fractional; divided, broken; split

**дробовой** *adj.* shot

**дробь** *f.* fraction (*math.*); **дроби в разных системах счисления** system fractions (*math.*); **несократимая д.** simplified fraction; **обратная д.** reciprocal (*math.*); **элементарная д.** partial fraction

дрогнуть *pf. of* дрожать

дрожание *n.* shake, shaking, vibration, shivering, tremor, jar, jarring, bounce, bouncing, jitter, trembling; flutter, flickering; bobbing; **д. от кадра к кадру** frame-to-frame jitter; **д. отметки** bobbing; **д. отметки дальности вдоль линии развёртки** time jitter; **д. развёртки** time-base flutter; **д. стрелки** flicker of a pointer; **д. строк** interline flicker, line flicker, line bounce

дрожать, дрогнуть *v.* to shake, to vibrate, to shiver, to quaver, to tremble, to bounce, to jar, to chatter; to flicker

дрожащий *adj.* vibrating, shivering, trembling

дрожжевой *adj.* yeast

дрозометр *m.* drosometer

дромометр *m.* dromometer

дросселирование *n.* installation of chokes; throttling, choking; **д. линии** line filtering

дросселировать *v.* to equip with chokes; to block by means of chokes, to throttle; to stop; to baffle

дросселирующий *adj.* throttling

дроссель *m.* choke, throttle, throttle governor; choke coil, reactive coil, reactor, impedance coil, inductance coil; **д. без стального сердечника** air-core choke; **бифилярный д.** bifilar inductor; **выпускной д.** bleed choke; **выходной д.** output choke, output inductance coil, output impedance coil; **д. для отфильтровки свистов** whistle box; **д. для сглаживания пульсации** ripple-filter choke; **д. для устранения помех** anti-interference impedance, noise-suppressor choke; **звуковой д.** tone-control choke; **корректирующий д.** anti-resonant coil; **манипуляторный д.** key-click choke; **д. насыщения** direct-current controllable reactor, direct-current presaturated reactor, direct-current saturable reactor; **ограничивающий д.** current limiting reactor; **отсасывающий д.** coil of a series resonant circuit; compensating choke (of a polyphase rectifier); **д. переменной индуктивности** swinging choke; **питающий д.** feed coil, feed retardation coil; **путевой д.** impedance bond; **разделительный д.** retardation coil, retarder; **регулируемый д.** variable inductor; **д. с галетной намоткой** pie-wound choke; **д. с переменной индуктивностью** swinging choke; **сглаживающий д.** smoothing coil, smoothing choke, filter choke, ripple-filter choke, hum filter; **сглаживающий д. с ответвлением в средней точке** interphase reactor; **сетевой д.** power choke, suppressor choke; **стыковой д.** impedance bond; **ударный д.** ringing choke; **управляемый д. с железным сердечником** controlled magnetic core reactor; **усилительный д.** saturable reactor; **д. фильтра источника питания** power-supply filter reactor

дроссельно-ёмкостный *adj.* choke-capacitance, impedance-capacitance

дроссельно-конденсаторный *adj.* choke-condenser

дроссельно-фланцевый *adj.* choke-flange

дроссельный *adj.* throttling, throttle, choke, choking; choke coil

другой *adj.* other, another, different

друза *f.* druse, node, nodule; **друзы в кристаллах** cluster crystals

Друммондов *poss. adj.* Drummond

дряхление *n.* aging

дряхлость *f.* aging

ДТРД *abbr.* (двухконтурный турбо-реактивный двигатель) turbofane engine, ducted-fan engine

дуализм *m.* dualism

дуальность *f.* duality

дуальный *adj.* dual

Дуан Duane

дуант *m.* dee, duant

дублёр *m.* doubler

дублет *m.* doublet, dipole; duplicate

дублетный *adj.* doublet

дублирование *n.* duplicating, duplication, doubling

дублированный *adj.* doubled

дублировать *v.* double, duplicate; fold

дублировочный *adj.* duplicating; doubling

дублирующий *adj.* duplicate, duplicated

дубль-дублет *m.* double doublet

дубляж *m.* dubbing

Дубровин Dubrovin

дуга *f.* arc, arch, bow, curve; gap; rib; bow collector, trolley bow; **д. высокополярного излучения** high field emission arc; **говорящая световая д., звучащая д.** sounding arc; **обратная д.** arc-back; **д. Петрова** electric arc, voltaic arc; **д. при размыкании** break arc; **ртутная д. с жидкостным катодом** mercury pool arc; **д. с термионным катодом** thermionic arc; **световая д.** arc, voltaic arc; **сквозная д.** arc-through

**дуговой** *adj.* arc, arch, arched, curve, curved

**дугогаситель** *m.* arc extinguisher, arc arrester, arc shield, (magnetic) blow-out

**дугогасительный, дугогасящий** *adj.* arc-extinguishing, arc-suppressing, arc-damping

**дугообразный** *adj.* quasi-arc, arched, curved, crooked, bow-shaped, hooked

**дугоотводящий** *adj.* arc-removing; arcing

**дугостойкий** *adj.* arc-resistant, non-arcing

**дугостойкость** *f.* arc resistance

**дугоустойчивый** *adj.* arc-resistant

**дудка** *f.* pipe

**дужка** *f.* handle, bar; arch; shackle; **д. для короткого замыкания** U-link (*tphny.*); **падающая д.** chopper bar, clutch mechanism; **перекрёстная д.** cross-over bend

**дуктилометр** *m.* ductilimeter, ductilometer

**дуодекальный** *adj.* duodecal

**дуодецима** *f.* duodecimo-tone

**дуодинатрон** *m.* duodynatron

**дуоль** *f.* duplet, couplet (*mus.*)

**дуплекс** *m.* twinplex (system)

**дуо-схема** *f.* lead-lag circuit

**дуплекс** *m.* duplex, twin; duplex telegraphy, duplex working; duplexing, two-way operation, full-duplex operation; **д.-манометр** double-tube pressure gage; **обыкновенный д.** combination duplex; **д. с дифференциальной схемой** differential duplex system; **д. с последовательным занятием провода** half-duplex operation; **ступенчатый д.** echelon duplexing

**дуплекс-диплексный** *adj.* duplex-diplex

**дуплексер** *m.* duplexer, duplexing circuit

**дуплексный** *adj.* duplex, double; two-way

**дупликатор** *m.* duplicator; electric charge duplicating device

**дупло** *n.* hollow, cavity

**дуралюминий** *m.* dural, duralumin

**«дурафен»** *m.* Durophen

**дуроплазматронный** *adj.* duroplasmatron

**«дуропласты»** *pl.* thermosetting plastics

**дутьё** *n.* blowing, blow-out, blast, draft; **с поперечным дутьём** cross-blast; **магнитное д.** magnetic blowout

**«духи»** *pl.* "ghost" images; ghost lines

**духовой** *adj.* wind, air

**ДЦВ** *abbr.* (**дециметровые волны**) decimetric waves

**дч** *abbr.* (**действительная часть**) $\Sigma$ real part (of complex numbers)

**дым** *m.* smoke, fume

**дымка** *f.* haze, mist, fog

**дымовой** *adj.* smoke

**дымообнаружитель** *m.* smoke detector

**дымостойкий, дымоупорный** *adj.* fume-resistant

**дымчатый** *adj.* smoked, smoky

**дыра** *f.* hole, aperture, gap, perforation; (electron) vacancy

**дырка** *f.* hole, perforation, aperture; vacant electron hole

**дырко-промежуточный** *adj.* vacancy-interstitial

**дыркол** *m.* perforator, punch, puncher

**дыромер** *m.* internal caliper gage, hole gage

**дыропробивной** *adj.* punching, perforating, piercing

**дырочный** *adj.* hole; **с дырочной проводимостью** acceptor-type

**дырчатый** *adj.* perforated

**дыхание** *n.* breathing, respiration

**дыхательный** *adj.* breathing, respiratory

**Дьюар** Dewar

**ДЭМУП** *abbr.* (**дифференциальный электромашинный усилитель с подмагничиванием**) differential-motor amplifier with premagnetization

**ДЭС** *abbr.* (**дуговая электросварка**) electric arc welding

**Дэшман** Dushman

**дюбель** *m.* dowel (pin), plug, peg

**дюза** *f.* nozzle, spout

**дюйм** *m.* inch

**дюймовый** *adj.* inch

**Дюлонг** Dulong

**дюралевый** *adj.* dural
**дюраль** *m.*, **дюралюмин** *m.*, **дюралю-миний** *m.* dural, hard-aluminum
**дюран** *m.* Duran
**дюранойд** *m.* duranoid

**дюратрон** *m.* duratron
**дюрометр** *m.* durometer
**Дюфур** Dufour
**Дюхамель** Duhamel
**Дюэн** Duane

# E/Ё

Note: technically **e** and **ё** are different letters, but **ё** is so often printed in the form **e** that the foreign reader cannot distinguish between the two. For this reason **e** and **ё** are treated as one letter in this dictionary.

**e.** *abbr.* (**единица**) unit

**Е** *abbr.* (**ёмкость**) capacitance; volume, capacity

**е.в.** *abbr.* (**единица веса**) weight unit

**Евклидов** *adj.* Euclidean

**евколлоиды** *pl.* eucolloids

**евр.** *abbr.* (**европейский**) European

**е. вр.** *abbr.* (**единица времени**) time unit

**европейский** *adj.* European

**Егер** Jager

**его** 1. *genit. and acc. of* **он, оно** him, it; 2. *poss. adj.* his, its, of it

**ед.** *abbr.* 1 (**единица**) unit. 2 (**единица допуска**) tolerance unit. 3 (**единственный**) singular, single, only, unique

**ЕД.** *abbr.* (**единица действия**) active unit

**едва** *adv.* hardly, only, just, scarcely, barely, narrowly; scant

**е. дл.** *abbr.* (**единица длины**) length unit

**ед. изм.** *abbr.* (**единица измерения**) unit of measurement

**единение** *n.* unity, accord; union, unification, uniting

**единительный** *adj.* uniting, unitive

**единить** *v.* to unite, to unify

**единица** *f.* unit; one; unity; digit; **абсолютные электрические единицы** rationalized M.K.S. units; **английские единицы торгового веса** avoirdupois; **основные единицы измерения** primary standards; **единицы системы меткилограмм секунда** Georgi units; **в единицах** in units (of), in terms (of); **е. в двоичной системе счисления** binary unit; **е. Виоля** violle (unit); **графическая е. полного импеданса** (*or* **сопротивления**) chart unit of impedance; **е. громкости** volume unit; **е. давления** bar (unit of pressure); **(двоичная) е. информации** information bit, binit; **е. действия** unit of operation; **десятичная е. информации** decit, decimal digit; **е. допусков** tolerance unit; **е. запоминаемой информации** bit; **е. затухания, равная 1/10 непера** hyp; **е. звукопоглощения, равная 1 кв. футу** square-foot unit of absorption; **е. излучения** emissivity unit; **е. измерения ощущения** liminal unit (*tv*); **е. индуктивности** secohm; **е. информации** field, item, bit of information, Hartley; **е. (информации) в двоичной системе счисления** binary unit (of information); **е. количества сведений** information (quantity) unit; **е. количества электричества** unit electrostatic charge; **е. контроля по чётности** parity check unit; **логарифмическая е. затухания** bel; **е. магнитного потока** megaline; **е. магнитного потока в $10^8$ максвелов** volt-line; **е. магнитного сопротивления** oersted; **е. массы** slug; **е. метрической информации** metron; **е. начала зоны** start-of-block unit; **неосновная е.** derived unit; **е. облучения** rad unit; **е. освещённости** foot-candle; **е. переговора** message unit, unit call; **е. перекрёстной наводки**, **е. переходного разговора** crosstalk unit; **е. площади эмитирующей поверхности** unit emitting area; **приводимая е.** operated unit; **е. производительности** unit of power; **е. работы, принятая Британским министерством торговли** Board of Trade Unit; **е. равная 0,001 оборота земного шара в сутки** milli-earth rate unit; **е. разговора** unit call; **световая е. расстояния** electrical distance; **е. СГСМ (равная 10 ампер)** abampere; **е. СГСМ (равная $10^{-9}$ генри)** abhenry; **е. СГСМ (равная $10^9$ фарад)** abfarad; **е. силы света** candel; **е. силы тока**

154

chemic; **е. силы тяжести** unit of gravity; **е. скорости ионов** ionic mobility, migration speed of an ion; **е. скорости сигнализации** (*or* **передачи**) baud; **е. содержания структурной информации** logon; **е. сопротивления Сименса** Siemens, mercury unit; **е. счёта** tally; **тактовая е.** clock bit; **тарифная е.** charge unit (*tphny*.); **е. телесного угла** unit of solid angle; **е. теплоты** therm; **е. угловой скорости вращения земли** earth-rate unit; **управляемая е.** operated unit; **е. учёта (трафика)** call unit; **централизационная е.** interlocking unit; **е. электрического сопротивления** mil-foot; **е. электрической энергии (Британского департамента)** Board of Trade Unit

**единичность** *f.* singleness

**единичный** *adj.* unit, unitary, one, single, only

**едино-** *prefix* uni-, mono-

**единовременно** *adv.* once only

**единовременный** *adj.* only once, once, one-time; non-recurring

**единообразие** *n.* uniformity

**единочный** *adj.* single; universal

**единственно** *adv.* only; solely

**единственность** *f.* singularity, singleness, oneness

**единственный** *adj.* singular, single, sole, unique

**единство** *n.* harmony, unity

**единый** *adj.* unified, united, uniform, common; single, sole; integral

**едкий** *adj.* corrosive, caustic

**едкость** *f.* corrosiveness, causticity

**её** 1. *genit. and acc. of* **она** her. 2. *poss. adj.* her, hers; its, of it

**ежег.** *abbr.* (**ежегодный**) yearly, annual

**ежегодно** *adv.* yearly, annually, per year

**ежегодный** *adj.* yearly, annual

**ежедн.** *abbr.* (**ежедневный**) daily

**ежедневно** *adv.* daily, per diem

**ежедневный** *adj.* daily, diurnal

**ежели** *conj.* if; in case, as soon as

**ежемесячно** *adv.* monthly, every month

**ежемесячный** *adj.* monthly, every month

**ежеминутно** *adv.* continually; every minute

**ежеминутный** *adj.* continual; occurring every minute

**ежечасно** *adv.* hourly

**ежечасный** *adj.* hourly

**ежовый** *adj.* hedgehog

**ёж-трансформатор** *m.* hedgehog transformer

**ездить** *v.* to drive, to ride

**ей** *dat. and instr. of* **она** her, to her, with her; it, to it, with it

**еле** *adv.* hardly, scarcely; only just

**еле-еле** *adv.* hardly

**ёлочный** *adj.* fishbone (design), herringbone (pattern); Christmas-tree

**ёмк.** *abbr.* (**ёмкость**) capacitance; capacity, volume

**ёмкий** *adj.* capacious; large-capacity

**ёмкостно** *adv.* capacitatively

**ёмкостнолинейный** *adj.* linear capacity

**ёмкостно-резистивный** *adj.* capacitance-resistance

**ёмкостно-связанный** *adj.* capacitance coupled

**ёмкостный** *adj.* capacitive; capacity; capacitance(-type); condensive

**ёмкость** *f.* capacitance; capacity, volume, cubic content; vessel, tank; tankage; charge capacity, capacity constant of a line; size; capacitor; condensance; space, spacing; **ё. аккумулятора при его прерывчатом разряде** ignition capacity of accumulator; **ё. анодного колебательного контура** tank capacity; **ё. блока памяти** size of memory; **ё. в арифметических знаках** digit capacity; **взаимная обратная ё.** mutual stiffness; **ё., вносимая человеческим телом** body capacitance; **вредная ё.** stray capacitance; **ё. вывода** lead capacitance; **действующая ё. между электродом** $j$ **и катодом** effective capacitance between the $j$th electrode and cathode; **дифференциальная ё.** incremental capacitance; **добавочная ё.** series capacitor; **ё. за счёт неоднородности в коаксиальной линии** discontinuity capacitance; **задействованная ё.** equipped capacity (of an exchange), fitted capacity; **ё. запирающего слоя** barrier layer capacitance, junction capacitance, transition capacitance; **заряжающая ё.** storage capacity; **ё. земли** earth capacity, wire-to-earth capacity; **ё. колебательного контура** tank capacitance; **компенсирующая ё.** balancing capacitance;

ё. конденсатора с воздушным диэлектриком air capacitance; ё. контура сетки grid-circuit capacitance; ё. корпуса shield capacity, stray capacity between component parts and enclosure; ё. ламповой панели socket capacity; ё. линии line shunt capacity; ё. магнита magnetic volume; ё. между двумя проводниками, ё. между жилами mutual capacitance; ё. между жилами основной цепи wire-to-wire capacity; ё. между жилой и металлической оболочкой shunt capacitance; ё. между проводами pair-to-pair capacity (cables); ё. между проводами основной цепи wire-to-wire capacity; ё. мужду проводниками (фантомной цепи) lateral capacity, pair-to-pair capacity; междувитковая ё. обмотки winding capacity; междувитковая шунтирующая ё. turn-to-turn shunt capacity; междупроводная ё. wire-to-wire capacity; междуэлектродная ё. лампы tube capacity; ё. модулятор-катод grid-cathode capacitance; ё. монтажа stray capacitance of mounting, wiring capacitance; монтированная ё. станций fitted capacity; ё. на единицу площади анода unit-area capacitance (of an electrolytic capacitor); ё. на стороне потребления demand-side capacity; ё. накопительного устройства, ё. накопителя memory size, storage capacity; начальная ё. (переменного конденсатора) primary capacitance; ё. неоднородности (в коаксиальной линии) discontinuity capacity (of a coaxial line); незадействованная ё. (станции) unequipped capacity, marginal capacity (of an office); нейтродинная ё. neutralizing capacitance, ё. нити накала filament capacitance; номинальная ё. rated capacity; обратная ё. reciprocal capacity; stiffness; обратная ё. цепи branch stiffness; ё. обратной связи feedback capacitance; ё. оперативной памяти internal storage capacity; ё. оператора body capacitance; ё. относительно земли wire-to-earth capacity, earth capacity; body capacity; ё. относительно катода cathode capacitance; ё. относительно крыши top loading capacity;

паразитная ё. stray capacitance, spurious capacitance, shunt capacitance, strays; ё. первичной обмотки primary capacitance; ё. перехода barrier(-layer) capacitance, junction capacitance, transition capacitance; ё. петли связи loop capacitance; полная ё. с учётом влияния земли grounded capacitance; поразрядная ё. digit capacity; постоянная ё. конденсатора fixed capacity; предельная ё. end capacity; ё. при зарядке loading capacity; ё. при прерывчатом разряде ignition capacity; проходная ё. transfer capacitance, direct capacitance; рабочая ё. mutual capacitance; ё. рассеяния stray capacity; ё. регулируемого объекта capacity; ё. с учётом влияния земли grounded capacitance; свободная ё. idle capacity; ё. связи crosstalk coupling-capacity; ё. сетка-анод grid-plate capacitance; собственная ё. self-capacity, natural capacity, intrinsic capacitance; собственная обратная ё. self-stiffness; сосредоточенная ё. lumped capacitance; ё. станции central office capacity; статическая ё. (между двумя проводниками) direct capacitance; ё. счётчика counter capacity, register length; удельная ё. capacitivity, specific capacity; укорачивающая ё., включённая последовательно с антенной antenna series capacitor; ё. установки fitted capacity; установленная ё. (станции) installed capacity (of an office); ё. цепи развёртки charge-discharge condenser in sweep circuits; ё. цепи смещения biasing capacitance; цифровая ё. digit spacing; частичная ё. к земле direct capacitance to ground; шунтирующая ё. shunt capacity; shunting condenser; эквивалентная ё. пространства взаимодействия effective gap capacitance; эксплуатационная ё. service capacity; электрическая ё. permittance; ё. (электро-)монтажа wiring capacitance; ё. эмиттерного перехода emitter transition capacitance

**ему** *dat. of* **он, оно,** him, to him; it, to it

**е.п.** *abbr.* (**единица площади**) area unit

**ёрш** *m.* ragbolt, jag; jagged rod; broach; brush, wire brush; split-lug; **ё. для заделки в стену** split-lug, expansion anchor

**если** *conj.* if, in case, as long as; **е. бы** if; **е. бы не** if not for, but for; were it not for; **е. не** unless, if not, but for; **е. только** provided, if only; **е. только вообще** if at all; **е. только не** unless; **е. уже** if anything

**ест.** *abbr.* (**естественный**) natural

**естественно** *adv.* naturally, of course; it is natural

**естественность** *f.* (true) fidelity; naturalness

**естественный** *adj.* natural, innate, inherent, inborn, native; spontaneous

**е.т.** *abbr.* (**единица тепла**) thermal unit, caloric unit

**ЕУ** *abbr.* (**ежедневный уход**) daily servicing, daily maintenance (of machines), daily care (of a car)

**ехать, поехать** *v.* to drive, to ride, to go

**ещё** *adv.* still, yet, but, again, else, more, any more

**ЕЭС** *abbr.* (**единая энергетическая система**) unified power system, unified energy system

**ею** *instr. of* **она**; *see also* **ней, нею,** with her, by her; with it, by it

# Ж

ж. *abbr.* 1 (жидкий) liquid; fluid; 2 (жидкость) liquid; fluid; 3 (журнал) journal

Ж *abbr.* 1 (жирный) rich; fat; greasy; 2 (журнал) journal

ЖАД *abbr.* (жидкостный аккумулятор давления) liquid fuel pressure accumulator

жалоба *f.* complaint, grievance

жалобный *adj.* sad, mournful

жалюзи *n. indecl.* louver, jaws, Venetian blind, jalousie; grating; **ж. системы терморегулирования** heat-regulation shutters; **ж. терморегулирования** thermal control vanes

жалюзийный *adj.* louver, jalousie, Venetian blind; grating

Жамэн Jamin

жар *m.* glow, heat; **краснокалильный ж.** red heat

жаркий *adj.* hot

жарко *adv.* hot; it is hot

жаровня *f.* devil, fire devil, brazier; chafing dish

жаровой *adj.* heat, fire

жаропрочность *f.* resistance to heat, high-temperature strength

жаропрочный, жаростойкий *adj.* refractory; heat-resisting, heatproof, high-temperature, temperature-resistant

жаростойкость *f.* thermal stability

жароупорность *f.* heat-proof quality; heat resistance

жароупорный *adj.* heatproof, heat-resistant, heat-resisting, fire-resistant

ЖАХ *abbr.* (Журнал аналитической химии) Journal of Analytical Chemistry

ж.-б *abbr.* (железобетонный) ferro-concrete, reinforced concrete

ЖГГ *abbr.* (жидкостный газогенератор) liquid-gas generator

жгут *m.* plait, band, braid; gasket; **ж. из проводов** bunched conductors

жгучесть *f.* causticity, corrosiveness

жгучий *adj.* burning, caustic, corrosive

ж.д., ж/д *abbr.* (железная дорога) railway, railroad

ж.-д *abbr.* (железнодорожный) railway, railroad

ждущий *adj.* waiting; expecting; slave

жезл *m.* staff, rod

жезловой *adj.* staff, rod

жезлообмениватель *m.* staff catcher

желаемый *adj.* desired, wished

желание *n.* desire, wish

желаный *adj.* desired, wished (for)

желательно *adv.* it is desirable

желательность *f.* desirability

желательный *adj.* desirable, desired

желатин *m.* jelly, gelatin

желатинация *f.* gelatinization, gelation

желатинизирующий *adj.* gelatinizer

желатинирование *n.* gelatinization, gelation

желатинированный *adj.* gelatinized, gelated.

желатинировать *v.* to gelatinize, to gelate

желатинный, желатиновый *adj.* gelatin, gelatinous, jelly(-like)

желатинозный *adj.* gelatinous, jelly-like

желатинообразный, желатиноподобный *adj.* gelatinous, gelatiniform, gelatinoid

желать, пожелать *v.* to wish, to desire, to want

желающий *adj.* wishing, desiring, wanting

жел. бет. *abbr.* (железобетонный) ferro-concrete, reinforced concrete

железисто- *prefix* iron, ferrous, ferro-

железистость *f.* ferruginosity

железистый *adj.* iron, ferrous, ferrierous, ferruginous

железнение *n.* iron plating; **гальваническое ж.** iron plating

железно- *prefix* iron, ferric, ferro-

железнодорожный *adj.* railway, railroad

железный *adj.* iron, ferric, ferrous

железняк *m.* iron ore; **магнитный ж.** magnetite, $Fe_3O_4$

**железо-** *prefix* iron, ferro-; ferri-, ferric

**железо** *n.* iron, Fe; **ж. Армко** mu-metal; **карбонильное ж. со связкой** ferrotron; **квадратное ж.** square bar iron; **ковкое ж.** forging grade steel, wrought iron; **корытное ж.** channel iron; **легированное ж.** steel alloy, stalloy; **ленточное ж.** hoop iron; **листовое динамное ж.** dynamo (iron) sheets; **обручное ж.** hoop iron; **оцинкованное ж.** galvanized iron; **полосовое ж.** flat iron, hoop iron; **прутковое ж.** bar iron; **рифленое ж.** corrugated sheet iron; **уголковое монтажное ж.** mounting angle; **фасонное ж.** profile steel, section(al) steel; **швеллерное ж.** channel iron; **шипное ж.** hoop iron

**железобетон** *m.* ferroconcrete, reinforced concrete

**железобетонный** *adj.* ferroconcrete, reinforced concrete

**железо-водородный** *adj.* hydrogen-iron

**железомагнитный** *adj.* ferromagnetic

**железоникелевый** *adj.* iron-nickel, nickel-iron

**железоподобный** *adj.* iron-like, ferruginous

**железосодержащий** *adj.* iron-containing, ferriferous, ferruginous

**желеобразный** *adj.* gelatinous, jelly-like

**жёлоб** *m.* groove, channel, furrow, notch; canal, conduit, gutter, chute, trough, troughing; slot; tray; **батарейный ж.** battery chute; **ж. для гашения дуги** arc chute; **ж. для проводов** trunking; **ж. для пропускания монеты** coin collecting chute; **контактный ж. со щелью** slot conduit; **направляющий ж.** guide slot

**жёлобистый** *adj.* grooved, channeled

**желобить, выжелобить** *v.* to groove, to notch, to slot, to channel, to flute

**жёлобоватый** *adj.* grooved, channeled

**жёлобление** *n.* grooving, channeling, fluting

**жёлобоватый** *adj.* trough-shaped, U-shaped

**желобок** *m.* groove, slot; key groove; channel; ball race; **с желобком** grooved, fluted, slotted; **ж. на шейке** neck groove

**жёлобообразный** *adj.* trough-shaped, U-shaped

**желобчатый** *adj.* grooved, channeled, fluted, slotted; corrugated; hollow, concave

**жёлтый** *adj.* yellow

**желудёвый** *adj.* acorn

**жёлудь** *m.* acorn

**жемчуг** *m.* pearl; bead

**жемчужный** *adj.* pearl, pearly; beaded

**жердевой** *adj.* pole

**жердь** *f.* pole, perch, rod

**жеребьёвка** *f.* sorting, allotment; draw, toss-up

**жёсткий** *adj.* hard; rigid, stiff, inelastic, inflexible, tough, stable; severe; **акустически ж.** sound-hard

**жёстко** *adv.* hard, rigidly, stiffly

**жестковатый** *adj.* stiff, hard, somewhat hard

**жёсткость** *f.* hardness; stiffness, rigidity, inflexibility; severity; **ж. воды** calcium carbonate content of water; **ж. мембраны** membrane tension; **ж. на изгиб** flexural rigidity; **ж. на кручение** torsional rigidity; **ж. при изгибе** flexural stiffness, bending strength; **удельная электрическая ж.** elasticity; **ж. формы** inherent rigidity, natural rigidity; **цилиндрическая ж.** flexural rigidity; **электрическая ж.** elastance

**жёстче** *comp. of* **жёсткий, жёстко**

**жестчение** *n.* aging; stiffening, hardening; **ж. (газоразрядных приборов)** clean-up

**жестчить** *v.* to age; to harden, to stiffen

**жесть** *f.* tin, tin plate, sheet iron, sheet (metal); **белая ж.** tin plate; **ж. для кожухов** shell plate; **легированная ж.** dynamo sheet; **ж. сердечника** core lamination; **чёрная ж.** black plate: sheet iron

**жестянка** *f.* box, tin box, can

**жестяной** *adj.* tin

**жетаватор** *m.* jetavator

**жечь, сжечь** *v.* to burn; to corrode

**жжение** *n.* burning, firing, calcinating

**жжёный** *adj.* burned, fired, calcined; burnt

**живой** *adj.* live, living, alive, animate; vivid, rich

**живость** *f.* liveliness, animation; vividness; **ж. воспроизводимых цветов** vividness of hues

**животное** *n. adj. decl.* animal; **подопытное ж.** test animal

**животный** *adj.* animal

**животрепещущий** *adj.* actual; of vital importance

**живучесть** *f.* survival, probability of survival; viability, tenacity; **ж. цели** target vulnerability

**живучий** *adj.* hardy, of great vitality

**живущий** *adj.* living

**ЖИГ** *abbr.* (железо-иттриевый гранат) yttrium-iron garnet

**жидкий** *adj.* liquid, fluid, water; thin

**жидко** *adv.* in liquid form

**жидко-** *prefix* liquid

**жидкометаллический** *adj.* liquid-metal

**жидкоплавкий** *adj.* liquid, fluid

**жидкоплавкость** *f.* fluidity

**жидкостно-воздушный** *adj.* liquid-pneumatic, pneudraulic

**жидкостно-заполненный** *adj.* liquid-filled

**жидкостность** *f.* liquidity, fluidity

**жидкостный** *adj.* fluid; hydraulic; fluid-flow (pump)

**жидкость** *f.* liquid, fluid; liquidity, fluidity; **Ньютоновы вязкие жидкости** Newton liquids; **горючая ж.** flammable fluid; **отработанная ж.** discharge liquid; **охлаждающая ж.** cooling fluid, liquid coolant; **охлаждающая ж. при сверлении** cutting solution, diluted soluble oil; **ж. применяемая для модулятора системы Эйдофор** Eidophor liquid; **ж. с большой плотностью** dense liquid

**жидкотекучесть** *f.* fluidity; fluid flow

**жидкотекущий** *adj.* fluid; fluid-flow

**жидкофазный** *adj.* liquid-phase

**жизнедеятельность** *f.* activity, active life

**жизнедеятельный** *adj.* active, vital

**жизненность** *f.* vitality, life

**жизненный** *adj.* life's, vital

**жизнеспособность** *f.* vitality, viability

**жизнь** *f.* life, existence; **ж. за пределами земной атмосферы** extraterrestrial life

**жиклёр** *m.* jet, jet nozzle, jet discharge

**жила** *f.* vein; filament, core (of cable), wire, conductor; **внутренняя (проводящая) ж.** inner conductor, internal conductor; **заземляющая ж.** supplementary earth wire; **измерительная ж.** test wire, second core; **изолированная ж.** core wire; **испытательная ж.** test wire; **контрольная ж.** pilot wire; **разговорная ж.**

telephone line, speaker wire

**жилка** *f.* fiber, vein, nerve; wire

**жилой** *adj.* residential, inhabited; habitable

**жир** *m.* grease, fat, lubricant; oil; **густой ж.** cup grease

**жираторный, жирационный** *adj.* gyrating, gyratory, gyration

**жирный** *adj.* fat, oily, greasy; rich; heavy

**Жиро** Girot

**жиро-** 1. *see* **гиро**; 2. *prefix* fat, fatty, lipo-

**жирование** *n.* lubrication, greasing

**жировать** *v.* to lubricate, to grease, to oil

**жироклинометр** *m.* gyrolevel

**жиромер** *m.* butyrometer

**жирометр** *m.* gyrometer

**жиронепроницаемый** *adj.* greaseproof

**жироскоп** *m.* gyro, gyroscope, gyrostat; *see also* **гироскоп**

**жироскопический** *adj.* gyroscopic, gyro-

**жиростатика** *f.* gyrostatics

**жиростатический** *adj.* gyrostatic

**ЖК** *abbr.* (жирные кислоты) fatty acids, aliphatic acids

**ЖМГ** *abbr.* (жидкометаллическое горючее) liquid-metal fuel

**ЖНХ** *abbr.* (Журнал неорганической химии) Journal of Inorganic Chemistry

**жокей-реле** *n. indecl.* Jockey relay

**жолоб** *see* **жёлоб**

**жолудь** *see* **жёлудь**

**ЖОХ** *abbr.* (Журнал общей химии) Journal of General Chemistry

**ЖР** *abbr.* (железнодорожная радиостанция) railway radio station

**ЖРД** *abbr.* (жидкостный ракетный двигатель) liquid-propellant unit, liquid-propellant rocket engine

**жребий** *m.* draw, lot

**ЖРФО** *abbr.* (Журнал Русского физического общества (1873-1930)) Journal of the Russian Physical Society

**ЖРФХО** *abbr.* (Журнал Русского физико-химического общества (1879-1930)) Journal of the Russian Physical and Chemical Society

**ЖРХО** *abbr.* (Журнал Русского химического общества (1869-1930)) Journal of the Russian Chemical Society

**ж. тех. физ., ЖТФ** *abbr.* (Журнал Технической Физики) Journal of Technical Physics

Жубер Joubert

жужжание *n.* hum, humming, buzz, buzzing, droning

жужжать, прожужжать *v.* to hum, to buzz, to drone, to burr, to rattle

жужжащий *adj.* humming, buzzing, droning

жук (в разматываемом проводе) kink

жук-дровосек *m.* wood beetle capricorne

журавль *m.* lever, arm, sweep; microphone boom

журнал *m.* journal; log; **аппаратный ж.** message log, message registry; **бортовой ж.** log book; **вахтенный ж.** station log, duty log; **ж. горения радиоламп** tube operating log; **ж. записи работы навигационной радиостанции** aeronautical communications log; **ж. нарушений связи** irregularity log, trouble (record) log; **ж. состояния оборудования** equipment performance log; **ж. учёта работы радиоламп** tube operation log; **ж. учёта работы электросиловых агрегатов** power-supply equipment log; **ж. учёта радиосредств** radio communications record log

журчание *n.* murmur, purl, babbling; splash, splatter, spatter; **модуляционное ж.** modulation splatter

журчать, зажурчать *v.* to ripple, to purl, to murmur, to babble

**ж. физ. хим., ЖФХ** *abbr.* (Журнал физической химии) Journal of Physical Chemistry

**ЖФХО** *abbr.* (Журнал физико-химического общества) Journal of the Physical and Chemical Society

**ЖХО** *abbr.* (Журнал химического общества) Journal of the Chemical Society

**ЖЭТФ** *abbr.* (Журнал экспериментальной и теоретической физики) Journal of Experimental and Theoretical Physics

Жюрен Jurin

# З

**з.** *abbr.* 1 (**западный**) western; 2 (**зенитный**) zenithal; antiaircraft

**З** *abbr.* (**земля**) earth, ground, land

**за** *prep. acc. and instr.* beyond, behind, after, out of; as, for; during; per

**за-** *prefix* beyond, trans-; behind

**заатмосферный** *adj.* extra-atmospheric, beyond the atmosphere

**забег** *m.* overshoot, overshooting, overswing, overswinging

**забегающий** *adj.* leading

**забивание** *n.* jamming; stopping up; driving (in); choking

**забивать, забить** *v.* to jam; to swamp; to choke up, to fill in, to block up, to obstruct; **з. помехами экран индикатора** to swamp the indicator screen

**забираемый** *adj.* input: **з. от** drawn from

**забитый** *adj.* plugged up, stopped up; driven in

**забить** *pf. of* **забивать**

**заблаговременно** *adv.* early, in advance

**заблаговременный** *adj.* early, done in time

**заблокированный** *adj.* interlocked

**забой** *m.* driving in; **з. ошибок** erasure of errors

**заболевание** *n.* disease, illness, sickness

**забор** *m.* fence, enclosure; intake

**заборный** *adj.* intake; fence, enclosure; partition

**забракованный** *adj.* rejected

**забраковать** *pf. of* **браковать**

**забронированный** *adj.* iron-clad, armoured, shielded

**заброс** *m.* throwing; **максимальный з.** maximum overshot

**забрызгивание** *n.* sputtering

**завал** *m.* obstruction; **з. верхних частот** de-emphasis; **з. частотной характеристики** steep slope of a frequency response curve

**заваленный** *adj.* clogged up, filled, heaped up, loaded

**заваливать, завалить** *v.* to clog (up), to choke, to heap up, to fill, to block up, to load, to overload; to compress; to cover up

**завалка** *f.* charging, charge

**завальцованный** *adj.* rolled

**заваренный** *adj.* sealed, welded; **з. в стекло** glass-sealed

**заваривать, заварить** *v.* to seal, to weld

**заварка** *f.* welding, weld, sealing, seal; **з. в бусинку** beading; **з. в стекло** beading

**заварочно-откачной** *adj.* sealing-exhaust

**заварочный** *adj.* sealing, sealing-in

**заведение** *n.* winding; **з. провода (под контакт) петлей** looping-in (a wire)

**заведование** *n.* administration

**заведомо** *adv.* deliberately, certainly, with knowledge

**заведующий** *m. adj. decl.* administrator, manager, director

**завернуть** *pf. of* **завёртывать**

**завёртка** *f.* button, knob, latch; wrapping up

**завёртывать, завернуть** *v.* to wrap up, to cover, to envelope; to screw up

**завёртывающий** *adj.* covering, enveloping; screwing-up

**завершать, завершить** *v.* to complete, to consumate

**завершающий** *adj.* closing, final, concluding

**завершение** *n.* completion, ending, accomplishment

**завершённый** *adj.* completed, final, accomplished

**завершить** *pf. of* **завершать**

**завеса** *f.* curtain, screen

**завесочный** *adj.* curtain, screen

**завести** *pf. of* **заводить**

**завивание** *n.* coiling action, winding, twisting, curling, folding, convolution

**завивать, завить** *v.* to twist, to curl

**завивающийся** *adj.* twisting, curling

**завивка** *see* **завивание**

**завизировать** *pf. of* **визировать**
**завинтить** *pf. of* **завинчивать**
**завинченный** *adj.* screwed
**завинчивать, завинтить** *v.* to screw up
**зависеть** *v.* to depend (on)
**зависимость** *f.* dependence, dependency, relation; characteristic; function (*math.*); *oft.* not translated, e.g.: **кривая зависимости усиления от высоты** height-gain curve, (*lit.* curve of the dependence of gain on height); **в зависимости от** depending on, in relation to; **з. выходной величины от входной** input-output characteristic; **з. звукового давления громкоговорителя от частоты** speaker pressure-frequency response; **з. коэффициента дифракции** (*or* **направленности**) **от частоты** pattern factor characteristic; **з. между энергией и диапазоном её действия** energy-range (inter)relationship; **з. мельканий** (*or* **мерцания**) **от яркости** flicker dependence on brightness, flicker-brightness performance; **з. модуля от магнитного состояния** ΔE effect; **з. оптимальной частоты от расстояния** range-frequency characteristic; **з. от времени** time dependence; **з. от частоты** frequency response; **прямая з.** ordinal response, ordinal relation; **прямолинейная з. фазы от частоты** straight phase-versus-frequency curve; **з. сигнал-свет** current-output-versus-light-input relationship; **з. скорости счёта от напряжения** counting-rate versus voltage characteristic; **з. тока от длины волны** current-wave-length characteristic; **з. физических эталонов от скорости** influence of velocity upon physical standards; **з. фототока от светового потока** photo-current-versus-light intensity; **з. хода характеристики от напряжения накала** influence of filament voltage on tube characteristic; **з. электрического** (*or* **светового**) **выхода от сеточного напряжения** grid-drive characteristic; **з. ядерных сил от спина** spin relationship of nuclear forces
**зависимый** *adj.* dependent, subordinate, on-line
**зависящий** *adj.* depending

**завитой** *adj.* twisted, curled
**завиток** *m.* spiral, curl, scroll; coil; helix (of ear); **з. в первой бороздке, входной з. первой бороздки** dog leg; **з. ушной раковины** helix of ear
**завитый** *see* **завитой**
**завить** *pf. of* **завивать**
**завихрение** *n.* eddy, eddying, vortex, curling, convolution, turbulence, whirl; whirlwind
**завихренность** *f.* vorticity
**завод** *m.* starter, winding mechanism; plant, factory, mill, works; **автоматический з.** self-starter
**заводить, завести** *v.* to wind, to wind up, to crank, to start; to establish, to set up; **з. петлю** to loop (in)
**заводка** *f.* winding up; starting; **з. петли** looping-in
**заводной** *adj.* starting, winding, cranking
**заводский** *adj.* factory, plant, mill
**заворот** *m.* turn, turning, recursion
**завывание** *n.* howl, howling, wail; warble
**завывать, завыть** *v.* to howl, to wail
**завышенный** *adj.* too high; overstated; oversized, excessive
**завязанный** *adj.* tied, bound
**завязать, завязнуть** *v.* to stick; to sink
**завязнуть** *pf. of* **вязнуть, завязать**
**завязывание** *n.* binding, tying
**загасить** *pf. of* **гасить**
**загаснуть** *pf. of* **гаснуть**
**загиб** *m.* bend, crease, fold, edge; **з. характеристики** knee (of B/N curve)
**загибание** *n.* bend, bending, twist, folding; curvature
**загибать, загнуть** *v.* to bend, to fold, to crease, to turn back, to turn in
**загибной** *adj.* folding, folded
**загибочный** *adj.* bending, creasing
**заглушать, заглушить** *v.* to attenuate, to damp, to jam, to blackout, to bar, to choke; to suppress, to deafen, to deaden, to silence
**заглушающий** *adj.* damping; blanketing; smothering, choking
**заглушение** *n.* damping, muffling, choking, drowning (out), blocking deafening; silencing, jamming, quenching, blackout, barrage
**заглушённый** *adj.* damped, choked; dead, dead-ended; suppressed; opacified; obtuse; anechoic

**заглушить** *pf. of* **глушить, заглушать**
**заглушка** *f.* plug, earplug, stopper, choke, end cap; dead end; silencer; **з. головного телефона** earphone cushion; **з. провода у оконечного изолятора** dead end
**загнутый** *adj.* curved, hooked, crooked, arched; bent, folded
**загнуть** *pf. of* **загибать, гнуть**
**заголовок** *m.* heading, title; **з. блока** block head; **з. цикла** for clause (in Algol language)
**загораживание** *n.* blocking, stopping, shutting; enclosure, enclosing
**загораживать, загородить** *v.* to block, to lock, to bolt, to interlock, to shut, to stop, to arrest, to latch; to cut off
**загорание** *n.* ignition, firing
**загораться, загореться** *v.* to ignite, to catch fire; to light
**загорающийся** *adj.* igniting, inflamable; ignition
**загореться** *pf. of* **загораться**
**загоризонтный** *adj.* beyond-the-horizon
**загородить** *pf. of* **загораживать**
**заготовить** *pf. of* **заготовлять**
**заготовка** *f.* blank, bar, billet; store, stock; **восковая з.** cake wax
**заготовлять, заготовить** *v.* to prepare, to provide; to store, to stock up
**заградитель** *m.* wavetrap, rejector, stopper, band-elimination filter, suppressor, barrier, trap; **вильчатый з. эхо** terminal echo suppressor; **з. обратной связи** feed-back suppressor
**заградительный** *adj.* jamming, obstructing; barrier, barrage
**заградить** *pf. of* **заграждать**
**заграждаемый** *adj.* obstructed; blocked
**заграждать, заградить** *v.* to block, to blockade, to obstruct; to shut in, to dam
**заграждающий** *adj.* obstructing, blocking, choking, suppressing; eliminating; restricting, restrictive; barrier
**заграждение** *n.* barrage, obstruction, rejection, stopping, blocking
**заграждённый** *adj.* blocked, obstructed, stopped
**загреметь** *pf. of* **греметь**
**загромождённый** *adj.* busy; blocked up, encumbered
**загрохотать** *pf. of* **грохотать**
**загрубление** *n.* ruggedization; desensitization

**загружать, загрузить** *v.* to load, to charge, to fill; to feed
**загруженность** *f.* overload; **з. использования линии связи** line utilization
**загруженный** *adj.* charged, filled, loaded, fed
**загрузить** *pf. of* **загружать, грузить**
**загрузка** *f.* charge, load, loading, filling, charging, priming; duty; firing; **весовая з. электропечи** weight of materials to be treated in the furnace; **з. по времени** duty; usage count; **з. тракта программы передач** program-circuit loading
**загрузочный** *adj.* charging, loading, feeding
**загрязнение** *n.* contamination, poisoning, pollution; impurity; soiling
**загрязнённый** *adj.* contaminated, polluted, impure; clogged up
**загрязнить** *pf. of* **загрязнять, грязнить**
**загрязнять, загрязнить** *v.* to contaminate, to pollute, to poison; to choke
**загрязняющий** *adj.* contaminating, polluting
**загудеть** *pf. of* **гудеть**
**задавать, задать** *v.* to give, to assign; to set, to drive; **з. начальное значение** to prestore
**задавить** *pf. of* **давить, задавливать**
**задавка** *f.* **задавливание** *n.* peening
**задавливать, задавить** *v.* to crush
**задание** *n.* task, mission, assignment; drive
**заданный** *adj.* assigned, given, set, fixed, preset, specified, prescribed; designated
**задатчик** *m.* control point adjustment, setter, controller
**задать** *pf. of* **задавать**
**задача** *f.* problem; task, mission; **краевые задачи, задачи при краевых условиях** limit-value problems; **з. в истинном масштабе времени** real-time problem; **з. выбора кратчайшего маршрута** shortest route problem; **з. нахождения собственных значений** eigenvalue problem; **з. о блуждании в области с экранами** barrier problem (theory of info); **з. поиска информации** information retrieval problem; **з. присвоения** assignment problem; **пробная з.**

для локализации неисправностей trouble-location problem; з. целераспределения target-assignment problem

задающий *adj.* master, driving; giving, assigning

задвигать, задвинуть *v.* to bolt, to bar, to shut, to push in

задвижка *f.* gate, slide gate, slide valve, slide plate; damper, bolt, shutter, catch, bar, locking bar, latch; быстродействующая з. quick-opening gate valve; вращающаяся з. rotary barrel throttle, rotary valve; з. для отсечки воздуха air throttle; з. люка chute gate

задвижной *adj.* sliding, drawable

задвинутый *adj.* shut, bolted, barred

задвинуть *pf. of* задвигать

задел *m.* (semi-)production lot

заделанный *adj.* embedded, fixed, built-in; sealed, stopped up, closed; terminated, dead-ended; clamped

заделать *pf. of* заделывать

заделка *f.* closing up, sealing, sealing-off, stopping up; seal; building-in, embedding; clinching; з. входа канала duct sealing; оконечная з. termination, terminal tie, sealing off; оконечная з. для испытания кабеля под напряжением live cable test cap; оконечная з. провода dead-end tie

заделочный *adj.* sealing

заделывание *see* заделка

заделывать, заделать *v.* to embed, to build-in; to fix, to fasten; to terminate, to seal, to close up, to stop up

задемпфированный *adj.* damped

задёргивать, задёрнуть *v.* to draw, to pull in

задержание *n.* delay, arrest, detention, holding back, trapping; retardation, inhibition; з. фигуры persistence of pattern

задержанный *adj.* delayed, stopped, inhibited, retarded, lagged, time-lagged; retained

задержать *pf. of* задерживать

задержающий *adj.* inhibitory, inhibiting, retarding, restraining; retaining, retentive; intercepting; holding back; stop(ping); check(ing)

задерживание *see* задержание

задерживать, задержать *v.* to delay, to detain, to retard, to hold back, to inhibit, to impede, to stop, to arrest, to check, to block, to suppress, restrain, to lag; to entrap, to keep, to retain; to intercept; з. вызов to hold a call in abeyance

задерживающий *adj.* delaying, holding back, restraining, retarding, inhibiting, retaining, blocking; intercepting; lock (mechanism); check (valve); choking; trapping; retentive

задержка *f.* delay, lag, lagging, retardation; stop, check, catch; entrapment; impediment; з. в один такт one-pulse delay; з. в ответчике beacon delay; з. в отпускании slow-release; з. (в цепи) передачи transmission delay; з. во времени time lag; з. (времени) с настройкой большими ступенями coarse delay; з. (времени) с точной регулировкой fine delay; з. вспышки ignition delay, starting delay; з. выборки access delay; з. заднего фронта trailing edge delay; з. на время между двумя соседними импульсами one-pulse delay; з. на длительность (одного) импульса one-pulse time delay, pulse time delay; з. на каскад stage delay; з. на несколько разрядов digits delay; з. на один разряд digit delay, one digit delay; з. начала развёртки sweep delay; з. освобождения линии extended hold (*tphny.*); плавно регулируемая з. fine delay; з. повторного сигнала multipath delay; з. пробега transit-time delay; произвольная з. (импульса) во времени arbitrary coding delay; з. пуска, з. пускового импульса trigger delay; з. с грубо регулируемыми ступенями coarse delay; з. селекторного импуьса trigger gate delay; стабильная з. во времени static time delay; з. яркостного сигнала luminance delay; з. яркостного сигнала относительно сигнала цветности luminance-versus-chroma delay

задёрнуть *pf. of* задёргивать

задненавесной *adj.* rear-mounted

задненередний *adj.* posterior-anterior

задний *adj.* back, rear, end, tail

задник *m.* heelpiece (of relay)

задолжавший *adj.* indebted, in debt

задувание *n.* blowout, blowing out, extinguishing

**задувать, задуть** *v.* to blow out, to extinguish, to put out

**задувающий** *adj.* blowing (out)

**задувка** *see* **задувание**

**задувочный** *adj.* blowing; blow-in

**задутый** *adj.* blown out, extinguished

**задуть** *pf. of* **задувать**

**заедание** *n.* jam, jamming, catching, sticking, seizing, binding

**заедать, заесть** *v.* to stick, to grip, to catch, to jam, to seize to hook in, to bind in, to dig in

**заём** *m.* borrow, loan; **круговой з.** end-around borrow

**заёмный** *adj.* borrowed; loan

**заесть** *pf. of* **заедать**

**зажатие** *n.* squeezing, pressing

**зажатый** *adj.* squeezed, pressed; clamped, held, fixed, fastened

**зажать** *pf. of* **зажимать**

**зажечь(ся)** *pf. of* **зажигать(ся)**

**зажжённый** *adj.* ignited, lit

**зажигание** *n.* firing, striking; ignition; kindling, lighting, setting fire; priming; **вторичное з. (счётной трубки)** re-ignition (of a radiation counter tube); **з. выпрямителя путём образования ртутного контакта между полюсами** spray ignition, jet ignition (of a mercury-pool cathode rectifier); **з. дуги погружным электродом** dipper-anode-type ignition; **з. (дуги) посредством наклона** mechanical ignition of mercury arc rectifier; condenser-type starter for cold cathode tubes; **неисправное з.** misfire, defective ignition; spark failure; **неконтактное з.** proximity fuse; **неконтролируемое прямое з.** arc-through loss of control; **обратное з. (дуги)** flash-back, backfire, arcing back; **отстающее** (*or* **позднее**) **з.** late firing; late ignition; **преждевременное з.** pre-ignition; **прямое з.** arc-through; **з. с опережением** advanced firing; advanced ignition

**зажигатель** *m.* firing electrode, grid, ignitor, starter; lighter

**зажигательный** *adj.* ignition, igniting

**зажигать, зажечь** *v.* to fire; to set fire (to); to ignite, to light

**зажигаться, зажечься** *v.* to fire, to catch fire, to ignite; to light

**зажигающий** *adj.* igniting, starting

**зажим** *m.* terminal, binding post; clamp, clip, grip, fastener, clutch, chuck, clamping device, gripping device, (crimped) lock; connector; adapter; cleat; conductor joint; **зажим (на приборе), зажим (с отверстием для провода)** binding post, binder; **натяжные зажимы** terminal clamp, anchor clamp, straining clamp; **анкерный з. для контактного провода** anchor ear; **выпускающий з.** free-center-type clamp; **з. для крепления** terminal screw, clamping screw; **з. для подключения антенны** antenna connection; **з. для присоединения к броне** armour clamp; **з. для присоединения к заземлителю** ground clamp; **з. для прямых участков контактного провода** straight-line ear; **з. для сеточного колпачка** grid clip; **з. для скрещённого присоединения проводов** four-wire connector; **затяжной з.** pressure connector; **кнопочный з.** turret terminal; **концевой з.** terminal connector; **концевой линейный з.** dead ending; **крепёжный з.** terminal screw, clamping screw; **магнитный з.** magnetic chuck; **з. на перемычке** bridging connector; **надёжный з.** pressure connector; **невибрирующий з., недребежащий з.** antirattle clip; **обжимной з.** clinch ear; **ответвительный з.** branch terminal, branch joint, distribution connector, line connecting terminal, tee connector; **прикрепляющий з.** screw terminal, binding clip; **присоединительный з.** binding post, circuit terminal, terminal block, accessible terminal; **пружинный (схватывающий) з.** spring clip, alligator clip; **сдвоенный концевой з.** twin lug; **скрепляющий з.** screw terminal, binding clip; **соединительный з.** connector, through joint; **сращивающий з., стыковой з.** splicing ear; **тройниковый соединительный з.** three-way connector; **холостой з.** vacant terminal; **щипковый з.** alligator clip

**зажимание** *n.* clamping, gripping, pressing

**зажимать, зажать** *v.* to clamp, to clip, to catch, to clutch, to grasp, to grip, to fasten, to fix, to press; to jam

**зажимающий** *adj.* gripping, clamping

**зажимный** *adj.* binding, tightening; gripping, clamping; terminal, grip

**зажурчать** *pf. of* журчать

**заземление** *n.* grounding, ground, ground connection, ground system; глухое з., жёсткое з. dead ground; з. на корпус hull return circuit; з. одного полюса внутренней проводки interior wiring system ground; з. подогрева катода filament ground; з. сети grounding of the (power) distribution system; з. средней точки одной фазы трансформатора mid-phase grounding; з. угловой точки трансформатора в треугольник corner-of-delta grounding

**заземлённый** *adj.* grounded, ground-connected, earth-shielded; ground

**заземлитель** *m.* grounding electrode, ground plate; ленточный з. (metal) strip earth conductor, (iron) strip earth conductor; лучевой з. radial grounding-system; пластинный з. ground plate; стержневой з. earth rod, ground rod

**заземлительный** *adj.* grounding

**заземлять, заземлить** *v.* to ground, to earth

**заземляющий** *adj.* grounding, earthing, earth, ground

**зазор** *m.* clearance, gap, airgap, space, spacing, margin; play, backlash, slack; slot, slit; tolerance; з. в подшипнике play of a bearing, bearing slackness; воздушный з. в магнитной цепи magnetic discontinuity; воздушный з. ярма magnetic gap; з. входного резонатора "buncher" gap; з. выходного резонатора "catcher" gap; з. между двумя группами информации block gap; з. отлипания residual gap (of a relay); з. при отпущеном якоре released gap (of a relay); з. при притянутом якоре operated gap (of a relay)

**зазубренный** *adj.* notched, toothed, cogged, jagged, indented, serrated; pointed

**заимствование** *n.* consumption (*elec.*); derivation; borrowing

**заимствованный** *adj.* borrowed; derived

**зайчик** *m.* reflection of a sunbeam, light spot, luminescent spot

**заказ** *m.* order; call; аннулированный з. на переговор cancelled call; внеочередной з. top-priority call; з. на определённое время fixed time call; з. на очереди call on hand; call on waiting list; з. на переговор по авансу call with indication of charge; з. на переговор со справкой о его стоимости advise duration and charge call; срочный з. rush order (*tphny.*)

**заказать** *pf. of* заказывать

**заказной** *adj.* made to order; order; registered (letter)

**заказчик** *m.* customer, client

**заказывать, заказать** *v.* to order, to book

**закал** *see* закаливание

**закалённый** *adj.* tempered, hardened, hard, quenched

**закаливаемость** *f.* hardness capacity, hardenability

**закаливание** *n.* hardening, tempering, temper, quenching, annealing

**закаливать, закалить** *v.* to harden, to temper, to quench

**закаливающий** *adj.* hardening, tempering

**закалить** *pf. of* закаливать

**закалка** *see* закаливание

**закалочный** *adj.* hardening, tempering

**заканчивание** *n.* termination

**заканчивать, закончить** *v.* to complete, to end, to finish, to conclude

**закись** *f.* oxide

**заклёпанный** *adj.* riveted

**заклепать** *pf. of* заклёпывать, клепать

**заклёпка** *f.* rivet, clinch, clincher, cramp; з. с потайной головкой countersunk rivet

**заклёпочный** *adj.* riveting; riveted; rivet

**заклёпывание** *n.* riveting

**заклёпывать, заклепать** *v.* to rivet, to clinch

**заклинивание** *n.* keying; blocking, wedging; jamming, sticking, catching

**заклинивать, заклинить** *v.* to brace, to block, to prop, to wedge (up); to key

**заклинивающийся** *adj.* wedge-on, wedged; jammed

**заклинить** *pf. of* заклинивать

**заключать, заключить** *v.* to include, to enclose, to contain, to comprise;

to conclude; to shut in, to confine; **з. в коробку** to encase; **з. в себе** to imply; to house

**заключаться, заключиться** *v.* to consist (of), to lie (in); to be contained; to end, to finish, to result (in)

**заключение** *n.* conclusion, deduction, inference; closing, finishing up; inclusion; occlusion; enclosure; **з. в капсулу** encapsulation, encapsulating; **з. в ящик** casing; **з. экспертизы** expert opinion

**заключённый** *adj.* enclosed, included; concluded; occluded; embedded; **з. в кожух** cased

**заключительный** *adj.* final, closing, terminal, conclusive

**заключить(ся)** *pf. of* **заключать(ся)**

**закодированный** *adj.* coded; **з. по десятичному коду** decimal coded

**заколебать** *pf. of* **колебать**

**закон** *m.* law, rule, principle; relationship; **законы сохранения энергии в газодинамике** energy theorems of gas-dynamics; **по случайному закону** at random; **з. действующих масс** law of mass-action; **з. изменения ёмкости конденсатора** law of condenser; **з. индукции** Faraday's law of induction, law of electromagnetic induction; **з. кратных отношений** law of simple proportions; **з. куба температуры Дебая** Debey's $T^3$-law; **з. лучеиспускания Кирхгофа** Kirchhoff's law of spectral radiation; **з. напряжения сдвига** (*or* **среза**) law of shearing stress, law of torsion stress; **з. нарастания габаритов с увеличением мощности** law of growth; **з. необходимого разнообразия** law of requisite variety; **з. обратной квадратичной пропорциональности** inverse-square law; **з. полного тока** Biot-Savart's law; **з. поражения** pattern damage function; **з. постоянства площади фазовой характеристики** phase area law; **з. предельной синусоиды** sinusoidal limit theorem; **з. развёртывания** law of development; **з. сохранения количества движения** momentum conservation law; **з. сохранения энергии** theorem energy; **степенной з.** power law; **з. Столетова** the proportionality between the number of escaping photoelectrons and the luminous flux; **з.**

**трёх вторых** three-halves power law, Child-Langmuir law, "$\frac{3}{2}$" power law; **усиленный з. больших чисел** strong law of large numbers

**закономерность** *f.* regularity; law, rule; principle, mechanism; **з. голоса** voice mechanism

**закономерный** *adj.* regular

**законченный** *adj.* completed, finished, final, terminated; dead

**закончить** *pf. of* **заканчивать**

**закопанный** *adj.* buried

**закорачивание** *n.* shorting, short-circuiting

**закорачиватель** *m.* short-circuiter; plunger (of a waveguide)

**закорачивать, закоротить** *v.* to short-circuit, to short out

**закорачивающий** *adj.* shorting, short-circuiting

**закоротить** *pf. of* **закорачивать**

**закоротка** *f.* short-circuiting jumper; "short"

**закороченный** *adj.* shorted-out

**закорочено наглухо** (*or* **намертво**) dead short

**закорочивающий** *adj.* short-circuiting, shorting

**закраина** *f.* edge, border, ledge, flange, rim; tip

**закраска** *f.* marking; **з. гнезд** jack marking; multiple marking (*tphny.*)

**закреп** *m.* fastener, clip, catch, tack; dowel

**закрепитель** *m.* fixer, fixing agent

**закрепить** *pf. of* **закреплять**

**закрепка** *see* **закреп**

**закрепление** *n.* clamping, fastening, holding, gripping, fixing, securing, attaching; safety device, securing device, attachment; fixation, fixing (*phot.*); anchoring; **з. под шайбу** clamp mounting; **з. проводов** binding of wires, termination of wires

**закреплённый** *adj.* fixed, secured, clamped, fastened, mounted, attached; fast, tight; **наглухо з.** dead-ended

**закреплять, закрепить** *v.* to fasten, to secure, to fix, to mount, to attach, to clamp, to tighten; to bind, to lace, to tie off; to anchor; **з. оплётку на концах проводов** to tie up, to lace

**закрепляющий** *adj.* fastening, fixing, clamping

**закругление** *n.* curvature, curve, bend, rounding, camber; **з. изгибов** corner bevelling

**закруглённый** *adj.* rounded (off), truncated

**закружить(ся)** *pf. of* кружить(ся)

**закрутить(ся)** *pf. of* крутить(ся), закручивать

**закрутка** *f.* capstan; torsion

**закрученный** *adj.* twisted, curled

**закручивание** *n.* torsion; coiling; curling, twisting, involution

**закручивать, закрутить** *v.* to twist, to curl, to crimp

**закручивающийся** *adj.* twisting, curling

**закрывание** *n.* closure, cutoff; **з. диафрагмы** iris out

**закрывать, закрыть** *v.* to shut, to close, to shut (off *or* down), to stop, to cut off, to turn off; to cover, to house, to shelter, to enclose

**закрылок** *m.* flap, flange

**закрытие** *n.* closing, shutting; cover; enclosure; close; **з. связи** end of service, end of work sign

**закрытый** *adj.* closed, sealed, locked, shut; potted, enclosed, housed, shielded, indoor, sheltered; **з. с внешним обдувом** enclosed fan-cooled; **з. с самоохлаждением** enclosed self-cooled; **частично з. на сетке** phased back

**закрыть** *pf. of* закрывать

**закупоренный** *adj.* sealed, plugged, stopped up

**закупоривать, закупорить** *v.* to seal, to stop up, to plug, to plug up, to choke, to obstruct

**закупоривающий** *adj.* choking, blocking

**закупорить** *pf. of* закупоривать

**закупорка** *f.* slug, sealing, packing, chocking; jam

**зал** *m.* room, hall; **автоматный з.** auto-room, automatic switch room, apparatus room (*tphny.*); **автоматный з. междугородной станции** trunk switchroom; **аппаратный з.** instrument room; **з. искателей** instrument room, apparatus room (*tphny.*); **коммутаторный з.** operating room, manual switch room (*tphny.*); **з. кроссов и стативов** terminal room (*tphny.*); **линейно-аппаратный з.** line equipment room; **машинный з. с расположенными в одну линию**

**турбогенераторами** in-line turbine house

**залегание** *n.* bed, stratification; seam

**заливать, залить** *v.* to seal, to seal in, to cast; **з. наглухо** to seal off

**заливающий** *adj.* filling, flooding

**заливка** *f.* filling, seal(ing), sealing-in, priming; potting; **з. вкладыша подшипника** lining of the bearing; **з. компаундом** filling with compound; potting (of capacitors or coils)

**заливочный** *adj.* flooding, pouring; sealing; casting

**залипание** *n.* sealing

**залипающий** *adj.* sticking

**залитый** *adj.* flooded; filled; spread over; potted

**залить** *pf. of* заливать

**заложенный** *adj.* embedded

**замагнетизировать** *pf. of* магнетизировать

**замазать** *pf. of* замазывать, мазать

**замазка** *f.* cement, putty, paste, plaster, sealing compound; **цокольная з.** cap cement

**замазывание** *n.* blurring (*tv*); cementing, plastering, puttying

**замазывать, замазать** *v.* to seal (up), to putty, to cement, to plaster (up), to fill up, to stop

**замаскированный** *adj.* masked, disguised, camouflaged

**замаскировать** *pf. of* замаскировывать, маскировать

**замаскировывать, замаскировать** *v.* to mask, to disguise, to camouflage; to screen

**замедление** *n.* delay, retardation, time lag, slowing-down, slowdown, deceleration; **з. отпускания якоря реле** releasing delay, slow releasing; **з. скорости** deceleration

**замедленный** *adj.* slow, slowed, delayed, retarded, deferred-action, lagged, time-lag(ged), time-limit, time-element, deferred action; **замедленного действия** delayed(-action)

**замедлитель** *m.* decelerator, delay element, retarder; delay line; moderator; **с графитовым замедлителем** graphite-moderated; **вихревоточный з.** eddy current brake; **з. времени в записи речи** sonastretcher; **катодный з.** cathodic inhibitor; **пьезоэлектрический з.** piezoelectric delay line; **з. травления** pickling inhibitor

**замедлять, замедлить** *v.* to slow down, to decelerate, to retard, to delay, to defer, to reduce, to ease, to moderate, to inhibit

**замедляющий** *adj.* decelerating, delaying, retarding, inhibiting; timing (relay)

**замена** *f.* replacement, replacing, substitution, substitute, equivalent, change, exchange, interchange, interchanging; transposing (*math.*); **круговая з.** cyclic replacement; **з. наименований** name replacement **з. переменной** change of variable; **з. проводки** rewiring

**заменённый** *adj.* substituted

**заменено** *adv.* unplugged

**заменимость** *f.* interchangeability

**заменимый** *adj.* interchangeable, exchangeable, replaceable, detachable, renewable, commutative

**заменитель** *m.* substitute, substitute material, spare; eliminator

**заменить** *pf. of* **заменять**

**заменяемость** *see* **заменимость**

**заменяемый** *see* **заменимый**

**заменять, заменить** *v.* to replace, to substitute, to renew, to exchange, to interchange, to change, to supersede

**заменяющий** *adj.* replacing, substituting, substitution, interchangeable

**замер** *m.* measuring, measurement; test, probe, sample, sampling; **з. дальности** ranging; **радиолокационный з. дождя** radar rainfall measurement

**замера** *f.* test point

**замереть** *pf. of* **замирать**

**замерзание** *n.* freezing, chilling, solidification

**замерять, замерить** *v.* to meter, to measure, to gage

**заместить** *pf. of* **замещать**

**заметить** *pf. of* **замечать**

**заметка** *f.* note, notice

**заметный** *adj.* noticeable, marked, observable, visible, conspicuous

**замечать, заметить** *v.* to notice, to spot; to remark

**замечение** *n.* noticing, spotting

**замешивание** *n.* mixing; injection; **перекрёстное з.** cross-mixing

**замещ** *abbr.* (**замещённый**) substituted

**замещаемый** *adj.* replaceable, displaceable

**замещать, заместить** *v.* to substitute, to replace, to renew, to supersede; to change, to convert

**замещающий** *adj.* replacing, substituting

**замещение** *n.* replacement, substitution, displacement; change, conversion

**замещённый** *adj.* replaced, substituted, displaced

**замирание** *n.* fade, fading; dying away, dying down; going out; decay; **ближнее з.** interference-type fading; **з. в вертикальной плоскости** vertical fading; **з. сигнала звукового сопровождения** sound fading; **з. сигналов изображения** vision fading; **з. слышимости при разговоре** sudden volume decrease because of contact resistance (*tphny.*)

**замирать, замереть** *v.* to fade, to die, to die away

**замирающий** *adj.* fading, dying

**замкнутость** *f.* closed condition, closure

**замкнутый** *adj.* closed, locked, self-contained; **з. накоротко** short-circuited, shorted-out

**замкнуть** *pf. of* **замыкать; з. накоротко** to short, to short-circuit

**замок** *m.* lock; catch; hinge; scarf, scarf joint; **оконечный ригельный з.** facing point lock; **стрелочный ригельный з.** bolt lock

**замолкать, замолкнуть, замолчать** *v.* to fall silent, to become silent

**замолкнуть** *pf. of* **замолкать**

**замолчать** *pf. of* **замолкать, молчать**

**замороженный** *adj.* frozen; congealed; trapped

**заморский** *adj.* overseas

**замочный** *adj.* lock

**замыкание** *n.* closing, closure; locking; completion, termination; closing the current, closing the circuit; fastening; bridge; **с замыканием до размыкания** make-before-break; **быстрое з.** quick-make; **глухое з. на землю** dead earth; **двойное з. на землю** double-ground fault, double-line-to-neutral fault; **з. двух контактов (реле) подвижным контактом** bridging of contacts (of a relay); **длительное з. на землю** continuous earth, continuous grounding; **короткое з.** short-circuit(ing); **междупостовое з.** check locking; **много-**

кратное з. на землю polyphase earth; **з. на землю** contact to ground, ground connection, earth terminal, ground fault; **з. на корпус** contact to frame; ironwork fault, chassis fault, body contact; **з. накоротко** short-circuiting, short-circuit; **неполное з. на землю** partial ground; **з. от защёлки** latch locking; **з. от рельсовой цепи** track-circuit locking; **перемежающееся з. на землю** intermittent earth; **повторное з.** reclosure, reclosing; **предварительное з.** approach locking; **предварительное з. на подходе к централизации** approach locking the interlockings; **предварительное з. рычага** preliminary locking; **предмаршрутное з.** approach locking; **з. сообщения** routing the traffic; **установившееся короткое з.** sustained short circuit; **з. через железо** magnetic shunt; **электрическое з. нецентрализованной стрелки** outlying switch lock; **электрическое з. приближения** electric approach locking; **электрическое стрелочное з.** electric switch lever locking

**замыкатель** *m.* switch, contactor, closer, contact maker; locking mechanism; **плавкий з. тревожной системы** alarm fuse

**замыкать, замкнуть** *v.* to lock, to close, to make; to terminate; to shut; **взаимно з.** to interlock

**замыкающий** *adj.* locking, closing; **з. накоротко** short-circuiting

**замычка** *f.* lock, catch, bolt; **з. на линейке** locking dog (of an interlocking machine); **з. рычага управления стрелочным приводом** switch machine lever lock; **электрическая з. стрелочной рукоятки** electric switch lever lock

**занавес** *m.* curtain, screen

**заниженный** *adj.* undersized; too low, understated

**занимание** *n.* borrow (*math.*)

**занимать, занять** *v.* to hold, to engage, to tie (up), to busy, to take up; to seize; to assign, to allocate (a line); to borrow

**зануление** *n.* neutralizing, connection to the neutral conductor

**занулять** *v.* to ground, to connect to the neutral wire

**зануляющий** *adj.* nullifying, neutralizing

**занумерованный** *adj.* numbered

**занумеровать** *pf. of* **занумеровывать, нумеровать**

**занумеровывать, занумеровать** *v.* to number

**занятие** *n.* busy condition, busying, holding; work; occupation; **двухминутное з.** equated call; **искусственное з.** backward busying; **предварительное з. (линии)** advance assignment, advance hold (of a trunk) (*tphny.*)

**занято** *adv.* busy, engaged (*tphny.*); **з. междугородным переговором** busy on a long-distance call; **з. по входящим линиям** incoming engaged; **з. по исходящим линиям** outgoing engaged

**занятой** *adj.* busy

**занятость** *f.* busy condition, engaged condition, occupancy (*tphny.*); **з. всех соединительных** (*or* **шнуровых**) **линий** all trunks busy, ATB (*tphny.*); **междугородная з.** trunk busy (*tphny.*); **з. последней соединительной** (*or* **шнуровой**) **линии** last trunk busy, LTB (*tphny.*)

**занятый** *adj.* busy, engaged

**занять** *pf. of* **занимать**

**заодно** *adv.* together, at the same time

**заострение** *n.* point, tip; sharpening, tapering; **з. контакта вследствие наплавки** (*or* **нагорания**) coning of contact; **з. (проволочки) электрическим путём** electropointing (of a wire)

**заострённый** *adj.* sharp-edged, pointed, tapered, taper, acute

**заострять, заострить** *v.* to point, to sharpen, to taper down

**заостряющий(ся)** *adj.* peaking, tapering

**Зап** *abbr.* (**запад**) west

**зап** *abbr.* 1 (**западный**) western, westerly; 2 (**запасный**) auxiliary, alternate, reserve, replacement

**запад** *m.* west

**западание** *n.* attenuation; **з. частотной характеристики** frequency attenuation

**запаечно-откачной** *adj.* sealing-exhausting

**запаечный** *adj.* sealing

**запаздывание** *n.* delay, lag, retardation, time lag, time delay, lagging, hysteresis; **з. в переходном процессе** transfer lag, dynamic lag; **ёмкостное з. подачи** supply side capacity lag; **ёмкостное з. потребления** demand side capacity lag; **з. начала обратного хода** fly-back delay; **обусловленное ёмкостью з.** capacity lag; **з. показания измерительного элемента** measuring lag; **постоянное з.** fixed time lag; **з. распространения** propagation time lag; **з. с показательной характеристикой** exponential time lag; **з. смещения и скорости** distance-velocity lag; **чистое з.** pure time delay

**запаздывать, запоздать** *v.* to lag, to delay, to retard, to creep

**запаздывающий** *adj.* lagging, sluggish; retarded, delayed

**запаивание** *n.* sealing off

**запаивать, запаять** *v.* to solder, to weld, to seal up; **з. наглухо** to seal off

**запайка** *f.* soldering, welding; seal, sealing; sealed end; **вращающаяся з.** rotating bearing seal; **стыковая з.** butt seal; **фасонная з.** molded seal

**запал** *m.* fuse, lighter, cap, primer; ignition; firing; **преждевременный з.** pre-ignition

**запальник** *m.* ignition device, igniter, ignition chamber; blasting fuse

**запальный** *adj.* igniting, firing, ignition; starter

**запараллеленный** *adj.* in parallel, multiple

**запараллелить** *v.* to connect in parallel, to multiple

**запас** *m.* store, supply, reserve, storage, stock; margin, provision, allowance; **з. в верхней части** ceiling margin; **з. в нижней части** floor margin; **ведущий з.** guide margin; **з. до зуммирования** (*or* **самовозбуждения**) singing margin; **з. по ёмкости (станции)** marginal capacity (of an office); **з. по нагрузке** marginal load capacity; **з. по усилению** gain margin; **з. прочности** factor of safety, coefficient of safety; **з. прочности на излом** breakage safety factor; **з. усиления допустимый зуммированием** singing margin of amplification

**запасание** *n.* storage, accumulation; **з. программы при помощи записи на магнитную ленту** magnetic-tape storage

**запасать, запасти** *v.* to store, to accumulate, to reserve

**запасающий** *adj.* storing

**запасённый** *adj.* stored

**запасной, запасный** *adj.* emergency, auxiliary, spare, reserve, duplicate, stand-by, alternate

**запасти** *pf. of* **запасать**

**запаянный** *adj.* soldered, welded; sealed, sealed-off, closed; **з. в стекло** glass-sealed

**запаять** *pf. of* **запаивать, паять**

**запекание** *n.* baking

**запеленговать** *pf. of* **пеленговать**

**запереть(ся)** *pf. of* **запирать(ся)**

**заперто** *adv.* it is closed; **з. автоматическим смещением** self-biased off

**запертый** *adj.* closed, locked; blocked, trapped, barred, cutoff, blanked, blacked out, ungated

**запечатанный** *adj.* sealed

**запечатать** *pf. of* **запечатывать**

**запечатывание** *n.* sealing

**запечатывать, запечатать** *v.* to seal, to seal up

**запечённый** *adj.* baked

**запил** *m.* notch, slot, groove

**запиленный** *adj.* notched

**запираемый** *adj.* locked, closed, bolted

**запирание** *n.* closing, locking, shutting, suppression, blanketing, blacking-out, blocking, barring, cutoff, blanking, squegging, wipe-out; biasing-off, bias-off; **з. луча** blanking, blackout, beam suppression; **сеточное з.** arc suppression, grid extinguishing; **з. строк** line blanking

**запирательный** *adj.* locking, lock

**запирать, запереть** *v.* to lock, to close, to shut, to seal, to block, to interlock, to bar, to cut off, to blank, to disable, to squegge, to black out, to wire out, to ungate; to stop; to make busy

**запирающий** *adj.* stopping, blocking, locking; cut-off, blanking, blackout, disabling; inhibitory

**записанный** *adj.* recorded, written-in, written; record

**записать** *pf. of* **записывать**

**записка** *f.* note, report, memorandum; **служебная з.** service record

**записывание** *n.* recording, putting down, transcribing, inscribing

**записывать, записать** *v.* to record, to transcribe, to register, to put down; to enter; to read in; to pick up; **з. в виде уравнения** to equate

**записывающий** *adj.* recording, writing; **автоматически з.** self-recording, self-registering

**запись** *f.* recordance, recording, record, transcription, writing, note, entry, write, write-in; memory; **с шестью записами** six-trace; **з. белым по чёрному** white-on-black writing; **бесскобочная з.** Lukasiewicz's notation, parenthesis-free notation; **з. в журнал** logging; **з. в неравновесном режиме потенциалов** non-equilibrium writing; **з. в режиме равновесия** equilibrium writing; **з. во всю ширину ленты** full track recording; **глубинная з.** hill-and-dale recording, vertical recording; **глубокая механическая з.** depth recording; **з. для непосредственного** (*or* прямого) **воспроизведения** instantaneous recording; **долговременная з.** nonvolatile recording; **з. звукового сопровождения кинофильма** sound-on-film recording; **з. изображения на магнитную ленту** video tape recording; **интенсивная з.** variable density recording; **з. колебаний телефонной мембраны** telephone record; **з. колебания в виде кривой** variable area film recording; **контрольная з.** supervisory record; reference recording; **з. корреляции** correlatogram; **магнитная з. с перемещаемой границей** boundary-displacement magnetic recording; **з. магнитографа** magnetogram; **з. методом нагрева выше точки Кюри** Curie point writing; **з. модулированием обеих сторон звуковой дорожки** bilateral recording; **з. на киноплёнку телевизионных программ** kinescope recording, motion-picture recording; **з. на копирку** carbon-pressure recording; **з. на ленте самописца** record chart; **з. на плёнку телевизионных передач, з. на плёнку телепередач** television recording, teletranscription; **з. на ультразвуковых частотах** ultrasonography; **з. наведённой проводимостью** induced-conductivity writing; **з. переходного процесса** tran-sient recording; **з. по методу наведённой проводимости** induced-conductivity writing; **з. по методу перераспределения электронов** redistributing writing; **з. показания** record of reading; **полночастотная з.** high-fidelity recording; **поперечная з.** lateral recording, perpendicular recording, variable-area recording; **пробная з. (звука)** reference (sound) recording; **з. прямого воспроизведения** instantaneous recording; **размноженная з.** serial copy; **з. распылённой струёй чернил** ink-vapor recording; **рельефная з.** embossed-groove recording; **з. рядом** juxtaposition; **з. с зазорами** non-contact recording; **з. с переменной глубиной** hill-and-dale recording; **з. с приёмной трубки для цветного телевидения** color kinescope recording; **з. с промежутками между знаками** (*or* цифрами) return recording; **телевизионная з.** canned television, bottled televison; **з. телевизионных программ на две киноплёнки с чередованием полей** alternate field recording; **темновая з.** dark trace; **ультразвуковая з.** ultrasonography; **з. фотовозбуждённой проводимостью** photoconductivity writing; **з. чёрным по белому** "black on white" writing; **широкополосная з.** full-frequency range recording; **з. электронно-возбуждённой проводимостью** bombardment-conductivity writing

**запитать** *pf. of* **запитывать**

**запитывание** *n.* feed, feeding, supply; **з. (антенны) в центре** apex drive

**запитывать, запитать** *v.* to feed, to supply

**заплавка** *f.* puddling, seal; **з. верхнего вывола** top puddling; **з. вывода по типу глазка** eyelet-type lead seal

**запланированный** *adj.* preplaned

**заплата** *f.* patch, piece

**запломбировать** *pf. of* **пломбировать**

**заподлицо** *adv.* flush (with); **з. с поверхностью** flush-type

**запоздавший** *adj.* late, retarded, deferred

**запоздать** *pf. of* **запаздывать**

**заполнение** *n.* filling, filling up, charging, priming; **с оксидным запол-**

нением oxide-impregnated; **з. об-мотки** closeness of winding

**заполненный** *adj.* filled, charged; occupied

**заполнитель** *m.* filler, filling; bolus material

**заполнять, заполнить** *v.* to fill, to fill up, to charge; to make up

**заполняющий** *adj.* filling

**запоминаемый** *adj.* stored

**запоминание** *n.* memorizing, remembering; storage (of data), registering; **з. местоположения (цели)** position memory; **з. на короткие промежутки времени** short-term memory; **з. на многопозиционных элементах** multiple-stable-state storage

**запоминать, запомнить** *v.* to memorize, to remember; to store; **предварительно з.** to prestore

**запоминающий** *adj.* memorizing; memory; storage

**запомнить** *pf. of* **запоминать**

**запор** *m.* fastener, bolt, lock, bar, latch, catch, stop; seal(ing); locking; **з. рельсов разводного моста** drawbridge rail lock; **з. рычага с защёлкой** lever latch lock

**запорно-выпускной** *adj.* shut-off

**запорный** *adj.* locking, closing, shut-off, blocking; barrier

**заправить** *pf. of* **заправлять**

**заправка** *f.* leading-in; threading; charging, priming; setting, preparing; servicing, repairing; **з. станков (для скрутки)** thread-up (of twisting machines)

**заправлять, заправить** *v.* to load, to fill, to fill up, to charge, to prime; to set, to prepare; to service, to repair

**заправочный** *adj.* filling, charging, priming; setting, preparing; repairing; servicing

**запрашиваемый** *adj.* interrogated

**запрашивать, запросить** *v.* to demand, to interrogate, to inquire

**запрессованный** *adj.* pressed, press-fitted, potted, embeded

**запрессовка** *f.* pressing, pressing process

**запрет** *m.* inhibition, prohibition; **з. включения** ungated reset

**запретительный** *adj.* restrictive, prohibitive, inhibitory

**запретить** *pf. of* **запрещать**

**запретный** *adj.* forbidden, prohibited

**запрещаемый** *adj.* inhibited

**запрещать, запретить** *v.* to restrict, to forbid, to inhibit, to prohibit

**запрещающий** *adj.* inhibiting, inhibitory

**запрещение** *n.* inhibition, prohibition, suppression; **местное з. печатания** localized print suppression; **з. переноса** carry suppress(ing)

**запрещённый** *adj.* forbidden, prohibited, restricted, inhibited; trapped

**запрограммированный** *adj.* programmed

**запроектировать** *pf. of* **проектировать**

**запрос** *m.* interrogation, challenge; demand; inquiry, request; **з. в боковом лепестке** side-lobe interrogation; **з. в диапазоне частот** multiple-frequency interrogation; **з. корреспонденту** question to the called party; **з. на многих частотах** multiple-frequency interrogation

**запросить** *pf. of* **запрашивать**

**запросный** *adj.* request; inquiry

**запросчик** *m.* interrogator, interrogator-responser, challenger; inquiry station; **з.-ответчик** interrogator-responser, challenger; **з. «свой или чужой»** interrogator "friend or foe"

**запруда** *f.* dam, dike

**запружать, запрудить** *v.* to dam

**запружение** *n.* damming, diking

**запруженный** *adj.* dammed, retained

**запруживание** *n.* damming, diking

**запруживать** *see* **запружать**

**запуск** *m.* run, start, triggering, trigger action; **с запуском от внешнего импульса** signal-triggered; **з. селектора строк** line selector trigger; **з. схемы от импульса** pulse triggering

**запускаемый** *adj.* launched; driven

**запускать, запустить** *v.* to start, to launch, to activate, to trigger

**запускающий** *adj.* starting, initiating, firing, triggering

**запустить** *pf. of* **запускать**

**запутывание** *n.* fouling, complication

**запылённый** *adj.* dusty

**запятая** *f. adj. decl.* comma; point; **з. в позиционном представлении числа** radix point; **нефиксированная з.** floating point; **отделяющая з.** decimal point; **числовая з.** radical point

**запятовидный** *adj.* comma, comma-shaped

**заражение** *n.* contamination; infection, contagion
**заражённый** *adj.* contaminated; infected
**заранее** *adv.* beforehand, early, previously; pre-; **з. накапливать** to prestore; **з. расчитанный** precomputed
**зарегистрированный** *adj.* registered, recorded
**зарегистрировать** *pf. of* **регистрировать**
**зарезервированный** *adj.* reserved
**зарница** *f.* sheet lightning
**зародить(ся)** *pf. of* **зарождать(ся)**
**зарождать, зародить** *v.* to generate, to produce
**зарождаться, зародиться** *v.* to be generated; to arise, to set in
**зарождающийся** *adj.* incipient, nascent, commencing, initial
**зарождение** *n.* origin; onset; generation, formation; **з. колебаний** onset of oscillations
**зарождённый** *adj.* produced, conceived
**зарубка** *f.* notch, slot, nick, cut, notching, mark, incision; indentation, dent
**зарытие** *n.* digging in, burying in
**зарытый** *adj.* buried
**заряд** *m.* charge, charging, load, loading; cartridge; blasting charge (of explosive); **з. взрывчатого вещества** explosive charge; **поверхностный з. (обкладки конденсатора)** charge of capacitor; **з. при постоянной силе тока** constant-current charge; **резонансный з. от источника постоянного тока** d-c resonance charging; **удельный з. электрона** electron-charge mass ratio; **з. ядра** nuclear-charge number
**зарядить** *pf. of* **заряжать**
**зарядка** *f.* threading up (the film); charging, charge, loading; **з. батареи через постоянное сопротивление при постоянном напряжении питания** modified constant-voltage charge; **буферная з.** feeding power by means of floating-batteries; **з. на расстоянии** remote charging; **настроенная з.** resonant charging; **непрерывная з.** trickle charge; **з. постоянным током через выпрямитель со сглаживающим дросселем** d-c choke rectifier charging; **з. при**

**постоянной величине тока** constant-current charge; **з. через постоянное сопротивление при постоянном напряжении** modified constant voltage charge
**зарядник** *m.* battery charger
**зарядно-разрядный** *adj.* charge-and-discharge, charge-discharge
**зарядный** *adj.* charging, charge, loading
**зарядово-инвариантный, зарядово-независимый** *adj.* charge-independent
**зарядовый** *see* **зарядный**
**зарядообменный** *adj.* charge-exchange
**заряжаемый** *adj.* chargeable, charged
**заряжание** *n.* loading
**заряжатель** *m.* charger; **буферный з.** trickle charger
**заряжать, зарядить** *v.* to charge, to electrify; to load (a film or tape); **повторно з.** to recharge
**заряжающий** *adj.* charging
**заряжение** *n.* charging, charge, loading
**заряженный** *adj.* charged, loaded, fed; contaminated; live (*elec.*)
**засасывать, засосать** *v.* to suck in, to draw in
**засасывающий** *adj.* suction
**засветить** *pf. of* **засвечать, засвечивать**; **з. изображение** to flare the picture
**засветиться** *pf. of* **засвечиваться**
**засветка** *f.* flare spot, light bias, bias lighting, illumination; gating, strobing; irradiation, exposure; **внешняя з.** ambient light; **з. края** edge flare (*tv*); **з. от постороннего источника** ambient light illumination; **з. по краям** edge flare; **з. снизу** bottom flare; **узкая з.** narrow gate; **з. экрана** flare spot (*tv*)
**засвеченный** *adj.* light-struck
**засвечать, засветить** *v.* to light, to kindle; to gate, to strobe
**засвечивать** *see* **засвечать**
**засвечиваться, засветиться** *v.* to be lighted, to light up, to flash
**засев** *m.* charging
**засекать, засечь** *v.* to determine by insertion, to intersect; to notch; to locate; to take the bearing, to take a radio bearing
**засекретить** *pf. of* **засекречивать**
**засекреченный** *adj.* confidential, classified, security-restricted, secret

**засекречивание** *n.* classifying, imposing secrecy

**засекречиватель** *m.* speech inverter, scrambler

**засекречивать, засекретить** *v.* to make secret, to hush up, to restrict, to classify

**засечённый** *adj.* notched, cut, marked

**засечка** *f.* cross bearing, fix, fixing; notch, cut, mark; intersection; **з. времени** timing, clocking; **звуковая з.** sound ranging; **з. направления** getting a fix

**засечь** *pf. of* засекать

**засиять** *pf. of* сиять

**заскок** *m.* blocking, catch; overrun

**заслонение** *n.* shielding, screening; **з. источников космического высокочастотного излучения** occultations of discrete sources of cosmic radiation

**заслонка** *f.* damper; door; vane, valve, choke, restrictor; baffle; flap; shutter, drop, lid; **возвратная з.** flap valve; **дифференцирующая з.** derivative restrictor; **дроссельная з.** butterfly valve, throttling butterfly valve; **затемняющая з.** dark screening slide; **круглая скользящая з.** circular slide damper; **обратная з.** backpressure valve, return valve; **падающая з.** drop indicator; **противопожарная з.** fire shutter; **з. типа жалюзи** louvre

**засов** *m.* bolt, locking bar, latch

**засорение** *n.* impurity, contamination, soiling; choking, plugging, obstruction, stoppage

**засорённый** *adj.* contaminated, soiled; choked, plugged up, clogged up, stopped

**засос** *m.* intake, inflow

**засосанный** *adj.* sucked in, drawn in, suction

**засосать** *pf. of* засасывать

**заставить** *pf. of* заставлять

**заставка** *f.* cue, cue mark; fuse

**заставлять, заставить** *v.* to force, to compel, to make; to block, to bar

**застёжка** *f.* fastening, fastener, hook, hasp, clasp

**застеклённый** *adj.* glass-fronted, glassed-in; glazed, vitrified

**застой** *adj.* stagnation, standstill

**застопоривание** *n.* stop, stopping, locking, choking, clogging

**застопоривать, застопорить** *v.* to stop, to lock, to plug, to clog

**застопорившийся** *adj.* choked, clogged, plugged

**застопорить** *pf. of* застопоривать, **стопорить**

**застревание** *n.* sticking, jamming; **з. асинхронного электродвигателя** crawling of an asynchronous machine

**застревать, застрять** *v.* to stick, to jam, to get stuck; to crawl

**застревающий** *adj.* sticking, lock(ed)-in, stuck

**застрявший** *adj.* stuck, clogged

**застрять** *pf. of* застревать

**заступ** *m.* spade

**застывание** *n.* solidification, congealing, freezing

**засурдиненный** *adj.* stopped

**засыпка** *f.* backfilling, filling up; filling, charging; covering

**засыпной** *adj.* charging

**затвердевание** *n.* solidification, solidifying, hardening, congealing; freezing

**затвердевать, затвердеть** *v.* to harden, to set, to solidify, to bind, to grow hard, to congeal; to fix

**затвердение** *see* затвердевание

**затвердеть** *pf. of* затвердевать, **твердеть**

**затвор** *m.* shutter, gate, valve; trap, plug, stopper; closing device; bar, bolt, lock, closure, seal(ing); flood gate; **безопасный з.** safety lock; **взаимный з.** interlock; **взвешенный световой з.** suspension light valve (*tv*); **жидкостный з.** liquid valve, mercury cut-off; **опускающийся** (*or* **опускной**) **з.** drop shutter; **ртутный з. с магнитным управлением** magnetic mercury cut-off; **секторный з.** radius gate; **штыковый з.** bayonet joint

**затворенный** *adj.* shut, closed

**затворять, затворить** *v.* to lock, to shut, to close

**затемнение** *n.* black-out, darkening, shading, dim-out, shadow, blanking, blanketing; **оптическое з. края** optical border obscurity; **з. по углам** corner shading

**затемнённый** *adj.* darkened, blacked, blackened, blacked out

**затемнитель** *m.* dimmer, shade; **дроссельный з.** inductor dimmer, reactance dimmer; **реостатный з.** resistor-type dimmer; **сценический з.** theater dimmer

**затемнять, затемнить** *v.* to darken, to black out, to obscure, to shade

**затемняющий** *adj.* blanking (*tv*)

**затенение** *n.* shadow(ing), shading, adumbration; shield

**затенённый** *adj.* shaded

**затенивать** *see* **затенять**

**затенитель** *m.* shade

**затенять, затенить** *v.* to shade, to shadow, to darken, to black out

**затеняющий** *adj.* shadowing, blanking

**затирание** *n.* smoothing out, rubbing over; binding

**заткнутый** *adj.* plugged, stopped up

**затмевать, затмить** *v.* to darken, to obscure, to overshadow; to cover

**затмевающий** *adj.* occulting

**затмение** *n.* eclipse

**затменный** *adj.* eclipsing

**затмить** *pf. of* **затмевать**

**заткнуть** *pf. of* **затыкать**

**затормаживание** *n.* "slewing"

**заторможённый** *adj.* braked; retarded, restrained, deferred, delayed; damped

**затормозить** *pf. of* **тормозить**

**заточка** *f.* edging, pointing, sharpening, rounding off; groove; *pl.* **заточки** polished sections, ground and polished surfaces

**затравка** *f.* seed; fuse, primer, priming device

**затрагивающий** *adj.* affecting, touching upon

**затрата** *f.* expense, outlay, expenditure; consumption; **з. на сооружение** cost of installation, investment

**затраченный** *adj.* consumed; expended

**затрачиваемый** *adj.* expended, spent; consumed

**затрачивание** *see* **затрата**

**затрещать** *pf. of* **трещать**

**затруднение** *n.* difficulty, trouble, inconvenience; **з. пеленгации** measures to prevent the use of radio signals for direction-finding purposes

**затухание** *n.* attenuation, damping, decay, decaying, loss, fading, extinction, build-down, subsidence, weakening, dying-out, decrement; **на границе затухания, с критическим затуханием** critically damped; **с малым затуханием** underdamped; **з. ближнего эха** local echo attenuator;

**з. в воздушной камере** air-vane damping; **з. в децибелах** db-loss; **з. в заграждаемой полосе, з. в полосе заграждения** (*or* **подавляемых частот**) stop band attenuation; **з. в пределах пропускаемой полосы фильтра** pass loss of a band pass; **вносимое з.** insertion loss, insertion transmission loss; **з. волн над плоской земной поверхностью** plan earth attenuation; **з. (волны) при распространении в трубе** tube attenuation; **з. вследствие отражений** balance attenuation; **з. вследствие отражений в активном четырёхполюснике** active return loss; **з. вследствие рассогласования** mismatch attenuation, regularity return-loss; **действующее з.** transducer loss; transmission efficiency; net loss; **з. зуммирования** through singing point (*tphny*.); **интерференционное з.** selective fading; **километрическое з. (высокочастотного луча) при дожде** kilometric attenuation due to absorption by rain; **з. на выходе** leakage damping; **з. на единицу длины линии** attenuation factor, attenuation constant; **з. на участке радиорелейной линии** path attenuation (of radio links); **з. нелинейного искажения** natural logarithm of the reciprocal of the distortion (factor); **неправильное з.** balance return loss, balance attenuation; **нормированное з.** standard trunk-loss; **з. обратного течения** return loss; **остаточное з.** overall circuit attenuation, net loss, effective equivalent; **з. от взаимных помех между каналами** crosstalk transmission equivalent, crosstalk damping; **з. от влияния экрана** natural logarithm of the reciprocal of the shielding factor; **з. от действия активного сопротивления** ohmic component of the attenuation constant; **з. от диафонии** near end crosstalk attenuation; **з. от дождя** raindrop attenuation; **з. от отражений в активном четырёхполюснике** active return loss; **з. от потерь** dissipative attenuation; **з. от шунтов (на линии)** (line) bridging loss; **перекрёстное з.** cross-attenuation; **переходное з.** crosstalk attenuation, iterative attenuation, crosstalk transmission equivalent; **переходное з. на отправном** (*or* **передающем**) **конце** near-end crosstalk attenuation; **переходное з. на приёмном конце** far-end crosstalk attenuation; **з. по весовому закону** weight law attenuation; **полезное з.** effective transmission equivalent; **ползучее з.** aperiodic damping; **полное з. системы передачи** net transmission equivalent; **з. полосы пропускания** attenuation in the pass-band; **з. при передаче** sending allowance, transfer attenuation-constant; **рабочее з.** effective attenuation, overall loss; **з. разветвительного контура** effective attenuation of a termination circuit; **з. расстройки** off-resonance attenuation; **сверхкритическое з.** overdamping; **з. симметрирующей схемы** balancing-network attenuation; **собственное з.** regularity attenuation, attenuation equivalent, non-reflection attenuation; **собственное з. четырёхполюсника** image attenuation constant of a network; **з. ударного звука** impact loss; **з. частоты сигнала звукового сопровождения** sound fading; **з. частоты сигналов изображения** vision fading; **чрезмерное з.** overdamping; **эквивалентное з. для артикуляции** articulation reference equivalent; **эквивалентное з. на конце междугородной связи** toll terminal equivalent of attenuation; **эквивалентное з. при передаче** (effective) transmission equivalent, sending attenuation; **эквивалентное з. при приёме** receiving equivalent, receiving attenuation; **з. эквивалентной схемы** balancing-network attenuation

**затухательный** *adj.* attenuation

**затухать, затухнуть** *v.* to attenuate, to damp, to die, to die away, to fade, to decay, to go out, to be extinguished

**затухающий** *adj.* dying, decaying, damping, fading; transient; damped

**затухнуть** *pf. of* **затухать**

**затушить** *pf. of* **тушить**

**затыкать, заткнуть** *v.* to jam, to choke up, to obstruct; to stop, to plug

**затычка** *f.* plug, stopper; duct plug

**затягивание** *n.* pulling, pull-in, ziehen

effect, overlap, backlash, coupling-hysteresis effect; oscillation hysteresis phenomenon; tightening; **з. строк** trailing; **з. частоты магнетрона** pulling of magnetron

**затягивать, затянуть** *v.* to tighten, to screw down, to fasten down; to delay; to draw out

**затяжной** *adj.* clamping, tightening

**затянуть** *pf. of* **затягивать**

**Заундерс** Saunders

**заунывный** *adj.* mournful

**заусенец** *m.* wire-edge, burr, barb, ridge, seam

**зафиксированный** *adj.* fixed

**зафиксировать** *pf. of* **фиксировать**

**захват** *m.* hold, holding, clamp, locking, clip, grip, catch, tappet, clutch; capture, trap, trapping, entrapment, seizing; **з. дырок** hole trapping; **радиолокационный з. низколетящих целей** target reconnaissance for low-level flight; **з. с гамма-излучением** radiative capture; **з. частоты** frequency sticking; **з. ядра** nuclear attraction

**захватить** *pf. of* **захватывать**

**захваты** *pl.* tongs

**захватывание** *n.* catching, seizing, holding, locking, lock-in, trapping, entrapment, capture, entrainment; **з. частоты генератора** pulling of oscillator

**захватыватель** *m.* grip, fastener, clamp

**захватывать, захватить** *v.* to take, to catch, to seize, to trap, to cover, to lock, to capture; to enclose; to interlock, to engage

**захватывающий** *adj.* engaging, gripping, catching

**захваченный** *adj.* trapped, captured, entrapped; gripped

**захлёстывание** *n.* whipping, lashing; sagging

**захлопывание** *n.* implosion (of sound); closure; **з. пузырка** bubble collapse

**заход** *m.* stopping (at); run, approach; **з. за шкалу** overshoot, overswing (of a pointer); **з. на посадку управляемый с земли** ground-controlled approach; **з. развёртки за пределы изображения** overscanning

**зацепить(ся)** *pf. of* **зацеплять(ся)**

**«зацепка»** *f.* catchword (*tgphy.*)

**зацепление** *n.* linkage, meshing, catching (of tooth), hooking, gearing, engagement; **зубчатое з.** gear coupling

**зацеплять, зацепить** *v.* to mesh, to catch, to gear, to engage, to hook, to lock, to bite

**зацепляться, зацепиться** *v.* to gear

**зацепляющий** *adj.* hooking, engaging

**зацикливание** *n.* ringing

**зачеканенный** *adj.* calked

**зачистить** *pf. of* **зачищать**

**зачистка** *f.* clearing, stripping, skinning; trimming

**зачищать, зачистить** *v.* to skin, to strip, to pencil down (the insulation), to scrape, to clean, to bare; **з. провод** to skin the wire, to strip the wire

**зашифрованный** *adj.* codified, ciphered, inverted

**зашифровать** *pf. of* **зашифровывать, шифровать**

**зашифровка** *f.* code, cipher

**зашифровывание** *n.* enciphering

**зашифровывать, зашифровать** *v.* to code, to codify, to cipher, to encode

**зашнуровать** *pf. of* **шнуровать**

**заштемплевать** *pf. of* **штемплевать**

**заштенгелеванный** *adj.* tabulated

**заштрихованный** *adj.* shaded, cross-hatched

**заштриховать** *pf. of* **штриховать**

**зашуметь** *pf. of* **шуметь**

**зашунтированный** *adj.* shunted, switched, connected across

**зашунтировать** *pf. of* **шунтировать**; **з. накоротко** to short out

**зашуршать** *pf. of* **шуршать**

**защёлка** *f.* catch, stop, cam, latch, detent, pawl, holding pawl, trigger, click, detainer, arrester, clip, stop, trip; lock, clamp, clamping ring; **з. для быстрой установки объектива** bayonet lock (*tv*); **расцепляющая з.** release pawl; **з. рычага** lever latch; **з. с быстрым размыканием** quick-release catch; **фиксирующая з.** fixed pawl

**защёлкивание** *n.* latching, locking, bolting, interlocking

**защёлкивать, защёлкнуть** *v.* to latch, to snap, to fasten

**защемить** *pf. of* **защемлять**

**защемление** *n.* jamming, choking

**защемлённый** *adj.* choked; fastened, pinched, entrapped

**защемлять, защемить** *v.* to jam, to choke; to bite, to fasten, to pinch

**защита** *f.* shield, shielding, screen, screening, guard, safeguard; proofing; relaying; wiping; protection; **балансная з. трёхфазных кабелей** core balance protective system; **з. второй линии, вышестоящая з.** back up protection; **дифференциальная з. фидеров с работой по принципу уравновешенных напряжений** biassed differential protective system; **зависимая от крутизны фронта з.** surge protection, rate-of-change protection; **з. контрольными проводами, з. (линии) с контрольными проводами** pilot wire protection, pilot protection, pilot relaying; **з. минимального напряжения** undervoltage protection; **з. минимальной частоты** underfrequency protection; **з. на пересечении путей** crossing protection, highway-railroad-crossing protection; **направленная з. энергии** directional power protection; **з. непрерывной кривой «расстояние-время»** continuous curve distance time protection; **з. нулевой последовательности** zero sequence protection; **з. от влияния линий передач** protection against (metallic-circuit) induction; **з. от выпадения из синхронизма** out-of-step protection; **з. от замыкания на «землю»** earth fault protection; **з. от климатических воздействий** weather proofing; **з. от коронирования** (*or* **короны**) corona prevention; **з. от неправильных коммутационных операций** switching error protection; **з. от несогласования фаз** phase comparison protection; **з. от обрыва поля** field-loss protection, loss-of-field protection; **з. от обрыва фазы** phase failure protection; **з. от пеленгации** anti-DF, antidirection finding; **з. от перемены направления фаз** phase-reversal protection; **з. от плесни** fungus proofing; **з. от потери возбуждения** field-loss protection, loss-of-field protection; **з. от превышения частоты** over frequency protection; **з. от преобразователя** phase balance protection; **з. от прикосновения к токоведущим частям** protection against electric-shock hazard; **з. от разговорных токов** voice-frequency trap; **з. от размыкания**

**фазы** open-phase protection; **з. от разъедания грибковой плесенью** fungus proofing; **з. от сообщения проводов** cross protection; **з. от утечки** frame leakage protection; **з. полного сопротивления** impedance protection, reactance protection; **з. при нормальной работе двигателя** motor-running protection; **з. против увеличения нагрузки сверх установленной** overpower protection; **противоветровая з.** windscreen; **резервная з.** back up protection; **релейная з. с вч блокировкой** carrier-current relaying; **роговая з., з. роговыми разрядниками** horn arrester, horn gap; **з. с вч телеблокировкой** carrier-current pilot protection; **з. серводвигателей** pilot protection; **сетевая з.** protection of power-distribution systems; **з. слуха** acoustic shock absorber; **з. электрических машин от прикосновения** (shock-hazard) protection of electrical machinery

**защитить** *pf. of* **защищать**

**защитник** *m.* protector

**защитный** *adj.* protecting, protective, guard, guarding; shield, shielding; proofing; relaying; suppressing

**защищаемый** *adj.* protected, guarded, screened, shielded, enclosed, sheltered

**защищать, защитить** *v.* to protect, to defend, to guard, to screen, to cover, to shield, to shelter; to proof; to relay, to relay out

**защищающий** *adj.* protecting

**защищённость** *f.* protection

**защищённый** *adj.* protected, guarded, screened, enclosed, sheathed, armoured, shielded, sheltered; proofed; -proof; **з. от внешних воздействий** environment-proofed; **з. от перегорания** antiburn-out; **з. от попадания водяных брызг** splash-proof; **з. от попадания капель** drip-proof; **з. от прикосновения** shockproof; **з. с вентиляцией** protected-enclosed; **з. футляром** encased

**заявленный** *adj.* declared

**заякоривание** *n.* anchoring

**звезда** *f.* star; star navigation system; Y-system; **соединённый звездой** star-connected; **простая з.** spiral quad, spiral four; **з.-треугольник**

stardelta; **якорная з.** spider armature

**звёздный** *adj.* star, stellar, starry; sidereal

**звездовидный, звездообразный** *adj.* star-shaped, star-like, stellate

**звездо-четверочный** *adj.* star-quad

**звёздочка** *f.* asterisk; spider; turnstile; little star

**звездчатка** *f.* star wheel, sprocket; **лентопротяжная з.** sprocket feed, pinfeed

**звено** *n.* section, link, ring, device, element, unit, member, component; monomer unit; **демпфирующие звенья для волноводов** attenuators for waveguides; **з. в виде L** mid-shunt termination; **з. в виде T** mid-series termination; **ведомое з.** driven member, slave unit; **Г-образное з.** L-network; **дифференцирующее з.** differentiating circuit; **измерительное з.** instrument movement, instrument link; **коммутационное з.** contact assembly; **копирующее з.** slave unit; **открытое цепное з.** swivel plus keystone link; **отрабатывающее з.** resetter; **з. по контрольному проводу** pilot wire link; **П-образное з.** split-capacitor section, P-network, $\pi$-section; **П-образное оконечное з.** mid-shunt termination (of a filter); **полосовое фазосдвигающее з.** restricted lag network; **поперечное з. сети** shunt arm (of a network); **предохранительное з.** fuse socket; **простое фазосдвигающее з.** simple lag network; **з. с запаздыванием** delay component; **з. с индуктивностью и ёмкостью** LC-circuit, LC-member; **з. связи** coupling member, link coupling; **сглаживающее з.** steadying circuit, smoothing filter; **симметрирующее з.** balancing network; **следящее за скоростью з.** speed follower; **Т-образное з.** T-network (of a filter); **Т-образное з. фильтра низких частот** middle-condenser circuit; **Т-образное оконечное з.** mid-series termination (of a filter); **управляющее з.** variable attenuator; **фазоопережающее з.** lead network; **з. фильтра нижних частот** low-pass filter section; **з. фильтра типа M** m-derived half section; **з. RC-цепи** RC network; **з.**

**четырёхполюсника** partial four-terminal network, partial fourpole network

**звенящий** *adj.* ringing, jingling

**звон** *m.* ring, ringing, peal; **з. на заднем фронте** trailing ringing (of a pulse)

**звонить, позвонить** *v.* to ring; to phone up, to call up

**звонкий** *adj.* ringing, clear, resounding, sonorous; voiced

**звонковый** *adj.* bell; ringing

**звонкость** *f.* clearness, sonorousness; reverberation

**звонок** *m.* bell, ring, ringer; **з. в номернике** indicating bell; **вызывной з. с управлением от реле подводного кабеля** submarine call bell; **дифференциальный з.** differential-wound bell; **дребезжащий з.** vibrating bell, trembler bell; **з.-колокол** circular bell; **з. с двумя чашками** double bell; **з. с длительными интервалами между ударами** pulsed ringer; **з. с нумератором** indicating bell; **з. с падающими клапанами** drop indicator bell; **з.-трещотка** trembling bell; **условный з.** code ringing; **электрический з. с прерыванием тока** buzzer

**звук** *m.* sound, tone; tune; **звуки в пределах слышимости** audio sound; **взрывные звуки** explosive (*or* stop) sounds; **глухие согласные звуки** voiceless consonants; **высокий з.** high-pitched sound; **гласный з.** vowel sound; **глухой короткий з.** dead short sound; **двугласный з.** diphthong; **засурдиненный з.** stopped note; **звонкий взрывной согласный з.** voiced stop; **нетональный з.** unpitched sound; **параллельный з.** separate sound (*tv*); **з., передаваемый методом частотной модуляции** frequency-modulated sound; **полугласный з.** semi-vowel sound; **порывистый з.** rushing sound; **пронзительный з.** squeal, squawk; **протяжный з. речи** continuant; **распространяющийся в воде з.** waterborne sound; **з. с недостатком верхних частот** no top sound; **з. с преобладанием верхних (*or* высоких) частот** all-top sound; **з. с преобладанием низких частот** all-bottom sound; **ударный з. от головной волны** sonic boom; **шумящий з.** sibilant

sound, hiss sound; **шумовой з.** unpitched sound

**звуко-** *prefix* phono-, phon-, sound

**звукоанализатор** *m.* sound analyzer

**звукоблок** *m.* sound reproducer; **з. кинопроекционной установки** optical sound reproducer, photographic sound reproducer

**звуковидение** *n.* ultrasonoscopy

**звуковой** *adj.* acoustic(al), sound, sonic, voice; audio; audible, aural; auditory

**звуковоспроизведение** *n.* sound reproduction, sound-reproducing, audio reproduction

**звуковоспроизводящий** *adj.* sound reproducing

**звукогенератор** *m.* audio-frequency generator

**звукоглушитель** *m.* silencer, muffler, sound damper; vibration damper

**звукозаписыватель** *m.* sound recorder, transcriber, voice recorder; phonograph recorder

**звукозаписывающий** *adj.* recording, sound recording

**звукозапись** *f.* sound recording, transcription; phonogram; **з. в фонотеке** canned music; **з. до съёмки** prescoring; **многозубчатая з. (на плёнку)** multi-track variable-area sound recording (on film); **з. на пластинку** disk recording, sound-on-disk recording; **з. на пластинку по способу Берлинера** lateral recording; **з. на плёнку** sound-on-film recording, magnetic tape recording; **з. (по методу) переменной ширины** variable-area recording; **з. с обесшумливанием** noiseless recording; **трансверсальная з.** variable-area sound recording on film; **фотографическая з.** variable-density sound recording (on film)

**звукозащитный** *adj.* sound-resisting

**звукозондаж** *m.* echo-sounding

**звукоизлучатель** *m.* acoustic generator, acoustic radiator

**звукоизлучение** *n.* sound generation

**звукоизмерение** *n.* sound ranging; **з. дальности** sound ranging

**звукоизолирование** *n.* sound deafening, deafening

**звукоизолировать** *v.* to deafen

**звукоизолирующий, звукоизоляционный** *adj.* sound-insulating, sound-resisting, soundproof

**звукоизоляция** *f.* soundproofing, sound insulation, sound deafening; draping (of a room); **з. заглушек головного телефона** earphone cushion attenuation

**звуколента** *f.* audio tape

**звуколокатор** *m.* sonar, asdic, audio-locator, aerial sounding line; **миниатюрный з.** standing wave area motion indicator

**звуколокация** *f.* sound fixing-and-ranging; sonar

**звуколюминесценция** *f.* sonoluminescence

**звукомаяк** *m.* aural radio range, aural-type beacon

**звукомер** *m.* phonometer, sound level meter

**звукометрический** *adj.* sound-ranging

**звукометрия** *f.* sound ranging, audibility test

**звуконепроницаемость** *f.* soundproofing

**звуконепроницаемый** *adj.* soundproof, free of resonance, non-echoic

**звуконоситель** *m.* sound carrier

**звукооператор** *m.* recordist, sound recordist, audio operator, soundman, monitor man, audio control engineer

**звукоотражение** *n.* sound reflection, acoustical reflection

**звукоощущение** *n.* sound sensation, tone perception

**звукопеленгатор** *m.* direction-listening device

**звукопеленгация** *f.* sound bearing, sound ranging

**звукопередача** *f.* sound transmission, speech transmission

**звукопередающий** *adj.* sound-transmitting, acoustic transmission

**звукописец** *m.* sound recorder, voice recorder; phonograph recorder

**звукопоглотитель** *m.* sound absorber, sound damper, silencer, sound-absorbing material, filterplexer; **подвесной з.** functional sound absorber

**звукопоглощательный, звукопоглощающий** *adj.* sound-absorbing, sound-proof

**звукопоглощение** *n.* sound absorption

**звукоподобный** *adj.* sound-like

**звукоподражнение** *n.* sound effects; onomatopoeia

**звукоподражательный** *adj.* onomatopoeic

**звукопреломление** *n.* acoustical refraction

**звукоприёмник** *m.* sound receiver, acoustic detector; microphone

**звукоприёмный** *adj.* sound-pickup, sound-picking

**звукоприставка** *f.* sound pickup, sound head

**звукопровод** *m.* acoustic line, sound duct

**звукопроводимость, звукопроводность** *f.* sound-transmission, sound conduction, sound conductivity

**звукопроводный, звукопроводящий** *adj.* sound-conducting, sound-transmitting

**звукопроекция** *f.* sound projection

**звукопрожектор** *m.* sound projector

**звукопроникаемость, звукопроницаемость** *f.* sound transmission, acoustical transmission

**звукопроницаемый** *adj.* sound-transmitting

**звукопропускание** *n.* sound transmission, acoustical transmitivity

**звукорадиобуй** *m.* sonoradio buoy, radio sonobuoy; **з. пускаемый в расход** expendable sonoradio buoy

**звукорассеивающий** *adj.* sound-scattering

**звукорепродукция** *f.* sound reproduction

**звукосигнал** *m.* sound signal, audio signal, audible signal

**звукосниматель** *m.* sound pickup, phonograph pickup, adapter, gramophone adapter; **керамический з. со сдвоенной иглой** dual-stylus ceramic pickup; **з. механической записи** mechanical reproducer; **з. с массивной головкой** bulkhead pickup; **угольный з.** carbon-contact pickup

**звукосочетание** *n.* sound combination

**звукостирание** *n.* sound erasure

**звукостирающий** *adj.* sound erasing

**звукосъёмник** *see* звукосниматель

**звукосъёмочный** *adj.* sound-pickup

**звукотень** *f.* sound shadow

**звукотехника** *f.* phonics

**звукоулавливание** *n.* sound ranging, sound locating

**звукоулавливатель, звукоуловитель** *m.* sonic detector, sound-locator, sound-ranger, phonozenograph; sound receiver

**звукоусиление** *n.* sound amplification

**звукоусилитель** *m.* sound amplifier

**звукоусилительный** *adj.* sound-amplifying

**звукофикационный** *adj.* public-address

**звукофикация** *f.* public address system

**звукочастотный** *adj.* voice-frequency, audio-frequency

**звукочувствительный** *adj.* sound-sensitive

**звукоячейка** *f.* sound cell

**звучание** *n.* sound(ing), vibration, sonorousness, phonation; ringing; **з. от релаксационных колебаний** motor-boating; **з. свирели** harmonic tones; **з. твёрдого тела** sound in solids

**звучать, прозвучать** *v.* to sound, to resound; to ring

**звучащий** *adj.* sounding, vibrating, reverberating, sonorific; ringing

**звучно** *adv.* loudly, sonorously

**звучность** *f.* richness of sound, sonority, sonorousness, resonance

**звучный** *adj.* sonorous, loud, sonant, resonant

**звякание** *n.* tinkling

**звякать, звякнуть** *v.* to tinkle, to jingle

**ЗГ** *abbr.* (**звуковой генератор**) audio oscillator

**з-д** *abbr.* (**завод**) plant, factory

**здание** *n.* building; **з. распределительного устройства** switch-house

**ЗДМ** *abbr.* (**закон действующих масс**) law of mass action

**зев** *m.* opening, jaw, mouth

**Зегер** Seger

**Зеебек** Seebeck

**Зееман** Zeemann

**зеемановский** *adj.* (of) Zeemann, Zeemann's

**Зейберт** Seiberth

**Зейбт** Seibt

**зелёный** *adj.* green

**зем.** *abbr.* (**земельный; земля**) ground, soil

**земле-** *prefix* ground, earth, land

**землекоп** *m.* excavator, digger

**землепроводный** *adj.* earth-conduction

**землесос** *m.* dredge pump, suction dredge

**землетрясение** *n.* earthquake

**землечерпательный** *adj.* dredging

**земля** *f.* ground, earth, soil; land; ground fault; **з. в качестве обратного провода** earth return circuit; **з.-воздух** ground-to-air; **з.-корпус** rack-ground; **«з.» на линии** line-to-ground fault; **перемежающаяся з.** swinging ground, intermittent ground; **полная «з.»** solid ground, dead ground

**земляной** *adj.* earth, earthen; ground

**земной** *adj.* ground; earth, earthen, earthly, terrestrial

**земномагнитный** *adj.* earth-magnetic

**Зенер** Zener's, (of) Zener

**зенеровский** *adj.* Zener's

**зензубель** *m.* tooth plane, molding plane

**зенит** *m.* zenith

**зенитный** *adj.* zenith; anti-aircraft

**зенкер** *m.* vertical drill

**зеркало** *n.* mirror; speculum (for optical instruments); surface (of liquid); reflector; **скрещённые светоделительные зеркала** crossed dichroics; **з., входящее в сечение светового пучка** incoming mirror; **з., выходящее из сечения светового пучка** outgoing mirror; **зелёное светоделительное з.** green-reflecting dichroic; **интерференционное светоделительное з.** color selective mirror, dichroic mirror; **красное светоделительное з.** red-reflecting dichroic; **противо-параллаксное з.** parallax-free mirror; **расщепляющее з.** beam-splitting mirror; **з. с наружной металлизацией** front-surface mirror; **светоделительное з.** color selective mirror, dichroic mirror; **синее светоделительное з.** blue-reflecting dichroic; **цветоизбирательное з.** dichroic filter, dichroic mirror

**зеркальноотражённый** *adj.* specularly-reflected, mirror-reflected

**зеркальный** *adj.* mirror; specular, mirror-like; reflecting; shining; optical (isomerism); image

**зеркальце** *n.* mirror

**зерненный** *adj.* granulated

**зернистость** *f.* grain, granularity; graininess, seediness; grain-size; mesh; **з. светочувствительного слоя** film graininess

**зернистый** *adj.* granular, grained, granulated

**зерно** *n.* grain, granule, kernel; nodule; **з. малого размера** fine grain; **з. угольного порошка** carbon granule

**зерновой** *adj.* grain, granular

**Зиверт** Sievert

**зигзаг** *m.* zigzag

**зигзагообразный** *adj.* zigzag, zigzag-shaped, crisscross, staggered; saw-like, notched, serrated

**зимозиметр** *m.* zymometer

**зимоскоп** *m.* zymoscope

**злободневность** *f.* actuality, actualness

**злободневный** *adj.* topical; on issues of the day

**злостный** *adj.* malicious

**змеевидный** *adj.* serpentine, sinuous, zigzag; coil, spiral

**змеевик** *m.* coil, coil pipe; **подогревающий з.** reheating coil

**змеевиковый** *adj.* coil, spiral, worm

**змейковый** *adj.* kite(-shaped); sinuous

**зн.** *abbr.* (знак) mark, sign, symbol; signal

**знак** *m.* mark, sign, symbol, character; signal; digit; **неверные знаки на ленте** tailings (*tphny.*); **з. бекар** natural (*mus.*); **выдавленный з.** embossed character; **выжженный з.** marking sign, branding symbol; **з. выноски** asterisk (*math.*); **двоичный з.** bit, binary digit, binit; **з. исключения** erase character; **з. испытательного напряжения** test voltage symbol; **з. корня** radical, radical sign (*math.*); **з. молнии** danger arrow; **наземный опознавательный з.** ground indicator; **з., нанесённый магнитными чернилами** magnetic-ink character; **отличительный з.** dot mark; **з. перехода на верхний регистр** shift-out character; **з. перехода на нижний регистр** shift-in character; **последовательный з. за знаком** serial by character; **потерянный двоичный з.** dropped bit; **предохранительный з.** security signal; **з. пропуска** ignore; **разрушенный з.** destroyed digit; **рельефный з.** embossed character; **з. смены регистра** escape character; **з. смены регистра без блокировки** nonlocking shift character; **з. смены регистра с блокировкой** locking shift character; **з. смены типа шрифта** font-change character; **з. сноски** asterisk (*math.*); **з. управления передачей** transmission control character; **управляющий з.** control character, functional character

**знаковыбиратель** *m.* sign selector

**знаковый** *adj.* sign, symbol; marking

**знакоизбирающий** *adj.* sign-selecting

**знакоинвертор** *m.* sign-inverter

**знакообразование** *n.* sign-formation, character-formation

**знакопеременный** *adj.* sign-changing; alternate, alternating

**знакопечатающий** *adj.* symbol-printing, sign-printing

**знаменатель** *m.* nominator, denominator; **з. относительного отверстия** aperture number

**знаменательный** *adj.* significant, important; denominative

**значащий** *adj.* significant; meaningful; meaning

**значение** *n.* value, magnitude; significance, meaning, sense, importance, import; (*oft.* omitted in translation, e.g.: **з. частоты** = frequency) magnitude of the frequency = frequency); **комплексные дискретные значения** complex samples; **(переключаемый) на два номинальных значения, с двумя номинальными значениями** double-rated; **средние значения величин переменного тока** mean values of alternating current quantities; **з. в наинизшей точке кривой** valley value; **з. в системе с избытком три** excess-three value; **выделенное дискретное з. (из непрерывного) сигнала** sampled signal, separated signal; **выделенное мгновенное з. сигнала** instantaneous sample; **действующее з.** effective value; **дифференциальное з. гаммы** point gamma; **долевое з. общего реактанса цепи возбуждения** per unit total field reactance; **долевое з. переходной реактивности** per unit transient reactance; **заданное з.** predetermined value, desired value, index value, reference quantity; **заданное з. регулируемой величины** control point; **идеальное з. величины** ideal value; **измеренное з. величины** test value, measured value; **иррациональное з.** incommensurate value; **з. команды** controlling-pulse value; **комплексное з. величины** complex value (*math.*); **з. контраста мелких деталей изображения** detail contrast ratio; **контрольное з.** reference quantity; **максимальное мгновенное з. тока** peak current; **максимальное отрица-**

тельное з. negative peak; **макси-мальное считываемое з.** full-scale value; **мгновенное з. напряжения** (instantaneous) voltage value; **з. модуля упругости при постоянном электрическом смещении** constant electrical displacement value of the elastic modulus; **номинальное з. влияющей величины** rating (in terms of a limiting quantity); **ориентиро-вочное з.** direction esteem; **пиковое з. плотности магнитного потока** peak flux density; **пиковое з. уровня белого** peak white; **з. по шкале серых цветов** grey-scale value; **по-следнее вычислительное з.** latest entry; **предельное з. шкалы** infinity; **предписанное з.** rated value, nominal level; **регулируемое измеренное з.** typical meter reading; **з. сопро-тивления** withstand value; **средне-взвешенное з.** weighted average (math.); **среднее з. безразмерного акустического импеданса** average acoustic impedance ratio; **среднее з. мощности флуктуации** mean fluctuation power; **среднее з. сопротивле-ния** average impedance; **среднее арифметическое з. квадрата ошибки** square of error (math.); **среднее по множеству з. функции** assembly average (math.); **среднеквадратичное з.** RMS value, root-mean-square value, effective value; **текущее з. суммы** running sum; **установив-шееся з.** potential value; **установив-шееся з. регулируемого параметра** (or **регулируемой величины**) final controlled condition; **частное з. на-вигационной координаты** element of a fix; **эффективное з. обратного анодного напряжения** RMS inverse voltage rating; **эффективное з. тока** RMS current value

**значимость** f. significance
**значимый** adj. significant
**значительно** adv. considerably, significantly, much
**значительность** f. significance, magnitude
**значительный** adj. considerable, important, significant, notable, sizable, marked
**значить** v. to mean, to imply, to signify
**значность** f. valency, atomicity

**значный** adj. marking, significant, valued; (as suffix) -character, -symbol
**значок** m. designation, marking, distinctive mark, sign
**золотник** m. slide valve, slide, gate valve; **вращающийся з.** rotary barrel throttle; **плоский з.** slide valve; **распределительный з. воздухопо-дачи** air slide valve
**золотниковый** adj. valve, slide-valve
**золото** n. gold, Au
**золотой** adj. golden, gold
**золочёный** adj. gold-plated, gilded
**Зоммерфельд** Sommerfeld
**зона** f. zone, field, area, spot, region, section; band, belt, range; sector; **мёртвые зоны вследствие отраже-ния от земли** ground-reflection nulls; **ближняя з.** close range; local zone; **дальняя з.** night reception zone, telephone trunk zone, telephone toll zone; **двусторонняя з. видимости (радиолокатора)** two-way coverage pattern; **з. действия** (zone of) cover-age; (maximum) range of radio waves; (maximum) transmission range (of telephone systems); service area; **з. действия по вертикали** (or **углу места**) vertical coverage; **з. действия централизационной уста-новки** interlocking limits; **заполнен-ная з. (полупроводника)** space charge region (of a semiconductor); **запретная** (or **запрещённая**) **з.** energy gap, forbidden band; **между-городная з.** telephone trunk zone, telephone toll zone; **з. мерцания** flutter fading area; **з. непрозрач-ности (фильтра)** (filter) attenuation band; **з. неуверенного приёма** secondary service area; **з. неустой-чивости слышимости** freak zone, freak range; **з. нечувствительности** dead zone; **з. нормального приёма** primary service area; **з. обзора на больших высотах** high-altitude coverage; **з. обзора на малых высо-тах** low-altitude coverage; **з. облу-чения** coverage, scanned area, gate of entry; **з. обмотки у вершины пазов** top of slot belt; **з. обнаруже-ния** coverage pattern; **односторон-няя з. видимости радиолокатора** one-way coverage pattern; **первая з. молчания** primary skip zone; **з. повреждения помехам** interference

range; **поисковая з. действия** search coverage; **з. поражения прямым ударом молнии** rideflashing zone, rebounding zone; **пригородная з.** close range; local zone; **з. прозрачности (фильтра)** (filter) transmission band; **пролётная з.** drift region; **з. прямой видимости** radio-optical line of distance, radio-optical range; **рабочая з.** occupied area, occupied space; **з. размытости** width of blurring, width of confusion (*tv*); **з. размытости по вертикали** vertical width of confusion (*tv*); **районированная з.** area of a multi-office exchange; **з. сумерек** twilight zone; **тарифная з.** exchange area, service zone (*tphny.*); **теневая з.** shadow, shadowed receiving site; **з. уверенного приёма** primary service area; **угловая з. нечувствительности** angular backlash

**зональный** *adj.* zonal, zone, regional

**зонд** *m.* sonde, probe, sound; sounding balloon; **акустический з.** probe microphone; **выходной з. высокой частоты** radio-frequency output probe; **з. для нахождения неисправностей** tracing probe; **защищённый термоизмерительный з.** pyrometer tube; **измерительный з. с движением по вертикали** hunting probe; **регулируемый з.** tuning probe; **термоизмерительный з.** pyrometer probe, thermometer probe

**зондаж** *m.*, **зондирование** *n.* sounding, probing; **з. вразброс** scatter sounding

**зондировать, позондировать** *v.* to probe, to sound, to search, to explore

**зондировочный, зондирующий** *adj.* sounding, exploring, probing

**зондовый** *adj.* sonde, sonde-type

**зонированный** *adj.* zoned

**зонный, зоновый** *adj.* zone, zonal

**зонолит** *m.* zonolite mica

**зонтик** *m.* umbrella; cover; cupola; **металлический з.** rain cone

**зонтичный** *adj.* umbrella, umbrella-shaped

**зонтообразный** *adj.* umbrella-shaped

**ЗПУ** *abbr.* (**заданный путевой угол**) given course angle

**зрачок** *m.* pupil

**зрение** *n.* vision, eyesight; **дневное з.** anoptic vision, photopic vision; **ночное з.** scotopic vision

**зритель** *m.* viewer; observer

**зрительный** *adj.* visual, optic, optical; signal (communication)

**зуб** *m.* tooth; dent, claw, cam; **конические зубья** vertical hub (of a Strowger switch); **стопорный з.** locking cog

**зубец** *m.* tooth, cog, cam, catch, lug; dent, notch; (drill) bit; **з. дискового разрядника** stud of a disk discharger

**зубило** *n.* chisel, firmer chisel, cutting chisel, bit, pinch; **з. с черенком** rod chisel

**зубильный** *adj.* chisel, bit, punch

**зубомер** *m.* gear gage, tooth gage

**зуборезный** *adj.* gear-milling, tooth-cutting

**зубострогальный** *adj.* gear-planning

**зубчатка** *f.* rack-wheel, tooth pinion, gear, gear wheel; cam; **импульсная з. (номеронабирателя)** pulsing cam (of a dial); **подъёмная з.** vertical hub (of a Strowger switch)

**зубчато-дисковый** *adj.* toothed disk

**зубчатость** *f.* serration

**зубчатый** *adj.* toothed, geared, gear, serrated, serrate, cogged, dented, indented, notched, jagged

**зубчик** *m.* kink; ripple; *pl.* **зубчики** serrations, ripple(s); blurs

**зуммер** *m.* buzzer, hummer, vibrator, howler, ticker, phonic ringer; tone signal, audio signal; **з. высокого тона** high-frequency buzzer; **з. для фонического вызова абонента** howler; **з. занятости** busy tone (*tphny.*); **з. изменения номера абонента** changed-number tone (*tphny.*); **измерительный з.** audio-frequency (signal) generator; **з. контроля посылки вызова** audible ringing signal, ringing tone (*tphny.*); **ламповый з.** (electron-)tube buzzer, vacuum-tube-type audio generator; **з. окончания разговора** clearing signal buzzer (*tphny.*); **з. ответа станции** dial tone (*tphny.*); **з. с изменяющимся тоном** variable-note buzzer; **з. с лампой тлеющего разряда** glow-discharge tube audio generator

**зуммерный** *adj.* humming; buzzer

**зуммирование** *n.* buzzing, howling, humming, singing

**зуммировать** *v.* to buzz

**зуммирующий** *adj.* buzzing, humming, singing

**зухтон** *m.* search tone
**зыбкий** *adj.* unsteady, vacillating, unstable
**зыбкость** *f.* vacillation, fluctuation

**зыбчатость** *f.* ripple factor, percent ripple
**зычность** *f.* loudness
**зычный** *adj.* loud, sonorous

# И

**и** *conj. and adv.* and, as well as; also, too, even; indeed . . . *as in* **вот и проблема** (or omit in transl.); **и... и...** both . . . and . . .

**ИА** *abbr.* (**истинный азимут**) true azimuth

**иатрон** *m.* iatron tube

**ИВ** *abbr.* (**искатель вызовов**) finder switch

**Ивенс** Evans

**ИВ-ИВ** *abbr.* (**искатель вызова-искатель вызова**) line selector with line finder

**ИВС** *abbr.* (**истинная воздушная скорость**) true airspeed

**ИГВФ** *abbr.* (**Институт гражданского воздушного флота**) Institute of the Civil Air Fleet

**«игедур»** *m.* igedur

**игла** *f.* stylus, needle; pivot; **и. Вика** test-probe for heat tests on solid insulating materials; **и. для выдавливания звуковой дорожки** embossing stylus; **и. для звукозаписи** recording stylus; **и. накопления** dip needle; **z-образная и.** shank needle; **перегнутая и.** bent-shank needle; **полировочная и.** burnishing tool; **чертёжная и.** scratch awl

**иглистый** *adj.* needle-shaped

**игловидный, иглообразный** *adj.* pin-shaped, needle-shaped, needle

**игнайтер** *m.* ignitor; igniter

**игнитор** *m.* ignitor

**игниторный** *adj.* ignitor

**игнитрон** *m.* ignitron, ignitron rectifier; **разборочный и., и. с непрерывной откачкой** pumped ignitron

**игнитронный** *adj.* ignitron

**игнорирование** *n.* ignoring

**игнорировать** *v.* to ignore, to disregard

**иголка** *f.* needle; **и. для отыскивания жил** test point needle, pricker, testing spike, test pick

**игольный** *adj.* needle

**игольчатый** *adj.* needle, needle-shaped, spiny, spicular

**игра** *f.* play, backlash, looseness, freedom, slackness; clearance; game; **безобидная и.** fare game; **и. более двух участников** more-than-two-person game; **и. валков** backlash; **и. двух участников** two-person game; **и. на выживание** game of survival; **и. на ускользание** eluding game; **и. нескольких участников** multi-person game; **и. Ним** game of Nim; **продольная и.** axial clearance; **прямоугольная и.** matrix game; **и. с блефом** bluffing game; **и. с накопленной суммой** general-sum game; **и. с ограничениями** constrained game; **и. с одним участником** solitaire; **и. с полной информацией** perfect-information game; **и. с фиксированным объёмом выборки** fixed sample size game; **и. N участников** N-person game

**играть, сыграть** *v.* to play

**играющий** *adj.* playing

**игровой** *adj.* game

**игрок** *m.* player

**ИДД** *abbr.* (**измеритель динамических деформаций**) measuring equipment for dynamic deformations

**идеализировать** *v.* to idealize

**идеализированный, идеализованный** *adj.* idealized, theoretical

**идеально** *adv.* perfectly, ideally

**идеальнорассеянный** *adj.* uniformly diffuse(d)

**идеальный** *adj.* ideal, perfect, theoretical, optimum

**идентификатор** *m.* identifier; **и. массива** array identifier; **и. переменной** variable identifier

**идентификационный** *adj.* identification

**идентификация** *f.* identification

**идентифицировать** *v.* to identify, to determine

**идентифицируемый** *adj.* identifiable, identified

**идентичность** *f.* identity

**идентичный** *adj.* identical

**идея** *f.* idea, notion, conception

идиома *f.* idiom
идиоматический *adj.* idiomatic(al)
идиометр *m.* idiometer
идиоморфный *adj.* idiomorphic
идиоплазма *f.* germ plasma
идиостатический *adj.* idiostatic
идиохроматический *adj.* idiochromatic
идиоэлектрический *adj.* idioelectric(al)
идти, пойти *v.* to go; to run, to work, to operate; to proceed
идущий *adj.* going, running, working, operating; reaching; **и. в положительном направлении** positive-going; **и. вверх** rising; **и. вниз** descending, falling
иенский *adj.* Jena
иерархия *f.* hierarchy; **и. запоминающего устройства** hierarchy of storage
из *prep. genit.* from, out of, with; of (*as in* **один из этих...**)
избавить(ся) *pf. of* избавлять(ся)
избавление *n.* release
избавленный *adj.* released, freed, rid
избавлять, избавить *v.* to free (of), to rid (of), to eliminate, to release
избавляться, избавиться *v.* to eliminate, to get rid of
избежание *n.* escape, avoidance, avoiding
избираемый *adj.* selected, chosen
избиратель *m.* selector
избирательность *f.* selectivity, selectance, selection, discrimination, sharpness of tuning; **логарифмическая и.** log-log selectivity; **и. по высокой частоте** radio-frequency selectivity; **и. по зеркальному каналу** image selectivity, image suppression; **и., повышенная за счёт боковых частот** skirt selectivity
избирательный *adj.* selective, discrimination
избирать, избрать *v.* to select, to choose
избрание *n.* selection
избранный *adj.* selected, selective, chosen
избрать *pf. of* избирать
избытный *adj.* excess
избыток *m.* excess, surplus; plenty, abundance, profusion; **и. трафика** detour traffic for peak loads (*tphny.*)
избыточность *f.* redundance, redundancy
избыточный *adj.* excessive, surplus, overflow, redundant, superfluous

извержение *n.* ejection, emission, outbreak, effusion, discharge, eruption
изверженный *adj.* ejected, emitted
известитель *m.* signaling device; indicator; **вспомогательный пожарный и.** auxiliary fire-alarm box; **пожарный и. с разбиваемым стеклом** break-glass station
известительный *adj.* indicating
известить *pf. of* извещать
известковать *v.* to lime(-coat)
известковый *adj.* lime, calcium, calciferous
известный *adj.* certain; well-known, famous
известь *f.* lime
извещатель *m.* alarm, signaling device
извещать, известить *v.* to inform, to notify, to let known; to indicate
извещение *n.* notification, information, notice, message, report; **и. приближения к заграждающему сигналу** restricting indication
извилистый *adj.* serpentine, sinuous, winding, meandering
извлекаемый *adj.* expendable, extractable
извлекать, извлечь to extract, to draw, to draw out, to draw off, to withdraw, to recover, to extricate; to isolate; to derive
извлечение *n.* extracting, extraction, recovery; taking (the root of); abstract; **и. влаги** dehydrating; **и. из ванн** "pulling" of tanks (*elec.-chem.*); **и. квадратного корня** square rooting, root-squaring; **и. корня** evolution, rooting, extracting; **электрохимическое и. металлов** electroextraction
извлечённый *adj.* extracted, drawn off; derived
извлечь *pf. of* извлекать
извне *adv.* from without
извращать, извратить *v.* to misinterpret, to misrepresent, to distort
извращение *n.* distortion; inversion
извращённый *adj.* distorted
изгиб *m.* bend, dending, curvature, curve, crook, flexure, arc, kink, deflection; fold; winding; offset; elbow (of a pipe); **и. в данной точке** slope (of a string); **и. в плоскости E** E bend; **и. в плоскости H** H bend, edgewise bend; **и. волновода в плоскости H** H-plane waveguide

bend; **малый и.** easy bend; **продольный и.** buckling, lateral flexure; **угловой и. волновода** elbows for coaxial transmission-lines; **угловой и. в плоскости** E E corner, E plane bend; **угловой и. в плоскости** H H corner, H plane bend

**изгибаемость** *f.* deflectivity

**изгибание** *n.* bending, curvature, curve, curving, deflection, buckling; **и. в нижних слоях атмосферы** low-temperature bending; **и. под углом** cambering

**изгибать, изогнуть** *v.* to bend, to curve to deflect, to distort, to deform

**изгибаться, изогнуться** *v.* to bend, to curve, to buckle, to sag

**изгибающий** *adj.* bending, deflecting, flexural

**изгибающийся** *adj.* wriggling

**изгибный** *adj.* flexural

**изглаживание** *n.* fading out, dying out; smoothing out, ironing out

**изготавливать, изготовить** *v.* to make, to manufacture, to produce, to fabricate; to carry out, to execute

**изготовитель** *m.* manufacturer, producer

**изготовить** *pf. of* **изготавливать**

**изготовление** *n.* preparation, manufacture, production; carrying out

**изготовленный** *adj.* prepared, manufactured, produced

**изготовлять** *see* **изготавливать**

**изделие** *n.* work piece, make, product, article, manufactured product; item; **промежуточное и.** semi-manufactured article, semi-finished product

**издержки** *pl.* expenditure, costs, outlay

**из-за** *prep. genit.* because of, through, on account of; from; from behind

**излишек** *m.* excess, surplus; **и. хода** overstroke

**излишне** 1. *adv.* in excess, too much, too many; 2. *pred. adj.* it is unnecessary

**излишний** *adj.* excessive, superfluous, unnecessary; redundant

**изложенный** *adj.* stated, presented, expounded, exposed, developed

**изложница** *f.* mold, casting mold, pan

**излом** *m.* fracture, break, fissure; breakdown; bend; **верхний и.** upper characteristic bend

**изломанный, изломленный** *adj.* fractured, broken

**излучаемость** *f.* radiance, radiancy, radiating power, emissivity

**излучаемый** *adj.* radiated, emitted, outgoing, transmitting

**излучатель** *m.* radiator, emitter, radiating element; source; projector (*acoust.*); active antenna; **вибраторный и.** dipole radiator; **гидроакустический и.** sonar transmitter, asdic transmitter; **и. гидролокатора** sonar projector, sonar transmitter; **диэлектрический стержневой и.** rod radiator (*ant.*); **и. звука** sound source, acoustic source, acoustic radiator; **идеальный и.** blackbody; **изогнутый и.** folded dipole, constrained radiator; **обтекаемый и.** streamlined radiator; **отклоняющий и.** collimating reflector; **полный и.** blackbody, complete radiator, plankian radiator; **и. продольных колебаний** longitudinal vibrator; **и., работающий в установившемся режиме** steady source, steady radiator; **рамочный и.** loop radiator; **рупорно-линзовый и.** horn-and-lense radiator; **рупорный и.** horn-type source, horn-type antenna; **и. с излучением перпендикулярно к оси** broadside aerial; **и. с осевой симметрией** axially symmetrical radiator; **и. Сен-Клера** St. Clair sound generator; **и. синусоидального тона** pure tone source; **и. со щелевой воздушной полостью** slotted cylindrical antenna; **совершенный и.** perfect radiator, blackbody; **стандартный и.** standard illuminant; **стержневой и. из полистирола** polyrod antenna; **трёхвибраторный и.** tridipole radiator; **трубчатый щелевой и.** slotted-tube radiator; **ферритовый стержневой и.** ferrite-filled rod radiator; **и. Фессендана** Fessenden's oscillator

**излучательность** *f.* radiant emittance

**излучательный** *adj.* radiating, emissive, emitting

**излучать, излучить** *v.* to radiate, to emit, to ray, to beam; **обратно и., повторно и.** to re-radiate

**излучаться, излучиться** *v.* to emanate, to radiate

**излучающий** *adj.* radiating, radiant, emitting, transmitting

**излучение** *n.* radiation, emission, emittance, emanation; projection (of sound); beam, evolution; **и. абсолютно чёрного тела** black-body radiation; **боковое и.** lateral radiation, fringe radiation; **и. в видимой части спектра** visible light; **и. в невидимой части спектра** dark light; **и. в ультрафиолетовой части спектра** ultraviolet radiation; **и. внеземного происхождения** extra-terrestrial radiation; **и. волн одной частоты** monofrequent radiation; **и. вразброс** scattered radiation; **высокочастотное и. межзвёздного дейтерия** r-f radiation of interstellar deuterium; **граничное и.** grenz rays, Bucky rays; **неиспользуемое и.** stray radiation; **немонохроматическое и.** heterogeneous radiation; **ненаправленное и.** ring radiation; **и. нормальных частот** transmission of standard frequency; **обратное и. (в антенну)** reradiation, reaction; **оконечное и.** end fire; **и. от медного антикатода** Cu-radiation, copper radiation; **и. отдачи** repulse radiation; **отражённое и.** indirect radiation; **и. перпендикулярно к оси** broadside radiation; **побочное и.** indirect radiation, stem radiation, spurious emission; **и. помимо основного излучателя** leakage radiation; **прямонаправленное и.** head-on radiation; **собственное и.** characteristic radiation, fluorescent radiation; **тормозное и.** bremsstrahlung; **тормозное и. протонов** proton-retardation radiation; **и. фона, фоновое и.** background radiation

**излучённый** *adj.* radiated, emitted, transmitted

**излучить(ся)** *pf. of* **излучать(ся)**

**измельчение** *n.* crushing, breaking up, grinding; refinement

**измельчённый** *adj.* powdered, ground, crushed, pulverized

**изменение** *n.* variation, change, alteration, modification, transformation, fluctuation, deviation; correction; reversal; pulling (*rad.*); **беспорядочные изменения величины** statistical variations; **быстрые изменения слышимости сигналов** swinging of signals; **заблаговременные изменения** pre-modifications; **кратковременные изменения качества детали** short-term drift; **изменения остаточной** высоты с истинной высотой mushing error; **предварительные изменения** pre-modifications; **и. в дальности** alternation in the range; **и. в поперечном направлении** transverse variation; **и. вида действия** mode change; **и. глубины модуляции** change of modulation-degree; **и. глубины эхолотом** echo-sounder work; **и. градации** variation of gamma, change of gamma; **и. затухания в зависимости от амплитуды** nonlinear amplitude distortion; **и. знака на обратный** sign reversal; **и. знака потока** flux reversal; **координатное и.** spatial variation; **и. коэффициента трансформации под нагрузкой** load ratio control; **кубообразное и.** cube translation; **и. линейности** rheolinearity; **и. масштаба времени** memomotion; **медленное и. частоты** slow frequency drift; **и. нагрузочного момента в двигательный** overdrive; **и. направления** change of direction, deviation, diversion; **и. направления стрелки** change of deflection; **и. настройки** tuning variation, tuning drift; **и. обозначений** re-indexing; **и. оптических свойств в электрическом поле** electrostatic optical stress; **относительное и. частоты** fractional frequency change; **паразитное и. полутонов** spurious halftone variation; **периодическое и. фазы цветовой поднесущей (частоты)** oscillating color sequence; **плавное и. величины зазора** tapered gap; **и. подачи** feed change; **и. полутонов из-за помех** spurious half-tone variation; **и. полюсов диода** reversal of diode; **и. порядка чередования фаз** phase reversal; **и. проницаемости** variation of inverse amplification factor; **и. регулируемой величины** change of desired value; **скачкообразное и. частоты** frequency jumping; **скачкообразное и. яркости** jump in brightness; **и. сопротивления порошка микрофона** breathing of microphone; **и. требований** re-retaing; **и. фазы на 180°** phase reversal; **и. фазы на обратную** phasing back; **и. фазы на противоположную** phase inversion; **и. фазы поднесущей на обратную** reversal of subcarrier phase; **и.**

цветового тона hue shift; **и. ширины канала** channel-width variation
**изменённый** *adj.* changed, converted
**изменитель** *m.* changer, converter; **и. знака** inverter
**изменить(ся)** *pf. of* изменять(ся)
**изменчивость** *f.* variability, impermanence
**изменчивый** *adj.* changeable, variable, irregular, floating, inconstant
**изменяемость** *f.* variability, changeability, alterability
**изменяемый** *adj.* variable, changeable, convertible; adjustable
**изменять, изменить** *v.* to alter, to change, to modify, to vary, to alternate, to variate; **и. масштаб** to scale, to change the scale
**изменяться, измениться** *v.* to change; to fluctuate, to vary; **постепенно и.** to fade
**изменяющийся** *adj.* variable, alternating, varying, fluctuating; **постепенно и.** tapered
**измерение** *n.* measurement, measuring, measure, metering, gaging; survey; size, dimension; determination; sounding, fathoming; sampling; ranging; test; **маятниковые измерения ускорения силы тяжести** pendulum gravity measurements; **оптические измерения времени жизни носителей** optical life time measurements; **автоматическое и. угла возвышения** automatic elevation measurement; **акустическое и. расстояний** sound ranging, sound spotting; **аудиометрическое и. порога слышимости** threshold audiometry; **и. без непосредственного контакта** isolation metering; **и. высоты уровня жидкости** liquid-level measurement; **и. глубин эхолотом** echo sounder work, echo sounding; **и. глубины модуляции** modulation measurement; **и. глубины по отражению (ультра)звука, и. глубины ультразвуком** reflection sounding; **и. дальности** distance measurement, ranging, range measurement; **и. дальности полёта снаряда** missile-range measurement; **и. дальности с помощью эха** echo ranging; **и. диаграмм направленности** antenna-pattern measurement; **дистанционное и.** telemetering, remote metering,

remote measuring; **и. дифференциального усиления** differential-gain test; **дуговое и.** circular measure, radian; **и. ёмкостей постоянным током** d.c. charging test method; **и. зависимости усиления от высоты** height-gain measurement; **и. затухания на участке линии** transmission efficiency test; **и. затухания нелинейного искажения** distortion measurement; **и. излучательной способности** emissivity measurement; **и. количества протекающей жидкости** rate of flow measurement; **косвенное и. коэффициента полезного действия** indirect measurement of efficiency; **и. коэффициента трансформации** transformer ratio test; **и. крутизны характеристики компенсационным методом** null measurement of transconductance; **и. (линии) на прямую** straightway measurement (of a line), go-and-return measurement; **магнитное и. ускорения силы тяжести** pendulum gravity measurement; **и. морских глубин эхолотом** marine echo sounding; **и. мостиком, и. мостом** bridge measurement, bridge test; **и. мощности высокой частоты** R-F power measurement; **и. на малой мощности** low-power measurement; **и. на мостовой схеме** bridge measurement; **и. напряжённости поля** field-intensity measurement; **непосредственное и. коэффициента полезного действия** direct measurement of efficiency; **и. неправильного затухания** balance attenuation measurement; **и. нечётных гармоник** test odd, measurement of odd harmonics; **и. нулевым методом** false zero test; **объективное и. громкости** measurement of volume by means of indicators; **и. остаточных помех** interference elimination measuring; **и. по методу сравнения** comparison measurement; **и. повреждения** fault location test; **и. полного сопротивления (импеданса) цепи заземления** earth loop impedance test; **и. пробелов** range measurement; **и. проницаемости** measurement of inverse amplification factor; **и. размеров пятна** spot measurement; **и. расстояния методом эха** echo ranging;

и. расходов пульсирующих потоков pulsating-flow measurement; **и. с обоих концов между кабелем и землёй** overlap test; **и. свойств магнитной ленты** tape measurement; **и. сопротивлением** thermoelectric measurement; **и. сопротивления высокой частоты** R-F resistance measurement; **и. угла места** measurement of elevation; **и. участка** end-to-end measurement; **и. частного клирфактора** measurement of partial distortion, measurement of nonlinear distortion; **и. шлейфа, и. шлейфом** loop measurement, go-and-return measurement; **и. эхолотом** sounding

**измеренный** *adj.* measured, gaged
**измеримость** *f.* measurability
**измеримый** *adj.* measurable, mensurable
**измеритель** *m.* measurer, meter, gage, counter movement; pickup, feeler; index; **и. акустической мощности** sound radiometer; **бесконтактный и. толщины** non-contacting thickness gage; **и. величины излучения** fallout meter; **и. внутренней ёмкости** capacitance meter for measurement of effective interelectrode capacitance; **и. вращательного движения** angular accelerometer; **и. времени прохождения звука** period meter; **вч и. ёмкости** radio-frequency capacitometer; **и. высоты облачности** nephoscope; **и. выходного уровня** output level meter; **и. выходной мощности** output-power meter; **и. глубины модуляции** percentage modulation meter; **и. давления радиации звуковой волны** acoustic(al) radiometer; **динамометрический и. хода двигателя** dynamometer motor meter; **дистанционный и. напряжённости поля** field intensity meter; **дифференциальный и. мощности** Johnson power meter; **и. добротности** Q-meter; **и. жёсткости лучей** radiochronometer; **и. заземления** earth tester, ground tester; **и. затухания** attenuation meter, transmission measuring set, decremeter; **и. затухания эха** return loss measuring set; **и. звука** acoustimeter; **и. колебаний высоты тона** pitch-variation indicator; **и. кратковременных интерва-**

лов по Бему Behm microchronometer; **и. кривизны линий** rotameter; **и. крутизны (характеристики лампы)** slope meter; **логарифмический и. излучений** logarithmic ratemeter; **и. магнитного потока** fluxmeter; **и. (магнитной) индукции** gaussmeter; **магнитный и. удлинения** magnetic strain gage; **и. максимума** demand meter, maximum-demand meter; **и. модуляции** level indicator, volume indicator, tone control; **и. мощности дозы** dose rate meter; **и. мощности, передаваемой в прямом направлении** forward-power meter; **и. напора** head meter; **и. напряжённости поля** field-intensity meter, field strength meter; **и. напряжённости поля в волноводе** traveling detector; **и. неоднородности цепи** reflection measuring set, return-loss measuring set; **и. неправильного затухания** unbalance attenuation measuring set; **и. нулевого напряжения** voltmeter with extended zero range; **и. объёмных единиц** VU meter; **и. освещённости** luxmeter, illumination meter; **и. остроты слуха** acuity meter; **и. отношения** ratiometer; **и. отношения токов** logometer, quotient meter; **и. отражений** backscatter meter; **и. перекрёстной наводки** crosstalk measuring set; **переносный и. освещённости** foot-candle meter; **и. переходного затухания, и. переходных разговоров** crosstalk meter, crosstalk measuring set; **и. полного сопротивления** Z meter; **и. полных сопротивлений нулевого типа** null-type impedance meter; **полупроводниковый и. крутящего момента** transistor torquemeter; **и. поля** radar densiometer; **и. помех** interference measuring apparatus; **и. постоянного напряжения** direct current voltmeter, direct voltage meter; **и. потери слуха** audiometer; **и. потерь от небаланса линии** impedance unbalance measuring set; **и. потока излучения** fluxmeter; **и. продувочной воды** blow-down meter; **и. пройденного расстояния** cyclometer; **и. проникающей силы** (*or* **способности**) penetrometer; **и. разности давлений** venturi system; **и. рас-**

согласования error meter, error sensor; **и. расхода по перепаду давлений** pressure differential meter; **и. реактивных вольт-ампер** idle-current wattmeter; **и. реактивных киловольтампер** kilovar instrument; **самопишущий и. уровня передачи** recording transmission measuring set; **сверхзвуковой и. толщины металла** sonigage; **и. светового потока** lumenmeter; **и. силы звука** acoustic(al) radiometer, acousti-meter, sound intensity meter; **и. силы поля** densiometer; **и. сколжения** slip meter; **и. скороподъёмности** rate-of-climb meter; **и. скорости вращения** gyrometer; **и. скорости протекания (жидкости)** flowmeter; **и. скорости распространения** velocity-of-propagation meter, VP-meter; **и. скорости снарядов** counter chronograph; **и. скорости счёта** counting-rate meter; **и. содержания влаги** moisture meter; **и. сопротивления заземления** earth resistance meter, groundmeter; **и. сопротивления изоляции** insulation tester; **и. среднеквадратичного значения** root-mean-square meter; **и. температуры накала по цвету** color (temperature) meter; **и. тока нагрузки** load-current meter; **и. толчков** jerk meter; **и. толщины линии** line-width meter; **и. толщины листового материала** pachymeter; **и. угла сноса** drift-meter; **и. удлинения** strain gage; **и. уклонов** gradiometer; **ультразвуковой и. толщины металла** sonigage; **и. уровня** decibelmeter; **и. уровня громкости** volume meter (of a signal); **и. уровня передачи** transmission measuring set; **и. усиления** gain measuring set; **и. утечки** electrical leakage tester; **и. фазного напряжения** phase-to-neutral voltmeter; synchroscope; **и. формы кривой** ondometer; **фотоэлектрический и. твёрдости (металлов)** photoelectric scleroscope; **и. частотной характеристики** frequency-response display set; **и. числа витков** coil-turn counter; **и. шумов контура** circuit noise meter; **электронный и. времени** electronic timer; **эталонный и. мощности** reference power meter

**измерительный** *adj.* measuring, gaging, metering; measurement
**измерить** *pf. of* **измерять**
**измеряемый** *adj.* measurable, measuring
**измерять, измерить** *v.* to meter, to measure, to gage, to size; to sound, to determine, to rate; **и. глубину** to sound; **и. на расстоянии** to tele-meter; **и. чётные гармоники** to test even
**измеряющий** *adj.* measuring
**изморозь** *f.* hoarfrost, rime
**изнашиваемость** *f.* wearability
**изнашивание** *n.* wear and tear, wear, wear-out, wearing out; deterioration
**изнашивать, износить** *v.* to wear out, to wear away
**изнашиваться, износиться** *v.* to wear away, to wear out, to deteriorate
**износ** *m.* wear and tear, wear, wearing away, abrasion, deterioration; consumption; depletion; shaving; **волнистый и. рельсов** fluting of rails, rail corrugation; **и. от действия дуг** sparkwear
**износить(ся)** *pf. of* **изнашивать(ся)**
**износовый** *adj.* wear-out
**износостойкий** *adj.* long-lasting, durable, resistant to wear
**износостойкость, износоупорность** *f.* durability, resistance to wear
**износоупорный** *adj.* long-lasting, durable, resistant to wear
**износоустойчивый** *adj.* ruggedized
**изношенный** *adj.* worn, worn out; used up, exhausted
**изнутри** *adv.* from within, from inside
**изо** *see* **из**
**изо-** *prefix* iso-
**изобара** *f.* isobar, constant-pressure line
**изобата** *f.* isobath, depth counter
**изображаемый** *adj.* imaginary
**изображать, изобразить** *v.* to picture, to depict, to represent, to describe
**изображающий** *adj.* representative
**изображение** *n.* picture, image; representation, description, pattern; layout; **искажающие изображения компоненты** distortion terms (*tv*); **побочные паразитные изображения** multipath effect, ringing effect, fold-over; **аффинное и.** affine transformation; **бледное неконтрастное и.** washed-out picture; **и. в виде заря-**

дов charge image; **и. в виде полос**
bar pattern; **и. в виде цветных**
**полос** color bar pattern, color bar
test pattern; **и. в виде шахматного**
**поля** checkerboard pattern; **и. в**
**одном из основных цветов** primary
image; **и. в отражённых лучах**
reflected image; **видимое и. спектров**
**речи** visible speech pattern; **воспро-**
**изводимое и.** display image; **и.,**
**воспроизводимое чересстрочным**
**растром** vertically interlaced image;
**гипсометрическое и.** elevated projection; **глубокое и.** deep dimension
picture; **и. градка на радиолока-**
**ционном индикаторе** masking effect
of hail in radar display; **движущееся**
**паразитное и.** travel ghost image;
**двоичное и. цифры** pulse code;
**двойное и.** echo image, fold-over,
multipath effect, double image, split
image; **и. дерева предложения** tree
diagram; **и. десятичных чисел дво-**
**ичным кодом** binary-coded decimal
representation; **и. живого объекта**
live image; **и. задачи в виде блок-**
**схемы** block representation; **заряд-**
**ное и.** picture charge, electrical
charge; **зеркальное и. антенны**
image antenna; **зеркальное и. источ-**
**ника** image source; **и. зернистой**
**структуры** grain pattern; **и. катода**
focal spot, cross-over, intermodulation; **контрольное и.** resolution
chart; **мозаичное радиолокационное**
**и.** radar mosaics; **и. на индикаторе**
**высоты** oscilloscope patterns of radar
altimeter; **и. на индикаторе круго-**
**вого обзора** stretched PPI; **и. на**
**индикаторе типа G** G-pattern on
radar scope; **и. на сетчатке глаза**
retinal image; **нефокусируемое и.**
extra-focal image; **и., образуемое**
**сигналом генератора цветных полос**
color bar generator pattern; **одно-**
**цветное и. с полной чёткостью** full-
resolution monochrome picture; **ос-**
**таточное и.** afterimage, burn, image
retention, sticking image; **паразит-**
**ное и. на обратном ходу** retrace
ghost image; **параллельное одно-**
**цветное и.** "by-passed monochrome"
image; **перемежающееся по гори-**
**зонтали и.** horizontally interlaced
image; **и. по Лапласу** Laplace transform; **повторное и.** echo image,

multi-image; **полное и. высокой чёт-**
**кости** complete high-resolution picture; **и., поражённое шумами** noisy
reproduction; **и. при студийной**
**передаче** live image; **прозрачное и.**
diaphanous view; **прямое и.** erect
image; **расплывчатое и.** soft picture;
**и. с плохими полутонами** high-
contrast image; **и. с рваными краями**
ragged picture; **светлое тоновое и.**
high key (*tv*); **светотеневое и.** black-
and-white picture; **тёмное тоновое**
**и.** low key (*tv*); **теневое и.** skiagraph;
**и. точечной цели** point-source image;
**фотографическое и. на непрозрач-**
**ном материале** top view image;
**частичное и.** frame; **штриховое и.**
facsimile copy, recorded copy; **экран-**
**ное и. типа L** L-pattern on radar
scope; **G-экранное и.** G-pattern on
radar scope

**изображённый** *adj.* represented, described

**изобразительный** *adj.* graphic; descriptive, imitative

**изобразить** *pf. of* изображать

**изобрести** *pf. of* изобретать

**изобретатель** *m.* inventor

**изобретательство** *n.* invention, development of inventions

**изобретать, изобрести** *v.* to invent, to
devise, to contrive, to develop

**изобретание** *n.* invention, device

**изобретённый** *adj.* invented, devised,
developed

**изогипса** *f.* isohypse, structure contour

**изогнутость** *f.* curvature, flexion,
camber

**изогнутый** *adj.* bent, curved, arched,
hooked, cranked; folded; camber
(beam)

**изогнуть(ся)** *pf. of* изгибать(ся)

**изогона** *f.* isogonic, isogonic line;
**нулевая и.** agonic line

**изогональный** *adj.* isogonal, equiangular

**изогонический** *adj.* isogonic

**изограмма** *f.* isogram

**изограф** *m.* isograph; **и. на сельсинах**
magslip isograph

**изогрива** *f.* connecting lines (on maps)
of equal grid-net error

**Изод** Izod

**изодин** *m.* isodyne

**изодинама** *f.* isodynamic line

**изодинамический** *adj.* isodynamic

**изодинный** *adj.* isodyne
**изодоза** *f.* isodose
**изодоп** *m.* isodop; **изодопы от наземных предметов** ground clutter isodops
**изодромы** *pl.* isodromous curves
**изойти** *pf. of* **исходить**
**изокинетический** *adj.* isokinetic
**изоклин** *m.*, **изоклина** *f.* isocline, isoclinal line, isoclinic line; **нулевая и.** aclinic line
**изоклиналь** *see* **изоклина**
**изоклинальный, изоклинический** *adj.* isoclinal, isoclinic
**изокон** *m.* isocon; **и. с переносом изображения** image isocon, super-image isocon
**изолак** *m.* isolac, insulating varnish
**изолантайт** *m.* isolantite
**изолента** *f.* insulating tape, friction tape
**изолирование** *n.* insulating, insulation; isolating
**изолированность** *f.* insulativity
**изолированный** *adj.* insulated, sealed; isolated; single
**изолировать** *v.* to insulate, to seal; to separate, to eliminate, to isolate
**изолировка** *f.* insulation
**изолировочный** *adj.* insulating, insulation
**изолируемый** *adj.* insulated
**изолирующий** *adj.* insulating, insulation, nonconducting
**изолит** *m.* isolit
**изолюкса** *f.* isolux (curve)
**изолятор** *m.* insulating material, insulator, insulant; isolator; **и. Брэдфилда** Bradfield leading-in insulator; **вводный и.** lead-in insulator, wall entrance insulator, bushing, inlet bell; terminal insulator; **вводный и. высокого напряжения** high-voltage feed-through insulator; **гирляндный и.** link insulator, chain insulator; **двухшейковый и.** double-groove insulator; **двухюбочный и. с массивным сердечником** solid-core-type insulator; **дистанционный и.** stand-off insulator; **и. для ввода через крышу** roof insulator; **и. для воздушных линий** outdoor insulator, overhead insulator, line insulator; **и. для линий связи** communication insulator; **и. для транспозиции проводов** transposition insulator; **за-**жимной и. cleat insulator, split knob insulator; **костылеобразный и.** spur insulator, line insulator; **кровельный и.** roof insulator; **крючный и.** pin-type insulator, swan-neck insulator; **маслонаполненный проходной и.** oil insulator; **натяжной и.** strain insulator, terminal insulator, guy strain insulator; **натяжной гирляндный и.** link strain insulator; **однокостыльный и.** line insulator, spur insulator; **опорный и.** base insulator, bracket insulator, stand-off insulator, support insulator; **опорный шинный и.** bus support; **орешковый и.** egg insulator; **и. под четвёртым рельсом** fourth rail insulator; **подвесной гирляндный и.** link suspension insulator; **подкладочный (под сосуды) и.** accumulator insulator; **подстелажный и.** battery rack insulator; **проходной и.** wall tube, partition insulator, insulating bead, wall entrance insulator, bush, through insulator; **проходной и. антенны** antenna trunk; **проходной через пол и.** floor insulator; **прочный на пробой и.** puncture-proof insulator; **разделительный и.** double-groove insulator, test insulator **распорный и.** stand-off insulator; **и. с двойной юбкой** double-petticoat insulator, double-shed insulator; **и. с конической заделкой** cone-head insulator; **и. с носком сбоку** spur insulator, side-knob insulator; **и. с одной юбкой** single-shed insulator; **и. с предохранителем** fuse insulator; **и. с шаровой заделкой** ball ring insulator; **сдвоенный и.** spacer insulator; **седловидный антенный и.** "shell" antenna insulator; **слаботочный и.** electro-communication-type insulator, sound insulator; **составной и.** multiple piece insulator; **и., стойкий против воздействия солей** salt-resistive insulator; **такелажный и.** strain insulator, guy insulator; **угловой и. в обойме** shackle; **шарнирный подвесной и.** cap-and-pin suspension insulator; **широкоюбочный и.** umbrella type insulator, mushroom insulator

**изоляторный** *adj.* insulator; isolator
**изоляционный** *adj.* insulating, insulation; insulated

**изоляция** *f.* insulation, insulating, sealing; isolation, segregation; **с неорганической изоляцией** mineral-insulated; **и. входа (схемы) от выхода** input-output isolation; **дополнительная и.** super insulation; **и. коллекторной втулки** commutator shell insulation; **и. между листами, и. между пластинками** laminated insulation, lamination insulation; **междусигнальная и.** cut section; **и. направляющего стакана** skirt insulation; **особо теплостойкая и.** class-H insulation; **и. от воздействия внешних условий** environmental insulation; **и. от воздушного звука** airborne sound insulation; **и. от конструкционного звука, и. от корпусного звука** solid-borne sound insulation; **и. пазов** slot lining; **и. пропиткой** varnished insulation; **и. путей** track circuiting; **и. с помощью шайб** beaded insulation; **эластичная поверхностная защитная и.** flexible blanket insulation

**изомер** *m.* isomer

**изомеризация** *f.* isomerization, isomeric heat

**изомеризм** *m.* isomerism

**изомеризованный** *adj.* isomerized

**изомеризовать** *v.* to isomerize

**изомерия** *f.* isomerism

**изомерный** *adj.* isomeric

**изометрический** *adj.* isometric, isometrical

**изоморфизм** *m.* isomorphism

**изоморфность** *f.* isomorphism

**изоморфный** *adj.* isomorphic, isomorphous

**изоорты** *pl.* curves of equal accuracy

**изопараметрический** *adj.* isoparametric

**изопикна** *f.* isopycn, equal-density line

**изопикнический** *adj.* isopycnic, isotermic

**изоплера** *f.* isopleric line

**изоплета** *f.* isopleth

**изополиморфизм** *m.* isopolymorphism

**изопотенциальный** *adj.* unipotential

**изопрен** *m.* isoprene

**изосейсма** *f.* isoseism, isoseismal line

**изосейсмический** *adj.* isoseismal, isoseismic

**изосейста** *f.* isoseism, isoseismal line

**изоскоп** *m.* isoscope

**изосмотический** *adj.* isosmotic, isotonic

**изосоединение** *n.* isocompound

**изостазия** *f.* isostasy

**изостата** *f.* isostath, isostatic curve

**изостатический** *adj.* isostatic

**изостера** *f.* isostere

**изостерический** *adj.* isosteric

**изостерия** *f.* isosterism

**изотерма** *f.* isotherm, isothermal curve

**изотермический** *adj.* isothermal, isothermic

**изотон** *m.* isotone

**изотонический** *adj.* isotonic, isosmotic

**изотония** *f.* isotonicity

**изотонный** *adj.* isotonic, isosmotic

**изотоп** *m.* isotope; **индикаторный радиоактивный и.** tracer element

**изотопический** *adj.* isotopic, isotope

**изотопия** *f.* isotopy

**изотопный** *adj.* isotopic, isotope

**изотрансформатор** *m.* isotransformer

**изотрон** *m.* isotron

**изотропический** *adj.* isotropic

**изотропия** *f.* isotropy, isotropism

**изотропность** *f.* isotropy, isotropism

**изотропный** *adj.* isotropic

**изоупругий** *adj.* isoelastic

**изофазный** *adj.* isophase

**изофазы** *pl.* lines of equal phase relations

**изофота** *f.* isophot

**изохор** *m.*, **изохора** *f.* isochore, isochore curve

**изохорный** *adj.* isochoric

**изохроматический, изохроматичный** *adj.* isochromatic, orthochromatic

**изохроматы** *pl.* isochromatic curves

**изохромный** *adj.* isochromatic, orthochromatic

**изохрона** *f.* isochrone

**изохронизм** *m.* isochronism

**изохронический** *adj.* isochronous, isochronal

**изохронность** *f.* isochronism

**изохронный** *adj.* isochronous, isochronal

**изоцентр** *m.* isocenter

**изоцианатный, изоциановый** *adj.* isocyanate

**изоциклический** *adj.* isocyclic

**изоэлектрический** *adj.* isoelectric

**изоэлектронный** *adj.* isoelectronic

**изоэнергета** *f.* constant-energy line

**изоэнергетический** *adj.* isoenergetic

**изоэхо** *n.* isoecho contour

**из-под** *prep. genit.* from under

**израсходованный** *adj.* spent, used

**изречение** *n.* utterance

**изрытый** *adj.* pitted; dug up

**изучать, изучить** *v.* to study, to investigate, to research

**изучение** *n.* study, investigation, research; analysis

**изученный** *adj.* studied, learned, investigated

**изучить** *pf. of* **изучать**

**изъеденный** *adj.* pitted, corroded

**изъявление** *n.* pitting

**изъявлённый** *adj.* pitted

**изъятие** *n.* removal, elimination

**изъятый** *adj.* removed, withdrawn, off

**изымать, изъять** *v.* to remove, to eliminate, to withdraw

**изыскание** *n.* finding, procuring; research, investigation; prospecting; survey; exploring; search

**изыскательный** *adj.* exploratory

**изэнтропа** *f.* isentrope, isentropic line

**изэнтропический** *adj.* isentropic

**ик** *abbr.* (**инфракрасный**) infrared

**ИК** *abbr.* 1 (**испытательный комплект**) test kit; 2 (**истинный курс**) true course

**Икклз** Eccles

**ИКЛ** *abbr.* (**инфракрасные лучи**) infrared rays, infrared beam(s)

**ИКМ** *abbr.* (**импульсно-кодовая модуляция**) pulse-code modulation

**ИКО** *abbr.* (**индикатор кругового обзора**) plan position indicator, PPI, J-scope

**иконометр** *m.* iconometer

**иконорама** *f.* iconorama

**иконоскоп** *m.* iconoscope; **безынерционный и.** nonstorage camera tube; **и. для передачи кино** movie pick-up iconoscope; **и. с умножителем** multiplier-iconoscope; **и. с электрическим переносом изображения** electrostatic image iconoscope

**иконоскопия** *f.* iconoscopy

**иконофон** *m.* iconophone

**икосаэдр** *m.* icosahedron

**икосаэдрический** *adj.* icosahedral

**икоситетраэдр** *m.* icositetrahedron; **пентагональный и.** icositetrahedron, leucitohedron, gyrohedron

**икс** *m.* x; x-unit

**икс-лучи** *pl.* X-rays

**ил** *m.* mud, silt, slime, sediment

**«илеско»** *n.* elesco

**или** *conj.* or, either

**ИЛИ-ИЛИ** OR ELSE; **ИСКЛЮЧА-**

**ЮЩЕЕ ИЛИ** OR ELSE; **ИЛИ-ячейка** OR-gate, OR-circuit

**иллий** *m.* illium

**иллитрон** *m.* illitron

**иллюзия** *f.* illusion, delusion

**иллюминатор** *m.* illuminator

**иллюминационный** *adj.* illumination

**иллюминация** *f.* illumination, decorative lighting

**иллюминированный** *adj.* illuminated, lit

**иллюминировать, иллюминовать** *v.* to illuminate, to light

**иллюминометр** *m.* illuminometer, photometer

**иллюстративный** *adj.* illustrative

**и. л. с.** *abbr.* (**индикаторная лошадиная сила**) indicated horsepower

**Ильгнер** Ilgner

**ИМ** *abbr.* (**импульсная модуляция**) pulse modulation

**именной** *adj.* nominal; name

**именованный** *adj.* named, called

**именовать, наименовать** *v.* to name, to label, to designate

**иметь** *v.* to have, to possess; **и. результатом** to result in; **и. следствием** to imply

**имеющий** *adj.* having; **и. два значения** double-valued, bivalent; **и. несколько значений** many-valued; **и. размерность** dimensional

**имеющийся** *adj.* available

**имитатор** *m.* imitator, simulator; **и. замираний** fading machine

**имитационный** *adj.* imitation

**имитация** *f.* imitation, simulation; **и. отражений от моря** sea clutter simulation

**имитирование** *n.* simulation

**имитированный** *adj.* simulated

**имитировать** *v.* to imitate, to copy, to simulate, to reproduce

**имитируемый** *adj.* imitated, imitating, false

**имитирующий** *adj.* imitative, simulative

**иммерсионный** *adj.* immersion

**иммерсия** *f.* immersion

**иммитанс** *m.* immittance

**иммитансный** *adj.* immitance

**иммунность** *f.* immunity

**иммунный** *adj.* immune

**имп.** *abbr.* (**импульсы**) impulses, pulses

**импеданс** *m.* impedance, apparent resistance; **импедансы рассечённого**

четырёхполюсника bisection impedance; **безразмерный и.** impedance ratio; **восстанавливающийся и.** recovery impedance; **«закороченный» и.** short-circuit impedance; **нагрузочный и.** load impedance, terminal impedance, terminating impedance; **номинальный и. говорителя** speaker rating impedance; **обратный и.** looking-back impedance; **отнесённый к первичной обмотке вторичный и.** reflected secondary impedance; **и. отправительного** (*or* **отправного**) **конца** (**линии**) second-end impedance (of a line); **и. параллельного резонансного контура** antiresonant impedance; **параллельный и.** leak impedance; shunt impedance; **и. при закороченном** (*or* **разомкнутом**) **выходе** open-circuit impedance (of a network); **продольный и.** series impedance; **прямой и.** looking-in impedance; **и. связи** reflected secondary impedance; **и. холостого хода, холостой и.** open-circuit impedance; **цепочный и.** iterative impedance

**импеданс-годограф** *m.* transmission-line calculator

**импеданс-метр** *m.* performance index meter

**импедансный** *adj.* impedance

**импеданц** *see* **импеданс**

**импеллер** *m.* impeller, blade

**импидор** *m.* impedor

**имплантат** *m.* implant

**имплантация** *f.* implantation

**импликативный, импликационный** *adj.* implicative, implicational

**импликация** *f.* implication

**имплицировать** *v.* to imply

**имплозивный** *adj.* implosive

**имплозия** *f.* implosion, bursting inwards

**имп/мин** *abbr.* (**импульсов в минуту**) pulses per minute, counts per minute

**импортация** *f.* importation

**импрегнация** *f.* impregnation

**импрегнированный** *adj.* impregnated

**импрегнировать** *v.* to impregnate

**имп/сек** *abbr.* (**импульсов в секунду**) pulses per second, counts per second

**импульс** *m.* impulse, pulse, impetus, impact; momentum; **импульсы ведомой станции, импульсы ведущей станции** slave pulses, master pulses;

**выделяющие импульсы** sorting pulses; **импульсы для оптического указателя номера** call-indicator pulses; **задние уравнивающие импульсы** post-equalizing pulses; **многократно сосчитанные импульсы (в счётной трубке)** multiple tube counts (in a radiation counter tube); **импульсы (напряжения) обратного хода строк** line fly-back pulses; **обратные кадровые импульсы** inverted field pulses; **импульсы одной полярности** single-polarity pulses, unidirectional pulses; **импульсы от обратного хода развёртки** fly-back pulses; **перевёрнутые импульсы частоты полей** inverted frame pulses, inverted field pulses; **передние уравнивающие импульсы** pre-equalizing pulses; **импульсы, перекрывающие друг друга во времени** overlapping pulses; **импульсы с крутыми фронтами** rise-fall pulses; **импульсы с переменной полярностью** bidirectional pulses; **импульсы с преобладанием** biased reversals; **импульсы сдвига** advance pulses; **сосчитанные импульсы фона** background counts; **управляющие цветовой синхронизацией импульсы** burst flag pulses; **и. большой длительности** very wide pulse, lengthened pulse; **ведущий и. частоты кадров** frame drive pulse; **ведущий и. частоты полей** field driving pulse; **и. возврата в исходное положение** reset pulse; **возмущающий и.** actuating pulse; **и. вращения** moment of momentum; **входной единичный и.** step-function input signal; **входной ударный и.** output stroke; **и. выброса** peaking pulse; **выделяющий и.** sampling pulse; **и. высокочастотной энергии** radio pulse; **и. вычитания** subtract pulse; **гасящий и. отрицательной полярности** negative blanking pulse; **и. гашения** quench pulse; **и. двойной строчной частоты** half-line pulse; **двухсторонний и.** unit doublet impulse; **деблокирующий и.** enabling pulse; **и. для отсчёта расстояния** range step; **и. добавления** correcting pulse; **и. закрытия строба** gate-closing pulse; **и. замыкания** make pulse; **и. занятия** inceptive pulse, incipient pulse, initiating pulse; **и. запирания приёмника** receiver disabling pulse; **и. запуска развёртки** sweep-initiating pulse; **и. засветки** gate (pulse); **затемняющий и.** blanking pulse, blanking gate; **и. индикации (истинной) дальности** true-range-indicating pulse; **и. коммутации цвета** color(-indexing) pulse; **ложный считанный и.** spurious tube count; **и. малой длительности** narrow pulse; **маркёрный и.** selector pulse; **мешающий и.** noise spike, noise pulse; **и. многократного эха** re-echoing pulse; **и. на обратном ходу развёртки** fly-back pulse, flyback kick, kick-back pulse; **и. на обратном ходу строчной развёртки** line fly-back pulse, horizontal flyback pulse; **накопленный и.** pile-up pulse; **и. напряжения помехи** noise pulse; **и. напряжения с плоской вершиной** flat-topped voltage pulse; **и. небольшой длительности** narrow pulse; **и. облака пространственного заряда** cloud pulse; **и. обратного хода** kick-back pulse; **и. обратного хода кадровой развёртки** vertical flyback pulse (*or* voltage); **и. обратного хода луча** flyback pulse; **и. обратного хода развёртки по строке** horizontal flyback pulse; **и. обратного хода строк** line flyback pulse; **и. обратной полярности** inverted pulse; **обратный и. при занятости** busy indication, engaged indication (*tphny.*); **и. обращённой полярности** inverted pulse; **общий и.** total impulse, sum of momenta; **объёмный удельный и.** density specific impulse; **однополярный и.** unidirectional pulse; **и. окончания переноса** end carry pulse; **опорный и. засветки** pedestal gating pulse; **и. основания** pedestal pulse; **основной и. засветки** main gating pulse; **и. от генератора импульсов** clock pulse; **и. от замыкания** make pulse; **и. от перерыва тока** break impulse; **и. от посылки тока** make pulse; **и. от размыкания** break impulse; **и. отметки** pip pulse, notch pulse; **отпирающий и.** trigger pulse, enabling pulse, unblocking pulse, gate pulse; **и. отсчёта** sampling pulse; **паразитный считанный и.** spurious tube count; **первоначальный и.** prefix signal; **и. переключе-**

ния в исходное положение set pulse; **перемещающийся селекторный и.** walking strobe pulse; **и. перенапряжения** surge; **и. переноса из старшего разряда** end carry pulse; **П-образный и.** square(d) pulse, square-topped pulse; **подготавливающий и.** priming pulse; **и. подсветки, и. подсвечивания развёртки** intensifier pulse, unblanking pulse, sensitizing pulse, brightening pulse; **и. (подсвечивания) сигнал-селектора** signal-selector patch; **подсвечивающий и.** *see* **и. подсветки**; **и. полувозбуждения при записи** half writing pulse; **и. полувозбуждения при считывании** half reading pulse; **и. последней ступени** final impulse; **и. последовательности** sequencing pulse; **поступающий и.** incoming pulse; **предварительный и.** pre-pulse; **«предваряющий» звуковой и.** "pretrigger"; **и. при включении** transient impulse; transient current; **и. при выключении** circuit-breaking, transient current; **и. при размыкании** break impulse; **пробельный и.** space pulse, spacing pulse; **и. пробивки** punch pulse; **пропускающий и.** gate pulse, gating pulse; **пропущенный и.** lost pulse; **и. пространственного заряда** cloud pulse; **прошедший и.** transmitted pulse; **и. пускового тока** starting impulse; **пусковой и. нулевой дальности** zero-range trigger; **пусковой селекторный и.** trigger gate; **и. работы** marking impulse; **разрезанный и.** serrated pulse; **разрешающий и.** enabling pulse; **и. разрушения перед тактом считывания** pre-read disturb pulse; **и. разрушения после такта записи** post-write disturb pulse; **расчётный и.** computed impulse; **и. регулируемой длительности** variable-length pulse; **и. реперной высоты** pulse of reference height; **и. с крутым фронтом** rise-fall pulse, steep-sided pulse, sharp-edged wave; **и. с опорным уровнем** pulse of reference height; **и. с плоской верхушкой** square-topped pulse; **и. сброса** reset pulse; **и. света** pulsed light; **селекторный и.** strobe pulse, gating pulse; dialing impulse; selector pulse; **селекторный и. дальности** range pip, range gate;

**селекторный и. отметки нулевой дальности** zero-range selector waveform; **селекторный и. подсветки индикатора** indicator gate; **селекторный сеточный и.** grid gate; **и. сигнала ошибки** error pulse; **и. «синхровспышки»** burst pulse; **и. синхронизации коммутатора** sampling synchronization pulse; **и. синхронизации по кадрам** vertical synchronizing pulse; **синхронизирующий и. кадровой частоты** frame synchronizing pulse; **синхронизирующий и. строчной частоты** line synchronizing pulse; **синхронизирующий и. частоты полей** field synchronizing pulse; **и. скорости сотрясения, проходящий через фундамент** foundation-induced velocity shock; **и. сложения** add pulse; **смешанный и. гашения** mixed suppression pulse; **смещающий и.** bias pulse; **сосчитанный трубкой и.** tube count; **«столбообразный» и.** plateau pulse; **и. строчного обратного хода** horizontal flyback pulse; **и. строчной синхронизации, и. строчной частоты** strip pulse, line synchronizing pulse; **и., сформированный посредством линии задержки** delay-line-shaped pulse; **считанный и.** tube count; **и. считывания** read pulse, full read pulse; **тактирующий и., тактовый и.** clock pulse; shift pulse; **и. тяги двигателя** engine pulse; **удельный и. на единицу объёма** density impulse, volume impulse; **удельный и. на уровне моря** sea-level specific impulse; **удельный и. силовой установки** power-plant specific impulse, system specific impulse; **управляющий и. привязки** clamp keying pulse; **управляющий пропусканием и.** gate pulse, gating pulse; **и., управляющий схемой фиксации** clamp pulse, clamp drive pulse; **установочный и.** setting pulse, set pulse; **и. частичного ввода** partial-write pulse; **и. частичного считывания** partial reading pulse; **и. частичной выборки** partial read pulse; **и. частичной записи** partial-write pulse, partial writing pulse; **и. частоты полей** field pulse; **и. чередования** sequencing pulse; **и. электрона** momentum of electron; **и.**

**эхолота** sounding pulse
**импульсатор** *m.* impulsator
**импульсер** *m.* impulse starter
**импульсивный** *adj.* impulsive
**«импульсмессер»** *m.* sound level meter
**импульсметр** *m.* pulse meter, impulse meter
**импульсник** *m.* key pulser, key sender
**импульсно-возбуждаемый** *adj.* pulse-actuated
**импульсно-временной** *adj.* pulse-time
**импульсно-допплеровский** *adj.* pulsed-doppler
**импульсно-кодовый** *adj.* pulse-code
**импульсно-модулированный** *adj.* pulsed, pulse-modulated
**импульсно-модуляторный** *adj.* pulse modulator
**импульсно-образующий** *adj.* pulse-forming
**импульсно-приложенный** *adj.* step-function (voltage)
**импульсно-регенеративный** *adj.* pulse-regenerating
**импульсный** *adj.* impulse, impact, momentum; pulse, pulsed, impulsive, pulsating; push-pull; flash (bulb); sampled (data)
**импульсованный** *adj.* pulsed
**импульсовидный** *adj.* pulse-like, pulse-wise
**импульсограф** *m.* pulse recorder
**импульсозаострённый** *adj.* pulse-sharpening
**импульсо-образующий** *adj.* pulse-forming
**импульсостойкий** *adj.* pulse-proof; surge-proof
**имущество** *n.* property; goods
**ин.** *abbr.* (**интенсивный**) strong
**инвар** *m.* invar
**инвариант** *m.* invariant
**инвариантность** *f.* invariance
**инвариантный** *adj.* invariant
**инварный** *adj.* invar
**инвентаризационный** *adj.* inventory
**инвентаризация** *f.* inventory control
**инвентаризировать** *v.* to take inventory
**инвентарный** *adj.* inventory
**инверсионный** *adj.* inversion
**инверсирование** *n.* inversion
**инверсированный** *adj.* inverted
**инверсия** *f.* inversion, inverse; **и. входов** input inversion; **и. градиента скорости** inversion in velocity gradient; **и. двоичных цифр** binary

inversion; **и. относительно сложения** additive inversion; **и. относительно умножения** multiplicative inverse; **и. переменных** variables inversion, inverse of variables
**инверсный** *adj.* inverted, inverse
**инвертер** *m.* inverter, inverted rectifier; **и. пускового импульса** trigger inverter; **и. с регулируемым напряжением** voltage-regulated inverter
**инвертирование** *n.* inversion, inverting, reversal
**инвертированный** *adj.* inverted
**инвертировать** *v.* to invert, to reverse
**инвертирующий** *adj.* inverting
**инвертный** *adj.* invert
**инвертор** *m.* inverter, inverted rectifier, inverted converter, inverting amplifier, complementing amplifier, sign-changing amplifier, polarity-inverting amplifier; **и. со схемой совпадения** gated inverter
**инверторный** *adj.* inverter
**инволютный** *adj.* involute, coiled
**ингибирование** *n.* inhibition
**ингибитор** *m.* inhibitor, arrester
**нигибиторный** *adj.* inhibitor, arrester
**ингибиция** *f.* inhibition
**индекс** *m.* index; subscript, suffix; **верхний и.** superscript; **классификационный и.** accuracy index for electrical instruments; **подстрочный и.** subscript; **и. температурной уставки** temperature setting index; **численный и. системы включения счётчика** hour-meter circuit code number
**индексация** *f.* indexing; numbering
**индекс-карта** *f.* guide card
**индексный** *adj.* indexing, index; subscript
**индексовый** *adj.* index
**индекс-регистр** *m.* B-register, base register, index register
**индекс-сигнал** *m.* index signal
**индивид** *m.* individual
**индивидуальный** *adj.* individual, personal, peculiar; independent, single, separate, self-contained
**индивидуум** *m.* individual
**индий** *m.* indium, In
**индийский** *adj.* Indian; India
**индикатор** *m.* indicator; indicator tube; indicator device, display unit, scope, marker; enhancer; tracer; detecting head; **и. азимут-угол места** azimuth-elevation indicator; **акустический и.**

**эффекта Допплера** audible Doppler enhancer; **визуальный вибрационный и.** visual reed indicator; **и. включения программы на спуск** retro sequence indicator; **вынесенный и. кругового обзора** plan position indicator repeater, remote plan indicator; **выносной и.** repeater scope; **выносной и. азимут-дальность** R-Theta repeater; **и. высокого напряжения** corona feeler; **высокочастотный и. скорости автомобилей** microwave vehicle-speed indicator; **и. дальности гидроакустической станции** sound range recorder; **и. дальности сопровождения цели** radar-coverage indicator; **и. дальность-угол места** elevation-position indicator; **двойной и. настройки** double-range tuning-indicator tube; **двойной оптический и.** double-range (visual-)tuning indicator tube; **двустрелочный и.** cross-pointer indicator; **двухкоординатный и.** two-dimensional display; **двухкоординатный и. с яркостной отметкой** two-dimensional intensity-modulated display; **и. диапазона настройки** sector-alignment indicator; **и. для грубого определения дальности** coarse-range scope; **добавочный и.** repeater scope, repeater-indicator; **дополнительный и. кругового обзора** plan repeater indicator; **допплеровский и. космической цели** space target Doppler indicator; **и. замыкания на землю** earth detector, ground indicator, leakage indicator; **и. ИКЛ** heat-sensitive device; **и. интервала разрыва** burst-distance indicator; **и. инфракрасных лучей** Golay cell; **и. коммутационной панели** switchboard lamp; **и. короткозамкнутых витков** shorted-turn indicator; **и. кругового обзора** plan-position indicator; **и. кругового обзора с отражающей оптикой** reflection plotter, virtual PPI reflectoscope; **и. курсового угла цели при торпедной стрельбе** torpedo target bearing indicator; **и. лучистой энергии** photon detector; **и. максимума** demand attachment, maximum pointer; **машинный и.** low frequency ringer, signaling set, power ringing generator; **и. местоположения отно-**

сительно **Земли** earth path indicator; **накопительный и.** indicator employing a signal storage-tube; **и. наличия цели** on-target detector; **и. напряжения ошибки** error-indicating light; **и. настройки с двумя секторами** double-range tuning-indicator tube; **и. облучения** ray-tracer; **и. одного из видов излучения** passive detector; **однокоординатный и. с амплитудной отметкой** one-dimensional deflection-modulated display; **и. останова при делении** divide stop light; **панорамный и. с наложением изображения** superimposed panoramic radar display; **и. переключения диапазонов** band-selector indicator; **периодически действующий и. максимума** restricted-hour maximum demand meter; **и. пиковых значений шума** peak-noise indicator; **и. повреждения солнечных элементов** solarcell-damage indicator; **и. подводных лодок** antisubmarine attack detector; **и. положения управляемой диафрагмы** iris setting indicator; **и. порогового значения** sector-alignment indicator; **и. пропуска импульсов** missing-pulse detector; **и. растянутой развёртки** selected range indicator; **и. с амплитудной отметкой** variable-displacement indicator, variable displacement instrument; **и. с круглой шкалой, и. с лимбом** dial-type indicator, dial indicator, dial gage; **и. с масштабированием развёртки** expanded partial-indication display; **и. с отсчётом азимута от направления на север** north-stabilized indicator; **и. с прямоугольной системой координат** rectangular-coordinate display; **и. с радиальной развёрткой** radial-time-base indicator; **и. с раздвоенной точкой** "doublet-dot" display; **и. с растянутой развёрткой** expanded indicator, expanded scope; **и. с частично растянутой развёрткой** expanded partial-indication display; **и. с яркостной модуляцией, и. с яркостной отметкой** intensity-modulated display, intensity-modulated indicator, variable-intensity instrument, intensity-modulation scan; **световой и. «безопасно»** safety light; **и. секторного обзора** (or

поиска), **секторный и. кругового обзора** sector scan indicator, sector plan position indicator, sector PPI; **и. сигналов от всенаправленного маяка** omnibearing indicator; **и. состояния средств системы** status indicator; **и. текущих координат цели** present-position indicator; **и. теплового излучения** heat-sensitive device, heat cell; **и. типа А** (or **типа B, C, D, E, F, G, H, I, J, К, L, M, N, R**) A-scope, A-indicator, A-display, range-amplitude display, range-height marker; **и. типа B с растянутой развёрткой** micro-B-scope; **тлеющий и. колебаний** ondoscope; **точечный и.** point source radiator; **точечный и. рассогласования** spot error indicator; **трёхкоординатный и.** perspective three-dimensional display; **и. углового положения** range-angle indicator; **и. угловой скорости орбиты** gyro orbit indicator; **и. узлов** nodalizer; **и. уровня** vu-meter, volume indicator, level meter; **фотопроводниковый и. из германия, легированного цинком** zinc impurity photoconductor; **и. частоты вибраций** reed-frequency detector; **электроннолучевой и.** indicator tube, (shadow) tuning indicator, cathode-ray indicator; **электронно-оптический и. настройки** magic eye, shadow-tuning eye, shadow-tuning indicator; **электронно-световой и.** electron-ray indicator tube, tuning indicator

**индикаторный** *adj.* indicating; indicated; indicator

**индикатороподобный** *adj.* indicator (-like)

**индикатрис** *see* **индикатриса**

**индикатриса** *f.* indicatrix

**индикация** *f.* indication; presentation, display; imagining; tracing; **и. биений** beat-frequency indication; **и. горизонтальной дальности** ground plan plot; **и. дистанции пролёта** missdistance indication; **и. кругового обзора** PPI display; **и. мерцания с плёнки** presentation of flicker film; **и. на осязание** tactile presentation; **наглядная и.** display presentation, pictorial indication; **и. по звуковому нулю, и. по отсутствию слышимости** aural-null presentation; **и. по**

прибору meter display; **и. при равносигнальной зоне** radio range indication, split presentation; **и. при точном наведении антенны на цель** on-target indication; **пространственная и.** three-dimensional display; **равносигнальная и. пеленга** split indication of bearing; **радиолокационная и. в полярных координатах** radar-scope for PPI representation; **и. с воспроизведением вида местности** realistic display of terrain; **и. с использованием скиатрона** skiatron display; **и. с показом характера местности** realistic display of terrain; **и. с помощью формированного электронного пучка** shaped-beam display; **и. синхронизации по частоте** frequency-lock indication

**индилитан** *m.* indilatans
**индистор** *m.* indistor
**индитрон** *m.* inditron tube
**индифферентный** *adj.* indifferent, passive, neutral, inert
**индицирование** *n.* indexing; indicator test; indication, indicating
**индицируемый** *adj.* indicated
**индоевропейский** *adj.* Indo-European
**индокс** *m.* indox
**индуктивно** *adv.* inductively
**индуктивно-ёмкостный** *adj.* inductance-capacitance
**индуктивно-импульсный** *adj.* inductive-impulse
**индуктивностный** *adj.* inductance
**индуктивность** *f.* inductance, inductivity, inductive reaction, coefficient of induction, self-inductance; **и. вторичной обмотки** secondary inductance; **динамическая и.** (self-) inductance under working conditions; **и. жилы** series inductance; **и. катушки с незамкнутым железным сердечником** aero-ferric inductance; **и. колебательного контура** tank (circuit) inductance; **подстроечная и.** trimming inductance; **поперечная и.** shunt inductance; **и. рассеяния** leakage inductance, stray inductance; **и. соединительных проводов** inductance of connections; **удлиняющая и.** antenna tuning coil, antenna inductance; **и. управляющего провода** control inductance; **и. холостого хода** no-load inductance

**индуктивный** *adj.* inductive; induced

**индуктирование** *n.* induction, inducing
**индуктированный** *adj.* induced
**индуктировать** *v.* to induce
**индуктируемый** *adj.* induced; generated
**индуктирующий** *adj.* inducing: inductive
**индукто-** *prefix* inducto-, induction
**индуктомер** *m.* gaussmeter
**индуктометр** *m.* induction meter, inductometer
**индуктор** *m.* inductor, induction coil, work coil; magneto generator, hand generator, field magnet; **безобмоточный и.** inert inductor; **вызывной и.** magneto ringer; **и. для последовательного включения** series ringer; **искровой и.** induction coil, magneto generator, inductor; **молоточковый и.** trembler coil; **и. нагрева** work coil; **ручной и.** calling magneto, hand ringing generator; **телефонный и.** generator, magneto ringer; **униполярный и.** homopolar field magnet
**индукторный** *adj.* inductor, induction (coil); magneto
**индуктосин** *m.* inductosyn
**индуктроник** *m.* inductronic
**индукционный** *adj.* induction, inductive
**индукция** *f.* induction, electric field; density; **внутренняя магнитная и. насыщения** saturation inductance; **действительная и.** actual flux density; **действительная и. в зубцах** actual tooth density; **магнитная и. насыщения** saturation flux density; **и. на частном цикле** incremental induction; **остаточная магнитная и.** remanence, remanent induction, residual flux density; **и. от линий сильного тока** power induction, power line interference; **и. от потока рассеяния** stray induction; **и. от энерголиний** power line interference, power induction; **содержательная и.** informal induction; **средняя и. на поверхности магнитного зазора** specific magnetic loading; **средняя магнитная и. под полюсом** mean pole-face density
**индустрия** *f.* industry
**индуцированный** *adj.* induced
**индуцировать** *v.* to induce
**иней** *m.* rime, frost, hoarfrost
**инертин** *m.* inerteen

инертность *f.* inertance, inertia, inertness, passiveness, inactivity, sluggishness (of a phototube); aftereffect; afterglow, persistence, lag; **и. счётной трубки** hysteresis of a radiation counter tube
инертный *adj.* inert, inactive, passive; sluggish
инерциально-астрономический *adj.* inertial-celestial
инерциально-доплеровский *adj.* inertial-doppler
инерциальный *adj.* inertial
инерционно-гравитационный *adj.* inertial-gravitational
инерционность *f.* inertness; afterglow, lag, aftereffect, persistence (of a screen), sluggishness (of a phototube), retentivity, time lag, time delay; **большая и.** slow response; **и. глаза, и. зрительного восприятия** visual persistence, retentivity time of eye, vision retentivity; **и. экрана** screen speed
инерционный *adj.* inertia, inertness, lag, reluctance; sluggish
инерция *f.* inertia, inertness, lag, time delay, reluctance; **по инерции** under one's own momentum; mechanically, automatically; **и. зрительного восприятия** *see* **инерционность глаза**; **световая и.** afterglow
инж. *abbr.* (**инженерный**) engineering
инжектированный *adj.* injected
инжектировать *v.* to inject
инжектор *m.* injector; accelerator
инжекторный *adj.* injector; injection
инжекционный *adj.* injection
инжекция *f.* injection; **и. носителей** carrier injection; **и. носителей тока** charge carrier injection
инженер *m.* engineer; **и. для научно-исследовательских работ** research engineer; **и. по вопросам защиты электрических установок** protection engineer; **и. по разработке новых конструкций** development engineer; **и. по технике безопасности** safety engineer; **и. по управлению видеопередачами** video engineer; **и. по цифровым вычислительным машинам** digital engineer; **и. по экспериментальным исследованиям** experimental engineer
инженер-диспетчер *m.* control engineer
инженер-звукооператор *m.* audio engineer
инженер-конструктор *m.* design engineer
инженер-монтажник *m.* installation engineer
инженер-проектировщик *m.* design engineer
инженерно-технический *adj.* engineering and technical
инженерный *adj.* engineering, technical
инженер-электрик *m.* electrical engineer
инженер-электроакустик *m.* audio engineer
инзоляция *see* инсоляция
ИНИ *abbr.* (**измеритель нелинейных искажений**) distortion meter
инициатор *m.* initiator, starter
инициирование *n.* initiation, starting
инициировать *v.* to initiate, to start, to trigger
инициирующий *adj.* initiating, starting
инклинатор, инклинометр *m.* inclinometer, deep needle, inclination compass
инкредуктор *m.* increductor
инкремент *m.* increment, growth
инкрементальный, инкрементный *adj.* incremental
инородный *adj.* foreign, extraneous, heterogeneous
инсоляция *f.* insolation, sun's radiation
инспектор *m.* inspector; **и. счётчиков и установок** meterman
инспекторский *adj.* inspectorial; inspection; survey
инспекция *f.* inspection, examination
инсталляция *f.* installation
институт *m.* institute, institution; **Ленинградский электротехнический и. сигнализации и связи** the Leningrad Electrotechnical Institute of Signals and Communications; **Московский электротехнический и. связи** the Moscow Electrotechnical Institute of Communication; **научно-исследовательский и.** research institute; **Одесский электротехнический и. связи** the Odessa Electrotechnical Institute of Communications; **Центральный научно-исследовательский и. связи** the Central Scientific Research Institute of Communications
инструктаж *m.* instruction, direction
инструктивный *adj.* instruction

инструктирование *n*. instruction, instructing

инструктировать *v*. to instruct, to direct, to advise

инструкционный *adj*. instruction

инструкция *f*. instruction; manual, handbook; specification; **временная и.** interim instruction; **и. по аварийной эксплоатации** emergency operating order; **и. по обслуживанию** service manual; **и. по применению** instruction for use, directions for application; **и. по производству полёта** flight instruction; **и. по техническому обслуживанию** maintenance instruction, engineering instruction; **производственная и.** operating instructions, operating rules; **рабочая и. по обслуживанию** service manual

инструмент *m*. instrument, tool, implement; instrument (*mus*.); **музыкальные инструменты со стержнями** musical instruments employing excited bars; **наладочные инструменты** aligning tools; **и. для выравнивания плоскости** planing tool, grading tool; **и. для подрезки слюды** mica undercutter; **и. для работы под током** hot-line tool; **и. для разметки** marker; **нулевой и.** null detector; **пистолетный и. для обмотки якорей** winding gun; **расточный и. на сверло** auger; **и. с малой собственной ёмкостью** neutralizing tool

инструментальный *adj*. instrumental; instrument, tool

инструментодержатель *m*. tool holder

инсулат *m*. insulate

интеграл *m*. integral; **и. вдоль замкнутого контура** contour integral; **и. квадрата ошибки** integrated square error; **«n»-кратный и.** n-fold multiple integral; **неопределённый и. функции** antiderivative of a function; **и. от квадрата** integrated square; **и. ошибок** error function integral; **«перекрёстный» и.** transfer integral; **и. по замкнутой поверхности** closed surface integral; **пространственный и.** volume integral; **и. с бесконечным пределом** infinite integral; **и. свёртывания** superposition integral, convolution integral; **и. состояния** phase integral

интегральный *adj*. integral, whole, cumulative; (fully-)integrated

интегратор *m*. integrator, integrating instrument; integraph; integrating circuit; **моделирующий и.** analog integrator; **и. на плоскостных кристаллических триодах** junction-transistor integrator; **и. непрерывного действия** analog integrator; **решающий и.** decision integrator; **и. с визуальным считыванием типа солиона** visual readout integrator; **и. с положительной обратной связью** regenerative integrator, regeneration integrator; **и. со следящей лампой** bootstrap integrator; **фрикционный и.** ball-and-disk integrator; **центробежный и.** gyroscopic integrator; **шаровой центробежный и.** flyball integrator

интеграторный *adj*. integrating

интеграф *m*. integraph

интеграция *f*. integration

интегрирование *n*. integration, integrating

интегрированный *adj*. integrated

интегрировать *v*. to integrate

интегрируемость *f*. integrability

интегрируемый *adj*. integrable; integrated

интегрирующий *adj*. integrative, integrating

интегрирующийся *adj*. integrand

интегродифференциальный *adj*. integrodifferential

интеллект *m*. intelligence

интеллектуальность *f*. intelligence

интенсивность *f*. intensity, intensiveness; density; rate; magnitude; strength; **и. астатического действия** floating rate; **и. отражений от дождя** rain-echo intensity; **и. отражённых от цели сигналов** target-echo intensity; **и. падающего потока облучения** intensity of incident radiation; **и. потока** radiation flux; **и. сигналов, отражённых от объекта** target-echo intensity; **и. скачка уплотнения** shock-wave intensity; **и. сообщения** traffic load; **спектральная и. величины излучения** spectral concentration of a radiometric quantity; **и. эхо-сигналов от дождя** rain-echo intensity

интенсивный *adj*. intensive, intense; high; heavy

интенсиметр *m.* ratemeter, counting-rate meter
интенсификатор *m.* intensifier
интенсификация *f.* intensification
интенсифицировать *v.* to intensify; to increase the capacity
интенсифицирующий *adj.* intensifying
интенция *f.* intension
интервал *m.* interval, space, spacing, gap, interspace, breach, interruption, pause, separation, distance, range; **и. включения** key-in region; **и. выключения** key-out region; **и. дискретности** sampling time; **запертый и.** ungated period; **и. измерения** sampling interval; **конечный и. времени** definite time interval; **концевой и.** back space; **и. между зонами** interrecord gap; **и. между кривыми коэффициента сигнала** signal differential interval; **и. между посылками импульсов** transmission interval; **и. между сигналами** momentum range; **и. между смежными передними фронтами (импульсов)** pulse spacing; **и. между строками развёртки** scanning line separation; **и. между точечными сигналами** dot keying; **и. между фокусирующими импульсами радиолокационной станции** radar repetition interval; **и. между штриховыми сигналами** dash keying; **междуимпульсный и.** off period; **и. отрицательного потенциала** inverse period; **и. протяжки (ленты)** pulldown interval; **и., равный девяти строкам** "nine-line keyout"; **разделительный и.** spacer interval; **растянутый и. дальности** expanded range interval
интервалометр *m.* intervalometer, timer; **и. для аэрофотосъёмки** photogrammetric intervalometer
интервальный *adj.* interval
«интерлесинг» *m.* interlacing
интерлингва *f.* interlingua
интерлокер *m.* interlocker
интерметаллид *m.* intermetallic semiconductor
интерметаллический *adj.* intermetallic
интермиттирующий *adj.* intermittent
интермодуляционный *adj.* intermodulation
интерполирование *n.* interpolation
интерполировать *v.* to interpolate, to intercalate, to collate

интерполирующий *adj.* interpolating
интерполятор *m.* interpolator, interpolater
интерполяционный *adj.* interpolation
интерполяция *f.* interpolation; **двумерная и.** bivariate interpolation; **кратная и.** oscillatory interpolation; **и. назад** regressive interpolation
интерпретатор *m.* interpreter; interpretive program, interpretive routine
интерпретация *f.* interpretation; **и. Гейзенберга** Heisenberg representation
интерпретирование *n.* interpretation
интерпретировать *v.* to interpret
интерпретируемый *adj.* interpretive
интерпретирующий *adj.* interpretative, interpretive; interpreting
интерсектинг *m.* intersecting
интерсептор *see* интерцептор
интерференциальный, интерференционный *adj.* interfering, interference
интерференционно-поляризационный *adj.* interferential-polarizational
интерференция *f.* interference, interfering; **и. высокочастотных сигналов** continuous wave interference; **и. каналов с близкими частотами** common-channel interference; **и. незатухающих колебаний** continuous wave interference; **ослабляющая и.** destructive interference; **усиливающая и.** constructive interference
интерферировать *v.* to interfere
интерферирующий *adj.* interfering
интерферограмма *f.* interferogram, interference pattern
интерферометр *m.* interferometer; **и. Клифа** Cliff interferometer, Loyd's mirror system; **и. Лобе** Lobe-switching interferometer; **и. с качающейся частотой** swept-frequency interferometer; **и. со слежением по фазе** phase-tracking interferometer
интерферометрический *adj.* interferometric; interferometer
интерферометрия *f.* interferometry
интерфлекс *m.* interflex
интимный *adj.* intimate, close
интрамолекулярный *adj.* intramolecular
интранзитивный *adj.* intransitive
интрафакс *m.* intrafax
интуиционизм *m.* intuitionism

**интуиционистический** *adj.* intuitionistic

**интуиция** *f.* intuition, insight

**инфазный** *adj.* in-phase

**инфекция** *f.* infection, contagion

**инфильтрация** *f.* infiltration, seepage

**инфлектор** *m.* inflector, deflector

**информационно-логический** *adj.* information-logical

**информационный** *adj.* information

**информация** *adj.* information; data, report; signal complex; intelligence; **и. в виде световых импульсов** optical-information pulses; **и. в виде электрических импульсов** electrical-information pulses; **визуальная и. о цели** visual target data; **дискретно-непрерывная и.** sampled analog data; **и. для составления программы** reference record; **и., зашифрованная двоичным кодом** binary-coded information; **коммутированная и.** gated information; **и. на один символ** information content (entropy); **наглядная и.** pictorial information; **накопленная в памяти и.** memory contents; **и. о градациях яркости** half-tone information; **и. о мелких деталях изображения** fine-detail information; **и. о местонахождении** bearing information, radar information; **и. о положении луча** indexing information; **и. о полутонах яркости** half-tone information; **и. о яркости изображения** luminance information; **и. об опорной фазе** phase reference information, reference-phase information; **и. одного поля разложения** field information; **и. от удалённого источника** remote information; **и., передаваемая на цветовой поднесущей** color subcarrier information; **и. системы самонаведения** homing intelligence; **сложная и. о цветовом оттенке и насыщенности** combined hue and saturation information; **и., содержащаяся в одном кадре разложения** frame information; **и., содержащаяся в разностном цветовом сигнале** color-difference information; **и., содержащаяся в форме сигнала** shape information; **угломестная и.** E information; **усреднённая передаваемая и.** average transinformation; **и. цветного сигнала в ЦТ** chromaticity information

**информировать** *v.* to inform, to advise

**инфра-** *prefix* infra-

**инфрадин** *m.* infradyne

**инфрадинный** *adj.* infradyne

**инфразвук** *m.* infra-audible sound, infrasound, infrasonics

**инфразвуковой** *adj.* subsonic, infrasonic, subaudio

**инфракрасный** *adj.* infrared

**инфралюминесценция** *f.* infraluminescence

**инфрачёрный** *adj.* infrablack

**инъектировать** *v.* to inject

**инъекция** *f.* injection

**иод** *see* йод

**ион** *m.* ion; **быстрый и. отдачи** fast recoil ion; **и. индикатора** tracer ion; **исходный и.** parent ion

**Иондрелль** Jodrell

**ионизатор** *m.* ionizer

**ионизационный** *adj.* ionization

**ионизация** *f.* ionization, electrolytic dissociation; **и. атома при соударении** collision ionization; **и. в газах** electromerism; **и. в области полярного сияния** aural ionization; **вторичная и. счётной трубки** re-ignition of a radiation counter tube; **дополнительная и. в счётчиках излучений** re-ignition of a radiation counter tube; **лавинная и.** cumulative ionization; **и. от полярного сияния** auroral ionization; **и. под воздействием света** photo-ionization

**ионизированный** *adj.* ionized, dissociated

**ионизировать** *v.* to ionize

**ионизируемый** *adj.* ionizable

**ионизирующий** *adj.* ionizing

**ионизирующийся** *adj.* ionizable

**ионизованный** *adj.* ionized

**ионизовать** *v.* to ionize

**ионно-геттерный** *adj.* getter-ion

**ионно-нагревный** *adj.* ionic-heated

**ионно-обменный** *adj.* ion-exchange

**ионно-фокусированный** *adj.* ion-focused

**ионный** *adj.* ionic; ion

**ионоген** *m.* ionogen

**ионогенный** *adj.* ionogenic

**ионограмма** *f.* ionogram

**ионозонд** *m.* ionosonde

**ионоизбирательный** *adj.* ion-selective

**ионоизлучающий** *adj.* ion-emitting

**ионо-импульсный** *adj.* ion-pulse

**ионолюминесценция** *f.* ionolumines-
cence
**иономер** *m.* pH meter
**ионометр** *m.* ionometer
**ионообмен** *m.* ion exchange
**ионообменитель** *m.* ion exchange
**ионообменник** *m.* ion exchanger
**ионообменный** *adj.* ion-exchange
**ионообразование** *n.* ion formation
**ионопауза** *f.* ionopause
**ионосфера** *f.* ionosphere, Heaviside
layer
**ионосферный** *adj.* ionospheric; ionosphere
**ионотрон** *m.* ionotron
**ионофон** *m.* ionophone, ionic loudspeaker
**ионофорез** *m.* ionophoresis
**Ионсон** Johnson
**ионтофорез** *m.* iontophoresis
**Иордан** Jordan
**ИП** *abbr.* (**истинный пеленг**) true
bearing
**ИПО** *abbr.* (**истинный пеленг ориен-
тира**) true bearing of a landmark
**ИПП** *abbr.* (**инструкция по производству полёта**) flight instruction
**и пр.** *abbr.* (**и прочее**) and so forth
**ИПР** *abbr.* (**истинный пеленг радиостанций**) true bearing of a radio
station
**ИПС** *abbr.* (**истинный пеленг самолёта**) true bearing of aircraft
**ИПУ** *abbr.* (**истинный путевой угол**)
true course angle
**иразер** *m.* iraser
**иридиевый** *adj.* iridic; iridium
**иридий** *m.* iridium, Ir
**ирис** *m.* iris
**ирисовый** *adj.* iris
**Ирли** Early
**ирнег** *m.* yrneh
**ИРП** *abbr.* (**истинный радиопеленг**)
true radio bearing
**иррадиация** *f.* irradiation
**иррадиировать** *v.* to irradiate
**иррациональный** *adj.* irrational, surd;
incommensurate
**иррегулярный** *adj.* irregular, uneven
**иррефлексивный** *adj.* irreflexive
**ИСЗ** *abbr.* (**искусственный спутник
земли**) artificial earth satellite
**искажать, исказить** *v.* to distort, to
deform, to alter, to mutilate
**искажающий** *adj.* distorting
**искажение** *n.* distortion, deformation,

alteration, mutilation; perturbation;
misrepresentation; **искажения, вносимые второй гармоникой** second
harmonic distortions; **градационные
искажения** tonal distortions; **искажения за счёт биений** beat-note distortions; **перекрёстные искажения**
crosstalk interference, cross modulation, intermodulation distortions;
**искажения при переходном режиме**
transient distortions; **искажения,
связанные с частичным подавлением одной боковой полосы** vestigial
sideband distortions; **трапецеидаль-
ные искажения по кадру** field keystone distortions, frame keystone
distortions; **фоновые искажения
изображения** hum interference in
picture-tube display; **и. в направлении строк** line distortion; **и. в оконечной лампе** volume distortion; **и.
в целях засекречивания** speech inversion; **и. вида колебаний** mode
distortion; **и., вносимое обтекателем**
radome distortion; **вносимое усилителем и.** amplifier distortion; **волнообразное и. (растра)** ripple effect,
weave; **и. времени нарастания** rise
time distortion; **и. времени пролёта**
transit time distortion; **и. вследствие
огибания** (*or* **рефракции**) tracing
distortion; **и. вследствие отражений**
echo distortion; **и., вызванное временем прохождения сигнала** transient time distortion; **и., вызванное
фоном питающего напряжения** hum
distortion; **и. гармонического состава сигналов звуковой частоты**
audio-frequency harmonic distortion; **и. диаграммы излучения** pattern distortion; **и. динамических
оттенков** volume-range distortion,
companding; **дугообразное и. растра**
bend (*tv*); **и. за счёт времени пролёта электронов** transit-time distortion; **и. за счёт отражений** echo
distortion; **и. за счёт сеточного тока**
grid-current distortion; **и. «зайчика»**
spot distortion; **и. из-за огибания**
tracing distortion; **и. изображения
типа «вспышки»** blooming (*tv*); **и.
импульсов набора** dialing distortion;
**и. крутизны** transconductance distortion; **кубическое и.** distortion of
third order; **и. на выходе** output
distortion, volume distortion; **и. на**

телевизоре slipping (*tv*); **и. неустановившегося процесса** transient distortion; **и. «нумерации»** quantization distortion; **и. обратной связи** distortion due to feedback; **и. огибающей** envelope distortion; **и. от затухания** amplitude distortion, frequency distortion; **и. от недостаточной крутизны фронта сигналов** underthrow distortion; **и. от нечётных гармоник** odd distortion; **и. от перегрузки** overshot distortion; **и. от разбивки (волны) на ступени** quantization distortion (of waves); **и. от разветвления** bifurcation distortion, multipath distortion; **и. от раскрыва** aperture-effect distortion; **и. от чётных гармоник** even distortion; **и. первого рода** amplitude distortion; **перегрузочное и.** overshot distortion; **предварительное и.** pre-distortion; **и. при отклонении** deflection distortion; **и. при передаче** sending-end distortion, transmitter distortion; **и. при переходном режиме** transient distortion; **и. при распространении по нескольким путям** multipath distortion; **и. проницаемости** distortion due to variation of inverse multiplication factor; **и. развёртки** pattern distortion; **реостатно-ёмкостное и.** resistance-capacitance distortion; **и. степенного закона** power law distortion; **и. тембра** audio-frequency harmonic distortion; **и. фазы сигнала цветности, и. фазы цвета** color-phase error; **и. формы волны и амплитуды** waveform-amplitude distortion; **и. формы сигнала** wave distortion, waveform distortion

**искаженность** *f.* distortion, deformation

**искажённый** *adj.* distorted, disfigured, altered, mutilated, abnormal

**исказить** *pf. of* **искажать**

**искание** *n.* search, searching, finding, hunting; selection; dialing, keying; **и. в нескольких декадах** level hunting, discrimination on several levels; **и. в одной декаде** rotary hunting, selection on one level; **и. вызова, вызывное и.** finding action, hunting action; **вынужденное и.** impulse action; **и. выхода** trunk hunting; **дальнее и.** long-distance dialing, toll line dialing;

**дальнее и. импульсами переменного тока** alternating current dialing; **дальнее и. импульсами подтональной частоты** subfrequency dialing; **дальнее и. импульсами тональной частоты** voice-frequency dialing; **дальнее и. с устранением искажений** distortionless long-distance dialing; **и. импульсами постоянного тока** d.c. dialing, d.c. selection; **и. индуктивным путём** inductive selection; **и. и сигнализация на тональных частотах** audio-frequency dialing and signaling; **и. на расстоянии** dialing (*tphny.*); **и. на тональных частотах** audio-frequency dialing; **и. номера по шлейфной системе** loop dialing; **обратное и.** register-type dialing; call-finding; **и. по шаговой системе** step-by-step selection; **и. по шлейфной системе** loop dialing; **и. под контролем регистра** register-controlling selection; **и. подтональной частотой** low frequency dialing; subaudio dialing; **последующее и.** post-selection; **предварительное и.** forward selection, preselection; **промежуточное и.** tandem selection; **ручное дальнее и.** operator's toll-line dialing, operator's long-distance selection; **свободное и. (вывода), и. свободной междугородной линии** trunk hunting; **серийное и.** (trunk) hunting; **и. с помощью тастатуры, тастатурное и.** key sending position, key pulsing, key sending, keying; **и. тональной частотой** voice-frequency dialing; **и. через транзитную станцию** through dialing; **шаговое и.** step-by-step selection

**искатель** *m.* selector, finder, switch; searcher, seeker; locator, localizer; scanner; telescope director; **и. абонентского номера без кода станции** "numerical" selector; **и. без исходного положения** non-homing finder; **и. без серийного искания** controlled switch; **вращательно-подъёмный и.** vertical and rotary selector, two-motion selector; **вращательный и. с исходным положением** homing finder, homing-type switch; **второй и. вызовов** second line finder; **и. второй и третьей буквы** B- and C-digit selector; **и. входящих соединений** call finder; **и. вызова-искатель**

**вызова** line selector with line finder; **групповой и. для транзитной связи** tandem selector; **групповой и. оконечной станции** terminal office group selector, central exchange frequency group-selector plug; **дальний и. соединительных линий** long-distance switch; **дальний групповой и.** trunk group selector; **двухмагнитный (подъёмно-вращательный) и.** tetragonal selector, two-magnet (two-motion) switch, square selector; **и. для предложения междугородного переговора** trunk offering selector; **и. для проглатывания цифр** digit absorber, digit absorbing selector; **и. для соединительных линий** hunting switch; **и. жил** wire finder; **избирательный и.** hunting switch, tapper; **коммутаторный линейный и.** rotary hunting connector; **и. контактных щёток** trip spindle; **координатный и.** crossbar switch, crossbar selector; **кулисный и.** rotary plunger selector, Hasler's selector; **линейный и.** connector, final selector, line selector, junction finder; **линейный и. больших серийных номеров, линейный и. для абонента, имеющего несколько линий, линейный и. для (больших) коммутаторов Р. В. Х.** final selector, rotary hunting connector; **машинный и.** motor-operated selector employing group drive; **междугородный групповой и.** trunk offering selector; **междугородный линейный и.** trunk connector, toll connector; **местный и междугородный линейный и.** long-distance and local connector, final selector common to local and trunk lines; **моторно-вращательный и. с палладиевым контактом** rare metal machine rotary switch; **моторно-координатный и. с драгоценным металлом** rare metal machine crossbar switch; **накопительный и.** register-type selector; **обходный и.** direct selector; **и. оптимального режима** optimalizing input drive; **оптический и.** optical view finder; **и. панельной системы** panel switch; **и. первой буквы** A-digit selector; **первый и. вызовов** first line finder; **первый групповой и.** district selector; **пиковый и. вызовов** peak-load

call finder; **поворотный и.** rotary switch, uniselector; **поглощающий и.** digit absorber, digit absorbing selector; **подъёмно-вращательный и.** two-motion switch, Strowger switch, two-motion selector; **и. посадочной полосы** zero reader; **предварительный и.** subscriber's selector; **проходной и.** tandem selector; **районный и.** discriminating selector; **регистровый и.** register finder, A-digit selector, sender selector; **релейный и. вызова** line-finder relay, call-finder relay; **и. с вынужденным движением** pulse-actuated selector, numerical connector switch; **и. с двумя движениями** vertical and rotary selector; **и. с исходным положением** homing finder, homing-type switch; **и. с машинным приводом** motor-uniselector; **и. с питающим мостиком** selector with feeding bridge (*tphny.*); **и. с положениями ожидания** register-type selector; **свободный и.** hunting selector; **служебный и.** auxiliary selector, special code selector, selector repeater; **и. служебных линий** order wire selector; **и. телефонистки** call distributor; **и. телефонистки стола заказов** order-wire distributor, operator's position distributor; **транзитный и.** tandem selector; **и. третьей цифры** C-digit selector; **универсальный линейный и.** local and trunk connector; combination connector; **и. через транзитную станцию** through-dialing connector; **шаговый и.** step-by-step switch, stepping switch, stepping counter, stepping relay, crossbar switch, multicontact switch

**искательный** *adj.* seeking, searching

**искать, поискать** *v.* to look (for), to search, to find, to seek; to select

**исключать, исключить** *v.* to eliminate, to discard, to exclude, to except; to release, to discharge, to eject; to shut out; to deduct

**исключающий** *adj.* excluding, exclusive, lock-out; **взаимно и.** conflicting

**исключение** *n.* screening out, elimination, exclusion, exception, expulsion, ejection; **и. двухзначности** sense finding, sense research; **и. незначащих нулей** zero suppression; **и. подстановкой** elimination by substitution

**исключённый** *adj.* excluded, excepted; ejected, eliminated, expelled

**исключительно** *adv.* exceptionally; solely, exclusively

**исключительный** *adj.* exceptional, exclusive, unusual

**исключить** *pf. of* **исключать**

**искорка** *f.* scintillation; spark

**искра** *f.* spark, flash; **и. вспомогательного зажигания** trigger spark; **звучащая и.** musical spark, singing spark; **неполностью затухающая и.** arcing spark; **поджигающая и.** trigger spark, pilot spark; **и. прерывания** break spark; **и. при прикосновении** touch spark; **и. при размыкании** break spark; **и. при разрыве индуктивной цепи** wipe spark; **разрядная и.** disruptive spark; **синхронная и.** timed spark

**искрение** *n.* sparking, flashing; arcing; **круговое и.** commutator sparking

**искривить** *pf. of* **искривлять**

**искривление** *n.* distortion, deformation, warp, warping, twisting, twist; curvature, curve, bend, bending, flexure; **и. звукового луча** sound-ray refraction; **и. изображения** curvature of the image field; **и. изображения вдоль строк** line bend, line tilt; **и. изображения по кадру** frame bend, field bend; **и. прямых линий изображения (наружу)** barrel distortion; **и. прямых линий изображения (внутрь)** pillow distortion; **и. растра** curvature of the image field

**искривлённый** *adj.* off-set, cranked; distorted, warped, twisted; bent, curved

**искривлять, искривить** *v.* to distort, to deform, to twist; to bend, to curve

**искривляющий** *adj.* bending

**искристый** *adj.* scintillating, flashing

**искрить** *v.* to spark, to flash

**искриться** *v.* to spark, to flash, to scintillate, to sparkle

**искробезопасный** *adj.* spark-proof

**искровой** *adj.* spark, spark-like

**искрогаситель** *m.* spark-quench, spark extinguisher, arc arrester; quench circuit, spark blowout; key filter (*tgphy.*); **и. ключа** key filter, key spark suppressor; **роговой и.** arcing horn

**искрогасительный, искрогасящий** *adj.* spark-quench, spark-quenching, spark-extinguishing, blow-out, spark-quenched

**искрогашение** *n.* spark quenching, spark suppression, arc suppression

**искроловитель** *m.* spark catcher, spark arrester

**искромер** *m.* spark meter, scintillometer

**искрообразование** *n.* sparking, spark formation

**искростойкий** *adj.* nonsparking, non-arcing

**искростойкость** *f.* arc resistance

**искротушитель** *m.* spark arrester, spark quencher, spark extinguisher

**искроудержатель** *m.* spark arrester

**искроуказыватель** *m.* spark indicator, spark detector

**искроуловитель** *m.* spark trap, spark catcher, spark arrester

**искрящий** *adj.* sparking

**искусственно** *adv.* artificially, synthetically

**искусственный** *adj.* artificial, synthetic; false, dummy

**искусство** *n.* art, craft; skill, workmanship; practice

**искушенность** *f.* sophistication

**ИСЛ** *abbr.* (**искусственный спутник Луны**) artificial Moon satellite

**исландский** *adj.* Icelandic; Iceland

**испарение** *n.* evaporation, evaporating, vaporization, sublimation; fume, vapor

**испаритель** *m.* evaporator, vaporizer, vapor source; evaporant cooler; **кольцевой и.** ring (evaporation) source; **и., нагреваемый джоулевым теплом** resistance-heated boat; **и. с индукционным нагревом** induction-heated boat; **тепловой и.** thermal evaporation source

**испарительный** *adj.* evaporative

**испаритель-сгуститель** *m.* evaporator condenser

**испарить(ся)** *pf. of* **испарять(ся)**

**испаряемый** *adj.* volatile

**испарять, испарить** *v.* to evaporate, to vaporize, to volatilize; to steam

**испаряться, испариться** *v.* to evaporate, to vaporize; to vanish

**испаряющий** *adj.* evaporating, vaporizing

**испаряющийся** *adj.* volatile, evaporating

**исполнение** *n.* performance, execution, completion, fulfillment, accomplishment; **и. аккумулятора по видам применения** types of storage battery according to use; **закрытое и.** enclosed type, enclosed structure; **защищённое и.** protected type, semienclosed type; **и. команды** execute part of cycle

**исполненный** *adj.* complete, fulfilled, accomplished

**исполнимость** *f.* feasibility, practicability

**исполнимый** *adj.* feasible, practicable

**исполнитель** *m.* performer, executor

**исполнительный** *adj.* executive, execution; actuating, driving; effective

**исполнить** *pf. of* **исполнять**

**исполняемый** *adj.* performed, executed, carried out

**исполнять, исполнить** *v.* to do, to perform, to accomplish, to carry out, to execute, to fulfill

**исполняющий** *adj.* carrying out, fulfilling, performing

**использование** *n.* utilization, use, employment, application, exploitation, consumption; recovery; development; **с использованием в обе стороны** worked alternately; **и. в полевых условиях** field use; **и. внутри** internal use; **и. девиации** usable (frequency) swing; **и. международной цепи** allocation of an international circuit; **многократное и. линии** multiplexing (of lines or cables); **и. одной из компаний-участиц** intercompany use; **повторное и.** re-use; **и. принципа многократности** multiplexing; **уплотнённое и. линии** multiplexing (of lines or cables); **и. хвостового оперения для передачи** empennage excitation

**использованный** *adj.* used, used up, spent, utilized, consumed

**использовать** *v.* to utilize, to use, to make use (of), to employ, to consume

**используемый** *adj.* used, employed, utilized

**испортить** *pf. of* **портить**

**испорченный** *adj.* faulty, defective, damaged; spoiled

**исправитель** *m.* corrector, corrective network

**исправительный** *adj.* corrective

**исправить** *pf. of* **исправлять**

**исправление** *n.* correction, correcting, adjustment, readjustment, readjust-

ing, fixing, reparation, repair, rectification; improvement; **и. двойных ошибок** double-error correction; **и. на сдвиг** drift correction; **и. одиночных ошибок** single-error correction

**исправленный** *adj.* corrected, fixed, adjusted, repaired; improved

**исправлять, исправить** *v.* to correct, to adjust, to readjust, to fix, to repair, to rectify; to revise, to amend

**исправляющий** *adj.* correcting, corrective; regenerative (circuit); corrector

**исправность** *f.* soundness, normalcy; continuity (of circuit); operable condition; good condition

**исправный** *adj.* in good working order, in good shape, serviceable, efficient, fit, operable, sound

**испускаемый** *adj.* ejected, emitted, given off

**испускание** *n.* emission, emanation, exhalation; emergence, egress; ejection; **и. моноэнергетического излучения** mono-energetic emission

**испускатель** *m.* radiator, emitter

**испускательный** *adj.* emitting, emissive, radiating

**испускать, испустить** *v.* to emit, to radiate, to give off; to eject; to expire; to utter

**испускающий** *adj.* emitting, emissive

**испустить** *pf. of* **испускать**

**испущенный** *adj.* emitted

**испытание** *n.* test, testing, probe, trail, experiment; examination, analysis, analysing, checking; research, investigation; **испытания в динамическом режиме** dynamic tests; **испытания в импульсном режиме** impulse tests; **испытания в статическом режиме** static tests; **испытания на микрофонный эффект** microphony tests; **испытания на окружающие условия** acceptable environmental range test(s); **очередные эксплуатационные испытания** routine tests; **испытания по оценке качества изображения** picture-appraisal tests; **полигонные испытания** field tests; **приёмо-сдаточные испытания** acceptance tests; **продолжительные циклические испытания на влагостойкость** prolonged humidity cycling tests; **циклические испытания подогревателя** heater-cycling life tests; **баллистическое и. (иконоскопа) на фоточувствительность** ballistic sensitivity test; **и. без разрушения образца** nondestructive testing; **и. в искусственно созданном режиме** staged test; **и. в непрерывном режиме** continuous test; **и. в нормальную величину** full-scale test; **и. в период эксплуатации** maintenance test; **и. в полевых условиях** field test; **и. в прерывистом режиме** intermittent test; **и. в рабочих условиях** performance test, operation testing; **и. в составе целого комплекса** on-line test; **и. в условиях повышенных температуры и влажности** tropicalization test; **и. волной с крутым фронтом** front-of-wave test; **всестороннее климатическое и.** pan-climatic testing; **и. вхолостую** no-load test; **и. вызовов** signalling test; **и. (гальванического элемента) на саморазряд** shelf test; **и. действием плавящейся проволоки** fuse wire test; **и. деталей без их повреждения** non-destructive inspection; **и. для локализации повреждения по Блавье** Blavier's test; **и. для определения к. п. д.** efficiency test; **и. допустимым напряжением** withstand test; **и. звуковым сигналом** audible test; **и. изделий без их разрушения** nondestructive inspection; **и. (изоляции) между слоями** layer test; **и. изоляции проводов** circuit (insulation) testing; **(импульсное) и. волной с крутым фронтом** front-of-wave test; **и. кабеля на плотность** cable pressure test; **контрольное и.** monitoring test; check routine test; **и. линейного искателя** final selector test; **и. линейности развёрток передающей камеры** camera linearity test; **и. линейности растра видеоконтрольного устройства** monitor linearity test; **и. (материала) на обгорание** charring test; **и. материалов без их разрушения** non-destructive material testing; **и. места спайки** soldering test; **и. методом взаимной нагрузки** back-to-back test; **и. методом выбега** retardation test; **и. методом качающейся частоты** sweep check; **и. методом моделирования** simulation test; **и. мето-**

**дом наблюдения изображения** picture viewing test; **и. методом самоторможения** retardation test; **и. на большой прерывистой нагрузке** heavy intermittent test; **и. на брызгоустойчивость** spray test; **и. на вибрационную усталость** fatigue-vibration test; **и. на внезапные волны с крутым фронтом** impulse test, surge test; **и. на выдерживание высокого напряжения** high-voltage holding test; **и. на выносливость** stability test; **и. на грибостойкость** fungus test; **и. на дифференциальные фазовые искажения** differential-phase test; **и. на дрейф нуля** drift test; **и. на живучесть** survival-rate test; **и. на изгиб в холодном состоянии** cold bend test; **и. на к. п. д.** efficiency test; **и. на линии** field test, line test; **и. на месте установки** site test; **и. на механическую прочность** mechanical robustness testing; **и. на осадкообразование** sludge test; **и. на перекрытие высоким напряжением** spark-over test; **и. на предупреждение столкновений** collision test; **и. на пробой** breakdown test, disruptive test, puncture test; **и. на проведение разговора** voice-ear measurement; **и. на проход** continuity test; **и. на прочность ударной нагрузкой** toughness test; **и. на разбег** running-up test; **и. на разговор** speech test, voice-ear test; **и. на разрыв** tension test; **и. на растяжение** tensile test; **и. на сброс нагрузки** load-dropping test; **и. на сохранность** storage test, shelf test; **и. на сплошность (оболочки)** pinhole test; **и. на стабильность параметров** stability life test; **и. на теплоустойчивость** temperature check; **и. на ударную нагрузку** drop test; **и. на утечку** leakage test; **и. на хаотические вибрации** "white noise" vibration test; **и. на холостом ходе** no-load test, running-light test; **и. на целость (линии)** continuity test (of a line); **и. напряжением с выдержкой времени** time-voltage test; **и. обливанием солёной водой** salt-spray test; **и. обратной полярностью** reverse battery test; **и. падением напряжения** drop test, fall-of-potential test; **и. перед запуском**

в производство preproduction-type test; **и. перед установкой** pre-installation test; **и. переполюсовкой батареи** reverse battery test; **петлевое и. по Варлею** Varley loop test; **и. петлей Морея** Murray loop test; **и. по методу взаимной нагрузки** opposition test; **и. по методу возвратной работы** feedback test; **и. по требованию заказчика** customer request test; **и. повышенным напряжением** pressure test; overpotential test; induced potential test; **и. под большой прерывистой нагрузкой** heavy instrument test; **и. под искусственным дождём** spray test; **и. под напряжением** full-scale test; **и. под небольшой прерывистой нагрузкой** light intermittent test; **и. при большой прерывистой нагрузке** heavy intermittent test; **и. при заторможённом роторе, и. при остановленном роторе** blocked rotor test; **и. при прерывистой нагрузке** intermittent test; **и. при разработке** development test; **и. провеса** dip test, sag test; **и. провода на перегиб** wrapping test; **и. прохождением вызова** ringing test, signaling test; **и. прочности на раздирание** tearing strength test; **и. путей соединения** testing the transmission paths; **и. путём наблюдения изображения** viewing test; **и. путём наружного осмотра** appearance test; **и. с помощью ультразвука** silent-sound testing; **и. сжатым воздухом** air pressure test, cable pressure test; **и. сигнальной лампой** visual engaged test; **и. слухового восприятия** auditory perception test; **и. срока службы** aging test, life test; **и. сростков** joint testing, splice testing; **сухоразрядное и.** dry test; **и. торможением** braking test; **и. тормозом Прони** Prony brake test; **ударное и. на изгиб** impact bending test; **ударное и. надрезанного образца** notched bar impact bending test; **и. усечённой волной** chopped-wave test; **и. через посредника** off-line test; **эксплуатационное и.** field test, service test, performance test, maintenance test, actual test; **и. эмиссии** filament activity test

**испытанный** *adj.* tested, tried, in-spected, examined; **и. временем** time-tested

**испытатель** *m.* testing apparatus, tester; investigator, research man; analyst; **и. дерева столбов** pole tester, timber tester, pole prod; **и. ламповых панелей в работе** electron tube tester, free-point tester; **и. переходной характеристики** transient analyzer; **и. переходных процессов в видеосигнале** transient video analyzer; **и. прохождения сигнала** signal-tracing instrument; **и. стыков** bond tester; **и. цепей** circuit tester; **и. эмиссии ламп** emission tester, emission-control tube tester

**испытательный** *adj.* testing, test, experimental, proving

**испытать** *pf. of* испытывать

**испытуемый** *adj.* testing, tested, experimental

**испытующий** *adj.* testing

**испытываемый** *adj.* tested

**испытывать, испытать** *v.* to test, to try, to put to test; to investigate, to check, to examine, to analyze; to experience, to undergo

**исследование** *n.* study, research, investigation, survey; analysis, examination, analysing; searching, exploration; inquiry; **исследования с помощью индикаторов** tracer studies; **кинематографическое и.** time magnifying study; **и. кривых** curve tracing; **и. «лупой времени»** time magnifying study; **и. на месте установки** field investigation; **и. на минимум** minimization (*math.*); **и. отдельных частей** partial system test; **и. с помощью замедленной кинопроекции** time magnifying study; **и. с помощью моделирования** analog study; **и. (функции) на минимум** minimization

**исследованный** *adj.* studied, examined, tested, investigated

**исследователь** *m.* researcher, investigator

**исследовать** *v.* to inquire into, to research, to study, to investigate, to examine; to test, to try; to analyse; **и. на ощупь** to sense

**истекание** *n.* efflux

**истекать, истечь** *v.* to elapse, to expire; to escape, to emanate, to run out

**истекший** *adj.* elapsed, expired; escaped

**истереть** *pf. of* **истирать**

**истечение** *n.* emission, emanation, escape; expiration, lapse; flow, discharge, outflow, efflux, effluve, effluvium, flux; **и. заряда с острия** effluvium, point effect, needle point effect; **и. несжимаемого газа** exhaustion of an incompressible gas; **и. с большим током** effluve; **и. сжимаемого газа** exhaustion of a compressible gas; **и. электрического заряда из острий** point effect

**истечь** *pf. of* **истекать**

**истина** *f.* true, verity, fact, truth

**истинностный** *adj.* true

**истинность** *f.* truth, validity

**истинный** *adj.* true, real, actual, proper, veritable; intrinsic

**истираемость** *f.* wearing, wear, wearability

**истирание** *n.* abrasion, wear, wearing away, abrading, deterioration; grinding, crushing; attrition

**истиратель** *m.* abrasive; grinder, pulverizer

**истирать, истереть** *v.* to wear down, to abrade; to grind, to crush

**исток** *m.* source, issue, outflow, discharge, flowing

**истолковать** *pf. of* **истолковывать, толковать**

**истолковывать, истолковать** *v.* to interpret, to expound; to reason out

**история** *f.* history

**источник** *m.* source, origin, cause; supply, resource; principle; generator; well, fountain; **источники излучения, распределённые на бесконечной полосе** infinite strip sources; **стандартизованные ахроматические источники излучения** specified achromatic lights; **стандартные источники света для колориметра** illuminants for colorimetry; **стандартные источники света МКО** CIE standard illuminants; **и. вихрей** vortex source, eddy source; **и. внешнего сигнала** remote video source; **и. внешней программы** remote program source; **и. возбуждения** excitation source; field power supply; **воображаемый и.** image source; **высоковольтный и. питания, использующий импульсы обратного** хода развёртки fly-back EHT supply, kickback power supply; **и. высокочастотных шумов** microwave noise supply; **газоразрядный и. шума** gas discharge noise source; **и. гармоник** harmonic generator; **и. Гюйгенса** Huygens source; **двойной и.** dipole; **дистанционный и. сигналов изображения** remote video source; **добавочный и.** booster; **и. звука прямого излучения** direct radiator source; **и. звуковых колебаний** tone source, acoustic source; **зенеровский эталонный и. напряжения** Zener reference unit; **идеальный и. тока** pure current source; **и. импульсного освещения** pulsed light generator; **импульсный и. питания** kickback type of supply; **искуственный и. звука, вмонтированный в муляж головы** artificial head source; **и. местных колебаний** local oscillator; **мнимый и. звука** acoustical image; **и. напряжений смещения** multibias box; **незаземлённый и. питания** insulated supply system; **и. незатухающих колебаний** continuous wave source; **и. неисправностей** source of failure; **и. непрерывного звука** steady source; **и. нулевого порядка** source of zero order; **объединённый стабилизованный и. питания** stacked regulated power supply; **опорный и. питания** reference supply; **ПВО и. получения данных** data-gathering source; **и. питания** power supply, power plant, power source, power pack; **и. питания анода** plate-power supply, B-power supply, B-source; **и. питания блока дальности** range power supply; **и. питания генератора** oscillator supply; **и. питания накалов** A-power supply; **и. питания с большим внутренним сопротивлением** high-resistance power supply; **и. питания с вибропреобразователем** vibrator power supply; **и. питания сварочной дуги** welding source, welding power supply; **и. питания сверхвысокого напряжения** EHT supply; **и. питания сеток** C-power supply; **и. питания цепи накала** A-power supply; **и. постоянной мощности** constant-output source; **и. проблескового освещения** flasher; **и. пусковых**

**сигналов** triggering source; **и. равных энергий** equal-energy source; **и. распыляемого материала** vapor source; **и. рассеянного излучения** beam-antenna with poor directivity; **и. с большим внутренним сопротивлением** high impedance source; **и. с интенсивностью, равной единице** unit source; **и. с конечным значением объёмной скорости** source of zero order; **и. с малым внутренним сопротивлением** low impedance source; **и. с поперечной поляризацией** transversely polarized source; **и. с продольной поляризацией** longitudinally polarized source; **и. с равномерным распределением света** uniform light source; **и. света с изменяющейся интенсивностью** intermittent light source; **и. света с постоянной интенсивностью** constant-intensity light source; **и. сеточного напряжения** grid bias supply; **и. сеточного смещения** grid-voltage supply; **и. сигналов информации** information generator; **спокойный и. тока** quiet power supply, quiet battery; **и. стабилизированного напряжения** voltage-stabilized source, voltage standardizer; **стабилизированный и. постоянного тока** regulated direct-current source; **и. тактовых импульсов** clock, impulser; cadence tapper; **и. тока накала** filament generator; **точечный и. звука** focused sound source; **точечный и. света** point (source) lamp, spot lamp; **и. удалённой программы** remote program source; **и. ультразвуковых колебаний** supersonic source; **универсальный и. питания** multiple power source; **элементарный и. колебаний** Huygens source; **и. энергии неподвижных звёзд** source of energy of the stars

**истощать, истощить** v. to exhaust, to drain, to waste, to deplete, to spend, to wear out, to impoverish

**истощение** n. exhaustion, depletion, impoverishment, draining

**истощённый** adj. depleted, exhausted, drained, spent, impoverished

**истощившийся** adj. exhausted

**истощить** pf. of **истощать**

**истребитель** m. fighter aircraft, interceptor

**истребительный** adj. destructive, destroying

**исход** m. outcome, result, issue, outlet

**исходить, изойти** v. to issue (from), to come (from), to originate, to proceed (from), to emanate, to emerge, to radiate

**исходный** adj. reference, base, original, initial, first, primary; launched; starting

**исходящий** adj. outgoing, issuing, emanating, coming (from)

**исчезание** see **исчезновение**

**исчезать, исчезнуть** v. to disappear, to vanish, to dissipate, to fade away; to die out, to become extinct; to merge (into)

**исчезающий** adj. disappearing, vanishing, fading away, dying out

**исчезновение** n. fading, disappearance; loss; **плавное и. изображения** fade out (of a picture); **подлинное плавное и.** true fade out (of a signal)

**исчезнуть** pf. of **исчезать**

**исчерпанный** adj. exhausted, spent

**исчерпать** pf. of **исчерпывать**

**исчерпывание** n. exhaustion

**исчерпывать, исчерпать** v. to exhaust, to exhause, to drain, to empty

**исчерпывающий** adj. exhaustive, comprehensible; draining

**исчисление** n. calculus; calculation, estimation, computation; enumeration; rating; **и. бесконечно-малых** infinitesimal calculus; **вариационное и.** calculus of variations; **и. высказываний** propositional calculus, sentential calculus; **интуиционистское и. высказываний** intuitionistic propositional calculus; **и. комплексных величин** algebra of complex quantities; **и. предикатов высших порядков** higher predicate calculus; **узкое и. предикатов** restricted predicate calculus

**исчисленный** adj. calculated, estimated; enumerated

**исчислимый** adj. computable

**исчислять, исчислить** v. to number, to enumerate; to compute, to calculate, to estimate

**итеративный, итерационный** adj. iterative

**итерация** f. iteration

**итерированный** adj. iterated, repeated

**итерировать** v. to iterate

**итог** *m.* sum, summation, total, total amount, result; **частичный и.** minor total

**итоговый** *adj.* final, summarized, summary; sum, total

**иттербий** *m.* ytterbium, Yb

**иттербовый** *adj.* ytterbium, ytterbic

**иттриевый** *adj.* yttrium; yttric, yttriferous

**иттрий** *m.* yttrium, Y, Yt

**иттровый** *adj.* yttrium; yttric, yttriferous

**ИФХА** *abbr.* (**Институт физико-химического анализа**) Institute of Physico-chemical Analysis

**ИЧХ** *abbr.* (**измеритель частотной характеристики**) frequency-response display set

# Й

йод *m.* iodine
йодистый *adj.* iodic
йодный *adj.* iodine

Йордан Jordan
Йэйтс Yates

# К

К *abbr.* 1 (компас) compass; 2 (компрессор) compressor; 3 (коэффициент) coefficient, factor, ratio

к *abbr.* (кулон) coulomb

к *prep. dat.* to; towards; by; for; against

ка *abbr.* (килоампер) kiloampere

КА *abbr.* (компасный азимут) compass azimuth

каб. *abbr.* (кабель, кабельный) cable

кабелеискатель *m.* cable detector

кабелек *m.* service cable, duplex wire; pigtail

кабелепровод *m.* cable conduit, cable duct, cable tube

кабелепроводка *f.* cable network; cabling

кабелепрокладочный *adj.* cable laying

кабелеукладчик *m.* cable layer, cable laying-out machine

кабелеукладыватель *m.* cable layer

кабель *m.* cable, service cable; **абонентский к.** subscriber's cable, internal cable, service cable; **к. без поясной изоляции** separately lead-covered cable; **береговой к.** shore-end cable; **ведущий к.** leader cable, guide cable, pilot cable; **внутрирайонный к.** trunk zone cable, district cable; **вч к.** h-f cable, r-f cable; **к. дальней связи** trunk cable; **к. двузвёздной скрутки, двузвёздно-скрученный к.** spiral-eight cable, quad pair cable; **к. для контактных систем** bank-to-bank cable; **к. для неоновых вывесок** neon sign cable; **к. для телефонирования на двух полосах частот** twin-band telephone cable; **к. для электровоспламенителей** shot-firing cable; **ёмкий к.** large-capacity cable, large-sized cable; **к. звёздной скрутки, звёздно-четверочный к.** spiral-quad cable; **кабинный к.** trailing cable, elevator electric traveling cable; **ковровый к.** braid-covered cable, braided cable; **комбинированный к.** composite cable; **комнатный к.** house cable, internal cable; **концентрический к.** coaxial cable; **крарупизированный к., краруповский к.** continuously loaded cable, Krarup cable; **магистральный к.** main cable; **междугородный к. быстрой связи** intercity cable for non-delay service; **межстанционный к.** interoffice cable; **многожильный пучковый к.** bunched cable; **к. многократного поля** bank cable; **многопарный к.** large capacity cable; **морской к.** deep-sea cable, submarine cable; **морской к. для средних глубин** shallow water cable; **мультипльный к.** bank cable; **освинцованный голый к.** plain lead-covered cable; **ответвительный к.** branching cable, stub cable; **отсасывающий к.** return feeder cable; **к. парной скрутки, парно-скрученный к.** paired cable, non-quadded cable; **к.-план** cable layout; **к. подачи сигнала на громкоговоритель** (*or* **микрофон**) feeder line; **поперечный соединительный к.** cross-connection cable, tie cable; **проходческий к.** trailing cable (for use in mines); **разнопарный к.** composite cable, compound cable; **районный к.** local cable; exchange cable (*tphny.*); **к. с воздушно-бумажной изоляцией** air-space cable, dry-core cable; **к. с восьмерками** quadruple pair cable; **к. с незащищённой свинцовой оболочкой** plain lead-covered cable; **к. с отдельно освинцованными жилами** separately lead-covered cable; **к. с противосыростной разделкой** terminal cable; **к. с пупинизированной фантомной цепью** composite loaded cable; **к. с пучком жил** bank cable; **к. с трёхслойной оболочкой** separate-lead type cable, S.L.-type cable; **к. с четвёрками** quadded cable, multiples twin cable; **к. с четвёрками звездой** spiral-four cable; **к. с экранированными изолированными жилами** screened cable; **к. с экранированными изолированными** проводами screened conductor cable; **к. с эмалевой и хлопчатобумажной изоляцией** varnished cambric cable; **скрученный к.** stranded cable; **к., скрученный двойной звездой** quad pair cable; **скрученный четвёрками к.** multiple twin cable, quadruplex cable; **слаботочный к.** communication cable, weak current cable; **к. со сложной скруткой** rope-lay cable; **к. со смешанными парами жил** combined cable; **к. со стироловой изоляцией** styrene-dielectric cable; **к. со штепсельным приспособлением** patching cable, patch-cord; **станционный к.** office cable; internal cable, switchboard cable; **к. статива искателей** bank cable (*tphny*.); **к. ТГ** plain-lead covered cable; **шланговый к.** cab-tire cable

**кабельный** *adj.* cable

**кабельщик** *m.* cable man, cable splicer, jointer

**кабина** *f.* cabin, cage, cab, booth; **дикторская к.** announcer booth; **к. для прослушивания** sound box, sound booth

**кабинный** *adj.* cabin, cage, booth

**каблирование** *n.* cabling

**каблировать** *v.* to cable

**каблограмма** *f.* cablegram

**кавалерийский** *adj.* cavalry

**Кавальери** Cavalieri

**кавитационный** *adj.* cavitation

**кавитация** *f.* cavitation

**кавитирующий** *adj.* cavitating

**кавычки** *pl.* quotation-marks, quotes

**каданс** *m.* cadence, close (*муз.*)

**каденция** *f.* cadence, close (*муз.*)

**кадмиево-никелевый** *adj.* cadmium-nickel

**кадмиевый** *adj.* cadmium

**кадмий** *m.* cadmium, Cd

**кадмирование** *n.* cadmium plating

**кадмированный** *adj.* cadmium-plated

**кадмировать** *v.* to plate with cadmium

**кадр** *m.* frame, framework, skeleton; outline; frame, picture, still, field, exposure; **по кадрам** framewise; **к. изображения** picture area (in facsimile); **к. плёнки** exposure

**кадрирование** *n.* framing

**кадровый** *adj.* frame, framework, skeleton; outline; frame, picture

**Кается** Cailletot

**кажущийся** *adj.* apparent, seeming

**казеиновый** *adj.* casein

**кайма** *f.* edging, edge, fringe, flange, border, rim

**кактусовый** *adj.* cactus

**кал** *abbr.* (**калория**) calorie

**каландр** *m.* calender

**калевка** *f.* molding plane; channel, channel molding

**калевочный** *adj.* channel; molding plane

**калейдофон** *m.* kaleidophone

**Календар** Callendar

**каление** *n.* incandescence

**калёный** *adj.* red hot

**калеометр** *m.* caleometer

**калиберный** *adj.* caliber, gage, standard

**калибр** *m.* caliber, gage, standard, size, bore; jig; **вставной к.** jack gage; **к. длины** guard-ring; **к. для измерения диаметра струн** string gage; **к. для листового металла** sheet gage; **к. для отверстий** internal calliper gage; **к.-нутромер** internal cylindrical gage, plug gage; **предельный к.** limit gage, go-no-go gage; **приёмочный к. заказчика** purchase inspection gage; **к.-пробка** see **к.-нутромер**; **раздвижной к.** calliper gage; **резьбовой к.** thread gage; **резьбовой к. для наружной резьбы** gage nut; **к.-скоба** snap gage, calliper gage; **к. со сферическими концами** spherical end measuring rod

**калибратор** *m.* calibrator; **маятниковый к.** physical pendulum calibrator

**калибрационный** *adj.* calibrating

**калибрирование** *n.* calibration, standardization; grooving

**калибрировать** see **калибровать**

**калибрирующий** *adj.* calibrating

**калиброванный** *adj.* calibrated, gaged, standardized, graduated, sized

**калибровать** *v.* to calibrate, to gage, to standardize, to graduate, to adjust, to test

**калибровка** *f.* calibration, standardization, matching; grooving; **к. в замкнутой камере** pressure calibration; **к. в свободном поле** field calibration; **к. действительной чувствительности** real-voice calibration; **дополнительная к.** recalibration; **к. нуля дальности** range-zero calibration; **к. радиокомпаса** radio-direction-finder

calibration; **к. трубочек катода** cathode sizing; **к. шкалы** graduation
**калибровочный** *adj.* calibrating; caliber, calibration
**калибровый** *adj.* gage
**калибромер, калиброметр** *m.* gage
**калиевый** *adj.* potassium, potash
**калий** *m.* potassium, K; **углекислый к.** potassium carbonate, potash
**калийный** *adj.* potassium, potash
**калильный** *adj.* incandescent, glowing
**калифорний** *m.* californium, Cf
**каллиротрон** *m.* kallirotron, negative-resistance tube
**каломельный** *adj.* calomel
**калори-** *prefix* calori-, heat
**калориметр** *m.* calorimeter; **дифференциальный паравой к.** twin calorimeter
**калориметрический** *adj.* calorimetric, caloric
**калориметрия** *f.* calorimetry
**калорифер** *m.* heat radiator, heater, heat dissipator, heating element
**калория** *f.* calorie
**калутрон** *m.* calutron
**Кальдерон** Calderon
**Калье** Callier
**калька** *f.* tracing paper, tracing cloth
**калькирование** *n.* tracing; calking
**калькированный** *adj.* traced; calked
**калькировать, скалькировать** *v.* to trace; to calk
**калькулирование** *n.* estimation
**калькулировать, скалькулировать** *v.* to calculate, to figure out, to estimate
**калькулятор** *m.* calculating machine; calculator
**калькуляционный** *adj.* calculating; calculation
**калькуляция** *f.* calculation, estimate
**кальций** *m.* calcium, Ca
**кальциметр** *m.* calcimeter
**калютрон** *m.* calutron
**Кампбелл** Campbell
**камедь** *f.* gum, resin
**каменный** *adj.* stone; jewel (bearing)
**камень** *m.* stone, rock; jewel (bearing)
**камера** *f.* chamber, cell, compartment; office, room; booth; camera (*photo*.); vault; vessel; **воздухоэквивалентная ионизационная к.** air wall ionization chamber; **гасительная к.** explosion chamber, arc chamber; **глухая к.** anechoic chamber; **гулкая к.** reverberation room, reverberation cham-

ber; **диодная к.** crystal mount; **к. для испытания под давлением** pressure tank; **к. для калибровки** (*or* **испытания**) test cavity; **к. для калибровки методом взаимности** reciprocity calibration cavity; **к. для подсчёта кровных телец** hemocytometer; **дуговая к.** arc-chute; **дугогасительная к.** arc-control device; **замкнутая к. связи** closed coupler; **звукомерная к.** sound chamber; **звукосъёмочная к.** sound (pick up) camera; **к. измерения звукоизоляции** transmission room; **ионо-импульсная к. высокого давления с газообразным радиатором** high-pressure gas-recoil ion-pulse chamber; **к. искусственного уха** earphone coupler, "artificial ear" coupler; **контрольная ионизационная к.** monitor ionization chamber; **многопластинчатая ионизационная к. для регистрации гамма-лучей** multiplate gamma-ray ionization chamber; **к. на видиконе с медленной развёрткой** slow-scan vidicon camera (*tv*); **к., обитая войлоком** felt-lined box; **к. одновременной системы цветного телевидения** simultaneous camera (*tv*); **к.-приставка** camera attachment; **к. с точечным отверстием** pinhole camera; **к. сгорания реактивного двигателя** jet settling chamber; **телевизионная к. для передачи крупным планом** close-up camera (*tv*); **телевизионная к. на трубке с накоплением** storage camera (*tv*)
**камерный** *adj.* chamber, chambered
**камертон** *m.* tuning fork; tonometer; **задающий к.** master tuning fork, mechanical master oscillator; **эталонный к.** standard fork, tuning fork standard
**камертонный** *adj.* tuning-fork
**кампометр** *m.* kampometer
**камуфлированный** *adj.* camouflaged
**камуфляж** *m.* camouflage
**канава** *f.* ditch, trough, trench, channel, tunnel, canal; groove, notch, slot
**канавка** *f.* groove, furrow, incision, cut; slot; **внутренняя концентрическая к.** guard circle; **к. для выхода воздуха** escape groove; **замкнутая к.** locked groove, eccentric circle (of a record); **концевая к.** guard circle;

**немая к.** blank groove, unmodulated groove; **переходная к.** lead-over groove, crossover spiral; **соединительная к.** fast groove; **шпоночная к.** keyway; **эксцентрическая к.** eccentric circle

**канавокопатель** *m.* ditching machine, trench digger, ditcher

**канал** *m.* canal, channel, link, conduit, passage, duct, path; bore; **к. административной связи** administration link; **блокировочный к.** carrier-blocking channel; **к. видеосигнала основного цвета** primary-color video-signal channel; **внутренний к.** bore; **волновой к.** wave duct; **к. восстановления поднесущей** subcarrier regeneration channel; **вч к. (с несущей)** carrier channel, high frequency channel; **к. вч блокировки защити** carrier-pilot relaying channel; **дежурный к.** guard channel; **к. звукового сопровождения** sound channel; **идеализированный к. с прямоугольной частотной характеристикой** theoretical rectangular frequency channel; **к. кабельной канализации** cable duct, cable way; **контрольный к.** pilot channel; **к. на несущей частоте** carrier-derived channel; **к. на сантиметровых волнах** microwave channel; **приповерхностный звуковой к.** surface wave duct, shallow sound channel; **к. промежуточной частоты сигналов изображения** picture i-f channel; **разговорный к.** voice-frequency channel; **к. разрешения многозначности** ambiguity resolving channel; **к. с частотной характеристикой, поднимающейся к верхним частотам** peaked channel; **к. связи** communication channel, transmission channel, link, traffic route; **к. связи взаимодействия** liaison link; **к. связи взаимодействия с авиацией** air co-operation link; **к. связи с командованием** command and report link; **к. сигнала цветности** chromaticity channel, chrominance channel; **к. сигналов зелёного (красного) цветоделённого изображения** green (red) channel; **симплексный к.** one-way channel; **служебный к. в системах направленного радио** service channel in radio relay systems; **слуховой к.** acoustic duct; **смежный к.** adjacent channel; **совместно используемый к.** shared channel; **телевизионный к. с передачей изображения в обоих направлениях** reversible television channel; **к. формирования сигнала синхронизации цветов** color-sync processing channel; **к. цветности с постоянной яркостью** constant luminance chromaticity channel

**канализационный** *adj.* canalization

**канализация** *f.* duct, duct bank, conduit; channeling; canalization; sewerage; **кабельная к.** conduit, duct, duct bank; **многоотверстная к.** multiple duct or conduit; **одноотверстная к.** single cut conduit, single duct

**канализированный** *adj.* channelized

**канализировать** *v.* to canalize

**каналовый, канальный** *adj.* canal, channel(ing)

**канальчатый** *adj.* grooved, channeled; tubular

**канат** *m.* cable, rope, cord, line, stranded cable; **к. спиральной свивки** twisted rope

**канатик** *m.* stranded conductor; **гибкий к. из тонких эмалированных жилок** litzendraht wire

**канатный** *adj.* rope, cord, line, cable, funicular

**кандела** *f.* candela

**кандолюминесценция** *f.* candoluminescence

**канифоль** *f.* colophony, white resin

**канон** *m.* canon

**канонический** *adj.* canonical

**кантовать** *v.* to turn on edge, to tilt, to cant

**Каньяр-Латур** Cagniard de la Tour

**капа** *f.* sealed end

**кападайн** *m.* capadyne

**капасуич** *m.* Capaswitch

**капать, накапать** *v.* to drip, to drop, to fall (in drops)

**капельный** *adj.* drop, dropping, trickling

**капилляр** *m.* capillary

**капиллярноактивный** *adj.* capillary active

**капиллярность** *f.* capillarity

**капиллярный** *adj.* capillary

**капицевский** *adj.* (of) Kapitza, Kapitza's

**каплевидный** *adj.* drop-shaped, drop

**каплезащищённый, капленепроницаемый** *adj.* drip-proof, driptight

**каплестойкий, каплеупорный** *adj.* drip-proof

**капля** *f.* drop; a bit, a grain

**капок** *m.* kapok

**капоковый** *adj.* kapok

**каппа-мезон** *m.* K-meson

**капремонт** *m.* overhauling

**капрон** *m.* cupron

**капсула** *f.* capsule; chamber; casing, case, shell; **к. кристалла** crystal protection tube

**капсюловать** *v.* to enclose

**капсюль** *m.* percussion cap, cap, cartridge primer; **к. звукоснимателя** pickup cartridge; **микрофонный к.** diaphragm case, telephone transmitter

**капсюльный** *adj.* percussion cap, cap; capsule

**капсюля** *see* капсула

**карабин** *m.* shackle, swivel plus keystone link, snap hook

**карабинный** *adj.* snap hook

**«карактрон»** *m.* charactron

**карандаш** *m.* pencil

**карандашный** *adj.* pencil-type, pencil

**карат** *m.* carat

**карбамид** *m.* carbamide, urea

**карбамидный** *adj.* carbamide

**карбид** *m.* carbide

**карбидный** *adj.* carbide

**карбидокремниевый** *adj.* silicon-carbide

**карболинеум** *m.* Carbolineum

**карбонат** *m.* carbonate

**карбонизационный** *adj.* carbonization

**карбонизация** *f.* carbonization

**карбонизирование** *n.* carbonization

**карбонизированный** *adj.* carbonized

**карбонизировать** *v.* to carbonize

**карбонизованный** *adj.* carbonized

**карбонизовать** *v.* to carbonize

**карбонильный** *adj.* carbonyl

**карборунд** *m.* carborundum

**карборундовый** *adj.* carborundum

**кардан** *m.* joint, link, universal joint, Cardan joint

**карданный** *adj.* cardan, cardanic

**кардинальный** *adj.* cardinal, principal

**кардиограмма** *f.* cardiogram

**кардиограф** *m.* cardiograph

**кардиоида** *f.* cardioid

**кардиоидный** *adj.* cardioid

**кардиология** *f.* cardiology

**кардиометр** *m.* cardiometer

**кардиоскоп** *m.* cardioscope

**кардиотахометр** *m.* cardiotachometer

**кардиционер** *m.* card conditioner

**кардоматик** *m.* card-o-matic

**Карей** Carey

**каретка** *f.* carriage, carrier; **к. для бумажной ленты** paper carriage; **к. печатного колёсика** print wheel carriage; **к. с управлением от перфоленты** tape-controlled carriage; **к. щёткодержателя** wiper carriage

**Кари** Carey

**кариллон** *m.* carillon

**каркас** *m.* framework, frame, skeleton, stand, end-frame; chassis; body, form, shell, hull, casing, housing, carcass; support, supporting; **к. катушки** coil form, coil holder; spool, bobbin; **к. катушки возбуждения** field pole, coil box, loading coil case; **к. ряда стативов** end-frame of racks

**карлик** *m.* dwarf, dwarf signal

**карликовый** *adj.* dwarf, dwarfish, diminutive

**карман** *m.* pocket; bin; container; housing

**карманный** *adj.* pocket; bin; container; housing

**карматрон** *m.* carmatron

**карнаубский** *adj.* carnauba

**Карнаф** Karnaugh

**Карно** Carnot

**кароттаж** *m.* logging, well logging, coring; **электрический к. скважин** electrical bore-hole prospecting

**Карпентер** Carpenter

**Карре** Carre

**карректор** *m.* currector

**карсинотрон** *m.* carcinotron; **к. с электронной перестройкой** voltage-tuned carcinotron

**карта** *f.* map, chart; card; sheet; **к. (глубин) в горизонталях** contour map; **к. передачи управления** transfer card; **прокладочная к.** padded card; **к. радиосредств** radio-facility chart; **к. с отрывным корешком** stub card; **титульная к.** guide card

**картина** *f.* picture; pattern, image, figure; diagram; **записанная к. интерференции** interferogram; **к. земных отражений** ground-echo pattern; **к. мнимых источников** picture of image sources; **к. обстановки в воздухе** (electric) air-situation map;

к. **стоячей волны** standing-wave pattern; к. **трещин** failure pattern
**картограмма** *f.* cartogram
**картографический** *adj.* cartographic
**картография** *f.* cartography, mapping
**картон** *m.* carton, cardboard, paper board
**картотека** *f.* card file, card index; filing cabinet; record system (*tphny.*)
**картотечный** *adj.* file
**«картофелетёрка»** *f.* "potato masher" (jamming antenna)
**карточка** *f.* card, index card; к.-**паспорт цепи** circuit layout card; к. **пневматической почты** pneumatic tube ticket; к. **регулировки реле** relay adjustment sheet
**карточный** *adj.* card, index card
**карусельный** *adj.* carrousel; rotary, revolving; turret-type
**карцинотрон** *m.* carcinotron
**касание** *n.* contact, touching, tangency, osculation
**касательная** *f. adj. decl.* tangent (*math.*)
**касательно** *prep. genit.* concerning, relative to, about, touching
**касательность** *f.* relation, connection
**касательный** *adj.* tangential, tangent; concerning; touching, grazing
**касаться, коснуться** *v.* to touch (upon), to concern, to relate (to); to be in contact (with), to touch
**касающийся** *adj.* tangent; touching; concerning
**каскад** *m.* cascade, tandem (*elec.*); stage, step; **буферный к.** buffer stage, isolator, separator; к. **введения постоянной составляющей** d-c inserter; **входной к.** pre-stage; к. **выключения канала цветности** color killer stage; **выходной к. видеоусилителя** final video amplifier, video-output stage; **выходной к. кадровой развёртки** vertical output circuit; **выходной к. развёртки** scanning output stage; **выходной к. строчной развёртки** horizontal-deflection circuit, line output stage; к. **вычитания** subtracter; **генераторный к., гетеродинный к.** oscillator stage; **детекторно-усилительный к.** directional amplifier; **добавочный к. усилителя** additional multiplier stage, intensifier; **задающий к.** master stage; **кварцевый к. управления** crystal-controlled stage; **модулирующий к.**

**(усилителя мощности)** power-amplifier stage; **мощный к. задающего генератора** master oscillator-power amplifier; **мощный выходной к. видеоусилителя** image power amplifier; **нейтродинный к. усиления вч** neutralized radio-frequency stage; **обострительный к.** peaking circuit; **ограничительный к. видеоусилителя** video limiter (*tv*); **ограничительный к. катодного повторителя** cathode-follower clipper (*tv*); **ограничительный к. с катодным смещением** cathode-bias clipper (*tv*); **оконечный к. канала изображений** (final) video output stage; к. **предварительного усиления** preamplifier cascade; **предварительный ультракоротковолновый к.** VHF R-F stage; **предоконечный к.** penultimate stage, driver stage; к. **преобразования частоты** conversion detector; к. **промежуточной частоты** i-f stage; **пушпульный выходной к.** push-pull power stage; к. **разделения цепей** OR-gate; **реостатный усилительный к. на пентоде** resistance-coupled pentode amplifier stage; к. **с катодным выходом** cathode-loaded stage, cathode follower (stage); к. **с отрицательной обратной связью по цепи катода** cathode-degenerated stage, negative-feedback stage; **селективный к. предусиления** preselector stage; **селекторный к. предварительного пускового импульса** pre-trigger time selector; **смесительный к.** adder stage; к. **совпадения** AND-gate; к. **стробированного усилителя** amplitude-gating circuit; **суммирующий к.** counter stage; **управляемый к. усиления вспышки, управляемый усилительный к. вспышки** keyed burst-amplifier stage; к. **усиления** amplifier stage, gain stage; к. **усиления видеосигнала цветности** color-video stage; к. **усиления мощности** power amplifier stage; к. **усиления сигнала цветности, к. усиления цветового сигнала** chroma amplifier stage; **усилительный к. с заземлённой сеткой** grounded grid amplifier, cathode-input amplifier; **усилительный к. с катодным входом** cathode-input amplifier; **фазоинверсный к. в канале сигнала I**

"I" phase inverter; **к. формирования сигнала** wave-shaping stage
**каскадирование** *n.* cascading, cascade operation
**каскадно** *adv.* in cascade, in series
**каскадно-соединённый** *adj.* tandem
**каскадный** *adj.* cascading, cascade, tandem
**каскод** *m.* cascode
**каскодный** *adj.* cascode
**Кассгрен, Кассегрен** Cassegrian(ian)
**кассегреновский** *adj.* Cassegrianian
**кассета** *f.* magazine, cassette; cartridge; spool, reel; adapter; **подающая к.** feed reel; **приёмная к.** takeup spool, takeup reel
**Кассини** Cassini
**кассирование** *n.* collecting (of coins); **к. монет (в таксофоне)** coin collection (in a telephone)
**кассировать** *v.* to collect (coins); to annul, to void, to reverse
**кассирующий** *adj.* collecting
**кастаньеты** *pl.* castanets, bones
**катадиоптрика** *f.* catadioptrics
**катализатор** *m.* catalyst, catalyzer
**каталог** *m.* catalog
**катанка** *f.* rolled wire
**катаракт** *m.* dashpot, damper; cataract
**катарометр** *m.* katharometer
**катафорез** *m.* cataphoresis
**катафорезный** *adj.* cataphoretic
**катафот** *m.* rear red reflex reflector
**категориальный** *adj.* categorial
**категорический** *adj.* categorical
**категория** *f.* category
**катена** *f.* catena
**катенарный** *adj.* catenary
**катеноидальный** *adj.* catenoidal
**катетометр** *m.* cathetometer
**катетрон** *m.* cathetron
**катион** *m.* cation, positively charged ion
**катионный** *adj.* cationic
**катод** *m.* cathode, filament, negative electrode; **к. воспроизводящего прожектора** viewing gun cathode; **действующий к.** virtual cathode; **к. дугового разряда** arc cathode; **жидкий к.** pool cathode; **к. записывающего прожектора** writing gun cathode; **к. из губчатого никеля** nickel-matrix cathode; **известковый к.** lime-spot cathode; **изопотенциальный к.** unipotential cathode, indirectly heated cathode; **калиль-**

**ный к.** hot cathode; **металлизованный к.** sintered cathode; **металлокапиллярный к., металлопористый к.** dispenser cathode; **накалённый к., накаливаемый к.** hot cathode, incandescent cathode, glowing cathode; **оксидный к.** oxide-coated cathode; **ореольный к.** halo-cathode; **подогревный к.** indirectly-heated cathode, heater cathode, hot cathode; **к. подстроечного устройства** tuner cathode; **к. покрытый стекловидными пятнами** glossy cathode; **к. прямого накала, прямонакальный к.** filamentary cathode, filament; **распределённый к.** dispensed cathode; **к. с напылением эмиттирующего слоя** cathode oxide coated by evaporation; **к. с отдельным выводом** separated cathode; **к. с приваренным хвостом** cathode with tab; **к. с экранировкой для уменьшения потерь тепла** heat-shielded cathode; **свёрнутый к.** Juno cathode, wrapped cathode; **свёрнутый к. со швом взамок** Juno locked-seam cathode; **сложный к.** coated cathode; **к. со сжатым верхом, к. со сплющенным верхом** pinched-top cathode; **к. со швом взамок** lock-seam cathode; **к. со швом внахлёстку** lap-seam cathode; **к. стирающего прожектора** erasing gun cathode; **экономичный к.** dim filament
**катоднолучевой** *adj.* cathode-ray
**катодный** *adj.* cathode, cathodic
**катодолюминесцентный** *adj.* cathodoluminescent
**катодолюминесценция** *f.* cathodoluminescence
**каток** *m.* roller, roll
**католит** *m.* catholyte
**катоптрика** *f.* catoptrics
**катоптрический** *adj.* catoptric
**катушечный** *adj.* reel-type, reel, roll, spool, bobbin; coil (*elec.*)
**катушка** *f.* reel, roll, spool, bobbin; coil (*elec.*); **встречно-включённые катушки** opposing coils; **катушки всыпной обмотки** "random"-wound coils, mush-wound coils; **выключающие катушки минимального напряжения** low-voltage tripping coils; **катушки для смещения орбиты электронов** orbit shift coils; **отклоняющие катушки с малым полным**

сопротивлением low-impedance deflection coils; **катушки с неперекрещивающимися лобовыми соединениями** concentric coils; **катушки секции переноса изображения** image section coils; **уравнительные катушки заземления** ground equalizer inductors; **к. анодного контура** plate coil; **антенная удлинительная к.** antenna loading coil; **к. без внешнего поля** fieldless coil; **к. без сердечника, безжелезная к.** air-core coil; **блокирующая к.** holding coil; **к. в схеме моста** bridging coil; **включённая навстречу к.** bucking coil; **к., возбуждаемая током рабочей частоты** driving coil; **к. возбуждения** energizing coil, magnet coil; **воздушная к.** air-core coil; **вставная к.** plug-in coil, pluggable coil, plug-in inductor; **выключающая к.** trip coil; **галетная к.** disk coil, pancake coil; **к.-датчик** sensing coil; **дисковая к.** spiderweb coil; **к. для компенсации внешних полей** field-neutralizing coil; **к. для корректирования сигналов развёртки** peaking coil; **к. для магнитного дутья, к. для магнитного срывания дуги** blow-out coil, magnetic blow-out coil; **к. для смещения центра растра** off-centering yoke; **к. для увеличения крутизны заднего фронта импульса** tail-sharpening inductor; **дугогасительная к.** blow-out coil; **закрытая стаканчиком к.** potted coil; **занимающая пару пазов к.** single coil; **заостряющая к.** peaking coil; **защитная к.** drain(age) coil; **измерительная к.** pickup coil; **индуктивная к. с раздвинутыми витками для уменьшения собственной ёмкости** air-spaced coil; **индуктивная заземлительная к.** earth-leakage coil, Peterson coil; **к. индуктивности для увеличения крутизны заднего фронта импульса** tail-sharpening inductor; **индукционная к.** induction coil, spark inductor; microphone transformer (*tphny.*); **испытательная к.** explorer, exploring coil; **кадровая отклоняющая к.** vertical deflection coil; **к. колебательного контура** oscillating circuit coil, tank inductance; **корзиночная к. с намоткой через две спицы** double basket coil;

**к. крестовой мотки** cross coil; **ленточная к.** strap coil; **к. магнитного дутья** blow-out coil; **к. магнитного сведения лучей** magnetic convergence coil; **к. на салазках** sledge coil; **наматывающая к.** take-up reel; **к. настройки с двумя движками** two-side receiving coil; **к. обмотки возбуждения** field coil; **к. обмотки якоря** armature coil; **к. обратной связи** feedback coil, reaction coil, tickler coil; **отклоняющая к. без сердечника** air-core(d) deflection coil; **отклоняющая к., включённая по трансформаторной схеме** transformer-fed deflector coil; **параллельно включённая к.** bridging coil; **переменная к.** variable inductor; **питающая к.** battery supply coil (*tphny.*); **плавно регулируемая к. индуктивности** continuously adjustable inductor, variable inductor; **к. по типу сотовой** duolateral coil; **поворотная к., подвижная к.** moving coil, voice coil, signal-current coil, speech coil; **к., поддерживающая положение включения** holding-on coil; **к. подмагничения** polarizing coil; **подогнанная к.** matched coil; **последовательно включаемая к.** boosting coil; **последовательно включённая к.** series coil; **противофоновая к.** hum-bucking coil; **пупиновская к.** loading coil, Pupin coil; **пупиновская к. основной цепи** side circuit loading coil; **реактивная к. с выведенной средней точкой** T-reactance coil; **к. регулировки линейности по горизонтали** horizontal linearity coil; **к. с зазором** air-gap coil; **к. с магнитной настройкой** permeability-tuned coil; **к. с насыщением** saturable inductor; **к. с ответвлениями, к. с отводами** tapped coil, bleeding coil; **к. с отогнутыми лобовыми частями** short-type coil; **к. с перекрещивающимися в лобовых частях витками** lap coil; **к. с проводом** wire-laying reel; **к. с фабричной настройкой** pretuned coil; **к. со сплошным сердечником** coil with powdered core, cable loading-coil with powdered core; **к. со штырьками** plug-in coil, plug-in conductor; **к. связи** coupling coil, pickup coil; jigger; **к. секции откло-**

нения scanning section coil; **сек-ционированная к.** tapped coil; **секционированная пополам к.** bisected coil; **сменная к.** plug-in coil, unit coil, plug-in inductor; **сотовая к.** honeycomb coil; **к. строчной отклоняющей системы** line-scan coil system; **сформированная с одного конца к.** hairpin coil; **к. схемы регулировки чистоты цвета** color purity coil, purifying coil; **к., удерживающая на положении включения** holding-on coil; **удлинительная к.** load(ing) coil, antenna loading coil, extension coil, lengthening coil; **удлинительная к. для настройки** lengthening-coil tuner; **к. указателя действия** target coil; **фасонная к.** preformed coil; **центрирующая изображение к.** picture control coil; **шаблонная к.** form-wound coil; **шунтирующая к.** bridging coil

**катушкодержатель** m. coil holder
**катэлектронус** m. catelectronus
**каузальный** adj. causal
**каустик** m. caustic
**каустика** f. caustic (optics)
**каутеризация** f. cautery
**каучук** m. rubber, caoutchouc
**каучуковый** adj. rubber, caoutchouc
**Кауэр** Cauer
**кафель** m. tile, stone tile
**кафельный** adj. tile, tiled
**качание** n. swing(ing), rocking, tilting, swaying, flutter, hunting, shaking, oscillation, vibration; fluctuation; wobbling, wobbulation; sweep; cycling; pumping; **к. изображения по горизонтали** horizontal hunting; **к. несущей** carrier flutter, carrier swinging; **к. частоты** frequency variation, wobbulation, wobbling
**качательный** adj. nodding
**качать, качнуть** v. to swing, to rock, to sway; to oscillate, to vibrate, to shake; to pump; **к. насосом** to pump
**качаться, качнуться** v. to rock, to swing, to tilt, to vacillate, to sway; to oscillate, to vibrate; to fluctuate, to wobble, to shake; to pump
**качающийся** adj. rocking, swinging, oscillating, oscillatory; fluctuating, jigging, shaking, wobbling, vibrating, vibratory; tilting; sweep(ing); swept
**качественный** adj. qualitative, fine; observational

**качество** n. quality, grade; nature, character, property; **к. воспроизведения формы импульса** indicial response; **высокое к. (звука)** high fidelity; **к. обработанной поверхности** surface finish; **к. обслуживания** quality of service, grade of service; **к. широкополосности** band merit
**качка** f. roll, rolling, tossing; looseness, free play; swing
**качнуть(ся)** pf. of качать(ся)
**«каша»** f. random noise, mush
**кашета** f. mat, mask, matte
**каюта** f. cabin, stateroom; store-room; small room
**КБВ** abbr. (**коэффициент бегущей волны**) traveling-wave ratio
**кбм** abbr. (**кубический метр**) cubic meter
**кв** abbr. 1 (**квадрат**) square; 2 (**квадратный**) square; 3 (**киловольт**) kilovolt
**КВ** abbr. 1 (**короткие волны**) short waves; 2 (**коротковолновый**) short-wave
**ква** abbr. (**киловольт-ампер**) kilovolt-ampere
**квадрант** m. quadrant; **к. централизационного аппарата** machine quadrant
**квадрантальный, квадрантный** adj. quadrantal, quadrant
**квадрат** m. square
**квадратирующий** adj. squaring
**квадратический** adj. quadratic
**квадратично-косекансный** adj. cosecant-squared
**квадратичный** adj. quadratic, quadrature; square, square-law
**квадратность** f. squareness
**квадратный** adj. square; quadratic
**квадратор** m. square-law function generator, squarer, square-law storage
**квадратрон** m. quadratron
**квадратура** f. quadrature, squaring
**квадратурно-модулированный** adj. quadrature-modulated
**квадратурный** adj. quadrature, squaring
**квадрикоррелятор** m. quadricorrelator
**квадрирование** n. squaring
**квадруплекс** m. quadruplex, quadruple telegraph system
**квадруплексный** adj. quadruplex

квадруполь *m.* quadrupole
квадрупольный *adj.* quadrupole
квази- *prefix* quasi-
квазиактивный *adj.* quasi-active
квазианизотропный *adj.* quasi-anisotropic
квазибистабильный *adj.* quasi-bistable
квазивременной *adj.* quasi-time
квазивырожденный *adj.* quasi-degenerated
квазигармонический *adj.* quasi-harmonic
квазидвусопряжённый *adj.* quasi-biconjugate
квази-двуустойчивый *adj.* quasi-bistable
квази-диагональный *adj.* quasi-diagonal
квазидифференцирующий *adj.* quasi-differentiating
квазиизолятор *m.* quasi-insulator
квазиимпульс *m.* crystal momentum
квазикритический *adj.* quasi-critical
квазилинейный *adj.* quasilinear
квазимоностабильный *adj.* quasi-monostable
квазимонохроматический *adj.* quasi-monochromatic
квазиодиночный *adj.* quasi-single
квазиоднородный *adj.* quasi-homogeneous
квазиодноустойчивый *adj.* quasi-monostable
квазиоптический *adj.* quasi-optical
квазипериод *m.* quasi-period
квазипериодический *adj.* quasiperiodic
квазиплоский *adj.* quasi-plane
квазипоперечный *adj.* quasi-transverse
квазипродольный *adj.* quasi-longitudinal
квазипроизвольный *adj.* quasi-random
квазирешётка *f.* quasi-array
квазисвязанный *adj.* quasibound
квазистатический *adj.* quasi-static
квазистационарный *adj.* quasistationary, quasistable
квазиупругий *adj.* quasi-elastic
квазиустановившийся *adj.* quasi-steady
квазиустойчивый *adj.* quasistable
квазицифровой *adj.* quasidigital
квалиметр *m.* qualimeter
квалификация *f.* qualification, ability
квалифицированный *adj.* qualified; skilled
квант *m.* quantum; bit (of information)

квантизатор *m.* quantizer
квантизация *f.* quantization
квантиль *m.* quantile
квантификатор *m.* quantifier
квантификация *f.* quantification
квантифицировать *v.* to quantify
квантование *n.* quantization, sampling
квантованный *adj.* quantized; quantum
квантователь *m.* quantizer
квантовать *v.* to quantize
квантовомеханический *adj.* quantum-mechanical, wave-mechanical
квантовый *adj.* quantum
квантометр *m.* quantometer
квантор *m.* quantifier
квантующий *adj.* quantizing
кварта *f.* fourth (*mus.*); quart
квартиль *m.* quartile
квартирный *adj.* apartment, residence, home
кварц *m.* quartz, crystal; **к., возбуждаемый на гармониках** harmonic mode crystal; **к. с воздушным зазором с тыльной стороны** air-backed quartz; **к.-тень** ghost quartz
кварцевый *adj.* quartz, quartzitic, crystal
кварцедержатель *m.* crystal holder; **к. с воздушным зазором** air gap mount
кварцованный *adj.* crystal-coated
квасцы *pl.* alum
кватернарный *adj.* quaternary
кватернион *m.* quaternion
Квинке Quincke
квинта *f.* quint, fifth; **чистая к.** perfect fifth
квитанция *f.* receipt, acknowledgment; (ac)quittance, finished reading
квитирование *n.* receipting, acknowledgment of receipt
квитирующий *adj.* acknowledgment
КВП *abbr.* (**коаксиально-волноводный переход**) waveguide-to-coaxial adapter
квт *abbr.* (**киловатт**) kilowatt
квтч, квт-ч *abbr.* (**киловатт-час**) kilowatt-hour
кг *abbr.* (**килограмм**) kilogram
КГ *abbr.* (**кратная горизонтальная**) multiband antenna, (center-fed) two-wave horizontal broadside array
КГЛ *abbr.* (**копировально-гибочный станок для листового материала**) plate copy bending machine
кг/л. с. ч. *abbr.* (**килограмм/лошадиная**

**сила, час)** kilograms per horsepower hour

**кгм** *abbr.* **(килограммометр)** kilogram-meter

**кг-моль** *abbr.* **(килограмм-молекула)** kilogram molecule

**кгс** *abbr.* **(килограмм-сила)** kilogram weight

**кгц** *abbr.* **(килогерц)** kilocycle per second, kilohertz

**кдж** *abbr.* **(килоджоуль)** kilojoule, large joule

**КДП** *abbr.* **(командо-диспетчерский пункт)** control tower

**Кейстон** Keystone

**Кейт** Keith

**Кельвин** Kelvin

**кембриджский** *adj.* Cambridge

**кембрик** *m.* cambric (tape); cambric (textile)

**кембриковый** *adj.* cambric

**Кеннели** Kennelly

**кенопдиотрон** *m.* kenopliotron

**кенотрон** *m.* kenotron, vacuum tube rectifier, valve tube

**кенотронный** *adj.* kenotron

**к. е. о.** *abbr.* **(коэффициент естественной освещённости)** daylight factor

**керамика** *f.* ceramics; **к. из титаната бария** titanate ceramics; **сегнето-электрическая к.** ferroelectric ceramics

**керамиковый, керамический** *adj.* ceramic, earthenware; clay

**кератометр** *m.* keratometer

**керит** *m.* kerite

**кермет** *m.* cermet, metal-ceramic

**керн** *m.* core, center; base; pivot; leg (of a transformer); core bearing; **к. нити** filament base

**керновый** *adj.* core

**Керр** Kerr

**Кёртис** Curtis

**кессон** *m.* caisson

**кессонный** *adj.* caisson

**кибернетика** *f.* cybernetics

**кибернетический** *adj.* cybernetic

**Кикухи** Kikuchi

**килеобразный** *adj.* fin-shaped

**кило-** *prefix* kilo-

**килобит** *m.* kilobit

**киловар** *m.* kilovar; **к.-час** kilovar-hour

**киловатт** *m.* kilowatt; **к.-час** kilowatt-hour

**киловольт** *m.* kilovolt

**киловольтампер** *m.* kilovolt-ampere

**киловольтметр** *m.* kilovoltmeter

**килогаусс** *m.* kilogauss

**килогерц** *m.* kilohertz, kilocycle per second

**килограмм** *m.* kilogram

**килоджоуль** *m.* kilojoul, large joule

**килокалория** *f.* kilocalorie

**килокюри** *m.* *indecl.* kilocurie

**килолиния** *f.* kiloline

**килолюмен** *m.* kilolumen

**киломега** *f.* kilomega

**киломегагерц** *m.* kilomegacycle

**киломегом** *m.* kilomegohm

**километр** *m.* kilometer

**километраж** *m.* distance in kilometers

**километрический, километровый** *adj.* kilometric

**киломоль** *f.* molecule kilogram

**килоом** *m.* kilo-ohm, kilohm

**килопонд** *m.* kilopond

**килопондметр** *m.* kilopondmeter

**килопондсекунда** *f.* kilopondsecond

**килоцикл** *m.* kilocycle

**килоэлектронвольт** *m.* kiloelectron-volt

**кильватер** *m.* ship's wake, dead water

**кильватерный** *adj.* wake

**КИМ** *abbr.* **(кодово-импульсная модуляция)** pulse-code modulation

**кимограф** *m.* kymograph

**кимография** *f.* kymography

**Кинбек** Kienboeck

**кинеграфический** *adj.* kinegraphic

**кинематика** *f.* kinematics

**кинематический** *adj.* kinematic

**кинематограф** *m.* motion picture, cinematograph

**кинематографический** *adj.* cinematographic, motion-picture

**кинемометр** *m.* kinemometer

**«кинеплекс»** *m.* Kineplex system

**кинескоп** *m.* kinescope, picture tube; **совмещённые кинескопы** registered kinescopes; **однопрожекторный к. для цветного телевидения** single-gun color tube; **плоский к.** flat TV screen; **к. с неалюминированным экраном** unfilmed tube; **к. синего канала** blue kinescope; **цветной к. с маской, использующий параллакс электронного луча** parallax mask color tube; **цветной к. с проникающим лучом** penetration-type kinescope; **цветной к. типа венецианской решётки** Venetian-blind color tube

«кинескор» *m.* Kinescore
кинетика *f.* kinetics
кинетический *adj.* kinetic, motional
кинетофон *m.* kinetophone
кинеч. *abbr.* (кинетический) kinetic
кино *n. indecl.* motion picture, movie, cinema; звуковое к. sound motion picture, talking movie
киногородок *m.* film city
кинозапись *f.* motion-picture recording
кинокадр *m.* moving-picture frame
кинокамера *f.* movie camera, picture camera
кинокамал *m.* film-channel
кинокартина *f.* motion picture, movie
кинолампа *f.* cinema lamp
кинолента *f.* film
киномеханик *m.* movie technician, projectionist
кинооборудование *n.* film equipment
кинооператор *m.* camera man
киноорган *m.* cinema organ
кинопередвижка *f.* portable film projector
киноплёнка *f.* film, moving picture
кинопроектор *m.* movie projector, film projector, movie-picture projector; к. для рир-проекции background film projector
кинопроекционная *f. adj. decl.* telefilm room
кинопроекционный *adj.* projection, film projection
кинопроекция *f.* film projection, motion-picture projection; к. фона при передаче из студии background projection, rear projection
кинопромышленность *f.* moving picture industry
киностудия *f.* moving-picture studio, film studio
киносъёмка *f.* motion-picture filming
кинотеатр *m.* movie theater, cinema
кинотеодолит *m.* cinetheodolite
кинотехника *f.* cinematograph engineering
киноустановка *f.* movie projector
кинофильм *m.* cinema picture, motion picture
кинофотомикрография *f.* cinephotomicrography
кинофотопулемёт *m.* camera gun
кинохроника *f.* newsreel
Кинсбэри Kingsbury
киоск *m.* kiosk, booth, stand, cabin; vault; звуконепроницаемый к.

silence cabin (*tphny.*)
кип *m.* kip
КИП *abbr.* (контрольно-измерительные приборы) control and measuring instruments
кипение *n.* boiling, gassing, ebullition; bubbling
киперный *adj.* heringbone; twilled
кипп-генератор *m.* kipp oscillator, single flip-flop oscillator
кипятильный *adj.* boiling
кипячение *n.* boiling
кипячёный *adj.* boiled
кипящий *adj.* boiling; bubbling
кирка *f.* pick, pickax; scraper
кирпич *m.* brick; к. сильного обжига clinker, hard burned brick
кирпичный *adj.* brick; masonry
Кирхгоф Kirchhoff
кислород *m.* oxygen, O
кислородный *adj.* oxygen, oxygenous
кислота *f.* acid; соляная к., хлористоводородная к. hydrochloric acid, muriatic acid
кислотность *f.* acidity
кислотный *adj.* acid; sour
кислотомер *m.* acidimeter, acidometer; сифонный к. hydrometer syringe
кислотостойкий, кислотоупорный, кислотоустойчивый *adj.* acidproof, acidresistant, acid-resisting
кислый *adj.* acid; sour
Кисслинг Kissling
кистевой *adj.* brush; cluster, bunch
кисть *f.* brush; cluster, bunch
китайский *adj.* Chinese
киттелевский *adj.* (of) Kittel, Kittel's
кишка *f.* hose, tube
ккюри *abbr.* (килокюри) kilocurie
клавиатура *f.* fingerboard, keyboard, keyframe, key motion
клавиатурный *adj.* keyboard
клавикорды *pl.* clavichord, harpsichord
клавиолин *m.* clavioline
клавиш *m.*, клавиша *f.* key, tapper (*tgphy.*); stop; type, letter; push button; bar; к. вычитания minus bar; к. гашения (счётчика) clearing key; микрофонный к. push-to-talk key, talk-listen switch, flap (of a handset); к. обратного перемещения resetting key; к. переводного механизма shift key; к. сложения plus bar
клавишный *adj.* key; keyboard
кладка *f.* laying; spotting; каменная к. brickwork, masonry

**кладоискание** *n.* treasure finding

**клаймэкс** *m.* climax (alloy)

**клаксон** *m.* Klaxon horn

**клапан** *m.* finger hole, stop (of a mus. instrument); valve, vent; flap (of a handset); drop (*tphny.*); lid, shutter; **воздушный нагруженный пружиной к.** spring-opposed air valve; **к. выдачи (результатов вычисления), к. выдачи суммы** read-out gate, sum read-out gate; **вызывной к.** drop-indicator, calling drop, shutter, (call) indicator; **гидравлический предохранительный к.** hydraulic back-pressure valve; **к. добавочного воздуха с автоматическим управлением** automatic supplementary air valve; **дроссельный к.** throttling valve, butterfly valve; **игольчатый нагруженный пружиной к.** spring-loaded needle valve; **микрофонный к.** push-to-talk key, talk-listen switch; **многоходовой поворотный обратный к.** multiple-swing check valve; **отбойный к.** clearing indicator, ringoff indicator (*tphny.*); **падающий к.** *see* вызывной к.; **переключающий к.** switch valve; two-way valve; **к. повторного вызова** plug-restored indicator shutter (*tphny.*); **предохранительный к. (давления)** pressure relief valve, blow-off valve; **распределительный к. автоматической подачи** automatic delivery control valve; **к. с электроприводом** electrically operated valve; **самозакрывающийся вызывной к.** self-restoring indicator, plug-restored indicator; **сильфонный управляющий к.** bellows-operated pilot valve; **соленоидальный к. для отбора проб** solenoid-actuated sampling valve; **управляемый нечётными числами к.** odd-controlled gate

**клапанный** *adj.* valve; flap; stop; drop

**Клапейрон** Clapeyron

**Клапп** Clapp

**Кларк** Clark

**кларнет** *m.* clarinet

**класс** *m.* class, grade, sort; category; **к. волнового сопротивления** classification of characteristic impedances; **к. перегрузки трансформатора тока** overcurrent class of a current transformer

**классификатор** *m.* classifier

**классификационный** *adj.* indexing; taxonomic

**классификация** *f.* classification, arrangement, grading, sorting; sizing; grouping; **к. по двум признакам** two-way classification

**классифицированный** *adj.* classified, graded, sorted

**классифицировать** *v.* to classify, to class, to sort, to arrange, to grade, to size; to tabulate

**классический** *adj.* classic, classical

**Клаудиус** Claudius

**клаузиус** *m.* clausius (unit of entropy)

**Клаузиус** Clausius

**клевание** *n.* flutter (of an automatic device), hunting, pumping

**клевер** *m.* clover

**клеверный** *adj.* clover

**клеёнка** *f.* wax-cloth, oilcloth; linoleum

**клеёный** *adj.* glued, cemented, pasted

**клеивающий** *adj.* adhesive

**клеить, склеить** *v.* to glue, to cement, to paste

**клей** *m.* glue, gum, paste, cement, putty, bond; adhesive

**Клейден** Clayden

**клейкий** *adj.* gummy, adhesive, gluey, viscous

**клеймо** *n.* brand, mark, seal, stamp, branding iron; **выжженное к. (на столбе)** branding (on a pole); **железное к. с номером** letter nail; number nail

**Клейн** Klein

**клейстер** *m.* paste, filling, sizing

**Клеман** Clément

**клемма** *f.* terminal; clamp, clip, binding post; **к. для проводов** wire grip; **коленно-рычажная к.** draw tongs; tensioning device for light wires

**клемник** *m.* terminal block

**клемный** *adj.* terminal; clamp, clip

**клепать, заклепать** *v.* to rivet

**клёпка** *f.* riveting

**Клеро** Clairaut

**клетка** *f.* cell; casing, box; cage; mesh (of screen); square, cubicle; **беличья к.** squirrel-cage (winding); **внутренняя волосковая к.** internal hair cell; **к. Гензена** Hensen's cell; **чувствительная нервная к.** sensory cell

**клеточка** *f.* cell

**клеточкообразный** *adj.* cellular

**клеточный** *adj.* squirrel-cage (type); cage; cell; cellular

**клетчатый** *adj.* square, squared; meshed; graph (paper)

**клеть** *f.* cage; housing

**клещевой** *adj.* tongs, forceps, pincers

**клещи** *pl.* pliers, cutting pliers, pincers, nippers, tongs, drawing pliers, forceps; **длинноклювые к.** long nose pliers; **к. для контактных винтов** dowel screw pliers; **к. для скручивания** twist pliers, twist clamp; **к. для юстировки** adjusting pliers; **токоизмерительные к.** hook-on meter; **трансформаторные к.** split-core type transformer, split-wire type transformer

**клидонограф** *m.* clydonograph

**клиент** *m.* client, user, customer, patron

**клизеометр** *m.* cliseometer

**климат** *m.* climate

**климатический** *adj.* climatic

**клин** *m.* wedge, key; cleat; cone; tapered bar; **клинья для проверки разрешающей способности** definition wedges; **клинья для проверки разрешающей способности по вертикали (по горизонтали)** vertical-(horizontal-)resolution wedges; **звукопоглощающий к.** wedge-shaped absorber

**клинический** *adj.* clinical, clinic

**клинкер** *m.* clinker, brick

**клино-** *prefix* wedge-, spheno-; clino-

**клиновидный** *adj.* wedge, wedge-shaped, V-shaped, tapered

**клиновой** *adj.* wedge, cotter, key; wedge-shaped, V-shaped

**клинок** *m.* blade

**клинометр** *m.* clinometer

**клинообразный, клинчатый** *see* клиновидный, клиновой

**клиппированный** *adj.* limited

**клиранс, клиренс** *m.* clearance

**клирфактор** *m.* klirr-factor, non-linear distortion factor

**клирфакторметр** *m.* distortion factor meter

**клистрон** *m.* klystron, Shepard tube, reflection oscillator, transit time tube; **двухконтурный к., двухполостный к., двухрезонаторный к.** double-cavity klystron; **пролётный к.** floating(-drift)-tube klystron; **к. с двойным пролётом** double-transit klystron, double-transit oscillator; **к. с конечным пучком** finite beam klystron; **к. с ленточным пучком** Heil tube; **к. с многократным отражанием (электронного) потока** multi-reflex klystron, multiple reflection klystron

**клистронный** *adj.* klystron

**Клиф** Cliff

**клица** *f.* cleat, insulating clamp

**клишограф** *m.* scan-a-graver

**клм** *abbr.* (килолюмен) kilolumen

**Клогстон** Clogston

**Клод** Claude

**клопфер** *m.* acoustic dial, sounder, tapper, Morse sounder; acoustic telegraph; **к. для двухполюсного телеграфирования** double-current sounder; **трансляционный к.** repeating sounder

**клопферный** *adj.* sounder, tapper

**клофен** *m.* clophen, diphenylchloride

**Клузиус** Clusius

**клупп** *m.* jointing clamp, sleeve twister, screwstock, screwdie; **рычажный к.** sleeve twisters, jointing clamp; **шарнирный винторезный к.** shears vice

**ключ** *m.* key; clef, key; key, code; switch; **азимутальный к. походного положения** azimuth stowing switch; **альтовый к.** alto clef; **английский к.** monkey wrench; **басовый к.** bass clef; **к. без арретира** non-locking key; **к. в схеме моста** bridge key; **к. внутристанционных переговорных линий** transfer key; **к. возвратного вызова** ring-back key (*tphny.*); **возвратный к.** restoring key; **возвращающийся к. (без арретира)** non-locking key (*tphny.*); **к. вызова, вызывной к.** call key, signaling key, ringing key; **гаечный к.** wrench, spanner; **двусторонний к., двухполюсный к., двухтоковый к., к. для двухполюсного телеграфирования** double-current key, double tapper key; **к. для включения приёмника** break-in key; **к. для избирательного вызова** party line ringing key; **к. для исправления ошибок** erase key; **к. для мостика сопротивления** bridge key; **к. для соединения рабочих мест телефонисток** position coupling key, position grouping key; **закорачивающий к.** short-circuit(-ing) key; **к. избирательного вызова** selector key, party-line ringing key (*tphny.*); **к. к мостику сопротив-**

ления с двумя парами изолированных контактов double contact key; **квадруплексный к.** reversing key; **коммутаторный к.** rocking key; **к. контрольной лампочки** pilot lamp key (*tphny.*); **к. контроля** monitoring key; **к. концентрации** operator-position switch (*tphny.*); **многоконтактный к.** multiple key, multipoint switch; **многодырочный к.** combination wrench; **направляющий к.** aligning key; **невозвращающийся к.** locking key (*tphny.*); **к. номеронабирателя** dialing key; **к. обратного вызова** ring-back key, ringing key (*tphny.*); **к. обрыва** cutoff key, interruption key; **общий к.** passing key; **к. одиночного выполнения команд** execute console instruction key; **опросный к.** talking key, answering key, listening key (*tphny.*); **отбойный к., к. отбоя** release key, clearing key, splitting key; **переговорно-вызывной к.** speak-buzz key; **плоскопружинный к.** strap key (*tphny.*); **к.-ползун (у струнного моста)** sliding-contact key; **полудуплексный к.** break-in key; **к. поправки** reset key; **к. потактовой работы** single step key; **к. с арретиром** locking key; **скрипичный к.** soprano, treble clef; **тенорный к.** tenor clef; **торцевой к.** socket wrench; **торцевой трубчатый гаечный к.** pipe wrench, stillson wrench; **к. управления несколькими цепями** multiple key; **к. усиленного тока** incrementing key; **к. электролиза** (shot-)firing key

**ключевать** *v.* to key
**ключевой** *adj.* key
**клякса** *f.* blurr, blurring
**клямера** *f.* bracket, clamp, clip
**КМ** *abbr.* (**киломега**) kilomega
**км** *abbr.* (**километр**) kilometer
**КНД** *abbr.* (**коэффициент направленного действия**) directivity factor
**кнезеровский** *adj.* Kneser
**книгообразный, книжкообразный** *adj.* book-form, book-shaped
**кнопка** *f.* knob, button, push button, key, push; clasp, fastener, snap; **к. без фиксации** non-locking key, "spring-opened" push button; **блокирующая к.** holding key (*tphny.*); **к. ввода (перфо)карты** load card button; **к. внутристанционных переговорных линий** assignment key, allotter key (*tphny.*); **к. выключения** cancelling key; **двойная к.** combined blocking-and-releasing switch; **к. заказной линии** order-wire speaking key, recording-trunk key; **к. «заряд конденсатора»** charge button; **звучащая к.** sounder pushbutton; **изолирующая к. пружины** nipple stud; **к. исходного положения** reset pushbutton; **к. исходного режима** stop pushbutton; **коридорная к. автоматически останавливающая кабину** elevator hall stop button; **коридорная к. лифта** elevator hall button; **коридорная к. посылки сигнала в кабину** elevator hall signal button; **к. монетного автомата** pay button; **общая к.** *see* **двойная к.**; **к. объединения нескольких рабочих мест в одно** position-grouping key, concentration key, concentration switch, day-night transfer key; **к. перевода сигналов на автоматическое действие** stick-nonstick button; **к. переговорной линии** press-to-talk button; **к. переключения шнуров** splitting key (*tphny.*); **плоскопружинная к.** strap key; **к. принудительного разъединения** cutoff key, cancel key; **к. пуска двигателя регистрирующего устройства** recorder-motor pushbutton; **к. «разряд конденсатора»** discharge button; **к. режима фиксации** hold pushbutton; **к. решения** compute pushbutton; **к. с арретиром** locking key, locking button; **к. с прилипанием** *see* **удерживающая к.**; **к. сбрасывания** cutoff key, cancel key; **служебная к.** order wire button; **к. счётчика** meter key; **к. тастатуры** pulser key; **удерживаюшая к.** locking pushbutton with magnetic release

**кнопочный** *adj.* knob, button, pushbutton, clasp
**Кнудсен** Knudsen
**кнудсеновский** *adj.* Knudsen's
**коагулирование** *n.* coagulation
**коагулированный** *adj.* coagulated
**коагулировать** *v.* to coagulate
**коагулируемый** *adj.* coagulable
**коагулирующий** *adj.* coagulating
**коагулометр** *m.* coagulometer
**коагуляционный** *adj.* coagulation, coagulating

**коагуляция** *f.* coagulation; **к. под действием ультразвуковых волн** ultrasonic coagulation
**коаксиал** *m.* coaxial
**коаксиально-тороидальный** *adj.* coaxial-torus
**коаксиальный** *adj.* coaxial
**коалиционный** *adj.* coalitional
**кобальт** *m.* cobalt, Co
**ковалентность** *f.* covalence
**ковалентный** *adj.* covalent
**Кован** Cowan
**ковар** *m.* kovar
**ковариантный** *adj.* covariant
**ковариационный** *adj.* covariance
**ковариация** *f.* covariance, covariation
**коваровый** *adj.* kovar(-glass)
**ковёр** *m.* rug, carpet
**ковкий** *adj.* forgeable, malleable, flexible, ductile
**ковочный** *adj.* forge, forging; forged
**коврик** *m.* mat; **к. с контактами** electrical floor matting
**ковровый** *adj.* rug, carpet; braided (cable)
**ковшеобразный** *adj.* pitcher-shaped
**когерентно** *adv.* coherently
**когерентноимпульсный** *adj.* coherent-impulse, coherent-pulse
**когерентность** *f.* coherency, coherence
**когерентный** *adj.* coherent
**когерер** *m.* coherer, sensitive tube; **встряхивающий к.** tapper; **к.-предохранитель** coherer protector, coherer-type acoustic shock reducer
**когерировать** *v.* to cohere
**кограница** *f.* coboundary
**коградиентный** *adj.* cogradient
**когти** *pl.* pole climbers, climbers, grapplers, spurs, gaff, clevice; **монтёрские к.** lineman's climbers
**код** *m.* code, key, cipher; **по десятичному коду** decimal coded; **к. без запятой** comma-free code; **буквенный к., образованный из проводящих точек** conductive-dot code; **к. в остатках** residue code; **двоичный к. с избытком три** excess-three binary code; **к. для буквопечатающего аппарата** printer telegraph code; **к. запроса** interrogation code; **к. интерпретирующего устройства** interpreter code; **к. команд части программы** coding section; **к. команды на входном языке** input obstruction code; **обратный к. числа**

base minus one code, base minus one complement, diminished radix complement; **к. опознавания** identity code; **опознавательный к.** authentication code; **перемещённый к., перестановочный к.** permuted code, permutation code; **к. позывных** authentication code; **рефлексный перестановочный к.** cyclic permuted code; **к. с единичным переходом** unit-distance code; **к. с избытком три** excess-three code; **к. с контролем ошибок** error checking code; **к. с простым контролем по чётности** simple parity-check code; **к. с равными основаниями** nonconsistently based code; **к. с расстоянием единица** continuous progressive code, unit-distance code; **к. с самопроверкой** error detecting code; **к. с элементами, отличающимися на единицу** continuous progressive code, unit-distance code; **сокращённый к.** brevity code; **станционный буквенный к.** office code; **к. «три в избытке»** excess-three code; **циклический двоичный к.** reflected binary code; **циклический перестановочный к.** cyclic permuted code
**«кодан»** *m.* CODAN (carrier operated device antinoise)
**кодатрон** *m.* kodatron
**кодер** *m.* coder, encoder, pulse coder
**кодирование** *n.* coding, encoding; **маркёрное к.** label coding; **местное к.** specific coding; **оптимальное к.** minimum access programming, optimum programming; **к. ответными импульсами** range coding; **«разрывное» к.** gap coding; **к. с меткой кода** label coding; **к. с применением обратной связи** feed-back encoding
**кодированно-десятичный** *adj.* coded decimal
**кодированный** *adj.* coded, in code
**кодировать** *v.* to code, to encode
**кодировщик** *m.* code clerk, encoder
**кодирующий** *adj.* coding, encoding
**кодово-импульсный** *adj.* pulse-code
**кодовый** *adj.* code
**кодопреобразователь** *m.* code converter
**коэрцитивный** *adj.* coercive
**кожа** *f.* skin; film, coating; leather, hide
**кожаный** *adj.* leather; skin

**кожно-гальванический** *adj.* cutaneo-galvanic

**кожный** *adj.* skin

**кожух** *m.* housing, casing, case, cover, covering, sheath, sheathing; sleeve; jacket, mantle, hood, bonnet, dome, shield; shell, envelope, enclosure; **к. антенны** radome, radar dome, blister, antenna fairing; **заземлённый к. (трансформатора)** ground shield; **защитный к.** jacket, shield, protective housing, protecting housing

**козлы** *pl.* pedestal, trestle jack, bench

**козырёк** *m.* hood, camera hood; visor; baffle plate, deflector; **затеняющий к.** viewing hood

**кокатегория** *f.* cocategory

**кокиль** *m.* shell (electro-typing)

**Кокле** Coquelet

**коклюшечный** *adj.* bobbin, spindle

**коклюшка** *f.* bobbin, spindle

**кол** *m.* stake, pole

**колба** *f.* shell, envelope; bulb, glass envelope; flask; **к. без носика** pipless bulb; **заштенгелеванная к.** tabulated bulb; **зеркальная к.** metal coated (mirror) bulb; **к. лампы** envelope of a tube, bulb, shell; **ртутная к.** pool-type tube; **к. с внутренней химической матировкой** satin-etched bulb

**колбасный** *adj.* sausage

**колбомоечный** *adj.* bulb-washing

**колбосниматель** *m.* bulb remover

**колбочка** *f.* bulb; cone

**колебание** *n.* variation, oscillation, fluctuation, vacillation, flutter, wavering; range; wave; sweep; vibration, swing(ing); **колебания во время обратного хода** back oscillations; **колебания во время переходного процесса** transient oscillations; **колебания времени пробега** transit-time oscillations; **временные колебания около номинала** short-term drift; **колебания второго рода** class B amplifier oscillations; **колебания генератора с обратной волной** backward-wave oscillations; **колебания изгиба, изгибающие колебания, изгибные колебания** flexural vibrations; **модулированные ключом незатухающие колебания** telegraph-modulated waves; **модулированные речью незатухающие колебания** continuous waves modulated by

speech; **колебания первого рода** class A amplifier oscillations; **колебания по направляющей** electron-orbit oscillations; **колебания по толщине** thickness mode; **колебания с нарастающей амплитудой** diverging oscillations; **колебания связанных систем** coupled oscillations; **колебания силы приёма** swinging (of reception); **служащие для отсчёта времени колебания** timing oscillations; **случайные колебания с широкой полосой частот** broadband random vibrations; **колебания строчной развёртки** line-scan ringing (*tv*); **колебания уровня сигнала при бортовой качке** roller fading; **устанавливающиеся колебания** transient oscillations, swing oscillations; **к. величины отметки** bounce; **к. громкости** variation in signal strength, fading; **к. длительности импульса** pulse-time jitter; **затухающее собственное к.** damped free vibration; **к. (кристалла) на срез** shear mode of vibration; **к. кристаллической решётки** lattice vibration; **к. луча пеленгатора** bearing fluctuation; **к. напряжённости поля** variation of field strength; **к. при самовозбуждении** self-excited vibration; **к. пузырька** bubble variation; **к. растяжения** longitudinal vibration; **к. сложного вида** multimode propagation; **собственное к.** free vibration, eigentone

**колебательно-вращательный** *adj.* vibrational-rotational

**колебательный** *adj.* oscillatory, oscillating, fluctuating; vibrating, vibration, vibrational, vibratory

**колебать, заколебать, поколебать** *v.* to swing, to agitate, to shake, to vibrate

**колебаться, поколебаться** *v.* to swing, to oscillate, to fluctuate, to vary, to range; to vibrate; to flicker; to wobble, to sway; to waver, to falter, to vacillate, to fluctuate, to wiggle; **к. туда и обратно** to surge back and forth; to swing back and forth

**колеблющийся** *adj.* oscillating, oscillatory, fluctuating, alternating, variable; vibrating, vibratory; unsteady, wavering

**коленкор** *m.* calico

**колено** *n.* elbow, bend, knee, joint, link;

curvature; **к. (волновода) в плоскости E** E-(plane) bend; **к. (волновода) в плоскости H** H-(plane) bend; **к. обратного замыкания** transformer-core yoke; **шарнирное к. волновода** waveguide hinge joint

**коленчатый** *adj.* knee, knee-like, elbowed, bent; off-set, cranked, crank elbow; jointed, articulate

**Колер** Kohler

**колёсико** *n.* little wheel, caster, roller; **красящее к., пишущее к.** inking wheel, inking disk

**колёсный** *adj.* wheel, wheeled

**колесо** *n.* wheel; drum; roller; disk; **лентопротяжное к.** feed wheel; **маховое к.** flywheel; **плоское лобовое зубчатое к.** studded disk

**колея** *f.* track; (railroad) line

**колинейный** *adj.* collinear

**количественный** *adj.* numerical, quantitative

**количество** *n.* quantity, number, amount, rate; **к. движения** momentum; **ничтожное к.** minute quantity; **подкоренное к.** radical (*math.*); **к. цепей** number of circuits, circuit capacity

**коллапс** *m.* collapse

**Коллац** Collatz

**коллективизированный** *adj.* collectivized

**коллективный** *adj.* collective

**коллектор** *m.* commutator (*elec.*); collector, accumulator, receiver, sampler; header, catcher; collecting agent; **к. с затяжкой по конусу** arch-bound commutator; **к. с затяжкой по цилиндрической поверхности** drum-bound collector; **к. с ловушкой** hook collector; **сетчатый к.** collector grid; **умножающий ток к.** P-N hook

**коллекторный** *adj.* collector; commutator; collecting

**Колли** Colley

**коллиматор** *m.* collimator

**коллиматорный** *adj.* collimator, collimating

**коллимационный** *adj.* collimation

**коллимация** *f.* collimation

**коллимированный** *adj.* collimated

**коллимировать** *v.* to collimate

**коллимирующий** *adj.* collimating

**коллинеарность** *f.* collinearity

**коллинеарный** *adj.* collinear

**коллинеация** *f.* collineation

**коллоид** *m.* colloid

**коллоидальный, коллоидный** *adj.* colloidal, colloid

**коловорот** *m.* drill bore; **к. для сверления в углах, к. с зубчатой передачей** angular borer

**коловратный** *adj.* circular, rotary, rotating, revolving

**кологарифм** *m.* cologarithm

**колода** *f.* deck, stack; trough

**колодец** *m.* well, pit, shaft; manhole; **вводный к.** cable duct; **вспомогательный к.** inspection shaft (for cable channels); **входной к.** intake manhole; **кабельный к. под тротуаром** sidewalk manhole; **муфтовый к.** splicing manhole; **к. на проезжей части улицы, к. под мостовой** manhole under the pavement; **к. под тротуаром** sidewalk manhole

**колодка** *f.* shoe (*mech.*), block, check, backing; bit, jaw; **зажимная к.** terminal box; **клемная переходная к.** terminal block for light-fixtures; **переходная к.** adapter, adapter socket; **переходная ламповая к.** tube adapter; **присоединительная к.** terminal block, connecting block; **к. с гнёздами** receptacle block, jack block, female receptacle; **к. с зажимами** terminal block, terminal strip, connection strip; **штампованная переходная к.** wafer adapter; **штепсельная к.** jack plug; **штепсельная разветвительная к.** current tap

**колодочка** *see* колодка

**колодочный** *adj.* shoe, block, backing, check

**колок** *m.* peg, (tuning) pin

**колокол** *m.* bell; bell jar; **переездной сигнальный к.** highway crossing bell

**колокола** *pl.* carillon

**колоколообразный, колоколоподобный** *adj.* bell-shaped, bowl(-shaped), funnel-shaped, funneled, dome-shaped

**колокольный** *adj.* bell, bell-shaped

**колонка** *f.* column; core; pillar; **направляющая к. для ленты** tape guide; **распределительная к.** pillar type switchgear

**колонковый** *adj.* column

**колонкообразный** *adj.* columnar

**колона** *f.* column, pillar; tower; pile (*elec.*); gang, crew; **громкоговорящая к.** sound column

**колонный** *adj.* columnar, column
**колоратура** *f.* coloratura (*mus.*)
**колоратурный** *adj.* coloratura
**колордаптер** *m.* colordapter
**колориметр** *m.* colorimeter, tintometer
**колориметрический** *adj.* colorimetric
**колориметрия** *f.* colorimetry
**колориндекс** *m.* color index
**колорит** *m.* coloring
**колорплексер** *m.* colorplexer
**колортрон** *m.* colortron
**колосниковый** *adj.* grate
**колотушка** *f.* sounder, tapper
**колотый** *adj.* split, cleaved
**колоть, наколоть, расколоть** *v.* to split, to chop, to hack, to cleave; to pierce, to prick
**колпак** *m.* cap, cover, hood, bell; cupola, dome; lid; bell jar
**колпачковый, колпачный** *adj.* cap, cover, hood, cap-type; top; lid
**колпачок** *m.* cap, bonnet; bubble cap; **выводной к.** lead cap, boot; **к. для коммутаторных ламп** signal-light cap; **к. для спуска воздуха** air bleeder cap; **крыльевой к. антенны** wing-cap antenna; **хвостовой к. антенны** tail-cap antenna
**Колпитс** Colpitts
**колчедан** *m.* pyrite
**колышек** *m.* peg, prop, picket
**кольманит** *m.* colemanite
**Кольпиц** Colpitts
**кольцевание** *n.* completion of circuit
**кольцевой** *adj.* ring, ring-shaped, annular, circular, circumferential; cyclic; orificed
**кольцеобразный** *adj.* annular, ring-shaped, circular; cyclic
**кольцо** *n.* ring, annulus, collar; coil; loop; band; circuit (*elec.*,); reel (of tape); **защитные кольца для подавления взаимных помех** cross-talk damping rings; **вращающееся к.** race, ball race; **к. для центровки чашечки** ferrule-centering ring; **защитное резиновое к. (телефона)** ear muff (of a telephone receiver); **курсовое к.** grid ring; **отталкивающее к.** cracking-off ring; **поддерживающее к.** cable ring; suspender; **проволочное к.** annular wire loop; **распорное к.** spacer ring, lock washer, locking washer; **токосъёмное к.** slip ring, collecting ring
**колющий** *adj.* cleaving; piercing

**ком** *abbr.* (**килоом**) kilohm
**кома** *f.* coma (aberration)
**команда** *f.* command, order, instruction, word; crew, gang, team; **к. автоматического изменения масштаба** automatic scale command; **к. автоматического подведения итога** automatic tally order; **к. без указания адресов** zero-address instruction; **к. блокировки** ignore instruction; **к. в машинном коде** computer instruction; **двухадресная к. с адресом следующей команды** one-plus-one instruction; **к. заполнения индексного регистра** load index register instruction; **к. заполнения счётчика повторений** load repeat counter instruction; **к. инвертирования** convert instruction; **к. контрольного останова** breakpoint instruction; **машинная к.** computer instruction; **начальная к.** key instruction; **к. «обзор таблицы»** table lock-up; **к. обращения к барабану** drum instruction; **к. обращения к ленте** tape order; **одиночная к.** coding line; **к. останова** breakpoint instruction; **к. передачи управления, к. перехода** jump instruction, control transfer instruction, transfer instruction; **к. перехода к выполнению следующей команды** skip instruction; **к. перехода по знаку регистра множителя-частного** multiplier-quotient (register) sign jump instruction; **к. перехода (по значению кода) со сдвигом** shift-jump instruction; **к. перехода при переполнении** overflow jump instruction; **к. по установке передаточных коэффициентов** gear ratio order; **к. подведения итога** tally order; **предварительно набранная к.** prewired instruction; **к. пропуска** blank instruction; **к. «сложить-выдать признак»** add index instruction; **к. сортировки** merge command; **текущая к. перехода** current transfer order; **условная к. передачи управления, к. условного контрольного останова (or перехода)** conditional breakpoint instruction; **к. условного перехода, к. условной передачи управления** conditional instruction, conditional transfer of control; **к. установки коэффициента передачи** gear ratio order; **цикли-**

ческая к. iterative instruction; **четы-рёхадресная к. с адресом следующей команды** three-plus-one address instruction

**командирский** *adj.* commander
**командный** *adj.* command, control, pilot; master
**командоаппарат** *m.* signalling device
**командо-контроллёр** *m.* master controller, master switch
**комбинанс** *m.* combinance
**комбинатор** *m.* combiner; combinator
**комбинаторика** *f.* combination theory
**комбинаторный** *adj.* combinatory; combiner
**комбинационный** *adj.* combination, combinative, combinational; combined
**комбинация** *f.* combination; **импульсная к.** pulse mode
**комбинирование** *n.* combination; rearrangement
**комбинированный** *adj.* combination (-type), combined; composite, compound; multiple-unit
**комбинировать, скомбинировать** *v.* to combine; to rearrange
**комиссия** *f.* commission, committee
**комитет** *m.* committee, bureau; **Международный консультативный к. по радио (по телеграфии, по телефонии)** International Radio (Telegraph, Telephone) Consultative Committee; **к. по промышленной стандартизации** Engineering Standards Committee; **Подготовительный к. по радиочастотам** CPF Provisional Frequency Board
**комма** *f.* comma
**коммерческий** *adj.* commercial, business
**коммунальный** *adj.* communal, municipal, public, public service
**коммуникационный** *adj.* communication
**коммуникация** *f.* communication
**коммутативный** *adj.* commutative
**коммутатор** *m.* collector, commutator; distribution board, switchboard (*tel.*); sampling switch, change-over switch; interphone control box; switch; **А-к., к. А-телефонистки** A-switchboard, A-position, outgoing position, outward operator's position; **Б-к.** B-switchboard, B-position, incoming position; **к. без многократного поля**

nonmultiple switchboard; **к. Б-телефонистки** inward operator's position, B-position; **к. входящей связи** incoming (trunk) position; **двухполюсный элементный к.** double pole battery regulating switch, double cell switch, double regulating switch; **директорский к.** chief operator's turret, chief operator's switchboard; **к. диспетчерской связи** intercommunicating plug switchboard; **к. для внутренней связи** dispatch switchboard; inter-through switch; **к. для индукторного вызова** magneto switchboard; **к. для линий коллективного пользования** party-line switch; **добавочный к.** trunk junction position; **зарядный к.** battery switch, cell switch; **к. исходной связи** outgoing position (*tphny.*); **ключевой к.** key box, keyboard (*tphny.*); **ламельный к.** cross-connecting board; **к. междугородной связи** trunk exchange; **междугородный к. (для транзитного трафика)** toll switchboard, long-distance switchboard; **междугородный контрольный к.** trunk test switchboard, trunk test panel; **номерной к.** dial switch; **ночной к.** concentration position (*tphny.*); **одноместный ручной к.** one-position manual switchboard; **полевой к. с клапанами** field-telephone exchange, portable switchboard; **пригородный к.** suburban toll board (*tphny.*); **к. с клапанами** telephone switchboard; **к. с лампами накаливания** lamp switchboard; **к. с местным полем** A-position, answering position; **к. с отпадающими клапанами** drop-indicator panel, indicator switchboard, drop type switchboard; **к. с поворотными включателями** rotary-switch manual switchboard; **к. с радиальным пучком** radial-beam tube; **станционный вызывной к.** dialtone switching relay; **телеграфный к.** teletype exchange; **телефонный клапанный к.** telephone switchboard; **к. транзитной связи, транзитный междугородный к.** through switchboard; **штепсельный к.** cross-connecting board, plug selector, plug connector panel, jack unit switchboard; **штепсельный ламельный к.** arc plug-switchboard;

**элементный к.** cell switch, accumulator switch, battery (regulating) switch

**коммутаторный** *adj.* commutator; switchboard, switchboard-type

**коммутационный** *adj.* commutation; switching; switchboard

**коммутация** *f.* switching, switching process, establishing a connection; commutation; sampling; **к. в системе с постоянной яркостью** constant-luminance sampling; **к. коллекторной электрической машины** commutation in a commutator machine; **плавная к. телевизионных каналов (с наплывом изображений)** cross-fade; **прямолинейная к.** resistance commutation; **к. сигналов цветного изображения** color-picture sampling; **к. частот** frequency-shift keying; **к. шаговым методом** step-by-step action

**коммутирование** *n.* commutation; gating; switching; **симметричное к.** symmetrical sampling (*tv*)

**коммутированный** *adj.* commutated; switched; commutator

**коммутировать** *v.* to commutate, to commute, to change over; to switch, to reverse

**коммутируемый** *adj.* commutated; switched; keyed

**коммутирующий** *adj.* commutating; switching; commutative

**комната** *f.* room, chamber; apartment; **к. со множественным звукоотражением** echo room

**комнатный** *adj.* room; indoor

**комол** *m.* Comol (alloy)

**компактность** *f.* compactness, density

**компактный** *adj.* compact, solid, dense, massive; tight; packaged

**компандирование** *n.* companding

**компандор** *m.* compandor (*tphny.*)

**компаратор** *m.* comparator, differential element, comparator circuit; **фазовый к.** phase-comparison circuit, raydist

**компас** *m.* compass; **магнитостробный к.** flux-gate compass, saturable-reactor type compass

**компасный** *adj.* compass

**компаунд** *m.* compound, filler, saturant; insulation compound; **компаунды для печатных схем** printed-circuit polyesters; **к.-генератор** compound-wound dynamo, compound-wound generator; **к.-машина** compound-wound machine; **к.-ядро** intermediate nucleus

**компаундирование** *n.* compounding

**компаундированный** *adj.* compound-wound, compounded, compound

**компаундировать** *v.* to compound

**компаундировка** *f.* compounding

**компаундирующий** *adj.* compounding

**компаундный** *adj.* compound; compound-wound

**компенсатор** *m.* compensator, equalizer, balancer; expansion piece; **к. искажений в линии** line equalizer; **к. перекоса кадра** tilt mixer (*tv*); **к. полного сопротивления для фильтров** filter impedance compensator

**компенсационный** *adj.* compensation, compensating, compensatory, balancing, equalizing; expansion (piece); canceller

**компенсация** *f.* compensation, balancing, balance, equalizing, equalization; neutralization cancellation, bucking-out; **к.влияния окружающей температуры** ambient-temperature compensation; **к. влияния соединительных проводов** lead-wire compensation; **к. влияния температуры холодного спая** cold-junction compensation; **к. выходных концов термопары** reference-junction compensation; **к. завала частотной характеристики** stagger damping **к. затухания** attenuation equalization; **к. звуковой ряби** flutter compensation; **к. квадратичной зависимости** square-root compensation; **к. паразитного сигнала передающей трубки** shading compensation; **перекрёстная к.** transposition balancing; **к. перекрёстных искажений** cross-talk compensation; **к. сдвига фаз** phase compensation, correction of phase

**компенсирование** *n.* compensation, balancing, equalizing

**компенсированный** *adj.* compensated

**компенсировать** *v.* to compensate, to balance, to equilibrate, to make up, to equalize; to cancel out

**компенсирующий** *adj.* compensating, balancing, equalizing; compensation

**компетентность** *f.* expert knowledge, competence

**компетентный** *adj.* expert, competent, able, experienced

**компилирование** *n.* compilation, compiling

**компилировать** *v.* to compile, to collect

**компилирующий** *adj.* compiling, compiler

**компилятор** *m.* compiler; compiler program

**компиляция** *f.* compilation, collection

**комплекс** *m.* complex; group; assemblage

**комплексный** *adj.* complex, unitized, integral; multiple

**комплект** *m.* assembly, series, kit, set, unit, suit, outfit; both terminal sets; complement; **абонентский к. (на станции)** subscriber's line equipment (*tphny.*); **вентильный к. из ряда столбиков** metallic rectifier stack assembly; **вентильный к. с одним полупроводниковым вентилем в плече** basic metallic rectifier; **к. деталей** kit; **к. деталей для сборки анализатора сигнала звуковой частоты** audio analyzer kit; **к. деталей для сборки испытателя конденсаторов** condenser checker kit; **к. деталей для сборки усилителя** amplifier kit; **к. деталей для сборки устройства связи с антенной** antenna coupler kit; **к. для включения дальних линий** long-distance calling equipment; **к. для внутреннего соединения** internal selector set; **к. катушек** coil assembly, coil set; **к. контактных пружин** contact spring set; **к. контакто-держателей** wiper set, wiper assembly; **к. кулачков** battery of cams; **линейный к.** line switch, terminal line equipment; **к. междугородных РСЛ** trunk block connector relay; **к. накала, анода и сетки** heater, plate and grid, ABC; **переносный к. приборов для кабельных измерений** portable cable measuring set; **переходный (дифференциальный) к.** hybrid set (*tphny.*); **к. пружин (реле)** spring pile-up (of a relay), spring assembly, spring set; **к. пупинизации, к. пупиновских катушек** loading (coil) unit, loading coil set; **к. реле вызова** ringing-current bypass relay set; **к. реле телеуправления** supervisory control relay; **соединительный к.** relay group for internal connections (*tphny.*); **«управляющий к.»** master switch; **центральный к.** common selector equipment; **шунтовый к. пружин** off-normal spring set; **к. эталона частоты** standard-frequency assembly

**комплектный** *adj.* complete; unitized

**комплектование** *n.* completion, making up a set

**комплектовать, укомплектовать** *v.* to complete (a set), to replenish; to supply

**комплементарность** *f.* complementarity

**композитный** *adj.* composite

**композиция** *f.* composition

**компонент** *m.* component, constituent; component force; **искажающие изображения компоненты** distortion terms (*tv*); **к., изменяющийся синусоидально в зависимости от пеленга** semicircular component

**компонента** *f.* component, constituent; component force; **к. напряжения прямой последовательности фаз** positive phase-sequence voltage component; **к. операции** operand; **проникающая к.** hard component (*nucl. physics*); **к. ряда Фурье** Fourier component

**компоновка** *f.* assembling, arrangement, grouping, composition; packaging; **к. схем** circuitry

**компонующий** *adj.* assembling; assembler

**компоситрон** *m.* compositron

**компостер** *m.* spot punch

**компрессионный** *adj.* compressional, compression

**компрессия** *f.* compression; **к. относительного размаха синхронизирующих импульсов** synchronization compression

**компрессометр** *m.* compressometer

**компрессор** *m.* compressor; **к. (диапазона) громкости, к.-сжиматель динамического диапазона** volume compressor

**компрессорный** *adj.* compressor

**компромиссный** *adj.* compromise

**комптометр** *m.* comptometer, adding machine

**Комптон** Compton

**комптоновский** *adj.* (of) Compton, Compton's

**компутрон** *m.* computron

**компфнеровский** *adj.* Kompfner, Kompfner's

**компьютрон** *m.* computron

**конвейер** *m.* conveyer, elevator

**конвейерный** *adj.* conveyer, conveying

**конвективно** *adv.* convectively

**конвективный** *adj.* convective, convection, convectional

**конвектрон** *m.* convectron

**конвекционный** *adj.* convection, convectional, convective

**конвекция** *f.* convection

**конвенциональный** *adj.* conventional

**конвенция** *f.* convention; **международная к. по электрорадиосвязи, Международная Конвенция Связи** International Telecommunication Convention, I.T.C.

**конвергентный** *adj.* convergent

**конвергенция** *f.* convergence

**конвергировать** *v.* to converge

**конверсионный** *adj.* conversion

**конверсия** *f.* conversion

**конвертер** *m.* converter; dynamotor

**конвертерный** *adj.* converter

**конвертирование** *n.* conversion

**конвертированный** *adj.* converted

**конвертировать** *v.* to convert

**конвертор** *m.* converter; dynamotor

**конверторный** *adj.* converter; conversion

**конграмматический** *adj.* congrammatical

**конграмматичность** *f.* congrammaticality

**конгруэнтность** *f.* congruence

**конгруэнтный** *adj.* congruent, congruous, corresponding

**конгруэнция** *f.* congruence

**конденсат** *m.* deposit, moisture, sediment, condensate

**конденсатор** *m.* capacitor (*elec.*); condenser; **анодный блокировочный к.** plate bypass capacitor; **блокировочный к.** blocking capacitor, stopping capacitor; bridging capacitor, shunt capacitor; **к. блокирующий вызывной ток** capacitor for blocking signal current; **блочный к.** multiple capacitor; **большой вакуумный к.** coconut capacitor, vacuum capacitor; **вводной втулочный к.** bushing capacitor; **к. воющего тона** warble tone condenser; **вч к. связи** carrier coupling capacitor; **выравнивающий к. гетеродина** padding capacitor,

oscillator padder; **двухстаторный к.** tandem capacitor; **к. для устранения помех** anti-interference capacitor; **заградительный к.** countercurrent capacitor; **залитый в стаканчик к.** potted capacitor; **зарядный к. в схеме развёртки** time-base charging capacitor; **к. из посеребренных пластин** silvered capacitor; **к. колебательного контура** tank capacitor; **коммутируемый накопительный к.** switched memory capacitor; **к. коррекции нуля** zero-adjusting capacitor; **логарифмический переменный к.** midline plate capacitor; **многосекционный к.** multiple unit capacitor; **многосекционный к. переменной ёмкости** gang(ed) capacitor, gang tuning capacitor; **накопительный к.** charging capacitor; **накопительный к. (питающий магнетрон)** reservoir capacitor (supplying a magnetron); **намотанный к.** roll (type) capacitor, wound capacitor; **к. настройки антенного контура** antenna tuning capacitor; **к. настройки передатчика** transmitter tuning capacitor; **нейтродинный к.** neutralizing capacitor; **к. обратной связи** feedback capacitor, reaction capacitor; **опрессованный в пластмассу к.** molded-insulated capacitor; **к. переменной ёмкости** rotary capacitor, variable capacitor; **перестраиваемый постоянным током к.** direct-current tuned capacitor; **переходный к.** block(ing) capacitor, isolating capacitor, coupling capacitor; **подгоночный к., подстроечный к.** trimmer, padder, "permalinery", passing capacitor, trimmer capacitor, compression-type capacitor, adjustment capacitor; **полупеременный к.** padding capacitor, padder, trimming capacitor, trimmer; **помехоподавляющий к.** anti-interference capacitor; **проходной к.** duct capacitor; **прямоёмкостный к.** straight-line capacitor, straight-line-capacity capacitor; **развязывающий к.** ground capacitor; **разделительный к.** coupling capacitor; **к. реактивного тока** static phase advancer, phase advancing capacitor; **регулируемый к.** rotary capacitor; **рулочный к.** tubular capacitor, Mansbridge capacitor;

к. с воздушным диэлектриком air capacitor; к. с ёмкостью, регулируемой по напряжению voltage sensitive capacitor; к. с закруглёнными углами bathtub capacitor; к. с защитным кольцом guard ring-type capacitor; к. с кремнеорганической заливкой silicon-filled capacitor; к. с надвигающимися обкладками sliding capacitor; к. с отводами multiple-unit capacitor; к. с твёрдой пропиткой jelly-filled capacitor; к. связи с осветительной сетью lighting network coupling capacitor; antenna eliminator, lamp-socket antenna; сглаживающий к. tank capacitor of a rectifier filter; сдвоенный к. переменной ёмкости two-gang variable capacitor; сдвоенный поворотный к. с общим управлением gang capacitor; слюданой опрессованный к. molded-mica capacitor; сопрягающий к. tracking capacitor, padding capacitor; стеклянный подстроечный к. glass trimmer capacitor; к. типа «стабиль» metallized mica capacitor; тонально-калиброванный к. incremental pitch capacitor; к. точной настройки fine-tuning capacitor, bandspreader, vernier capacitor; тройной переменный к. three-gang capacitor; укорачивающий к. антенны, к. укорочения антенны antenna shorting capacitor, antenna series capacitor; форсирующий к. speed-up capacitor; шунтированный сопротивлением к. shunted capacitor

конденсаторный *adj.* capacitor; condenser

конденсационный *adj.* condensation, condensing

конденсация *f.* condensation

конденсер *see* конденсор

конденсирование *n.* condensation, condensing

конденсированный *adj.* condensed

конденсировать *v.* to condense

конденсирующий *adj.* condensing

конденсит *m.* condensite

конденсор *m.* condenser; condenser lens, condensing lens

конденсорный *adj.* condenser

кондиционер *m.* conditioner, air conditioning plant

кондиционирование *n.* conditioning, humidity control

кондиционированный *adj.* conditioned

кондицировать *v.* to condition

кондуктивность *f.* conductivity

кондуктивный *adj.* conductive

кондуктометрический *adj.* conductometric

кондуктор *m.* conductor; jig

кондукторный *adj.* conductor, guide; jig

кондукционный *adj.* conduction, conductive

кондукция *f.* conduction

конель *m.* Konel

конец *m.* end, termination, close, consummation; tip; terminal; outflow, issue, discharge; distance; lead, jumper; length (of wire, rope); батарейный к. battery lead; к. вторичной обмотки out secondary; к. выводной спирали throw-out tail; высокопотенциальный к. отклоняющей системы hot side of yoke; заделанный к. capped end, sealed end; заправочный к. ленты leader tape; зарядный к. leader; к. катушки coil out; к. нерегулируемой вилки butt end; к. ножки с тарелкой flare end; отправной к. sending end; near end; к. первичной обмотки out primary; к. работы связи end of communication; режущий к. инструмента tool bit; смысловой ведущий к. logical leading end

конечноразностный *adj.* finite-difference

конечный *adj.* finite; final, ultimate, end, terminal; top

кониметр *m.* konimeter

конический *adj.* conical, conic, cone; tapered, tapering; beveled

конкатенация *f.* concatenation

конкорданция *f.* concordance

конкретность *f.* concreteness

конкретный *adj.* concrete

конкурентноспособный *adj.* cost-competitive

коноид *m.* conoid

коноидальный *adj.* taper

конопля *f.* hemp

коноскоп *m.* conoscope

конперник *m.* Konpernik

консервант *m.* preservative

консервативный *adj.* conservative

консерватор *m.* conserver; expansion tank of transformers

**консервация** *f.* conservation; preservation

**консервирование** *n.* conservation; preservation

**консервированный** *adj.* conserved; preserved, canned

**консервировать** *v.* to conserve; to preserve, to can

**консервирующий** *adj.* conserving; preserving

**консистентный** *adj.* consistent

**консистенция** *f.* consistency, consistence, density

**консистометр** *m.* consistometer

**«Консоль»** Sonne (long-range radio-navigation system)

**консоль** *f.* angle bearer, bracket, bracket support, arm, ancon truss, cantilever; console; **к. для оттяжки** guy attachment; **к. для укрепления кабелей** cable cleat, cable clamp, cable support; **оконечная к.** terminal bracket; **угловая к.** angle bracket

**консольный** *adj.* angle bearer, cantilever, bracket, arm; console, console-type

**консонанс** *m.* consonance

**конспект** *m.* abstract, summary, compendium

**константа** *f.* constant; **к., характеризующая группу цифр** block constant

**константан** *m.* constantan

**константановый** *adj.* constantan

**константировать** *v.* to construe, to establish, to state; to ascertain

**конституировать** *v.* to constitute

**конституэнт** *m.* constituent

**конструирование** *n.* design, designing, construction, development, shaping, forming

**конструировать, сконструировать** *v.* to design, to construct, to develop; to form

**конструктивность** *f.* designability

**конструктивный** *adj.* constructive, constructional

**конструктор** *m.* constructor, designer

**конструкционный** *adj.* constructional, structural, construction

**конструкция** *f.* make, design; construction, installation, erection, formation; structure; assembly, framework, setup; phrase; **модульные конструкции** stacked modules; **блочная к.** plug-in assembly; **звукопоглощающая к. стального пере-**крытия sound-absorptive steel roof-deck assembly; **к. лампы, собранная на керамических кольцах** stacked ceramic tube; **сдвоенная к.** tandem assembly; **упрочнённая к. ножки** rugged mount structure

**консультативный** *adj.* consultative, advisory

**контакт** *m.* contact, connection, terminal; touching; tip; **вращательные контакты** off-normal spring (of a Strowger switch); **контакты вращения искателя** rotary off normal contacts; **временно-замыкающие контакты** wiper-type contacts; **дугогасительные контакты** arcing contacts; **запараллеленные контакты** contacts in parallel, multiple contacts; **контакты коммутаторных линий** hunting contacts; **кососоединённые контакты поля** slipped banks; **контакты обгорания (выключателя)** arcing tips (of circuit breakers); **контакты подъёмного движения (искателя)** vertical contacts (of a switch); **разрывные контакты выключателя** arcing tips of a circuit breaker; **акустический к.** echo contact; **к. вкладыша штепсельного разъёма** male contact; **вращающийся нормальный к.** rotary off-normal contact, shaft contact; **временно замыкающий к.** impulse contact; **к. гнезда штепсельного разъёма, гнездовой к.** female contact; **двойной к. коммутаторной линии** muster number double contact, double hunting contacts; **к. для проверки величин** check point; **к. для провёртывания искателя** overflow contact; **добавочный к.** trailing contact, sequence contact, consecutive contact; **добавочный к. для монетного автомата** coin-feeler contact; **к. импульсного номеронабирателя** dial pulse-contact; **кнопочный к.** finger contact; **линейный к.** line contact; ohmic contact; **к. между головкой и корпусом штепселя** tip and sleeve contact; **к. между петлями** constant loop, contact between two wires of one circuit; **мостящий к.** make-before-break contact; **набирательный к.** hunting contact; **к. начала подъёма** vertical off-normal contact, head contact; **к. начального положе-**

ния normal contact; **начальный к. подъёмного движения** vertical off-normal contact; **невыпрямляющийся к.** ohmic contact; **неплотный к.** intermittent contact; **непосредственный гидролокационный к.** close sonar contact; **нижний рабочий к. ключа** anvil (*tphny*.); **ножевой к.** knife-type contact, jack-in contact; **нормально разомкнутый к.** operating contact, make-contact; **к., образованный контактной пружиной** whisker contact; **к. основного носителя** majority-carrier contact; **отбойный к.** break contact, spacing contact, space contact, release contact; **отрывной к.** arcing contact, breaker point; **охватываемый к.** male contact; **охватывающий к.** female contact; **педальный к.** floor contact; **перекидной к.** break-before-make contact; **переключающий к.** double-throw contact; **перекрывающий к.** $\lambda/4$-joints for waveguides; **переходный к.** make-before-break contact; **плоский к.** area contact; **к. подъёмного движения** head contact; **к. подъёмного электромагнита** vertical interrupter contact; **к. покоя вращающего электромагнита** rotary interrupter contact; **к. покоя номеронабирателя** N.C. dial contact (*tphny*.); **последовательно включённый к.** sequence contact, trailing contact; **последовательный к. с замыканием после размыкания** break-before-make contact; **последовательный к. с размыканием после замыкания** make-before-break contact; **приёмно-передающий к.** send-receive contact; **рабочий к. номеронабирателя** N.O. dial contact; **рубящий к.** knife blade contact, blade contact; **к. с головкой** butt contact; **сварной к., образованный контактной пружиной** welded whisker contact; **к. серийного искания** collective contact (of a selector); **спокойный к. вращающего электромагнита** rotary interrupter contact; **укороченный к. распределителя** signal segment of a distributor; **к. усиленного типа** heavier duty contact; **формованный к., образованный контактной пружиной** formed whisker contact

**контактно-модулированный** *adj.* contact-modulated

**контактно-разрывной** *adj.* make-and-break

**контактный, контактовый** *adj.* contact, contacting

**контактор** *m.* contactor, switch; **к. бдительности** acknowledging contactor; **кулачковый к.** cam contactor; **к. с реле** automatic tripping contactor, magnetic full-voltage starter

**контакторный** *adj.* contactor, switch

**контаминация** *f.* contamination

**контейнер** *m.* container, jar, vessel, tank

**контекст** *m.* context

**контекстуальный** *adj.* contextual

**контингенция** *f.* contingency

**континентальный** *adj.* continental

**континуум** *m.* continuity, continuum

**континюанта** *f.* continuant

**контора** *f.* office; station, exchange, central office; **монтажная к.** installer firm

**конторский** *adj.* office

**контр-** *prefix* counter-

**контрабас** *m.* double bass, contrabass, bass viol, violoncello

**контравалентность** *f.* contravalence

**контравариантный** *adj.* contravariant

**контраградиентный** *adj.* contragradient

**контрадикторный** *adj.* contradictory

**контрадикция** *f.* contradiction

**контракт** *m.* contract, agreement

**контральто** *n. and m. indecl.* contralto

**контрапункт** *m.* counterpoint

**контрарный** *adj.* contrary

**контраст** *m.* contrast; **к. яркостей** variation in light intensity, picture contrast

**контрастирование** *n.* superelevation, lift (*tv*)

**контрастировать** *v.* to contrast, to compare

**контрастность** *f.* contrast (range), fine detail resolution; **к. изображения** image contrast range, picture contrast; **к. мелких деталей** detail contrast ratio; **к. сцены** scene contrast range

**контрастный** *adj.* contrast, contrasting

**контрафагот** *m.* double bassoon

**контрбатарея** *f.* counter-cell

**контргайка** *f.* counter-nut, check-nut, locknut

**контргруз** *m.* counterweight, balance weight

**контркалибр** *m.* control gage, master gage, countergage

**контркомпаунд** *m.* differential compound

**контрмеры** *pl.* countermeasures

**контrobмотка** *f.* opposing winding, bucking coil

**контр-октава** *f.* contraoctave

**контролёр** *m.* monitor; controller; checker, inspector, controller, supervisor, verifier operator; **к. радиомаяка, к. радиостанции** radio-range monitor; **к., собранный частично из магнитных элементов** semi-magnetic controller; **стрелочный к.** switch box; **к. точности синхронизма** scrutinizer, scrutineer; **фотоэлектрический к. пламени** photoelectric flame-failure detector

**контролирование** *n.* monitoring; control, controlling, supervision

**контролированный** *adj.* controlled, control

**контролировать, проконтролировать** *v.* to control, to check, to inspect; to monitor; to verify

**контролировка** *see* **контролирование**

**контролируемость** *f.* controllability

**контролируемый** *adj.* controllable; controlled; monitored

**контролирующий** *adj.* controlling; monitoring

**контроллер** *m.* controller; **барабанный к.** starter drum; **к. линейки ящика зависимости** circuit tapper controller; **регулирующий барабанный к.** drum starter and controller

**контроллерный** *adj.* controller

**контроль** *m.* check, checking, control, controlling, inspection, supervision, monitoring; standard, blank test; service observation; survey; verification gaging; **автоматический к. размеров** autosizing; **к. «в две руки»** duplication check; **выборочный к.** spot check, random inspection; **выборочный к. по количественным признакам качества** method of variable; **двукратный выборочный к.** double sampling; **к. на приёмнике** radio watching; **к. передачи по модулю *n*** modulo-*n* check; **к. пере-**

полнения exceed capacity check; **к. по избыточности** redundant check; **к. по контрольной сумме** echo checking; **к. по разности** differencing check; **к. по сумме** sum check; **к. по увеличенной тени** shadowgraphing; **к. по чётности** even-parity check, odd-even check; **к. подслушиванием** blind supervision, monitoring; **проекционный к.** shadowgraphing; **профилактический к.** marginal checking; **к. совмещения** register (photoelectric) control; **к. считыванием после записи** read-after-write check; **к. у рычага** lever indication; **к. частоты передатчика** supervision of transmitter operation; **к. эксплоатации междугородных линий** trunk line observation, toll service observation; **электронный к. размеров** electronic autosizing

**контрольно-измерительный** *adj.* control and measurement

**контрольно-планшетный** *adj.* monitor charting

**контрольный** *adj.* supervisory; control, check, test, pilot; regulating; monitor, monitoring; master; reference (point)

**контрольщик** *m.* card-proof punch; verifier

**контроттяжка** *f.* antiflex flying wire

**контрпривод** *m.* countershaft

**контрпружина** *f.* opposing spring, reacting spring

**контррефлектор** *m.* primary radiator, subdish

**контрсистема** *f.* negative phase-sequence system

**контрспираль** *f.* double helix

**контрфланец** *m.* counterflange

**контрход** *m.* countermove

**контур** *m.* contour, outline, form, sketch, shape, profile; section; loop, circuit, boundary; circumference, periphery; tank; **многополосные колебательные контуры** multiband tank-circuits; **с настроенным анодным контуром** tuned-anode; **с расстроенными контурами** stagger-tuned; **агрегатированный к.** gang circuit; **анодный к. (передатчика)** tank circuit, plate circuit, anode circuit; **балансный к. с приближённой настройкой** compromise balancing network; **вспомогательный хо-**

лостой к. idling circuit; **выравнивающий амплитудный к.** attenuation equalizer; **высокочастотный запирающий к.** high-pass selective circuit; **горшковый к.** pot resonator; **к. для придания формы** shaping network; **добавочный балансный к. (для пупинизированной линии)** building-out section, building-out network; **запирающий к.** rejective circuit, trap circuit, anti-resonance circuit; **запирающий к. в цепи анода** anode rejector circuit; **звенящий к., выполненный на кристалле, звенящий к. с кварцевой стабилизацией** crystal ringing circuit; **к. исправления сигнала** signal shaping network; **коаксиальный к. УВЧ** coaxial butterfly circuit; **колебательный к. в аноде** plate tank; **колебательный к. последовательного включения** series-resonant circuit; **объёмный к.** cavity circuit; resonant cavity; **объёмный к. с магнитной настройкой** ferrite-tuned resonant cavity; **отпирающий к.** gate opener; **периодический к.** tuned circuit; **к. подавления зеркального канала** image suppression circuit; **последовательный к. для выделения сигнала определённой частоты** acceptor circuit, series resonance circuit; **последовательно включённый корректирующий к.** series equalizer; **противоперерегулировочный к.** snubbing circuit; **к. развёртывающего устройства** time-base circuit (tv); **разъединяющий к.** network-decoupling transformer; **режекторный к.** band-elimination circuit; **к.-резервуар** tank (circuit); **резонансный к. генератора** resonance oscillator circuit; **к. с затуханием выше критического** overdamped circuit; **к. с затуханием ниже критического** underdamped circuit; **к. с отрицательной обратной связью** degenerative circuit; **к. с положительной обратной связью** regenerative circuit; **к. с резонансом тока** parallel-resonance circuit; **к. с сосредоточенными постоянными** lumped-constant circuit; **самовозбуждаемый колебательный к.** autodyne; **к. смесителя** injection circuit; **специальный объёмный к.** pot tank cir-

cuit; **срезывающий к.** attenuation equalizer; **к. типа «бабочка»** coaxial butterfly circuit; **увч колебательный к. из параллельных стержней** parallel-rod uhf tank circuit; **ударный к.** impulsing circuit, impulse circuit, surge-voltage test circuit; **устраняющий радиопомехи к.** noise silencer

**контурный** adj. contour, outline

**контуроподобный** adj. circuit-wise

**конус** m. cone, taper; jaw; **плавкие конуса** fusion cones; **мёртвый к., к. молчания** cone of nulls, cone of silence; **пироскопический к.** Seger cones, fusion cones

**конусность** f. conicity, angle of taper

**конусный** adj. cone, conic, conical; taper, tapered; bevel

**конусовидность** f. conicity, cone shape, taper

**конусовидный** adj. cone-shaped, conical; taper, tapered

**конусообразность** f. conicity, cone shape; taper

**конусообразный** adj. cone-shaped, conical; taper, tapered

**конферирование** n. presentation

**конфигурационность** f. projectivity

**конфигурационный** adj. projective, configuration, profile, contour

**конфигурация** f. configuration, profile, contour; layout; shape; gulp; **к. пучка излучения** beam pattern; **к. системы проводов** conductive system; **к. схемы** network topology

**конфликт** m. conflict

**конфликтный** adj. conflict; conflicting

**конфлюэнтность** f. confluence

**конфлюэнтный** adj. confluent

**конфокальный** adj. confocal

**конформный** adj. conformal

**конхоида** f. conchoidal curve

**концевик** m. microswitch

**концевой** adj. end, terminal; edge

**концентрат** m. concentrate

**концентратор** m. concentrator; concentration switch

**концентрационный** adj. concentration, concentrating

**концентрация** f. concentration, concentrating, focussing; density; **к. неосновных носителей-дырок** minority electron density; **объёмная к. ионов** volume ionization density; **к. основных носителей-электронов** majority electron density; **к. приме-**

**сей** impurity concentration; **к. собственных носителей** intrinsic concentration

**концентрированный** *adj.* concentrated, strong

**концентрировать, сконцентрировать** *v.* to concentrate, to lump, to focus, to direct

**концентрирующий** *adj.* concentrating, focussing

**концентрический** *adj.* concentric; co-axial

**концентричность** *f.* concentricity

**концентричный** *adj.* concentric

**концертино** *n. indecl.* concertina

**концертный** *adj.* concert; musical

**концовка** *f.* tail-piece; runout (of a film)

**кончать, кончить** *v.* to finish, to end, to complete, to terminate

**конченный** *adj.* finished, completed

**кончик** *m.* tip, point, end

**кончить** *pf. of* **кончать**

**конъюнктивный** *adj.* conjunctive

**конъюнкция** *f.* conjunction

**кообъём** *m.* co-content

**кооперативный** *adj.* cooperative

**координата** *f.* coordinate; fix; **координаты отсчёта полных сопротивлений** impedance coordinates; **пространственные координаты** solid axes; **пространственные координаты оси** space axes (*math.*); **прямоугольные координаты** Cartesian coordinates; **координаты цвета** tristimulus values (of a light); **координаты цветности** chromacity coordinates; **координаты цели** target position data; **к. в цифровой форме** digital coordinate; **к. времени** time base; **пространственная к.** spatial value

**координатник** *m.* coordinate spacer

**координатный** *adj.* coordinate, coordinated; spatial

**координатограф** *m.* coordinatograph

**координатор** *m.* coordinator; homer

**координационный** *adj.* coordination

**координация** *f.* coordination

**координирование** *n.* coordination

**координированный** *adj.* coordinated, coordinate

**координировать** *v.* to co-ordinate (with)

**координирующий** *adj.* co-ordinating

**копал** *m.* copal

**копаловый, копальный** *adj.* copal

**копать, выкопать** *v.* to dig, to trench, to excavate

**копёр** *m.* pile driver, ram, drop hammer; paving hammer; impact tester

**копилка** *f.* coin box; receptacle; **денежная к.** coin box, collection receptacle (of a telephone)

**копир** *m.* copy, master form; cam

**копировальный** *adj.* duplicator, duplicating, copying, tracing; printer

**копирование** *n.* copying, duplication, reproduction; printing; tracing; **масштабное к.** scaling; **масштабное к. магнетрона** scaling of a magnetron

**копировать, скопировать** *v.* to copy, to duplicate; to imitate; to trace

**копирующий** *adj.* duplicating, copying, reproducing; master-slave (manipulator)

**копирэффект** *m.* magnetic transfer, crosstalk magnetic printing, spurious printing, magnetic printing

**копить, накопить** *v.* to accumulate, to store up, to pile up, to lay up; to build up

**копия** *f.* copy, duplicate, transcript, replica, counterpart; **к. решётки** replica grating

**копланарный** *adj.* coplanar

**кора** *f.* bark, rind, cortex

**корабельный** *adj.* ship, vessel; marine; ship-based, ship-borne

**корабль** *m.* ship, vessel, boat; **к. с радиолокаторами наведения истребителей** fighter direction ship

**Корбино** Corbino

**кордерит** *m.* corderite

**кореллограмма** *f.* correlatogram

**коренной** *adj.* root; fundamental, main, original, radical

**корень** *m.* root; radical (*math.*); radix, root (*math.*); **первообразный к.** generator (*math.*); **к. третьей степени** cubic root

**корзина** *f.* basket, crate; bucket

**корзиночный, корзинчатый** *adj.* basket, basket-like

**коридор** *m.* corridor, passage (way); switchbay

**коридорный** *adj.* corridor, passage, hall

**Кориолис** Coriolis

**корка** *f.* crust, scale; bark

**кормовой** *adj.* stern (*naut.*)

**корневой** *adj.* root, radical

**«корнеквадратный»** *adj.* square-root

**корнет** *m.* cornet (*mus.*); **к.-а-пистон** cornet-a-piston, key bugle

**Корню** Cornu

**короб** *m.* basket; duct, ducting; box; **изоляционный к.** wooden raceway

**коробка** *f.* box, chest, case; can; housing; body; chamber; **абонентская к.** terminal block; **гнездовая к.** jackbox; **к. для многократного включения** terminal block; **золотниковая к.** valve cage; **кабельная концевая к.** cable-head distribution rack; **кабельная ответвительная к.** cable branch-joint box; **оконечная кабельная к.** cable end box, cable head, sealing chamber; **ответвительная к.** branching box, connection box, dividing box, distribution box; **к. передач** reduction gear box; **переходная к.** *see* **ответвительная к.**; **к. пневматической почтовой установки** pneumatic dispatch carrier; **поверочная зажимная к.** test terminal box; **присоединительная к.** connection box, terminal box, pothead; **разветвительная к.** splitter box, distributor box, multiple joint, distributing cable connector, distributing cable muff; **разъединительная к.** terminal block; **к. с зажимами** connection box, terminal box; **к. скоростей** gear box; **телефонная к.** plug-and-socket; **к. управления модульной конструкции** modular control box

**коробление** *n.* buckling, warping, deformation, distortion; shrinkage

**коробочка** *f.* box

**коробочный** *adj.* box, case

**коробчатый** *adj.* box, box-type; cellular

**коромысло** *n.* yoke, beam, balance beam; rocking shaft

**корона** *f.* corona, crown, corona discharge, brushing

**коронал** *m.* coronal

**коронирование** *n.* corona, corona discharge

**коронировать** *v.* to have corona

**коронка** *f.* crown; **к. из контактов** group of contacts; **к. пробки** screw-plug cartridge fuse carrier

**коронный** *adj.* corona, crown

**коронограф** *m.* coronograph

**коронстойкий** *adj.* corona-proof

**корончатый** *adj.* corona, crown

**короткий** *adj.* short, brief

**коротко** *adv.* shortly, briefly

**коротко-** *prefix* short-

**коротковолновик** *m.* short-wave radioham

**коротковолновый** *adj.* short-wave

**короткодействующий** *adj.* short-range

**короткозамкнутый** *adj.* short-circuited, short-circuit, shorted

**короткозамыкатель** *m.* shorting device, shorting plug, short-circuiting device; u-link

**короткозамыкающий** *adj.* short-circuiting

**короткопериодический** *adj.* short-term

**короткопробежный** *adj.* short-range

**короткость** *f.* shortness, brevity

**короткофокусный** *adj.* short-focus

**короткоходовой** *adj.* short-stroke

**корпорация** *f.* corporation

**корпрен** *m.* corprene

**корпус** *m.* chassis, body, framework, frame, rack, carcass, shell, hull; box, housing, case, casing; tank; **к. подшипника** bearing box, bearing shell; **к. штепселя** sleeve (of a plug); S-wire

**корпускула** *f.* corpuscle

**корпускулярный** *adj.* corpuscular

**корпусный** *adj.* chassis, body, framework, frame, carcass, shell, hull; housing, case, casing

**корректирование** *n.* correction; **к. искажений с применением перфорированной ленты** equalization of telegraph-pulses by means of the recorder method; **к. трапецеидального искажения** keystone correction

**корректированный** *adj.* corrected, adjusted, compensated

**корректировать, прокорректировать** *v.* to correct, to adjust, to rectify, to eliminate distortion, to tailor, to update

**корректировка** *f.* adjustment, updating, equalization; **к. (огня) при помощи акустического измерения расстояния** sound-ranging adjustment

**корректировочный** *adj.* corrective, correcting, adjusting

**корректирующий** *adj.* correcting, adjusting, corrective; compensating, equalizing, error-correcting

**корректный** *adj.* correct, proper

**корректор** *m.* corrector, corrective device, corrective network; equalizer;

reader, proof-reader; **к. градации яркости** gamma corrector; **к. затухания** attenuation compensator; **к. импульсов тока** impulse corrector; **регулируемый групповой фазовый к.** variable group-delay equalizer; **к. фазовых искажений** phase compensator, phase equalizer, transit time equalizer

**коррекционный** *adj.* correction

**коррекция** *f.* correction, compensation, normalization, equalization, equalizing; **к. бочкообразных искажений растра** antibarrelling; **к. в катодной цепи** cathode compensation, cathode correction; **к. времени нарастания** rise time correction; **к. высоких частот параллельным контуром** shunt peaking correction; **к. геометрического искажения кадра** frame keystone correction; **к. градации яркости** gamma correction, brightness correction; **к. для работы с фокусировкой на бесконечность** infinite focus correction; **к. для уравнивания углов** equi-signal correction; **к. для устранения влияния нагрузки на потенциометр** potentiometer-loading correction; **к. искривления изображения вдоль строк** line bend correction; **к. искривления изображения по кадру** frame bend correction; **к. кадровой динамической сходимости** vertical dynamic convergence correction; **к. нелинейности** linearity correction; **обратная к.** de-emphasis; **к. паразитного сигнала передающей трубки** shading correction; **к. по входу** pre-accentuation; **к. по выходу** postequalization; **к. по дальности** range normalization; **к. по закону, обратному закону корректируемой цепи** inverse-power correction; **к. по закону, обратному квадратичному** inverse square-law correction; **поперечная к.** shunt admittance type equalization; **к. последовательного включения** equalization in series; **последующая к.** de-emphasis, postequalization, post-accentuation; **к. послесвечения в канале зелёного (красного, синего) цветоделённого изображения** green (red, blue) afterglow correction; **к. послесвечения люминофора** phosphor decay correction; **предва-**

**рительная к.** pre-accentuation, pre-equalization, pre-emphasis, pre-compensation, pre-distortion; **предварительная к. нелинейности** gamma pre-correction; **к. строчной динамической сходимости** horizontal dynamic convergence correction; **к. тёмного пятна** shading correction *(tv)*; **к. трапецеидального искажения строк** line keystone correction; **к. трапецеидальных искажений** keystone correction, trapezium correction; **к. фазовых искажений** transit time correction; **к. фединга** automatic volume control, automatic gain control; **частотная к. усилителя** accentuation in an amplifier; **к. частотной характеристики видеодетектора в области верхних частот** video-detector peaking; **к. частотной характеристики видеоусилителя в области верхних частот** video(-amplifier) peaking; **к. частотных искажений** response equalization

**коррелированный** *adj.* correlated

**коррелировать** *v.* to correlate

**коррелограмма** *f.* correlogram

**коррелятивный** *adj.* correlative

**коррелятограф** *m.* correlatograph

**коррелятор** *m.* correlator; **к. для анализа речи** speech-waveform correlator; **к. с большим временем корреляции** long-time correlator; **к. с малым временем корреляции** short-time correlator

**корреляционный** *adj.* correlation

**корреляция** *f.* correlation; **к. вдоль потока** downwind correlation; **взаимная к.** cross-correlation; **к. между членами совокупности** intercorrelation; **к. по ветру** downwind correlation; **схоластическая к.** nonsense correlation

**корреспондировать** *v.* to correspond

**корреспондирующий** *adj.* corresponding

**коррозиестойкий** *adj.* corrosion-resising, rustproof; stainless (steel)

**коррозиеустойчивость** *f.* corrosion resisting quality

**коррозиеустойчивый, коррозийностойкий** *adj.* rustproof, corrosion-resisting; stainless (steel)

**коррозийный** *adj.* corrosion; corrosive

**коррозионно-** *prefix* corrosion-

**коррозионный** *see* **коррозийный**

**коррозия** *f.* corrosion; **точечная к.** pitting

**корток** *abbr.* (**ток короткого замыкания**) short-circuiting current

**корытный** *adj.* trough, trough-shaped; channel

**корытообразный** *adj.* trough-shaped, trough

**косвенно** *adv.* indirectly, obliquely

**косвенно-возбуждаемый** *adj.* indirectly-excited

**косвеннодействующий** *adj.* indirect

**косвенный** *adj.* indirect, oblique; cross

**косеканс** *m.* cosecant

**косекансно-квадратичный** *adj.* cosecant-squared

**косекансный** *adj.* cosecant

**косинус** *m.* cosine; **к.-компенсатор** aperture corrector with delay-line; **к. фи** power factor

**косинусный** *adj.* cosine

**косинусоида** *f.* cosine wave, cosine curve

**косинусоидальный** *adj.* cosinusoidal

**космический** *adj.* cosmic

**космогонид** *m.* cosmogonid

**космогония** *f.* cosmogony

**космонавигация, космонавтика** *f.* astronautics, interplanetary navigation

**космотрон** *m.* cosmotron

**косность** *f.* inertness, inertia, persistence

**коснуться** *pf. of* **касаться**

**косо** *adv.* obliquely, slantwise, sideways, on a bias, askew, askance

**косозаглушённый** *adj.* transversally-damped

**косой** *adj.* inclined, slanting, sloping; splayed; diagonal, transverse; oblique; skew, sidelong

**косо-отражённый** *adj.* oblique-incidence

**кососвет** *m.* angle lighting fitting, corner reflector

**кососиметричный** *adj.* skew-symmetric, antisymmetrical

**косоугольный** *adj.* bevel, beveled, canted; oblique-angled

**Коста Рибейро** Costa Ribeiro

**косточка** *f.* (small) bone; bushing (*elec.*); **слуховые косточки** oscicles; **к. на ведущей пружине (реле)** lever bushing (of a relay); **к. на пружине (реле)** spring bushing (of a relay); **к. поднимающая пружину (реле)**

spring operating bushing (of a relay); **к. с обозначением (у зажимов)** marking bush

**костылеобразный** *adj.* crutch-shaped

**костыль** *m.* peg, pivot, spike, pin, cotter; crutch

**костыльный** *adj.* cotter, pin, spike, cramp; crutch

**кость** *f.* bone

**костяной** *adj.* bone

**косяк** *m.* shoal (of fish)

**котангенс** *m.* cotangent

**котёл** *m.* kettle, boiler; reactor (*nucl.*); **к.-утилизатор** waste-heat boiler

**котельный** *adj.* kettle, boiler

**который** *rel. and interr. pron. adj.* which?, what?, which, that, who

**коуш** *m.* thimble, iron thimble; eye, eye ring; **к. для оттяжки** stay thimble, guy thimble

**Кох** Koch

**Кохенбургер** Kochenburger

**кохерер** *see* **когерер**

**кохлеарный** *adj.* cochlear

**Коши** Cauchy

**кошка** *f.* car, trolley, carriage (of crane); grab, grapnel

**кошки** *pl.* pole climbers

**коэлостат** *m.* coelostat

**коэнергия** *f.* co-energy

**коэрциметр** *m.* coercimeter, coercive force meter

**коэрцитивность** *f.* coercivity

**коэрцитивный** *adj.* coercive

**коэф.** *abbr.* (**коэффициент**) coefficient, factor, ratio

**коэффициент** *m.* coefficient, factor, ratio, multiplier; **трихроматические коэффициенты цветового уравнения** tristimulus values; **удельные коэффициенты равноэнергетического монохроматического излучения** distribution coefficients; **к. амплитуды импульса** crest factor of a pulse; **к. влияния земли на диаграмму антенны** pattern propagation factor; **к. возврата реле** drop-off-to-pick-up ratio; **к. восприимчивости** incremental response index, incremental response characteristic; **к. выпрямления** rectification factor, coefficient of detection; value ratio; **к. выпрямления кристалла** crystal ratio; **к. глубины модуляции видеосигнала** picture modulation percentage; **к. готовности** in-commission rate; к. действующей вторичной эмиссии collected-current ratio; к. деления частоты scaling ratio; диффузный к. поглощения random-incidence absorption coefficient; к. естественной освещённости daylight factor; к. загрузки duty factor; к. занятия call fill (*tphny.*); к. запаса (прочности) factor of assurance, safety factor; к. заполнения space factor, fill factor, stacking factor, activity coefficient; mark-to-space ratio; к. заполнения импульсного сигнала mark-to-space ratio; к. защищённости front-back ratio; к. звукоизоляции acoustical reduction coefficient, sound reduction factor; к. звукопоглощения, измеренный методом гулкой камеры reverberation sound absorption coefficient; к. зеркальных помех image interference ratio; к. изменения ёмкости конденсатора от частоты capacity frequency factor; к. импульсного цикла pulse duty factor; к. ионного усиления (газового фотоэлемента) gas amplification factor (of a gas phototube); к. искривления рупора flare factor; к. использования utilization factor; fill capacity factor; к. использования полосы частот канала wide-band ratio; к. использования ρ радиолампы оконечного усилителя plate efficiency in power amplifier stage; к. использования установки coefficient of utilization; к. компрессии полной проводимости transadmittance compression ratio; к. концентрации load factor, capacity factor, demand factor; к. лобового сопротивления drag coefficient; к. лучепоглощения immissivity; к. лучистого отражения radiant reflectivity, radiant reflectance; к. наблюдения observed differential; к. нагрузки demand factor; relative severity factor; к. направленного действия directive gain, directivity factor; к. направленного действия антенны front-to-rear ratio; к. направленности directivity index, directivity factor, direction gain, space factor; front-to-rear ratio; к. направленности антенны оптического телефона reflector gain of an optophone; к. направленности для полосы излучения частот в заданном диапазоне частот transmitting band directivity factor; к. направленности по полю pattern-propagation factor; к. нелинейных искажений distortion factor; к. неравномерности variation factor; к. несогласованности mismatching factor, transition factor, reflection factor; к. неустойчивости wobble factor; к. обратного рассеяния backscattering coefficient; к. обратной связи reaction coefficient, feedback factor; к. одновременности (нагрузки) demand factor; к. опрессовки bulk factor; к. ослабления по зеркальному каналу image attenuation coefficient; к. ослабления промежуточной частоты intermediate-frequency rejection factor; к. отбора вторичной эмиссии collected-current ratio; относительный к. направленности по полю pattern-propagation factor; к. отражения баланса balance return loss; к. отражения вследствие неоднородности линии irregularity return loss; к. отражения для видимой части спектра luminous reflectivity; к. отражения дна bed reflection coefficient; к. отражения на выходе четырёхполюсника reflection factor on the fourpole network output; к. отражения оконечной аппаратуры terminal return loss; к. отражённого рассеяния backscattering coefficient; к. отрицательной обратной связи degeneration factor; к. первичного распределения тока primary current ratio; к. передачи напряжения transmission gain; к. передачи преобразователя transconductance of a converter; к. перекрёстных искажений cross-talk factor; к. питания многократной антенны feed ratio of a multiple-tuned antenna; к. планирования drag-lift ratio; повторный к. затухания звена iterative attenuation constant per section; к. поглощения для свободной волны free-wave absorption coefficient; к. поглощения поверхности стен wall coefficient; к. покрытия полюса ratio pole arc, pole pitch; к. полезного действия efficiency factor, efficiency output; полный к. излучения total emissivity of a thermal radiator; к. **понижения частоты (импульсов)** scaling ratio; к. **поправки на дифракцию** pattern factor; **поправочный к. на пульсацию** pulsating-correction factor; к. **последействия** residual-induction coefficient; к. **потерь в диэлектрике** dielectric power factor; к. **потерь на гистерезис** hysteresis constant; к. **потокосцепления** (Hopkinson's) linkage coefficient; к. **преломления в области инфракрасного излучения** infrared refraction index; к. **приспособления** accommodation coefficient; к. **проницаемости электрода** electron-stream transmission efficiency; к. **простоя** down time ratio; к. **пульсирующего напряжения** percent ripple voltage; к. **разветвления** pyramiding factor; к. **разложения** resolution ratio; к. **разновременности** diversity factor; к. **разностного тона** intermodulation factor; к. **раскрыва рупора** flare factor; к. **реверберации в заданном диапазоне частот** band reverberation factor; к. **связи группирователя** buncher coupling factor; к. **скоса пазов** skew factor; к. **скрутки** lay ratio; к. **слоистости** lamination factor; к. **смешанной корреляции** coefficient of determination; **спектральный к. излучения** transmitting spectrum factor, transmitting spectrum index, spectral emissivity of a thermal radiator; **суммарный к. расхода установившегося потока** overall steady-flow coefficient; **температурный к. диэлектрической проницаемости** temperature coefficient of permittivity; к. **трансформаторный связи** coefficient of indirect magnetic coupling; **угловой поправочный к.** phase-angle correction factor; к. **укорочения шага** pitch factor; к. **уменьшения дальности** range-attenuation factor; к. **умножения электронного умножителя** multiplication factor of a multiplier type of tube; к. **уравнения возмущённого движения** perturbation coefficient; к. **усиления** gain, gain factor, amplification factor; к. **у. антенны** gain of an antenna; к. **у. контура** magnification factor of a circuit; к. **у. по** замкнутому контуру loop gain; к. **у. по мощности при согласовании (на выходе)** matched(-output) power gain; к. **у. по току в режиме короткого замыкания в схеме с общей базой (с общим эмиттером)** collector-to-emitter (collector-to-base) short-circuit current amplification factor; к. **у. электронного умножителя** multiplier gain; к. **утолщения диполей** reduction-factor of dipole elements; к. **ухудшения надёжности** degradation factor; к. **формы кривой** form factor; к. **формы поперечного сечения** cross-sectional shape factor; к. **частотного отклонения** deviation ratio (of a modulation frequency); **шаговый обмоточный к.** pitch factor; к. **шумов** noise factor; **электроакустический к. полезного действия** acousto-electric index; к. **электроакустической связи** electro-acoustical force factor; к. **электронной связи** beam-coupling factor

**КП** *abbr.* 1 (**командный пункт**) command post; 2 (**компасный пеленг**) compass bearing; 3 (**коробка передач**) gear box, gear housing

**КПД, к.п.д.** *abbr.* (**коэффициент полезного действия**) efficiency, effectiveness; **к.п.д. электрического преобразователя** projector efficiency

**КПУ** *abbr.* (**квантовый парамагнитный усилитель**) paramagnetic amplifier

**краевой** *adj.* edge, rim, fringe, margin; end (effect); contact; outer; marginal; fringing

**кразер** *m.* craser

**край** *m.* edge, ledge, rim, border, margin, boundary; periphery; tip, end, fringe, side, lip; **заострённый к.** sharp-edged lip; **мешающий к. полосы** interference fringe

**крайне** *adv.* extremely, very, highly

**крайний** *adj.* extreme, utmost; last, end, terminal, ultimate; outer, outside; urgent

**Крамер** Cramer

**крамеровский** *adj.* (of) Kramers, Kramers'

**Крамерс** Kramers

**кран** *m.* cock, tap, faucet, stopcock; crane, dolly; к. **на два направления** two-way cock; **к.-штатив** camera crane

**краник** *m.* cock
**краниометр** *m.* craniometer
**кранный** *adj.* cock, stopcock, faucet
**крановый** *adj.* crane
**крапинка** *f.* tracer, identification thread; speck, spot
**крарупизация** *f.* Krarup loading, continuous loading, distributed inductance; **клинообразная к.** taper loading
**крарупизированный** *adj.* continuously loaded
**крарупизировать** *v.* to load continuously
**краруповский** *adj.* Krarup's, (of) Krarup
**краситель** *m.* dye, dyestuff; pigment, colorant, coloring material
**краска** *f.* paint, dye, color, pigment; ink
**краскораспылитель** *m.* paint sprayer
**красно-оранжевый** *adj.* red-orange
**красный** *adj.* red
**красочный** *adj.* color; inking
**красящий** *adj.* coloring, dyeing; inking
**кратер** *m.* crater; target
**кратерный** *adj.* crater; target
**краткий** *adj.* short, short-form, brief
**кратковременный** *adj.* short, of short duration, short-lived; temporary, momentary, transient, transitory
**краткосрочный** *adj.* short, short-term, short-range
**кратное** *n. adj. decl.* multiple (*math.*); **общее наименьшее к.** least common multiple, lowest common multiple
**кратность** *f.* multiplicity factor; rate, ratio; **к. насыщения** overcurrent factor; **к. резервирования** redundancy rate; **к. тока выплавления** fusing factor
**кратный** *adj.* multiple; *n*-к. *n*-uple
**крафт-бумага** *f.* Kraft paper
**крацевание** *n.* cutting down, polishing
**крашеный** *adj.* dyed, colored, painted; sprayed
**креативный** *adj.* creative
**крейсерский** *adj.* cruising
**крейцкопф** *m.* cross-head
**крейцмейсель** *m.* cross-cut chisel, groove chisel
**крейцшпуля** *f.* cross coil
**кремальера** *f.* rack, rack and pinion, rack gear; lens mount
**кремень** *m.* flint, flint nodule; silica, silex

**Кремер** Krämer
**кремне-, кремнево-** *prefix* silico-
**кремнёвый** *adj.* silicon, silicic, siliceous
**кремнекислота** *f.* silicic acid
**кремнекислый** *adj.* silicate (of); silicic acid
**кремнеорганика** *f.* organosilicon; **к. для печатных схем** printed-circuit silicon
**кремнеорганический** *adj.* organosilicon
**кремниево-карбидный** *adj.* silicon carbide
**кремниевый** *adj.* silicon, silic, siliceous
**кремний** *m.* silicon, Si; pebble, flint; **к.-органический материал** silicon
**кремнисто-** *prefix* silico-
**кремнистый** *adj.* siliceous, flinty; silicide (of)
**креномер** *m.* inclinometer
**креозот** *m.* creosote
**креозотовый** *adj.* creosote
**крепёж** *m.* bracing, bracketing, fastening; brackets; attachments
**крепёжный** *adj.* fastening, holding, mounting; reinforcing
**Крепен** Crepin
**крепить, подкрепить** *v.* to strengthen, to fasten, to fix, to brace; **к. раскосами** to stay, to strut
**крепкий** *adj.* strong, firm, solid, hard, tough, sturdy
**крепление** *n.* mount, mounting, bracketing, bracing, strengthening; attachment, linkage; support, reinforcement; **жёсткое к.** stiffening; **к. на шайбах** beaded support; **к. распорками** strut(ting); **к. (тела преобразователя) в точке узла колебаний** nodal mounting
**крепость** *f.* strength, stability, toughness
**крепящий** *adj.* strengthening, supporting, bracing
**крест** *m.* cross, four-way piece; **визирный к.** cross lines; **к.-накрест** crisscross; **к. нитей** cross spiderline, cross lines
**крестатрон** *m.* crestatron
**крестовидный** *adj.* cross-shaped, cruciform
**крестовина** *f.* cross, cross piece, cross beam, crossover, crossing-over, spider, center-piece; **воздушная к.** stay clamp, overhead crossing; **к. ротора** field spider
**крестовый** *adj.* cross

**крестообразно** *adv.* crosswise
**крестообразный** *adj.* cross-shaped, cruciform
**крешендо** *adv.* crescendo
**кривая** *f. adj. decl.* curve, line, plot; characteristic; **кривые равной освещённости** isolux curves; **кривые цветового возбуждения** tristimulus value curves; **к. бинаурального порога слышимости** binaural curve; **к. в форме розетки** rosette (electron) orbits; **к. верности воспроизведения** fidelity curve; **к. видности** luminance function; **к. входящего тока** arrival curve; **к. выбега** deceleration curve, coasting curve; **градуировочная к.** calibration curve; **к. зависимости амплитуды от частоты** amplitude-versus-frequency curve; **к. зависимости магнитной проницаемости от температуры** permeability-temperature curve (*see* **зависимость**); **к. зависимости натяжения от стрелы провеса** sag-tension curve; **к. зависимости натяжения от стрелы провеса при разных нагрузках** pull-up curve; **к. изменения тока в зависимости от времени** current-time curve; **импульсная посадочная к.** pulsed glide path, pulsed glide slope; **к. нарастания тока на конце кабеля** arrival curve (of current); **к. остаточного намагничивания** residual flux density curve; **к. ответвления** branch-line curve; **к. относительного спектрального распределения величины излучения** relative spectral distribution curve of a radiometric quantity; **к. относительного спектрального распределения фотометрической величины** relative spectral distribution curve of a photometric quantity; **к. погони** dog-leg path; **посадочная к.** glide slope, glide path, landing beam; **к. предельных значений сеточного напряжения** critical-grid-voltage curve; **к. пути по времени** time-displacement curve; **к. равной освещённости** isolux line, isophot; **к. равной силы света** isocandela curve; **к. распределения излучения в функции угла места** angular intensity curve; **к. распределения энергии излучения по спектру** spectroradiometric curve;

**резонансная к. полосового фильтра** band-pass resonance curve, band-pass response; **сложная к. тока** total complex current wave; **спектральная к. относительной видности** luminosity curve; **к. спектрального распределения величины излучения** spectral distribution curve of a radiometric quantity; **к. срока службы** life curve; **к. трафика** incidence of traffic curve; **тупая к. резонанса** broad resonance curve; **к. цветности внутри цветового треугольника** Planckian locus
**кривизна** *f.* curvature, curve, camber, flexure; warping, buckling; bend, bending, slope
**кривить, покривить, скривить** *v.* to bend, to curve, to flex; to distort
**кривой** *adj.* bent, crooked, buckled, curved, curve
**криволинейный** *adj.* curvilinear, curved, curvilineal
**кривошип** *m.* crank, crankshaft; **ручной к.** crank handle
**кривошипный** *adj.* crank, crankshaft
**Крид** Creed
**криогеника** *f.* cryogenics
**криозистор** *m.* cryosistor
**криолит** *m.* cryolite
**криометр** *m.* cryometer
**криосар** *m.* cryosar
**криоскоп** *m.* cryoscope
**криоскопический** *adj.* cryoscopic
**криоскопия** *f.* cryoscopy
**криостат** *m.* cryostat
**криотрон** *m.* cryotron; **входной к.** read-in cryotron; **выходной к.** read-out cryotron; **плёночный к.** planar cryotron; **плёночный к. без сверхпроводящего экрана** open-field cryotron; **плёночный к. со сверхпроводящим экраном** closed-field cryotron; **поперечный плёночный к.** crossed film cryotron, cross-strip cryotron; **проволочный к.** wire-wound cryotron; **продольный к.** in-line cryotron; **разрешающий к.** enable cryotron
**криотроника** *f.* cryotronics
**криотронный** *adj.* cryotron
**криохирургия** *f.* cryosurgery
**криптанализ** *m.* cryptanalysis
**крипто-** *prefix* crypto-
**криптовалентность** *f.* cryptovalence
**криптограмма** *f.* cryptogram

**криптографический** *adj.* cryptographic
**криптодетерминированный** *adj.* crypto-deterministic
**криптометр** *m.* cryptometer
**криптон** *m.* krypton, Kr
**криптоскоп** *m.* kryptoscope
**криптоцианиновый** *adj.* cryptocyanine
**кристадин** *m.* oscillating crystal receiver
**кристалл** *m.* crystal; **двойниковые кристаллы скольжения** twin crystal resulting from gliding; **германиевый к. с собственной проводимостью** intrinsic germanium crystal; **двупреломляющий к.** uniaxial crystal; **к., (колеблющийся) на гармонике** overtone crystal; **к.-призрак** phantom crystal; **к., работающий с одной стороны в воздух** air-backed crystal; **к. с постоянной точкой** fixed crystal; **к. х-среза** X-cut crystal
**кристаллизационный** *adj.* crystallizing, crystallization
**кристаллизация** *f.* crystallization, crystallizing
**кристаллик** *m.* dice, chip
**кристаллический** *adj.* crystalline, crystallized, crystal
**кристалловый** *adj.* crystal
**кристаллограф** *m.* crystallograph
**кристаллографический** *adj.* crystallographic
**кристаллография** *f.* crystallography
**кристаллодержатель** *f.* crystal holder, crystal mount
**кристаллолюминесценция** *f.* crystalloluminescence
**кристаллометр** *m.* crystallometer
**кристаллометрический** *adj.* crystallometric
**кристаллообразование** *n.* yielding of crystals
**кристаллооптика** *f.* crystal optics
**кристаллофосфоры** *pl.* crystal phosphors
**кристаллохимия** *f.* crystal chemistry
**кристальный** *adj.* crystal, crystalline
**Кристоффель** Christoffel
**критерий** *m.* criterion; measure; number, group; **квадратичный к. ошибок** error-squared criterion; **к. на основе двойной выборки** two-sample test
**критически** *adv.* critically
**критический** *adj.* critical, crucial
**критичность** *f.* criticality

**КРМ** *abbr.* (**курсовой радиомаяк**) radio-range beacon
**«крокодил»** *m.* alligator clip
**кролит** *m.* crolite
**кромаг** *m.* cromag
**кромка** *f.* edge, border; rim, brim; bead, shoulder; porch; **к. колбы** bulb shoulder; **режущая к.** (steel) cutting edge
**кромочный** *adj.* edge, border, hem; rim, brim; edging
**Кронекер** Kronecker
**кронциркуль** *m.* outside calipers
**кронштейн** *m.* bracket, arm, cantilever, outrigger; stand, truss, hanger; **двойной к. для изолятора** double insulator spindle; **к. для скрещивания** transportation bracket; **оконечный кабельный к.** cable support rack
**Кросби** Crosby
**кросс** *m.* distributing frame; terminal room; **к. для искателей** selector distributing frame; **трансляционный к.** repeater distributing frame (*tphny*.)
**кроссбар** *m.* crossbar system; **к.-искатель** crossbar selector
**кроссировка** *f.* cross-connections, cross-cuts, cross-connection field
**кроссировочный** *adj.* cross-connection, jumper
**кросскорреляция** *f.* crosscorrelation
**кроссмодуляция** *f.* cross-modulation
**кроссовер** *m.* crossover, crossing over
**кроссовый** *adj.* cross, cross-connection, jumper
**Кроу** Crowe
**кроулей** *m.* crowley
**кроющий** *adj.* covering; coating
**КРП** *abbr.* (**компасный радиопеленг**) compass radio bearing
**круг** *m.* circle, ring, disk, reel, wheel; circumference; range, scope; period, cycle; **круги в качестве линий основания** circular lines of position; **к. телеграфной ленты** telegraph reel
**круглогубцы** *pl.* round nosed pliers
**круглополяризованный** *adj.* circularly polarized
**круглопроволочный** *adj.* round-wire
**круглосуточно** *adv.* round-the-clock
**круглосуточный** *adj.* continuous, twenty-four-hour, round-the-clock
**круглый** *adj.* circular, round, spherical, annular, globular
**круговой** *adj.* circular, round, ring;

circling, cyclic; continuous, endless; circumferential

**круговорот** *m.* rotation, circular motion; turn over, cycle; circulation

**круговращательный** *adj.* rotary, circular, circulatory

**круговращение** *n.* rotation; circulation, circular motion

**кругозор** *m.* horizon, scope; range, range of vision

**кругом** *adv.* round, around, about; in a circle

**кругонаправленный, кругообзорный** *adj.* omnidirectional

**кругооборот** *m.* circuit, cycle; circulation

**кругообразный** *adj.* round, circular

**кругообращение** *n.* rotation, circular motion; circulation

**кружение** *n.* eddy, turning, spinning

**кружить, вскружить, закружить** *v.* to turn, to spin around, to rotate, to wheel

**кружиться, закружиться** *v.* to revolve, to rotate, to turn, to spin; to circulate

**кружок** *m.* small circle; disk; **к. рассеяния** figure of confusion

**Крукс** Crookes

**круксов** *poss. adj.* Crookes

**крупитчатость** *f.* noise effects on radar scopes

**крупнозернистый** *adj.* coarse-grained, coarse

**крупноклетчатый** *adj.* wide-mesh(ed)

**крупнокристаллический** *adj.* macro-crystalline

**крупномасштабный** *adj.* large-scale; broad-scale

**крупносерийный** *adj.* large-scale (production)

**крупный** *adj.* big; large-scale, vast; heavy-duty (machine)

**крутизна** *f.* steepness, sharpness, slope; transconductance, mutual conductance; **с переменной крутизной** variable-mu; **к. анодного тока по напряжению управляющей сетки** control grid gate transconductance; **к. анодносеточной характеристики** grid-plate transconductance, mutual conductance; **боковая к. приёмника** steepness of receiver response curve; **к. катодного тока** ratio of cathode current change to grid voltage change at constant plate voltage; **комплексная к.** transadmittance; **к. наклона** характеристики затухания cut-off attenuation rate; **настроечная к. линий** slope of the wavelength curves of transmission-lines; **обратная к.** transimpedance, inverse transconductance, mutual resistance; **к. передаточной характеристики собственно транзистора** intrinsic transconductance; **переменная к.** variable mu; **к. перепада** jump steepness; **к. преобразования преобразователя** transconductance of converter; **к. ската кривой затухания (фильтра)** cut-off attenuation rate; **к. характеристики анодного тока в функции экранного напряжения** screen-plate transconductance; **к. характеристики при начале колебаний** minimum transconductance required to start oscillations

**крутильный** *adj.* torsion, torsional, twisting, torque

**крутить, закрутить, скрутить** *v.* to turn, to twist, to whirl, to ring; to reverse

**крутиться, закрутиться** *v.* to turn, to spin, to gyrate, to eddy, to whirl

**крутой** *adj.* steep, sudden, sharp, abrupt, fast; severe

**крутящий** *adj.* torsion, torsional, twisting; torque; **к. момент** torque

**кручение** *n.* torsion, twisting; distortion

**крылатка** *f.* vane; wing nut

**крыло** *n.* blade, vane; wing; **пригласительное к. семафора** slow-speed arm of a semaphore

**крыловидный, крылообразный** *adj.* wing-shaped, alar

**крылышко** *n.* vane; wing

**крыльевой** *adj.* wing

**крыльчатка** *f.* vane wheel, blade wheel

**крыльчатый** *adj.* wing, vane; blade

**крыть, покрыть** *v.* to cover, to coat

**крышевидный** *adj.* roof-shaped

**крышевый** *adj.* roof

**крышечный** *adj.* cover, cap, lid, hood

**крышка** *f.* cap; cover, lid, hood, top; **вторая к. колодца** catch pan; **к. зажимной колодки счётчика** meter terminal cover; **к. колодца** cover disk, cover slab; **к. люка** chute gate; **нарезная к. пробки** screw-plug cartridge fuse carrier

**крышный** *adj.* roof

**Крюгер** Krueger

**крюйс-пеленг** *m.* cross-bearing

**крюк** *m.* hook, crook; hook-shaped bracket, cramp iron; pick, pickax; **к. для изолятора** swan-neck spindle, cupholder pin, hook-shaped pin; **к. для открывания люков** manhole hook, manhole bar; **к. с пружинным замком** snap hook

**крючки-подвески** *pl.* cable suspenders

**крючкообразный** *adj.* hooked, hook-shaped, hook-like, crooked

**крючный** *adj.* hook

**крючок** *m.* hook, catch; lifting handle; **карабинный к.** snap hook, swivel; **к.-переключатель** switch-hook (*tphny.*); **спусковой к.** trigger

**ксатрон** *m.* xatron

**КСВ** *abbr.* (**коэффициент стоячей волны**) standing wave ratio

**КСВН, к.с.в.н.** *abbr.* (**коэффициент стоячей волны напряжения**) (*or* **по напряжению**) voltage standing-wave ratio

**ксеноморфный** *adj.* xenomorphic

**ксенон** *m.* xenon, Xe

**ксеноновый** *adj.* xenon

**ксерографический** *adj.* xerographic

**ксерография** *f.* xerography; xeroprinting

**ксерорадиография** *f.* xeroradiography

**ксилен** *m.* xylene

**ксилол** *m.* xylene

**ксилометр** *m.* xylometer

**ксилофон** *m.* xylophone

**КТО** *abbr.* (**контрольно-технический осмотр**) control inspection

**КУ** *abbr.* 1 (**курсовой угол**) course angle; 2 (**квантовый усилитель**) quantum mechanical amplifier

**куадрадар** *m.* quadradar

**куб** *m.* cube; vat; still; stack; **в кубе** cubed; **к. с центрированными гранями** face-centered cube

**кубатура** *f.* cubic content, volume; cubature

**кубит** *m.* cubit

**кубический, кубичный** *adj.* cubic, cubical

**кубовидный** *adj.* cubical, cuboid, cube-shaped

**кубовый** *adj.* cube; vat; still

**кубо-пирамидальный** *adj.* cubical pyramid

**кувалда** *f.* sledge hammer

**кувыркающийся** *adj.* tumbling

**Куде** Coudé

**кузов** *m.* basket; body (of a vehicle); hood (of a car); **антенный к.** antenna trailer

**кул.** *abbr.* (**кулон**) coulomb

**кулак** *m.* clamp, terminal, binding post, cam, lug

**кулачковый** *adj.* cam, cog, lug, pin, tooth, finger; catch, detent; tappet

**кулачкообразный** *adj.* cam-shaped, jaw-shaped

**кулачок** *m.* cam, cog, lug, pin, finger; pawl, catch, detent, checking device; tappet; bit, jaw; **к. в переключателе** "cat-head"; **двухдисковый к.** double profile cam; **к. для регулирования давления** compression relief cam; **зажимной к.** clamping jaw; **к. компенсации квадратичной зависимости** square root compensating cam; **обкатывающий к.** cam follower pin; **плоский к.** two-dimensional cam; **пространственный к.** three-dimensional cam; **регулирующий к. для понижения давления** half-compression relief cam; **к. токораспределения** sequence switch cam; **к. холостого хода** slipping cam (in a dial), delayed pulse tripping cam

**Кулидж** Coolidge

**кулиса** *f.* link, connecting link, crank; slot, hole, slide(way); **шумовая к.** background effect source

**кулометр** *m.* coulometer

**кулон** *m.* coulomb

**Кулон** Coulomb

**кулонметр** *m.* coulometer

**кулоновский** *adj.* coulombian; Coulomb

**кулонометрический** *adj.* coulometric

**кульминация** *f.* culmination

**культурно-просветительный** *adj.* educational

**кумароновый** *adj.* cumar, coumaric

**куметр** *m.* Q-meter (*elec.*)

**куметрический** *adj.* Q-metric

**кумулятивный** *adj.* cumulative

**«кунайф»** *m.* cunife

**Кундт** Kundt

**куниаль** *m.* kunial

**кунико** *n.* cunico

**Кунсмэн** Kunsman

**купол** *m.* dome, cupola, bowl

**куполообразный** *adj.* dome-shaped, arched

**купольный** *adj.* cupola, dome

**купрокс** *m.* copper oxide rectifier; cuprous oxide

**купроксный** *adj.* cuprous oxide
**купрон-элемент** *m.* caustic soda cell, copper-oxide cell, cupron cell
**КУР** *abbr.* (**курсовой угол радиостанции**) radio station angle of approach
**курбель** *m.* knob
**курвиметр** *m.* curvometer, opisometer, rotameter
**курковый** *adj.* snapping; (gun) cock
**курс** *m.* course; track, path; **к. по координатной сетке** grid course; **к. по маяку** beacon course; **к. по неизменному азимуту** directional homing
**курсовой** *adj.* course
**курсограф** *m.* course recorder, avigraph
**курсомер** *m.* course calculator
**курсоуказатель** *m.* course indicator
**Курц** Kurtz
**КУС** *abbr.* (**комбинированный указатель скорости**) combined airspeed indicator
**кусачки** *pl.* cutting pliers, nippers, wire cutter
**кусок** *m.* piece, fragment, block, bar, chunk; length
**кусочечный** *adj.* piecewise
**кусочно-линейный** *adj.* line-segment

**кусочно-ломаный** *adj.* broken-line
**кусочный** *adj.* piecewise
**кустарный** *adj.* homemade
**кустование** *n.* interconnection, interconnecting
**кустованный** *adj.* interconnected; bank (*tphny.*)
**Кутта** Kutta
**кухня** *f.* kitchen, range
**кучность** *f.* cluster, concentration
**кучный** *adj.* cluster
**Кью** Kew
**КЭ** *abbr.* (**кинетическая энергия**) kinetic energy
**кэб-индикатор** *m.* cab indicator
**Кэв** *abbr.* (**килоэлектрон-вольт**) kiloelectron volt
**Кэди** Cady
**Кэмпбелл** Campbell
**«кэпстен»** *m.* capstan
**кэрбинг** *m.* curbing (*tgphy.*)
**Кэри** Carey
**кюбит** *m.* cubit
**кювета, кюветка** *f.* cell, container, vessel, tank; bulb; tray
**Кюне** Kuene
**кюри** *m. indecl.* curie
**кюрий** *m.* curium, Cm
**кюритерапия** *f.* Curie therapy

# Л

л. *abbr.* (левый) left; counterclockwise

лаб. *abbr.* (лаборатория) laboratory; test field

лабиальный *adj.* labial

лабиринт *m.* labyrinth, maze

лабиринтный, лабиринтоый *adj.* labyrinth, maze

лабиринтообразный *adj.* labyrinthine

лаборант *m.* tester, research man; laboratory worker

лаборатория *f.* laboratory; гидроакустическая л. ВМС Naval Underwater Sound Laboratory; л. с электронно-вычислительным оборудованием computerized laboratory

лабораторный *adj.* laboratory

лава *f.* lava, lavarock

лавина *f.* avalanche; л. заряжённых частиц electron avalanche

лавинный *adj.* avalanche

лавинообразный *adj.* avalanche-type, avalanche-like

лавиннопролётный *adj.* avalanche-and-transit

лаг *m.* chip log, log, sillometer; механический вертушечный л. patent log, taffrail log

Лагерр Laguerre

Лагранж Lagrange

лад *m.* mode, fret, stop (*mus.*); harmony, accord; way, manner; плагальные лады plagal modes; мажорный л. натурального строя major scale of just temperament; мажорный л. равномерно темперированного строя major scale of equal temperament; пифагорейский л. Pythagorean scale

ладить, поладить *v.* to adjust, to fit, to adapt; to tune; to repair, to prepare

ЛАЗ *abbr.* (линейно-аппаратный зал) line equipment room

лаз *m.* manhole opening; л. колодца manhole chimney

лазейка *f.* opening, gap, loophole

лазер *m.* laser; газовый л. в режиме непрерывного излучения CW gas laser; газовый л. подкачки gas injection laser; л., генерирующий видимое излучение visible beam laser; л. для работы в полевых условиях field operation laser; л. для слежения за космическими объектами space tracking laser; импульсный л. на твёрдом теле pulsed solid-state laser; л. подкачки на твёрдом теле solid-state injection laser; л. с замкнутым контуром closed-circuit laser; л. с мегаватной мощностью в импульсе megapulse laser; л. с накачкой солнечными лучами sun-pumped laser; л. с непрерывным излучением continuously-operating laser, continuous-wave laser; л. с несколькими последовательно расположенными кристаллами many-element laser; л. с одним видом колебаний single mode laser; л. с оптической подкачкой optically-pumped laser; л. со слабо выраженной дифракцией diffraction limited laser

лазерный *adj.* laser

лай *m.* barking, bark; «собачий л.» monkey chatter (interference)

Лайман Lymann

лак *m.* lacquer, varnish; lac; л. воздушной сушки, высыхающий на воздухе л. air-drying varnish; высыхающий при нагреве в печи л., л. горячей сушки baking varnish; л. для производства лакоткани baking cloth varnish; печной пропиточный л. для обмоток baking coil varnish; л. печной сушки baking varnish; пропиточный л. dipping varnish; л. с большим количеством наполнения heavily loaded varnish; л. с большим содержанием масла long oil varnish; л. с малым содержанием масла short oil varnish; термореактивный л. thermosetting varnish; тяжёлый покровный л. spar varnish; упругий

л. для отделки elastic finishing varnish; л. холодной сушки air-drying varnish
лакирование *n.* varnishing; lacquering
лакированный *adj.* varnished; lacquered
лакировать, отлакировать, полакировать *v.* to varnish; to lacquer
лакировка *f.* varnishing; lacquering; enamelling
лаковый *adj.* varnished, lacquered, japanned; varnish; lacquer
лаколента *f.* varnish tape
лакотканевый *adj.* cambric
лакоткань *f.* varnished cambric
лактобутирометр *m.* lactobutyrometer
лактометр *m.* lactometer
Лаланд Lalande
ламберт *m.* lambert
ламбда-мезон *m.* λ-meson
Ламе́ Lamé
ламель *m.* lamella, lamina; sheeting; commutator bar, commutator segment; л. коллектора, л. переключателя commutator bar, commutator segment
ламельный *adj.* lamellar, lamellate, scale-like
ламинарный *adj.* laminar, laminal
ламинирование *n.* lamination
ламинированный *adj.* laminated, laminary, lamellar
ламинировать *v.* to laminate
Ламонт Lamont
лампа *f.* lamp; bulb; tube; light; лампы с модуляцией скорости с коаксиальным резонатором coaxial-line velocity modulated tubes; лампы с модуляцией скорости электронов velocity-modulated tubes; лампы с мягкими выводами wired-in tubes; электронные лампы с дрейфовым пространством drift tubes; л. автоматической регулировки усиления automatic gain-control tube; амплитудная л. glow-tube amplitude indicator; л. АРУ automatic gain-control tube; батарейная л. dry-cell tube; л. бегущей волны traveling wave tube; безэлектродная газонаполненная л. electrodeless tube, nullode, spark-gap tube; бесцокольная пальчиковая л. baseless subminiature vacuum tube; л. бесшумной АРГ, л. бесшумной регулировки усиления squelch tube; л.-

бутафория tube simulator; быстро коммутируемая л. rapidly switched tube; вакуумная л. накаливания vacuum filament tube; л., включённая по схеме триода triode connected tube; л. возбуждающего каскада drive tube; вспыхивающая л. flash tube; л. выдержки времени time-delay tube, phantastron; л.-выключатель switch tube; выпрямительная л. rectifier tube; выпрямительная л. высокого вакуума high-vacuum tube (*or* rectifier); выпрямительная ртутная л. ignitron; высоковольтная л. для схем развёртки high-voltage scanning tube; высокочастотная усилительная л. high frequency tube; выходная л. видеоусилителя video output tube; выходная л. кадровой развёртки frame-output tube; газонаполненная л. дугового разряда gas-filled arc tube; газонаполненная счётная л. gas-tube counter; газосветная л. gas-discharge lamp, luminous gas lamp; luminous discharge tube; л.-генератор с обратной волной backward-wave oscillator; л. генератора высоковольтного выпрямителя high-voltage drive tube; генераторная л. transmitting tube, oscillator tube, generator tube, electron power tube; генераторная трёхэлектродная л. oscillion; генерирующая л. (self-) oscillating tube; двусветная л. bilux light bulb; двухэлектродная л. с накалённым катодом two-element hot-cathode tube; девятиэлектродная л. enneode; декадная л. decade-counter tube; л.-делитель частоты modulator divider; делительная л. scaling tube; л. десятичного счёта decade counting tube; дисковая л. disk-seal tube; л. для автоматической подстройки частоты transitrol; л. для незатухающих колебаний continuous-wave tube; л. для переключательных схем switch(ing) tube; л. для преобразования частоты mixing tube; л. для проверки счётчиков meter lamp; л. для сантиметровых волн microwave triode; л. для сверхвысоких частот very high frequency tube, disk-seal tube; л. для уничтожения насекомых insecticide bulb; л. для уравнивания тока current balance tube; дуговая л. в замкнутом сосуде enclosed-arc lamp; дуговая л. с вертикальными углями Debrun candle; дуговая л. с закрытой дугой enclosed-arc lamp; л. дугового разряда arc-discharge tube; л.-жёлудь peanut tube, acorn tube; л., забракованная по шуму noise-rejected tube; л. задающего каскада drive tube; закрыта л. the tube is cut off; л.-заменитель replacement tube; л. замешивания гасящих импульсов blanking-mixer tube; л. занятости busy lamp, visual engaged lamp; записывающая л. recording lamp; л. «застревающего» типа lock-in tube, loctal tube; л.-зонд Farnsworth tube; измерительная л. electrometer tube; изнутри матированная л. inside-frosted bulb; л., имеющая характеристику с острой отсечкой sharp cut-off tube; импульсная газоразрядная л. flash-discharge tube; л.-индикатор настройки tunoscope, electron-ray tube; л. каскада ограничения clipper tube; каскадная фазовращающая л. octode-type phase inverter; л.-квадратор square-law tube; керамическая л. штабельной конструкции stacked ceramic tube; л. кода «нумерации» quantizing-coding tube; комбинированная л. multiple-unit tube, multisection tube; коммутирующая л. switching tube, keyer(-type) tube; контрольная л. рабочего места position pilot lamp; контрольная л. счётчика register pilot lamp; контрольная вызывная л. ringing pilot lamp; л. контроля времени time-check lamp; л. косвенного накала cathode-type tube, indirectly heated tube; лучевая л. с поперечным полем beam-deflection tube; люминесцентная л. с мгновенным зажиганием rapid start lamp; л. манипулированного усилителя вспышки burst gate tube; матричная газоразрядная л. grid switching gas tube; л. местного гетеродина local oscillator tube; металлическая л. с пространством дрейфа metallic drift tube; микролучевая генераторная л. microray oscillator tube; миниатюрная бесцокольная л. для

УВЧ door-knob tube; многоанодная триггерная л. polyanode flip-flop tube; многодисковая л. с бегущей волной disk-on-rod-type tube; многоэлементная л. для счётных устройств computron; мощная л. бегущей волны high-power TWT; мощно-усилительная л. high-power amplifier tube; л. на выходе output tube; надёжная л., л. надёжной серии premium tube, reliable tube; л. накаливания с газовым наполнением gas-filled lamp; неоновая л. в качестве делителя напряжения glow gap divider tube; л. непрерывного генерирования continuous-wave tube; л., не требующая накала heaterless tube; л. обратной волны carcinotron, backward-wave tube; общевызывная л. ringing pilot tube; л.-ограничитель limiter tube; ограничительная л. clipper tube, threshold tube; одноконтурная пролётная л. single-circuit drift tube; оконечная л. output tube; оконечная пушпульная л. push-pull output tube; л. опорного напряжения voltage reference tube; опытная л. developmental tube; орбитальная л. orbital-beam tube; л. освещения безопасности emergency lamp; л. освещения прибора instrument lamp; отбойная л. clearing lamp, supervisory lamp; л. открывается the tube conducts; открытая л. conducting tube; отпаянная л. со связью через спираль helix-coupled sealed-off tube; пальчиковая л. bantam tube, small-button glass tube, miniature tube; паросветная л. vapor-discharge lamp, metal-vapor lamp; первая л. усилителя first amplifier tube; перекидная л. relay tube; л.-переключатель switch-tube; переключающая л. key(ing) tube, switch(ing) tube; переключающая электронная л. nomotron; л. переменной крутизны variable-mutual conductance tube, variable-mu tube, remote-cut-off tube; переносная л. inspection lamp, test lamp, land lamp; плоскоэлектродная тлеющая л. flat-plate glow tube; подопытная л. test lamp; подсвечивающая л. exciter lamp, bright-up lamp; позиционная л. position-

voltage tube; **полностью открытая л.** fully conducting tube; **поперечно-лучевая л. бегущей волны** transverse-current TW tube; **предоконечная л.** driver tube; **предоконечная л. развёртки** sweep driver tube; **предоконечная л. схемы кругового отклонения** circular-deflection driver tube; **л.-преобразователь** transducer tube, inverted tube, converter tube; **л. привязки** clamper tube; **приёмноусилительная л.** radio-receiving tube; **л. продолжительности разговора** paid time lamp, chargeable time lamp; **промежуточная л.** intertube; **пропускающая л.** gate tube; **просвечивающая л.** exciter lamp; **простая дуговая угольная л.** carbon arc lamp; **прочная л.** ширпотреба commercial ruggedized tube; **л. прямого накала** directly heated tube, battery tube, filament-type tube; **пуговичная л.** doorknob tube; **пушпульная усилительная л.** five-electrode push-pull amplifier tube, pentatron; **рабочая сигнальная л.** operating lamp, pilot lamp; **разборная л.** assembled parts tube, demountable tube; **разборная л. Хольвека** Holweck tube; **разделительная л.** pulse separator tube; **реактивная л. для подстройки** reactance tube for frequency control; **реактивная л. модуляторная** reactance tube modulator; **л.-реактор** reactance tube; **л. регулирования уровня опорных импульсов** pedestal control tube; **л. регулируемого усиления** fading tube, fading hexode; **л., регулирующая смещение развёртки** off-centering control tube; **ртутная л. с катодом и сеткой** grid-pool tube; **ртутная л. с кварцевой колбой** silica lamp; **ртутная л. с магнитным запуском** hodectron; **ртутная газосветная л.** mercury (vapor) lamp; **л. с автоматическим смещением** self-biased tube; **л. с большим коэффициентом усиления** high-mu tube, high-amplification-factor tube; **л. с большим сроком службы** long-life tube; **л. с большой крутизной характеристики** high-transconductance tube; **л. с быстрой коммутацией** rapidly-switched tube; **л. с вариацией по скорости** velocity-

modulated tube; **л. с верхним выводом** double-ended tube; **л. с взаимно заслоняющими сетками** aligned grid tube; **л. с водородным заполнением** hydrogen tube; **л. с двумя комплектами катода** multitube; **л. с дрейфовым пространством** drift tube; **л. с заземлённой базой, л. с заземлённой сеткой** grounded-grid tube; **л. с запирающимся цоколем** lock-in tube; **л. с изменением скорости** velocity-variation tube; **л. с катодной сеткой** space-charge grid tube; **л. с коммутируемым лучом** beam switching tube; **л. с ленточным пучком** sheet-beam tube; **л. с магнитным управлением** permatron; **л. с малой межэлектродной ёмкостью** low-capacitance tube; **л. с мгновенной вспышкой (для фотосъёмок)** photoflash bulb; **л. с мгновенным загоранием** instant-start tube; **л. с наружной сеткой** outer-grid tube; **л. с настроенной сеткой** aligned grid tube; **л. с объёмным зарядом** transit-time tube, traveling-wave tube; **л. с ограниченным объёмным зарядом** space-charge-limited tube; **л. с острой отсечкой характеристики** sharp cut-off tube; **л. с отклоняемым лучем** beam-deflection tube; **л. с отрицательным сопротивлением** negative-resistance tube, kallirotron; **л. с «памятью»** storage tube, memory tube; **л. с переменной крутизной** variable-mu tube, exponential tube, remote cut-off tube, controlled tube, supercontrol tube, tube with variable slope; **л. с переменной крутизной характеристики** super-control tube; **л. с плохим вакуумом** soft tube; **л. с подогревным катодом** cathode-heater tube, separate heater tube; **л. с положительной сеткой** brake-field tube; **л. с поперечным управлением** balitron tube; **л. с пространственным зарядом** space-charge tube; **л. с ртутным катодом и сеткой** grid-pool tube; **л. с сеточным управлением** grid-control tube; **л. с совмещёнными сетками** aligned-grid tube; **л. с тормозящим полем** brake-field tube, retarding-field oscillator; **л. с тусклым накалом** dim lamp; **л. с удаленной отсечкой** remote cut-off tube; **л. с ударным возбуждением** shock tube; **л. с универсальным питанием** A.C.-D.C.-tube; **л. с управлением по сетке** grid-controlled tube; **л. с усиленной жёсткостью** ruggedized tube; **л. с характеристикой в форме экспоненты** exponential tube; **л. с экономичным катодом** dull emitter tube; **л. с ярким накалом** bright-emitter tube; **сверхминиатюрная л. прямого накала** filamentary subminiature tube; **л. свободной линии** idle indicating lamp; **л. сигнала окончания операции** finish light; **л. сигнализации повреждения** alarm lamp, fault lamp; **сигнальная л. занятости всех регистров** all-sender-busy lamp; **сигнальная л. свободной линии** free line signal, idle indicating signal; **сигнальная вызывная л.** direction-indicator light; **л.-смеситель (сигналов)** mixer tube; **смесительная л. на триодепентоде** triode-heptode tube; **смесительная л. с двойным управлением** double-control converter tube; **смесительная лучевая л. с поперечным полем** beam-deflection mixer tube; **л. со скоростной модуляцией** velocity-modulated tube; **л. со слабым накалом** dull emitting tube; **л. со стальным баллоном** steel tube; **л. со стеклянными колбой и ножкой** all-glass tube; **л. со стробированным лучом** gated-beam tube; **л. со штыковым цоколем** bayonet tube; **л. «совмещения»** coincidence tube, "and tube"; **л., создающая опорное напряжение** voltage reference tube; **спектральная эталонная л.** standard light-source; **л.-стабилизатор** regulator tube, stabilizing tube; **л.-стабилизатор напряжения** voltage-regulator tube, voltage stabilizing tube; **л.-стабилизатор напряжения на принципе коронного разряда** corona regulator tube; **л. стабилизатора, стабилизирующая л.** regulator tube, voltage regulator tube, regulation tube; **стабилизованная л.** stabilizer tube; **стробированная л., стробирующая л.** gate tube; **л. схемы выключения канала**

**цветности** color killer tube; **л. схемы растягивания синхроимпульсов** sync stretch tube; **л. схемы фиксации уровня** clamper tube; **счётная л. Гейгера-Мюллера** Geiger-Müller counter tube; **термоэлектронная л.** thermionic tube; **тлеющая газосветная л.** glow (discharge) lamp; **л. тлеющего разряда** glow-discharge lamp (or tube), negative glowlamp; **л. тлеющего разряда с сеточным управлением** grid-glow tube; **л. тлеющего света** Moore lamp; **трёхэлектродная газосветная л.** gas-discharge strobo-triode lamp; **трубчатая электронная л.** rod-shaped tube; **тусклогорящая л.** dull emitter; **угольная л. накаливания** carbon filament lamp; **л., указывающая продолжительность разговора** chargeable time lamp, time check lamp; **л.-умножитель** photo-multiplier tube; **л.-умножитель функции** square-law tube; **л.-умножитель частоты** modulator multiplier; **л., управляемая флаг-импульсом** burst gate tube; **управляющая л.** control tube, driver tube, key(ing) tube, regulation-control tube; **л., управляющая пропусканием сигналов** gate tube; **усилительная л. с поглощающими стенками** resistance-wall tube; **усилительная л. тлеющего разряда** glow-discharge amplifier tube; **усилительная лучевая л. с поперечным полем** beam-deflection amplifier tube; **л. усилителя вспышки** burst gate tube; **л. усилителя промежуточной частоты** intermediate-frequency tube; **л., устойчивая против сотрясений** rough-service lamp (or tube); **л. фазирования по кадрам** vertical phasing tube; **л. формирования** driver tube; **л. формирования прямоугольных импульсов** squaring tube; **«черепаховая» л.** doorknob type tube; **читающая л.** readout tube, exciter lamp; **широкополосная приёмно-передающая л.** broadband TR tube; **эквивалентная л. при обратной связи** replacement tube in case of regenerative feedback; **экономичная л.** low-filament drain tube; **экранированная л.** screened-grid tube, shielded-grid tube, screen-grid tetrode; **экранированная л. с переменной крутизной** multimu screen-grid tube, variable-mu screen-grid tube; **электрическая л. накаливания для фотосъёмок** photoflood lamp; **электроизмерительная л., электрометрическая л.** electrometer tube; **электронная л. переменной крутизны** variable-mutual conductance tube, remote cut-off tube; **э. л. повышенной прочности** rugged electron tube; **э. л. постоянного тока** vacuum tube for D.C. filament supply; **э. л. преобразователя** transducer tube; **э. л. с дисковыми впаями** disk-seal tube; **э. л. с поперечным управлением** beam deflection tube; **э. л. с универсальным питанием** A.C.-D.C. tube; **э. л. со вторичной эмиссией** electron multiplier; **электроннолучевая переключательная л. с перекрещивающимися полями** electron-beam switch tube with cross fields; **электроннолучевая переключательная л. с трохоидальным лучом** electron-beam switch tube with trochoidal beam; **яркогорящая л.** bright emitter tube

**ламповый** *adj.* lamp; tube; light bulb

**ламповыниматель** *m.* lamp extractor, tube extractor

**ламподержатель** *m.* lamp holder, lamp socket, lamp jack; tube holder

**лампоизмеритель, лампоиспытатель** *m.* tube tester, tube checker, tube analyzer

**лампочас** *m.* tube-hour; lamp-hour

**лампочка** *f.* tube; bulb; lamp; **вызывная л. для междугородных вызовов** trunk calling lamp; **л. для осмотра пластин аккумулятора** cell inspection lamp; **л. (источника) питания** power-supply signal lamp, power light; **контрольная стрелочная л.** switch machine lever light; **л. накаливания для иллюминационных гирлянд** festoon lamp; **л. освещения приборов** instrument light; **л. подсветки** dial-lighting lamp; **сигнальная л. «в эфире»** "on-air" light; **служебная вызывная л.** order wire lamp

**Ланак** Lanac

**Ланде** Lande

**Ланжевен, Ланжевэн** Langevin

**лантан** *m.* lanthanum, La

**Ланчестер** Lanchester

**лапка** *f.* draw-vice, eccentric clamp, toggle; **л. для заделки в стену** expansion anchor, split-lug; **контактная л.** contact blade

**лапки** *pl.* draw-tongs, draw-vice

**Лаплас** Laplace

**лапласиан** *m.* Laplacian, Laplacian operator; buckling

**лапласовский** *adj.* Laplacian, Laplace

**лапчатый** *adj.* paw-shaped; finger-shaped

**ларингофон** *m.* laryngophone, throat microphone

**Ларк** Lark

**Лармор** Larmor

**Ларсен** Larsen

**латекс** *m.* latex

**латентный** *adj.* latent, dormant

**латеральный** *adj.* lateral

**латероид** *m.* fish paper

**латинизировать** *v.* to Romanize

**латинский** *adj.* Latin; Roman

**латунный** *adj.* brass

**латунь** *f.* brass; **л. для вкладышей** bearing brass

**Латур** la Tour

**лаyталь** *m.* Lautal

**Лауэ** Laue

**лауэграмма** *f.* Laue pattern, Laue diffraction pattern

**лацкан** *m.* lapel

**лацканный** *adj.* lapel

**ЛБВ** *abbr.* (**лампа бегущей волны**) traveling wave tube, TWT; **двухлучевая ЛБВ** double-stream amplifier; **ЛБВ с внутренним фокусированием** Estiatron; **ЛБВ с низкими шумами** double-stream amplifier; **ЛБВ со скрещёнными полями** crossed-field TWT

**ЛГУ** *abbr.* (**Ленинградский государственный университет**) Leningrad State University

**Лебег** Lebesgue

**лебёдка** *f.* winch, hoist, (lifting) jack, windlass; **л. с грейфом** bucket hoist

**лебедочный** *adj.* winch, windlass

**Лебедь** Cygnus

**Леблан** Le Blanc

**Леблановский** *adj.* Le Blanc

**левер** *m.* syphon, syringe

**лево-** *prefix* levo-, left

**левовращающий, левовращающийся** *adj.* left-handed, left-running, levorotatory

левозаходный *adj.* left-handed
левополяризованный *adj.* left-handed polarized
леворучной, левосторонний *adj.* left-handed, left-hand
левый *adj.* left, left-handed, left-hand; counterclockwise, levo-
легенда *f.* legend
легирование *n.* alloying, doping
легированный *adj.* alloyed, doped; alloy
легировать *v.* to alloy, to dope, to compound
легирующий *adj.* alloying
лёгкие *pl.* lungs
лёгкий *adj.* light-weight, light; easy, simple; thin, slight
легко *adv.* easily, readily; lightly; it is easy
легковесность *f.* lightness, light weight
легковесный *adj.* light-weight, light
легкоплавкий *adj.* semi-fluid; fusible, low-melting, easily melted
легкоподвижный *adj.* mobile
легкоходовой *adj.* free-running, smooth-running
лёд *m.* ice
леддик *m.* laddic
ледостойкий *adj.* sleetproof
Ледюк Leduc
ледяной *adj.* ice
лежалка *f.* couch
Лежандр Legendre
лежачий, лежащий *adj.* lying, recumbent, horizontal
лежень *m.* stay block, stay tightener; якорный л. anchor log
лезвие *n.* edge, blade; bit; л. резца steel edge
Лейбниц Leibniz
лейденский *adj.* Leyden
Лейденфрост Leidenfrost
лейцитоэдр *m.* leucitohedron
лекало *n.* caliber rule; curve templet; form, gage, pattern, standard
лекальный *adj.* curve; mold; gaged
Лекланше Leclanché
лексикографический *adj.* lexicographic(al)
лексикография *f.* lexicography
лексикон *m.* dictionary
лексикостатистика *f.* glottochronology
лексический *adj.* lexical
лектрон *m.* Lectron
лекционный *adj.* lecture
лекция *f.* lecture, discourse

лемма *f.* lemma
Леммес Lemmens
лемниската *f.* lemniscate
лен. *abbr.* (ленинградский) Leningrad
Ленар, Ленард Lenard
Ленгмюр Langmuir
Ленинградский *adj.* Leningrad
лента *f.* tape, band, ribbon, strip, string, strap, lace; belt; film; **бесконечная магнитная л.** endless magnetic tape loop, magnetic belt; **бесшовная косая л.** seamless bias tape; **бронирующая л.** restricting tape; **бумажная л. с отпечатанными данными** printed paper tape; **л. ввода программы** input program tape; **л. внешнего запоминающего устройства** file tape; **л. входных данных** input tape; **л. выходных данных** output tape; **дегтярная л.** tarred tape; **л. для вклейки** timing tape; **л. для записи показаний прибора** record chart; **л. для заправки** leader tape; **л. для изготовления сопротивлений** resistor tape; **л. для обмотки** swaddling tape; **л. для печатания** printing-out tape; **л. для проверки** (*or* **наладки**) **головок звукозаписи** audio head aligning tape; **л. для регистрирующих приборов** (perforated) recorder tape; **л. для хранения и электронно-оптического воспроизведения изображения** electro-optical imaging and storage tape; **л. из асбестовой пряжки** asbestos woven tape; **л. из джутовой ровницы** hessian tape; **испытательная л. для юстировки магнитных головок** head aligning tape; **магнитная л. с записанным телевизионным изображением** recorded video tape; **магнитная л. с пластмассовой подложкой** plastic-backed magnetic tape; **магнитная л. со стёртой записью** erased tape; **металлизированная л., металлизующая л.** bonding jumper, bonding strip; **недоперфорированная л.** chadless tape; **некачественная магнитная л.** imperfect magnetic tape; **неразмеченная л.** virgin tape; **однослойная однородная магнитная л.** magnetic powder impregnated tape; **парусиновая л.** duct tape; **поддерживающая л.** securing strip; **л., пред-**

варительно пропитанная связующим веществом pre-impregnated tape; **(предварительно) размеченная л.** pre-addressed tape; **промежуточная л.** change tape; **л. с косым резом** bias tape; **л. с намагничиваемым слоем, л. с нанесённым магнитным слоем** coated plastic recording-tape, magnetic powder-coated tape; **л. с подпрограммой** subroutine tape; **л. с программой наведения** guidance tape; **л. с распылённым ферромагнитным порошком** magnetic powder-coated tape; **л. с шестью рядами отверстий** six-hole tape; **л. со стёртой записью** erased tape; **склеенная петлёй магнитная л.** endless tape, tape loop; **транспортёрная л.** belt conveyor; **л. управления** steering tape; **л. управления намоткой нити** filament winding control tape; **фольговая л.** chaff, window; **цветная л.** ink ribbon

**лентообмоточный** *adj.* taping

**лентообразный** *adj.* band-shaped, ribbon

**лентопротяжка** *f.* tape drive system, tape transport

**лентопротяжный** *adj.* feed, tape-winding, tape-moving, tape transport, tape-drive; (telegraph) paper-drive; film-transport, film-traction

**ленточка** *f.* strip, tape, ribbon, band, strap, string, lace

**ленточно-спиральный** *adj.* tape-helix

**ленточный** *adj.* band, tape, ribbon, belt-type, strip, ribbon-type

**Ленц** Lenz

**Леонард** Leonard

**Лепель** Lepel

**лепестковый** *adj.* lobed; petal-shaped, petaled, leaf-shaped

**лепесток** *m.* lobe, leaf; **ближние боковые лепестки** inner side lobes; **задние боковые лепестки** outer side lobes; **побочные лепестки** unwanted lobes; **боковой л. в дифракционной картине** diffraction side lobe; **л. зеркального отражения** specular reflection lobe; **л. излучения с учётом влияния земли** ground-deflection lobe; **л. ламповой панели** tube-socket lug; **основной л. диаграммы излучения** main lobe of rabiation; **побочный л.** parasite lode

**лепидолит** *m.* lithia mica, lepidolite

**Леппель** Leppel

**лептон** *m.* lepton

**леса** *f.* scaffold, rack, framework, supporting (structure)

**лесоматериал** *m.* timber, lumber

**лестница** *f.* staircase; ladder; scale; **приставная л.** ladder; **самодвижная л.** escalator

**лестничный** *adj.* ladder(-type); staircase; scale

**летание** *n.* flying, flight

**летательный** *adj.* flying

**летать** *v.* to fly

**летающий** *adj.* flying

**летероид** *m.* fish-paper insulation, leatheroid

**лететь, полететь** *v.* to fly, to be flying; to volatize

**летний** *adj.* summer

**лётный** *adj.* flying; flight

**летучесть** *f.* volatility

**летучий** *adj.* volatile; flying

**летучка** *f.* harness, breadboard hookup; leaflet

**лётчик** *m.* pilot

**летящий** *adj.* flying

**Лехер** Lecher

**лехеровский** *adj.* Lecher('s), (of) Lecher

**лечебный** *adj.* medical, medicinal, medicine

**лечение** *n.* medical treatment; **л. фарадизацией** faradism; **л. электричеством** electrotherapy

**ЛЗП** *abbr.* (линия заданного пути) specified track

**ЛИ** *abbr.* (линейный искатель) connector (*dts.*)

**Либен** Lieben

**Либенов** Lieben's

**ливень** *m.* shower

**ливневый** *adj.* shower

**лига** *f.* tie, slur; bind (*mus.*)

**лигнолит** *m.* plywood

**Лиенард** Liénard

**ЛИИ** *abbr.* (Лётно-испытательный институт) flight-test institute

**ЛИК** *abbr.* (линейный искатель для коммутаторов) rotary hunting connector (*dts.*)

**ЛИКБ** *abbr.* (линейный искатель для больших коммутаторов) rotary hunting connector (*dts.*)

**ликвидация** *f.* elimination, liquidation; **л. остановки** restoring of service

**ликвидирование** *n.* liquidation

**ликвидировать** *v.* to liquidate, to eliminate, to remove, to stop; to dispose

**ликоподий** *m.* lycopodium powder

**Лилл** Lill

**ЛИМ** *abbr.* (**международный линейный искатель**) trunk connector (*dts.*)

**лимб** *m.* limb, dial, graduated circle; **л. антенны** antenna repeat dial

**лимниметр** *m.* limnimeter

**лингвистика** *f.* linguistics

**лингвистический** *adj.* linguistic

**Линдеманн** Lindemann

**линеаризатор** *m.* linearizer

**линеаризация** *f.* linearity, linearization; **л. нарастания** slope linearization

**линеаризировать** *v.* to linearize

**линеаризирующий** *adj.* linearizing; linearity-control

**линеаризованный** *adj.* linearized

**линеаризовать** *v.* to linearize; to equalize

**линейка** *f.* rule, straight edge, ruler, gage; **визирная л.** aiming rule; **высотная л.** altitude slide; **дешифраторная л.** code bar; **запорная л.** locking bar; **л. к ригельному замку** lock rod; **масштабная л.** measuring rule, plotting scale; scale, rule, rate rule; **мерная л.** gage; **многошкальная счётная л.** multiscale rule; **наборная л.** combination bar; selector bar; **навигационная л. для прокладки курса** dead-reckoning ruler, navigators ruler; **навигационная счётная л.** navigation computer; **печатающая л.** printer bar; **подвижная счётная л.** slide rule; **радиолокационная л.** radar "ruler"; **счётная л.** slide rule; **л. усиления промежуточной частоты** intermediate-frequency strip; **л. централизационного аппарата** locking bar, tappet

**линейно** *adv.* linearly

**линейно-логарифмический** *adj.* linear-to-log

**линейно-поляризованный** *adj.* linearly polarized

**линейно-ступенчатый** *adj.* linear-staircase

**линейность** *f.* linearity; **л. амплитудной характеристики** amplitude linearity; **л. кадровой развёртки** vertical sweep linearity; **л. переда-** точной характеристики transfer linearity; **л. развёртки по вертикали** vertical-sweep linearity; **л. развёртки по строкам** line (*or* horizontal) linearity

**линейный** *adj.* linear, lineal; line, run, straight-line; one-dimensional, unidimensional; circuit; in-line

**линейчатотость** *f.* lineation

**линейчатый** *adj.* line; linear; ruled; bright-line (spectra); phosphorline

**линейщик** *m.* lineman, wireman

**линза** *f.* lens; lamp cap; button (*tbs.*); **л. для ландшафтной съёмки** landscape lens; **л. для увеличения поля зрения** field lens; **л. для фокусировки магнитного поля** magnetic lens; **л. «катраль»** special spectacle glass to replace the lens of the human eye; **конденсаторная л.** illuminating lens; **металлопластинчатая л.** metal-plate lens; **оборачивающая л.** relay lens; **однопотенциальная л.** univoltage lens; **л. пасивного усиления радиолокационного эхо-сигнала** passive augmentation lens; **первая электродная л.** cathode lens; **л. под нулевым потенциалом** equipotential lens; **просветлённая л.** coated lens; **простая металлическая л.** unzoned metal-plate lens; **рассеивающая л.** spreading lens, diverging lens, concave lens; **решетчатая л.** lattice lens; **рифлёная л.** stippled lens; **л. с меняющимся фокусом** zoom lens; **л. с «седлообразным» полем** saddle-field lens; **л., состоящая из двух цилиндров** two-tube lens; **тормозящая л.** delay-type lens

**линзово-растровый** *adj.* lenticulated

**линзовый** *adj.* lens

**линзовыниматель** *m.* cap extractor, lens extractor

**линзообразный** *adj.* lenticular

**линие-выбиратель** *m.* selecting switch

**линия** *f.* line, mark; path, direction; link; **визирные линии** cross lines; **граничные линии равного потенциала** equipotential boundaries; **двухпроводные открытые линии** parallel wires; **зеркальные линии** image lines; **крестообразные линии** cross lines; **однотипные линии** duplicate lines; **линии связи, основанные на прямом рассеянии** forward scatter circuits; **линии уровня** contour lines (*math.*); **абонентская л. с одним аппаратом** direct line (*tphny.*); **автоматическая сквозная производственная л.** automatic flow-through production system; **л. автоматической передачи речевой команды** automatic voice data link; **акустическая л. задержки на проволоке** wire-type acoustic delay line; **акустическая длинная л.** acoustic transmission line; **л. акустического пеленга** sound line; **балансная л.** balancing network, compensation circuit; **л. без ответвлений** fixed network; **л. без потерь** loss-free line, zero-loss line; **безупречная л.** line in good order; **бесконечно-длинная л.** infinite line; **блокировочная л.** lock-out circuit; **л. в спектре испускания** emission line; **ведущая л.** directrix; **л. визирования** line of sight, sight line; **л. визирования светила** stellar line; **вихревая л.** vortex line; **внутрирайонная л. коллективного пользования** rural party line; **внутристанционная л.** trunk (between two switching devices), link; **воздушная л. на столбах** open-wire pole line; **воздушная л. связи** overhead telecommunication line; **воздушная л. сильного тока** overhead power line; **л. времени** base line, time base, time vector; **вторичная л. передачи** subtransmission circuit; **вызываемая л.** called line; **л. вылета** line of departure; **«выравненная л.»** loaded line; **высокочастотная телефонная л. электростанции** electric-power station telephony; **главная л. питания вторичной распределительной системы** secondary distribution trunk line; **главная л. питания первичной распределительной системы** primary distribution trunk line; **главная л. питания распределительной системы** distribution trunk line; **л. главной международной станции** main international office trunk; **горбылёвая л.** slab line; **городская л. (коммутатора)** PBX trunk; **двусторонняя соединительная л.** both-way trunk; **двухканальная л.** two-way circuit; **двухпроводная л.** two-wire line, two-wire circuit, metallic circuit, Lecher wire; **л. двухстороннего действия** two-way circuit; **л. действия силы** line of application; **директорская л.** intercepting line (*tphny.*); **диэлектрическая плоская л. передачи** dielectric-sandwich line; **л. для внешних переговоров** outward line (*tphny.*); **л. для входящей связи** incoming one-way circuit; **л. для исходящей связи** outgoing one-way circuit; **л. для радиотрансляции** program line; **дуплексная телевизионная радиорелейная л.** two-way television relaying; **единичная силовая л. электростатического поля** unit electrostatic flux; **жёлобообразная передающая л.** trough line; **жидкая акустическая л. задержки** liquid medium sonic delay line; **л. жирного шрифта** extra-heavy line; **л. заданного пути** specified track; **л. задержки** delay line, lag line, delay circuit, time delay line, retardation line; **л. задержки с сосредоточенными параметрами** lumped-parameter delay line; **л. задержки с сосредоточенными постоянными** lumped constant delay line; **л. задержки сигнала цветности** chrominance delay line; **заказная л.** recording trunk, record operator's line, call circuit; **заказная л. с использованием для междугородных переговоров, заказная-переговорная л.** recording-completing trunk, CLR trunk; **идущая к потребителю л.** service main; **л. из скрученных проводов** skew-wire line; **л. из четырёхполюсников** circuit network; **л. изгиба** curvature; **измерительная л.** measuring line, Lecher wires, slotted line; **искусственная л.** artificial line, phantom line; Guillemin circuit, pulse-forming circuit, attenuation network; **л. источников векторного поля** source line; **калибровочная л. для определения азимута** lubber line; **л. коллективного пользования** party line; **л. коллективного пользования в окружной связи** rural party line; **командная проводная л.** wire command link; **комбинированная л. связи** composite communication link; **л. коммутаторной установки**

PBX line; **коническая винтовая л.** tapered helix (line); **коническая полосковая л.** tapered strip line; **контрольная л.** monitoring line; **короткозамкнутая на конце л.** short-ended line; **крарупизированная л.** continuously loaded circuit; **л. кривых** graph, chart, curve sheet; **круговая л. развёртки, л. круговой развёртки** circular trace; **ленточная л.** ribbon feeder; **ленточная полосковая л.** microstrip; **магическая л.** magic strip; **магнитная л.** isoclinic line, unit magnetic flux; **магнитострикционная л. с регулируемой задержкой** variable magnetostrictive delay line; **л. манипулятора** keying circuit; **междугородная л.** toll line, long-distance line, trunk line; **междукоммутаторная л.** tie trunk; **междустанционная соединительная л.** inter-office trunk; **местная соединительная л.** inter-office trunk cable; **местная соединительная л. двухстороннего действия** both-way junction; **местная соединительная л. симплексной связи** two-way junction, duplex junction; **л. местонахождения, л. местоположения** line of position, base line; **микроволновая л. связи с промежуточными станциями** microwave repeater system; **модуляционная л.** program line; **л. на двойных опорах** H-pole line; **л. на металлических изоляторах** stub-supported (transmission) line; **л. на токах несущей частоты** carrier-frequency line; **л. на шлейфах** stub-supported (transmission) line; **л. нагрузки задающего каскада** driver load line; **назначенная позиционная л.** precomputed line of position; **л. направленной передачи** beam transmission link; **настроенная л.** resonant line; **л. начала отсчёта** reference line, datum line; **нейтральная л. (коммутации)** neutral axis; **нейтральная л. с остаточным затуханием** zero-loss circuit; **несекционированная л.** single-link line; **несиметричная полосковая л.** single-ground-plane line; **нижняя л.** lower trace; **л. низовой связи** rural line; **л. нулевого реактивного сопротивления** zero-reactance line; **л. нулевых магнитных**

склонений agonic line; **л. обмена (мощностью)** tie line, branch line, interconnector, interconnection; **л. обратного хода** return trace, return line, retrace line; **обходная л.** by-pass, by-pass line; **общая соединительная л.** concentration line, collecting line; **л. общего пользования** associated line (*tphny.*); **объединяющая л.** concentration line, interconnection line, tie line; **л. одинакового уровня** equipotential line; **однопроводная л. с поверхностной волной** Goubou line; **однопроводная фидерная л.** single-wire transmission line, guide wire; **л. одностороннего действия** straightforward line; **л. одностороннего использования** one-way circuit; **оконечная магистральная л.** final telephone trunk line; **л. окружной связи** rural line, rural subscriber's line; **л. основного аппарата** main line, direct exchange line; **л. от транзитной станции к оконечной** tandem-completing trunk; **отвесная л.** plumb line; **отдельная л.** single wire circuit, single wire line, earth return circuit; individual line; **откачная л.** vacuum pump line; **отходящая л.** output circuit; **л. падения** lines of dip; **л. параллельных вибраторов, соединённых в противофазе** end-fire array; **ПВО л. радиолокационных станций** radar fence, belt; **ПВО вынесенная в море л. радиолокационных станций** advanced radar fence, advanced coast belt; **переговорная л.** talk-back circuit; **передаточная л.** transfer circuit, lending circuit, interposition trunk; **л. передачи звуковой программы** audio program link; **л. передачи команд по проводам** wire command link; **л. передачи с согласованной нагрузкой** matched transmission line; **передающая л. с потерями** lossy transmission line; **передающая л. типа «провод над землёй»** strip-above-ground line; **л. пересечения** intersection; **петлеобразная л.** folded line; **питаемая с двух концов л.** two-terminal line; **плавнорегулируемая л. задержки** continuously-variable delay line; **пластинчатая л.** slab line; **л. повышенной индук-**

тивности inductively loaded circuit; л. подачи питания feed line; подключиваемая л. transfer circuit; л. покоя neutral position (of a selector); поперечная соединительная л. inter-switchboard line, junction line; последняя соединительная л. декады (искателя) late choice trunk (of a switch); л. постоянной скорости isovel; поточная л. production line; л. потребителя service line; л. предупреждения о появлении подводных лодок противника electronic warning line; присоединительная л. direct line, subscriber's line; прямая л. inter-office line, beeline; л. прямого абонента toll terminal; л. прямого абонента междугородной станции direct trunk connection line; л. прямого и обратного хода go and return line; л. прямого соединения standard routing of calls; л. прямого хода outgoing line; л. прямой видимости line of sight; пунктирная л. точкатире dot-and-dash line; пупинизированная л. coil-loaded circuit; л. пурпурных цветов purple boundary; л. равной слышимости equal-loudness contour; л. равных радиопеленгов locus of equal radio bearings; л. равных скоростей isotach; радиопроволочная л. combined radio and wire link; л. радиосвязи за счёт рассеянного распространения ionoscatter circuit; л. развёртки sweep trace, scanning trace, scanning line, time axis; разомкнутая л. open-circuit line; разомкнутая на конце л. задержки open-circuited delay line; расплывчатая л. blurred trace; распределительная л. service cable, distributing line, current distributor; регулируемая л. задержки variable delay line; релейная л. радиосвязи radio-relay link, radio-relay system; л. речевой связи voice communication link; руководящая междугородная л. controlling trunk line; л. с воздушным заполнением airspace line; л. с двумя экранирующими плоскостями triplate line; л. с диэлектрическим заполнением solid-dielectric sandwich; л. с землёй в качестве обратного провода earth-return line; л. с ответвлениями

branched distributor, multi-ended line; л. с проводами различного назначения joint line; л. с «пучковыми» проводами для каждой фазы multiple conductor line; л. с разомкнутым концом open-ended line; л. с распределёнными постоянными distributed-parameter line; л. с рассеянием dissipative line; л. с согласованной нагрузкой match-terminated line; л. с сосредоточенными постоянными lumped-parameter line; л. с турникетным расположением проводов transposed transmission line; л. со смешанной связью line with mutual coupling; сборная заказная л. split order-wire; л. связи sound line, (electro) communication line, trunk line, telegraph and telephone line; link, tie, circuit, link circuit; л. связи всеобщего европейского телевизионного вещания Eurovision link; л. связи между студией и радиопередатчиком studio-to-transmitter link; л. связи на сантиметровых волнах microwave radio link; л. связи с тылом rear link; л. связи системы наведения управляемых ракет guided missile fire control circuit; секущая л. secant; сельская л. коллективного пользования rural party line; л. симплексной связи both-way circuit; слаботочная воздушная л. overhead telecommunication line; слоистая л. solid-dielectric sandwich; служебная л. order circuit, order wire, recording trunk, call circuit, interposition trunk; смонтированная на шайбах л. bead-supported line; л., соединение с которой выполняется вызовом ringdown trunk; соединительная л. junction line, connecting line, trunking circuit, connecting circuit, interconnection line, interconnection trunk; с. л. в группе районных сетей interconnection line between exchange and network center; с. л. к междугородному коммутатору trunk junction circuit; с. л. между ГТС и МТС toll switching trunk; с. л. одностороннего действия single-action trunk; с. л. от междугородной к районной городской телефонной станции toll switching trunk; с. л.

промежуточной станции tandem trunk; л., соединяющая точки с равными магнитными склонениями isogonic line; спектральная л. полярного сияния auroral line; л. спектральных цветов spectral locus, spectrum locus; справочная л. inquiry circuit, information circuit, intercepting trunk; л. средней нагруженности medium-haul circuit; среднепупинизированная л. medium loaded circuit; л., стойкая против атмосферных влияний weather-resistant insulation line; твёрдая акустическая л. задержки solid-medium sonic delay line; л. телефона-автомата coin box circuit; транзитная л. through line, through circuit; транзитная междугородная л. through toll line; транзитная соединительная л. built-up trunk, tandem trunk; трансляционная проволочная л. pick-up line; л. трансляционных радиостанций radio relay system; трансформирующая л. matching stub; трёхплоскостная л. с воздушным заполнением air-spaced triplate line; тройниковая л. three-terminal line; укреплённая на шайбах л. bead-supported line; уложенная на землю кабельная л. earth line; ультразвуковая жидкостная л. задержки ultrasonic trunk; ультразвуковая керамическая л. задержки ultrasonic vitreous silica delay line; уплотнённая л. связи shared line, multiplex line; л. управления вычислительной машиной computer control link; учетверённая четырёхпроводная л. quadruple phantom circuit; характеристическая л. с перегибом bent characteristic, buckled characteristic; л. цветов абсолютно чёрного тела achromatic locus, Plankian locus; цепная л. ladder circuit, recurrent circuit, catenary circuit; чёрная вертикальная л. испытательной таблицы flagpole (tv); чистая л. silent circuit (tphny.); шнуровая л. trunk, link; л. шунтированная на конце bridged-end line; эквивалентная л. со включённым посредине конденсатором middle-condenser circuit; л. электролиза, л. электролитических ванн potline (elec.-chem.); л. электротяги electrified track; эталонная л. высокой частоты high-frequency reference circuit; л. эталонной задержки standard lag line

**линолеум** m. linoleum
**лиофильный** adj. lyophilic
**липкий** adj. adhesive, sticky
**Липовиц** Lipowitz
**лира** f. lyre, harp
**Лиссажу** Lissajous
**лист** m. sheet, lamina, lamination; scale; foil; leaf; plate, blade; **выштампованные листы якоря** armature stampings; **перфорированные листы сердечника** pierced core disks; **л. для перекрытия** cover sheet; **л. жёсткости** stiffening plate; **корпусный л.** shell plate; **магнитный л.** magnetic lamella; **л. накладки** fish-plate; **л. обшивки корпуса** hull plate; **освинцованный железный л.** terne plate; **л. поверхности** sheet of surface (math.); **л. с отверстиями, выполненными травлением** etched sheet; **улавливающий л.** propping sheet; **л. чертёжной бумаги** drawing sheet; **штампованный л.** punching; **штампованный л. в форме буквы « E »** E-punching; **экранирующий л.** baffle shield
**лиственный** adj. leaf, leafy
**листовой** adj. lamellar; leaf; sheet; flake
**листок** m. leaf; sheet; **л. плёнки** film wafer
**листообразный** adj. lamellar
**листорезный** adj. plate-shearing, plate-cutting, shearing
**литавра** f. kettledrum
**литейная** f. adj. decl. foundry shop, casting shop
**литейный** adj. casting, founding; foundry
**литера** f. letter (of alphabet), type
**литература** f. literature
**литиево-аммониевый** adj. lithium-ammonium
**литиево-дрейфовый** adj. lithium-drifted
**литиево-натриевый** adj. lithium-sodium
**литиевый** adj. lithium
**литий** m. lithium, Li
**литографический** adj. lithographic
**литография** f. lithography

литографский *adj.* lithographic
литой *adj.* cast; poured; molten
литр *m.* liter
Литров Littrow
литцендрат *see* лицендрат
литый *adj.* cast; poured
лить, полить *v.* to pour; to cast
литьё *n.* casting, founding, molding; pouring; casts, castings; красное л. red brass; латунное л. cast brass; л. пластмасс plastic molding; л. под давлением die casting
ЛИУ *abbr.* (универсальный линейный искатель) combination connector, local and trunk connector
Лиувилль Liouville
лифт *m.* elevator, lift; pump; грузовый л. без проводника service elevator; л. с кнопочным управлением automatic elevator; управляемый лифтёром л. attendant-operated elevator
лифтёр *m.* elevator operator; lifter
лифтовый *adj.* elevator, lift
Лихтенберг Lichtenberg
лица *f.* cord, strand, stranded wire, lace, tinsel
лицевой *adj.* face, facial, front, front-face
лицендрат *m.* litz wire, strand wire, r-f cable; л. из тонких проволок fine strand
лицензия *f.* license, permit, concession
лицо *n.* face, side; person; обслуживающее л. operator, serviceman
личной *adj.* face, facial
личный *adj.* personal, private, individual, particular
лишать, лишить *v.* to deprive (of ), to cut off, to eliminate, to remove; л. напряжения, л. энергии to de-energize
лишение *n.* deprivation, loss
лишённый *adj.* deprived (of ), devoid of
лишить *pf. of* лишать
лишний *adj.* superfluous, excessive, unnecessary; spare, odd
лишь *adv.* only, but, even; as soon as; л. бы provided
лк *abbr.* (люкс) lux
ЛКАО *abbr.* (линейная комбинация атомных орбит) linear combination of atomic orbits, LCAO
Ллойд Lloyd
лм *abbr.* (люмен) lumen
лмб *abbr.* (ламберт) lambert

лм-с *abbr.* (люмен-секунда) lumen-second
лм-ч *abbr.* (люмен-час) lumen-hour
л.н.с. *abbr.* (линия наименьшего сопротивления) line of least resistance
лоб *m.* front, face, head, crown
лобзик *m.* fret-saw, jig saw
лобный *adj.* frontal, forehead
лобовой *adj.* frontal, front, face; head-on; end
ЛОВ *abbr.* (лампа обратной волны) backward-wave tube, reflected wave tube
Лове Love
ловитель *m.* catcher, safety gear; catch, stop
ловкость *f.* skill, dexterity, cleverness; л. рук operating skill
ловушка *f.* trap, snare; decoy
логарифм *m.* logarithm, log; десятичный л. common logarithm, Brigg's logarithm, denary logarithm; интегральный л. logarithmic integral; пятизначный л. five-place logarithm
логарифмирующий *adj.* logarithm-taking
логарифмический *adj.* logarithmic
логарифмовать *v.* to take the logarithm
логатомный *adj.* syllabic (articulation)
логика *f.* logic
логистика *f.* logistics
логистический *adj.* logistic
логицизм *m.* logicism
логически *adv.* logically
логический *adj.* logical
логично *adv.* logically
логичный *adj.* logical
логометр *m.* ratiometer, logometer, quotient meter; магнитоэлектрический л. с подвижной катушкой moving-coil quotient meter with permanent magnet; л. с подвижным (мягким) железом moving iron quotient meter; ферродинамический л. с замкнутой магнитной системой electrodynamic quotient meter with closed iron circuit; цифровой л. digital ratiometer
логон *m.* logon
логон-мера *f.* logon content
логопериодический *adj.* log-periodic
Лодж Lodge
лодка *f.* boat; подводная л. с действующим радиолокатором radar-using submarine; подводная л. с

**неработающим радиолокатором** non-radar-using submarine

**лодочка** *f.* boat; combustion boat; **л., нагреваемая джоулевым теплом** resistance-heated boat; **л. с индукционным нагревом** induction-heated boat

**лодочный** *adj.* boat

**ложечный** *adj.* spoon(-shaped); concave

**ложкообразный** *adj.* spoon-shaped, hollow, concave

**ложность** *f.* invalidity

**ложный** *adj.* false, erroneous, untrue, fallacious; spurious; dummy; pseudo-

**ложь** *f.* falsehood

**локализатор** *m.* localizer, detector, finder; **л. повреждений** fault finder, fault detector; **световой л.** light beam localizer

**локализация** *f.* localization; trapping

**локализированный** *adj.* localized

**локализировать** *v.* to localize, to locate

**локализованный** *adj.* localized

**локализовать** *v.* to localize, to locate

**локально** *adv.* locally

**локальный** *adj.* local

**локатор** *m.* locator, position finder; location, station; **звуковой л. ближнего действия** "sonicator"

**локация** *f.* location; **гидроакустическая л.** subaqueous sound ranging; **инфракрасная л.** infrared range and detection equipment

**локомотив** *m.* locomotive

**локомотивный** *adj.* locomotive

**локотной** *adj.* elbow

**локсодрома** *f.* loxodrome, loxodromic curve

**локтальный** *adj.* loctal

**локтевой** *adj.* elbow

**локус** *m.* locus; **л. излучения абсолютно чёрного тела** Plankian locus

**лом** *m.* junk, scrap; crowbar, digging bar; **л. для вытаскивания гвоздей** nail puller, crowbar

**ломано-линейный** *adj.* dog-leg

**ломаный** *adj.* broken, fractured; irregular

**ломать, сломать** *v.* to break (up), to crush

**ломаться, сломаться** *v.* to break, to get out of order

**ломик** *m.* pinch bar, small crowbar; **л. для открывания крышки колодцев** manhole hook

**ломка** *f.* breaking, demolishing

**ломкий** *adj.* brittle, fragile; inflexible

**лопастеобразный** *adj.* paddle-shaped, beavertail(-shaped)

**лопастный** *adj.* blade, vane, paddle

**лопасть** *f.* blade, vane, fan, paddle; **л. семафорного крыла** semaphore blade

**лопата** *f.* spade, shovel

**лопатка** *f.* shovel; blade, vane, paddle; stem press; **двойная л.** blade thickened on back side; **л. ломаного профиля** broken blade; **настроечная л.** tuning paddle; **л. ножки** stem press; **л., отлитая вместе с ротором** cast-in blade; **поворотная л.** hinged guide blade; **л. рабочего колеса (турбины)** float board; **л. с утолщённой спиной** blade thickened on back side

**лопаткообразный** *adj.* spade-shaped

**лопаточка** *f.* spade

**лопаточный** *adj.* shovel; scapular

**Лопиталь** L'Hospital

**Лорак** Lorac

**Лоран** Loran

**лорановский** *adj.* (of) Loran

**Лордотический** *adj.* Lordotic

**Лорент** Laurent

**Лоренц** Lorenz, Lorentz

**лоренцов** *poss. adj.* (of) Lorenz, (of) Lorentz

**лот** *m.* plumb line, plumb bob, depth finder

**лотерея** *f.* lottery

**лотковый** *adj.* tray, trough, pan

**лоткообразный** *adj.* trough-shaped

**лоток** *m.* tray, trough, pan; **л. для колб, л. для сборки ламп** bulb tray, tube tray; **загрузочный л., сборочный л.** assembling tray, assembly tray

**Лоу** Law

**Лофтин** Loftin

**« лохис »** *m*, lohys

**лоцман** *m.* navigator, pilot; **л. точной радиолокационной системы** PAR controller

**лошадиный** *adj.* horse

**Лошмидт** Loschmidt

**лощение** *n.* burnishing, glossing, polishing; **л. шариками** ball burnishing

**лощёный** *adj.* polished, glossy; glazed

**лощило** *n.* trowel, slicker; burnisher

**лощильный** *adj.* burnishing, glossing, polishing; burnisher

**лощить, налощить** *v.* to polish, to burnish, to gloss, to smooth

**ЛП** *abbr.* (линия прицеливания) sight line, aiming line

**ЛПС** *abbr.* (линия положения самолёта) line of position, LOP

**ЛРА** *abbr.* (линия равных азимутов) line of equal azimuth

**ЛРРП** *abbr.* (линия равных радиопеленгов) locus of equal radio bearings

**л.с., ЛС** *abbr.* (лошадиная сила) horsepower, hp

**л.с.-ч, л.с.-час** *abbr.* (лошадиная сила-час) horsepower-hour, hph

**лудильный** *adj.* tinning

**лудить, полудить** *v.* to tin

**Лудольф** Ludolf

**лужение** *n.* tin-plating, tinning

**лужёный** *adj.* tinned, tin-plated

**лужица** *f.* pool, puddle

**лукалокс** *m.* lucalox

**Луммер** Lummer

**луна** *f.* moon

**лунно-солнечный** *adj.* lunisolar

**лунный** *adj.* lunar, moon

**лупа** *f.* magnifier, magnifying glass, lens; **л. времени** time magnifier; time magnifying study

**Лупанов** Lupanow

**луч** *m.* ray, beam, pencil (of light), shaft (of light); leg (of a beacon); path; **лучи-вита** vita-rays; **асимметрично расходящийся л.** asymmetrically flared beam; **бегающий л., бегущий л.** scanning beam, scanning pencil of light, flying spot; **боковой л.** marginal ray; **ведущий л.** localizer beam; **веслообразный л.** beavertail beam; **запертый л.** cut-off beam; **импульсный л. наведения** pulsed guidance beam; **коммутированный л.** sampled beam; **лопастеобразный л., ножевой л.** beavertail beam; **остронаправленный л.** pencil beam; **л., отклоняемый по кругу** circulating beam; **л. пеленгагатора, пеленгаторный л.** bearing ray, localizer beam; **плоский л.** fan beam; **побочный ведущий л.** side lobe beam; **поддерживающий л.** holding beam; **прерывистый л.** chopped beam; **прижатый л.** low-altitude beam; **прожекторный л.** searchlight beam; **прямолинейный л.** thread beam; **л. радиомаяка** radio-range beam; **л., регулируемый по углу возвышения, л., регулируемый по углу места** vari-

able-elevation beam; **л. с огибающей формы косеканс в квадрате** cosecant squared beam; **слабо фокусированный л.** low-resolution beam; **снимающий л. (в иконоскопе)** scan-off beam, play-off beam (in an icono-scope); **сопровождающий л.** locked-on beam; **стирающий л.** play-off beam, scan-off beam; **сформированный для воспроизведения знака электронный л.** character-shaped beam; **считывающий л.** scan-off beam, viewing beam; **убивающий л.** death ray; **узкий л. в виде полоски** ribbon beam; **л. улавливания снаряда** gathering beam; **фиксирующий л.** holding beam, test pattern beam; **широкораскрытый л.** wide-angle beam; **л. электронно-лучевой трубки** cathode-ray beam

**лучевод** *m.* light guide, light-pipe

**лучевой** *adj.* ray, radial; beam

**лучезвуковой** *adj.* radiophonic

**лучеиспускаемость** *f.* emissivity, emittance

**лучеиспускание** *n.* radiation, emission, emissivity; irradiation

**лучеиспускательный** *adj.* radiating

**лучеиспускать** *v.* to radiate, to beam, to emit rays

**лучеиспускающий** *adj.* radiating

**лученепроницаемый** *adj.* radiopaque

**лучеобразный** *adj.* radial, ray-like; radiating

**лучеобразующий** *adj.* beam-forming

**лучеотражаемость** *f.* radiant reflectivity, radiant reflectance

**лучепоглощение** *n.* absorption of exterrestrial radiation

**лучепреломление** *n.* refraction, refringence

**лучепреломляющий** *adj.* refractive

**лучепроницаемый** *adj.* radioparent

**лучистость** *f.* radiance, radiancy

**лучистый** *adj.* radiant, radiating, radial, radiated

**лыжа** *f.* collecting shoe (*elec.*); **л. на бюгеле** collector strip, bow collector

**льготный** *adj.* reduced; favorable

**льдоудалитель** *m.* de-icer

**Лэмб** Lamb

**Лэнгмюр** Langmuir

**лэнгмюров** *poss. adj.* Langmuir's, (of) Langmuir

**любитель** *m.* amateur, fan, ham, layman

любительский *adj.* amateur, fan
Люилье L'Huilier
люк *m.* manhole, hatch, trap; л. колодца manhole cover frame
люковый *adj.* manhole, hatch, trap
люкс *m.* lux, meter-candle
люксембургский *adj.* Luxembourg
люксметр *m.* luxmeter, illuminometer, lightmeter
люкс-секунда *f.* lux-second
люлька *f.* cable car, chair, bosun's chair; cage, cradle; bucket
люмарит *m.* lumarith
люмен *m.* lumen
люменметр *m.* lumenmeter, integrating photometer
люмен-секунда *f.* lumen-second
люмен-час *m.* lumen-hour
люмерг *m.* lumerg
люметр *m.* lumeter
люметрон *m.* lumetron
люмикон *m.* lumicon
люминер *m.* luminaire
люминесцентный *adj.* luminescent; scintillation (counter); fluorometric
люминесценция *f.* luminescence; л. в переменном электрическом поле electroluminescence; звуковая химическая л. sonic chemiluminescence
люминесценц-микроскоп *m.* luminescence microscope
люминесцировать *v.* to luminesce
люминесцирующий *adj.* luminescent
люминоген *m.* phosphorogen
люминоскоп *m.* luminoscope
люминофор *m.* luminophor; phosphor; л. белого свечения white phosphor;

инерционный л. persistent phosphor; многокомпонентный л. composite phosphor; л., обладающий послесвечением persistent phosphor; л., обладающий средней длительностью послесвечения medium-persistence phosphor; л. с быстрым затуханием short-persistence phosphor, phosphor of rapid decay; л. с коротким послесвечением, л. с малой длительностью послесвечения rapid-decay phosphor, short-lag phosphor, short-persistence phosphor; л. с равноэнергетической спектральной характеристикой свечения equal-energy phosphor; л., светящийся зелёным (красным) цветом green (red) (emitting) phosphor; цветной точечный л. color phosphor dot, color-emitting phosphor dot
Люнеберг Luneberg
Люссак Lussac
лютеций *m.* lutecium, Lu
лютня *f.* lute (*муз.*)
люфт *m.* backlash, free play, freedom; gap, clearance
лягушачий, лягушечий *adj.* frog; grip, toggle, (draw) vice, (draw) tongs
лягушка *f.* frog; grip, toggle, (draw) vice, (draw) tongs; зажимная л. (draw) tongs, draw vice, toggle, grip; cable grip
Лямэ Lame
Ляпунов Liapunoff

# M

**M** *abbr.* (*Max*) Mach
**м** *abbr.* (**метр**) meter
**МА** *abbr.* (**магнитный азимут**) magnetic azimuth
**ма** *abbr.* (**миллиампер**) milliampere
**мавар** *m.* mavar
**магазин** *m.* store, warehouse, storeroom, magazine; box; stack (of perforated cards); unit; **декадный м. ёмкостей** decade capacitor, decade capacitance box; **нагрузочный м. сопротивлений, м. нагрузочных сопротивлений** impedance-matching load box; **рычажный м. ёмкостей** switch capacitance box; **рычажный м. сопротивлений с дополнительным сопротивлением** Thompson-Varley coils; **м. с замещающими декадами** box containing two identical sets of resistors; **м. сопротивлений с многопозиционными переключателями** dial resistance box; **штепсельный м. ёмкостей** capacitance box with plugs, plug capacitance box
**магазинный** *adj.* magazine; box; store, warehouse; unit
**магистраль** *f.* main line, main route, artery, trunk line; intertoll trunk
**магистральный** *adj.* main-line, artery, trunk(-line)
**магический** *adj.* magic
**магметр** *m.* magmeter
**магнадур** *m.* magnadur
**магнакард** *m.* magnacard
**магналий** *m.* magnalium
**магнальный** *adj.* magnal
**магнаскоп** *m.* magnascope
**магнесин** *m.* magnesyn
**магнескоп** *m.* magnescope
**магнестат** *m.* magnestat
**магнетизёр** *m.* magnetizer
**магнетизёрский** *adj.* magnetizer
**магнетизёрство** *n.* magnetizing
**магнетизирование** *n.* magnetization
**магнетизированный** *adj.* magnetized
**магнетизировать, замагнетизировать, намагнетизировать** *v.* to magnetize

**магнетизм** *m.* magnetism; magnetics; **естественный м.** spontaneous magnetism; **м., обусловленный движением по орбите** orbital magnetic moment
**магнетит** *m.* magnetite, lodestone
**магнетитовый** *adj.* magnetite
**магнетический** *adj.* magnetic
**магнето** *n. indecl.* magneto (generator)
**магнетограф** *m.* magnetograph
**магнетометр** *m.* magnetometer; **м. с большим динамическим диапазоном** wide-range magnetometer; **м. с насыщенным сердечником** flux-gate magnetometer
**магнетон** *m.* magneton
**магнетопроводимость** *f.* magnetoconductivity
**магнетор** *m.* magnettor
**магниторезистор** *m.* magnetoresistor
**магнитоэлектрический** *adj.* magnetoelectric
**магнетрон** *m.* magnetron; **двухцокольный м.** double-ended magnetron; **одноцокольный м.** single-ended magnetron; **разнополостный м., разнорезонаторный м.** "rising-sun" magnetron; **м. с беличьим колесом** donutron, squirrel-cage magnetron; **м. с неразрезным анодом** diode magnetron; **м. с электронной подстройкой частоты** voltage-tunable magnetron; **м. со связками** strapped magnetron; **сверхмощный м.** multimegawatt magnetron; **стабилизованный по методу захвата м.** injection-locked magnetron **стержневой м.** interdigital magnetron
**магнетронный** *adj.* magnetron
**магниевый** *adj.* magnesium
**магний** *m.* magnesium, Mg
**магнистор** *m.* magnistor
**магнит** *m.* magnet; **м. возврата стрелки** release magnet, resetting magnet; **м. для ступенчатого запуска** stepping magnet; **заграждающий м.** application magnet; **нажимной м.** printing magnet; **оттормаживающий**

277

**путевой м.** reset magnet; **м. подъёма** vertical magnet; **путевой м.** application magnet; **тактовый м.** cadence magnet, time tapper

**магнитизм** *m.* magnetism, magnetics

**магнитно-дипольный** *adj.* magnetic-dipole

**магнитно-дисковый** *adj.* magnetic-disk

**магнитножёсткий** *adj.* magnetically hard, hard-magnetic

**магнитно-ионный** *adj.* magneto-ionic

**магнитномягкий** *adj.* magnetically soft, soft-magnetic

**магнитно-ртутный** *adj.* magnetic-mercury

**магнитность** *f.* magnetizability

**магнитный** *adj.* magnetic, magnetical, magnet

**магнито-** *prefix* magneto-

**магнитоакустический** *adj.* magneto-acoustic

**магнитогазодинамика** *f.* magnetogasdynamics

**магнитогидродинамика** *f.* magnetohydrodynamics

**магнитогидродинамический** *adj.* magnetohydrodynamic, hydromagnetic

**магнитограмма** *f.* magnetogram

**магнитограф** *m.* magnetograph

**магнитография** *f.* magnetography

**магнитодвижный** *adj.* magnetomotive

**магнитодержатель** *m.* magnet support, magnet cradle, magnet carrier, magnet holder

**магнитодиод** *m.* magnetodiode

**магнитожёсткий** *see* **магнитножёсткий**

**магнитозвуковой** *adj.* magneto-acoustic; magneto-ionic

**магнито-ионный** *adj.* magneto-ionic

**магнитомер** *m.* magnetometer, magnet tester; **астатический м. компенсационного типа** null astatic magnetometer; **м. с компенсационной обмоткой** null coil magnetometer; **м. с насыщением** saturable magnetometer; **м. с насыщенным сердечником** flux-gate magnetometer

**магнитометрический** *adj.* magnetometric; magnetometer

**магнитометрия** *f.* magnetometry

**магнитомеханический** *adj.* magnetomechanic; magnetomotive

**магнитомоторный** *adj.* magnetomotive

**магнитооптика** *f.* magneto-optics

**магнитооптический** *adj.* magneto-optic

**магнитоплазмадинамика** *f.* magnetoplasmadynamics

**магнитоплазменный** *adj.* magnetoplasma

**магнитопровод** *m.* magnetic circuit, leg

**магнитопроводимость** *f.* magnetoconductivity

**магниторезистивность** *f.* magnetoresistivity

**магниторезистивный** *adj.* magnetoresistive

**магнитоскоп** *m.* magnetoscope

**магнитосопротивление** *n.* magnetoresistance; magnetoresistor

**магнитостабильный** *adj.* magnetically-stable

**магнитостатика** *f.* magnetostatics

**магнитостатистический** *adj.* magnetostatic

**магнитостриктор** *m.* magnetostrictor

**магнитострикционный** *adj.* magnetostrictive, magnetostriction

**магнитострикция** *f.* magnetostriction

**магнитосфера** *f.* magnetosphere

**магнитотеллурический** *adj.* magnetotelluric

**магнито-тепловой** *adj.* magneto-thermal

**магнитотранзисторный** *adj.* transistor-magnetic

**магнитоуловитель** *m.* magnetic detector

**магнитоупругий** *adj.* magnetoelastic

**магнитоупругость** *f.* magnetoelasticity

**магнитофон** *m.* tape recorder, magnetic sound recorder; magnetophone; **м. воспроизведения, воспроизводящий м.** magnetic tape recorder; **м. для совещаний** conference recorder; **стереофонический м.** binaural tape recorder

**магнитофонный** *adj.* magnetophone; magnetic (tape) recorder, tape recorder

**магнитохимия** *f.* magnetochemistry

**магнитоэлектрический** *adj.* magnetoelectric

**магнитоэлектричество** *n.* magnetoelectricity

**магнитоякорный** *adj.* magnetic-armature

**магнон** *m.* magnon

**магслип** *m.* magslip, magnetic slip-ring

**Мадер** Mader

**маджикор** *m.* magicore
**мадистор** *m.* madistor
**мажор** *m.* major key (*mus.*)
**мажоритарный** *adj.* majority
**мажорный** *adj.* major
**мазать, замазать, намазать, помазать**
 *v.* to smear, to grease
**мазер** *m.* maser; **двухуровневый м.**
 **генераторного типа** two-level maser
 oscillator; **односторонний м.** unilateral maser amplifier; **м. с циркуля-**
 **торным устройством** circular maser;
 **трёхуровневый м. твёрдого состоя-**
 **ния** three-level solid-state maser
**мазня** *f.* blur, blurring, smear
**мазок** *m.* smear
**мазурий** *m.* masurium, Ma
**мазь** *f.* salve, ointment; grease
**майер** *m.* mayer
**Майкельсон** Michelson
**майкротерм** *m.* microtherm
**майлар** *m.* milar, mylar
**майларовый** *adj.* mylar, milar,
**майорана** Majorana
**майорановский** *adj.* Majorana
**майоранта** *f.* series consisting of positive terms only
**майснеровский** *adj.* Meissner
**макет** *m.* model, mock-up, dummy,
 simulator, prototype; working diagram
**макетирование** *n.* breadboarding
**макетный** *adj.* breadboard, model,
 protoype, dummy
**МакЛеод** McLeod
**Маклорен** Maclaurin
**Мак-Мас** McMath
**макро-** *prefix* macr-, macro-; large-
 scale, broad-scale
**макровозмущение** *n.* broadscale perturbation
**макро-генератор** *m.* macro-generator
**макро-генерирующий** *adj.* macro-
 generating
**макрография** *f.* macrography
**макроиндикатор** *m.* high-speed indicator
**макроклимат** *m.* macroclimate
**макрометеорит** *m.* macrometeorite
**макрометр** *m.* macrometer
**макроосциллографический** *adj.* macro-
 oscillographic
**макропрограмма** *f.* macroprogram,
 automation plan
**макропрограммирование** *n.* macropro-
 gramming

**макроскопический** *adj.* macroscopic,
 macroscopical
**макроскопия** *f.* macroscopy
**макростатистика** *f.* macrostatistics
**макроструктура** *f.* macrostructure
**макросхема** *f.* macrocircuit
**макрочастица** *f.* macroparticle
**макроэлемент** *m.* macrocell
**макс.** *abbr.* 1 (**максимальный**) maximum; 2 (**максумум**) maximum
**максвелл** *m.* maxwell
**максвеллвиток** *m.* maxwell-turn
**максвеллметр** *m.* maxwellmeter
**максвелловский** *adj.* maxwellian, Maxwell's
**максимально** *adv.* maximally, maximum, as much as possible, at most
**максимально-возможный** *adj.* maxi-
 mum-available
**максимально-допустимый** *adj.* maximum-permissible
**максимальный** *adj.* maximum, highest,
 top, peak, greatest, optimum
**максимизироваться** *v.* to maximize
**максимизирующий** *adj.* maximizing
**максимин** *m.* maximin, peak
**максимум** *m.* maximum, upper limit,
 highest quantity, peak; **м. белого**
 peak white (*tv*); **м. пеленгации** loop-
 position of maximum intensity; **м.**
 **чёрного** peak black (*tv*)
**маленький** *adj.* small, little; slight;
 diminutive
**мало** *adv.* little, few, not enough
**малоактивный** *adj.* low-activity; low-
 level
**малоамперный** *adj.* low-current, low-
 amperage
**малоблагоприятный** *adj.* scarcely con-
 ducive, scarcely favorable
**маловажный** *adj.* unimportant
**маловаттный** *adj.* low-watt
**маловероятный** *adj.* unlikely, hardly
 probable
**маловысотный** *adj.* low-level
**малогабаритный** *adj.* midget, miniature, small-scale; bantam
**малодействительный** *adj.* ineffective
**малодостоверный** *adj.* unlikely, not
 well-founded
**малодоступный** *adj.* inaccessible
**малое** *n. adj. decl.* little; **малые**
 **высшего порядка** high-order small
 parameters; **бесконечно м.** infinitesimal
**малоёмкий, малоёмкостьный** *adj.*

small-capacity, low-capacity; small-sized

**малозначный** *adj.* unimportant, of little significance

**малоиндукционный** *adj.* low-induction

**малоинерционный** *adj.* low-inertia, quick-response

**малоинтенсивный** *adj.* low-intensity; low(-flux)

**малокалиберный** *adj.* small-caliber

**малоконтрастный** *adj.* soft (picture)

**малолитражный** *adj.* low-powered

**маломощный** *adj.* low-powered, low-capacity, low-level, low-duty, low-yield

**малонаправленный** *adj.* semidirectional

**малорадиоактивный** *adj.* weakly radioactive

**малосигнальный** *adj.* small-signal

**малосильный** *adj.* low-powered; weak

**малоскоростной** *adj.* low-velocity, low-speed

**малостойкий** *adj.* not stable

**малострочный** *adj.* coarse, low-definition

**малость** *f.* smallness, littleness; trifle

**малоупотребительный** *adj.* rare, rarely used

**малоценный** *adj.* inferior, of inferior quality

**малочисленность** *f.* small number

**малочисленный** *adj.* scanty, few, not numerous

**малочувствительный** *adj.* insensitive, low-sensitivity

**малошумный, малошумовой, малошумящий** *adj.* low-noise, quiet; noise-less

**малтиар** *m.* multiar

**малый** *adj.* small, little, minor; fine; light; weak; shallow; low; **малой энергии** low-energy; **бесконечно м.** infinitesimal

**мальтаза** *f.* maltase

**мальтийский** *adj.* Maltese

**манганат** *m.* manganate

**манганин** *m.* manganin

**манганиновый** *adj.* manganin

**манганит** *m.* manganite

**мандолина** *f.* mandoline

**манёвр** *m.* maneuver

**манёвренность** *f.* maneuverability

**манёвренный** *adj.* maneuvering, maneuver

**маневрирование** *n.* maneuvering

**маневрировать** *v.* to maneuver, to manipulate, to operate; to shunt

**маневрирующий** *adj.* maneuverable

**маневровый** *adj.* maneuvering; shunting

**манжета** *f.* baffle, shield; collar, flap, sleeve

**манильский** *adj.* (of) Manilla; manilla

**манипулирование** *n.* handling, manipulation, operation; keying

**манипулированный** *adj.* manipulated; keyed

**манипулировать** *v.* to manipulate, to handle, to operate; to key; to treat

**манипулируемый** *adj.* manipulated; keyed

**манипулятивный** *adj.* manipulation

**манипулятор** *m.* manipulator, manipulating key, signaling key, keyer; operator; handler; **м. для испытания магнитных сердечников** memory core handler; **импульсный м.** impulse sending key; **м. по длительности импульсов** pulse-width keyer; **частотный м.** frequency-shift keyer

**манипуляторный** *adj.* manipulator, keyer; keying

**манипуляционный** *adj.* manipulation, handling; keying

**манипуляция** *f.* manipulation, handling, operation; keying, key modulation, on-off modulation; **м. в первичной цепи** primary keying; **высоковольтная м.** plate keying (*tphny.*); **м. обрывом** open-circuit keying; **м. обычного типа** on-off signaling; **м. по защитной сетке** suppressor-grid keying; **м. сдвигом фазы** phase-shift keying; **м. смещением** blocked-grid keying

**мановакуумметр** *m.* vacuum manometer

**манограф** *m.* manograph

**манокриометр** *m.* manocryometer

**манометр** *m.* pressure gage, manometer; vacuum gage; **альфаионизационный м.** alphatron gage; **двухжидкостный дифференциальный м.** differential multiplying manometer; **дифференциальный жидкостный м.** liquid level manometer; **м. для измерения давления питания** boost pressure gage; **жидкостный м.** mercury (vacuum) gage, liquid (vacuum) gage; **U-образный батарейный**

**м.** multiple U-gage, U-tube manometer; **однотрубный м.** single column manometer; **поворотный м.** tilting manometer; **поршневой м. с весовой нагрузкой** dead-weight piston gage; **радиевый м.** radium-type vacuum gage; **м. с безнулевой шкалой** suppressed zero pressure gage; **м. с воздушным пузырьком в капиллярной трубке** air-bubble capillary tube manometer; **м. с прерывателем** click gage; **тепловой м.** hotwire gage

**манометрический** *adj.* manometric, manometer

**маностат** *m.* manostat

**Манселл** Munsell

**мантисса** *f.* mantissa

**марганец** *m.* manganese, Mn; **с присадкой марганца** doped with manganese

**марганцевый** *adj.* manganese

**марганцово-цинковый** *adj.* manganese-zinc

**марганцовый** *adj.* manganic, manganese

**мареограф** *m.* marigraph, tide-gage

**«маримба»** *f.* marimba

**Мариотт** Mariotte

**марка** *f.* mark, sign, stamp, brand, make, trade mark, designation strip; grade, sort, quality; post stamp; **м. кабеля** cable make-up; **м. лампы** tube symbol, tube type designation

**маркёр** *m.* marker, label, tag; selector position-marker; **z-м.** zone marker; **м. с конусом молчания** cone-of-silence marker

**маркёрный** *adj.* marker, marking, label

**маркирование** *n.* marking, labelling, characterization

**маркированный** *adj.* marked, labelled, stamped

**маркировать** *v.* to mark, to indicate, to stamp; to gage

**маркировка** *f.* marking, stamping, designation, labelling, tagging; **ведущая м., направляющая м.** radar markers

**маркировочный** *adj.* marking, stamping, labelling, tagging

**маркировщик** *m.* marker

**маркирующий** *adj.* marking, stamping, labelling, tagging

**марковский** *adj.* Markovian; Markoff's, (of) Markoff

**Маркони** Marconi

**маркшейдер** *m.* surveyor

**маркшейдерский** *adj.* surveying

**марлекс** *m.* marlex

**марлинь** *m.* marline, houseline

**марля** *f.* gauze

**мартеновский** *adj.* Martin's

**маршрут** *m.* route, course

**маршрутный** *adj.* route, course

**ма. сек.** *abbr.* (миллиампер-секунда) milliampere-second

**масер** *see* мазер

**маска** *f.* mask, mat, matte; **блуждающая м.** traveling matte; **м. для использования параллакса электронного луча** parallax mask; **теневая м.** aperture mask, shadow mask, aperture plate; **м. цветной приёмной телевизионной трубки** color-selecting mask

**маскирование** *n.* masking; shadowing

**маскированный** *adj.* masked, disguised, camouflaged; concealed

**маскировать, замаскировать** *v.* to mask, to disguise, to camouflage

**маскировка** *f.* masking, disguise, camouflage; concealment; screening; **звуковая м., слуховая м.** aural masking

**маскировочный** *adj.* masking, disguising, camouflaging; concealing

**маскируемый** *adj.* masked

**маскирующий** *adj.* masking, disguising, camouflaging; concealing

**маслёнка** *f.* grease cup; oil can; lubricating valve

**масло** *n.* oil; **м. из дёгтя** tar oil

**маслогасимый** *adj.* oil-quenched

**масломер** *m.* oil gage

**маслонаполненный** *adj.* oil-filled

**маслонепроницаемый** *adj.* oiltight, oilproof, greaseproof

**маслополный** *adj.* oil-filled

**маслопровод** *m.* oil-piping layout; oil pipeline, oil line

**маслопроводный** *adj.* oil-piping

**маслопропитанный** *adj.* oil-impregnated

**маслостойкий** *adj.* oilproof

**масляно-водяной** *adj.* oil-water

**масляно-воздушный** *adj.* oil-air

**масляно-пневматический** *adj.* oleo-pneumatic

**масляный** *adj.* oil, oily, greasy

**масочный** *adj.* mask

**масса** *f.* mass; bulk, volume; shoal, swarm; block; substance, compound, composition, matter; body (of casting); medium; **активная м.** active material, filling paste; **заливочная м.** filling compound, fill-in paste; **приведённая м.** reduced mass; **присоединённая м.** added mass; **соколеблющаяся м.** added medium; added mass

**массив** *m.* body, solid mass; block, group, array; file (of data)

**массивный** *adj.* massive, bulky, solid, heavy, substantial, sturdy, compact

**массовый** *adj.* mass, bulk

**масс-параметр** *m.* mass parameter

**масс-сепаратор** *m.* mass separator

**масс-спектр** *m.* mass spectrum

**масс-спектрограф** *m.* mass spectrograph

**масс-спектрометр** *m.* mass spectrometer

**масс-спектрометрический** *adj.* mass-spectrometer

**масс-спектрометрия** *f.* mass spectrometry

**мастер** *m.* expert, master, skilled workman; **м. по гальваноклише** electrotyper, electrotypist

**мастерская** *f. adj. decl.* workshop, shop

**мастика** *f.* putty, paste, cement, composition

**мастикатор** *m.* masticator

**мастичный** *adj.* mastic, cement, paste, putty

**масштаб** *m.* scale, gage, rule, measuring rule; rate; **в истинном масштабе времени** real-time; **большой м.** high range; **логарифмический м. по обеим осям** log-log plot; **логарифмический м. по одной из осей** semilog plot; **меньший м.** lower range; **полулогарифмический м.** semilog plot; **раздвижной м.** slide gage; **условный м.** representative scale

**масштабирование** *n.* scaling; multiplexing

**масштабный** *adj.* scale, gage, rule; rate

**мат** *m.* mat; matte finish

**маттаух** Mattauch

**математик** *m.* mathematician

**математика** *f.* mathematics

**математический** *adj.* mathematical

**материал** *m.* material; data; cloth; fabric; **намазываемые акустические материалы, наносимые акустические материалы** troweled-on acoustical materials; **амортизующий м.** resilient material; **м. без потерь** nonlossy material; **м. вызывающий потери** lossy material; **вязочный м.** binder, ties; **дефицитный м.** critical material; **магнитный м. с большой остаточной индукцией** highly remanent magnetic material; **м., обладающий гистерезисными свойствами** hysteretic material; **м. с избирательным поглощением** frequency-selective damping material

**материализация** *f.* materialization

**материаловедение** *n.* science of material

**материальный** *adj.* material, physical

**материя** *f.* matter, substance; material, fabric

**матерчатый** *adj.* fabric, cloth

**матированный** *adj.* frosted, depolished, ground; mat, dull

**матировать** *v.* to frost, to grind; to dull, to deaden; to tarnish

**матовый** *adj.* dull, lusterless, mat; frosted; ground

**матрица** *f.* matrix (*math.*); stamper; die; case (*electro-typing*); array; plane; **м. для изготовления пластинок звукозаписи** record stamper; **м. для оптического вывода данных** graphical plotting matrix; **дуплексная м.** backed stamper; **запоминающая м. на ферроэлектриках** ferroelectric memory matrix; **м. запоминающего устройства** memory bit; **м. комплексной проводимости** admittance matrix; **м. кристаллового сопротивления** crystal-resistor matrix; **накопительная сердечниковая м.** memory core plane; **м. неопределённой проводимости** indefinite-admittance matrix; **м. памяти «двойная точка»** double dot pattern; **м. памяти «точка-окружность»** dot-circle pattern; **м. памяти «точка-ореол»** dot-blur pattern; **м. памяти «точка-тире»** dot-dash pattern; **переключательная м. на неоновых лампах и фотосопротивлениях** neon-photoconductive switching matrix; **м. полной проводимости** admittance matrix; **положительная**

м. mold; **присоединённая м.** adjoint matrix; **м. разряда** bit plane; **м. с переключающимися выводами диодов** trunking matrix
**матрицирование** *n.* matrixing
**матрицированный** *adj.* matrixed
**матричный** *adj.* matrix; stamper; die
**Матьё** Mathieu
**Max** Mach
**мах** *m.* motion, move, stroke, wave; oscillation, vibration
**Maxe** Mache
**маховик** *m.* flywheel; **ручной м.** handwheel
**маховичный, маховой** *adj.* flywheel
**мачта** *f.* mast, post, pole, tower, support, column; **анкерная м.** dead-end tower, strutted pole, stayed pole; **м. на пересечении** transposition pole; **м. с оттяжками** guyed mast; **свободнонесущая м.** pylon, guyed mast
**мачтовый** *adj.* mast, mast-type, post, tower, pole, support; *also as suffix:* двухмачтовый two-pole, double-pole
**Машерон** Masceroni
**машина** *f.* machine, engine; mechanism; computer; **м.-аналог** analog computer; **батарейная м. вызывного тока** battery ringer; **вентилируемая электрическая м. с воздушным охлаждением** ventilated radiator machine; **взрывонепроницаемая электрическая м. для подземных выработок** fire-damp proof machine; **вибрационная м. с механическим** (*or* **кривошипным**) **приводом** brute force vibration machine; **вольтодобавочная м. для зарядки** battery charging booster; **вольтодобавочная электрическая м.** positive booster; **вольтопонижающая электрическая м.** negative booster; **выдающая м.** paying-out machine; **вызывная м., м. вызывного тока** ringing set, ringing generator, ringing dynamo, ringer; **м. высокой частоты** high frequency alternator; **вычислительная м.** computer; **в. м. для определения производительности** machinability computer; **в. м. для расчёта (атомного) реактора** reactor computer; **в. м. для расчётов службы тыла** logistic computer; **в. м. для решения задач дальней навигации** long-range navigation computer; **в.**

**м. импульсного типа** alternating-current computer; **в. м. малого быстродействия** low-speed computer; **в. м. насветоводах** optical path computer; **в. м. ограниченных возможностей** small-scale computer; **в. м. огромных возможностей** giant-scale computer; **в. м. потенциального типа** direct-current computer; **в. м., работающая в двоичной системе счисления** radix two computer; **в. м., работающая в истинном масштабе времени** real-time computer; **в. м. с автоматической регулировкой десятичной запятой** automatic decimal point computer; **в. м. с запоминаемой программой** stored program computer; **в. м. с запоминающим устройством на тонких плёнках** thin-film memory computer; **в. м. с наборной программой** plugged program computer, wired program computer; **в. м. с периодизацией решения** repetitive computer; **в. м. с последовательно выполняемой программой** sequenced computer; **в. м. с программным управлением** sequence-controlled computer; **в. м. с программным управлением от перфокарт** card-programmed calculator, card-programmed computer; **в. м. с совмещением операций** simultaneous computer; **в. м. сверхвысокой производительности** giant-powered computer; **в. м., собранная из различных функциональных блоков** multi-unit computing machine; **в. м. средних возможностей** medium-scale computer; **дифференциальная вольтодобавочная м.** differential booster; **м. для автоматического контроля перфокарт** punched hole verifier; **м. для заварки смонтированной ножки** mount sealing-in machine; **м. для зачистки изоляции проводов** wire-stripping machine; **м. для обработки информации с внешним накопителем** file processor; **м. для опрыскивания кратеров** target spray machine; **м. для приварки горловины колбы** bulb neck splicing machine; **м. для сборки ножки** stem machine; **м. для скрутки пар** twinning machine; **м. для скрутки четвёрток** quadding machine; **м. для спуска кабеля**

paying-out machine; **м. для штенгелевки колб** bulb tabulating machine; **каплезащищённая электрическая м. с защитной сеткой** dripproof screen-protected machine; **карусельная м.** turret type machine; **клавишная м.** key-actuated machine, unityper; **крутильная м.** twisting machine; stranding machine, quadding machine; **обдувочная м.** bulbblowing machine; **обратная вольтодобавочная м.** reversible booster; **м. ответа голосом** machine answering device; **раскладочная м.** collator; **рулевая м.** control actuator; **синусокосинусная м.** resolver; **м. со случайным выходом** chance machine, randomizer; **собственно вычислительная м.** central computer; **сортировально-подборочная м.** collator, interpolator; **счётная м. последовательного действия** serial computing machine; **счётная м. с задаваемой последовательностью действия** selective sequence calculator; **счётно-аналитическая м.** punched card machine; **теленаборная м.** teletypesetter; **цифровая вычислительная м. параллельного действия** parallel digital computer; **цифровая вычислительная м. последовательного действия** serial digital computer; **цифровая интегрирующая м.** incremental computer; **электрическая м. открытого исполнения** open type machine; **электрическая м. с охлаждаемой ребристой поверхностью** ventilated ribbed surface machine; **электрическая м. с разомкнутой системой вентиляции** machine with open-circuit ventilation; **электрическая м. со смешанным возбуждением** compound-wound machine, compound generator

**машинист** *m.* engineer, engineman

**машинка** *f.* small machine, device; **м. для зачистки концов проводов** wire skinner, wire stripper

**машинный** *adj.* machine, engine; power (driven); mechanical, automatic; computer

**машиноведение** *n.* engineering

**машинописный** *adj.* typewritten; typewriter

**машиностроение** *n.* mechanical engineering; machine building

**маяк** *m.* beacon, signal tower, lighthouse; **аэропортовый м. для ориентировки самолётов вслепую** blind approach (airport) beacon; **главный линейный м.** principal airway beacon; **глиссадный м.** landing beam beacon, glide slope facility; **допплеровский м. для определения скорости управляемых снарядов** Doppler velocity and position, Dovar; **звуковой м.** aural-type beacon, talking radio beacon, aural radio range; **линейный м.** airway beacon; **непрерывно излучающий радиолокационный м.** radar marker, Ramark; **м. опознавания целей** target identification beacon; **опознавательный м.** land mark beacon; **м.-ответчик** responder beacon, transponder beacon; **отличительный световой м.** land mark beacon; **м. посадочной дорожки** runway-localizing beacon; **м. предупреждения об опасности** hazard beacon; **приводной м.** homing beacon; **приводной маркёрный м.** approach-marker beacon-transmitter; **радиогидроакустический м.** radiosonic ranging system; **радиолокационный м. для вывода в заданную точку** radar homing beacon; **радиосигнальный м.** radio-range beacon, visual radio range; **рамочный м.** rotating-loop beacon; **м.-указатель** position indicator

**маятник** *m.* pendulum, bob, balance; **секундный м.** length of seconds pendulum

**маятниковый** *adj.* pendular, pendulum, swing; floating

**маятникообразный** *adj.* pendular

**маячковый** *adj.* lighthouse

**маячный** *adj.* beacon

**МБ** *abbr.* (**местная батарея**) local battery

**мб** *abbr.* (**миллибар**) millibar

**МБР** *abbr.* (**межконтинентальная баллистическая ракета**) intercontinental ballistic missile

**МБС** *abbr.* (**межконтинентальный баллистический снаряд**) intercontinental ballistic missile

**МВ** *abbr.* (**магнит вращения**) rotary magnet

**мв** *abbr.* (**милливольт**) millivolt

**мвт** *abbr.* (**милливатт**) milliwatt
**мг** *abbr.* (**миллиграмм**) milligram
**мгвт** *abbr.* (**мегаватт**) megawatt
**мгвт-ч** *abbr.* (**мегаватт-час**) megawatt-hour
**мггц** *abbr.* (**мегагерц**) megacycle per second, megahertz
**мгдж** *abbr.* (**мегаджоуль**) megajoule
**мгн** *abbr.* (**миллигенри**) millihenry
**мгновение** *n.* instant, moment
**мгновенность** *f.* instantaneousness
**мгновенный** *adj.* momentary; instantaneous, instant, prompt; flash
**мгом** *abbr.* (**мегом**) megohm
**Мд** *abbr.* (**масштаб дальности**) range scale
**м.д.с.** *abbr.* (**магнитодвижущая сила**) magnetomotive force
**меандеровый, меандерообразный** *adj.* meander
**меандрический** *adj.* meander
**Мёбиус** Möbius
**мег-, мега-** *prefix* meg-, mega-
**мегабар** *m.* megabar
**мегабит** *m.* megabit
**мегаварметр** *m.* megavarmeter
**мегаватт** *m.* megawatt; **м.-час** megawatt-hour
**мегавольт** *m.* megavolt
**мегагаусс** *m.* megagauss
**мегагерц** *m.* megahertz, megacycle per second
**мегагерцный** *adj.* megahertz, megacycle
**мегаджоуль** *m.* megajoule
**мегадина** *f.* megadyne
**мегалайн** *m.* megaline
**мегалиния** *f.* megaline
**мегамега** *f.* megamega
**мегамегагерц** *m.* megamegacycle
**мегаметр** *m.* megameter
**мегампер** *m.* megampere
**меганод** *m.* meganode
**мегарад** *m.* megarad
**мегасвип** *m.* megasweep
**мегатрон** *m.* megatron, lighthouse tube
**мегафарада** *f.* megafarad, macrofarad
**мегафон** *m.* megaphone, speaking trumpet
**мегафонный** *adj.* megaphone
**мегацикл** *m.* megacycle
**мегаэрг** *m.* megaerg
**мега-электрон-вольт** *m.* mega-electron volt

**меггер** *m.* megger, "meg" insulation tester, megohmmeter, earth tester
**мегеварметр** *m.* megavarmeter
**Мёгель** Moegel
**мегом** *m.* megohm
**мегометр** *m.* megohmmeter, megger
**мегомит** *m.* megomit
**мегомметр** *m.* megohmmeter, megger
**мегомный** *adj.* megohm
**меготальк** *m.* megotalc
**мегэрг** *m.* megerg
**медиан** *m.*, **медиана** *f.* median
**медианный** *adj.* median
**медианта** *f.* mediant
**медиатор** *m.* plectrum
**медицинский** *adj.* medical
**медленно** *adv.* slowly; **м. срабатываемый** slow operating
**медленновращающийся** *adj.* slow-speed
**медленно-действующий** *adj.* slow, slow-acting, slow-operating
**медленно-отпускающий** *adj.* slow-release, slow-releasing
**медленно-потухающий** *adj.* long-persistence
**медленно-притягивающий** *adj.* slow-operating (relay)
**медленность** *f.* slowness
**медленный** *adj.* slow, sluggish, low-speed
**медлительность** *f.* sluggishness
**медлительный** *adj.* sluggish, slow
**медно-** *prefix* cupri-, cupric, copper
**меднозакисный** *adj.* cuprous-oxide, copper-oxide
**медноконстантановый** *adj.* copper-constantan
**меднокупоросный** *adj.* copper-sulphide
**медноокисный** *adj.* copper-oxide
**медноплакированный** *adj.* copper-plated
**медносульфидный** *adj.* copper-sulphide
**медный** *adj.* copper, cupric, cuprous
**медь** *f.* copper, Cu; **с перерасходом меди** overcoppered; **основная уксуснокислая м.** verdigris, copper rust; **сернокислая м.** copper sulphate; blue vitriol
**медь-константан** *m.* copper-constantan
**меж-** *prefix* inter-
**межатомный** *adj.* interatomic
**межгрупповой** *adj.* between-group
**междолинный** *adj.* intervalley
**междоузельный** *adj.* interstitial; internodal

**междоузлие** *n.* internode; interstice, interstitial, interstitial site

**междоузловой** *see* **междоузельный**

**между** *pre. instr. and genit.* between, among; **м. зажимами** between the terminals; (connected) across; **м. сети** grid-to-grid

**между-** *prefix* inter-, between

**междуатомый** *adj.* interatomic

**междублочный** *adj.* interunit

**междуведомственный** *adj.* interdepartmental

**междувитковый** *adj.* turn-to-turn, interturn

**междугородный** *adj.* intercity, interurban, long-distance, trunk, toll (*tphny.*)

**междузубье** *n.* tooth jag

**междукаскадный** *adj.* interstage

**междукоммутаторный** *adj.* interswitchboard

**междукристаллический** *adj.* intercrystalline

**междуламповый** *adj.* intertube

**междулинзовый** *adj.* between-the-lens

**междумолекулярный** *adj.* intermolecular

**международный** *adj.* international, standard

**междуобмоточный** *adj.* interwinding

**междупарный** *adj.* pair-to-pair

**междуполюсный** *adj.* interpolar

**междупроводной** *adj.* wire-to-wire

**междупутный** *adj.* intertrack

**междупутье** *n.* intertrack space, track spacing

**междурельсовый** *adj.* intertrack

**междусегментный** *adj.* intersegmental

**междуслойный** *adj.* interlaminar, interlayer, interleaving

**междустанционный** *adj.* interoffice

**междустрочный** *adj.* interlinear

**междуточечный** *adj.* interdot

**междуузлие** *n.* internode; interstice

**междуштырьковый** *adj.* inter-pin

**междуэлектродный** *adj.* interelectrode

**междуэтажный** *adj.* interfloor

**межевание** *n.* survey, surveying

**межевой** *adj.* boundary

**межзвёздный** *adj.* interstellar

**межзвеньевый** *adj.* interlinkage

**межзернистый** *adj.* intergranular

**межзонный** *adj.* interband

**межионный** *adj.* inter-ion

**межканальный** *adj.* interchannel

**межкаскадный** *adj.* interstage

**межконтинентальный** *adj.* intercontinental

**межкристаллитный, межкристаллический** *adj.* intercrystalline

**межлабораторный** *adj.* interlaboratory

**межобмоточный** *adj.* interwinding

**межпланетный** *adj.* interplanetary

**межрадиальный** *adj.* interradial

**межразрядный** *adj.* intercolumnar

**межрайонный** *adj.* "inter-raion"

**межремонтный** *adj.* between repairs

**межрешёточный** *adj.* interlattice

**межслойный** *adj.* interlamination

**межсоединение** *n.* interconnection

**межстанционный** *adj.* interoffice, interexchange

**межстрочный** *adj.* interline

**межточечный** *adj.* interdot

**межфазный** *adj.* interphase

**межъязыковый** *adj.* interlanguage

**межэлектродный** *adj.* interelectrode

**межэлементный** *adj.* interelement, interunit

**мезадиод** *m.* mesa diode

**мезаструктура** *f.* mesa structure, mesa design

**мезатранзистор** *m.* mesa transistor; **эпитаксиальный м. с двойной диффузией** double-diffused epitaxial mesa transistor; **эпитаксиальный м. с диффузионной базой** diffused-base epitaxial mesa transistor

**мездровый** *adj.* hide

**мезон** *m.* meson

**мезонный** *adj.* meson, mesonic

**мезопауза** *f.* mesopause

**мезопик** *m.* mesopeak

**мезопрограммирование** *n.* mesoprogramming

**мезоскопический** *adj.* mesoscopic

**мезосфера** *f.* mesosphere

**мезотрон** *m.* mesotron

**Мейдингер** Meidinger

**мейсель** *m.* chisel

**Мейснер** Meissner

**мейснеровский** *adj.* Meissner's

**мел** *m.* 1. mel; 2. chalk

**мелалит** *m.* melalith

**мелалитовый** *adj.* melalith

**меламиновый** *adj.* melamine

**мелизма** *f.* melisma (*mus.*)

**мелкий** *adj.* fine, small, minute; shallow

**мелко** *adv.* fine, very small

**мелкозернистость** *f.* fineness, fineness of grain

**мелкозернистый** *adj.* fine, fine-grained, close-grained

**мелкоклетчатый** *adj.* close-mesh, fine-mesh

**мелкокристаллический** *adj.* fine-crystalline, finely crystalline

**мелкомасштабный** *adj.* small-scale

**мелкопетлистый** *adj.* close-meshed, fine-meshed

**мелкопористый** *adj.* fine-pored, finely porous

**мелкосерийный** *adj.* small-scale

**мелкосетчатый** *adj.* close-meshed, fine-meshed

**мелкослоистый, мелкослойный** *adj.* thinly laminar; fine-grained

**мелкоструктурный** *adj.* fine-structure

**мелкоячеистый** *adj.* fine-mesh, close-mesh

**Меллен** Melin

**мелмак** *m.* melmak

**мелодически** *adv.* melodiously

**мелодический** *adj.* melodious, melodic, tuneful

**мелодичность** *f.* melodiousness

**мелодичный** *see* **мелодический**

**мелодия** *f.* melody, tune

**меломайн** *m.* melomine

**мелохорд** *m.* melochord

**«мелт-квенч»** *m.* melt-quench method

**мелькание** *n.* flicker, flickering, flashing, sparkling

**мембрана** *f.* membrane, diaphragm, film; plate; **м. бесконечной протяжённости** infinite membrane; **возбуждённая ударом м.** struck membrane; **возбуждённая щипком м.** plucked membrane; **подвижная м.** vibrating diaphragm; **угольная м.** carbon microphone, carbon transmitter

**мембранный** *adj.* membrane, membranous, diaphragm

**мемистер** *m.* memister

**мемистор** *m.* memistor

**мемноскоп** *m.* memnoscope

**мемоскоп** *m.* memoscope

**мемотрон** *m.* memotron

**мензура** *f.* ratio of pipe diameter to pipe length

**Мени** Mesny

**мениск** *m.* meniscus

**менисковый** *adj.* meniscus

**менискообразный** *adj.* meniscus-shaped

**Менсон** Munson

**Менье** Meusnier

**меньший** *comp. of* **малый**, lesser, smaller, least, minor; inferior, lower

**менять, поменять** *v.* to alter, to alternate, to change, to vary, to shift

**меняться, поменяться** *v.* to change, to vary, to shift, to fluctuate

**меняющийся** *adj.* changing, varying, fluctuating, alternating, intermittent

**мера** *f.* measure, size, dimension, rate; degree, extent; caliber, gage, standard; **меры против глушения** anti-jamming measures; **м. информации ёмкости по Хартли** Hartley's measure; **м. против управляемых снарядов** guided missile countermeasure

**меридиан** *m.* meridian

**меридианный** *adj.* meridian

**меридиональный** *adj.* meridional; meridian

**мерило** *m.* standard, measure, gage, scale

**меритель** *m.* measurer

**мерительный** *adj.* measuring

**мерить, померить, смерить** *v.* to measure, to gage, to meter

**мерка** *f.* measure, gage; measuring rod

**мерник** *m.* metering vessel, measuring tank, calibrated tank

**мерный** *adj.* measured, uniform, rhythmic; measuring; dimensional

**мероморфный** *adj.* meromorphic, fractional

**мероприятие** *n.* countermeasure, measure, action; practice

**Меррей** Murray

**мёртвый** *adj.* dead; idle

**Мерц** Merz

**мерцание** *n.* blinking, flicker, flickering, glimmer, glitter, scintillation, shimmer, winking, twinkling, flashing, gleam; flutter

**мерцать** *v.* to blink, to flicker, to glimmer, to glitter, to shimmer, to flash, to gleam

**мерцающий** *adj.* flickering, glimmering, glittering, scintillating

**месмеризм** *m.* mesmerism, electro-biology

**месдоза** *f.* load cell

**месскоффер** *m.* measuring set

**местность** *f.* locality, region, place, site; area

**местный** *adj.* local; domestic, native, home; partial; site

**место** *n.* place, spot, site, locality, location, seat, point, position; room, space; blank; **А-м.**, **м. абонентских телефонисток** A-position, outgoing position; **Б-м.** B-position, incoming (trunk) position; **белое м. оригинала** picture white (in facsimile); **геометрическое м.** locus; **геометрическое м. спектральных цветов на цветовом графике** spectral locus, spectrum locus; **геометрическое м. точек полной проводимости** admittance locus; **м. замыкания силовых линий** vanishing point; **м., засечённое звуковой разведкой** sound ranging location; **контрольное м.** supervisor's position, observation desk, observing desk; **ночное рабочее м.** night service position, concentration position; **опросное (рабочее) м.** answering position, outgoing position, A-position; **м. отпайки** tip-off point; **м. плохой слышимости** dead spot; **пустое м.** blank, blank space; **рабочее м.** switchboard position; **р. м. быстрой связи** toll switching position; **р. м. мгновенного включения** demand position; **р. м. при комбинированной эксплуатации** combination of recording and long-distance position; **р. м. с указателем номеров** call indicator position; **р. м. стола заказов** record position, trunk record position; **р. м. телефонистки для международного сообщения** international position; **р. м. телефонистки на соединительных линиях** toll switching position, junction switching position; **р. м. телефонисток на входящих соединительных линиях** incoming position, B-position; **р. м. телефонисток на соединительных линиях без шнуров** cordless B-position; **р. м. телефонисток на соединительных линиях с импульсником** key-sending B-position; **р. м. транзитной связи** through switchboard; **м. сварки ввода** lead wire weld; **свободное м. в кристаллической решётке** lattice vacancy; **м. соединения кабеля с воздушными проводами** wire lead-in point of a cable; **соединительное м. ручного обслуживания** Telex switchboard; **справочное м. для междугородных станций** toll information desk (*tphny.*); **справочное рабочее м.** information position, inquiry position; **м. сращивания** cable splice; **счислимое м.** dead reckoning position; **тастатурное рабочее м.** key sending position, key pulser position; **м. телефонисток на соединительных линиях** *see* **Б-место**; **транзитное рабочее м.** through position, tandem position, lending position; **транзитное междугородное рабочее м.** through switching position; **м. установки изолирующих стыков (за крестовиной)** fouling point; **м. хорошей слышимости** loud spot; **чёрное м. оригинала** picture black in facsimile

**местоназначение** *n.* destination

**местонахождение** *n.* position, location, spot, site, seat, localization

**местоопределение** *n.* fixing, position finding

**местоположение** *n.* location, locality, spot, site, position, seat; station; situation, locus

**месторасположение** *n.* location, situation; locus

**метаболон** *m.* metabolon

**метагалактика** *f.* metagalaxy

**метадин** *m.* metadyne

**метадинный** *adj.* metadyne

**метакоманда** *f.* meta-instruction

**метакриловый** *adj.* methacrylic

**металингвистика** *f.* metalinguistic

**металингвистический** *adj.* metalinguistic

**металл** *m.* metal; **вентильный м.** valve metal, barrier-layer metal; **м. с электрохимическими вентильными свойствами** electrochemical valve metal

**металлизация** *f.*, **металлизирование** *n.* metallization, metallizing, plating, metallic coating; bonding (of metallic parts); **горячая м.** metal spraying; **м. самолёта** aircraft bonding

**металлизированный** *adj.* metallized, plated

**металлизировать** *v.* to metallize, to plate, to coat with metal

**металлизованный** *adj.* metallized, plated

**металлизовать** *see* **металлизировать**

**металлизующий** *adj.* bonding

**металлический** *adj.* metallic, metal

**металло-** *prefix* metallo-, metal

**металлобумажный** *adj.* metallized paper, metallic-paper

**металловолокно** *n.* metal wool

**металлографический** *adj.* metallographic

**металлография** *f.* metallography

**металлоид** *m.* metalloid

**металлоискатель** *m.* metal locator, metal detector; **м. на полупроводниковых триодах** transistor locator, transistor metal locator

**металлокерамика** *f.* metal ceramics, sintering, cermet

**металлокерамический** *adj.* ceramic-metal, ceramic-to-metal, metal-ceramic

**металломагнитный** *adj.* metallomagnetic

**металлометр** *m.* metallometer, metal tester

**металлометрический** *adj.* metallometric

**металлообрабатывающий** *adj.* metalworking

**металлоорганический** *adj.* metalloorganic, organometallic

**металло-отражатель** *m.* speculum metal

**металлопластинчатый** *adj.* metalplate

**металло-полупроводниковый** *adj.* metal-semiconductor

**металлосодержащий** *adj.* metal-containing, metalliferous

**металлостеклянный** *adj.* metal-glass

**металлургический** *adj.* metallurgic, metallurgical

**металлургия** *f.* metallurgy

**метаматематика** *f.* metamathematics

**метаматематический** *adj.* metamathematical

**метанометр** *m.* methanometer

**метасистема** *f.* metasystem

**метаскоп** *m.* metascope

**метастабильность** *f.* metastability

**метастабильный** *adj.* metastable

**метатеория** *f.* metatheory

**метатитанат** *m.* metatitanate

**метахимический** *adj.* metachemical

**метацентр** *m.* metacenter

**метаязык** *m.* metalanguage

**метеозонд** *m.* meteo sounding balloon

**метеор** *m.* meteor

**метеорадиолокатор** *m.* weather radar

**метеорит** *m.* meteorite

**метеоритика** *f.* meteoritics

**метеоритный, метеоритовый** *adj.* meteoritic

**метеорический, метеорный** *adj.* meteor, meteoritic

**метеорограф** *m.* meteorograph, radiometeorograph

**метеорографический** *adj.* meteorographic

**метеорография** *f.* meteorography

**метеороид** *m.* meteoroid

**метеорологический** *adj.* meteorologic, meteorological, weather

**метеорология** *f.* meteorology

**метеосигнал** *m.* meteorological signal

**метеошар** *m.* meteo balloon

**метилметакрилат** *m.* methylmetacrylate

**метиловый** *adj.* methylated, methyl

**метить, наметить** *v.* to mark, to indicate, to label, to tag

**метка** *f.* mark, marking, sign, notch, marker; tag, label; tracer; score; **длинные метки шкалы** major graduations; **мелкие метки шкалы** minor graduations; **граничная м.** sentinel; **кольцевая калибрационная м.** range marks, distance marks; **м. конца массива** end-of-file mark; **м. на машинной ножке** stem mark; **м. положения** sprocket bit; **м. рабочего положения** position mark **м. частоты** frequency marker, frequency pip

**Меткаф** Metcalf

**метод** *m.* method, technique, process, way, procedure, mode; **по методу** by (*oft.* omitted in translation: e.g. **аудиометр по методу маскировки** screening audiometer); **по импульсному методу** pulse-wise; **методы « длинных линий »** transmission-line methods; **методы повторения решений** resetting methods; **амплитудный м.** amplitude-difference method; **м. амплитудных сеток** (*or* **решёток**) amplitude grating method; **астатический м. регулирования** floating control method; **м. Брэгга** method of crystal analysis; **м. быстрейшего спуска** method of steepest descent; **м. взаимной нагрузки** back-to-back method; **м. вплавления-диффузии** post alloy diffusion method; **м. вращающегося изогнутого образца** rotating cantilever beam method; **м. вращающегося**

прерывателя contactor method; м. встречного включения opposition method; м. выбега retardation method, deceleration method; м. вызова с двойным опросом direct trunking, ring down connection; м. геометрического места корней root-locus method; гиперболический м. hyperbolic (radio) navigation method; м. горизонтально перемежающейся развёртки horizontal-interlace technique; м. горячей запрессовки heated-die pressing process; дальний радиолокационный м. измерения radar lorar system, long-range accuracy radar system; м. двойного сдвига double rank technique; м. двухцветного воспроизведения two-color process; м. двухчастотной тональной модуляции double-tone system; м. зернового растра «агфаколор» Agfacolor granulated screen printing-method; м. измерения звукового давления во внешнем слуховом канале outer-ear canal method; м. измерения давления на барабанной перепонке eardrum pressure method; м. измерения искажения гармоник harmonic distortion method; м. измерения (мощности звука) в слабореверберирующем помещении semireverberant-field method; м. измерения сопротивлений по величине падения напряжения fall-of-potential method; м., использующий перемежение вдоль строки horizontal-interlace technique; м. калибровки в замкнутой камере closed-chamber method; м. калибровки в свободном поле free-field method; м. компоновки (программ) assembly method; м. линзового растра «агфаколор» Agfacolor lenticular screen printing-method; м. магнитного перешейка Ewing method, isthmus method; м. междугородной линейной связи trunk-offering method; м. мыльных пузырей soap-bubble method; м. накалённого зонда search-filament method, exploring-filament method; м. наплыва wipe screen effect; м. наращивания rate growing; м. негативной модуляции "CR-burn" method; м. неопределённых множителей Лагранжа Lagrange-

multiplier method; м. обнаружения подводных лодок с помощью гидроакустических приборов asdic method; м. обратного плавления melt-quench method, meltback method; ограниченно адитивный м. pseudo-additive combination of overtones; однозондовый м. исследования потенциального поля single-probe-potential profile method; м. одноразового покрытия single-coat method; операторный м. symbolic(al) method (math.); м. определения разборчивости по баллам rating-scale method; м. отдельных потерь loss summation method; м. отношения напряжений voltage-ratio method; м. отражённых волн в сейсморазведке reflection seismograph method; м. оценки по баллам rating-scale method; м. оценки сравнения rank-order method; м. перевала method of steepest descent; м. передачи с разделением каналов по времени time sharing method; м. переключения диаграммы направленности beam switching method; м. перемагничивания method of reversals; м. переменной плотности movietone; м. подбора trial and error method; м. поля кодовых комбинаций coded pattern method; м. последовательной передачи кадров field sequential method (tv); м. последовательных интервалов, м. последовательных операций (or приближений) step-by-step method, method of successive approximations, approximation method, cut-and-try method; м. постоянных приращений static increment method; м. построения кривой по выборочным точкам method of selected point; м. прерывания пучка chopped beam method; м. приведения к нулю null method, zero balance; м. присвоения (адресов) assemble method; м. пристрелки target method (math.); м. проб и ошибок try-and-error method, cut-and-try method; м. проводимости (при анализе цепей) node-pair method (during analysis of circuits); м. работы method (of operation); м. работы по принципу совпадения токов с подмагничиванием biased coincident-current mode of operation; **м. работы с управлением от вычислительной машины** computer control mode of operation; **м. равносигнальной зоны** pip matching method, split-beam method; **равносигнальный м.** lobeswitching method; **радиолокационный м. построения плана местности** radar plotting method; **м. разборчивости сравнением** rank-order method; **м. разделения по цветовому тону** chroma-key method; **м. раздельного квантования сигнала по участкам спектра** hyperquantizing method (*tv*); **м. раздельной передачи информации о яркости и цветности** shunted monochrome technique (*tv*); **м. разности несущих частот, м. разности несущей** intercarrier system, carrier-difference system (*tv*); **растровый м.** beam-scanning method; **м. расчёта полей по распределению тока** current-distribution method; **м. расчёта по распределению поля в раскрыве антенны** aperture-field method; **м. регистрации излучения при помощи фотопластинок** photographic plate detection; **м. с повторением импульса** impulse sender method; **м. светового блика** Buchmann-and-Meyer pattern; **м. свилей** Schlieren method; **м. сдвига поднесущей** offset subcarrier method; **м. сдвига узлов** nodal-shifting method; **м. «седловины»** crevass method; **м. сложных сочетаний** method of multiplexing; **м. смещения несущей** offset carrier method; **м. спадания заряда** loss-of-charge method; **м. сравнения громкости** equal-loudness method; **м. стоячих волн** standing wave analysis; **м. «строка на строку»** row-by-row method (*math.*); **ступенчатый м. (измерений)** step-by-step method; **суммирующий импульсный м. набора** ditigal-total pulse system (*tphny.*); **м. теней** direct-shadow method; **фазовый м.** phase-difference method; **м. характеристики** method of determining flow lines by means of Mach characteristics; **м. электронных видеовставок, основан-**

ный на разделении по цветовому тону chroma-key electronic video insert technique; **м. ярма** yoke method of permeability test

**методика** *f.* technique, method(s); procedure; **м. расчёта** design procedure

**методический, методичный** *adj.* methodical, orderly

**метр** *m.* meter; **усадочный м.** shrinkage gage; measure of shrinkage, measure of contraction

**метр-ампер** *m.* meter-ampere

**метрамперы** *pl.* radiation constant

**метрекон, метрехон** *m.* metrechon. half-picture storage tube

**метрика** *f.* metrics

**метрический** *adj.* metric

**метро-ампер** *m.* meter-ampere

**метровый** *adj.* meter, metric

**метрон** *m.* metron

**метроном** *m.* metronome, rhythmometer

**метчик** *m.* tap, screw tap; cutting gimlet, twist drill; **м. для первого прохода** entering tap

**мех.** *abbr.* 1 (**механизированный**) mechanized; 2 (**механический**) mechanical

**меха, мехи** *pl.* bellows

**механизация** *f.* mechanization

**механизированный** *adj.* mechanized, power; aided

**механизировать** *v.* to mechanize

**механизируемость** *f.* mechanizability

**механизм** *m.* mechanism, gear, works, device; movement; machine, machinery; **астатический исполнительный м.** floating power unit; **м. ввода перфокарт** card feed; **верньерный м.** vernier drive, mechanical bandspread; **возвращающий м. молоточка (фортепиано)** repeating mechanism, release mechanism (of a piano); **м. вращения антенны** antenna rotator; **двойной лентопротяжный м.** twin tape transporter; **м. для прокладки и подъёма подводных кабелей** submarine cable gear; **м. звёздоискателя** star following mechanism; **исполнительный м.** power unit; **исполнительный м. клапана** valve actuator; **кассирующий м.** coin collector, collecting device; **м. качания зеркала оптического компенсатора** mirror-tilting mechanism; **колесный м. счётчика**

meter clockwork; **коммутационный м.** switch (mechanism); **контактно-разрывной м.** make-and-break mechanism; **кривошипный исполнительный м.** crank-type power unit; **лентопротяжный м.** spooling mechanism, tape winder; (automatic) paper winder; paper guide; chart mechanism; **переводный м.** shift mechanism; **м. передачи к индикатору** indicator gear; **подающий м.** feeder; **м. полуавтоматического сопровождения** aided tracking mechanism; **м. поправок на параллакс** parallax offset mechanism; **прямоходный исполнительный м.** straight-type regulating cylinder, push-stem power unit; **м. работы нервной сети** neural net mechanism; **м. с коленчатым валом** toggle-lever mechanism; **м. с разомкнутой обратной связью** open-cycle mechanism; **м. с целенаправленным поведением** teleological mechanism; **м. смены диафрагм** iris change mechanism; **соединительный м. без начального положения** stay-put switch; **соединительный м. с начальным положением** restore-to-normal switch; **статический исполнительный м.** proportional power unit; **м. сцепления грейфера** in-and-out movement; **сцепляющий м. на оси крыла** spindle slot; **счётный м. счётчика** counting mechanism of a meter, register of a meter; **м. установки антенны** antenna mount; **м. установки высоты** height servo; **м. установки на нуль** zero adjusting device; **м. установки на регулируемое значение** set point adjuster; **часовой м. для включения передатчика** pip squeak (mechanism); **электрический м. подачи карт** electric chart drive

**механик** *m.* mechanic, operator; **станционный м.** exchange maintenance man

**механика** *f.* mechanics; mechanism, movement, machinery; **клавишная м.** piano action

**механико-акустический** *adj.* mechanical-acoustic

**механико-электрический** *adj.* mechanical-electrical

**механически** *adv.* mechanically

**механический** *adj.* mechanical, power-driven, machine

**механорецептор** *m.* mechanoreceptor

**меховой** *adj.* bellows

**меццо-сопрано** *n. indecl.* mezzo-soprano

**мечение** *n.* labeling, tagging

**меченный** *adj.* labeled, tagged, marked; tracer

**мешалка** *f.* agitator, mixer, stirring rod

**мешание** *n.* agitation, mixing, stirring

**«мешанина»** *f.* hash, mush (*rdr.*)

**мешать, помешать, смешать** *v.* to hinder, to hamper, to impede, to interfere, to disturb; to stop, to clog; to mix, to stir, to blend, to agitate

**мешающий** *adj.* mixing, stirring, disturbing; interfering; preventing; inhibiting

**мешок** *m.* bag, sack

**мешочный** *adj.* bag, bag-type, sack

**Ми** Mie

**мигание** *n.* blinking, winking, twinkling, flickering, flicker

**мигатель** *m.* blinker, flasher; **терморелейный м.** thermal flasher

**мигательный** *adj.* blinking, blinker, flickering; nictational, nictating; pulsed

**мигать, мигнуть** *v.* to blink, to flicker, to twinkle, to wink; to flash, to glitter, to shimmer

**мигающий** *adj.* blinking, flickering, flashing; nictating; pulsed

**мигнуть** *pf. of* **мигать**

**миграционный** *adj.* migration

**миграция** *f.* migration

**мигрировать** *v.* to migrate

**мидель** *m.* mid-ship line

**мидоп** *m.* Midop

**микалекс** *m.* Mycalex

**микалента** *f.* mica tape

**миканит** *m.* micanite, built-up mica

**миканитовый** *adj.* micanite

**микарта** *f.* micarta

**микро-** *prefix* micro-; small; microscopic

**микроампер** *m.* microampere

**микроамперметр** *m.* microammeter

**микроатмосфера** *f.* microatmosphere

**микробар** *m.* microbar

**микробарограф** *m.* microbarograph

**микроблок** *m.* microassembly

**микроватт** *m.* microwatt, micropower

**микроваттный** *adj.* microwatt

**микровесы** *pl.* microbalance
**микроволна** *f.* microwave
**микроволновый** *adj.* microwave
**микровольт** *m.* microvolt
**микровольтметр** *m.* microvoltmeter
**микровольтовый** *adj.* microvolt
**микровплавной** *adj.* micro-alloyed
**микровыключатель** *m.* microswitch
**микрогенри** *m. indecl.* microhenry
**микроглоссарий** *m.* microglossary
**микрогологрáмма** *f.* microhologram
**микрограф** *m.* micrograph
**микрографический** *adj.* micrographic
**микрография** *f.* micrography
**микродвигатель** *m.* micromotor, subfractional motor
**микроденситометр** *m.* microdensitometer
**микроденситометрический** *adj.* microdensitometer
**микродеталь** *f.* micro-element
**микродиод** *m.* microdiode
**микродозировка** *f.* micro-metering
**микрозапись** *f.* microgroove, fine groove
**микроизмерение** *n.* micro-metering
**микрокалориметр** *m.* microcalorimeter
**микрокамера** *f.* Sievert chamber
**микроканонический** *adj.* microcanonical
**микрокварц** *m.* microquartz
**микроколебание** *n.* micro-oscillation
**микрокоманда** *f.* micro-order
**микроконденсатор** *m.* microcapacitor
**микроконтакт** *m.* microswitch
**микроконтроллер** *m.* microcontroller
**микрокристаллический** *adj.* microcrystalline
**микрокулон** *m.* microcoulomb
**микрокюри** *m. indecl.* microcurie
**микролебёдка** *f.* microlevelling
**микролучевой** *adj.* microray
**микроманипулятор** *m.* micromanipulator
**микроманометр** *m.* micromanometer, micropressure gage
**микромасштаб** *m.* microscale
**микромация** *f.* micromation
**микрометеорит** *m.* micrometeorite
**микрометод** *m.* micromethod
**микрометр** *m.* micrometer, micrometer gage; **м. для измерения биения конца оси гироскопического ротора** gyro motor end play micrometer; **нитяной м.** crossbar micrometer

**микрометрический** *adj.* micrometrical; micrometer
**микрометрия** *f.* micrometry
**микромикро-** *prefix* micromicro-
**микромикроамперметр** *m.* micromicroammeter
**микромикрофарада** *f.* micromicrofarad
**микроминиатюризация** *f.* microminiaturization
**микроминиатюрный** *adj.* microminiature
**микромо** *n. indecl.* micromho
**микромодуль** *m.* micromodule
**микромодульный** *adj.* micromodular, micromodule
**микромощный** *adj.* microenergy
**микромсантиметр** *m.* microhm-centimeter
**микрон** *m.* micron
**микронеоднородность** *f.* microinhomogenity
**микрообъектив** *m.* micro objective
**микрообъёмный** *adj.* microvolumetric
**микроом** *m.* microhm
**микроомметр** *m.* microhmmeter
**микроорганизм** *m.* micro-organism
**микроосмометр** *m.* microosmometer
**микроосциллограф** *m.* microoscillograph
**микроосциллографический** *adj.* microoscillograph(ic)
**микропервеанс** *m.* microperveance
**микропередатчик** *m.* miniature transmitter
**микропереключатель** *m.* microswitch
**микроплазма** *f.* microplasma
**микрополосковый** *adj.* microstrip
**микропористый** *adj.* microporous
**микропорошки** *pl.* micropowders
**микропочта** *f.* V-mail
**микропрерыватель** *m.* microchopper
**микропрограммирование** *n.* microgramming; **м. на прошитых магнитных сердечниках** threaded-core type microprogramming
**микропрограммный** *adj.* microprogramming
**микропроектирование** *n.* micro-projection
**микропульсация** *f.* micropulsation
**микрорадиография** *f.* microradiography
**микрорадиометр** *m.* microradiometer
**микрорезерфорд** *m.* microrutherford
**микрореле** *n. indecl.* microrelay

**микрорентгеноанализ** *m.* microradiography

**микросвитч** *m.* microswitch

**микросгусток** *m.* microclot; packet

**микросейсмический** *adj.* microseismic

**микросейсм** *m.* microseism

**микросекунда** *f.* microsecond

**микросекундный** *adj.* microsecond

**микросекундомер** *m.* microdial

**микросемантика** *f.* microsemantics

**микросин** *m.* microsyn

**микросинтаксический** *adj.* microsyntactic

**микросканинг** *m.* microscanning

**микроскоп** *m.* microscope; **м. для наблюдений в поляризованном свете** polarizing micrometer (*tv*); **м. с бегущим лучом** flying-spot micrometer; **спектрозональный телевизионный м.** color translating microscope; **спектрозональный ультрафиолетовый телевизионный м. с воспроизведением изображения в условных цветах** ultraviolet color-translating television microscope; **телевизионный м., работающий в ультрафиолетовом свете** ultraviolet television microscope

**микроскопический, микроскопный** *adj.* microscopic, microscopical

**микроскопия** *f.* microscopy; **телевизионная м. в ультрафиолетовых лучах** televised ultraviolet microscopy

**микроскэннинг** *m.* microscanning

**микроснимок** *m.* micrograph

**микроспектроскоп** *m.* microspectroscope

**микросплавление** *n.* microalloying

**микросплавной** *adj.* micro-alloy, micro-alloyed

**микроструктура** *f.* microstructure

**микросферический** *adj.* microspheric

**микросхема** *f.* microcircuit

**микросхемотехника** *f.* microcircuitry

**микросъёмка** *f.* microfilming

**микротасиметр** *m.* microtasimeter

**микротвёрдость** *f.* microhardness

**микротелефон** *m.* microtelephone, hand set

**микротелефонный** *adj.* microtelephonic, microtelephone

**микротермометр** *m.* microthermometer

**микроторр** *m.* microtorr

**микротранзистор** *m.* microtransistor, transistor microelement

**микротрон** *m.* microtron

**микроузел** *m.* microassembly

**микрофарада** *f.* microfarad

**микрофарадметр, микрофарадометр** *m.* microfarad meter

**микрофильм** *m.* microfilm

**микрофон** *m.* microphone, mike, telephone transmitter, microphone transmitter; **безбатарейный м.** self-powered microphone; **биградиентный м.** second-order pressure-gradient microphone; **м. ближнего действия** close-talking microphone; **волноводный м.** tubular microphone; **гарнитурный м.** breastplate microphone; **градиентный м.** velocity microphone, (pressure-) gradient microphone; **дикторский м.** speaker microphone; **м. для имитации бинаурального приёма звука** stereocephaloid microphone; **м. для прослушивания пульса** sphygmophone; **м. для работы на открытом воздухе** open-air microphone; **измерительный м.** standard microphone; **м., использующий костную проводимость** osteophone; **капсюльный м.** button microphone; **конденсаторный м. с акустической резонансной полостью перед диафрагмой** resonator-coupled condenser microphone; **конденсаторный м. с плоской диафрагмой** microphone with a plate diaphragm; **концертный м.** music-pickup microphone; **кристаллический м. с диафрагмой** diaphragm actuated crystal microphone; **лацканный м.** lapel microphone; **ленточный м.-приёмник давления** pressure-actuated ribbon microphone; **ленточный м.-приёмник колебательной скорости, ленточный градиентный м.** velocity ribbon microphone; **молчащий м.** idle microphone; **м. на укосине, м. на штанге** boom microphone; **м. переговорного устройства** talk-back microphone; **м. переменного магнитного сопротивления** variable-reluctance microphone; **петличный м.** lapel microphone; **м.-приёмник градиента высокого порядка** high-order gradient microphone; **м.-приёмник градиента давления** pressure-gradient microphone, velocity-ribbon microphone; **м.-приёмник колеба-**

тельной скорости velocity microphone; пьезоэлектрический м. с диафрагмой diaphragm actuated crystal microphone; м. с акустической фазосдвигающей цепочкой phase-shift microphone; м. с двумя угольными колодочками double-button carbon microphone; м. с кристаллом сегнетовой соли Rochelle salt microphone; м. с нагретым проводником hot-wire microphone, thermal microphone; м. с параболическим (звуко)отражателем parabolic-reflector microphone; м. с угольной колодочкой solid-back microphone; м. с шумоподавлением noise-cancelling microphone, antinoise microphone; сверхнаправленный м. ultradirectional microphone; м. слухового протеза hearing-aid receiver; м., соединённый с головным телефоном detectophone; суперградиентный м. high-order gradient microphone; суперградиентный м. второго порядка second-order pressure-gradient microphone; телефонный м. telephone transmitter; эталонный м. связи post office pattern microphone; эталонный м. с плоской частотной характеристикой flat microphone

микрофонирующий adj. microphonic

микрофония f. microphonism, microphonicity, microphonic effect

микрофонный adj. microphone, microphonic

микрофонящий adj. microphonic

микрофотограмма f. microphotogram

микрофотографирование n. photomicrography

микрофотографический adj. photomicrographic

микрофотография f. photomicrography; photomicrograph, micrograph

микрофотометр m. microphotometer

микроэлектроника f. microelectronics, microsystem electronics

микроэлектронный adj. microelectronic

микроэлемент m. microcell

микшер m. mixer, fader, mixing unit, dramatic control unit; monitor man; м. звукового сопровождения центральной аппаратной central sound mixer; м.-коммутатор mixer-switch unit; ламповый м. electron fader,

(electron-)tube mixer; «немой» м. dummy fader; одноручечный м. fader potentiometer

микшерно-линейный adj. mixer-line

микшерный adj. mixer

микшерский adj. mixing

микширование n. mixing; м. наплывом fade-over, lap dissolve; плавное м. телевизионных каналов cross-fade; м. фона background fade-in

мил m. mil (unit of length)

милар m. mylar, milar

миларовый adj. mylar, milar

милли- prefix milli-

миллиампер m. milliampere

миллиамперметр m. milliammeter; м. с непосредственной записью direct-reading recording milliammeter

миллиамперный adj. milliampere

миллиампер-секунда f. milliampere-second

миллиард m. billion, milliard

миллибар m. millibar

милливатт m. milliwatt

милливаттный adj. milliwatt

милливольт m. millivolt

милливольтамперметр m. millivolt-ammeter

милливольтметр m. millivoltmeter

миллигал m. milligal

миллигенри m. indecl. millihenry

миллиграмм m. milligram

милликюри m. indecl. millicurie; м. полного распада на расстоянии 1 см millicurie destroyed at one cm

миллиламберт m. millilambert

миллиметр m. millimeter

миллиметровый adj. millimetric, millimeter; graph (paper)

миллимикро- prefix millimicro-

миллимикромикроамперметр m. millimicromicroammeter

миллимикрон m. millimicron

миллимикросекунда f. millimicro-second

миллимикросекундный adj. millimicro-second

миллиомметр m. milliohmmeter

миллион m. million

миллирадиан m. milliradian

миллирентген m. milliroentgen

миллисекунда f. millisecond

миллисекундомер m. millisecond timer

миллстонский adj. Millstone

миль m. mil (unit of length)

Мильс Mills

**миля** *f.* mile
**мимический** *adj.* mimic
**мин.** *abbr.* 1 (**минимум**) minimum; 2 (**минута**) minute
**мина** *f.* mine; torpedo
**минерализованный** *adj.* mineralized
**минеральный** *adj.* mineral
**миниатюризация** *f.* miniaturization
**миниатюризировать** *v.* to miniaturize
**миниатюрный** *adj.* miniature, midget, tiny, micro-
**миникард** *m.* Minicard system
**минимакс** *m.* minimax
**минимаксный** *adj.* minimax
**минимально** *adv.* minimally
**минимально-обнаружимый** *adj.* minimum-detectable
**минимальный** *adj.* minimum, minimal, smallest, least
**минимизация** *f.* minimization, minimizing
**минимизировать** *v.* to minimize
**минимизирующий** *adj.* minimizing
**минимодуль** *m.* minimodule
**минимум** *m.* minimum
**минировать** *v.* to mine
«**минископ**» *m.* minescope
**министак** *m.* ministac
«**минитрак**» *m.* Minitrack (system)
**минитрон** *m.* minitron
**минитрэк** *m.* Minitrack (system)
**миноискание** *n.* mine detection
**миноискатель** *m.* mine detector, mine locator; metal locator
**минор** *m.* minor key (*mus.*); minor, minor determinant (*math.*)
**миноранта** *f.* infinite series containing only positive numbers
**минорантный** *adj.* minorant
**минорный** *adj.* minor
**Минтроп** Mintrop
**минус** *m.* minus; defect, drawback, shortcoming; **м. батареи** B-negative (of the battery)
**минусовый** *adj.* minus; negative
**минута** *f.* minute; moment; **м. на развёртку** minutes per scan
**минуто-занятие** *n.* call-minute (*tphny.*)
**минующий** *adj.* passing, bypassing; bypassed
**миньон** *m.* minion
**мираж** *m.* mirage, optical illusion; **верхний м.** superior mirage, blooming; **нижний м.** inferior mirage
**миран** *m.* Miran
**мирар** *m.* Mirar camera

**мириа-** *prefix* myria-
**мириаватт** *m.* myriawatt
**мириаметр** *m.* myriameter
**мириаметровый** *adj.* myriametric
**миска** *f.* basin, pan, dish
**мискообразный** *adj.* dish-shaped, pan-shaped, dished
**мистор** *m.* mistor
**миткаль** *m.* calico; cambric
**митрон** *m.* mitron
**мицеллярный** *adj.* micellar
**Мичэм** Meacham
**мишенный** *adj.* target
**мишень** *f.* target; target assembly; target electrode; **двусторонняя м.** double-sided mosaic; **м. калибратора дальности** range-calibrator target; **мозаичная м. с накоплением зарядов** charge storage mosaic; **м. москопа** picture plate; **накапливающая м.** storage plate; **однородная м.** isotropic target
**мишура** *f.* tinsel; metallic thread, patchcord wire
**мишурный** *adj.* tinsel
**МК** *abbr.* 1 (**магнитный курс**) magnetic course; 2 (**маяк**) beacon
**мк** *abbr.* 1 (**микро-**) micro-; 2 (**микрон**) micron; 3 (**милликулон**) millicoulomb
**мка** *abbr.* (**микроампер**) microampere
**мкб** *abbr.* (**микробар**) microbar
**мкв** *abbr.* (**микровольт**) microvolt
**мквт** *abbr.* (**микроватт**) microwatt
**мкгн** *abbr.* (**микрогенри**) microhenry
**мкк** *abbr.* (**микрокулон**) microcoulomb
**МККТ** *abbr.* (**Международный Консультативный Комитет по Телефонии**) Comité Consultatif International Téléphonique
**мккюри** *abbr.* (**микрокюри**) microcurie
**мкмк** *abbr.* (**микромикро**) micromicro
**мкмкф** *abbr.* (**микромикрофарада**) micromicrofarad
**мком** *abbr.* (**микроом**) microhm
**мкр** *abbr.* (**микрорентген**) microroentgen
**мкс** *abbr.* (**максвелл**) maxwell
**МКС** *abbr.* 1 (**метр-килограмм-секунда**) meter-kilogram-second; 2 (**многократный координатный соединитель**) cross-bar (selector) switch
**МКСА** *abbr.* (**метр, килограмм, секунда, ампер**) MKSA (system of units)
**мксек** *abbr.* (**микросекунда**) microsecond

**мкф** *abbr.* (**микрофарада**) microfarad

**мкюри** *abbr.* (**милликюри**) millicurie

**младший** *adj.* minor; junior; lower (part)

**ММ** *abbr.* (**маркерный маяк**) marker beacon

**мм** *abbr.* (**миллиметр**) millimeter

**ммк** *abbr.* (**миллимикро-**) millimicro-

**ммксек** *abbr.* (**миллимикросекунда**) millimicrosecond

**мм рт. ст** *abbr.* (**миллиметр ртутного столба**) millimeter of mercury column

**МН** *abbr.* 1 (**магнитное насыщение**) magnetic saturation; 2 (**минута, минуты**) minute, minutes

**мнемоника** *f.* mnemonics

**мнемонический** *adj.* mnemonic, memory

**мнимый** *adj.* imaginary, supposed, simulated, virtual; false

**многие** *pl. adj.* multiple; many, great many

**много** *adv.* many, much, considerably

**много-** *prefix* poly-, multi-, many, multiple

**многоадресный** *adj.* multiple-address, multi-address

**многоамперный** *adj.* heavy-current

**многоанодный** *adj.* multianode

**многовалентность** *f.* multivalence

**многовалентный** *adj.* multivalent, polyvalent

**многоваттный** *adj.* high-watt

**многовибраторный** *adj.* multidipole

**многовидность** *f.* multimoding

**многовидный** *adj.* multimode

**многовитковый** *adj.* multiturn, multiloop, multiple-turn

**многоволнистость** *f.* multimoding

**многоволновой** *adj.* multiwave

**многогнёздный** *adj.* multijack; multislide

**многогранный** *adj.* polyhedral; many-sided, varied

**многогрупповой** *adj.* multigroup

**многодвигательный** *adj.* multiple-generator, multimotor

**многодекадный** *adj.* multiple-decade

**многодетальный** *adj.* rich in detail and contrast

**многодиапазонный** *adj.* multiband, multirange, multiscale

**многодипольный** *adj.* multidipole

**многодолинный** *adj.* multivalley, many-valley

**многодоменный** *adj.* multidomain

**многодорожечный** *adj.* multipath, multi-track

**многодырчатый** *adj.* perforated

**многодырочный** *adj.* multiple-aperture, multi-aperture

**многоёмкостный** *adj.* multicapacity

**многожильный** *adj.* multicored, multiple-core, polycore; multiwire, multiple (cord), multiple-strand, multiconductor

**многозажимный** *adj.* multiterminal

**многозазорный** *adj.* multigap

**многозарядный** *adj.* multicharge

**многозаходный** *adj.* multiple-thread (screw)

**многозвенный** *adj.* multisectional, multipart, ladder-type; iterated

**многозвучный** *adj.* polyphonic

**многозначительный** *adj.* significant

**многозначность** *f.* multivalence, polyvalence; ambiguity, multiple meaning, polysemanticism

**многозначный** *adj.* multivalent, polyvalent; many-valued, multiple-value, multiple-valued; ambiguous; multidigit

**многозональный, многозонный** *adj.* multizone, multirange, multiregion

**многоигольный** *adj.* multistylus

**многоинерционный** *adj.* multilag

**многокамерный** *adj.* multibarrel, multibarreled; multichambered, multiple-chamber, multicellular

**многоканальность** *f.* multichanneling, multiple channels, channeling, channel multiplexing

**многоканальный** *adj.* multi-channel, multi-way, multi-duct; polysleeve

**многокаскадный** *adj.* multistage, multiple-stage

**многокатодный** *adj.* polycathode, multicathode

**многоквартирный** *adj.* community

**многокерновый** *adj.* multicore

**многоклеточный** *adj.* many-celled, multicellular

**многокомпонентный** *adj.* multicomponent

**многоконтактный** *adj.* multicontact, multiple-contact

**многоконтурный** *adj.* multicircuit, multiloop

**многокорпусный** *adj.* multiple, multiple-unit

**многокрасочный** *adj.* multicolor, polychromatic, many-colored

**многократ** *m.* multiple (*tphny.*)

**многократно** *adv.* repeatedly, many times

**многократность** *f.* multiplex(ing), frequency, multiplicity; **м. видами импульсов** pulse multiplex, pulse-mode multiplex

**многократный** *adj.* multiple, numerous, frequent, compound, manifold, repeated, multi-, multiplex, multistage

**многокурсовой** *adj.* multiple-course, multiple-track

**многоламповый** *adj.* multitube; multiple-bulb, multi-lamp

**многолепестковый** *adj.* multilobe

**многолинейный** *adj.* multicircuit

**многолопастный** *adj.* multiblade

**многолучевой** *adj.* many-pronged; multibeam; multi-wire; multi-wave

**многомерный** *adj.* multidimensional; multivariable

**многоместный** *adj.* multiplace

**многомодовый** *adj.* multimode, multimodal, polymodal

**многомоторный** *adj.* multiple-motor, multi-engine

**многонакальный** *adj.* multifilament

**многонаправленный** *adj.* polydirectional, multidirectional

**многонитный** *adj.* multifilament

**многониточный** *adj.* multiple-thread, multiple

**многообмоточный** *adj.* multiwinding

**многооборотный** *adj.* of high revolutions, multiturn

**многообразие** *n.* variety, diversity; multiplicity (*math.*)

**многообразный** *adj.* varied, diverse, manifold, multiform

**многообъективный** *adj.* multiple-lens

**многоосколковый** *adj.* multichip

**многоотверстный** *adj.* multiple-duct

**многоотражательный** *adj.* multireflex

**многопальцевый** *adj.* multifinger

**многопарный** *adj.* multi-pair, multi-paired

**многоперьевой** *adj.* multi-pen

**многопетлевый** *adj.* multiple-loop; compound (cycle)

**многопластинный, многопластинчатый** *adj.* multiplate

**многоплечий** *adj.* multiple-arm, multiple

**многоплоскостный** *adj.* polyhedral

**многопозиционный** *adj.* multi-position; multipoint

**многополосный** *adj.* multichannel, multiband

**многополостный** *adj.* multicavity

**многопольный** *adj.* multiple-field

**многополюсник** *m.* multiterminal network, multiport device, multiport (network); **разветвлённый м.** transfer tree

**многополюсный** *adj.* multiple-pole, multiterminal, multipolar

**многопредельный** *adj.* multirange

**многопримесный** *adj.* multidopant

**многопроводниковый, многопроводный** *adj.* multiple-conductor, multiconductor, multiwire, multiple-wire, multiple (line)

**многопроволочный** *adj.* stranded

**многопрожекторный** *adj.* multigun

**многопутевой, многопутный** *adj.* multipath

**многоразрезной** *adj.* multislot; multisegment, multisegmental

**многоразрядный** *adj.* multidigit; multiple-order

**многорежимный** *adj.* multimode

**многорезонаторный** *adj.* multiresonator, multicavity, multiple-cavity

**многорезцовый** *adj.* multiblade, multicutting

**многорупорный** *adj.* multiple-horn

**многорядный** *adj.* polyserial, multiseries; stacked

**многосегментный** *adj.* multisegment, multisegmental

**многосекционный** *adj.* multisection, multisectional, multi-tap(per)

**многосеточный** *adj.* multigrid

**многоскачковый** *adj.* multishock

**многоскоростный** *adj.* multivelocity, multiple-speed

**многосложный** *adj.* complex, complicated, intricate

**многослойный** *adj.* multilayer, laminated, sandwich-type

**многоспиральный** *adj.* multispiral

**многосрезный** *adj.* multiple-shear

**многоставочный** *adj.* multirate

**многостандартный** *adj.* multistandard

**многостанционный** *adj.* multi-office

**многостолбовой** *adj.* multiple-pole

**многосторонний** *adj.* polygonal, multilateral, many-sided; versatile

**многострендовый** *adj.* multiple-strand

**многоступенный** *see* **многоступенчатый**

**многоступенчатость** *f*. multistaging

**многоступенчатый** *adj*. multistage, multiple-stage, polystage, multilevel, multistep, multiple-phase; compound; cascade

**многотарифный** *adj*. multiple-tariff, multirate

**многотоковый** *adj*. multicurrent

**многотональный** *adj*. polytonic

**многотонный, многотоновый** *adj*. "multitone", heavy; multitone

**многоточечный** *adj*. multiple-point, multipoint

**многотрубный** *adj*. multitubular

**многоугольник** *m*. polygon

**многоугольный** *adj*. polygonal; multi-V

**многоуровневый** *adj*. multilevel

**многофазный** *adj*. polyphase, multiphase, polycyclic

**многофакторный** *adj*. multiple-factor, complex

**многофидерный** *adj*. multiple-feed, multifeed

**многофонический** *adj*. multiphonic

**многофункциональный** *adj*. multifunction, multifunctional

**многоходовой** *adj*. multipath, multiway, multipass, multiple-pass; multiworm (screw), multithread (screw), multiple-thread

**многоцветность** *f*. polychromy

**многоцветный** *adj*. polychromatic, polychrome, multicolor(ed)

**многоцелевой** *adj*. multipurpose, multifunction

**многоцепный** *adj*. multicircuit, multiway

**многоцикловый** *adj*. multicycle

**многоцилиндровый** *adj*. polycylindrical, multicylindrical

**многочастотный** *adj*. multifrequency

**многочисленность** *f*. multiplicity

**многочисленный** *adj*. multiple, numerous, manifold

**многочлен** *m*. multinomial, polynomial

**многочленный** *adj*. multinomial, polynomial

**многошкальный** *adj*. multiscale, multidial; multirange

**многошлейфный, многошлейфовый** *adj*. branched-line

**многошпиндельный** *adj*. multiple-spindle

**многоштанговый** *adj*. multibar

**многоштифтовый** *adj*. multiple-pin

**многоштыревой** *adj*. multiple-rod, polyrod

**многощелевой** *adj*. multislot, multisegment

**многощёлочной** *adj*. multi-alkali

**многоэлектродный** *adj*. multi-electrode

**многоэлементный** *adj*. multiple-unit, multi-element, multispot

**многоэмиттерный** *adj*. multiple-emitter

**многоэтажный** *adj*. many-storied, multistage, multiple-stage; stacked

**многоязычковый** *adj*. multiple-reed

**многоярусный** *see* **многоэтажный**; stacked

**многоячеечный, многоячеистый, многоячейковый** *adj*. multicellular, multicell, multiple-cell; multisection

**многоячейный** *adj*. multichambered, multiple-chamber

**множественно-параллельный** *adj*. multiple-parallel

**множественность** *f*. multiplicity, plurality

**множественный** *adj*. multiple, plural

**множество** *n*. great number, numbers, multitude; multiplicity; set; mass; assemble (*math*.), population (*math*.); **выборочное м.** sampling population (*math*.); **общерекурсивное м.** general recursive set; **рекурсивно перечислимое м.** recursively enumerable set; **специально подобранное м.** artificial population (*math*.); **м. элементарных событий** aggregate of simple events; **м. электронных состояний** manifold of electronic states

**множимое** *n. adj. decl.* multiplicand

**множитель** *m*. factor, coefficient, multiplier, multiple; **масштабный м. времени** time-scale factor; **угловой поправочный м.** phase-angle correction factor; **м. частоты** thickness-frequency ratio; **м. шунта, шунтовый м.** multiplying power (of a galvanometer shunt)

**множительно-делительный** *adj*. multiplication-division

**множительый** *adj*. multiplying, multiplier, multiplication

**множить, помножить, умножить** *v*. to multiply

MO *abbr.* (магнит отбоя) release magnet

мо *n. indecl.* mho

моавр Moivre

мобиланс *m.* mobilance

мобилометр *m.* mobilometer

мобильность *f.* mobility

мобильный *adj.* mobile

мода *f.* mode; style; запрещённые моды колебаний trapped modes; изгибная м. колебаний flexural mode; м. колебаний mode, mode of vibration; м. колебаний по толщине thickness mode; м. колебаний «расширение-сжатие» extensional mode; м. колебаний сдвига shear mode; «критическая» м. колебаний cutoff mode; м. натуральных колебаний natural mode; незатухаюшая м. колебаний undamped mode; м. объёмных колебаний extensional mode; сдвиговая м. колебания по толщине thickness shear mode; собственная м. колебаний, м. собственных колебаний eigen mode

модальный *adj.* modal

моделирование *n.* simulating, simulation, imitation; modeling; analog computation, analog formation; м. в истинном масштабе времени real-time simulation; м. зоны нечувствительности backlash simulation; комбинированное аналого-цифровое м. analog-digital simulation; м. линии balancing network, artificial (balancing) line; м. с дискретными отсчётами sampling simulation; м. с параметрическим квантованием sampling parametric computation; м. явлений распространения радиоволн radio-wave propagation simulation

моделировать *v.* to model, to shape, to simulate; to imagine

моделирующий *adj.* simulating; analogous, analog

модель *f.* model, mock-up; pattern, form, shape; standard; make, type; analog; simulator; м. в истинном масштабе времени real time simulator; м. голосового аппарата voice operation demonstrator; дебаевская м. колебаний решётки в кристаллах Debye model; капельная м. drop-model of the nucleus; м. пространственного явления three-axis

simulator; расчётная м. сети network analyzer, calculating board

модельный *adj.* model, mock-up; pattern; form, shape; make, type

модем *m.* modem

модератор *m.* moderator

модернизация *f.* modernization, updating

модернизировать *v.* to modernize, to update

модификатор *m.* modifier, transformer; B-register, index register

модификация *f.* modification; modifier

модифицированный *adj.* modified

модифицировать *v.* to modify

модифицирующий *adj.* modifying, modification

модулирование *n.* modulation

модулированный *adj.* modulated; м. по длительности width-modulated; м. по частоте frequency-modulated, warbled; м. полностью fully modulated; м. речью speech-modulated, voice-modulated

модулировать *v.* to modulate, to control

модулирометр *m.* modulation meter

модулируемость *f.* modulability

модулируемый *adj.* modulated

модулирующий *adj.* modulating

модулирующийся *adj.* modular; modulating

модулометр *m.* modulation meter

модуль *m.* modulus, coefficient, magnitude; module; standard; м. гибкости modulus of compliance; объёмный м. bulk modulus, volume modulus; м. проводимости (scalar) admittance; м. растяжения modulus of elasticity, coefficient of elasticity; сегментный м. для визуального вывода цифр N-segment numeric display module; м. скольжения modulus of rigidity; м. согласования index of cooperation; м. сопротивления scalar impedance

модульный *adj.* modular, modularized, module

модулярный *adj.* modular

модулятор *m.* modulator; buncher (of a klystron); keyer; driving device; control efficiency; control grid; м. без подмодулятора line-pulsing modulator; временной м. с колебаниями пилообразной формы voltage-sawtooth time modulator; магнитный м. на второй гармонике magnettor; **минующий** м. bypassed luminance signal (*tv*); **минующий м. на стороне передачи и демодулятор на стороне приёма** bypassed monochrome signal (*tv*); **многократный м. по ширине импульса** pulse-width multiplexer; **м. на ключах** switch-type modulator; **м. по длительности импульса** PDM keyer; **потенцирующий** м. antilogarithmic circuit, logarithmic-to-linear converter; **м., работающий от формирующей линии** line-pulsing modulator; **м. с возбуждением в сеточной цепи** grid-circuit modulator; **м. с модуляцией импульсов по длительности** pulse-width modulator; **м. с подмодулятором** driver-power-amplifier modulator; **м. сигналов звукового сопровождения** aural modulator; **м., состоящий из подмодулятора и усилителя мощности** driver-power-amplifier modulator; **м. устройства многократной связи** multiplex modulator; **цветовой м. передатчика** transmitter sampler (*tv*)

**модуляторный** *adj.* modulator

**модуляционный** *adj.* modulation; modulator

**модуляция** *f.* modulation, tone control; keying; **с амплитудной модуляцией** amplitude-modulated; **с импульсной модуляцией** pulse-modulated; **с модуляцией на анод** plate-modulated; **с недостаточной модуляцией** undermodulated; **амплитудная м. несущей импульса** pulse-amplitude modulation, PAM; **взаимная м. (составляющих сложной волны)** intermodulation; **м. в каскадах предварительного усиления** pre-stage modulation; **м. в предварительном каскаде** low-level modulation; **м. временем нарастания (импульса)** rise-time modulation; **м. выключением** on-off modulation; **глубокая м.** hard driving modulation; **м. дефазированием** outphasing modulation; **м. защитной сетки** suppressor-grid modulation; **квадратурная м. цветового поднесущей** color multiplexing; **кодовоимпульсная м.** quantized pulse modulation; **кодовоимпульсная м. с передачей**

изменений модулированного сигнала "servo" modulation; **м. конического сканирования** lobing modulation; **многократная м. с разделением по частоте** frequency-division multiplex modulation; **м. на антидинатронную сетку** grid modulation; **м. на большой мощности** high-level modulation; **м. на защитную сетку** suppressor grid modulation; **м. на малой мощности** low-level modulation; **м. на сигнальную пластину** backplate modulation; **м. напряжением на коллекторе** collector-voltage modulation; **м. напряжения на сигнальной пластине** backplate-voltage modulation; **м. напряжения на теневой маске** aperture-mask voltage modulation; **м. несущей в цепи антенны** antenna modulation; **м. несущей частоты помехи** jamming modulation; **м. по интервалам между импульсами** pulse-interval modulation; **м. по крутизне импульса** pulse-slope modulation; **м. по крутизне наклона импульса** variable-slope pulse modulation; **м. помехой** interference modulation; **прерывистая м.** chopper modulation; **м. при незатухающих колебаниях** continuous-wave modulation; **м. противофазой** phase-reversal modulation; **прямоугольная м.** square-wave modulation; **м. пучка первичных электронов** primary-current modulation; **м. с автоматической стабилизацией глубины** controlled-carrier modulation; floating-carrier modulation; **м. с двумя боковыми полосами** double-sideband modulation; **м. с « качающейся » частотой** sweep-frequency modulation; **м. с подавлением несущей (в пробелах)** quiescent-carrier modulation; **м. с постоянной глубиной** controlled-carrier modulation; floating-carrier modulation; **м. с предусилением н. ч.** (a-f) amplifier modulation, low-frequency pre-emphasis modulation; **м. с фазным смещением** outphasing modulation, Chireix modulation; **м. с частично подавленной боковой полосой частот** vestigial sideband modulation; **м. сеточным смещением** grid modulation, grid-bias modulation, direct current grid modulation; **м. сигналами цветности** chrominance modulation; **скоростная м. развёртки** scanning-velocity modulation; **м. смещением, м. смещения** grid-bias modulation; **м. тока луча** beam-current modulation; **фазоимпульсная м.** pulse-time modulation; **м. частотой импульсов, м. частоты следования импульсов** pulse-frequency modulation, pulse repetition-rate modulation; **м. шагом импульсов** pulse-interval modulation

**мозаика** *f.* mosaic; mosaic electrode; **м. из изолированных элементов** insulating mosaic; **светочувствительная м.** mosaic electrode

**мозаичный** *adj.* mosaic, inlaid

**мозг** *m.* brain; cerebrum

**Мозли** Moseley

**Мозотти** Mosotti

**Мокрушин** Mokrushin

**мокрый** *adj.* wet, moist, damp

**мол.** *abbr.* (**молекулярный**) molecular

**мол. в.** *abbr.* (**молекулярный вес**) molecular weight

**молектроника** *f.* moletronics, molecular electronics

**молектронный** *adj.* molecular

**молекула** *f.* molecule

**молекулярность** *f.* molecularity

**молекулярный** *adj.* molecular, molar

**молибден** *m.* molybdenum, Mo; **м.-пермаллой** molybdenum-permalloy

**молибденистый** *adj.* molybdenous; molybdenum

**молибденит** *m.* molybdenite

**молибденовый** *adj.* molybdenum

**молнеотвод** *see* **молниеотвод**

**молниевидный** *adj.* lightning-like

**молниезащита** *f.* lightning protection

**молниеотвод** *m.* lightning conductor, lightning arrester, lightning rod; **безвоздушный м.** vacuum lightning protector; **м. с ножевыми контактами** knife-edge lightning protector

**молниеуловитель** *m.* lightning rod

**молния** *f.* lightning; "lightning" priority

**молотковый** *adj.* hammer

**молоткообразный** *adj.* hammer-shaped

**молоток** *m.* hammer; **веретачный м.** fitter's hammer; **контактный м.** trembler

**молоточек** *m.* hammer; clapper (of a bell); tapper; **м. звонка** bell clapper; **м.-прерыватель** hammer break, hammer interrupter

**молоточковый, молоточный** *adj.* hammer

**молочный** *adj.* milky, frosted, opal

**молчание** *n.* silence

**молчать, замолчать, помолчать** *v.* to be silent, to keep silent

**молчащий** *adj.* silent, quiescent; idle

**моль** *m.* mole, gram-molecule

**Мольвейде** Mollweide

**Молье** Mollier

**молярность** *f.* molecular concentration

**молярный** *adj.* molar, gram-molecular; molal

**мом** *abbr.* (**мегом**) megohm

**момент** *m.* moment, instant (time), instance; momentum; feature, factor, point; torque; monent (mechanical, electrical); **вращающий м.** torque, torsional moment, twisting moment, deflecting torque; **втягивающий (в синхронизм) м.** pull-in torque (of a synchronous motor); **м. выпадения из синхронизма** pull-out torque; **м. количества движения** momentum of momentum, angular momentum; **крутящий м., м. кручения** torsion torque, torque; **минимальный пусковой м.** pull-up torque; **м. начала отсчёта** reference time; **м. начала решения** zero computing time; **номинальный м. вращения** rated torque; **опрокидывающий м.** tilting moment; pull-out torque; **м. пары сил** moment of couple; **поверхностный м. инерции** polar moments inertia, moment of inertia of a plane surface; **м. попадания электронов на анод** moment of interception; **предельный перегрузочный м. синхронного электродвигателя** pull-out torque of a synchronous motor; **м. при номинальной нагрузке (электродвигателя)** torque at rated load (of a motor); **противодействующий м.** restoring torque; **смешанный м.** product moment (*math.*); **смешанный м. второго порядка** covariance (*math.*); **м. срабатывания** speed releasing; **м. тока антенны** radiation constant; **м. трогания** stalled torque; **угловой орбитальный м.** orbital moment of momentum; **удельный м.** torque-

weight ratio (of a meter); **м. успокоения** damping torque; **устанавливающий м.** controlling torque
**моментальный** *adj.* instantaneous
**моментный** *adj.* quick
**моментомер** *m.* torque meter
**Мон** Mohn
**монауральный** *adj.* monaural
**монель-металл** *m.* Monel metal
**монета** *f.* coin
**монетный** *adj.* monetary, coin
**моникон** *m.* monicon
**« монимакс »** *m.* monimax
**монитор** *m.* monitor; picture monitor, television (viewing) monitor; executive program; **м. камерного канала** camera control monitor
**монитрон** *m.* monitron
**моно-** *prefix* mono-, single, one
**моноатомный** *adj.* monoatomic, monatomic, monovalent
**моноблок** *m.* monoblock, single block
**моноблочный** *adj.* monoblock
**монодия** *f.* monody (*mus.*)
**моноимпульс** *m.* giant-pulse radiation
**моноимпульсный** *adj.* monopulse
**монокок** *m.* monocoque
**монокристалл** *m.* monocrystal, single crystal
**монокристаллический** *adj.* monocrystalline, single-crystal
**монолитный** *adj.* monolithic
**моном** *m.* monomial (*math.*)
**мономер** *m.* monomer
**мономерный** *adj.* monomeric
**мономолекулярный** *adj.* monomolecular
**моноокись** *f.* monoxide
**моноплан** *m.* monoplane
**монорельс** *m.* monorail
**монорельсовый** *adj.* monorail
**моноскоп** *m.* monoscope, phasmajector
**моноспираль** *f.* single-coil filament
**моноспиральный** *adj.* single-coil
**моностабильный** *adj.* monostable
**моностатический** *adj.* monostatic
**монотонный** *adj.* monotonic, monotone; monotonous
**монотрон** *m.* monotron
**монотропный** *adj.* monotropic
**монофонический** *adj.* monophonic, monaural
**монофония** *f.* monophony
**моноформер** *m.* monoformer
**монохлордифторметан** *m.* monochlorodifluoromethane

**монохорд** *m.* monochord
**монохроматизация** *f.* monochromatization
**монохроматизировать** *v.* to monochromize
**монохроматический** *adj.* monochromatic
**монохроматичность** *f.* monochromaticity
**монохроматор** *m.* monochromator
**монохромный** *adj.* monochrome
**моноциклический** *adj.* monocyclic
**моноэнергетический** *adj.* monoenergetic
**монтаж** *m.* mounting, fitting, installing, installation, mount, assembly, assembling, erection, erecting, setting up, rigging; wiring; adjustment; **м. в секциях** cell mounting; **м. в стойке** rack mounting; **м. в ячейках** cell mounting; **выдвижной м.** telescopic mounting, draw-out mounting; **м. голой проволокой** piano wiring strapping, piano wiring, bare wiring; **горизонтальный телефонный м. кабельной проводки** straight telephone wiring; **м. кабельной проводки** cabling; **м. на панели** panel mounting; **м. на раме** frame mounting; **наращённый м.** plated wiring; **петлеобразный м. проводов** looping-in the wire; **м. спереди панели** front-of-panel mounting; **м. струнного поля** piano wiring; **м. схемы** wiring, joining up, connecting up; **травленный оловянный м.** tinned etched mounting
**монтажная** *f. adj. decl.* assembly room
**монтажник** *m.* installer, mounter, assembly-man, rigger; mechanic, repairman; adjuster
**монтажный** *adj.* assembling, assembly, fitting, erecting
**монтёр** *m.* fitter, mounter; adjuster; assembler, rigger; service man, repairman, mechanic; **кабельный м.** cable splicer, cable man; **линейный м.** lineman; **м. по уходу** service man, maintenance man; **м.-регулировщик** adjuster; **м. связи** communication technician
**монтёрский** *adj.* fitter's, lineman's
**монтирование** *n.* assembling, assembly, mounting, erection
**монтированный** *adj.* mounted, assembled, erected

монтировать, смонтировать *v.* to assemble, to mount, to erect, to install, to set (up), to arrange, to rig up, to fit (up), to build up

монтировка *see* монтирование

монтируемый *adj.* mounted, installed, erected

Мор Mohr

мордент *m.* mordent; верхний м. inverted mordent

море *n.* sea

Морей Murray

Морелль, Морель Morell

Морера Morera

Морзе Morse

морозостойкий *adj.* frostproof, non-freezing, winter-proof

морозостойкость *f.* resistance to frost, winter-proof feature

морозоустойчивый *see* морозостойкий

Моррей Murray

Моррел Morrel

морской *adj.* sea, marine, maritime; nautical; naval

Мортон Morton

морф *m.* morph

морфема *f.* morpheme

морфемный *adj.* morphemic

морфологический *adj.* morphological

морфология *f.* morphology

морфонология *f.* morphophonemics

морщинистый *adj.* wrinkled, creased; rugose

Мос Mohs

мост *m.* bridge; м. для измерения активных проводимостей conductance bridge; м. для измерения акустических сопротивлений acoustic impedance bridge; м. для измерения добротности Jaumann's differential bridge; м. для измерения коэффициента взаимоиндукции Felici balance, mutual induction bridge; м. для измерения крутизны transconductance bridge; м. для измерения удельных проводимостей conductivity measuring bridge; м. для определения места повреждения кабеля fault localizing bridge; м. для сравнения коэффициентов само- и взаимоиндукции Heaviside-Campbell bridge; измерительный м. с калиброванной проволокой slide-wire bridge; контрольный м. limit bridge; линейный м. slide-wire bridge; магазинный м.

box bridge; магнитный м. Юинга Ewing permeability bridge; м. Максвелла для сравнения коэффициентов само- и взаимоиндукции Maxwell M-L bridge; одинарный м. Wheatstone bridge; питающий м. battery supply bridge, battery supply circuit, battery supply feed, transmission bridge; м. с круговым коммутатором (*or* переключателем) dial bridge; м. с магазином сопротивлений box bridge; м. с метровой струной meter slide-wire bridge; м. с реохордом, струнный м. slide-wire bridge; тензометрический м. strain gage measuring bridge

мостик *m.* bridge, circuit, feed; cross bar, cross link; м. для проверки допусков limit bridge; закорачивающий м. контура сетки grid-shorting bar; съёмный м. plug-in bridge

мостовой *adj.* bridge

мостящий *adj.* bridging; make-before-break

моталка *f.* rewinder, winder, reel(er), coiler, coiling machine

мотальный *adj.* reeling, winding

мотанный *adj.* coiled, wound

мотать, намотать *v.* to wind (up), to spool, to coil, to reel

мотив *m.* tune; theme; motive, ground, reason, cause

мото- *abbr.* 1 (моторизованный) motorized; 2 (моторный) motor, engine

мото- *prefix* motor, power; mechanical

мотовило *n.* reel, reeling frame; coiler

мотовильный *adj.* reeling

моток *m.* bundle, hank, skein; link

мотонейрон *m.* neuromotor unit

мотор *m.* motor; engine; асинхронный м. induction motor; м. больше 1 л.с. integral-hp motor; вентильный м. thyratron motor; конденсаторно-расщеплённый м. capacitor-start motor; конденсаторный м. capacitor motor; м. постоянно-переменного тока ac/dc motor, universal motor; м. привода сигнала signal motor; м. с конденсаторным пуском capacitor-start motor

моторгенератор *m.* motor-generator set; dynamotor, dynamo

моторизация *f.* motorization

моторизованный *adj.* motorized

**моторно-вращательный** *adj.* machine-rotary

**моторный** *adj.* motor

**моторчик** *m.* small motor, fractional-horsepower motor; **м., приводимый в действие часовой пружиной** clock-spring motor

**мотоциклетный** *adj.* motorcycle

**мощностный** *adj.* power

**мощно-усилительный** *adj.* high-power amplifier

**мощность** *f.* power, horsepower, force; potency, strength, energy; output, efficiency, capacity; capability; duty (of engine); **с двумя номинальными мощностями** dual-rated; **м. анодного рассеяния генераторной лампы** anode dissipation of an oscillator tube; **м. анодного рассеяния электронной лампы** anode dissipation of a vacuum tube; **м. в согласованной нагрузке** available power; **м. возбуждения (цепи) сетки** grid-driving power; **выходная м. на несущей частоте** carrier output power; **выходная м. передатчика изображения** visual-transmitter power; **выходная м. помехи** noise output (power); **м. дозы в заданном месте** local doze rate; **допустимая м. волновода** power-carrying capacity of a waveguide; **забираемая м.** input power; **импульсная м. на выходе** peak output (power); **м. критерия** strength of a test; **м. лучистой энергии** luminous power; **максимально-возможная м.** maximum available power; **мгновенная м. при толчках нагрузки** instantaneous) swing capacity; **м. модулирующих колебаний** speech output, speech power; **м. на входе антенны** antenna input power; **м. на согласованной нагрузке** matched-load power; **номинальная выходная м. на несущей частоте** carrier power output rating; **отдаваемая м.** power output; **отдаваемая м. на несущей частоте** carrier output power; **м. передатчика изображения** visual transmitter power; **пиковая м. импульса несущей частоты** carrier-frequency peak pulse power; **м. питания анода** plate-supply power, B-power, anode input power; **м. питания сетки** grid driving power;

**подводимая м.** power input, energy input; **подводимая постоянная м.** constant available power; **м. подвозбуждения** priming power; **м. подключаемых установок** installed load, connected load, connected value; **м. подогрева** heater power; **полная м. на входе, полная входная м.** total power input; **поступающая в антенну м.** antenna input power; **м. потерь** rate of energy loss; **потребляемая м.** intake power; **м. потребляемая цепью сетки** grid input power; **предельно-допустимая м. рассеяния** maximum permissible dissipation, rated dissipation; **м. при коротком замыкании** short-circuiting power; **м. при продолжительной работе** continuous output, continuous rating; **м. при холостом ходе** no-load input power; **пробивная м. в импульсе** pulse-power breakdown; **проходная м.** actual transformer rating; **м. раскачки на сетке** grid driving power; **м., рассеиваемая в анодном контуре** plate dissipation, anode dissipation; **м. рассеяния на аноде** D.C. rating of anode; plate dissipation; **теоретическая м. трафика** theoretical efficiency of a line; **удельная м.** power density; **установочная м.** relayed capacity; **м. электронной лампы** tube output

**мощный** *adj.* powerful, vigorous; high-power(ed), high-duty, heavy-duty; power; sturdy

**мощь** *f.* power, force

**мп** *abbr.* (**магнит подъёма**) vertical magnet

**МП** *abbr.* 1 (**магнитный пеленг**) magnetic bearing; 2 (**мёртвое пространство**) dead space

**МПР** *abbr.* (**магнитный пеленг радиолокатора**) magnetic bearing of a radio range station

**МПС** *abbr.* (**магнитный пеленг самолёта**) magnetic bearing of aircraft

**МПУ** *abbr.* (**магнитный путевой угол**) magnetic course angle

**МПЧ** *abbr.* (**максимальная применимая частота**) maximum usable frequency

**мр** *abbr.* (**миллирентген**) milliroentgen

**мрамор** *m.* marble

**мраморный** *adj.* marble

**мрезерфорд** *abbr.* (**миллирезерфорд**) millirutherford

**МРМ** *abbr.* (**маркёрный радиомаяк**) radio marker beacon

**МРП** *abbr.* 1 (**магнитный радиопеленг**) magnetic radio bearing; 2 (**маркерный радиоприёмник**) marker receiver

**МРУ** *abbr.* (**межрайонный узел**) "interraion" center, toll center

**мсек** *abbr.* (**миллисекунда**) millisecond

**МСК** *abbr.* (**московское время**) Moscow time

**МТ** *abbr.* (**мёртвая точка**) dead center

**МТР** *abbr.* (**магнитный термоядерный реактор**) magnetic thermonuclear reactor

**МТС** *abbr.* (**междугородная телефонная станция**) long-distance office, toll center, long-distance exchange; **координатная МТС** crossbar toll office; **межрайонная узловая МТС** "interraion" center, toll center (*tphny.*)

**МУ** *abbr.* 1 (**магнитный усилитель**) magnetic amplifier; 2 (**моторная установка**) power system, power plant

**муар** *m.* moire

**мувитон** *m.* movieton

**«муза»** *f.* musa, multiple-unit steerable antenna

**музыка** *f.* music

**музыкальный** *adj.* musical

**мульти-** *prefix* multi-, poly-

**мультивариантный** *adj.* multivariant

**мультивибратор** *m.* multivibrator, relaxation generator, flip-flop; **м. в режиме свободных колебаний** free-running multivibrator; **ведомый м.** slave flip-flop; **м. временной развёртки** time base sweep multivibrator; **м. генератора отметок** marker flip-flop; **м. длительной засветки** wide-gate multivibrator; **м. длительности пусковых импульсов** trigger gate width multivibrator; **ждущий м.** Eccles-Jordan circuit, flip-flop circuit, biased flip-flop, biased multivibrator, monostable multivibrator; **ждущий м. с непосредственным соединением** direct-coupled flip-flop; **м. задерживающий узкие селекторные импульсы** narrow-gate delay multivibrator; **м. задержки пусковых им-**пульсов trigger delay multivibrator; **моностабильный м.** univibrator; **м. на газовых разрядниках** VR-tube flip-flop; **м. на полупроводниковых триодах** transistor multivibrator; **одноразовый м., однотактный м.** start-stop multivibrator, one-kick multivibrator, one-shot multivibrator; **одноходовой м.** monostable flip-flop, biased multivibrator; **пентодный ждущий м. с непосредственной связью** direct-coupled pentode flip-flop; **м. развёртки дальности** range multivibrator; **м. с анодной связью** plate-coupled multivibrator; **м. с двумя квазиустойчивыми состояниями** astable multivibrator; **м. с двумя устойчивыми положениями** bistable multivibrator, phase-bistable flip-flop, flip-and-flop generator; **м. с независимым возбуждением** slave flip-flop; **м. с одним устойчивым положением** mono-stable multivibrator, flip-flop multivibrator; **м. с положительным смещением** positive-bias multivibrator; **м. с регулировкой длительности импульса, м. с регулировкой ширины** width multivibrator; **м. с симметричным выходом** push-pull multivibrator; **м. со связью на сопротивлениях** resistance-coupled flip-flop; **свободноидущий м.** astable multivibrator; **м. строчной частоты** horizontal multivibrator; **м. строчных гасящих импульсов** horizontal blanking multivibrator; **м. узких селекторных импульсов** narrow-gate multivibrator; **узкостробно-задерживающий м.** narrow-gate delay multivibrator; **м. формирующий строчные гасящие импульсы** horizontal blanking multivibrator; **м. цепи временной задержки** time-delay flip-flop; **м. частоты перемены фазы цвета** color phase alternation multivibrator

**мультигармониграф** *m.* multiharmoniograph

**мультикардиотрон** *m.* multicardiotron

**мультикон** *m.* multicon tube

**мультиметр** *m.* multimeter

**мультимодальный** *adj.* multimodal

**мультимодный** *adj.* multimode

**мультипакторный** *adj.* multipactor

**мультиплексер** *m.* multiplexer, traffic pilot; multiplex switch
**мультиплексный** *adj.* multiplex
**мультиплет** *m.* multiplet
**мультиплетный** *adj.* multiplet
**мультипликативный** *adj.* multiplicative
**мультипликатор** *m.* multiplier, intensifier; multi-lens camera
**мультипликационный** *adj.* multiplication
**мультипликация** *f.* multiplication; animated cartoon
**мультиплиситет** *m.* multiplicity
**мультиполь** *m.* multipole
**мультипольность** *f.* multipolarity
**мультипольный** *adj.* multipole
**мультиполюсный** *adj.* multipolar, multipole
**мультипрограммирование** *n.* multiprogramming
**мультипрограммный** *adj.* multiprogramming
**мультипроцессорный** *adj.* multiprocessing
**мультирадиксный** *adj.* multiradix
**мультисистема** *f.* multisystem
**мультископ** *m.* multiscope
**мультитон** *m.* multitone
**мультитрак** *m.* multitrack (hyperbolic radionavigation system)
**мультиустойчивость** *f.* multistability
**мультиустойчивый** *adj.* multistable
**мультиэкран** *m.* multiple screen
**муляж** *m.* modeling, molding, casting, moulage
**муметалл** *m.* Mu metal
**муметаллический** *adj.* mumetal
**мундштук** *m.* mouthpiece, bit; jet, nozzle, tip
**Мунзель** Munsell
**Мунсон** Munson
**Мур** Moore
**мусковит** *m.* muscovite
**мускульный** *adj.* muscular, muscle
**муслиновый** *adj.* muslin
**мутатор** *m.* mutator, grid-controlled mercury rectifier
**мутационный** *adj.* mutation
**мутация** *f.* mutation
**мутность** *f.* turbidity, dullness, cloudiness; opacity
**мутный** *adj.* turbid, dull, cloudy, hazy, dim
**муфельный** *adj.* muffle
**муфта** *f.* muff; clutch; coupling, con-

duit-coupling, connecting piece, connection, sleeve, sleeve pipe, joint; junction box (for cables); socket, socket joint, plug; **винтовая м.** screwed socket; **дисковая м.** floating ring clutch, disk clutch, flange coupling, plate coupling; **зажимная соединительная м.** clamp sleeve; **кабельная м.** jointing, sleeve; cable joint; **кабельная концевая м.** cable-head distribution rack; **м. канализационной трубы** spigot (of a conduit); **конденсаторная м.** cable coupling box for installation of balancing capacitors; **концевая кабельная м.**, **оконечная кабельная м.** cable terminal, cable head, cable shoe, end sleeve, pothead, cable terminal box; **ответвительная м.** parallel jointing sleeve, cable jointing sleeve, joint box, conduit-tee; **переходная м.** adapter coupling; **продольно-свёртная м.** shaft coupling, split coupling; **м. с кожаным упругим звеном** leather coupling; **м. с резьбой** screwed socket; **соединительная м.** joint, cable joint, connecting sleeve, jointing sleeve, coupling box; **соединительная м. для труб** pipe joint, pipe socket; **м. сцепления** clutch; **угловая м.** conduit-elbow; **электромагнитная м. на принципе вихревых токов** dynamic clutch
**муфтовый** *adj.* muff; clutch; coupling, connection, joint, sleeve
**муфточка** *f.* socket, sleeve, small coupling
**мф** *abbr.* (**миллифарада**) millifarad
**мч** *abbr.* (**мнимая часть**) $\Sigma$ imaginary part (of complex numbers)
**мысль** *f.* thought, idea, notion
**мыслящий** *adj.* reasoning, thinking, intellectual
**мышечный** *adj.* muscle, muscular
**мышьяковистый** *adj.* arsenous, arsenide, arsenic
**мэв** *abbr.* (**мегаэлектронвольт**) megaelectron-volt
**МЭИС** *abbr.* (**Московский электротехнический институт связи**) the Moscow Electrotechnical Institite of Communications
**мэл** *m.* mel (*acoust.*)
**МЭТИ** *abbr.* (**Московский электротехнический институт**) Moscow Electro-Technical Institute

мю *n. indecl.* mu ($\mu$)

**Мюллер** Mueller

**мюльтипльный** *adj.* multiple, bank

**мюметалл** *m.* Mu metal

**Мюрхед** Muirhead

**мягкий** *adj.* soft, mild, smooth, mellow, pliant, flexible; **акустически м.** sound-soft

**мягко** *adv.* softly, mildly

**мягкоотожжённый** *adj.* soft-annealed

**мягкотянутый** *adj.* soft-drawn

**мягчитель** *m.* softening agent

**мятина** *f.* hollow, dent, gouging

# Н

**Н** *abbr.* 1 (**нижний**) lower. 2 (**новый**) new

**Н-** *abbr.* (**норма**) standard

**н.** *abbr.* 1 (**нормальность**) normality. 2 (**нормальный**) normal. 3 (**ньютон**) newton

**на** *prep. acc. and prepos. case* at, by, on, upon; in, into; toward, to; over; for, per (in phrase, often translated by adj.): **несущая на выходе** incoming carrier (*lit.* carrier at the entrance); **на слух** orally

**набег** *m.* incursion, inroad; **н. фазы** phase change

**набегание** *n.* running on; flowing in; creeping; buildup; **н. (звуков)** tailing (of counds); **н. (знаков)** tailing (*tgphy.*); piling-up

**набегать, набежать** *v.* to run against, to dash against, to hit, to encounter; to creep, to climb; to run in, to flow in; to accumulate

**набегающий** *adj.* leading; creeping, climbing; incoming, inflowing, entering; incident

**набежать** *pf. of* **набегать**

**набивание** *n.* stuffing, filling, filler, gasket, packing; padding

**набивать, набить** *v.* to stuff, to fill, to pack; to pad

**набивка** *f.* filling, filler, packing, stuffing, gasket; pad, padding, lining; **сальниковая н.** packing; **теплоизолирующая н.** insulating packing

**набирательный** *adj.* dialing

**набирать, набрать** *v.* to dial; to collect, to gather, to accumulate; **н. номер** to select a number, to dial, to keysend; **н. номер на импульснике** to keysend

**набирающий** *adj.* dialing; collecting

**набитый** *adj.* padded, packed

**набить** *pf. of* **набивать**

**набла** *f.* nabla

**наблюдаемый** *adj.* observable

**наблюдатель** *m.* observer; plot observer; lookout, spotter; recipient; **вероятностный н.** success-run ob-

server; **н. самолётов в системе ПВО** airplane spotter; **средний н. МКО** standard observer; **стандартный н. МКО** ICI standard observer

**наблюдательный** *adj.* observation; observant; supervisory; lookout, viewing

**наблюдать, наблюсти** *v.* to watch, to observe, to study; to sight; to follow, to spot; to supervise; to inspect

**наблюдающий** *adj.* observing; supervisory

**наблюдение** *n.* observation, supervision, control, controlling, superintendence; inspection, study; **накопительные наблюдения** cumulative observations; **воздушное н. оповещение и связ** civil defense warning network, intelligence net, reporting chain; **н. за линиями** line controlling; **н., независимое от времени** timeless observation; **предварительное н.** preview; **н. снаряда** tracking the shell

**наблюдённый** *adj.* observed

**наблюсти** *pf. of* **наблюдать**

**набор** *m.* set, kit, outfit, assembly, gang, collection, stack; bank (of cells); gage (of wire); setup, suit; dialing; **наборы из модулей** stacked modules; **батарейный н.** battery dialing; **вынужденный н. номера** numerical selection; **н. генов** gene pattern; **дальний н.** long-distance dialing, toll line dialing; **н. деталей для быстрой сборки схемы** breadboard kit; **н. деталей (для) детекторного приёмника** crystal receiver kit; **н. задач** set-up; **н. задач на моделирующей установке** setting-up the computer; **н. инструментов** assembly, outfit kit; **искажённый н. номера** distorted dialing; **клавишный н.** push-button dialing; push-button type range switch; **н. коллекторных пластин** commutator bar assembly; **междугородный абонентский н.** subscriber's trunk dialing;

**многочастотный н. номера** multi-frequency dialing; **незаконченный н.** incompleted call; **неправильный н.** faulty selection; **н. номера** dialing, numerical selection, pulse stepping, impulse action; **н. номера на постоянном токе** direct current keying; **н. номера по шлейфной системе** loop dialing; **н. пластин сердечника** core-laminated stack; **н. подтональной частотой** low-frequency dialing, subaudio dialing; **последующий н.** extension dialing from central-office lines; **н. приборов** assembly, kit; **н. программы** programming; **н. рисованных студийных декораций** cartoon set, abstract set, studio decoration; **н. символов** language (of a computer); **сквозной н.** (single-) through dialing; **ступенчатый н. номера телефона** automatic toll system employing dial registers; **н. схемы** set-up; **тастатурный н.** key sending, key pulsing, keying; **телеграфный н.** telegraphic typesetting; **н. тональной частотой** voice-frequency dialing, voice-frequency keying; **транзитный н. номера** tandem dialing; **н. уравнений** equation set-up; **н. шестерен** set of wheels

**наборка** *f.* assembly, gathering
**наборный** *adj.* dialing; selector
**набранный** *adj.* collected, gathered
**набрасывание** *n.* sketching; throwing on
**набрасывать, набросать, набросить** *v.* to sketch, to draw, to outline; to throw on; to hook up; **н. схему** to sketch, to figure out, to draw
**набрать** *pf. of* **набирать**
**набросанный** *adj.* sketched, outlined; thrown on
**набросать** *pf. of* **набрасывать**
**набросить** *pf. of* **набрасывать**
**набросок** *m.* sketch, draft, layout, rough copy; outline
**набрызг** *m.* sputtering; **н. (в изготовлении прессового оригинала)** sputtering, cathode sputtering (in production of a metal master)
**набрызгивание** *n.* sputtering
**набухание** *n.* swelling
**наваглайд** *m.* Navaglide
**наваглоб** *m.* Navaglobe
**навамандер** *m.* Navamander

**навар** *m.* Navar
**наваренный** *adj.* welded (on), built up, deposited; tipped, faced
**наваривание** *n.* burning-on; **н. сталью** acierage
**наваривать, наварить** *v.* to weld on, to build up; to face, to tip
**наварить** *pf. of* **наваривать**
**наварка** *f.* welding (on), building up; weld bead
**наварной** *adj.* weld, welded
**наваскоп** *m.* Navascope
**наваскрин** *m.* Navascreen
**наваспектор** *m.* Navaspector
**наведение** *n.* induction; leading, guiding, bringing on, homing action; ground control of interception; pointing (of rocket); application, coating; vectoring; **автоматическое н. на цель** automatic homing, passive homing; **автономное н.** precomputed course, preset control; onboard guidance; **автономное н. ракеты** missile-borne guidance; **астроинерциальное н. по двум звёздам** two-star stellar inertial guidance; **н. в вертикальной плоскости** up-down guidance; **н. в двух координатных плоскостях** two-dimensional guidance; **н. в полёте** airborne guidance; **н. в точку, образованную пересечением двух окружностей** circular guidance system; **н. в трёх координатных плоскостях** three-dimensional guidance; **н. в упреждённую точку** predicted point guidance; **н. для встречи** rendezvous guidance; **н. для обеспечения стыковки** docking guidance; **н. из задней полусферы** tail-chase guidance; **инерциальное н. с коррекцией по вертикали** inertial-gravitational guidance; **командное н. в упреждённую точку** collision-course command guidance; **командное н. по информации радиолокационных станций сопровождения** track-command guidance; **командное н. по методу наведения самолётов-мишеней** drone-type command guidance; **командное н. ракеты по лучу** beam-rider command guidance; **командное н. с оптическим сопровождением снаряда и цели** optical track command guidance; **командное н. с радиолокационным сопровождением (сна-**

ряда и цели) radar-track command guidance; **комбинированное н. по лучу радиолокатора и оптическому прицелу** optical beam-riding system; **н. методом параллельного сближения** constant-bearing guidance; **н. методом совмещения точек кривой** matching-curve guidance; **н. методом совмещения трёх точек** line-of-sight guidance; **н. на активном участке траектории по командам с земли** ground controlled boost guidance; **н. на визуально ненаблюдаемую цель** blind guidance; **н. на встречно-пересекающихся курсах** collision-course guidance; **н. на конечном участке маршевого полёта** preterminal guidance; **н. на конечном участке при стыковке** terminal docking guidance; **н.на конечном участке траектории входа в плотные слои атмосферы** re-entry corridor trajectory guidance; **н. на маршевом участке** incourse guidance, midcourse (-phase) guidance; **н. на начальном участке** initial guidance, primary guidance; **н. на РЛС** anti-radar guidance; **н. на участке спуска** descent guidance; **н. на цель во время пуска** launching phase guidance; **наземное н. по информации бортовой РЛС** radar repeat-back guidance; **н. наземных объектов** land vehicle guidance; **ПВО н.** vectoring; **н. по азимуту** aiming azimuth; **н. по геофизическим параметрам** terrestrial reference guidance; **н. по кривой погони** pursuit-course homing, curved-course guidance; **н. по кривой погони с постоянным углом упреждения** deviated pursuit guidance; **н. по лучу** beam rider radar, beam-follower guidance, beam-climber guidance, beam guidance; **н. по методу погони без упреждения** pure pursuit guidance; **н. по «предвычисленной» траектории** plotting trajectory guidance; **н. по радиолучу** beam rider guidance; **н. по тепловому излучению** heat-seeking guidance; **н. при догоне** tail-chase guidance; **пространственное н.** three-dimensional guidance; **н. противоракеты на МБР** anti-ICBM guidance; **н. путём распознавания местности**

terrestrial-reference guidance; **радиолокационное н. с визуальным слежением** direct radar guidance; **н. (ракеты) вручную по командной (радио)линии** manual command-line guidance; **н. ракеты по лучу** beam-riding missile guidance; **н. ракеты по лучу в упрежденную точку** beam-rider collision course guidance; **н. ракеты по лучу, осуществляющему программированное движение** programed beam riding guidance; **резервное автономное н.** backup guidance; **н. с помощью РЛС непрерывного излучения** constant-wave radar guidance; **н. с распознаванием радиолокационной карты местности** map-matching guidance; **слежечное н.** track homing; **н. снарядов на цель при запуске** launching-phase guidance; **н. шумов** noise induction

**наведённый** *adj.* induced; led, guided, directed; brought on
**наверх** *adv.* up, upward
**наверху** *adv.* above, at the top, aloft
**навесить** *see* **навешивать**
**навеска** *f.* hinge; button; fitting; hanging
**навесной** *adj.* hook-on, hinged; mounted; attached; inserted
**навести** *pf. of* **наводить**
**навешать** *pf. of* **навешивать**
**навешенный** *adj.* suspended, hung
**навешивание** *n.* hanging up; suspension; **н. бирок** tagging; **н. квантов** quantification
**навешивать, навешать** *v.* to hand up; to suspend; **н. кванторы** to quantify
**навивальный** *adj.* winding, coiling
**навивание** *n.* winding on
**навивать, навить** *v.* to wind, to wind on, to reel, to coil, to roll on; to spool
**навивка** *f.* winding on, rolling on, twisting on; **н. сеток** grid spiraling
**навивной, навивочный** *adj.* winding, coiling
**навигатор** *m.* navigator
**навигационный** *adj.* navigation, navigating, navigational
**навигация** *f.* navigation, guidance; **н. по приборам** blind navigation; **н. по счислению пути** navigation by dead reckoning; **н. с постоянным углом упреждения** fixed-lead navigation

**навинтить** *pf. of* **навинчивать**

**навинченный** *adj.* screwed on

**навинчивать, навинтить** *v.* to screw on, to screw to; to clasp

**навить** *pf. of* **навивать**

**навитый** *adj.* wound on, rolled on

**наводимый** *adj.* inducible; **н. при движении** motionally induced

**наводить, навести** *v.* to direct, to aim (at), to set, to lay, to level; to guide, to lead; to induce; to home; to control the interception; to spot

**наводка** *f.* aiming, directing, pointing, setting, laying; sighting; homing; induction; ground control of interception; pick-up; **внятная н.** intelligible crosstalk; **невнятная н., неразборчивая н.** inverted crosstalk; **н. одной цепи на другую** cross coupling; **окончательная н.** final laying; **орудийная н. береговой артилерии** coastal gunfire control; **н. от многих (телефонных) каналов** (telephone) babble, induction from many (telephone) channels; **перекрёстная н.** cross fire *(tgphy.)*, crosstalk; **перекрёстная н. антенны** antenna crosstalk; **перекрёстная н. на передающем конце** near-end crosstalk; **перекрёстная н. на приёмном конце** far-end crosstalk, receiving-end crosstalk; **перекрёстная н. с основной цепи на фантомную** side-to-phantom crosstalk; **полуавтоматическая н. орудий** aided gun laying; **н. помех** noise induction, stray pickup; **прямая н. (помех)** direct pickup, stray pickup; **разборчивая н.** intelligible crosstalk; **самолётная орудийная н.** aircraft gun laying; **фоновая н.** hum pickup

**наводчик** *m.* director, spotter; gunner, sighter

**наводящий** *adj.* inducing; aiming, directing, guiding; bringing on

**навстречу** *adv.* toward

**Навье** Navier

**навьючиваемый** *adj.* burdened, packed

**нагар** *m.* scale, deposit

**нагель** *m.* peg, pin, dowel, axle

**наглухо** *adv.* tightly, hermetically; permanently; **н. закорочено** dead short; **н. закреплённый** dead-ended

**наглядный** *adj.* pictorial, display, visual; graphic, descriptive

**наглянцевать** *pf. of* **глянцевать**

**нагнести** *pf. of* **нагнетать, гнести**

**нагнетатель** *m.* (force) pump, blower; booster, supercharger

**нагнетательный** *adj.* pressure, force

**нагнетать, нагнести** *v.* to press, to pressurize, to squeeze; to pump, to feed

**нагнетающий** *adj.* pressure, forcing

**нагнетённый** *adj.* pressed, forced

**наголовник** *m.* head, cap; **поворотный н. троллея** swivelling trolley head

**наготове** *adv.* ready, in readiness; at the point

**нагрев** *m.* heat, heating, warming; **н. в высокочастотном поле** radio-frequency heating; **высокочастотный ёмкостный н.** heat by induced high-frequency capacitive current; **н. джоулевым теплом** Joule heating, resistance heating; **обратный н.** backheating; **н. по ребру изделия** edge heating; **повторный н.** reheat; **предельный н.** maximum permissible temperature rise; **н. при помощи джоулева тепла** resistance heating; **н. токами высокой частоты** (*or* **радиочастоты**) high-frequency heating, radio-frequency heating

**нагреваемый** *adj.* heated

**нагревание** *n.* heating, warming

**нагреватель** *m.* heater, resistive heater; **н. лучеиспусканием** radiant heater; **погружаемый н.** immersion heater, cartridge heater; **проточный н.** flow heater, flow-type hot-water supply

**нагревательный** *adj.* heating, warming

**нагревать, нагреть** *v.* to heat, to warm

**нагреваться, нагреться** *v.* to become hot, to get warm, to run hot, to overheat

**нагревающий** *adj.* heating, warming

**нагревный** *adj.* heating, warming

**нагретый** *adj.* hot, heated, warmed

**нагреть(ся)** *pf. of* **нагревать(ся), греть**

**нагрудный** *adj.* breast, chest; breastplate

**нагружаемость** *f.* load-carrying capacity, load capacity

**нагружаемый** *adj.* loaded, chargeable

**нагружать, нагрузить** *v.* to load, to charge; to burden

**нагружающий** *adj.* loading, charging; burdening

**нагружение** *n.* load, loading, charge,

stress, strain; **параллеьное н. линии** shunt loading; **н. по заданному графику** pre-set loading; **последовательное н. линии** series loading
**нагруженность** *f.* loading
**нагруженный** *adj.* loaded, charged
**нагрузить** *pf. of* **нагружать, грузить**
**нагрузка** *f.* load, loading, charge; burden, strain, stress; weight; filling; **нагрузки, возникающие при столкновении** crash loads; **балластная н.** dummy load; **безындуктивная и безъёмкостная н.** non-reactive load; **большая н.** heavy traffic *(tphny.)*; heavy load; **н. большой мощности** high-power load; **бытовая н.** appliance load; **воздушная н.** aerodynamic load; **волноводная н. с поглощающими стенками** dissipative-wall waveguide load; **вторичная н.(измерительного трансформатора)** burden (of an instrument transformer); **высокоомная н.** high-resistance load; **высшая н.** peak load, maximum load; **н. генератора** oscillator loading; **действительная н.** net load; **динамическая н.** live load; transient load; **длительная н.** permanent load; **допустимая н.** safe load, allowable load; current-carrying capacity; load-carrying capacity; **допустимая н. искателей** selector carrying capacity; **импедансная н.** reactive load; **инерционная н.** mass load; acceleration load; inertial load; **н. искателей** selector carrying capacity; **квартирная н.** residential load; **крупная н.** block load; **линейная токовая н.** flux density per peripheral unit length; **максимальная расчётная н. рентгеновской трубки** maximum radiographic ratings; **максимальная допустимая н. по моменту** torque capacity; **маломощная н. волновода** low-power waveguide termination; **меняющаяся н.** discontinuous load; **н. на конце линии** line terminating network; **н. на пучок (в клистроне)** beam loading; **н. на сетку** grid dissipation; **настроенная н. в анодной цепи** tuned plate load; **неподвижная н.** continuous load; **неполная н.** fractional load, underload; **номинальная вторичная н.** maximum permissible loading of

instrument transformers; **оконечная н.** termination; **оконечная н. волноводов** terminal resistance of waveguides; **оконечная н., равная волновому сопротивлению** characteristic-impedance termination; **омическая н.** resistive load; **опрокидывающая н.** stalling load; **отрицательная н.** overhauling load; **повторно-кратковременная н.** intermittent load; **полезная н.** additional load; actual load; pay load; live load; **полная н. измерительного устройства** burden of an instrument; **предварительная н.** preloading; **предельная н.** ultimate load; stalling load; **н., равная волновому сопротивлению** characteristic-impedance termination; **н. с одним скосом** single-taper load; **н. с опережающим током** leading load; **н. с отстающим током** lagging load; **н. с током, опережающим по фазе** leading load; **н. с тяжёлыми условиями пуска** hard-starting load; **сетевая н.** maximum permissible power-line load; **сеточная н.** grid dissipation; **н. со стороны среды** load of medium; **согласованная оконечная н.** matched termination, non-reflecting termination; **сосредоточенная н.** single load; **суммарная н. на единицу длины провода** conductor loading; **толчковая н.** shock load; **толчкообразная н.** fluctuating load; **удельная н.** loading capacity; **удельная токовая н.** specific current density per unit area; **частичная н.** fractional load; **электронная н. резонатора вторичными электронами** secondary electron gap loading, multipactor; **электронная н. резонатора первичными электронами** primary transit-angle gap loading; **эталонная оконечная н.** reference termination
**нагрузочный** *adj.* load, loading, charge; freight
**над** *prep. instr.* over, above, on, upon
**над-** *prefix* over-, super-, hyper-, above
**надводный** *adj.* overwater, above-water, emergent; surface
**надвое** *adv.* in two, in half; ambiguously
**наддув** *m.* pressure, charging, pressurization; boost
**надеваемый** *adj.* hook-on; slip-on
**надевание** *n.* putting on

**надевать, надеть** *v.* to put on, to slip on

**надёжность** *f.* reliability, safety, security, dependability; serviceability, maintainability; margin of safety; **н. аппаратуры общего применения** commercial reliability; **н. в работе** reliability of operation; **н. в эксплуатации** reliability of service, reliability of operation; **гарантированная н.** inherent reliability; **н. детали** component and part reliability; **н. и контроль качеств продукции** reliability and quality control; **н. конструкции устройства** built-in reliability; **н. против приёма промежуточных частот** safety against direct reception of the I.F. section; **рабочая реальная н.** operational reliability; **н. соединения скруткой** wire-wrap reliability; **н. суммы всех элементов устройства** overall reliability **эксплуатационная н.** serviceability, use reliability, operational reliability; **н. элемента** component and part reliability

**надёжный** *adj.* reliable, foolproof, safe, dependable, trustworthy; secure

**надеть** *pf. of* **надевать**

**надзвуковой** *adj.* supersonic, superaudio

**надземный** *adj.* overhead, aerial, elevated, overground

**надзирать** *v.* to supervise, to control, to inspect, to look after

**надзор** *m.* supervision, control, inspection; servicing; **дистанционный производственный н.** trunk line observation, trunk operating supervision; **н. за предохранителями** open fuse alarm; **производственный н.** operating observation

**надзорный** *adj.* supervising

**надир** *m.* nadir

**надлежащий** *adj.* proper, fit

**надмодовый** *adj.* supermode

**надпанель** *f.* built-up panel, vertical instrument section

**надпись** *f.* inscription, superscript; legend; notice

**надрез** *m.* notch, cut, incision, gash, groove

**надрезанный** *adj.* notched

**надрезать** *pf. of* **надрезывать**

**надрезка** *f.* notches

**надрезывание** *n.* cutting in

**надрезывать, надрезать** *v.* to cut in, to notch, to incise

**надсинхронный** *adj.* hypersynchronous, supersynchronous

**надслуховой** *adj.* supersonic, ultrasonic

**надсмотрщик** *m.* supervisor, overseer, surveyor; **участковый н.** section lineman

**надставка** *f.* extension, extension rod, extension piece; adapter, built-up part; **н. на траверзу** arm extension bracket; terminal iron; **полюсная н.** pole shoe

**надставленный** *adj.* extended, added on, lengthened

**надставной** *adj.* extension

**надстраивать, надстроить** *v.* to build upon, to build on

**надстроенный** *adj.* built-on

**надстроивать** *see* **надстраивать**

**надстроить** *pf. of* **надстраивать**

**надстройка** *f.* superstructure; building on

**надтепловой** *adj.* epithermal

**надтональный** *adj.* supertonic, hyperacoustic, supersonic, ultrasonic, superaudio

**надувание** *n.* inflation

**надувной** *adj.* inflatable; air-inflated

**надутый** *adj.* inflated, blown up

**надфиль** *m.* broaching file, needle file

**наезд** *m.* incursion; inroad, irruption; **н. камерой** track in, dolly in

**наездник** *m.* clamp-on

**наём** *m.* hire, hiring, rent

**наёмный** *adj.* hired; rented

**нажатие** *n.* pressure, impression, depression; **н. клавиша** type printing; **н. рукоятки бдительности** acknowledging, acknowledgement

**нажать** *pf. of* **нажимать**

**наждак** *m.* emery

**наждачный** *adj.* emery

**нажим** *m.* pressure, stress, thrust, push; clamp; **н. Прони** Prony brake

**нажимать, нажать** *v.* to press, to depress, to push, to contract, to force (against); to clamp, to pinch; **н. рукоятку бдительности** to acknowledge

**нажимной** *adj.* pressure; clamp, clamping; pressing, thrust, push

**нажимно-отжимный** *adj.* push-and-pull

**назад** *adv.* back, backwards

**название** *n.* name, designation

**названный** *adj.* named, called

**назвать** *pf. of* **называть**

**наземный** *adj.* ground, land, terrestrial; land-based

**назначать, назначить** *v.* to assign, to allot, to allocate, to fix, to set; **н. соединительную линию** to assign a trunk (*tphny.*)

**назначение** *n.* function, scope; purpose, destination; designation, allocation, assignment; **двойного назначения** double-duty, dual-purpose; **комплексное н.** integral use; **н. мощности (электродвигателя)** sizing

**назначенный** *adj.* assigned, set, designated, allotted, appointed

**назначить** *pf. of* **назначать**

**называемый** *adj.* named, called

**называть, назвать** *v.* to name, to call, to designate, to describe

**наи-** *prefix* the most

**наиболее** *adv.* utmost, the most; above all; most of all

**наибольший** *adj. superl.* the greatest, maximum, extreme, peak

**наивероятнейший** *adj. superl.* most probable

**наивыгоднейший** *adj. superl.* most advantageous, most favorable, best

**наивысший** *adj. superl.* highest, maximum, ceiling

**наилучший** *adj. superl.* the best; optimal

**наименование** *n.* designation, marking; lettering; name; **н. абонента** listing, address name; **н. адресата** name of the addressee; **н. отправителя** sender's name, originator's name

**наименовать** *pf. of* **именовать**

**наименьший** *adj. superl.* the least, minimum

**наинизший** *adj. superl.* the lowest

**наихудший** *adj. superl.* the worst

**Найквист** Nyquist

**найлон** *m.* nylon

**найти(сь)** *pf. of* **находить(ся)**

**накал** *m.* incandescence, intense heat, red heat, white heat, glow; heating, filament, heater; **прямого накала** directly-heated; **отделный н.** individual filament heating; **последовательный н.** series filament supply

**накалённость** *f.* heat, incandescence

**накалённый** *adj.* heated, incandescent, glowing

**накаливание** *n.* incandescing, incandescence, heating, glowing

**накаливать, накалить** *v.* to heat, to glow, to incandesce

**накаливаться, накалиться** *v.* to get hot, to heat, to incandesce

**накаливающий** *adj.* incandescent

**накалить(ся)** *pf. of* **накаливать(ся)**

**накальный** *adj.* filament, heating

**накалять** *see* **накаливать**

**накапать** *pf. of* **капать**

**накапливаемый** *adj.* cumulative

**накапливание** *n.* storage

**накапливать, накопить** *v.* to store, to register, to accumulate; **заранее н.** to prestore

**накапливающий** *adj.* accumulating, storing, registering

**накапливающийся** *adj.* cumulative

**накатать** *pf. of* **накатывать**

**накатка** *f.* knurl, knurling tool

**накатный** *adj.* knurled

**накатывать, накатать** *v.* to wind, to coil

**накачать** *pf. of* **накачивать**

**накаченный** *adj.* pumped up, inflated

**накачивание** *n.* pumping, inflation

**накачивать, накачать** *v.* to pump, to inflate

**накачка** *f.* pumping; pump

**накидывать, накидать, накинуть** *v.* to throw (on), to cast on; to slip on; to hook up

**накинуть** *pf. of* **накидывать**

**накипь** *f.* scum; scale, boiler scale, incrustation; sediment, deposit

**накладка** *f.* lap, splice, cover plate; facing; butt-strap; **н. (у стыка рельсов)** fishplate; **лицевая н.** flush plate; **н. на траверзу** arm extension bracket; **н. рельсового стыка** joint bar; **соединительная н.** connecting link; **н. утопленного выключателя** switch plate

**накладной** *adj.* put-on, superposed, laid on, applied

**накладываемый** *adj.* superposable

**накладывание** *n.* superimposing, superposition, laying on; plating; **н. высокой частоты** radio frequency biasing

**накладывать, наложить, накласть** *v.* to put on, to lay on, to lay over, to superimpose, to superpose, to heterodyne; to plate, to coat, to apply; **н. колебания** to heterodyne

**накладывающийся** *adj.* superposable

**накласть** *pf. of* **накладывать**

**наклейка** *f.* gluing, pasting; patch, label, sticker

**наклёп** *m.* riveting; cold working, cold hardening

**наклёпанный** *adj.* riveted; cold-worked

**наклёпывать, наклепать** *v.* to rivet, to clench, to strain harden

**наклон** *m.* slope, incline, inclination, dip, incidence, slant, grade, gradient, pitch; tilt, deviation; drag; fall; droop; **н. (от вертикальной линии)** rake; **н. антенны** antenna tilt; **н. верха импульса** pulse tilt; **н. вершины импульса** pulse droop; **н. волны, н. вперёд фронта волны** wave tilt; **н. горизонтальной площадки импульса** pulse tilt; **н. дорожки** track pitch; **единичный н.** unity slope; **н. кривой намагничивания** differential permeability; **н. опоры** rake of a pole; **н. пологого участка** plateau slope; **н. ряда** array pitch, row pitch; **н. статической характеристики** offset ratio; **н. строки** array pitch, row pitch; **фиксированный н. (антенны)** fixed elevation; **н. фронта волны** waveform angle; **н. фронта импульса** wave tilt; **н. фронта сигнала** wavefront angle; **н. характеристики, равный 1** unity slope

**наклонение** *n.* inclination, dip, tilting, pitch; obliquity

**наклонённый** *adj.* inclined, dipped, canted; biased

**наклонить(ся)** *pf. of* **наклонять(ся)**

**наклонно** *adv.* obliquely, aslant

**наклонность** *f.* obliquity; inclination, leaning, bent, propensity

**наклонный** *adj.* oblique; inclined, sloping, slope, slanted, slanting, tilting; truncated

**наклонять, наклонить** *v.* to tilt, to slant, to incline, to slope, to tip; to decline, to depress; to bias

**наклоняться, наклониться** *v.* to slope, to incline, to lean

**наковальня** *f.* anvil; **верстачная н., настольная н.** bench anvil; **однорогая н.** beak iron; **правильная н.** elevating anvil

**наколоть** *pf. of* **колоть**

**наконечник** *m.* tip, point; adapter;

head, cap; terminal, pole terminal, clip; tag, burr, lug, ferrule terminal, (pole) shoe, nosepiece; thimble; banana pin; ramp; **аксиальный н.** axial-tab terminal; **алюминиевый кабельный н.** aluminum cable lug; **н. для гибкого провода** spade tag; **н. для двух проводов** twin lug; **н. для крепления в нескольких точках** multimount lug; **н. для многопроволочного канатика** spade terminal; **н. для одного провода** single lug; **н. для паяльника** soldering pencil; **зажимной н.** bus-bar clamp lug; **заклинивающийся н.** wedge-on lug; **н. кабеля** tag terminal; **контактный н.** contact joint; **металлический н.** metal end-cup; **н. на питательной линии** feed nozzle; **н. на штыре** thimble; **напаиваемый н., напайный н.** sweating thimble, soldering lug; **полюсный н.** pole terminal, pole shoe, pole piece; **н. провода** tag terminal; **радиальный полюсный н.** radial pole piece; **соединительный н.** connector lug; **н. стойки** finial; **н. шнура** cord terminal

**накопитель** *m.* storage, storage element, storage plate, circuit; tank; accumulator; memory, register; **адресный н.** address file; **н. в счётных устройствах** counter tank; **внешний н.** secondary storage; **входной буферный н.** input store, input block, input buffer; **главный н.** main store; **дополнительный н.** backing store, backing memory; **н. имеющий адрес для каждого накопленного элемента информации** addressed memory; **н. конфликтов** hindsight pool; **магнитный н. с многократным совпадением** multiple coincidence magnetic memory; **матричный н. из сердечников и диодов** core-diode storage matrix; **н. на люминофорах** phosphor storage; **н. на магнитной плёнке** tape file; **н. на магнитных плитах** magnetic plate memory; **н. на ртутных линиях задержки** mercury memory, mercury store; **н. не требующий восстановления** non-volatile memory, non-volatile storage, permanent memory, permanent store; **оперативный н.** internal memory, internal storage; **оперативный быстродейст-**

вующий н. rapid memory; **н. ошибок округления** round-off accumulator; **перевозимый н. информации** transport memory; **периодический н.** cycle storage; **н. с акустической линией задержки** acoustic memory, acoustic store; **н. с быстрой выдачей данных** fast-access memory; **н. с катушками плёнки** reel-type transport; **н. с квазипроизвольной выдачей данных** quasi-random access memory; **н. с линией задержки** delay-line memory, delay-line store; **н. с малым временем выборки** (or **обращения**) rapid access memory, rapid access store; **н. с нулевым временем обращения** zero-access storage; **н. с произвольным временем выборки** (or **обращения**) random access memory, random access store; **н. с ферритовыми сердечниками** ferrite-core memory; **н. со средним временем выдачи информации** medium-access time memory; **н., требующий восстановления** volatile memory, volatile storage; **н. управляющего устройства** control memory, control store; **электрический н.** electrostatic storage

**накопительный** *adj.* memory, storage, accumulator

**накопить** *pf. of* **накапливать, накоплять, копить**

**накопление** *n.* store, storage, accumulation, congestion, pile-up, build-up, cumulation, gathering; handling memory; **н. в линии задержки** delay-line storage; **дискретное н. на ленте** digital tape storage; **н. за период строки** line storage; **н. информации на микроплёнке** micro-image data; **н. на сеточном электроде** mesh-screen storage; **н. неосновных носителей** minority-carrier storage; **н. носителей(тока)** carrier storage; **н. от импульса к импульсу** pulse-to-pulse integration; **н. отклонения** accumulated deviation; **н. отметок (сигналов)** track storage; **н. ошибок округления** round-off accumulating; **н. произведения** product accumulation; **селективное н.** selective localization; **н. с произвольной выборкой** random access storage

**накопленный** *adj.* stored, accumulated, cumulative

**накоплять, накопить** *v.* to store, to accumulate, to collect, to register, to cumulate, to gather, to stock up

**накопляющий** *adj.* storing, accumulating

**накрест** *adv.* cross, crossways, traverse

**накрывание** *n.* covering

**накрывать, накрыть** *v.* to cover, to lay over

**накрытый** *adj.* covered

**накрыть** *pf. of* **накрывать**

**нактоуз** *m.* binnacle

**налагаемый** *adj.* congruent

**налагать, наложить** *v.* to superimpose, to impose, to lay on, to put on

**налагающийся** *adj.* overlapping

**наладить** *pf. of* **налаживать**

**наладка** *f.* alignment, adjusting, adjustment, tuning up, setting up; fixing; repairing; presetting; arrangement; debugging; **эксплоатационная н.** operating adjustment

**наладочный** *adj.* aligning

**налаженный** *adj.* fixed, repaired; set, tuned up, adjusted

**налаживание** *n.* adjusting, adjustment, tuning up; repair, repairing, fixing; **предпусковое н. и испытание системы управления** prelaunching adjustment

**налаживать, наладить** *v.* to set, to tune up, to adjust, to align, to debug; to fix, to mend, to repair, to put right; **повторно н.** to readjust

**наледь** *m.* sleet, layer of ice

**налёт** *m.* deposit, coating; film, bloom, tarnish

**налетающий** *adj.* incident

**налётный** *adj.* fallen on, fallen upon; flying on

**наливание** *n.* filling, pouring in

**наливать, налить** *v.* to pur in, to fill (up)

**наливной** *adj.* pouring, filling

**налитый** *adj.* poured in, filled

**налить** *pf. of* **наливать**

**наличие** *n.* presence; availability; occurrence; **при наличии** in the presence (of); **н. гистерезиса** hysteretic nature; **н. деталей** stock of material; **н. «кнопки несчастья» в управлении** deadman's feature; **н. пятен на мишени** target blemishing

**наличник** *m.* finishing stile, stile strip

**наличный** *adj.* available, on hand, present; effective

**налог** *m.* tax, duty, imposition; impost

**наложение** *n.* superposition, superimposing, superimposure, imposition, laying on, overlap, overlapping, overlay, overlaying, pile-up; **н. бандажа** banding; **н. изоляции** lapping; **н. порядков** overlapping of order; **н. тонов** modulation at audible frequency: **н. характеристик** retrace

**наложенный** *adj.* superimposed, superposed, laid on

**наложить** *pf. of* **накладывать, налагать**

**налощить** *pf. of* **лощить**

**намагнетизировать** *pf. of* **магнетизировать**

**намагнитить** *pf. of* **намагничивать**

**намагничение** *n.* magnetization, magnetizing; excitation; **мгновенное н., поверхностное н.** flash magnetization; **поперечное н.** transverse magnetization

**намагниченность** *f.* magnetization, intensity of magnetization; **н. насыщения** saturation intensity, specific magnetization; **н. ферромагнитных областей** domain magnetization

**намагниченный** *adj.* magnetized

**намагничиваемость** *f.* magnetizability

**намагничиваемый** *adj.* magnetizable

**намагничивание** *n.* magnetization, magnetizing (process); **н. в одном и том же направлении** unidirectional magnetization; **неправильное н.** anomalous magnetization; **остаточное н.** retentivity; **н. пластинки, плоское н.** lamellar magnetization; **поперечное н.** cross magnetization, perpendicular magnetization; **предварительное н.** premagnetization, magnetic bias, polarization; **н. с направлением силовых линий по окружности** circular magnetization; **н. стального бруска двумя стержневыми магнитами** divided touch magnetization; **н. стального бруска натиранием его полюсами двух стержневых магнитов** double touch magnetization; **н. стержня двумя постоянными магнитами** magnetization by separate touch; **н. током** excitation

**намагничивать, намагнитить** *v.* to magnetize

**намагничивающий** *adj.* magnetizing, magnetization, magnetomotive

**намагничивающийся** *adj.* magnetizable

**намазанный** *adj.* smeared, coated, daubed

**намазать** *pf. of* **намазывать, мазать**

**намазка** *f.* filler; smearing; **н. цоколей** base filling

**намазываемый** *adj.* troweled-on

**намазывать, намазать** *v.* to cover, to paste, to coat, to smear, to daub

**наматывание** *n.* winding, reeling, coiling; lapping

**наматывать, намотать** *v.* to wind, to reel, to coil, to spool; to wind up, to take up; **н. в нахлёстку** to wind with overlapping, to overlap; **н. на катушку** to spool

**наматывающий** *adj.* winding, reeling, coiling; take-up (reel)

**намеренный** *adj.* intentional, deliberate; premeditated

**намертво** *adv.* dead

**наметить** *pf. of* **метить**

**намётка** *f.* presetting

**намотанный** *adj.* wound, coiled, reeled; **н. на ребро** edgewise-wound; **н. по шаблону** form-wound

**намотать** *pf. of* **наматывать, мотать**

**намотка** *f.* winding, coil, coiling, spooling; **со спиральной намоткой** helically wound; **беспорядочная н., н. беспорядочными витками** random winding, haphazard winding, scramble winding, mush winding; **н. внахлёстку** pile winding; **н. вперекидку, н. в перекрышку** banked winding; **«дикая» н.** random winding; **ёмкостная н.** slab winding; **н. катушки** coil winding; **кучная н.** *see* **беспорядочная н.; малоёмкостная н.** low-capacitance winding; **неплотная н.** loose winding, space winding; **простая н.** unifilar winding, banked winding; **рядовая н.** layer winding; **н. с чередованием витков в каждой паре слоёв** pile winding; **сотовая н.** honeycomb winding

**намоточный** *adj.* winding, spooling

**нанесение** *n.* drawing, plotting, marking, charting, tracing; heaping, drifting; **н. (слоя)** spraying; **н. люминофора** settling of phosphor; **н. люминофора методом печатания** phosphor printing; **н. люминофора на экран ЭЛТ** pouring of phosphor; **н.**

люминофора путём печатания phosphor printing; **н. маркировочного знака** labelling; **н. металлических покрытий** metallization; **н. многоцветного экрана фотоспособом** photoprinting process; **н. на карту** mapping, plotting; **н. покрытия** plating; **н. покрытия струйным методом** jet plating; **н. тонкой нерефлектирующей плёнки** blooming; **н. тонкой плёнки** sputtering; **н. точек на график** plotting; **н. траектории** tracing; **н. трассы на карте** track charting; **частичное н. покрытия** partial plating; **н. экрана осаждением** screen condensation; **н. экранов электростатическим методом** electrostatic printing; **н. экранов ЭЛТ** gravity pouring

**нанесённый** *adj.* coated, applied; drawn, plotted; entered (on chart); etched, inscribed; deposited

**нанести** *pf. of* **наносить**

**нано-** *prefix* nano-

**наноампер** *m.* nanoampere

**наноамперметр** *m.* nanoammeter

**нановаттный** *adj.* nanowatt

**нанограф** *m.* Nanograph

**наносекунда** *f.* nanosecond

**наносекундный** *adj.* nanosecond

**наносимый** *adj.* troweled-on; applied

**наносить, нанести** *v.* to mark, to plot, to map, to draw, to mark down; to insert; to heap, to drift; to deposit; to apply, to coat, to superimpose; **н. деления** to graduate, to make divisions on a scale

**наноскоп** *m.* Nanoscope

**наносхема** *f.* nanocircuit

**нанофарада** *f.* nanofarad

**напад** *m.* attack, invasion

**нападать, напасть** *v.* to attack, to invade

**нападение** *n.* attack, invasion; **скрытое н.** sneak attack

**напаиваемый** *adj.* soldered on, fused on

**напаивать, напаять** *v.* to solder on, to fuse on, to seal on; to burn on

**напайка** *f.* sealing on, building up

**напайный** *adj.* soldering

**напасть** *pf. of* **нападать**

**напаянный** *adj.* soldered on, welded on

**напаять** *pf. of* **напаивать**; **н. (лампу) в вакуумной системе** to seal in a vacuum system

**напёрстковый** *adj.* thimble

**напёрсток** *m.* thimble

**напечатанный** *adj.* printed, published

**напечатать** *pf. of* **печатать**

**напильник** *m.* file; **н. для заточки пил** saw file; **н. для зачистки контактов (or для обработки мест контакта)** contact file; **драчевой н.** bastard file, coarse file; **игольчатый н.** broaching file; **круглый н.** rat-tail file; **личной н., мелкий н.** smooth-cut file; **ножовочный н.** knife file, edged file; **обдирочный н.** rought-cut file; **плоский остроносный н.** taper file; **проволочный н.** broaching file; **н.-рифлуар** bow file; **н. с грубой насечкой** rasp, rough-cut file; **н. с перекрёстной насечкой** double cut file; **тонкий шлифной н.** dead smooth file; **узкий н.** feather edge; **шлифной н.** smooth-cut file

**напитать** *pf. of* **питать**

**наплавить** *pf. of* **наплавлять**

**наплавка** *f.* beading, weld seam; cladding, surfacing

**наплавление** *n.* building up

**наплавленный** *adj.* welded (on), fused on, built up; deposited

**наплавлять, наплавить** *v.* to fuse, to melt, to smelt; to build up, to fuse on

**наплыв** *m.* wipe, wipe screen effect, wipe superimposure, fade over, dissolve; influx, abundance; **боковой н.** lateral dissolve

**наплывать, наплыть** *v.* to fade up and down; to float, to drift

**наполнение** *n.* charge, filling; packing; admission; filing; **с кремноорганическим наполнением** silicone-bonded; **аргоновое н.** argon filling; **неоновое н.** neon filling

**наполненный** *adj.* filled, full

**наполнитель** *m.* filler, filling, stuff; feeder, charger; binder

**наполнительный** *adj.* filling; feeding, charging

**наполнять, наполнить** *v.* to fill, to stuff; to charge; to file; to impregnate; to top up (cells)

**наполняющий** *adj.* filling; charging

**наполовину** *adv.* half, by halves, semi-

**напольный** *adj.* outdoor; ground-type, floor

**напор** *m.* pressure, compression, head, thrust, rush; **динамический н.** kinetic energy head; impact pressure напорный *adj.* pressure, force, head

напр. *abbr.* (например) for example

направительный *adj.* guiding, directing

направить *pf. of* направлять

направление *n.* direction, sense, heading, bearing, way, run, course, route; alignment; guidance; trend; tendency; **н. (на цель)** line of sight; **по направлению** parallel, in the direction (of), toward; **н. антенны, при котором рассогласование равно нулю** apparent radar center; **блокируемое н.** inverse direction, reverse direction; **н. в сторону от принятого курса** off-course direction, deviation from course; **н. волны** wave line; **н. воспроизведения ленты** tape travel; **встречное н.** opposite direction, reverse direction; **главное н. приёма** main receiving direction; **н. движения** running direction, direction of motion, sense direction; traffic route; **запасное н.** reserve route, sidetrack; **запертое н., запирающее н., запорное н.** *see* **блокируемое н.; н. избираемое набором кодовой цифры** dial prefix routing; **истинное н. полёта** true heading; **исходное н.** reference direction; **н. каната** rope guide; **н. лёгкого намагничения** easy direction of magnetization; **н. на базу** directional homing; **на одно н.** single-throw; **н. намотки витков трансформатора** polarity of transformer; **неправильное н.** misrouting (*tphny.*); **н. неправильных соединений к телефонистке** denied-service routing, operator intercepting; **непроводящее н.** inverse direction; **н. обмотки** sense of winding; **обходное н.** bypass routing, alternative trunking; **опорное н.** reference direction; **определяющее н. движения** electric traffic locking; **оптическое н.** Y-direction; **н. оптической оси антенны радиолокатора** radar center; **н. оси X(Y, Z)** X-(Y-, Z-) direction; **н. отражения** directional reflectance; **ошибочное н.** misalignment; **н. полёта** heading, flight direction; **н. полёта цели** target flight direction; **правильное н. полёта** correct flight direction; **н. пропускания тока** forward direction (of current);

**пропускное н.** forward direction; **пьезоэлектрическое н.** direction in a crystal; **н. распространения** direction of propagation; **н. распространения волны** wave line; **н. связи** route, routing; **н. синхронизирующих импульсов** sync direction; **н. скрутки** direction of lay; **смещённое н.** offset direction; **трудное н.** hard direction; **эталонное н.** reference bearing

направленность *f.* direction; directivity, directionality, directional effect; **н. в широком диапазоне частот** broadband directivity; **н. громкости** loudness directivity index; **регулируемая н. антенны** controlled directivity (of an antenna)

направленный *adj.* directed, set; directional, directive, oriented, orientational, headed, guided, beam(ed); direct (current); point-to-point; **н. в одну сторону** unidirectional

направляемый *adj.* directed, guided

направлять, направить *v.* to direct, to guide, to lead; to set, to adjust, to aim; to head; to beam; **н. в обход** to reroute, to divert the traffic

направляющая *f. adj. decl.* guide, rail, slide, runner; **н. клапана** valve guide; **н. пишущего штифта** pencil guide; **н. штока клапана** valve stem guide

направляющий *adj.* guide, guiding, leading, directing, directive, director; control(ling), regulating; aligning; pilot

напротив *adv.; prep. genit.* on the contrary; counter, opposite, facing

напрягать, напрячь *v.* to strain, to force

напряжение *n.* voltage, potential, tension; stress, strain, straining; pressure; exertion; **два номинальных напряжения** double voltage rating; **напряжения, единаковые по величине и полярности** push-pull voltage; **флуктуационные напряжения, возникающие в антенной цепи** antenna noise; **азимутальное управляющее н.** azimuth-control voltage; **н. азимутальной развёртки** azimuth-sweep voltage; **анодное н. зажигания** anode breakdown voltage (of a discharge tube); **н. блокировки (люминесцентного экрана)** sticking voltage (of a luminescent screen); **большое н. отсечки** remote cut-off; **н. в начале линии** sending-end voltage; **н. в ненагретом состоянии** maximum permissible plate voltage for electron tubes at cold filament condition; **н. в первичной цепи (обмотки)** primary voltage; **н. в точке схождения пучка** crossover voltage; **вводимое н.** injecting voltage; **ведущее н.** slaving voltage; pilot voltage; **н. видеосигнала, подводимого к «зелёному» («красному», «синему») прожектору трубки** green (red, blue) video voltage; **вносимое н.** insertion voltage; **н. во вторичной цепи** secondary voltage; **возбуждающее н.** exciting voltage, driving voltage, keep-alive voltage, excitation drive; **н. возбуждения** driving voltage, excitation voltage; **н. возбуждения альтернатора** alternator field voltage; **н. возврата** drop-out voltage; **возвращающееся н.** recovery voltage; **н. возникающее при изменении потока от состояния насыщения до остаточного состояния того же знака** flyback voltage; **н. возникновения электрического разряда** breakdown voltage (between two electrodes); **н. вольтовой дуги** arc voltage, arc potential; **вольтодобавочное н.** boost voltage; **восстанавливающееся н.** restriking-voltage; recovery voltage, restoring voltage; **выключающее н.** release voltage; **н. выключения** interrupting voltage; **выпрямленное н. шумов** clipped noise; **выходное н. компенсационного устройства** canceller output; **выходное н. помех** noise output voltage, noise power, noise output; **выходное н. схемы «привязки»** clamp output voltage; **н. гашения луча** spot cut-off voltage; **н. генератора качающейся частоты** sweep oscillator voltage; **н. генератора развёртки** sweep oscillator voltage; **главное н.** line voltage; **н. «горба» частотной характеристики** "hump" voltage; **н. горения (тиратрона)** sustaining voltage; **группирующее н.** bunching voltage; **н. датчика** cue voltage; **н. двоичного сигнала** binary voltage; **действующее н.**

effective voltage; **длительное н.** steady stress; **н. для выборки мёртвой зоны** ditcher voltage; **добавочное н., дополнительное н.** boost(ing) voltage, surplus voltage, additional voltage; **допустимое н. по нагреву** temperature-rise voltage; **н. дробового шума** shot-noise voltage; **н. единичного перепада** unit voltage; **задающее н.** driving voltage, lumped (control) voltage; **н. задержки** delay voltage; threshold voltage (*tv*); **н. зажигания** firing voltage, critical grid voltage, striking potential; **н. замедляющей системы** circuit voltage; **н. замыкания магнитного контактора** pick-up voltage; **н. запирания** cut-off bias, pinch-off voltage; **н. запирания луча** spot cut-off voltage; **запирающее н.** blackout voltage, blanking voltage, nonconducting voltage, cutoff voltage; **запирающее н., определённое методом экстраполяции** extrapolated cut-off voltage; **запирающее н. смещения** cut-off bias; **н. заскока** blocking potential; **затемняющее н.** blanking voltage; **н. затягивания** locking voltage; **знакопеременное н.** alternate stress; **идеальное выпрямленное н. холостого хода** ideal no-load D. C. voltage; **н., имеющее наклонный перепад** sloping voltage; **н. импульса выброса** peaking voltage; **н. импульса обратного хода** flyback voltage; **н. импульса обратного хода строчной развёртки** horizontal fly-back voltage; **импульсно-приложенное н.** step-function voltage; **н. импульсов прямоугольной формы** square-wave voltage; **индуктируемое н. при номинальной нагрузке** virtual voltage; **испытательное н. изоляции, испытательное н. пробоя диэлектрика** dielectric test voltage; **н. источника питания** supply voltage; **н. касания** pickup voltage; **н. качания частоты, качающееся н.** sweep voltage; **н. квадратурной составляющей поднесущей** quadrature subcarrier voltage; **н. коллектора в рабочей точке** quiescent collector voltage; **коммутирующее н.** switching voltage; **комплексное синусоидальное н.** complex harmonic voltage; **конечное н.** end-point voltage, cut-off voltage; **н. конца разряда** final voltage, terminal voltage, end-point voltage; **н. короткого замыкания** short circuit voltage, impedance voltage; **корректирующее н. подстройки** tuning-correction voltage; **критическое коронное н.** initial voltage; **критическое сеточное н.** critical grid voltage (of a thyratron); **линейное н. (в трёхфазной системе)** delta voltage; **максимальное н. без перекрытия** maximum permissible A.C. voltage; **максимальное мгновенное обратное н.** maximum peak inverse voltage; **максимальное положительное н.** peak blocked voltage; **н. между двумя смежными ламелями коллектора** commutator-segment voltage; **н. между двумя смежными линиями многофазной системы** mesh voltage; **н. между двумя смежными линиями шестифазной системы** hexagon voltage; **н. между диаметрально-противоположными проводами в симметричной шестифазной системе** diametrical voltage; **н. между контактными кольцами** slip ring voltage, ring-to-ring voltage; **н. между пластинами** bar-to-bar voltage; **н. между проводами** circuit voltage; **н. между проводами в вершинах треугольника в симметричной шестифазной системе** delta voltage; **н. между фазой и нейтралью** star voltage; **междуламельное н.** bar-to-bar voltage; **междулинейное н.** circuit voltage; **механическое н. в диэлектрике от разности потенциалов** electric stress; **(механическое) н. в диэлектрике при переходных режимах** transient stress; **механическое н. электрострикции** electric stress; **мешающее н.** noise voltage; **н. модулируемых импульсов** sampling voltage; **мокроразрядное н.** wet flash-over potential; **н. на активном сопротивлении** resistive voltage; **н. на вторичной обмотке трансформатора** transformer-secondary voltage; **н. на вторичной цепи** secondary voltage; **н. на выходе генератора кадровой развёртки** image output, picture output, video time base; **н. на выходе генератора строчной раз-**

**вёртки** line output; **н. на выходе фиксирующей схемы** clamp output voltage; **н. на защитном кольце** guard-ring voltage; **н. на конце линии** received-end voltage; **н. на омическом сопротивлении** resistive voltage; **н. на остановленном роторе** locked rotor voltage; **н. на резонаторе** cavity voltage; **н. на сетке относительного катода** grid-to-cathode voltage; **н. на сетке стабилизирующего усилителя** regulator-amplifier-grid voltage; **н. на ускоряющем электроде** image accelerator voltage; **наведённое н. (при перемагничивании)** switching voltage; **наибольшее обратное н.** inverse voltage; **наивысшее н.** ceiling voltage; **наименьшее линейное н.** mesh voltage of a polyphase system; **н. начала коронирования** initial voltage; **н. начала повреждения изоляции** tracking voltage; **н. начала разряда** stopping voltage of discharge; initial voltage; **невыпрямленное н. шумов** unclipped voltage; **неполное н.** undervoltage, lower voltage; **непрерывно меняющееся н.** analog voltage; **низшее н.** under-tension, undervoltage, low voltage; **н. нового зажигания (дуги)** restriking voltage; **номинальное н.** rated voltage, standard voltage, normal voltage; **номинальное н. прибора (or цепи)** normal circuit voltage; **номинальное рабочее н.** service voltage; **нормальное номинальное н.** normal voltage, rating; **н. обратного зажигания** flashback voltage, reignition voltage; **обратное н.** reverse voltage, return voltage, kick-back voltage, back voltage, inverse voltage, counter voltage; **н. обратной связи** feedback voltage; **н. ограничения** cut-off bias; **ограниченное н. шумов** clipped noise; **однономинальное н.** single-voltage rating; **опасное н.** excess voltage, over-voltage; **опорное н.** reference voltage, Zener voltage, reference potential; **н. опорных импульсов** pedestal voltage; **остаточное н. (после сварки)** residual stress; **н. от нагрева** thermal stress; **н. от остаточного магнетизма** residual voltage; **отводящее н.** "seeker" voltage; **н. отпадания** drop-out voltage;

отпирающее н. trigger voltage; н. отпускания drop-out voltage; н. отражателя (в клистроне) repeller voltage, reflector voltage; отрицательное н. смещения back bias voltage, negative bias voltage; отрицательное сеточное н. отсечки negative cut-off grid voltage; паразитное н. stray voltage; перекрывающее н. sparking voltage, spark-over voltage; н. перекрытия flashover voltage, break-down voltage; н. перемещения изображения positioning voltage; пиковое переменное н. в зазоре peak alternating gap voltage; н. питания управляющей сетки control-grid supply voltage; н. плотного прижатия sealing voltage; н. площадки гасящего импульса относительно уровня чёрного pedestal voltage; н. повторного зажигания re-stricking voltage; н. погасания extinction voltage; н. поджигания, поджигающее н. firing voltage, ignition voltage, keep-alive voltage, critical grid voltage; н. поднесущей, сдвинутое по фазе на 90° quadrature subcarrier voltage; н. подогрева, н. подогревателя, подогревное н. heater voltage; н. подсветки intensifier potential; н. помехи от токов частичной выборки disturb voltage; пониженное н. undervoltage, low voltage; поперечное н. shearing stress, torsion voltage; постоянное н. constant voltage, continuous voltage, direct voltage, direct-current potential, fixed potential; постоянное н. для центрирования растра shift voltage; н., постоянное по величине constant voltage; н. превышающее расчётное overrating voltage; предварительное н. prestress, pre-load; предельно допустимое пиковое обратное н. peak inverse voltage rating, peak reverse voltage rating; предельное н. заряда cut-off voltage; предельное н. поля change-over field strength, change-over field intensity; предельное импульсное н. (insulation) test voltage; предельное обратное н. peak inverse voltage; предельное переменное синусоидальное н. maximum permissible A. C. voltage: н. предыонизации keep-alive volt-

age; н. при несимметричной нагрузке biasing voltage; н. при переходном процессе transient voltage; н. при растяжении tensile stress; н. при сжатии compression stress, crushing stress; н. при соединении треугольником delta voltage, mesh voltage; н. при установившемся режиме steady stress; н. при холостом ходе floating voltage; приведённое н. lumped voltage, reduced voltage; приложенное н. applied voltage, impressed voltage; принятое для сравнения постоянное н. reference voltage; н. пробоя breakdown voltage, peak inverse voltage; н. пробоя при дожде rain flash-over voltage (of insulators); н. пробоя через масло dielectric-strength of oil; н. проводника при перемещении в магнитном поле e.m.f. induced by motion; н. произведения (двух величин) product voltage; н., пропорциональное заданному числу analog voltage; н., пропорциональное скорости velocity voltage; противодействующее н. counter voltage, bucking voltage, inverse voltage, backlash potential; н. прямого пробоя punch-through voltage; н. развёртки sweep voltage, sawtooth voltage, scanning voltage; time-base waveform; н. развёртки дальности range-sweep voltage; раздвигающее н. spread voltage; н. разностного цветного сигнала color-minus-difference voltage; разрядное н. игольного разрядника needle-point voltage; н. реагирования minimum operating voltage; реактивное н. reactance voltage, imaginary voltage; реактивное н. искрения sparking voltage; н. регулировки синхронизации locking control voltage; н. регулировки уровня гасящих импульсов pedestal control voltage; регулировочное н. calibration voltage; результирующее н. total voltage; релаксационное н. sweep voltage, sawtooth voltage; решающее н. computing voltage; н. с жёсткой характеристикой, н. с пологой характеристикой flat voltage; н., сдвинутое по фазе out-of-phase voltage; н. секции обмотки y-voltage, star voltage; селектирующее н. молекулярного усилителя maser voltage; сетевое н. supply voltage, supply pressure; н. сеточного смещения grid priming voltage, grid-bias voltage; сеточное запирающее н. cutoff bias; н. (сигналов) питания power-supply voltage; н. силовой сети mains voltage, power line voltage; слабо пульсирующее н. ripple electromotive force, ripple voltage; смещающее н. bias(ing) voltage, смещающее н. детектора detector polarizing voltage; смещающее н. на сетке grid leak; смещающее сеточное н. (direct) grid bias; н. смещения (на сетке) bias (ing) voltage, biasing potential, grid bias, priming grid voltage; собственное н. natural stress, residual stress, inherent tension; совпадающее по фазе н. inphase voltage; содействующее н. additional voltage of same direction as the existing voltage; н., соответствующее вращению элемента motion-derived voltage; н. срабатывания pick-up voltage; н. статора primary voltage; н. столба дуги arc-stream voltage; н. строчной развёртки horizontal sweep voltage, line-sweep voltage; н. ступенчатой формы staircase voltage; сухоразрядное н. dry flash-over potential; тактовое н. timing voltage; тренировочное н. burning voltage, aging ·voltage; н. трогания pickup voltage (of a relay); н. у нагрузки receiver voltage; н. у начала линии sending-end voltage; н. у потребителя, н. у приёмника utilization voltage; удерживающее н. (в реле) voltage restraint; управляющее н. на сетке grid drive voltage; управляющее переменное н. scan-control voltage; уравновешивающее н. balancing waveform; н. ускоряющего электрода image accelerator voltage; ускоряющее н. (в фотоэлементе) external voltage; ускоряющее н., при котором экран заряжается отрицательно sticking voltage; ускоряющее н. пучка beam voltage; устойчивое н. flat voltage; фазное н., фазовое н. star-voltage, y-voltage; phase voltage; н. фиксирования clamp voltage; н. Холла, холловское н. Hall voltage; холод-

ное н. cold straining, cold working; н. холостого режима (or хода) no-load voltage, open-circuit voltage; н. холостого хода двухполюсника no-load voltage of a two terminal network; н. шума покоя quiescent noise voltage; н. шумов гидроакустического преобразователя transducer noise voltage; н. эквивалентного диода composite controlling voltage; эквивалентное н. lumped voltage; н. энергоснабжающей системы system voltage; эталонное н. зенер-диода zener voltage

**напряжённость** *f.* tension, intensity, intensiveness, strain, stress; strength; коэрцитивная н. поля coercive force intensity, coercivity; магнитная н. magnetizing force; н. нагрузки, н. обмена density of traffic (*tphny.*); н. отраженного сигнала echo strength; н. поля высокой частоты radio field intensity; н. поля по Бриллюэну Brillouin flux density; н. поля при приёме intensity of received field, received field strength; н. поля при пробое disruptive electric field strength; н. поля радиовещательной станции radio field intensity; предельная н. поля insulating strength; статическая пороговая н. магнитного поля DC threshold magnetizing force

**напряжённый** *adj.* intense, intensive; stressed; tense, taut, strained

**напрямую** *adv.* straight, straight-way

**напрячь** *pf. of* напрягать

**напуншированный** *adj.* punched

**напыление** *n.* deposition, sputtering; dusting spraying; evaporation; вакуумное н. vacuum evaporation, thermal evaporation in vacuum

**напылённый** *adj.* spray-coated, sprayed; evaporated; н. в вакууме vacuum-deposited

**напыливать, напылить** *v.* to spray, to dust; to sputter, to shadow

**напылить** *pf. of* напыливать, пылить

**напыляемый** *adj.* sprayed-on

**наработка** *f.* non-failure operating time; заданная н. mission time; машинная н. serviceability; н. между отказами time between failures; н. на отказ mean time between failures; суммарная н. accumulated operating time

**наравне** *adv.* on a level (with), flush (with), like

**нарастание** *n.* growing, growth, rise, accumulation, build-up; intensification, increment; **н. импульса** building-up of pulse; **н. какой-либо величины во времени** time rise; **мгновенное н. амплитуды** overshoot; **постепенное н. напряжённости намагничения в железе** creeping, magnetic creeping

**нарастать, нарасти** *v.* to grow on, to be formed on; to rise, to increase, to accumulate, to build up

**нарастающий** *adj.* growing, increasing, rising, incremental; positive-going

**нарасти** *pf. of* **нарастать**

**нарастить** *pf. of* **наращивать**

**наращение** *n.* growing, increment, building up; accumulation, augmentation

**наращенный** *adj.* built-up, accumulated

**наращиваемый** *adj.* built-up; continuous

**наращивание** *n.* building-up; **электролитическое н.** plating

**наращивать, нарастить** *v.* to build up, to accumulate; to plate

**нарегулированный** *adj.* regulated

**нарез** *m.* cut, incision; notch; thread (of screw), worm; threading, cutting

**нарезание** *n.* cutting; **н. резьбы** threading, cutting thread

**нарезанный** *adj.* cut, threaded

**нарезать** *pf. of* **нарезывать; н. зубья** to cog, to tooth

**нарезка** *f.* cut, incision; thread (of screw), worm; cutting, threading; **н. болта** male thread; **внутренняя н.** female thread; **н. Эдисона** Edison screw

**нарезной** *adj.* cut, threaded

**нарезывать, нарезать** *v.* to cut, to notch; to thread (screw)

**нарисовать** *pf. of* **рисовать**

**нарост** *m.* growth, excrescence

**нарочный** 1. *adj.* intentional, designed; 2 (*noun*) *m. adj. decl.* express messenger

**НАРС** *abbr.* (**неуправляемый авиационный ракетный снаряд**) unguided aircraft rocket

**наружно** *adv.* externally; apparently

**наружность** *f.* exterior, appearance, outside, outward aspect

**наружный** *adj.* external, outward, outer, out, exterior, outside; outdoor; surface; extraneous

**наружу** *adv.* out, outside, outwards

**нарушать, нарушить** *v.* to break, to infringe, to violate, to transgress; to disturb, to disrupt

**нарушающий** *adj.* impairing; disturbing

**нарушение** *n.* disturbance; perturbation, perturbance; violation; infringement, infraction, breaking; irregularity; failure; upset; abnormality; **н. калибра** overage; **н. контакта** contact fault; **н. ортогональности** nonorthogonality; **н. режима** irregularity, trouble, disturbance; **н. связи** communication irregularity, signal irregularity, service irregularity; **н. симметрии** asymmetry; **н. синхронизации** desynchronizing, lock-out, out-of-lock, out-of-step; **н. служебного порядка** service irregularity; **н. сообщения** interruption of traffic, interruption of communication; **н. сходимости** misconvergence; **н. фокусировки** defocusing

**нарушенный** *adj.* disturbed, upset, disrupted, perturbed; broken, infringed, faulty

**нарушить** *pf. of* **нарушать**

**наряд** *m.* order; detail; **н. на исправление повреждения** repair order; "bad" order; **н. на отключение цепи** clearance order; **н. на работу в (междугородной) цепи** (toll) circuit order; **н. на работу у абонента** service order

**насадить** *pf. of* **насаживать**

**насадка** *f.* nozzle, mouthpiece; cap, capping; putting on, load; building up; built-up part; adapter, attachment; setting, fitting on; **н. для гудения** additional equipment for aural indication of Doppler effects in radar systems; **концевая н.** end closer; **концевая н. муфты** threaded end-fitting; **органная н.** acoustic funnel; **поглощающая н.** sand load, absorption cell, absorption pad; power termination; dissipative element; **полюсная н.** pole shoe

**насадок** *see* **насадка**; **н. Вентури** Venturi; **н. статического давления** static tube

**насадочный** *adj.* packed
**насаженный** *adj.* built-up; set, put on
**насаживать, насадить** *v.* to put on, to fit (on), to set (on), to mount, to slip over; **н. в горячем состоянии** to warmdraw
**насекальный** *adj.* cutting, notching
**насекательный** *adj.* pruning, cutting
**население** *n.* population
**населённый** *adj.* populated, inhabited; residential
**насеченный** *adj.* cut, incised
**насечка** *f.* notch, groove, slot, incision, cut
**насквозь** *adv.* through, right through
**наслаивание** *n.* stratification; superposition, overlapping
**наслаивать, наслоить** *v.* to stratify, to laminate; to overlap, to superpose
**наслоение** *n.* stratification, lamination; layer; **н. импульсов** pulse interleaving
**наслоенный** *adj.* stratified, laminated; superposed
**наслоить** *pf. of* **наслаивать**
**насос** *m.* pump; **под насосом** on the pump; **водоструйный воздушный н.** water-jet air pump; **вязкостный вакуумный н.** molecular pump; **герметический н. с электроприводом** canned motor pump; **грязевой н.** mud pump; **н. для откачки газа** off-gas pump; **н. для предварительной откачки** forepump; **ионный н.** evapor-ion pump; **ионный вакуумный н.** ion-type pump; **конденсационный вакуумный н.** vapour pump; **пароструйный н.** air-ejector, steam ejector, diffusion pump; **н. предварительного вакуума** forepump; **н. предварительного разрежения** high-pressure vacuum pump; **спиральный н.** volute pump; **струйный вакуумный н.** ejector; **тепловой н.** Peltier heat pump; **форвакуумный н.** forepump
**насосно-вакуумный** *adj.* pumped-storage
**насосный** *adj.* pump
**наставление** *n.* manual; direction, instruction
**наставления** *pl.* service instruction(s); operating instruction(s); specifications, practice
**наставной** *adj.* added, set on
**насталенный** *adj.* steel-faced

**насталивание** *n.* steel facing, acierage
**насталивать** *v.* to steel, to plate with steel
**настенный** *adj.* wall, wall-type
**настил** *m.* frame; trestles; layer; floor; deck; **половой н.** floor covering
**настильность** *f.* flatness
**настольный** *adj.* table, table-type, desk; cabinet-type
**настраиваемый** *adj.* tunable, adjustable
**настраивание** *n.* adjust(ing), align(ing), tune (up), syntony
**настраивать, настроить** *v.* to adjust, to tune, to tune up, to attune, to set, to tune in, to syntonize; to control, to regulate; to reproduce, to imitate, to simulate, to construct; **н. на резонанс** to tune in resonance
**настраивающий** *adj.* adjusting, tuning, syntonizing
**настроенный** *adj.* adjusted, tuned; resonant; constructed, built (on); in tune; **заранее н.** pretuned
**настроечный** *adj.* adjusting, tuning, aligning, alignment
**настроить** *pf. of* **настраивать**
**настройка** *f.* adjusting, tuning, adjustment, accordance, setting, syntonization, syntony, alignment; **с анодной настройкой** anode-tuned; **с магнитной настройкой** ferrite-tuned; **с тупой настройкой** flatly tuned; **автоматическая н. нуля** automatic balancing; **н. в сторону повышения (понижения)** tuning upward (downward); **н. величины давления** pressure adjustment; **н. верньером** fine tuning, sharp tuning; **н. выдержки** timing; **грубая н. (нуля)** coarse balance; coarse tuning, flat tuning, rough adjustment; **двухконтурная н.** simultaneous tuning of two tunable circuits; **н. десятичной запятой** decimal point alignment; **н. диапазона частот** band spreading; **заранее установленная н.** pretuning, preset tuning; **кнопочная ручная н.** push-button selector; **н. колебательных контуров на одну частоту** coincidence tuning; **н. линии** balancing of a circuit, line balance; balancing, equalization, compensation; **н. магнетрона окном** iris-coupled magnetron tuning; **н. магнетрона при помощи проводящего**

сердечника sprocket tuning; **н. на одну боковую полосу частот** single-sideband tuning; **н. на приём направленной передачи** tuning for radiobeam; **н. на резонанс** syntonization; **н. нажимной кнопкой** push-button tuning; **н. не на середину полосы частот** off-center tuning; **неточная н. последовательных каскадов** staggered tuning; **н. нуля** balancing; **н. объёмного контура** tuning of a cavity resonator; **одновинтовая н. волновода** single-screw waveguide tuner; **н. одного контура на две мало различающиеся частоты** double-hump tuning; **н. одной кнопкой (or ручкой)** ganged control, single-dial control, single-dial tuning, single-span tuning; **окончательная н.** debugging; **оптическая н.** visual tuning; **очень острая н.** hair-breadth tuning; **плохая н.** maladjustment; **н. по визуальному индикатору** visual tuning; **н. по высоте звука и по длине волны** sound and wave tuning; **н. по низким частотам** bass tuning; **полосовая н.**, **н. полосового фильтра на заданную полосу частот** band-pass tuning, band-pass transmission; **последовательная н.** series tuning, stagger tuning; **постоянная н. антенного контура** constant antenna tuning; **н. при помощи кольца, вставляемого между связками** cookie-cutter tuning; **н. при помощи линии** line tuning; **н. при помощи пластинки** spade tuning; **н. при помощи прослушивания щелчков** click method tuning; **н. регистрирующего устройства** recorder adjustment; **н. реле времени** timing; **н. с помощью магнитных сердечников** permeability tuning; **н. с пониженной громкостью** quiescent tuning; **н. с растяжкой диапазона** band-spread tuning; **н. сердечником** slug tuning; **н. смежных контуров на разнесённые частоты** double-hump tuning; **совпадающая н. нескольких каскадов на одну частоту** coincidence tuning; **согласная н.** accordance; **темперированная н.** tempered music-pitch; **точная н.** tight alignment, sharp tuning, fine tuning, vernier tuning, coincidence tuning, fine adjustment;

**точная н. (нуля)** fine balance; **тупая н.** flat tuning, broad tuning; **физическая н.** physical standard pitch; **четырёхкратная н.** quadruple tuning; **н. электронного прожектора** gun alignment

**настройщик** *m.* tuner, adjuster

**наступать, наступить** *v.* to set in; to come, to approach; to advance

**наступательный** *adj.* offensive, aggressive

**наступающий** *adj.* approaching, advancing

**наступить** *pf. of* **наступать**

**наступление** *n.* onset, approach, advance, coming

**насыпка** *f.* filling

**насытимый** *adj.* saturable

**насытить** *pf. of* **насыщать**

**насыщаемость** *f.* saturability

**насыщаемый** *adj.* saturable

**насыщать, насытить** *v.* to saturate, to impregnate

**насыщающий** *adj.* saturating

**насыщение** *n.* saturation, impregnation; purity; **(глубокое) н.** bottoming; **н. анодного тока** plate saturation, plate-current; **н. бромидами** bromiding; **н. в «белом»**, **н. в области белого** white saturation; **н. в области чёрного**, **н. в «чёрном»** black saturation; **н. лампы** plate saturation, saturation voltage; **н. по напряжению** voltage saturation effect

**насыщенность** *f.* saturation, degree of saturation, impregnation; purity

**насыщенный** *adj.* saturated, impregnated; deep (sound); **н. красный** saturated red

**наталкиваться, натолкнуться** *v.* to come across, to meet, to encounter

**натачивать, наточить** *v.* to sharpen, to grind

**натекание** *n.* leakage, inleakage, soaking-out; **н. (в конденсаторе)** residual charge (of a condenser); soaking out, after effect; **н. через цоколь** base leakage

**натекать, натечь** *v.* to leak; to run in, to flow in; to accumulate

**натекающий** *adj.* gassy, leaky

**натереть** *pf. of* **натирать**

**натечь** *pf. of* **натекать**

**натирать, натереть** *v.* to rub, to rub on; to grate

**НАТО** *abbr.* (научно аэро-техническое общество) Scientific Aeronautical Society

**натовский** *adj.* (of) Nath

**натолкнуться** *pf. of* **наталкиваться**

**наточить** *pf. of* **натачивать, точить**

**натр** *m.* soda, natron

**натриевый** *adj.* sodium, soda

**натрий** *m.* sodium

**натронный** *adj.* soda

**натура** *f.* nature

**натурально** *adv.* naturally

**натуральность** *f.* naturalness

**натуральный** *adj.* natural; real, pure, simple; actual; **в натуральном масштабе времени** real-time

**натурный** *adj.* live; full-scale

**натяг** *m.* hoop cramp; negative allowance; interference; tightness

**натягивание** *n.* stringing, spanning, pulling, tightening; **н. провода** spanning the wire, stringing the wire, pulling (up), tightening (up)

**натягивать, натянуть** *v.* to stretch, to draw, to pull; to tighten, to span, to string; to fix, to fasten; to bend, to tension

**натягивающий** *adj.* tightening, drawing, pulling

**натяжение** *n.* tension, stress, strain, draw, pull(ing), tightening; **н. ввода** lead strain; **предварительное н.** prestress, pre-load; **н. провода** pull of wire, stress exerted by wire

**натяжка** *f.* tightening

**натяжной** *adj.* tension(ing), strain, tightening; pull; stretching

**натянутый** *adj.* drawn, tight, tense; strained

**натянуть** *pf. of* **натягивать**

**наугад** *adv.* by guess, by rule of thumb, at random, haphazardly

**наука** *f.* science; knowledge; **прикладная н.** applied science

**научно-исследовательский** *adj.* scientific-research

**научно-консультативный** *adj.* scientific-advisory

**научный** *adj.* scientific

**наушники** *pl.* earpieces, ear muffs; earphones, headgear, head receiver, headset; **заглушающие н.** ear muffs; **н. пульта** turret headgear; **н. с держателем** head receiver

**НАФА** *abbr.* (ночной аэрофотоаппарат) night aerial camera

**нафталин** *m.* naphthalene

**нахлёст** *m.*, **нахлёстка** *f.* lap, lap joint, overlap, overlapping

**нахлестать** *pf. of* **нахлёстывать**

**нахлёстывание** *n.* lapping, overlapping

**нахлёстывать, нахлестать** *v.* to lap, to overlap

**находить, найти** *v.* to find, to come upon, to detect, to locate; to arrive at, to determine

**находиться, найтись** *v.* to be, to occur, to exist; to be found, to turn up

**находящий** *adj.* finding

**находящийся** *adj.* being, occurring; **н. в контуре, н. в схеме** circuital; **н. в устойчивом состоянии** steady-state; **н. в фазе** in-phase; **н. в ходе** in progress; **н. в цепи** circuital; **н. на хозрасчёте** self-supporting; **н. под током** current carrying, live, alive

**нахождение** *n.* finding, locating, location, detecting; being (at); occurrence; calculation; **н. места повреждения** fault localization; **н. неисправности** failure detection; **н. обрывов цепи** continuity test; **н. повреждений** location of faults

**нацеливание** *n.* aiming, pointing

**нацеливать, нацелить** *v.* to guide; to point, to aim, to level

**нацелить** *pf. of* **нацеливать, целить**

**национальный** *adj.* national

**нач.** *abbr.* (**начальный**) initial

**начало** *n.* beginning, commencement, start, onset, inception; principle, basis; origin, source; **н. автоматического сопровождения цели** lock-on; **н. блока** block head; **н. вторичной обмотки (трансформатора)** secondary (of a transformer); **н. действия при ненагретом состоянии лампы** cold starting of fluorescent lamps; **н. катушки** coil-in; **н. координат** origin of coordinates, zero; **н. линии** sending end; **основное н. теории теплоты** fundamental laws of thermodynamics; **н. отсчёта** origin; **н. отсчёта времени** zero time reference, reference time; **н. первичной обмотки (трансформатора)** in primary (of a transformer); **повторное н. цикла** recycle; **н. пробоя** prebreakdown; **произвольное н. отсчёта** arbitrary origin; **н. с задержкой** delayed start

**начальник** *m.* head, master, chief, superior, officer; **н. радиосвязи** radio officer

**начальный** *adj.* initial, first, original, inceptive; rudimentary, elementary; reference

**начать** *pf. of* **начинать**

**начернить** *pf. of* **чернить**

**начерно** *adv.* in the rough, roughly, coarse(ly)

**начертание** *n.* sketch, plan, outline

**начертить** *pf. of* **начерчивать, чертить**

**начерченный** *adj.* traced, outlined

**начерчивать, начертить** *v.* to draw, to sketch, to trace, to draft, to outline

**начинание** *n.* beginning, start

**начинатель** *m.* initiator, trigger, originator

**начинать, начать** *v.* to begin, to start, to initiate, to set in, to originate

**начинающий** *adj.* beginning, starting, initiating, initial

**начисто** *adv.* clean(ly); thoroughly

**нач. ск.** *abbr.* (**начальная скорость**) initial velocity; muzzle velocity

**нашатырь** *m.* ammonium chloride, sal ammoniac

**нашелушить** *pf. of* **шелушить**

**нашуметь** *pf. of* **шуметь**

**наэлектризование** *n.* electrification

**наэлектризованный** *adj.* electrified

**наэлектризовать** *pf. of* **электризовать**

**нвг** *abbr.* (**навигация**) navigation

**НГ** *abbr.* (**нейтральный газ**) neutral gas, inert gas

**не-** *prefix* im-, in-, un-, non-, dis-, mis-

**не** *adv.* not, no, none; **не в фазе** out of phase; **не и** NAND, NOT AND; **не или** NOR, NOT OR

**неавтоматический** *adj.* non-automatic, manual

**неавтономный** *adj.* nonautonomous, co-operative

**неаддитивный** *adj.* non-additive

**неадиабатический** *adj.* nonadiabatic

**неактивированный** *adj.* non-activated

**неактивность** *f.* inactivity

**неактивный** *adj.* inactive, inert, passive; idle

**неакустический** *adj.* nonacoustical

**неалгебраический** *adj.* nonalgebraic

**неалюминированный** *adj.* nonaluminized, unfilmed

**неаналитический** *adj.* non-analytic

**неарифметический** *adj.* nonarithmetical

**неассоциированный** *adj.* nonassociated

**небаланс** *m.* unbalance

**небесный** *adj.* sky, celestial

**неблагоприятный** *adj.* unfavorable, adverse, disadvantageous

**неблагородный** *adj.* ignoble, common, base, low

**неблокирующий** *adj.* nonlocking

**нёбо** *n.* palate

**небольшой** *adj.* small, little, slight, light, low

**небосвод** *m.* sky, the firmament

**небронированный** *adj.* unarmored

**небьющийся** *adj.* safety (glass), unbreakable

**неведение** *n.* ignorance

**неверно** *adv.* incorrectly, wrong; mis-

**неверность** *f.* invalidity, inaccuracy, falsity, deception, inexactness

**неверный** *adj.* incorrect, inaccurate, wrong, erroneous; false, untrue; inexact; poor

**невесомость** *f.* zero gravity, weightlessness; imponderability

**невесомый** *adj.* agravic, weightless; imponderable

**невзаимный** *adj.* nonreciprocal, unilateral; unidirectional

**невзвешенный** *adj.* nonweighted

**невибрирующий** *adj.* antirattle

**невидимый** *adj.* invisible, imperceptible

**невихревой** *adj.* irrotational, noncircuital

**невключённый** *adj.* off (duty)

**невнятность** *f.* indistinctiveness, inarticulateness, unintelligibility, inaudibility

**невнятный** *adj.* inarticulate, indistinct, unintelligible, inaudible; scrambled

**невозбуждаемый** *adj.* non-excited

**невозбуждённый** *adj.* unexcited; unperturbed

**невозвратный** *adj.* nonreturn

**невозвращаемый** *adj.* nonrecoverable

**невозможность** *f.* impossibility

**невозможный** *adj.* impossible

**невозмущённый** *adj.* unperturbed, undisturbed; stagnant

**невозрастающий** *adj.* non-increasing, stable

**невоспламеняемый** *adj.* flame-proof, nonflammable

**невоспламеняющийся** *adj.* flame-proof, nonflammable, noninflammable, incombustible

**невосприимчивость** *f.* lack of receptivity; non-susceptibility, immunity, insensibility

**невосприимчивый** *adj.* non-susceptible

**невосстанавливаемый** *adj.* unrestorable

**невосстанавливающийся** *adj.* nonregenerative

**невосстановленный** *adj.* unreduced

**невращающийся** *adj.* non-spinning, non-rotating, non-rotary, irrotational

**неврит** *m.* neuritis

**неврубленный** *adj.* noncut-in

**невыбалансированный** *adj.* unbalanced, non-balanced

**невыбранный** *adj.* unselected, non-selected

**невыверенность** *f.* misalignment

**невыгода** *f.* disadvantage

**невыгодно** *adv.* disadvantageously, unprofitably

**невыгодный** *adj.* disadvantageous, unprofitable, unfavorable

**невыключаемый** *adj.* permanent

**невыполнение** *n.* failure, non-performance

**невыполнимость** *f.* impracticability

**невыполнимый** *adj.* impracticable

**невыпрямленный** *adj.* unrectified

**невыпрямляющий** *adj.* nonrectifying

**невырожденный** *adj.* non-degenerate

**невысокий** *adj.* low, not high

**невыступающий** *adj.* non-projecting, non-protruding

**невыясненный** *adj.* unexplained, obscure

**невязка** *f.* discrepancy, lack of coordination, maladjustment, failure

**негадин** *m.* negadyne circuit

**негармонический** *adj.* nonharmonic, inharmonic; anharmonic *(math.)*

**негармоничность** *f.* anharmonicity; inharmoniousness

**негармоничный** *adj.* inharmonious

**негатив** *m.* negative, negative proof, negative matrix

**негативность** *f.* negativeness

**негативный** *adj.* negative

**негатон** *m.* negative electron

**« негатор »** *m.* negator

**негатрон** *m.* negatron; negative electron

**негерметизированный** *adj.* unsealed

**негерметический** *adj.* non-hermetic

**негерметичность** *f.* leaking, seepage

**негерметичный** *adj.* non-hermetic

**негибкий** *adj.* rigid, inflexible, stiff, stationary

**негибкость** *f.* rigidity, inflexibility, stiffness

**негигроскопический** *adj.* nonhygroscopic

**негладкий** *adj.* pitted, jagged, uneven, rough

**неглубокий** *adj.* shallow, not deep; light

**неглубоко** *adv.* shallowly

**негодный** *adj.* improper, useless, unfit, unsuitable, worthless, faulty; waste

**негомогенный** *adj.* non-homogeneous, heterogeneous

**негорючесть** *f.* incombustibility

**негорючий** *adj.* noncombustible, fire-resistant, fire-proof; safety

**неготовый** *adj.* not ready, unprepared, unfinished

**неграмматический** *adj.* nongrammatical

**негранёный** *adj.* unfaced, uncut, rough

**негэнтропия** *f.* negentropy

**недвижущийся** *adj.* stationary

**недвоичный** *adj.* non-binary

**недвусмысленный** *adj.* unambiguous

**недействительность** *f.* ineffectiveness, inefficiency; invalidity

**недействительный** *adj.* ineffective, inefficient, inoperative; invalid

**недействующий** *adj.* inactive, non-active, passive, idle, inert; inoperative

**неделимость** *f.* indivisibility

**неделимый** *adj.* indivisible

**недельный** *adj.* week, weekly

**неделящийся** *adj.* non-fissionable

**недемпфированный** *adj.* undamped

**недеструктивный** *adj.* nondestructive

**недетерминированный** *adj.* non-determinate

**недетерминистический** *adj.* non-deterministic

**недеформированный** *adj.* undistorted

**недеформируемость** *f.* nondeformity, form persistence

**недеятельность** *f.* inactivity, passivity, inertness; ineffectiveness

**недеятельный** *adj.* inactive, passive, inert; inoperative, idle

**недиагональный** *adj.* off-diagonal

**недискретный** *adj.* indiscrete

**недиспергирующий** *adj.* dispersionless

**недисперсионный, недисперсный** *adj.* nondispersive

**недо-** *prefix* under-, incompletely

**недоброкачественный** *adj.* inferior, of inferior quality

**недовес** *m.* underweight, shortweight

**недовозбуждать, недовозбудить** *v.* to under-excite

**недовозбуждение** *n.* underexcitation

**недовозбуждённый** *adj.* underexcited

**недогрев** *m.* underheating

**недогруженный** *adj.* underloaded

**недогрузка** *f.* underload(ing), marginal load capacity

**недогрузочный** *adj.* underload

**недогруппирование** *n.* underbunching

**недоделанный** *adj.* incomplete, unfinished

**недоделка** *f.* non-completion, bug

**недодемпфирование** *n.* underdamping

**недодемпфированный** *adj.* underdamped

**недодержка** *f.* underexposure

**недокал** *m.* underheating

**недокальный** *adj.* underheating

**недокомпаундированный** *adj.* undercompound

**недокомпенсация** *f.* undercompensation

**недокомпенсированный** *adj.* undercompensated

**недоконченный** *adj.* unfinished, incomplete

**недолговечный** *adj.* temporary, transient, short-lived

**недолёт** *m.* short, short round, falling short, undershot

**недомодулированный** *adj.* undermodulated

**недонапряжение** *n.* undervoltage

**недонасыщенный** *adj.* undersaturated

**недоопределённый** *adj.* underdetermined

**недоохлаждённый** *adj.* undercooled

**недопустимость** *f.* inadmissibility

**недопустимый** *adj.* inadmissible, intolerable, objectionable, illegal

**недоразряд** *m.*, **недоразрядка** *f.* undercharge, undercharging

**недорегулирование** *n.* underregulation, undershooting, undercontrol

**недорегулированный** *adj.* undercompensated

**недорез** *m.* undercutting (of groove)

**недосвязан** *pred. adj.* undercoupled

**недосмотр** *m.* oversight, slip, error

**недоставать, недостать** *v.* to be insufficient, to run short, to be wanting, to miss, to lack

**недостаток** *m.* lack, shortage, deficiency; defect, flaw, fault, imperfection; drawback; **недостатки изображения** image distortion(s), image defects; **н. верхов** no top *(acoust.)*; **н. конструкции** fault of construction; **н. низов** no bottom *(acoust.)*; **н. разрешающей способности, н. разрешающей чёткости** lack of resolution

**недостаточно** *adv.* insufficiently

**недостаточность** *f.* insufficiency, inefficiency, deficiency; failure; imperfection; shortage; **н. ширины пятна** underlap

**недостаточный** *adj.* insufficient, inadequate, deficient, scanty, meager, poor, bad; defective, faulty, inefficient, imperfect; under-sized

**недостать** *pf. of* **недоставать**

**недостающий** *adj.* missing, lacking, deficient

**недостоверность** *f.* uncertainty

**недостоверный** *adj.* uncertain, doubtful; unauthentic

**недостроенность** *f.* incompleteness

**недостроенный** *adj.* unfinished, incomplete

**недоступно** *adv.* inaccessibly; difficult

**недоступность** *f.* inaccessibility; unavailability; remoteness

**недоступный** *adj.* inaccessible, impenetrable; isolated; remote; unavailable

**недра** *pl.* bosom, depths, interior (of the earth)

**недребезжащий** *adj.* antirattle

**недрейфующий** *adj.* nondrifting

**Неель** Neel

**неестественный** *adj.* unnatural, abnormal

**нежелательность** *f.* undesirability

**нежелательный** *adj.* undesirable, unwanted, objectionable; parasitic

**нежелезный** *adj.* non-ferrous

**нежёсткий** *adj.* nonrigid, flexible; loose, slack

**незавершённость** *f.* incompleteness

**незавершённый** *adj.* unfinished

**независимо** *adv.* independently

**независимо-возбуждаемый** *adj.* separately-excited

**независимость** *f.* independence; **н. частоты** independence of frequency response, fidelity of frequency response

**независимый** *adj.* independent, separate, free, isolated, individual; insulated; arbitrary

**независящий** *adj.* independent, free

**незаглушённый** *adj.* not suppressed, not drowned out; live

**незагруженный** *adj.* uncharged, unloaded; idle

**незадействованный** *adj.* spare, vacant

**незадемпфированный** *adj.* undamped

**незадержанный** *adj.* undelayed

**незаедающий** *adj.* nonsticking

**незаземлённый** *adj.* ungrounded, off-ground(ed), earth-free

**незаконченный** *adj.* unfinished, incomplete

**незакорачивающий** *adj.* nonshorting

**незакреплённый** *adj.* loose, unmounted, unfastened; mobile, floating; unsupported

**незамерзающий** *adj.* frostproof, antifreezing, non-freezing; uncongealable

**незаметный** *adj.* unnoticeable, imperceptible

**незамкнутость** *f.* open condition (of a circuit), discontinuity

**незамкнутый** *adj.* open, unlocked; incomplete; unenclosed

**незанумерованный** *adj.* unnumbered

**незанятость** *f.* availability, free condition (of a trunk)

**незанятый** *adj.* not busy, not engaged, free, available, idle, clear, vacant, unoccupied; disengaged

**незапертый** *adj.* noncut-off

**незаполненный** *adj.* blank, not filled, vacant, empty, unoccupied; raw

**незарегистрированный** *adj.* unlicensed

**незаряженный** *adj.* uncharged, unloaded

**незасекреченный** *adj.* unclassified, nonsecret

**незатвердевший** *adj.* nonhardened

**незатухающий** *adj.* continuous, sustained, persistent, undamped

**незащищённый** *adj.* unguarded, unshielded, unprotected, exposed

**незеркальный** *adj.* nonspecular, nonreflecting

**незернистый** *adj.* grainless

**незнание** *n.* ignorance

**незначащий** *adj.* nonsignificant

**незначимость** *f.* insignificance

**незначительность** *f.* insignificance

**незначительный** *adj.* insignificant, trivial, negligible, minute; light, small, little, low

**неидеальный** *adj.* imperfect

**неизбежно** *adv.* inevitably, of necessity

**неизбежность** *f.* inevitability

**неизбежный** *adj.* inevitable, unavoidable; unintentional

**неизбирательный** *adj.* non-selective

**неизбыточный** *adj.* irredundant

**неизвестное** *n. adj. decl.* the unknown, unknown quantity

**неизвестный** *adj.* unknown, uncertain, obscure

**неизлучающий** *adj.* nonradiating, nonradiative

**неизменность** *f.* stability, constancy, invariability, inalterability, immutability

**неизменный** *adj.* invariable, immutable, unalterable; fixed, constant, permanent, stationary, stable, steady, continuous

**неизменяемость** *f.* stability, constancy, inalterability

**неизменяемый** *adj.* invariable, invariant, immutable, constant, unalterable, permanent; **н. по времени** time-invariant

**неизменяющийся** *adj.* invariable, unchangeable, permanent

**неизмеримый** *adj.* immeasurable, bottomless

**неизолированный** *adj.* uninsulated, bare

**неизотермический** *adj.* nonisothermal

**неизотропность** *f.* anisotropy, anisotropism

**неизотропный** *adj.* anisotropic

**неизъязвляющийся** *adj.* nonpitting, noncracking (material)

**неиндексируемый** *adj.* non-indexed

**неиндуктивный** *adj.* noninductive

**неинжектирующий** *adj.* noninjecting

**неионизирующий** *adj.* nonionizing

**неионизованный** *adj.* nonionized

**неионный** *adj.* nonionic

**неискажающий** *adj.* distortionless, nondistorting

**неискажение** *n.* nondistortion, nondeformation

**неискажённый** *adj.* undistorted, distortionless, undisturbed, unbiased

**неискрящий(ся)** *adj.* nonsparking, sparkless, nonarcing

**неиспользование** *n.* disuse

**неиспользованный** *adj.* blank, vacant; empty; unused

**неиспользуемый** *adj.* spare; dead; unusable

**неисправленный** *adj.* defective

**неисправность** *f.* fault, faultiness, defect, disrepair, trouble, failure, breakdown; interference; inaccuracy; **н. аппаратуры наведения** (*or* **управления**) guidance malfunction; **н. в течение гарантийного срока** in-warranty failure

**неисправный** *adj.* inoperable, defective, out of repair, faulty, deficient, in bad order; inaccurate, incorrect, unsatisfactory

**неисчезающий** *adj.* non-vanishing, non-zero

**ней** *instr. of* **она** with her, by her; with it, by it

**нейзильбер** *m.* argentan, German silver

**нейлон** *m.* nylon

**нейлоновый** *adj.* nylon

**Нейман** Neumann

**нейристор** *m.* neuristor

**нейрон** *m.* neuron; nerve cell, neuron

**нейрохирургия** *f.* neurosurgery

**нейроэлектричество** *n.* neuroelectricity

**нейтр.** *abbr.* 1 (**нейтральный**) neutral; 2 (**нейтрон, нейтронный**) neutron

**нейтрализационный** *adj.* neutralization

**нейтрализация** *f.* neutralization, neutralizing, balancing out

**нейтрализованный** *adj.* neutralized

**нейтрализовать** *v.* to neutralize

**нейтрализующий** *adj.* neutralizing

**нейтраль** *f.* neutral, neutral conductor, neutral wire, neutral feeder, zero line; **н.** (**в многофазных коллекторных электрических машинах**) neutral plane (on a polyphase commutator machine); **н.** (**в однофазных коллекторных машинах**) neutral plane (on a single-phase commutator machine); **н.** (**в электрической машине постоянного тока**) neutral plane (on a direct current machine); **геометрическая н.** normal neutral plane; **н., заземлённая в нескольких точках** multiple earthed neutral

**нейтрально** *adv.* neutrally

**нейтральность** *f.* neutrality; **н. шкалы серых тонов** gray-scale balance

**нейтральный** *adj.* neutral, unpolarized; inert, indifferent

**нейтретто** *n.* neutretto

**нейтреттовый** *adj.* neutretto

**нейтринный** *adj.* neutrino

**нейтрино** *n.* neutrino

**нейтродин** *m.* neutrodyne

**нейтродинирование** *n.* neutralization, neutralizing, neutrodyne; **анодное н. по схеме Хазелтайна** Hazeltine neutralization

**нейтродинный** *adj.* neutralized, neutrodyne

**нейтродиновый** *adj.* neutrodyne

**нейтрозонный** *adj.* neutrosonic

**нейтрон** *m.* neutron

**нейтронный** *adj.* neutron

**нейтронограмма** *f.* neutron diffraction pattern

**нейтронограф** *m.* neutron diffraction camera

**нейтронография** *f.* neutron radiography

**нейтрононепроницаемый** *adj.* neutron-tight

**нейтронотерапия** *f.* neutron therapy

**неканонический** *adj.* noncanonical

**неквадратный** *adj.* nonsquare

**неквалифицированный** *adj.* unqualified

**неквантованный** *adj.* non-quantized, unquantized

**некоаксиальность** *f.* misalignment

**некоаксиальный** *adj.* misaligned

**некогерентность** *f.* incoherence

**некогерентный** *adj.* incoherent, non-coherent

**некодированный** *adj.* uncoded

**неколебательный** *adj.* nonoscillating, nonoscillatory

**некомпенсированный** *adj.* uncompensated

**некомплект** *m.* shortage, deficiency

**некомплектный** *adj.* partial, deficient

**неконтактный** *adj.* noncontact, non-contacting; proximity, influence

**неконтрастный** *adj.* washed-out

**неконтролируемый** *adj.* uncontrolled

**неконцентрический** *adj.* non-concentric

**некооперативный** *adj.* noncooperative

**некорректированный** *adj.* uncompensated, uncorrected

**некоррелированный** *adj.* uncorrelated

**некоторый** *adj.* some, certain

**некритический** *adj.* non-critical

**некруглый** *adj.* out of round, untrue

**нелатинский** *adj.* non-Roman

**нелинеарность** *f.* nonlinearity

**нелинейность** *f.* nonlinearity; **действующая н.** effective gamma; **н. модуляционной характеристики в области «чёрного»** black saturation; **общая н. системы** overall gamma
**нелинейный** *adj.* nonlinear
**нелокализованный** *adj.* nonlocal
**Нель** Néel
**немагнитный** *adj.* nonmagnetic
**нематематический** *adj.* nonmathematical
**немедленный** *adj.* quick, no-delay, instant, immediate, immediate-action
**немелодичный** *adj.* tuneless
**неметалл** *m.* non-metal
**неметаллический** *adj.* non-metallic
**немеханический** *adj.* nonmechanical
**немодифицируемый** *adj.* unmodified; non-indexed
**немодулированный** *adj.* unmodulated
**немой** *adj.*, *(noun)* *m.* *adj.* *decl.* mute; dumb, mute; silent; blank, dummy
**немонокристалл** *m.* nonsingle crystal
**немонотонный** *adj.* non-monotonous
**немонохроматический** *adj.* polychromatic
**ненагретый** *adj.* unheated, cold
**ненагружённый** *adj.* unloaded, not loaded, empty, idle
**ненадёжность** *f.* equivocation, uncertainty, unreliability, insecurity; **эксплуатационная н.** unserviceability
**ненадёжный** *adj.* uncertain, unreliable, troublesome, insecure, unsafe, untrustworthy; **н. в эксплуатации** unserviceable
**ненакалённый** *adj.* cold, unheated
**ненамагниченный** *adj.* unmagnetized
**ненаправленный** *adj.* omnidirectional, undirected, nondirectional, astatic
**ненапряжённый** *adj.* dead; without tension; relaxed, slack
**ненастроенный** *adj.* untuned, unadjusted; nonresonant
**ненасыщающийся** *adj.* unsaturable, antisaturation
**ненасыщенность** *f.* non-saturation
**ненасыщенный** *adj.* unsaturated, non-saturated, diluted, desaturated; **н. красный** unsaturated red
**ненатяжной** *adj.* nonstrain, nontension
**ненатянутость** *f.* looseness, slack(ness)
**ненатянутый** *adj.* loose, slack

**ненормальность** *f.* abnormality
**ненормальный** *adj.* abnormal, irregular
**ненужный** *adj.* unnecessary, waste, needless, useless; idle
**ненулевой** *adj.* non-zero
**необделанный** *adj.* unwrought, non-machined, rough, unfinished
**необитаемый** *adj.* unmanned; uninhabited
**необнаруживаемый, необнаружимый** *adj.* undetectable
**необновляемый** *adj.* nonerasable; non-renewable
**необработанный** *adj.* raw, crude, untreated; unfinished, rough
**необратимость** *f.* irreversibility
**необратимый** *adj.* irreversible, non-reversible; nonreciprocal; nonconservative
**необращающий** *adj.* noninverting
**необслуживаемый** *adj.* unattended
**необтекаемый** *adj.* blunt
**необученный** *adj.* untrained, unskilled
**необходимо** *adv.*, *pred.* *adj.* necessarily; necessary
**необходимость** *f.* need, necessity, indispensability
**необходимый** *adj.* necessary, needed, indispensable, required, requisite, essential
**необъявленный** *adj.* unlisted
**необъяснимость** *f.* inexplicability
**необъяснимый** *adj.* inexplicable, unaccountable
**необыкновенно** *adv.* unusually
**необыкновенность** *f.* unusualness, singularity
**необыкновенный** *adj.* unusual, extraordinary, rare, uncommon, singular
**необязательный** *adj.* optional, not obligatory
**неограниченный** *adj.* unlimited, unrestricted, unbounded, indefinite, absolute
**неодим** *m.* neodymium, Nd
**неодинаковый** *adj.* not uniform, different; unequal
**неоднозначность** *f.* nonuniqueness, ambiguity
**неоднозначный** *adj.* nonunique, ambiguous
**неоднократный** *adj.* multiple, manifold, repeated
**неоднообразный** *adj.* irregular

**неоднородность** *f.* inhomogeneity, discontinuity, heterogeneity, irregularity, nonuniformity; **н. в распределении массы по колеблющейся системе** mass perturbation; **электрическая н.** point of reflection (transmission lines)

**неоднородный** *adj.* nonuniform, heterogeneous, nonhomogeneous, inhomogenous; dissimilar; not uniform

**неожиданно** *adv.* suddenly, unexpectedly

**неожиданность** *f.* suddeness, unexpectedness

**неожиданный** *adj.* sudden, unexpected

**неокардиограф** *m.* Neocardiograph

**неокисляющий** *adj.* non-oxidizing

**неокисляющийся** *adj.* non-oxidizable

**неокруглённый** *adj.* unrounded

**неомический** *adj.* non-ohmic

**неон** *m.* neon, Ne

**« неоника »** *f.* neon tube; neon lamp

**неоновый** *adj.* neon

**неопёртый** *adj.* unsupported

**неопознанный** *adj.* unidentified

**неопределенно** *adv.* indefinitely

**неопределённость** *f.* indeterminancy; vagueness, uncertainty, indefinitiveness; ambiguity, equivocation; **априорная н.** prior uncertainty

**неопределённый** *adj.* indeterminate, indefinite, not fixed, vague, uncertain; undefined, undetermined

**неопределимость** *f.* indefinability

**неопределимый** *adj.* indefinable, indeterminable

**неопределяемый** *adj.* undefined

**неопрен** *m.* neoprene

**неопреновый** *adj.* neoprene

**неопровержимость** *f.* irrefutability

**неопровержимый** *adj.* irrefutable, undeniable, incontestable, indisputable

**неоптимальный** *adj.* nonoptimal

**неоптический** *adj.* anoptic, nonoptical

**неорганический** *adj.* inorganic

**неориентированный** *adj.* unoriented

**неосвещённый** *adj.* dark, unlighted

**неослабленный** *adj.* unattenuated, unattenuating

**неослабляющий** *adj.* unattenuating

**неослабность** *f.* unremittance

**неослабный** *adj.* unremitting

**неосмысленный** *adj.* nonsensical

**неосновной** *adj.* minority

**неострый** *adj.* imperfect, not sharp

**неосциллирующий** *adj.* non-oscillating

**неотбелённый** *adj.* unbleached

**неотвеченный** *adj.* unanswered

**неотделанный** *adj.* rough, unfinished; raw

**неотделимый** *adj.* inseparable

**неотклонённый** *adj.* non-deflected

**неотложность** *f.* urgency

**неотложный** *adj.* urgent, pressing

**неотражающий** *adj.* antireflection, nonreflective, reflection-free

**неотражённый** *adj.* nonreflected

**неотрицательный** *adj.* nonnegative

**неотрон** *m.* neotron

**неотчётливый** *adj.* indistinct, vague, indefinite, inarticulate, illegible

**неотъемлемый** *adj.* inherent

**неощущаемый** *adj.* nonsensory

**непарный** *adj.* unpaired

**непер** *m.* neper

**неперекрывающийся** *adj.* non-overlapping

**непересекающийся** *adj.* non-intersecting

**непереставляемый** *adj.* non-interchangeable

**непериодический** *adj.* nonperiodic, nonperiodical, noncyclic, acyclic; transient

**непериодичность** *f.* aperiodicity

**неперметр** *m.* neper meter, transmission measuring set

**неперов** *poss. adj.* Napierian

**непилотируемый** *adj.* unmanned

**неплакированный** *adj.* unclad

**непланарный** *adj.* nonplanar

**неплоский** *adj.* nonplane, nonplanar

**неплотность** *f.* looseness; leakage

**неплотный** *adj.* loose; leaky, not tight; low-density

**неповоротимый** *adj.* irrotational

**неповреждаемый** *adj.* nondestructible; shakeproof; -proof; **н. пламенем** flameproof; **н. при тряске** shakeproof

**неповреждённый** *adj.* intact, sound, unimpaired

**непоглощающий** *adj.* nonabsorbing, nonabsorptive, nonacoustical

**неподавленный** *adj.* unsuppressed

**неподавляемый** *adj.* unjammable

**неподатливый** *adj.* inflexible, rigid; unyielding

**неподвижно** *adv.* immovably, securely

**неподвижность** *f.* immobility, immovability

**неподвижный** *adj.* immovable, motionless, inflexible, fixed, still, stationary, constant, at rest, resting; rigid; standing, dead, quiescent; static

**неподобранный** *adj.* unmatched

**неподходящий** *adj.* unsuitable, inadequate, unfitted, inappropriate

**неполадка** *f.* defect, disorder, disturbance, trouble, failure, bug, maladjustment; shortcoming

**неполированный** *adj.* mat

**неполно** *adv.* incompletely

**неполнота** *f.* incompleteness, imperfection

**неполный** *adj.* incomplete, partial, fractional, imperfect, defective

**неполяризованный** *adj.* nonpolarized

**неполярный** *adj.* nonpolar, apolar

**непомеченный** *adj.* unmarked, unlabelled

**непопадение** *n.* no-hit

**непористый** *adj.* nonporous, dense, compact

**непосредственно** *adv.* immediately, directly, next; **н. (на одном валу)** directly coupled (to the same shaft)

**непосредственность** *f.* spontaneity; immediateness

**непосредственный** *adj.* direct, immediate; instantaneous, spontaneous; close; straight, straight-forward

**непостоянный** *adj.* nonuniform, changeable, variable, unstable, nonconstant, unsteady

**непостоянство** *n.* impermanence, inconstancy, instability, mobility, variability; **н. положения изображения по вертикали, н. положения кадров (в кадровом окне)** vertical judder; **н. скорости** flutter flicker

**неправильно** *adv.* incorrectly

**неправильность** *f.* irregularity; incorrectness, inaccuracy; anomaly; infidelity

**неправильный** *adj.* incorrect, inaccurate, erroneous, wrong, false, untrue; defective; irregular, abnormal, anomalous; **с неправильными очертаниями** ungeometrical

**непредвиденный** *adj.* unexpected, unforeseen

**непреднамеренный** *adj.* unintentional

**непредсказываемый** *adj.* unpredictable

**непрекращающийся** *adj.* incessant

**непреложный** *adj.* immutable, unalterable; unbiased

**непрерываемый** *adj.* unchopped, uninterrupted

**непрерывно** *adv.* continuously, uninterruptedly

**непрерывность** *f.* continuity, regularity, persistence; **н. (магнитной цепи)** magnetic continuity; **н. состояния** continuity of state

**непрерывный** *adj.* continuous, unbroken, uninterrupted, constant, ceaseless, incessant, steady, permanent

**неприводимый** *adj.* irreducible

**непригодность** *f.* unfitness, uselessness

**непригодный** *adj.* unfit (for use), useless, unsuitable, ineffective

**неприемлемый** *adj.* unsuitable, unacceptable

**неприжатый** *adj.* open, not pressed down

**непринятие** *n.* rejection, non-acceptance, refusal

**неприспособленность** *f.* maladjustment

**неприспособляемость** *f.* inadaptability

**неприспособляемый** *adj.* inadaptable, inapplicable

**неприступность** *f.* inaccessibility

**неприступный** *adj.* inaccessible

**неприятность** *f.* annoyance, trouble, nuisance, unpleasantness

**неприятный** *adj.* unpleasant, troublesome, disagreeable, objectionable

**непробиваемый** *adj.* puncture-proof

**непроводник** *m.* nonconductor, insulator, dielectric medium

**непроводящий** *adj.* nonconducting, nonconductive

**непрограммируемый** *adj.* nonprogrammable

**непродолжительный** *adj.* brief, short; intermittent, discontinuous

**непродуктивный** *adj.* nonproductive; no-charge

**непрозрачность** *f.* opacity

**непрозрачный** *adj.* opaque, nontransparent, impervious; **н. для излучений** radiopaque

**непроизводительно** *adv.* unproductively

**непроизводительность** *f.* unproductiveness

**непроизводительный** *adj.* unproductive, barren

**непроизвольно** *adv.* involuntarily, unintentionally

**непроизвольный** *adj.* unintentional, involuntary, nonrandom

**непромокаемость** *f.* impermeability, imperviousness

**непромокаемый** *adj.* impermeable, watertight, waterproof

**непроницаемость** *f.* impermeability, impenetrability, tightness; opacity

**непроницаемый** *adj.* impermeable, impenetrable, impervious, tight; proof; opaque; **н. для дождя** raintight

**непропаянный** *adj.* solderless, dry

**непропитанный** *adj.* unimpregnated, untreated

**непропускаемый** *adj.* impervious; suppressed

**непропускающий** *adj.* impervious, tight

**непросвечивающий** *adj.* opaque

**непротиворечность** *f.* consistency

**непротиворечный** *adj.* consistent

**непроходимость** *f.* impenetrability, impassability

**непроходимый** *adj.* impenetrable, impassable, impervious

**непрохождение** *n.* nonpassage of signals, no communication

**непрямой** *adj.* indirect

**непрямолинейность** *f.* misalignment

**непрямоугольный** *adj.* nonsquare

**нептуний** *m.* neptunium, Np

**непупинизированный** *adj.* nonloaded

**непьезоэлектрик** *m.* nonpiezoelectric lattice

**неработающий** *adj.* inoperative, nonoperative, idle, standing

**нерабочий** *adj.* idle, not working, inactive, inoperative, unoperated; **в нерабочем положении** inoperative

**неравенство** *n.* inequality, disparity, nonequivalence; **н. Бьенэмэ-Чебышева** Bienaymé-Chebyshev inequality

**неравновесие** *n.* unbalance; bias

**неравновесный** *adj.* non-equilibrium

**неравномерно** *adv.* irregularly

**неравномерность** *f.* irregularity, unevenness, discontinuity, nonuniformity, inequality, disproportion; variation; unbalance; **остаточная н.** offset; **н. преломления волн** irregular refraction

**неравномерный** *adj.* uneven, irregular, nonuniform, erratic, unequal, discontinuous, disproportionate, nonequilibrium

**неравный** *adj.* unequal, uneven

**нерадиоактивный** *adj.* nonradioactive

**неразборный** *adj.* solid

**неразборчивость** *f.* illegibility, unintelligibility

**неразборчивый** *adj.* illegible, unintelligible, undecipherable, unreadable, undiscriminating

**неразделимый** *adj.* indivisible, inseparable

**неразделительный** *adj.* inclusive

**нераздельный** *adj.* inseparable, indivisible; undivided

**неразжижённый** *adj.* undiluted

**неразличимый** *adj.* indistinguishable, indiscernible, undecipherable

**неразмеченный** *adj.* unmarked, virgin

**неразобщающий** *adj.* nontrip(ping)

**неразориентирующий** *adj.* nondisorienting

**неразрезной** *adj.* solid, continuous

**неразрешенный** *adj.* unresolved; forbidden

**неразрешимость** *f.* unsolvability, insolubility, undecidability

**неразрешимый** *adj.* insoluble, unsolvable, undecidable

**неразрушающий** *adj.* nondestructive

**неразрушающийся** *adj.* nonvolatile

**неразрушенный** *adj.* undisturbed

**неразрушимость** *f.* indestructibility

**неразрушимый** *adj.* indestructible

**неразрывно** *adv.* inseparably

**неразрывность** *f.* continuity; indissolubility

**неразрывный** *adj.* continuous; indissoluble

**неразъедаемый** *adj.* corrosion-resistant, noncorrodible; **н. ртутью** mercury-proof

**неразъедающий** *adj.* noncorrosive, noncorroding

**неразъёмный** *adj.* solid, single, one-part, permanent, nondetachable

**нераспределенный** *adj.* concentrated

**нерастворенный** *adj.* undissolved

**нерастворимость** *f.* insolubility

**нерастворимый** *adj.* insoluble

**нерастущий** *adj.* non-increasing; not growing

**нерастяжной** *adj.* non-expandable, non-tension, nonspread

**нерасходуемый** *adj.* nonconsumable

**нерасцепляющий** *adj.* nontrip(ping)

**нерв** *m.* nerve

**нервно-мышечный** *adj.* neuromuscular, neuromyal

**нервный** *adj.* nerve, nervous
**нереагирующий** *adj.* nonreacting
**нереактивный** *adj.* nonreactive
**нереальный** *adj.* nonreal, fictitious, nonphysical; unrealizable
**нереверсивность** *f.* irreversibility
**нереверсивный** *adj.* nonreversible, non-reversing, irreversible; unidirectional
**нереверсируемый** *adj.* nonreversible, irreversible
**нерегламентированный** *adj.* unrestricted
**нерегулируемый** *adj.* unregulated; solid, fixed
**нерегулярно** *adv.* irregularly
**нерегулярность** *f.* irregularity
**нерегулярный** *adj.* irregular, sporadic, random, uneven, casual
**нерезервированный** *adj.* irredundant
**нерезкий** *adj.* soft (sound)
**нерезкость** *f.* haziness, out-of-focus; **н. изображения подвижного объекта** motional distortion, blur caused by moving object
**нерезонансный** *adj.* nonresonant, non-resonating
**нерезонирующий** *adj.* nonresonating
**нерекомбинационный** *adj.* nonrecombining
**нерекурсивный** *adj.* nonrecursive
**нерелейный** *adj.* direct-acting
**нерелятивистский** *adj.* nonrelativistic
**нерефлексивный** *adj.* nonreflexive
**нерешённый** *adj.* unsolved
**нержавеющий** *adj.* noncorrosive, stainless, rustproof
**Нернст** Nernst
**неровно** *adv.* unevenly, irregularly
**неровность** *f.* unevenness, irregularity, roughness; **н. границ, н. краёв** edge scalloping, jagged edges
**неровный** *adj.* uneven, irregular, rough, ragged, jagged; odd; unequal
**нертилька** *f.* marking awl
**несамовоспламеняющийся** *adj.* non-self-igniting, nonhypergolic
**несамостоятельный** *adj.* nonself-maintained, semi-self-maintained
**несбалансированный** *adj.* unbalanced
**несвариваемый** *adj.* unweldable
**несваривающийся** *adj.* nonwelding
**несверхпроводящий** *adj.* nonsuperconducting
**несветящийся** *adj.* nonluminous
**несводимый** *adj.* irreducible

**несвязанный** *adj.* uncoupled, uncombined, unbound, free, loose, slack; incoherent
**несвязно** *adv.* incoherently
**несвязность** *f.* incoherence
**несвязный** *adj.* incoherent, disconnected
**несгораемость** *f.* incombustibility
**несгораемый** *adj.* fireproof, incombustible, refractory
**несгорающий** *adj.* nonburning, incombustible
**несекретный** *adj.* unclassified
**несекционированный** *adj.* single-link
**неселективный** *adj.* nonselective
**несжимаемость** *f.* incompressibility
**несжимаемый** *adj.* incompressible, non-condensable
**несимметрический** *adj.* unsymmetrical, asymmetric(al), unbalanced; irregular
**несимметричность** *f.* asymmetry, dissymmetry, unbalance; **допустимая н. нагрузки** maximum permissible unbalance; **н. распределения вероятностей** skewness
**несимметричный** *adj.* unsymmetrical, asymmetrical, asymmetric, dissymmetrical, unbalanced; irregular; single-ended
**несимметрия** *f.* asymmetry, dissymmetry, asymmetrical effect; unbalance; **н. в линии** unbalance of lines
**несинусоидальный** *adj.* nonsinusoidal, nonsine, distorted
**несинхронизированный, несинхронизованный** *adj.* free-running, self-running, out-of-lock, out-of-step; a-stable, nonsynchronized
**несинхронный** *adj.* nonsynchronous, asynchronous
**несистематический** *adj.* nonsystematic
**несквозной** *adj.* blind (passage)
**несколько** *adv.* several, some; somewhat, slightly
**нескомпенсированный** *adj.* uncompensated; uncancelled
**неслепящий** *adj.* antidazzle
**несложный** *adj.* simple, plain; single
**неслучайный** *adj.* non-incidental; assignable
**неслышимость** *f.* silence
**неслышимый, неслышный** *adj.* inaudible, unheard
**несмещённый** *adj.* unbiased

**несмонтированный** *adj.* unmounted, unassembled; unwired

**несобранный** *adj.* unassembled, dismantled

**несобственный** *adj.* improper; extrinsic

**несовершенный** *adj.* imperfect, incomplete, inadequate, defective

**несовершенство** *n.* imperfection, irregularity

**несовместимость** *f.* inconsistency, incompatibility

**несовместимый** *adj.* inconsistent, incompatible

**несовместность** *f.* inconsistency, incompatibility

**несовместный** *adj.* inconsistent, incompatible, antithetic

**несовмещение** *n.* nonregistration (of colors); **н. (цветов)** nonregistration (of colors), fringe, misregistration, misregistry; **н. отдельных цветных изображений на экране кинескопа** misregistration

**несовмещённость** *f.* misconvergence (of colors); **н. цветоделённых изображений** color misconvergence

**несовпадать, несовпасть** *v.* to mismatch

**несовпадающий** *adj.* anticoincidence

**несовпадение** *n.* discrepancy, variance, disagreement, noncoincidence, anticoincidence; misalignment; mismatching; offset; **н. осей** misalignment (of axes); **н. фазы** out-phase, out-of-phase, noncoincidence of phases; **н. чётности** parity failure

**несовпасть** *pf. of* **несовпадать**

**несогласие** *n.* discord, variance, difference, disagreement, unconformity

**несогласно** *adv.* at variance (with), in disagreement (with)

**несогласность** *f.* disagreement

**несогласный** *adj.* discordant, disagreeing, differing

**несогласованность** *f.* mismatching, mismatch, discord, disagreement, maladjustment, inconsistency; **н. сопротивлений** impedance mismatch

**несогласованный** *adj.* unmatched, mismatched, nonmatched, not in agreement, incoordinate, uncoordinated; disorganized

**несогласующийся** *adj.* incompatible

**несозвучный** *adj.* dissonant (to), inconsonant (with, to); out of tune (with)

**несоизмеримость** *f.* incommensurability

**несоизмеримый** *adj.* incommensurate, incommensurable

**несоосность** *f.* malalignment

**несоответственный** *adj.* incongruous, conflicting, contrary, inappropriate, inadequate; discordant, discrepant; unsuited

**несоответствие** *n.* discrepancy, disparity, inequality, inadequacy, nonconformity, noncorrespondence, conflicting, mismatch(ing), shortcoming; dissonance

**несоответствовать** *v.* to mismatch

**несоответствующий** *adj.* unsuitable, inadequate, ineffective, inappropriate, contrary, incongruous, inexpedient

**несоразмерность** *f.* disproportion, inadequacy

**несоразмерный** *adj.* disproportional

**несостоявшийся** *adj.* uncompleted, not taken place

**несостоятельность** *f.* incompetence, failure, unfitness

**несостоятельный** *adj.* incompetent, unfit

**несохранение** *n.* nonconservation

**неспадающий** *adj.* nondecreasing

**неспаренный** *adj.* unpaired

**неспасаемый** *adj.* nonrecoverable, expendable

**неспаяный** *adj.* solderless

**неспектральный** *adj.* nonspectral

**неспециальный** *adj.* nontechnical, nonspecialized, general-purpose, universal

**неспирализированный** *adj.* noncoiled, straight

**несплошной** *adj.* scattered

**неспокойный** *adj.* fluctuating, restless, erratic

**несрабатывание** *n.* nonoperation (of a relay), failure to operate

**несравненный, несравнимый** *adj.* perfect, matchless, incomparable, unequalled

**нестабилизированный** *adj.* unstabilized, unregulated

**нестабилизируемый** *adj.* unstabilized, unregulated

**нестабилизованный** *adj.* unstabilized, uncontrolled, unregulated

**нестабильность** *f.* instability, unstable state; **н. изображения** smudging; **н. поверхности ускоряемой жидкости** Taylor instability

**нестабильный** *adj.* unstable

**нестандартный** *adj.* nonstandard, irregular, nontypical; off-gage; nonstock; optional

**нестареющий** *adj.* non-aging

**нестатистический** *adj.* nonstatistical

**нестационарный** *adj.* transient, transitory, transitional, nonsteady, nonstationary, unsteady; portable, field

**нести, понести** *v.* to carry, to bear; to perform; to sustain

**нестираемый** *adj.* nonerasable, indelible

**нестирающийся** *adj.* nonerasable

**нестойкий** *adj.* unstable; nonpersistent

**нестойкость** *f.* instability

**нестробный** *adj.* ungated

**неступенчатый** *adj.* nonstepped, stepless

**несуммирующий** *adj.* non-additive

**несуразный** *adj.* incoherent, absurd; irregular

**несущая** *f. adj. decl.* carrier; **автоматическая регулируемая н.** floating carrier; **введённая н.** injected carrier; **восстановленная н.** reinserted carrier, exalted carrier; **вспомогательная н.** subcarrier, intermediate carrier; **н. высокой частоты** radio-frequency carrier; **детектированная н.** discriminating carrier; **н. звукового сопровождения** sound carrier; **контрольная н.** pilot carrier; **н., модулированная шумами** randomly-modulated carrier; **остаточная н. модуляционного устройства** residual carrier, white-level reference; **н. светлотного сигнала** luminance carrier; **сдвинутая н.** offset carrier; **н. сигнала звукового сопровождения** aural carrier, sound carrier, voice carrier; **н. сигнала изображения чёрно-белого телевидения** monochrome carrier; **н. сигнала цветности** chrominance carrier; **н. сигналов изображения** television picture carrier; **срединная н.** mean carrier frequency, center carrier frequency, resting frequency; **управляющая н.** pilot carrier

**несущественный** *adj.* unessential, unimportant, immaterial

**несуществование** *n.* nonexistence

**несуществующий** *adj.* non-existing

**несущий** *adj.* carrier, carrying, supporting, bearing

**несфокусированность** *f.* out-of-focus

**несфокусированный** *adj.* out-of-focus

**несходимость** *f.* divergence, nonconvergence

**несходимый** *adj.* divergent

**несходный** *adj.* diverse, dissimilar, unlike

**несходство** *n.* dissimilarity, difference, discrepancy; disparateness

**несчастный** *adj.* unfortunate

**несчётный** *adj.* uncountable, countless, innumerable, numberless, incalculable

**несчитающий** *adj.* noncounting

**нет** *adv.* not, no, not any; **н. на месте** (desired party) is not available *(tphny.)*; **н. тока** "no power", power outage

**нетелефонизируемый** *adj.* no-telephone

**нетемнеющий** *adj.* nondarkening, nonbrowning

**нетепловой** *adj.* nonthermal

**нетеплопроводный** *adj.* non-heat-conducting

**нетерминальный** *adj.* nonterminal

**нетеряемый** *adj.* nonlosable

**нетехнический** *adj.* nontechnical

**нетипичный** *adj.* foreign

**нетональный, нетоновый** *adj.* unpitched

**неточно** *adv.* inaccurately, incorrectly

**неточность** *f.* inaccuracy, incorrectness, inexactitude, error, discrepancy, infidelity; *pl. oft.* misalignments

**неточный** *adj.* inaccurate, inexact, incorrect; coarse, rough, poor; (*+noun*) mis-: **неточная настройка** mistuning

**нетронутый** *adj.* untouched, intact, whole, virgin

**нетто** *adv.* net

**неуверенность** *f.* uncertainty

**неуверенный** *adj.* uncertain, unsure

**неувязка** *f.* discrepancy, lack of coordination, maladjustment, failure, trouble

**неугрожаемость** *f.* safety

**неугрожаемый** *adj.* safe

**неудача** *f.* failure

**неудачно** *adv.* unsuccessfully

**неудачный** *adj.* unsuccessful, unfortunate

**неудобство** *n.* inconvenience; difficulty, disadvantage, drawback; shortcoming

**неудовлетворительный** *adj.* inadequate, imperfect, insufficient, unsatisfactory, poor, bad

**неудовлетворяющий** *adj.* unsatisfactory, substandard

**неуместный** *adj.* misplaced, irrelevant, inappropriate, out of place, superfluous

**неумышленный** *adj.* unintentional, inadvertent

**неупорядоченный** *adj.* disordered

**неуправляемый** *adj.* uncontrolled; unstabilized; unguided, random

**неупругий** *adj.* inelastic, rigid

**неупругость** *f.* anelasticity

**неуравновешенность** *f.* unbalance, balance error

**неуравновешенный** *adj.* unbalanced, out of balance, out of alignment

**неусреднённый** *adj.* unaveraged

**неустановившийся** *adj.* unsettled, irregular, interrupted, unsteady; transient, transitory, transitional

**неустановленный** *adj.* unestablished, unstated; not fixed, unmounted

**неустойчивость** *f.* instability, unsteadiness, fluctuation; wobble, swinging, jitter; **н. во времени** time jitter; **н. развёртки** time flutter; **н. усилителя** near singing condition, tendency to sing; **н. чересстрочной развёртки** interlace jitter

**неустойчивый** *adj.* unstable, unsteady, astable, labile, transient, shifting, drifting; fluctuating, changeable

**неустройство** *n.* disorder, disorganization

**неутопленый** *adj.* unmounted

**неучитываемый** *adj.* nonmetered, nonregistered, negligible

**неучтённый** *adj.* unaccounted-for

**неуязвимость** *f.* survival, invulnerability, invulnerableness

**нефелометр** *m.* nephelometer, turbidimeter

**нефиксированный** *adj.* floating

**нефильтрованный** *adj.* unfiltered

**нефокусированный** *adj.* nonfocused

**нефоскоп** *m.* nephoscope

**нефтеденсиметр** *m.* oleometer

**нефть** *f.* oil, petroleum, rock-oil; **натуральная н., сырая н.** crude oil, crude petroleum

**нефтяной** *adj.* oil, petroleum

**нехватать, нехватить** *v.* to be insufficient, to be wanting, to lack

**нехватка** *f.* deficiency, lack, shortage, absence, scarcity

**нехлопающий** *adj.* nonpopping, nonmotorboating

**нецентрализованный** *adj.* noncentralized, decentralized

**нецентральный** *adj.* noncentral; off-center, side

**нецентрированность** *f.* eccentricity, misalignment

**нецентрированный** *adj.* eccentric

**нециклический** *adj.* acyclic; loop-free

**нецилиндрический** *adj.* noncylindrical

**нечаянно** *adv.* accidently, unexpectedly

**нечаянность** *f.* unexpectedness

**нечаянный** *adj.* unexpected, sudden; unintentional, accidental, inadvertent, incidental

**нечерненый** *adj.* clear

**нечёрный** *adj.* nonblack

**нечет** *m.* odd number

**нечёткий** *adj.* illegible, undecipherable; difficult, indistinct, defective

**нечётно-гармонический** *adj.* odd-harmonic

**нечётно-нечётный** *adj.* odd-odd

**нечётно-симметричный** *adj.* odd-symmetric

**нечётность** *f.* oddness

**нечётно-чётный** *adj.* odd-even

**нечётный** *adj.* odd, uneven; odd-numbered

**нечисленный, нечисловой** *adj.* nonnumerical

**нечувствительность** *f.* insensibility, insensitivity

**нечувствительный** *adj.* insensitive, insensible; dead; **н. к изменениям фазы, н. к фазе** phase-insensitive; **н. к цвету** colorblind

**нешаблонный** *adj.* random

**нешунтированный** *adj.* unshunted

**неэвклидов** *poss. adj.* non-Euclidean

**неэквивалентность** *f.* non-equivalence

**неэквивалентный** *adj.* non-equivalent

**неэквидистантный** *adj.* unequally spaced

**неэкранированный** *adj.* unshielded

**неэластичность** *f.* inelasticity, rigidity

**неэластичный** *adj.* inelastic, rigid

**неэлектризуемый** *adj.* anelectric

**неэлектрифицированный** *adj.* unelectrified

**неэлектрический** *adj.* non-electric

**неэффективность** *f.* ineffectiveness

**неэффективный** *adj.* ineffective, inefficient

**нею** *see* **ней**

**неявка** *f.* non-appearance, absence; **н. на связь** failure to report to communication work

**неявный** *adj.* implicit

**неядерный** *adj.* nonnuclear

**неядовитый** *adj.* nontoxic, nonpoisonous

**неясно** *adv.* vaguely

**неясность** *f.* vagueness, obscurity, uncertainty, confusion

**неясный** *adj.* vague, obscure, indistinct, unclear, confused; inarticulate, illegible, indefinite, undistinguishable

**НЕ-ячейка** *f.* NOT-circuit, NOT-gate

**НИ** *abbr.* (**научный институт**) scientific institute

**НИАИ** *abbr.* (**Научно-исследовательский авиационный институт**) Scientific Research Aviation Institute

**НИАТ** *abbr.* (**Научно-исследовательский институт авиационной технологии**) Scientific Research Institute of Aviation Technology

**нивелир** *m.* level, leveling instrument, surveyor's level

**нивелирование** *n.* grading, leveling; **барометрическое н.** barometric measurement of altitude

**нивелировать** *v.* to level, to grade

**нивелировочный** *adj.* level; leveling

**нивелирующий** *adj.* leveling

**ниже** *compar. of* **низкий, низко** lower

**ниже** *prep. genit.* below, beneath, under

**нижележащий** *adj.* underlying, sublayer

**нижепропускающий** *adj.* low-pass(ing)

**нижний** *adj.* lower, bottom; inferior; under

**низ** *m.* bass (*mus.*); bottom, base, lowest part; **низы** *pl.* lows, low tones, bass, bottom

**низкий** *adj.* low; base; deep (sound); « **низкие** » *pl.* lows

**низко** *adv.* low

**низковольтный** *adj.* low-voltage

**низковязкий** *adj.* low-viscosity

**низкокачественный** *adj.* low-definition, coarse

**низколетящий** *adj.* low-flying

**низкоомный** *adj.* low-resistance, low-impedance

**низкопроходящий** *adj.* low-pass(ing)

**низкорасположенный** *adj.* low-sited, low-lying

**низкотемпературный** *adj.* low-temperature

**низкоуглеродистый** *adj.* low-carbon

**низкофонный** *adj.* low-background

**низкочастотный** *adj.* low-frequency, subsonic, audiofrequency; low-speed

**низкоэнергетический** *adj.* low-energy

**низовой** *adj.* bottom

**низший** *adj.* lower, inferior

**НИИ** *abbr.* (**научно-исследовательский институт**) scientific research institute

**НИИ ВВС** *abbr.* (**Научно-исследовательский институт военно-воздушных сил**) Scientific Research Institute of the Air Force

**никалой** *m.* nicaloi

**никелевый** *adj.* nickel

**никелин** *m.* nickeline

**никелиновый** *adj.* nickeline

**никелирование** *n.* nickel facing, nickel plating

**никелировать, отникелировать** *v.* to nickel-plate

**« Никелодеон »** *m.* Nickelodeon

**никель** *m.* nickel, Ni

**никель-молибденовый** *adj.* nickel-molybdenum

**никель-цинк-феррит** *m.* nickel-zinc-ferrite

**Николь** Nicol

**николь** *m.* Nicol prism; **скрещённые николи** crossed nicols

**НИКФИ** *abbr.* (**Научно-исследовательский кино-фото институт**) Motion Picture and Photography Scientific Research Institute

**нильвар** *m.* nilvar

**Ним** Nim (game)

**нимб** *m.* halation, nimbus, halo

**ниобат** *m.* niobate, columbate

**ниобатовый** *adj.* niobate, columbate

**ниобиевый** *adj.* niobium

**ниобий** *m.* niobium, Nb

**ниппель** *m.* nipple, adapter; sleeve; contact block; **переходный н.** adapter (nipple), diminishing socket; **шнуровой н.** cord grip

**Нипков** Nipkow

**НИР** *abbr.* (**научно-исследовательская работа**) research work

**нисходящий** *adj.* descending, down, downward

**нисхождение** *n.* descent

**нитевидный** *adj.* filament, filamentary, filiform, filar, threadlike, capillary

**нитевой** *adj.* thread

**нитеобразный** *see* **нитевидный**

**нитка** *f.* thread, yarn, twine, fiber, filament

**НИТО** *abbr.* (научное инженерное техническое общество) Scientific Engineering and Technical Society

**нитон** *m.* niton, Nt, radon, Rn

**ниточка** *f.* thread, fine thread, filament

**нитрат** *m.* nitrate

**нитро-** *prefix* nitro-

**нитробензол** *m.* nitrobenzene

**нитробензольный** *adj.* nitrobenzene

**нитрометр** *m.* nitrometer, azotometer

**нитрон** *m.* nitron (plastic)

**нитроплёнка** *f.* nitrate film

**нитросоединение** *n.* nitro compound

**нитрофильм** *m.* nitrate film

**нитроцеллюлоза** *f.* nitrocellulose

**нитроцеллюлозный** *adj.* nitrocellulose, cellulose nitrate

**нитчатый** *adj.* filamentous

**нить** *f.* filament, thread, fiber; **биспиральная н. (накала)** coiled coil filament; **биспиральная намотанная н.** coiled coil; **визирная н. на экране электронно-лучевой трубки** bearing cursor on oscilloscope screen; **витая н. накала** cabled filament, coiled filament; **дважды спирализированная н. накала** coiled-coil filament; **н. доказательства** proof thread; **зигзагообразная н.** straight up-and-down filament; **мишурная н.** patch cord wire; **н. накала** glow lamp filament, filament, glower, heater; **н. накала, имеющая намотку с малой индуктивностью** non-inductively wound filament; **н. накаливания** heated filament, incandescent filament; **неспирализированная н. накала** straight filament; **оксидированная н.** oxide (coated) filament; **отличательная н.** colored tracer thread; **прессованная н.** colloidal filament; **прямолинейная н. накала** line filament; **н., расположенная в одной плоскости** monoplane filament; **секционированная н.** tapped filament; **угольная н.** carbon filament; **угольная н. накала** Helion filament; **укреплённая на крючках н. накала** anchored filament; **условная н.** colored tracer thread

**нитяный** *adj.* thread, filament, fiber, filar

**нихром** *m.* nichrome

**нихромовый** *adj.* nichrome

**ничтожно** *adv.* insignificantly

**ничтожный** *adj.* insignificant, minute, slight, negligible, infinitesimal, faint; worthless

**ниша** *f.* recess, niche, housing, bay

**НЛ** *abbr.* (навигационная линейка) navigation computer

**НМТ** *abbr.* (нижняя мёртвая точка) bottom dead center

**нобатрон** *m.* Nobatron

**Нобили** Nobili

**нов.** *abbr.* (новый) new

**нова** *f.* nova

**новейший** *superl. adj.* up-to-date, newer; newest, latest, modern

**новый** *adj.* new, modern, recent

**нога** *f.* leg; foot; brace, stand; **н. опоры** pole footing, pole pedestal, pole butt

**нож** *m.* knife; blade; **н. для свинца** chipping knife; **монтерский н.** lineman's knife, hack knife; **насекательный н.** pruning knife, pruning hook; **подающий н.** picker knife; **н. рубильника** switch blade; **складной кабельный н.** cable splitting pocket knife; **н. с ломающимся шарниром** broken-back blade; **снимающий н.** shaving knife, scraping knife

**ножевой** *adj.* knife; knife-type, blade

**ножка** *f.* leg; stem, shank; shoe (*mech.*); jaw (of measuring instrument); arm, tine (of tuning fork); prong; mount, pinch; **гребенчатая н.** pinched base; **заштенгелеванная н.** tabulated flare stem; **н. кристаллического триода** transistor mount; **н. лампы** stem, pinch assembly, mount; **монтированная н.** mount; **направляющая н.** aligning plug, aligning key, positioning pin; **Н-образная монтированная н.** H-mount; **н. прожектора** gun stem; **расплющенная н.** compression socket; **н. с тарелкой** flare (*tbs.*); **смонтированная н.** lamp mount; **спрессованная н.** pressed-glass base; **стеклянная н.** press; **стеклянная н. металлической лампы** all-metal tube base; **н. транзистора** transistor mount; **четырёхштырьковая н.** four-prong base; **штампованная стеклянная н.** pressed glass base

**ножкодержатель** *m.* mount support

**ножницеобразный** *adj.* scissors-type

**ножницы** *pl.* scissors, shears, cutter; **двухрычажные** н. alligator shears, crocodile shears; **дисковые** н. circular shears, slitter with reel; **н. для обрезки ветвей** pruning hook, tree pruner; **н. для подрезки щёток** brush trimmer; **н. для резки листового железа** (*or* **металла**) plate shears, plate cutter, cutting shears; **н. для резки углового железа** angle iron shears; **листорезные** н. plate-shears, sheet-iron shears

**ножной** *adj.* foot, pedal; floor

**ножовка** *f.* hack saw; hand saw; **узкая (прорезная)** н. keyhole saw, panel saw

**ножовочный** *adj.* knife; edged

**ножовый** *see* **ножевой**

**ноздреватость** *f.* porosity, sponginess

**ноздреватый** *adj.* porous, spongy

**нокс** *m.* nox (unit)

**ноктовизор** *m.* snooperscope

**ноль** *m.* zero; *see* **нуль**

**« номаг »** *m.* nomag

**номенклатура** *f.* nomenclature

**номер** *m.* number; size, gage (of wire); caliber; item; mesh (of screen); **абонентский** н. call(ing) unit; **н. волны разгруппировки** debunching wave number; **индивидуальный телефонный** н. standard subscriber connection; **исходящий** н. consecutive number, current number; **коллективный** н. directory number; **н. накопителя** address (*computing*); **направляющий** н. dial prefix; **н. не работает** not a working number, number unobtainable; **несуществующий** н. "no such number"; **н., определяющий вид трафика** dial prefix; **н. по абонентскому списку** directory number; **порядковый** н. **элемента** atomic number; **н. проволоки** gage of wire; **серийный** н. collective number; **н. ступени** quantization level; **н. шифра** box-number; **н. ячейки памяти** address (*computing*)

**номератор** *m.* automatic sender

**номерник** *m.* annunciator board, drop annunciator, drop indicator; number switch; switchboard, wall-pattern switchboard; **н. с вызывными клапанами** drop indicator

**номерной** *adj.* number, numerical

**номеронабиратель** *m.* (telephone) dial, dial switch, selecting mechanism; **автоматический** н. punch-card-operated register selector; **н. с холостым ходом** telephone dial with pulse-spacing device

**номероуказатель** *m.* call indicator; **оптический** н. automatic number-annunciator panel

**номинал** *m.* rating; nominal value, rating value, set point; **длительный** н. continuous-duty rating; **повторно-кратковременный** н., **прерывистый** н. intermittent-duty rating

**номинальный** *adj.* nominal; rated

**номограмма** *f.* nomogram, nomograph, chart, alignment chart, computer board; **диагональная** н. nomogram of three variables; **н. для перевода в масштаб децибел** decibel chart; **н. для расчёта ...** calculator; **н. для расчёта диэлектрических свойств смешаных диэлектриков** dielectric mixture chart; **н. для расчёта сопротивления термистора в зависимости от температуры** thermistor monogram; **н. Крепена** Crépin monographic chart; **прямолинейная** н. nomogram, nomographic chart; **н. с параллельными шкалами** parallel-nomogram; **сетчатая** н. bunch graph

**номограф** *m.* nomograph

**номографический** *adj.* nomographic, nomograph, nomogram

**номография** *f.* nomography

**номотрон** *m.* Nomotron

**нона** *f.* ninth

**нониус** *m.* nonius, vernier

**нониусный** *adj.* nonius, vernier

**нонод** *m.* nonode

**норатор** *m.* norator

**нория** *f.* bucket elevator; **н. с редкими (частными) ковшами** intermittent (continuous) bucket elevator

**норм** *m.* norm, standard; rate; quota

**норм.** *abbr.* (**нормальный**) normal

**норма** *f.* standard; rate, *pl. oft.* limits; **нормы Британского Департамента Торговли** (British) Board of Trade Standards; **нормы Международного Консультативного Комитета по Радиосвязи** CCIR standards; **нормы нагрузки для ламп** tube ratings; **н. отказов (детали)** (part) failure rate; **н. профилактики** maintenance ratio;

н. рабочего времени устройства, н. эксплуатационной готовности in-commission rate

**нормализация** *f.* normalization, normalizing, standardization

**нормализованный** *adj.* normalized, standardized

**нормализовать** *v.* to normalize, to standardize

**нормализующий** *adj.* normalizing, standardizing

**нормаль** *f.* normal, standard, standard specification; **н. оконечной нагрузки** terminating unit

**нормально** *adv.* normally; **н. закрыт** normal danger (*rr appl*); **н. открыт** normal clear (*rr appl*)

**нормальный** *adj.* normal, standard, regular; rated

**норматив** *m.* norm, standard

**нормативный** *adj.* norm, standard; normative

**норматрон** *m.* normatron

**нормирование** *n.* normalization, standardization, normalizing; rate fixing, rating

**нормированный** *adj.* normalized; fixed, set

**нормировать** *v.* to normalize, to standardize

**нормировка** *f.* normalization, standardization; rate fixing

**нормирующий** *adj.* normalizing; normalization

**Норт** North

**нортовский** *adj.* North, North's

**Нортон** Norton

**нос** *m.* point, forepart; lug; nose

**носик** *m.* tip; bill; spout

**носилки** *pl.* handbarrow; **н. с тамбуром** wire barrow

**носимый** *adj.* portable

**носитель** *m.* carrier, bearer; carrier vehicle; medium; **н. вводимых данных** input medium; **н. выводных данных** output medium; **«господствующий» н.** majority carrier; **н. информации в машине** machinable medium; **многократно используемый н. информации** erasable memory, erasable store, erasable storage; **неосновной н.** minority carrier; **неравновесный н.** excess carrier; **однократно используемый н.** non-erasable storage; **основной н. (заряда)** majority

carrier; **н. положительного заряда** positive carrier; **сегнетоэлектрический н.** ferroelectric carrier; **тепловой н.** thermal acceptor; **«угнетённый» н.** minority carrier

**носить** *v.* to carry, to bear

**носовой** *adj.* nasal, nose; bow, fore; forward

**нота** *f.* note (*acoust.*); **ноты** *pl.* notes, music; keys (on keyboard); **гармонически несвязанные ноты** remote keys; **гармонически связанные ноты** related keys; **н. в одну восьмую** eight note; **н. в одну тридцать вторую** thirty-second note; **н. в одну четвёртую** quarter note; **н. в одну шестнадцатую** sixteenth note; **н. в одну шестьдесят четвёртую** sixty-fourth note; **восьмая н.** quaver, eighth note; **основная н. ключа** keynote; **тридцать вторая н.** demisemiquaver, thirty-second note; **целая н.** whole note, whole tone, whole step, semibrave; **шестнадцатая н.** semiquaver, sixteenth note

**нотация** *f.* notation; tabulature (*mus.*)

**нотный** *adj.* note, music

**нотоносец** *m.* stave, staff (*mus.*)

**ночник** *m.* night light

**ночной** *adj.* night, nightly

**НП** *abbr.* (**наблюдательный пункт**) observation post, observation station

**НПЧ** *abbr.* (**наименьшая применимая частота**) lowest usable frequency

**НР** *abbr.* 1 (**нормаль**) standard (specification). 2 (**ночная разведка**) night reconnaissance

**НРС** *abbr.* (**наземный радиолокационный запросчик**) ground-based interrogator

**НТО** *abbr.* (**научно-технический отдел**) Technological Department

**НТС** *abbr.* (**научно-технический совет**) Scientific Technical Council

**Нуаре** Noiré

**нувистор** *m.* nuvistor

**нужда** *f.* need, necessity, want

**нуклеарный** *adj.* nuclear

**нуклеиновый** *adj.* nuclein, nucleinic

**нуклеоника** *f.* nucleonics

**нуклид** *m.* nuclide

**нуклон** *m.* nucleon, nuclear particle

**нуклонный** *adj.* nucleon(ic)

**нулевой** *adj.* zero, null, neutral

**нуллаторно-нораторный** *adj.* nullator-norator

**нуллоид** *m.* electrodeless tube, nullode, spark-gap tube

**нуллор** *m.* nullor

**нуль** *m.* zero, null; zero point, null point; cipher; **внешкальный н.** inferred zero; **н.-орган** null-element; **н., разрушенный током записи** write-disturbed zero; **н., разрушенный током считывания** read-disturbed zero; **сдвинутый н.** false zero; **н., стоящий слева** left-hand zero; **фиксирующий н.** leading zero

**нуль-адресный** *adj.* zero-address

**нуль-детектор** *m.* null detector, balance detector

**нуль-индикатор** *m.* null indicator; **н.-и. полёта** flight director

**нумератор** *m.* annunciator, indicator board; numerator, ratemeter; **блинкерный н.** drop annunciator; **н. с электромагнитным подъёмом выпавших номеров** throw-back indicator; **световой н.** annunciator; **сигнальный н.** teleseme

**нумерация** *f.* numeration, numbering; quantization, quantizing; numbering scheme; **н. по амплитуде** quantization of amplitude

**нумерикорд** *m.* numericord control system

**нумерирование** *see* **нумерация**

**нумерованный** *adj.* numbered; quantized

**нумеровать, занумеровать** *v.* to number; to index

**нумероскоп** *m.* numeroscope

**НУП** *abbr.* (**необслуживаемый усилительный пункт**) unattended repeater station

**нутационный** *adj.* nutation

**нутация** *f.* nutation

**нутромер** *m.* (inside) calipers, inside micrometer, hole-gage

**НЧ** *abbr.* 1 (**низкая частота**) audio frequency, low frequency. 2 (**низкочастотный**) low-frequency

**нырнуть** *pf. of* **нырять**

**ныряло** *n.* plunger, ram

**нырять, нырнуть** *v.* to plunge, to immerse, to dip, to dive, to pitch

**Ньюман, Ньюмэн** Newmann

**ньютон** *m.* newton (unit of force)

**Ньютон** Newton

**ньютонов** *poss. adj.* Newton's

**ньютоновский** *adj.* Newton(ian)

**нэгит** *m.* naegite

**Нэттол** Nuttall

**Нюквист** Nyquist

# O

о *prep. acc. and prepos. case* against; about, concerning, of, on, upon

о. *abbr.* (область) region, oblast

об *see* о

об. *abbr.* 1 (оборот) revolution; 2 (объёмный) volume, volumetric

оба *num.* both

об. вес *abbr.* (объёмный вес) density

обвивать, обвить *v.* to wind around, to twist around

обвитый *adj.* wound around, wrapped around, overspun

обвить *pf. of* обвивать

обвод *m.* bypass; mesh, loop; enclosing, surrounding

обводка *f.* looping

обводный *adj.* by-pass; surrounding, encircling

обволакивать, обволочить *v.* to coat, to cover, to envelop, to wrap

обволочить *pf. of* обволакивать

обвязка *f.* binding, tying, strapping, bandage

обвязочный *adj.* binding, fastening

обвязывание *n.* binding, tying

обгонять, обогнать *v.* to gain upon, to outstrip, to pass, to overtake, to outdistance

обгорание *n.* burning, pitting; spark-wear; scorching

обгорать, обгореть *v.* to burn around; to scorch, to char

обгорающий *adj.* sparking; burning around

обгорелый *adj.* burnt, scorched, charred

обгореть *pf. of* обгорать

обдирать, ободрать *v.* to strip, to peel; to skin, to tear off, to separate

обдирка *f.* stripping; skinning; грубая о. burnishing

обдирочный *adj.* stripping

обдуваемый *adj.* ventilated

обдувание *n.* blowing; о. смонтированной ножки blowing the mount

обдувать, обдуть *v.* to blow, to blow out, to blow off

обдувка *f.* blow, blow out

обдувочный *adj.* blowing; bulb-blowing

обдуть *pf. of* обдувать

обе *see* оба

обедневший *adj.* impoverished, poor

обеднение *n.* depletion, exhaustion, impoverishment; depopulation

обеднённый *adj.* depleted, impoverished; stripped

обеднять, обеднить *v.* to deplete, to impoverish; to depopulate

обезводить *pf. of* обезвоживать

обезвоженный *adj.* dehydrated

обезвоживание *n.* dehydration, desiccation; draining

обезвоживатель *m.* dehydrator, desiccator

обезвоживать, обезводить *v.* to dehydrate; to drain

обезвоживающий *adj.* dehydrating

обезгаженный *adj.* evacuated, gas-free

обезгаживание *n.* degassing, degasification, de-aeration, outgassing, exhaustion, evacuation

обезгаживать, обезгазить *v.* to degas, to outgas; to evacuate

обезжиренный *adj.* degreased

обезжиривание *n.* degreasing; cleaning

обезжиривать, обезжирить *v.* to degrease

обезжиривающий *adj.* degreasing

обезжирить *pf. of* обезжиривать

обезьяний *adj.* monkey

обёрнутый *adj.* wrapped, enveloped

обернуть *pf. of* обёртывать

обёртка *f.* envelope, cover, covering, casing, case, jacket, sheath, wrapper, wrapping; liner

обертон *m.* overtone, (higher) harmonic; самый громкий на слух о. physiologically loudest overtone

обёрточный *adj.* wrapping, casing

обёртывание *n.* lapping, taping

обёртывать, обернуть *v.* to invert, to turn; to wrap up, to envelope, to cover

обёртывающий *adj.* wrapping, enveloping

**обеспечение** *n.* securing, security, provision, providing; maintenance; **о. разборчивости речи** extraction of intelligence

**обеспеченный** *adj.* secured

**обеспечивать, обеспечить** *v.* to protect, to secure; to guarantee

**обеспечивающий** *adj.* guaranteeing; safety

**обеспечить** *pf. of* обеспечивать

**обеспыление, обеспыливание** *n.* removal of dust, dust elimination

**обесточение** *n.* de-energizing, killing

**обесточенный** *adj.* dead, de-energized, no-current, currentless; flared; floating

**обесточивать, обесточить** *v.* to de-energize, to cut off current

**обесцветить** *pf. of* обесцвечивать

**обесцвечивание** *n.* fading, decoloration, discoloration; decolorization

**обесцвечивать, обесцветить** *v.* to decolorize, to decolor; to discolor

**обесценение** *n.* depreciation

**обесцененный** *adj.* depreciated

**обесшумливание** *n.* quieting, noise reduction

**обечайка** *f.* purfling (of a violin); shell, rib

**обжатие** *n.* strain; pressing, squeezing

**обжечь** *pf. of* обжигать

**обжиг** *m.*, **обжигание** *n.* burning, roasting; firing, annealing, baking

**обжигательный** *adj.* annealing; roasting, calcining

**обжигать, обжечь** *v.* to bake; to fire, to calcinate; to anneal, to kiln

**обжимка** *f.* strain, pressing, compression, squeezing

**обзор** *m.* scan, scanning field; field of vision, field of view, coverage, surveillance; survey, review, synopsis; **о. по азимуту** azimuth coverage; **о. по углу места** elevation coverage, vertical coverage

**обзорный** *adj.* scanning, search; surveillance, coverage

**обивка** *f.* casing; upholstery; **внутренняя о.** lining

**обитаемый** *adj.* inhabited; manned

**обитый** *adj.* lined, padded, covered, upholstered

**обкладка** *f.* lining, coating, covering; (capacitor) plate; armature

**обл.** *abbr.* (**область**) region, district, oblast

**облагораживание** *n.* precipitation hardening; improvement

**обладание** *n.* possession

**обладающий** *adj.* possessing, having

**облако** *n.* cloud; swarm

**облакомер** *m.* ceiling-height indicator

**областной** *adj.* region, regional, district, territorial, "oblast"

**область** *f.* region; field, domain, sphere, zone, area, spot, realm; range, band; space; scope; "oblast"; **ближняя инфракрасная о.** fundamental vibration-rotation region; **о. более сильного приёма** beaming area; **о. возникновения метеоров** M-region; **вторая катодная тёмная о.** Crookes dark space, Hittorf dark space, cathode dark space; **о. дырочной проводимости** p-region; **о. заглушения** blackout area, wipe-out area; **о. заграждения** key-out region; **о. затягивания частоты, о. захватывания** lock-in range, locking range, pull(-in) range (*tv*); **о. избытка электронов** n-region (in a semiconductor); **о. изменения переменной** variable range; **о. недостатка электронов** p-region (in a semiconductor); **о. неразрешаемости сигнала** confusion region; **о. определения функции** domain of function; **о. приёма помех** interference area; **о. примесной проводимости** extrinsic range; **о. пропускания** key-in region, free transmission range; **рабочая о. тока** current margin (*tgphy.*); **о. связи** association region (*math.*); **о. синхронизации** hold-in range, insync range, lock-in range, locking range, retention range, pull-in range; **о. слышимости** auditory sensation area; **о. слышимости на данной частоте** range of audibility; **о. смешанного приёма** mush area; **собственная беспримесная о.** intrinsic range; **о. собственной проводимости** intrinsic region; **спирально намагниченная о.** spirally polarized magnetic zone; **средняя о. цели** radar center; **о. срыва колебаний** oscillation stop; **о. точечного излучения** Fraunhofer region; **о. устранения искажений** frequency range of equalization

**облегать, облечь** *v.* to encircle, to enclose, to encompass, to surround

**облегающий** *adj.* encircling, encompassing

**облегчать, облегчить** *v.* to ease, to discharge, to lighten, to facilitate, to relieve, to alleviate

**облегчающий** *adj.* relieving, alleviating

**облегчение** *n.* relief, alleviation, mitigation, mollification, facilitation, lightening, easing

**облегчённый** *adj.* relieved, alleviated, eased, facilitated, lightened

**облегчить** *pf. of* **облегчать**

**облечь** *pf. of* **облегать**

**облитерация** *f.* obliteration

**облитый** *adj.* covered, drenched

**облицевать** *pf. of* **облицовывать**

**облицованный** *adj.* lined, faced, coated

**облицовка** *f.* revetment, facing, lining, coat, coating, covering, jacketing, casing; finish

**облицовочный** *adj.* facing, lining, coating, covering, casing; finish, finishing

**облицовывать, облицевать** *v.* to revet (with), to face (with), to line, to coat, to cover

**облический** *adj.* oblique

**обложка** *f.* cover, envelope, casing; wrapper

**обломанный** *adj.* broken off; truncated

**обломок** *m.* piece, chip, fragment, splinter; *pl.* **обломки** debris, rubble, bits, scrap

**облучаемый** *adj.* irradiated; swept

**облучатель** *m.* irradiator, primary radiating element, exciter; feed; **о. антенны** antenna feed; **о. в виде линейной решётки вибраторов** dipole-array extended feed; **излучающий вперёд сквозной о.** front waveguide feed; **излучающий назад сквозной о.** rear waveguide feed; **о. с бегущим лучом** flying-spot illuminator

**облучать, облучить** *v.* to irradiate; to sweep; to expose

**облучение** *n.* irradiation, raying, illumination, exposure; floodlighting; **конусное о.** gable illumination; **кратковременное о.** acute exposure; **многопольное перекрёстное о.** cross firing

**облучённость** *f.* irradiancy, irradiance

**облучённый** *adj.* irradiated

**облучить** *pf. of* **облучать**

**обмазанный** *adj.* coated, covered, smeared, greased

**обманка** *f.* blende (*min.*); **урановая смоляная о.** pitchblende

**обматывание** *n.* lapping, taping, covering, wrapping, coiling; winding; sheath, cover

**обматывать, обмотать** *v.* to lap, to cover, to wrap, to tape, to wrap around, to wind, to coil; to encircle, to sheath

**обмахивать, обмахнуть** *v.* to sweep (off); to fan; to brush away

**обмеднение** *n.* coppering, copper plating

**обмеднённый** *adj.* coppered, copper-plated

**обмеднять, обмеднить** *v.* to copper, to copper plate

**обмен** *m.* exchange, interchange, change; communication; traffic; **о. малой плотности** slack traffic; **о. по дальним линиям** long-distance traffic (*tphny.*); **о. с внешним устройством** peripheral transfer

**обмена** *f. see* **обмен**

**обменивать, обменить, обменять** *v.* to exchange, to interchange

**обмениваться, обмениться, обменяться** *v.* to exchange, to interchange

**обменник** *m.* exchanger, interchanger

**обменно-связанный** *adj.* exchangeable

**обменный** *adj.* interchangeable; exchange; converted; composite (wave)

**обменять(ся)** *pf. of* **обменивать(ся)**

**обмер** *m.* measuring, measurement

**обмеренный** *adj.* measured

**обмерять, обмерить** *v.* to rate, to determine, to proportion; to meter, to measure

**об/мин.** *abbr.* ([число] **оборотов в минуту**) rpm, revolutions per minute

**обмотанный** *adj.* covered, taped, wound, coiled, wrapped

**обмотать** *pf. of* **обматывать**

**обмотка** *f.* winding, turn, coil, coiling, covering, taping, wrapping, lapping; braiding, braid; cover, sheath; **дисковые чередующиеся обмотки трансформатора** sandwich windings of a transformer; **обмотки, намотанные взаимно перпендикулярно** quadrature windings; **о. в полюсных наконечниках** pole face winding; **о. внахлёстку** pile winding; **о. возбуждения** excitation winding, field winding, magnet winding; field coil, magnetizing coil; **о. возврата в исходное положение** reset winding; **всыпная о.** mush winding, random winding; **вторичная о. секционированная пополам** center-tapped secondary winding, split secondary winding; **о. выборки** drive winding, drive wire; **выходная о.** sense winding; **галетная о.** pie winding; **гемитропическая о.** half-coiled winding, hemitropic armature winding; **диаметрическая о.** full-pitch winding; **дисковая о.** pie winding; **дробношаговая о.** fractional-pitch winding; **одноходовая о.** simplex winding; **о. переключения в исходное** (*or* **заданное) положение** set winding; **о. подмагничения** magnetization winding, DC winding, field winding; **протяжная о.** pull-through winding; **размагничивающая о.** bucking winding; **разрядная о.** digit winding; **разрядная о. записи** bit-write winding, digit-write winding; **разрядная о. запрета** bit-plane winding, digit-plane winding; **разрядная о. считывания, разрядная сигнальная о.** bit-sense winding, sense-digit winding; **о. ровными слоями** layer-by-layer winding; **о. с дробным числом катушек на полюс и фазу** fractional-slot winding; **о. с изоляцией на полное напряжение** fully insulated winding; **о. с косыми лобовыми соединениями** skew-coil winding; **о. с левым обходом** retrogressive winding; **о. с правым обходом** progressive winding; **о. с регулируемым импульсным напряжением** stress-control winding; **о. с укороченным шагом** short-pitch winding; **о. с целым числом пазов** (*or* **шагом катушек) на полюс и фазу** integer-slot winding, integral-slot winding; **сеточная о. обратной связи** grid-feedback winding; **сигнальная о.** sense winding; **согласная о.** cumulative winding; **о. считывания** sense winding; **о. у якоря** inside end winding (of a relay); **о. у ярма** heel end winding (of a relay); **управляющая о.** advance winding; **шаблонная о.** preformed winding, diamond winding; **шаблонная о. с рассыпными катушками** armature winding for semiclosed slots; **эталон-**

**ная о. возбуждения** reference field winding

**обмоточный** *adj.* winding, taping, coiling, covering, wrapping

**обмурованный** *adj.* brick-lined

**обмуровка** *f.* brickwork, masonry, covering, lining; casing

**обнаружение** *n.* detection, finding, location, acquisition, sighting, uncovering, discovery, disclosure; appearance, development; display; **дальнее о.** early warning; **о. на слух** aural detection

**обнаруженный** *adj.* located, discovered, exposed, uncovered; bare

**обнаруживаемость** *f.* detectability

**обнаруживаемый** *adj.* detectable

**обнаруживание** *see* обнаружение

**обнаруживать, обнаружить** *v.* to detect, to discover, to disclose, to reveal, to uncover, to find, to locate, to trace, to track; to develop, to display; to identify

**обнаруживающий** *adj.* detecting, locating, identifying

**обнаружимость** *f.* detectability

**обнаружимый** *adj.* detectable

**обнаружитель** *m.* finder, detector, locator; feeler; **о. раковин** (*or* **трещин)** crack detector; **о. частоты вибраций** reed-frequency detector

**обнаружительный** *adj.* detector, locator, finder; feeler; warning

**обнаружить** *pf. of* обнаруживать

**обновитель** *m.* regenerator, restorer

**обновление** *n.* renewal, regeneration, restoration, innovation; change

**обновлённый** *adj.* renewed, restored

**обновляемость** *f.* renewability

**обновляемый** *adj.* renewable

**обо** *see* о

**обобщать, обобщить** *v.* to generalize, to theorize; to correlate

**обобщение** *n.* generalization; correlation

**обобщённый** *adj.* generalized; unified

**обобществленный** *adj.* socialized; business

**обобщить** *pf. of* обобщать

**обогатительный** *adj.* enriching

**обогащение** *n.* enrichment, enriching; concentration

**обогащённый** *adj.* enriched; concentrated; dressed

**обогнать** *pf. of* обгонять

**обогнуть** *pf. of* огибать

**обогрев** *m.* heating, warming
**обогревание** *n.* heating, warming
**обогреватель** *m.* heater
**обогревать, обогреть** *v.* to heat, to warm
**обогретый** *adj.* heated, warmed
**обогреть** *pf. of* **обогревать**
**обод, ободок** *m.* ring, hoop; rim
**ободрать** *pf. of* **обдирать**
**Обое** Oboe, Observed Bombing of Enemy
**обожжённый** *adj.* annealed; calcinated, burnt, roasted
**обозначать, обозначить** *v.* to designate, to denote, to indicate, to mark, to label, to define; to determine, to characterize; to mean, to represent
**обозначающий** *adj.* denoting, designating
**обозначение** *n.* designation, mark, label, identification, marking, notation, denotation; symbol, sign, character, indication; index, characteristic; definition, determination; **бесскобочное о.** Lukasiewicz's notation, Polish notation, parentheses-free notation; **о. зажимов** terminal marking; **о. запоминающего регистра** absolute address; **часовое о. векторной группы соединения** clock hour figure of the vector group
**обозначенный** *adj.* designated, marked; declared; specified
**обозначить** *pf. of* **обозначать**
**обозревать, обозреть** *v.* to survey, to review, to inspect
**обозрение** *n.* survey, review, visualization; observation
**обозреть** *pf. of* **обозревать**
**обойдённый** *adj.* by-passed
**обойма** *f.* collar, ring, iron ring, girdle, clip, clamp, band, yoke, shackle; **о. ролика** trolley harp
**обойти** *pf. of* **обходить**
**оболочечный** *adj.* shell
**оболочка** *f.* cover, covering, jacket; coat, coating, proofing, film; envelope, sheath, sheathing; shell; case, casing; capsule; membrane; mantle; bulb; **о. аксона** membrane; **радужная о.** iris (of the eye)
**оболочный** *adj.* shell
**оборачивать, оборотить** *v.* to turn, to revert; to change, to transform
**оборачивающий** *adj.* inverting

**оборванный** *adj.* torn, broken, cut short
**оборвать** *pf. of* **обрывать**
**оборона** *f.* defense
**оборонительный** *adj.* defensive, defense
**оборонный** *adj.* defense
**оборот** *m.* revolution, rotation, turn; direction; turnover; convolution; phrase
**оборотить** *pf. of* **оборачивать**
**оборотный** *adj.* reversible, reverse, back; circulating
**оборудование** *n.* equipment, equipping, machinery, outfit, outfitting, fitting, gear, apparatus; instrumentation; arrangement, system; plant; facility, facilities, appliance; **о.** ATC dialing equipment; **гидроакустическое о.** sound gear, underwater sound equipment; **гидролокационное о.** sonar detection gear; **о. для работы с одноколонным пенсовым кодированием** pence conversion equipment **коммутационное о.** switching equipment; **о. ЛАЗ** trunk-line equipment (*tphny.*); **приборное о.** instrumentation; **о. пути рельсовыми цепями** track circuiting; **о., работающее в истинном масштабе времени** on-line equipment; **о. работающее вне основного времени работы вычислительной машины** off-line equipment; **связное о. с подавлением помех от многократного распространения** anti-multipath equipment
**оборудованный** *adj.* equipped
**оборудовать** *v.* to equip, to fit out; to arrange
**оборуча** *f.* hoop
**обоснование** *n.* proof, basis, ground
**обоснованность** *f.* validity
**обособить(ся)** *pf. of* **обособлять(ся)**
**обособление** *n.* setting apart, isolating, isolation, separation, insulation
**обособленность** *f.* isolation, insulation
**обособленный** *adj.* individual, single, solitary, isolated, insulated; detached, set off; self-contained
**обособлять, обособить** *v.* to isolate, to insulate, to keep apart
**обособляться, обособиться** *v.* to keep separate, to isolate oneself
**обострение** *n.* enhancement, sharpening, accentuation; peaking

**обострённый** *adj.* aggravated, strained; sharp

**обостритель** *m.* peaker, peaking circuit, sharpener

**обострительный** *adj.* peaking

**обострять, обострить** *v.* to intensify, to sharpen, to peak; to aggravate; to increase

**обостряющий** *adj.* sharpening, peaking (circuit)

**обоюдно** *adv.* mutually, reciprocally

**обоюдность** *f.* reciprocity, mutuality, correlation, reciprocation; ambiguity

**обоюдный** *adj.* reciprocal, mutual, alternate

**обоюдовогнутый** *adj.* concavo-concave, biconcave

**обоюдовыпуклый** *adj.* convexo-convex, biconvex

**обоюдоострый** *adj.* double-cutting, two-edged, double-edged

**обр.** *abbr.* (**образец**) sample

**обрабатываемость** *f.* workability; machinability

**обрабатываемый** *adj.* workable, processable; machinable

**обрабатывание** *see* **обработка**

**обрабатывать, обработать** *v.* to work, to process; to machine, to tool; to handle; to develop, to elaborate; to adapt, to fashion; to manufacture; to interpret (data), to reduce

**обрабатывающий** *adj.* working, processing, treating; machining; process

**обработанный** *adj.* worked, processed, treated; machined, tooled; finished, trimmed, dressed; precision

**обработать** *pf. of* **обрабатывать**

**обработка** *f.* working, processing, treatment, treating; machining, tooling; handling; preparation, manufacture; adaptation; conditioning; evaluation, analysis; reduction; milking; **о. данных в истинном масштабе времени** real-time data processing; **о. данных параллельно с их поступлением** on-line data reduction; **микромеханическая о.** micromachining; **о. поглотителем** absorptive treatment

**образ** *m.* transform; image, picture; shape, form; **о. действия** procedure, modus operandi

**образец** *m.* standard, model, sample, specimen, example, pattern, test piece, copy; form, shape, type; original; **отобранный о.** selection, sample

**образный** *adj.* graphic, figurative; descriptive; *suffix* -shaped, -formed, -form, -oid

**образование** *n.* formation, production; generation, evolution; development, origination; **о. «барашков»** (*or* **«жуков»**) kinking; **о. каналов** channeling; **о. кистевого разряда** brushing; **о. очередей** queueing; **о. слышимости** rendering audible; **о. центров кристаллизации** nucleation

**образованный** *adj.* formed, produced; generated, evolved

**образовать(ся)** *pf. of* **образовывать(ся)**

**образовывать, образовать** *v.* to form, to make, to produce; to generate, to evolve; to construct, to organize, to establish; to constitute, to make up

**образовываться, образоваться** *v.* to form, to appear, to be generated, to develop, to originate

**образуемый** *adj.* formed, generated, produced

**образующая** *f. adj. decl.* generant (*math.*), generatrix

**образующий** *adj.* forming, producing, generating

**образцовый** *adj.* model, sample, exemplary, standard, master, original, test; reference

**образчик** *m.* specimen, sample, pattern

**обрамить** *pf. of* **обрамлять**

**обрамление** *n.* framework, framing; mask; **затеняющее о.** shadow box (*tv*); **светозащитное о.** framing mask; **о. экрана** decorative mask

**обрамлённый** *adj.* framed

**обрамлять, обрамить** *v.* to frame; to set off

**обрастание** *n.* treeing, overgrowing

**обратимость** *f.* reversibility, inversibility, convertibility, reciprocity

**обратимый** *adj.* reversible, reciprocal, bilateral

**обратитель** *m.* reverser; inverter

**обратить(ся)** *pf. of* **обращать(ся)**

**обратно** *adv.* back, conversely, inversely, reversibly, counter-, anti-

**обратнозависимый** *adj.* inverse-relation

**обратноидущий** *adj.* retrogressive; returning

**обратно-магнитострикционный** *adj.* converse magnetostrictive

**обратнонаправленный** *adj.* antagonistic

**обратнооплавленный** *adj.* melt-back

**обратнопьезоэлектрический** *adj.* converse piezoelectric

**обратносинхронный** *adj.* reverse synchronous

**обратносмещённый** *adj.* reverse biased, back-biased

**обратноступенчатый** *adj.* step-back (welding)

**обратный** *adj.* reverse, reversible, reversed, return, back, revertive, backward; reciprocal; inverse, inverted, converse, counter, opposite, anti-; **о. ход** flyback

**обращать, обратить** *v.* to turn, to change, to convert, to transform; to reverse, to invert; to reduce

**обращаться, обратиться** *v.* to refer, to reference; to handle, to treat, to manipulate; to return, to revert; to circulate; to turn, to rotate; **о. к запоминающему устройству** to (make) reference to storage

**обращающий** *adj.* reversing

**обращающийся** *adj.* rotating, circulating

**обращение** *n.* revolution, rotation, turn; handling, care, treatment, usage, manipulation; application; inverse; conversion, reduction; inversion; reversal; access; reference; **о. изображения по горизонтали** lateral inversion (*tv*); **о. к запоминающему устройству** storage access; **неправильное о.** maloperation; **о. трезвучия** inversion of triad (*mus.*); **тройное о.** triple access

**обращённый** *adj.* backward, inverted, inverse; reversed; turned (to), exposed, facing

**обрезание** *n.* clipping, cutting, trimming; cut-off

**обрезанный** *adj.* clipped, truncated, cut-off

**обрезатель** *m.* chopper

**обрезать** *pf. of* обрезывать

**обрезка** *f.* clipping, trimming, cutting; suppression

**обрезной** *adj.* cut, trimming

**обрезок** *m.* cut, piece, length

**обрезывание** *see* обрезание

**обрезывать, обрезать** *v.* to clip, to cut, to cut short, to cut off

**обруч** *m.* collar, hoop, ring, band, clamp

**обручный** *adj.* hoop, ring, collar

**обрыв** *m.* opening, break, breaking away, breakage, fracture, discontinuity; cut-off, premature disconnection, interruption, abruption

**обрывание** *n.* opening, breaking; cutting

**обрывать, оборвать** *v.* to tear, to break, to tear off; to intercept, to cut off, to quench

**об/сек** *abbr.* ([число] **оборотов в секунду**) revolutions per second

**обсерватория** *f.* observatory

**обсерваторский** *adj.* observatory

**обсервационный** *adj.* observation; observatory

**обсервация** *f.* observation

**обследование** *n.* search, scan; survey, inspection, examination

**обследованный** *adj.* inspected

**обследовать** *v.* to survey, to inspect, to examine; to scan

**обследуемый** *adj.* surveyed, inspected; scanned

**обслуживаемый** *adj.* attended, operated, serviced

**обслуживание** *n.* service, servicing, maintenance, working, care, attendance, tending, handling, operation, manipulation; **о. и текущий ремонт** operating maintenance; **комбинированное о.** combined line and recording service (*tphny.*); **ручное о. с вызовом от абонента на оптический номероуказатель** non-coded call display working (*tphny.*); **о. с номероуказателем** call display working (*tphny.*); **о. с номероуказателем комбинированного вызова** coded call display working (*tphny.*); **о. через промежуточную АТС** dial system tandem operation; **о. через промежуточную станцию** tandem working (*tphny.*)

**обслуживать, обслужить** *v.* to serve, to bring service, to service, to maintain, to handle, to attend; to manipulate, to operate

**обслуживающий** *adj.* operating, servicing; auxiliary

**обслужить** *pf. of* обслуживать

**обстановка** *f.* condition(s), situation, circumstance(s); setting, arrangement, background

**обстоятельство** *n.* circumstance; modifier (*lang.*)

**обсуждение** *n.* consideration, discussion

**обтачивание** *n.* turning, rounding off, machining

**обтачивать, обточнть** *v.* to turn, to round off, to machine

**обтекаемый** *adj.* streamlined, streamline; **о. волной** pressing the wave, conducting the wave

**обтекание** *n.* passing around, flowing around; **о. полутела** semi-infinite body-vortex

**обтекатель** *m.* streamlined unit, streamlined housing, dome; deflector; shield; (*aero.*) fairing; **о. антенны** radar dome, blister, radome; **о. гидролокатора** sonar housing, sonodome, sonar dome; **носовой о.** bow dome

**обтекать, обтечь** *v.* to flow around, to by-pass, to circumvent

**обтекающий** *adj.* flowing around, circumvent, ambient

**обтереть** *pf. of* обтирать

**обтечь** *pf. of* обтекать

**обтирать, обтереть** *v.* to wipe, to wipe over, to rub round; to rub away, to wear away

**обтирка** *f.* wiper; wiping, rubbing

**обтирочный** *adj.* cleaning

**обточенный** *adj.* turned, machined, rounded off

**обточить** *pf. of* обтачивать

**обточка** *f.* facing, dressing; turning, rounding off, machining; **о. на станке** facing

**обтюратор** *m.* obturator, seal, cut-off; shield, baffle plate; diaphragm, stop; shutter, tube shutter, flicker shutter

**обтюраторный** *adj.* obturator

**обтюрация** *f.* obturation, sealing; masking

**обтюрировать** *v.* to screen off, to dim, to stop down

**обугленный** *adj.* carbonized

**обуглероживание** *n.* carbonization

**обугливаемый** *adj.* carbonizable

**обугливание** *n.* carbonization

**обугливать, обуглить** *v.* to carbonize

**обугливающий** *adj.* carbonizing

**обуглить** *pf. of* обугливать

**обуславливающий** *adj.* determinant

**обусловить** *pf. of* обусловливать

**обусловленный** *adj.* stipulated, specified, dependent (on), conditional; explained, determined (by)

**обусловливаемый** *adj.* determined, caused

**обусловливать, обусловить** *v.* to make reservations, to make conditions, to specify, to stipulate; to determine, to cause, to effect

**обучающий** *adj.* teaching

**обучение** *n.* teaching, training, education, instruction; habituation, learning

**обхват** *m.* circumference, volume, compass; perimeter

**обхватывать, обхватить** *v.* to embrace, to clasp, to seize, to catch, to envelope, to surround, to girth

**обход** *m.* by-pass, by-passing; passing around; alternate path, alternate route; alternative trunking; diversion; patrolling; round; **включено обходом** in by-pass, in bridge; **о. вызова** ringing-current bypass; **о. полюсов** circuit of the poles (*math.*)

**обходить, обойти** *v.* to by-pass, to go round; to turn; to spread (all over); to patrol

**обходный** *adj.* by-pass; roundabout, alternate, indirect, circuitous

**обходчик** *m.* patrol lineman

**обхождение** *n.* conduct, ways, dealings

**обшивание** *see* обшивка

**обшивать, обшить** *v.* to encase, to board, to cover; to face, to line, to sheath

**обшивка** *f.* sheath, sheathing, sheeting, lining, facing, covering, casing; paneling, planking, boarding, trimming; plating; poling

**обшивочный** *adj.* facing, covering, sheathing, lining, coating

**обшить** *pf. of* обшивать

**обще-** *prefix* general, generally, widely

**общедоступность** *f.* accessibility

**общедоступный** *adj.* accessible

**общеизвестный** *adj.* popular, well-know, reputed

**общепринятый** *adj.* generally accepted, universal, standard, conventional

**общепромышленный** *adj.* general-purpose

**общерекурсивный** *adj.* general recursive

**общественный** *adj.* social, public, common

**общеупотребительный** *adj.* of general use, customary

**общеязыковый** *adj.* nontechnical

**общий** *adj.* general, common; total, aggregate, joint, mutual; miscellaneous; net; average; gross

**общипать** *pf. of* **щипать**

**общность** *f.* generality, totality; joint, community

**объединение** *n.* merging, grouping, combination, interconnection, union, joining; **о. рабочих мест (кнопкой)** grouping of positions (with a key) (*tphny.*)

**объединённый** *adj.* united, joint, integrated, integral, interconnected; associate; **с объединёнными эмиттерами** emitter-coupled

**объединять, объединить** *v.* to unite, to combine, to assemble, to join, to pack, to unify, to integrate, to merge

**объединяющий** *adj.* interconnecting; interconnection

**объездной** *adj.* patrol

**объект** *m.* object, member, item; objective, target; entity, substance; **объекты связи** communication facilities, communication installations; **о. действия** operand; **контрастный о.** high-contrast scene

**объектив** *m.* objective; lens, objective lens; **о. с большой светосилой** high-power objective; **светосильный о.** fast lens, high-aperture lens

**объективность** *f.* objectivity

**объективный** *adj.* objective, unbiased; physical

**объём** *m.* volume, circumference, size, dimension, capacity, bulk, space, contents; extent, amplitude, compass; amount; extension; **связанный о.** covolume; **о. станционного оборудования** amount of switching equipment (*tphny.*)

**объёмистый** *adj.* voluminous, bulky

**объёмлющая** *f. adj. decl.* envelope

**объёмлющий** *adj.* enveloping, convolute

**объёмно** *adv.* volumetrically, in volume

**объёмность** *f.* voluminosity; **о. звучания** stereophonism; **о. звучания при двух источниках** (*or* **каналах**) **воспроизведения** diphonic stereophonism

**объёмно-управляемый** *adj.* volume-controlled

**объёмноцентрированный** *adj.* body-centered

**объёмный** *adj.* volume, volumetric, bulk; three-dimensional

**объёмометр** *m.* volumometer

**объявить** *pf. of* **объявлять**

**объявление** *n.* announcement, announcing

**объявленный** *adj.* announced, propagated

**объявлять, объявить** *v.* to announce, to advertise, to propagate; to declare, to state; to set, to signify

**объяснение** *n.* interpretation, explanation, comment

**объяснимый** *adj.* explicable

**объяснительный** *adj.* explanatory

**объяснять, объяснить** *v.* to explain, to interpret, to define, to state, to expound, to clear up, to elucidate, to illustrate

**обыкновенная** *f. adj. decl.* ordinary priority (telegram)

**обыкновенно** *adv.* ordinarily, usually, as a rule, habitually

**обыкновенный** *adj.* usual, normal, customary; ordinary, simple, plain

**обычно** *adv.* usually, generally

**обычный** *adj.* routine, usual, ordinary, normal, regular; ordinary, common, plain

**ОБЭ** *abbr.* (**относительная биологическая эффективность**) relative biological effectiveness, RBE

**обязанность** *f.* duty, obligation, obligement

**обязанный** *adj.* obliged, indebted

**обязательно** *adv.* without fail, certainly, surely

**обязательный** *adj.* obligatory, binding, compulsory

**овал** *m.* oval

**овальный** *adj.* oval, oval-shaped

**Оверхаузер** Overhauser

**овладевать, овладеть** *v.* to master; to seize

**овладение** *n.* mastering; seizing

**овладеть** *pf. of* **овладевать**

**огарок** *m.* cinder, ash

**огибание** *n.* diffraction; rounding; **искажённое о.** tracing distortion (*acoust.*)

**огибать, обогнуть** *v.* to round, to bend round, to turn; to envelope

**огибающая** *f. adj. decl.* envelope; enveloping

**огибающий** *adj.* enveloping; creeping

**ОГИЗ** *abbr.* (Объединение государственных издательств) Unified State Publishing House

**оглавить** *pf. of* **оглавлять**

**оглавление** *n.* index, table of contents

**оглавлять, оглавить** *v.* to index, to make an index

**оглавленный** *adj.* indexed

**оглушать, оглушить** *v.* to deafen, to stun

**оглушающий** *adj.* deafening

**оглушённый** *adj.* deafened, stunned

**оглушительный** *adj.* deafening, stunning

**оглушить** *pf. of* **оглушать**

**огне-** *prefix* fire, pyro-

**огнебезопасный** *adj.* fireproof

**огневидный** *adj.* fire-like, igneous

**огневой** *adj.* fire

**огнегаситель** *m.* fire extinguisher

**огнегасительный** *adj.* fire-extinguishing

**огнезадерживающий** *adj.* fire-retardant

**огнезащитный** *adj.* fireproof, fireproofing

**огнеопасность** *f.* fire hazard, inflammability

**огнеопасный** *adj.* inflammable

**огнепостоянный** *adj.* fire-resistant, incombustible, refractory

**огнестойкий** *adj.* fireproof, fire-resistant, heat-resistant, flame-proof, nonignitable

**огнестойкость** *f.* heat proof quality, resistance to fire

**огнестрельный** *adj.* firing

**огнетушитель** *f.* fire extinguisher

**огнетушительный** *adj.* fire-extinguishing

**огнеупорность** *f.* heat proof quality, fire-proofness, refractoriness, resistance to fire

**огнеупорный** *adj.* fireproof, refractory, non-ignitable

**оголённый** *adj.* bared, bare, exposed, uncovered

**оголовье** *n.* headband (*tphny.*)

**огонь** *m.* fire, flame; light, lamp; **входные и ограничительные огни** threshold lights; **рулёжные огни гидроаэродрома** taxi-channel lights; **бакбортный о.** port position light;

**вспышковый проблесковый о.** flashing light; **входной пограничный о.** approach light (*av.*); **глисадный о.** angle-of-approach light; **дополнительный о. на сигнале при тяжелом профиле** additional light by hard grade; **затмевающийся проблесковый о.** occulting light; **линейный о.** airway light; **о. приземления** contact light (*av.*); **проблесковый о.** rhythmic light; **ходовой о.** navigation light

**оградитель** *m.* guard, protector

**оградительный** *adj.* obstruction; protecting, guarding; enclosing

**ограждать, оградить** *v.* to enclose, to fence, to protect; to guard

**ограждающий** *adj.* enclosing; safety, security

**ограждение** *n.* guard, guarding, safety device, guardrail, fender, protection; barrier, enclosure, fence, fencing; locking

**огранённый** *adj.* faced

**ограничение** *n.* limiting, limitation, delimitation, restriction, termination, restricting, clamping, constraint; clipping, chopping, cutting; localization; threshold; **о. записанием анодного тока** plate-current cut-off clipping; **о. по объёму памяти** storage limitation; **о. разрешающей способности** resolution limiting

**ограниченно-линейный** *adj.* linear-limited, limited-linear

**ограниченный** *adj.* limited, bounded, bound, restricted, restrained, narrowed, confined, finite; clipped

**ограничивать, ограничить** *v.* to limit, to restrict, to restrain, to bound, to narrow, to define, to confine; to terminate; to gate; to clip

**ограничивающий** *adj.* confining, confined, restricted, limiting, clipping

**ограничитель** *m.* limiter, killer, clipper, clipper tube, threshold tube; restrictor, chopper, slicer, delimiter, debooster, suppressor; limiter circuit; **безынерционный о. пиков помех** instantaneous noise-peak limiter; **о. в частотной модуляции** cycle-by-cycle device; **о. выбросов** overshoot clipper; **о. выходного напряжения звуковой частоты** audio output limiter; **двухсторонний о.** clipper-limiter; **о. запросов** demand limiter, overinterrogation gate; **о. наклона** elevation stowing switch; **о. помех с отрицательной обратной связью** degenerative noise limiter; **регулируемый о. хода** adjustable stop, adjustable dog; **о. угла места** elevation stowing switch; **о. усилия** force limiting device; **о. шумов** noise suppressor, noise silencer, noise limiter

**ограничительный** *adj.* restrictive, restricting, limiting, arresting; limiter

**ограничить** *pf. of* **ограничивать**

**огрубевать, огрубеть** *v.* to roughen, to grow rough, to grow coarse

**огрубелый** *adj.* coarse

**огрубеть** *pf. of* **грубеть, огрубевать**

**огрубить** *pf. of* **огрублять**

**огрубление** *n.* desensitization

**огрублять, огрубить** *v.* to desensitize; to roughen

**одежда** *f.* clothing; insulation, jacket, lining; **о. пола** floor covering

**Одесский** Odessa

**одеяло** *n.* blanket, quilt

**один** *num.* one; single, only, alone; a, an

**одинаково** *adv.* equally, alike, in like manner

**одинаковый** *adj.* identical, same, the same, duplicate, equal

**одинарный** *adj.* single; single-play; **с одинарной точностью** single-precise

**одиннадцатиплоскостной** *adj.* hendecahedral

**одиннадцатиугольник** *m.* hendecagon

**одиннадцатиугольный** *adj.* hendecagonal

**одиннадцатиштырьковый** *adj.* eleven-pronged, magnal

**одинокий** *adj.* solitary, single, unique, only

**одиночный** *adj.* lone, single, individual, separate, self-contained; single-cell

**одиометр** *m.* odiometer

**одна, одно** *see* **один**

**одно-** *prefix* one, mono-, uni-, single

**одноадресный** *adj.* one-address, single-address

**одноаксиально-двоякопреломляющий** *adj.* uniaxially birefringent

**одноанодный** *adj.* single-anode

**одноатомный** *adj.* monatomic

**однобаковый** *adj.* single-tank

**однобарабанный** *adj.* single-drum

**однобатарейный** *adj.* single-battery

**одноблочный** *adj.* single-block

**одновалентность** *f.* univalence

**одновалентный** *adj.* univalent, monovalent

**одновальный** *adj.* single-shaft

**одновариантный** *adj.* univariant, monovariant

**одновибратор** *m.* single flip-flop oscillator, kipp oscillator, monovibrator, univibrator

**одновинтовой** *adj.* single-screw

**одновитковый** *adj.* single-turn, single-coil

**одноволновой** *adj.* single-wave

**одновременно** *adv.* simultaneously, at the same time, coincidently

**одновременность** *f.* synchronism, coincidence, simultaneousness, time-coincidence; **о. поступления вызовов** coincidence of calls, density of traffic (*tphny.*)

**одновременный** *adj.* simultaenous, concurrent, synchronous, isochronous

**одногнёздный** *adj.* one-celled, unicellular

**одногрупповой** *adj.* one-group

**однодекадный** *adj.* single-decade, one-digit decimal

**однодетекторный** *adj.* single-detector

**однодиапазонный** *adj.* single-band, single-range

**однодисковый** *adj.* single-disk

**однодиффузорный** *adj.* single-diffuse

**однодоменный** *adj.* single-domain

**однодорожечный** *adj.* single-track

**однодуантный** *adj.* one-dee

**одноёмкостный** *adj.* single-capacity

**одножелобчатый** *adj.* single-groove

**одножидкостный** *adj.* one-fluid

**одножильный** *adj.* single-core, one-cable

**однозазорный** *adj.* single-gap

**однозаконченный** *adj.* singly terminated

**однозамещённый** *adj.* monosubstituted

**однозарядный** *adj.* single-charged; single-level

**однозаходный** *adj.* single-cut (screw); unifilar

**однозвенный** *adj.* single-mesh, single-section

**однозвучный** *adj.* monotonous

**однозначащий** *adj.* synonymous, identical

**однозначность** *f.* univocacy; **о. направления пеленга** sense (*nav.*)

**однозначный** *adj.* unique, single-valued; simple, clear, univocal, unequivocal, unambiguous, well-defined; unilateral; one-figure; one-digit **взаимно о.** one-to-one, one-one

**однозондовый** *adj.* single-probe

**одноимённополюсный** *adj.* homopolar

**одноимённость** *f.* equivalence, material equivalence

**одноимённый** *adj.* analogous, like, homologous, similar

**одноимпульсный** *adj.* monopulse

**однокадровый** *adj.* single-frame

**однокалиберный** *adj.* of the same caliber

**однокамерный** *adj.* single-camera; single-chambered, single-barreled

**одноканальный** *adj.* single-channel, single-duct

**однокапсюльный** *adj.* single-button

**однокаркасный** *adj.* monoskeleton

**однокаскадный** *adj.* single-stage, one-stage, single-step

**однокатушечный** *adj.* monocoil, single-coil, single-spool, single-capacitor

**однокачественный** *adj.* isomorphic

**одноквадрантный** *adj.* one-quadrant

**одноквантовый** *adj.* one-quantum

**однокерновый** *adj.* unipivot

**одноклапанный** *adj.* single-valved, one-valve

**одноклеточный** *adj.* one-celled, unicellular

**одноклиномерный** *adj.* monoclinic

**однокнопочный** *adj.* single-button, single-dial

**одноколейный** *adj.* single-track, one-gage

**одноколенчатый** *adj.* single-jointed, one-jointed

**одноколонный** *adj.* single-column

**однокольчатый** *adj.* monocyclic

**однокомпонентный** *adj.* single-component, monocomponent

**одноконечный** *adj.* single-ended

**одноконтактный** *adj.* single-contact

**одноконтурный** *adj.* single-circuit, one-circuit, single-loop

**однокоординатный** *adj.* one-coordinate

**однокорпусный** *adj.* single-unit

**однокрасочный** *adj.* monochromatic

**однократнозамкнутый** *adj.* single re-entrant

**однократный** *adj.* single, single-stage, simplex, unitary, one-; **однократного действия** single-acting

**однокристалловый, однокристальный** *adj.* single-crystal

**однокрылый** *adj.* single-blade

**одноламповик** *m.* one-tube receiver

**одноламповый** *adj.* single-tube, one-tube

**однолепестковый** *adj.* single-lobe

**однолинейный** *adj.* unilinear, one-line

**однолопастный** *adj.* unilobed, single-lobed; single-blade

**однолучевой** *adj.* single-beam

**одномачтовый** *adj.* single-tower

**одномашинный** *adj.* single-machine

**одномегагерцный** *adj.* one-megacycle

**одномерный** *adj.* one-dimensional, uni-dimensional, univariate

**одноместный** *adj.* one-position, one-place

**одномодовый** *adj.* unimodal, unimode

**одномолекулярный** *adj.* monomolecular

**одномоторный** *adj.* single-motored, single-motor

**однонаправленный** *adj.* unidirectional, oneway, irreversible, unicursal; unilateral; uniaxial

**однонитный, однониточный** *adj.* unifilar, single-thread; single-rail

**однообмоточный** *adj.* single-coil, single-winding

**однооборотный** *adj.* single-revolution; single-cut, single-thread

**однообразие** *n.* uniformity, similarity; monotony

**однообразный** *adj.* uniform, homogeneous, identical, alike, monotonous; monotonic (function)

**однообратный** *adj.* single-revolution

**одно-однозначный** *adj.* one-to-one

**однооскольковый** *adj.* single-chip

**однооосновный** *adj.* monobasic

**однооосность** *f.* uniaxiality

**однооосный** *adj.* uniaxial, mono-axial

**однооотверстный** *adj.* single-duct

**однооотсчётный** *adj.* single-reading

**однопальцевой** *adj.* finger-tip (control)

**однопеременный** *adj.* one-variable; unit-distance

**однопереходный** *adj.* unijunction, single-junction

**однопериодный** *adj.* single-shot

**однопламенный** *adj.* single-flame

**одноплечный** *adj.* one-armed

**одноповерхностный** *adj.* single-surface

**однопозиционный** *adj.* single-position; monostatic

**однополосный** *adj.* single-band; single-channel; single-sideband

**однополостный** *adj.* single-cavity

**однополупериодный** *adj.* half-wave, one-half period

**однополюсник** *m.* unipole

**однополюсность** *f.* unipolarity

**однополюсный** *adj.* unipolar, single-pole, monopolar, one-poled; single-throw (switch); line-to-ground

**однополярный** *adj.* unipolar, unidirectional

**однопотенциальный** *adj.* univoltage

**однопоточный** *adj.* single-flow

**однопредельный** *adj.* single-range

**однопреломляющий** *adj.* singly refracting

**одноприборный** *adj.* single-instrument

**однопроблесковый** *adj.* single-flash

**однопроводный, однопроволочный** *adj.* single-wire, one-line, unifilar

**однопрограммный** *adj.* one-program, spot-program

**однопрожекторный** *adj.* single-gun, single-projector

**однопроточный** *adj.* single-flow

**однопроходный** *adj.* single-pass

**однопутный** *adj.* single-track, one-way, single-line

**одноразовый** *adj.* single, one-time; single-action, single-shot, one-kick; single-swing; start-stop; expendable; nonrecoverable

**одноразрывный** *adj.* single-break; one-column

**одноразрядный** *adj.* one-column, one-digit; single-order

**однорельсовый** *adj.* single-rail, monorail

**однородность** *f.* homogeneity, uniformity, unity, similarity; **о. размерностей** dimensional homogeneity

**однородный** *adj.* homogeneous, uniform, similar, isotropic

**одноручечный** *adj.* one-handed; single-dial, single-knob

**однорядный** *adj.* single-row, one-row, one-range, one-tier; uniserial, unilinear, single

**односедельный** *adj.* single-seat

**односеточный** *adj.* single-net, single-grid; single-input

**односкоростный** *adj.* single-speed

**однослойный** *adj.* single-layer, one-layer, single-play, single-braid; single-wound

**односрезный** *adj.* single-shear

**одностворчатый** *adj.* univalve

**одностержневой** *adj.* single-bar, single-rod

**одностоечный** *adj.* single-column; open-side (machine)

**односторонний** *adj.* unilateral, one-sided, single-sided; single, single-ended; unidirectional, one-way; biased, prejudiced; double-track(ed) (*rr.*)

**односторонно** *adv.* unilaterally

**однострелочный, однострельчатый** *adj.* single-needle

**однострочный** *adj.* one-line

**одноступенный, одноступенчатый** *adj.* one-stage, single-stage, single-step, single-level; simple (process)

**односуставный** *adj.* single-jointed, one-jointed

**однотактный** *adj.* single-cycle, one-cycle, single-phase; single-shot

**однотарифный** *adj.* one-rate, single-fee

**однотипичный** *adj.* monotypic

**однотональный, однотонный** *adj.* single-tone, monophonic

**одноточечный** *adj.* single-point

**однотрубный** *adj.* single-column; single-funnel

**одноугольный** *adj.* one-angled

**одноударный** *adj.* single-stroke

**одноухий** *adj.* monaural

**однофазный** *adj.* monophase, uniphase, single-phase

**однофокусный** *adj.* single-focusing

**одноходовой** *adj.* single-throw, single-pass, one-pass, one-shot, simplex, straight-through; single-thread (screw); monostable

**одноцветный** *adj.* monochrome, monochromatic, unicolor, one-colored

**одноцентровый** *adj.* concentric

**одноцепный** *adj.* single-circuit; single-chain

**одноцилиндровый** *adj.* single-cylinder, monocylindrical; single-casing

**одноцокольный** *adj.* single-base, single-ended

**одночасовой** *adj.* one-hour, hourly

**одночастичный** *adj.* one-particle, single-particle

**одночастный** *adj.* one-part

**одночастотный** *adj.* single-frequency, monofrequent

одночлен *m.* monomial
одночленный *adj.* monomial
одношаговой *adj.* single-step
одношарнирный *adj.* single-joint, single-link; unipivot
одношкальный *adj.* single-scale
одношлейфный, одношлейфовый *adj.* single-stub
одношнурный, одношнуровый *adj.* single-cord
одношпиндельный *adj.* single-spindle
однощелевой *adj.* single-hole, single-slot
одноэлектронный *adj.* single-electron
одноэлементный *adj.* single-element, single-element type, single-unit
одноэтажный *adj.* single-tier, single-stage, single-deck, one-story
одноядерный *adj.* mononuclear
одноякорный *adj.* single-armature
одноярусный *adj.* single-tier, single-stage
одноячейковый *adj.* unicellular
одометр *m.* odometer, cyclometer
одориметр *m.* odorometer
одориметрия *f.* odorometry
одряхление *n.* aging
Оже Auger
оживальный *adj.* ogive
оживить *pf. of* оживлять
оживление *n.,* оживлённость *f.* resuscitation, regeneration, revival, vivification; animation; liveliness
оживлённый *adj.* revived, vitalized; animated, lively; brisk; о. током (current) excited
оживлять, оживить *v.* to activate, to animate, to enliven; to revive
ожидаемость *f.* expectancy
ожидаемый *adj.* expected, expectant
ожидание *n.* expectation, expectancy; waiting; anticipation; математическое о. universal mean, assembly mean
ожижитель *m.* refrigerator
озвучание, озвучение *see* озвучивание
озвученный *adj.* sonicated; sound
озвучивание *n.* scoring (process or result), sonication; distribution of sound; последующее о. postscoring
озвучивать, озвучить *v.* to wire for sound, to sonicate, to irradiate (with ultrasonic waves)
означать, означить *v.* to indicate, to denote; to mean, to stand (for)

озокерит *m.* ader-wax, mineral wax, lignite wax, ozocerite
озон *m.* ozone
озонометр *m.* ozonometer
озонометрический *adj.* ozonometric
озонометрия *f.* ozonometry
озоноскоп *m.* ozonoscope
озоносфера *f.* ozonosphere
ОИП *abbr.* (обратный истинный пеленг) true bearing
ОИРП *abbr.* (обратный истинный радиопеленг) reciprocal true radio bearing
ОИЯИ *abbr.* (Объединённый институт ядерных исследований) United Institute of Nuclear Research
оказываться, оказаться *v.* to prove to be; to appear, to occur; о. из to result from
окаймить *pf. of* окаймлять
окаймление *n.* fringe effect; bordering, edging, flange; световое о. halation, halo, halation ring
окаймлённый *adj.* edged
окаймлять, окаймить *v.* to flange, to edge, to border
окаймляющий *adj.* fringing
окалина *f.* scale, sinter, cinder
окантовать *pf. of* окантовывать
окантовка *f.* contour accentuation, edging, fringe effect, fringing, ending; передняя белая о. leading white
окантовывать, окантовать *v.* to frame
оканчивать, окончить *v.* to terminate, to end, to finish
окарина *f.* ocarine
ОКГ (оптический квантовый генератор) laser
окисел *m.* oxide
окисление *n.* oxidation
окислённый *adj.* oxidized, oxidated
окислитель *m.* oxidizer
окислить(ся) *pf. of* окислять(ся)
окислять, окислить *v.* to oxidize, to oxidate; to acidify; о. электрически to anodize
окисляться, окислиться *v.* to oxidize, to be oxidized; to become sour
окиснортутный *adj.* mercury-mercurous oxide
окисносеребряный *adj.* silver-oxide
окисный *adj.* oxide
окись *f.* oxide
окклюдирование *n.* occlusion
окклюдированный *adj.* occluded

**окклюдировать** *v.* to occlude

**окклюзия** *f.* occlusion

**окно** *n.* window; opening, rift, aperture; slot; window space

**ОКО** *abbr.* (**отметчик кругового обзора**) plan position indicator

**около** *prep. genit. and adv.* by, about, around; near, toward; approximately

**около-** *prefix* near, para-, peri-, circum-

**окологоризонтальный** *adj.* circumhorizontal

**околозвуковой** *adj.* transonic

**околозенитный** *adj.* circumzenithal

**околополюсный, околополярный** *adj.* circumpolar

**окольный** *adj.* oblique, indirect

**оконечник** *m.* terminator

**оконечность** *f.* end, tip, extremity

**оконечный** *adj.* terminal; end, end output, final, terminating

**окончание** *n.* termination, finishing, end, ending, completion, conclusion, consummation, closing; desinence; **чувствительное о.** sensory ending

**окончательный** *adj.* ultimate, terminal, final, finishing, closing, definitive, target

**оконченный** *adj.* finished, completed

**окончивающий** *adj.* terminating

**окончить** *pf. of* оканчивать

**окошко** *n.* window, aperture; **мало ослабляющее о.** low-absorption window

**ОКП** *abbr.* (**обратный компасный пеленг**) reciprocal compass bearing

**окр.** *abbr.* (**округ**) district

**ОКР** *abbr.* (**опытно-конструкторская работа**) development work, development effort

**окрасить** *pf. of* окрашивать

**окраска** *f.* paint, coat of paint; painting, color, tint, coloration

**окрашенный** *adj.* painted; dyed; colored, tinted

**окрашиваемость** *f.* colorability

**окрашиваемый** *adj.* colorable

**окрашивание** *n.* painting, staining; dyeing; coloring, coloration, tinting; tint, tinge

**окрашивать, окрасить** *v.* to color, to tint; to paint, to stain; to dye

**округление** *n.* rounding, round-off

**округлённость** *f.* roundness, rotundity

**округлённый** *adj.* rounded (off), round, round-off, spherical; blunt

**округлить** *pf. of* округлять

**округлый** *adj.* round, rounded, curved, rotund, spherical, orbicular

**округлять, округлить** *v.* to round, to round off

**окружать, окружить** *v.* to surround, to enclose, to compass, to ring; to encircle

**окружающий** *adj.* surrounding, encircling, circumfluent, circumjacent; fringing, peripheral; environmental; ambient

**окружение** *n.* environment, surroundings; background; enclosing, encircling, surrounding

**окружённый** *adj.* surrounded

**окружить** *pf. of* окружать

**окружной** *adj.* peripheral, circumferential, circuitous

**окружность** *f.* circumference, periphery, circle; circuit (*math.*); district, neighborhood; annulus; shell; **масштабные окружности дальности** range rings; **пилообразные масштабные окружности дальности** serrated range rings; **о. соприкасания** bevelling circle

**окружный** *adj.* circumferential, peripheral, circling; surrounding; district, neighboring

**оксалат** *m.* oxalat

**оксид** *m.* emissive material, oxide

**оксидирование** *n.* oxidizing, oxidation

**оксидированный** *adj.* oxidized, oxide-coated

**оксидировать** *v.* to oxidize

**оксидировка** *f.* oxidizing, oxidation; oxide coating

**оксидный** *adj.* oxide; oxide-coated

**оксиферовый** *adj.* ferrocart

**оксихлоридный** *adj.* oxychloride

**октава** *f.* octave

**октавный** *adj.* octave; octave-band

**октальный** *adj.* octal, scale-of-eight

**октанальный** *adj.* octanal

**октант** *m.* octant

**октантальный, октантный** *adj.* octantal

**октаэдр** *m.* octahedron

**октаэдрический** *adj.* octahedral

**октод** *m.* octode

**октоплексный** *adj.* octuplex

**ОКУ** *abbr.* (**оптический квантовый усилитель**) laser

**окуляр** *m.* eyepiece, ocular; eyeglass

**олеометр** *m.* oleometer

**олеорефрактометр** *m.* oleorefractometer

**олифа** *f.* varnish, drying oil

**Оллендорф** Ollendorf

**олово** *n.* tin, Sn; **твёрдое о.** pewter

**оловянный** *adj.* tin, stannic

**олометр** *m.* holometer, altitude gage

**олотропный** *adj.* aeolotropic

**ольфактометр** *m.* olfactometer

**ольфактометрия** *f.* olfactometry

**ом** *m.* ohm

**омад** *m.* ohmad

**омбилический** *adj.* umbilical (*math.*)

**омброграф** *m.* hyetograph, ombrograph, pluviograph

**омбpoметр** *m.* hyetometer, ombrometer, pluviometer

**омега** *f.* omega ($\omega$)

**омега-непротиворечность** *f.* omega-consistency

**омегатрон** *m.* omegatron

**омеднение** *n.* copper plating, coppering

**омеднённый** *adj.* copper-plated, copper-clad

**омеднять, омеднить** *v.* to copper plate, to copper, to clad with copper

**омический** *adj.* ohmic, ohm; low-resistance

**« омлак »** *m.* ohmlac

**омметр** *m.* ohmmeter

**омниграф** *m.* omnigraph

**омниматик** *m.* Omnimatic

**ОМО** *abbr.* (**отдел механо-сборных и сварочных работ**) department for mechanical assembly and welding

**омовский, омовый** *adj.* ohmic, ohm

**омограф** *m.* homograph

**омографически** *adv.* homographically

**омография** *f.* homography

**омонимичный** *adj.* ambiguous

**омонимия** *f.* ambiguity

**ОМП** *abbr.* 1 (**обратный магнитный пеленг**) reciprocal magnetic bearing; 2 (**оружие массового поражения**) weapon of mass destruction

**ОМРП** *abbr.* (**обратный магнитный радиопеленг**) reciprocal magnetic radio bearing

**ОН** *abbr.* (**основное направление**) main direction

**онгстрем** *m.* angstrom

**ондограф** *m.* ondograph

**ондоскоп** *m.* ondoscope

**ондулировать** *v.* to undulate

**ондулятор** *m.* undulator, telegraphic register, siphon-recorder

**онозот** *m.* onozote

**ономатика** *f.* onomatics

**онтология** *f.* ontology

**ООД** *abbr.* (**оконечное оборудование данных**) data terminal equipment

**ОП** *abbr.* 1 (**обратный пеленг**) reciprocal bearing; 2 (**огневая позиция**) launching site, firing position; 3 (**отдел перелётов**) air traffic control section

**опал** *m.* opal

**опалесценция** *f.* opalescence

**опаливание** *n.* burning, singeing

**опаловый** *adj.* opal, opaline, opalescent

**опалубка** *f.* poling (of a trench), poling boards, casing, lining, sheathing

**опасиметр** *m.* opacimeter

**опасность** *f.* hazard, danger, risk, emergency

**ОПБ** *abbr.* (**оптический прицел для бомбометания**) optical bombsight

**опера** *f.* opera

**операнд** *m.* operand (*math.*)

**оперативно-технический** *adj.* operating-technical

**оперативный** *adj.* operative, operating, operational, operation(s)

**оператор** *m.* operator; statement (in ALGOL lang.); telephone operator; camera man; **о. комплексной проводимости** admittance operator; **о. по измерению дальности** range operator; **о. по измерению пеленга** bearing operator; **о. полной проводимости** admittance operator; **о. ПУАЗО** director operator; **сложный о.** molecular statement; **о. сопровождения цели** target tracker

**операторный** *adj.* operational, operating

**операционный** *adj.* operation, operational, operating

**операция** *f.* operation, working, step; operator; **о. безусловного перехода** unconditional jump; **о. в накопителе, о. в памяти машины** memory operation; **о. взятия производной** derivation; **о. выборочного слияния** match-merge operation; **о. гашения переносов** carry clearing operation; **о. над двоичными числами** binary operation; **о. обращения** storage operation; **о. перехода** jump; **о. с значительной затратой времени**

time consuming operation; **о. условного перехода** conditional jump

**опережать, опередить** *v.* to outstrip, to lead, to advance, to leave behind; to anticipate

**опережающий** *adj.* leading

**опережение** *n.* lead, leading, outrunning, outstripping; advance, advancing; anticipation

**оперение** *n.* feathering, feathers, plumage

**опереть** *pf. of* **опирать**

**оперирование** *n.* handling, operation, operating, manipulation

**оперировать** *v.* to handle, to operate; to work, to do, to perform

**опёртый** *adj.* supported

**опечатываемый** *adj.* sealable

**опиливание** *n.* filing

**опиливать, опилить** *v.* to file

**опиловочный** *adj.* filing

**опирать, опереть** *v.* to lean, to prop, to rest; **о. поворотно** to pivot

**описание** *n.* description, characterization, report; declaration; **о. блока** block head

**описанный** *adj.* described; circumscribed

**описатель** *m.* declarator

**описательный** *adj.* descriptive

**описывать, описать** *v.* to describe, to depict, to report; to delinate; to circumscribe

**оплавка** *f.* glazing

**оплавление** *n.* glazing, fire-polishing (of glass); flash, flashing off; fusion, fusing

**оплавленный** *adj.* fused; clad

**оплата** *f.* rate; payment; **о. по времени** charging by time (*tphny.*); **о. по дальности** charging by distance (*tphny.*); **повременная о.** flat rate; **поразговорная о.** message rate

**оплаченный** *adj.* paid, postpaid

**оплачиваемый** *adj.* payable, paid, paying

**оплетать, оплести** *v.* to entwine, to braid

**оплетённый** *adj.* braided, covered, insulated

**оплётка** *f.* covering, sheathing, braid, braiding, sleeving; armour

**оплёточный** *adj.* braiding, covering

**оповеститель** *m.* reporter, annunciator; signal device

**оповестительный** *adj.* annunciator

**оповещать, оповестить** *v.* to inform, to notify

**оповещение** *n.* notification, reporting, warning; announcement

**опознавание** *n.* identification, identity, recognition; sensing; spotting; **о. «свой-чужой»** IFF, identification friend or foe

**опознаватель** *m.* recognizer

**опознавательный** *adj.* identifying, identification, recognition, authentication; marker

**опознавать, опознать** *v.* to identify, to recognize; to sense, to spot

**опознание** *n.* identification

**опознанный** *adj.* identified, recognized

**опознать** *pf. of* **опознавать**

**оползание** *n.* creep, creeping, sliding

**оползать, оползти** *v.* to creep, to slide, to slip

**опора** *f.* support, rest, mast, post, pillar, prop, carrier, bearer, bottom bearing, bearing, foot, footing; pole, standard; mounting, mount, bracket, fixture; frame; pin, rod; **вводная о.** office pole (*tphny.*); **о. для перехода** terminal pole; **массивная о.** mass-controlled support; **о. на катках** roller bearing; **о. на цапфах** pivot suspension; **натяжная о. на уклоне** strain take-off tower; **ножевая о.** steel-prism bearing; **о. с кронштейнами** bracket pole; **о. с оттяжкой** stayed pole, guyed pole; **угловая промежуточная о.** angle suspension tower

**опоражнивание** *n.* discharging, discharge, evacuation, emptying, draining, dumping

**опоражнивать, опорожнить** *v.* to discharge, to drain, to empty, to evacuate, to dump

**опорный** *adj.* backing, supporting, bearing, seating; reference, guide, index, key; pedestal

**опорожнить** *pf. of* **опоражнивать**

**опосредствованный** *adj.* mediate

**опоясывать, опоясать** *v.* to gird, to girdle, to encircle, to tie, to span

**опоясывающий** *adj.* encircling

**оппанол** *m.* Oppanol

**Оппенгеймер** Oppenheimer

**оппозитный** *adj.* opposite, contrary

**оппозитрон** *m.* oppositron

**оппозиция** *f.* opposition

**оправа** *f.* case, holder; mounting, setting; instrument case; mandrel; **о. для стекла** bezel

**оправка** *f.* holder, mandrel, arbor, mounting, chuck, jig, setting; form; **о. для растяжки сетки** grid stretcher

**опрашивание** *n.* interrogation

**опрашивать, опросить** *v.* to answer a call, to accept a call (*tphny*); to inquire, to interrogate; to request

**опрашивающий** *adj.* interrogating; interrogation

**определение** *n.* determining, determination, identification; (radio-)localization, finding, detection, location, spotting; definition; measurement, computation, reckoning, calculation, estimation; test, analysis; rating; ranging; modification (*lang.*); **о. дальности, о. дистанции** ranging, range finding; **о. коэффициента полезного действия по суммарным потерям** determination of efficiency by total losses; **о. места** position fixing, localization; **о. места повреждения** fault location; localization of transmission-line mismatch; **о. местонахождения, о. местоположения** location, fixing, localization, position finding, taking of bearings; **о. местоположения расчётным путём** dead reckoning; **о. направления по сверхвысокочастотным ориентирам** VHF direction finding; **предварительное о.** predetermination; **о. расстояния по отражённому звуку** sound ranging technique; **о. угла места** elevation position finding; **о. цели гидролокатором** asdic spotting; **о. числа повторений** repetition count test

**определённость** *f.* precision, definiteness, definitiveness; accuracy, exactitude

**определённый** *adj.* defined, definite, fixed, certain, specific, given; determined; distinct, concrete, absolute, sharp

**определимость** *f.* definability

**определимый** *adj.* definable, determinable

**определитель** *m.* determinant (*math.*); detector, finder, locator; key, guide, index; finding; recognizer; **о. номера** identifier, discriminator; **о. смеси** mixture analyzer

**определить(ся)** *pf. of* **определять(ся)**

**определяемый** *adj.* defined

**определять, определить** *v.* to determine, to define, to ascertain, to establish, to identify; to detect, to sense, to locate; to fix, to position, to specify, to state; to modify; to assign, to settle; **о. пеленг** to take bearing

**определяться, определиться** *v.* to be defined; to be determined; **о. по пеленгу и углу** to fix by bearing and angle

**определяющий** *adj.* determinant, determining, defining; identifying

**опрессованный** *adj.* molded, pressed; press-fitted, pressurized; **о. оболочкой** sheathed

**опрессовка** *f.* molding, pressing; pressurization; pressure test; **о. под давлением** injection molding

**опрессовочный** *adj.* molding

**опробование** *n.* test, testing, trying out; sampling

**опровержение** *n.* contradiction, disproof, refutation, denial

**опровержимый** *adj.* refutable

**опрокидывание** *n.* turning over, turnover, overthrow, tipping, tipping over, tilting, changeover, upsetting, tumbling; stalling (of a motor); reversal (of phase), inversion (of phase); commutation failure

**опрокидыватель** *m.* tipper, tippler, tipple, dumper, trip, tripper, inverter, reverser, tipover device

**опрокидывать, опрокинуть** *v.* to turn over, to overturn, to overthrow, to upset, to tip over, to tip, to trip, to tilt, to invert; to reverse

**опрокидываться, опрокинуться** *v.* to capsize, to tip over

**опрокидывающий** *adj.* tipping, tilting, changeover; dump, dumping; **опрокидывающая схема** flip-flop

**опрокидывающийся** *adj.* tipping, tilting, dumping, dump; tipover; flip-flop

**опрокинутый** *adj.* turned over, overturned, tipped, tilted; inverted, reversed

**опрокинуть(ся)** *pf. of* **опрокидывать (-ся)**

**опрос** *m.* interrogation, inquiry, challenge, request; **повторный о.** checkback, return question

**опросить** *pf. of* **опрашивать**

**опросно-вызывной** *adj.* listening-and-speaking

**опросно-переговорный** *adj.* answering-and-talking

**опросный** *adj.* answering; interrogatory, inquiry; speaking

**опросчик** *m.* interrogator; inquiry station

**опрысканный** *adj.* sprayed, sprinkled

**опрыскивание** *n.* spray, spraying, sprinkling

**опрыскиватель** *m.* sprayer, spray machine

**опрыскивать, опрыснуть** *v.* to spray, to sprinkle

**опрыскивающий** *adj.* spraying, spray

**опрыснутый** *adj.* sprayed

**опрыснуть** *pf. of* **опрыскивать**

**ОПС** *abbr.* (**одноперекидная схема**) single flip-flop

**оп. ст.** *abbr.* (**опытная станция**) experiment station

**опт.** *abbr.* 1 (**оптика**) optics; 2 (**оптический**) optical

**оптар** *m.* optar

**оптика** *f.* optics; **о. проекционного телевидения, о. Шмидта** reflective optics (*tv*)

**оптико-тепловой** *adj.* optitherm

**оптико-электронный** *adj.* opto-electronic

**оптимальный** *adj.* optimum, best, most favorable; optimal; optimized

**оптиметр** *m.* telescope caliper

**оптимизация** *f.* optimizing, optimization

**оптимизированный** *adj.* optimized

**оптимум** *m.* optimum, the best, ideal

**оптифон** *m.* optiphone

**оптически** *adv.* optically

**оптический** *adj.* optical, optic, visual

**оптовый** *adj.* wholesale

**оптометр** *m.* optometer

**оптотехника** *f.* technical optics

**оптотранзистор** *m.* optotransistor

**оптофон** *m.* optophone

**оптоэлектронный** *adj.* opto-electronic

**опуск** *m.* omission

**опускаемый** *adj.* dipping, dip; droppable

**опускание** *n.* lowering, letting down, dip, dipping, drop, dropping, sinking, settling, descent, plunging; deposition; omission; **допускающий о.** droppable

**опускать, опустить** *v.* to dip, to lower, to let down, to dip, to drop, to sink, to immerse; to suppress, to omit; to deposit

**опускаться, опуститься** *v.* to sink, to fall, descend, to drop, to subside

**опускающийся** *adj.* descending, dropping, lowering

**опускной** *adj.* drop, lowering

**опустить(ся)** *pf. of* **опускать(ся)**

**опустошение** *n.* emptying; release

**опущение** *n.* omission; lowering, letting down

**опущённый** *adj.* missing; omitted; lowered, dropped, depressed, sunk

**опыт** *m.* experiment, test, trial; experience, practice; **о. эксплоатации** operating experience

**опытно** *adv.* experimentally; expertly

**опытно-конструкторский** *adj.* developmental

**опытно-показательный** *adj.* experimental-demonstrative

**опытность** *f.* experience, proficiency

**опытный** *adj.* experienced, expert, competent; research, experimental, developmental; pilot (plant); empirical; production; trail

**ор** *abbr.* (**орудие**) 1. gun, cannon. 2. tool, instrument

**ораз** *abbr.* (**отдельная разведывательная авиационная эскадрилья**) detached reconnaissance air squadron

**оранжево-голубой** *adj.* orange-cyan

**оранжевый** *adj.* orange, orange-colored

**оранж-циан** *m.* orange-cyan

**орбита** *m.* orbit; race-track

**орбитально** *adv.* orbitally

**орбитальный** *adj.* orbital

**орбитер** *m.* orbiter

**орган** *m.* 1. element, unit; member, organ; instrument, tool, device. 2. organ (*mus.*); **органы чувств** sensory organs; **большой о.** grand organ; **индукторный о. настройки** inductuner; **кнопочный о. настройки** pushbutton tuner; **настроечный о., о. настройки** control device, control unit, tuner, tuning control; **о. подстройки частоты строк** horizontal hold control (*tv*); **регулируемый о. связи** variocoupler; **о. связи** coupler; **о. слуха** auditory system; **о. управления** control unit. controller

**организация** *f.* organization

**организм** *m.* organism, life

**огранизованный** *adj.* organized, arranged, managed

**организовывать, организовать** *v.* to organize, to arrange

**организующий** *adj.* organizational, organizing

**органически** *adv.* organically

**органический** *adj.* organic

**органный** *adj.* organ, organpipe

**оргасвязь** *f.* data transmission

**оргтехника** *f.* management technique

**ординальный** *adj.* ordinal

**ординарный** *adj.* ordinary, common; single, plain

**ордината** *f.* ordinate, axis of the ordinates, Y-axis

**ордир** *m.* ORDIR (omnidirectional digital radar)

**ордрат** *m.* Ordrat (ordnance dial recorder and translator)

**ореол** *m.* halation, halo, corona, bloom (-ing), aureole

**ореольный** *adj.* halation, halo

**орешковый** *adj.* egg, egg-shaped; nut

**оригинал** *m.* original, subject copy, master copy; **прессовый о.** master stamper, master

**оригинальный** *adj.* original; live; singular; source

**ориентация** *f.* orientation

**ориентир** *m.* orienting point, guiding line, description point; indicator; landmark, mark, reference point

**ориентирование** *n.* orientation

**ориентированный** *adj.* oriented, orientational; stabilized

**ориентировать** *v.* to orient, to orientate, to direct

**ориентироваться** *v.* to orientate, to get one's bearings

**ориентировка** *f.* orienting, orientation; finding

**ориентировочный** *adj.* directional, direction, course

**ориентирующий** *adj.* directing, orienting, directive

**ориентирующийся** *adj.* orienting

**ОРК** *abbr.* (**отсчёт радиокомпаса**) radio compass reading

**оркестр** *m.* orchestra, band; **духовой о.** brass band

**оркестровый** *adj.* orchestral, orchestra

**орнитоптер** *m.* ornithopter

**ОРП** *abbr.* (**обратный радиопеленг**) reciprocal radio bearing

**орт** *m.* crosscut, cross drift

**ортикон** *m.* orthicon; **о. с переносом** image orthicon

**ортоводород** *m.* ortho-hydrogen

**ортогелий** *m.* ortho-helium

**ортогонализация** *f.* orthogonalization

**ортогональность** *f.* orthogonality

**ортогональный** *adj.* orthogonal, right-angled

**ортографический** *adj.* orthographic

**ортодин** *m.* orthodyne

**ортодихлорбензол** *m.* orthodichlorobenzene

**ортодрома** *f.* orthodrome

**ортомагнитный** *adj.* orthomagnetic

**ортомод** *m.* Orthomode

**ортоник** *m.* orthonik

**ортоноль** *m.* orthonol

**ортонормальный** *adj.* orthonormal

**ортонормированный** *adj.* ortho-normalized

**ортоось** *f.* ortho-axis, ortho-diagonal

**орторадиоскопия** *f.* orthoradioscopy, orthodiagraphy

**орторомбический** *adj.* orthorhombic, rhombic, prismatic

**ортосиликат** *m.* orthosilicate

**ортоскоп** *m.* orthoscope

**ортоскопический** *adj.* orthoscopic

**ортоскопия** *f.* orthoscopy

**ортосоединение** *n.* ortho compound

**ортотелефонный** *adj.* orthotelephonic

**ортотест** *m.* measurer

**ортотрон** *m.* orthotron

**ортоферрит** *m.* orthoferrite

**ортофонический** *adj.* orthophonic

**ортохроматический** *adj.* orthochromatic, isochromatic

**ортоцентр** *m.* orthocenter

**ОРТП** *abbr.* (**ортодромический пеленг**) orthodromic bearing

**орудие** *n.* gun, cannon; instrument, tool; **совмещённое с радиолокатором зенитное о.** self-aiming AA gun

**орудийный, оружейный** *adj.* gun, armament

**оружие** *n.* weapon; arms

**осадитель** *m.* precipitant, precipitator

**осадительный** *adj.* precipitating, precipitation; settling

**осадить(ся)** *pf. of* **осаждать(ся)**

**осадка** *f.* sag, sagging; setting, settling, set; sinking, immersion

**осадок** *m.* sediment, deposition, deposit, residue; fallout; **о. в эле-**

**ментах** battery mud; **компактный о.** reguline; **контактный о.** strike (deposit)

**осадочный** *adj.* settling; sedimentary; precipitation; upsetting

**осаждаемость** *f.* precipitability

**осаждаемый** *adj.* precipitable

**осаждание** *n.* precipitation

**осаждать, осадить** *v.* to precipitate, to deposit, to set down, to settle

**осаждаться, осадиться** *v.* to precipitate, to settle, to be deposited, to fall

**осаждающий, осаждающийся** *adj.* precipitating, settling

**осаждение** *n.* deposition, settling, precipitation, precipitating; plating, coating, coat; deposit, sediment; **мгновенное о. металла** striking, deposition of a strike; **электролитическое о. с использованием подвижных катодов** mechanical electroplating

**осаждённый** *adj.* deposited, precipitated, settled; plated; sedimentary

**осаживание** *n.* bucking effect, pressing back; upsetting, jumping up; strain

**осведомлённость** *f.* expert knowledge, information

**осведомлённый** *adj.* expert, experienced, informed

**осветитель** *m.* lighter, illuminator

**осветительный** *adj.* lighting, illuminating, illumination

**осветить** *pf. of* **освещать**

**освещаемый** *adj.* lighted

**освещать, осветить** *v.* to light, to illuminate, to throw light upon, to irradiate; to expose (to light)

**освещение** *n.* light, lighting, illumination; exposure; **о. заливающим светом** floodlighting; **не слепящее о.** "black" lighting; **о. предметного поля** bright-field illumination

**освещённость** *f.* illumination, illuminance, illuminancy, illumination intensity; **о. ярких мест изображения** highlight illumination

**освещённый** *adj.* lighted, lit, illuminated

**освинцование** *n.* lead plating, lead lining

**освинцованный** *adj.* lead-covered

**освинцовывать, освинцовать** *v.* to lead-coat, to lead

**освободительный** *adj.* freeing, liberating

**освобождать, освободить** *v.* to release, to set free, to free, to disconnect, to clear, to disengage; to relieve; to loose, to loosen; to eliminate, to clear

**освобождающий** *adj.* releasing, setting free, liberating; clearing

**освобождение** *n.* clearing, disengagement, release, unblocking, uncaging; relief; retrapping

**освобождённый** *adj.* released, freed, liberated

**осевой** *adj.* axial; axle; (of a) beam

**оседаемость** *f.* precipitability

**оседаемый** *adj.* precipitable

**оседание** *n.* landing (of electrons); shrinking, shrinkage; sag, sagging, collapse; settling, precipitation; fallout; depression, subsiding, subsidence, sinking, lowering, settlement

**оседать, осесть** *v.* to sag, to yield; to set, to sink, to settle, to settle down, to subside

**осекаться, осечься** *v.* to fail, to refuse, to miss-fire

**осесимметричный** *adj.* axisymmetric, axially-symmetric

**осесть** *pf. of* **оседать**

**осечённый** *adj.* shortened, truncated

**осечься** *pf. of* **осекаться**

**ослабевание** *n.* weakening

**ослабевать, ослабеть** *v.* to weaken, to grow weak, to slacken, to relax, to fall off, to diminish, to abate, to fade away, to decline

**ослабевший** *adj.* weakened, weak, feeble

**ослабеть** *pf. of* **ослабевать**

**ослабитель** *m.* attenuator, attenuation network, pad, attenuating pad; reducer, isolator; resistance network; **о. в виде петли связи** loop-coupled-mode attenuator; **о. волн длиннее критической** beyond-cutoff attenuator; **о. измерения уровня звукового давления** sound-level meter attenuator; **о. на критической волне** cutoff attenuator; **нерегулируемый о.** pad, attenuator; **поршневой о. для поперечной магнитной волны ТМ** mode piston attenuator; **предельный о.** cut-off attenuator

**ослабить(ся)** *pf. of* **ослаблять(ся)**

**ослабление** *n.* weakening, slackening, slack, relaxation; loosening; reduction, decrease, loss, abatment, attenuation, decay; extinction; isolation; rejection; de-amplification; dilution; cable reduction-factor; **o. зеркального канала** image attenuation

**ослабленный** *adj.* weakened, decreased, reduced, attenuated; loose, lax, relaxed

**ослаблять, ослабить** *v.* to weaken; to reduce, to quench, to decrease, to attenuate, to fall, to abate; to relax, to slacken, to ease, to loosen

**ослабляться, ослабиться** *v.* to die away

**ослабляющий** *adj.* weakening; reducing; loosening

**ослепительный** *adj.* glaring, blinding, dazzling

**ослепить** *pf. of* **ослеплять**

**ослеплённый** *adj.* blinded, dazzled

**ослеплять, ослепить** *v.* to blind, to dazzle

**ослепляющий** *adj.* blinding, dazzling

**осмаливать, осмолить, осмолять** *v.* to pitch, to tar

**осматривать, осмотреть** *v.* to inspect, to survey, to examine, to look over

**осмиевый** *adj.* osmium, osmic

**осмий** *m.* osmium, Os

**осмоз** *m.* osmosis, osmose

**осмолённый** *adj.* pitched, tarred

**осмолить** *pf. of* **осмаливать**

**осмолка** *f.* tar coating, tarring

**осмолять** *pf. of* **осмаливать**

**осмометр** *m.* osmometer

**осмос** *m.* osmosis, osmose

**осмотически** *adv.* by osmosis

**осмотический** *adj.* osmotic

**осмотр** *m.* inspection, examination; maintenance; survey, search, review; **наружный о.** appearance test, visual inspection

**осмотреть** *pf. of* **осматривать**

**осмысленный** *adj.* meaningful, intelligent

**осн.** *abbr.* (**основанный**) based

**оснастить** *pf. of* **оснащать**

**оснастка** *f.* attachment(s), equipage, equipment, instrumentation, hardware, rigging; ropes, cordage, tackle; **о. столбов скобками** stepping of poles

**оснащать, оснастить** *v.* to equip, to rig, to fit out

**оснащение** *n.* instrumentation; **о.** (**контрольно-измерительными**) **приборами** instrumenting

**основа** *f.* base, basis, foundation; principle, element; framework, back, backing; origin, starting point; host, host crystal; stem; pedestal base, base plate; bottom, ground; **катодная о.** starting sheet; **о. крыла** (**семафора**) casting arm (of a semaphore)

**основание** *n.* foundation, base, basis, stand, body, mounting, pedestal, footing, foothold, bed; origin; principle; establishment, founding; radix, root; modulus; motive, reason; **о. внутренней системы счисления** internal number base; **о. пуговичной ножки** button header; **о. системы счисления** radix, number base; **специальное о.** nonclassical radix; **фольгированное о.** metal-clad base material; **о. шкалы** dial

**основанный** *adj.* based

**основать** *pf. of* **основывать**

**основной** *adj.* fundamental, basic, base, cardinal, principal, main, leading, major, primary, chief, essential, master; ultimate; elemental, dominant; side

**основный** *adj.* basic

**основывать, основать** *v.* to base; to construct, to establish, to found, to erect, to set up, to constitute

**особенно** *adv.* especially, specifically, particularly, unusually

**особенность** *f.* feature, property, characteristic, specialty, peculiarity, singularity

**особенный** *adj.* singular, peculiar, special, specific, particular

**особо** *adv.* extra; apart, separately; especially, particularly

**особочувствительный** *adj.* hypersensitive

**особый** *adj.* special, particular, specific; separate, singular

**ософон** *m.* osophone

**ОСП** *abbr.* (**оборудование слепой посадки**) instrument-landing equipment

**осреднение** *n.* averaging; **о. по времени** time-averaging operation

**осреднённый** *adj.* averaged

**осреднитель** *m.* averager

**осредняющий** *adj.* averaging

**ОСТ** *abbr.* (**общесоюзный стандарт**) national standard

**оставить** *pf. of* **оставлять**
**оставленный** *adj.* left, abandoned
**оставлять, оставить** *v.* to leave, to abandon, to desert
**оставшийся** *adj.* residual
**останавливать, остановить** *v.* to stop, to shut down, to put out of action, to lock, to discontinue, to close; to check, to arrest
**останавливаться, остановиться** *v.* to stop, to rest, to stay
**останавливающий** *adj.* stopping
**останов** *m.* caging; stop, break, arrester, safety gear; stopping, halting
**остановить(ся)** *pf. of* **останавливать (-ся)**
**остановка** *f.* stop, halt, cessation; stopping, arresting, shutdown; intermission, pause; interruption, disturbance; **о. счётчиков** nonmetering (*tphny.*)
**остановленный** *adj.* stopped, standing
**остановочный** *adj.* stop, stopping, check
**остаток** *m.* residue, remainder, remnant, surplus, rest, balance; excess; vestige; tail
**остаточный** *adj.* residual; remanent; rest, remainder; after-
**остающийся** *adj.* permanent, lasting, durable, stable; residual, remaining, remanent; staying, stopping
**Оствальд** Ostwald
**остеклованный** *adj.* vitrified, glazed, fused; vitreous; glass ambient
**остеофон** *m.* bone-conductor receiver
**остирование** *n.* standardization
**остировать** *v.* to standardize
**остов** *m.* frame, body, carcass, shell, casing; core; **о. для нагревательных элементов** element carrier
**остойчивость** *f.* stability (of a ship)
**острие** *n.* edge, point, cutting point; spike, peak, cusp; pivot; tracer; **контактное о.** test point needle, battery spear
**острийный** *adj.* point; peak
**остро** *adv.* sharply
**островной** *adj.* island, insular
**острогубцы** *pl.* nippers, cutting pliers, pincers, tapernose pliers
**остроконечный** *adj.* peaked; sharp, pointed, acute, tapered, tapering
**остролучевой** *adj.* pencil-beam
**остронаправленный** *adj.* highly-directional, high-directional, narrow

(-beamed), pointed; beamed, pencil-beam
**острота** *f.* sharpness; acuity, acuteness, fineness
**остроугольный** *adj.* acute-angled
**острофокусный** *adj.* sharp-focused
**острый** *adj.* sharp, acute, keen, fine, pointed, edged; peaked; narrow
**остряк** *m.* switch blade; point
**осушение** *n.* drainage; dehumidification
**осушённый** *adj.* drained; dried, dry
**осушитель** *m.* dehydrator, drier, desiccant
**осушительный** *adj.* drying, desiccating; draining
**осуществимость** *f.* feasibility, practicability
**осуществимый** *adj.* feasible, practicable, realizable
**осуществить** *pf. of* **осуществлять**
**осуществление** *n.* realization, realizing, accomplishment, actuation, actuality, achievement; **о. соединения абонентов** trunking scheme
**осуществлённый** *adj.* realized, accomplished
**осуществляемый** *adj.* realizable
**осуществлять, осуществить** *v.* to effect, to cause, to bring about, to realize, to carry out, to accomplish
**осуществляющий** *adj.* bringing about
**осциллатор** *see* **осциллятор**
**осциллирование** *n.* oscillation
**осциллировать** *v.* to oscillate
**осциллистор** *m.* oscillistor
**осциллограмма** *f.* oscillogram, oscilloscope record, oscillograph trace; cathode-ray tube trace; cyclogram; **о. дефектоскопа** reflectogram
**осциллограф** *m.* oscillograph; synchroscope; cyclograph; **контрольный о.** picture signal monitor, waveform monitor; **магнитоэлектрический о. с подвижным шлейфом** Duddell oscillograph; **многошлейфовый о.** multichannel recording oscillograph; **о. с выделением строки** line strobe monitor; **о. с магнитоэлектрическим петлевым вибратором** oscillograph with bifilar suspension **о. с электромагнитным вибратором** soft iron oscillograph; **тепловой о.** hot-wire oscillograph; **универсальный о.** multi-element oscillograph

осциллографический *adj.* oscillographic

осциллография *f.* oscillography

осциллометр *m.* oscillometer

осциллосинхроскоп *m.* oscillosynchroscope

осциллоскоп *m.* oscilloscope, ondoscope; двухлучевой о. double-trace oscilloscope; о. на лампе тлеющего разряда ondoscope; о. с выделением строки line-selector oscilloscope; о. с двумя линиями развёртки double-trace oscilloscope

осциллоскопия *f.* oscilloscopy

осциллятор *m.* oscillator; sinusoid(al) generator, wave generator; о. с потерями damped oscillator

осцилляторный *adj.* oscillator

осцилляция *f.* oscillation

ось *f.* axis; axle, spindle, pin, shaft, pivot; о. азимутального привода azimuth-drive shaft; ведущая о. capstan; о. вещественных частот real frequency axis; о. звукового канала axis of the sound channel, level of minimum sound velocity; о. зетов Z-axis; о. V кварца V-axis, mechanical axis of a quartz crystal; о. крутящего момента axis of torque; лёгкая о. axis of easy magnetization, easy direction; о. отметки времени, о. развёртки timing axis; трудная о., о. трудного намагничивания hard direction, axis of hard magnetization

осязаемость *f.* tangibility

осязаемый *adj.* tangible, tactile

осязание *n.* touch, feel, sense of touch, feeling, touching; tickle

осязательный *adj.* tactile, palpable, tangible; sensitive

осязать *v.* to touch, to feel

от *prep. genit.* from, out of, off, of; for

ОТ *abbr.* (огневая точка) firing point

отбелённый *adj.* bleached

отбелка *f.* bleaching

отбивание *n.* repulsion, repelling; beating off

отбираемый *adj.* withdrawing

отбирание *n.* withdrawal, taking away; sorting, selecting, picking

отбирать, отобрать *v.* to withdraw, to take away, to remove; to select, to pick, to sort, to single out, to choose

отбирающий *adj.* collecting

отблеск *m.* reflectoin, gleam

отбой *m.* release, clearing, ring-off, clearing signal; backsweep; repulsion, repulse, repelling; о. во время набора номера incompletely dialed call; двусторонний о. last-party ring-off, last-party release (*tphny.*); односторонний о. first-party release (*tphny.*)

отбойный *adj.* clearing, ring-off, release; repelling, recoil; supervisory

отбор *m.* selection, choice; separation, separating; picking up, sampling; taking, take-off, drain; выборочный о. образцов random sampling; о. и контроль шумов noise selection; о. импульсных посылок quantization, quantizing; о. образцов для испытания sampling, taking of samples; о. по скорости velocity sorting; о. проб наугад random sampling; о. тока drawing of current

отборник *m.* separator, sifter, sorter; sampler; (power) take-off; selection, choice; о. синхроимпульсов clipper, synchronizing separator, amplitude separator (*tv*)

отбракованный *adj.* rejected

отбраковка *f.* rejection

отбраковщик *m.* quality checker; scanner

отбрасываемый *adj.* rejected, discarded; ejected

отбрасывание *n.* discarding, rejection, throwing away, casting out; truncation; kick; spattering

отбрасывать, отбросить *v.* to throw away, to reject, to discard, to throw back; to reflect; to repel, to repulse; to spatter; to kick; to truncate

отброс *m.* deflection, maximum deflection; waste, residue

отброшенный *adj.* thrown away, discarded, rejected; repulsed

отбросить *pf. of* отбрасывать

отбросный *adj.* waste

отброшенный *see* отбросанный

отбытие *n.* departure

отведение *n.* leading away, removal, elimination, carrying off, drawing off

отведённый *adj.* removed, drained off, drawn off

отвердевание *n.* congealing, hardening, setting

отвердевать, отвердеть *v.* to harden, to grow hard, to congeal, to set

**отверделый** *adj.* hardened

**отвердеть** *pf. of* **отвердевать**

**отверждение** *n.* crosslinking

**отверстие** *n.* opening, hole, orifice, aperture, iris; perforation, port, mesh (of screen); mouth, vent, inlet, outlet, passage, access; gap, break; bore; signal hole; **выводное о.** fairlead; **о. для развёртки** exploring aperture (*tv*); **заборное о.** intake; **маркёрное о.** feed hole; **о. сетки** grid mesh; **угловое о. (антенны)** flare angle, half-power width (of an antenna)

**отвёртка** *f.* screw-driver

**отвес** *m.* plumb, plumb line, bob, perpendicular

**отвесность** *f.* perpendicularity, verticality, steepness

**отвесный** *adj.* plumb, vertical, perpendicular, upright

**отвести** *pf. of* **отводить**

**ответ** *m.* answer, reply, response

**ответвитель** *m.* coupler; **о. Бете** Bethe-hole coupler; **двуслойный направленный о.** back-to-back directional coupler; **направленный о. с противофазным возбуждением** reverse-coupling directional coupler; **направленный о. с регулируемой связью** Transvar coupler; **шлейфовый о.** branched-guide coupler

**ответвительный** *adj.* branching, distributing, tapping; distributor

**ответвить(ся)** *pf. of* **ответвлять(ся)**

**ответвление** *n.* offshoot, offset, branch, branching, tap, tapping, arm, parting, tap-off; spur line; derivation, derivation wire; shunt, shunting, by-pass; **абонентское о.** private connection, residence telephone; **о. звукового тракта** audio channel take-off connection; **о. на ответвлении** bridged tap; **передвижное о. от линии** wire-netting tapping

**ответвлённый** *adj.* branched, branch; tapped, tapped off; derived; shunt, shunted

**ответвлять, ответвить** *v.* to branch off, to tap, to derive, to take off, to tee off; to shunt

**ответвляться, ответвиться** *v.* to offshoot, to branch off, to branch out, to bifurcate

**ответвляющий** *adj.* branching; shunting

**ответить** *pf. of* **отвечать**

**ответный** *adj.* answer, response, reply, answering; return, repeating

**ответственный** *adj.* essential, critical; responsible

**ответчик** *m.* responder, echo enhancer, radar booster; transponder; **двухчастотный о.** crossband beacon; **радиолокационный о.** tracking beacon

**отвечать, ответить** *v.* to answer, to reply, to respond

**отвинтить** *pf. of* **отвинчивать**

**отвинченный** *adj.* unscrewed

**отвинчивать, отвинтить** *v.* to unscrew, to screw off, to take down, to dismantle

**отвлекать, отвлечь** *v.* to draw off, to withdraw; to divert to deflect, to turn aside

**отвлекающий** *adj.* distracting

**отвлечение** *n.* diversion, diverting, distraction; withdrawing; discharge, removal

**отвлечённый** *adj.* abstract, abstracted; removed; distracted

**отвлечь** *pf. of* **отвлекать**

**отвод** *m.* bend, elbow; tap, tapping, drain, outlet, by-pass, run-off; branch, branching, offset; drawing-off, take-off; removal, extraction, withdrawal, discharge; diversion; **о. в землю** drainage to ground, shunting off to ground; **о. сетки** grid return

**отводимый** *adj.* withdrawable; outgoing, exit

**отводитель** *m.* outlet; baffle

**отводить, отвести** *v.* to lead away, to lead off, to derive, to drain off; to remove, to divert, to deflect; to shunt; to discharge

**отводка** *f.* shifting device

**отводный** *adj.* drain(age), outlet; by-pass; branch

**отводящий** *adj.* deflecting, diverting; discharge, outlet; draining

**отвозка** *f.* haul

**отгибание** *n.* turning back; bending away, bending aside; diffraction; deflection

**отгонка** *f.* rectification; elimination, driving off

**отдаваемый** *adj.* output

**отдавать, отдать** *v.* to return, to deliver, to yield; to give away; to give back; to rebound, to reverberate, to recoil

**отдаваться, отдаться** *v.* to resound, to echo, to ring

**отдаление** *n.* distance, remoteness; removal

**отдалённый** *adj.* distant, remote, far, far out

**отданный** *adj.* returned; yielded, given off

**отдать(ся)** *pf. of* **отдавать(ся)**

**отдача** *f.* efficiency, output, performance, delivery, yield; emission, evolution (of heat); gain, return, giving back; recoil, kick, rebound, spring back; deflection; side tone; **о. излучателя, о. излучающей системы** source-system rating; **квантовая о. (фотоэлемента)** quantum efficiency (of a phototube); **лучеиспускательная о. экрана** screen radiant efficiency; **максимально возможная о. мощности** maximum available power gain; **о. по аноду** plate efficiency; **о. по напряжению** volt efficiency; **о. при постоянной подводимой мошности** constant-power available response; **световая о.** photometric radiant equivalent; **условная о. в процентах** percentage of available power efficiency

**отдел** *m.* department, division, section, branch; step, stage; class

**отделанный** *adj.* finished, trimmed, dressed

**отделать** *pf. of* **отделывать**

**отделение** *n.* separation, segregation, separating; compartment; section, branch, partition, department, division; evolution, emission; detaching, detachment

**отделённый** *adj.* isolated, separated, segregated, eliminated; detached

**отделившийся** *adj.* separated

**отделимость** *f.* separability

**отделимый** *adj.* separable; detachable

**отделитель** *m.* separator; eliminator; divider

**отделительный** *adj.* separating

**отделить** *pf. of* **отделять**

**отделка** *f.* finish, finishing, dressing, trimming; after-treatment; framing; **о. верхушки опоры** finial of a pole; **о. кромки** edging; **монолитная акустическая о. потолка** monolithic acoustical ceiling; **о. опоры** lagging

**отделочный** *adj.* finishing, trimming

**отделывать, отделать** *v.* to plane, to smooth; to finish, to trim, to dress, to clean; to align, to adjust; **о. резьбу** to chase (a screw thread)

**отдельный** *adj.* separate, discrete, individual; independent, isolated, single, detached; divided, partial

**отделяемый** *adj.* detachable

**отделять, отделить** *v.* to separate, to detach, to disjoin, to severe, to cut off, to disengage, to release; to isolate, to section out, to part, to sort, to eliminate, to segregate; to divide, to part

**отдулина** *f.* bulging, bulge

**отдушина** *f.* air vent, air hole, ventilator; hole, vent, vent hole

**отечественный** *adj.* home, native

**отжатие** *n.* release

**отжатый** *adj.* pulled, detached; pressed out, squeezed out

**отжечь** *pf. of* **отжигать**

**отжиг** *m.*, **отжигание** *n.* annealing, bake-out; **предварительный о.** pre-firing

**отжигательный** *adj.* annealing

**отжигать, отжечь** *v.* to anneal; to burn off; to roast

**отжимный** *adj.* pull

**отзвук** *m.* echo, repercussion, reverberation, reflected sound; response; audibility

**отзвучать** *v.* to echo, to resound

**отзыв** *m.* echo, report; response

**отзывающийся** *adj.* resonant

**отзывчивость** *f.* responsiveness; response, effect

**отзывчивый** *adj.* responsive

**ОТК** *abbr.* (**отдел технического контроля**) quality control section

**отказ** *m.* failure, fault, non-operation; rejection, refusal; cancellation of a subscriber's agreement; natural (*mus.*); **о. в работе** failure, breakdown; **конструкционный о.** design-error failure; **о. от разговора** cancellation of a call, withdrawal of a call (*tphny.*); **постепенный о.** degradation failure, gradual failure, deterioration failure; drift failure

**отказанный** *adj.* refused

**отказать(ся)** *pf. of* **отказывать(ся)**

**отказывать, отказать** *v.* to fail; to reject, to refuse

**отказываться, отказаться** *v.* to fail,

to get out of order, to break down, not to respond; to refuse

**откатка** *f.* haul, haulage

**откачанный** *adj.* evacuated; exhaust

**откачать** *pf. of* **откачивать**

**откачиваемый** *adj.* pumped

**откачивание** *n.* exhaustion

**откачивать, откачать** *v.* to pump out, to evacuate, to exhaust

**откачка** *f.* pumping (out), exhaustion, evacuation, exhaust, degassing; **о. на заварочно-откачном автомате** sealex exhaust; **о. на передвижных постах, о. на подвижных постах** trolley exhaust

**откачной** *adj.* exhaust, exhausting; pump

**откидной** *adj.* folding, collapsible; drop, flap, hinged, throw-over; tipping, dumping; reversible

**откладывать, отложить** *v.* to put aside, to lay aside, to set aside; to plot (curve), to lay off, to measure off; to postpone, to delay, to defer

**отклик** *m.* response

**отклонение** *n.* deviation, deflection, declination, excursion, diffraction, divergence, diversion, digression, aberration; departure, drift, discrepancy, variance, variation, error, fluctuation; derivation; oscillation, swing, sweep; distance; inclination, bending; clearing; **допустимое о. частоты** frequency tolerance; **о. на большой угол** wide-angle deflection; **о. на всю шкалу** fsd, full scale deflection; **остаточное о.** zero variation; **о. от номинала вниз** undershoot, undershooting, underswing; **о. от установленного значения** deviation from the index value; **о. по кадру** vertical deflection, field deflection, frame deflection; **о. по оси времени** time-base deflection

**отклонённый** *adj.* deflected, divergent, difracted, swept

**отклонитель** *m.* deflector

**отклонить(ся)** *pf. of* **отклонять(ся)**

**отклонять, отклонить** *v.* to deflect, to swing, to decline, to throw, to turn aside, to deviate, to divert; to alter; to difract

**отклоняться, отклониться** *v.* to diverge, to deviate, to decline, to deflect, to digress, to depart; to slant, to tilt

**отклоняющий** *adj.* deflecting, deflection, diverting, sweeping; diffracting

**отклоняющийся** *adj.* divergent, deviating

**отключаемый** *adj.* interrupting, breaking, detachable

**отключатель** *m.* circuit breaker, cut-out siwtch

**отключать, отключить** *v.* to open, to cut off; to cut out, to disconnect, to switch off, to detach, to isolate; to de-energize

**отключающий** *adj.* cut-off, tripping, breaking, interrupting; isolating

**отключение** *n.* cut-off, cutting-off, separation, disconnection, disconnecting, tripping, switching off, detachment, isolating; killing, de-energizing, powercut

**отключённый** *adj.* disconnected, switched off, dead, cut off; **о. от сети** off the line

**отключить** *pf. of* **отключать**

**отковка** *f.* swaging

**открутить** *pf. of* **откручивать**

**открутка** *f.* untwisting

**откруточный** *adj.* untwisting

**откручивать, открутить** *v.* to untwist, to untwine

**открываемый** *adj.* detectable

**открывание** *n.* opening

**открыватель** *m.* opener

**открывать, открыть** *v.* to open, to uncover, to reveal, to disclose, to discover, to detect; to turn on; to unlock, to clear

**откр.-закр.** *abbr.* (**открыто-закрыто**) open-closed

**открытие** *n.* opening, aperture, hole; discovery; detection

**открытый** *adj.* open, exposed; clear, accessible

**открыть** *pf. of* **открывать**

**отлагать, отложить** *v.* to deposit; to delay, to put off

**отладить** *pf. of* **отлаживать**

**отладка** *f.* check-out, debugging; **о. программы** program check-out

**отлаженный** *adj.* checked-out; running

**отлаживание** *n.* debugging, checking-out

**отлаживать, отладить** *v.* to debug

**отлакировать** *pf. of* **лакировать**

**отлив** *m.* reflux, return flow, discharge; ebb, ebb-tide; swell; hue, shade (of color)

**отливать, отлить** *v.* to cast, to found; to pour, to pour off

**отливка** *f.* casting, molding, founding; cast, ingot; pouring off

**отливной** *adj.* founding, casting; cast, molded, founded

**отливочный** *adj.* casting

**отлипание** *n.* peeling off, loosening

**отлипать, отлипнуть** *v.* to come loose, to come off, to peel off

**отлипающий** *adj.* peeling off, coming loose

**отлипнуть** *pf. of* **отлипать**

**отлитый** *adj.* cast, founded; poured off

**отлить** *pf. of* **отливать**

**отличать, отличить** *v.* to distinguish, to discern, to differ, to discriminate

**отличаться, отличиться** *v.* to be distinguished (by), to differ, to be characterized (by)

**отличающийся** *adj.* differing, (being) characterized (by); **о. по фазе** dephased

**отличие** *n.* distinction, difference; discrimination, discriminateness; contrast; nature, feature

**отличительный** *adj.* distinctive, distinguishing, characteristic, peculiar, special

**отличить(ся)** *pf. of* **отличать(ся)**

**отличный** *adj.* different (from), distinct; excellent, perfect

**отлог** *m.* inclination, slope

**отложение** *n.* deposit, deposition, sedimentation; postponing; sediment, precipitate, precipitation

**отложенный** *adj.* deferred, delayed, postponed, put off; deposited, precipitated

**отложить** *pf. of* **откладывать, отлагать**

**отматывать, отмотать** *v.* to wind off, to unwind

**отмена** *f.* knock down (of a coded transmission); cancelling, cancel, cancellation, abrogation, annulment, revocation; abolition

**отменённый** *adj.* cancelled, abrogated, annulled

**отменить** *pf. of* **отменять**

**отменный** *adj.* superior, excellent, exquisite; different

**отменять, отменить** *v.* to cancel, to revoke, to repeal, to annul, to abolish, to quash, to suppress

**отмер** *m.*, **отмеривание** *n.* measuring off

**отмеривать, отмерить** *v.* to measure off, to measure out, to mark off

**отмерять** = **отмеривать**

**отметина** *f.* mark, marking, notch, nick

**отметить(ся)** *pf. of* **отмечать(ся)**

**отметка** *f.* note, mark, marker, notch, nick; sign, indication, index, indice criterion; marking, indexing; label; blip, pip, radar blip; **о. времени** time marker; time marking, timing, clocking, **о. группы записи** end-of-file mark; **калибровочная о.** marker pip; **контрольная о.** reference mark; **о. масштаба, масштабная о.** range marker, range-marker pip; **о. механического нуля** mechanical zero; **о. о категории телеграммы, о. об особом виде телеграммы** prefix of a radiotelegram; **подвижная масштабная о.** movable range marker; **пусковая о.** trigger pip; **селекторная о.** blip, notch, pip

**отметчик** *m.* marker, indicator; **о. времени** time marker, timer; time-sweep unit; **вынесенный о. кругового обзора** plan position repeater; **о. кругового обзора** plan position indicator; **поверочный о. дальности** reference range marker

**отмечать, отметить** *v.* to mark, to note, to record, to register; to plot; to indicate; to take (readings)

**отмечаться, отметиться** *v.* to be denoted, to be annotated; to register

**отмечающий** *adj.* recording, registering, marking

**отмеченность** *f.* well-formedness; **грамматическая о.** grammaticality

**отмеченный** *adj.* marked, indicated, recorded, registered; plotted; **грамматически о.** grammatical

**отмотать** *pf. of* **отматывать**

**отмыкание** *n.* unlocking, opening

**отмыкать, отомкнуть** *v.* to unlock, to open

**ОТН** *abbr.* (**отделение технических наук**) Division of Technical Science

**отн. ед.** *abbr.* (**относительная единица**) relative unit

**отнесённый** *adj.* referred (to); carried away

**отнести(сь)** *pf. of* **относить(ся)**

**отникелировать** *pf. of* **никелировать**

**отнимать, отнять** *v.* to deduct, to take away, to take off; to carry away; to cut off

**относ** *m.* release range (*av.*); delivery; deviation

**относительно** *adv. and prep. genit.* relatively, comparatively; concerning, relative to, regarding, with regard to, with reference to

**относительность** *f.* relativity, relativeness

**относительный** *adj.* relative, comparative, reference

**относить, отнести** *v.* to remove, to carry away, to deliver, to take; to refer; to deviate; to attribute (to)

**относиться, отнестись** *v.* to refer, to report; to relate, to pertain, to concern

**относящийся** *adj.* concerning, pertaining; acting

**отношение** *n.* relation, relationship, connection; ratio, proportion; behavior, attitude; reference, rate; equation; **о. амплитуд несущих изображения и звука** picture-to-sound ratio; **о. вращающего момента к моменту инерции** torque-to-inertia ratio; **о. вращающего момента при номинальной мощности к весу подвижной части** torque-weight ratio; **о. времени движения к времени остановки** drive-rest ratio; **о. времени обратного хода по кадру к периоду кадровой развёртки** vertical retrace ratio; **о. выходного сигнала единицы к выходному сигналу при частичной выборке** one-output-to-partial-output ratio; **о. (выходного) сигнала единицы к сигналу помехи при частичной выборке** one-to-partial select ratio; **о. выходного тока к току на выходе** current transfer ratio; **о. длительности импульса к периоду следования** mark(-to)-space ratio; **о. длительности импульса к частоте повторения** pulse duration ratio; **о. единичного выходного сигнала к нулевому выходному сигналу** one-to-zero ratio; **о. единичного выходного сигнала к частично селективному выходному сигналу** one-to-partial-select ratio; **о. мощностей, несущих изображения и звуки** picture(-to)-sound ratio; **о. наиболь-**

**шей яркости к наименьшей** highlight-to-low-light ratio; **о. параметра возврата к параметру срабатывания (реле)** pick-up-to-drop-off ratio; **о. плеч рычага** lever transmission, leverage; **о. показателей преломления** relative refractive index; **о. прямой скорости кадра к обратной** vertical retrace ratio (*tv*); **о. прямой скорости (луча) к обратной** retrace ratio (*tv*, *rdr*.); **о. прямой скорости строк к обратной** horizontal retrace ratio; **о. размаха видеосигнала от белого к размаху сигнала от чёрного в объекте** white-to-black-amplitude range; **о. размаха (собственно) видеосигнала к размаху импульсов синхронизации** picture to synchronizing ratio; **о. распространённости** abundance ratio; **о. сигнал с помехой/помеха** signal-plus-noise-to-noise ratio; **о. сигнала единицы к сигналу нуля** one-to-zero ratio; **о. сопротивлений при гелиевой и комнатной температурах** residual resistance ratio; **о. частот модулирующей к несущей** modulation-frequency-to-carrier ratio; **о. частот основного и зеркального каналов** channel ratio (of frequencies); **о. числа витков вторичной обмотки к первичной** secondary-to-primary-turn ratio

**отнятие** *n.* elimination, removal, taking away

**отнятый** *adj.* removed, eliminated, taken away

**отнять** *pf. of* **отнимать**

**ото** *see* **от**

**отображать, отобразить** *v.* to reflect; to represent; to map (*math.*)

**отображающий** *adj.* reflecting; representing; configuration

**отображение** *n.* reflection; representation; image, echo-image; mapping (*math.*); **биполярное о.** bipolar transformation, bipolar mapping (*math.*); **обратное о.** reciprocal representation (*math.*)

**отображённый** *adj.* reflected

**отобразить** *pf. of* **отображать**

**отобранный** *adj.* selected, chosen

**отобрать** *pf. of* **отбирать**

**отогнутый** *adj.* bent-up, bent, bent-back

**отождествить** *pf. of* **отождествлять**

отождествление *n.* identification
отождествлять, отождествить *v.* to identify
отожжённый *adj.* annealed
отойти *pf. of* отходить
отолог *m.* otologist
отомкнутый *adj.* open, unlocked
отомкнуть *pf. of* отмыкать
отопительный *adj.* heating, heat
отопление *n.* heating
отопленный *adj.* heated
отослать *pf. of* отсылать
отофон *m.* otophone
отпадание *n.* dropout (of a relay), drop-back, fall-back, release, releasing, separation
отпадать, отпасть *v.* to fall off, to drop off, to release; to take out
отпадающий *adj.* falling, dropping, dropout
отпадение *n.* falling off, falling away, dropping
отпаечный *adj.* tipping, sealing; exhaust
отпаивать, отпаять *v.* to seal off; to unsolder
отпайка *f.* seal(ing) off, tip-off, tip, pip, tap, tapping; unsoldering
отпасть *pf. of* отпадать
отпаянный *adj.* sealed-off
отпаять *pf. of* отпаивать; о. (лампу) от насоса to seal off the pump
отпереть *pf. of* отпирать
отпертый *adj.* unblocked
отпечатание *n.* imprint, impression; printing
отпечатанный *adj.* imprinted, stamped; printed
отпечатать *pf. of* отпечатывать
отпечаток *m.* replica; print, impression, imprint, stamp, seal; пробный о. proof (*tgphy.*); о. резца bit replica
отпечатывание *see* отпечатание
отпечатывать, отпечатать *v.* to print, to imprint, to impress, to stamp
отпирание *n.* unlocking, unblocking, trigger, triggering, trigger action, firing; breakdown (of a thyratron); unblanking; о. импульсами на анод plate-pulse triggering; о. тиратрона conduction of a thyratron, triggering, firing, breakdown of thyratron
отпирать, отпереть *v.* to unlock, to open, to unblock, to trigger; to unbolt; to enable

отпирающий *adj.* unblanking, unblocking, unlocking; enabling; trigger
отполировать *pf. of* полировать
отправитель *m.* sender, transmitter
отправительный *adj.* transmitting, sending, transmission
отправить *pf. of* отправлять
отправка *f.* dispatch, dispatching, sending; setting
отправление *n.* transmitting, transmission; dispatching, departure, sending
отправленный *adj.* sent, dispatched
отправляемый *adj.* transmitted
отправлять, отправить *v.* to transmit, to send, to dispatch, to forward
отправной *adj.* sending; starting
отпрессованный *adj.* pressed
отпуск *m.* release; delivery; tempering, annealing
отпускание *n.* release, releasing, dropout
отпускать, отпустить *v.* to release, to let go; to quench (of metal); to loosen, to ease, to unfasten; to temper, to draw, to anneal
отпускающий *adj.* releasing
отпускной *adj.* releasing, release; tempering
отпустить *pf. of* отпускать
отпущенный *adj.* released
отработанный *adj.* used up, spent, exhausted, depleted; waste; foul
отработка *f.* performance; finishing (of work); о. аппаратуры final adjustment of equipment; о. волны netting
отравление *n.* poisoning
отравляющий *adj.* poisoning
отражаемость *f.* reflectivity
отражаемый *adj.* reflected
отражание *n.* bouncing
отражатель *m.* reflector; deflector, repeller; dish, dishpan; reflector plate; baffle; многорядный о. mattress reflector; равноизлучающий о. equal-energy dish; светоделительный о. dichroic reflector; «совковый о.» snow-shovel reflector
отражательно-детекторный *adj.* reflex-detector
отражательный *adj.* reflective, reflecting, reflection, reflex; reverberatory; baffle, deflecting; reflector
отражать, отразить *v.* to reflect, to reverberate; to repel, to repulse;

to echo, to re-echo; to rebound; to indicate

**отражаться, отразиться** *v.* to reflect, to reverberate; to be repulsed; to be reflected, to rebound, to impinge; to echo

**отражающий** *adj.* reflecting, reverberatory; deflecting; echo(ing)

**отражение** *n.* reflectance, reflection, reverberation; echo, echoing; image; return, rebound, impingement; hop (of a wave); repulsion; reradiation; **катадиоптрическое о.** reflex reflection, retroreflection; **о. от грозовых образований** precipitations echo; **о. от дна** bottom bounce, bottom reflection; **о. от местности** terrain echo; **о. от местных предметов** clutter, ground-clutter return; **последовательное зеркальное о.** specular reflection

**отражённый** *adj.* reflected, reverberated, echoed; indirect

**отразитель** *m.* reflector

**отразить(ся)** *pf. of* **отражать(ся)**

**отрасль** *f.* branch

**отрастить** *pf. of* **отращивать**

**отращивание** *n.* tapping

**отращивать, отрастить** *v.* to tap, to branch off

**отрегулирование** *n.* adjustment

**отрегулированный** *adj.* adjusted, set

**отрегулировать** *pf. of* **регулировать**

**отрез** *m.* section, stretch; piece; cut

**отрезанный** *adj.* cut, cut off, severed

**отрезать** *pf. of* **отрезывать, отрезать**

**отрезать** = **отрезывать**

**отрезок** *m.* length, piece; section; remnant; segment; distance, stretch, tract

**отрезывать, отрезать** *v.* to cut, to clip, to cut off

**отремонтировать** *v.* to patch up

**отрицание** *n.* negation; negative; non-; **о. дизъюнкции** nondisjunction; **о. конъюнкции** nonconjunction

**отрицательно** *adv.* negatively; adversely

**отрицательно-ионный** *adj.* negative-ion

**отрицательный** *adj.* negative; bad, deleterious, unfavorable

**отросток** *m.* tap, tapping, spur, branch piece; extension; outlet; **боковой о. рукоятки** tail lever; **о. линии** spur line, "T"-line

**отруб** *m.* butt

**отрыв** *m.*, **отрывание** *n.* detachment, removal; break, breaking away, break-off, separation

**отрывистость** *f.* abruptness, suddenness

**отрывистый** *adj.* abrupt, sudden

**отрывной** *adj.* loose; detachable; (contact-)breaking

**отрывок** *m.* piece, fragment

**отрывочный** *adj.* interrupted; fragmental

**ОТС** *abbr.* (**оконечная телефонная станция**) terminal community office

**Отс** Oates

**отсасывать, отсосать** *v.* to suck away, to draw off; to suck out; to filter by suction

**отсасывающий** *adj.* outgoing, outlet; suction, sucking; absorption; draining, absorbing; trap

**отсвечивание** *n.* brilliancy, glare; reflection

**отсев** *m.* sift, sifting; selection

**отсек** *m.* compartment; cabinet, cubicle, part, section, bay; **о. модулятора** modulator section, bay; **о. радиооборудования** radio bay; **о. телеметрического оборудования** telemetry bay

**отсекание** *n.* cutting off, interception; cleaving, splitting off

**отсекать, отсечь** *v.* to cut off, to sever, to chop off, to clip, to detach; to intercept

**отсекающий** *adj.* pick-off, intercepting, cut-off, cutting-off

**отсечение** *see* **отсечка**

**отсечённый** *adj.* cut off, intercepted; closed; split off

**отсечка** *f.* cutoff, closing; clipping; interception, cutting off; margin; pinch-off

**отсечь** *pf. of* **отсекать**

**отскакивание** *n.* rebound, kick, recoil, ricochet, jumping back, bounce; breaking away, coming off

**отскакивать, отскочить** *v.* to bounce, to spring back, to jump back, to kick, to recoil; to break away, to come off

**отскок** *see* **отскакивание**

**отскочить** *pf. of* **отскакивать**

**отслаивание** *n.* peeling-off, peeling; stripping, separation

**отслаиваться, отслоиться** *v.* to scale

(off), to peel off, to flake

**отслоение** *see* **отслаивание**

**отслоиться** *pf. of* **отслаиваться**

**отсоединение** *n.* disconnecting, disconnection, breaking (away), separation, isolation, cutting, switching off

**отсоединённый** *adj.* separated, disconnected; floating

**отсоединять, отсоединить** *v.* to cut, to cut off, to switch off, to disconnect; to isolate

**отсоединяющий** *adj.* isolation

**отсортированный** *adj.* selected

**отсос** *m.* drain; pulling

**отсосанный** *adj.* sucked off, drawn off

**отсосать** *pf. of* **отсасывать**

**отсроченный** *adj.* delayed

**отсрочивать, отсрочить** *v.* to defer, to delay, to put off, to retard, to suspend, to postpone, to prolong

**отсрочка** *f.* suspension, postponement, delay, deferment

**отставание** *n.* lag, lagging, creep; peeling off, detachment; delay, retardation; hysteresis; **о. в переходном процессе** transfer lag; **о. по времени** time-delay, time lag; **о. по скорости** velocity lag

**отставать, отстать** *v.* to lag, to fall behind, to delay, to retard; to peel; to creep

**отстаивание** *n.* deposition

**отстать** *pf. of* **отставать**

**отстающий** *adj.* lagging, retarding, slow, late; **о. на 90°** quadrature-lagging

**отстойник** *m.* thickener; settling tank, settler

**отстой** *m.* deposit, residue, sediment

**отстойный** *adj.* settling

**отстояние** *n.* clearance, distance, space, interval

**отстраивать, отстроить** *v.* to tune out, to tune away from

**отстраиваться, отстроиться** *v.* to tune away from, to tune out from

**отстроить(ся)** *pf. of* **отстраивать(ся)**

**отстройка** *f.* tuning out, detuning, mistuning; **о. от смежного канала** adjacent channel selectivity

**отступающий** *adj.* retrograde, retrogressive, receding; divergent

**отступление** *n.* deviation, shift, departure, divergence, variation; falling back, regression, retrogression, recession

**отсутствие** *n.* nonexistence, absence, lack; shortage, deficiency; **о. переноса** not-carry; **о. связанности** incoherence

**отсутствующий** *adj.* absent

**отсчёт** *m.* reading, indication; metering, registering, registration; count; read-out; **о. времени** timing, time reading; **многократный о.** multimetering, multiple registration; **о. напряжения** voltage sample; **о. обратным током** reverse battery metering (*tphny.*); **о. перенапряжением** booster battery metering (*tphny.*); **о. переполюсовкой батареи** reverse battery metering (*tphny.*); **о. по дальности** range reading; **о. по зонам и времени** zone and overtime registration (*tphny.*); **о. по планшету** plot reading; **о. повышением напряжения** booster battery metering (*tphny.*); **о. продолжительности** time metering

**отсчётный** *adj.* reading; count; metering

**отсчитанный** *adj.* read off; counted off

**отсчитать** *pf. of* **отсчитывать**

**отсчитываемый** *adj.* counted-off, read-off

**отсчитывание** *n.* reading, reading off; counting (off), reckoning

**отсчитывать, отсчитать** *v.* to read, to read off, to take a reading; to count off, to reckon

**отсылать, отослать** *v.* to refer to; to forward (to), to transmit, to relay over, to convey, to transport; to send off; **о. за информацией** to refer to information

**отсылающий** *adj.* sending (off), dispatching; refering back

**отсылка** *f.* dispatch, dispatching; sending off; reference back

**оттаивание** *n.* thawing, thawing out, defrosting

**отталкивание** *n.* repulsion, repelling

**отталкиватель** *m.* repeller

**отталкивать, оттолкать** *v.* to repel, to push back, to thrust away, to drive back

**отталкивающий** *adj.* repelling, pushing back; repulsive; cracking-off

**оттенение** *n.* adumbration

**оттенивать, оттенить** *v.* to shade, to shadow, to tint, to blend

**оттенок** *m.* shade, hue, tint, tinge, tone, contrast; **о. цвета** tint

**оттенять** *see* **оттенивать**

**оттиск** *m.* pressing; mold; impress, print, copy; mark; stamp, impression, imprint; printing

**оттолкать** *pf. of* **отталкивать**

**оттормаживающий** *adj.* lifting, releasing (the brake)

**отточенный** *adj.* edged, pointed, sharpened, sharp

**оттягивание** *n.* drawing off; delaying, prolonging

**оттягивать, оттянуть** *v.* to retract, to pull back; to stay, to tie, to span, to tension, to anchor, to guy; to stretch, to extend, to expand

**оттягивающий** *adj.* drawing out; retractile; restraining

**оттяжка** *f.* guy, boom guy, stay, anchor rope, stay wire; anchoring, staying, guying; rigging, span; drawing out, pointing (of wire); **о. к кольцу** (*or* **столбу**) stub guy; **о. пересекающая дорогу** over-road stay; **продольная о.** head guy

**оттяжной** *adj.* guy, stay; pull, pull-off, drawing out

**оттянутый** *adj.* drawn (out)

**оттянуть** *pf. of* **оттягивать**

**отфильтрованный** *adj.* filtered

**отфильтровать** *pf. of* **отфильтровывать**

**отфильтровка** *f.* filtering

**отфильтровывать, отфильтровать** *v.* to filter, to filter off, to filter out

**отформованный** *adj.* molded

**отфрезеровать** *pf. of* **фрезеровать**

**отход** *m.* drop-out (of a relay); tailing(s); waste; departure, start; removal, withdrawal

**отходить, отойти** *v.* to fall back; to depart, to leave, to vacate, to abandon

**отходящий** *adj.* outgoing, exit; waste; output

**отцепить** *pf. of* **отцеплять**

**отцепка** *f.* uncoupling

**отцепляемый** *adj.* detachable

**отцеплять, отцепить** *v.* to detach, to release, to disengage, to uncouple, to disconnect, to unhook, to unhitch

**отчеканить** *pf. of* **чеканить**

**отчёт** *m.* account; report, record; reference

**отчётливость** *f.* clearness, distinctness, intelligibility

**отчётливый** *adj.* clear, distinct; sharp; comprehensive

**отчётность** *f.* accounting, bookkeeping

**отчисление** *n.* deduction, reckoning off

**отчислять, отчислить** *v.* to deduct, to reckon off

**отчуждаемость** *f.* alienability

**отчуждаемый** *adj.* alienable

**отчуждать, отчуждить** *v.* to alienate, to estrange

**отчуждение** *n.* exclusion; alienation

**отчуждить** *pf. of* **отчуждать**

**отшлифованный** *adj.* polished, ground

**отшлифовать** *pf. of* **шлифовать**

**отшнуровывание** *n.* pinch, pinching

**отштамповать** *pf. of* **штамповать**

**отъединение** *n.* isolation; disconnecting, switching off

**отъезд** *m.* departure, leaving

**отъём** *m.* take-off, removal, tripper

**отъёмный** *adj.* detachable, removable

**отыскание** *n.* finding, localization; detection; tracing; **о. максимума** maximizing; **о. обрывов цепи** break localization, continuity test; **о. производной** differentiation

**отыскать** *pf. of* **отыскивать**

**отыскивание** *n.* seeking, searching, looking

**отыскивать, отыскать** *v.* to retrieve; to trace, to locate, to detect, to discover, to find; to seek, to search for

**отяжелевший** *adj.* heavy, leaden

**отяжеление** *n.* loading, weighting; growing heavy

**отяжелять, отяжелить** *v.* to weight

**ОУ** *abbr.* (**областной узел**) oblast center, sectional center, principal outlet (*tphny.*)

**Оуен, Оуэн** Owen

**ОУП** *abbr.* (**обслуживаемый усилительный узел**) attended repeater station

**офитрон** *m.* Ophitron

**офицер** *m.* officer; **о.-связист** signal officer

**официальный** *adj.* official, formal

**оформить** *pf. of* **оформлять**

**оформление** *n.* design, appearance; shaping, forming; mounting; housing

**оформлять, оформить** *v.* to shape, to form; to mold; to design

**офсет** *m.* offset
**офсетный** *adj.* off-set
**офтальмометр** *m.* ophthalmometer
**офтальмоскоп** *m.* ophthalmoscope
**охват** *m.* scope, range, reach, coverage, compass; overlapping; envelopment; **малый о.** low coverage; **о. материала** coverage
**охватить** *pf. of* **охватывать**
**охватываемый** *adj.* male (contact)
**охватывать, охватить** *v.* to envelope, to compass, to encompass, to embrace; to cover
**охватывающий** *adj.* female (contact)
**охваченный** *adj.* covered, enveloped, encompassed, embraced, seized
**охладитель** *m.* cooler, refrigerator; coolant, cooling agent; condenser
**охладительный** *adj.* cooling
**охладить** *pf. of* **охлаждать**
**охлаждаемый** *adj.* cooled; **о. горючим** fuel-cooled
**охлаждать, охладить** *v.* to cool, to chill; to condense
**охлаждающий** *adj.* cooling, refrigerating, refrigerant; condensing
**охлаждение** *n.* cooling, coolness, refrigeration; condensation; **с водяным (воздушным) охлаждением** water-(air-)cooled; **масляное о. с наружным обдувом** oil-immersed air-blast cooling
**охлаждённый** *adj.* cooled, chilled, refrigerated; condensed
**охрана** *f.* protection, guard
**охранение** *n.* guarding
**охранительный** *adj.* protective, guarding, guard
**охранить** *pf. of* **охранять**
**охранный** *adj.* guard, protective
**охранять, охранить** *v.* to guard, to protect
**оцарапать** *pf. of* **царапать**
**оцелит** *m.* ocelit
**оценивание** *n.* evaluation
**оценивать, оценить** *v.* to estimate, to evaluate, to value, to rate, to appraise; to gage, to grade
**оценка** *f.* estimation, estimate, evaluation, appraisal; study, analysis, estimate; rate, rating; criterion, definition; rank; measure; estimator; **о. быстродействия** representative-circulating time; **о. по ускорению** acceleration figure of merit; **о. сверху** upper estimate (*math.*)

**оцепить** *pf. of* **оцеплять**
**оцепление** *n.* encompassing, surrounding
**оцеплять, оцепить** *v.* to encircle, to surround, to encompass
**оцепляющий** *adj.* encircling
**оцинкование** *n.* galvanizing, zinc plating
**оцинкованный** *adj.* galvanized, zinc-plated
**оцинковать** *pf. of* **оцинковывать, цинковать**
**оцинковка** *f.* galvanizing, galvanization, zinc plating
**оцинковывать, оцинковать** *v.* to galvanize, to zinc
**оцифровать** *v.* to digitize
**очаг** *m.* range
**очевидный** *adj.* obvious, evident, manifest, patent
**очередной** *adj.* next, next in turn; consecutive, routine, regular, usual
**очерёдность** *f.* sequence, regular succession, order, order of priority; queue
**очередь** *f.* turn, series, row, course, range, rank; queue
**очертание** *n.* outline, contour, profile, form, configuration, cutout; ambit (*tv.*)
**очерчивать, очертить** *v.* to ouline, to trace round, to sketch, to define
**очиститель** *m.* cleaner, cleanser, purifier; rectifier; separator; clarifier
**очистительный** *adj.* cleaning, purifying
**очистить** *pf. of* **очищать**
**очистка** *f.* cleaning; refinement, rectification, refining, purification, purifying; clarification; reset, resetting
**очистной** *adj.* purging, cleaning
**очищать, очистить** *v.* to clean; to clear, to clarify, to free, to refine, to purify; to rectify; to empty; to reset; to strip
**очищающий** *adj.* purifying, cleaning; sweeping (electrode)
**очищение** *see* **очистка**
**очищенный** *adj.* purified, cleaned, clarified; decontaminated
**очки** *pl.* glasses; eyepieces; goggles; **сигнальные о. (семафора)** spectacle (of a semaphore)
**очко** *n.* mesh (of screen); eye
**очувствитель** *m.* activator
**очувствить** *pf. of* **очувствлять**

**очувствление** *n.* sensitization, activation, sensitizing, photoactivating

**очувствлённый** *adj.* sensitized, activated; **о. примесями** impurity sensitized, impurity activated

**очувствлять, очувствить** *v.* to photoactivate, to activate, to sensitize

**ошелушить** *pf. of* **шелушить**

**ошибка** *f.* mistake, error, inaccuracy, fault, failure, blunder, oversight; malfunction; **о. в определении цвета** color-index error; **о. в отработке** error of behavior; **о. в считывании** read-out error; **о. выборочного обследования** ascertainment error, sampling error; **о. за счёт разрешающей способности** resolution error; **о. из-за дрейфа нуля умножителя** multiplier zero error; **о. индикации** datum error; **итоговая о.** resultant error; **о. монтажа** connection error, wiring error; **о. наложения записей** overwriting error; **неслучайная о.** response error; **о., обусловленная программой** program-sensitive error; **о. от подключения нагрузки** loading error; **о. от подключения нагрузки к потенциометру** potentiometer loading error; **о., портящая программу** program-sensitive error (*math.*); **предвнесённая о.** inherited error; **о. представ-**

ления truncation error; **о. рассогласования** following error; **сознательно допускаемая о.** conscious error; **строго возрастающая о.** ever-increasing error; **фазовая о. опорного напряжения** reference phase error

**ошибочно** *adv.* by mistake, erroneously; (with verbs) mis-: **о. направлять** to misroute, to misdirect

**ошибочность** *f.* inaccuracy; fallibility

**ошибочный** *adj.* erroneous, inaccurate, incorrect, faulty, mistaken, wrong

**ошивка** *f.* casing

**ошиновать** *v.* to bus

**ошиновка** *f.* leads (*elec.*)

**оштукатуренный** *adj.* plastered

**оштукатуривание** *n.* plastering

**оштукатурить** *pf. of* **штукатурить**

**ощипать** *pf. of* **щипать**

**ощупание** *n.* sounding, probing, feeling

**ощупывать, ощупать** *v.* to sound, to probe, to feel, to sense

**ощутимый** *see* **ощущаемый**

**ощутительность** *f.* perceptibility

**ощутить** *pf. of* **ощущать**

**ощущаемый** *adj.* perceptible, sensible, apparent

**ощущать, ощутить** *v.* to feel, to perceive

**ощущение** *n.* sensation, feeling, feel, perception

# П

П *abbr.* (**плоскопроволочная броня**) flat-wire armoring

п. *abbr.* 1 (**правый**) right-hand, clockwise; 2 (**пункт**) point

ПА *abbr.* 1 (**полевой аэродром**) field airdrome; 2 (**полевой аэростат**) field balloon

ПАД *abbr.* (**пороховой аккумулятор давления**) cartridge pressure accumulator, solid fuel pressure accumulator

падать, пасть *v.* to fall, to drop, to decline, to decrease, to dip

падающий *adj.* falling, dropping; incident, impinging; drooping, decrescent; incoming

падение *n.* incidence (of waves or rays), fall, drop, dip, decrease, reduction, lowering; gradient, slant, slope, incline, descent, grade, inclination; collapse; impact; **активное п. выпрямленного напряжения** resistive D.C. voltage drop; **анодное п. напряжения** anode drop, plate drop; **внутреннее п. напряжения** tube voltage drop, anode voltage drop, internal drop; **наклонное п.** oblique incidence; **п. напряжения на электровакуумном приборе** tube voltage drop (in an electronic tube); **отвесное п.** axial incidence; **п. по касательной** grazing incidence; **п. по нормали** normal incidence; **скользящее п.** grazing incidence

паечный *adj.* soldering

ПАЗ *abbr.* (**противатомная защита**) atomic defense, antinuclear defense

паз *m.* groove, slot, channel, flute, chase; slit, notch, recess, gap; channeling; **наклонный п.** helical armature slot; **направляющий п. (искателя)** selector shaft guide

пазный *adj.* grooved, slotted

пазовать *v.* to groove, to notch, to slot

пазовик *m.* groover, grooving plane

пазовочный *adj.* grooving, mortising

пазовый *adj.* grooving, groove, slot, mortise

пайка *f.* soldering, solder, brazing; **п. крепким припоем**, **п. медью** brazing; **п. погружением** solder dip

пайлистор *m.* pylistor

пайрекс *m.* pyrex

пайросерам *m.* pyroceram

пак *m.* pack; **п. питания** power pack

пакет *m.* packet, package, pile, stack, pack; **п. роторных пластин** rotor plate assembly; **п. сердечника** core-lamination stack; **симметричный пластинчатый п.** symmetrical stack

пакетированный *adj.* packaged; stacked

пакетный *adj.* package, packaged

пакля *f.* oakum, tow

паление *n.* ignition; burning; discharging, firing

палец *m.* finger; pin, peg, bolt, stud; guard, cam, cog, catch, tooth; **контактный п.** wiper; **поводковый п.** keeper pin; **п. подачи, транспортный п.** spacing cam

палисадный *adj.* palisade

палка *f.* rod, stick

палладиевый *adj.* palladic, palladium

палладий *m.* palladium, Pd

палладированный *adj.* palladized, palladium-coated

палочка *f.* stick, small rod, rod; **дирижёрская п.** baton (*med.*)

палочный *adj.* rod, stick

пальцевой *adj.* finger, finger-type, pin

пальчиковый *adj.* finger-type; bantam

память *f.* memory, recollection; storage, store (of data); **быстродействующая п.** quick-access memory, short-access memory; **внешняя п. с пополняемым** (*or* **обновляемым**) **массивом** file memory; **магазинная п.** stack memory; **магнитная п. с выборкой при совпадении нескольких токов** multiple-coincidence magnetic memory; **п. на элементах с захваченным потоком** persistent current memory; **оперативная п.** rapid storage, rapid memory, high-speed store, internal store, immediate-access memory; **постоянная п.** read-only memory; **п. с плоской выборкой слова** two-dimensional word selection; **п. с поразрядной выборкой** bit-organized memory; **п. с пренебрежно малым временем выборки** zero-access memory; **п. с прямой выборкой** switch-driven memory, switch-organized memory; **п. сверхбольшой ёмкости** mass memory; **п. со средней скоростью выборки** medium-speed access memory; **п. типа Z** linear-selection memory

**паналарм** *m.* Panalarm

**панель** *f.* panel, board section, shelf, bay, (switch)board, desk; socket, receptacle; base; **акустическая облицовочная п.** acoustical form board; **вводная п., п. вводных гребёнок** terminal panel (*tphny.*); **ключевая п., п. ключей** keyshelf; **п. коммутатора с гнездом** jack panel **коммутационная п.** patch board, patch panel, peg board, problemboard, switchboard panel; **п. лабораторной схемы** breadboard; **ламповая п.** tube socket, tube holder, bank of lamps, lamp panel; **малоёмкостная ламповая п.** anticapacitance tube socket; **междугородная испытательная п.** toll test panel; **п. многократного поля** section of multiple, bank; **монтажная п.** subpanel; **морская ламповая п.** navy socket; **наборная коммутаторная п.** patch panel; **неакустическая облицовочная п.** nonacoustical form board; **п. ожидания** call-storing panel; **п. опросных гнёзд** answering jack panel; **отделочная п.** finishing board; **п. передачи данных** data panel; **переходная ламповая п.** tube adapter; **п. переходных трансформаторов** repeating coil panel; **печатная п.** printed-circuit board; **распределительная п. передачи данных** data panel; **п. распределительного щита** switchboard section; **п. с предохранителем** fuse panel; **п. с расположением монтажа на обратной стороне** back-wired panel; **сменная наборная коммутационная п.** removable patch panel; **смесительная п.** audio mixer (*tv*); **п. шнуровых соединений** patch board (*tphny.*);

**штампованная ламповая п.** wafer socket; **электролюминесцентная п. визуального вывода (данных) типа сильватрон** sylvatron EL display panel

**панелька** *f.* socket

**панельный** *adj.* panel, board-section, shelf; (switch) board

**панидиоморфный** *adj.* panidiomorphic

**панорама** *f.* panorama

**панорамирование** *n.* scanning, panoraming, panning; panorama; follow shot, tracking shot (*tv*)

**панорамировать** *v.* to scan, to pan; to panoram

**панорамный** *adj.* panoramic, panorama

**пантелеграф** *m.* pantelegraph, phototelegraph, facsimile transmission

**пантелеграфия** *f.* pantelegraphy, facsimile telegraphy

**пантелефон** *m.* pantelephone, microtelephone

**пантограф** *m.* pantograph, pantograph collector

**пантодрель** *f.* pantodrill

**пантометр** *m.* pantometer

**пантоскоп** *m.* pantoscope

**панхроматический** *adj.* panchromatic

**панцирный** *adj.* metal-sheathed, metalclad, armorclad, armored

**панцирь** *m.* armor, shell

**пар** *m.* vapor, steam, fume

**пара-** *prefix* para-

**пара** *f.* pair, couple, dyad; **коаксиальная п.** coaxial line; **первая п. боковых частот** (*or* **полос модуляции**) first side pair modulation product; **транзитная шнуровая п.** through switching cord circuit; **универсальная шнуровая п.** universal cord circuit

**парабаллун** *m.* paraballoon

**парабола** *f.* parabola; **кадровая п.** vertical parabola; **строчная п.** horizontal parabola

**параболический** *adj.* parabolic

**параболоид** *m.* paraboloid; parabolic dish; **несимметричный усечённый п.** Cindy antenna; **п. передающей антенны** transmitting paraboloid; **п., питаемый не в фокусе** offset paraboloid section

**параболоидный** *adj.* paraboloid(al)

**параболо-цилиндрический** *adj.* parabolic-cylinder

**парадигма** *f.* paradigm

**парадигматический** *adj.* paradigm, paradigmatic

**парадокс** *m.* paradox

**парадоксальный** *adj.* paradoxical

**паразит** *m.* parasitic oscillation, stray(s), bug(s), spurious effect(s); idler, idle gear

**паразитический, паразитный, паразитовый** *adj.* stray, spurious, parasitic; idle

**паракаучук** *m.* para rubber

**параксиальный** *adj.* paraxial

**парализованный** *adj.* paralyzed, paralytic

**парализовать** *v.* to paralyze

**парализующий** *adj.* blocking

**параллакс** *m.* parallax

**параллаксный, параллактический** *adj.* parallactic, parallax

**параллелепипед** *m.* parallelepiped

**параллелизм** *m.* parallelism

**параллелограм** *m.* parallelogram

**параллель** *f.* parallel

**параллельно** *adv.* parallel (with); in parallel, in bridge

**параллельновключённый** *adj.* by-pass, connected in parallel

**параллельно-последовательный** *adj.* multiple-series, parallel-serial, parallel-series, series-parallel

**параллельность** *f.* parallelism

**параллельный** *adj.* parallel, multiple, shunt; cocurrent; collateral; **п. по разрядам** parallel by bit

**парамагнетизм** *m.* paramagnetism

**парамагнетик** *m.* paramagnet

**парамагнитный** *adj.* paramagnetic

**параметр** *m.* parameter; **зеркальный п.** image parameter; **п. коректирования рассогласования по скорости** velocity smoothing constant; **п. кривизны** buckling

**параметрика** *f.* parametrics

**параметрический** *adj.* parametric

**параметрон** *m.* parametron, parametric oscillator, phase-locked oscillator

**параметронный** *adj.* parametron

**парасоединение** *n.* para combination

**парасостояние** *n.* para-state

**парафазный** *adj.* paraphase

**парафин** *m.* paraffin

**парафинированный** *adj.* paraffined, waxed with paraffin

**парафиновый** *adj.* paraffin

**парафонический** *adj.* close-talking

**парахор** *m.* parachor

**парацентрический** *adj.* paracentric

**парашют** *m.* parachute

**парашютный** *adj.* parachute

**параэлектрический** *adj.* paraelectric

**паргелий** *m.* parhelium

**паритет** *m.* parity, equality

**паркеризировать** *v.* to phosphate-coat

**парно-пятиричный** *adj.* biquinary

**парный** *adj.* twin, pair, paired, double, dual; conjugate

**паро-** *prefix* steam, vapor

**паровоз** *m.* locomotive, steam engine

**паровозный** *adj.* locomotive, steam-engine

**паровой** *adj.* steam; vapor, vaporous

**паровпуск** *m.* steam admission

**паровыпускной** *adj.* (steam-)exhaust

**парогенератор** *m.* steam generator

**парогидравлический** *adj.* steam-hydraulic

**парозапорный** *adj.* steam cut-off

**паромер, парометр** *m.* steam meter, steam flow meter

**паронепроницаемый** *adj.* steamproof; vaportight

**парообразный** *adj.* steamlike; vaporous

**парообразование** *n.* steaming, steam formation, steam generation; vaporization

**парообразовательный** *adj.* steam-generating

**пароперегреватель** *m.* steam super-heater

**пароподвод** *m.* steam supply

**паропроизводитель** *m.* steam generator

**паропроизводительность** *f.* steam power, steam output, evaporative value

**паропроизводительный** *adj.* steam-producing; evaporating

**паросветный** *adj.* vapor-discharge

**паросиловой** *adj.* steam-power

**паростойкий** *adj.* vaporproof

**пароструйный** *adj.* steam-ejecting, steam-jet

**паротурбина** *f.* steam turbine

**паротурбоагрегат** *m.* steam-electric generating set

**паротурбогенератор** *m.* steam-turbine-driven alternator

**парофазный** *adj.* vapor-phase

**пароэлектростанция** *f.*, **пароэлектраль** *f.* steam power plant

**парсек** *m.* heliocentric parallax, parsec

**партер** *m.* (orchestra) pit

партитура *f.* score (*mus.*)

партия *f.* part (*mus.*); batch; lot, set; party, crew, group

парц. *abbr.* (парциальный) partial

парциальный *adj.* partial, fractional, divided

пасик *m.* little drive belt; belt

Паскаль Pascal

пасма *f.* filler

паспорт *m.* record; certificate, rating, rating plate, name plate; nominal data, rated values; **п. искателя** switch adjustment sheet; **п. реле** relay adjustment sheet; **п. цепи** summarized circuit details, circuit layout record

паспортизация *f.* rating; conditioning

паспортный *adj.* record; certificate, rating-plate, name-plate

пассаметр *m.* indicating snap gage

пассивация *f.*, **пассивирование** *n.* passivation

пассивированный *adj.* passivated, passive

пассивировать *v.* to passivate

пассивирующий *adj.* passivating

пассивность *f.* passivity

пассивный *adj.* passive, inert, parasitic

пассиметр *m.* indicating plug gage

паста *f.* paste; paint

пастирование *n.* pasting

пастированный *adj.* pasted

пасть *pf. of* падать

патефон *m.* gramophone, phonograph

патефонный *adj.* gramophone, phonograph

Патли Patley

патрон *m.* cartridge, case; (lamp) socket, lamp holder, lamp receptacle; holder, chuck, jig, mount, socket; adapter; **наполненный предохранительный п.** filled fuse unit; **плавкий п.** cartridge fuse; **поводковый п.** driving disk, tappet, driver, dog, catch; **п. с винтовой нарезкой** Edison screw holder; **самоцентрирующийся п.** centering head; **четырёхкулачковый п.** four-jaw chuck

патронный *adj.* lamp-holder, lamp-socket; socket, holder, chuck, jig; cartridge(-type)

патрончик *m.* cartridge

патрубок *m.* nipple, nozzle, outlet; socket, sleeve, connection; connecting pipe (*or* piece); branch pipe, socket pipe; **выпускной п. (висячей антенны)** fairlead (of a trailing antenna); **нарезной п.** nipple; **переходный п.** pipe adapter, tube adapter

патруль *m.* patrol

патрульный *adj.* patrol

пауза *f.* rest, break, interval, quiescent interval, pause, stop, spacing, interruption, paralysis; blank; **п. в измерении** plotting interval

паузовый *adj.* rest, break, interval, pause, stop

Паули Pauli

Паулинг Pauling

Паульсен Poulsen

паульсеновский *adj.* Poulsen's

паундаль *m.* poundal

паутинный *adj.* spiderweb

паучок *m.* spider collar

« пауэр-пэк » *m.* power pack

пачка *f.* block, pack, packet, batch; pile, deck, stack, unit

Пашен Paschen

паяльник *m.* soldering iron, soldering bit

паяльный *adj.* soldering

паяльщик *m.* solderer, soldering iron; tinsmith

паяние *n.* soldering

паяный *adj.* soldered

паять, запаять *v.* to solder

ПБМ *abbr.* (переключатель блокировки магнетрона) antitransmit-receive tube, ATR

ПБО *abbr.* (провод с бумажной обмоткой) see wire, single cotton covered wire

ПВ *abbr.* (противовоздушный) anti-aircraft

ПВД *abbr.* 1 (приёмник воздушных давлений) pilot-static tube 2 (прямоточный воздушно-реактивный двигатель) ramjet-engine

ПВНОС *abbr.* (пост воздушного наблюдения, оповещения и связи) aircraft warning service post

ПВО *abbr.* (противовоздушная оборона) antiaircraft defense

ПВП *abbr.* (полоса воздушных подходов) lane of approach

ПВРД *abbr.* (прямоточный воздушно-реактивный двигатель) ramjet engine

ПВС *abbr.* (пост воздушной связи) ground-air liaison post

пг *abbr.* (**пикограмм**) micromicrogram
**ПГАП** *abbr.* (**пневмогидравлический автопилот**) pneudraulic autopilot
**ПГГ** *abbr.* (**парогазогенератор**) steam-gas-generator
**п. д.** *abbr.* (**парциальное давление**) partial pressure
**ПД** *abbr.* (**поршневой двигатель**) piston engine
**ПДК** *abbr.* 1 (**потенциометрический дистанционный компас**) telecompass with potentiometer tap; 2 (**предельно допустимая концентрация**) maximum permissible concentration
**певотрон** *m.* pevotron
**певучий** *adj.* melodious
**певческий** *adj.* singing
**пегельмессер** *m.* level meter
**педаль** *f.* pedal, treadle, foot lever; **рельсовая п.** track instrument
**педальный** *adj.* pedal, foot-operated
**леддинговый** *adj.* padding
**педион** *m.* single-surface crystalline structure
**педометр** *m.* pedometer
**Пекле** Peclet
**пекулярный** *adj.* peculiar
**пелена** *f.* fog, set-up (*tv*); film
**пеленг** *m.* bearing, direction; **безопасный п.** clearing bearing; **п. засечкой** cross-bearing; **п. на большое расстояние** long-path bearing; **п. по засечке из двух точек** two-point bearing; **п. по координатной сетке** grid bearing; **п. по углу места** elevation bearing
**пеленгатор** *m.* finder, DF, direction finder, course-and-bearing indicator; **п. Адкока с разнесёнными вертикальными диполями** Adcock elevated-type direction finder; **п. Адкока с трансофматорной связью** Adcock coupled direction finder; **п. атмосферных разрядов** static direction finder; **п. высотного угла** radio theodolite; **однозначный п.** sense direction finder; **п. по нулю звука** aural-null direction finder; **п. со скрещёнными катушками** crossed loop direction finder
**пеленгаторный** *adj.* direction-finder, localizer
**пеленгация** *f.* direction finding, position fixing, goniometry; **п. с наземных станций** ground-station direction finding

**пеленгирование, пеленгование** *n.* direction finding; **п. направленного действия** radio directional bearing; **однозначное п.** sense finding; **п. по минимуму** direction finding system using minimum intensity as indication; **штриховое п.** radio-direction finder with oscilloscope-type bearing indicator
**пеленговать, запеленговать** *v.* to bear, to take a bearing, to set, to take a course
**пеленговый** *adj.* bearing, direction
**Пёлер** Poehler
**пельвиметр** *m.* pelvimeter
**пелькомпас** *m.* azimuth compass
**Пельтье** Peltier
**пемза** *f.* pumice
**пена** *f.* foam, froth
**пенетрация** *f.* penetration
**пенетрометр** *m.* penetrometer, qualimeter, radiochronometer
**пенетрон** *m.* penetron
**пение** *n.* singing; pipe; **п. гребного винта** propeller singing
**пенистый** *adj.* foam, foamy, frothy
**пенокерамический** *adj.* foam(ed)-ceramic
**пенопласт** *m.* foam material, polyfoam
**пенополистирол** *m.* styrofoam, polyfoam
**пенорезина** *f.* foam rubber
**пенсовый** *adj.* pence
**пента-** *prefix* penta-, five
**пентагональный** *adj.* pentagonal
**пентагрид** *m.* pentagrid, heptode
**пентагридный** *adj.* pentagrid
**пентакарбонил** *m.* pentacarbonyl
**пентатоника** *f.* pentatonic scale
**пентатоновый** *adj.* pentatonic
**пентатрон** *m.* pentatron
**пентахорд** *m.* pentachord (*mus.*)
**пентациклический** *adj.* pentacyclic, five-membered
**пентод** *m.* pentode; **высокочастотный п. с переменной крутизной** variable-mu R.F. pentode; **двойной п. с раздельным катодом** double output pentode; **оконечный п. горизонтальной развёртки** line output pentode; **п. переменной крутизны** variable-mu tube, variable mutual conductance tube; **п. с короткой характеристикой, п. с крутым изгибом характеристики** sharp-cutoff pen-

tode, gated-beam tube; **п. с малой крутизной, п. с низкой крутизной характеристики** low-slope pentode; **п. с переменной крутизной (характеристики)** variable-mu (high-frequency) pentode; **п. с удлинённой характеристикой** remote cutoff pentode

**пентодный** *adj.* pentode

**пенька** *f.* hemp

**пеньковый** *adj.* hemp

**пер.** *abbr.* 1 (**период**) period, cycle; 2 (**периодический**) periodic

**пералуман** *m.* peraluman

**перв.** *abbr.* (**первичный**) primary

**первеанс** *m.* perveance

**первично** *adv.* first, primarily, initially

**первичный** *adj.* primary, original, initial, first, fundamental, primordial; parent

**первоисточник** *m.* original, origin, primary source

**первоначальный** *adj.* primary, primitive, elementary; original, initial, incipient

**первообразный** *adj.* primitive, original, protoplastic

**первоочерёдность** *f.* first priority, high priority

**первоочередной** *adj.* first-priority, first

**первопричина** *f.* source, origin, original cause

**первосортный** *adj.* high-class, first-rate, top-grade

**первостепенный** *adj.* foremost, paramount, chief

**первый** *adj.* first; main; former; raw, starting

**пергаментирование** *n.* parchmentizing

**переадресация** *f.* address substitution, address modification, readdressing

**переадресовывать, переадресовать** *v.* to re-address, to modify

**перебазирование** *n.* relocation

**перебалансировка** *f.* retrimming

**перебег** *m.*, **перебегание** *n.* crossing, running over, overrunning

**перебивание** *n.* jamming, breaking in

**перебивать, перебить** *v.* to break in, to interrupt; to break up

**переблокировать** *v.* to interlock

**перебой** *m.* disturbance, trouble, outage, hitch, failure, bug; erasure; cuts-out; interruption, intermission, delay, stop; missing (of motor); **п. в зажигании** spark failure

**перебойный** *adj.* erasure; trouble, failure

**перебор** *m.* gear, compound gear, shift; excess

**переборка** *f.* overhaul, overhauling; baffle, partition, wall, diaphragm; sorting

**перебрасывать, перебросать, перебросить** *v.* to flip, to throw over; to transfer

**переброс** *m.* throw-over, change-over, shifting, transfer; splashing over; surge; flip-flop; **п. головки тумблера** flipping the switch; **п. реактивной мощности** power-factor swing

**перебросать, перебросить** *pf. of* **перебрасывать**

**переброшенный** *adj.* thrown-over; transferred

**перевал** *m.* pass, passing, crossing, passage; descent

**перевальный** *adj.* passing, passage

**переведённый** *adj.* transferred, shifted; translated; switched, shunted; operated, reverse; **п. на транзисторы** transistorized; **п. на холостой ход** off the line

**перевезти** *pf. of* **перевозить**

**перевёрнутый** *adj.* inverted, reverse, upset, turned over

**перевёртывать, перевернуть** *v.* to turn; to turn over, to change, to invert, to overturn, to trip, to upset; to reverse

**перевес** *m.* 1. overweight, overbalance; 2. bias

**перевеска** *f.* restringing

**перевести** *pf. of* **переводить**

**перевод** *m.* transformation, conversion, reduction, recalculation; interpretation; translation, translating; transfer, shift, shifting, change-over; shifter; operation; **взаимный п.** interconversion

**переводимый** *adj.* operated; transferable; convertible; translatable

**переводить, перевести** *v.* to change, to change over, to transfer, to shift; to translate; to switch, to shunt; to operate, to reverse; to store (a number); to conduct; to gear; to convert, to reduce; to carry over; to convey; **п. (схему) на транзисторы** to transistorize

**переводник** *m.* adapter

**переводной, переводный** *adj.* reduc-

ПЕРЕВОДЧИК 388 ПЕРЕГРУППИРОВЫВАТЬСЯ

tion, conversion; reversing; transfer, shift, shifting, switch; translation

**переводчик** *m.* translator

**перевозбуждать, перевозбудить** *v.* to overdrive, to over-excite

**перевозбуждение** *n.* overexcitation, overdrive, overswing, superexcitation

**перевозбуждённый** *adj.* overdriven, overexcited

**перевозимый** *adj.* transportable

**перевозить, перевезти** *v.* to convey, to transport, to carry, to remove, to move

**перевозка** *f.* conveyance, transport, transportation, transfer, transmission

**переворачивание** *n.* overturn

**переворачивать, переворотить** *v.* to invert, to turn, to turn over, to overturn; to change, to revert, to reverse

**перевулканизация** *f.* overcuring

**перевязка** *f.* staggering; lacing, binding, bandage; **п. линий** termination of wire, dead-end tie

**перевязочный** *adj.* binding, bandage; bond

**перевязывание** *n.* bandage

**перевязь** *f.* bandage, tie

**перегиб** *m.* bend, twist, fold, folding, kink, inflection, knee (of curve), recurvature; slope, crest; discontinuity

**перегибный** *adj.* inflectional

**перегнать** *pf. of* **перегонять**

**перегнутый** *adj.* bent, folded

**переговор** *m.* call, conversation, message; **п. между районными АТС** interoffice local call; **местный п. по соединительной линии** local trunk call (*tphny.*); **п. оплачиваемый вызываемым абонентом** transfered charge call, collector call; **отменённый п.** cancelled call; **п. с вызовом через нарочного** messenger call

**переговорно-вызывной** *adj.* speak-buzz

**переговорный** *adj.* telephone (booth); call; talking, speaking; intercommunication

**перегонка** *f.* rectification; distillation; sublimation; fractionation

**перегонять, перегнать** *v.* to outdistance, to outrun, to surpass, to outdrive; to rectify

**перегорание** *n.* burning out, burning, burn-out, burning off; combustion

**перегорать, перегореть** *v.* to burn out; to blow (a fuse), to melt

**перегоревший, перегорелый** *adj.* burnt out

**перегореть** *pf. of* **перегорать**

**перегородка** *f.* partition, baffle plate, baffle, barrier, screen, wall, dividing wall; membrane, diaphragm, iris; compartment; baffler; deflector; lamina; septum; **п. из пустотелой черепицы** hollow tile panel; **костная п., идущая вдоль спирального хода улитки** lamina spiralis, limbus; **разделительная п.** baffle-board; **спиральная п.** spiral septum

**перегородочный** *adj.* dividing; partition; septal

**перегороженный** *adj.* septate; partitioned, divided

**переградуировка** *f.* recalibration

**перегрев** *m.*, **перегревание** *n.* overheating, overheat, superheating; baking-out; overtemperature

**перегреватель** *m.* superheater

**перегревать, перегреть** *v.* to overheat, to superheat; to overrun, to run hot; to heat again; to burn out, to fuse

**перегретый** *adj.* overheated, superheated

**перегреть** *pf. of* **перегревать**

**перегружаемость** *f.* overload capacity

**перегружать, перегрузить** *v.* to overload, to overburden; to overdrive, to overtax; to overcharge

**перегруженность** *f.* overload, overwork

**перегруженный** *adj.* overdriven, overworked; overladen, overloaded

**перегруживать** *see* **перегружать**

**перегрузить** *pf. of* **перегружать**

**перегрузка** *f.* overload, overloading, excess load, overburden, overcharging, overdriving, straining (of machine); transfer, reloading

**перегрузочный** *adj.* overload; transfer, transport; loading

**перегруппированный** *adj.* rearranged

**перегруппировать(ся)** *pf. of* **перегруппировывать(ся)**

**перегруппировка** *f.* regrouping, reshuffle, rearrangement; migration

**перегруппировывать, перегруппировать** *v.* to rearrange, to regroup

**перегруппировываться, перегруппи-**

**роваться** *v.* to migrate; to regroup (itself)

**перед** *prep. instr. and acc.* before, in front of; against; to

**передаваемый** *adj.* delivered; transferable; transmitted

**передавать, передать** *v.* to transfer, to transmit, to relay, to signal, to key, to send, to convey, to deliver; to broadcast; to retransmit, to report, to communicate; to direct, to gear

**переданный** *adj.* transmitted, conveyed

**передаточность** *f.* transmissibility

**передаточный** *adj.* transfer, conveying, transmitting, transmission, transmissive, carrier; gear; driving

**передатчик** *m.* transmitter, transferrer, carrier; sender, sending set, transmitter, transmitting station; translator; **ведомый п.** repeater transmitter, slave transmitter; **п. звукового сопровождения** aural transmitter; **искровой п. затухающих колебаний** quenched spark transmitter; **искровой п. с ударным возбуждением** shock excitation type spark transmitter; **механический п.** omnigraph (*tgphy.*); **п. по схеме импульсной автомодуляции** self-pulsed transmitter; **п. помех** jammer, jamming transmitter; **посадочный п.** PAR, precision approach radar; localizer; **п. радиолинии звукового сопровождения внестудийных передач** audio program link transmitter; **п. с двумя боковыми полосами** double-sideband transmitter; **п. с задающим генератором, стабилизованным кварцем** crystal-controlled transmitter; **п. с одной боковой полосой** single-sideband transmitter; **п. с частотой непосредственно задаваемой (пьезо-)кристаллом** crystal-controlled transmitter; **телевизионный п. прямого видения** direct pickup scanner; **п. умышленных помех** jammer, jamming transmitter

**передать** *pf. of* **передавать**

**передача** *f.* sending, transmission, transfer; passage, conveyance, delivery, delivering; communication; gear, gearing, driving gear, drive; instruction, assignment; broadcast; program; pickup; modulation; **ази-**

мутальная **п.** azimuthgear; **п. без постоянной составляющей** alternating current transmission; **п. в зоне прямой видимости** horizon transmission; **п. видеосигнала без постоянной составляющей** alternating current transmission; **внестудийная п.** field pickup; **п. двумя боковыми полосами** double-sideband transmission; **п. депеши через цепочку связных** messenger relay system; **дифференциальная п.** equalizer; **м. заказов сериями** sequence calling (*tphny.*); **п. звука в разветвлённом трубопроводе** branch transmission; **зубчатая п.** gear, gearing; **п. из студии** studio pickup, direct pickup; **п. изображений** facsimile transmission; image transmission; **п. импульсов закорачиванием линии** shortcircuit working; **п. кинофильма** film pickup; **п. клиновым ремнём** vee-velt drive; **п. коническими зубчатыми колесами** bevel gearing; **п. крупным планом** close-up, close-up view; **мальтийская п.** Geneva movement; **механизированная п. данных о дальности** aided range gearing; **многократная п. с разделением каналов во времени** time division multiplex (transmission); **многократная п. (с разделением каналов) по частоте** frequency-division multiplex (transmission); **многократная п. с уплотнением во времени** time (-division) multiplex transmission; **многочастотная п. импульсов** multifrequency pulsing; **п. наклона** elevation drive gear; **п. неподвижных изображений** facsimile transmission, phototelegraphy, facsimile telegraphy; **п. пищиком** buzzer signaling; **п. по разветвлённому каналу** branch transmission; **п. под углом** angle drive; **полуавтоматическая п. данных о дальности** aided range gearing; **п. полуволнами переменного тока** a-c half-cycle transmission; **п. понятий** intelligence transmission; **последовательная п. цветовых элементов** dotwise color switching (*tv.*); **п. постоянной составляющей** d-c transmission; **п. прямого видения** floodlight scanning (*tv.*); **п. регистрирующего пера** recording pen linkage; **рычажная п.** leverage; **п. с киноплёнки** motion-picture pickup; **п. с подавленной боковой частотой** vestigial-sideband transmission (*tv.*); **сегментноцепная п.** chain-and-segment linkage; **сквозная п.** rippling through; **п. соединения** transfer of a call; **п. сообщений по радиосвязи** communications traffic; **суммирующая дифференциальная зубчатая п.** differential gear adder; **театральная п.** electrophone call, program supply (*tphny.*); **п. телевидения по эфиру** radiated television transmission; **п. угла места** elevation drive gear; **узконаправленная п.** beam transmission; **п. управления и переадресации** branch and index (transmission); **фототелеграфная п. неподвижных изображений** facsimile picture transmission; **червячная п.** worm gear drive; **п. яркостного сигнала в обход канала цветности** by-passed monochrome transmission

**передающий** *adj.* transmitting, sending, radiating

**передвигать, передвинуть** *v.* to migrate; to shift, to displace, to move, to remove

**передвигаться, передвинуться** *v.* to move, to migrate, to travel

**передвигающий** *adj.* moving, shifting

**передвижение** *n.* migration, travel, movement, moving, traction; removal, transfer, transportation; shifting, displacement; **п. на новую строку** line feed

**передвижка** *f.* move, moving, shifting, transfer; mobile unit, mobile equipment, mobile set; field equipment

**передвижной** *adj.* movable, mobile, vehicular, portable, traveling; sliding, adjustable; shifting

**передвинуть(ся)** *pf. of* **передвигать (-ся)**

**передел** *m.* conversion; process stage

**переделанный** *adj.* converted, changed; transplanting

**переделать** *pf. of* **переделывать**

**переделка** *f.* alteration, remodeling, redesigning, repairs; **п. схемы** redesigning of a circuit; reconnection

**переделывать, переделать** *v.* to convert, to change, to alter, to remodel, to remake, to redesign; **п. монтажную**

**схему** to change the wiring diagram; to rewire

**передельный** *adj.* remade, changed, altered; limit

**передемпфирование** *n.* overdamping

**передемпфированный** *adj.* overdamped

**передержанный** *adj.* burned up, overexposed

**передерживать, передержать** *v.* to overexpose, to burn up; to hold over

**передержка** *f.* overexposure

**переднезадний** *adj.* anterior-posterior

**передний** *adj.* leading, front, fore, forward, anterior

**передовой** *adj.* forward, advanced, leading, foremost, front, fore

**передувание** *n.* overblowing

**передувать, передуть** *v.* to overblow

**передувка** *f.* overblowing

**передутый** *adj.* overblown

**передуть** *pf. of* **передувать**

**переезд** *m.* crossing, passage, transit; moving

**переездной** *adj.* crossing

**переезжающий** *adj.* transportable, mobile

**пережечь** *pf. of* **пережигать**

**пережжённый** *adj.* burnt, burned; died away; cooled down

**пережигание** *n.* burnt-out; burning, burning off, combustion

**пережигать, пережечь** *v.* to burn out; to burn, to consume

**пережигающий** *adj.* blowing

**пережог** *m.* burn-out; overheating

**перезаделка** *f.* repair

**перезаписывать, перезаписать** *v.* to rewrite, to regenerate

**перезапись** *f.* re-recording, dub, dubbing, transcription, duplication; overwriting; recirculation (of information), regeneration; rewrite, rewriting; **п. с редактированием** dubbing combined with editing

**перезапрос** *m.* overinterrogation

**перезапуск** *m.* restart

**перезаряд** *m.* overcharge, overcharging; recharge

**перезарядить** *pf. of* **перезаряжать**

**перезарядка** *f.* recharging, reloading; overloading; supercharging, overcharge, overcharging; refilling; **п. частиц** charge-exchange phenomenon

**перезаряжать, перезарядить** *v.* to recharge, to reload; to overcharge

**переизлучение** *n.* re-radiation

**перейти** *pf. of* **переходить**

**перекал** *m.* overheating, overrunning

**перекалённый** *adj.* overheated, burnt

**перекаливать, перекалить** *v.* to overheat, to overrun

**перекаточный** *adj.* converting

**перекатывающийся** *adj.* rocking

**перекачка** *f.* pumping back; **взаимно-обратная п.** pumpback

**перекашиваться, перекоситься** *v.* to warp, to twist, to get out of alignment

**перекидка** *f.* jumper, (temporary) overhead span (of wire or cable); throwing over; **кабельная п.** aerial repair cable, aerial breakdown cable

**перекидной** *adj.* flip-flop, double-throw, two-way, reversible, reversing, throw-over, break-before-make; tipping

**перекидывать, перекинуть** *v.* to throw, to throw over; to tilt; to turn over; to span

**перекись** *f.* peroxide

**перекладина** *f.* crossbeam, cross bar, brace, tie beam

**переклейка** *f.* plywood

**переключаемый** *adj.* switchable, reversible; exchangeable

**переключатель** *m.* selector, selector knob, selector switch, switch, throw-over switch, change-over switch, double-throw switch, two-way switch, commutator, duplexer, reverser, key; **на переключателях** switch-type; **барабанный п. настройки** turret tuner (*tv.*); **п. бдительности** acknowledging switch, forestalling switch; **п. блокировки магнетрона** ATR, antitransmit-receive tube; **п. возбуждения на шунтирующее сопротивление** field break switch, field discharge switch; **п. выбора длины развёртки** sweep-length selector switch; **головной п.** vertical off-normal assembly, vertical off-normal spring; **голосовой п. приём-передача** vodas, voice-operated device anti-singing; **п. грубой и тонкой настройки** standby-tune switch; **п. делителя напряжения видеосигнала** video attenuator switch; **п. диапазонов** band selector, range selector, wave selector, band-switch; **конечный реверсивный п.**

travel-reversing switch; **п. контрольного останова** breakpoint switch; **лампа-электронный п.** beam switching tube; **матричный п. на сердечниках** selection core matrix; **п. масштаба дальности** range-scale selector; **п. «маяк-телефон»** range-both-voice switch; **многопозиционный п.** uniselector; **п. на две ступени нагрева** two-heat control switch; **п. на несколько направлений** multiple-way switch; **п. направления** direction-control switch, reverser; **п. направления тока** current reverser, reversing key; **п. ограничения ширины полосы пропускания** passband limiting switch; **п. ответвлений** tap changing device; **п. ответвлений для полного отклонения мёртвых витков** second section switch; **п. отметок** peg switch; **п. оттормаживания** reset switch; **п. полосы пропускания** bandwidth switch; **п. пределов измерения** range control switch; **п. приём-передача** TR switch, ATR switch, send-receive switch; polyplexer; **пусковой п. с звезды на треугольник** Y-delta starter; **п. рабочих мест** position switching key; **п. режимов работы вычислительной машины** compute-hold-reset switch; **п. рода работы** automatic-manual switch; **рычажный п. с перекидной головкой** jack switch, toggle switch; **п. с малым временем включения** low-duty-cycle switch; **п. с остановкой на каждой ступени** multi-position switch; **п. с перекидной головкой** trigger, switch, toggle switch; **п. со звезды на треугольник** star-delta switch; **п. стрелочного привода** switch machine pole changer; **п. угла места** elevation commutator; **п. шкалы дальности** range selector, range switch, range-changing switch

**переключательный** *adj.* switching, switch-type, commutation

**переключать, переключить** *v.* to key, to switch, to commutate, to commute; to reverse, to change poles; to shift; to change over; to change connections

**переключаться, переключиться** *v.* to switch

**переключающий** *adj.* switching; reversing; throw-over, change-over, two-way, double-throw; commutation; read-in

**переключающийся** *adj.* shifting

**переключение** *n.* reset, resetting, switching, change-over, changing, commutation, throw-over; patching; shift, shifting; reversing, reversion; dynamic remagnetization process; transferring; **п. аппарата абонента** transfer of a subscriber's set (*tphny.*); **п. аппарата на цифры** figure shift; **п. в наносекундном диапазоне** milli-microsecond switching; **п. голосом** voice-operated switching (*tphny.*); **обратное п. (схемы)** reconnection (of a circuit); **п. передач** gear change; **п. станции** cutting-in of an office (*tphny.*)

**переключённый** *adj.* reconnected, shifted, switched, changed over

**переключить(ся)** *pf. of* **переключать (-ся)**

**перекодировать** *v.* to convert

**перекоммутация** *see* **переключение**

**перекомпаундирование** *n.* overcompounding

**перекомпаундированный** *adj.* overcompound(ed)

**перекомпенсация** *f.* overcompensation, overcorrection; overneutralizing; overdoping

**перекомпенсировать** *v.* to overcompensate

**перекорректированный** *adj.* overcompensated, overcorrected; overpeaked

**перекорректировать** *v.* to overcompensate, to overcorrect

**перекоррекция** *f.* overcorrection, overcompensation; overpeaking

**перекос** *m.* bias, misalignment; squint, skew, skewing, slant, angularity, bending, curving, obliquity, tilt, sag, sagging; **п. плоской части импульса** tilt distortion

**перекоситься** *pf. of* **перекашиваться**

**перекошенный** *adj.* biased, out of alignment; warped, skew, twisted; drunken (screw)

**перекрестие** *n.* cross lines

**перекрестить(ся)** *pf. of* **перекрещивать(ся)**

**перекрёстно-включённый** *adj.* cross-connected

**перекрёстный** *adj.* cross, crossing, crossed; lattice, cross-talk

**перекрёсток** *m.* (road) crossing, cross-over

**перекрестье** *n.* cross; crosshair

**перекрещённый** *adj.* crossed

**перекрещивание** *n.* cross, crossing, intersecting, traverse; **п. колебаний различных видов** mode crossing

**перекрещивать, перекрестить** *v.* to cross, to intersect, to traverse; to transpose

**перекрещиваться, перекреститься** *v.* to cross, to intersect

**перекрещивающийся** *adj.* crossing, intersecting, crisscross, overlapping

**перекристаллизация** *f.* recrystallization, regrowth

**перекрутка** *f.* torsion

**перекрывание** *n.* overlap, overlapping, lap

**перекрывать, перекрыть** *v.* to overlap, to lap, to superimpose; to span, to bridge over; **п. внакрой** (*or* **внахлестку**) to overlap; **п. накладкой** to splice

**перекрывающий** *adj.* overlapping, covering

**перекрытие** *n.* cover, covering, coverage, overlap, overlapping, lap; floor, flooring; roof, ceiling; flashover, arc-over, spark-over; span; **без перекрытия** non-shorting; **п. зон действия сигналов** overlap (*rr. appl.*); **п. изолятора** flashing over; **п. импульсами** impulse flashing over; **искровое п., п. искрой** sparkover

**перекрытый** *adj.* covered, bridged, spanned; overlapped

**перекрыть** *pf. of* **перекрывать**

**перекрышка** *f.* overlap, overlapping, lap; cover, ceiling

**перелёт** *m.* overshot, overreach; flight, passage; **обратный п. (луча)** retrace, back swing, flyback (of a beam) (*tv*)

**перелив** *m.* modulation (of sound) overflow

**переливание** *n.* overflow, overflowing; pouring over

**переливаться, перелиться** *v.* to overflow, to run over

**переливчатость** *f.* iridescence

**переливчатый** *adj.* modulating; bubbling; iridescent, chatoyant, opalescent

**перелиться** *pf. of* **переливаться**

**перелом** *m.* discontinuity, fracture, break

**переломлять** *v.* to diffract

**перемагнитить** *pf. of* **перемагничивать**

**перемагничение, перемагничивание** *n.* magnetic polarity reversal, magnetic pulsing, magnetization process, remagnetization, reversal, flux reversal, magnetization reversal; **п. импульсами тока постоянной амплитуды** constant-current switching; **п. однородным вращением** coherent flux reversal, uniform rotational reversal

**перемагничивать, перемагнитить** *v.* to reverse the magnetism

**перематывание** *n.* rewinding

**перематывать, перемотать** *v.* to rewind, to roll back, to reel

**перемежать, перемежить** *v.* to alternate, to interleave

**перемежаться, перемежиться** *v.* to intersperse, to intermit, to be interrupted; to alternate; to interleave, to interlace

**перемежающийся** *adj.* alternating, alternate, discontinuous, intermittent, unstable, swinging; interlaced, interleaved; permutation

**перемежение** *n.* interlace, interlacing, interleave, interleaving; sharing; **п. частотных спектров** band-sharing; **чересстрочное п.** line interlacing; **чересстрочное п. с чётным числом строк** even-line interlacing

**перемежить(ся)** *pf. of* **перемежать(ся)**

**перемежный** *adj.* intermittent, intermissive

**перемена** *f.* alternation; changeover, change, alteration, transformation, inversion, reversal, mutation, variation; shift, move; interval; **п. знаков** shift (*tgphy.*); **п. направления** reversal; **п. направления радиосигналов** wandering of radio signals; **п. строки** line feed (*tgphy.*)

**переменить** *pf. of* **переменять**

**переменная** *f. adj. decl.* variable, variant (*math.*); **п. величина** variable quantity; **п. в моделирующих устройствах** analog variable; **п. в цифровой форме** digital variable; **п. с индексами** subscripted variable

**переменно** *adv.* alternately; **п. действующий** alternating; **п.-замедленный** variable-decreasing; **п.-поляр-**

ный heteropolar; **п.-ускоренный** variable-increasing

**переменность** *f.* changeability, variability, versatility, inconsistency, fluctuation, instability, mutability

**переменный** *adj.* variable, varying, changeable, interchangeable; alternating, alternate

**переменять, переменить** *v.* to change, to modulate, to shift, to vary, to modify; to interchange

**переместительный** *adj.* commutative (*math.*); transposable, transposing

**переместить(ся)** *pf. of* **перемещать(ся)**

**перемешанный** *adj.* mixed, stirred; mixed up, confused; inverted; scrambled

**перемешать** *pf. of* **перемешивать**

**перемешивание** *n.* mixing, agitation, stirring; intermixing; confusion; randomization

**перемешиватель** *m.* scrambler

**перемешивать, перемешать** *v.* to mix, to stir, to blend, to intermix; to mix up, to confuse; to agitate

**перемешивающий** *adj.* mixing, stirring; confusing

**перемещаемый** *adj.* movable, mobile; transportable

**перемещать, переместить** *v.* to move, to transfer, to transpose, to transport, to shift, to displace; to drive, to convey; to migrate; to relay over; to push, to slide

**перемещаться, переместиться** *v.* to move, to slide, to slip; to migrate, to travel

**перемещающий** *adj.* moving, motive

**перемещающийся** *adj.* migrating, walking, traveling; moving, movable, mobile, adjusting, sliding; traversing, shifting; permutation

**перемещение** *n.* transference, transfer, translation; shift, shifting, moving, motion, move, travel, migration, drift, wandering; displacement, dislocation, sliding; adjustment; permutation; conveying, conveyance, transportation, transposition; **п. запятой** arithmetical shift; **поперечное п.** lateral motion; **угловое п. вала** angular shaft motion

**перемещённый** *adj.* moved, shifted, transferred; dislocated

**перемкнутый** *adj.* by-passed, jumpered

**перемкнуть** *pf. of* **перемыкать**

**перемножатель** *m.* multiplier

**перемножать, перемножить** *v.* to multiply

**перемножение** *n.* multiplication; **асинхронное п. сигналов** asynchronous multiplexing; **п. непрерывных величин** analog multiplication

**перемножить** *pf. of* **перемножать**

**перемодулировать** *v.* to overmodulate

**перемодуляционный** *adj.* overmodulation

**перемодуляция** *f.* overmodulation, overcutting

**перемонтаж** *m.* remounting, repackaging; rewiring

**перемотать** *pf. of* **перематывать**

**перемотка** *f.* rewind, rewinding, reroll, rollback; reconnecting; wind

**перемоточная** *f. adj. decl.* rewinding room

**перемоточный** *adj.* rewinding, reroll

**перемыкание** *n.* bridging, shunt

**перемыкать, перемкнуть** *v.* to bridge; to place a jumper, to tie, to shunt, to span

**перемыкающий** *adj.* bridging

**перемычка** *f.* jumper, connecting link, crossbar, cross connection, tie, bond, crosspiece, connector, bridge, bridle, bridling, patch cord, wire, plug wire, cross wire, by-pass; leg; **аварийная п.** emergency cable, emergency wire; **гибкая п.** flexible bond, pigtail; **дугообразная п.** U-link; **кабельная п.** breakdown cable, jumper cable, bridle cable; **разгруженная п.** nontension bow; **сдвижная п. (на клемнике)** offset link (on a terminal block)

**перенапрягать, перенапречь** *v.* to overstrain

**перенапряжение** *n.* overstrain, overexertion, overdrive, supertension, overstress, overstressing; excess voltage, overvoltage; surge; **п. при резонансе** resonance elevation

**перенапряжённый** *adj.* overstrained, overdriven

**перенаселённость** *f.* overpopulation

**перенастройка** *f.* retuning, readjustment; change-over

**перенасыщать, перенасытить** *v.* to oversaturate, to supersaturate

**перенасыщение** *n.* oversaturation, supersaturation

**перенасыщенный** *adj.* oversaturated, supersaturated

**перенесение** *n.* transference, transportation, translation, removal; bearing, endurance; postponement

**перенести(сь)** *pf. of* **переносить(ся)**

**перенос** *m.* transfer, transfering, transport, transportation, shifting; transmission; migration (of ions); carry, carry-over; carry pulse; transposition, translation (*math.*); **п. из предыдущего разряда** previous carry; **п. передаточной функции** translation of transfer function; **сквозной п.** ripple through carry, high-speed carry; **сквозной п. через девятки** standing-on-nines carry; **п. станции** cutting-in of an office (*tphny.*); **управляемый п.** separately instructed carry

**переносимый** *adj.* transferable; endurable; bearable

**переноситель** *m.* transmitter, transporter, transferrer; carrier

**переносить, перенести** *v.* to transfer, to carry, to carry over, to convey, to take, to transmit, to relay; to shift, to transport; to transpose; to endure; to superimpose; to migrate

**переноситься, перенестись** *v.* to migrate; to shift

**переносный** *adj.* transportable, portable, movable, moving; translational; figurative; applicable

**переносчик** *m.* carrier

**перенумерация** *f.* relabeling

**перенумеровывать, перенумеровать** *v.* to enumerate, to number

**переопределённость** *f.* overidentification

**переориентация** *f.* reorientation

**переотклонение** *n.* overshooting, overswing

**переотправление** *n.* retransmission

**переохлаждение** *n.* supercooling, subcooling

**переохлаждённый** *adj.* supercooled, undercooled

**перепад** *m.* drop, jump; gap; step (of voltage); difference; **крутой п.** steep edge; **п. напряжения** step(ped) signal; **п. (потенциала) с предыдущей лампы** preceding tube swing; **п. яркости** brightness transient

**перепаивать, перепаять** *v.* to resolder

**переписать** *pf. of* **переписывать**

**переписывание** *n.* rewriting

переписывать, переписать *v.* to re-write, to copy; to transcribe, to type

переписывающий *adj.* transcribing

перепись *f.* transcription

переплести(сь) *pf. of* переплетать(ся)

переплёт *m.* binding; lacing

переплетать, переплести *v.* to plait, to interlace, to interweave

переплетаться, переплестись *v.* to inlerlace, to interleave, to interweave, to intertwine, to interlock

переплетающий *adj.* interlacing; inter-locking

переплетающийся *adj.* interlocked, in-terlaced, interwoven

переплетение *n.* interlace, interlacing, interleave, intertwining, interlock-ing, interweaving; interlock

переподъём *m.* overhoisting, overwind

переполнение *n.* overflow, overflowing; redundancy

переполюсование *n.* changing of poles, polarity reversal, poling

переполюсовать *v.* to change the poles, to reverse the polarity

переполюсовка *f.* polarity reversal, poling

перепонка *f.* membrane, diaphragm, lamella; septum, pellide; **барабан-ная п.** ear diaphragm, tympanum, ear-drum; **вышибаемая п.** knockout diaphragm; **кортиева п., покровная п.** tectorial membrane

переправлять, переправить *v.* to relay, to cross over; to convey

переприём *m.* re-sending (*tgphy.*)

переприёмник *m.* translator (of tele-graph systems), transfer, network, transducer

перепроверять, перепроверить *v.* to recheck

перепрограммирование *n.* reprogram-ming

перепроявленный *adj.* overdeveloped

перепуск *m.* by-pass

перепускание *n.* underrunning

перепускной *adj.* by-pass; relief; trans-fer; passage

перепутать *pf. of* перепутывать

перепутывание *n.* change; confusion; entanglement

перепутывать, перепутать *v.* to con-fuse, to mix up, to muddle up; to entangle; **п. полюса** to connect with the wrong polarity

переработка *f.* processing, conversion; recovery; reduction (of data)

переразмечать, переразметить *v.* to relable

перераэряжать, перераэрядить *v.* to run down (a battery)

перераспределение *n.* redistribution, redistribution effect

перераспределять, перераспределить *v.* to redirect; to redistribute

перерегистрация *f.* rewrite, rewriting, reregistration

перерегистрировать *v.* to rewrite, to reregister

перерегулирование *n.* overcontrol, overregulation, overcorrection; over-shooting, overswing, hunting, over-travel, overshoot, override

перерегулированный *adj.* overcom-pensated

перерегулировать *v.* to overregulate

перерегулировка *f.* retrimming; **п. на отставание фазы** "overlagging"

перерез *m.* crosscut, cutting through

перерезание *n.* overcut, overcutting

перерезать *pf. of* перерезывать

перерезка *f.* overcutting

перерезывать, перерезать *v.* to cut, to divide; to crosscut, to intersect

перерезывающий *adj.* intersecting; transverse

перерыв *m.* interruption, break, dis-continuity, disturbance; gap, inter-val, stop; **с перерывами** interrupted, intermittent; **п. линии у парного изолятора** circuit loop break

пересекать, пересечь *v.* to cross, to cut, to intersect; to interlace; to traverse

пересекающий *adj.* secant, crossing, intersecting; transverse (axis)

пересекающийся *adj.* intersecting; in-terlaced; crossed

пересечение *n.* intersection, cross, crossing, crossover, traversal, tra-verse; interlacing; interception

пересеченный *adj.* intercepted; cros-sed

пересечь *pf. of* пересекать

пересиливать, пересилить *v.* to over-power

перескакивающий *adj.* jumping over; skipping

перескок *m.* skip effect, skipping, omitting; jump, jump-over, hopping, overreach; transfer, passage

**пересмотр** *m.* overhauling, inspection; **п. номинала** rerating

**пересоединённый** *adj.* reconnected

**пересоединять, пересоединить** *v.* to interchange, to change the connections

**переставить** *pf. of* **переставлять**

**переставка** *f.* reset

**переставленный** *adj.* permuted

**переставляемый** *adj.* resettable, adjustable

**переставлять, переставить** *v.* to transfer, to shift; to alternate; to reset, to readjust, to regulate, to transpose, to rearrange; to commute, to invert

**переставной** *adj.* adjustable, sliding, shifting, resettable; reversible

**перестанавливать** *pf. of* **переставлять**

**перестановка** *f.* transposition, change, displacement, replacement, reset; transfer, relocation; permutation; readjustment, rearrangement; interchange, interchanging, exchange; inversion; **п. хода** stroke adjustment

**перестановочный** *adj.* permutation; adjustable

**перестраиваемый** *adj.* tunable, tuned; adjustable

**перестраивать, перестроить** *v.* to rebuild, to reconstruct; to change over; to tune, to attune; to rearrange; to switch over

**перестраиваться, перестроиться** *v.* to switch over

**перестраивающийся** *adj.* tunable; turnable

**перестроить(ся)** *pf. of* **перестраивать (-ся)**

**перестройка** *f.* retuning; rebuilding, reconstruction; rearrangement

**переступать, переступить** *v.* to overstep, to exceed; to transgress; to cross, to step over

**пересчёт** *m.* conversion, translation, scaling; recalculation; matrix, matrixing, matrix operation; counting; **взаимный п.** interconversion; **п. кода** coder, decoder, discriminator; **п. номера (в регистровых системах)** decoding, translating (in register networks)

**пересчётный** *adj.* conversion, translation; recalculation; scaling; counting

**пересчётчик** *m.* computer, alotter; director, register; translator

**пересчитанный** *adj.* recounted; converted

**пересчитать** *pf. of* **пересчитывать**

**пересчитываемый** *adj.* countable

**пересчитывание** *n.* recounting, counting over; conversion

**пересчитывать, пересчитать** *v.* to count over, to count; to scale; to convert; to recalculate

**пересыщать, пересытить** *v.* to supersaturate

**пересыщение** *n.* supersaturation

**пересыщенный** *adj.* supersaturated

**перетаскивание** *n.* haul

**перетягивать, перетянуть** *v.* to overdrive, to overpower; to overhaul; to overbalance; to draw, to tighten

**перетягивающий** *adj.* overhauling

**перетянутый** *adj.* tightened

**перетянуть** *pf. of* **перетягивать**

**переуспокоение** *n.* overdamping

**переуспокоенный** *adj.* overdamped

**переустроенный** *adj.* modified

**переустройство** *n.* modification, change-over, reconstruction, rebuilding

**перефокусировка** *f.* overfocusing

**переформирование** *n.* reforming, reshaping

**переформированный** *adj.* reshaped

**перехват** *m.* wiretapping, interception, pickup; intercept (*math.*); constriction; intake; **п. вслепую** blind interception

**перехватить** *pf. of* **перехватывать**

**перехватный** *adj.* intercept

**перехватчик** *m.* interceptor

**перехватывание** *n.* interception

**перехватывать, перехватить** *v.* to intercept, to pickup; to overshoot the mark

**перехваченный** *adj.* intercepted

**переход** *m.* transit, transition, crossover, spill-over, passing over, changing over, switching over, conversion, transformation; transfer; passage, migration, crossing; junction (in a semiconductor device); jump; overstroke, overrun, overtravel; transmutation; taper; adapter; **п. в генерацию, п. в режим генерации** spillover; **выращенный п.** rate-grown junction; **п. за крайнее положение** overrun, overstroke, overtravel, overshoot; **конический п. в круглом волноводе** circular waveguide taper; **п. на многократность** multiplexing, multiplication; **п. на одинаковом уровне** level crossing; **п. от полярных координат к прямоугольным** resolution of polar to cartesian (*math.*); **п. от симметричного к несимметричному устройству** balanced-to-unbalanced unit; **плавный п.** graded junction, gradient junction; **п. по знаку плюс** positive jump; **п. полученный изменением скорости выращивания кристалла** rate-grown junction; **предельный п.** transition to a limit; **п. при переполнении** jump on overflow; **резкий п.** step junction; **п. с одного аппарата на другой** fade-over; **п. с треугольника на звезду** star-mesh transformation; **сварной п.** pressure-welded junction; **п. сверху** overhead crossing; **тянутый п.** rate-grown junction; **условный п.** branch operation

**переходить, перейти** *v.* to pass, to pass over, to be converted, to change; to cross; to shunt; to develop, to proceed; **п. с одного аппарата на другой** to fade up and down; to pass from one apparatus (or device) to another

**переходка** *see* **переход**

**переходновременной** *adj.* transit-time

**переходный** *adj.* transient, transitional, transit, transitory, transition, crossover, crossing, passing; connecting; transitive; reversible; reduction

**переходящий** *adj.* transient, transitory; erratic; passing over, changing (into)

**перецоколовка** *f.* rebasing

**перечень** *m.* enumeration; schedule, inventory, docket; list; sum, total

**перечерчивание** *n.* replotting, redrafting

**перечисление** *n.* enumeration; transfer, translation

**перечисленный** *adj.* enumerated

**перечислимый** *adj.* enumerable

**перечислять, перечислить** *v.* to enumerate, to number, to add up; to list; to transfer

**перешеек** *m.* isthmus; vent

**переэкспонированный** *adj.* overexposed, burned up

**переэкспонировать** *v.* to overexpose, to burn up

**пери-** *prefix* peri-

**перигелий** *m.* perihelion

**перидин** *m.* peridyne

**перидинамический** *adj.* peridynamic

**перикон** *m.* pericon

**периконовый** *adj.* pericon

**перилимфа** *f.* perilymph

**периметр** *m.* perimeter

**период** *m.* period, stage, phase; cycle; interval, time, date; spacing; **п. временного ряда** return period; **п. выборки информации** access cycle; **п. двоичных разрядов** bit interval; **п. обратного хода по строкам** horizontal retrace period, horizontal flyback period, line return period; **п. отпертого состояния лампы** "on" period (of a tube); **п. памяти** storage cycle period; **п. полуведения** biological half-life; **п. полураспада** half-life period, transformation period; **п. приостановки (действия)** down time; **п. развёртки** action period, action phase; **п. свободного колебания системы, п. свободных колебаний, п. собственного свободного колебания** natural period, free period; **п. точки** dot cycle; **п. числа** one-number time

**периодизация** *f.* repetitive operation

**периодически** *adv.* periodically

**периодический** *adj.* recurrent, periodic(al); alternating, intermittent, cyclic(al), circulating

**периодичность** *f.* periodicity, noncontinuity; (feed) rate; commutating number; frequency; **п. посылки вызова** ringing cycle

**периодметр** *m.* periodmeter

**периодный** *adj.* period

**периодограмма** *f.* periodogram

**периодопреобразователь** *m.* frequency changer

**перископ** *m.* periscope

**перископический** *adj.* periscopic

**периферийный, периферический** *adj.* peripheral, circumferential

**периферия** *f.* periphery, circumference

**пёрка** *f.* pointed end drill, drill point, bit, cutter; flat drill; **червячная п.** twisted auger; spiral drill

**пермаллой** *m.* permalloy

**перматрон** *m.* permatron

**пермафил** *m.* permafil

**пермеаметр** *m.* permeameter, ferrometer

**пермендур, пермендюр** *m.* permendur

**пер/мин** *abbr.* (периоды в минуту) cycles per minute

**перминвар** *m.* perminvar

**пермиссивный** *adj.* permissive

**пермитол** *m.* permitol

**пермутация** *f.* permutation; transmutation

**Перо** Pérot

**перо** *n.* pen, stylus

**перовой** *adj.* pen; pointed

**перовскит** *m.* perovskite

**перпендикуляр** *m.* perpendicular

**перпендикулярно** *adv.* perpendicularly, perpendicular (to)

**перпендикулярность** *f.* perpendicularity

**перпендикулярный** *adj.* perpendicular, vertical; normal (*geom.*)

**пер/сек** *abbr.* (периоды в секунду) cycles per second

**персептрон** *m.* perceptron

**персистатрон** *m.* persistatron

**персистор** *m.* persistor

**персонал** *m.* personnel, staff

**персональный** *adj.* person-to-person, personal

**перспекс** *m.* perspex; **п. с наполнением** expanded perspex

**перспективный** *adj.* perspective

**пертинакс** *m.* pertinax

**пертурбационный** *adj.* disturbance, perturbation

**пертурбация** *f.* perturbation, disturbance

**перфокарта** *f.* punched card, perforated card; **п. со считываемыми метками** mark-sensing punch card

**перфолента** *f.* punched tape, perforated tape

**перфоратор** *m.* perforator, punch, puncher; drill, drilling machine; punch-card machine; **итоговый п.** summary card punch

**перфораторщик** *m.* puncher

**перфорационный** *adj.* punching; punched

**перфорация** *f.* perforation; drilling, boring; (key) punching; punched hole

**перфорирование** *n.* perforation; drilling, boring; punching, keypunching

**перфорированный** *adj.* perforated, punched; drilled, bored

**перфорировать** *v.* to perforate, to punch; to drill, to pierce, to bore

**перфорирующий** *adj.* perforating

**перхэпсатрон** *m.* perhapsatron

**перцептрон** *m.* perceptron

**перцепция** *f.* perception

**перчатка** *f.* glove; sleeve; **кабельная п.** multiple cable joint

**перьевой** *adj.* pen

**пескодувка** *f.* sandblast unit

**пескоструйный** *adj.* sandblast

**песочный** *adj.* sand, sandy

**пестик** *m.* pin, tamping bar, pestle

**песчатый** *adj.* sand, sandy

**петарда** *f.* petard; firecracker

**петардонакладыватель** *m.* torpedo placer

**петелька** *f.* eyelet, mesh

**петельный** *adj.* loop

**петлевание** *n.* looping

**петлевидный** *adj.* loop, loop-shaped

**петлевой** *adj.* lap, bifilar, double-wound; loop; bent, folded

**петлеобразный** *adj.* loop, loop-shaped, looped; folded, bent

**петлистый** *adj.* looped, loop

**петличный** *adj.* button-hole; lapel; loop

**петля** *f.* loop, stub, kink; slip knot, noose; buttonhole; eye, lug, ear; collar; loop circuit, loop; hair spring; hinge; cusp; **п. в первой бороздке** dog leg; **п. держателя** pigtail; **п. динамического гистерезиса** flux-current loop; **оконечная п. связи** halo-coupled loop, segment-fed loop; **предельная п. гистерезиса** full hysteresis loop, major hysteresis loop, saturated hysteresis loop; **п. с незатухающим током** persistent current loop, storage loop; **п. связи** loop coupler, pickup loop, coupling loop

**петтикот** *m.* petticoat

**петушок** *m.* commutator riser, commutator lug, riser

**Петцваль** Petzval

**печатание** *n.* printing; imprint; reproduction; **автоматическое п. билетов** (*or* **ярлыков**) automatic ticketing (*tphny.*); **п. в десятичных знаках** decimal printing; **п. схем** etching

**печатанный** *adj.* printed

**печатать, напечатать** *v.* to print; to type; to publish; to copy; to impress, to imprint

**печатающий** *adj.* printing

**печатный** *adj.* printed, print; stamped, marked

**печать** *f.* printing, print; seal, stamp
**печь** *f.* furnace, oven, stove
**пещерка** *f.* shell; cave, cavern
**ПИ** *abbr.* (**предыскатель**) preselector
**пианино** *n. indecl.* upright piano; **п. с косорасположенными струнами** oblique pianoforte
**пианиссимо** *adv.* pianissimo
**пиано** *adv.* piano
**Пик** Peek
**пик** *m.* peak, apex, cusp, crest, surge; maximum amplitude; spike
**Пикар** Picard
**пикинговый** *adj.* peaking
**пикнометр** *m.* pycnometer
**пико-** *prefix* pico-, micromicro-
**пиковый** *adj.* peak, crest, spike
**пикоограничитель, пикосрезатель** *m.* peak limiter
**пикосрезающий** *adj.* peak-clipping, peak-limiting, despiking
**пикотранзистор** *m.* pico-transistor
**пикофарада** *f.* picofarad, micromicrofarad
**пиктограмма** *f.* pictogram
**пикфактор** *m.* peak factor
**пила** *f.* saw; file
**пилбокс** *m.* pillbox
**пилить, выпилить** *v.* to saw; to file
**пиловидный, пилозубный, пилозубчатый** *adj.* sawtooth
**пилонный** *adj.* pylon
**пилообразный** *adj.* sawtooth, saw-like, saw-toothed, notched, serrate
**пилот** *m.* pilot
**пилотаж** *m.*, **пилотирование** *n.* pilotage, piloting
**пилотировать** *v.* to pilot
**пилотируемый** *adj.* manned
**пилотирующий** *adj.* guided
**пильный** *adj.* saw, sawing; file
**пильчатый** *adj.* saw-tooth, saw-like, serrate, notched
**пинакоид** *m.* pinacoid
**пинакоидальный** *adj.* pinacoid, pinacoidal
**пинта** *f.* pint
**пинцет** *m.* pincers, tweezers, large forceps, nippers; **п. для ламп** lamp extractor (*tphny.*); **п. для линз** cap extractor (*tphny.*)
**пинчевый** *adj.* pinch
**пинч-эффект** *m.* pinch effect
**пипетка** *f.* pipette, dropper; **п. для электролита** battery syringe
**пирамида** *f.* pyramid

**пирамидальный** *adj.* pyramidal, tapered
**Пирани** Pirani
**пиранол** *m.* pyranol
**пиранометр** *m.* pyranometer
**пиргелиометр** *m.* pyrheliometer
**пиргеометр** *m.* pyrgeometer
**пирекс** *m.* pyrex
**пирит** *m.* iron pyrite; copper pyrite
**пиритный, пиритовый** *adj.* pyritic, pyron
**пирогелиометр** *m.* pyroheliometer
**пирограф** *m.* pyrograph
**пирокатехин** *m.* pyrocatechin, pyrocatechol
**пиролитический** *adj.* pyrolytic
**пиролюзит** *m.* pyrolusite
**пиромагнетизм** *m.* pyromagnetism
**пиромагнитный** *adj.* pyromagnetic
**пирометр** *m.* pyrometer; **оптический п.** disappearing filament pyrometer
**пирометрический** *adj.* pyrometric
**пирометрия** *f.* pyrometry
**пиропроводимость** *f.* pyroconductivity
**пироскоп** *m.* fusion cone, Seger cone
**пиротрон** *m.* pyrotron
**пироэлектрический** *adj.* pyroelectric
**пироэлектричество** *n.* pyroelectricity
**Пирс** Pierce; Pearce
**писк** *m.* squeal, squeaking; peep
**пискать** *or* **пищать, пискнуть** *v.* to squeak, to pipe
**пискливый** *adj.* squeaky, squeaking
**пискнуть** *pf. of* **пискать, пищать**
**пистолет** *m.* pistol
**пистолетный** *adj.* pistol
**пистон** *m.* interface connection; piston; percussion cap; **п. с гальваническим покрытием** plated-through interface connection
**пистонный** *adj.* piston; stack-mounting
**пистонофон** *m.* pistonphone, pistophone (sound generator employing a piston in a closed cylinder)
**пистончик** *m.* eyelet
**пистофон** *m.* pistophone
**письменный** *adj.* writing; written
**письмо** *n.* letter
**питаемый** *adj.* fed, driven, energized
**питание** *n.* feed, feeding, power supply, current supply; supply, delivery; charging, loading; drive; **с автоматическим питанием** self-powered; **п. антенны качающимся рупором** Robinson feed; **п. в пучности тока** current feed; **вч п.** radio-frequency

power supply; **п. (сеточного) сме-щения)** grid voltage supply

**питатель** *m.* feeder, feed wire

**питательный** *adj.* feeder, feeding, feed, power-supply; driving

**питать, напитать** *v.* to feed, to deliver, to supply; to run; to energize

**питающий** *adj.* feeding, feed, supplying, power-supplying

**Пито** Pitot

**питометр** *m.* pitometer

**пифагорейский** *adj.*, **пифагоров** *poss. adj.* Pythagorean, Pythagoric

**пицеин** *m.* picein wax

**пичковый** *adj.* spiking

**Пише** Piché

**пишущий** *adj.* writing; recording; printing

**пищалка** *f.* tweeter

**пищать, пискнуть, пропищать** *v.* to squeak, to peep, to cheep, to pipe; to buzz, to hum, to howl

**пищик** *m.* buzzer, howler, ticker; (car) horn; reed(-type) rectifier

**ПК** *abbr.* (**поправка курса**) course correction

**ПКБ** *abbr.* (**проектно-конструкторское бюро**) project and design office

**ПКО** *abbr.* (**противокосмическая оборона**) astro-defense

**ПЛ** *abbr.* (**подводная лодка**) submarine

**плавание** *n.* floating, swimming; navigation; **п. тона** wow, wow-wows (*acoust.*)

**плавательный, плавающий** *adj.* floating, swimming

**плавень** *m.* soldering flux, brazing flux, flux

**плавильный** *adj.* melting

**плавить, расплавить** *v.* to melt, to fuse

**плавиться, расплавиться** *v.* to melt, to fuse; to blow out (*as:* a fuse)

**плавка** *f.* melting, fusion; **бестигельная зонная п.** floating zone melting

**плавкий** *adj.* melting, fusible, meltable, liquefiable; fused; fuse

**плавкость** *f.* fusibility, liquefaction

**плавление** *n.* melting, fusion, liquefaction

**плавленный** *adj.* fused, melted; smelted

**плавниковый** *adj.* finline

**плавно** *adv.* smoothly

**плавнорегулируемый** *adj.* continuously adjustable, continuously variable

**плавность** *f.* continuity, smoothness, eveness

**плавноуправляемый** *adj.* continuously-controlled

**плавный** *adj.* smooth, even, jerkless; continuous, stepless; tapered; liquid (*linguist.*)

**плавучесть** *f.* buoyancy

**плавучий** *adj.* buoyant, floating

**плавящий, плавящийся** *adj.* fusing, melting

**плагальный** *adj.* plagal

**плазма** *f.* plasma

**плазматический** *adj.* plasmatic, plasmic

**плазматрон** *m.* plasmatron

**плазменный** *adj.* plasma

**плазмоид** *m.* plasmoid, donut-shaped plasma blob

**плазмон** *m.* plasmon

**плакат** *m.* poster, placard

**плакирование** *n.* cladding, metal-cladding, plating

**плакированный** *adj.* clad, plated

**плакировать** *v.* to clad, to plate, to electroplate

**плакировка** *f.* cladding, plating

**пламенный** *adj.* flame, flaming

**пламестойкий** *adj.* flameproof

**пламечувствительный** *adj.* flame-sensitive

**план** *m.* plan, scheme, project; layout, design; projection; plane, surface; device; schedule, program; **сопряжённые планы** associate designs; **задний п.** background; **крупный п.** close-up, close-up view (*tv*); **«перевёрнутый» п. эксперимента** crossover design (of data processing); **передний п.** foreground; **п. расположения оборудования** layout of equipment; **п. съёмки** camera shot

**планарный** *adj.* planar

**планвектор** *m.* plane vector

**планёр** *m.* glider

**планета** *f.* planet

**планетарный, планетный** *adj.* planetary, planet

**планетоид** *m.* planetoid

**планиграфия** *f.* planigraphy

**планиметр** *m.* planimeter

**планиметрический** *adj.* planimetric

**планиметрия** *f.* planimetry

**планирование** *n.* planning, projecting, designing; smoothing, planing, leveling; systematization

**планировать, распланировать** *v.* to program; to plan, to design, to project

**планировка** *f.* planning, laying-out; lay-out

**планирующий** *adj.* gliding

**Планк** Planck

**планка** *f.* strip, plank; baffle, baffle plate; **нажимная п.** cover plate; **п. с наконечниками** (solder) terminal strip

**планово-предупредительный** *adj.* routine-preventive

**Планте** Planté

**планшайба** *f.* face-plate, bulb face, face glass

**планшет** *m.* plotter, plotting board, drawing board, board, table, plane table; topographic map; plot; **гидроакустический п.** sonar plot; **п. для построения функции от двух независимых переменных** X-Y plotter (*math.*); **п.-преобразователь высоты** altitude-conversion table

**планшетист** *m.,* **п.-наблюдатель** *m.* plot observer

**планшетный** *adj.* plotter, board, plotting-board, table

**пласт** *m.* layer, sheet, lamina, lamination, wafer

**пластик** *m.* plastic

**пластика** *f.* 1. plastic. 2. contour accentuation, plasticity of TV pictures, plastic effect; **п. изображения** ghost image, ghost picture, double image, distortion of television image in form of multiple contours of diminishing intensity

**пластикон** *m.* plasticon

**пластина** *f.* plate, flake, tablet, lamina, lamination, sheet, sheeting, panel, vane, leaf, lamella, wafer; membrane; blade, disk, (phonograph) record; fin; plate (of an accumulator); **п. отлипания** strip of nonmagnetic material to prevent armature sticking; **п. пакетного типа** pockets-type plate; **п., работающая на расширение** expander bar; **сигнальная п.** backplate; signal plate, pickup plate; **спечённая п.** sintered plate

**пластинка** *f.* (phonograph) record; **контактная п. цоколя** base eyelet; **чистая п.** recording blank; *see also* **пластина**

**пластинка-основание** *n.* plate base

**пластинный** *adj.* plate, laminar

**пластинообразный** *adj.* lamellar, plate-like, tabular

**пластинчато-роликовый** *adj.* roller-leaf

**пластинчатость** *f.* lamination

**пластинчатый** *adj.* sheet, sheet-like; laminated, lamellar, laminar, lamellate, flaky, flaked; plate, plate-like; finned

**пластификатор** *m.* softener; plasticizer

**пластический** *adj.* plastic, pliable

**пластичность** *f.* plasticity, pliability; ductility; **п. изображения** *see* **пластика изображения**

**пластичный** *adj.* plastic

**пластмасса** *f.* plastic, plastic material, composition material; moulding material

**пластмассовый** *adj.* plastic

**пластование** *n.* wafering

**пластометр** *m.* plastometer

**пластообразный** *adj.* sheet, sheet-like

**плат** *m.* section

**плата** *f.* fee, toll, charge, pay; (mounting) plate; frame, plane, board; sheet, panel; base; wafer; **зажимная п.** terminal plate, terminal box; **основная п. аппарата** mounting bedplate of a device; **печатная п.** printed-circuit board; **повызывная п., поразговорная п.** call fee, call charge, message rate

**платёж** *m.* payment, pay

**платёжный** *adj.* paid, chargeable, billing, payment

**платиметр** *m.* platymeter

**платина** *f.* platinum, Pt

**платинит** *m.* platynite, copper-clad wire

**платиновый** *adj.* platinum, platinic

**платиноид** *m.* platinoid

**платинородиевый** *adj.* platinorhodium

**платинотрон** *m.* platinotron tube

**платион** *m.* plation tube

**платный** *adj.* paid, chargeable; toll; coin-payment

**плато** *n. indecl.* plateau, table land

**платформа** *f.* stage, platform; flatcar

**платформенный** *adj.* platform, platform-like, stage

**плашка** *f.* die, cutter, screw die, cutting die

**плашмя** *adv.* flat, flatwise

**плашмя-направленный** *adj.* broadside-directional

**плевание** *n.* spluttering (of an arc)

**Плейел** Pleijel

**плексиглас** *m.* plexiglass

**плектр** *m.* plectrum, pick

**плёнка** *f.* film; sheet, layer, pellicle, pellide; coating; strip; **двусторонняя п.** double emulsion film, double coated film; **п. для транспарантных съёмок** background film; **односторонняя п.** single emulsion film, single coated film

**плёночный** *adj.* film, pellicular; tape; planar

**плеоморфизм** *m.* pleomorphism

**плеоморфный** *adj.* pleomorphic

**плеохроизм** *m.* pleochroism

**плеохроический, плеохроичный** *adj.* pleochroic, pleochromatic

**плесень** *f.* mold, must; mildew; **грибковая п.** fungus, molds

**плессиметр** *m.* plessimeter

**плетёный** *adj.* woven, braided; wicker; mesh

**плетизмограмма** *f.* plethysmogram

**плетисмограф** *m.* plethysmograph

**плечо** *n.* shoulder; lever (arm), lever, arm, crank (arm); leg (of cathode, circuit); branch; side; **плечи моста, пропорциональные плечи** ratio arms; **п. ножки** stem shoulder; **п. триггера** side of flip-flop

**плинт** *m.* distributing block, terminal block, baseboard, base molding, plinth; connecting strip, terminal strip (*tphny.*)

**плиодинатрон** *m.* pliodynatron

**плиотрон** *m.* pliotron

**плита** *f.* plate, slab, tile; panel, board; base, table; (gas) range, stove; **выверочная п.** surface plate; **мезонитовая п.** fiberboard; **правильная п.** aiming plate, surface plate

**плитка** *f.* slab, block, cake; tile; **древесноволокнистая п.** cellulose fiber tile, fissured cellulose fiber tile; **п. из минерального волокна с гибкой облицовкой** membrane-faced mineral-fiber tile

**плитковатый** *adj.* laminated

**плиткообразный** *adj.* plate-like, tubular; batch

**пловец** *m.* float, floater

**пловучий** *adj.* floating, buoyant

**пломатрон** *m.* plomatron

**пломба** *f.* seal, stamp

**пломбирование** *n.* sealing, stamping

**пломбированный** *adj.* sealed, stamped

**пломбировать, запломбировать** *v.* to seal, to stamp

**плоский** *adj.* flat, square-topped, flat-topped, plane, planar, level, horizontal; flush

**плоско** *adv.* flatly, flat

**плоскобокорезцы** *pl.* sidecutting pliers

**плосковерхний** *adj.* flat-top

**плосковогнутый** *adj.* plano-concave

**плосковыпуклый** *adj.* plano-convex

**плоскогубцы** *pl.* pliers, flat-nosed pliers; **п.-бокорезцы** sidecutting pliers

**плоскодонный** *adj.* flat-bottomed

**плоско-компаундированный** *adj.* flat-compounded, level-compounded

**плоскопараллельный** *adj.* plane-parallel; parallel-plate (electrode)

**плоскополяризованный** *adj.* plane-polarized

**плоскопроволочный** *adj.* flat-wire

**плоскорасстроенный** *adj.* flat-staggered

**плоскослоистый** *adj.* plane-layered, plane-stratified

**плоскостной** *adj.* plane, planar; junction, junction-type

**плоскостность** *f.* planeness, smoothness

**плоскость** *f.* plane, level, surface, flat, flatness; face; pad, sheet, tier; **п. кривизны** levelling plane (*math.*); **отображающая п.** configuration plane; **п. отсчёта** reference plane; **передняя п.** face-plate; **п. разряда** bit plane

**плоскоцилиндрический** *adj.* plane-cylindrical

**плоскоэлектродный** *adj.* flat-electrode

**плотина** *f.* dam, dike

**Плоткин** Plotkin

**плотномер** *m.* densitometer, densimeter

**плотнопрочный** *adj.* composite

**плотностный** *adj.* dense, solid

**плотность** *f.* density, thickness, consistency; tightness, solidity, impenetrability, compactness, massiveness; bunching; opacity; extinction coefficient; **п., измеренная в направленном свете** specular density; **линейная п. ионизации** specific ionization; **мгновенная п. потока звуковой энергии** instantaneous acoustic power per unit area; **п. облучения** irradiancy, irradiance, intensity of illumination; **погонная п.** linear density; **п. потока звуковой энергии** sound energy flux density; **п. приложенного заряда** impressed-charge density; **пространственная п. лучистого потока** radiant intensity; **п. пространственного объёмного заряда** space-charge density; **п. расположения ответчиков** beacon density

**плотный** *adj.* dense, compact, consistent, close, thick, solid; compressed; impermeable, tight, leakproof; massive, closegrained; intimate

**плотовой** *adj.* thump (noise)

**плохой** *adj.* bad, poor; faulty, imperfect, defective

**площадка** *f.* platform, stand, stage, floor; site; plateau (of graph); area; porch; **п. гасящего импульса** pedestal, blanking pedestal; **п. гасящего импульса перед синхроимпульсом** front porch, front porch of pedestal; **п. гасящего импульса позади синхроимпульса** back porch, back porch of pedestal; **п. кода цифры** digit area

**площадь** *f.* area, surface, section, space; (city) square; **выходная п. сопла** nozzle exit area; **концевая п.** annulus; **п. отверстий в распределительной решётке** (*or* **маске**) open aperture of mask; **п. поперечного сечения** cross-sectional area; **п. рассеяния антенны** scattering cross-section of an antenna; **п. рассеяния цели** scattering cross-section of a target; **эффективная п. отражения** specular cross-section; **эффективная п. рассеяния** cross section of target

**плужок** *m.* plow

**плунжер** *m.* plunger, piston, ram; plug

**плунжерный** *adj.* plunger, piston, ram

**плутон** *m.* penumbra

**плутониевый** *adj.* plutonium

**плутоний** *m.* plutonium, Pu

**плутонический** *adj.* plutonic

**плювиограф** *m.* hyetograph, pluviograph

**плювиометр** *m.* hyetometer, pluviometer

**плювиометрический** *adj.* pluviometric

**плюс** *m.* plus; advantage; positive element (of a primary cell)

**плюсовой** *adj.* plus, positive

**п.н.** *abbr.* (**порядковый номер**) ordinal number; index number

**пневматика** *f.* pneumatics

**пневматический** *adj.* pneumatic, pressure-type; air

**пневматолитический, пневматолитовый** *adj.* pneumatolytic

**пневматология** *f.* pneumatology

**пневмогидравлический** *adj.* pneumatic-hydraulic, pneu(mohy)draulic

**пневмодатчик** *m.* pressure gage, pressure pickup

**по** *prep.* 1. *dat.* on, by, at, over, through, in, to, according to, conforming to; (*in phr. oft. transl. by Eng. adj.*): **нормализованный по дальности** range-normalized (*lit.:* normalized by range). 2. *acc.* as far as, up to, to, till. 3. *prepos. case* on, after, for

**побежалость** *f.* iridescence

**побежать** *pf. of* **бежать**

**побелеть** *pf. of* **белеть**

**побережье** *n.* coast, shore

**побить** *pf. of* **бить**

**побледнеть** *pf. of* **бледнеть**

**побороться** *pf. of* **бороться**

**побочие** *n.* leak to ground

**побочный** *adj.* secondary, by-; subsidiary, subordinate, incidental; indirect; supplementary, additional; accessory, extraneous; parasitic, spurious; lateral; reduced

**п-образный** *adj.* bent, folded; p-shaped

**побудитель** *m.* stimulus; stimulator, booster

**побудительный** *adj.* stimulating, inciting; boosting

**побудить** *pf. of* **побуждать**

**побуждаемый** *adj.* stimulated, impelled

**побуждать, побудить** *v.* to stimulate, to impel, to induce, to urge; to boost

**побуждающий** *adj.* stimulating, impelling

**побуждение** *n.* stimulation, drive, boosting; motive, motivation, impulse

**повалить** *pf. of* **валить**

**поведение** *n.* behavior, conduct; procedure; response; **п. в неустановившемся режиме** transient response

**поверить** *pf. of* **поверять**

**поверка** *f.* checking, check-up, verification, calibration; control, control test, test; proof (*math.*)

**повернутый** *adj.* turned

**повернуть** *pf. of* **поворачивать**

**поверочный** *adj.* testing, checking, check, verifying; reference

**повертеть(ся)** *pf. of* **вертеть(ся)**

**поверх** *prep. genit.* over, above, upon

**поверхностноактивный** *adj.* surface-active

**поверхностно-барьерный** *adj.* surface-barrier

**поверхностно-волновой** *adj.* surface-wave

**поверхностно-сплавной** *adj.* surface-alloy

**поверхностноуправляемый** *adj.* surface-controlled

**поверхностный** *adj.* superficial; surface, face; exposed (wire); open, accessible

**поверхность** *f.* surface, area, plane, face, interface; **по всей поверхности** unrestricted, overall; **по наружной поверхности** external; **по части поверхности** restricted; **контактная п. раздела** contact surface; **лицевая п.** front-face area; **лобовая п.** end face; **накапливающая п. из диэлектрика** insulating storage surface; **облучаемая п.** X-ray coverage; **очищенная контактная п.** blank (surface); **плоская п. раздела** plane junction; **рабочая п. полюса** pole face; **п. раздела** interface; **п., сглаживающая канавку** burnishing surface; **смачиваемая п.** hydrophilic surface; **фотометрическая п.** surface of intensity distribution; **частично отражающая п.** surface of separation

**поверять, поверить** *v.* to check, to verify

**повесить** *pf. of* **вешать**

**повести** *pf. of* **вести**

**повешение** *n.* hang-up

**повив** *m.* lay, layer

**повлечь** *pf. of* **влечь**

**повлиять** *pf. of* **влиять**

**повод** *m.* occasion, cause, ground, reason; **п. к ионизации** ionizing event

**поводковый** *adj.* driving, guide; carrier

**поводок** *m.* driver, catch, dog, tappet, lathe carrier

**повозка** *f.* vehicle, car, cart

**поволочить** *pf. of* **волочить**

**поворачиваемый** *adj.* rotatable, reversible

**поворачивание** *n.* turning, slewing, angulation, semirotary motion

**поворачивать, повернуть** *or* **поворотить** *v.* to rotate over, to hunt over a complete level; to turn, to swing; to turn over, to tilt; to reverse; to divert; to cock

**поворачивающий** *adj.* reversing, inverse, inversion

**поворачивающийся** *adj.* rotating, running; reversible; swinging

**поворот** *m.* turn, turning, bend, curve, angulation, sluing, winding; swinging, slewing, reversal; rotation; **п. лёгкой оси намагничивания** anisotropy rotation

**поворотить** *pf. of* **поворачивать**

**поворотно-нажимный** *adj.* turn-push

**поворотно-отжимный** *adj.* turn-pull

**поворотный** *adj.* rotary, rotatable, rotable, rotating, revolving, steerable; turning, steering; tilting, slewing, swing, swivel; reversing, reversible

**повредить(ся)** *pf. of* **повреждать(ся), вредить**

**повреждаемость** *f.* fault rate, failure rate, occurrence of faults; probability of faults

**повреждать, повредить** *v.* to damage, to impair, to cause trouble

**повреждаться, повредиться** *v.* to fail, to get out of order

**повреждение** *n.* damage, breakage, injury, harm; failure, trouble; defect, fault, impairment; accident, disturbance, interference

**повреждённый** *adj.* faulty, out of order, inoperable, defective, damaged, disturbed

**повременный** *adj.* periodic, periodical; flat (rate)

**«повтор»** *m.* echo; ghost, double image, echo image, fold-over, multipath effect

**повторение** *n.* repetition, recurrence, iteration, duplication; rerun

**повторитель** *m.* repeater; (cathode) follower, translator (*tgphy.*); amplifier; repeater-indicator; **анодный п.** anode follower, grounded cathode amplifier; **импульсный п.** transponder; **катодный п.** cathode follower, grounded anode amplifier; **катодный п. в канале цветного разностного сигнала** color difference cathode follower; **регенеративный п.** interpolator (*tgphy.*); **регенеративный п.-синхронизатор** synterpolator (*tgphy.*); **п. сигналов данных** data repeater

**повторительный** *adj.* repeater, repeating, reiterative

**повторить** *pf. of* **повторять**

**повторно** *adv.* repeatedly, over again

**повторно-кратковременный** *adj.* intermittent

**повторный** *adj.* repeated, reiterated, repetitive, iterative; several; duplicate, multiple; re-: **повторная градуировка** recalibration

**повторяемость** *f.* recurrence, repetition rate, frequency; resettability

**повторять, повторить** *v.* to repeat, to iterate; to follow

**повторяющийся** *adj.* repetitive, repeating, iterated, recurrent, reiterative; loop(ed)

**повызывный** *adj.* call, message

**повыситель** *m.* augmenter

**повысительный** *adj.* increasing, boosting; step-up

**повысить(ся)** *pf. of* **повышать(ся)**

**повышать, повысить** *v.* to raise, to heighten, to increase; to boost; to promote, to advance, to step up; to enliven

**повышаться, повыситься** *v.* to increase, to rise

**повышающий** *adj.* step-up; boosting

**повышающийся** *adj.* increasing, rising; upsweep

**повышение** *n.* rise, rising, increase, boost, boosting; stepping up; improvement, accentuation

**повышенный** *adj.* raised, increased; stepped up

**повязка** *f.* tape, bandage, band

**погасание** *n.* extinction, extinguishment, quenching

**погасать, погаснуть** *v.* to go out, to be out, to be extinguished

**погасить** *pf. of* **погашать, гасить**

**погаснуть** *pf. of* **погасать, гаснуть**

**погасший** *adj.* extinguished; extinct

**погашать, погасить** *v.* to extinguish, to quench, to put out; to cancel, to liquidate

**погашение** *n.* amortization; extinction, extinguishment; depreciation; cancellation

**погашенный** *adj.* extinguished, out; cancelled; quenched

**поглотитель** *m.* absorber; absorbent, sorbent; damper; **двухчастотный резонансный п.** two-circuit resonant absorber; **единичный подвешиваемый п.** unit absorber; **коаксиальный п. мощности** coaxial dry load; **п.**
колебаний с двумя резонансными частотами two-circuit damper; **п. мощности коаксиального кабеля** coaxial dry load

**поглотительный** *adj.* absorbing, absorbent, absorption; sorptive

**поглотить** *pf. of* **поглощать**

**поглощаемость** *f.* absorbability; absorptivity

**поглощаемый** *adj.* absorbed, absorbable

**поглощательный** *see* **поглощающий**

**поглощать, поглотить** *v.* to absorb, to pick up, to take up, to capture; to occlude; to resorb, to suck up, to consume

**поглощающий** *adj.* absorbing, absorbent, absorptive; damping

**поглощение** *n.* absorption, sorption, resorption; consumption (of power), input; attenuation; capture; **п. звука в сэбинах** Sabine absorption; **п. окном счётчика** window absorption

**поглощённый** *adj.* absorbed, taken up

**поглушить** *pf. of* **глушить**

**погнать** *pf. of* **гнать**

**погнить** *pf. of* **гнить**

**погнуть** *pf. of* **гнуть**

**поговорить** *pf. of* **говорить;** to have a talk (or a conversation)

**погода** *f.* weather

**погодостойкий, погодоустойчивый** *adj.* weatherproof

**погонный** *adj.* linear, length, run

**погонять** *pf. of* **гонять**

**пограничный** *adj.* boundary, bordering

**погреб** *m.* basement, cellar

**погремушка** *f.* rattle, rattler

**погрешность** *f.* error, mistake; defect; inaccuracy; fault; **абсолютная п. отсчёта** inaccuracy of reading; **п. визирования** pointing error; **п. за счёт системы** systemic error; **п. за счёт установки рамки** loop alignment error; **п. напряжения трансформатора напряжения** ratio error of a voltage transformer; **п. обтекателя** boresight slope; **п., обусловленная угловой погрешностью** phase displacement error; **п. от земляного пути** groundpath error; **п. от преходящих явлений** transient error; **п. по расположению** site error; **п. у нуля** origin distortion

**погружаемый** *adj.* submergible, submersible; immersed, immersion

**погружать, погрузить** *v.* to dip, to immerse, to plunge, to submerge, to sink; to load

**погружающийся** *adj.* merging, sinking, plunging

**погружение** *n.* sinking, immersion, dip, submersion; plunging, dipping, insertion

**погружённость** *f.* submergence

**погружённый** *adj.* sunk, submerged, immersed, plunged; buried

**погружной** *adj.* immersible, immersion

**погрузить** *pf. of* **погружать**

**погрузочный** *adj.* loading

**под** *prep. acc. and instr.* under; to, toward, near

**подаваемый** *adj.* delivered, supplied, fed

**подаватель** *m.* feeder

**подавать, подать** *v.* to give; to convey, to feed, to supply, to apply, to deliver; to enter; **п. в разрыв** to feed in series-opposition, to buck out

**подаваться, податься** *v.* to yield, to give way

**подавитель** *m.* attenuator (of oscillations); suppressor, remover, killer; **п. обратной связи** reaction suppressor

**подавить** *pf. of* **подавлять**

**подавление** *n.* suppression, suppressing, inhibition, repression; quenching, jamming; depression, cancelling, elimination, rejection, nullification; **п. зеркального канала, п. несущей изображения** image rejection, image suppression; **п. обратного хода по кадрам** frame flyback suppression; **п. обратной связи** feedback suppression, antifeedback, regeneration suppression; **п. побочных колебаний** spurious-mode suppression; **п. сигналов звукового сопровождения в канале сигналов изображения** sound-on-vision rejection; **п. слабых сигналов ЧМ** blanket effect (*rad.*); **п. фона переменного тока** hum suppression, hum bucking

**подавленный** *adj.* suppressed; depressed

**подавляемый** *adj.* suppressed; quiescent

**подавлять, подавить** *v.* to suppress, to repress, to crush, to smother; to depress, to kill, to choke; to jam; to inhibit

**подавляющий** *adj.* suppressing

**податливость** *f.* compliance; pliancy, pliability, yielding, elastance; admittance; **полная механическая п.** mechanical admittance

**податливый** *adj.* yielding, pliable, pliant, nonrigid

**податчик** *see* **датчик, передатчик**

**подать(ся)** *pf. of* **подавать(ся)**

**подача** *f.* giving; supply, feed, feeding, delivery, conveyance, inflow, input; sending, transmission; pickup; approach, motion, travel; supply(ing), furnishing; filing, handing-in (of a message); application; entering; entry; **с автоматической подачей** self-feeding; **п. импульса на сетку** grid pulsing; **п. лицевой стороной вверх** face-up feed; **п. лицевой стороной вниз** face-down feed; **п. обратных импульсов** revertive pulsing; **п. опережающих импульсов** prepulsing; **п. смещения** biasing; **п. углей** clockwork feed of carbons

**подающий** *adj.* feed, feeding, conveying, supplying; lead

**подбирать, подобрать** *v.* to select, to match, to fit, to assort; to collate, to merge

**подбор** *m.* fitting, matching, match, selection, sampling; set, assortment; **п. масштаба** scale factoring

**подборка** *f.* merging; set, selection; **п. информации** collating, collation

**подвал** *m.* basement, cellar

**подведение** *n.* supply, supplying, feeding; sluing, slewing; **п. итога** summation, tally

**подведённый** *adj.* supplied, fed, conducted

**подвергать, подвергнуть** *v.* to subject, to submit, to treat (with), to machine

**подвергнутый** *adj.* subjected

**подвергнуть** *pf. of* **подвергать**

**подверженность** *f.* susceptibility; liability

**подверженный** *adj.* subjected (to), exposed; subject (to), susceptible

**подверхностный** *adj.* subsurface

**подвес** *m.* suspension; suspender, carrier arm, dropper, clip, sling, hanger, pendant; **растяжной п.** cord-absorber pendant

**подвесить** *pf. of* **подвешивать**

**подвеска** *f.* suspension, suspension device, suspension support, suspension arm, carrier arm, suspended span, dropper, suspender, slinging wire, sling, clip; hanger; mounting; **демпфирующая п.** shock-absorbing mounting; **продольная** (*or* **цепная**) **п.** catenary suspension

**подвесной, подвесный** *adj.* suspension, suspended, hanging, hanger; mounted, attached; swinging; pendant; hook-up; overhead, overhung, aerial

**подвести** *pf. of* **подводить**

**подвешенный** *adj.* suspended, pendent

**подвешиваемый** *adj.* hung, suspended

**подвешивание** *n.* suspension, hanging; stringing

**подвешивать, подвесить** *v.* to suspend, to hang up, to hang

**подвигать, подвинуть** *v.* to advance, to approach, to move forward, to push

**подвигающий** *adj.* advancing, moving

**подвижной** *adj.* mobile; moving, movable; loose, free, active, live; portable, transportable; sliding, traversing; migrating

**подвижность** *f.* mobility; transportability, portability; fluidity; maneuverability; liveliness; **п. носителей тока** drift mobility (in a homogeneous semiconductor)

**подвижный** *adj.* motional, active, live; moving, movable; floating; transient

**подвинуть** *pf. of* **подвигать**

**подвинчивать, подвинтить** *v.* to screw up, to retighten, to tighten

**подвод** *m.* admission, supply, feed, delivery; lead, feeder; feed line

**подводимый** *adj.* fed, supplied; input

**подводить, подвести** *v.* to supply, to feed; to link up with, to run to, to lead up to; **п. баланс** to balance

**подводка** *f.* supply lead; leading in; feed, supply

**подводнозвуковой** *adj.* submarine-sound, underwater-sound

**подводный** *adj.* submarine; submerged, sunk, sunken, underwater; subaqueous

**подводящий** *adj.* conducting, supplying, feeding, conveying, delivery, leading in

**подвозбудитель** *m.* pilot exciter, subexciter

**подвозбуждение** *n.* priming

**подгонка** *f.* adjustment, fitting, alignment, matching, padding, trimming;

**п. контуров** alignment of tuned circuits; **п. частоты кварцевых пластин** close etching of crystals

**подгоночный** *adj.* adjusting, trimming

**подгонять, подогнать** *v.* to drive on; to adjust, to adapt, to align, to match, to suit, to fit; to start up, to speed up, to bring up to speed

**подгорать, подгореть** *v.* to burn in; to burn, to be burned; to catch fire

**подгорелый** *adj.* burnt, scorched

**подготавливать** = **подготовлять**

**подготавливающий** *adj.* priming

**подготовительный** *adj.* preparatory, preliminary

**подготовить** *pf. of* **подготовлять**

**подготовка** *f.* preparation; prime; **п. к полёту** preflight action; **п. режима** pre-conditioning

**подготовленный** *adj.* prepared

**подготовлять, подготовить** *v.* to prepare, to make ready; to prime

**подграф** *m.* subgraph

**подгрифок** *m.* tailpiece

**подгруппа** *f.* subgroup, subunit, subset; **п. информации** sublock, blockette

**поддавательный** *adj.* yielding, yield

**поддаваться, поддаться** *v.* to yield, to give in; to give way

**поддающийся** *adj.* yielding (to); conformable; **не поддающийся** resisting; **п. прокатке** able to be rolled; **п. управлению** controllable

**поддержание** *n.* maintenance, keeping, upkeep; holding, support, holder, rest, prop; **п. постоянства частоты** frequency maintenance

**поддержать** *pf. of* **поддерживать**

**поддерживаемый** *adj.* holding; supported, sustained

**поддерживание** *n.* prolongation; support

**поддерживать, поддержать** *v.* to support, to hold up, to hold, to bear, to carry; to sustain, to keep; to maintain, to feed

**поддерживающий** *adj.* supporting, carrying, holding, sustaining; carrier

**поддержка** *f.* support; prolongation

**поддетерминанта** *f.* sub-determinant

**поддиапазон** *m.* sub-band, subrange

**поддон** *m.* tray, trough, sump, drip pan

**подействовать** *pf. of* **действовать**

**поделённый** *adj.* shared, distributed

поделить(ся) *pf. of* делить(ся)

подёргивание *n.* bounce; twitching, jerking, jerk, jitter

подёргиваться, подёрнуться *v.* to bounce; to be covered (with)

подержать(ся) *pf. of* держать(ся)

поджечь *pf. of* поджигать

поджигание *n.* ignition, firing; **п. разряда** pre-ignition

поджигатель *m.* igniter

поджигать, поджечь *v.* to fire, to set on fire; to light

поджигающий *adj.* starting; incendiary; trigger

подзаголовок *m.* subheading, subtitle

подзаряд *m.* additional charge, boost charge, replenishing; **непрерывный п.** trickle charge

подзарядить *pf. of* подзаряжать

подзарядка *f.* recharging, boost charge, quick charge, additional charge

подзаряжать, подзарядить *v.* to recharge, to charge additionally; to buffer

подзванивание *n.* (tube) ringing

подзвуковой *adj.* subsonic, subaudio

подземный *adj.* underground, subterranean, below ground; buried

подканал *m.* subchannel, subcarrier channel

подкасательная *f. adj. decl.* subtangent

подкачка *f.* pumping

подкладка *f.* lining, liner, backing, back, distance piece, packing; block; pad; base, base plate, support

подкладной *adj.* backing, put under

подкладочный *adj.* lining

подкласс *m.* subclass

подключаемый *adj.* connected; connecting

подключать, подключить *v.* to connect, to switch (on)

подключение *n.* connection, coupling; **п. для предварительного просмотра** preview switching (*tv*); **п. линий к рабочему месту** operator-position wiring (*tphny.*)

подключённый *adj.* coupled, connected

подключить *pf. of* подключать

подковообразный *adj.* horseshoe, horseshoe-shaped, U-shaped

подконтрольный *adj.* (under) inspection, (under) control

подкос *m.* strut, brace, prop, stay; **п. тваверсы** cross-arm brace

подкрепить *pf. of* подкреплять, крепить

подкрепление *n.* reinforcement, fortification

подкреплять, подкрепить *v.* to reinforce, to strengthen, to fortify; to support

подкрепляющий *adj.* reinforcing, strengthening, fortifying; supporting

подкритический *adj.* subcritical

подкузовный *adj.* underframe

подлёдный *adj.* underice

подлинник *m.* script; original

подлинный *adj.* true, original, real, authentic

подложка *f.* base, backing; foundation, base layer, support; core; sublayer, substrate; header

подлунный *adj.* sublunar

подмагнитить *pf. of* подмагничивать

подмагничение, подмагничивание *n.* magnetic biasing, magnetic bias, field, excitation, superposed magnetization; **вч п., п. переменным полем** a-c magnetic biasing; **п. постоянным током** (*or* **полем**) d-c magnetic biasing

подмагничивать, подмагнитить *v.* to magnetize, to bias, to polarize

подмеднённый *adj.* copper plated

подмешивание *n.* admixture

подмножество *n.* subset (*math.*)

подмодулятор *m.* submodulator, driver circuit, driver device; **п. с управляемым линией блокинг-генератором** line-controlled blocking oscillator driver

подмодуляторный *adj.* submodulator

подначальный *adj.* subordinate

поднесущая *f. adj. decl.* subcarrier; **п. со сдвинутой фазой** phase-shift subcarrier

поднимание *n.* lifting, vertical motion, raising, hoisting

поднимать, поднять *v.* to lift, to raise, to hoist; to step up; to pick up; to turn up; to remove, to take up, to lift off; **п. воротом** to reel

поднимающий *adj.* lifting, raising

поднимающийся *adj.* rising, ascending

подножие *n.* footing, pedestal

подножка *f.* footstep, footboard, running board

поднормаль *m.* subnormal (*math.*)

поднятие *n.* lifting, vertical motion, rise, rising, elevation, ascent; boost

**поднятый** *adj.* raised, elevated, hoisted; boosted

**поднять** *pf. of* **поднимать**

**подобие** *n.* likeness, similarity, similitude, resemblance, comparison, equality; **по подобию** in the image (of), resembling

**подобранность** *f.* match, matching

**подобранный** *adj.* matched, selected, assorted; picked, gathered

**подобрать** *pf. of* **подбирать**

**подогнанный** *adj.* matched; adjusted

**подогнать** *pf. of* **подгонять**

**подогрев** *m.*, **подогревание** *n.* warming up, heating; preheating

**подогреватель** *m.* heating element, heater, heat booster; preheater

**подогревательный** *adj.* heating; preheating

**подогревать, подогреть** *v.* to warm up, to heat up; to preheat

**подогревный** *adj.* heating; heater, heater-type; heated

**подогретый** *adj.* heated, warmed up; preheated

**подогреть** *pf. of* **подогревать**

**подойти** *pf. of* **подходить**

**подопытный** *adj.* test, experimental, tentative

**подотдел** *m.* subsection, subdivision, branch

**подпанель** *m.* subpanel, subbase

**подпёртый** *adj.* strutted, supported, propped

**подпилок** *m.* file

**подпирание** *n.* bracing, strutting, propping, supporting

**подписать** *pf. of* **подписывать**

**подписка** *f.* subscription

**подписчик** *m.* subscriber

**подписывать, подписать** *v.* to sign

**подпитка** *f.* field current, injection, magnetization current; feed maintenance; make-up

**подпиточный** *adj.* interspersing; make-up

**подпитывание** *see* **подпитка**

**подподпрограмма** *f.* sub-subroutine

**подпольный** *adj.* underfloor; underground

**подпора, подпорка** *f.* prop, support, brace; stand; pole brace, carrier brace, push brace, abutment, bearer, bracket, rest; foundation

**подпорный** *adj.* support; booster

**подпорядок** *m.* suborder

**подпоследовательность** *f.* subsequence

**подпочва** *f.* subsoil, underground

**подпрограмма** *f.* subroutine, subprogram; **п. для вычислений в десятичной системе с учётом порядков** floating decimal subroutine; **п. для вычислений с учётом порядков** floating subroutine; **п. одного уровня** first order subroutine, one-level subroutine; **п. первого уровня** first remove subroutine; first order subroutine

**подпружиненный** *adj.* spring-opposed

**подпятник** *m.* step bearing, thrust bearing, meter bottom bearing, bearing; jamb

**подраздел** *m.* subsection

**подразделение** *n.* subdivision, section; **концентрическое п. провода** concentric stranding

**подразделённый** *adj.* divided, subdivided, split; graduated; tapped

**подразделять, подразделить** *v.* to divide, to subdivide; to class, to classify

**подрегулирование** *n.* adjusting, adjustment, readjustment

**подрегулировать** *v.* to adjust, to readjust

**подрегулировка** *f.* readjusting, readjustment, adjustment

**подрезание** *n.* chopping, clipping; undercutting; trimming, cutting

**подрезанный** *adj.* undercut; trimmed, cut

**подрезать** *pf. of* **подрезывать**

**подрезающий** *adj.* clipping

**подрезка** *f.* clipping, cutting, trimming

**подрезной** *adj.* cutting, trimming; undercutting

**подрезывание** *n.* clipping

**подрезывать, подрезать** *v.* to cut, to clip, to trim; to undercut

**подрешётка** *f.* sublattice

**подробность** *f.* detail

**подробный** *adj.* detailed, minute; explicit

**подрыв** *m.* harm, injury; destruct, destruction

**подсборка** *f.* subassembly

**подсветить** *pf. of* **подсвечивать**

**подсветка** *f.* light bias, bias lighting, lighting up, brightening, intensity gate; illumination; blanker; blanking, strobing; **вспомогательная п.** bias lighting; **п. обратного хода**

(луча) retrace brightening (of light); **п. по угловой координате** angle blanking; **п. сзади** back light

**подсвечивание** *n.* brightening, intensification, unblanking

**подсвечиватель** *m.* sweep magnifier, sweep intensifier

**подсвечивать, подсветить** *v.* to flash on; to intensify; to strobe

**подсвечивающий** *adj.* brightening; bright-up, illuminating; intensifying, intensifier

**подсинхронный** *adj.* subsynchronous, hyposynchronous

**подсистема** *f.* subsystem

**подслой** *m.* sublayer, substratum; **разделяющий п.** stripping compound

**подслушать** *pf. of* **подслушивать**

**подслушивание** *n.* listening, monitoring, interception; wire tapping

**подслушиватель** *m.* monitor, monitoring device

**подслушивать, подслушать** *v.* to intercept, to listen in, to tap the wire, to eavesdrop; to monitor

**подсобный** *adj.* subsidiary, auxiliary; secondary, by-

**подставить** *pf. of* **подставлять**

**подставка** *f.* pedestal, rest, base, stand, support; bracket, bearer; socket; block, chock; bridge (of a violin); **изолирующая п.** bench insulator

**подставляемый** *adj.* substituted

**подставлять, подставить** *v.* to substitute, to replace (*math.*); to place under; to hold up, to lift up; to extract

**подстановка** *f.* substitution, replacement (*math.*); resubstitution

**подстанция** *f.* substation; suboffice, subcontrol office, branch (dial) office, satellite exchange; **п. без районных выключателей** satellite office without switching selector repeaters; **вспомогательная п.** satellite office (*tphny.*); **помещаемая в отдельной точке сети п.** spot-network type substation; **фидерная п.** distributing office, junction office (*tphny.*)

**подстилка** *f.* mat, flooring; lining

**подстраиваемый** *adj.* tuneable

**подстраивать, подстроить** *v.* to tune (up), to trim, to align, to readjust

**подстроечник** *m.* tuner

**подстроечный** *adj.* trimming, aligning; pre-set

**подстроить** *pf. of* **подстраивать**

**подстройка** *f.* fine tuning, tiny adjustment, very small tuning; alignment, trimming, frequency trim, readjusting, readjustment, control; **с подстройкой** tunable; **п. контуров** alignment of tuned circuits; **п. по частоте** frequency control

**подстрочный** *adj.* under the line; interlinear

**подсушивать, подсушить** *v.* to dehumidify, to dry a little

**подсушка** *f.* dehumidification

**подсхема** *f.* subcircuit

**подсчёт** *m.* estimation, calculation, computation, count, determination, rating; tabulation; metering; **п. нагрузки** summary of traffic; **п. числа возведений в квадрат, п. числа квадрирований** squaring count

**подсчитывать, подсчитать** *v.* to calculate, to compute, to count up, to tally, to estimate; to tabulate

**подсчитывающий** *adj.* counting

**подталкивание** *n.* jogging

**подтангенс** *m.* subtangent

**подтверждать, подтвердить** *v.* to acknowledge, to confirm

**подтверждение** *v.* acknowledgement, affirmation

**подтональный** *adj.* infrasonic, low-frequency, subsonic, subaudio, infra-acoustic

**подточка** *f.* dubbing; sharpening; groove, recess

**подтравленный** *adj.* undercut

**подтягивание** *n.* inching, jogging

**подтягивать, подтянуть** *v.* to readjust, to reset; to retighten, to pull up, to tighten, to screw up

**подуровень** *m.* sublevel; substate

**подушечка, подушка** *f.* cushion, pillow, (small) pad, bearer, bearing, chock, block; bedding

**подушковидный, подушкообразный** *adj.* pillow-like, pillow-type, pillow, pincushion

**подфидер** *m.* subfeeder

**подхват** *m.* pickup; catching up, picking up; cross support

**подхватить** *pf. of* **подхватывать**

**подхватчик** *m.* pickup

**подхватывать, подхватить** *v.* to intercept, to pick up; to take up, to catch up

**подход** *m.* approach, approaching, access, entrance; manner, way

**подходить, подойти** *v.* to fit, to suit, to match; to approach, to come near, to draw near, to arrive (at)

**подходный** *adj.* approach

**подходящий** *adj.* suitable, fitting, appropriate, adequate, proper, right; incoming, approaching

**подчёркивание** *n.* accentuation, underlining, emphasis, preemphasis, stress; boosting

**подчёркиватель** *m.* accentuator, emphasizer

**подчёркивать, подчеркнуть** *v.* to underline, to accentuate, to stress, to emphasize

**подчёркивающий** *adj.* emphasizing

**подчеркнуть** *pf. of* **подчёркивать**

**подчинённый** *adj.* subordinate, inferior; slave

**подчищать, подчистить** *v.* to retouch, to finish, to dress, to trim; to clean, to erase

**подшасси** *n. indecl.* subchassis

**подшипник** *m.* bearing, bush, bushing, collar; chock; **коренной п.** crankshaft bearing; **опорный п.** bottom bearing; **п. с кольцевой смазкой** ring-oiling bearing, oil-ring bearing, self-oiling bearing; **упорный п.** thrust bearing; **п. шейки вала** journal bearing, collar bearing

**подшипниковый** *adj.* bearing, bush, bushing, collar; chock

**подштукатурный** *adj.* under-plaster

**подщит, подщиток** *m.* branch switchboard

**подъём** *m.* lift, rise; boosting; hoisting, lifting, vertical motion, raising, elevation; pitch; build-up, building-up; stroke; **п. высоких частот при сохранении линейности фазовой характеристики** linear-phase high-frequency peaking; **п. частотной характеристики подбором элементов цепи катода** cathode peaking

**подъёмник** *m.* elevator, hoist, hoisting jack

**подъёмно-вращательный** *adj.* two-motion, vertical-and-rotary

**подъёмно-транспортный** *adj.* hoisting-and-transport

**подъёмный** *adj.* hoisting, lifting, raising; vertical; swing

**подынтегральный** *adj.* subintegral

**подытоживать, подытожить** *v.* to sum up, to summarize, to add up, to total

**поезд** *m.* train; **шарнирно сочленённый п.** articulated train

**поездной** *adj.* train

**поездограф** *m.* train recorder, railway traffic recording apparatus

**поехать** *pf. of* **ехать**

**пожар** *m.* fire

**пожарно-известительный** *adj.* fire-alarm

**пожарный** *adj.* fire

**пожелать** *pf. of* **желать**

**позволять, позволить** *v.* to permit, to allow, to let

**позволяющий** *adj.* permitting, permissive

**позвонить** *pf. of* **звонить**

**поздний** *adj.* late, tardy, retarded

**позитив** *m.* positive, positive image, positive picture

**позитивный** *adj.* positive

**позитрон** *m.* positron, positive electron

**позитронный** *adj.* positron

**позиционер** *m.* positioner

**позиционный** *adj.* position, positioning, positional

**позиция** *f.* position, site, station; attitude; item

**познание** *n.* knowledge, perception; conception

**позолоченный** *adj.* gold-plated, gilded

**позондировать** *pf. of* **зондировать**

**позонный** *adj.* zone

**позывной** *adj.* call; call signal, call letter, call sign, station identification letter(s)

**поиск** *m.* search, searching, seeking, hunt, hunting, scan, scanning, sounding; investigation; pinging, sweep; look-up; retrieval; **гидроакустический п.** sonar search, asdic search; **гидролокационный п.** investigation by echo; **дежурный п.** raster scan

**поискать** *pf. of* **искать**

**поисковый** *adj.* search, searching, scanning; scanner; survey

**Пойнтинг** Poynting

**пойти** *pf. of* **идти**

**показ** *m.* presentation, show

**показание** *n.* indication, presentation, showing, exhibiting, display; reading, readout; registration; aspect; **минимальное п. шкалы** minimum scale value; **открытое п.** clear aspect

**показатель** *m.* index, exponent, value, factor (*math.*); merit; property, characteristic; rate; indicator, pointer; **показатели работы радиолокатора** radar performance; **п. адиабаты** specific-heat ratio, adiabatic index; **видоизмененный п. преломления** modified index of refraction; **п. корня** radical sign; **п. « мощность-полоса пропускания »** power-band merit; **п. направленности в заданном диапазоне частот** band directivity index; **п. НС-структуры** phrase marker; **относительный п. преломления** relative refractive index; **п. станции** code number, code letter, index, numerical office code; **п. степени** exponent sign, exponent, power exponent, index; **п. степени со знаком** signed exponent; **п. степени характеристики передачи** gamma exponent; **п. « усиление-полоса пропускания »** gain-band merit

**показательный** *adj.* exponential; demonstrative, model; significative

**показать(ся)** *pf. of* **показывать(ся)**

**показной** *adj.* display

**показывать, показать** *v.* to display, to exhibit, to show, to demonstrate; to indicate, to point, to register; to signal

**показываться, показаться** *v.* to display; to show itself, to emerge, to appear, to seem

**показывающий** *adj.* indicating

**покаскадный** *adj.* staggered, by stages

**покатость** *f.* slope, sloping, incline, inclination, declivity, slant, descent, grade, pitch, tilt; droop; bias

**покатый** *adj.* sloping, slope, inclining, slanting

**покачать** *pf. of* **покачивать**

**покачивание** *n.* tilting, tilt, swinging; pumping

**покачивать, покачать** *v.* to tilt; to shake, to swing, to waver

**покидать, покинуть** *v.* to leave, to vacate, to abandon

**Поккельс** Pockels

**покой** *m.* rest(ing), standstill, peace, quiet, quiescence

**покойный** *adj.* quiet, quiescent, calm, restful

**поколебать(ся)** *pf. of* **колебать(ся)**

**покомпонентно** *adv.* componentwise

**покоющийся** *adj.* stationary

**покоящийся** *adj.* quiescent

**покраска** *f.* paint, coat of paint; painting, dyeing

**покривить** *pf. of* **кривить**

**покров** *m.* cover, sheath, sheathing, mantle, coat, coating; sheet; deposit

**покровный** *adj.* cover, covering; masking; tectorial

**покрываемый** *adj.* covered; coated

**покрывание** *n.* covering, coating

**покрывать, покрыть** *v.* to cover, to coat, to overlay, to overspread, to deposit, to plate; to house, to roof; to blanket, to envelope, to sheath; to span, to bridge

**покрывающий** *adj.* covering

**покрытие** *n.* covering, sheathing, envelope, enclosure, coating, coverage, lining; layer, film, blanket; insulation (of wire); plating, galvanizing; coat, facing; deposition; **дуговое п.** arc-cover; **звукопоглощающее п. самолётного типа** absorbing airplane blanket; **п. пиковой нагрузки** load factoring; **противогидролокационное п.** anti-asdic coating

**покрытый** *adj.* clad, covered, coated, plated; insulated

**покрыть** *pf. of* **покрывать, крыть**

**покрышка** *f.* cover, lid; case, hood, jacketing, mantle; casing; cover-plate

**пол** *m.* floor, ground

**полагать, положить** *v.* to put, to place, to lay down, to lay on

**поладить** *pf. of* **ладить**

**полакировать** *pf. of* **лакировать**

**полдеровский** *adj.* polder

**поле** *n.* field, ground; bank; margin; frame; **прямо-соединённые контактные поля** straight banks; **п. в относительных единицах** relative field; **вихревое (векторное) п.** circuital vector field, rotational field; **высокочастотное п. « откачки »** RF pump field; **п. гнёзд, гнездовое п.** jack panel (*tphny.*); **п. зрения камеры** camera coverage; **п. искателя** terete field (*tphny.*); **контактное п.** contact bank, link frame (*tphny.*); **контактное п. искателя** line bank, selector bank (*tphny*.); **контактное п. испытательных линий** test bank, private bank (*tphny*.); **п. ламп** lamp panel (*tphny*.); **линейное контактное п.** line bank (*tphny*.); **п. лобового магнитного рассеяния** face-ring stray loss (of armatures); **местное п. коммутатора** extension jacks, answering jackfield (*tphny*.); **местное п. с сигналами** display panel (*tphny*.); **многократное п. системы ЦБ** common battery multiple (jackfield); **п. на выходе** exit field; **наборное п.** patch bay, patch board, jackfield (*tphny*.); **п., направленное по винтовой линии** helical field; **п. начала образования зародышей доменов** nucleation field; **противодействующее п.** chocking field, controlling field; **п. распределителя вызовов** master bank, traffic-distributor bank; **п. рассеяния** stray field, leakage field, extraneous field; **п. рассеяния вобулирующего устройства** stray spot-wobble field; **сетчатое п.** crosshatch pattern; **струнное п.** piano (in a rotary system); **п. чересстрочной развёртки** interlaced field

**полевой** *adj.* field
**полезный** *adj.* useful, helpful, serviceable, beneficial; active, effective; profitable; available; net; **полезная нагрузка** payload
**полёт** *m.* flight, flying
**полететь** *pf. of* **лететь**
**полётный** *adj.* flight
**ползание** *n.* crawling, creeping, sliding
**ползать, ползти** *v.* to creep, to crawl, to slide; to spread
**ползун** *m.* runner, slide, slider, slide bar, slipper, guide shoe; crosshead
**ползунковый** *adj.* sliding, slider-type
**ползунок** *m.* sliding contact, slider, slide; **п. потенциометра** potentiometer movable arm
**ползучесть** *f.* creep, creeping, creepage; efflorescence
**ползучий** *adj.* creeping, crawling; viscous
**ползущий** *adj.* creeping, crawling
**поли-** *prefix* poly-, many, multiple
**полиадический** *adj.* polyadic
**полиакрилнитрил** *m.* polyacryle-nitrile
**полиамид** *m.* polyamide
**поливать, полить** *v.* to water, to pour on

**полигон** *m.* proving ground, firing ground, range; polygon
**полигональный** *adj.* polygonal
**полигонный** *adj.* firing-ground, firing-range, ordnance practice ground
**полиграф** *m.* polygraph
**полиграфический** *adj.* polygraphic
**полигубчатый** *adj.* polyfoam
**полидиод** *m.* polydiode, multi-element diode
**полиизобутилен** *m.* poly-isobutylene
**поликонденсат** *m.* polycondensate
**поликонденсация** *f.* polycondensation
**поликристалл** *m.* polycrystal
**поликристаллический** *adj.* polycrystalline, polycrystal
**полимер** *m.* polymer
**полимеризат** *m.* polymerization product
**полимеризация** *f.* polymerization, crosslinking
**полимеризованный** *adj.* polymerized
**полимеризовать(ся)** *v.* to polymerize
**полимерный** *adj.* polymeric
**полиметр** *m.* polymeter
**полимонохлоротрифторэтилен** *m.* polymonochlorotrifluoroethylene
**полиморф** *m.* polymorph
**полиморфизм** *m.* polymorphy, polymorphism
**полиморфический** *adj.* polymorphic, polymorphous
**полиморфия, полиморфность** *f.* polymorphy, polymorphism
**полиморфный** *adj.* polymorphic, polymorphous
**полином** *m.* polynomial
**полиномиальный** *adj.* polynomial, multinomial
**полипласт** *m.* polyplastic material
**полипропилен** *m.* polypropylene
**полирование** *n.* polishing, burnishing, buffing
**полированный** *adj.* polished, burnished; bright
**полировать, отполировать** *v.* to polish, to burnish, to buff, to brighten
**полировка** *f.* polishing, buffing; polish, finish, gloss; etching; **п. дисками** buffing
**полировочный** *adj.* polishing, buffing, burnishing
**полирующий** *adj.* polishing
**полисемия** *f.* polysemanticism
**полиспаст** *m.* polyspast, tackle pulley, compound pulley

**полистиреновый** *adj.* polystyrene
**полистирол** *m.* polystyrene
**полистироловый** *adj.* polystyrene
**политен** *m.* polythene
**политеновый** *adj.* polythene
**политетрафторэтилен** *m.* polytetra-fluor-ethylene
**политональность** *f.* polytonality
**политрифтормонохлорэтилен** *m.* poly-trifluormonochlor-ethylene
**политрон** *m.* polytron, electromeson
**политропа** *f.* polytropics
**политропический, политропный** *adj.* polytropic
**политура** *f.* polish; varnish
**полить** *pf. of* **поливать, лить**
**полиуретан** *m.* polyurethane
**полиуретановый** *adj.* polyurethane
**полифония** *f.* polyphony
**полихлорвинил** *m.* polyvinyl chloride
**полихлорнафталин** *m.* chloro-naphthalene wax, halowax
**полихлоропрен** *m.* polychloroprene
**полихроматический** *adj.* polychromatic, polychromic, multicolored
**полихромия** *f.* polychromy
**полицейский** *adj.* police
**полициклический** *adj.* polycyclic, polynucleated
**полицилиндрический** *adj.* polycylindrical
**полиэдр** *m.* polyhedron
**полиэдральный, полиэдрический** *adj.* polyhedral
**полиэлектрод** *m.* polyelectrode, multiple electrode
**полиэнергетический** *adj.* polyenergetic
**полиэтилен** *m.* polyethylene
**полиэтиленовый** *adj.* polyethylene
**полиэфир** *m.* polyester, polyether
**полиэфирный** *adj.* polyester, polyether
**полиядерный** *adj.* polynuclear
**полка** *f.* shelf, rack
**полковой** *adj.* regimental
**полноавтоматический** *adj.* full-automatic
**полногабаритный** *adj.* full-sized
**полнозанятый** *adj.* fully-occupied, occupied
**полнозначный** *adj.* meaning-bearing
**полномерный** *adj.* full-sized
**полномочие** *n.* power, authority; commission
**полнопериодный** *adj.* full-wave; full-period

**полноразмерный** *adj.* full-size, full-scale
**полностью** *adv.* fully, utterly, totally, completely
**полнота** *f.* volume (of sound); amplitude; fullness; completeness, absoluteness, exhaustiveness; **п. звука** speech volume, "fullness" of sound
**полноцветный** *adj.* full-color, in full color(s)
**полночастотный** *adj.* full-frequency, high-frequency, high-fidelity
**полношаговый** *adj.* full-pitch
**полный** *adj.* full, complete, total, complex, exhaustive; overall, net, integral; solid; deep (sound)
**половина** *f.* half; **нижняя п. колбы** bulb side, lower half of the bulb
**половинный** *adj.* half
**половой** *adj.* floor
**пологий** *adj.* slanting, sloping, gently sloping
**положение** *n.* position, posture, location, situation, place, station; state; condition; stand; aspect; conclusion; present position (of a target) **без исходного положения** nonhoming; **в нерабочем положении** inoperative; **исходное п.** homing position, home position; initial position; **мгновенное п. цели** target present position; **начальное п.** homing position, home position; **обратное п. покоя** reversed normal position (*tgphy.*); **п. ожидания** stop position (of a selector); **основное п. трезвучия** root position of a triad; **п. повторного опроса** operating condition for PBX equipment; **п. покоя** normal position of relay contacts; **походное п.** stowed position; **п. разряда** digit layout; **п. спектра в частотном диапазоне** frequency location of spectrum; **п. цели** present position of a target
**положительно** *adv.* positively, decidedly
**положительный** *adj.* positive, plus; absolute; affirmative; favorable
**положить** *pf. of* **полагать**
**полоз** *m.* way, runner, slide
**полознозаграждающий** *adj.* band-elimination
**поломка** *f.* breaking, breakdown; fracture
**полоний** *m.* polonium, Po

**полопас** *m.* pollopas

**полоса** *f.* strip, stripe, bar, streak, strap, border; band, bandwidth, wideband; zone, region, belt; stretch, width; margin; **встречные полосы частот** crossbands (in a two-band carrier system) (*tphny.*); **цветные полосы** color bar(s); **п. валентных связей** valence-bond band; **п. заграждения** (*or* **задержания, затухания**) **(фильтра)** (filter) attenuation band, filter stop band; **критическая п. слуха** aural critical band; **межевая п.** interference guard band; **п. непрозрачности** (*or* **непропускания**) **фильтра** filter attenuation band, filter stop band; **п. ограждения от взаимных интерференционных помех** interference guard band; **п. отчуждения** wayleave; **п. переменной частоты** swept band; **подавленная боковая п. частот** vestigial side band; **п. помех** fringe interference; **п. пропускания** bandwidth, passband; **п. пропускания звука** sound bandwidth; **п. пропускания по промежуточной частоте** intermediate-frequency bandwidth; **п. пропускания с перерывами** discontinuous passband; **разговорная п. частот** voice-frequency band; **п. сигнала I** orange-cyan wideband; **п. сигнала Q** red-blue wideband; **сильно выраженная боковая п.** strong sideband; **п. частот** frequency band, wave range, band, bandwidth; **п. частот заглушения** barrage width; **п. частот модулированного по фазе колебания при одной боковой частоте** bandwidth of single side-band phase-modulated oscillation; **п. частот яркостного видеосигнала** monochrome vision band; **п. частот яркостной информации** luminance band, monochrome band; **чёрная п.** flagpole (*tv*); **эквивалентная п. пропускания шумов** noise-equivalent passband

**полосатый** *adj.* banded, stripped, ribbed, streakly, striated; band

**полоска** *f.* strip, band, streak; **п. с ёмкостным тензометрическим датчиком** capacitance strain gage; **п. с номерами** designation strip; **п. со штифтами** strip of tags, terminal strip

**полосковый** *adj.* strip, strip-line

**полосно-заграждающий** *adj.* band-elimination, band-exclusion, band-rejection

**полосной** *adj.* band, passband; strip, bar

**полосно-пропускающий** *adj.* bandpass

**полосный, полосовой** *adj.* band, passband; bar, strip; strap

**полостной, полостный** *adj.* cavity-type, cavity; hole

**полость** *f.* cavity, hollow, void; housing, enclosure, chamber; **барабанная п.** tympanum, tympanic cavity; **п. глотки и рта** oral cavity; **коаксиальная п. с перепонкой** septate coaxial cavity; **п., образуемая входящим в воду телом** water-entry cavity; **п. рта** oral cavity

**полотнище** *n.* cloth, blanket; panel; blade (of a saw); breadth, width

**полотно** *n.* cloth, linen; curtain

**полотяный** *adj.* cloth, linen

**полочка** *f.* shelf

**полпериод** *m.* half-period

**полу-** *prefix* semi, demi, hemi-, half

**полуавтомат** *m.* semi-automatic device, semi-automatic machine; **п. для установки контактов** semi-automatic terminal inserter

**полуавтоматический** *adj.* semi-automatic, aided

**полуактивный** *adj.* semi-active

**полуанкерный** *adj.* half-anchor

**полуапериодичный** *adj.* semi-aperiodic

**« полубабочка »** *f.* semi-butterfly

**полубесконечный** *adj.* semi-infinite

**полуватт** *m.* half watt

**полуваттный** *adj.* half-watt

**полувитковый** *adj.* half-turn, half-turn coil

**полувиток** *m.* half turn; **п. обратной связи** feedback loop; **п. связи** pick-up loop

**полувозбуждённый** *adj.* half-excited

**полуволна** *f.* half-wave; **п. синусоиды** lobe (*math.*)

**полуволновой, полуволновый** *adj.* half-wave

**полувращающийся** *adj.* semirotary

**полувыбранный** *adj.* half-selected, partially selected

**полувыступающий** *adj.* quasi-flush (-mounted)

**полувычитатель** *m.* half subtractor, two-input subtractor

полугоночный *adj.* half-racer
полугоризонтальный *adj.* inclined, sloping
полугруппа *f.* semigroup
полудиполь *m.* half-dipole
полудить *pf. of* лудить
полудрагоценный *adj.* semiprecious
полудуплекс *m* .half-duplex, single-band duplex; одночастотный п. single-band duplex
полудуплексный *adj.* half-duplex, semiduplex
полужёсткий *adj.* semirigid
полузакрытый *adj.* semi-enclosed
полузапись *f.* half-write
полузащищённый *adj.* semiprotected
полузвено *n.* half-section
полуизбирательный *adj.* semiselective
полуизображение *n.* one-field picture
полуизолированный *adj.* semi-insulated
полуизолирующий *adj.* semi-insulating
полуинтервал *m.* half-gap
полуистина *f.* half-truth
полукадр *m.* field, frame field (*tv*)
полуклассический *adj.* semiclassical
полукомплект *m.* subassembly; one terminal set
полуконус *m.* half-angle cone
полукоронирующий *adj.* semicorona
полукорпус *m.* half-yoke
полукристаллический *adj.* semicrystalline, hemicrystalline
полукруглый, полукруговой, полукружный *adj.* semicircular, half-round; semispherical; hemicyclical; crescent; hollow, concave
полукубический *adj.* semicubical
полулогарифмический *adj.* semilogarithmic
полуматовый *adj.* flat-finish, dull-finish
полумаячный *adj.* half-beacon
полумеханизированный *adj.* semiautomatic, aided
полумикро-калориметр *m.* semimicro calorimeter
полумуфта *f.* half-coupling
полунатяжной *adj.* semistrain
полунепрерывный *adj.* semicontinuous
полунепроницаемый *adj.* semitight, semipermeable
полуоборот *m.* half turn
полуокружность *f.* semicircle, semicircumference
полуоктава *f.* semi-octave, half-octave

полуоктавный *adj.* semi-octave, half-octave
полуопределённый *adj.* semidefinite
полуось *f.* semiaxis; большая п. semimajor axis
полуотвод *m.* half-normal bend
полуотражающий *adj.* semireflecting
полуотражённый *adj.* semi-indirect
полупереведённый *adj.* half-reversed
полупеременный *adj.* semivariable
полупериод *m.* half-period, half-cycle, alternation, semiperiod; pulse; semioscillation; п. жизни, п. распада half-life
полупериодный *adj.* half-period
полуплоскость *f.* half-plane, semiplane
полупостоянный *adj.* semipermanent
полупотенциометр *m.* semipotentiometer
полупроводник *m.* semiconductor; п. с дырочной проводимостью P-type semiconductor; п. с примесями extrinsic semiconductor; п. с собственной проводимостью pure semiconductor, intrinsic semiconductor
полупроводниковый *adj.* semiconductor; semiconducting
полупроводящий *adj.* semiconducting, semiconductive
полупрозрачный *adj.* semitransparent, semitranslucent, translucent, semitransmitting
полупроизвольный *adj.* semirandom
полупроницаемый *adj.* semipermeable
полупропускающий *adj.* semipermeable, semitransparent
полупросвечивающий *adj.* semitranslucent
полупространство *n.* half space
полупрямая *f. adj.decl.* half line (*math.*)
полуразмах *m.* semispan, amplitude
полураспределённый *adj.* semidistributed
полусамостоятельный *adj.* semi-half-maintained
полусекундный *adj.* half-second
полусинусоида *f.* half-sinusoid
полусинусоидальный *adj.* half-sine
полуслово *n.* half-word
полусосредоточенный *adj.* semilumped
полустандартный *adj* semistandard
полустатический *adj.* semistatic
полустационарный *adj.* semiportable
полустроб *m.* gate; задний п. late gate; передний п. early gate

полустрочный *adj.* half-line
полустык *m.* half joint
полусумма *f.* half-sum; **п. кратных значений** midrange
полусумматор *m.* half-adder
полусухой *adj.* semiarid; moist
полусфера *f.* hemisphere
полусферический *adj.* hemispherical, semispherical
полусходящийся *adj.* semiconvergent
полутарифный *adj.* half-rate
полутвёрдый *adj.* semisolid, medium-hard
полутеневой *adj.* penumbra
полутень *f.* penumbra, shading
полутолщина *f.* half-thickness
полутон *m.* half-tone, semitone, half-step, penumbra
полутоновый *adj.* half-tone, semitone
полутрубчатый *adj.* semitubular
полуугловатый *adj.* subangular
полуутопленный *adj.* semiflush, semi-sunk, semirecessed
полуфабрикат *m.* semifinished product, semimanufactured article
полухордовый *adj.* half-span (of an arc)
полуцелый *adj.* half-integral, half-integer
полуцентр *m.* half center
полуцилиндрический *adj.* semicylindrical
полуциркуль *m.* semicircle
получаемый *adj.* resulting, resultant; received
получатель *m.* recipient, receiver
получать, получить *v.* to receive, to get, to obtain, to secure; to derive; to produce, to generate
получение *n.* receipt, reception, receiving, getting; derivation; production, output, generation; **п. каналов** channeling; **п. шкалы серых тонов** gray scale tracking
полученный *adj.* received, obtained, resulting
получить *pf. of* получать
полушаг *m.* half-step
полуширина *f.* half width; half thickness
полущелевой *adj.* half-slot
полуэлемент *m.* half-cell
полуэмпирический *adj.* semi-empirical
полуэталонный *adj.* semistandard
полуячейка *f.* half-cell
полый *adj.* hollow, tubular; bare, open

полыхание *n.* smudging
пользование *n.* use, utilization, treatment; **п. цепью** allocation of a circuit
пользователь *m.* user
пользоваться, воспользоваться *v.* to use, to employ, to apply, to make use (of)
полюс *m.* pole, terminal; **полюса возбуждения** field poles; **дополнительный п.** commutating pole; interpole
полюсный *adj.* polar, pole
полюсоопределитель *m.* polarity indicator
поляра *f.* polar (*math.*)
поляренциальный *adj.* polar-differential, polarential
поляризатор *m.* polarizer, polarized disk
поляризационный *adj.* polarization, polarizing
поляризация *f.* polarization, polarizing; **при наличии остаточной п.** on remanence
поляризируемый *adj.* polarizable
поляризование *n.* polarization, polarizing
поляризованность *f.* polarization
поляризованный *adj.* polarized, electropolar, polar-sensitive
поляризовать *v.* to polarize
поляризуемость *f.* polarizability, polarization capacity
поляризуемый *adj.* polarizable
поляризующий *adj.* polarizing
поляризующийся *adj.* polarizable
поляриметр *m.* polarimeter
поляриметрия *f.* polarimetry
полярископ *m.* polariscope
полярископия *f.* polariscopy
полярность *f.* polarity; **обратная п. трансформатора** additive polarity of a transformer; **п. сигнала на входе передатчика** transmitter input polarity
полярный *adj.* polar; arctic
полярограмма *f.* polarogram
полярограф *m.* polarograph
полярографический *adj.* polarographic
полярография *f.* polarography
поляроид *m.* polaroid, polaroid filter
поляроидный *adj.* polaroid
полярон *m.* polaron
поляроскоп *m.* polaroscope

**помазать** *pf. of* **мазать**

**помеднённый** *adj.* copper plated

**поменять(ся)** *pf. of* **менять(ся)**

**померить** *pf. of* **мерить**

**поместить** *pf. of* **помещать**

**пометить** *pf. of* **помечать**

**пометка** *f.* mark, note

**помеха** *f.* interference, disturbance; sturb(s), stray(s); noise, hum, clutter, static; jamming; hindrance, obstacle, impediment; difficulty, trouble, kink; restriction; bug; bloop; *pl.* stray, noise, statics; **не дающий телефонных помех** telephonically silent; **атмосферные помехи** statics, disturbances, atmospheric noise; **зеркальные помехи** image interference; **интерференционные помехи** beat interference; **искусственные помехи** jamming; **местные реверберационные помехи** clutter; **помехи, наводимые по линиям питания** conducted interference; **помехи от газоуловителей** precipitator noise; **помехи от гармоник гетеродина** oscillator harmonic interference; **помехи от зеркального канала** second-channel interference; **помехи от обтекателя, помехи от огибания** translation noise; **помехи от пылеуловителей** precipitator noise; **помехи от своего** (*or* **от совпадающего**) **канала** co-channel interference; **помехи от утечки** leakage-path noise; **помехи от характера местности** site noise; **перекрёстные помехи** crosstalk; **узкополосные помехи** selective interference; **заградительная п.** barrage jamming; **зеркальная п.** image interference; **прицельная п.** selective jamming **прямая п.** forward noise

**помеховосприимчивость** *f.* sensibility of disturbances

**помехоглушитель** *m.* noise silencer

**помехозащитный** *adj.* noise-suppressing

**помехозащищённость** *f.* noise immunity, anti-interference

**помехозащищённый** *adj.* antistatic; jamproof, anti-interference

**помехоограничитель** *m.* noise limiter, noise silencer, noise suppressor; **п. с отрицательной обратной связью** degenerative noise limiter

**помехоподавляющий, помехостойкий** *adj.* anti-interference, staticproof, noiseproof, antijamming, interference-free

**помехоустойчивость** *f.* noise immunity, interference immunity, noise stability, noiseproof feature, interference-killing feature, anti-jamming ability

**помехоустойчивый** *adj.* staticproof, noiseproof, jamproof, antijamming, interference-free, unjammable

**помехофильтр** *m.* noise filter

**помечать, пометить** *v.* to mark, to label, to denote, to indicate; to date

**помеченный** *adj.* marked, labeled

**помешать** *pf. of* **мешать**

**помещать, поместить** *v.* to set, to place, to insert, to install; to locate, to set up, to establish; to stow away; to store (a number); **п. рядом** to juxtapose

**помещение** *n.* room, apartment, lodging; housing, space, chamber, compartment; premises; installation, insertion; **гулкое п.** live room; **п. для тонмейстра** mixing room; **заглушённое п.** anechoic room; **п. режиссуры** master-control room (in a broadcast studio); **п. с релейными устройствами** apparatus room (*tphny.*)

**помещённый** *adj.* housed

**помножать, помножить** *v.* to multiply

**помноженный** *adj.* multiplied

**помножить** *pf. of* **помножать, множить**

**помолчать** *pf. of* **молчать**

**помост** *m.* stage, dais, scaffold, bridging

**помощь** *f.* help, assistance, aid; service

**помпа** *f.* pump

**помпаж** *m.* hunting, flutter, cycling; pumping motion, pumpage

**помутнение** *n.* bloom(ing); darkening, turbidity, dimness, blur; **п. плёнки** blooming

**пондеромоторный** *adj.* pondermotive

**понести** *pf. of* **нести**

**понижать, понизить** *v.* to lower, to reduce, to decrease; to depress; to step-down

**понижаться, понизиться** *v.* to fall, to go down, to drop, to lower, to diminish

**понижающий** *adj.* reducing, depressing, dropping; step-down

**понижающийся** *adj.* falling, dropping

**понижение** *n.* fall, drop, dropping, decrease, lowering, settling; reducing, reduction, decay; depression (of constant); stepping down, bucking out (*elec.*); degradation

**пониженный** *adj.* lowered, reduced, depressed

**понизитель** *m.* reducer; **п. выбросов** despiker; **п. частоты** scaler, scaling circuit

**понизительный** *adj.* reducing; step-down

**понизить(ся)** *pf. of* **понижать(ся)**

**понудительный** *adj.* impellent, pressing, compelling, coercive

**понятие** *n.* idea, concept, notion, intelligence

**понятность** *f.* intelligibility, clearness

**понятный** *adj.* intelligible, clear, understandable, comprehensible

**пооперационный** *adj.* step-by-step

**поочерёдно** *adv.* in turn, by turns, alternately

**поочерёдный** *adj.* alternate, by turn, successive, sequential; time-shared

**попадание** *n.* hit; entry

**попадать, попасть** *v.* to hit, to strike; to fall, to land

**попарно** *adv.* in pairs

**попарно-сопряжённый** *adj.* conjugate-pair

**попарный** *adj.* by pairs, paired

**попасть** *pf. of* **попадать**

**поперек** *adv.* crosswise, cross, transverse, transversal

**попеременно** *adv.* alternately, by turns, in turn

**попеременно-полюсный** *adj.* heteropolar

**попеременный** *adj.* alternate, alternative

**поперечина** *f.* tie, crossarm, cross piece, cross member, crossbar, crossbeam, traverse, transom

**поперечно** *adv.* transversely

**поперечно-лучевой** *adj.* transverse-current, transverse-beam

**поперечно-намагничивающий** *adj.* cross-magnetizing

**поперечнополяризованный** *adj.* cross-polarized

**поперечный** *adj.* transverse, transversal, cross, diametrical; lateral, perpendicular; cross-cut, cross, cross-sectional

**поплавковый** *adj.* floating, float

**поплавок** *m.* float, buoy

**поплавочный** *adj.* floating, float

**пополам** *adv.* in two, half-and-half, by halves

**пополнение** *n.* completion, completing, addition, replenishment; supplement; replacement, reinforcement

**пополнять, пополнить** *v.* to augment, to add, to supplement, to refill, to complete, to replenish; to enlarge, to widen, to enrich

**поправимый** *adj.* reparable, remediable

**поправить** *pf. of* **поправлять**

**поправка** *f.* correction, rectification; repair; readjustment; alteration, modification; allowance; correlation; **с поправкой на вакуум** reduced to vacuum; **индивидуальная п. на наблюдателя** personal equation

**поправление** *n.* correction; restoration; recovery

**поправлять, поправить** *v.* to correct, to rectify; to readjust; to set right; to repair, to mend

**поправочный** *adj.* correction, correctional, corrective, equalizing

**попробовать** *pf. of* **пробовать**

**популяция** *f.* population (*math.*)

**попуск** *m.* blank

**попутно** *adv.* along the way, at the same time, in passing; incidentally

**попутный** *adj.* passing

**попытка** *f.* effort; trial, test, experiment, venture, attempt

**пора** *f.* pore

**поработать** *pf. of* **работать**

**поравнять(ся)** *pf. of* **равнять(ся)**

**поражение** *n.* loss; destruction

**поразговорный** *adj.* message (rate)

**поразрядный** *adj.* digit, step-by-step

**порванный** *adj.* broken, torn

**порвать** *pf. of* **рвать**

**пористость** *f.* porosity

**пористый** *adj.* porous, spongy, cellular, pitted

**порог** *m.* threshold; baffle, baffle plate; cut-off; borderline, margin; level; **п. болевого ощущения** threshold of pain, pain threshold; **болевой п. интенсивности** threshold intensity for pain; **визуальный п. запирания** visual cut-off; **п. возникновения генерации** point of self-oscillation; **п. контрастной чувствительности** brightness difference threshold; **п. неприятного ощущения** threshold of discomfort; **п. осязания** threshold of tickle; **п.**

**осязания вибраций** vibrotactile threshold; **п. отпирания** trigger level; **п. различимости сигнала** signal threshold; **п. слышимости** threshold of audibility, borderline of audibility; **п. срабатывания регулятора** control resolution; **стандартный нормальный нулевой п.** standard reference normal threshold; **п. улучшения (ЧМ)** sputter point (FM); **п. фотоэффекта** photoelectric threshold; **п. чувствительности** threshold of response

**пороговый** *adj.* threshold; cut-off; baffle; fringe

**порождать, породить** *v.* to generate, to produce, to give rise (to)

**порождающий** *adj.* generating, generative

**порождение** *n.* result, outcome; generation, production

**порожек** *m.* string support; **верхний п.** nut (of a violin); **п. на конце** end support (of a string)

**порожний** *adj.* empty, void, vacant, evacuated

**порозиметр** *m.* porosimeter

**порок** *m.* flaw, defect, imperfection, blemish, fault, failure, crack

**порометр** *m.* porometer, porosimeter

**пороховой** *adj.* solid-fuel

**порошковый** *adj.* dust, powder(ed)

**порошкообразный** *adj.* powder-like, powdery, powdered

**порошок** *m.* powder, dust; granule

**порт** *m.* port, harbor

**портативность** *f.* portability

**портативный** *adj.* portable, handy

**портик** *m.* porch, portico

**портить, испортить** *v.* to damage, to spoil, to waste, to impair, to deteriorate

**портовый** *adj.* port, harbor

**портящий** *adj.* damaging, spoiling

**поручение** *n.* commission, assignment; message, order

**поручень** *m.* handrail, railing, guardrail; handle

**порционный** *adj.* batch, in batches, portion

**порция** *f.* batch, portion, lot, share

**порча** *f.* ill effect; damage, injury, breakage, waste; trouble; defect, flaw, impairment; failure

**порченный** *adj.* damaged, injured

**поршень** *m.* plunger, piston; **бескон-**

**тактный п.** choke piston; **п. в экране** baffled piston; **п., нагруженный пружиной** spring-opposed piston; **п. с экраном с фланцем** piston and baffle

**поршневой, поршневый** *adj.* piston, piston-type

**порыв** *m.* gust, puff, inrush

**порывистый** *adj.* gusty, violent, rushing; irregular, jerky, percussive

**порядковый** *adj.* ordinal, serial, consecutive, order

**порядок** *m.* order, rank; form; procedure, sequence, series, array, arrangement, course; degree; exponent; ordering; **п. величины** order of magnitude; **п. следования** sequence; **п. числа** exponent, exponent part of cycle

**ПОС** *abbr.* (**потенциометр обратной связи**) feedback potentiometer

**посадить** *pf. of* **сажать**

**посадка** *f.* fit, fitting, setting; landing; dip, bucking; **п. 2-го класса точности** fine fit; **п. напряжения** voltage dip; **напряжённая п. 2-го класса точности** socket seat, wringing fit, fine fit; **п. по 4-му классу точности** coarse fitting; **п. по приборам** instrument landing; **п. по указаниям с земли** talk-down landing; **п. под давлением** force fit; **ходовая п.** running fit

**посадочный** *adj.* landing, glide; setting

**посаженный** *adj.* fitted, potted

**посветить** *pf. of* **светить**

**посезонный** *adj.* seasonal

**посеребрение** *n.* silver-plating

**посеребрённый** *adj.* silver-plated, silvered, silver-jacketed

**посеребрить** *pf. of* **серебрить**

**посистемный** *adj.* system

**послание** *n.* message; sending

**послать** *pf. of* **посылать**

**после** *prep. genit.* after, since; *adv.* later (on), afterwards

**после-** *prefix* post-, after, secondary, subsequent, supplementary

**последействие** *n.* aftereffect, residual effect, residual induction; fatigue; secondary action, reaction, remanence

**последетектирование** *n.* postdetection

**последетекторный** *adj.* postdetection

**последний** *adj.* resent, latest; final, last, finishing; late; ultimate

**последование** *n.* following
**последовательно** *adv.* in succession, in sequence, in turn; in series, serially; tandem
**последовательно-параллельный** *adj.* series-parallel
**последовательностный** *adj.* consecutive
**последовательность** *f.* sequence, sequencing, succession, order, series, continuity, successiveness; train; consistency; graduation; **временная п.** time-sequential routine; **п. клемм и направление вращения** standard terminal-sequence and direction of rotation; **курсовая п.** homing
**последовательный** *adj.* successive, consecutive, sequential; gradual, step-by-step, cascade; consistent, systematic; series, in series, serial; logical; progressive
**последовать** *pf. of* **следовать**
**последствие** *n.* result, consequence, after-effect
**последующий** *adj.* following, subsequent, posterior, sequent, consecutive, succeedent; consequent (*math.*); after, post; future
**послезвучание** *n.* reverberation, echo
**послеизображение** *n.* burn, image retention, sticking, afterimage, retained image
**послеимпульс** *m.* afterpulse
**послеотклонение** *n.* post-deflection
**послесветить** *v.* to persist, to afterglow
**послесвечение** *n.* afterglow, persistence, phosphorescence afterglow, decay, lag; **с послесвечением** persistent (phosphor); **с длительным послесвечением** long-persistence, long-lag; **с коротким послесвечением** short-persistence, rapid-decay; **п. люминофора** phosphorescent decay; **п. экрана** screen persistence
**послесплавной** *adj.* post-alloy
**послеускорение** *n.* post-deflection acceleration, post-acceleration
**послеускоряющий** *adj.* post-acceleration, post-accelerating, after-accelerating; post-deflection
**послефокусировка** *f.* secondary focusing
**пословный** *adj.* word-by-word
**послойный** *adj.* layer-by-layer, in layers; section

**послушать** *pf. of* **слушать**
**пособие** *n.* manual, textbook; help, assistance
**посодействовать** *pf. of* **содейстовать**
**поспособствовать** *pf. of* **способствовать**
**посреди** *adv. and prep. genit.* in the middle, among
**посредине** *adv.* in the middle
**посредничать** *v.* to intervene
**посреднический** *adj.* referee; intervening, interceding
**посредственность** *f.* mediocrity
**посредственный** *adj.* moderate, middling; mediocre, fair
**посредство** *n.* means, medium; **при посредстве** by means of, thanks to; **через п.** owing to, by, through, thanks to
**посредством** *prep. genit.* by means of, through; **п. которого** by means of which, by virtue of which; whereby; **п. этого** thereby
**посредствующий** *adj.* intermediate
**пост** *m.* post, station; **гидроакустический п.** listening post (*hydroacoustics*); **п. обработки информации гидроакустической аппаратуры** sonar evaluation station; **п. централизации** interlocking tower; **централизационный п.** signal box, shunting box
**пост.** *abbr.* (**постоянный**) constant
**поставить** *pf. of* **поставлять, ставить**
**поставка** *f.* supply, delivery; procurement
**поставленный** *adj.* placed, set
**поставлять, поставить** *v.* to supply, to deliver; to put, to place, to set; to set up, to erect
**постановка** *f.* erection, raising; arrangement, set-up; statement; show; **п. измерительных приборов** instrumentation
**постановление** *n.* regulation, decree, resolution, decision
**постановщик** *m.* producer
**постаревший, постарельнй** *adj.* aged, grown old
**постареть** *pf. of* **стареть**
**постель** *f.* bed; bottom
**постепенно** *adv.* step-by-step, in steps, gradually, by degrees, little by little
**постепенный** *adj.* gradual, progressive, step-by-step; fractional

**пост-контроль** *m.* post-mortem checking

**постовой** *adj.* indoor

**посторонний** *adj.* foreign, strange, extraneous, outside; separate

**постоянная** *f. adj. decl.* constant; **большая п. времени** slow response, long time constant; **восстанавливающая п.** control constant; **п. времени** time constant; **п. времени обмотки статора при короткозамкнутой цепи возбуждения** short-circuit time constant of armature winding; **п. времени цепи сетки** grid time constant; **п. выдержки времени** constant time lag; **диэлектрическая п.** inductivity, electronic inductivity, relative inductivity; **зеркальная п. затухания** image attenuation constant; **п. колебаний, колебательная п.** vibration constant; **п. колебательного контура** resonance constant; **п. магнитоупругой связи** magnetoelastic coupling constant; **малая п. времени** fast time constant, short time constant; **п. передачи четырёхполюсника** image transfer constant; **переходная п. трёхфазного короткого замыкания при включении на данное сопротивление или при отключении внешней цепи** transient time-constant on three-phase short-circuit, or on a given impedance, or on open-circuit; **повторная п. затухания звена** iterative attenuation per section; **поперечная переходная п. времени при короткозамкнутом статоре** quadrature-axis transient short circuit time constant; **п. постоянного затухания** iterative attenuation factor; **п. потерь кабеля** attenuation constant of a cable; **продольная ударная п. времени цепи возбуждения при разомкнутом статоре** direct-axis subtransient open-circuit time constant; **п. пространственного заряда** perveance (of a diode); **п. пространственного заряда многоэлектродной лампы** perveance of a multi-electrode tube; **сверхпереходная п. времени трёхфазного короткого замыкания при включении на заданное сопротивление или при размыкании цепи** subtransient time-constant on three-phase short-cir-cuit, or on a given impedance, or on open-circuit; **частотная п. радиальных колебаний** radial mode frequency constant

**постоянно** *adv.* constantly, continually, always, permanently; uniformly

**постоянно-направленный** *adj.* constant (field)

**постоянный** *adj.* constant, invariable, permanent, consistent, steady, stable, fixed, stationary, dead (center); perpetual, uniform, lasting, continuous; **по существу п.** sensibly constant

**постоянство** *n.* constancy, steadiness, stability, consistency, uniformity; continuity, continuance, persistence, permanence; **п. отношений** constant proportions

**постоять** *pf. of* **стоять**

**пост-программа** *f.* post mortem, post mortem program

**постранично** *adv.* page by page, a page at a time

**постраничный** *adj.* per page, paginal, page-by-page

**постредактирование** *n.* postediting

**построение** *n.* construction, building up, formation, designing, developing; structure; (curve) plot, plotting; **п. кривых** plotting; **п. кривых по точкам** discrete plotting; **п. поля** field mapping; **п. потока** flux plot; **п. схемы** designing the circuit, developing the circuit; nature of a circuit, configuration of a circuit

**построенный** *adj.* built, constructed; plotted; composite

**построитель** *m.* plotter; **п. графика выходных функций** output table; **п. кривых** plotter

**построить** *pf. of* **строить**

**постройка** *f.* building, structure, construction

**построчный** *adj.* linewise, line, sequential, line-at-a-time, line-by-line

**постукивание** *n.* rapping, light knocking, tapping

**постукивать** *v.* to tap, to knock, to rap

**постулат** *m.* law, postulate, hypothesis

**поступательно-возвратный** *adj.* reciprocating

**поступательный** *adj.* progressive, forward, advancing; step; translational (motion)

**поступать, поступить** *v.* to proceed, to commence, to do; to go in, to enter; to act, to deal, to treat

**поступающий** *adj.* entering, incoming; feeding (into)

**поступить** *pf. of* **поступать**

**поступление** *n.* entrance, entering, entry, arrival, ingress, inflow, intake, admission; return

**постучать** *pf. of* **стучать**

**посылатель** *m.* transmitter, sender

**посылать, послать** *v.* to send, to dispatch, to transmit; to refer

**посылающий** *adj.* sending

**посылка** *f.* sending, transmission, dispatching; sample; package, parcel; message; premiss, premise; **автоматическая п. вызова (абоненту)** automatic ringing (to a subscriber); **бесключная п. вызова** keyless ringing (*tphny.*); **большая п.** sumption; **п. вызова от токовращателя** dc ringing, pole-changer ringing; **п. гидролокационной станции** sonar transmission; **п. избирательного вызова** selective ringing, party line ringing; **импульсная п. модулирования частотой (повторения) импульсов** rate-modulated pulse chain; **кодовая п. вызова** code ringing; **машинная п. вызова** power ringing, machine ringing (*tphny.*); **п. наложенного вызова** superposed ringing (*tphny.*); **одновременная п. вызова** composite through-ringing; **однонаправленная п. импульсов** unidirectional pulse train; **п. первого вызова** first ringing impulse, primary call; **п. полуизбирательного вызова** semiselective ringing (*tphny.*); **прерывистая п. вызова, п. прерывчатого вызова** interrupted ringing; **ручная п. вызова** hand-generator ringing (*tphny.*); **п. частотного избирательного вызова** harmonic selective ringing (*tphny.*)

**потайной** *adj.* countersunk, flush, sunk; secret

**потамометр** *m.* potamometer

**поташ** *m.* potash, potassium carbonate

**поташный** *adj.* potash, potassic

**потащить** *pf. of* **тащить**

**потемневший** *adj.* dim, grown dark

**потемнение** *n.* dimness, dullness, darkening blackening, shading; fogging

**потемнять, потемнить** *v.* to darken, to dim, to obscure

**потенциал** *m.* potential, voltage; potential; **одинакового потенциала** midpotential; **п. анода эквивалентного диода** equivalent diode voltage; **п. Гальвини** absolute electrode potential; **п. гашения разряда** extinction voltage; **действующий п.** equivalent diode voltage; **максимальный п. анода** peak forward anode voltage; **п. нулевого заряда** electrocapillary maximum; **п. обратного зажигания** flash-back voltage; **п. повторного зажигания разряда** re-striking voltage; **пробивной п.** anode breakdown voltage

**потенциалоскоп** *m.* beam storage tube, storage tube, memory tube, picture storage tube; **п. прямого видения** Tonotron; **п. с видимым изображением** image storage tube; **п. с видимым цветным изображением** multi-color storage tube; **п. с длительным сохранением записи** recording storage tube

**потенциальность** *f.* potentiality

**потенциальный** *adj.* potential

**потенциограф** *m.* potentiograph

**потенциометр** *m.* potentiometer, potentiometer-type rheostat, bleeder; **п. выдачи данных** data potentiometer; **п.-датчик дальности** range-transmitting potentiometer; **дающий п.** data potentiometer; **п.-делитель** scale-factor potentiometer; **п. для установки коэффициентов вручную** hand(-set coefficient) potentiometer; **контрольный п. для сигнальных ламп** "lamp check" (*tphny.*); **п. коррекции по дальности** normalization potentiometer; **левый п.** left-hand taper; **п. отработки** "follow-up" potentiometer; **п. передачи данных о дальности** range-transmitting potentiometer; **п. регулировки смещения** bias control potentiometer; **п. с малым моментом** low-torque potentiometer; **п. с нелинейным изменением сопротивления** tapered function potentiometer; **п. с неполным уравновешиванием** deflection potentiometer; **п. с плавным переключением** continuous potentiometer; **п. с реохордом** slide-wire potentiometer; **самопишущий**

**п. для непрерывной записи двух кривых** two-pen potentiometer continuous recorder; **п. смещения растра по кадру** vertical centering potentiometer; **п. смещения растра по строке** horizontal centering potentiometer; **п. со ступенчатым переключением** step potentiometer; **п. тонмейстера** fade unit, fader; **п. угла** bearing potentiometer; **п. угла места** elevation potentiometer; **п. установки крутизны** slope potentiometer; **п. центрирования растра по кадру** vertical centering potentiometer; **п. центрирования растра по строкам** horizontal centering potentiometer

**потенциометрический** *adj.* potentiometric, potentiometer(-type)

**потенцировать** *v.* to raise to a higher power (*math.*)

**потенцирующий** *adj.* antilogarithmic, antilogarithm-taking

**потереть** *pf. of* **тереть**

**потеря** *f.* loss, waste, disappearance, escape, drop; **потери активной мощности** loss of watt; **активные потери** ohmic loss, resistance loss; **вносимые потери** insertion loss; **потери мощности на рассеяние в преобразователе** transducer dissipation losses; **потери на кистевой разряд** brush discharge loss; **потери на неогибание** translation loss, playback loss; **потери на преобразование в кристаллическом преобразователе частоты** conversion loss of a frequency-changer crystal; **потери на расширение фронта** spreading loss (of a wave); **потери от вихревых токов** eddy-current loss; **потери преобразования в кристаллическом преобразователе частоты** conversion loss of a frequency-changer cyrstal; **потери при распространении по воздуху** airborne transmission losses; **потери энергии на вихревые токи** eddy-current energy losses; **с потерями** lossy; **временная п. эмиссионной способности** black-out effect; **п. выхода на рыскание** output hunting loss; **п. значащих цифр** loss of accuracy; **п. на лучеиспускание** radiation loss; **п. разрешающей способности** loss of resolution; **угловая п.** angular deviation loss

**потерявший** *adj.* having lost, lost

**потерянный** *adj.* lost, dropped

**потерять** *pf. of* **терять**

**потечь** *pf. of* **течь**

**поток** *m.* stream, streaming, current, flow, flux; torrent; shower (of electrons); run, duct, race; **п. в струе реактивного двигателя** jet flow; **п. двоичной информации** bit traffic; **п. количества движения** flow of momentum; **косой п. (электронов)** skew congruence; **краевой п.** fringing; **лучистый п.** radiant power, radiant flux; **магнитный п. в зазоре, магнитный п. зазора** air-gap flux; **магнитный п. сцепления** magnetic linkage; **мгновенный п. звуковой энергии** instantaneous acoustic power across a surface element; **обратный п.** reflux; **общий п. энергии** net intensity; **п. рассеяния** leakage flux; **п. рассеяния в лобовых частях** coil-end leakage; **световой п.** luminous flux; **п. свободной струи** free-jet stream

**потокосцепление** *n.* interlinkage, linkage, flux linkage

**потолок** *m.* ceiling

**потолочный** *adj.* ceiling, overhead

**поточечный** *adj.* point-by-point

**поточный** *adj.* stream, current, flow; continuous, mass

**потребитель** *m.* customer, consumer, patron, user

**потребить(ся)** *pf. of* **потреблять(ся)**

**потребление** *n.* consumption, use, demand, drain (of current); **п. тока нитью накала** filament current consumption, heater current consumption

**потребляемый** *adj.* consumed; intake, input

**потреблять, потребить** *v.* to consume, to use, to expend

**потребляться, потребиться** *v.* to be consumed, to be used

**потребность** *f.* demand, need, requirement, want, necessity

**потребный** *adj.* required, needful, necessary

**потребовать** *pf. of* **требовать**

**потрескавшийся** *adj.* cracky, flawy

**потрескивание** *n.* crackle, crackling, frying, decrepitation, scratching; engaged click

**потрескивать, потрещать** *v.* to crackle, to decrepitate, to crack, to burst

потрескивающий *adj.* crackling

потрещать *pf. of* потрескивать

потрясение *n.* shock; disturbance

потушить *pf. of* тушить

Потье Potier

потянуть(ся) *pf. of* тянуть(ся)

пофразовый *adj.* sentence-by-sentence

походный *adj.* stowed; traveling

похолодеть *pf. of* холодеть

похолодить *pf. of* холодить

поцифровой *adj.* digit

почва *f.* soil, earth, ground

почвенный *adj.* soil, earth, ground

« почерк » *m.* mannerism (of an operator)

почерневший, почернелый *adj.* blackened, grown black

почернение *n.* blackening, growing black

почернеть *pf. of* чернеть

почернить *pf. of* чернить

починить *pf. of* починять, чинить

починка *f.* repair, repairing, mending, fixing

починять, починить *v.* to repair, to mend, to fix

почистить *pf. of* чистить

почленно *adv.* term-by-term, termwise (*math.*)

почленный *adj.* term-by-term

почта *f.* post office; mail

почтамт *m.* main post office, postmaster office

почти *adv.* nearly, almost; *prefix* near-, almost-, quasi-

почтовый *adj.* post, postal, mail

почувствовать *pf. of* чувствовать

пошаговый *adj.* marching; per step

пошуметь *pf. of* шуметь

пощадить *pf. of* щадить

пощёлкивание *n.* cracking noise

пощупать *pf. of* щупать

поэлементно *adv.* elementwise

поэлементный *adj.* elementwise, element

поэтому *conj.* therefore, consequently

поющий *adj.* singing, humming

появиться *pf. of* появляться

появление *n.* appearance, emergence; onset, advent; lineage; **п. биения** electric dissonance, appearance of beats; **внезапное п.** percussive onset; **п. земли** ground indication; **плавное п. изображения, постепенное п. (изображения)** fade-in (of a picture); **резкое п.** percussive onset

появляться, появиться *v.* to display; to emerge, to originate, to appear, to show itself

появляющийся *adj.* appearing, emerging

пояс *m.* belt, band; zone, region; **п. с инструментами** tool belt

пояснение *n.* explanation, elucidation

пояснительный *adj.* explanatory

поясный *adj.* zone, zonal; belt

ПП *abbr.* (**полупроводник**) semiconductor

ППЛ *abbr.* (**посадочная площадка**) landing field

ППТ *abbr.* (**плоскостный полупроводниковый триод**) junction transistor

ПР *abbr.* (**передвижная радиостанция**) mobile radio station

пр. *abbr.* (**прочее, прочий**) the rest

правдоподобие *n.* likelihood, probability

правдоподобный *adj.* likely, probable, verisimilar

правило *n.* rule, maxim, principle; regulation, procedure; guide bar, straightener; **правила обозначения векторов и стрелок в схемах** directivity rules for vectors; **правила слепого полёта** instrument-flight rules;**Правила технической эксплоатации средств радиосвязи** Radio Communication Practices; **п. буравчика** right-hand screw rule, corkscrew rule; **п. гидроакустического поиска** echo-sweeping procedure; **п. пробочника** corkscrew rule, right-hand screw rule; **п. трёх пальцев** Fleming's rule; **п. штопора** right-hand screw rule, corkscrew rule

правильно *adv.* correctly, properly, accurately, right

правильность *f.* correctness, accuracy; regularity

правильный *adj.* normal, regular, sound; accurate, correct, proper; true; straightening

правительственный *adj.* governmental, government

правка *f.* correcting, correction; straightening; trimming, setting; **п. трапеции** keystone correction, trapezium correction

правовращающий *adj.* right-hand(ed), detrorotatory

правописание *n.* spelling

**право-поляризованный** *adj.* right-polarized

**правосторонний** *adj.* right-hand

**правый** *adj.* right, right-hand(ed), clockwise; rightful; dextro-; **с правым ходом** clockwise, right-turning

**прагаусс** *m.* pragauss

**прагматика** *f.* pragmatics

**прагматический** *adj.* pragmatic

**празеодим** *m.* praseodymium, Pr

**« прайминг »** *m.* priming

**Прайс** Price

**практика** *f.* practice; operating experience

**практически** *adv.* practically

**практический, практичный** *adj.* practical, useful

**прамаксвелл** *m.* pramaxwell

**Прандтль** Prandtl

**превосходить, превзойти** *v.* to dominate; to exceed, to surpass, to excel, to gain on

**превосходно** *adv.* excellently, superiorly

**превосходный** *adj.* excellent, superior, first-class

**превратить** *pf. of* превращать

**превращаемый** *adj.* convertible, transformable

**превращать, превратить** *v.* to convert, to transform, to change, to transmute, to turn; to reduce (*math.*)

**превращение** *n.* conversion, transformation, change, transmutation; reduction; transition

**превращённый** *adj.* converted, transformed, changed

**превышать, превысить** *v.* to exceed, to surpass, to excel, to outdo

**превышающий** *adj.* exceeding

**превышение** *n.* excess, exceeding, surpassing

**преграда** *f.* obstruction, obstacle, barrier, block, bar, fender; interception

**преграждать, преградить** *v.* to bar, to block, to obstruct; to intercept, to interrupt

**преграждающий** *adj.* blocking, obstructing; intercepting

**предвакуум** *m.* initial vacuum, rough exhaust

**предварение** *n.* precedence, preact; advance, advancing, lead; forewarning

**предварительно** *adv.* preliminarily, first; pre-, fore-; as a preliminary

**предварительный** *adj.* preliminary, previous, initial, advance, prime, tentative, pre-

**предварять, предварить** *v.* to prevent; to precede, to anticipate, to forestall; to warn

**предваряющий** *adj.* preventing; preceding, anticipating; warning

**предвключённый** *adj.* superimposed

**предвнесённый** *adj.* inherited

**предвычисление** *n.* prediction, precomputation

**предвычисленный** *adj.* precomputed

**преддиссоциация** *f.* predissociation

**предетекторный** *adj.* predetector

**предел** *m.* limit, limitation, bound, boundary, end; capacity, extent, compass; point; range; margin of error; **пределы измерения по шкале прибора** full-scale readings; **пределы рабочего диапазона счётчика по мощности** limits of power range for accuracy of a meter; **пределы рабочего диапазона счётчика по силе тока** limits of effective current range; **пределы регулировки источника сигнала** source regulation; **пределы сканирования по углу места** elevation scan angle; **верхний п. измерения** rating; **верхний п. измерения (в отношении измеряемой величины)** rating (in terms of the quantity measured); **верхний п. по углу места** upper elevation limit (*rad.*); **п. естественной устойчивости системы передачи электроэнергии** natural stability limit of a transmission system; **п. естественной устойчивости системы передачи электроэнергии в переходном режиме** natural transient stability limit of a transmission system; **п. измерений** effective range, measurement range; **п. исправления** margin (*tphny.*); **нижний п. угла места** low(er) elevation range; **п. прочности** breaking strength; **п. прочности на сжатие** compressive strength; **п. прочности при изгибе** bending strength; **п. прямой видимости** radio-optical line of distance; **п. статической устойчивости системы передачи электроэнергии при неизменном возбуждении** steady-state stability limit, with constant flux, of a transmission system; **п. усталости** fatigue strength; **п. устойчивости**

**системы передачи электроэнергии в переходном процессе при неизменном возбуждении** transient stability limit with constant flux of a transmission system

**предельно** *adv.* extremely; **п. допустимый** maximum rated

**предельно-допустимый** *adj.* permissible, maximum-permissible, maximum-rated

**предельный** *adj.* limiting, limit, boundary, cut-off, threshold; end, terminal, extreme, utmost; ultimate, full, maximum, overall; resolvable

**предзажигаемый** *adj.* pretriggered, triggered

**предзажигание** *n.* prestrike (of an arc), prefire, keep-alive discharge; **п. сигналов** approach lighting

**предзажигатель** *m.* excitation electrode, trigger electrode, keep-alive electrode

**предзажигающий** *adj.* triggering

**предизбиратель** *m.* preselector

**предизбирательность** *f.* preselection

**предикат** *m.* predicate

**предикативный** *adj.* predicative

**предикатный** *adj.* predicate

**предиктивный** *adj.* predictive

**предиктор** *m.* predictor

**предионизатор** *m.* keep-alive electrode

**предионизация** *f.* pre-ionization

**предиссоциация** *f.* predissociation

**предколебательный** *adj.* pre-oscillating

**предконечный** *adj.* penultimate

**предкоррекция** *f.* pre-emphasis, precompensation, precorrection, predistortion

**предложение** *n.* suggestion, proposition, offering; statement, sentence, clause; **рационализаторское п.** efficiency suggestion

**предложительный** *adj.* propositional

**предложный** *adj.* prepositional

**предмет** *m.* object; subject, topic; objective

**предметный** *adj.* objective, object

**предмодулятор** *m.* submodulator

**предмодуляторный** *adj.* submodulator

**предмонтажный** *adj.* pre-installation

**преднагрузка** *f.* preload

**преднастройка** *f.* pretuning

**предоконечный** *adj.* penultimate

**предоминирующий** *adj.* predominant

**предоплата** *f.* prepayment

**предоставление** *n.* allocation, reservation; giving, leaving to

**предостерегающий** *adj.* warning, caution

**предостережение** *n.* warning, caution, notice

**предосторожность** *f.* precaution, caution, safeguard

**предотвратить** *pf. of* **предотвращать**

**предотвращаемый** *adj.* preventable

**предотвращать, предотвратить** *v.* to avert, to ward off, to prevent, to avoid

**предотвращающий** *adj.* averting, preventing

**предотвращение** *n.* averting, warding off, prevention; preclusion

**предохранение** *n.* protection, guard, prevention; safety; proofing, conservation, preservation; **п. от пеленгации** anti DF, anti-direction-finding

**предохранённый** *adj.* protected, shielded, preserved

**предохранитель** *m.* safety lock, safety catch, safety stop, safety device, protector; safeguard, guard; cutout, fusible cutout, fuse; monitor; preserver, preservative; **п. без выдержки времени** non-delayed fuse, quick-action fuse; **п. Бозе** small glass-tube fuse (*tphny*.); **быстродействующий плавкий п.** quick-break fuse; **п. в цепи накала** filament fuse; **п. для носика** tip protector; **контрольный п.** pilot fuse (*tphny*.); **п. кристалла** crystal shutter; **п. многопетлевой сети** fuse for intermeshed networks; **п. от сверхтока и перенапряжения** combined protector and fuse; **плавкий п.** fuse, safety fuse, safety cutout, plug fuse, wire fuse; **плавкий п. обратного тока** heat-coil-type fuse, solder fuse; **плавкий п. с сигнализацией** alarm fuse, grasshopper fuse; **розеточный п.** box fuse; **п. с указателем срабатывания** indicating fuse; **трубчатый п. ножевого типа, трубчатый ножевой п.** bridge fuse; **трубчатый плавкий п.** tubular fuse, tube fuse

**предохранительный** *adj.* protective, protecting, protection; safety, guard; fuse; precautionary; preservative

**предохранять, предохранить** *v.* to safeguard, to guard, to protect, to insulate; to prevent; to conserve, to preserve; to monitor

**предохраняющий** *adj.* protective, protecting, protection

**предписание** *n.* instruction, order, rule, regulation, direction

**предписанный** *adj.* specified, prescribed

**предплазменный** *adj.* pre-plasma

**предплата** *f.* prepayment

**предполагаемый** *adj.* probable, supposed, conjectural, reputed, prospective

**предполагать, предположпть** *v.* to suppose, to presuppose, to presume, to surmise, to conjecture, to guess; to propose, to contemplate

**предполётный** *adj.* preflight

**предположение** *n.* assumption, supposition, theory, hypothesis, guess, guesswork

**предположенный** *adj.* supposed, assumed; proposed

**предположительно** *adv.* hypothetically, presumably, supposedly

**предположить** *pf. of* **предполагать**

**предпоследний** *adj.* penultimate

**предпосылка** *f.* premise, prerequisite; ground, reason

**предпочтение** *n.* advantage; preference

**предпочтительно** *adv.* preferably, rather

**предпочтительность** *f.* preferableness

**предпочтительный** *adj.* preferential, preferable

**предпраздничный** *adj.* week-end

**предприятие** *n.* undertaking, enterprise, business, agency; **п. связи** electrocommunication agency

**предпробивной** *adj.* pre-breakdown

**предпрограммный** *adj.* pre-programed

**предпусковой** *adj.* pre-operational

**предразрядный** *adj.* preconduction

**предраспределительный** *adj.* predistributing

**предредактирование** *n.* pre-editing

**предсказание** *n.* prediction, forecast

**предсказанный** *adj.* predicted

**предсказатель** *m.* predictor

**предсказать** *pf. of* **предсказывать**

**предсказуемостный** *adj.* predictive, predicative

**предсказуемость** *f.* predictability

**предсказывать, предсказать** *v.* to predict, to foretell

**предсказывающий** *adj.* predictive; predictor

**представить** *pf. of* **представлять**

**представление** *n.* representation; idea, notion, theory, concept, formulation; presentation; introduction; notation; picture; **в представлениях геометрической оптики** in (a) geometric optics approximation; **п. в непрерывной форме** analog representation; **п. десятичных чисел в двоичном коде** binary-coded decimal notation; **комбинированное п. чисел** mixed number representation; **п. с учётом порядков** floating-point representation; **п. чисел** notation; **п. чисел без учёта порядков** fixed-point notation; **п. чисел в двенадцатеричной системе** duodecimal notation; **п. чисел в двоично-десятичной системе** binary-coded decimal notation; **п. чисел в системе со смешанным основанием** mixed base notation; **п. чисел с учётом порядков** floating-point notation

**представленный** *adj.* presented, introduced; represented

**представлять, представить** *v.* to present, to offer, to introduce; to represent, to describe

**представляющий** *adj.* representative

**предстартовый** *adj.* prelaunch, prestart

**предупредительный** *adj.* precautionary, preventive; alert, alerting, warning; advance; anticipating; approach

**предупредить** *pf. of* **предупреждать**

**предупреждаемый** *adj.* preventable

**предупреждать, предупредить** *v.* to prevent, to forestall; to warn, to caution; to anticipate, to notify

**предупреждающий** *adj.* preventing; warning

**предупреждение** *n.* prevention, forestalling; warning

**предупреждённый** *adj.* prevented; warned

**предусиление** *n.* pre-amplification

**предусилитель** *m.* preamplifier, prime amplifier; preliminary intensifier

**предускорение** *n.* preacceleration

**предусматривать, предусмотреть** *v.* to foresee, to provide for, to specify

**предусмотренный** *adj.* provided for, specified

**предусмотреть** *pf. of* **предусматривать**

**предусмотрительный** *adj.* prudent, longsighted, foreseeing

**предустановленный** *adj.* predetermined, pre-established

**предшественник** *m.* forerunner, precursor

**предшествовавший** *adj.* preceding, foregoing, antecedent, prior

**предшествование** *n.* precedence, priority, antecedence

**предшествующий** *adj.* predecessor; precedent, preceding, leading, precursory, antecedent, foregoing, older, previous, prior

**предыдущий** *adj.* preceding, foregoing, former, previous

**предызбиратель** *m.* preslector

**предыонизатор** *m.* keep-alive electrode

**предыонизация** *f.* pre-ionization

**предыскажающий** *adj.* predistorting

**предыскажение** *n.* predistortion

**предыскание** *n.* preselection

**предыскатель** *m.* preselector, line switch; subscriber's selector (*tphny.*); **групповой п.** master switch; **обратный п.** line selector with line finder (*tphny.*)

**предыстория** *f.* prehistory, past

**предэкспоненциальный** *adj.* pre-exponential

**преемник** *m.* successor

**преждевременность** *f.* prematurity

**преждевременный** *adj.* premature, early

**прежний** *adj.* previous, former, preceding, prior, earlier

**презентация** *f.* presentation

**преимущество** *n.* advantage; preference

**прейскурант** *m.* catalog, price list

**прейти** *pf. of* преходить

**прекращать, прекратить** *v.* to stop, to cease, to discontinue, to break off, to cut off, to end, to terminate, to finish, to suspend; to shut down

**прекращаться, прекратиться** *v.* to end, to cease, to stop; to die away (of oscillations)

**прекращение** *n.* stopping, stop, cessation, ceasing, interruption, discontinuance, finishing, cutting off; extinction; closing down, shutting down; **п. выборочного обследования** curtailment of sampling

**прекращённый** *adj.* severed, cancelled, discontinued, stopped, cut off, finished; shut down

**преломимый** *adj.* refractable, refrangible

**преломитель** *m.* refractor

**преломить(ся)** *pf. of* **преломлять(ся)**

**преломление** *n.* refraction, deflection; breaking; interception (of beams)

**преломлённый** *adj.* refracted

**преломляемость** *f.* refractivity, refractability, refrangibility

**преломляемый** *adj.* refractable, refrangible

**преломлятель** *m.* refractor

**преломлять, преломить** *v.* to refract; to diffract, to deflect; to break

**преломляться, преломиться** *v.* to be refracted, to be deflected; to be broken

**преломляющий** *adj.* refracting, refractive

**пренебрегаемый** *adj.* negligible

**пренебрегать, пренебречь** *v.* to neglect, to disregard, to ignore; to omit

**пренебрежение** *n.* neglect, disregard

**пренебрежимый** *adj.* negligible

**пренебрежительный** *adj.* neglectful

**пренебречь** *pf. of* пренебрегать

**преобладание** *n.* predominance, prevalance, preponderance, dominance; bias; **п. низов** all bottom (*acoust.*); **п. при посылке пробелов** spacing bias (*tgphy.*); **п. при посылке сигналов** marking bias (*tgphy.*)

**преобладать** *v.* to predominate, to dominate, to prevail

**преобладающий** *adj.* dominant, predominant, preponderant, prevailing

**преобразающая** *f. adj. decl.* transform (*math.*)

**преобразование** *n.* transformation, conversion; transforming, transduction, alteration; variation; translation; transform (*math.*); reforming (*math.*); reduction, reducing; generation; interconversion, reconversion, inversion; rearranging; **п. временных интервалов в код** analog-to-digital time conversion; **п. данных из непрерывной формы в цифровую** analog-digital data interconversion; **п. данных курса** yaw data reduction; **п. из двоичной системы (счисления) в восьмеричную** binary-to-octal conversion; **п. непрерывной информации в дискретную** analog-digital conversion; **п. непрерывных**

данных в цифровые и цифровых в **непрерывные** analog-digital data interconversion; **промежуточное п.** buffering; **функциональное п. с неявным заданием функции** implicit-function generation; **цифро-аналоговое п. напряжения сложением токов** current summation digital-to-voltage conversion

**преобразованный** *adj.* transformed, converted; reduced

**преобразователь** *m.* transducer (*elec. comm.*); converter, transformer; inverter; converter tube, heterodyne conversion transducer, translator; changer; quantizer, generator; reducer; transcriber; adapter; processor; **аналого-цифровой п. времени** time-to-number converter; **аналого-цифровой п. положения** position-to-number converter; **аналого-цифровой п. углового положения(вала)** angle-to-digit converter; **п. в цифровую форму методом сравнения** comparison coder; **п. величины в последовательную форму** analog-to-serial converter; **п. вида колебаний** mode transducer, wave converter; **п. временных интервалов в код** time-to-number converter; **п. вызывного тока звуковой частоты** audio-frequency ringing current converter; **п. выходного сигнала** monitoring element; **гармонический (конверторный) п. (частоты)** harmonic conversion transducer; **гидроакустический п.** sonar transducer; **гидролокационный п.** echo-ranging transducer; **п.-гидрофон** underwater transducer; **двоичный п. (чисел)** binary quantizer; **двусторонний** (*or* **двунаправленный**) **п.** bilateral transducer; **диафрагменный п.** baffle-plate converter; **п. дискретной информации в непрерывную** digital-to-analog converter; **п. записи на перфокартах в запись на магнитную ленту** punched card-to-magnetic tape converter; **п. из цифровой формы в непрерывную** digital-to-analog converter; **п. изображений** image processing system, image translator, image converter tube; **п. изображений в цветные** color converter (*tv*); **п. импульсных сигналов** wave

sampler; **п. информации непрерывной в дискретную и снова в непрерывную** analog-digital-analog converter; **ионный п.** ionized fluid-flow transducer; **ионный п. частоты для питания действующей сети переменного тока** cycloinverter; **ионный п. частоты для работы в двух направлениях** cycloconverter; **ионный п. частоты для работы в одном направлении** cyclorectifier; **п. кода в интервал времени** number-to-time converter; **п. кода в напряжение с промежуточным преобразованием в положение** number-to-position-to-voltage converter; **п. кода в напряжение с промежуточным преобразованием во временные интервалы** number-to-time-to-voltage converter; **п. кода в положение с промежуточным преобразованием в напряжение** number-to-voltage-to-position converter; **коллекторный п. частоты** frequency converter; **логарифмический п. напряжений в цифровую форму** logarithmic voltage quantizer; **п. малых механических смещений в изменения электрической ёмкости** capacitance displacement transducer; **п. механических перемещений в электрическое напряжение** mechanical-electrical transducer; **механотронный п.** vacuum-tube transducer; **п. напряжения в цифровую форму** voltage digitizer; **п. непрерывной информации в дискретную** digitalyzer; **п. непрерывной информации в дискретную типа магнитоэлектрического прибора** low-current analog-to-digital converter; **п. непрерывных данных в дискретные** (*or* **цифровые**) quantizer; **пластинчатый п.** laminated transducer; **п. полного сопротивления** impedance inverter circuit; **п. положения вала в цифру** shaft position digitizer; **п. с вогнутой излучающей поверхностью** concave transducer; **п. с переменным реактивным сопротивлением** variable-reactance transducer; **п. с переменным сопротивлением** variable-resistance transducer; **п. с потоком ионизированной жидкости** ionized fluid-flow transducer; **п. с прямоугольной излучаю-**

щей поверхностью rectangular plate transducer; **п. сигналов** signal data converter; **симметрированный п. для перехода от симметричной линии к несимметричной** symmetrical balance-to-unbalanced converter; **п. системы счисления** radix converter; **п. скорости вращения** rate-of-turn transducer; **сложнопараллельный п.** multiple converter; **тензометрический п.** strain-gage transducer; **п. типа вал-число с фазовращателем** phase-shift coder; **п. типа напряжение-число** voltage coder; **п. типа число-вал с фазовращателем** phase-shift decoder; **п. углового положения(вала)в код** angle-to-digit converter; **функциональный п. на переключателях** switch-type function generator; **цифро-аналоговый п. времени** number-to-time converter; **цифро-аналоговый п. напряжения** number-to-voltage converter; **цифро-аналоговый п. частоты** number-to-frequency converter; **цифровой п.** digitizer; **п. частоты на повышение** up-converter (*rad.*); **п. чисел двоичной системы в восьмиричную** binary-to-octal converter; **п. чисел из одной системы счисления в другую** radix converter; **электроннолучевой функциональный п. с переменной плотностью светового тока** variable-density-type cathode-ray function generator; **электроннооптический п.** image converter tube, optical-electronic transducer

**преобразовательный** *adj.* converting, conversion, transforming, transformative, transformation

**преобразовывать, преобразовать** *v.* to transform, to convert, to modify, to alter, to turn, to converse, to reduce, to change; to regenerate; to transpose; to transcribe; **п. в цифровую форму** to digitize; **масштабно п.** to scale

**преобразующий** *adj.* transforming, inverting; transfer

**преодолевание, преодоление** *n.* overcoming, surmounting; discrimination

**препарат** *m.* preparation, preparative, compound

**препарировать** *v.* to prepare, to make

**препозициональный** *adj.* prepositional

**препятствие** *n.* obstacle, obstruction, difficulty, hindrance, barrier, impediment, block; drawback, stop; restriction

**препятствовать, воспрепятствовать** *v.* to prevent, to hinder, to interfere, to interrupt, to impede, to obstruct, to inhibit; to block, to oppose, to cross; to stop

**препятствующий** *adj.* obstructing, interfering, stopping, preventing, impeding; **п. радиопомехам** anti-interference

**прерванный** *adj.* interrupted, discontinuous, cut-off

**прервать** *pf. of* **прерывать**

**прерыв** *see* **прерывание**

**прерываемый** *adj.* interrupted, discontinued

**прерывание** *n.* interruption, interception, break, breaking, cutting off, chopping; discontinuity; **п. от внешних устройств** peripheral interruption; **п. по времени** clock interruption

**прерыватель** *m*, interrupter, breaker, circuit breaker, contact breaker, cut-out, shut-down switch, make-and-break device; chopper; buzzer, vibrator, ticker, trembler

**прерывать, прервать** *v.* to interrupt, to intercept; to stop, to break off, to discontinue; to break contact, to break, to chop, to disconnect, to open; to shut

**прерывающий** *adj.* interrupting, intercepting; breaking; chopping; disruptive

**прерывающийся** *adj.* discontinuous, intermittent, interrupting

**прерывистость** *f.* discontinuity, intermittence, brokeness

**прерывистый** *adj..* discontinuous, broken, interrupted, intermittent, chopped

**прерывность** *f.* discontinuity, brokeness

**прерывно-управляемый** *adj.* intermittently-controlled

**прерывный** *adj.* discontinuous, interrupted, intermittent, broken

**прерывчатость** *f.* discontinuity, brokenness

**прерывчатый** *see* **прерывистый**

**преселектор** *m.* preselector

преселекция *f.* preselection
пресипитрон *m.* precipitron
преследование *n.* pursuing, pursuit, chasing
преследующий *adj.* pursuing, chasing
пресный *adj.* fresh
пресс *m.* press, pressing machine, punch, punching machine; **много-пуансонный п.** multislide machine; **п. с револьверной головкой и панто-графическим управлением по ша-блону** template-guided turret press
прессбоард *m.* pressboard
прессование *n.* compression, pressing, compressing; moulding
прессованный *adj.* pressed, compressed, squeezed; compacted
прессовать, спрессовать *v.* to press, to compress, to squeeze
прессовка *f.* pressing, compression, squeezing; moulding; **п. в формах** percussion pressing
прессовочный *adj.* molded
прессовый *adj.* press; force
прессуемый *adj.* molded, moldable; pressed
прессующий *adj.* pressing, press
прессующийся *adj.* compressible
прессшпан *m.* pressboard, fullerboard, press-spahn
претерпевать, претерпеть *v.* to undergo, to bear, to endure, to be subjected to
преувеличение *n.* exaggeration
преувеличенно *adv.* exaggeratedly
преувеличенный *adj.* exaggerated
преходить, прейти *v.* to pass, to cross
преходящий *adj.* passing, transient, temporary, short-lived
прецессионный *adj.* precessional
прецессировать *v.* to precess
прецессия *f.* precession
прецизионность *f.* precision, accuracy
прецизионный *adj.* precision
« преципитрон » *m.* precipitron
« преэмфазис » *m.* pre-emphasis
при *prep., prepos. case* by, in, at, near; attached (to); during; in the presence of; about, with
прибавить *pf. of* прибавлять
прибавка *f.,* прибавление *n.* addition, additive, increase, supplement, adding; allowance
прибавлять, прибавить *v.* to augment, to add, to enclose, to increase

прибавляемое *n. adj. decl.* addend (*math.*)
прибавочный *adj.* additional; excessive; after-
прибивать, прибить *v.* to nail, to fasten
приближать, приблизить *v.* to approach, to approximate, to bring near
приближаться, приблизиться *v.* to approach, to draw near, to approximate, to converge
приближающий *adj.* approximate, approximating
приближающийся *adj.* approaching, incoming
приближение *n.* approach, approaching; approximation; proximity; **п. к решению** trial (*math.*); **п. с весом** weighted approximation
приближённость *f.* nearness, proximity
приближённый *adj.* approximate, rough, coarse; close
приблизительно *adv.* approximately, roughly
приблизительность *f.* approximateness, approximation
приблизительный *adj.* approximate, rough
приблизить(ся) *pf. of* приближать(ся)
прибор *m.* apparatus, device, instrument; gage; set, outfit; gear; equipment; unit; meter; **безнулевой п.** suppressed-zero instrument; **беспод-шипниковый п.** taut band meter; **вентильный п.** rectifier instrument; **п. выводящий на станцию помех** jamming homing set; **двусторонний п.** center zero instrument; **двустре-лочный измерительный п.** cross-pointer instrument; **детекторный п.** rectifier instrument; **дистанционный п.** remote acting device; telemeter, remote indicating instrument; **п. для диэлектрических измерений** dielectrometer; **п. для записи звука на фотоплёнке** phonodeik; **п. для записи рельефа** talysurf; **п. для за-писи штриховых изображений** facsimile recorder; **п. для звукозаписи** sound-track engraving apparatus; **п. для измерения группового вре-мени распространения** envelope-delay meter; **п. для измерения кривизны роговой оболочки** kerato-meter; **п. для измерения рас-стояния и угла места** apomecometer;

п. для измерения силы увеличения линзы auxiometer; п. для отмета (*or* отсчёта) времени timer, timing device; п. для отсчёта (заданных) интервалов времени intervalometer; п. для решения дифференциальных уравнений digital differential analyzer; п. для счисления dead reckoning tracer; п. для точного определения местоположения (самолёта) accurate reporting device; п. для тугоухих hearing aid, osophone, deaf aid; п. для увеличения дальности эхо-измерений echo-ranging booster; п. звукового определения глубины sonic depth-finding instrument; измерительный п. с безнулевой шкалой set-up-scale meter, suppressed-zero meter; измерительный п. с непосредственным отсчётом direct-reading instrument, direct-reading meter; измерительный п. с нулём в середине шкалы zero-center meter; клещевой п. split electromagnet instrument; комбинированный п. для испытания передатчиков ЧМ panalyzor; п. коммутационного щита panel instrument; контролирующий п. monitor; контрольный п. check meter; купроксный п. rectifier instrument; ламповый (измерительный) п. (electron-) tube instrument, electronic instrument; магнитоэлектрический п. permanent-magnet moving coil instrument; многошкальный п. unimeter, onemeter, multimeter, multirange instrument, multiscale instrument; multitester; п.-образец prototype; осветительный п. luminaire; панорамный п. panadapter, magic box; п. пересчёта координат на заданный курс offset-course computer; пересчётный п. scaler (in radiation counting); пиковый п. peak (-reading) meter; п. питания от сети power supply unit; полупроводниковый п. с точечным контактом point-to-point transistor; поляризованный электромагнитный п. permanent-magnet moving-iron instrument; потенциальный п. voltage-operated device; проекционно-контрольный п. shadowgraph; регистрирующий п. дальности range recorder; п. с безнулевой шкалой suppressed-zero instrument; п. с задним присоединением, п. с зажимами на задней стороне back connected instrument; п. с индикацией тенью токопроводящей струны string-shadow instrument; п. с клещевым магнитопроводом split-electromagnet instrument; п. с непосредственным отсчётом показаний direct-reading instrument; п. с подвижной шкалой projected-scale instrument; п. с теневым указателем shadow column instrument; самопишущий п. registering apparatus, recorder, recording meter; п. со световым указателем instrument with optical pointer; п. со шкалой без нуля set-up-scale instrument; среднеквадратичный измерительный п. square-law instrument, mean-square indicating instrument; счётно-решающий п. computer; счётно-решающий п. выдачи дальности ranging computer; тепловой п. с выдержкой времени time-lagged thermal instrument; тепловой п. с нагреваемой нитью hot wire instrument; тепловой п. сопротивления bolometric instrument; туннельный п. tunnel effect device; туннельный п. на напылённой плёнке evaporated film tunneling device; п.-указатель detecting instrument; универсальный п. multimeter; п. управления зенитным огнём antiaircraft director, AA director; усредняющий измерительный п. average indicating instrument; щитовой п. switchboard instrument, back connected instrument; щитовой п. с удлинённой шкалой long-scale panel indicator; электромагнитный поляризованный п. permanent-magnet moving-iron instrument

**приборный** *adj.* instrument, instrumental

**прибыль** *f.* gain, profit; increase, rise

**прибытие** *n.* arrival

**приваренный** *adj.* welded; п. встык butt-welded

**приваривать, приварить** *v.* to weld (on), to braze

**приварка** *f.* welding, soldering; sealing; п. глазка eyelet welding; п. к стакану цоколя shell weld; п. штенгеля tabulating

**приварной** *adj.* welding, welded

**приведение** *n.* bringing; reduction; putting (in order); setting (in motion); adducing, adduction; **п. в действие** driving; **п. в соответствие** matching

**приведённый** *adj.* corrected; reduced, lumped; referred; quoted; presented; equated

**привесить** *pf. of* **привешивать**

**привести** *pf. of* **приводить**

**привешивать, привесить** *v.* to hang on, to append, to suspend

**привилегированный** *adj.* privileged

**привилегия** *f.* privilege, preference, priority

**привинтить** *pf. of* **привинчивать**

**привинченный** *adj.* screwed on

**привинчивание** *n.* screwing on

**привинчивать, привинтить** *v.* to screw (on, to); to clamp on

**привнесённый** *adj.* assignable; introduced

**привносить, привнести** *v.* to introduce, to import

**привод** *m.* drive, gear, driving gear, drive mechanism, actuator; servo, operation, movement; bringing; homing; approach; **с приводом** driven by; **с пневматическим приводом** air-powered; **п.-замыкатель** switch-and-lock movement; **комбинированный п. регулятора** combination actuator; **п. нецентрализованной стрелки** outlying switch-and-lock movement; **п. от двигателя** power drive; **п. пера** pen-driving mechanism; **п. с земли** ground-controlled approach, GCA; **п. с коническими зубчатыми колесами** bevel gear drive; **п. тормозного крана машиниста** brake valve actuator; **п. угла места** elevation drive; **п. управляемый с земли** ground-controlled approach, GCA; **п. шагового искателя** stepping selector drive mechanism, stepping relay drive mechanism; **п. эксцентрика** tapped actuator

**приводимость** *f.* reducibility

**приводимый** *adj.* driven; reducible; cited; **п. в действие** operated, driven; **п. от батареи** battery-driven; **п. снаружи** externally operable

**приводить, привести** *v.* to bring, to lead; to drive; to equate, to reduce; to adduce, to present; to set (iu motion); to put (in practice); to cite, to quote; **п. в движение, п. в действие** to drive, to bring into motion, to set in motion, to actuate, to put into operation; **п. в порядок** to order, to put in order, to arrange; to fix; **п. в соответствие** to match

**приводно-замыкающий** *adj.* switch-and-lock

**приводной** *adj.* driving, drive, actuating; power-driven, power-operated; approach; guidance; homing

**приводный** *adj.* homing; brought; quoted, cited

**приводящий** *adj.* operating, driving; reducing; leading; bringing

**привыкание** *n.* habituation, acclimatization

**привычный** *adj.* conservative, usual, habitual, customary

**привязанный** *adj.* tied, attached

**привязать(ся)** *pf. of* **привязывать(ся)**

**привязка** *f.* clamp, clamping; clamping operation, locking; tying; **кадровая п.** vertical clamping; **управляемая п.** keyed clamp; **п. уровня** clamping

**привязной** *adj.* tethered, held down, captive

**привязываемый** *adj.* clamped

**привязывать, привязать** *v.* to tie, to attach, to bind, to fasten

**привязываться, привязаться** *v.* "to lock on"

**привязывающий** *adj.* clamping, catching

**приглаживающий** *adj.* smoothing

**пригласительный** *adj.* call-on, calling-on (of signal)

**приглушать, приглушить** *v.* to damp down, to fade down; to muffle, to deaden

**приглушающий** *adj* muting

**приглушение** *n.* muting, deadening

**приглушить** *pf. of* **приглушать**

**пригнанность** *f.* fitting together, matching

**пригнанный** *adj.* adjusted, matched, fitted

**пригнать** *pf. of* **пригонять**

**пригодиться** *pf. of* **годиться**

**пригодность** *f.* adaptability; suitableness, usefulness, fitness, serviceability; validity; approval

**пригодный** *adj.* fit, suitable, useful, good, adaptable, adequate; **п. для необученного персонала** foolproof

**пригонка** *f.* adjustment, adjusting, alignment; fitting, matching, jointing, fit; processing; **п. свинцовой перчатки (муфты)** beating-in (of the end of a lead sleeve)

**пригоночный** *adj.* fitting, adjusting, matching

**пригонять, пригнать** *v.* to adjust, to adapt; to fit, to join, to joint; to work in

**пригорание** *n.* burning; sticking

**пригоревший, пригорелый** *adj.* burnt, scorched

**пригородный** *adj.* suburban; local; short-distance

**приготавливать** *see* **приготовлять**

**приготовить** *pf. of* **приготовлять**

**приготовление** *n.* preparation; arrangement

**приготовленный** *adj.* prepared

**приготовлять, приготовить** *v.* to prepare, to arrange, to provide

**придавать, придать** *v.* to assign to, to adjoin, to attach; to add, to give, to impart; to increase, to augment

**придавить** *pf. of* **придавливать**

**придавленный** *adj.* pressed, squeezed; jammed

**придавливать, придавить** *v.* to press, to squeeze; to jam

**придание** *n.* giving, imparting; communication; **п. жёсткости** stiffening; **п. морозостойкости** winterization; **п. прямоугольной формы** squaring

**приданный** *adj.* given, added, imparted; attached, associated

**придаток** *m.* attachment, gadget, appendage, addition, additional part; **набегающий наклонный п.** leading ramp

**придаточный** *adj.* additional, accessory

**придать** *pf. of* **придавать**

**придающий** *adj.* adding, imparting, giving

**приделывать, приделать** *v.* to fit, to adapt; to attach, to join

**приём** *m.* receiving, reception; pickup; acceptance; admission, intake; mode, way, process, method, procedure; device (*math.*); **п. заказов** booking of calls, filing of calls; **клопферный п.** sound reading (*tgphy.*); **п. на копиру** carbon recording; **п. на**

**нулевых биениях** zero-beat reception, homodyne reception; **п. на основе опыта** rule-of-thumb method; **п. на разнесённые антенны** diversity reception; **п. на прямую** direct pickup; **п. на слух** sound reading, audible code reception (*tgphy.*); **пишущий п.** visual reception; **п. по методу нулевых биений** zero-beat reception; **п. с диаграммой направленности в виде «восьмёрки»** "figure-of-eight" reception; **п. с записью** visual reception; **п. с чернильным распылением** ink-vapor recording; **п. сигналов в зоне прямой видимости** line-of-sight reception

**приёмка** *f.* accepting, acceptance, taking over, reception; inspection

**приемлемость** *f.* acceptability, admissibility

**приемлемый** *adj.* acceptable, admissible

**приёмник** *m.* radio (set); receiver, receiving vessel, tank, receptacle, collector, container, bin; transducer (*elec.*); (radiation) detector; catcher, hopper; magazine, receptor; recipient; **автодинный п.** autoheterodyne, autodyne, self-heterodyne; **п. гидроакустической станции** sonar-listening set; **п. гидролокатора кругового обзора** sonar receiver scanner; **п. для приёма сигналов с круговой поляризацией волны** right-hand receiver; **инфракрасный п. прямого видения** image-forming infrared receiver; **контрольный п.** monitor(ing) receiver; **п. неподвижных изображений** facsimile receiver; **п. ответов** responser; **п. переменно-постоянного тока** ac/dc receiver, universal receiver; **п. прямого усиления** tuned radio frequency receiver; **регенеративный п.** regenerative receiver; autodyne detector, oscillating detector; **рулочный п.** page printer; **п. с двумя антеннами, нейтрализующий помехи** spaced antenna receiver, barrage receiver; **п. с раздельными каналами усиления по промежуточной частоте для изображения и звука** split sound receiver; **п. с трубкой, имеющей модуляционную характеристику** n-й **степени** nth law receiver, nth power-law receiver; **сетевой п.** socket-powered receiver, socket-powered set, mains(-operated) receiver; **п. слепой посадки** glide-path receiver, glide-slope receiver; **страничный п.** page printer; **фототелеграфный** (*or* **штриховой**) **п. с открытой записью** facsimile printer; **п. цветного телевидения с трёхлучевой** (*or* **трёхпрожекторной**) **трубкой** three-gun display color receiver, three-gun color television receiver

**приёмно-передатчик** *m.* transceiver

**приёмно-передающий** *adj.* send-receiver, transmitting-and-receiving, two-way

**приёмно-усилительный** *adj.* receiver-amplifier

**приёмный** *adj.* receiving, reception, collecting; take-up; pickup, catcher

**приёмозаписывающий** *adj.* receiving-recording

**приёмоответчик** *m.* transmitter-responder, transponder

**приёмопередаточный** *adj.* two-way (*rad.*)

**приёмопередатчик** *m.* transceiver, transmitter-receiver, transponder; **импульсный п.** transponder; **совмещающий п. ответов на запрос** coincident transponder

**приёмо-передающий** *adj.* send-receive, transmitting-and-receiving; transmitter-receiver

**приёмоуказатель** *m.* finder

**приёмочный** *adj.* reception, acceptance

**прижатый** *adj.* low-altitude

**прижать** *pf. of* **прижимать**

**прижигание** *n.* cauterization, searing

**прижизненный** *adj.* in one's lifetime, while alive

**прижим** *m.* clip

**прижимание** *n.* pressing, squeezing; clamping, tightening

**прижимать, прижать** *v.* to press (to), to clasp (to); to tighten, to clamp, to screw

**прижимающий** *adj.* clamping, tightening; pressing, squeezing

**прижимной** *adj.* pressure, pressing; tightening, clamping

**прижимный** *adj.* pressure

**приземление** *n.* landing

**приземляться, приземлиться** *v.* to land

**призма** *f.* prism; **п. из крона** crown-glass prism

**призматический** *adj.* prismatic

**признак** *m.* sign, character, indication, mark, index, criterion; distinctive feature, feature, attribute, characteristic; marker, tag; sentinel; **для нескольких признаков** multiple; **п. группы** block constant; **импульсный п.** mode of a pulse

**признаковый** *adj.* tag; sign, indication, mark; characteristic

**прийти** *pf. of* **приходить**

**приказ** *m.* order, command

**приказание** *n.* direction, instruction; order, command

**приказывать, приказать** *v.* to order, to command; to direct, to instruct

**прикасаться, прикоснуться** *v.* to contact, to touch, to adjoin

**прикатодный** *adj.* cathodic, at the cathode; catholyte (layer)

**прикладной** *adj.* applied

**прикладывание** *n.* application

**прикладывать, приложить** *v.* to apply, to join, to affix, to add; to enclose; to switch to

**приклеенный** *adj.* pasted (on), glued

**приклеивать, приклеить** *v.* to glue, to attach, to paste, to stick

**приключать, приключить** *v.* to connect up

**приключение** *n.* connection; terminating

**приключить** *pf. of* **приключать**

**прикосновение** *n.* contact, touching, touch

**прикоснуться** *pf. of* **прикасаться**

**прикрепить** *pf. of* **прикреплять**

**прикрепление** *n.* attachment, fastening, strengthening

**прикреплённый** *adj.* mounted, attached, fastened; **п. основанием** base-mounted

**прикрепляемый** *adj.* fastened, attached, connected

**прикреплять, прикрепить** *v.* to fasten, to secure, to attach, to affix, to fix, to connect, to anchor; to fit, to stick

**прикрепляющий** *adj.* fastening, attaching, securing, connecting

**приладить(ся)** *pf. of* **прилаживать(ся)**

**прилаживание** *n.* fitting, adapting, adjusting, adjustment

**прилаживать, приладить** *v.* to fit together, to adjust, to adapt

**прилаживаться, приладиться** *v.* to fit, to apply; to be adjusted

**прилегающий, прилежающий** *adj.* adjoining, adjacent, neighboring, abutting

**прилёт** *m.* arrival, flight

**прилив** *m.* tide; flow, influx; cleat; boss, lug, tongue, rib

**приливающий** *adj.* inflowing

**приливомер** *m.* tide-gage

**приливообразующий** *adj.* tide-generating

**прилипаемость** *f.*, **прилипание** *n.* adherence, adhesion, sticking, attachment; trapping

**прилипать, прилипнуть** *v.* to adhere, to stick, to cling, to cohere, to hold

**прилипающий** *adj.* adhering, sticking, clinging

**прилипнуть** *pf. of* **прилипать**

**прилипший** *adj.* stuck; adhesive, adherent

**приложение** *n.* application; supplement, appendix, addition; enclosure

**приложенный** *adj.* applied; impressed

**приложить** *pf. of* **прикладывать**

**прим** *m.* prime (*math.*)

**прим.** *abbr.* (**примечание**) note

**прима** *f.* tonic, prime (*mus.*); first string; first violin; **п. аккорда** root (*mus.*)

**применение** *n.* application, use, employment, adaptation, adaptability; **п. рефлексных схем** reflexing; **п. теории реляксации к рассеянию звука в газах** relaxation theory treatment of sound dispersion in gases; **п. транзисторов** transistorization

**применённый** *adj.* applied, adapted, employed, utilized

**применимость** *f.* applicability, adaptability

**применимый** *adj.* applicable, suitable, adaptable, usable, practicable. appropriate; available

**применительный** *adj.* applicable, adaptable, suitable

**применить** *pf. of* **применять**

**применяемость** *f.* applicability, adaptability

**применяемый** *adj.* employed, used, applied; accepted

**применять, применить** *v.* to apply, to employ, to use, to adapt, to practice

**пример** *m.* example, instance, sample, model, specimen

**примерный** *adj.* apparent

**примесный** *adj.* impurity, extrinsic, doping; impure

**примесь** *f.* impurity, contaminant, doping material; addition, admixture; alloy

**примета** *f.* identification mark, sign, characteristic, feature, indication, criterion

**примечание** *n.* note, comment, annotation; footnote

**примитивный** *adj.* primitive; simple, rough

**примкнуть** *pf. of* **примыкать**

**примораживающий** *adj.* freezing

**приморский** *adj.* maritime

**примыкать, примкнуть** *v.* to join, to adjoin, to border; to shut; to fasten

**примыкающий** *adj.* adjacent, adjoining, abutting

**принадлежность** *f.* accessory, fixture, fitting(s), adjunct(s), implement, gadget(s), outfit, attachment

**принимаемый** *adj.* received; acceptance

**принимать, принять** *v.* to receive, to take, to accept; to take up, to take on; **п. за нуль** to take as a datum; **п. значение** to possess the value (*math.*)

**принимающий** *adj.* receiving

**принтер** *m.* printer; teleprinter receiver

**принтоскоп** *m.* printoscope

**принудительно** *adv.* compulsorily

**принудительный** *adj.* compulsory, forced, coercive; positive

**принуждать, принудить** *v.* to force, to compel, to constrain, to impel, to coerce, to enforce

**принуждение** *n.* compulsion, constraint, coercion, forcing

**принуждённый** *adj.* forced, constrained

**принцип** *m.* principle; mode (of operation); concept; **в принципе** in the main; theoretically; as a matter of principle; **п. близости** short-path principle; **п. восстановления (координат цепи)** regenerative principle (restoring the coordinates of a circuit); **п. искусственного воспроизведения низких тонов** synthetic-bass principle; **мелкоблочный п.** building block; **п. неопределённости** indeterminateness relation (*math.*); **п. огибающей импульсов** pulsed-envelope principle; **п. отбора по скоростям** velocity selector principle; **п. «параллельного одноцветного изображения»** "by-passed monochrome" principle; **п. передачи и приёма на разных частотах** cross-band principle (*tphny.*); **п. смеси высоких частот** mixed highs principle

**принципал** *m.* principal (*mus.*)

**принципально** *adv.* as a principle, on principle

**принципальный** *adj.* principle, basic

**принятие** *n.* acceptance, adoption, admission, reception, assumption; **п. решения** decision-making

**принятый** *adj.* accepted, admitted, assumed; established

**принять** *pf. of* **принимать**

**приобрести** *pf. of* **приобретать**

**приобретание** *n.* acquisition, obtaining, gain, gaining, acquirement, acquiring

**приобретать, приобрести** *v.* to acquire, to obtain, to gain, to get, to take on

**приобретение** *see* **приобретание**

**приобретённый** *adj.* acquired

**приоритет** *m.* priority

**приоритетный** *adj.* priority

**приостановка** *f.* suspension, stoppage, stopping, cessation; pause, stop, rest; turning off, cutting off

**приостановленный** *adj.* stopped, closed

**припаечный** *adj.* soldering

**припаиваемый** *adj.* soldering, solderin; solderable

**припаивание** *n.* soldering

**припаивать, припаять** *v.* to solder to, to fix on

**припайка** *f.* soldering; **п. к контактной пластинке** eyelet solder; **п. т стакану цоколя** shell solder; **п. штенгеля к колбе** tubulating

**припасованный** *adj.* form-fit, fitted lined-up, aligned

**припасовать** *pf. of* **припасовывать**

**припасовка** *f.* adjusting, fitting, matching; alignment

**припасовочный** *adj.* fitting

**припасовывать, припасовать** *v.* to fit together

**припаяный** *adj.* soldered-to, brazed-on; **п. индием** indium-soldered

**припаять** *pf. of* **припаивать**

**приписной** *adj.* assigned, added

**приплющивать, приплюснуть** *v.* to flatten (out)

**приповерхностный** *adj.* shallow, surface; surface-bounded

**приподнимать, приподнять** *v.* to boost, to raise; to lift, to take up

**приподнятие** *n.*, **приподнятость** *f.* elevation

**приподнятый** *adj.* raised

**припой** *m.* solder; braze; **п. в палочках** solid wire solder; **легкоплавкий п.** quick solder; **п. с канифольным фитилем, трубчатый п. (с канифолью), фитильный п.** rosin-core solder

**приполярный** *adj.* circumpolar

**припуск** *m.* allowance, margin; over-measure

**прирабатываться, приработаться** *v.* to run in, to work in (of bearing)

**приработавшийся** *adj.* worn in

**приравнивание** *n.* equating, matching

**приравнивать, прировнять** *v.* to equate (*math.*); to level, to equalize, to adapt, to adjust; to set (to), to set equal to; to compare

**прирастать, прирасти** *v.* to increase, to grow, to accrue; to adhere

**приращать, прирастить** *v.* to increase, to augment, to enlarge, to extend; to attach

**приращение** *n.* increment, increase, gain, difference; **п. входной ёмкости** input capacitance increase, Miller effect

**прировнять** *pf. of* **приравнивать**

**природа** *f.* nature

**прирост** *m.* increment, increase, gain, growth

**Прис** Preece

**присадка** *f.* addition, addition agent, dope, doping material, additive

**присадочный** *adj.* additional

**присасывание** *n.* sucking, suction, indraft

**присасывать, присосать** *v.* to suck, to suck in, to draw in, to pull; to surge

**присасывающий** *adj.* sucking

**присвоение** *n.* assignment, appropriation

**присвоенный** *adj.* assigned, authorized

**присмотр** *m.* maintenance, upkeep, care, attendance

**присоединение** *n.* addition, joining; connection, contact, connecting; terminating; **п. делителя напряжения** bridging; **п. коммутационного** (*or* **распределительного**) **щита** switchboard connection

**присоединённый** *adj.* added, joined; associated, connected, augmented, adjoint; **п. на прямую** direct-connected

**присоединительный** *adj.* binding, connecting, adjunctive; connection

**присоединять, присоединить** *v.* to join, to add, to attach; to connect, to joint, to jumper, to interconnect, to link up, to hook up; **п. к зажиму** to terminate

**присоединяющийся** *adj.* joining, adjoining; additive

**присосать** *pf. of* **присасывать**

**приспосабливание** *n.* adjustment

**приспосабливать** *see* **приспособлять**

**приспособить** *pf. of* **приспособлять**

**приспособление** *n.* attachment, device, appliance, apparatus, gadget, gear, fixture; adaptation, addition, adjustment, fitting, arrangement, accommodation; equipment, outfit; **визирное п.** (optical) finder, visor; **п. для выдержки времени** timing device; **п. для настройки** tuner; **п. для поддержки труб сигнальных тяг при пересечении полотна** transverse pipe carrier (*RR appl.*); **п. для покачивания (ртутника)** tilter (for a mercury-arc rectifier); **п. для прямления штырьков** pin straightener; **п. для точной установки** adjuster; **п. для установки камеры** camera mount; **противоугонное п.** rail anchor, rail brace; **регулировочное п.** adjuster, adjustment, control; **сцепляющее п.** clutch; slot (*RR appl.*)

**приспособленность** *f.* fitness, suitability

**приспособленный** *adj.* adjusted, fitted, adapted, suited

**приспособляемость** *f.* adaptability

**приспособляемый** *adj.* adjustable, adaptable, applicable

**приспособлять, приспособить** *v.* to adapt, to fit, to accommodate, to adjust, to arrange; to equip, to establish, to set up, to install, to mount

**приставание** *n.* adhesion, adherence, sticking, clinging

**приставать, пристать** *v.* to join, to attach oneself, to cling, to stick, to ahere; to stop

**приставка** *f.* adapter; attachment, gadget, attached piece, auxiliary piece, unit, device; derrick pole; **высокочастотная п.** radiofrequency head; **высотная п.** height warning device; **граммофонная п.** phonograph attachment; **звуковая п.** pickup head; **п. к радиоприёмнику** adapter; **противопомеховая п.** interference eliminator, static eliminator, antijamming unit

**приставной** *adj.* added, attached

**пристать** *pf. of* **приставать**

**пристающий** *adj.* sticking, adhering, adhesive

**Пристлей** Priestly

**пристли** *n. indecl.* priestly (unit)

**пристраивать, пристроить** *v.* to attach, to fix, to fasten; to build to, to add to; to settle, to establish

**пристрелка** *f.* target

**пристрастный** *adj.* biased, partial

**пристроенный** *adj.* built-on, added on

**пристроивать** *see* **пристраивать**

**пристроить** *pf. of* **пристраивать**

**присутствие** *n.* occurence, presence

**присущий** *adj.* inherent, innate, intrinsic

**присущность** *f.* inherence

**присчитывание** *n.* counting-out (*math.*)

**притёртый** *adj.* ground, ground-in; wiped

**притир** *m.* vise clamps

**притирание, притиранье** *n.* wipe, wiping; rubbing, friction; grinding; fitting in

**притирать, притереть** *v.* to grind, to polish, to rub; to fit

**притирка** *f.* bedding, seating; grinding, regrinding, abrasion, abrading, lapping; fitting

**приток** *m,* influx, inflow; admission, feed, intake

**приточный** *adj.* tributary; addition

**притупить** *pf. of* **притуплять**

**притупление** *n.* blunting; dulling, deadening, de-emphasis; **п. верхних частот** de-emphasis of higher audio frequencies

**притупленный** *adj.* blunted, dulled; deadened, damped

**притуплять, притупить** *v.* to blunt, to dull; to deaden, to damp, to quench

**притупляющий** *adj.* despiker; damping, deadening

**притягивать, притянуть** *v.* to pull up, to draw to, to attract; to brace; to tighten, to screw up

**притягиваться, притянуться** *v.* to take up, to be drawn to; to draw near, to come near, to approach

**притягивающий** *adj.* pulling, drawing, tractive

**притяжение** *n.* attraction, gravitation, pull; adhesion; pickup; **продольное п. в поле** longitudinal force of a (magnetic or electrical) field; **п. частиц** adhesion of particles

**притянутый** *adj.* pulled, attracted; tightened, screwed up

**притянуть(ся)** *pf. of* **притягивать(ся)**

**прифлянцованный** *adj.* flange-mounted

**приход** *m.* arrival, coming

**приходить, прийти** *v.* to come, to arrive

**приходящий** *adj.* arriving, incoming, incident

**прицел** *m.* sight, gunsight; (radar) eye; **п. с решающим устройством** computing gunsight

**прицеливание** *n.* aiming, sighting

**прицеливаться, прицелиться** *v.* to sight, to aim, to take aim

**прицелить** *pf. of* **целить**

**прицельный** *adj.* sighting; selective, spot

**прицеп** *m.* trailer; hook

**прицепка** *f.* hooking, hitching; trailer

**прицепление** *n.* hooking, hitching, connecting; coupling

**прицепленный** *adj.* hooked, connected; attached; coupled

**прицепной** *adj.* trail, trailing, hook-on; trailer

**причастность** *f.* implication

**причастный** *adj.* participial; participating, involved

**причина** *f.* cause, reason, origin, source, principle

**причинение** *n.* causing

**причинённый** *adj.* caused

**причинность** *f.* causality

**причинный** *adj.* causal, causative

**пришабренный** *adj.* scraped

**пришабривание** *n.* scraping off

**пришабривать, пришабрить** *v.* to scrape; to fit

**пришвартовать** *pf. of* **швартовать**

**пришлифованный** *adj.* ground, ground in, ground down

**пришлифовать** *pf. of* **пришлифовывать**

**пришлифовка** *f.* seating

**пришлифовывать, пришлифовать** *v.* to grind, to grind true

**ПРО** *abbr.* (**противоракетная оборона**) missile defense

**проба** *f.* test, testing, trial, probe, experiment, try(-out); sample, proof, specimen; analysis; trial slip (*tgphy.*); **п. на разрядку** discharge test; **п. « на язык »** galvanic taste

**пробег** *m.* run; running through, flow; passage, travel; range, path; rundown; **п.-энергия** range-energy relation

**пробегать, пробежать** *v.* to pass, to run, to run (by, through); to pass over; to travel, to traverse; to scan

**пробежно-ионизационный** *adj.* range-ionization

**пробел** *m.* blank, gap, omission; discontinuity, space, spacing, interval, interspace, separation; vacancy; dead interval

**пробельный** *adj.* spacing

**пробивание** *n.* breakdown (*elec.*); piercing, punching, puncture, perforation; **п. передачи (через другую)** breakthrough

**пробивать, пробить** *v.* to break through, to break down; to pierce, to punch, to puncture, to perforate, to make a hole; to clear

**пробивающий** *adj.* disruptive; piercing, punching, puncturing

**пробивка** *f.* punching, perforation, slotting; punch; **пробивки-признаки** control holes, designation punchings; **п. одиночных отверстий** spot punching; **п. с пантографическим управлением по шаблону** template-guided punch

**пробивной** *adj.* break-through; piercing, punching, puncturing; disruptive, breakdown

**пробирка** *f.* test tube

**пробитие** *n.* breaching, piercing

**пробитый** *adj.* pierced, punched, perforated, punctured; cleared

**пробить** *pf. of* **пробивать**

**пробка** *f.* cork; plug, stopper; fuse, screw-plug fuse; plug gage, internal gage; **предохранительная п.** cartridge plug; **резьбовая п.** screw plug; **п. со вставным плавким патроном** screw-plug cartridge fuse

**пробковый** *adj.* plug, stopper; cork, cork-like

**проблема** *f.* problem, task; **п. собственной величины** eigenvalue problem; **п. трёх тел** three-body problem

**проблеск** *m.* flash, gleam, spark

**проблесковый** *adj.* flashing, flickering, intermittent

**проблуждать** *pf. of* **блуждать**

**пробник** *m.* probe; sampler, sampling tube; test rod; gage cock; circuit tester; proof-plane, test-tool

**пробный** *adj.* trial, experiment, test; proof testing

**пробование** *n.* testing, trying

**пробовать, попробовать** *v.* to probe, to test, to try, to sample, to check

**пробой** *m.* breakdown, break, flashover, spark-over, spark discharge; disruption, puncture; breakthrough; failure; **повторный п. разряда** spark-gap restriking

**пробойник** *m.* punch, marker, puncher, piercer, drift; mandrel

**пробоотбиратель** *m.*, **пробоотборник** *m.* sampler, sampling apparatus

**пробочник** *m.* corkscrew

**пробочный** *adj.* cork; plug; screwplug

**пробуравить** *pf. of* **буравить**

**пробурить** *pf. of* **бурить**

**провал** *m.* failure; dip (of a curve); drop; valley, trough, crevasse; cave-in; undershoot; gap; **п. в чувствительности** dead spot (of a radio receiver); **п. волны** trough of a wave; **п. на горизонтальной площадке импульса, п. на импульсе** pulse valley; **резко выраженный п.** crevasse; **п. тока сетки** grid dip; **п. чувствительности** blind spot, dead spot

**проваливаться, провалиться** *v.* to collapse, to break down, to fall through; to fail

**провальный** *adj.* downcomer; slack

**проваривание** *n.* boiling; baking

**проведение** *n.* conveyance, conducting, leading; installation; execution, carrying out; construction, building

**провентилировать** *pf. of* **вентилировать**

**проверенный** *adj.* checked, screened

**проверить** *pf. of* **проверять**

**проверка** *f.* verification, verifying, check-up, checking, check, inspection; examination, testing, control; test, trial; proof (*math.*); **п.: годен-негоден** go-no-go check; **лётная п. правильности курса** flight check of course alignment; **п. линии радиоприцеливания** bore sighting test; **п. на занятость зуммером** buzzer test (*tphny.*); **п. на надёжность** marginal checking; **п. на нечётность** odd-parity check; **п. на чётность** even-parity check; **п. ошибок с помощью резервирования знаков** redundancy error check; **п. по внешнему виду** observation check; **п. по чётности** odd-even check; **п., предусмотренная программой** programmed checking; **п. прохождения сигналов** signal tracing; **п. совпадения результатов с заданным диапазоном значений** consistency check

**проверочный** *adj.* checking, verifying, examining, testing, calibrating; test, control

**провёртывание** *n.* level overflow motion (*tphny.*); **п. искателя** level overflow motion of a selector, full rotation

**проверяемый** *adj.* under examination

**проверять, проверить** *v.* to check, to verify, to calibrate; to test, to examine, to control, to monitor; to screen

**провес** *m.* slack, sag, sagging, dip (of wire); bending deflection

**провесить** *pf. of* **провешивать**

**провести** *pf. of* **проводить**

**проветриваемый** *adj.* ventilated

**проветривание** *n.* ventilation, airing

**проветривать, проветрить** *v.* to ventilate, to air

**провешивание** *n.* alignment

**провешивать, провесить** *v.* to align; to plumb, to make vertical; to weigh wrong

**провисание** *n.* sag, sagging, slackening, dipping, bending deflection

**провисать, провиснуть** *v.* to sag, to dip, to deflect

**провисающий** *adj.* slack, sagging; protruding

**провиснувший** *adj.* sagged; protruded

**провиснуть** *pf. of* **провисать**

**провисший** *adj.* sagged, slack

**провитамин** *m.* provitamin
**провод** *m.* wire, lead, conductor, cable, circuit, strand; duct, conduct, conduit; loop; **калиброванные провода** instrument leads; **п. большого сечения** heavy wire, heavy-gage wire; **п. в соединении с головкой штепселя** (*or* **пружиной гнезда**) tip wire (*tphny.*); **вентильный п.** gate; **выходной п.** pickup wire; **зануляющий п.** neutral wire, protective conductor; **п. квадратного сечения** square conductor, bus-bar wire; **корпусный п.** sleeve wire, S-wire; **п. круглого сечения** round conductor; **п. малого сечения** light gage wire; **п. многократного поля** bank wire (*tphny.*); **монтажный п.** wiring leads, hook-up wire, cross-connecting wire, jumper wire; **п. от головки штепселя** tip side, tip wire, T-wire; **п. от корпуса штепселя** S-wire, sleeve wire, third wire (*tphny.*); **п. от шейки штепселя** R-wire, ring wire, ring side (*tphny.*); **п. передачи запрещающего импульса** inhibit wire; **питательный п. направленной радиосети** beam aerial feeder; **п. ПР** rubber-insulated impregnated-cotton braided copper wire, black cable; **п. прямоугольного сечения** bus-bar wire; **разрядный п.** bit line, digit line, digital wire; **раскалённый п.** hot-wire; **п.-рельс с нижним контактом** under contact rail; **п. с пущенным током** live wire; **силовой п.** main lead; **п. считывания** sense wire, sensing wire; **третий п.** private wire (*tphny.*)
**проводимостный** *adj.* conduction
**проводимость** *f.* conduction, conductivity, conductibility; conducting power; conductance; admittance; **активная п.** conductance; **активная динамическая амплитудная п. одного электрода относительно другого** transconductance; **активная динамическая амплитудная п. электрода** electrode conductance; **безразмерная (активная) п.** specific conductance, conductance ratio; **безразмерная акустическая п.** acoustic admittance ratio; **безразмерная реактивная п.** specific susceptance, susceptance ratio; **ваттная п.** conductance; **п., вносимая цепью обратной связи** feedback admittance;

**волновая п.** characteristic admittance, natural admittance, surge admittance; **входная комплексная п. короткого замыкания** short-circuit input admittance; **входная комплексная п. (многополюсника)** driving-point admittance (of an n-terminal network); **входная комплексная п. электрода** electrode admittance; **п. для зеркальной частоты** image-frequency admittance; **дополнительная п. цепи сетки** feedback admittance; **ёмкостная п.** capacitive susceptance, permittance; **п. изоляции** shunt conductance, leakance; **индуктивная п.** inductive susceptance; **комплексная п.** (vector) admittance; **комплексная п., вносимая цепью обратной связи** susceptance; **комплексная п. у точки питания (многополюсника)** driving-point admittance (of an n-terminal network); **комплексная п. электрода** electrode admittance; **п. лампы** perveance; **магнитная п.** permeance; **п. нагрузки** load admittance; **обратная переходная (комплексная) п.** backward transfer admittance; **объёмная п.** cubic conductance; **однородная п.** isotropic conductivity; **переходная (комплексная) п.** transfer admittance, transadmittance, indicial admittance; **полная п.** (total) admittance; **п. п. зазора** circuit gap admittance; **п. п. при антирезонансе** antiresonance admittance; **п. п. при заторможённом выходе** clamped admittance; **п. п. при освобождённом выходе** free admittance; **п. п. при последовательном резонансе** series resonance admittance; **полная внесённая п.** motional admittance; **полная внутренняя выходная п.** internal output admittance; **полная входная п.** input admittance; **полная динамическая амплитудная п. одного электрода относительно другого** transadmittance; **полная динамическая п. электрода** electrode admittance; **положительная реактивная п.** positive susceptance; **п. при токе насыщения** perveance; **приведённая п.** transfer admittance; **проходная (комплексная) п. короткого замыкания** short-circuit transfer admittance; **прямая проходная п.** forward transadmittance; **реактивная п.** susceptance; **реактивная динамическая амплитудная п. электрода** electrode susceptance; **удельная п.** conductivity, specific conductivity; **п. электрода при выпрямлении** conductance for rectification; **эффективная входная полная п.** effective input admittance; **эффективная выходная полная п.** effective output admittance

**проводимый** *adj.* conducted

**проводить, провести** *v.* to lead, to conduct; to conduct, to carry (*elec.*); to install; to lay out, to mark; to draw, to trace; to run, to pass, to wire

**проводка** *f.* installation, wiring, wiring system; line, electric line, lead, lead wire, circuit; conduit, line, main; **временная п.** stop-gap wiring, haywire; **монтажная п.** wiring, harness

**проводник** *m.* conductor, guide; lead, wire; conductive body; conductive pattern; **впаянный в лопатку п.** seal wire; **п. под током** current-carrying conductor

**проводниковый** *adj.* conductive; guide, conductor

**проводной** *adj.* leading; wire

**првводность** *see* **проводимость**

**проводный** *adj.* conducting; wire, conductor

**провододержатель** *m.* hanger

**проводящий** *adj.* conducting, conveying, carrying, transmitting, conductive

**проволока** *f.* wire; **п. для подвески** slinging wire; **калиброванная п.** calibrated slide-wire, graduated slide-wire; **обмоточная п.** magnet wire; **п. с ферромагнитным покрытием** magnetic-plated wire

**проволокообразный** *adj.* wire-shaped, filiform

**проволочка** *f.* fine wire, short (piece of) wire

**проволочно-волочильный** *adj.* wire-drawing

**проволочнонамоточный** *adj.* wire-coiling

**проволочный** *adj.* wire; wire-wound

**проворачивать, проворотить** *v.* to rotate over, to hunt over a complete level (of a selector)

**провощённый** *adj.* waxed

**прогар** *m.* burn-out

**прогарина** *f.* burnt place, burn

**прогиб** *m.*, **прогибание** *n.* sag, sagging, deflection, caving in, depression, bending deflection, buckling, flexure; bend, camber

**прогибать, прогнуть** *v.* to deflect

**прогибаться, прогнуться** *v.* to sag, to yield, to give, to collapse; to deflect, to cave in

**прогибомер** *m.* deflectometer

**проглатываемый** *adj.* swallowable

**проглатывание** *n.* swallowing; absorbing

**проглатывать, проглотить** *v.* to swallow

**прогнать** *pf. of* **прогонять**

**прогноз** *m.*, **прогнозирование** *n.* forecast, forecasting, predicting, prediction; planning

**прогнозированный** *adj.* predictive

**прогнозировать** *v.* to predict, to forecast, to prognosticate

**прогнозирующий** *adj.* predictive

**прогностический** *adj.* prognostic; predicted, forecast, forecasting

**прогнуть(ся)** *pf. of* **прогибать(ся)**

**прогон** *m.* run, running, drive; span, girder; machine pass; **п. теста** running of test routine

**прогонка** *f.* pass, playing; die, screw die

**программа** *f.* instruction, program, schedule; scheme, plan, project; routine; **п. ввода** bootstrap; **восстанавливающая п.** rerun routine; **жёсткая п.** wired-in program; **п. контроля программы** tracing routine; **наборная п.** plugged program; **общая п. вывода содержимого накопителя** general post-mortem program; **оптимально кодированная п.** minimum access routine, minimum latency routine; **отладочная п.** error routine, diagnostic routine; **отсылающая п.** calling order; **переключающая п.** reference order; **покомандная компонующая п.** one-to-one assembler; **п. прерывания по адресу блока** block address interrupt routine; **п. проверки** diagnostic routine; **п. с минимальным временем выборки** minimum access routine; **п. субпорядков** breakout program; **управляемая п. вывода**

**после просчёта ленты** controlled post-mortem program; **п. условного перехода** branching program

**программирование** *n.* programming, compiling, coding; **п. в кодах команд машин** object coding; **п. на наборном поле** pinboard programming; **п. с минимальным временем выборки** (*or* **ожидания**) minimum access programming, minimum latency programming, least wait programming

**программированный** *adj.* programmed

**программировать** *v.* to program; to compile

**программируемый** *adj.* programmed

**программирующий** *adj.* programming, programmer; compiling

**программист** *m.* programmer, coder

**программно-радиовещательный** *adj.* radio-program-broadcasting

**программно-управляемый** *adj.* program-controlled

**программный** *adj.* program

**прогрев** *m.*, **прогревание** *n.* warming-up, initial heating, preheating; bakeout, baking

**прогревать, прогреть** *v.* to warm up, to heat; to bake

**прогрессивно** *adv.* progressively, little by little, gradually, by degrees

**прогрессивный** *adj.* progressive, progressional, progressing, gradual

**прогрессировать** *v.* to progress; to improve

**прогрессирующий** *adj.* progressing, progressive, gradual

**прогрессия** *f.* progression, series (*math.*)

**прогреть** *pf. of* **прогревать**

**продажный** *adj.* commercial; selling; market

**продалбливать, продолбить** *v.* to chisel through, to make a hole in

**продвигаться, продвинуться** *v.* to advance, to move forward

**продвигающий** *adj.* advance, advancing, shift

**продвижение** *n.* advance, advancement, progress, movement

**продвинуться** *pf. of* **продвигаться**

**продевание** *n.* threading, passing through

**продевать, продеть** *v.* to thread, to pass through, to run through, to insert

**продёргивание** *n.* threading

**продёргивать, продёрнуть** *v.* to thread, to put through

**продетектированный** *adj.* rectified

**продеть** *pf. of* **продевать**

**продление** *n.* prolongation, lengthening, extension; retardation, delay, protracting; **п. (времени) ожидания** prolongation of delay; **п. занятия** extended hold

**продлиться** *pf. of* **длиться**

**продолбить** *pf. of* **продалбливать, долбить**

**продолговатость** *f.* oblong form, oblongness

**продолговатый** *adj.* elongated, extended; oblong

**продолжать, продолжить** *v.* to continue, to go on, to proceed, to persist; to prolong, to extend; to lengthen, to elongate, to broaden; to pursue, to carry on; to resume

**продолжаться, продолжиться** *v.* to be prolonged, to be extended, to be lengthened; to continue, to last; to keep on; to persist, to endure

**продолжающийся** *adj.* lasting, continuing

**продолжение** *n.* continuation, continuance, duration; prolongation, extension; interval, course, space (of time); streaking, trailing; resumption; **белое п.** pulling on whites, white smear; **чёрное п.** black after white, pulling-in on blacks, black smear

**продолжительность** *f.* duration, continuance, length, period, time, date, cycle; endurance; performance; **п. вызова** ring time (*tphny.*); **п. действия, п. жизни** life; **п. занятия** holding time, holding period; **п. занятия цепи в процентах** percentage circuit occupation (*tphny.*); **п. измерения** sampling time; **п. использования** demand time; **относительная п. работы** service life; **п. работы, п. службы** life, age

**продолжительный** *adj.* long, of long duration, long-term, lasting, continuous, prolonged, sustained; exhaustive

**продолжить(ся)** *pf. of* **продолжать(ся)**

**продольно** *adv.* longitudinally, lengthwise

**продольно-перфорированный** *adj.* lengthways perforated

**продольно-фрезерный** *adj.* planer-type milling, plano-milling

**продольный** *adj.* longitudinal, lengthwise, linear; extentional, drawn out, direct-axis; **в продольном направлении** lengthwise

**продораживание** *n.* recessing; grooving

**продуваемый** *adj.* air-blast, blow-through

**продувание** *n.* blowing off, blowing through; scavenging; purging, purge, drainage; **п. воздуха** draught (of air); **п. воздуха вверх (вниз)** up-(down-)draught

**продувательный** *adj.* blowing through, blow-through, blow off, blow-down; drain

**продувать, продуть** *v.* to blow, to blow out, to blow through; to remove, to exhaust, to scavenge

**продувка** *f.* blow, blowing through, blowing down

**продувной, продувочный** *adj.* blow-off; drain

**продукт** *m.* product

**продуктивность** *f.* productivity, productiveness, efficiency

**продуктивный** *adj.* productive, producing, efficient

**продукция** *f.* production, output, yield, productive capacity

**продутый** *adj.* blown, blown through, blown out

**продуть** *pf. of* **продувать**

**продуцирующий** *adj.* producing, productive

**продырявить** *pf. of* **продырявливать**

**продырявленный** *adj.* perforated, pierced

**продырявливание** *n.* perforation, hole injection

**продырявливать, продырявить** *v.* to perforate, to punch, to pierce

**проезд** *m.* transit, passage, passing; **проездом** in transit, in passing

**проект** *m.* project, projection, plan, design, scheme, device, layout, blueprint

**проективность** *f.* projectivity

**проективный** *adj.* projective

**проектирование** *n.* projecting, planning, designing, design; **п. на экран** visual projection; **п. с учётом наихудшего варианта** worst-case design

**проектировать, запроектировать,**

**спроектировать** *v.* to project, to plan, to design

**проектировка** *see* **проектирование**

**проектировочный** *adj.* designing

**проектировщик** *m.* designer, projector, design engineer, planner

**проектируемость** *f.* designability

**проектно-расчётный** *adj.* preliminary-estimation

**проектный** *adj.* project, plan, design; designed

**проектор** *m.* projector, projection machine

**проекционный** *adj.* projection

**проекция** *f.* projection

**проём** *m.* opening, aperture, embrasure

**прожектор** *m.* projector, searchlight, floodlight, spotlight; (electron) gun; **п. для фотозаписи** photographic recording gun; **п. с теневой перегородкой** shadow-bar floodlight; **считывающий п.** viewing gun

**прожекторный** *adj.* searchlight; beam; projector

**прожечь** *pf. of* **прожигать**

**прожигание** *n.* burn, burn-in, burning through

**прожигать, прожечь** *v.* to burn (through)

**прожилок** *m.* filament, fiber

**прожужжать** *pf. of* **жужжать**

**прозвонить** *v.* to ring out

**прозвонка** *f.* ringing out, ringout, sounding, signaling test; wire identification

**прозвучать** *pf. of* **звучать**

**прозвучивание** *n.* sounding

**прозрачность** *f.* transparence, transparency, transmittance, pellucidity, transmission, penetrability

**прозрачный** *adj.* transparent, clear, translucent, transmissive, pellucid; **п. для электронов** electron-transmissive

**проиграть** *pf. of* **проигрывать**

**проигрывание** *n.* play, playing, playback; **п. с одной дорожки** half-track play

**проигрыватель** *m.* turntable, record player, sound reproducer

**проигрывать, проиграть** *v.* to play, to play over, to play back

**произведение** *n.* product (*math.*); work, production; derivation, origination; **векторное п.** crossproduct; **п. из величины индукции на максималь-**

**ную напряжённость поля** energy product; **п. из дальности** range product

**произведённый** *adj.* produced, manufactured; generated; derived

**произвести** *pf. of* **производить**

**производимый** *adj.* producible

**производитель** *m.* producer, generator; generant

**производительность** *f.* production, productivity, productiveness, output, yield, delivery; capacity, capability, duty, efficiency, effect, performance; rating; **большой производительности** heavy-duty; **п. насоса** pumpage; **п. точечного источника** strength of simple source

**производительный** *adj.* productive, efficient, producing, production

**производить, произвести** *v.* to make, to produce, to manufacture; to create; to exert; to generate; to drive; to perform, to effect; to do; **п. отсчёт** to take the reading; **п. подставку** to substitute (*math.*); **п. подстановку** to insert (*math.*)

**производная** *f. adj. decl.* derivative, differential coefficient; **производные по координатам** spatial derivatives; **п. по нормали** normal derivative; **п. слева** derivative on the left, left-hand derivative; **п. справа** derivative on the right, right-hand derivative; **частная п.** partial; **частная п. высших порядков** higher partial derivative

**производный** *adj.* derived, derivative

**производственный** *adj.* production, industrial, manufacturing

**производство** *n.* making, preparation, production, manufacture, manufacturing; generation; execution, effecting; derivation; industry; **п. измерений** metering; **опытное п.** pre-production; **п. партиями** (*or* **сериями**) batch production

**производящий** *adj.* producing, forming, yielding, generating; productive

**произвольно** *adv.* arbitrarily, voluntarily

**произвольность** *f.* randomness

**произвольный** *adj.* arbitrary, voluntary, spontaneous; any, random

**произносить, произнести** *v.* to pronounce, to utter

**произношение** *n.* pronunciation, artic-

ulation, utterance, delivery (of speech), enunciation; speech

**произойти** *pf. of* **происходить**

**проинтегрированный** *adj.* integrated

**проистекать, проистечь** *v.* to result, to ensure

**проистекающий** *adj.* resulting, resultant

**проистечь** *pf. of* **проистекать**

**происходить, произойти** *v.* to originate, to arise, to spring, to proceed, to come from, to emanate; to stem (from), to issue; to happen, to occur, to come about, to take place

**происхождение** *n.* origin, nature; derivation; emanation; **воздушного происхождения** airborne

**происшествие** *n.* accident; incident; event, occurrence

**пройденный** *adj.* run through, passed, looked over

**пройти** *pf. of* **проходить**

**прокалённый** *adj.* calcinated, roasted; annealed, sintered; cooled down

**прокаливание** *n.* calcination, roasting; annealing; **п. в водороде** air calcination

**прокаливать, прокалить** *v.* to calcinate, to calcine, to roast, to bake, to fire; to harden; to anneal

**прокалка** *f.* annealing, tempering; lighting

**прокалывать, проколоть** *v.* to pierce, to puncture, to punch

**прокатать, прокатить** *pfs. of* **прокатывать**

**прокатка** *f.* rolling, flattening

**прокатный** *adj.* rolled, rolling

**прокатчик** *m.* roller

**прокатывание** *n.* rolling, flattening, lamination

**прокатывать, прокатать, прокатить** *v.* to roll, to flatten, to mill, to draw out, to laminate

**прокладка** *f.* packing, pad, padding, lining, layer, strip, slab, filler; washer, gasket, bearing disk; separator (of battery); plotting; spacer, distance piece, shim; laying, running; step, socket; interlayer; cross-over block; **п. гибкой тяги** wire run; **п. из шлаковой ваты** slag-wool slab; **п. кабеля** cabling, cable laying; **п. курса** plotting; **п. поверх штукатурки** surface wiring

**прокладной, прокладочный** *adj.* packing, lining, stuffing; interleaving; padded; gasket

**прокладывать, проложить** *v.* to lay, to run (a wire); to interlay; to build, to construct; to lay, to lay out

**прокладывающий** *adj.* laying

**проклейка** *f.* sizing; pasting, glueing

**проключение** *n.* switching through (*tphny.*)

**прокол** *m.* pinhole; needle hole

**проколоть** *pf. of* **прокалывать**

**проконтролировать** *pf. of* **контролировать**

**прокорректировать** *pf. of* **корректировать**

**пролёт** *m.* transit; spacing, span, bay, arch, aperture, opening; flight; **п. в свету** span

**пролётный** *adj.* transit, drift, floating

**проложенный** *adj.* laid; sandwiched

**проложить** *pf. of* **прокладывать**

**пролом** *m.* gap, break, breach, split

**проломанный** *adj.* broken breached

**проломка** *f.* break, breakage, breakdown

**промасленный** *adj.* oiled, oil-(impregnated)

**промасливать, промаслить** *v.* to oil, to lubricate; to grease

**промедление** *n.* delay

**промежуток** *m.* interval, space, span, interspace, clearance, gap, distance; pause, period, time interval; interim, interspace; discontinuity; lapse; **п. между перегородками** interbaffle space; **пробивной п.** spark gap

**промежуточночастотный** *adj.* i-f, intermediate-frequency

**промежуточный** *adj.* intermediate, discontinuous, interrupted; intervening, intermedial, middle, mediate, interim; interstitial; interjacent

**промер** *m.* survey; sounding; measuring, test, testing; error in measurement

**промеривать** *or* **промерять, промерить** *v.* to measure, to survey; to sound; to make a mistake in measurement

**промерный** *adj.* measure, measuring, surveying

**промерять** = **промеривать**; **п. шумом** to feel, to gage, to prod, to probe

**прометий** *m.* promethium, Pm

**промпомехи** *pl.* industrial noise, man-made noise

**промчастотный** *adj.* i-f, intermediate-frequency

**промщит** *m.* IDF, intermediate distributing frame; **входящий п.** incoming intermediate distributing frame (*tphny.*)

**промыватель** *m.* washer, purifier

**промывательный, промывной, промывочный** *adj.* washing, wash-out, flushing, rinsing

**промышленность** *f.* industry

**промышленный** *adj.* industrial, commercial

**пронзительность** *f.* shrillness, sharpness, acuteness, piercingness

**пронзительный** *adj.* shrill, sharp, piercing, penetrating, acute, strident

**Прони** Prony

**пронизывать, пронизать** *v.* to strike; to thread; to pierce, to perforate

**проникание** *n.* penetration, permeation, pervasion, infiltration, infusion

**проникать, проникнуть** *v.* to penetrate, to go through, to pass through, to pervade, to infiltrate, to permeate, to impregnate; to pierce, to sink, to bore

**проникающий** *adj.* penetrating, thorough; entrant

**проникновение** *n.* penetration, permeation, infusion, infiltration, pervasion; link; transmission

**проникнутый** *adj.* penetrated, permeated

**проникнуть** *pf. of* **проникать**

**проницаемость** *f.* penetrability, permeability, permittivity, perviousness, penetrance, penetration factor, porosity; transmission, transmissivity, transmittancy; reciprocal of amplification factor, grid through, specific inductive capacity; **внутренняя (магнитная) п.** intrinsic permeability; **диэлектрическая п.** permittivity, electronic inductivity, specific inductive capacitance; **диэлектрическая п. зажатого кристалла** clamped permittivity, clamped dielectric constant; **диэлектрическая п. свободного кристалла** free dielectric constant; **магнитная п.** magnetic inductivity, magnetic permeability, magnetic inductive capacity, permeance; **магнитная п. на частном цикле** incremental permeability; **нормальная п.** cyclic permeability;

**обратная п.** reciprocal of transmissivity

**проницаемый** *adj.* permeable, pervious, permeant, penetrable; passable

**проницание** *n.* permeation, infiltration

**проницательность** *f.* penetrability; penetration

**проницательный** *adj.* penetrating

**пронумерованный** *adj.* numbered

**пропадание** *n.* drop-out, fade; **п. следов (сигналов)** death path

**пропадать, пропасть** *v.* to fade; to be lost, to disappear, to vanish

**пропаянный** *adj.* soldered

**пропеллер** *m.* propeller, fan, airscrew

**пропеллерный** *adj.* propeller, fan, screw-type, airscrew

**пропечатывание** *n.* accidental printing, spurious printing

**пропионат** *m.* propionate

**пропитанность** *f.* impregnation

**пропитанный** *adj.* treated, impregnated, permeated, saturated

**пропитать** *pf. of* **пропитывать**

**пропитка** *f.* impregnation, steep, steeping, soaking, saturation, treatment, impregnating; saturant; **с твёрдой пропиткой** jelly-impregnated; **п. по способу полного поглощения** full-cell treatment

**пропиточный** *adj.* impregnating

**пропитывание** *see* **пропитка**

**пропитывать, пропитать** *v.* to impregnate, to saturate, to soak, to steep, to imbue

**пропитывающий** *adj.* impregnating

**пропищать** *pf. of* **пищать**

**пропозициональный** *adj.* propositional; sentential

**пропорционально** *adv.* in proportion (to), proportionally

**пропорционально-навигационный** *adj.* proportional-navigation

**пропорциональность** *f.* proportionality, proportion, symmetry, uniformity

**пропорциональный** *adj.* proportional, proportionate, in proportion

**пропорционирующий** *adj.* proportioning; proportioner

**пропорция** *f.* proportion, ratio; degree, rate

**пропуск** *m.* lapse, omission; admission, passing, passage; blank, gap, vacancy; leak; gating; skip, omission, miss, lapse; **п. зажигания**

misfire; **п. импульсов** misfiring (in a magnetron); **п. поля** (*or* **полукадра**) field suppression, frame suppression

**пропускаемость** *f.* transmissivity, transmittance, transmittivity

**пропускаемый** *adj.* transmitted

**пропускание** *n.* passing, passing through, passage, filtering, transmission; plying; gating

**пропускатель** *m.* band-pass filter

**пропускать, пропустить** *v.* to pass, to let pass, to conduct; to ignore, to omit, to skip, to miss; to leak; to filter; to slip; to gate, to gate through

**пропускающий** *adj.* passing, conducting, allowing passage, transmitting; carrying; gating; **п. ток** current-carrying

**пропускной** *adj.* allowing passage; permeable; carrying; forward

**пропустить** *pf. of* **пропускать**

**пропущенный** *adj.* passed through; filtered; lost; rejected, suppressed

**прорез** *m.* slot, slit; notch, recess, cut, nick; groove; aperture, perforation; section

**прорезанный** *adj.* slotted, slit; notched, cut

**прорезать** *pf. of* **прорезывать**

**прорезиненный** *adj.* rubber-treated, rubberized; gummed

**прорезинивание** *n.* rubber-proofing, rubberizing

**прорезывать, прорезать** *v.* to slot, to slit, to notch, to cut through

**прорезь** *see* **прорез, щель**

**прорыв** *m.* breakthrough, breaking, break, breach, gap; rupture, outbreak, outburst, eruption, blowout

**просадка** *f.* sag, sagging

**просадочный** *adj.* sagged

**просачивание** *n.* leakage, leak, leaking, escape, seepage; infiltration, permeation; soaking; percolation

**просачиваться, просачиться, просочиться** *v.* to leak, to seep, to escape, to ooze, to exude; to filter (through), to penetrate, to infiltrate; to soak, to impregnate

**просачивающий(ся)** *adj.* leaking, leakage, escaping; percolating

**просочиться** *pf. of* **просачиваться**

**просверлённый** *adj.* perforated, drilled

**просверлить** *pf. of* **сверлить**

**просвет** *m.* gap, airgap, clearance, open

space, opening, chink, slit; gate; **п. гортани** straight slit

**просветить** *pf. of* **просвечивать**

**просветлающий** *adj.* transmitting

**просветление** *n.* transmission augmentation; acoustic penetration; translucence

**просветлённый** *adj.* cleared, clarified; coated

**просветный** *adj.* translucent

**просвечиваемость** *f.* transparency, translucence

**просвечивание** *n.* transillumination, translucence, radioscopy

**просвечивать, просветить** *v.* to X-ray; to be translucent

**просвечивающий** *adj.* translucent, transparent, diaphanous; transmitting, transmission-type

**просвинцованный** *adj.* lead-treated

**просеивание** *n.* screening, sifting, riddling

**просеивать** *pf. of* **просеять**

**просеивающий** *adj.* screening, sifting, riddling

**просеянный** *adj.* screened, sifted, riddled

**просеять, просеивать** *v.* to screen, to sift, to riddle, to size, to sieve, to bolt

**просигналить** *pf. of* **сигналить**

**проскакивание** *n.* overrun

**проскакивать, проскочить** *v.* to slip, to spring, to jump, to get through

**проскакивающий** *adj.* jumping

**проскальзывание** *n.* slippage, slipping, slip

**проскальзывать, проскользнуть** *v.* to slip (past *or* through)

**проскальзывающий** *adj.* slipping

**проскок** *m.* overswing, overshoot, overthrow, overtravel; passage; breakthrough; getting through

**проскользнуть** *pf. of* **проскальзывать**

**проскочить** *pf. of* **проскакивать**

**проследить** *pf. of* **прослеживать**

**проследование** *n.* passing, going; sequence

**прослеживание** *n.* tracing; hunting

**прослеживать, проследить** *v.* to trace, to track, to follow

**прослоек** *m.* streak, layer, band

**прослоённый** *adj.* interstratified, interbedded

**прослой** *m.,* **прослойка** *f.* layer, lamina, sheet; interlayer; padding, stuffing

**прослужение** *n.* service for some time

**прослуженный** *adj.* served

**прослуживать, прослужить** *v.* to serve

**прослушать** *pf. of* **прослушивать**

**прослушиваемый** *adj.* audible; listened through

**прослушивание** *n.* listening, listening through, audition; playback; crosstalk; **контрольное п.** monitoring; **п. междукадровых штрихов** frame noise; **п. на приёмнике** radio watching; **п. переходных разговоров** overhearing the crosstalk (*tphny.*); **п. перфорации** sprocket hum; **п. (подтонального) телеграфа** telegraph crosstalk; **сквозное п.** break-in (*tphny.*)

**прослушивать, прослушать** *v.* to hear; to listen

**просмолённый** *adj.* tarred; treated with resin; friction (tape)

**просмотр** *m.* pass, scan, scanning; examination, preview, review, survey; omission, oversight

**просмотренный** *adj.* scanned, swept; looked over, examined, checked, revised; reviewed

**просмотровый** *adj.* viewing, preview

**просовывание** *n.* pushing through, forcing through; **п. палок, п. суставной штанги** rodding

**просовывать, просунуть** *v.* to push through, to shove through, to force through

**просочиться** *pf. of* **просачиваться**

**проспрягать** *pf. of* **спрягать**

**простейший** *adj.* the simplest

**простереть(ся)** *pf. of* **простирать(ся)**

**простирание** *n.* spread, spreading

**простирать, простереть** *v.* to stretch, to extend, to hold, to reach out

**простираться, простереться** *v.* to stretch, to extend, to range, to reach, to spread

**простирающийся** *adj.* stretching, extending; extended

**просто** *adv.* simply, plainly; just, merely

**простой** 1. *adj.* simple, common, ordinary, plain, single. 2. *m. adj. decl.* inactivity, idle period, standstill, outage, down time, dead time; **взаимно п.** coprime

**пространно** *adv.* extensively; in detail

**пространность** *f.* extensiveness

**пространный** *adj.* extensive, spacious, vast, expansive

**пространственно-гармонический** *adj.* space-harmonic

**пространственно-зарядный** *adj.* space-charge

**пространственно-нарастающий** *adj.* spatially-growing

**пространственно-однородный** *adj.* spatially homogeneous

**пространственноподобный** *adj.* space-like

**пространственный** *adj.* spatial, space, three-dimensional, steric; solid (angle); roomy; directional; **по пространственным координатам** spacewise

**пространство** *n.* space, spacing, expanse, gap, range, extent, expansion, amplitude; area, field; room, volume; distance; spread; **между-железное п.** air-gap clearance, entrefer; **п., попадающее в угол зрения** angular field of view; **пустое п.** void; **рабочее п. ванны** cell activity

**простукивание** *n.* tapping

**простукивать, простучать** *v.* to tap, to rap

**просунуть** *pf. of* **просовывать**

**просушенный** *adj.* dried

**просушивание** *n.* drying, dehumidifying, desiccating

**просушивать, просушить** *v.* to dry, to desiccate

**просушка** *f.* drying, drying-out, desiccation; baking

**просчёт** *m.* checking; check; counting loss, miscalculation, error in counting

**просчитывать, просчитать** *v.* to check; to count; to count over again, to recount

**протактиний** *m.* protactinium, Pa

**протаскивание** *n.* pulling-in, pull-in; pulling through

**протаскивать, протащить** *v.* to pull, to draw in, to pull through, to drag through; to run (cable)

**протез** *m.* prosthesis, artificial limb

**протезный** *adj.* prosthetic

**протеин** *m.* protein

**протеиновый** *adj.* protein

**протекание** *n.* flow, flowing; conveyance, course, progress; passing

**протекать, протечь** *v.* to run, to flow, to pass through, to pass by; to take

place, to proceed; to leak; to elapse

**против** *adv. and prep. genit.* against, versus; facing; contrary (to); to, as against, in comparison with; **п. часовой стрелки** counter-clockwise

**противник** *m.* adversary, enemy

**противо-** *prefix* counter-, anti-

**противобаллистический** *adj.* antiballistic

**противовариантный** *adj.* contravariant

**противовес** *m.* counterpoise, back balance, counterweight, counterbalance, balancing weight; balancing capacitance, lower capacitance; earth screen, capacity earth

**противовибрационный** *adj.* antivibration, vibration-damping

**противовключение** *n.* opposition balancing, opposition circuit, feedback circuit

**противовключённый** *adj.* back-to-back

**противовоздушный** *adj.* anti-aircraft

**противогидролокационный** *adj.* anti-asdic

**противоглушитель** *m.* antijamming device

**противоглушительный** *adj.* antijamming

**противогнилостный** *adj.* antirot, antiputrefactive; preservative; aseptic

**противогрибковый** *adj.* fungisized, fungicidal

**противодавление** *n.* antivacuum; back pressure, reaction pressure, counterpressure, resistance

**противодействие** *n.* reaction, counterreaction, opposition, resistance; countermeasure, jamming; countercheck; buckling effect; **п. колебаниям** oscillation impedance; **электронное п.** electronic jamming

**противодействовать** *v.* to counteract, to react; to oppose, to resist, to buck; to countercheck

**противодействующий** *adj.* counteractive, reactive, reactionary; opposing, antagonistic; restoring

**противодетонирующий** *adj.* antiknock

**противоёмкостный** *adj.* anticapacitance

**противоизносный** *adj.* antiwear

**противоион** *m.* counterion, gegenion

**противоионный** *adj.* counterionic

**противоискровой** *adj.* antispark

**противоколебательный** *adj.* antihunt

**противокомпаундированный** *adj.* counter compound-wound

**противокомпаундный** *adj.* differentially compounded

**противокорабельный** *adj.* antiship

**противокражный** *adj.* burglar-proof, theft-proof, tamper-proof

**противолежащий** *adj.* contrary, opposite, lying opposite

**противолодочный** *adj.* antisubmarine

**противолокационный** *adj.* antiradar

**противомагнитный** *adj.* counter-magnetic

**противоместный** *adj.* antisidetone (*tphny.*)

**противоминный** *adj.* antimine

**противоминомётный** *adj.* counter-mortar

**противомодуляция** *f.* counter-modulation

**противообледенитель** *m.* de-icer

**противообледенительный** *adj.* anti-icing

**противо-ореольный** *adj.* antihalo

**противоотклонение** *n.* counter-deflection

**противопараллаксный** *adj.* antiparallax

**противопараллельный** *adj.* antiparallel

**противопереходный** *adj.* anti-induction; anticrossfire

**противоподлодочный** *adj.* antisubmarine

**противопожарный** *adj.* fireproof; fire-fighting

**противополе** *n.* counter field

**противоположение** *n.* contrast, contradistinction, antithesis

**противоположно** *adv.* contrarily, in contrast, oppositely

**противоположность** *f.* opposition, contrast, counteraction, difference; **прямая п.** the exact opposite (to)

**противоположный** *adj.* adverse, contrary, opposed, opposite, contradictory; reverse, inverse; different; counter-

**противопомеховый** *adj.* antijamming, anti-interference

**противопомпажный** *adj.* antihunting

**противопоставление** *n.* contraposition

**противопыльный** *adj.* antidust, dustproof, dusttight

**противорадиолокационный** *adj.* antiradar, radarproof

**противорадиолокация** *f.* radar countermeasure, counter radiolocation

**противоракета** *f.* antimissile

**противоракетный** *adj.* antimissile, antiballistic

**противорегулирование** *n.* counter regulation

**противоречащий** *adj.* contradictory

**противоречивость** *f.* inconsistency

**противоречивый** *adj.* contradictory, inconsistent

**противоречие** *n.* contradiction, inconsistency, conflict, variance, discrepancy; violation (*math.*)

**противосвязь** *f.* negative feedback

**противосовпадение** *n.* anticoincidence

**противоспутниковый** *adj.* antisatellite

**противостать** *pf. of* **противостоять**

**противостояние** *n.* opposition, resistance

**противостоять, противостать** *v.* to resist, to stand against, to withstand, to oppose, to face

**противостоящий** *adj.* opposed, resisting; opposite

**противосыростный** *adj.* antimoisture, dampproof, moisture-resistant

**противотанковый** *adj.* antitank

**противотечение** *n.* counterflow

**противоток** *m.* opposing current, countercurrent, counterflow, inverse current, bucking current; reflux

**противоторпедный** *adj.* antitorpedo

**противоточный** *adj.* counterflow, countercurrent; reflux

**противотуманный** *adj.* antifog

**противоугонный** *adj.* anticreepage

**противоударный** *adj.* shockproof

**противофаза** *f.* antiphase, opposite phase, reverse phase; phase opposition; **в противофазе** antiphase

**противофазность** *f.* push-pull

**противофазный** *adj.* antiphase, reverse-phase, inversely phased, out-of-phase

**противофединговый** *adj.* antifading

**противофоновый** *adj.* antihum; antibackground; hum-bucking

**противоходный** *adj.* counter-current

**противошумный, противошумовой** *adj.* antinoise

**противошунтовый** *adj.* antishunt

**противоэлектрод** *m.* counter-electrode, back plate

**противоэлектродвижущий** *adj.* counter-electromotive

противоэлемент *m.* counter e.m.f. cell, counter-electromotive force cell, counter cell

противоэховый *adj.* anti-echo, anti-reflection

проток *m.* channel, canal, tube

протокол *m.* register, record, official report; **п. испытаний** test record sheet

протокольный *adj.* protocol

протон *m.* proton

протонный *adj.* proton, protonic

протопирамида *f.* protopyramid

протопризма *f.* protoprism

прототип *m.* prototype, primary standard

проточный *adj.* flowing, flow, flow-type; circulating, continuous; production (line)

протрава *f.* dip, pickle, pickling

протравка *f.* etching, etch, pickling; **грубая п.** mass etch; **точная п.** close etching

протравленный *adj.* pickled; etched

протравливание *n.* pickling, dipping, corroding off; etching

протравливать, протравить *v.* to pickle, to dip; to etch; to scour, to corrode

протравлять *see* протравливать

протравляющий *adj.* corrosive

протуберанц *m.* protuberance, prominence

протягивание *n.* stretching forth, drawing, pulling through, pull-down, pull-in; traction, transport, forward wind

протягивать, протянуть *v.* to stretch out, to draw out, to extend, to spread; to prolong

протягивающий *adj.* extending, stretching; overhauling

протяжение *n.* extent, stretch, expanse; dimension, expansion, extension, spread, space; range, amplitude; length; field; run

протяжённость *f.* span, extent, length; extension, expansion, spread; range

протяжка *f.* pull-down (*tv*); drawing; broach

протяжный *adj.* continuous, long, drawn (out), lengthy, extended; slow; pull-through

протянутый *adj.* stretched, extended; prolonged; pulled through

протянуть *pf. of* протягивать

проушина *f.* ear, eye, lug

профессиональный *adj.* professional, occupational

профилактика *f.* preventive maintenance, maintenance, marginal checking; preventive measure

профилактический *adj.* preventive, protective, prophylactic

профилирование *n.* installation of sound-diffusers; profiling, shaping

профилированный *adj.* contoured, shaped, profiled, profile

профилограф *m.* surface analyzer

профилометр *m.* profilometer

профиль *m.* profile, outline, contour, shape, design; section, cross section, side view; elevation

профильно-лучевой *adj.* shaped-beam

профильно-отражательный *adj.* shaped-reflector

профильный *adj.* profile, contour; shaped; section, sectional

профильтрованный *adj.* filtered through

профильтровать *pf. of* фильтровать

профрезерованный *adj.* milled

проход *m.* way, passage, canal, conduit; pass; passing, transit; inlet, aperture, opening, orifice; port; **внутренний слуховой п.** internal auditory canal; **слуховой п.** auditory canal, acoustic duct

проходимость *f.* permeability, penetrability; passability

проходимый *adj.* permeable, pervious; passable

проходить, пройти *v.* to pass, to elapse, to expire; to go, to go through, to pass through, to run through, to proceed, to travel; to cover (a distance); to propagate; **п. по всей декаде** to hunt over a complete level (of a selector)

проходная *f. adj. decl.* clockhouse

проходной *adj.* going through, passing through; leading-in; through; straight-way; transfer; double-ended; re-entrant; feedthrough

проходческий *adj.* trailing; sinking

проходящий *adj.* passing, (going) through; transient

прохождение *n.* passing, passage, crossing, going through; flow; conveyance, transmission, transit; advancing; propagation; **п. обходным путём** flanking transmission; **п. кривой** rounding the curve; **п. соединения** advancement of a call, completion of a call, advancing the call (*tphny.*); **п. трафика** handling of traffic (*tphny.*)

**прохрипеть** *pf. of* **хрипеть**

**процедить(ся)** *pf. of* **процеживать(ся)**

**процедура** *f.* procedure

**процеживать, процедить** *v.* to filter, to strain

**процеживаться, процедиться** *v.* to creep; to be filtered; to filter

**процент** *m.* percentage, per cent, rate (per cent); **п. замедленных соединений** delayed calls proportion, probability of delay (*tphny.*); **п. искателей** selector ratio

**процентиль** *m.* percentile

**процентно-дифференциальный** *adj.* percentage-differential

**процентный** *adj.* percentage, per cent

**процесс** *m.* process, act, operation; procedure, method, practice; **п. многократной передачи с разделением по времени** time-division multiplex process; **п. нарастания** transient effect; **п., обратный по отношению к процессу печатания** de-printing operation; **переходный п.** transient, transient state, transient effect; **п. проектного осаждения люминофора** three-step phosphor-settling process; **п. сжатия-расширения** companding

**процессор** *m.* processor

**процитировать** *pf. of* **цитировать**

**прочерчивание** *n.* tracing

**прочерчиватель** *m.* tracer; **п. формы кривой** ondometer

**прочесть, прочитать** *pfs. of* **читать**

**прочистка** *f.* cleaning, cleansing

**прочность** *f.* durability, strength, toughness, endurance, permanence; stability, sturdiness, firmness, rigidity, solidity; safety, reliability, dependability; ruggedness; **вентильная п.** peak inverse anode voltage; **п. на изгиб** bending strength; **п. на износ** wearability, resistance to wear; **п. на пробой** disruptive strength; **п. на разрыв** breaking strength, tearing strength, tensile strength; **п. на удар** resistance to impact; **п. при переменных нагрузках** vibration strength

**прочный** *adj.* firm, durable, strong; stable, tough, sturdy, solid; lasting,

permanent; ruggedized; **п. на пробой** puncture-proof

**прошедший** *adj.* past, previous, last; transmitted

**прошивать, прошить** *v.* to interweave; to sew through, to enlace; to pierce; to broach

**прошивка** *f.* weaving process; sewing through; stitching; braid; insertion; piercing; broaching; reamer, broach

**прошивной, прошивочный** *adj.* piercing; broaching

**прошипеть** *pf. of* **шипеть**

**прошить** *pf. of* **прошивать**

**прошнуровать** *pf. of* **шнуровать**

**прошуметь** *pf. of* **шуметь**

**прощупать** *pf. of* **прощупывать**

**прощупывание** *n.* probing

**прощупывать, прощупать** *v.* to probe, to feel through; to sound

**проэкзаменовать** *pf. of* **экзаменовать**

**проявитель** *m.* developer

**проявительный** *adj.* developing

**проявить** *pf. of* **проявлять**

**проявление** *n.* display, exhibition, manifestation; appearance; development, developing process, bringout, point-up; **п. плёнки** photographic development

**проявленный** *adj.* developed; shown, manifested

**проявлять, проявить** *v.* to show, to exhibit, to manifest, to display, to exert, to give rise (to); to develop, to expose; to point up

**проявочный** *adj.* development

**ПРС** *abbr.* (**приводная радиостанция**) homing station

**пружина** *f.* spring, coil; **п. выборки люфта** tension spring; **движущая п.** lever spring, pressure spring; **контактная п. покоя (реле)** back contact spring; **оттягивающая п.** antagonistic spring, restoring spring, opposing spring; release spring, return spring; **п. против мёртвого хода** antibacklash spring; **п. работающая на растяжение** tension spring

**пружинение** *n.* flexibility, resilience, elasticity, cushioning; spring, springing; **п. нитей** spring-suspension of filament, filament tensioning support

**пружинистость** *f.* springiness

**пружинистый** *adj.* flexible, springy, elastic

**пружинить** *v.* to spring, to be elastic, to yield
**пружинка** *f.* small spring, catwhisker
**пружинно-торсионный** *adj.* spiral-torsion
**пружинный** *adj.* spring, spring-loaded
**пружиновыгибатель** *m.* spring adjuster
**пружинодержатель** *m.* spring holder
**пружинящий** *adj.* elastic, flexible, springy, springing
**Пруст** Proust
**прут** *m.* stick, rod, bar
**прутковый** *adj.* stick, rod, bar, rod-shaped
**пруток** *m.* wand, rod, bar
**прыгание** *n.* jumping, jerking, skipping
**прыгать, прыгнуть** *v.* to jump, to spring, to leap, to jerk; to knock (of valve)
**прыгающий** *adj.* jumping, jerking, skipping
**прыгнуть** *pf. of* прыгать
**пряжа** *f.* yarn, thread, twine
**пряжечный** *adj.* buckle, clasp
**пряжка** *f.* 1. yarn. 2. buckle, clasp
**прямая** *f. adj. decl.* straight line; **п. проницаемости** straight-line curve of the reciprocal of amplification factor
**прямление** *n.* straightening
**прямо** *adv.* straight, directly, straight-ways; uprightly; immediately
**прямо-** *prefix* straight, rect-, recti-
**прямо-возбуждаемый** *adj.* directly excited
**прямоволновый** *adj.* straight-line-wavelength
**прямодействующий** *adj.* direct-action, direct-acting, direct
**прямоёмкостный** *adj.* straight-line-capacitance
**прямозависимый, прямозависящий** *adj.* direct-relation
**прямозвенный** *adj.* straight-link
**прямозубый** *adj.* straight-toothed
**прямоизлучающий** *adj.* direct-radiating
**прямоимпульсный** *adj.* direct-pulse
**прямой** *adj.* direct, straight; right; erect, upright; real; forward; instantaneous
**прямолинейно** *adv.* rectilinearly; straight-forwardly
**прямолинейно-возвратный** *adj.* reciprocating, moving to and fro
**прямолинейность** *f.* rectilinearity

**прямолинейный** *adj.* rectilinear, straight, linear, rectilineal, straight-line(d)
**прямонакальный** *adj.* directly-heated
**прямонаправленный** *adj.* head-on
**прямопоказывающий** *adj.* direct-indicating; direct-reading
**прямосмещённый** *adj.* forward-biased
**прямо-соединённый** *adj.* directly connected
**прямоточность** *f.* direct flow, straight-through feed
**прямоточный** *adj.* direct-flow; once-through, straight-through; concurrent; uniflow
**прямоугольник** *m.* rectangle, square
**прямоугольность** *f.* rectangularity, squareness
**прямоугольный** *adj.* rectangular, right-angled, square, quadrate; orthogonal; square-wave
**прямочастотный** *adj.* straight-line-frequency
**прямошумовой** *adj.* direct-noise
**псалтерион** *m.* bell harp, psaltery
**псевдо-** *prefix* pseud-, pseudo-
**псевдовектор** *m.* pseudovector
**псевдогармонический** *adj.* pseudoharmonic
**псевдодиэлектрик** *m.* pseudodielectric
**псевдозатухание** *n.* pseudodamping
**псевдокод** *m.* pseudo-code, abstract code
**псевдокоманда** *f.* pseudo order, instructional constant
**псевдоморфоз** *m.* pseudomorphism
**псевдонасыщение** *n.* pseudosaturation
**псевдопассивный** *adj.* pseudopassive
**псевдопериодический** *adj.* pseudo-periodic
**псевдопрограмма** *f.* pseudoprogram
**псевдорегистр** *m.* pseudoregister
**псевдосимметричный** *adj.* pseudo-symmetric
**псевдоскаляр** *m.* pseudoscalar quantity
**псевдоскалярный** *adj.* pseudoscalar
**псевдослучайный** *adj.* pseudorandom
**психо-** *prefix* psycho-, psychological
**психоакустика** *f.* psychoacoustics
**психогальванический** *adj.* psychogalvanic
**психогальванометр** *m.* psychogalvanometer, pathometer
**психогенный** *adj.* psychogenic
**психосоматограф** *m.* psychosomato-graph

**психостатистический** *adj.* psychostatistical

**психофизика** *f.* psychophysics

**психофизиология** *f.* psychophysiology

**психофизический** *adj.* psychophysical

**психрометр** *m.* bulb thermometer, psychrometer

**психрометрический** *adj.* psychrometric

**псофометр** *m.* circuit noisemeter, psophometer

**псофометрический** *adj.* psophometric

**Пти** Petit

**птица** *f.* bird; **п. обладающая звуколокатором** bird sonar

**ПТО** *abbr.* 1 (**пункт технического обслуживания**) maintenance point. 2 (**противотанковая оборона**) antitank defense

**ПТОР** *abbr.* (**противотанковое орудие**) antitank gun

**ПТП** *abbr.* (**переключатель телевизионных программ**) tuner, switch tuner, television tuner

**ПТС** *abbr.* 1 (**передвижная телевизионная станция**) pick-up station, telecruiser, mobile television unit. 2 (**противотехнические средства**) pyrotechnical means

**ПТЭ** *abbr.* 1 (**правила технической эксплуатации**) technical operation instructions. 2 (**Приборы и техника эксперимента**) Instruments and Experimental Techniques (journal)

**ПУ** *abbr.* 1 (**пункт управления**) control post. 2 (**пусковая установка**) starting system; launching installation. 3 (**путевой угол**) track angle. 4 (**переговорное устройство**) interphone

**пуаз** *m.* poise (unit of viscosity)

**Пуазейль** Poiseuille

**ПУАЗО** *abbr.* (**прибор управления артиллерийским зенитным огнём**) AA-director, (antiaircraft fire) director

**Пуанкаре** Poincaré

**пуансон** *m.* punch knife, punch, die, stamp; pin-setting device; plunger

**Пуассон** Poisson

**публичный** *adj.* public, common

**ПуВРД** *abbr.* (**пульсирующий воздушно-реактивный двигатель**) pulse jet engine

**пугало** *n.* bird guard, scarecrow

**пуговичный** *adj.* button

**пуговка** *f.* button, stud; head, knob

**пузырёк** *m.* bubble, blister; bead; phial, vial

**пузырение** *n.* bubbling

**пузыристый** *adj.* bubbly; blistered, porous

**пузыриться** *v.* to bubble, to blister; to effervesce

**пузырчатый** *adj.* bubbly; blistered, porous

**пузырь** *m.* bubble, blister, air hole; blow hole; pocket, sac

**пузырьковый** *adj.* bubble, blister; pocket, sac

**пулемёт** *m.* machine gun

**пулемётный** *adj.* machine-gun

**пульверизатор** *m.* dust gun, pulverizer; atomizer, sprayer

**пульверизационный** *adj.* pulverization, pulverizing; atomizing, spraying

**пульверизация** *f.* pulverization; atomization, spraying; **п. сквозь пламя** flame-spraying of plastics

**пульверизирование** *n.* pulverization; atomization, spraying

**пульверизированный, пульверизованный** *adj.* pulverized; atomized, sprayed

**пульверизировать** *or* **пульверизовать** *v.* to pulverize; to atomize, to spray

**пульс** *m.* pulse

**пульсатор** *m.* pulser; pulsator

**пульсатрон** *m.* pulsatron

**пульсация** *f.* pulsation, pulsing, pulse; beat, pulsatance, ripple, fluctuation, flutter; surging, flicker, throb(bing), pulsative oscillation; **п. напряжения в линии** mains ripple; **п. напряжения источника питания** power supply flutter; **п. расхода** mass-flow rate perturbation; **п. силы звука** flutter, hearing effect

**« пульсескоп »** *m.* pulsescope

**пульсирование** *see* **пульсация**

**пульсировать** *v.* to expand and contract; to pulse, to pulsate, to beat, to fluctuate; to ripple, to flutter

**пульсирующий** *adj.* pulsating, pulsatory, fluctuating; undulating; variable, intermittent

**пульсометр** *m.* pulsometer; vacuum pump

**пульс-реле** *n. indecl.* relay-interrupter

**пульт** *m.* desk, stand, benchboard; panel, board; console; **п. без проёма** continuous-type benchboard; **п. вы-**

работки данных, дающий п. data panel; **п. с проёмом** open-type benchboard; **сдвоенный п.** duplex benchboard; **смесительный п.** master control, master control desk, mixer (*tv*); **п. управления** control booth; control desk; benchboard; lever machine

**Пунгс** Pungs

**пункт** *m.* point, spot; station, post, center; item, article, clause, paragraph; **органный п.** pedal point, pedal-note; **переговорный п.** public telephone, call office, pay station; **п. прокладки целей** battery plotting room

**пунктир** *m.* dotted line; **точечный п.** dotted line

**пунктирный, пунктированный** *adj.* dotted, punctate; stippled; spot (welding)

**пунктировать** *v.* to dot, to prick, to point; to stipple

**пунктировка** *f.* dotting; stippling

**пуншер** *m.* punch, puncher, perforator

**пунширование** *n.* punching

**пуншировать** *v.* to punch, to stamp, to perforate

**пуншировка** *f.* punching, perforation

**пупинизация** *f.*, **пипинизирование** *n.* coil loading, lumped loading, pupinization; **концертная п.** musical loading; **п. параллельным включением катушек** shunt loading

**пупинизированный** *adj.* coil-loaded

**пупинизировать** *v.* to coil-load, to load, to pupinize

**пупинизованный** *adj.* coil-loaded

**пупиновский** *adj.* Pupin; loading

**ПУПЧ** *abbr.* (**предварительный усилитель промежуточной частоты**) I. F. preamplifier, intermediate-frequency preamplifier

**пурка** *f.* hondrometer

**пурпурный** *adj.* purple, mauve

**ПУРС** *abbr.* (**противотанковый управляемый реактивный снаряд**) guided antitank missile

**пуск** *m.* start, starting, start-up, setting in motion, triggering, trigger action, initiation; take-off; **п. в обход** start, starting, starting up, setting in motion; bringing into service; **п. в эксплоатацию** putting in operation, set-up procedure, cut-over; **п. от полного напряжения**

across-the-line starting, full-voltage starting; **п. переключением со звезды на треугольник** star-delta starting; **п. по схеме без разрыва цепи** closed-transition starting

**пускание** *see* **пуск**

**пускатель** *m.* starter; **п. для прямого пуска от сети** across-the-line starter, direct-on starter; **многоступенчатый роторный п. с сопротивлениями** rotor-resistance starter; **многоступенчатый статорный п. с сопротивлениями** stator-resistance starter, stator-inductance starter; **п. с лицевой плитой** disk-type starter; **п. с приспособлением для медленного вывода сопротивлений** slow-motion starter; **статорный п.** reduced-voltage starter

**пускать, пустить** *v.* to let, to allow; to run, to start, to set in motion, to put in action, to trigger, to initiate; **п. в обход** to put in action, to put into operation, to put into service; to set in motion, to actuate, to drive, to start

**пусковой** *adj.* starting, start-up, actuating, triggering; initiating; initial; tripping; trigger, starter

**пустить** *pf. of* **пускать**

**пустота** *f.* void, emptiness, vacuum; blankness; hollow, cavity; vacancy

**пустой** *adj.* vacuous, empty, void; vacant, blank, bare; hollow, cored

**пустотелость** *f.* hollowness

**пустотелый** *adj.* hollow; vacuum

**пустотность** *f.* hollowness; vacuum

**пустотный** *adj.* hollow; vacuum, vacuous, evacuated; void

**путаница** *f.* confusion, maze, tangle; mix-up

**путанный** *adj.* confused, tangled; confusing

**путёвка** *f.* clearance card; permit

**путевой** *adj.* track; way, road; traveling

**путемер** *m.* distance gage, pedometer

**путепрокладчик** *m.* map tracer

**путь** *m.* way, road, track, path, coarse, route; passage; means, method; **п. (волны) за пределы горизонта** beyond-line-of-sight path; **п. замыкания (магнитного) потока** flux return path; **параллельный п. (для тока)** alternate path; **перевальный п.** passage path; **подкрановый п.** runway

**пучёк** *see* **пучок**

**пучкование** *n.* bunching

**пучкованный** *adj.* bunched

**пучкователь** *m.* buncher

**пучковатый** *adj.* bundle

**пучковидный** *adj.* clustered, tufted

**пучковый** *adj.* bunched, bundled, clustered

**пучность** *f.* antinode, loop; crest, peak (of a wave); velocity loop; **п. волны** oscillation loop; **п. давления** pressure antinode

**пучок** *m.* bundle, bunch, cluster, pack, band, pile; pencil (of rays), beam; group, bank (of outgoing trunks); **равнодоступные пучки** equally accessible trunk groups (*tphny.*); **выделенный п. соединительных линий** segregated trunk groups (*tphny.*); **п. лучей** beam, pencil of rays, ray bundle; **п. лучей звукового гидролокатора** sonar beam; **п. многократных проводов в одной секции** section multiple (*tphny.*); **п. порядков всех рабочих мест для исходящих разговоров** multiple junction (*tphny.*); **п. соединительных линий** trunk group. multiple group (*tphny.*); **фиксирующий п.** test pattern beam

**пушечный** *adj.* gun, cannon

**пушка** *f.* gun, cannon; irradiation unit; **п. с двойным кроссовером** double-crossover gun; **электронная п.** cathode-ray gun

**пушпулл, пушпуль** *m.* push-pull

**пушпульный** *adj.* push-pull

**пф** *abbr.* (**пикофарада**) picofarad, micromicrofarad

**ПЧ** *abbr.* (**промежуточная частота**) i.f. intermediate frequency

**пылеватый** *adj.* dusty, dust-like

**пылевой** *adj.* dust

**пылезащищённый** *adj.* dustproof

**пылеизмеритель, пылемер** *m.* dust meter, dust counter

**пыленепроницаемый** *adj.* dustproof, dust-tight

**пылеосадитель** *m.* dust extractor

**пылеосаждение** *n.* dust precipitation

**пылеотделитель** *m.* dust separator, dust remover

**пылеочиститель** *m.* duster

**пылесос** *m.* vacuum cleaner, dust cleaner

**пылестойкий** *adj.* dustproof

**пылеудаление, пылеудаливание** *n.* dust removal, elimination of dust, dust collecting, dust precipitation

**пылеуловитель** *m.* dust trap, dust collector, dust catcher; dust precipitator

**пылеуловительный** *adj.* dust-catching, dust-collecting

**пылить, напылить** *v.* to raise dust, to fill the air with dust

**пыль** *f.* dust, powder; spray; **неповреждаемый пылью** dustproof

**пыльный** *adj.* dusty; powdery

**пьедестал** *m.* pedestal, stand, base; pedestal pulse; porch

**пьеза** *f.* pieze (unit of pressure)

**пьезо-** *prefix* piezo-, pressure; piezoelectric

**пьезовибратор** *m.* piezoelectric resonator, piezoelectric vibrator

**пьезовосприимчивость** *f.* piezoelectric susceptibility

**пьезогромкоговоритель** *m.* crystal loudspeaker

**пьезодиффузия** *f.* pressure diffusion

**пьезодиэлектрический** *adj.* piezodielectric

**пьезоид** *m.* piezoid

**пьезокварц** *m.* piezoelectric crystal

**пьезокварцевый** *adj.* piezoelectric-crystal

**пьезокерамика** *f.* piezoelectric ceramics

**пьезокерамический** *adj.* piezoceramic

**пьезокристалл** *m.* piezocrystal, piezoelectric crystal; **п., срезанный под 35° к оси Z** AT-cut crystal

**пьезокристаллический** *adj.* piezoelectric-crystal

**пьезомагнетизм** *m.* piezomagnetism

**пьезоманометр, пьезометр** *m.* piezoelectric pressure gage, piezometer

**пьезометрический** *adj.* piezometric, piestic

**пьезомикрофон** *m.* piezoelectric microphone, crystal microphone

**пьезомодуль** *m.* piezoelectric-modulus

**пьезооптический** *adj.* piezooptic

**пьезорезец** *m.* crystal cutter

**пьезорезистивный** *adj.* piezoresistive

**пьезорезонатор** *m.* piezoelectric resonator

**пьезотелефон** *m.* piezoelectric receiver

**пьезотензометр** *m.* piezoelectric strain gage

**пьезотранзистор** *m.* piezotransistor

**пьезотропный** *adj.* piezotropic

**пьезохимия** *f.* piezochemistry

**пьезоэлектрик** *m.* piezoelectric, piezoelectric material
**пьезоэлектрический** *adj.* piezoelectric, ferroelectric
**пьезоэлектричество** *n.* piezoelectricity
**пьезоэффект** *m.* piezoelectric effect
**пьезоячейка** *f.* piezocell, sound cell
**пьеса** *f.* piece (*mus.*)
**ПЭШО** *abbr.* (провод эмалированный, шёлк, однослойный) enamelled single silk-covered wire
**пюпитр** *m.* desk, reading desk
**пята** *f.* bottom bearing, base, foot; heel; pivot, pivot journal, pin
**пятерично-двоичный** *adj.* quibinary
**пятеричный** *adj.* quinary, fivefold, quintuple
**пяти-** *prefix* penta-, quinque-, five
**пятиадресный** *adj.* five-address
**пятивалентность** *f.* pentavalence
**пятивалентный** *adj.* pentavalent
**пятигруппный** *adj.* pentode, pentodal
**пятизначный** *adj.* five-unit; five-place
**пятиконечный** *adj.* pentagonal, five-pointed
**пятикратный** *adj.* quintuple

**пятикупольный** *adj.* pentadome
**пятиламповый** *adj.* five-tube
**пятипроводный** *adj.* five-wire
**пятиричный** *adj.* quinary
**пятисекундный** *adj.* five-second
**пятислойный** *adj.* five-layer
**пятистрельчатый** *adj.* five-needle
**пятиступенный** *adj.* pentatonic
**пятиугольник** *m.* pentagon
**пятиугольный** *adj.* pentagonal
**пятифазный** *adj.* five-phase
**пятичастный** *adj.* five-part
**пятичленный** *adj.* five-membered
**пятиэлектродный** *adj.* five-electrode
**пятнадцатеричный** *adj.* quindenary
**пятнистость** *f.* blemish, blight, spot, spottiness, spotting-out; smudge, smear
**пятнистый** *adj.* spotty, spotted, blotched, mottled, stained, speckled, dappled; patchy
**пятно** *n.* spot, patch, blotch, blur; blot, smudge, smear, blemish; burn; **п. на солнце** sunspot
**пятнообразование** *n.* spotting
**пятый** *adj.* fifth

# Р

**Р** *abbr.* (**разведчик**) reconnaissance aircraft

**р** *abbr.* 1 (**радиан**) radian; 2 (**рентген**) roentgen

**РА** *abbr.* (**разведывательная авиация**) reconnaissance aviation

**Рабек** Rahbeck

**Рабиц** Rabitz

**работа** *f.* work; operation, working, service; performance, action, run, running, duty; **р. буфером** floating; **р. в граничном режиме** marginal operation; **р. в истинном масштабе времени** real time application, real time working; **р. в несколько смен** multishift operation; **р. в одном канале со смещёнными несущими** offset(-carrier) operation; **р. в режиме большого усиления** high-gain operation; **встречная р.** full-duplex operation (*tgphy.*); **р. вылета** work function (of an electron); **р. выхода** work-function; **генераторная р.** power station service; **р. для покрытия нагрузочных пиков** peak-load-service operation of storage batteries; **р. допускающая перебивание** break-in operation (*tgphy.*); **дуплексная р.** contraplex working (*tgphy.*); **дуплексная встречная р.** duplex working (*tgphy.*); **р. ключом** manipulation, keying; **малая р. выхода** low work function; **р. на высокой мощности** high-powered operation; **р. на одной волне** common frequency working; **р. на основной частоте** fundamental operation; **р. на перемежающихся частотах** frequency interlacing (*tv*); **неправильная р.** malfunction; **опытно-конструкторская р. на уровне целых систем** systems-level development work; **р. переговорно-справочной станции** direct record service; **р. по дифференциальной схеме** differential winding working (*tgphy.*); **р. по пригонке** fitting work; **повторно-кратковременная р. с прерывистой нагрузкой** continuously-running duty with intermittent loading; **р. ири наладочной скорости** threading operation; **р. при переменной длительности рабочего цикла** variable cycle operation; **р. при пиках** load factoring; **р. пуска** starting energy; **пятикратная р.** pentode working (*tgphy.*); **р. с дозвуковой струёй** unchoked operation; **р. с кондуктивной связью** D.C. operation; **р. с косвенным управлением** off-line working; **р. с подавлением гармоник** impeded harmonic operation; **р. с постоянной величиной скважности** constant d-c. operation; **р. с применением малых автоматических станций** U.A.X. working, unit automatic exchange working; **р. с током покоя** closed circuit operation; **р. со сверхзвуковой струёй** choked operation; **р. связистов** signalmen activities; **ступенчатая р.** echelon working; **р. схемы выключения канала цветности** color-killer operation; **р. схемы около точки лампы** bottoming; **термоэлектронная р. выхода** thermionic work function; **р. токами двух направлений** double current working; **р. током одного направления** single current working; **трёхкратная р.** triode working (*tgphy.*); **устойчивая р.** quiescent operation; **четырёхкратная р.** tetrode working (*tgphy.*); **шестикратная р.** hexode working (*tgphy.*)

**работать, поработать** *v.* to work; to run, to work, to operate, to function, to act, to perform; **р. ключом** to key; **р. на аппарате Морзе** to morse; **р. телеграфным ключом** to key, to control

**работающий** *adj.* functioning, operating, running; (often omitted in translation: **генератор, работающий на боковой полосе частот** sideband generator [*lit.* generator on a side band]); **бесперебойно р.** trouble-proof

**работоспособность** *f.* efficiency
**рабочий** 1. *adj.* operating, running, working, normal; actuating; effective; work; worker's, working; 2. (*noun*) *m. adj. decl.* worker, workman, laborer
**равенство** *n.* equality, parity; equation; matching
**равнина** *f.* plain, plane
**равно** *adv.* equally, alike
**равно-** *prefix* equi-, iso-, homo-
**равнобедренный** *adj.* isosceles (*math.*)
**равнобокий** *adj.* equilateral
**равновероятно** *adv.* equally probable
**равновероятный** *adj.* equally probable, equally likely
**равновесие** *n.* equilibrium, balance, equipoise, equibalance
**равновесный** *adj.* equilibrium
**равновзвешенный** *adj.* equal-weighted
**равновозможный** *adj.* equally-likely
**равновременность** *f.* isochronism
**равновременный** *adj.* isochronal, isochronous, simultaneous
**равнодействующая** *f. adj. decl.* resultant (force)
**равнодействующий** *adj.* equally effective, equal; resultant
**равноденственный** *adj.* equinoctial, equinox; equatorial
**равноденствие** *n.* equinox
**равнодоступный** *adj.* equally accessible
**равнозначащий** *adj.* equivalent, equipolent; tantamount
**равнозначность** *f.* equivalence
**равнозначный** *adj.* equivalent, equipolent; tantamount
**равноизлучающий** *adj.* equally radiating; equal-energy
**равноконтрастный** *adj.* equality-of-contrast, uniform chromacity
**равномерно** *adv.* equally, uniformly, evenly
**равномерно-распределённый** *adj.* uniformly distributed
**равномерность** *f.* flatness, evenness, uniformity, constancy; uniformity factor; quality, proportionality
**равномерный** *adj.* uniform, even, smooth, steady; equi-; equal, proportional; isometric
**равнонаправленный** *adj.* equidirectional, unidirected, rectified, in the same direction
**равноосный** *adj.* equiaxial

**равноотстоящий** *adj.* equidistant, equispaced
**равноплечий** *adj.* equal-arm
**равнопотенциальный** *adj.* equipotential, midpotential
**равнопромежуточный** *adj.* equidistant
**равносигнальный** *adj.* equisignal, equal-signal
**равносилие** *n.* equivalence
**равносильный** *adj.* equivalent, of equal strength, tantamount
**равносторонний** *adj.* equilateral
**равноугольный** *adj.* equiangular
**равноудалённый** *adj.* equidistant
**равноускоренный** *adj.* uniformly accelerated
**равнофазный** *adj.* equiphase
**равноцветный** *adj.* homochromatic, polychromatic
**равноценность** *f.* equivalence
**равноценный** *adj.* equivalent, tantamount; of equal worth
**равночастотный** *adj.* equifrequent
**равноэлементный** *adj.* equal-element
**равноэнергетический** *adj.* equal-energy, equi-energy
**равный** *adj.* equal, similar, like, alike; even, constant
**равнять, поравнять** *v.* to equate, to equalize; to smooth, to even, to flatten; to compare
**равняться, поравняться** *v.* to equal, to be equalized; to compare (to)
**рад** *m.* rad
**радар** *m.* radar; **р. управления зенитным огнём** antiaircraft fire control radar
**радарный** *adj.* radar
**« радасвип »** *m.* radasweep
**радекон, радехон** *m.* radechon (tube)
**радиально** *adv.* radially
**радиально-лучевой** *adj.* radial-beam
**радиально-однородный** *adj.* radial-homogeneous
**радиально-сверлильный** *adj.* radial-drilling
**радиальный** *adj.* radial
**радиан** *m.* radian, circular measure
**радианный** *adj.* radian
**радиант** *m.* radiant; radiant point
**радиатор** *m.* radiator, dissipator; cooling flange; heat sink
**радиационный** *adj.* radiation, radiative
**радиация** *f.* radiation, emission
**радиевый** *adj.* radium, radium-type

**радий** *m.* radium, Ra

**радийсодержащий** *adj.* radium-containing, radium-bearing

**радикал** *m.* radical

**радикальный** *adj.* radical; efficient, complete

**радио-** *prefix* radio-, radio

**радио** *n. indecl.* radio, wireless

**радиоавтограф** *m.* radioautograph

**радиоавтография** *f.* autoradiography

**радиоактивация** *f.* radioactivation

**радиоактивировать** *v.* to radioactivate

**радиоактивность** *f.* radioactivity

**радиоактивный** *adj.* radioactive

**радиоактиний** *m.* radioactinium, RaAc

**радиоакустический** *adj.* radioacoustic

**радиоальтиметр** *m.* radio altimeter, radio height finder, terrain-clearance indicator

**радиоантенна** *f.* radio antenna

**радиоаппарат** *m.* radio set, radio equipment

**радиоаппаратура** *f.* radio equipment

**радиоастроном** *m.* radio astronomer

**радиоастрономия** *f.* radio astronomy

**радиоатмосфера** *f.* radio atmosphere

**радиобашня** *f.* radio tower

**радиобиолог** *m.* radiobiologist

**радиобиологический** *adj.* radiobiological

**радиобиология** *f.* radiobiology

**радиоблок** *m.* radio component; electronic component

**радиобуквопечатающий** *adj.* radioteletype

**радиовахта** *f.* radio watch(ing)

**радиовещание** *n.* broadcast, broadcasting, radio broadcasting, telediffusion; **внутреннее р.** domestic broadcasting; **высокочастотное р. по проводам** wired wireless broadcast, carrier rediffusion; **р. по проводам** line radio, wire broadcasting; **р. через несколько станций на одной волне** common frequency broadcasting

**радиовещательный** *adj.* broadcasting, radio broadcast(ing)

**радиовзрыватель** *m.* radio detonator, radio fuze, proximity fuze

**радиовидение** *n.* radiovision, panoramic radar

**радиовилка** *f.* radio bifurcation

**радиовождение** *n.* radio aids to navigation

**радиоволна** *f.* radio wave, broadcast wave

**радиоволновод** *m.* horn antenna

**радиовсплеск** *m.* radio burst

**радиовызов** *m.* radio contact

**радиовысотомер** *m.* radio altimeter, radio height finder

**радиовыставка** *f.* radio show, radio exhibition

**радиовышка** *f.* radio tower

**радиогальванометр** *m.* radiomicrometer

**радиогеничный, радиогенный** *adj.* radiogenic

**радиогеоразведка** *f.* radioprospecting

**радиогидроакустический** *adj.* radiohydroacoustic, radiosonic

**радиоглушение** *n.* radio jamming, radio countermeasures

**радиогониометр** *m.* direction finder, radiogoniometer

**радиогониоскоп** *m.* radiogonioscope

**радиогоризонт** *m.* radio horizon

**радиограмма** *f.* radiogram; radiotelegram, wireless message; roentgenogram

**радиограммофон** *m.* electric phonograph

**радиограф** *m.* radiograph

**радиографический** *adj.* radiographic

**радиография** *f.* radiography, X-ray photography; **поточная р.** mass radiography

**радиодальномер** *m.* range only radar

**радиодевиационный** *adj.* radio-deviation, radio-calibration

**радиодевиация** *f.* direction finding correction, direction finder deviation, deviation of radio bearing, deviation of radio azimuth

**радиодело** *n.* radio business, radio techniques, radio art

**радиодеталь** *m.* radio part, radio component, electronic component

**радиодефектоскопия** *f.* radiomateriology, X-ray defectoscopy

**радиожурнал** *m.* radio station log; radio magazine

**радиозавод** *m.* radio manufacturing plant

**радиозаглушение** *n.* radio jamming, radio countermeasures

**радиозамирание** *n.* radio fadeout, fading

**радиозасечка** *f.* radio fix

**радиозвезда** *f.* radio star

**радиозеркало** *n.* radio mirror
**радиознание** *n.* radio knowledge
**радиозонд** *m.* radiosonde, radiometeorograph, radio sounding balloon
**радиозондаж** *m.*, **радиозондирование** *n.* radiosondage technique, radio-sounding
**радиозондовый** *adj.* radiosonde, radiosonic
**радиозритель** *m.* televiewer
**радиоизлучение** *n.* radio noise, radio wows; radio emission; radio waves; **рассеянное р. фона** diffuse background radiation; **р. холодного разряда** plasma noise
**радиоизмерительный** *adj.* radio-measuring
**радиоизотоп** *m.* radio isotope
**радиоимпульс** *m.* radio frequency pulse, wave impulse, wave packet
**радиоиндуцированный** *adj.* radiation-induced
**радиоинерциальный, радиоинерционный** *adj.* radio-inertial
**радиоинженер** *m.* radio engineer
**радиоинспекция** *f.* radio inspection
**радиоинтерференция** *f.* radio interference
**радиоинтерферометр** *m.* radio interferometer
**радиоинтерферометрия** *f.* radio interferometry
**радиоинформация** *f.* radio information
**радиоисточник** *m.* radiation source
**радиокабина** *f.* radio cabin, radio room
**радиоканал** *m.* radio channel, radio link, radio duct
**радиоклуб** *m.* radio club
**радиокод** *m.* radio code
**радиокомандный** *adj.* radio-command
**радиокомпас** *m.* radio compass, homing device, direction finder, bearing compass, azimuth compass, sense finder; **автоматический р.** automatically tuned direction finder
**радиоконтрмера** *f.* radio counter measure
**радиоконтрнападение** *n.* radio counter-offensive
**радиоконтроль** *m.* radio monitoring service
**радиоконтур** *m.* radio circuit
**радиокоордината** *f.* radio fix
**радиокристаллография** *f.* radio crystallography
**радиокружок** *m.* radio ham's group

**радиокурсоуказатель** *m.* radio direction finder
**радиола** *f.* radio-gramophone, phonograph
**радиолампа** *f.* radio tube
**радиолечение** *n.* radiotherapy, radio-praxis
**радиолиз** *m.* radiolysis
**радиолиния** *f.* radio link; wire-broadcasting line
**радиолог** *m.* radiologist
**радиологический** *adj.* radiological
**радиология** *f.* radiology
**радиолокатор** *m.* radio-locator, radio-detector, radar; **самолётный р. наводки (орудий)** airborne gun-laying radar; **р. управления прожектором** searchlight-control radar
**радиолокационный** *adj.* radio-homing, radio-locating, radiolocation, radar
**радиолокация** *f.* radiolocation, radar, radio direction finding, radio detecting and ranging, wireless bearing; **равносигнальная р.** lobing radar
**радиолуч** *m.* radio beam, radio frequency beam; **р. наведения** radio vector; **плоский р.** fanned beam
**радиолюбитель** *m.* radio fan, radio amateur, radio enthusiast
**радиолюбительский** *adj.* radio-fan, radio-ham
**радиолюбительство** *n.* amateur radio
**радиолюминесцентный** *adj.* radio-luminescent
**радиолюминесценция** *f.* radiolumin-escence
**« радиом »** *m.* radiom
**радиомагистраль** *m.* main (long-distance) radio
**радиомагнитный** *adj.* radio-magnetic
**радиомаркёр** *m.* radio marker, radio marker beacon; **р. с секретным полем** radio fan marker
**радиомастер** *m.* radio technician
**радиомачта** *f.* radio-mast, antenna mast, radio tower
**радиомаяк** *m.* radio(-range) beacon, marker beacon, bearing transmitter, direction finding transmitter, radiophare, A-N range, equisignal beacon; **аэродромный р. зоны ожидания** airport danger beacon; **ведущий р.** radio-range beacon; **визуально-звуковой р.** visual-aural radio range; **всенаправленный р. сверхвысокой частоты** VHF omnirange; **зональ-**

ный **р.** A-N radio range; **р. зоны ожидания** airport hazard beacon; **курсовой р.** radio-beacon, radio range beacon; **направленный четырёхкурсовой р.** AN-radio-range, radio range beacon; **р.-ответчик** transponder, responder beacon; **р.-ответчик по коду на импульсные запросы** transponder radar beacon; **р. со звуковым сопровождением** aural radio range; **управляемый импульсами посадочный р.** blind approach beacon system

**радиометалл** *m.* radiometal

**радиометаллография** *f.* radiometallography

**радиометаллоискатель** *m.* radio metal locator

**радиометеорограф** *m.* radiometeorograph

**радиометеорография** *f.* radiometeorography

**радиометеорологический** *adj.* radiometeorological

**радиометеорология** *f.* radiometeorology, radio weather-forecast service

**радиометр** *m.* radiometer; **р. с охлаждаемым приёмником излучения** cooled cell radiometer, cooled detector radiometer

**радиометрист** *m.* radar operator; radio operator

**радиометрический** *adj.* radiometer; radiometric

**радиометрия** *f.* radiometry; **р. на сантиметровых волнах** microwave radiometry

**радиомеханика** *m.* radiomechanics; **сверхточная р.** micronics

**радиомикрометр** *m.* radiomicrometer

**радиомиметрический** *adj.* radiomimetric

**радиомодулятор** *m.* radio modulator

**радиомолчание** *n.* radio silence

**радиомонтаж** *m.* radio review

**радиомонтёр** *m.* radio technician, radio electrician

**радионаведение** *n.* electronic guidance, radio guidance

**радионавигационный** *adj.* radio navigational

**радионавигация** *f.* radio navigation

**радионаушники** *pl.* earphones

**радионепроницаемый** *adj.* radiopaque

**радиообмен** *m.* radio traffic

**радиообнаружение** *n.* radar detection, radar warning; **р. и измерение дальности** radio detection and ranging, radar

**радиооборудование** *n.* radio equipment, wireless apparatus

**радиообслуживание** *n.* radio service

**радиооператор** *m.* radio operator, radio-man

**радиооптический** *adj.* radio-optical

**радиоориентировка** *f.* radio direction finding, radio fixing

**радиоотдел** *m.* radio division

**радиоотделение** *n.* radio section

**радиоотметчик** *m.* radio marker

**радиоотправитель** *m.* radio transmitter

**радиопатрульный** *adj.* radio-patrol

**радиопеленг** *m.* wireless (directional) bearing, radio bearing, observed radio bearing; **р. без поправки на девиацию** observed radio bearing

**радиопеленгатор** *m.* radio direction finder, radiogoniometer; **р. с визуальной индикацией** radiogonioscope; **р. с отсчётом пеленга по методу нулевой слышимости** aural-null direction finder

**радиопеленгаторный, радиопеленгационный** *adj.* radio direction-finding

**радиопеленгация** *f.* radio homing, radio direction finding, radio bearing, radiogoniometry

**радиопереводный** *adj.* radio-translator

**радиопереговор** *m.* radio traffic

**радиопередатчик** *m.* radio transmitter, transmitter, radio-transmitting station, localizer; **р. звукового сопровождения** aural transmitter, sound transmitter; **машинный р.** alternator-transmitter; **р. направленного действия** beam transmitter; **р. посадочной дорожки, посадочный р.** glide-slope localizer, glide-path localizer

**радиопередача** *f.* broadcast, radio transmission, translation, radio traffic, radio diffusion; **направленная р.** point-to-point radio transmission, beam transmission

**радиоперехват** *m.* radio interception

**« радиопилюля »** *f.* radio "pill"

**радиопозывной** *adj.* radio-call, radio call signal

**радиополе** *n.* radio field

**радиополукомпас** *m.* radio compass

**радиополяриметр** *m.* radio polarimeter

**радиопомеха** *f.* radio noise, radio interference, spurious response, receiving disturbance(s); **не создающий радиопомех** radio-free, radioquiet, antistatic; **заградительная р.** jamming, barrage

**радиопостановка** *f.* radio show

**радиопредупреждение** *n.* radiowarning

**радиоприбор** *m.* radio set, radio equipment

**радиопривод** *m.* radio coupler

**радиоприём** *m.* radio reception

**радиоприёмник** *m.* radio receiver, radio-receiving set; **разведывательный р.** passive detector

**радиоприёмный** *adj.* radio-receiving

**радиоприцел** *m.* radio sight

**радиоприцеливание** *n.* radar aiming, bore sighting

**радиопроводка** *f.* installation of radio

**радиопрогноз** *m.* radio forecast

**радиопрограмма** *f.* radio program, broadcasting program

**радиопрожектор** *m.* searchlight-control radar, radiophare

**радиопромышленник** *m.* radio manufacturer

**радиопромышленный** *adj.* radio-manufacturing

**радиопроницаемый** *adj.* radiotransparent

**радиопротиводействие** *n.* communication countermeasure(s)

**радиоразведка** *f.* radio reconnaissance, radio intelligence

**радиорезистентность** *f.* radioresistance

**радиореле** *n. indecl.* radio relay; **р. для создания помех радиолокации** anti-interception relay station

**радиорелейный** *adj.* radio-relay

**радиорубка** *f.* radio cabin, radio room

**радиосвечение** *n.* radioluminescence

**радиосвязь** *f.* radio communication, point-to-point radio communication, space radio; radio circuit, radio link, radio contact; **двусторонняя р.** transmitting and receiving radio service

**радиосекстант** *m.* radio sextant

**радиосенсибилизация** *f.* radiosensitization

**радиосеть** *f.* radio net, radio network, chain; antenna system; **р. антенна-противовес** double antenna; **горизонтальная р.** earth antenna; **р. из антенны и противовеса** double antenna; **направленная отражательная приёмная р.** beam reflector receiving aerial system; **р. с противовесом-экраном** screened antenna; **« сниженная р. »** earth antenna

**радиосигнал** *m.* wireless signal, radio signal; **р. бедствия** SOS-call, distress call

**радиосигнальный** *adj.* radio-signal

**радиоскоп** *m.* radioscope

**радиоскопический** *adj.* radioscopic

**радиоскопия** *f.* radioscopy

**радиослужба** *f.* radio service, wireless service

**радиослушание** *n.* listening in, radio listening

**радиослушатель** *m.* radio listener, broadcast listener

**радиоснимок** *m.* radiograph

**радиосообщение** *n.* telediffusion

**радиосопровождение** *n.* radio tracking

**радиоспектрометр** *m.* radio spectrometer

**радиоспектроскопия** *f.* radiospectroscopy

**радиоспециалист** *m.* radio specialist, radio expert

**радиосредство** *n.* radio facility, radio aid

**радиостанция** *f.* radio station, radio set; **главная р. сети** directing radio station; **передающая р. направленного действия** radio-beam transmitting station; **приводная р.** field localizer; **трансляционная р.** relay station, relay broadcasting station

**радиостат** *m.* radiostat

**радиостойкость** *f.* radioresistance

**радиосутки** *pl.* radio day

**радиосхема** *f.* radio circuit, electronic circuit

**радиотаблетка** *f.* radio pill

**радиотанк** *m.* signal tank

**радиотелеграмма** *f.* radiotelegram

**радиотелеграф** *m.* radiotelegraph

**радиотелеграфирование** *n.* wireless telegraphy

**радиотелеграфировать** *v.* to radiotelegraph

**радиотелеграфист** *m.* wireless operator, wireless telegraphist, radio telegraph operator

**радиотелеграфия** *f.* radiotelegraphy, continuous wave radio

**радиотелеграфный** *adj.* radiotelegraph, radiotelegraphic

**радиотелеизмерение** *n.* radar and radio-navigation

**радиотелеметрический** *adj.* radiotele-metering, radiotelemeter

**радиотелеметрия** *f.* radiotelemetry, radiotelemetering

**радиотелемеханический** *adj.* radio-telemechanical; radio-control

**радиотелескоп** *m.* radio telescope

**радиотелетайп** *m.* radioteletype(writer)

**радиотелеуправление** *n.* radio remote control, radio-telecontrol

**радиотелеуправляемый** *adj.* radio remote-controlled

**радиотелефон** *m.* radio telephone, radiophone set, aerophone

**радиотелефония** *f.* radiotelephony, radiophony

**радиотелефонный** *adj.* radio-telephone

**радиотень** *f.* radio shadow

**радиотеодолит** *m.* radiotheodolite

**радиотеплолокация** *f.* passive radar

**радиотерапевт** *m.* radiotherapist

**радиотерапевтический** *adj.* radiothera-peutic

**радиотерапия** *f.* radiotherapy, radiation therapy, radio-isotope therapy, radium therapy

**радиотермия** *f.* radiothermy, radio-thermics

**радиотехник** *m.* radio mechanic, radio technician, radio electrician, radio man; **р.-монтёр** radio repairman

**радиотехника** *f.* radio engineering, radio technology, radio-frequency engineering

**радиотехнический** *adj.* radio-engineer-ing, radio-technological

**радиотовар** *m.* radio merchandise

**радиоторий** *m.* radiothorium, RaTh

**радиоточка** *f.* "radio point"

**радиотрансляционный** *adj.* radio-relay (-ing), broadcasting

**радиотрансляция** *f.* rebroadcasting, reradiation, radio repeating, broad-cast transmission, wire broadcasting, line radio, radio relay system

**радиотрон** *m.* radiotron

**радиоугол** *m.* radio angle

**радиоузел** *m.* broadcasting center, radio center, radio relay center, re-diffusion station, "radio knot"

**радиоуправление** *n.* radio guidance, radio control

**радиоуправляемый** *adj.* radio-guided, radio-controlled

**радиоустановка** *f.* radio set; radio station, radio plant

**радиоустойчивость** *f.* radioresistance

**радиоустройство** *n.* radio installation

**радиофабрикант** *m.* radio manufacturer

**радиофизика** *f.* radiophysics

**радиофикатор** *m.* "radioficator"

**радиофикация** *f.* installation of radio

**радиофицированный** *adj.* radio-equipped

**радиофицировать** *v.* to install radio

**радиофон** *m.* radio telephone set

**радиофотография** *f.* radiophotography

**радиофотолюминесценция** *f.* radio-photoluminescence

**радиофотоснимок** *m.* radiophotogram

**радиофототелеграфия** *f.* wireless pic-ture transmission, radiotelegraphy, facsimile radio, facsimile transmis-sion, television, photoradio

**радиофургон** *m.* radio-truck, radio-van

**радиохимик** *m.* radiochemist

**радиохимический** *adj.* radiochemical

**радиохимия** *f.* radiochemistry

**радиоцентр** *m.* radio center, broad-casting center

**радиоцепь** *f.* radio circuit

**радиочастота** *f.* radio frequency

**радиочастотный** *adj.* radio-frequency

**радиочувствительность** *f.* radiosensi-bility

**радиочувствительный** *adj.* radiosensi-tive

**радиошум** *m.* radio noise

**радиоэкранирование** *n.*, **радиоэкрани-ровка** *f.* radio shielding

**радиоэлектрокардиограф** *m.* radioelec-trocardiograph

**радиоэлектромиограф** *m.* radioelectro-myograph

**радиоэлектроника** *f.* radio-electronics, electronics, electronic engineering; communications-electronics

**радиоэлектронный** *adj.* radio-electronic

**радиоэлектроэнцефалограф** *m.* radio-electroencephalograph

**радиоэхо** *n.* radio-echo, radio signal

**радиояркость** *f.* brightness

**радировать** *v.* to radio, to wireless, to wire, to radiotelegraph

**радируемый** *adj.* radioed, wired

**радист** *m.* radio operator, radioman, radio engineer, radio technician; telegraphist, radio telegraphist; **р. поста наблюдения и связи** resident radioman

**радиус** *m.* radius; range; **в радиусе действия** in range; **р. действия** range, zone, area, district; range of action, radius of service; **р. инерции** radius of gyration; **радиооптический р. действия** radio-optical line of distance

**радон** *m.* radon, Rn

**радуга** *f.* rainbow; iris

**радужный** *adj.* rainbow, iridescent, opalescent

**раз** *adv.* once, one time; one

**разбавитель** *m.* dilution agent, diluent

**разбавить** *pf. of* **разбавлять**

**разбавление** *n.* dilution

**разбавленный** *adj.* diluted, dilute; rare

**разбавлять, разбавить** *v.* to dilute; to thin, to rarefy

**разбаланс** *m.* unbalance

**разбалансированный** *adj.* unbalanced

**разбалансировка** *f.* unbalance; **р. по массе** mass unbalance

**разбалтывающийся** *adj.* loosening

**разбег** *m.* starting, start, acceleration, racing, warming up

**разбежка** *f.* race

**разбивать, разбить** *v.* to split, to divide; to break, to fracture, to rupture; to mark off, to lay out, to space; to align; **р. на части** (*or* **сегменты**) to segment

**разбивка** *f.* laying out, layout, marking off; division, splitting; dismantling; **р. диапазона** allotment of frequencies; **р. на части** segmentation; **р. по ступеням** grading, gradation, graduation; **р. сигнала по уровням** (*or* **ступеням**) quantizing, quantization

**разбивной** *adj.* separable

**разбивочный** *adj.* marking, spacing

**разбиение** *n.* decomposition, partition (*math.*); split, splitting; **р. на подгруппы** quantization

**разбирать, разобрать** *v.* to dismantle, to take down, to disassemble, to strip, to take apart, to dismount, to disjoint; to decipher; to analyze, to discuss

**разбитый** *adj.* broken; **р. на ступени** quantized

**разбить** *pf. of* **разбивать**

**разбланкирование** *n.* unblanking (*tv*)

**разблокировать** *v.* to unlock, to release, to clear

**разблокировка** *f.* unlocking, release, releasing

**разболтанность** *f.* looseness

**разболтанный** *adj.* loose, stirred; unbalanced

**разбор** *m.* analysis, examination; discretion; distinction; choice, selection; discernment, discerning

**разборка** *f.* separation, sorting out; dismantling, disassembling, disassembly, stripping, taking apart

**разборный** *adj.* collapsible, separable, dismountable, take-down, demountable

**разборчиво** *adv.* clearly, plainly

**разборчивость** *f.* intelligibility, legibility, distinctness; **р. фраз** sentence intelligibility

**разборчивый** *adj.* readable, clear, distinct, legible, intelligible; exacting, discriminating

**разбрасывание** *n.* dispersing, scattering, scattering effect; spreading

**разбрасыватель** *m.* scatterer, disperser, distributor, spreader

**разбрасывать, разбросать** *v.* to scatter, to disperse, to scatter about, to strew about, to throw about

**разбрасывающий** *adj.* scattering, dispersing

**разброс** *m.* diversity, divergence, variance, variation, spread, spreading, dispersal, scatter, scattering; splitting; fringe; straggling; **р. в величине амплитуды** spread in pulse height; **р. времени зажигания** jitter; **р. по длинам волн** wave spread; **р. рабочих характеристик** spread in performance; **стандартный р.** process standard deviation

**разбросанность** *f.* sparseness, dispersedness, dispersion; incoherence, disconnectedness

**разбросанный** *adj.* scattered, spread, dispersed; speckled; disconnected

**разбросать** *pf. of* **разбросить, разбрасывать**

**разбрызганный** *adj.* spattered, sprayed, sprinkled

**разбрызгать** *pf. of* **разбрызгивать**

**разбрызгивание** *n.* spraying, sprinkling, splashing, sputtering, spattering, spurting

**разбрызгивать, разбрызгать** *v.* to spray, to sprinkle, to spatter, to splash; to atomize

**разбрызгивающий** *adj.* spraying, sprinkling; spattering; atomizing

**разбудить** *pf. of* **будить, разбуживать**

**разбуживать, разбудить** *v.* to awaken, to awake; to ring

**развеваться, развеяться** *v.* to flutter, to wobble, to swing; to be dispersed

**разведать** *pf. of* **разведывать**

**разведка** *f.* exploration, prospecting, search; intelligence, reconnaissance; ranging; spotting; surveillance; survey, surveying; **р. средствами связи** signal intelligence, signal reconnaissance

**разведочный** *adj.* exploring, exploratory; intelligence, reconnaissance

**разведчик** *m.* seeker, selector level (*tgphy.*); prospector; « **морской р.** » mariner's pathfinder

**разведывание** *n.* reconnaissance; searching, exploring

**разведывательный** *adj.* intelligence, reconnaissance; searching, exploring

**разведывать, разведать** *v.* to spot; to explore, to investigate; to reconnoiter, to scout; to prospect

**развёрнутый** *adj.* unfolded, developed; extensive, large-scale; detailed

**развернуть** *pf. of* **развёртывать**

**развёрстка** *f.* division, repartition, distribution

**развёртка** *f.* scanning, scanning hunting, scan, trace, sweep, development; analyzing, exploring, exploration, resolution, analysis; time-base; presentation; deflection; development, evolvement, presentation, unfolding; evolute; evolvent; **р. бесконечным ремнем** belt scanning (*tv*); **временная р.** time-base; **временная р. дальности** range-time display; **ждущая р.** driven sweep, slave sweep, single sweep, ratchet time-base; **р. изображения на прямолинейные строки** rectilinear scanning; **кадровая р.** frame time base, frame scan; **линейная временная р.** linear time base; **р. лучом** beam deflection; **малострочная р.** coarse scanning; **многократная чересстрочная р.** multiple interlace-scanning (*tv*); **р. на обратном ходу** retrace scanning; **р. накопленного ранее потенциального рельефа** memory scanning; **несинхронизованная р.** flywheel time-base; **перемежающаяся р.**

interlacing; **р. перемещением изображения** image deflection scanning (*tv*); **р. по дальности** range scanning; **радиальная р.** radial scanning, radial time base; **растянутая р.** expanded time base; **р. с задержкой** ratchet time-base; **строчная р.** horizontal hold (*tv*); **чересстрочная р.** interlacing, interlaced scanning, line interlacing, alternate-line scanning, line jump scanning (*tv*); **чересстрочная р. кадра в четыре приёма** quadrupled scanning; **четырёхкратная перемежающаяся р.** quadruple scanning; **четырёхстрочная р. с результирующим ослаблением сигналов** subtractive type interlacing

**развёрточный** *adj.* scanning, sweep, sweeping

**развёртываемый** *adj.* scanned; scanning

**развёртывание** *n.* unwinding, unrolling; unfolding; development, evolution; scan, scanning, scanning hunting, exploring; rectification; **р. бегущим лучом** (*or* **пятном**) flying spot scanning; **р. без обратного хода** oscillatory scanning; **р. в ряд** series development; **р. световым бегущим лучом** (*or* **пятном**) floodlight scanning

**развёртыватель** *m.* scanner; **р. переднего сектора** forward-looking scanner; **следящий р.** (lock-and-)follow scanner

**развёртывать, развернуть** *v.* to unwind, to unroll, to uncoil; to display, to develop, to extend, to expand, to evolve, to spread, to unfold; to scan, to sweep; to analyze; to explore

**развёртывающий** *adj.* scanning, scan-off, sweeping; exploring; flaring

**развёртывающийся** *adj.* developable

**разветвитель** *m.* splitter, bifurcating device; branch(ing) box, connector box, coupler; **специальный волноводный р.** "magic tee"

**разветвительный** *adj.* branching, distributing, distribution, separating; splitter

**разветвить(ся)** *pf. of* **разветвлять(ся)**

**разветвление** *n.* branch, branching, ramification, bifurcation, fork, forking; splitting; node; fringing; **последовательное р. волноводов** waveguide hybrid junction; **р. с согласо-**

ванием matched junction; **р. тока** fringing of current

**разветвлённый** *adj.* ramified, far-flung, bifurcated, forked, divided, branched, branch; multi-path, multi-way; tapped off

**разветвлять, разветвить** *v.* to branch, to divide

**разветвляться, разветвиться** *v.* to branch, to branch out, to fork, to ramify; **р. надвое** to bifurcate

**развеяться** *pf. of* **развеваться**

**развиваемый** *adj.* developed, evolved, generated

**развивание** *n.* development, generation; untwining; uncurling

**развивать, развить** *v.* to develop, to generate, to evolve; to untwine, to unroll, to wind off, to untwist; to amplify, to build up, to increase

**развившийся** *adj.* developed

**развилистый** *adj.* forked, furcate

**развилка** *f.* dispatch tube branching; fork

**развильчатый** *adj.* forked

**развитие** *n.* development, development work, formation, evolution, expansion, growth; distribution, spread

**развитой, развитый** *adj.* developed, evolved

**развить** *pf. of* **развивать**

**разводка** *f.* separation, diverging; parting; **р. кабелей** cable laying, cabling; cable layout, cable allocation

**разводной** *adj.* separating

**развозбуждение** *n.* de-energization

**разворачивание** *n.* unwinding; unfolding, unwrapping; turning

**разворачивать, разворотить** *v.* to unwind; to unwrap, to unroll, to unfold; to split, to crack; to destroy, to upset

**разворачиваться, разворотиться** *v.* to run up

**разворот** *m.* turn; development

**разворотить(ся)** *pf. of* **разворачивать(ся)**

**развязать** *pf. of* **развязывать**

**развязка** *f.* decoupling, uncoupling; bypass, bypassing; isolation; padding; outcome, conclusion; **р. входа** input choke; **р. на землю** bypass to ground; **р. на корпус** bypass to chassis

**развязывание** *n.* decoupling, uncoupling, untying, unfastening; insulation

**развязывать, развязать** *v.* to isolate; to loosen, to untie, to undo, to unbind, to unfasten

**развязывающий** *adj.* decoupling, uncoupling; isolating; isolation

**разгерметизация** *f.* depressurization

**разглядывать, разглядеть** *v.* to scan; to view, to examine, to consider

**разговаривать, разговорить** *v.* to speak, to talk

**разговор** *m.* speech, conversation, talk; call, message (*tphny.*); **без переходного разговора** crosstalk-free; **р. быстрой связи** no-delay call; **внятный переходный р.** intelligible crosstalk, uninverted crosstalk; **невнятный переходный р.** unintelligible crosstalk, inverted crosstalk; **переходный р.** crosstalk; **п. р. между основными цепями** side-to-side crosstalk (*tphny.*); **п. р. на передающем конце** sending-end crosstalk, near-end crosstalk (*tphny.*); **п. р. на приёмном конце** far-end crosstalk, receiving-end crosstalk; **п. р. с основной цепи на основную** side-to-side crosstalk; **п. р. с основной цепи на основную на приёмном конце** side-to-side far-end crosstalk; **п. р. с основной цепи на фантомную** side-to-phantom crosstalk; **р. по абоненту** subscription call (*tphny.*); **р. по соединительной линии** trunk call; **р. с многократным отсчётом** multi-metered call; **р. с предварительным приглашением** avis d'appel call, messenger call; **р. с предупреждением о вызове** appointment call; **р. сверх абонента** extended subscription call; supplementary call; **р. шопотом** unvoiced speech

**разговорить** *pf. of* **разговаривать**

**разговорный** *adj.* conversational, speaking, speech; colloquial; verbal

**разгон** *m.* acceleration, racing, speed-up; start, starting, rise, run; launching; dispersion; hunt, hunting

**разгонный** *adj.* starting, start; racing, speed-up; launching

**разгоняемый** *adj.* accelerable; **быстро р.** quick-to-accelerate

**разгонять, разогнать** *v.* to accelerate, to speed up, to race; to disperse, to dissipate, to scatter; to hunt

**разгорание** *n.* build(ing)-up

**разграничение** *n.* differentiation, discrimination; delimination, demarcation

**разграниченный** *adj.* defined, demarcated, delimited, bound

**разгружатель** *m.* discharger

**разгружать, разгрузить** *v.* to unload, to dump, to discharge; to relieve

**разгружающий** *adj.* discharging, unloading, dumping

**разгруженный** *adj.* discharged, unloaded, dumped; balanced

**разгрузить** *pf. of* **разгружать**

**разгрузка** *f.* discharging, discharge, unloading, dumping; clearing, throwing off; relieving, relief

**разгрузочный** *adj.* discharging, discharge, unloading, dumping

**разгруппирование** *n.* debunching

**разгруппировать** *pf. of* **разгруппировывать**

**разгруппировка** *f.* debunching

**разгруппировывать, разгруппировать** *v.* to group, to divide into groups; to unpack (*math.*)

**раздавить** *pf. of* **раздавливать, давить**

**раздавленный** *adj.* crashed

**раздавливание** *n.* crushing, smashing

**раздавливать, раздавить** *v.* to crush, to smash, to squash

**раздваивать, раздвоить** *v.* to halve, to bisect, to bifurcate, to split

**раздвижение** *n.* spread; **р. выбросов** pip spacing

**раздвижка** *f.* separating, dispersing, parting, moving apart

**раздвижной** *adj.* slide, sliding, extension-type, extensible, telescopic, expanding, expansible; adjustable; flexible

**раздвинутый** *adj.* separated; extended

**раздвоение** *n.* bifurcation, branching, forking, fork; **р. азимутальной отметки** split-azimuth presentation, azimuth split-in

**раздвоенный** *adj.* split, bifurcated, bifurcate, forked; double

**раздвоить** *pf. of* **раздваивать**

**раздел** *m.* separation, division; partition; section, class

**разделать** *pf. of* **разделывать**

**разделение** *n.* separation, disconnecting, cutting-off; division, dissociation, segregation, parting, partition; splitting, split, fission, cleaving; unpack; crossover; sharing; distribu-

tion; indexing, classing; **р. каналов по времени** time division of channels; **р. отметок** echo separation, blib separation; **р. по форме сигналов** waveform separation; **р. по частоте** frequency separation, mode separation; **р. экрана** split screen effect

**разделённый** *adj.* split, divided, parted; graded, classed

**разделившийся** *adj.* separated; divided, fissioned

**разделимость** *f.* divisibility, separability

**разделимый** *adj.* divisible, separable; analyzable

**разделитель** *m.* separator; vent segment, vent spacer, spacer; separating agent; divider; **р. импульсов фазирования цветов** color-index pulse separator

**разделительный** *adj.* dividing, separating, separator, separation, separative; partitive; disconnecting, cut-off, trip (ping); partitioning

**разделить(ся)** *pf. of* **разделять(ся), делить(ся)**

**разделка** *f.* marking-off, fanning out, fan; dicing; **р. кабеля по четвёркам** quading of a cable

**разделывать, разделать** *v.* to fan out, to fan, to splice; to finish, to do; to dress

**разделять, разделить** *v.* to separate, to part, to portion, to split, to divide, to subdivide; to share; to distribute, to segregate; to disconnect, to cut off, to interrupt

**разделяться, разделиться** *v.* to split, to divide (into); to separate, to part, to branch

**разделяющий** *adj.* separating, dividing; isolating; crossover

**раздражать, раздражить** *v.* to irritate; to stimulate

**раздражение** *n.* stimulation; irritation; **длительное р.** steady stimulation

**раздражитель** *m.* stimulus, stimulating agent

**раздражительный** *adj.* stimulating; irritable

**раздражить** *pf. of* **раздражать**

**раздробить** *pf. of* **раздроблять, дробить**

**раздроблять, раздробить** *v.* to split (up), to smash; to disjoin, to disconnect; to reduce (*math.*); to shatter, to crush, to grind, to disintegrate

**разжижать, разжидить** *v.* to dilute; to thin, to rarefy

**разжижающий** *adj.* diluting; rarefying

**разжижение** *n.* dilution; rarefaction, thinning; liquefaction

**разжижённый** *adj.* diluted; thin, rare

**разжижитель** *m.* dilution agent, diluent

**разлагаемый** *adj.* decomposable; analyzable

**разлагатель** *m.* dissector, analyzer

**разлагать, разложить** *v.* to decompose, to separate, to analyse, to dissociate; to resolve; to explore, to scan; to corrupt; to dissect; to dissolve; to transform, to expand (*math.*); **р. на множители** to factor

**разлагаться, разложиться** *v.* to decompose, to separate, to disintegrate, to break up; to decay, to rot; to dissolve; to expend

**разлагающий** *adj.* sampling; scanning; decomposing

**разлагающийся** *adj.* decomposing; analyzable; decomposable

**разлаженность** *f.* maladjustment, misalignment

**разлёт** *m.* divergence; dispersion, scattering

**разлив** *m.* overflow; irruption

**разливать, разлить** *v.* to pour out, to pour; to found, to cast; to spill; to diffuse

**различать, различить** *v.* to discriminate, to distinguish, to differentiate, to discern; to recognize; to resolve

**различаться, различиться** *v.* to differ, to be distinguished, to be unlike

**различающий** *adj.* discriminatory; **р. по фазе** dephased

**различение** *n.* distinguishing, discerning, distinction, discrimination; recognition; difference; **р. изменений интенсивности** intensity discrimination; **р. по фазовому углу** phase angle discrimination

**различие** *n.* distinction, difference, discrimination; discrepancy, diversity; **р. акустических сопротивлений** acoustic mismatch

**различимость** *f.* discernability, definition, detectability; visibility; intelligibility

**различимый** *adj.* distinguishable, observable, discernible; resolvable

**различитель** *m.* discriminator; **р. для сантиметровых волн** microwave discriminator

**различительно** *adv.* in contra-distinction (to)

**различительный** *adj.* distinctive

**различить(ся)** *pf. of* **различать(ся)**

**различно** *adv.* differently

**различность** *f.* difference, unlikeness

**различный** *adj.* different, diverse, various, varied, dissimilar, unlike, distinct; **р. по фазе** dephased

**разложение** *n.* decomposition, separation, disintegration, dissociation, splitting, split-up; expansion; resolution; analysis, exploration; fission; scan, scanning, analyzing, exploring; decay; separation of reals and imaginaries; development (*math.*); dissection; interlace, interleave; **р. в гармонический ряд** harmonic expansion; **р. в ряд** expanding, expansion, transform, transformation; **р. в степенной ряд** power-series expansion, Taylor series expansion; **р. высокой чёткости** high resolution (*tv*); **р. диапозитивных изображений** slide scanning; **р. изображения** resolution, disintegration, scanning, dissecting, exploring, analyzing; **р. изображения на прямолинейные строки** rectilinear scanning; **р. на множители** factoring, factorization, product development; **р. на плоские волны** plane-wave resolution; **р. на элементарные операции** factorization; **р. по степеням...** expansion in powers of . . .; **р. с перемежением точек изображения** picture-dot interlacing; **р. с перемещающимися точками, р. с чередованием точек** interlaced spot-scanning system (*tv*); **р. световым бегущим лучом** (*or* **пятном**) floodlight scanning; **чересстрочное р.** interlacing

**разложенный** *adj.* analyzed; decomposed, dissociated; expanded (*math.*)

**разложившийся** *adj.* decomposed; putrid, rotten

**разложимость** *f.* decomposability

**разложимый** *adj.* decomposable, dissociable, separable; reducible; analyzable; **р. на множители** factorable

**разложить** *pf. of* **разлагать**

**разложиться** *pf. of* **разлагаться**

**размагнитить** *pf. of* **размагничивать**

**размагничение** *n.* demagnetization (phenomenon)

**размагниченный** *adj.* unmagnetized, neutral

**размагничивание** *n.* demagnetization, demagnetizing (action), degaussing; **p. до записи** recording demagnetization; **p. периодически меняющим направление магнитным потоком** demagnetization by continuous reversals

**размагничивать, размагнитить** *v.* to demagnetize, to degauss, to de-energize

**размагничивающий** *adj.* demagnetization, demagnetizing, degaussing

**размазанность** *f.* blur

**размазанный** *adj.* blurred, smeared (out)

**размазать** *pf. of* **размазывать**

**размазывание** *n.* blurring, smearing effect, smear

**размазывать, размазать** *v.* to blur, to smear, to spread

**разматывание** *n.* uncoiling, unreeling, unwinding, paying out

**разматыватель** *m.* recoiling machine

**разматывать, размотать** *v.* to unwind, to uncoil, to reel off, to unreel, to pay out, to unroll, to wind off, to run off

**разматывающий** *adj.* uncoiling, unwinding; take-off, pay-off

**размах** *m.* swing, swinging, sweep; scope, range; total amplitude, double-amplitude, peak-to-peak separation, peak-to-peak value; **p. входного сигнала** peak-to-peak signal input; **максимальный p. сигнала** peak-to-peak voltage; **p. напряжения помех** (*or* **шумов**) peak-to-peak noise; **полный p. колебаний** (*or* **сигнала**) peak-to-peak amplitude; **полный p. тока** peak-to-peak current; **p. сигнала** peak-to-peak signal; **p. сигнала изображения** picture-signal amplitude; **p. сигнала на входе** peak-to-peak signal input; **p. сигнала на сетке** grid swing; **p. сигнала, соответствующий перепаду с чёрного на белый уровень** peak black-to-white signal

**размахать** = **размахивать**

**размахивание** *n.* swinging

**размахивать, размахать, размахнуть** *v.* to swing, to sway; to wave

**размен** *m.* exchange, change

**разменный** *adj.* exchange, change

**размер** *m.* dimension, size, gage, caliber; degree, extent; scale; rate, measure; quantity, amount; yield; **размеры экранированного ленточной бронёй кабеля для укв-антенн** dimensions of twin lead for vhf antennas; **p. в свету** inner width, width in the clear; **p. кадра** picture segment, field of view; **p. кадра по горизонтали** horizontal frame dimension; **p. под ключ** width across flats

**размерение** *n.* measuring, measurement

**размеренный** *adj.* measured

**размеривание** *n.* measuring, measurement

**размеривать, размерить** *v.* to measure off, to partition, to proportion, to divide

**размерность** *f.* dimension, dimensionality; length; scale; **имеющий p.** dimensional; **p. массива** dimension of array; **p., обратная длине** inverse length

**размерный** *adj.* dimensional

**размерять** *see* **размеривать**

**разместить** *pf. of* **размещать**

**разметить** *pf. of* **размечать**

**разметка** *f.* marking, marking-out, laying out

**разметочный** *adj.* marking, layout

**разметчик** *m.* marker-off; **магнитный p. центров** magnetic center punch

**размечать, разметить** *v.* to label, to mark; to mark off, to mark out, to lay off; to trace; to locate, to space

**размеченный** *adj.* marked, marked out

**размещать, разместить** *v.* to group, to locate, to allocate, to arrange, to distribute, to place, to dispose; to interlace; to install

**размещение** *n.* arrangement, positioning, spacing, placing, location, disposal, disposition, allocation, distribution, placement, layout, accommodation; deployment; installation; permutation (*math.*)

**размещённый** *adj.* distributed, allocated; set, arranged

**разминовка** *f.* turnout, siding, by-pass

**размножать, размножить** *v.* to manifold, to copy; to multiply, to increase; to propagate

**размножение** *n.* multiplication, duplication, reproduction, increase, propagation

**размноженный** *adj.* multiplied; mimeographed

**размножить** *pf. of* **размножать**

**размол** *m.* milling; grinding, pulverization

**размотать** *pf. of* **разматывать**

**размотка** *f.* unwinding, uncoiling, unreeling

**размолчаленный** *adj.* frayed

**размолчаливание** *n.* fringing, fraying

**размывание** *n.* "erosion", flattening, leveling (of signal); blurring

**размывать, размыть** *v.* to blur; to wash off

**размыкание** *n.* break, breaking, disconnection, open-circuiting, separation, interruption, opening; release, releasing, tripping, trip, unlocking; **р. цепи на входе** alternating current interruption

**размыкатель** *m.* release, release gear, breaker, trigger; **р. дефектного напряжения** leakage protective system; **р. маршрута** route locking release; **р. с выдержкой времени** time release

**размыкать, разомкнуть** *v.* to release, to break, to disconnect, to cut off, to interrupt, to open; to unlock, to unlatch, to trip, to turn off

**размыкающий** *adj.* release, releasing, disconnecting, tripping, breaking, splitting, interrupting, cut-off

**размыкающийся** *adj.* unlocking

**размытость** *f.* spreading, spread; blur, fuzz

**размытый** *adj.* blurred; diffuse; washed-out

**размыть** *pf. of* **размывать**

**размягчать, размягчить** *v.* to soften, to mollify

**размягчающий** *adj.* softening

**размягчение** *n.* softening

**размягчить** *pf. of* **размягчать**

**разнесение** *n.* diversity; diversity spacing; separation

**разнесенность** *f.* diversity, spacing, separation

**разнесённый** *adj.* spaced; dispersed

**разнимать, разнять** *v.* to dismantle, to strip, to dismount, to take apart; to separate, to part, to disjoin, to dismember; to disengage

**разница** *f.* difference, disparity; contrast, distinction; **р. по высоте двух тонов** pitch interval

**разновидность** *f.* type, variety, variant, variation; **р. одновременной системы ЦТ** derived simultaneous system (*tv*)

**разновидный** *adj.* various, diverse, multiform

**разновремённость** *f.* time-diversity, diversification, diversity, non-coincidence

**разновременный** *adj.* alternative; (taking place) at different times

**разногласный, разноголосный** *adj.* discordant, conflicting

**разнозначащий** *adj.* having a different meaning

**разноимённость** *f.* nonequivalence

**разноимённый** *adj.* unlike, opposite, of different kinds, of different names; of opposite charge

**разномер** *m.* multimeter, onemeter, unimeter, multirange instrument, multiscale instrument; multitester

**разнообразие** *n.* diversity, variety, range; multiplicity

**разнообразить** *v.* to vary, to diversify, to deviate; to change, to alter

**разнообразность** *f.* variety, diversity; modification, variation

**разнообразный** *adj.* diverse, various, varied; modified; unlike, dissimilar; miscellaneous

**разнополярный** *adj.* heteropolar

**разнородность** *f.* heterogeneity, diversity

**разнородный** *adj.* heterogeneous, mixed, different, dissimilar, various, diversified, unlike, manifold

**разнос** *m.* separation, spacing, distance; diversity; carrying, delivery; overrun, overspreading; racing; dispersion; **р. видов колебаний по частоте** mode separation, mode frequency separation; **р. внутренних выводов** lead tip spacing; **р. выбросов** split separation; **р. между пиками** peak separation; «**р.**» несущих sound-to-picture separation; **р. отметок** blip spacing; **р. по частоте** frequency diversity, frequency separation, mode separation; **р. резонансных точек различителя** discriminator peak separation

**разностный** *adj.* different, difference; differential; subtractive

**разносторонний** *adj.* scalene, scalenous; many-sided; versatile

**разносторонность** *f.* versatility; multiplicity; many sidedness

**разность** *f.* difference, remainder (*math.*); variety, difference, diversity; **р. значений остаточной индукции** remanent induction difference, retentivity difference; **квадратурная р. фаз** quadrature (*tv*); **конечная р. первого порядка** first difference (*math.*); **р. остаточных (магнитных) потоков** remanent flux difference; **р. остаточных (магнитных) потоков по предельной петле гистерезиса** major remanent flux difference; **сигнальная р. тока** current margin (*tgphy.*); **фазовая р.** transient time difference; **р. хода (лучей)** propagation difference (of beams); **р. шагов нарезки** path difference, pitch of a thread

**разнотипность** *f.* diversity, different types

**разнохарактерный** *adj.* diverse, various, diversified

**разноцветный** *adj.* many-colored, multi-colored, variegated, polychromatic, heterochromatic

**разный** *adj.* different, diverse, various, unlike, miscellaneous

**разнять** *pf. of* **разнимать**

**разобранный** *adj.* dismantled

**разобрать** *pf. of* **разбирать**

**разобщать, разобщить** *v.* to disconnect, to release, to uncouple; to disengage, to separate, to disunite; to insulate; to isolate

**разобщающий** *adj.* releasing, release, cut-out, disconnecting, disengaging

**разобщение** *n.* disconnection, separation, disconnecting, release, releasing, dissociation, disengagement, disjunction, uncoupling; gating pulse; interruption; **свободное р.** trip free release

**разобщённо** *adv.* apart

**разобщённость** *see* **разобщение**

**разобщённый** *adj.* open, disconnected, disengaged, released; separate, discrete; insulated

**разобщитель** *m.* disconnector

**разобщительный** *adj.* disconnecting, releasing

**разобщить** *pf. of* **разобщать**

**разовый** *adj.* single

**разогнать** *pf. of* **разгонять**

**разогрев** *m.*, **разогревание** *n.* warming-up, heat-up, preheat; run-up

**разогревать, разогреть** *v.* to warm up, to heat up

**разогретый** *adj.* warmed-up, heated

**разогреть** *pf. of* **разогревать**

**разойтись** *pf. of* **расходиться**

**разомкнутость** *f.* open condition

**разомкнутый** *adj.* released, disconnected, broken, open, open-circuited, interrupted; clear

**разомкнуть** *pf. of* **размыкать**

**разорвать** *pf. of* **разрывать**

**разориентирующий** *adj.* disorienting

**разорять, разорить** *v.* to destroy, to ruin

**разрабатываемый** *adj.* workable, capable of development

**разрабатывать, разработать** *v.* to develop, to work out; to process, to treat

**разработанный** *adj.* developed; exploited; processed, treated

**разработка** *f.* development work, development, working out; working, exploitation; elaboration

**разработывать** *see* **разрабатывать**

**разрастание** *n.* extension, expansion, widening, growth; **р. аварии** dislocation of fault

**разрегулирование** *n.*, **разрегулированность** *f.* misalignment, maladjustment

**разрегулированный** *adj.* misaligned, maladjusted

**разрегулировка** *f.* maladjustment, misalignment

**разрежать, разредить** *v.* to evacuate, to exhaust, to rarefy

**разрежение** *n.* vacuum; exhaust, exhaustion, evacuation; rarefying, rarefaction; depression

**разрежённость** *f.* rarefaction, rarity, thinness; vacuity

**разрежённый** *adj.* rarefied, rare, reduced; vacuum, evacuated, exhausted; partially evacuated; **сильно р.** negligible

**разреживать** *see* **разрежать**

**разрез** *m.* slot, slit, cut; section, cross-section, profile, cut-away view, plan; **вертикальный р.** elevation; **горизонтальный р.** plan

**разрезание** *n.* cutting

**разрезанный** *adj.* serrated, cut

**разрезать** *pf. of* **разрезывать, резать**

**разрезной** *adj.* split, slot(ted), slit(ted), cut, serrated; sectional; laminated

**разрезывать, разрезать** *v.* to slit, to slot; to cut, to cut to size, to cut out; to cut up, to shred

**разрешаемый** *adj.* resolvable; permissive

**разрешать, разрешить** *v.* to resolve; to solve; to allow, to permit; to discriminate

**разрешающий** *adj.* resolving; allowing, permitting, enabling

**разрешение** *n.* grant, license, permit, permission, clearance, authorization; resolution, resolving; discrimination; solution; **р. на вход в зону посадки** clearance to enter the traffic pattern (*av.*)

**разрешённый** *adj.* authorized, assigned, permitted, allowed, permissible; resolved

**разрешимость** *f.* solvability

**разрешимый** *adj.* solvable, soluble

**разрешительный** *adj.* permissive, permitting; absolutory, absolving

**разрешить** *pf. of* **разрешать**

**разрушать, разрушить** *v.* to destroy, to demolish, to wreck, to break down

**разрушаться, разрушиться** *v.* to break down, to fail; to disintegrate, to decompose, to decay

**разрушающий** *adj.* breaking, destroying, disruptive, destructive

**разрушение** *n.* breakdown, destruction, demolition, crushing, shattering; decay; disturbance; disintegration; failure; **р. нити накала** disintegration of filament; **р. поля** collapse of field

**разрушенный** *adj.* disturbed; destroyed, ruined; decayed

**разрушитель** *m.* destructor

**разрушительный** *adj.* destructive, destroying

**разрушить(ся)** *pf. of* **разрушать(ся)**

**разрыв** *m.* break(age), break-up, breaking, gap, fracture, rupture; severance, fissure, void, space, clearance, crack, flaw; splitting; burst, explosion; discontinuity, disruption, interruption, disturbance; tearing; **р. непрерывности** discontinuity (*math.*)

**разрывание** *n.* breaking, tearing, rupturing

**разрыватель** *m.* breaker, breaking device, interrupting device

**разрывать, разорвать** *v.* to tear, to disrupt, to pull-down, to pull apart; to break; to blow up

**разрывающий** *adj.* bursting, rupturing, tearing; splitting, cleaving

**разрывной** *adj.* breaking, interrupting, tearing; bursting, rupturing, disruptive; discontinuous

**разрывность** *f.* discontinuity

**разрывомер** *m.* gap meter

**разряд** *m.* discharge; category, rank, sort, rating, class; column; rate (*math.*); order, place, position; bit, digit; decade; glow; **разряды** atmospherics; **р. в форме пробоя и истечения** disrupto-convective; **двоичный р.** bit, binit, binary digit, binary place; **дежурный р.** keep-alive discharge; **р. за разрядом** bit by bit; **кистевой р.** brushing; **коронный р.** corona; **младший р.** less (significant) digit, low-order digit; **наименьший значащий р.** least significant digit; **несамостоятельный р.** gas-discharge caused by external potentials; **несамостоятельный термический дуговой р.** externally heated arc; **поющий р.** musical spark gap; **предварительный газовый р.** keep-alive gas discharge; **промежуточный р.** sandwich digit; **р. с края** (*or* **с острия**) marginal discharge; **(самый) младший р.** least significant digit, lowest-order digit; **(самый) старший р.** higher-order digit, most significant digit; **самый старший двоичный р.** most significant bit; **следующий десятичный р. (чисел)** next decade; **старший р.** top digit, high-order digit, more (significant) digit; **старший значащий р.** most significant digit; **тёмный** (*or* **тихий, тлеющий**) **р.** glow discharge, corona discharge, dark discharge, first Townsend discharge, silent discharge; **р. через проводник** conductive discharge; **р. числа** digit order number

**разрядитель** *m.* discharger

**разрядить(ся)** *pf. of* **разряжать(ся)**

**разрядка** *f.* discharging, discharge; deexcitation

**разрядник** *m.* discharger, discharge switch, discharge lever, discharge rod, discharge arrester, lightning arrester; spark gap, arc gap, arrester, gap, surge absorber, excess voltage suppressor; surge diverter; aerial discharger; **алюминиевый р.** oxide film lightning arrester, aluminum arrester; **антенный р.** antenna lightning protector; **р. блокировки передатчика** ATR tube; **р. блокировки приёмника** cut-off tube, blocking tube; **вакуумный р.** vacuum lightning protector, vacuum arrester; **вентильный р.** non-linear-resistance arrester; **вилообразный р.** discharging rod; **р. Вина** quenched gap; **возбуждающий р., р. возбуждения** exciting spark gap; **газовый р.** gas-discharge arrester; **гребенчатый р.** comb lightning arrester; **грозовой р.** lightning arrester; **дисковый р.** Marconi's commutator; **р. для защиты от акустического удара** acoustic shock reducer (*tphny.*); **р. для защиты от перенапряжений** arrester, surge arrester; **дополнительный р.** outer gap; **защитный р.** lightning arrester, lightning protector, protective gap; **защитный р. передачи** anti-transmit-receive tube; **р. защиты приёмника** ATR tube, T. R. tube; **измерительный шаровой р.** sphere gap voltmeter; **импульсный р.** surge gap; **искровой р. с постоянным промежутком** fixed spark-gap modulator; **искрогасящий р.** quenched gap; **р. камеры « приём-передача »** ATR tube; **ламповый р.** discharger tube; **микрометрический р.** spark micrometer; **предзажигаемый р.** triggered spark gap; **предохранительный р.** protective gap, lightning arrester, lightning protector; **резонансный р.** soft rhumba-tron, R-T tube; **резонансный р. блокировки передатчика** TR tube; **роговой р.** horn-gap arrester; **р. с дополнительным фиксированным электродом (для) поджига** fixed spark gap; **р. с искровым промежутком между остриями** knife-edge lightning arrester; **р. с несколькими параллельными искровыми путями** multipath arrester; **р. с несколькими последовательными искровыми про-**межутками multigap arrester; **р. с последовательными искровыми промежутками** triggered gaps; **р. с предохранителем** combined protector and fuse; **р. с сопротивлением** resistance-type lightning arrester; **р. со вспомогательным зажиганием** impulse protective gap; **стреляющий р.** expulsion gap; **трубчатый р.** expulsion-type arrester; **шаровой р.** spark(ing) ball

**разрядность** *f.* capacity

**разрядный** *adj.* discharge, discharging; bit, digit, digital; disruptive

**разряжать, разрядить** *v.* to discharge; to unload; to space out

**разряжаться, разрядиться** *v.* to strike

**разряжающий(ся)** *adj.* discharging, discharge

**разряжение** *n.* discharging, discharge; unloading

**разряженный** *adj.* exhausted, run-down

**разумный** *adj.* sound, intelligent, reasonable, responsible

**разъедаемость** *f.* corrodibility

**разъедаемый** *adj.* corrodible

**разъедание** *n.* corrosion, pitting

**разъедать, разъесть** *v.* to corrode, to erode

**разъедающий** *adj.* corroding, corrosive

**разъеденный** *adj.* corroded, pitted, eroded

**разъединение** *n.* release, disconnection, separation, disconnecting, clearing, cutting, disjunction, disengagement, interruption, break, breaking, switching off; cancellation; busy-back release; **двустороннее р.** last-party release (*tphny.*); **р. местного разговора с междугородной станции** through clearing, breaking of local calls for trunk calls; **одностороннее р.** first party release; **одностороннее р. при отбое со стороны вызывающего абонента** calling-party release (*tphny.*); **посредственное р.** indirect tripping circuit control; **р. при двустороннем отбое** called and calling party release, last party release (*tphny.*); **р. при отбое со стороны вызывающего абонента** calling subscriber release (*tphny.*); **р. с вызываемой стороны** calling party release (*tphny.*)

разъединённый *adj.* separate; disconnected, cut off, disengaged; out of gear

разъединитель *m.* switch, disconnecting switch, interruption key, disjunctor, disconnector, isolator, isolating switch, sequential isolator, circuit breaker; load-isolating breaker; leakage-protection breaker; **выдвижной р.** (contact-chamber-type) disconnect breaker; **минимальный р.** undervoltage release; **предохранительный р.** disconnect switch with built-on fuses; disconnecting link, isolating link; **р. с поворотом ножей в вертикальной плоскости** vertical-break disconnecting switch; **р. с поворотом ножей в горизонтальной плоскости** side-break disconnecting switch; **складной одноколонковый р.** knife-blade-type disconnect switch; **участковый р.** sectionalizer

разъединительный *adj.* break, breaking, cut-off, isolating

разъединить(ся) *pf. of* разъединять(ся)

разъединять, разъединить *v.* to disconnect, to cut off, to switch off, to interrupt; to disengage, to break, to uncouple, to unlatch; to release, to clear, to clear out; to separate, to disjoin, to free, to detach, to dissociate

разъединяться, разъединиться *v.* to get disconnected

разъединяющий *adj.* tripping, releasing, disconnecting, isolating, disengaging; separating, dissociative; interruption

разъём *m.* connector, plug and socket; joint; **р. с запрессованными штепселями** molded-edge connector; **р. с печатным монтажем** printed-wire connector; **уголковый р.** right-angle connector; **фиксированный штепсельный р.** nonreversible plug, nonreversible connector, locking-type plug; **цилиндрический штепсельный р.** cannon plug; **штепсельный р.** plug-and-socket, plug(-type) connector, bullet connector; cannon plug, connector

разъёмный *adj.* dismountable, demountable, detachable, separable; split, divided, detached; reducible; release, disengaging

разъесть *pf. of* разъедать

район *m.* area, district, range, zone, locality, field, "raion"; **р. с лёгкими атмосферными условиями** light (mechanical) loading area (*tphny.*); **р. телефонной системы** central office area; **р. техобслуживания** maintenance area (*tphny.*)

районирование *n.* division into districts, zoning

районированный *adj.* divided into districts, zoned

районировать *v.* to district, to divide into districts

районный *adj.* area, district, zone, regional

Райт Wright

ракета *f.* rocket; flare; fuse; **осветительная р.** target-indicating flare, target indicator

ракетный *adj.* rocket, rocket-powered; rocket-borne; jet

ракетоплан *m.* rocket aircraft

раккорд *m.* leader tape

раковина *f.* shell; sink, pit; bandstand; flaw, crack; cavity; **р. (головного) телефона** earphone cushion, ear cap, earpiece; **слуховая р.** ear shell, earpiece, receiver cap; **ушная р.** helix, pinua; earflap

«ракон» *m.* Racon (radar beacon)

ракурс *m.* aspect

рама *f.* frame, framing; loop; stand, bay, rack, support, chassis, carriage, base of a device; framework; bed, bedplate; **двойная р.** double loop antenna; **катодная р.** plating rack; **р. колодца** manhole frame, manhole cover frame; **компасная р.** gimbal suspension; **р. люка** joint box channel; **р. многократного поля** multiple frame; **мультипльная р.** bank (*dts.*, *tphny.*); **р. с искателями** selector shelf; **сборная р.** split frame; **р. фортепиано** string plate; **р. централизационного аппарата** locking bed (*RR*); **р. штатива** rack

Раман Raman

Рамзай Ramsay

Рамзауэр Ramsauer

рамка *f.* gate; frame; loop, coil; strip, shelf (*tphny.*); picture segment, field of view; loop antenna; mask (*tv*); **с запертой рамкой** loop-locked; **р. ГИ** selector shelf; **гнездная (*or* гнездовая) р.** jack strip; **р. для оригинала** copy gate; **звуковая р.** sound gate; **р. ИВ** finder shelf; **р. кинопроектора** picture aperture; **ключевая р., кнопочная р.** keyshelf; **р. ЛИ** connector shelf; **магическая р.** luminous edge (*tv*); **ограничивающая р.** ellipse of essential information, framing mask (*tv*); **оконечная р.** terminal block (*tphny.*); **р. с гнёздами** jack strip (*tphny.*); **р. с ключами, р. с кнопками** keyshelf; **р. с лампами** lamp-socket mounting, signal-lamp strip (*tphny.*); **р. со штифтами** soldering terminal strip, tag strip, connecting strip; **холостая р.** jack spacer, end panel

**рамочный** *adj.* frame; loop, looped, loop-like; chassis

**рампа** *f.* footlights, studio light board(s); ramp, contact rail

**ранг** *m.* rank

**ранговый** *adj.* rank, class, grade

**рандомизация** *f.* randomization

**ранец** *m.* backpack, knapsack, pack, kit

**Ранкин** Rankine

**ранний** *adj.* early, previous, pre-

**рант** *m.* ledge, welt, shoulder, boss; **опорный р.** holding ledge

**ранцевый** *adj.* backpack, knapsack, pack, kit-bag

**« рапидин »** *m.* rapidyne generator

**раскалённость** *f.* incandescence, glow; **р. добела** white heat, incandescence; **р. докрасна** red heat

**раскалённый** *adj.* burning, scorching, hot; incandescent, glowing, red hot; **р. добела** white hot, incandescent

**раскаливать, раскалить** *v.* to incandesce, to make red-hot

**раскаливаться, раскалиться** *v.* to become red hot, to glow

**раскалываемость** *f.* cleavage; cleavability

**раскалываемый** *adj.* cleavable

**раскалывание** *n.* cleaving, cleavage, splitting; cutting, division, separation

**раскалывать, расколоть** *v.* to split (up), to cleave, to fissure, to crack; to slit, to cut off

**раскалять(ся)** *see* **раскаливать(ся)**

**раскатывать, раскатать** *v.* to unroll, to uncoil, to spool off; to flatten, to laminate

**раскачать** *pf. of* **раскачивать**

**раскачивание** *n.* build-up; oscillation, vibration; fluctuating, fluttering; swinging, swaying

**раскачиватель** *m.* driver, tilter

**раскачивать, раскачать** *v.* to swing; to drive; to shake, to loosen

**раскачивающий** *adj.* swinging, rocking; driving

**раскачка** *f.* build(ing)-up; drive, driving; swing, swinging, rocking; resonant step-up; **резонансная р. напряжения** resonant voltage stepup

**раскисление** *n.* deoxidation

**раскисленный** *adj.* deoxidized

**раскладка** *f.* allocation, division, allotment, distribution

**раскладочный** *adj.* distributing; interpolating; collator

**расковочный** *adj.* stretching

**раскодирование** *n.* decoding

**раскодировать** *v.* to decode, to decipher

**расколотый** *adj.* split, cleaved

**расколоть** *pf. of* **раскалывать, колоть**

**раскомпенсация** *f.* decompensation

**раскорачивать, раскоротить** *v.* to unshort

**раскос** *m.* diagonal, cross brace, pole brace, angle brace, cross stay, truss, strut, prop, diagonal strut

**раскрасить** *pf. of* **раскрашивать**

**раскраска** *f.* coloring, painting, tinting, coloration

**раскрашенный** *adj.* colored, painted

**раскрашивать, раскрасить** *v.* to color, to paint

**раскрепление** *n.* uncaging

**раскройка** *f.* cutting, cutting-out

**раскрыв** *m.* aperture, flare

**раскрывание** *n.* opening, uncovering, disclosure; **р. диафрагмы** iris-in

**раскрывать, раскрыть** *v.* to open, to expose, to lay bare, to reveal, to uncover, to disclose

**раскрывающийся** *adj.* opening; hinged

**раскрывный** *adj.* aperture

**раскрытие** *n.* opening, uncovering; exposure, disclosure; mouth; evaluation; expansion

**раскрыть** *pf. of* **раскрывать**

**распад** *m.* disintegration, decay, decomposition, dissociation, break-up; disruption; collapse; separation; resolution; destruction; **р. при обмене** degenerate state due to energy exchange

**распадаться, распасться** *v.* to disintegrate, to come apart, to decompose, to break down; to collapse; to dissociate

**распадающийся** *adj.* decaying, fissible, fissile; decomposing; disintegrating

**распаивание** *n.* unsoldering

**распаивать, распаять** *v.* to unsolder

**распайка** *f.* unsoldering

**распаковывать, распаковать** *v.* to unpack

**распасться** *pf. of* **распадаться**

**распаянный** *adj.* unsoldered

**распаять** *pf. of* **распаивать**

**распереть** *pf. of* **распирать**

**распиливать, распилить** *v.* to saw (apart)

**распирать, распереть** *v.* to brace, to prop; to thrust, to push apart

**расписание** *n.* schedule, timetable; **р. периодических измерений** routine test schedule

**расписать** *pf. of* **расписывать**

**расписывание** *n.* coding; unwinding (of a loop)

**расписывать, расписать** *v.* to unwind (a loop); to paint; to describe

**расплав** *m.* electrolyte, fused electrolyte, bath, fused salt; melt; fusion

**расплавить(ся)** *pf. of* **расплавливать (-ся), плавить(ся)**

**расплавление** *n.* melting, fusion; smelting

**расплавленный** *adj.* molten, melted, fused

**расплавливать, расплавить** *v.* to melt, to melt down, to fuse; to found

**расплавливаться, расплавиться** *v.* to melt; to be liquefied

**расплавлять(ся)** *see* **расплавливать (-ся)**

**расплавляющий** *adj.* fusing, melting

**распланировать** *pf. of* **планировать**

**расплывание** *n.* blurring, bloom, blooming; running, spreading, deliquescence, deliquium; **р. краски** blurring, bleeding

**расплываться, расплыться** *v.* to run, to spread, to dissolve, to deliquesce

**расплывающийся** *adj.* deliquescent

**расплывчатость** *f.* diffusion, diffuseness, diffusiveness, dimness, indistinctness

**расплывчатый** *adj.* diffuse, diffused, dim, indistinct, blurred

**расплыться** *pf. of* **расплываться**

**распознаваемый** *adj.* discernible, recognizable

**распознавание** *n.* recognition, discerning, discernment, discrimination, perception, distinction, clarification

**распознавать, распознать** *v.* to recognize, to discern, to discriminate, to distinguish

**распознание** *see* **распознавание**

**распознающий** *adj.* discriminatory, discriminative, discerning

**располагаемый** *adj.* available

**располагание** *n.* disposal, disposition, arrangement

**располагать, расположить** *v.* to array, to arrange, to dispose, to place, to locate, to lay out, to situate, to put, to set; to install; to intend; to group; **р. в случайном порядке** to randomize; **р. последовательно** to interlace; **р. симметрично** to symmetrize

**располагающий** *adj.* prepossessing, conductive (to)

**расположение** *n.* arrangement, layout, disposition, disposal, distribution, allocation, grouping, spacing, array; location, situation, position, positioning; installation; structure; trunking, transposing; **р. в случайном порядке** randomization; **р. в шахматном порядке** staggering; **заранее заданное р. цифр** preset digit layout; **р. проводов в шахматном порядке** wiring in diagonal pairs; **смежное р.** juxtaposition; **р. со взаимным сдвигом** (*or* **смещением**) staggering; **шкафное р. оборудования** bay of racks (*tphny.*)

**расположенный** *adj.* disposed, arranged, placed, set; situated, located

**расположить** *pf. of* **располагать**

**распор** *m.* thrust; sweaving

**распора** *f.*, **распорка** *f.* brace, cross bar, tie beam, tie rod, stay, strut; vent segment, vent spacer, spacer, distance piece, spreader; walling board; separator; **крестообразная р.** diagonal cross brace; **прямая р.** straight pole brace; **столбовая р.** pole brace

**распорный, распорочный** *adj.* spacer, distance (piece), spacing, stand-off; thrust, brace

**распорядительный** *adj.* administrative, control, controlling, commanding, efficient, active, capable

**распоряжаться, распорядиться** *v.* to order, to arrange, to dispose, to manage

**распоряжение** *n.* order, instruction, direction; arrangement, disposal, disposition

**распределение** *n.* allocation, allotment, distribution, division, dispensation, partition; sharing; handling (of traffic); pattern; shape; dispersion, diffusion, spreading, dissemination; **клапанное р.** valve control mechanism; **кулачковое р.** cam gear; **р. максимумов ускорения** acceleration peaks distribution; **р. нагрузки по хорде лопасти** blade chordwise load distribution; **несобственное р.** singular distribution (*math.*); **р. по дальности** range distribution; **р. по частоте** frequency distribution; **р. поля** field density; **р. поля в раскрыве антенны** aperture illumination, field strength distribution in the aperture, field distribution of the aperture; **р. поля по раскрыву** aperture-field distribution; **произвольное р.** unimodal distribution, arbitrary distribution; **равновесное р. бария** "freezing" distribution of barium; **равномерное р.** equi-partition; **р. скоростей** velocity shape; **р. собственных значений** distribution of eigenvalues; **р. спектра** spectrum conservation; **р. температуры по глубине** vertical temperature distribution; **р. частот по международному соглашению** internationally allocated band

**распределённый** *adj.* distributed, divided; dispensed; space; **р. во времени** timed

**распределитель** *m.* distributor, allotter, distributing frame, desampler; alloting switch; (cable) branch point; sampler (*tv*); spreader; manifold; **биметаллический р.** electrothermally-operated relay; **р. вызовов справочного стола** position distributor (*tphny.*); **р. искателей вызовов** assignment switch, assignment selector (*tphny.*); **контактный р.** tapper, hunting switch; **р. обратных вызовов** reverting call switch; **р. служебных линий** order wire distributor

**распределительный** *adj.* distributive, distributing, distribution, distributor; switching; regulating, control

**распределять, распределить** *v.* to distribute, to disperse, to allot, to allocate, to share, to assign; to segregate; to sequence; to sort

**распредщит** *m.* distributor, distributing frame, distributing board, switchboard

**распространение** *n.* propagation, spread, spreading, diffusion, dissemination; distribution; extension; enlargement, amplification; **р. в море** marine propagation; **р. в поверхностном слое** surface-duct propagation; **р. звука в водной среде** propagation of underwater sound; **касательное р.** grazing propagation; **р. на закрытой трассе** nonoptical propagation (of waves); **направленное р.** guided propagation; **перпендикулярное р.** normal propagation; **р. по цепочке** rippling down the chain; **р. радиоволн на закрытой трассе** nonoptical propagation; **сверхдальнее р. звука в подводном звуковом канале** sofar propagation; **скользящее р.** grazing propagation

**распространённость** *f.* prevalence, extent, frequency

**распространённый** *adj.* widespread, extended

**распространитель** *m.* spreader, diffuser, propagator

**распространительный** *adj.* extended; spreading, propagating

**распространять, распространить** *v.* to broadcast, to circulate, to spread, to diffuse, to propagate, to disseminate; to disperse, to distribute; to radiate, to emit

**распространяться, распространиться** *v.* to expand, to branch out, to spread, to enlarge; to propagate (waves); to travel; to radiate, to emit

**распространяющийся** *adj.* propagating; transmitted

**распускать, распустить** *v.* to dissolve, to solve, to melt, to liquefy, to deliquesce; to diffuse, to disperse

**распыление** *n.* dispersion, spraying, diffusion, atomization; splutter, sputtering, spattering; scattering; pulverization; evaporization; disintegration; **р. газопоглотителя, р. геттера** getter flash; **р. нити накала** disintegration of filament

**распылённый** *adj.* sprayed, atomized; pulverized, divided, powdered

**распылпвать** *see* **распылять**

**распылитель** *m.* dissipator, diffuser, sprayer, atomizer, pulverizer

**распылительный** *adj.* spraying, sprayer

**распылить** *pf. of* **распылять**

**распыляемый** *adj.* sprayed

**распылянный** *see* **распылённый**

**распылять, распылить** *v.* to spray, to atomize, to diffuse; to pulverize, to disperse, to scatter; to sputter

**распыляющий** *adj.* spraying, spray; pulverizing

**рассасывание** *n.* resorption

**рассверливание** *n.* drilling, boring

**рассверливать, рассверлить** *v.* to bore out, to drill out, to rebore, to widen

**рассевать** *see* **рассеивать**

**рассеиваемый** *adj.* dissipated

**рассеивание** *n.* dissipation, dispersion, dispersal, dispersing, diffusivity, diffusion, scattering

**рассеиватель** *m.* diffuser, disperser, scatterer

**рассеивать, рассеять** *v.* to diffuse, to disperse, to dissipate, to spread, to scatter, to disseminate; to leak; to spray, to atomize

**рассеиваться, рассеяться** *v.* to disperse, to dissipate, to scatter; to leak; to diverge

**рассеивающий** *adj.* dispersive, dispersing, dispersion, dissipative; diffusing, diffusion, scattering, spreading; diverging

**рассекаемый** *adj.* cleaved, cut

**рассекатель** *m.* dissector (*tv*)

**рассекать, рассечь** *v.* to cleave; to dissect; to cut apart, to cut up

**рассекречивать** *v.* to release

**Рассел** Russell

**рассечение** *n.* bisection, dissection

**рассечка** *f.* cutting, breaking (a circuit); **в рассечку** in series

**рассечь** *pf. of* **рассекать**

**рассеяние** *n.* scatter, scattering, dissipation, dispersion, diffusion, dissemination, diffusivity; spread; attenuation (of waves); leaking, leakage, leak; degradation; **магнитное р. в лобовых частях** flank dispersion; **р. на аноде** anode dissipation, plate dissipation; **р. на турбулентных неоднородностях** near-field correction; **р. на углы, близкие к нулю** near-

forward scattering; **р. на экране** screen dissipation; **номинальное р.** dissipation rating; **обратное р.** back-scattering; scattered ёcho; **р. под малыми углами вперёд** near-forward scattering; **р. полей** leakage (fields), stray magnetic fields

**рассеянность** *f.* dispersion

**рассеянный** *adj.* diffused, dispersed, disseminated, scattered, stray, dissipated; drifting; trace (element); sprinkled, strewn

**рассеять(ся)** *pf. of* **рассеивать(ся)**

**расслаивание** *n.* stratification, lamination, cleavage, separation; phase separation

**расслаивать, расслоить** *v.* to stratify, to divide into layers; to differentiate

**расслоение** *n.* stratification, lamination; **р. цветов** color break-up

**расслоённый** *adj.* stratified, laminated

**расслоить** *pf. of* **расслаивать**

**рассматривание** *n.* viewing; examination; consideration

**рассмотрение** *n.* consideration; examination, investigation, inspection

**рассогласование** *n.* error, disagreement, mismatch; mistuning, detuning; displacement; misalignment, unbalance; deviation; **вызванный рассогласованием** error-actuated

**рассогласованность** *f.* mismatch, mismatching fault

**рассогласованный** *adj.* mismatched

**рассогласовывать, рассогласовать** *v.* to mismatch

**рассоединять, рассоединить** *v.* to uncouple

**рассортировать** *pf. of* **сортировать**

**рассортировывание** *n.* sorting

**рассредоточение** *n.* spread, distribution, dispersal, dispersing

**расставлять, расставить** *v.* to array, to place, to set, to arrange; to move apart, to space

**расстановка** *f.* order, ordering, spacing, arrangement, disposition

**расстопоривание** *n.* release, unlocking, unlatching

**расстояние** *n.* range, distance; clearance, separation, span, space, spacing, interval; length, line, section, tract; width; **на расстоянии** at a distance, distant, remote; **р. в световых единицах** electrical distance (in microseconds); **р. в свету** inner width, width in the clear, clearance; **р. для передачи крупным планом** close-up distance; **р. до маяка** omnidistance; **еди́ничное р.** unit length; **р. между электродами по поверхности диэлектрика** creepage distance; **минимальное р. между проводами и поверхностью земли** minimum clearance of lines from ground; **р. от оси вращения** pivot distance; **р. перекрывающего разряда** jumping distance; **р. по касательной** grazing distance; **р. по курсу** longitudinal separation; **р. по прямой** air line distance; **р. (по частоте) между несущими сигналов изображения и звукового сопровождения** sound-to-picture separation; **предельное р. слышимости звука** hearing distance, hearing limit; **разрядное р.** arcing distance, striking distance; **р. рассматривания** viewing ratio; **р. скачка** skip distance, jump-over distance, hop distance; **угловое р. в одну минуту дуги** minute of arc; **фокусное р.** focal length, equivalent focus; **шумовое р.** signal-to-noise ratio

**расстраивание** *n.* detuning, mistuning

**расстраивать, расстроить** *v.* to detune, to untune, to mistune; to disturb, to disorder, to disarrange, to disrupt, to perturb, to unsettle, to upset, to derange, to unbalance; to put out; to mismatch

**расстроенность** *f.* detuning, mistuning; maladjustment, misalignment

**расстроенный** *adj.* untuned, detuned, off-tune, out of tune, stagger-tuned; dumb, mute; disturbed; unsettled; upset; **с расстроенными контурами** stagger-tuned; **сильно р.** overstaggered

**расстроить** *pf. of* **расстраивать**

**расстройка** *f.* mismatch(ing); mistuning, detuning; maladjustment, misalignment; frequency difference; **с сильной расстройкой** overstaggered; **(взаимная) р. контуров** staggering; **р. по частоте** frequency separation

**расстройство** *n.* disturbance, disorder, disorganization, disruption, derangement, confusion; abnormality; upset

**рассуждение** *n.* reasoning, deliberation, discussion, discourse, consideration

**рассчитанный** *adj.* intended (for), designed (for), meant; rated; calculated, computed; **р. для длительной нагрузки** continuously rated; **заранее р.** pre-computed; **р. на неквалифицированное обслуживание** fool-proof

**рассчитывать, рассчитать** *v.* to count, to calculate, to compute, to figure out; to rate; to design; to depend, to count (on), to expect, to reckon (on)

**рассылание** *n.,* **рассылка** *f.* distribution

**рассыльный** *m. adj. decl.* messenger, errand-man

**расталкивание** *n.* repulsion; pushing apart

**растачивать, расточить** *v.* to bore out, to drill out; to cut out, to chisel out

**раствор** *m.* opening, gap, aperture; flare (of a horn); mouth; solution, dissolution; **р. для амальгамирования** blue dip; **смоляной р.** dissolved resin

**растворение** *n.* solution, dissolution, dissociation

**растворённый** *adj.* dissolved

**растворимость** *f.* solubility; deliquescence

**растворимый** *adj.* soluble, dissoluble, liquefiable

**растворитель** *m.* solvent, menstrum; dissolver

**растворить(ся)** *pf. of* **растворять(ся)**

**растворяемый** *adj.* soluble; being dissolved

**растворять, растворить** *v.* to solve, to dissolve; to fuse; to open, to unfasten

**растворяться, раствориться** *v.* to dissolve, to be dissolved; to open, to be opened

**растворяющий** *adj.* dissolving, solvent, menstrum

**растекание** *n.* spreading out

**растекаться, растечься** *v.* to run, to spread, to flow, to spill

**растекающийся** *adj.* running, spreading; current

**растереть** *pf. of* **растирать**

**растечься** *pf. of* **растекаться**

**расти, вырасти** *v.* to grow, to spring up

**растирать, растереть** *v.* to rub, to pulverize, to grind; to spread

**расточить** *pf. of* **растачивать**

**расточка** *f.* cutting out, chiselling out; armature bore, armature chamber; bore, boring, drilling; **р. пространства между полюсами** bore of field

**расточный** *adj.* boring

**растр** *m.* frame, field, raster, trace, (scanning) pattern, scanning field; line raster; **линейный р.** lattice; **р. на мишени передающей трубки** camera-scanning pattern; **полутоновый р.** mesh half-tone; **р. с перемежением вдоль строк** horizontally interlaced pattern; **р. с погашенными обратными ходами** blank raster; **сетчатый р.** cross-hatch pattern; **точечный р.** horizontally interlaced pattern, dot pattern

**растрёпанный** *adj.* frayed

**растрёпывание** *n.* fraying, fringing

**растрескивание** *n.* crazing, bursting, cracking, fracture, splitting, fissuring

**растровый** *adj.* raster

**раструб** *m.* mouth, mouthpiece, bell, bell-mouth; trumpet horn, funnel, funnel-shaped; cable-trench nozzle, cable-trench snout; socket; **с раструбом** bell-mouthed; **рупорообразный р.** flared end (of a waveguide)

**растягиваемый** *adj.* stretchable, stretching

**растягивание** *n.* spread, spreading, expansion, extension, protraction, widening, stretching, pulling; drawl; **р. изображения** picture pull-down; **р. изображения по краям** edge pull-down

**растягиватель** *m.* stretcher

**растягивать, растянуть** *v.* to stretch, to extend, to expand, to elongate, to lengthen; to broaden; to strain

**растягиваться, растянуться** *v.* to expand, to stretch, to extend, to give

**растягивающий** *adj.* stretching, extensional, tensile

**растяжение** *n.* tension, extensional strain, stress-strain; expansion, extension, enlargement, stretch, stretching, dilatation, dilation, elongation; pull; **поперечное р.** sidewise curl

**растяжимость** *f.* tensility, stretchability, tensile strength, elasticity, extensibility; expansibility; ductility, ductibility

**растяжимый** *adj.* tensile, stretchable, elastic, extensible; expansible, extensive; ductile

**растяжка** *f.* stretching, extending, extension, lengthening out; **р. сеток** grid sizing

**растяжной** *adj.* stretching, extension

**растянутость** *f.* prolixity, amplification; lengthiness, extension; enlargement

**растянутый** *adj.* expanded, stretched, spread, elongated, distended, pulled; lengthy

**растянуть(ся)** *pf. of* **растягивать(ся)**

**расфазирование** *n.* misphasing

**расфазировка** *f.* out-of-phase

**расфокусирование** *n.* defocusing, misfocusing, debunching

**расфокусированный** *adj.* defocused, out-of-focus

**расфокусировать** *v.* to defocus

**расфокусировка** *f.* defocusing, defocus, out-of-focus; debunching

**расхлябанность** *f.* slack, slackness, looseness, shakiness

**расхлябанный** *adj.* slack, loose

**расход** *m.* consumption; input; expense, expenditure; flow, rate of flow; dissipation; **расходы на уход, расходы по содержанию** maintenance cost; **постоянный р.** fixed rate flow, steady flow; **средний по времени массовый р.** time-average mass-flow rate; **удельный р. тепла** heat rate

**расходимость** *f.* divergence, divergency; **р. электрических смещений** Laplace's equation, Laplace's law

**расходиться, разойтись** *v.* to dissolve; to diverge, to branch off; to radiate; to disperse; to drift apart, to break up, to part

**расходомер** *m.* flowmeter, rate-of-flow meter, flow gage; **дисковый р.** nutating-disk flowmeter; **магнитный р.** magnetic induction flowmeter; **р. по перепаду давлений** pressure differential meter; **р. поршневого типа с переменным проходным сечением** piston-type variable area flowmeter; **регистрирующий р. с двумя пределами измерений** double-range recording flowmeter; **регистрирующий р. типа кольцевых весов** tilting-manometer recording flowmeter; **регистрирующий колокольный р. для малых перепадов** low-pressure bell-type recording flowmeter; **р. с вращающимися лопастями** rotary

bucket-type flowmeter; **р. с компен-сацией квадратичной зависимости** square-root-compensated recording flowmeter; **р. с линейным вязким сопротивлением** linear-resistance flowmeter; **р. с переменной площа-дью проходного сечения** variable area flowmeter; **р. с переменным сечением** flowrator
**расходомерный** *adj.* flow-measuring
**расходящийся** *adj.* diverging, di-vergent, branching off; outgoing; dispersive; flared (beam)
**расхождение** *n.* separation, space, divergence, discrepancy, spread, spreading; error, disagreement; de-viation, deflection; misconvergence; **р. во времени** mistiming; **р. кон-тактов** contact separation; **р. линий тока** stream-line effect; **р. по фазе** phase split
**расцветить** *pf. of* **расцвечивать**
**расцветка** *f.* coloration, coloring, tint-ing, shading; colors, tint, hue; **услов-ная р.** color code
**расцвечение** *n.* tinting, coloring
**расцвеченный** *adj.* colored
**расцвечивание** *n.* coloring, tinting
**расцвечивать, расцветить** *v.* to color, to tint, to tone
**расценка** *f.* tariff; evaluation, appraisal, estimate, estimation
**расцентровка** *f.* decentering
**расцеп, расцепитель** *m.* trip, tripping device, release, release (gear), trigger
**расцепить** *pf. of* **расцеплять**
**расцепка** *f.* release, release catch
**расцепление** *n.* trip, tripping, release, uncoupling, releasing, disjunction, disengagement, disconnecting, un-linking, unhooking; clearing, un-blocking; decoupling, declutching; starting; **со свободным расцепле-нием** trip-free; **р. на холостом ходе** no-load release; **р. при сверхтоке** overload release; **свободное р.** trip free release
**расцепленный** *adj.* unhooked, dis-engaged, disconnected, out of gear
**расцеплять, расцепить** *v.* to trip, to release, to disengage, to decouple, to uncouple; to shut off, to unhook, to unlink, to disconnect, to declutch
**расцепляющий** *adj.* tripping, dis-connecting, releasing, release, dis-engaging

**расцепной** *adj.* disengaging, disengage-able, detachable
**расчёт** *m.* calculation, computing, com-putation, estimation, evaluation, cal-culus; design; accounting; account, bill, invoice; crew; **военнотыловые расчёты** logistics; **цветовые расчёты** computational colorimetry, indirect colorimetry; **р. в относительных единицах** (*or* **величинах**) per unit calculation; **р. вращающего мо-мента** torque calculation; **р. на (вычислительной) машине** machine calculation; **р. по контурам** loop analysis, mesh analysis; **р. по узлам** node analysis; **р. при конструирова-нии** design calculation; **р. цветов по спектральным кривым** indirect colorimetry
**расчётник** *m.* calculator
**расчётный** *adj.* calculated, rated, estim-ated, designed, design; calculation, calculating
**расчётчик** *m.* calculator, estimator; designer
**расчисление** *n.* calculation, computa-tion, reckoning
**расчислять, расчислить** *v.* to calculate, to compute, to reckon, to figure out
**расчитывать, расчитать** *v.* to calculate, to compute
**расчленение** *n.* articulation, arrange-ment, organization; decomposition, partition; differentiation (*math.*); separation, disintegration, disjunc-tion, breaking up
**расчленять, расчленить** *v.* to split, to separate, to disengage, to break up; to analyze, to dissect; to disjoint, to dismember, to disarticulate
**расшатанность** *f.* looseness, shakiness, instability
**расшатанный** *adj.* loose, unstable, shaky
**расшивать, расшить** *v.* to fan out
**расшивка** *f.* fanning out
**расширение** *n.* widening, broadening, extension, expansion, expending, spread(ing); enlargement, stretch-ing, dilatation, elongation; flare; tank (of a sound duct); completion; amplification, development, enlarge-ment; **автоматическое р. диапазона громкости** automatic volume ex-pansion; **р. пути тока в электролите** diffusion creep

**расширенный** *adj.* expanded, broadened, widened, dilated, enlarged; extensive

**расширитель** *m.* dilator, stretcher, widener, extender; expander

**расширительный** *adj.* expanding, broadening, widening; expansion

**расширить(ся)** *pf. of* **расширять(ся), ширить**

**расширяемость** *f.* expansibility, extensibility, dilatability

**расширяемый** *adj.* expansible, expansive, extensible, dilatant, dilatable

**расширять, расширить** *v.* to amplify, to widen, to enlarge, to expand, to extend, to broaden, to elongate, to dilate

**расширяться, расшириться** *v.* to expand, to stretch, to extend, to enlarge, to widen; to be widened, to be extended, to be dilated, to be enlarged

**расширяющий** *adj.* expanding, dilating

**расширяющийся** *adj.* flared; diverging, spreading out; extensible, expansible, dilatable, expanding

**расшить** *pf. of* **расшивать**

**расшифровать** *pf. of* **расшифровывать**

**расшифровка** *f.* deciphering; interpretation; **р. данных с быстрой подачей** on-line data reduction; **р. сигналов опознавания** IFF signal decoding

**расшифровывать, расшифровать** *v.* to decipher, to decode, to interpret

• **расшифровывающий** *adj.* deciphering, decoding

**расщелина** *f.* split, crack, chink, fissure, crevice, cleft, crevasse

**расщепитель** *m.* smasher; splitter; beam-splitting mirror

**расщепить(ся)** *pf. of* **расщеплять (ся)**

**расщепление** *n.* splintering, splitting, split, separating, separation, breaking up, disintegration, division, fission, smashing, parting, cleaving, cleavage; resolution (into); **со свободным расщеплением** trip-free

**расщеплённость** *f.* split

**расщеплённый** *adj.* split, divided, cleaved

**расщепляемость** *f.* fissility, fissionability; splitting, cleavage, cleavability

**расщепляемый** *adj.* fissionable; cleavable

**расщеплять, расщепить** *v.* to split, to splinter, to cleave; to break up, to disintegrate, to break down, to decompose; to crack, to laminate; to chip

**расщепляться, расщепиться** *v.* to split (up), to disintegrate, to break down; to splinter, to shatter; to crack, to cleave

**расщепляющийся** *adj.* fissionable; cleavable

**Раттен** Wratten

**Рауль** Raoult

**Раус** Routh

**Рафакс** Rafax

**Рафаэль** Raphael

**рафинирование** *n.* refining, purification

**рафинированный** *adj.* refined, purified

**рафинировать** *v.* to refine, to purify; to rectify

**рафинировочный** *adj.* refining

**рационализаторский** *adj.* rationalization; efficiency

**рационализация** *f.* rationalization; industrial efficiency

**рационализировать** *v.* to rationalize

**рациональность** *f.* rationality, reasonableness; efficiency

**рациональный** *adj.* rational, reasonable; efficient

**рация** *f.* portable radio set, radio station; **аварийная р.** emergency radio set; **р. взаимодействия** liaison radio set; **р. навьючиваемая на лошадь** packhorse radio station

**рацпредложение** *n.* efficiency suggestion

**рашпиль** *m.* rasp, rasp file

**РБП** *abbr.* (**радиолокационный бомбоприцел**) radar bombsight

**РВ** *abbr.* 1 (**радиоволна**) radio wave; 2 (**радиовысотомер**) radio altimeter; 3 (**распределитель вызовов**) call distributor

**рваный** *adj.* ragged, broken

**рвать, порвать** *v.* to tear, to disrupt; to break, to sever

**РГ** *abbr.* (**ромбическая горизонтальная**) rhombic (horizontal) antenna

**РГД** *abbr.* (**ромбическая горизонтальная двойная**) double rhombic (horizontal) antenna

**РД** *abbr.* 1 (**ракетный двигатель**) rocket engine; 2 (**реактивный двигатель**) jet engine; 3 (**регулятор давления**) pressure regulator, pressure control

**рд** *abbr.* (**резерфорд**) rutherford

**РДД** *abbr.* (**ракета дальнего действия**) long-range rocket

**реагирование** *n.* responsè, reaction, reacting

**реагированный** *adj.* reacted

**реагировать** *v.* to respond, to react; to couple back

**реагирующий** *adj.* reacting, reactive, responsive, sensitive

**реактанс** *m.* reactance; **общий р. в относительных единицах** per unit total field reactance; **переходный р., поперечный р.** quadrature-axis transient reactance; **поперечный ударный р.** quadrature-axis subtransient reactance; **продольный переходный р.** direct-axis transient reactance; **продольный синхронный р.** direct-axis synchronous reactance; **продольный ударный р.** direct-axis subtransient reactance

**реактатрон** *m.* reactatron

**реактивация** *f.* reactivation; **р. нити** flashing of filament, rejuvenation, reactivation

**реактивность** *f.* reactance; reactivity; **вносимая р.** insertion reactance; **долевая р.** per-unit reactance; **р. рассеяния в лобовых частях обмотки** end-connection reactance; **р. рассеяния между зубцами** tooth-tip reactance; **ударная р.** subtransient reactance

**реактивный** *adj.* reactive, reaction; wattless, reactive, reactance; jet (-driven), jet(-propelled); rocket

**реактиметр** *m.* reactimeter

**реактор** *m.* reactor (*nucl.*); reactance coil, reactor; **р.-двигатель** propulsion reactor; **защитный р.** choke coil, line choking coil; **путевой р.** choke coil; **разрядный р.** discharge coil; **р. с подмагничиванием** saturating core device

**реакторный** *adj.* reactor

**реактрон** *m.* reactron

**реакционный** *adj.* reaction

**реакция** *f.* reaction, reactive effect, counter-action, retroaction; response; reactance; **активная р.** re-sistive reaction; **граничная неспадающая ступенчатая р.** bonded nondecreasing step response; **р. излучения** radiation load; **локализованная р.** localized loading; **р. на единичное ступенчатое возмущение** unit-step response; **р. на импульсное возмущение** impulse response; **р. на синтез** fusion reaction; **р. на ступенчатое возмущение** step response; **однородная р.** uniform loading; **частотная р. нагрузки** pulling figure

**реализация** *f.* realization, substantialization

**реализировать** *or* **реализовать** *v.* to realize

**реализуемый** *adj.* available; realizable

**реалистический** *adj.* realistic

**реальность** *f.* reality

**реальный** *adj.* real, actual, tangible; practicable, workable

**реаэрация** *f.* reaeration

**ребатрон** *m.* relativistic electron, bunching accelerator, rebatron

**Ребекка** Rebecca (*rdr.*)

**Ребель** Roebel

**реборда** *f.* flange, collar rim (of pulley)

**ребристопризматический** *adj.* angular-prismatic

**ребристотубулярный** *adj.* fin-tube

**ребристый** *adj.* rib, ribbed, costate, costal; flanged; finned, corrugated

**ребро** *n.* rib, fin; edge, verge, border; flange; vane; cooling fin; **на ребро** edgewise, on edge, upright; **внутреннее р. жёсткости** bass-bram, sound-bar; **р. поглощающей конструкции** absorbing rib; **скошенное р. (кристалла)** chamfered edge

**ребром-направляемый** *adj.* end-on directional

**рёв** *m.* howling

**реверберационный** *adj.* reverberation, reverberatory

**реверберация** *f.* reverberation, reverberant sound

**реверберировать** *v.* to reverberate, to echo

**реверберирующий** *adj.* reverberating, reverberant

**ревербометр** *m.* reverberation meter, reverberometer

**реверс** *m.* backspacing (of tape); reversing, reverse; reversing gear, reverser

**реверсер** *m.* reverser
**реверсивность** *f.* reversibility
**реверсивный** *adj.* reversible, reversing; two-way
**реверсирование** *n.* reversion, reversing, reversal, reverse; change of direction, direction(al) control; **р. зубчатыми колесами** wheel reversing gear
**реверсированный** *adj.* reversed
**реверсировать** *v.* to reverse, to change direction
**реверсируемость** *f.* reversibility
**реверсируемый** *adj.* reversible
**реверсирующий** *adj.* reversing, reverse
**реверсия** *f.* reversion, reversal, throwback
**реверсор** *m.* reverser, reversing switch
**револьверный** *adj.* capstan; revolving; revolver
**ревун** *m.* howler
**регенеративный** *adj.* regenerative, regeneration, self-oscillating
**регенератор** *m.* regenerator, oscillating audion
**регенерация** *f.* regenerative action, regeneration, positive feedback; reactivation, retroaction, retroactive effect; self-oscillation; damping reduction; restitution (*tgphy.*); howling; reclamation, recovery; **р. ламп** tube reclamation; **р. ламповых цоколей** base reclaiming
**регенерирование** *see* **регенерация**
**регенерированный** *adj.* regenerated, restored, recovered; reclaimed
**регенерировать** *v.* to regenerate, to restore, to recover; to reclaim; to reprocess
**регенерирующий** *adj.* regenerating
**региональный** *adj.* regional, district
**регистр** *m.* register, organ stop (*mus.*); director, register, sender (automatic telephony); coder (*tphny.*); dial-pulse register; list; **р. второго слагаемого** addend register, augend register; **р., делящий пополам** halving register; **р. компоненты операции** operand register; **многочастотный р.** multifrequency sender; **р. множителя-частного** multiplier-quotient register; **р. на двухобмоточных сердечниках** two windings per core register; **органный р.** organ register, organ stop; beaked flute; **р. переадресации** index register, base register,

modifier register; **р. первого слагаемого** addend register; **пересчётный р.** computer, allotter; **р. с удвоенным количеством разрядов** double-ranked register; **самый верхний р. голоса** head tone; **сдвиговый р. частного** quotient shifter; **р. слагаемого произведения** addend-partial product register; **статический р.** staticizer; **р. транзитной связи** tandem sender (*tphny.*); **язычковый р.** reed stop
**регистратор** *m.* recording device, register, recorder; registering clerk; monitor; **р. времени** calculagraph; **р. времени движения повозки по инерции** coasting recorder; **р. импульсов** pulse storing device; **многоточечный р.** scanner recorder; **многоточечный р. для проволочных тензометров** strain gage scanner recorder; **прямопоказывающий р. высоты** direct-reading pitch recorder; **р. пульса** sphygmograph; **р.-регулятор** recorder-controller; **синхронный цифропечатающий р.** synchroprinter; **р. совпадений** coincidence counter
**регистрационный** *adj.* registration, recording, record
**регистрация** *f.* recording, record, registration, registering, writing, logging; memory; metering; **р. отметок сторожей** watchman's recording
**регистрирование** *see* **регистрация**
**регистрировать, зарегистрировать** *v.* to register, to record, to file; to receive; to meter; to scan
**регистрируемый** *adj.* registered; received
**регистрирующий** *adj.* registering, recording, writing; graphic
**регистровый** *adj.* register
**региструющий** *adj.* registering, recording, writing
**регламентный** *adj.* control, controlling
**регрессивный** *adj.* regressive
**регрессировать** *v.* to regress, to turn back
**регрессия** *f.* regression
**« регулекс »** *m.* regulex
**регулирование** *n.* control, controlling, governing, regulation, correction, adjustment, adjusting, setting; tuning, modulation, percentage modulation; resetting; correction;

tracking; **автоматическое р. несущей частоты** floating-carrier system; **быстрое р.** split-cycle control; **изодромное р.** proportional-plus-floating control; **р. на две ступени нагрева** two-heat control; **непрямое р.** power-operated control, servo-operated control; **р. нескольких взаимосвязанных величин** multivariable control; **р. обратной связи** reaction control; **р. от руки** manual control; **р. по времени** time-schedule control, time-variable control; **р. по интервалу** automatic reset; **р. по координате, производной и интегралу** derivative-proportional-integral control; **р. по косвенным параметрам** indirect control; **р. по нагрузке** load control; **р. по твёрдо заданной величине** constant adjustment, constant calibration; **р. подачи** (*or* **питания**) **с разветвлёнными потоками** split-feed control; **поправочное р. усиления** spotting gain control; **программное р.** time schedule control, time variable control; **р. процесса по анализу выходных данных** end-point control; **прямое р.** self-acting control, self-operated control; **р. с корректировкой** (*or* **упреждением**) predictor control; **р. с обратной связью** closed-loop control, feedback control; **р. с применением обратной связи** closed-cycle control, **р. с пропорциональным и интегральным управлением** proportional-plus-integral control; **р. с усилителем** servo-operated control; **следящее р.** follow-up control; **р. соотношения потоков** flow-ratio control; **упреждающее р. по производной** anticipatory control; **р. частоты и (обменной) мощности** load-frequency control; **р. числа оборотов двигателя** motor-speed control; **р. языком сопла** nozzle flap regulation

**регулированный** *adj.* regulated, adjusted

**регулировать, отрегулировать** *v.* to adjust, to regulate, to govern, to set, to control, to shift, to readjust; to manipulate, to operate, to handle; to align, to tune; to throttle; to trim; to line up

**регулировка** *f.* control, regulation, setting, readjustment, adjustment, adjusting, resetting, correction, alignment, line-up; harmonizing; tuning; **компенсирующие регулировки** compensating tabs (*tv*); **автоматическая р. контрастности изображения** automatic background control (*tv*); **автоматическая р. усиления изображения** automatic contrast control (*tv*); **автоматическая р. усиления с задержкой по времени** delayed automatic gain control; **р. амплитуд изображения** framing (*or* vertical deflection) control; **р. батарейным коммутатором** end cell control; **бесшумная р.** squelch system, mute control; **р. величины сводящего сигнала строчной частоты** horizontal dynamic amplitude control; **р. величины сигнала динамического сведения по вертикали** vertical dynamic convergence amplitude control; **р. величины сигнала динамического сведения по горизонтали** horizontal dynamic convergence amplitude control; **винтовая р.** screw adjustment; **р. возбуждения** field adjustment; **временная р. усиления** anticlutter gain control, fast time gain control, sensitivity-time control; **временная автоматическая р.** anticlutter gain control; **р. высоты изображения** height control (*tv*); **голосовая р. уровня громкости** vogad, voice-operated gain-adjusting device; **р. градации** gamma control; **р. громкости** volume control; **р. диапазона громкости** volume range control; **р. диапазона изменения цветового тона** hue range control; **р. динамического оттенка** control of volume range, control of volume contrast, companding, adjustment of contrast; **р. динамической сходимости по вертикали** vertical-frequency dynamic convergence control; **р. динамической сходимости по горизонтали** horizontal-frequency dynamic convergence control; **р. интенсивности отметок в результате изменения усиления** videogain control; **р. кадровой синхронизации** vertical hold control (*tv*); **р. линейности кадровой развёртки, р. линейности по кадрам** frame linearity control; **р. накала** filament control; **р. напряжения встречным**

включением элементов counter-cell control; **р. насыщенности (цвета)** chroma control, color control, saturation control; **неверная р.** misadjustment; **р. номинального значения частоты** steady-state frequency control; **р. обратной связи** regeneration control, feedback control; **плавная р.** stepless regulation; continuously variable control; **р. по дальности** range adjustment; **повторная р.** readjusting, readjustment; **р. « под шлиц »** screw adjustment; **р. положения растра по вертикали** vertical positioning control; **р. положения растра по горизонтали** horizontal positioning control; **р. полосы пропускания** bandwidth control; **поправочная р. усиления** spotting gain control; **р. равномерности цвета по полю** color shading control; **р. размера по вертикали** vertical-amplitude control; **р. размера по кадру** frame amplitude control; **р. режима камеры** camera line-up; **р. реле с преобладанием** bias of a relay (*tgphy.*); **р. сигнала динамического сведения по вертикали** vertical-frequency dynamic convergence control; **р. сигнала динамического сведения по горизонтали** horizontal-frequency dynamic convergence control; **р. синхронизации по вертикали** (*or* **кадрам**) vertical hold control; **р. синхронизации по строкам** horizontal hold control; **р. скорости двигателей последовательно-параллельным переключением обмоток возбуждения** series-parallel field control; **р. скорости двигателя перемежающимся включением и выключением сопротивления** notching-back control; **р. скорости каскадным включением** concatenation control, cascade control; **р. скорости прямым каскадным включением двух индукционных двигателей** direct-concatenation control; **р. скорости со стороны ротора** secondary speed control; **р. скорости сопротивлением в цепи ротора** secondary resistance control; **р. смещения на ЭЛТ** gun bias control (*tv*); **р. смещения по вертикали** vertical positioning control; **р. сопряжения** gang adjustment; **р. уровня**

**гасящих импульсов** pedestal control; **установочная р. средней яркости изображения** service background control (*tv*); **р. фазы напряжения динамического сведения лучей** convergence phase control; **р. фазы развёртки по вертикали** vertical phasing control; **р. формы сигнала сведения по вертикали** vertical convergence shape control; **р. формы сигнала сведения по горизонтали** horizontal convergence shape control; **р. цветового тона** color shading control, hue control; **р. частоты кадровой развёртки** field frequency control; **шунтовая р. усиления** automatic gain-control system employing an a.g.c. amplifier for simultaneous control; **р. яркости отметок от отражённых сигналов** video-grain control

**регулировочный** *adj.* regulating, regulation, adjusting, controlling

**регулировщик** *m.* control man

**регулируемость** *f.* adjustability, controllability

**регулируемый** *adj.* regulated, controlled; adjustable, regulable, regulatable, variable; steerable

**регулирующий** *adj.* adjusting, regulating, aligning, governing, setting, controlling; regulator

**регулирующийся** *adj.* adjustable; controllable

**регуляризация** *f.* regularizing

**регулярно** *adv.* regularly

**регулярность** *f.* regularity, order; **р. формы помещения** room regularity

**регулярный** *adj.* regular, routine

**регулятор** *m.* adjuster, regulator, setter, corrector, controller, control (device); governor; knob; **регуляторы** controls; **автоматический р. громкости** automatic fader; **автоматический р. фазы приёмника ЦТ** quadricorrelator (*tv*); **автоматический статический р. с воздействием по производной** proportional-plus-derivative controller; **р. амплитуды сигнала сходимости по кадру** vertical convergence amplitude control; **безверньерный р. настройки** direct selector; **р. возбуждения** field regulator, field rheostat; **р. выдержки времени** timer; **р. вызовов** ringing interrupter (*tphny*.); **p. высоты** vertical amplitude control, treble control; **p. громкости** volume control, volume adjuster, volume regulator, loudness control, fader, attenuator, gain control; speaker control, speaker pad; **p. громкости с коррекцией тембра** tone-compensated volume control; **грузовой p.** weight-loaded governor; **p. давления питания** boost-pressure controller; **двойной переменный p.** double-rotor induction regulator; **двухпозиционный p. уровня** high-low level control; **p. динамических оттенков** dynamic-range control means, volume-range control means, compandor; **p. для установки в рамку** framing device, framer (*tv*); **p. задаваемого режима** set point adjuster; **p. масштаба** range-marker control (*rdr*.); **p. накала** dimmer control; **нелинейный p.** dead-zone regulator; **p. непосредственного действия** self-actuated controller, self-operated controller; **p. непрямого действия** pilot actuated regulator; **p. обратной производной** inverse derivative controller; **p. остроты резонанса** quality corrector; **p. прерывистого действия** intermittent controller; **программный p.** time schedule controller; **p. размера по кадрам** frame amplitude control, vertical size control; **p. размера по строкам** horizontal size control, line amplitude control; **ручной p. громкости** manual volume control; **p. с дроссельной заслонкой** flapper-valve controller; **p. с неактивной зоной** dead-band regulator; **p. с падающей дужкой** hoop drop relay; **p. с перекатывающимися секторами** roller-type control; **p. с плоским ходом рукоятки** disk-type rheostat; **p. с противовесом** weight-loaded governor; **p. с управлением от измерительного элемента** self-actuated regulator; **p. сети** line-voltage regulator; **p. скорости** phonophone (*tgphy*.); **угольный p. напряжения** carbon-pile regulator; **p. уровня раздела** interface controller; **p. (уровня) с контрольным проводом** pilot-wire regulator (of level); **p.-усилитель** preamplifier; **p. установки кадра** framer; **p. установки**

**нуля** zero control; **p. устойчивости** hold control; **p. хода двигателя** engine governor; **центробежный p.** centrifugal pendulum; **p. центровки по вертикали** vertical positioning control; **p. частоты вибратора** gate governor, phonophone (*tgphy*.); **p. частоты кадровой развёртки** field frequency control; **p. частоты строк** horizontal hold control (*tv*); **p. четырёхполюсника скрещённого типа** lattice network-regulator, variable pad

**регуляторный** *adj.* regulator, adjuster, controller

**редактирование** *n.* editing

**редактировать** *v.* to edit

**редкий** *adj.* rare, uncommon, scarce; thin, sparse

**редко** *adv.* rarely, seldom

**редкоземельный** *adj.* rare-earth

**редуктор** *m.* reducing valve, reducer; reducing agent; gear, gear assembly, geared system, speed transformer, reduction gear; reductor; **с редуктором** geared; **с внутренним редуктором, с встроенным редуктором** internally geared; **с понижающим редуктором** back-geared, geared-down; **дифференциальный p.** mechanical differential; **однодекадный p.** tenth gearbox; **повысительный p.** speed increaser; **p. постоянного тока** glow-discharge tube used as current limiter; **p. числа оборотов** (*or* **скорости**) speed reducer

**редукторный, редукционный** *adj.* reduction, reducing

**редукция** *f.* reduction, reducing

**редуцирование** *n.* reduction

**редуцированный** *adj.* reduced

**редуцировать** *v.* to reduce

**редуцирующий** *adj.* reducing

**режектирование** *n.* rejection

**режектируемый** *adj.* rejected

**режектор** *m.* rejector

**режекторный** *adj.* rejective; rejector; rejection, elimination; notch

**режекция** *f.* rejection; trapping

**режим** *m.* regime, system, practice, process, method; behavior, mode; condition(s), state; rate; (operating) condition(s), service, operation; duty; modifying; schedule; *often not specif. translated:* **чувствительность в режиме приёма** receiving response, (*lit.* response in conditions of receiving); **номинальные режимы** standard service ratings for electric motors; **в импульсном режиме** pulsed; **с лёгким режимом** low-duty; **аварийный р.** emergency conditions; **безотражённый р.** reflectionless conditions; **буферный р.** buffer working, floating; **р. B** class B operation; **временной р. работы** time-cycle operation; **входной р.** input conditions; **вызывной р.** ring down operation; **двигательный р.** motoring, motorizing; **р. дозвуковой струи** unchoked operation; **р. запирания** cutoff condition; **р. записи** record mode; **р. заправки** threading operation; **контрольный р.** reference performance; **р. кратковременной нагрузки** short-time rating; **неустановившийся р.** transient (conditions); **номинальный р.** duty (of an apparatus); **номинальный р. работы** normal rating; **р. обжига** lighting-up schedule; **р. ограничения тока пространственным зарядом** space-charge-limited-current state; **р. откачки** exhaust schedule; **р. первого включения (для обжига)** flashing schedule (for firing); **р. переключения** change-over operation; **перемежающийся р. работы** intermittent operation; **р. «посылка-прослушивание-наведение»** ping-listen-train method; **р. предельной нагрузки** full-load conditions; **предельный (рабочий) р.** maximum permissible operation conditions, utmost permissible operation conditions; **прерывистый р.** switching mode; **произвольный нестационарный р.** arbitrary time behavior; **противоточный р.** plug-reversing service; **р. работы** operating conditions, duty; **р. работы гидроакустической станции на прослушивание** asdic silence; **р. работы при риёме слабых сигналов** fringe operation; **р. работы с перебоями** intermittent service, intermittent operation; **р. работы с положительным напряжением на сетке** grid positive-driven conditions; **рабочий р.** operating conditions, performance; **р. редкой работы** standby duty; **р. сверхзвуковой струи** choked opera-

tion; **р. связанных колебаний** coupled modes; **р. связи на одной боковой полосе частот** single-side-band system; **синхронный р.** in-sync operation, hold-in operation; **слепой р. (полёта)** instrument conditions (of a flight); **р. тока начальных скоростей** residual-current state; **р. фиксации решения** hold condition; **р. холостого хода** idling; **р. шумопеленгования** asdic silence

**режимность** *f.* modiness
**режиссёр** *m.* producer, program director
**режиссура** *f.* master control (*tv*)
**режущий** *adj.* cutting; sharp
**резак** *m.* torch, cutting torch, acetylene cutter; cutter, chopper
**резание** *n.* cutting; carving
**резанный, резаный** *adj.* cut
**резать, разрезать, срезать** *v.* to cut, to slice, to slit
**резерв** *m.* reserve, spare, stand-by
**резервирование** *n.* reservation, redundancy; **р. замещением** active redundancy, stand-by redundancy; **постоянное р.** passive redundancy, parallel redundancy
**резервировать** *v.* to reserve
**резервный** *adj.* spare, reserve, stand-by, emergency; off-duty; storage
**резервуар** *m.* tank, reservoir, vessel, receiver, container, basin, well, cistern; **напорный р.** overhead tank; **нижний р.** sump
**резервуарный** *adj.* reservoir, receiver, tank, vessel, container
**Резерфорд** Rutherford
**резерфорд** *m.* rutherford
**резерфордовский** *adj.* Rutherford's
**резец** *m.* bit, stylus, cutter, cutting stylus, recording stylus; blade, knife; (drill) bit; cutting tool; **р. быстрорежущей стали** high speed steel; **сапфировый р. с дюралевым черенком** dural-shank sapphire cutting stylus
**резидуум** *m.* residuum
**резина** *f.* rubber
**резиновый** *adj.* rubber
**резинокс** *m.* resinox
**резиноподобный** *adj.* rubber-like
**резист** *m.* resist
**резистер** *m.* resistor; resistance
**резистивно-ёмкостный** *adj.* resistance-capacitance, resistance-capacity, RC

**резистивность** *f.* resistivity, specific resistance

**резистивноустойчивый** *adj.* resistance-stable

**резистивный** *adj.* resistive, resistance

**резистор** *m.* resistor, nonohmic resistor; resistance

**резисторно-диодно-ёмкостный** *adj.* capacitor-resistor-diode

**резистрон** *m.* resistron

**резка** *f.* cutting, clipping, severing, severance; stem height

**резкий** *adj.* sharp, harsh, shrill, piercing, percussive; abrupt, sudden; brisk

**резко** *adv.* sharply; abruptly; clearly (defined)

**резкость** *f.* definition, resolution, sharpness; abruptness; shrillness; asperity

**резнатрон** *m.* resnatron, high-power tetrode

**резолвер** *m.* resolver

**резольвента** *f.* resolvent, resolvent equation

**резольвентный** *adj.* resolvent

**резонанс** *m.* resonance, syntony; echo, response; **р. на основной частоте** fundamental resonance; **р. напряжения** series (phase) resonance; **нелинейный р.** anharmonic resonance; **объёмный р.** cavity resonance effect; **р. (пластинки) по толщине** thickness resonance; **р. по фазе** phase resonance; **последовательный р.** acceptor resonance; **распределённый р.** space resonance; **р. с одной составляющей частотой** submultiple resonance; **собственный р.** natural resonance, periodic resonance; **р. субгармоники** submultiple resonance; **р. токов** parallel (phase) resonance, antiresonance, current resonance, inverse resonance, rejector resonance

**резонансный** *adj.* resonant, resonating, vibrating, resonance, vibration(al)

**резонатор** *m.* resonator, cavity, resonance cavity; sounding board; rod; ringing circuit; **входной р.** buncher, input resonator; **выходной р.** catcher, output resonator; **герметизированный объёмный р.** sealed cavity; **двусторонний р.** duplex cavity; **задающий р.** exciter; **р. клистрона** intercepting (*or* collecting) circuit; **контрольный р. с автоматической**

настройкой performeter; **нагруженный объёмный р.** loaded (resonant) cavity; **настраиваемый р. с двумя элементами настройки** two-way stretch resonator; **настраиваемый объёмный р.** tuneable cavity; **негерметизированный объёмный р.** unsealed cavity; **объёмный р.** resonant cavity, resonant tank, echo box, cavity resonator, cavity circuit; T-R box cavity; **объёмный р. в переключателях передача-приём** TR box cavity; **объёмный р. в цепи обратной связи** reaction cavity; **петлеобразный объёмный р.** folded cavity; **полый р.** hollow-space oscillator, endovibrator; **р. проходного типа** re-entrant cavity, re-entrant resonator; **расстроенный объёмный р.** off-resonance cavity; **реактивный р.** reaction cavity; **р. со спаем в узле колебаний** nodal seal resonator; **торцевой р.** end cavity; **эталонный р.** cavity meter

**резонаторный** *adj.* resonator, cavity

**резонировать** *v.* to resonate, to resound

**резонирующий** *adj.* resonating, resonant

**резоноскоп** *m.* resonoscope

**резотрон** *m.* resotron

**результат** *m.* result, outcome, effect; product, yield; **иметь результатом** to result in

**результатный** *adj.* resultant, resulting

**результирующая** *f. adj. decl.* resultant

**результирующий** *adj.* resulting, resultant; net, overall, combined

**резцедержатель** *m.* chisel holder, cutter holder, tool holder

**резцовый** *adj.* cutter, tool

**резьба** *f.* (screw) thread, screw; carving, engraving; **внутренняя р.** female thread; **двухниточная р., двухоборотная р., двухходовая р.** double thread; **многозаходная р., многооборотная р., многоходовая р.** multiplex thread; **р. с крупным шагом** coarse pitch thread; **стандартная винтовая р.** standard thread

**резьбовой** *adj.* thread, threaded, threading

**резьбомер** *m.* screw pitch gage, thread gage

**резьбонарезной** *adj.* thread-cutting, screw-cutting

резьбофрезерный *adj.* thread-milling

резьбошлифовальный *adj.* thread-grinding

резюме *n. indecl.* résumé, abstract, summary

резюмировать *v.* to summarize

рейд *m.* raid, mission (*av.*)

рейдер *m.* rider; **наводимый по радиолучу р.** beam rider

«**рейдист**» *m.* radist

рейдовый *adj.* raiding

рейзистор *m.* raysistor

рейка *f.* strip, ledge, border, rim; lath, stick, rod, staff, batten, cleat; gage; rack; **р. для измерения стрелы провеса** sag gage, dip gage; **р. для проводки** moulding; **зубчатая р.** rack and pinion, rack gear, pinion gear; **клемная р.** connecting block, terminal block, connection block, connection strip; terminal strip; **контактная р.** wiper, tapper, hunting switch; **лицевая р.** finishing stile; **направляющая р.** guide-hole strip; **облицовочная р.** finishing stile, stile strip

рейконарезной *adj.* rack-cutting

Реймер Reimer

Рейнарц Reinartz

Рейнольдс Reynolds

рейс *m.* trip, run; passage

Рейс Reis

рейсмас *m.* height gage; surface gage; shifting gage; marking gage

Рейснер Reissner

рейснеров *poss. adj.* Reissner's

рейстрек *m.* racetrack

рейсфедер *m.* drawing-pen, ruling-pen

рейсшина *f.* T-square, drawing rule

Рейхерт Reichert

рекалесценция *f.* recalescence

реклама *f.* advertisement, publicity

рекламация *f.* rejection, objection, complaint, protest, reclamation

рекламировать *v.* to advertise; to make a claim

рекламный *adj.* publicity, advertisement, advertising; commercial

рекламодатель *m.* advertiser, sponsor

рекомбинатор *m.* recombiner

рекомбинационный *adj.* recombination

рекомбинация *f.* recombination; **р. в объёме** initial recombination

рекомендация *f.* recommendation, reference, introduction

рекомендованный *adj.* recommended, introduced; preferred

рекомендовать *v.* to recommend, to introduce

реконструировать *v.* to reconstruct, to redesign, to remodel, to rebuild, to restore, to modernize

реконструктивный *adj.* reconstructive

реконструкция *f.* restoration, modernization; rearrangement

рекорд *m.* record

рекордер *m.* recorder, mechanical recording head; cutter head, cutter; **выносный р.** external recorder; **р. механической записи** cutting head; **р. с обратной связью** feedback cutter

рекордерный *adj.* recorder, recording

рекордный *adj.* record, record-breaking

рекордсмен *m.* record holder; **р.-слухач** champion of sound reading (*tgphy.*)

рекристаллизация *f.* recrystallization

рекристаллизованный *adj.* recrystallized

«**ректигон**» *m.* rectigon

ректификатор *m.* rectifier

ректификационный *adj.* rectification

ректификация *f.*, **ректифицирование** *n.* rectification

ректификованный, **ректифицированный** *adj.* rectified

ректифицировать *v.* to rectify

ректокс *m.* rectox

ректрон *m.* rectron

рекуперативный *adj.* recuperative, recovery; regenerative

рекуператор *m.* recuperator; regenerator

рекуперация *f.* recuperation, regeneration

рекупировать *v.* to recuperate, to regenerate; to recover

рекуррентный *adj.* recurrent, recurring, recursion

рекурсивность *f.* recursiveness

рекурсивный *adj.* recursive

рекурсионный *adj.* recurrence

рекурсия *f.* recursion

релаксатор *m.* relaxation oscillator, Eccles-Jordan circuit; toggle

релаксационный *adj.* relaxation

релаксация *f.* relaxation; **верхняя р.** overrelaxation (*math.*)

реле *n. indecl.* relay; **аварийное р.** power transfer relay; **асинхронное** р. времени asynchronous timer, self-synchronous timer; **р. бдительности** acknowledging relay; **биметаллическое р.** thermo relay, hot wire relay, bimetallic strip relay; **блокирующее р.** "compelling" relay; holding relay (*tphny.*); **вакуумное р. с гибким язычком** dry-reed relay; **включающее р.** cut-in relay, relay switch; **включающее вызывной ток р.** ringing relay; **р. включения** starting relay; **р. возврата в исходное положение** reset relay; **р. времени** timer, time relay, time-lag relay, time-delay switch; **р. времени с плавным регулированием** continuously variable electrotimer; **р. времени со ступенчатым регулированием** fixed-position variable electrotimer; **р. вступления** timing relay; **р. вызывной лампы** pilot relay, transmitting relay, supervisory relay; **р. выключения вызывного тока** tripping relay (*tphny.*); **р. выключения счётчика** nonmetering relay (*tphny.*); **вышестоящее р.** back-up relay; **газовое р.** gas-actuated relay; gas detector relay, gas-pressure relay, gas relay, gaseous relay; **дистанционное р.** ohm relay; **р. длины световой волны** color matcher, colorimetric relay; **р. для включения энергосети** mercury switch, mercury relay; **р. для смягчения толчка тока** antiplug relay; **дополнительное путевое р.** floater relay; **р.-дроссель** high impedance relay; **дроссельное р.** battery supply relay; **р. зависимого действия, зависимо-замедленное р.** inverse time relay; **замедленное р.** slow-acting relay, time-delay relay, time lag(ged) relay, time-limit relay; **р. замедленное на отпускание** slow-release relay; **р. замедленное на срабатывание** slow-operating relay; **р. замыкания на землю** earth-leakage relay, ground relay; **р. защиты от кругового огня и замыкания на землю** flash and ground protecting relay; **р. знака величины** selective relay, directional relay; **импульсное р. времени** pulse timer; **ионное р.** cold cathode thyratron, gas-filled relay; **исполнительное р.** individual point relay; **р. исходного**

режима reset relay; **р. катушки расцепления** trip relay, clear-out relay; **р.-клопфер** relaying sounder (*tgphy.*); **конечное р.** final impulse operating relay; **р.-контактор** closing relay, contactor-type relay; **контрольное р.** control relay; pilot relay, supervisory relay; leak relay (*tgphy.*); holding relay (*tphny.*); **р. контрольной лампы рабочего места** pilot relay; **коромысловое р.** beam relay; **р. крутизны фронта** surge relay; **максимальное р.** overhead relay, overcurrent relay, overload circuit breaker, current overload relay; **масляное р.** oil-filled contactor; **маятниковое р. (времени)** mechanical escapement (time) relay; **минимальное р.** minimum voltage relay, undervoltage relay, undercurrent relay; **р. миниполь** polarized miniature-relay; **многократное р. времени** multi-contact timer, definite time relay; **накальное р.** filament-circuit relay; **начальное р., начинающее р.** initiating relay, primary relay, trigger relay; **р. непрерывного согласования** rematching relay; **р. неправильной передачи** differential relay; **нулевое р.** no-voltage relay, no-volt release switch; **р. обратной мощности** reserve power relay; **р. ограничения нагрузки двигателя** step-back relay; **ограниченно-зависимое р. с постоянной минимальной выдержкой** definite minimum time-limit relay; **р. останова** plugging relay; zero-speed switch; **отбойное р.** cut-off relay, clear-out relay, clearing relay, separating relay, step-back relay; supervisory relay, monitoring relay; **откидное р.** automatic drop; **р. отношения (величины)** quotient relay; **р. переключения питания** power-transfer relay; **р. перепада давления** pressure difference switch; **питающее р.** battery supply relay (*tphny.*); **р. повторного включения** reclosing relay; **р. повышения частоты** overfrequency relay; **р. подачи возбуждения** field-application relay; **р. предварительной установки временных интервалов** preset time switch; **р. предзажигания сигналов, предзажигающее р.** approach lighting relay; **р.**

производной некоторой величины по времени rate-of-change relay; **процентно-дифференциальное р.** ratio balance relay; **разделительное р.** bridge cut-off relay (*tgphy.*); ringing-trip relay (*tphny.*); **р. режима решения** compute relay; **резонансное приёмное р.** harmonic receiver (*tgphy.*); **р. с втяжным сердечником, р. с втяжными контактами и с втяжным сердечником** plunger relay, dipper relay; **р. с выдержкой времени** time-delay relay, slow-releasing relay, time-limit relay, time lag(ged) relay; **р. с высоким сопротивлением на звуковой частоте** high-impedance relay; **р. с защёлкой** mechanically locking relay, latched relay; **р. с местным повышением чувствительности** (*or* **с местным огрублением**) biased relay; **р. с механической самоблокировкой** trigger relay; **р. с независимой выдержкой** definite time relay; **р. с преобладанием (чувствительности на одном из контактов)** biased relay (*tgphy.*); **р. с прилипанием** magnetic lock-in relay; **р. с притягивающимся якорем** clapper-type relay; **р. с пружинным опиранием** relay with flexible armature; **р. с регулируемой выдержкой времени** graded time-lag relay; **р. с шарнирным подшипником** Z-relay; **р. с язычковым якорем** clapper-type relay; **р. с якорем на призматической опоре** knife-edge relay, relay with knife-edge armature bearing; **секторное р.** vane-type relay; **селекторное р.** long impulse relay; selector-type relay; **сигнальное р.** transmitting relay, pilot relay, supervisory relay; **р. симметричной составляющей** phase-sequence relay; **р. соединительных линий** connector relay (*tphny.*); **р. составляющей нулевого следования** zero phase sequence relay; **спаренное р.** latch-in-type relay with electric interlock; **страховочное р.** back-up relay; **р. ступенчатого действия** stepping relay; **ступенчатое р.** relay with sequence operation; **р.-счётчик, р. счётчика** counting relay, metering relay (*tphny.*); **р. тлеющего разряда** cold cathode thyratron; трансля-

ционное р. autorelay (*tgphy.*); **удерживающее р.** magnetic lock-in relay; **р. фиксации решения** hold relay

**релеев** *poss. adj.* Rayleigh's
**релеевский** *adj.* (of ) Rayleigh
**Релей** Rayleigh
**релейно-контактный** *adj.* relay-switching, relay-contact
**релейный** *adj.* relay, relay-type, relay-operated
**релуктанц** *m.* reluctance, magnetic resistance
**рельеф** *m.* relief, contour; pattern; configuration; **зарядный р.** charge pattern, picture charge pattern, potential pattern; **отрицательный потенциальный р.** negative stored charge; **потенциальный р.** charge image, electrical image, image pattern, picture charge, potential hill, potential image, picture charge pattern
**рельефный** *adj.* relief, raised, bold, embossed, prominent
**рельс** *m.* rail, track, runway; bar; **р. в рельсовой цепи** block rail; **контактный р.** power supply bar, bus-bar; **рамный р.** stock rail; **третий р.** power supply bar, bus-bar; **третий р. с верхним токосниманием** overrunning third rail; **третий р. с нижним токосниманием** underrunning third rail
**рельсовый** *adj.* rail, track, runway
**рельсогибочный** *adj.* rail-bending
**рельсоправильный** *adj.* rail-straightening
**релэ** *n. indecl.* relay; *cf.* **реле**
**релюктанс, релюктанц** *m.* reluctance, magnetic resistance
**реляксатор** *m.* relaxation oscillator
**реляксационный** *adj.* relaxation
**релятивизм** *m.* relativity
**релятивистский** *adj.* relativistic, relativity
**релятивность** *f.* relativity
**релятивный** *adj.* relative
**ремалой** *m.* remalloy
**реманентность** *f.* remanence, retentivity
**реманентный** *adj.* remanent
**ремённый** *adj.* belt, strap
**ремень** *m.* belt, strap; **с ремнём** belted; **приводной р.** belting, driving belt
**ремесленник** *m.* mechanic, craftsman

**ремесленный** *adj.* industrial, trade

**ремесло** *n.* trade, craft, handicraft

**ремонт** *m.* repair(s), repairing, reconditioning, restoration; upkeep, maintenance; **текущий р.** maintenance

**ремонтирование** *n.* repairing, repair, reconditioning, overhauling

**ремонтировать** *v.* to repair, to refit, to overhaul, to recondition, to fix, to remedy

**ремонтник** *m.* repair man

**ремонтно-механический** *adj.* mechancal-repair

**ремонтный** *adj.* repair

**ремонтопригодность** *f.* serviceability, maintainability, repairability

**ремтрон** *m.* remtron

**рениевый** *adj.* rhenium

**рений** *m.* rhenium, Re

**ренистый** *adj.* rhenium

**рентабельно** *adv.* profitably; it is profitable

**рентабельность** *f.* profitableness, productiveness

**рентабельный** *adj.* profitable, commercial

**рентген** *m.* roentgen, X-ray

**рентгенизация** *f.* X-raying, roentgenization

**рентгенировать** *v.* to X-ray

**рентгенметр** *m.* roentgenometer

**рентгено-** *adv.*, *prefix* roentgen, X-ray

**рентгеноанализ** *m.* radio-examination

**рентгеновский, рентгеновый** *adj.* Roentgen, X-ray

**рентгенограмма** *f.* X-ray photograph, X-ray pattern, radiograph, skiagram, roentgenogram

**рентгенограф** *m.* roentgenograph, X-ray photograph

**рентгенографический** *adj.* radiographic (al)

**рентгенография** *f.* radiography, roentgenography, X-ray radiography

**рентгенодефектоскопия** *f.* radiomateriology, X-ray flow detection

**рентгенодиагностика** *f.* diagnostic roentgenology, X-ray diagnosis

**рентгенокинематография** *f.* roentgen cinematography

**рентгенокристаллография** *f.* X-ray crystallography

**рентгенолог** *m.* radiologist

**рентгенологический** *adj.* roentgenologic, X-ray

**рентгенология** *f.* radiology, roentgenology

**рентгенолюминесценция** *f.* roentgenoluminescence

**рентгенометр** *m.* roentgen meter, radiacmeter

**рентгенопроницаемый** *adj.* radiotransparent

**рентгенопросвечивание** *n.* roentgenoscopy, fluoroscopy

**рентгенопросвечивающий** *adj.* radiolucent

**рентгеносвечение** *n.* radioluminscence

**рентгеноскоп** *m.* roentgenoscope, fluoroscope

**рентгеноскопия** *f.* roentgenoscopy, fluoroscopy, radioscopy, X-ray examination

**рентгеноснимок** *m.* roentgenogram, X-ray photograph

**рентгеноспектральный** *adj.* X-ray spectral

**рентгеноструктурный** *adj.* X-ray diffraction

**рентгенотелевидение** *n.* X-(ray) television

**рентгенотерапия** *f.* roentgenotherapy, X-ray therapy

**рентгенотехник** *m.* X-ray technician, radiographer, roentgen ray technician

**рентгенотехника** *f.* radiology, X-ray technology

**рентгеночувствительность** *f.* radiosensibility

**рентгеночувствительный** *adj.* radiosensitive

**рентген-эквивалент** *m.* roentgen equivalent

**рео-** *prefix* rheo-

**реограф** *m.* rheograph

**реология** *f.* rheology

**реометр** *m.* rheometer, flow meter; **р. капиллярного типа** caplastometer

**Реомюр** Réaumur

**реопирометр** *m.* resistance pyrometer

**реоскоп** *m.* rheoscope, current detector

**реосплав** *m.* high-resistivity alloy

**реостат** *m.* rheostat, resistor, varistor, adjustable resistance, regulating resistance, variable resistor; **винтовой р.** spindle-operated rheostat; **р. возбуждения** field regulator, field rheostat; **р. для зарядки** charging resistor; **р. для шунтирования обмотки**

возбуждения field divertor field rheostat; **жидкостный пусковой р.** liquid starter; **р. накала** filament resistance; **плоский р.** face-plate rheostat, plate-type rheostat; **плоский р. с кольцевым контактным ходом** dial-type rheostat; **пусковой р.** rheostatic starter; **пусковой р. в цепи якоря** rotor starter; **пусковой р. с тремя зажимами** three-point starter; **пусковой р. с четырьмя зажимами** four-point starter; **путевой р.** track resistor; **регулировочный р.** governor; **р. с рукояткой** lever resistance-box; **струнный р.** rheochord

**реостатно-ёмкостный** *adj.* resistance-capacitance

**реостатно-транзисторный** *adj.* resistor-transistor

**реостатный** *adj.* rheostatic, rheostat, resistance

**реострикция** *f.* rheostriction, rheostriction effect, pinch effect

**реотаксиальный** *adj.* rheotaxial

**реотан** *m.* rheotan

**реотановый** *adj.* rheotan

**реотом** *m.* rheotome

**реотрон** *m.* rheotron, betatron

**реохорд** *m.* measuring wire, slide wire, rheochord

**реохордный** *adj.* slide-wire, rheochord

**репер** *m.* reference point, datum point, datum mark, **сборочный р.** location mark, matchmark; aiming spot

**реперный** *adj.* reference, reference point; datum

**репертуар** *m.* repertoire

**реперфоратор** *m.* reperforator, receiving perforator (*tgphy.*)

**репетиционный** *adj.* repeating; rehearsal

**репитер** *m.* repeater; **р. гидролокатора** asdic repeater

**репиторный** *adj.* repeater

**реплика** *f.* replica; remark; **вогнутая р.** negative replica

**репортаж** *m.* reporting

**репрезентивный** *adj.* representative

**репродуктивный** *adj.* reproductive

**репродуктор** *m.* loudspeaker, speaker, reed drive loudspeaker; reproducer; **перфорационный р.** reproducing puncher; **широкополосный коаксиальный р.** coaxial loudspeaker

**репродукция** *f.* reproduction

**репродуцировать** *v.* to reproduce

**репульсионный** *adj.* repulsion

**ресивер** *m.* receiver; storage receiver

**ресорбировать** *v.* to resorb, to reabsorb

**ресорбция** *f.* resorption, reabsorption

**респиратор** *m.* respirator, mouthpiece

**респираторный** *adj.* respiratory

**респирометр** *m.* respirometer

**Рессель** Russell

**рессора** *f.* spring

**рессорный** *adj.* spring

**реставратор** *m.* reconditioner

**реставрация** *f.*, **реставрирование** *n.* restoration, renovation

**реставрированный** *adj.* restored, repaired, renovated

**реставрировать** *v.* to restore, to repair, to renovate

**реституция** *f.* restitution, regeneration

**ретина** *f.* retina

**реторта** *f.* retort

**ретортный** *adj.* retort

**ретранслированный** *adj.* relayed, retransmitted

**ретранслировать** *v.* to retransmit

**ретранслирующий** *adj.* relaying, retransmitting

**ретранслятор** *m.* repeater, translator, two-way repeater

**ретрансляционный** *adj.* relay, relaying, repeating, retransmitting, retransmitter, rebroadcasting

**ретрансляция** *f.* retransmission, (radio) relaying, repeating relaying; **многократная р.** chain relaying

**ретрансмиссия** *f.* retransmission

**ретрансмиттер** *m.* retransmitter

**ретро-** *prefix* retro-

**ретрорефлектор** *m.* retroreflector

**ретрорефлексия** *f.* retroreflection

**ретрофлексия** *f.* retroflexion

**реферат** *m.* abstract; reference; paper

**реферирование** *n.* abstracting

**реферировать** *v.* to review; to report, to read a paper

**рефлекс** *m.* reflex

**рефлексивность** *f.* reflexive

**рефлексивный** *adj.* reflexive, reflected

**рефлексный** *adj.* reflex, reflected

**рефлексия** *f.* reflection, reflex action

**рефлексный** *adj.* reflex; reflected

**рефлективный** *adj.* reflective, reflex

**рефлектирующий** *adj.* reflecting

**рефлектограмма** *f.* reflectogram

**рефлектометр** *m.* reflectometer, reflection-coefficient meter

**рефлектор** *m.* reflector, mirror; reverberator

**рефлекторный** *adj.* reflector; reflex; reflectory

**рефлектоскоп** *m.* reflectoscope

**рефокусировать** *v.* to refocus

**рефрактометр** *m.* refractometer

**рефрактометрический** *adj.* refractometric

**рефрактор** *m.* refractor; refracting telescope

**рефракторность** *f.* refractoriness

**рефракторный** *adj.* refractory, refracting; refractor

**рефракционный** *adj.* refraction

**рефракция** *f.* refraction; **рефракции главной точки** principal-point diopters, principal-point refractive powers

**рефригератор, рефрижератор** *m.* refrigerator; condenser, cooler

**рецептор** *m.* receptor

**реципиент** *m.* container, lead container, receiver; recipient

**рециркулировать** *v.* to recirculate, to recycle

**рециркулятор** *m.* recirculator

**рециркуляция** *f.* recirculation, recycling

**речевой** *adj.* vocal, voice; speech

**речеобразование** *n.* speech production

**речной** *adj.* river, (sub) fluvial

**речь** *f.* speech; language; **зашифрованная р., искажённая р.** inverted speech; **невнятная р.** scrambled speech; **неразборчивая р.** babble

**решаемый** *adj.* solvable

**решать, решить** *v.* to decide, to determine, to settle, to fix, to conclude; to resolve, to solve

**решающий** *adj.* decisive, determinant, determining, determinative, conclusive; solving, resolving, computing

**решение** *n.* solution, solving (*math.*); decision, determination, conclusion, resolution; answer; derivation; **р. методом подбора** trial-and-error solution; **р. неявных функций** implicit solution; **одновременное р. нескольких задач** multishift operation; **р. подбором** solution by inspection; **р. с непрерывной выдачей данных** analogue computing

**решённый** *adj.* determined; solved

**решётка** *f.* lattice; grid, grate, grating, grill; framework; array (*ant.*); aperture plate; rack; cascade; **р. аккумуляторной пластины** plate grid; **антенная р. с излучением вдоль оси** end-fire array, end-on directional array; **дифракционная р., создаваемая сдвиговой волной** shear wave grating; **многорядная р.** mattress array; **многорядная синфазная р.** billboard array, broadside array; **перфорированная распределительная р.** apertured shadow-mask; **пластинчатая р.** sheet grating, sheet grid; **распределительная р.** shadow mask; **ступенчатая дифракционная р.** echelon grating; **«треугольная» р.** gabled array

**решёткообразный** *adj.* lattice-like, lattice

**решётный** *adj.* screen, sieve

**решето** *n.* sieve; coarse screen

**решёточный** *adj.* lattice, screen; network

**решётчатый** *adj.* lattice, latticed, lattice-type; grating, grate, grill, grid; screen; meshed

**решить** *pf. of* **решать**

**рея** *f.* yard; **мачтовая р. для крепления антенны** antenna yard

**ржавление** *n.* corrosion, rusting

**ржавленный** *adj.* corroded, rusted

**ржавость** *f.* rustiness

**ржавчина** *f.* rust

**ржавчиноустойчивый** *adj.* rust-resistant

**РИ** *abbr.* (**регистровый искатель**) register finder, A-digit selector

**Ривз, Ривс** Reeves

**Риггер** Riegger

**ригель** *m.* cross bar, collar beam; **р. ригельного замка** facing point lock plunger

**ригельный** *adj.* cross-bar, collar-beam

**Риги** Righi

**Рид** Reed; Read

**Ридберг** Rydberg

**«Ридикс»** *m.* Readix

**Ридли** Ridley

**Ридль** Riedl

**ризиметр** *m.* rhysimeter

**Рикати** Riccati

**Рике** Rieke

**рикотрон** *m.* rycotron

**Риман** Riemann

**риманов** *poss. adj.* Riemann's

**римлок** *m.* rimlock (base) tube

**римский** *adj.* Roman

**риометр** *m.* riometer

**риотрон** *m.* variable inductance cryogenic device, ryotron

**рир-проекция** *f.* background projection, rear(-screen) projection

**рис.** *abbr.* (**рисунок**) drawing, pattern, picture, design, illustration

**Рис** Riesz

**рисельиконоскоп** *m.* Rieselikonoscope

**риск** *m.* risk, hazard

**рисовать, нарисовать** *v.* to draw, to trace, to design

**риспота** *f.* answer, companion (*mus.*)

**рисунок** *m.* pattern, design, tracery; picture, drawing, figure, sketch, illustration; representation, diagram; **штриховой р.** facsimile picture, recorded copy

**ритм** *m.* beat, rhythm

**ритмический** *adj.* rhythmic(al)

**ритмично** *adv.* rhythmically

**ритмичность** *f.* rhythm, rhythmicality, evenness

**ритмичный** *adj.* rhythmic(al)

**ритрон** *m.* retron

**Ритц** Ritz

**рифлевать, рифлить** *v.* to channel, to groove, to riffle, to rib, to flute, to crimp, to corrugate

**рифление** *n.* corrugation, channeling, grooving, fluting

**рифлёный** *adj.* corrugated, channeled, fluted, grooved, rifled, ribbed; chequered

**рифлить** *pf. of* **рифлевать**

**Рихтер** Richter

**рихтовальный** *adj.* straightening; aiming

**рихтовать** *v.* to prepare, to arrange, to dress; to align; to level, to straighten

**Рич** Reech

**Ричардсон** Richardson

**Риччи** Ricci

**РК** *abbr.* (**радиокомпас**) radio compass

**РЛ** *abbr.* (**радиолиния**) radio link

**рлк** *abbr.* (**радиолокационный**) radar

**РЛС** *abbr.* (**радиолокационная станция**) radar station, radar set

**РМ** *abbr.* (**радиомаяк**) radio beacon

**РМЦ** *abbr.* (**радиометеорологический центр**) radio-meteorological center

**р-н** *abbr.* (**район**) region, district, area, zone

**РНС** *abbr.* 1 (**радионавигационная система**) radio navigation system; 2 (**радионавигационные средства**) radionavigational aids

**РНТ** *abbr.* (**радионавигационная точка**) radio check point, radionavigational point

**робот** *m.* robot

**роботестер** *m.* robotester

**«робурин»** *m.* roburine

**ровно** *adv.* smoothly, uniformly, steadily; exactly, just, equally

**ровность** *f.* flatness, evenness, smoothness; equality; uniformity

**ровный** *adj.* even, flat, plane, level; uniform, steady; equal

**ровнять, сровнять** *v.* to plane, to level, to even, to flatten; to align, to dress

**рог** *m.* horn; **рога разрядника** arcing horns

**рогач** *m.* crotch, butt-prop, deadman

**роговой** *adj.* horny, horn-type, horn, corneous

**рогообразный** *adj.* hornlike, horn-type, horn-shaped; horny

**род** *m.* sort, kind, type, nature; class, system; origin, family; stock, species

**родиевый** *adj.* rhodium

**родий** *m.* rhodium, Rh

**родистый** *adj.* rhodium

**родометр** *m.* rodometer

**родственный** *adj.* allied, related, relative

**рожковый** *adj.* horn-type, pronged

**рожок** *m.* horn, clarion; small horn; prong, catch; socket (of a lamp); **пастуший р.** alpenhorn

**роза** *f.* rose; **р. ветров** compass card

**розетка** *f.* rosette; electrical connector, socket; **р. для включения** connection rose; **осветительная штепсельная р.** light socket; **ответительная р.** branch(ing) box, connector box; **патронная штепсельная р.** lampholder plug, plug adapter; attachment plug; **подвесная штепсельная р.** hanging socket; **потолочная р.** ceiling block, ceiling rose, ceiling button; **приборная штепсельная р.** set (wall) socket; apparatus plug socket; **соединительная р.** wall socket; **соединительная штепсельная р.** socket coupler; **стенная штепсельная р.** wall socket, switch plug; **штепсельная р.** socket, wall socket, plug box, power outlet, socket outlet, plug receptacle; **штепсельная р. для вставки вилки сверху** top-entry socket; **штепсельная р. для вставки вилки спереди** front-entry socket; **штепсельная р. для открытой проводки** surface socket; **штепсельная р. для скрытой проводки** flush socket

**розеточный** *adj.* rosette

**Розинг** Rosing

**Роквелл** Rockwell

**рокот** *m.*, **рокотание** *n.* roar, rumble, murmur

**роксит** *m.* roxite

**ролик** *m.* castor, caster, roller; pulley; reel, roll; roller, cylinder; knob; sheave; porcelain insulator; **ролики на носу кабельного суда** bow sheaves; **наклонные холостые ролики** troughing idlers; **ведомый р.** idler; **ведущий р.** capstan; **р. для натяжки проводов** snatch block; **р. для подъёма кабеля** bow sheave (*tgphy.*); **р. для спуска кабеля** spider sheave (*tgphy.*); **кормовой р. для спуска кабеля** stern spider sheave (*tgphy.*); **круглый р. в обойме** shackle insulator; **р.-наездник** jockey roller; **р. направляющий ленту** paper-feed roll (*tgphy.*); **натяжный р.** iron thimble; **р. нижнего токоснимателя** underruning trolley; **поддерживающий р. (для гибких тяг)** wire carrier (*RR*); **прижимной р.** capstan idler; **р. толкателя** cam follower roller; **упорный р.** tappet roller; **холостой р.** idler

**роликовый** *adj.* roller, roll, drum-type

**Ролль** Rolle

**рольганг** *m.* table, mill table; roller conveyer; **приводящий р.** approach table

**ромб** *m.* rhomb, rhombus; diamond

**ромбический** *adj.* rhombic

**ромбовидный** *adj.* diamond-shaped, rhomboid, rhombiform

**ромбоид** *m.* rhomboid

**ромбоидальный** *adj.* rhomboid(al)

**ромбообразный** *adj.* diamond-shaped; rhomboid, rhombiform

**ромбоэдр** *m.* rhombohedron

**ромбоэдрический** *adj.* rhombohedral

**« рометалл »** *m.* rhometal

**Ронки** Ronchi

**росомер** *m.* drosometer

**рост** *m.* growth, increase, develop-
ment; height, size; **ростом** in height; **во весь р.** at full length

**рот** *m.* mouth

**рота** *f.* company; **кинофотосъёмочная р.** signal photographic company; **портовая р. связи** signal port service company; **р. связи** signal company; **р. склада связи** signal depot company

**ротаметр** *m.* rotameter, variable area flowmeter; **р. на обводной линии** by-pass rotameter; **р. с коническим осевым стержнем** tapered center-column rotameter

**ротари** *n. indecl.* rotary system; **р.-система** rotary dial system (*tphny.*)

**ротативный** *adj.* rotating, rotative, rotary

**ротатор** *m.* rotator

**ротационно-барабанный** *adj.* rotary-drum

**ротационно-молекулярный** *adj.* rotary-molecular

**ротационно-шунтовый** *adj.* rotary shunt-type

**ротационный** *adj.* rotation(al), rotary, rotatory

**ротация** *f.* rotation

**Роте** Rothe

**ротор** *m.* rotor, cursor; **р. вектора** curl of a vector; **высокостержневой р.** asynchronous motor with special rotor for starting current reduction; **глубокопазный р.** extended-groove-type rotor, eddy-current rotor; **р. с вытеснением тока** (current) throttling-type rotor; extended-groove-type rotor; **р. с двойной беличьей клеткой, р. с двойным пазом** double cage armature, double cage rotor, double-deck rotor, double-wound rotor; **р. с явновыраженными полюсами** pole wheel; **ступенчатый короткозамкнутый р.** skewed-winding-type rotor; **р. электродвигателя запускающего локатор** pulse-initiating rotor

**роторный** *adj.* rotor

**ротоскоп** *m.* rotoscope

**рототерапия** *f.* (radiation) rotation therapy

**« рототрол »** *m.* rototrol

**Роу** Rowe

**Роуланд, Роулэнд** Rowland

**рошеллев** *poss. adj.* Rochelle's

**рояль** *m.* piano, grand piano

**рояльный** *adj.* piano

**РП** *abbr.* 1 (**радиопеленг**) radio bearing; 2 (**радиопрожектор**) searchlight control radar; 3 (**распределительный пункт**) distribution point

**РПД** *abbr.* (**радиопротиводействие**) radio countermeasure, electronic countermeasure

**РПК** *abbr.* (**радиополукомпас**) fixed loop radio compass

**РПС** *abbr.* (**радиопеленгаторная станция**) radio direction-finding station

**РРЛ** *abbr.* (**радиорелейная линия**) radio relay line

**РРС** *abbr.* (**радиорелейная станция**) radio relay station

**РС** *abbr.* 1 (**радиосвязь**) radio communication; 2 (**реактивный снаряд**) rocket, rocket missile; 3 (**регулятор смеси**) mixture control; 4 (**ромбическая согнутая**) bent rhombic antenna

**РСЛ** *abbr.* (**реле соединительных линий**) connector relay

**РСТ** *abbr.* (**радиостанция**) radio station, radio set

**РТПД** *abbr.* (**радиотехническое противодействие**) radio countermeasure

**РТС** *abbr.* (**ручная телефонная станция**) manual telephone office

**ртуте-стойкий** *adj.* mercury-proof

**ртутистый** *adj.* mercurous, mercury

**ртутник** *m.* mercury-arc rectifier; **баковый р.** pool tube, pool tank; **металлический р.** metal-tank mercury-arc rectifier

**ртутно-дуговый** *adj.* mercury-discharge

**ртутно-индиевый** *adj.* indium-mercuric(-oxide)

**ртутно-кварциевый** *adj.* quartz-mercury

**ртутно-плунжерный** *adj.* mercury-plunger

**ртутно-таллиевый** *adj.* mercury-thallium

**ртутно-электролитический** *adj.* mercury-electrolyte

**ртутный** *adj.* mercury, mercuric

**ртуть** *f.* mercury, Hg

**РУ** *abbr.* 1 (**районный узел**) "raion" center, toll office (*tphny.*); 2 (**регулировка усиления**) gain control; 3 (**рентгеновская установка**) X-ray unit; 4 (**ручное управление**) manual control

**рубанок** *m.* plane (tool); **фигурный р.** moulding plane, fillet plane

**рубашка** *f.* shirt; jacket, housing; casing, lining; **р. цилиндра** case, casing, jacket, shell

**рубидиевый** *adj.* rubidium

**рубидий** *m.* rubidium, Rb

**рубильник** *m.* knife switch, blade switch, closing switch, lever switch, cut-off switch; plug-in strip; **двухножевой р.** tandem (or double throw) knife-switch; **р. на уравнительном проводе** equalizer switch

**рубин** *m.* ruby

**рубинный, рубиновый** *adj.* ruby

**рубка** *f.* chart house; house, building, construction; **акустическая р.** acoustic sound room; **гидроакустическая р.** antisubmarine cabinet; **кормовая р.** round house; **штурманская р.** pilot house

**рубрика** *f.* head(ing); rubric, column

**рубящий** *adj.* cutting; knife, blade

**рудиментарный** *adj.* rudimentary

**рудник** *m.* mine, pit

**рудничный** *adj.* mine, mining, mine-type

**Рузерфорд** Rutherford

**рузерфорд** *m.* rutherford

**рукав** *m.* tube, tubing, hose, flexible pipe; sleeve

**руководитель** *m.* director, manager, supervisor, leader

**руководительный** *adj.* guiding, guide, leading

**руководить** *v.* to manage, to run, to guide, to lead, to conduct; to rule, to direct

**руководство** *n.* management, guidance, lead, leadership, supervision, direction; handbook

**руководящий** *adj.* control, controlling, guide, guiding, leading, master

**рукоятка** *f.* handle, grip, shaft, lever, stem, arm, hand lever; crank, crank handle; **р. аварийного горочного сигнала** hump signal emergency lever (*RR*); **р. междупостового замыкания** check lock lever (*RR*); переводная **р.** control lever; **перекидная р.** capstan; **простая р. настройки** direct selector; **р. с накаткой** fluted knob; **р. стрелочных указателей** switch signal lever (*RR*); **р. указателей путей** trimmer signal lever (*RR*); **р. управления стрелоч-**

**ным приводом** switch machine lever (*RR*)

**рукоять** *f.* handle, crank

**рулевой** *adj.* rudder, steering

**рулёжный** *adj.* taxiing

**рулетка** *f.* tape measure, measuring tape; reel

**рулон** *m.* roll, roller, reel

**рулонный, рулочный** *adj.* page; tubular

**руль** *m.* (steering) wheel, rudder, helm, control

**румб** *m.* rhumb; bearing, line (of compass)

**румбовый** *adj.* rhumb; bearing, point

**румбатрон** *m.* rhumbatron

**Румкорф** Ruhmkorff

**румкорфовый** *adj.* Ruhmkorff's

**Румфорд** Rumford

**Рунге** Runge

**рупор** *m.* megaphone, mouthpiece, speaking-trumpet, loudspeaker; acoustic inlet (of a microphone); horn, horn radiator; funnel, flare; **вывернутый р.** rear-feed horn; **лабиринтный р.** folded horn; **мегафонный р.** mouthpiece; **р. с изогнутой осью** folded horn; **р. с отражателем** reflex horn; **сдвоенный р.** dual-aperture horn

**рупорно-зеркальный** *adj.* horn-reflector

**рупорно-линзовый** *adj.* horn-lens, horn-and-lens

**рупорнообразный** *adj.* flared

**рупорный** *adj.* horn, horn-type; funnel-shaped, cone; mouthpiece, speaker

**рупорообразный** *adj.* flared

**русло** *n.* channel, race, waterway

**Руссо** Rousseau

**Рут** Routh; Ruth

**рутениевый** *adj.* ruthenium, ruthenic

**рутений** *m.* ruthenium, Ru

**рутенистый** *adj.* ruthenium, ruthenious, ruthenous

**рутер** *m.* rooter

**рутил** *m.* rutile

**рутиловый** *adj.* rutile

**ручка** *f.* arm, hand; knob, handle, shaft, grip, holder; crank, crank handle; **р. управления с фиксированным числом положений** selector knob; **шаровая р.** knob

**ручной** *adj.* hand, arm; manual, hand-operated, manually-operated; portable

**ручьевидный** *adj.* stream-like

**РФС** *abbr.* (**радиально-фрезерный станок**) radial milling machine

**РЦ** *abbr.* (**районный центр**) "raion" central office (*tphny.*)

**рыбосчётчик** *m.* ichthyometer

**рывок** *m.* jerk, kick, shock, dash; **рывки** jerking; hunt effect, hunting; **рывки магнетрона** pulling of magnetron; **р. частоты** scintillation, momentary swing of frequency

**рыскание, рысканье** *n.* hunting, hunt, sounding; cycling

**рычаг** *m.* arm, crank, lever; crowbar; control, joystick; linkage; **р. для ручного перевода (стрелки)** dual selector lever (*RR*); **коленчатый р.** crank, toggle, toggle lever; **направляющий р.** switch lever; **р. переключения** contact lever; **приводной р.** rocket shaft arm; **угловой р.** radial arm; **р. управления с возвратно-поступательным движением** push-pull control; **четырёхплечий р.** cross lever

**рычажно-роликовый** *adj.* roller-lever

**рычажный** *adj.* lever, lever-type, lever-arm; needle-arm

**рычажок** *m.* lever, small lever; blade; rod

**рэдат** *m.* Radat

**« Рэйдист »** *m.* Raydist

**« Рэйсент »** *m.* Rayescent

**« Рэйсер »** *m.* Racer

**рэйтметр** *m.* ratemeter

**Рэйтон** Raytheon

**рэл** *m.* rel

**рэлеевский** *adj.* Rayleigh's

**Рэлей** Rayleigh

**рэлсин** *m.* ralsin

**Рэнд** Rand

**Рэнкин** Rankine

**Рэттен** Wratten

**Рюдберг** Rydberg

**рябь** *f.* ripple(s), ripple marks, rippled surface; **звуковая р.** flutter effect; **р. на фотографии** jitters in the received picture

**ряд** *m.* series, progression, line, train, row, bank; range, order, sequence, succession, continuum; rank; level; layer; set; **бесконечно-убывающий р.** harmonic progression (*math.*); **верхний р.** series containing only positive terms (*math.*); **р. выключателей** bank of keys; **р. напряжений** electric series, electrochemical

series, electromotive series; **р. откры-
тых ячеек** repeater bay, line of
repeater bays; **последовательный
р. (чисел)** series (*math.*); **р. стативов**
row of racks, train of racks, bay of

racks (*tphny.*); **частотный р.** pitch
continuum (*acoust.*); **р. штативов**
line of racks

**рядовой** *adj.* ordinary, common; layer;
consecutive, serial; in rows

# C

с 1 *prep. genit.* from, off, down from; on, over; since; for, of; *prep. acc.* as, about, approximately; for; 2 *prep. instr.* with, by, by means of, on; (often omitted in translation: **генератор с выступающими полюсами** salient-pole alternator [*lit.* generator with salient poles])

с *abbr.* 1 (**северный**) north, northern; 2 (**секунда**) second; 3 (**скорость света**) velocity of light

**Сабуро** Sabouraud

**Савар** Savart

**сагиттальный** *adj.* sagittal

**садка** *f.* charge; load

**сажа** *f.* soot; carbon black

**сажать, посадить** *v.* to seat; to buck; to set, to place, to put; **с. напряжение** to kill the voltage, to buck the voltage

**сайпак** *m.* cypak

**саксгорн** *m.* saxhorn

**саксофон** *m.* saxophone

**салазки** *pl.* carriage, slide, sliding block; slide rails, skids

**салинометр** *m.* salinometer

**сальник** *m.* stuffing box; gland, packing gland; seal, gasket; **с. с кожаной набивкой** leather stuffing box

**самарий** *m.* samarium, Sa

**само-** *prefix* self, auto-, automatic, spontaneous

**самоактивируемый** *adj.* self-activated

**самобалансирующийся** *adj.* self-balancing

**самоблокировка** *f.* self-holding, holding

**самоблокирующий(ся)** *adj.* interlocking, self-locking

**самовводящийся** *adj.* self-loading

**самовентилируемый** *adj.* self-ventilated, self-forced ventilated

**самовентиляция** *f.* self-ventilation

**самовзаимность** *f.* self-reciprocity

**самовключатель** *m.* recloser, circuit recloser

**самовключающий** *adj.* reclosing

**самовключение** *n.* reclosure, reclosing

**самовнушение** *n.* self-suggestion, autosuggestion

**самовождение** *n.* self-piloting

**самовозбуждающийся** *adj.* self-excited; self-driven, self-oscillating, self-oscillatory, self-energizing, self-activated, self-running, self-sustaining

**самовозбуждение** *n.* self-excitation, autoexcitation, direct-driven; self-oscillation; feedback; self-triggering; regeneration caused by reactive elements; **с самовозбуждением** self-excited, self-driven; **с. при отрицательной обратной связи** instability of amplifiers employing feedback

**самовозбуждённый** *adj.* self-excited

**самовозврат** *m.* self-restoring, self-reset; self-recovery

**самовозгорание** *n.* spontaneous ignition

**самовозникающий** *adj.* spontaneous

**самовоспламенение** *n.* spontaneous ignition, self-ignition

**самовоспламеняемость** *f.* spontaneous combustibility

**самовоспламеняющийся** *adj.* auto-igniting

**самовоспроизведение** *n.* self-reproduction, self-production

**самовосстанавливаемость** *f.* self-restorability

**самовосстанавливающийся** *adj.* self-restoring, self-healing

**самовосстановление** *n.* self-recovery

**самовыключатель** *m.* recloser

**самовыпрямляющий** *adj.* self-rectifying, self-righting

**самовыравнивание** *n.* self-regulation, self-recovery; inherent regulation

**самовыравнивающийся** *adj.* self-regulating, self-aligning

**самогасящий(ся)** *adj.* self-quenched, self-quenching

**самогашение** *n.* self-quenching

**самогерметизующийся** *adj.* self-sealing

**самодвижный** *adj.* power-driven

**самодвижущийся** *adj.* self-propelled, self-powered, automatic

**самодействующий** *adj.* self-acting, self-action, automatic
**самодельный** *adj.* home-made
**самодистрибутивность** *f.* self-distributiveness
**самодиффузия** *f.* self-diffusion
**самодополняющийся** *adj.* self-complementary
**самозавод** *m.* self-winding
**самозаводной** *adj.* self-winding
**самозаводящийся** *adj.* self-starting; self-winding
**самозагружающийся** *adj.* self-loading
**самозажигание** *n.* self-ignition
**самозажигающийся** *adj.* self-igniting
**самозакрепление** *n.* self-locking
**самозакрепляющийся** *adj.* self-locking
**самозакрывающийся, самозамыкающий** *adj.* self-closing, self-locking
**самозапирание** *n.* squegging; automatic closing, automatic locking; **с. (генератора)** squegging (of an oscillator)
**самозаписывающий** *adj.* recording, self-recording, graphic
**самозапуск** *m.* self-triggering, self-starting; self-running
**самозапускающийся** *adj.* self-starting
**самозарядный** *adj.* self-loading, autoloading
**самозатягивающийся** *adj.* self-sealing
**самозачищающийся** *adj.* self-cleaning
**самозащищённый** *adj.* self-protected
**самоиндукция** *f.* self-inductance, self-induction; electrodynamic capacity
**самоиндуцированный** *adj.* self-induced, self-stimulated
**самоисправляющийся** *adj.* self-healing
**самокомпенсированный** *adj.* self-compensated, autocompensated
**самокомпенсирующий** *adj.* self-compensating
**самоконтролирующийся** *adj.* self-supervisory, self-checking
**самоконтроль** *m.* self-test, automatic check; **с самоконтролем** self-monitoring
**самокорректирующийся** *adj.* self-correcting
**самолёт** *m.* aircraft, airplane
**самолётный** *adj.* airplane, aircraft; airborne
**самолётовождение** *n.* pilotage, air navigation
**самоликвидатор** *m.* safety exploder, self-destructor

**самомодуляция** *f.* self-modulation
**самонаведение** *n.* homing, self-induction, passive homing; **пассивное с. по радиоизлучению цели** RF passive homing guidance; **с. по тепловому излучению цели** heat passive homing guidance
**самонаведённый** *adj.* self-induced
**самонаводящий** *adj.* self-directing
**самонаводящийся** *adj.* self-guided, self-homing, homing
**самонакал** *m.* direct heating, self-heating
**самонакаливающийся** *adj.* self-heating
**самонастраивающийся** *adj.* self-organizing, self-aligning, self-tuning
**самонасыщающийся** *adj.* self-saturating, self-saturated
**самонасыщение** *n.* self-saturation
**самообращение** *n.* self-reversal; self-absorption
**самообучение** *n.* self-instruction, self-teaching
**самоорганизующийся** *adj.* self-organizing
**самоориентирующийся** *adj.* self-orienting
**самоотжиг** *m.* self-annealing
**самоотключение** *n.* automatic opening
**самоохлаждающийся** *adj.* self-cooled
**самоохлаждение** *n.* self-cooling; **закрытый с самоохлаждением** enclosed self-cooled
**самооценка** *f.* self-appraisal, self-scoring
**самоочищающийся** *adj.* self-cleaning, self-wiping
**самописец** *m.* self-recording unit, recorder, recording instrument, writer; curve-drawing instrument, contour plotter; data plotter; logger; **аналоговый с. функций** coordinate plotting table; **с. времени** calculagraph; **с. диаграмм направленности** pattern recorder; **координатный с.** plotter; **ленточный с.** strip chart instrument; **многократный с. без оси времени** multiple X-Y recorder; **с. на регулируемый диапазон (напряжений)** adjustable-span recorder; **непрерывный с.** curve-drawing recorder; **нерелейный с.** direct-acting recording instrument; **с. неустановившихся процессов** transient recorder; **релейный с. электрических величин** Everett-Edgcumbe recorder; **с. с**

одним пером varioplotter; **точечный с.** dot-dash recorder, intermittent contact recorder; **с. уровней** level recorder, automatic level recorder

**самопишущий** *adj.* (self-)recording, registering, self-registering, autographic, chart recording, graphic-recording

**самопоглощение** *n.* autoabsorption, self-absorption

**самоподавление** *n.* self-cancellation

**самоподача** *f.* self-feed, self-feeding

**самоподающий** *adj.* self-feeding

**самоподдержание** *n.* self-sustaining

**самопознание** *n.* self-knowledge

**самопрерывание** *n.* squegging

**самопрерывающий** *adj.* self-interrupting

**самопрерывающийся** *adj.* self-squegging

**самоприпаивающийся** *adj.* self-soldering

**самоприспосабливающийся** *adj.* self-adaptive, self-adapting

**самопроверка** *f.* self-control, self-checking, self-verifying

**самопроверяющийся** *adj.* self-checking, self-verifying

**самопрограммирующийся** *adj.* self-programming

**самопрогрев** *m.* self-heating

**самопроизвольно** *adv.* spontaneously

**самопроизвольность** *f.* spontaneity

**самопроизвольный** *adj.* spontaneous, self-arbitrary

**самопуск** *m.* self-starting; self-starter; **с самопуском** self-running

**самопускающийся** *adj.* self-starting

**саморазгружающийся** *adj.* self-dumping, self-discharging

**саморазмагничивание** *n.* self-demagnetizing, self-demagnetization

**саморазмагничивающий(ся)** *adj.* self-demagnetizing

**саморазмыкание** *n.* automatic opening

**саморазогрев** *m.* self-heating

**саморазогревный** *adj.* self-heating

**саморазряд** *m.* self-discharge, local action

**саморазряжающийся** *adj.* self-discharging, running down

**саморазряжение** *n.* self-discharge

**саморазъёмный** *adj.* self-release

**самораскачивание** *n.* cumulative hunting

**самораспространяющийся** *adj.* self-propagating

**саморегистрирующий** *adj.* self-recording, self-registering

**саморегулирование** *n.* self-regulation, regulation, inherent regulation, inherent stability, self-adjustment, self-alignment

**саморегулируемость** *f.* (self-)adaptability

**саморегулируемый** *adj.* self-controlled

**саморегулирующийся** *adj.* self-adjusting, self-correcting, self-regulating, adaptive

**саморегуляция** *f.* self-control

**саморемонтирующийся** *adj.* self-repairing

**самосветный, самосветящийся** *adj.* self-luminous, self-luminescent

**самосин** *m.* autosyn

**самосинхронизация** *f.* self-synchronization, self-locking; self-timing

**самосинхронизированный** *adj.* self-synchronous, autosynchronous, self-synchronized

**самосинхронизирующий(ся)** *adj.* self-synchronizing, self-synchronous; self-pulsing; self-locked

**самосинхронный** *adj.* self-synchronizing

**самосклеивающий** *adj.* self-adhering

**самослышимость** *f.* sidetone (*tphny.*)

**самосмазывающий(ся)** *adj.* self-oiling, self-lubricating

**самосмещение** *n.* self-bias, automatic bias

**самосовершенствование** *n.* self-improvement

**самосовместный** *adj.* self-consistent

**самосопряжённый** *adj.* self-conjugate; self-adjoint (equation)

**самоспадание** *n.* self-decay

**самоспекающийся** *adj.* self-baking; hardened

**самостановление** *n.* self-healing

**самостоятельно** *adv.* independently

**самостоятельность** *f.* independence

**самостоятельный** *adj.* independent, autonomous, self-contained, self-maintained; self-sustained; self-consistent

**самостягивание** *n.* self-pinching, pinch, pinching

**самотёк** *m.* drift; gravity feed; gravity flow, flow

**самотёком** *adv.* by gravity

**самоторможение** *n.* self-retardation, self-braking, self-locking, self-catching

**самотормозящийся** *adj.* self-braking, self-stopping; self-locking

**самоуничтожающийся** *adj.* self-destroying

**самоуплотняющийся** *adj.* self-sealing

**самоуправление** *n.* automatic control

**самоуправляемый** *adj.* self-guided, homing

**самоуравновешивающийся** *adj.* self-balancing

**самоустанавливание** *n.* self-adjustment

**самоустанавливающийся** *adj.* self-adjusting, self-aligning, self-fixing

**самофазировка** *f.* autophasing

**самофокусировка** *f.* self-focusing, self-trapping; filamenting

**самофокусирующий(ся)** *adj.* self-focusing

**самоход** *m.* motor vehicle, motor car; shunt running; creeping, power feed, self-feed; self-acting

**самоходный** *adj.* self-propelled, self-powered, automotive, power-driven

**самохронирующий** *adj.* self-pulsed

**самоцентрирующийся** *adj.* self-centering, self-aligning

**самоэкранирование** *n.* self-shielding, extinction

**санатрон** *m.* sanatron

**санафант** *m.* sanaphant

**сандвич** *m.* sandwich

**сани** *pl.* slide, carriage; sledge, sleigh

**санти-** *prefix* centi-

**сантибел** *m.* centibel (*acoust.*)

**сантиграмм** *m.* centigram

**сантиметр** *m.* centimeter

**сантиметровый** *adj.* centimeter, centimetric; microwave

**сантиоктава** *f.* centioctave

**сантипауз** *m.* centipoise

**сапфир** *m.* sapphire

**сапфировый** *adj.* sapphire

**саронг** *m.* sarong

**Саррюс** Sarrus

**сателлит** *m.* satellite

**сателлитный, сателлитовый** *adj.* satellite

**Сатче** Satche

**сахариметр** *m.* glucometer, saccharometer

**сб** *abbr.* (**стильб**) stilb

**сбалансирование** *n.* neutralization

**сбалансированность** *f.* balanced state

**сбалансированный** *adj.* balanced, neutralized, compensated

**сбалансировать** *pf. of* **балансировать**

**сбегающий** *adj.* leaving, running-off; trailing

**сбивание** *n.* churning; **с. пеленга** meaconing

**сбивать, сбить** *v.* to knock off, to knock down, to beat down; to throw down; to upset; to nail together, to put together

**сбитый** *adj.* biased; out of alignment, out of position

**сбить** *pf. of* **сбивать**

**сближать, сблизить** *v.* to approach, to converge, to bring together, to bring near; to bind, to connect; to compare

**сближение** *n.* approaching, closing, drawing together; proximity

**сближенный** *adj.* adjacent, contiguous; drawn together

**сблизить** *pf. of* **сближать**

**сблокированный** *adj.* interlocked, interlocking, interlinked, ganged

**сбой** *m.* failure, short duration failure, malfunction; reduction, lowering; **с. на контрольной сумме** check sum failure

**сбоку** *adv.* from one side, on the side, at the side (of)

**сбор** *m.* gathering, assemblage, assembly, collection, collecting; acquisition; toll, fee

**сборка** *f.* assembly, assembling, assemblage, mounting, putting together, installation, fitting, setting (up), erection, erecting, line-up, building up; **шинная с.** busduct, busway

**сборник** *m.* collection; stacker, hopper; accumulator, collector, receiver, sump, storage tank

**сборный** *adj.* collapsible; combined, composite; collecting; accumulative, aggregate; heterogeneous, miscellaneous

**сборочный** *adj.* assembly, assembling, gathering; erecting

**сборщик** *m.* adjuster; erector, fitter, assembler, mounter; collector

**сбрасываемый** *adj.* discarded, dropped, thrown (down), thrown out

**сбрасывание** *n.* drop, dropping (test); cutting off; through clearing (*tphny.*);

clearance; throwing-off; dropping, disposal, dumping, discarding

**сбрасыватель** *m.* tripper; pull-out; knock-off

**сбрасывать, сбросить** *v.* to drop, to throw down, to throw off, to dump, to discard, to dispose; to reset, to unset; to discharge; to clear

**сбрасывающий** *adj.* throwing down, dumping

**сброс** *m.* reset, resetting, unset; drop, dropping, shedding; trip-out; **с. регистра циклов** cycle reset

**сбросить** *pf. of* **сбрасывать**

**сброшенный** *adj.* thrown down, thrown off, dumped, discarded

**св** *abbr.* (**свеча**) candle

**СВ** *abbr.* 1 (**средневолновый**) medium wave; 2 (**средние волны**) medium waves

**свайка** *f.* spud; pole; **с. для сплетания нитей** marline spike

**Сван** Swan

**сваренный** *adj.* welded, fused

**свариваемость** *f.* weldability

**свариваемый** *adj.* weldable

**сваривание** *n.* welding

**сваривать, сварить** *v.* to weld, to fuse, to braze, to hard solder

**сваривающийся** *adj.* weldable

**сварить** *pf. of* **сваривать**

**сварка** *f.* welding, soldering, brazing; weld; **с. внахлестку** lap welding; **с. вполунахлёст** scarf welding; **с. впритык, с. встык** butt-welded joint; **с. встык с гребнем** upset welding; **контактная стыковая с. оплавлением** butt arc welding; **косая с.** scarf welding; **с. наплавлением** built-up welding; **с. оплавлением** flash welding; **с. оплавлением с осадкой** flash-upset welding; **роликовая с.** continuous welding; **с. с прихватыванием кромок** tack welding; **точечная с. с навариванием выступов** projection welding

**сварной** *adj.* welded, brazed, forged

**сварочный** *adj.* welding, weld

**сварщик** *m.* welder

**свая** *f.* pile, post, peg, stake, pole

**сведение** *n.* information, intelligence, knowledge; converging (process), convergence; reduction; **с. в таблицу** tabulation; **с. к арифметическим операциям** reducing (*math.*); **с. к нулю** nullification

**сведённый** *adj.* reduced; settled; **с. к нулю** nullified

**сверкание** *n.* sparkling, glitter, flash, flashing, twinkling; glare, flare, brilliancy

**сверкать, сверкнуть** *v.* to sparkle, to flash, to light up, to twinkle, to glitter; to glare; to gleam, to scintillate

**сверкающий** *adj.* sparkling, flashing, glittering, bright

**сверление** *n.* drilling, boring, piercing

**сверлильник** *m.* borer, driller

**сверлильный** *adj.* boring, drilling

**сверлильщик** *m.* borer, driller

**сверлить, просверлить** *v.* to bore, to drill, to perforate, to pierce; **с. начерно** to rough-drill; **с. начисто** to rebore, to widen

**сверло** *n.* drill bit, drill, borer, perforator, auger, gimlet; **винтовое с.** gimlet; **длинное перевое с.** auger; **с. для дюбеля** dowel borer; **с. для накладок** fish borer; **с. по дереву** wood boring drill, bit; **с. с крестообразным сечением** cross-mouthed chisel; **спиральное с.** twist drill; **screw tap; brace drill; чистовое с.** finishing bit

**свёрнутость** *f.* convolution, winding; twist

**свёрнутый** *adj.* convolute, folded, rolled; wrapped

**свернуть** *pf. of* **свёртывать**

**свёртка** *f.* convolution (*math.*)

**свёрточный** *adj.* convolutional

**свёртывание** *n.* convolution (*math.*); folding; rolling, coiling; «**с.**» **изображения** picture roll-over

**свёртывать, свернуть** *v.* to coil; to roll, to roll up, to wrap up; to turn (aside); to twist, to wrench; to envelope

**сверх** *prep. genit.* above, beyond, over, besides; in addition to

**сверх-** *prefix* super-, over-, hyper-, ultra-

**сверхбесшумный** *adj.* supersilent

**сверхбыстродействующий, сверхбыстроходный, сверхбыстрый** *adj.* superfast, ultrafast, superhighspeed

**сверхвысокий** *adj.* ultra-high, very high, superhigh, extremely high

**сверхвысоковольтный** *adj.* supervoltage

**сверхвысокочастотный** *adj.* super-high frequency, microwave

**сверхгенеративный** *adj.* superregenerative

**сверхгенерация** *f.* superregeneration

**сверхгруппа** *f.* supergroup

**сверхдальний** *adj.* long-distance

**сверхдальнобойный** *adj.* super-range

**сверхзатухание** *n.* overdamping

**сверхзвуковой** *adj.* superaudible, supersonic, hyperacoustic, ultrasonic, supertonic

**сверхзвукозапись** *f.* ultrasonography

**сверхзвукоскопия** *f.* ultrasonoscopy

**сверхизлучающий** *adj.* superradiant

**сверхизлучение** *n.* superradiation, superradiant radiation

**сверхкомпактный** *adj.* supercompact

**сверхкомплектный** *adj.* supernumerary

**сверхкритический** *adj.* supercritical

**сверхкритичность** *f.* supercriticality

**сверхлегирование** *n.* overdoping

**сверхлёгкий** *adj.* extra light, superlight

**сверхмашина** *f.* superengine

**сверхмикроскоп** *m.* electron microscope, super microscope

**сверхминиатюризация** *f.* subminiaturization

**сверхминиатюрный** *adj.* subminiature, microminiature

**сверхмодуляция** *f.* overmodulation

**сверхмощный** *adj.* superpower; supersound

**сверхнагрузка** *f.* surplus-load

**сверхнаправленный** *adj.* ultradirectional, superdirective

**сверхнасыщение** *n.* supersaturation

**сверхнизкий** *adj.* ultralow

**сверхникель** *m.* supernickel

**сверхобменный** *adj.* superexchange

**сверхпереходный** *adj.* subtransient

**сверхплоскость** *f.* hyperplane (*math.*)

**сверхплотный** *adj.* overdense

**сверхпреломление** *n.* superrefraction

**сверхпроводимость** *f.* superconductivity

**сверхпроводник** *m.* superconductor

**сверхпроводниковый** *adj.* superconductive

**сверхпроводящий** *adj.* superconducting

**сверхрадиолокатор** *m.* super radar

**сверхразмерный** *adj.* overdimensional

**сверхраскачка** *f.* overdrive, overswing

**сверхрасчётный** *adj.* overrating

**сверхрегенеративный** *adj.* superregenerative

**сверхрегенератор** *m.* superregenerator, Armstrong circuit

**сверхрегенерация** *f.* superregeneration

**сверхрезкий** *adj.* hyperabrupt

**сверхрезонанс** *m.* overresonance

**сверхрефрактивный** *adj.* superrefractive

**сверхрефракция** *f.* superrefraction; spill-over

**сверхсильный** *adj.* superpower; megagauss

**сверхсинхронный** *adj.* supersync

**сверхсинхротрон** *m.* supersynchrotron

**сверхскоростной** *adj.* superfast, superspeed

**сверхсогласование** *n.* matching for optimum power-transfer

**сверхстандартный** *adj.* superstandard

**сверхсходимость** *f.* dynamic focusing, overconvergence

**сверхсчётный** *adj.* odd

**сверхтермометр** *m.* hyperthermometer

**сверхток** *m.* excess current, overcurrent

**сверхтонкий** *adj.* ultrathin, microthin, hyperfine

**сверхточный** *adj.* overcurrent; ultraprecise

**сверхтяжёлый** *adj.* superheavy

**сверху** *adv.* from top, from above; on top, on, above, over, upon

**сверхурочный** *adj.* overtime

**сверхусиление** *n.* overamplification

**сверхфильтр** *m.* ultrafilter

**сверхчистый** *adj.* ultra-pure, high-purity

**сверхчувствительный** *adj.* supersensitive, ultrasensitive

**сверхширокополосный** *adj.* ultrabandwidth

**сверхэлектризация** *f.* hyperelectrification

**сверхэффективность** *f.* superefficiency

**свес** *m.* overhang, projection, extension

**свести** *pf. of* **сводить**

**свет** *m.* light; **ближний с.** passing light; **дальний с.** headlight, driving light; **заливающий с.** floodlight

**светило** *n.* celestial body, star, luminary

**светильник** *m.* lamp; luminaire, light fixture, lighting fitting, illuminant, illuminator; **с. для кодового огня** flash lamp

**светильный** *adj.* illuminating

**светимость** *f.* luminous density, luminance, luminosity; radiance; transmission (of spectrometer)

**светить, посветить** *v.* to light, to give light; to glow, to shine, to gleam, to scintillate; to glitter, to glisten

**светиться** *v.* to glow, to shine, to gleam, to glisten, to sparkle

**светло-лиловый** *adj.* mauve

**светлота** *f.* luminosity, subjective brightness; lightness, luminance

**светлотный** *adj.* luminous; whiteness

**светлый** *adj.* luminous, light, bright; clear, lucid

**светность** *f.* high light; luminous emittance, luminous emissivity

**свето-** *prefix* light, photo-

**светоблик** *m.* light pattern

**световод** *m.* optical path, light guide; beam (mode) waveguide

**световой** *adj.* light, luminous

**световыход** *m.* light output

**светоделитель** *m.* dichroic

**светоделительный** *adj.* light-dividing, dichroic

**светоизлучающий** *adj.* light-emitting

**светоизлучение, светоиспускание** *n.* radiation of light

**светоиспускающий** *adj.* luminous, light-emitting

**светоклапанный** *adj.* light-valve

**светокопия** *f.* phototype

**светокопировальный** *adj.* light-printing, blueprinting; tracing

**светокопирование** *n.* blueprinting, photostating

**светомаскированный** *adj.* blackout

**светомаскировка** *f.* blackout, dimout

**светомаяк** *m.* light beacon

**светомер** *m.* photometer

**светомерный** *adj.* photometric

**светомодулятор** *m.* light modulator, light cell

**светонепроницаемый** *adj.* light-tight, light-proof; opaque

**светоотдача** *f.* light efficiency, lumen efficiency, luminous efficiency, conversion efficiency, light output; **с. люминесценции** stimulation quantum efficiency

**светоотрицательный** *adj.* light-negative

**светопечатание** *n.* photographic tracing

**светопоглотитель** *m.* light trap

**светопоглощаемость** *f.* luminous absorptivity

**светопоглощение** *n.* luminous absorption, light absorption

**светоположительный** *adj.* light-positive

**светопреломляющий** *adj.* light-refracting

**светопровод** *m.* light guide; beam (mode) waveguide

**светопроекционный** *adj.* light-projecting; projector

**светопрозрачный** *adj.* translucent

**светопроницаемость** *f.* permeability to light, transparency; translucence

**светопроницаемый** *adj.* transparent; translucent

**светорассеивающий** *adj.* light-scattering, light-diffusing

**светорассеяние** *n.* diffusion of light, scattering of light

**светосигнал** *m.* light signal, visible signal

**светосигнализация** *f.* light signaling

**светосигнальный** *adj.* light-signal

**светосила** *f.* illumination, illuminating power; aperture ration, speed (of lens); transmission (of spectrometer)

**светосильный** *adj.* high-aperture (lens), fast (lens)

**светосостав** *m.* phosphor

**светостойкий** *adj.* lightproof, light-resistant, photostable

**светостойкость** *f.* light resistance, photostability

**светотеневой** *adj.* black-and-white, light-and-shadow

**светотень** *f.* light and shade, light and shadow

**светотехника** *f.* lighting engineering

**светофильтр** *m.* optical filter, light filter, filter, heliofilter, color frame, light trap, ray filter; **защитный с. для уменьшения внешней засветки** ambient light filter, optic light filter; **с. передающего устройства** color analysing filter

**светофор** *m.* light signal, traffic light

**светофорный** *adj.* light-signal

**светочувствительность** *f.* photosensitivity, photoelectric sensitivity, light sensitivity, speed (of light)

**светочувствительный** *adj.* light-sensitive, light-reactive, photoactive, photosensitive

**светоэлектрический** *adj.* photoelectric

**светящийся** *adj.* luminant, luminous, luminescent, glowing, lighted, illuminated; light-emitting; fluorescent; phosphorescent

**свеча** *f.* candle, candle-power; candle; light; taper; (spark) plug; **с. зажигания, зажигательная с., запальная с.** spark plug; **новая с.** candela

**свечение** *n.* light emission, luminescence, luminosity, glow, fluorescence, phosphorescence, brightness, glint, glimmer, radiance, shimmer; lighting; **белое с.** white heat, incandescence; **избыточно-яркое с.** blooming (in a television picture tube); **калильное с.** candoluminescence; **красное с.** red heat; **начальное с.** corona onset; **с. от трения** triboluminescence

**свечной** *adj.* candle

**свечность** *f.* candle power

**свешивающийся** *adj.* overhanging

**свивание** *n.* coiling, twisting, twist

**свивать, свить** *v.* to twist, to strand, to twine, to spin

**свивка** *f.* twisting, twist, coiling

**свидетельство** *n.* evidence; certificate, licence

**свидетельствующий** *adj.* indicating, significative

**свиль** *f.* curve, knurl, bend

**свинец** *m.* lead, Pb

**свинка** *f.* (mercury) switch; ingot, bar; **ртутная «с.»** mercoid, mercoid switch

**свинтить** *pf. of* **свинчивать**

**свинцово-кислотный** *adj.* lead-acid

**свинцово-оловянный** *adj.* lead-tin

**свинцово-селеновый** *adj.* lead-selenide

**свинцовый** *adj.* lead, leaden, plumbic

**свинчивать, свинтить** *v.* to screw, to screw together

**свирель** *f.* pipe, reed; reed-pipe, reed-organ; **пастушеская с.** Pandean pipes

**свисать, свиснуть** *v.* to hang down, to dangle, to droop, to sag; to trail; to overhang

**свисающий** *adj.* hanging, pendent; trailing

**свиснуть** *pf. of* **свисать**

**свист** *m.* whistle, singing; pipe, piping; hiss, hissing, squealing, swishes, whistler atmospherics, whistler-type noise; howling; **с. биений** heterodyne whistling; **с. на пороге гене-**рации fringe howling; **переливчатый с.** bubbling whistle

**свистание** *n.* whistling, whistle

**свистать** *see* **свистеть**

**свистение** *n.* hissing, whistling, whizzing

**свистеть, свистнуть** *v.* to whistle, to sing; to pipe; to hiss

**свисток** *m.* whistle

**свистящий** *adj.* whistling; hissing; sibilant

**свитый** *adj.* coiled, convoluted

**свить** *pf. of* **свивать, вить**

**свобода** *f.* freedom

**свободно** *adv.* freely, easily, loosely

**свободнобегущий** *adj.* free-progressing, free-running

**свободновисящий** *adj.* cantilever

**свободноидущий** *adj.* continuously running

**свободнонесущий, свободностоящий** *adj.* self-supporting

**свободность** *f.* free condition, availability (*tphny.*)

**свободный** *adj.* free, idle, available; blank, vacant; loose; spare; open; clear, not-busy, disengaged; floating; unsupported

**свод** *m.* dome, arch, arc, vault, cupola; code; summary

**сводимость** *f.* reducibility

**сводить, свести** *v.* to bring together, to converge; to reduce; to bank; to conduct, to lead, to take; **с. в таблицу** to tabulate

**сводка** *f.* résumé, summary; compendium

**сводный** *adj.* combined, compound, composite

**сводящий** *adj.* converging, convergent, convergence

**свойственно** *adv.* naturally; it is natural

**свойственность** *f.* peculiarity, singularity

**свойственный** *adj.* intrinsic, inherent, natural, indigenous; peculiar, characteristic, distinctive

**свойство** *n.* characteristics, quality, property, feature, attribute, aspect; effect; nature; condition; property; **свойства** properties; **свойства звукопроводности** sound transmission qualities; **индивидуальные свойства человека** personality factors

**сворачивание** *n.* deflection; shunting, turning; **с. спиралью** coiling

**СВЧ** *abbr.* 1 (**сверхвысокая частота**) SHF, superhigh frequency; 2 (**сверхвысокочастотный**) superhigh frequency

**связанность** *f.* coherence; contingency

**связанный** *adj.* bound, bounded, linked, tied, connected, bonded, coordinated, affiliated, coupled, associated; on-line; compounded, ganged; latent; coherent; (often omitted in translation: **искажения, связанные с квантованием** quantization distortions [*lit.* distortions connected with quantization]); **гальванически с.** d-c coupled; **с. по переменному току** a-c coupled; **с. по постоянному току** d-c coupled

**связать(ся)** *pf. of* **связывать(ся), вязать**

**связист** *m.* signal man, signaller, signal corps man, communication man, communicator

**связка** *f.* packet, pack, bundle, bunch, pile; tie, gang; chord, ligament; compound; binder, strap, strapping, band, belting; bond; connective; **связки внахлёстку** echelon strapping (of a magnetron); **голосовые связки** vocal cords; **двойная кольцевая с.** double-ring strapping; **ступенчатая с.** echelon strapping

**связной** *m. adj. decl.* messenger; communication

**связность** *f.* connection (*math.*); coherence, connectedness

**связный** *adj.* connected, coherent, cohesive; compendent

**связочный** *adj.* bundle, bunch; strapped

**связущий** *adj.* binding, connecting, connective, interconnecting, tying, tie, cementing; couplant

**связывание** *n.* coupling; tying, binding, bonding, linking, combining; strapping

**связывать, связать** *v.* to tie, to bind, to connect (with), to couple, to join, to link, to (inter)connect, to gang, to jumper; to cement, to bond; to strap; to brace, to stay

**связываться, связаться** *v.* to communicate (with), to contact by phone, to get in touch with; to be bound, to be combined

**связывающий** *adj.* linking, linkage, connecting, binding

**связь** *f.* coupling (*elec.*); tie, bond, connection, coupling, link, linkage, linking, jointing, joining, junction, binding, bonding; association, contact, coherence, continuity, closeness, relation(ship); (tele)communication service, telecommunication traffic, intercommunication; (telephone) service; signalling, signals; relaying; binder; connector, tie piece, stay; **без обратной связи** non-regenerative; **с обратной связью** back-coupled, regenerative; feedback; **автоматическая с.** automatic operation, dial working; **автоматическая междугородная с.** automatic long-distance service; **акустическая обратная с.** acoustical feedback; **буквопечатающая телеграфная с.** printing telegraphy, printing telegraph system; **быстрая с. через узловые станции** tandem toll circuit dialing (*tphny.*); **внешняя обратная с.** external feedback; **внутренняя с. для экипажа** crew intercommunication (*tphny.*); **внутренняя обратная с.** flux reset control; **с. внутриведомственного пользования** departmental (radio) service; **внутриобластная с.** "oblast"(-wide) communication service; **внутрипроизводственная с.** intercommunication at a production enterprise (*tphny.*); **внутрирайонная с.** "raion" (-wide) communication service; **вч с. по проводам** carrier communication, line radio, carrier telephony; **вч с. с несущей** carrier current communication; **гальванически-индуктивная с.** autocoupling; **с. (гидро-)акустическими средствами** sonar communication; **государственная с.** state communication service, national communication service; **гражданская с. на УКВ** citizen-radio; **двойная обратная с.** double feedback, double reaction; **двукратная с. по системе Пикара** superposed diplex-telegraphy; **двусторонняя оперативная с.** talkback circuit, intercommunication; **двухконтурная с.** double-tuned coupling; **двухпроводная разночастотная с.** two-frequency duplex operation; **двухсторонняя с.** intercommunications; two-way traffic, duplex traffic; **двухсторонняя груп-**

повая телефонная с. conference call, round call, conference circuit; десятикратная обратная с. times one-tenth feedback; дециметровая с. в службе электросвязи microwave (or uhf) radio links in telecommunications; директорская с. master station service (*tphny.*); дифференциальная с. derivative coupling; дроссельная с. inductor coupling, impedance coupling; ёмкостная обратная с. electrostatic feedback; жёсткая обратная с. follow-up direct feedback; с. за пределами оптического горизонта transhorizon communication; с. за счёт перекрёстных воздействии (or наводок) crosstalk coupling; комбинированная с. combined link, mixed coupling; контролирующая обратная с. monitoring feedback, primary feedback; магистральная с. national radio link; longline service (*tphny.*); nation-wide communication service; с. между неподвижными пунктами fixed (radio) service; с. между телефонными проводами phantom-to-side unbalance; междуламповая с. interstage coupling, interstage linkage; многовитковая обратная с. multiloop feedback; многократная с. multiplex transmission, multiplex telegraphy; multiplex telephony; многократная буквопечатающая телеграфная с. printing multiplex system; многократная обратная с. multiplex feedback; с. на выходе output coupling; с. на дросселях impedance coupling, inductor coupling; с. на ходу vehicle-to-vehicle communication; направленная с. beam communication; неполная с. undercoupling; с. несущими токами line radio; низовая с. rural service, peripheral service, lower-echelon service; обратная с. feedback (coupling), feedback path, back coupling, reaction, reaction coupling, regeneration, retroaction, tickling; ultraudition, ultramagnifier; Armstrong circuit; monitoring feedback; о. с. анодного повторителя anode-follower feedback; о. с. дискретного действия sampled-data feedback; о. с. от последнего каскада к первому envelope feedback

(around the transmitter); о. с. по анодному напряжению plate (or anode) voltage feedback; о. с. по напряжению parallel feedback, voltage feedback; о. с. по разности скоростей counting-rate-difference feedback; о. с. по току series feedback, current feedback; о. с. сопротивлением resistance feedback; о. с. через ёмкость capacity reaction, electrostatic reaction; о. с. через ёмкость сетка-анод grid to plate (capacity) feedback; с. окном iris coupling; оперативная с. talk-back circuit; отрицательная обратная с. negative feedback, degeneration, inverse back coupling; отрицательная обратная с. по анодному напряжению anode voltage degeneration; отрицательная обратная с. с катодом cathode negative feedback, cathode degeneration; с. перекрёстной наводкой cross-talk coupling; с. по переменной составляющей a-c coupling; с. по постоянной составляющей d-c coupling; с. по фронту lateral communication; подвижная с. mobile service; положительная обратная с. positive feedback, regenerative feedback loop; последовательная обратная с. series feedback; прямая с. connecting circuit, cross connection, internal circuit, tie line; радиорелейная с. relay repeater; распределённая с. space coupling; регенеративная обратная с. regenerative feedback; регулируемая с. variable coupling; рокадная с. lateral communication; с. с подразделением каналов во времени time dividing channeling; с. с помощью запирающего фильтра parallel resonance coupling, tuned coupling; сильная с. close coupling, tight coupling; симплексная с. intercommunication; служебная с. link between operators (*tphny.*); телеграфная с. teletypewriter service; токовая обратная с. current feedback, series feedback; трансляционная с. radio telephone circuit; с. фильтром низкой частоты low-pass coupling; с. через промежуточный виток link coupling; с. через районные переключатели reversing technique (toll system)

**Связьиздат** *m.* the State Publishing House of Electrocommunication

**СГ** *abbr.* 1 (**синфазная горизонтальная антенна**) cophasal horizontal antenna; 2 (**синхронный генератор**) synchronous generator; 3 (**стойка группового оборудования**) group terminal bay

**сгиб** *m.* flexure, flexion, bend, bending; joint, link; fold, crimp; ply

**сгибаемость** *f.* flexibility, pliability

**сгибаемый** *adj.* flexible, pliable, bendable; collapsible

**сгибание** *n.* flexure, lateral flexure, flection, buckling, deflection, bending

**сгибательный** *adj.* bending, flexing

**сгибать, согнуть** *v.* to bend, to flex; to curve, to crook; to deflect

**сгибающий** *adj.* bending, flexing

**сгибающийся** *adj.* bending, flexible, pliable

**сгладить** *pf. of* **сглаживать, гладить**

**сглажение** *n.* smoothing

**сглаженный** *adj.* smoothed-out

**сглаживание** *n.* smoothing, flattening; polishing; leveling (out); fitting (*math.*); filtering; despiking; cleaning

**сглаживатель** *m.* smoother, flattener; de-emphasizer, deaccentuator

**сглаживать, сгладить** *v.* to smooth out, to iron out; to calm, to stabilize; to even out, to adjust, to balance, to equilibrate; to filter; to offset; to deaccentuate, to level (down)

**сглаживающий** *adj.* smoothing, steadying, equalizing; burnishing; rectifier; equalizer; buffer

**СГН** *abbr.* (**стойка групповых несущих частот**) group carrier supply bay

**сгнить** *pf. of* **гнить**

**сгон** *m.* joining, fitting

**сгонный** *adj.* joined, driven together

**сгорание** *n.* burning, combustion; burn-out, burn-up; **с. предохранителя** blowout of a fuse

**сгорать, сгореть** *v.* to burn down, to burn out, to be consumed, to be used up; to fuse, to blow, to melt (of a fuse)

**сгореть** *pf. of* **сгорать, гореть**

**сграничить** *pf. of* **граничить**

**сгруппирование** *n.* grouping

**сгруппированный** *adj.* bunched; grouped, banked, coupled; integrated

**сгруппировать** *pf. of* **группировать**

**сгруппировывание** *n.* grouping

**СГС** *abbr.* (**сантиметр-грамм-секунда**) centimeter-gram-second

**сгуститель** *m.* thickener, condenser; tank

**сгустить** *pf. of* **сгущать**

**сгусток** *m.* cluster, bunch

**сгущаемость** *f.* condensability

**сгущаемый** *adj.* condensable; compressible

**сгущать, сгустить** *v.* to fix, to concrete, to condense, to thicken; to compress

**сгущающий** *adj.* condensing, concentrating

**сгущение** *n.* concentration, condensation, thickening; accumulation; bunching; compression; setting

**сгущённый** *adj.* concrete, condensed, thickened, concentrated; compressed

**сдавить** *pf. of* **сдавливать**

**сдавленный** *adj.* squeezed, pressed, compressed

**сдавливаемость** *f.* compressibility

**сдавливание** *n.* squeezing, compression, pressing; pinching

**сдавливать, сдавить** *v.* to press, to compress, to squeeze, to condense; to contract, to pinch; to crush; to throttle

**сдваивание** *n.* doubling, duplication, coupling

**сдваивать, сдвоить** *v.* to double, to duplicate, to couple; to bend

**сдваивающий** *adj.* doubling, coupling; joining

**сдвиг** *m.* shear, shearing, dislocation; displacement, shift, shifting, excursion, slip; deviation; drift, sliding; offset; frequency swing, frequency sweep, frequency deviation; upset; **с. во времени на 90°** time quadrature; **километрический с. фаз** phase constant, fractional wave length; **с. на 90°** quadrature; **с. на один разряд** shift of one position, single-place shift; **с. нуля за пределы шкалы** inferred zero; **с. по времени** time delay; **с. по времени на четверть периода** time quadrature; **с. по фазе на 90°** phase quadrature, opposition (*tv*), quadrature; **с. положения нуля** zero drift; **с. при вращении** tangential force

**сдвигание** *n.* shifting, displacement, displacing

**сдвигать, сдвинуть** *v.* to shift, to move, to displace, to offset, to remove, to slide; to push together, to cross interlace; to skip operation; to shear

**сдвигающий** *adj.* shearing, shifting

**сдвиговый** *adj.* shear, shift, displacement, deviation; drift

**сдвижение** *n.* shifting, shearing, displacement

**сдвижной** *adj.* offset, movable; slip; telescopic (mast)

**сдвинутый** *adj.* removed, moved, offset, shifted, displaced; out of alignment; **с. назад на четверь периода** quadrature-lagging; **с. по фазе** dephased, out of phase; **с. по фазе на 90°** quadrature-lagging, in-phase opposition

**сдвинуть** *pf. of* **сдвигать**

**сдвоение** *n.* coupling, doubling, duplication

**сдвоенный** *adj.* doubled, double, dual, duplex, tandem; binary, paired, coupled, matched; twin(ned)

**сдвоить** *pf. of* **сдваивать**

**сделать** *pf. of* **делать**

**сдерживать, сдержать** *v.* to stop, to sustain, to restrain, to hold in, to keep back, to contain; to support; to check

**сдерживание** *n.* check; restrain, suppression

**сдерживающий** *adj.* holding, restrictive

**сдирание** *n.* stripping, peeling

**сдирать, содрать** *v.* to strip (off), to peel, to skin

**СДМ** *abbr.* (**статистическая дельта-модуляция**) statistical delta-modulation

**СДС** *abbr.* (**стойка дифференциальных систем**) voice frequency terminating equipment bay

**сеанс** *m.* seance; **с. связи** communication

**себестоимость** *f.* working cost

**сев.** *abbr.* (**северный**) north, northern

**сев.-вост.** *abbr.* (**северо-восточный**) northeast

**север** *m.* north; **с. по координатной сетке** grid north

**северный** *adj.* north, northern; boreal

**сев.-зап.** *abbr.* (**северо-запад**) northwest

**сегмент** *m.* segment, section; **полукруг-**

**лый с.** binant; **сферический с., шаровой с.** calotte, spherical calotte

**сегментальный** *adj.* segmental

**сегментация** *f.*, **сегментирование** *n.* segmentation

**сегментировать** *v.* to segment

**сегментноцепный** *adj.* chain-and-segment

**сегментный, сегментообразный** *adj.* segment, segmental, segmentary

**сегнетоактивный** *adj.* ferroelectric, Seignette-electric

**сегнетовый** *adj.* Seignette

**сегнетоэлектрик** *m.* ferroelectric (material), segnetoelectric

**сегнетоэлектрический** *adj.* ferroelectric, segnetoelectric

**сегнетоэлектричество** *n.* ferroelectricity, ferroelectrics

**сегрегационный** *adj.* segregation, segregated

**сегрегация** *f.* segregation

**седельный** *adj.* dip (of a curve); saddle; cleat, clamp; pad; collarbeam

**седло** *n.* saddle; collar beam; **с. клапана** valve cone

**седловидный** *adj.* saddle-shaped, saddle

**седловина** *f.* saddle, valley, trough, crevasse, dip; hyperbolic point of a surface

**седловый** *adj.* saddle

**седлообразный** *adj.* saddle-like, saddle-shaped; dip (of a curve)

**Сезерланд** Sutherland

**сезонный** *adj.* seasonal

**сейсм** *m.* seism

**сейсмикрофон** *m.* seismicrophone

**сейсмический** *adj.* seismic

**сейсмичность** *f.* seismicity

**сейсмограмма** *f.* seismogram

**сейсмограф** *m.* seismograph

**сейсмография** *f.* seismography

**сейсмология** *f.* seismology

**сейсмометр** *m.* seismometer

**сейсмометрия** *f.* seismometry

**сейсморазведка** *f.* seismic prospecting

**сейсмостойкий** *adj.* earthquake-proof

**сейсмостойкость** *f.* seismic stability

**сек** *abbr.* (**секунда**) sec, second

**секанс** *m.* secant

**секвенция** *f.* sequence (*mus.*); sequent

**Секки** Secchi

**секретарский** *adj.* secretarial

**секретно** *adv.* secretly, covertly

**секретность** *f.* secrecy

**секретный** *adj.* secret, classified, confidential

**секторообразный** *adj.* sector-pattern

**сексагональный** *adj.* hexagonal

**секста** *f.* submediant, sixth (*mus.*)

**секстан, секстант** *m.* sextant; **с. с уровнем** bubble sextant

**секстет** *m.* sextet

**сектор** *m.* sector, segment, zone, arc; **с. гидроакустического поиска** sonar search arc; **с. ненадёжного определения пеленга** bad bearing sector; **теневой с.** shadow angle

**секторный** *adj.* sectoral, sectorial, sector

**секторообразный** *adj.* sector-shaped

**секунда** *f.* second; supertonic, second (*mus.*); **чрезмерная с.** sesquitone

**секундный** *adj.* second

**секундомер** *m.* stop-watch, timer, seconds counter, chronoscope; **с. для измерения долей секунд** split-second timer; **самопишущий с.** chronograph

**секундомерный** *adj.* timing

**секущая** *f. adj. decl.* secant

**секущий** *adj.* cutting, intersecting; secant

**секционирование** *n.* sectionalization, sectionalizing; **с. катушки** coil grading; **с. участков** sectioning

**секционированный** *adj.* split, tapped, sectional, sectionalized; subdivided; step, stepped, graded, graduated; multicellular

**секционировать** *v.* to sectionalize, to subdivide, to tap

**секционный** *adj.* section, sectional, subdivided, divided; unit-type

**секция** *f.* section, unit, cell; department, division; space; step, stage; **генераторная с.** oscillator section; **добавочная с. многократного поля** switch section of multiple (*tphny.*); **контактная с.** contact bank; **мёртвая с.** dead coil; **с. многократного поля** section of a multiple (*tphny.*); **с. переключателя** switch level; **с. переноса (изображения)** image section, image stage; **сопряжённая с.** tracking section; **холостая с.** dead coil, idle coil

**селективность** *f.* selectivity, discrimination

**селективный** *adj.* selective, discriminative, discriminatory, discriminating

**селектирование** *n.* strobing, gating

**селектированный** *adj.* selected

**селектировать** *v.* to select; to gate

**селектирующий** *adj.* selecting, sampling

**селектор** *m.* selector, selector switch, switch, auto-switch; separator, sorter; (time) gate; **с. выбора канала** channel selector; **с. кадровых синхроимпульсов** vertical separator, vertical-sync separator; **с. синхровспышки** burst separator; **с. строчных синхроимпульсов** horizontal separator, horizontal-sync separator

**селекторно-квадрантный** *adj.* quadrant-selector

**селекторный** *adj.* selector; gated, gating; strobe

**селектрон** *m.* selectron

**селекционный** *adj.* selection, selective

**селекция** *f.* selection, gating

**селен** *m.* selenium, Se

**селенистый** *adj.* selenium, selenious, selenous

**селеновый** *adj.* selenium, selenic; seleniferous

**селеноид** *m.* selenoid

**селенофон** *m.* selenophone

**селитра** *f.* saltpeter, niter

**сельсин** *m.* selsyn, synchro, autosyn, automatic synchronizer, driving synchro, synchrodrive, synchromotor, resolver; control transformer; **бесконтактный с.** magslip; **бесконтактный с. переменного тока** magnetic slip-ring indicator; **возбуждающий с.** transmitting selsyn; **вынесенный с. угла места** remote elevation selsyn; **с.-датчик** transmitting selsyn, transmitting synchro, synchro transmitter, control synchro, data synchro; **с.-датчик точного (отсчёта) азимута** fine-azimuth transmitting selsyn; **с.-датчик точного (отсчёта) угла места** fine-elevation transmitting selsyn; **с.-датчик угла наклона** elevation transmitting selsyn; **с.-индикатор** synchro indicator; **с.-приёмник** synchro receiver, synchro motor, mechanical repeater, synchro-follower; **с.-приёмник угла возвышения** (*or* подъёма), **с.-приёмник угла места** elevation selsyn transformer; **сверхточный с.** inductosyn; **с. точного отсчёта** fine selsyn; **с.-**

трансформатор synchro control transformer; **с. угла места** elevation selsyn, tilt autosyn

**сельсинный** *adj.* selsyn, self-synchronous

**сельский** *adj.* rural, country

**сем.** *abbr.* (**семейство**) family

**семантика** *f.* semantics

**семантический** *adj.* semantic

**семафор** *m.* semaphore, signaler

**семафорный** *adj.* semaphore, semaphoric

**семейство** *n.* family; assemblage; group; set; series; **с. колебательных характеристик** oscillator characteristics of a tube; **с. кривых** set of curves; **равновесное с.** equilibrium population; **с. траекторий** track population

**семема** *f.* seme, sememe

**семемический** *adj.* sememic

**семеричный, семерной** *adj.* septenary, septuple, sevenfold

**семи-** *prefix* seven-, of seven, hepta

**семи-** *prefix* semi-

**семизначный** *adj.* seven-element

**семиинвариант** *m.* semi-invariant (*math.*)

**семиинтерквартильный** *adj.* sem-interquartile (*math.*)

**семикратный** *adj.* sevenfold, septuple

**семиотика** *f.* semiotics

**семипозиционный** *adj.* seven-place, seven-point

**семипроводный** *adj.* seven-wire

**семистор** *m.* semistor

**семиугольник** *m.* heptagon

**семиугольный** *adj.* heptagonal

**семнадцатеричный** *adj.* septendecimal

**сендитрон** *m.* sendytron

**сензистор** *m.* sensistor

**Сен-Клер** St. Clair

**сенсибилизатор** *m.* sensitizer

**сенсибилизация** *f.* sensitization, sensitizing

**сенсибилизированный** *adj.* sensitized

**сенсибилизировать** *v.* to sensitize, to form, to activate

**сенситивный** *adj.* sensitive

**сенситометр** *m.* sensitometer

**сенситометрический** *adj.* sensitometric

**сенситометрия** *f.* sensitometry

**сенсорный** *adj.* sensory

**сент** *m.* cent (unit of pitch)

**сентибел** *m.* centibel

**сентрон** *m.* sentron

**сепарабельный** *adj.* separable

**сепаратный** *adj.* separate, separative, independent

**сепаратор** *m.* separator; **магнитный с.** magnetic cobbing machine

**сепарационный** *adj.* separation

**сепарация** *f.*, **сепарирование** *n.* separating, separation, segregation

**сепарировать** *v.* to separate, to segregate, to eliminate, to isolate

**септаккорд** *m.* seventh (*mus.*)

**септет** *m.* septet (*mus.*)

**септима** *f.* leading note, seventh (*mus.*)

**сер.** *abbr.* (**серийный; серия**) serial; series

**сера** *f.* sulphur, S; brimstone

**серводвигатель** *m.* servo-motor, brush-shifting motor, pilot motor; **с серводвигателем** servo-operated

**сервоинтегратор** *m.* servo integrator

**сервоканал** *m.* servo channel

**сервоклапан** *m.* servo valve

**сервоконтакт** *m.* servo contact

**сервоконтроллер** *m.* servo controller

**сервоконтур** *m.* servo-loop

**сервомеханизм** *m.* servomechanism, servo, servo unit, servosystem; **дискретный с.** digital servomechanism; **моделирующий с.** servo-simulator

**сервомодулятор** *m.* servomodulator

**сервомотор** *m.* servomotor, brush-shifting motor, actuating motor; **двухобмоточный сериесный с.** split series servomotor

**сервомоторный** *adj.* servomotor; servo-operated

**сервопотенциометр** *m.* servo-potentiometer

**сервопривод** *m.* servodrive

**серворегистратор** *m.* servo-recorder

**серворегулирование** *n.* servo control

**сервосистема** *f.* servosystem, servomechanism, servoloop

**сервоскоп** *m.* servoscope

**сервотайпер** *m.* servotyper

**сервоумножитель** *m.* servomultipler

**сервоуправление** *n.* servocontrol

**сервоуправляемый** *adj.* servocontrolled

**сервоусилитель** *m.* servo-amplifier; **предварительный с. на полу-проводниковых триодах** transistor servo preamplifier

**сердечник** *m.* core, center, pith, heart, slug; leg; mandrel; strand (of cable); **с сердечником** cored; **с воздушным**

**сердечником** air-cored; **с тремя сердечниками** triple-cored; **с четырьмя сердечниками** four-limbed; **с. барабана** powdered-iron core; **втяжной с.** plunger; slug; **с. из пластин** laminated iron core; **с. из феррокарта** dust core, powdered iron core; **настроенный с.** slug tuner; **оксиферовый с.** ferrocart core; **пластинчатый с.** laminated core, divided-iron core; **с. полюса** magnet leg; **прессованный с.** dust core; **ш-образный с.** three-leg core

**сердечниковый** *adj.* core, center

**сердцевидный** *adj.* heart-shaped

**сердцевина** *f.* heart, core, pith, center; **со стальной сердцевиной** steel-cored

**серебрение** *n.* silvering, silver plating

**серебристый** *adj.* silvery; silver

**серебрить, посеребрить** *v.* to silver-plate, to silver

**серебро** *n.* silver, Ag

**серебряно-кадмиевый** *adj.* silver-cadmium

**серебряно-цезиевый** *adj.* silver-cesium, cesium on silver

**серебряный** *adj.* silver

**середина** *f.* center, middle, midpoint; mean

**серединный** *adj.* middle, central, mean

**серёжка** *f.* lug; small link; loop, ring; shackle, fastening; **с. передачи тяги** switch-point lug (*RR*); **с. стрелочного контроллера** point lug (*RR*)

**сериальный** *adj.* serial

**сериесный** *adj.* series; series-wound

**серийно** *adv.* in series

**серийный** *adj.* series, serial; collective

**серия** *f.* row, rank, line, set, bank, series; turn; train, group (of waves); range, order, succession; **сериями** in series; **с. заказов (на переговоры)** sequence of calls, batch of calls (*tphny.*); **с. затухающих колебаний** jig; **с. типоразмеров** range of sizes; **с. тональных колебаний** tonic train

**«серкуитрон»** *m.* circuitron

**сернистокадмиевый, сернокадмиевый** *adj.* cadmium-sulfide

**сернокислый** *adj.* sulfuric acid; sulfate (of)

**серный** *adj.* sulfur, sulfury; (*chem.*) sulfuric

**сероуглерод** *m.* carbon disulfide

**серпент** *m.* serpent (*mus.*)

**серрасоидный** *adj.* serrasoidal

**серродинный** *adj.* serrodyne

**серьга** *f.* shackle, stirrup; sleeve piece, (connecting) link

**сетевой** *adj.* line, power-line, supply-line

**сетеобразный** *adj.* reticular, net-like

**сетка** *f.* net, netting; network; gauze; lattice; grate, grating, grid; mesh; cage; screen; cross-hatch pattern; grid electrode; mat; aperture plate; **с тремя сетками** triple-grid; **N-ая с.** grid number $n$; **антидинатронная с.** suppressor (grid), grid suppressor; **декартова координатная с.** Cartesian display; **с. для заземления** ground mat; **защитная с.** protective grating, screengrid; protective-wire network; **защитная жёлобообразная с.** guard cradle; **с. из сопротивлений** resistor matrix; **испытательная с.** checker board pattern; **катодная с.** space charge grid; **координатная с.** reference grid, reference frame; **модуляторная с.** intensity grid; **с. на земле** earth mat; **с. однотактного усилителя** single-ended grid; **проволочная с., охватывающая шар дугового фонаря** spark arrestor; **рассасывающая с.** space-charge grid; **с. с переменным шагом намотки** variable-pitch grid; **третья с.** suppressor grid; **управляющая с.** control gird of a variable mu tube; intensity grid, signal grid; **цилиндрическая с.** gauze cylinder; **экранирующая с.** anode-screening grid; screening grid

**сетко-анодный** *adj.* grid-anode

**сетковидный** *adj.* grid-like

**сеткообразный** *adj.* net-shaped, reticular, reticulated

**сеточно-анодный** *adj.* grid-plate

**сеточный** *adj.* net; grid; **с сеточным управлением** grid-controlled

**сетчатка** *f.* retina; meshwork

**сетчатый** *adj.* mesh, meshed, grating, netted, network, net-shaped, reticulate(d); latticed; wiregauze

**сеть** *f.* net, netting, mesh; network, mains system; row (of antennas); circuit; system; **автоматическая телефонная с.** automatic exchange, dial exchange; **антенная с.** billboard array, broadside array; **с. быстрой связи** no-delay telephone network; **воздушная с.** aerial; **луче-**

**вая с.** independent feeder, radial feeder; **междугородная с.** trunk network, long-distance network; **междугородная кабельная с.** toll cable network; **с. междугородной линии** trunk network, long-distance network; **местная с. из соединительных линий** junction line network; **многократно замкнутая с.** interconnecting mains; **с. многократной передачи** diversity transmission system; **многорядная вибраторная с.** beam array, stacked dipole array; **нерайонированная телефонная с.** single-office exchange; **осветительная с. в качестве антенны** lamp-socket antenna; **с. проводов** distributing network, main system wire netting; **радиотрансляционная с.** a-f rediffusion net; **районированная городская с.** multi-office city exchange (*tphny.*); **районированная телефонная с.** multi-office exchange; **районная с.** tandem toll circuit; **ручная телефонная с., с. с ручным обслуживанием** manual exchange, manual area; **с. сильного тока** power supply system, mains; **сложная петлевая с.** interconnecting mains; **с. соединительных линий** junction network; **с. станций(привода)** home chain (*rdr.*); **телефонная с.** telephone exchange; telephone system; **трансляционная с. (низкой частоты)** a-f rediffusion net; **с. ЦБ** common battery exchange (*tphny.*)

**сечение** *n.* section, cross section, cut, profile; gage, size; **большого сечения** heavy-gage (wire); **завышённого сечения** overcoppered (wire); **малого сечения** light-gage (wire); **измерительное поперечное с.** measuring cross section; **крыловидное с.** aerofoil section; **с. луча** beam crossover, area of a beam; **с. нетто** effective section; **поперечное с.** cross section; **поперечное с. пучка** aperture of the beam, beam width; **с. пучка** beam area; **с. рассеяния на связанных атомах** bound cross section; **эффективное с. цели** radar cross section

**сеченный** *adj.* chopped

**сечка** *f.* cleaver, chopping-knife, chopper

**сжатие** *n.* pressure, pressing; compres-

sion, compressional strain, condensation; contraction, reduction; collapse (of a bubble); striction; flattening; shrinkage, shrinking; jam; clutch; grip; pinch; **с. в области белого** white compression (*tv*); **с. динамического диапазона** volume compression; **с. относительного размаха синхронизирующих импульсов** synchronization compression; **с. по частоте** frequency compression; **с. при застывании** contraction of solidification

**сжатость** *f.* compression; compactness, conciseness

**сжатый** *adj.* compressed, condensed; compression; constricted, pinched; brief, concise, compact

**сжать(ся)** *pf. of* **сжимать(ся)**

**сжечь** *pf. of* **сжигать, жечь**

**сжигание** *n.* combustion, burning (up), consumption

**сжигать, сжечь** *v.* to burn, to burn down, to burn up, to consume

**сжижение** *n.* liquefaction; condensation

**сжиженный** *adj.* liquified

**сжим** *m.* connector, ferrule; clamp, grip, clip, forceps, tongs; **с. для присоединения к свинцу** lead grip; **с. ножки** stem press

**сжимаемость** *f.* compressibility, condensability; compliance, contractibility

**сжимаемый** *adj.* compression-type, compressible, squeezable; contractible

**сжимание** *n.* compression, condensation; shrinkage, contraction, constriction

**сжиматель** *m.* compressor; **с.-расширатель** compandor

**сжимать, сжать** *v.* to condense, to compress; to squeeze, to contract, to constrict, to press, to pinch; to grip

**сжиматься, сжаться** *v.* to shrink, to contract; to compress, to condense

**сжимающий** *adj.* compressive, compressing. condensing

**сжимающийся** *adj.* squeezable, contractile, constringent

**сжимный** *adj.* squeezable; squeeze

**сзади** *adv.* behind, from behind, at the back, at the rear

**СИ** *abbr.* (**смешивающий искатель**)

mixing selector, access switch, hunting switch

**« сигара »** *f.* high-speed strander

**сигнал** *m.* sign, signal, alarm; call; cue; wave, echo; flag; (often omitted in translation: **генератор сигнала цветных полос** color bar (*or* strip) generator; **генератор сигналов вертикальных полос** vertical bar generator); **с поиском сигнала** signal-seeking; **с прерывистым** (*or* **пульсирующим) сигналом** pulse-monitored; **сигналы изображения** video signals; **« нумерованные » сигналы** quantized signals; **сигналы с одинаковой загрузкой полупериодов** equal-alteration waves; **с. аварии** out of order call; **акустический с. посылки вызова** audible ringing signal; **амплитудно-импульсный модулированный с.** sampled signal; **с. бедствия** distress call; **с. белого прямоугольника** white-window signal; **с. в прямой цепи воздействия** action signal; **визуальный с. занятости** busy indicator; **с. включения** offering signal; **входной с. записи** write input; **входной с. цепи обратной связи** loop input signal; **с. вхождения в связь** netting call (*rad.*); **вч с. для распознавания пар** high-frequency tone; **с. выключенного номера** "no such number" tone; **выходной с. датчика вертикали** track vertical output; **выходной с. единицы** one input signal; **выходной с. со вторичной обмотки** secondary output; **выходной с. цепи обратной связи** loop output signal; **главный с. для прилёта** inner marker beacon; **горочный с.** hump signal; **с. готовности к набору номера** proceed to dial signal, dial tone; **дальний с.** transfer signal (*tphny.*); **диспетчерский с.** selective signal (*tphny.*); **с. для пробы занятости дальнего абонента** test-busy signal (*tphny.*); **единичный с.** step signal; **с. забивки** erasure signal (*tgphy.*); **с. задающего генератора строчной развёртки** horizontal-drive signal; **с. занятости** busy signal, busy tone, visual busy tone, engaged signal; **с. занятости междугородным разговором** sign for busy trunk line; **с. занятости пучка** group busy

signal; **с. индикации положения луча** beam indexing signal; **искусственный с. рассогласования** artificial error signal; **с. исправления ошибки** erase signal, rub-out, erasure; **исходящий с. окончания набора** forward end-of-selection signal; **квантованный по времени с.** sampled signal; **квитирующий с.** answer-back signal; **с. компенсации « чёрного пятна »** picture-shading signal; **контрольный с.** monitor signal; reference signal; answer-back signal; **корабельный звуковой с.** (ship's) foghorn; **с. коррекции тёмного пятна** shading-compensation signal (*tv*); **ламповый с.** time check lamp; **с. лампой занятости** busy flash signal (*tphny.*); **максимальный с., соответствующий разности между уровнями чёрного и белого** peak black-to-white signal; **мгновенно изменяющийся с.** ramp input signal; **модулированный с. с обеими боковыми полосами** double-sideband signal; **монохроматический с., передающий детали цветного изображения** mixed high frequency signal; **с. на выходе камеры** camera output (*tv*); **с. нажатия** marking signal (*tgphy.*); **с. накопления** trunk congestion signal; **нарастающий с.** step wave (*tv*); **с. незанятости** free line signal, ringing guard signal, audible ringing signal, ringing tone; **с. неисправности** alarm signal; **обламываемый с.** smashboard signal (*RR*); **с. обратной связи** feedback signal; **обратный с. « конец набора » (номера)** backward end-of-selection signal; **с. однозначности** sense signal; **оконечный с. набора номера** number received signal; **с. опорного генератора** bootstrap waveform; **оптический с.** visual signal, flag signal, lamp signal, light signal; **с. освобождения цепи** circuit available signal, reorder signal (*tphny.*); **основной с. точного времени** basic tune interval; **с. от датчика деформаций** strain gage signal; **с. от самого яркого места изображения** highlight signal; **отбойный с. вызванного (абонента)** clear-back signal (*tphny.*); **с. отбоя** ring-off signal, clearing signal; **с. ответа станции** dial tone; **с. отжатия** spacing signal (*tgphy.*); **с. открытия связи** commencing sign; **с. ошибки** erasure signal (*tgphy.*); **паразитный с. передающей трубки** dark-spot signal; **первоначальный с. переноса** carry initiating signal; **перебойный с., с. перебоя** erasure signal (*tgphy.*); **с. повторного вызова** rering signal; **с. подавления** blanketing signal (*tv*); **с. подготовки** prefix signal; **с. поднесущей с квадратурной фазой** QCW signal; **полный с.** composite signal (*tv.*); **полный (*or* сложный) с. синхронизации приёмника** composite synchronization; **с. полотнищем** panel signal (*av.*); **с. понижения давления в кабеле** contactor alarm (*tphny.*); **с. посылки вызова** line signal, calling signal, audible ringing signal, ringing tone; **с. потенциального типа** steady-state signal; **с. поясного времени** standard time signal; **предельный с.** saturation signal; **с. предложения переговора** offering signal (*tphny.*); **с. прерывателя** ticker signal; **с. продвижения** inner home signal; **с. прохода до ближайшего следующего сигнала** approach annunciator; **прямой с. окончания набора** forward end-of-selection signal; **разностный с. угла места** elevation difference signal; **с. разрешения переноса** carry clear signal; **ритмический с. точного времени** standard time interval; **с. с временным сдвигом** shifted signal; **с. с крылом вверх** upper quadrant signal; **с. с крылом вниз** lower quadrant signal; **с. с плоской верхушкой** flat-topped signal; **с. сетчатого поля** grating signal; **с. сложной формы** complex wave; **сложный с.** serrated signal (*tv*); **служебный вызывной с.** order wire signal; **слуховой с.** busy tone, ringing tone, dial tone; **с. такта** cadence (*tgphy.*); **телевизионный с. с введёнными гасящими импульсами** blanked picture signal; **с. «чёрного пятна »** shading signal; **яркостный с., передаваемый в обход канала цветности** by-pass monochrome signal

**сигнализатор** *m.* signalling apparatus, signal indicator, indicator; alarm,

alarm box; **ионизационный с.** ionization-type fire detector; **максимальный с.** thermostat-type fire detector; **с. номеров** automatic number-annunciator panel; **обратный с.** position indicator; **с. перегрева подшипников** bearing alarm; **с. потери скорости самолёта** airplane stall warning device

**сигнализационный** *adj.* signaling

**сигнализация** *f.* signaling, signal sending; **заградительная с.** fence alarm system; **охранная с.** burglar alarm, emergency alarm; **электрическая обратная с.** electrical position indication

**сигнализирование** *n.* signaling

**сигнализировать** *v.* to signal

**сигнализирующий** *adj.* signaling

**сигнализовать** *v.* to signal

**сигнализующий** *adj.* signaling

**сигналист** *m.* signalman

**сигналить, просигналить** *v.* to signal

**сигналообразующий** *adj.* signal-shaping

**сигнально-вызывной** *adj.* ringing-and-signaling

**сигнально-опросный** *adj.* ringing-and-answering

**сигнальный** *adj.* signal, signaling, indicating, alarm; warning; marking; pilot (lamp)

**сигнальщик** *m.* signaler, signalman; flagman

**сигнум** *m.* signum

**сидероскоп** *m.* sideroscope

**сиккатив** *m.* desiccator, desiccant, siccative

**сила** *f.* force, strength, power; intensity; energy; **возмущающая с.** driving force; **вынуждающая с.** forcing function (*math.*); **допустимая с. тока** current-carrying capacity; **наибольшая допустимая с. тока** maximum current carrying capacity; **ориентирующая с.** versorial force, directive force; **полная с., действующая на поверхности** circumferential force; **поперечная ударная (сверхпереходная) электродвижущая с.** $E_q''$ quadrature-axis subtransient internal voltage; **поперечно приложенная с.** transverse force; **постоянная с. тока** constant current; **с. привода** motive force; **с., приложенная в точке** concentrated force;

**с. притяжения в зазоре** air-gap pull; **с. притяжения в момент соприкосновения** sealing pull; **продольная ударная (сверхпереходная) электродвижущая с.** $E_d$ direct-axis subtransient internal voltage; **противодействующая с.** controlling force; **с. света в свечах** candle power; **с. тяги** intensity of draft; **с. тяги на крюке** draw-bar pull; **с. тяжести** gravitation, gravity; **с.-час** horse-power-hour; **с. шумов переходной наводки** crosstalk volume; **электродвижущая с. вращения** speed-induced voltage; **электродвижущая с. соприкосновения** contact potential

**силастик** m. silastic
**силикагель** m. silicagel
**силикат** m. silicate
**силикатный** adj. silicate
**силикон** m. silicone
**силиконовый** adj. silicone
**силит** m. silit
**силитовый** adj. silit
**силлабический** adj. syllabic
**силлабула** f. syllable
**силлабульный** adj. syllable
**силлогизм** m. syllogism
**силовой** adj. force, power
**силумин** m. silumin
**силуэт** m. silhouette
**силуэтный** adj. silhouette; masking
**сильванит** m. sylvanite
**сильватрон** m. sylvatron
**« сильверстат »** m. silverstat regulator
**« сильмалек »** m. silmalec
**сильно** adv. powerfully, strongly; greatly; heavily; vastly, highly
**сильнолегированный** adj. heavily-doped
**сильносвязанный** adj. overcoupled
**сильноточный** adj. heavy-current, heavy-duty
**сильнофокусирующий** adj. strong-focusing
**сильный** adj. strong, powerful; sharp, intense; potent; heavy; hard; high; tight
**сильфон** m. bellows, sylphon; **с. обратной связи** feedback bellows
**сильфонный** adj. bellows, bellows-sealed
**сим** abbr. (**сименс**) mho
**« симагал »** m. simagal
**символ** m. symbol, sign, mark; image; signal; figure; digit; character; note

**символизировать** v. to symbolize, to represent, to stand (for)
**символизм** m. **символика** f. symbolism
**символический** adj. symbolic
**сименс** m. mho, siemens
**симметрирование** n. balancing, rendering symmetrical, symmetrization
**симметрированный** adj. balanced, symmetrical
**симметрировать** v. to balance, to symmetrize
**симметрирующий** adj. balancing
**симметрический** adj. symmetric(al)
**симметрично** adv. symmetrically
**симметричность** f. symmetry, balance
**симметрично-цикличный** adj. symmetrically-cyclical
**симметричный** adj. symmetric(al), balanced; **с симметричным питанием** center-fed
**симметрия** f. symmetry, balance
**симплекс** m. single current working, transmission by simplex current; simplex (*math.*)
**симплексный** adj. one-way, single, single-channel; simplex (*math.*)
**симпьезометр** m. sympiesometer
**симулировать** v. to simulate, to feign
**симультанный** adj. simultaneous
**симультантный** adj. combined
**симулятор** m. simulator
**симуляция** f. simulation
**синапс** m. synapse
**синасин-двигатель** m. synduct motor
**сингония** f. syngony; **гексаэдрическая с., кубическая с.** cubic system
**сингулентный** adj. singulett
**сингулярность** f. singularity
**сингулярный** adj. singular
**синеватый** adj. bluish
**сине-зелёный** adj. cyan
**синекалильный** adj. (at) blue heat
**синий** adj. blue, dark blue
**синимакс** m. sinimax
**СИНК** abbr. (**стойка индивидуальных несущих и контрольных частот**) channel carrier and pilot supply bay (*tphny.*)
**синкопа** f. syncope (*mus.*)
**синкопирование** n. syncopation
**синоним** m. synonym
**синонимический** adj. synonymous
**синонимия** f. synonymy
**синтагматический** adj. syntagmatic
**синтаксис** m. syntax
**синтаксический** adj. syntactic

**синтактика** *f.* syntactics

**« синтамайка »** *f.* synthamica

**синтез** *m.* synthesis; design; **с. для случая вещественных корней** synthesis for real roots; **с. системы управления** control design

**синтезатор** *m.* synthesizer; **с. компиляторов** compiler generator

**синтезированный** *adj.* synthesized, synthetic

**синтезировать** *v.* to synthesize; to construct, to produce

**синтезируемый** *adj.* synthetic

**синтезирующий** *adj.* synthesizing

**синтерирование** *n.* sintering

**синтерированный, синтерованный** *adj.* sintered

**синтеровочный** *adj.* sintering

**синтетически** *adv.* synthetically

**синтетический** *adj.* synthetic(al)

**синтонизация** *f.* syntonization, tuning

**синтонизированный** *adj.* syntonized, tuned

**синтонизировать** *v.* to syntonize, to tune

**синтонический** *adj.* syntonic

**синтрон** *m.* syntron

**синус** *m.* sine; **с. версус** versed sine; **с.-генератор** function generator; **с. угла потерь** power factor (of a dielectric)

**синус-косинусный, синусно-косинусный** *adj.* sine-cosine

**синусный** *adj.* sine

**синусоида** *f.* sinusoid, sine curve, sinusoidal trace, harmonic curve; **затухающая с.** damped sinusoid; sinc function

**синусоидальный** *adj.* sinusoidal; sine, sine-shaped, sine-wave; alternating

**синфазирование** *n.* phasing-in, synphasing

**синфазировать** *v.* to bring into step

**синфазирующий** *adj.* phasing

**синфазно** *adv.* inphase

**синфазность** *f.* correct phase; coherence

**синфазный** *adj.* inphase, cophasal, cophased; coherent

**синхробетатрон** *m.* synchrobetatron

**синхровспышка** *f.* burst signal

**синхрогенератор** *m.* synchro generator, synchronization generator, waveform generator

**синхродин** *m.* synchrodyne

**синхродинный** *adj.* synchrodyne

**синхроимпульс** *m.* synchronizing pulse, lockout pulse

**синхрометр** *m.* synchrometer

**синхромотор** *m.* synchro motor

**синхронизатор** *m.* synchronizer, lock unit; **с. передачи сигналов цветности** chrominance synchronizer

**синхронизационный** *adj.* synchronizing

**синхронизация** *f.* synchronization, synchronizing, timing, hold-in, holding, hold, lock, locking, interlock; triggering; frequency sticking; syntonization; **ведомая с.** genlock operation; **с. вручную** manual synchronization; **с. кадровой развёртки от осветительной сети** mains hold (*tv*); **с. передачи сигналов цветности** chrominance synchronization; **принудительная с.** genlock operation; **с. развёртки** lock for the sweep (*tv*); **с. с сетью** mains hold, locking; **с. сетей, с. с частотой сети** (vertical) synchronization employing power-line frequency

**синхронизирование** *n.* synchronizing, synchronization, timing

**синхронизированный** *adj.* synchronized, timed, simultaneous; locked; tuned; **с. по фазе** phase-locked

**синхронизировать** *v.* to synchronize, to lock, to bring into step; to time; **с. сигнал** to lock-on

**синхронизируемость** *f.* lock-in feature

**синхронизируемый** *adj.* synchronized; triggered; driven

**синхронизирующий** *adj.* synchronizing, timing; triggering

**синхронизм** *m.* synchronism, ganging, unison; step; **в синхронизме** in step

**синхронизованный** *adj.* synchronized

**синхронизующий** *adj.* synchronizing, lock-in

**синхронистический** *adj.* synchronistic (al); synchronous

**синхронический, синхроничный** *adj.* synchronous, coincident; synchronistic(al)

**синхрония** *f.* synchronism

**синхронно** *adv.* in step, in synchronism

**синхронно-следящий** *adj.* synchronous-tracking

**синхронность** *f.* synchronism; tuning

**синхронный** *adj.* synchronous, synchronized, coincident

**синхронометр** *m.* synchronometer

**синхроноскоп** *m.* synchronoscope, synchroscope; phase detector

**синхрорезолвер** *m.* synchro resolver

**синхроридер** *m.* synchroreader

**синхросигнал** *m.* synchronizing signal, sync-signal

**синхроскоп** *m.* synchroscope

**синхротаймер** *m.* synchrotimer

**синхротектор** *m.* synchrotector

**синхротрон** *m.* synchrotron, synchrocyclotron; **с. с знакопеременным градиентом магнитного поля** alternating-gradient synchrotron

**синхротронный** *adj.* synchrotron

**синхрофазотрон** *m.* synchrophasotron

**синхроцепь** *f.* sync circuit

**синхроциклотрон** *m.* synchrocyclotron

**синхрочасы** *pl.* synchronous clock, synchroclock

**синька** *f.* phototype, blueprint; bluing

**сирена** *f.* siren, hooter; howler; **с. Каньяр-Латура** perforated-disk-type siren

**сиринкс** *m.* Pandean pipe, panpipe; syrinx

**« сиркуитрон »** *m.* circuitron

**Сирс** Sears

**сирутор** *m.* cuprous-oxide rectifier of small size

**« сируфер »** *m.* sirufer

**система** *m.* system, method, scheme, arrangement, organization; service; set; assembly; aid; **с. абонентской телеграфной связи** telex system; **абсолютная электромагнитная с. практических единиц** MKS electromagnetic system; **с. автоматического сопровождения цели радиолокатором** automatic tracking loop; **с. авторегулирования по положению** positioning system; **активная гидроакустическая с.** active sonar system; **безрегистровая с.** non-director system; **с. в режиме вынужденных колебаний** driven oscillator; **с. визуального вывода** display system; **с. внутренней автоматической связи** dictograph telephone; **восьмикратная с.** octuplex telegraphy; **вторая с. отклоняющих пластин** beam-positioning plates; **с. вывода (самолётов на аэропорт)** approach system (*nav.*); **гиперболическая с. местоопределения** distance-difference measurement, distance-difference fixing; **групповая с. ультразву-**ковых гидроакустических преобразователей ultrasonic transducer system; **групповая с. распределения** party-line system (*tphny.*); **двоичная с. выражения чисел** binary notation (*math.*); **двойная встречная телеграфная с.** quadruple telegraph system; **двукратная с.** multiplex diode system (*tgphy.*); **двухконсольная с. подвески** center pole suspension, center-bracket system; **двухфазная четырёхпроводная с. с соединением фаз звездой** four-phase system; **десятичная с., составленная по 3-коду** excess-3 coded decimal system; **с. дискретной телефонной связи** telephone transmission reference standard; **дискретная с.** sample-data system; **дуплексная с. с мостовой схемой** bridge duplex system; **с. единиц « метр-килограмм-секунда »** Georgi units; **заказная с. эксплоатации** delay-basis operation, delay working (*tgphy.*); **замедляющая с. ЛБВ** TW tube interaction circuit; **замедляющая с. « пальцы в пальцы »** interdigital structure; **замкнутая с. регулирования** closed cycle control system, error-actuated system; **замкнутая замедляющая с.** re-entrant circuit; **с. записи по двум уровням** two-level return system; **с. записи по трём уровням** three-level return system; **с. защиты от индуктивных влияний** anti-induction device; **с. звукофикации помещений** public address system; **земляная с.** three-wire system (*tphny.*); **импульсная с.** sampled-data system; **импульсная следящая с.** sampling servo-mechanism, sampled-data control system; **с. искателя с машинным приводом** motor-uniselector telephone system; **(колебательная) с., управляемая массой** mass-controlled oscillator; **кольцевая с. соединений** closed circuit arrangement; **комбинированная с. эксплоатации** combined line and recording operation; **с. координат** frame of axes; **координатная с. АТС, с. координатного искателя** crossbar system, crossbar exchange; **лехерова с.** Lecher wire; **линейная колебательная с. с затуханием** damped harmonic system; **маги-**

стральная с. backbone-type system; магнитная отклоняющая с. magnetic yoke (*tv*); мальтийская с. Geneva movement; с. «МБ» local-battery operation; с. Мейснера feedback-oscillator system; многоканальная с. уплотнения multiband carrier-current system, multiband carrier-telephone system; многоканальная с. частотной селекции frequency-division multiplex, FDM; многократная с. одновременной передачи coincidence multiplex system; многократная с. телеграфирования пульсирующим током различных частот phonoplex telegraphy; многократная радиорелейная с. на сантиметровых волнах multiplex microwave radio relaying; многочастотная с. frequency-division multiplex, FDM; с. монтажа с многократным сращиванием провода multi-splicing wiring system; с. на разностной несущей intercarrier system; с. набора задач setup system; с. наведения на среднем участке траектории midcourse guidance system; с. наведения по лучу beam-rider system; с. накопления одной единицы на две точки double-dot system; наружная с. рычагов house pole, roof standard; незамкнутая следящая с. open-cycle control system; с. нелинейных уравнений nonlinear simultaneous equations; немедленная с. эксплоатации "no-delay" service, combined line and recording operation (*tphny*.); непрерывно воспринимающая с. автоматического регулирования continuously sensing control system; неприведённая с. unprimed system; обобщённая следящая с. колебательного управления general oscillating control servomechanism; оборачивающая оптическая с. relay-lens system, relay optical system; с. одноканального приёма звукового сопровождения intercarrier sound system; оптическая с. из плавикового шпата microscope objective with fluorspar component lens; отклоняющая с. deflecting system; yoke, deflecting yoke, deflection yoke, scanning yoke, sweep yoke, mounting yoke; отклоняющая

с. при радиальной развёртке trace rotation system; отклоняющая с. с отогнутыми краями катушек, отклоняющая с. с «уширением» flared deflection yoke; с. отсчёта reference frame; с. «пальцы в пальцы» interdigital structure; с. передачи telephone system, communication system, overhead line system; с. передачи согласованного положения coincidence transmission system; с. переключения антенны transmitter blocker switch; с. питания переменным током с буферной аккумуляторной батареей alternating-current floating storage battery system; повторительная с., повторяющая с. follow-up device; с. подачи (*or* питания) самотёком gravity system; позиционная с. представления чисел, позиционная с. счисления radix notation, positional notation, radix scale; позиционная с. экстремального регулирования peak-holding optimalizing control; позиционная следящая с. zero-displacement-error servosystem; полюсная с. с чередующимися полюсами hemitropic field; полярная с. навигации omnibearing distance navigation, R-Theta; полярная отклоняющая с. deflection system of polar coordinate of tube; потенциальная с. level-signal system, direct-current system; с., преобразующая одну последовательность знаков в другую endomorphic system; с. прерывистого регулирования sampled-data control system; с. приёма звукового сопровождения на разностной несущей intercarrier sound system; с. принудительной (*or* ведомой) синхронизации genlock, general locking; с. проводки с местными распределительными щитами distribution board wiring system; с. проводов distributing network, main system; простая двоичная с. счисления pure binary number base; простейшая колебательная с. simple oscillator; пространственная с. индикации нагрева continuous fire detector; пространственная с. отсчёта reference frame; противолокационная с. radar jamming system; прямоимпульсная

**с. (без пересчёта)** nondirective system; **путевая блокировочная с.** railway block system; **пятикратная с.** multiplex pentode system; **с., работающая в истинном масштабе времени** real-time system; **разнорезонаторная с.** "rising sun" resonator; **разомкнутая с. регулирования** unmonitored control system; **распределительная с. для поразговорных ведомостей** ticket distributing system; **с. регулирования** sampling servomechanism; **с. регулирования с обратной связью** closed-loop control system; **с. рычагов** leverage; **с. с внешней синхронизацией** externally pulsed system (*rdr.*); **с. с восстановленной несущей** exalted-carrier system; **с. с временным разделением каналов** time-division system; **с. с запаздывающим аргументом** time lag system; **с. с обратной связью** feedback system, looped system; **с. с одновременной передачей (цветных) полукадров** field simultaneous (color) system (*tv*); **с. с освобождением промежуточных механизмов искания** common control dial telephone system; **с. с подавлением половины одной из боковых полос** sequi-side-band system; **с. с последовательной передачей (цветных) полукадров** field-sequential (color) system; **с. с последовательным чередованием цветов по точкам изображения** dot sequential system (*tv*); **с. с постоянной глубиной модуляции** floating carrier system; **с. с предгруппами** group modulation system; **с. с применением земли (для посылки импульсов)** earth (return) telegraph, earth-return automatic system, composite signaling; **с. с применением указателей** call-announcer system; **с. с раздельным квантованием сигнала по участкам спектра** hyper-quantizing system; **с. с упреждением** forward-acting system; **с. связи с упреждающими сигналами** predicted-wave signaling system; **сельсинная с. выдачи данных** selsyn-data system; **сквозная с.** backbone-type system; **сквозная с. управления** bushing control; **скоростная следящая с.** servo speed

control; **« скрученная » с.** warped normal mode; **следящая с.** servosystem, servo, servomechanism, coincidence transmission; **с. с. моделирующей установки** computer servomechanisms; **с. с. модульной конструкции** modular control system; **с. с. по частоте** frequency tracker; **с. с. прерывистого (*or* импульсного) действия** definite-correction servomechanism; **с. с. простейшего типа** zeroth-order servomechanism; **с. с. с контролем по скорости** zero-velocity-error servomechanism; **с. с. с несколькими датчиками на входе** multi-actuator system; **с. с. с обратной связью** feedback control system; **с. с. с одной постоянной времени** single-time-lag servo; **с. с. с управлением по скорости** rate servosystem; **следящая сельсинная с.** synchro-system; **смешанная с.** simultaneous-system (*tv*); **смешанная с. питания** a-c floating storage battery system; **совмещающая с. цветного телевидения** compatible color-television system; **с. сравнений** set of congruences; **с. сравнения лепестков характеристики направленности** lobe-comparison system; **с. станций с посторонней хронизацией** master-oscillator radar system; **с. счисления** scale of notation, notation, number system, numerical system; number language; **с. (счисления) в остатках** residue (number) system; **с. счисления с взаимно простыми основаниями** nonconsistently based number system; **с. счисления с отрицательным основанием** negative-base number-representation system; **с. телевидения с отвязкой от сети** free-running TV system; **телеграфная с. с двусторонним током** polar direct-current telegraph system; **телеграфная с. с односторонним током** neutral direct-current telegraph system; **телескопическая с. счётчиков** cosmic-ray telescope, counter telescope; **телефонная с. косвенной связи** indirect routing system; **тепловая с. самонаведения** heat passive homing guidance, infrared passive homing guidance; **с. трансляционного радиовещания** public address system; **с. трансформации переменного тока при постоянном напряжении в постоянный ток постоянной величины** mono-cyclic-square system; **трёхкратная с.** multiplex triode system (*tgphy*.); **трёхпроводная с. постоянного тока** Edison distribution system; **с. управления** monitoring system, control system; final value controller; **с. у. по квадратному корню из модуля рассогласования** sign error root-modulus error system; **с. у. полётом** flight director; **с. у. угловым положением самолёта с ручным наведением на цель** orientational control system with manual guidance; **с. у. станками** machine tool control system; **с. уравнений** set of equations, simultaneous equations; **с. учёта и отчётности** accounting system; **фазовая с. слепой посадки** lateral guidance (of airplanes); **с. формирователей запоминающего устройства** memory driver system; **фотоэлектрическая следящая с.** photoelectric follow-up; **с. цветного телевидения с последовательным чередованием цветов по полям** frame sequential system, field-sequential color-television system; **с. цветного телевидения с сокращённой полосой** band-saving color television system; **с. ЦТ с вычитанием двух цветов из белого цвета** subtractive color system (*tv*); **с. ЦТ с переплетением элементов** dot multiplex system (*tv*); **с. ЦТ с подразделёнными частотными диапазонами** band-shared system (*tv*); **с. ЦТ с точечным перемежением** dot multiplex system (*tv*); **с. ЦТ с чередующимися высокочастотными каналами** alternating highs system (*tv*); **с. чересстрочной развёртки** quadruple scanning; **четырёхзначная с. сигнализации** three-block scheme indication; **четырёхкратная с.** multiplex tetrode system; **шестикратная с.** multiplex hexode system (*tgphy*.); **широкополосная с.** broad-band carrier system; **с. электромеханической аналогии второго рода** electromagnetic analogy; **с. электромеханической аналогии первого рода** electrostatic analogy

**систематизация** *f.* systematization; filing, classification
**систематизировать** *v.* to systematize; to file, to classify, to arrange
**систематика** *f.* systematics, systematism; classification; systemization
**систематический** *adj.* systematic, regular
**систематично** *adv.* systematically
**сито** *n.* sieve, fine screen, screen; sifter, strainer; **ящичное с.** low-pass for waveguides
**ситовый** *adj.* screen
**ситуационный** *adj.* situation
**ситуация** *f.* situation
**сифон** *m.* syringe, syphon, siphon; **с.-рекордер** siphon recorder (*tgphy*.)
**сифонный** *adj.* syringe, siphon, syphon
**сияние** *n.* radiation, radiance, glow, shining, luminescence; aureala, aurora, halo; **полярное с.** aurora polaris **северное с.** aurora borealis; **южное с.** aurora australis
**сиять, засиять** *v.* to radiate, to shine, to beam, to eradiate
**сияющий** *adj.* radiant, beaming, shining
**СК** *abbr.* (**стойка каналов**) channel bay
**сказать** *pf. of* **говорить**
**«скайхук»** *m.* skyhook (*ant.*)
**скалка** *f.* plunger, piston plunger, ram; stem, rolling pin
**скалывание** *n.* cleaving
**скалькировать** *pf. of* **калькировать**
**скалькулировать** *pf. of* **калькулировать**
**скальчатый** *adj.* plunger
**скаляр** *m.* scalar, scalar quantity
**скалярный** *adj.* scalar
**скамейка** *f.* stool; bench
**скамья** *f.* bench; **светоизмерительная с.** bar photometer
**скандиеый** *adj.* scandium
**скандий** *m.* scandium, Sc
**сканирование** *n.* scan, scanning, analyzing
**сканировать** *v.* to scan
**сканирующий** *adj.* scanning, scan-off
**сканистор** *m.* scanistor
**скат** *m.* gradient, pitch, declivity, slope, descent, incline; slide, chute; fall, drop; fall of potential, potential gradient; driver; ramp; cut-off; **колесный с.** set of wheels; **набегающий с. (импульса)** leading edge

(of a pulse); **сбегающий с.** trailing edge

**скафандр** *m.* pressure suit; diving suit

**скачкообразно** *adv.* spasmodically

**скачкообразный** *adj.* uneven, spasmodic, intermittent; abrupt, sudden; step

**скачок** *m.* jump, hop, step, skip, leap, spring; transient (*tv*); disturbance, rapid change, drop; break, discontinuity; (potential) difference; **единичный с.** unit step input

**скашивание** *n.* beveling, chamfering, skewing, sloping

**скашивать, скосить** *v.* to bevel, to slope, to cut aslant, to chamfer, to cant

**скважина** *f.* hole, aperture, chink, slit, pore, gap, interstice, rift; drillhole; well

**скважистый** *adj.* porous; blown

**скважность** *f.* on-off time ratio, duty factor; porosity; **высокая с. импульсов** high pulsing ratio; **с. импульсов** duty cycle; **с. сигнала** mark-space ratio

**сквозной** *adj.* through, continuous; open, transparent; riser

**сквозь** *adv. and prep. acc.* through

**скелет** *m.* skeleton, frame

**скелетный** *adj.* skeleton, frame; skeletal

**скептрон** *m.* sceptron

**скиаграф** *m.* skiagraph

**скиаметр** *m.* skiameter

**скиаскоп** *m.* sciascope

**скиаскопический** *adj.* sciascopic

**скиатрон** *m.* skiatron, dark-trace tube

**скидка** *f.* reduction, deduction, discount, allowance

**скин-глубина** *f.* skin depth

**скинэффект** *m.* skin effect, surface effect

**скип** *m.* skip, charging skip

**скиповый** *adj.* skip

**склад** *m.* store, storeroom, warehouse; storage, store

**складка** *f.* fold, crease, lap, wrinkle, crimp, ply

**складной** *adj.* folding, collapsible; portable

**складный** *adj.* coherent, harmonious

**складской** *adj.* storage; warehouse

**складываемое** *n. adj. decl.* addend (*math.*)

**складывание** *n.* folding; storing, stowing; putting together

**складывать, сложить** *v.* to add, to sum, to summarize; to fold, to fold up; to put together; to accumulate, to store, to pile

**склеенный** *adj.* glued, cemented

**склеивание** *n.* gluing, cementing, pasting; patching

**склеивать, склеить** *v.* to splice, to glue (together), to patch, to cement, to paste (together), to stick (together); to adhere, to size

**склеивающий** *adj.* adhering, adhesive, sticking

**склеить** *pf. of* склеивать, клеить

**склейка** *f.* gluing together, pasting together; patch; splice; adhesive; curing

**склепанный** *adj.* riveted

**склепать** *pf. of* склёпывать

**скёпка** *f.* riveting, fastening

**склёпный** *adj.* riveted

**склёпывать, склепать** *v.* to rivet, to rivet together, to clench

**склёпывающий** *adj.* riveting

**склера** *f.* sclera

**склерометр** *m.* sclerometer

**склерометрический** *adj.* sclerometric

**склероскоп** *m.* scleroscope

**склон** *m.* fall, slope, slide, side, decline, descent, declivity

**склонение** *n.* inclination (*math.*); declination, deflection, variation; depression; dip, declivity, pitch; incline, inclination, slope; **магнитное с.** indication error, magnetic declination; **магнитное с. земли** ground-path error; **с. светила** declination (*astron.*)

**склонённый** *adj.* inclined, sloped; biased

**склонить** *pf. of* склонять

**склонность** *f.* inclination (to, for), tendency, disposition, propensity, bent, aptitude, affinity; **с. к холодному наклёпу** strain hardenability

**склонный** *adj.* inclined (to), disposed (to), given (to), prone, ready (to), susceptible (to), biased; **с. к зуммированию** tending to sing, near singing

**склонять, склонить** *v.* to bias, to incline, to bend

**склоняющийся** *adj.* dipping

**СКО** *abbr.* (**станция кругового обзора**) surveillance radar

**скоба** *f.* clip, spring, strap, bow, yoke; cramp, cleat, clamp, stay clamp; bracket, brace; staple; handle; catch, fastening hook, detent, detainer; **с. для крепления трубчатой проводки** pipe clamp, pipe clip, wall clamp; **измерительная с.** caliper gage; **индикаторная с.** dial snap gage; **калиберная с.** gap gage, caliper gage, snap gage; **предельная с.** limit gap gage; **скрепляющая с.** mounting bracket; **съёмная с. для крепления трубки к стене** spacer-bar saddle

**скобка** *f.* bracket(s), parenthesis; clamp, clip, bracket, loop; **скобки** lugs; parentheses; *see also* **скоба**

**скобочный** *adj.* parenthetic

**скол** *m.* cleaving

**скольжение** *n.* slippage, slide, sliding, glide, gliding; crawl, crawling; backlash; slip; stroke (of piston); yaw

**скользить, скользнуть** *v.* to slip, to skid, to slide; to sweep; to glide; to creep

**скользящий** *adj.* sliding, skidding, slipping, slipper-type, slip, gliding; creeping; sweeping, sweep, grazing

**скомбинированный** *adj.* combined

**скомбинировать** *pf. of* **комбинировать**

**скомпенсированный** *adj.* compensated

**скомпилированный** *adj.* compiled, collected

**скомплектованный** *adj.* ganged

**сконструированный** *adj.* engineered, constructed, designed

**сконструировать** *pf. of* **конструировать**

**сконцентрированный** *adj.* concentrated

**сконцентрировать** *pf. of* **концентрировать**

**скопировать** *pf. of* **копировать**

**скопление** *n.* accumulation, mass, aggregation; congestion, cluster, agglomerate, conglomeration, congregation

**скопленный** *adj.* accumulated, collected

**скопляемый** *adj.* accumulative, cumulative

**скопометр** *m.* scopometer

**скордатура** *f.* scordatura (*mus.*)

**скородействующий** *adj.* high-speed, quick-operating

**скороподъёмность** *f.* rate-of-climb, vertical speed

**скоростемер** *m.* velocity meter

**скоростной** *adj.* velocity; high-speed; rapid; current-type

**скорострельность** *f.* rate-of-fire

**скорострельный** *adj.* quick-firing, rate-of-fire

**скорость** *f.* velocity, speed, rate, rapidity, quickness; **с. в точке опрокидывания** stalling speed; **с. в четырёхмерной системе** quarternary velocity; **с. выбега** running down speed; **с. выборки (данных)** acess speed; **с. вывода** output speed; **с. (выхода газов) у среза сопла** muzzle velocity; **дозвуковая с. конца лопасти** subsonic tip speed; **заданная с.** datum speed; **с. изменения пеленга** angular rate; **с. изодрома** speed of reset; **истинная с. изменения пеленга** true bearing rate; **исходная с.** datum speed; **колебательная с.** a-c velocity; **колебательная с. звука** sound particle velocity; **начальная с. нарастания напряжения возбудителя** initial voltage response of an exciter; **номинальная с.** datum speed; **объёмная с. излучения** volume velocity of the source; **объёмная колебательная с.** volume velocity (across a surface); **с. объёмного истечения** volume emission rate; **относительная с. нарастания возбуждения в течение первой полусекунды** relative voltage response of the exciter during the first half-second; **с. передачи** key speed; speed of transmission; **с. по кругу** peripheral speed; **пониженная с.** underspeed; threading speed; **с. поступления перфокарт** card-handling speed; **с. протягивания киноленты** film velocity; **с. работы трубки** priming speed; **с. развёртки по дальности** range-scanning rate (*rdr.*); **с. разложения (изображения)** pick-up velocity; **сверхсветовая с.** super-velocity of light; **с. спуска** running down speed; **установившаяся с.** free-running speed; **фазовая с. волны в отсутствии тока электронов** interaction circuit phase velocity; **с. хода для использования шумопеленгатора** listening speed; **с. хода механизма**

running speed; **центростремительная с.** inward radical velocity; **шаговая с.** stepping pitch (of a selector); **с. электронов пучка в вольтах** beam voltage; **с. элемента лопасти** blade-section velocity

**скоротечность** *f.* transiency, transience, rapidity, short duration

**скоротечный** *adj.* transient, brief, fast, short-lived

**скорректированный** *adj.* corrected, self-equalized

**скорый** *adj.* rapid, fast, quick, speedy, swift

**скос** *m.* slant, chamfer, chamfering, bevel, rake, slope, taper, tapering, splay; **с. резца** dubbing

**скосить** *pf. of* **скашивать**

**скотография** *f.* scotography, skiagraphy

**скототрон** *m.* scototron

**скотофор** *m.* scotophor

**Скофони** Scophony

**скошённый** *adj.* tapered, chamfered, bevelled, canted; biased

**« скрамблер »** *m.* scrambler (*rad.*)

**скрап** *m.* scrap

**скребковый** *adj.* scraper, blade, scraping

**скребок** *m.* scraper, doctor blade; scratch brush; rake

**скрежещущий** *adj.* grinding, gnashing

**скрепа** *f.* brace, clamp, tie

**скрепить** *pf. of* **скреплять**

**скрепка** *f.* brace, clamp, fastener; splice; split pin

**скрепление** *n.* joint, fastening, tightening, attachment, clamping; bond, brace; splice; strengthening; **с. вмятиной** crimped lock

**скреплённый** *adj.* fastened, clamped; reinforced; cemented

**скреплять, скрепить** *v.* to fasten, to tie, to secure (to); to strengthen, to brace, to clamp, to bolt; to splice; to cement

**скрепляющий** *adj.* fastening, clamping; strengthening, reinforcing; cementing

**скрестить** *pf. of* **скрещивать**

**скрещение** *n.* intersecting, crossing, cross-over; junction; interlacing, twisting

**скрещённый** *adj.* crossed, crossed over; twisted; lattice

**скрещивание** *n.* crossing, transposition, intersecting; junction; twisting, interlacing; **с. проводов в пролёте** span transposition of wires; **с. проводов на опоре** point transposition of wires

**скрещивать, скрестить** *v.* to cross, to intersect, to traverse

**скривить** *pf. of* **кривить**

**скривлённый** *adj.* twisted, warped

**скрип** *m.*, **скрипение** *n.* squeaking, squeak, creaking, creak

**скрипеть, скрипнуть** *v.* to creak, to squeak, to grate

**скрипичный** *adj.* violin

**скрипка** *f.* violin

**скрипнуть** *pf. of* **скрипеть**

**скругление** *n.* rounding off, roundness, curvature

**скрутить** *pf. of* **скручивать, крутить**

**скрутка** *f.* twist, twisting, lay-up, spiraling, lay(ing); (unsoldered) splice; twisted joint; crossing, intercrossing, barreling (of a line); wire-wrapping, end fixture splice; transposition; junction; joint; joining; **с. восьмёркой** eight-fold twisting; **гильзовая с. (проводов)** twisted sleeve joint, McIntyre joint; **с. двойной звездой** spiral-eight twisting, quad pair formation; **звёздная с., с. звездой** spiral quad formation, spiral quad, spiral four; **с. кабеля четвёрками** quad formation; **с. парами** pairing, pair twisting; **простая с.** simple twist joint, Western Union splice; **с. с неодинаковым шагом** noncontinuous barreling (of a line); **с. с одинаковым шагом** continuous barreling (of a line); **ступенчатая волноводная с.** step-twist waveguide; **с. энерголиний** power line crossing

**скруточный** *adj.* twisting, stranding

**скрученный** *adj.* stranded, spliced, splayed, twisted; **с. четвёрками** quadded

**скручиваемый** *adj.* twisting, twisted

**скручивание** *n.* twisting, torsion, coiling action; transposition; twisting of cables, buckling, warping; **с. парами** pairing, twinning; **с. четвёрками** quadding, quad pairing

**скручивать, скрутить** *v.* to twist, to strand, to roll; to spin; to tie up, to bind

**скручивающий** *adj.* twisting, torsion, torsional

**скрытность** *f.* secrecy, secretiveness

**скрытный** *adj.* secretive, hidden, concealed

**скрытый** *adj.* secret, concealed, hidden, cryptic; buried; latent; stored, potential (energy)

**СЛ** *abbr.* (**соединительная линия**) trunk, trunk circuit (*tphny.*)

**слабина** *f.* slack, sag

**слабо** *adv.* faintly, feebly; weakly, mildly, slightly; poorly

**слабовозбуждённый** *adj.* feebly excited

**слабодистрибутивный** *adj.* weakly-distributive

**слабонаправленный** *adj.* low-gain

**слабонатянутый** *adj.* slack, loose

**слабонеоднородный** *adj.* weak(ly) non-homogeneous

**слабообусловленный** *adj.* ill-conditioned

**слабореверберирующий** *adj.* semi-reverberant

**слабосильный** *adj.* feeble, weak

**« слаботочник »** *m.* "milliampere man," weak-current-engineering man

**слаботочный** *adj.* weak-current, low-current; sound (cable)

**слабофокусирующий** *adj.* weak-focusing

**слабый** *adj.* weak, faint, slight, light; loose, slack, lax; poor, low, inefficient; soft, mild

**слаг** *m.* slug

**слагаемое** *n. adj. decl.* addend, augend, item, term (*math.*); summand; sum; **второе с.** augend; **первое с.** addend

**слагательный** *adj.* additive

**слагать, сложить** *v.* to put together, to add, to join; to clasp; to compose

**слагающая** *f. adj. decl.* component, constituent

**слагающий** *adj.* smoothing; component, constituent; cumulative

**сланец** *m.* shale; schist; slate

**сланцевый** *adj.* shale, shaly; flaky, scaly; slatelike; schistose

**слева** *adv.* on the left, from the left, to the left, left-wards, left

**след** *m.* trace, sign, track, mark, trail; wake; spur (of matrix); **с. матрицы** main-diagonal sum; **с. на горизонтальной плоскости** horizontal characteristic; **с. средней длительности** trace P-1

**следить** *v.* to control, to monitor, to observe, to supervise, to watch, to track, to trace, to follow, to hunt

**следование** *n.* sequence, succession, following; spacing; repetition, recurrence; movement

**следовать, последовать** *v.* to follow, to result, to succeed

**« следопыт »** *m.* path finder (*rdr.*)

**следствие** *n.* consequence, effect, result, conclusion

**следующий** *adj.* next, following

**следящий** *adj.* tracking, following; follow-up; control; tracer, servo (mechanism)

**слежение** *n.* following, tracing, tracking, hunting

**слёжечный** *adj.* track

**слёзка** *f.* insulating bead

**слепимость** *f.* glare, dazzle

**слепой** 1. *adj.* blind, sightless; 2. (*noun*) *m. adj. decl.* blind person

**слепота** *f.* blindness

**слесарь** *m.* locksmith; mechanic, fitter

**сливание** *n.* overflow; mixing; pouring off

**сливать, слить** *v.* to merge; to mix, to merge together; to pour off, to run off

**сливаться, слиться** *v.* to merge, to amalgamate, to blend, to interflow, to fuse, to run together, to combine; to overflow

**сливающийся** *adj.* flowing together, blending, fusing, interfluent, confluent

**сливной** *adj.* overflow, pouring; mixed; bleeder

**слизь** *f.* sludge, slime

**слипаемость** *f.* adherence

**слипание** *n.* adherence, adhesion, cohesion; merging; pairing; conglomeration, accumulation; peaking; **с. строк чересстрочного растра** twinning (*tv*); **с. угольного порошка** packing of granules

**слипаться, слипнуться** *v.* to adhere, to cohere, to stick together; to conglomerate

**слипшийся** *adj.* adhering; conglomerate

**слиток** *m.* ingot, pig, bar, rod

**слитый** *adj.* poured-off; joint

**слить(ся)** *pf. of* **сливать(ся)**

**сличать, сличить** *v.* to varify, to collate, to compare, to check

**сличение** *n.* comparison, collation, checking

**сличительный** *adj.* comparative

**сличить** *pf. of* **сличать**

**слияние** *n.* blending, merging, consolidation, coalescence; fusion, union; junction, confluence; **с. знаков** tailing (*tgphy*.); **с. зрительных ощущений** visual fusion

**СЛК** *abbr.* (**стойка линейной коммутации**) high frequency line and patching bay

**словарный** *adj.* dictionary, lexicon

**словарь** *m.* dictionary, lexicon

**словесный** *adj.* oral, verbal

**слово** *n.* word, term; address, say, speech; **составные слова** spondees; **с. с удвоенным числом разрядов** double-length word; **условное с.** keyword

**словоизменение** *n.* inflection, word-changing

**словоизменительный** *adj.* inflectional

**словообразование** *n.* derivation, word-building, word formation

**словораздел** *m.* word boundary

**словосочетание** *n.* phrase, combination of words

**слог** *m.* syllable, logatom; byte

**слоговой** *adj.* syllabic, syllabified; syllable-building

**слогосочетание** *n.* syllable

**слоеватость** *f.* lamination

**слоеватый** *adj.* laminated, stratified; scaly, flaky; slaty

**слоевой** *adj.* layer(-by-layer), bed

**сложение** *n.* addition, summation, adding (*math.*); composing; composition, build, constitution; integration; combining

**сложённый** *adj.* formed, built; folded

**сложимый** *adj.* summable (*math.*)

**сложить** *pf. of* **слагать**, **складывать**

**сложно** *adv.* complexly, complicatedly

**сложность** *f.* complexity, complication, intricacy; multiplicity

**сложный** *adj.* complicated, complex, intricate; involved; compound; multiple, multiplex; composite; accumulative; multi-element; reticular

**слоисто-неоднородный** *adj.* layered-inhomogeneous

**слоистость** *f.* lamination; stratification

**слоистый** *adj.* laminated, lamellar, lamelate, flaky; stratified, stratiated, layered, layer; sandwich-type

**слой** *m.* layer, ply, coat(ing), film, stratum, bed, band; lamina, lamination, sheet; blanket; barrier; **слоями** in layers; **вентильный с.** rectifying barrier, barrier layer, barrier film; **запирающий с.** barrier film, barrier region, back-biased barrier; depletion region, transition region; **поверхностный с. распространения** surface-duct propagation; **поверхностный волнопроводящий с.** surface duct; **с. половинного ослабления** half-value thickness, half-value layer; **с. скачка** layer of sharp density gradient; thermal barrier; **с. температурного скачка** thermal layer; **ударный с.** shock layer

**сломать(ся)** *pf. of* **ломать(ся)**

**служба** *f.* service; duty, job, work; **с. аэронавигации осуществляемой подвижными радиостанциями** aeronautical mobile radio service; **с. защиты от помех** radio interference suppression service; **с. испытания на качество** testing service; **коммутационная с.** telephone exchange service; **с. контроля** supervisory service, observation service; **материальная с. связи** ordnance department of communication; **Международная с. связи** International Telecommunication Service; **с. на коммутаторе** telephone exchange service; **с. предупреждения ПВО** Aircraft Warning System; **с. профилактического надзора** maintenance service; **с. сооружения дальней связи** construction service in telecommunication; **стационарная с. радионавигации** aeronautical fixed radio service; **телефонная заказная с.** absent subscriber service; **с. технического содержания** maintenance service; **с. устранения неисправностей** fault clearing service

**служебный** *adj.* official; auxiliary, ancillary; maintenance, service; order

**слух** *m.* hearing; **абсолютный с.** sense of absolute pitch

**слухач** *m.* sound-reading operator, sound-reader, listener (*tgphy*.)

**слуховой** *adj.* auditory, audio, acoustic, aural

**случай** *m.* case, occurrence, incident, circumstance, instance, opportunity; **частный с.** special case (*math.*)

случайно *adv.* accidentally, by chance, casually

случайность *f.* randomness, absence of pattern; chance, accident; contingency (*math.*)

случайный *adj.* random, incidental, stray, haphazard, arbitrary, accidental, casual, chance, occasional

слушание *n.* listening, hearing

слушатель *m.* hearer, listener

слушать, послушать *v.* to listen; to hear

слушающий *adj.* listening; listener

слышимость *f.* audibility, audibleness, hearing; quality of reception; understanding; с. переходных разговоров side-to-phantom crosstalk, overhearing; с. собственного аппарата sidetone generation

слышимый *adj.* audible

слышно *adv.* audibly

слышный *adj.* audible, heard

слэг *m.* slug

слюда *f.* mica, glimmer; индийская с. macallen mica; колотая с. block mica; комовая с. rough cobbed mica; прокалённая с. calcined mica; пятнистая с. с включениями spotted mica; смотровая с. skylight mica; с. со стеклянным наполнением glass-bonded mica; щипанная с. muscovite, mica splittings

слюдинит *m.* mica combination

слюдистый *adj.* micaceous

слюдной, слюдяной *adj.* mica, micaceous

сляб *m.* slab

слябинг *m.* slabbing mill

см *abbr.* (сантиметр) cm, centimeter

смазанный *adj.* blurred, smeared; lubricated, greased

смазать *pf. of* смазывать

смазка *f.* greasing; oiling, lubrication; grease, fat, lubricant, oil; adhesive; с. для скольжения cable grease; с. маслом под давлением forced feed lubrication

смазочный *adj.* lubricating, lubrication, grease

смазывание *n.* lubrication, greasing, oiling; slurring; smearing effect

смазывать, смазать *v.* to lubricate, to grease, to oil; to smear

смазывающий *adj.* lubricating, greasing, oiling; smearing

сматывание *n.* winding, reeling; unreeling, uncoiling; с. в бухту coiling

сматывать, смотать *v.* to wind, to reel (on); to unreel, to unroll, to uncoil, to wind off, to run off, to pay out

смачиваемый *adj.* wettable, hydrophilic

смачивание *n.* wetting, moistening

смачивать, смочить *v.* to wet, to moisten, to soak, to damp; to steep, to imbue

смачивающий *adj.* wetting

СМВ *abbr.* (сантиметровые волны) centimeter waves, microwaves

смежно *adv.* contiguously

смежность *f.* contiguity, adjacency, juxtaposition, proximity

смежный *adj.* adjacent, contiguous, neighboring, adjoining, proximate; allied; affine (*math.*)

смена *f.* changing, change, shift, interchange, replacement, replacing, renewing; reversal

сменить(ся) *pf. of* сменять(ся)

сменность *f.* interchangeability

сменный *adj.* changeable, interchangeable, renewable, exchangeable, removable, plug-in; change; shift; replaceable; noncontinuous; staggered

сменяемость *f.* removability, interchangeability

сменяемый *adj.* removable, renewable, interchangeable

сменять, сменить *v.* to change, to supersede; to replace, to remove, to renew, to exchange, to interchange; to shift

сменяться, смениться *v.* to take turns, to alternate; to exchange (for)

смерить *pf. of* мерить

смертельный, смертный *adj.* mortal, deadly, fatal, lethal; death

смеситель *m.* mixer, mixing tank, combiner unit; (frequency-)mixer tube; converter; adder; mixing pad; frequency changer; с. для сантиметрового диапазона microwave mixer; обычный с. unbalanced mixer; с. со связью окном iris-coupled mixer

смесительный *adj.* mixer, mixing

сместить *pf. of* смещать

смесь *f.* mixture, mix, composition, blend, conglomerate, compound; газовоздушная с. carburetted air

**смета** *f.* estimate, appraisal, estimated cost

**смешанный** *adj.* complex, composite, mixed, compound, combination, blended; hybrid; series-parallel

**смешать** *pf. of* смешивать, мешать

**смешение** *n.* mixing action, mixer action, mixing, blending, merging, combination; mixture; conversion; complication; **обратное с. при УКВ** re-conversion at VHF; **с. окрашенных световых потоков** mixing of colored lights

**смешивание** *n.* mixing, blending; stirring, agitation; jumpering

**смешивать, смешать** *v.* to mix, to blend, to merge, to combine; to lump together; to confuse; to compound; to stir, to agitate

**смешивающий** *adj.* mixing

**смещать, сместить** *v.* to displace, to offset, to remove, to shift, to dislodge; to bias; to reset

**смещающий** *adj.* shifting

**смещение** *n.* displacement, dislodgment, shift, shifting, reset, resetting; removal, dislocation, disturbance, distortion, offset; deviation; bias, biasing; parallax; drift; upset; **со смещением** biased; **со смещением для отсечки** biased to cut-off; **вч с. а-с** magnetic biasing; **с. нулевого положения** zero drift; **с. нулевой точки** zero error; **с. от цепи катода** cathode bias; **с. отметки** pip displacement (*rdr.*); **с. относительного центра** decentering; **относительное с. растров** pattern spacing; **с. по вертикали** vertical centering; **с. разговоров** co-ordination (*acoust.*); **с. растра** off-centering; **с. и стирание** bias-erase; **с. характеристики** zero drift; **с. частиц на поверхности** circumferential displacement; **электрическое с.** dielectric flux density

**смещённый** *adj.* shifted, displaced, dislocated, out of line; off, off-set, off-center; biased; skew; staggered; **с. до запирания** biased to cutoff; **с. по частоте** frequency-shift

**Смит** Smith

**смоделированный** *adj.* simulated

**смола** *f.* tar, pitch; resin, rosin, gum; **горная с.** bitumen; **кумароновая с.** indene resin

**смолевой** *adj.* tar, tarred; resin

**смолённый** *adj.* resined; tarred, pitched

**смолистость** *f.* resinousness; tarriness, pitchiness

**смолистый** *adj.* resinous, resin; tarry, pitch

**смолкать, смолкнуть** *v.* to grow silent, to fall into silence; to cease

**смоляной** *adj.* tar, tarry, pitch; resin, rosin, resinous; negative, resinous (*elec.*)

**смонтированный** *adj.* mounted, assembled, erected, built up, set up; **с. в стойке** rack-mounted; **вплотную с.** close-coupled; **с. заподлино** flush-mounted

**смонтировать** *pf. of* монтировать

**смотровой** *adj.* observation, sight, inspection

**смоченный** *adj.* moistened, humidified, wetted

**смочить** *pf. of* смачивать

**смыкание** *n.* closure, closing; pinch-off; joining; punch-through breakdown

**смыкать, сомкнуть** *v.* to shut, to close; to joint, to fit in; to link, to interlock, to couple; to clamp

**смысл** *m.* meaning, sense, significance

**смысловой** *adj.* semantic; logical

**смычковый** *adj.* bow, bowed

**смычок** *m.* bow

**смягчать, смягчить** *v.* to damp, to modify, to moderate; to soften, to mellow, to mollify

**смягчающий** *adj.* softening

**смягчение** *n.* softening; damping, modification, moderation

**смягчённый** *adj.* softened; modified, moderated; subdued

**смягчить** *pf. of* смягчать

**СН** *abbr.* (**самонаведение**) homing, homing guidance

**снаб** *abbr.* (**снабжение**) supply, provision

**снабдить** *pf. of* снабжать

**снабжаемый** *adj.* supplied, furnished, provided; **с. энергией** powered

**снабжать, снабдить** *v.* to supply, to provide, to feed, to deliver; to furnish, to equip; to store

**снабжающий** *adj.* supplying, supply, delivery, delivering

**снабжение** *n.* supply, provision, delivery, delivering, supplying, feed; **с. углом** (*or* **выступом**) angulation

**снабжённый** *adj.* supplied, provided; **с. транзисторами** transistorized

**Снайдер** Snyder

**снайперскоп** *m.* sniperscope

**снаряд** *m.* projectile, missile, shell; apparatus, instrument, gear, contrivance; **зенитный (управляемый) с.** ground-to-air missile, surface-to-air missile; **с. с крестообразным оперением** cruciform missile; **с. с подлодки по подводной цели** underwater-to-underwater missile; **с. с подлодки по самолёту** underwater-to-air missile; **телеуправляемый с.** **с системой самонаведения** radar guided missile with target homing; **управляемый с.** guided missile

**снаряжать, снарядить** *v.* to equip, to fit out, to supply, to provide, to furnish

**снаряжение** *n.* equipage, equipment, fittings, outfit; implements

**снаряжённый** *adj.* equipped, outfitted

**снег** *m.* snow

**снеговой** *adj.* snow

**Снелл** Snell

**снижать, снизить** *v.* to reduce, to build down, to lessen, to lower, to bring down, to cut, to abate

**снижающий** *adj.* lowering, reducing

**снижение** *n.* reduction, build(ing)-down, decrease, lowering, deterioration, drooping; drop, dropping, abatement; downlead (of an antenna); descent; **с. вибраций за счёт рассеяния** dissipative control of vibration; **с. механических напряжений** stress relief; **с. ранга** nullity of a matrix (*math.*); **с. скорости счёта** scaling of the counting rate

**сниженный** *adj.* reduced, decreased; geared-down

**снизить** *pf. of* **снижать**

**сниматель** *m.* pickup

**снимать, снять** *v.* to remove, to take off, to take down; to dump; to pick-off, to strip, to skin; to record, to take (a reading); to take a picture, to film, to photograph; to plot (a curve); **с. закоротку** to unshort; **с. кинофильм** to film; **с. показания** to take a reading; **с. с эксплоатации** to put out of service

**снимающий** *adj.* scraping, shaving, stripping; scan-off, play-off

**снимок** *m.* photograph, print, picture, shot; copy; **с. крупным планом** close-up, close-up view; **с., полученный по теневому методу** shadow photograph; **с. средним планом** medium shot; **фотографический с.** photograph

**снова** *adv.* anew, again, afresh; re-

**сноп** *m.* beam, shaft, cone (of light); shower (of sparks)

**снос** *m.* pulling-down; wrecking, demolition; drift, shift; wear

**сноска** *f.* note, footnote, reference

**СНР** *abbr.* (**станция наведения ракет**) missile guidance station

**Снук** Snook

**снуперскоп** *m.* snooper scope

**снятие** *n.* taking down, removal, dismantling; dump; erasing; taking (a reading); throwing-off; plotting (a curve); **быстрое с. возбуждения** quick-response de-energizing; **с. возбуждения** de-energization; **с. возбуждения противотоком** de-energizing of synchronous generators; quick-response de-energizing; **с. вольтсекундной характеристики** time-voltage test; **с. диаграммы излучения антенны** pattern measurement; **с. индикаторной диаграммы** indicating, indicator test; **с. карантина** pratique; **с. показаний (счётчиков)** reading, taking a meter reading; **с. характеристики антенны** antenna pattern measurement

**снятый** *adj.* removed, withdrawn, taken down, taken off; **при снятом кожухе** uncased

**снять** *pf. of* **снимать**

**со** *see* **с**

**собачка** *f.* clamp dog, dog, cam, catch, stop, detent; pawl; trigger, click, trip; **движущая с.** driving pawl, driving cam; **двойная стопорная с.** double detent, double dog, pair of pawls; **поворотная с.** catch hook; **храповая с., с. храповика** driving pawl, ratchet cam, ratchet mechanism

**собирание** *n.* collecting, collection, gathering, assembling

**собиратель** *m.* collector, gatherer; conductor

**собирательный** *adj.* collective, collecting; converging

**собирать, собрать** *v.* to gather, to collect, to accumulate, to aggregate,

to catch; to equip, to fit (out), to erect, to assemble, to mount, to install; **с. схему** to connect up, to hook-up, to patch-up

**собираться, собраться** *v.* to agglomerate, to gather, to collect, to congregate

**собирающий** *adj.* collecting; counting; converging

**собрание** *n.* bank, assembly; collection, accumulation, complex, congregation

**собранный** *adj.* assembled, gathered, collected; erected, built-up, mounted; **с. из стандартных узлов** modularized; **с. на магнисторах** magnistorized

**собрать(ся)** *pf. of* **собирать(ся)**

**собственно** *adv.* properly, correctly, strictly

**собственный** *adj.* natural, proper, inherent, fundamental, intrinsic; own; self-; eigen-; internal; i-type; **с собственным питанием** self-energizing

**событие** *n.* event, occurrence; **равновозможные события** equally likely possibilities (*math.*)

**совершать, совершить** *v.* to do, to perform, to accomplish, to effect, to achieve

**совершение** *n.* completion, fulfillment, performance, achievement, accomplishment

**совершенно** *adv.* quite, completely, entirely, fully, wholly, totally, absolutely, utterly, perfectly

**совершенный** *adj.* perfect, ideal, complete, absolute, thorough

**совершить** *pf. of* **совершать**

**совещание** *n.* conference; communication

**совковый** *adj.* scoop, shovel

**совместимость** *f.* compatibility; consistency

**совместимый** *adj.* compatible, combinable; consistent

**совместить** *pf. of* **совмещать**

**совместно** *adv.* jointly, in common, together, simultaneously

**совместно-используемый** *adj.* shared

**совместность** *f.* compatibility; consistency

**совместный** *adj.* common, joint, combined, team, simultaneous; composite; compatible; **сильно с.** super-consistent

**совмещаемость** *f.* congruence, congruency

**совмещать, совместить** *v.* to combine (with), to join; to overlap, to superimpose

**совмещающий** *adj.* compatible; coincident

**совмещающийся** *adj.* coinciding; superposable

**совмещение** *n.* superimposing, superimposition, superposition, coincidence; combination; matching, registry, registration, registering; convergence; fusion; overlapping; **с. изображения с картой** chart matching device; **оптическое с.** optical registering (*tv*); **с. цветов** registration of colors

**совмещённый** *adj.* in register, registered; superimposed, superposed; combined, integrated

**совокупно** *adv.* jointly, in common

**совокупность** *f.* set, series, aggregate, combination, conjunction, assembly, totality; part; population (*math.*); system; **выборочная с.** sample, selection; **генеральная с.** population (*math.*); **с. отметок шкалы** scale marks; **с. проводов одинакового назначения** lines of equal importance and/or equal properties; **с. рациональных чисел** rationality range (*math.*)

**совокупный** *adj.* joint, combined, aggregate, collective, cumulative

**совпадать, совпасть** *v.* to merge, to coincide (with), to conform, to agree

**совпадающий** *adj.* concurrent, coincidental, coincident, congruent, corresponding; **с. во времени** synchronized; **с. по фазе** in phase, co-phasal

**совпадение** *n.* coincidence, concurrence, congruence, concordance, conformity, matching, correspondence; accord; superposition

**совпасть** *pf. of* **совпадать**

**современный** *adj.* up-to-date, contemporary, modern, recent

**согласающий** *adj.* matching

**согласно** *adv.* according (to), in accordance (with), in conformity (with), accordingly, conformably, consistently

**согласно-компаундный** *adj.* cumulative-compound

**согласность** *f.* consistence; consonance, concord, harmony; concordance

**согласный** 1. *adj.* consonantal, consonant; concordant, harmonious, conforming (to), agreeable, consistent (with); 2 (*noun*) *m. adj. decl.* consonant

**согласование** *n.* match, matching, harmony, coordination, agreement, correspondence, accordance, adapting, adaptation, balance; register; coincidence; smoothing; termination; congruence (*math.*); **с. изображений** registration (*tv*); **оптическое с.** optical registering (*tv*); **с. с помощью пластины у вершины** vertex-plate matching; **шумовое с.** matching for minimum noise transfer

**согласованно** *adv.* in concord

**согласованность** *f.* co-ordination, agreement, coherence, concordance, accordance, harmony, match, compatibility, consistency

**согласованный** *adj.* co-ordinate(d), simultaneous, (pre)concerted, accordant; consistent; matched

**согласователь** *m.* matcher; waveguide stub; coupler

**согласовательный** *adj.* matching

**согласовать(ся)** *pf. of.* **согласовывать(ся)**

**согласовывать, согласовать** *v.* to match, to coordinate, to harmonize, to balance, to accord, to adapt; to correlate; to adjust; to comply, to accommodate; **с. по фазе, с. фазу** to bring in phase

**согласовываться, согласоваться** *v.* to agree, to conform, to comply; to coincide; to cohere

**согласовывающий, согласующий** *adj.* matching

**согласующийся** *adj.* consistent, conforming; compatible

**соглашение** *n.* agreement, understanding, arrangement, contract

**согнутый** *adj.* bent, curved; **с. на угол** angled

**согнуть** *pf. of* **сгибать, гнуть**

**согревание** *n.* heating, warming

**согреватель** *m.* heater

**согревательный** *adj.* heating

**согревать, согреть** *v.* to heat, to warm

**согретый** *adj.* heated, warmed

**согреть** *pf. of* **согревать, греть**

**« содар »** *m.* Sodar

**Содди** Soddy

**содействие** *n.* aid, cooperation, help, assistance, concurrence

**содействовать, посодействовать** *v.* to assist, to help, to cooperate, to concur; to contribute, to further, to promote

**содействующий** *adj.* assisting, helping; contributing, promoting; additional

**Содерберг** Soderberg

**содержание** *n.* content(s), capacity, volume; percentage; area; matter, substance; concentration; maintenance, upkeep; housing (system); **с. информации** signal complex; **кратное с.** summary

**содержательный** *adj.* pithy, informal

**содержать** *v.* to contain, to hold, to include, to comprise; to keep, to maintain; **с. в исправности** to keep in working order

**содержащий** *adj.* containing

**содержимое** *n. adj. decl.* contents

**содержимость** *f.* capacity, volume

**содрать** *pf. of* **сдирать**

**соединение** *n.* connection, connecting, interconnection, junction, joining, bond, join, joint(ing), communication, bonding; wiring; coupling, linkage, link, linking; seam; compound; combination, formation, union; splice, splicing; juxtaposition; call; grouping; **неперекрещивающиеся лобовые соединения** concentric connections; **без шунтового соединения** unshunted; **блокировочное с.** single-unit circuit; **с. взезду** star connection; **с. в параллелограмм** parallel wiring; **с. в прямоугольник** rectangular wiring; **вещательное с.** electrophone call, theatrophone call (*tphny.*); **взаимное с. (между двумя абонентами коллективной линии)** reverting call (between two parties on the same line); **с. вибраторов в одну линию** collinear array; **с. внапуск** lap joint; **с. впритык** butt joint, abutting joint; **входное с.** incoming trunk (*tphny.*); **с. двойным треугольником** hexagon connection, double-delta connection; **двухрядное заклепочное с.** double-line riveting; **с. для параллельной работы** position coupling; **замедленное с.** delayed call; **с. звезда-зигзаг** star-interconnected-star connection; **К-образное замкнутое с.** closed double-bevel butt weld; **лобовое с.** end winding, face-ring connection; **междугородное с.** trunk call, trunk connection, toll call; **междуламповое с. с двумя дросселями** double-impedance coupling; **с. многократных полей** interconnecting (*tphny.*) **с. многоугольником** mesh connection, ring connection; **с. муфтой** sleeve joint; **с. на пазах двух волноводов** slit coupling of wave-guides; **с. на шлифе** ground joint; **немедленное с.** no-delay call, effective call (*tphny.*); **неплотное с. в цепи** microphone joint (*tphny.*); **несостоящееся с.** ineffective call, uncompleted call, lost call, abandoned call; **с. перемычкой** bonding; **с. по обходным путям** rerouting service (*tphny.*); **последовательно-параллельное с.** series-multiple connection, series-parallel connection; **с. при помощи лепестков** stapling; **притёртое спаянное с.** wiped joint; **пробное с.** test call, tentative call; **с. проводом** wiring; **с. распорками** strutting; **резьбовое с. труб** bolted pipe joint; **с. с обратной последовательностью** cross-connection of wave windings; **свободное с.** idle trunk; **с. скруткой** twisted joint; **с. скручиваемой гильзой** twisted sleeve joint; **смешанное с.** series-multiple connection, series-parallel connection; **согласующее с.** mesh connection; **сомкнутое стыковое с. с двусторонним чашечным скосом** closed double-J butt weld; **состоявшееся с.** completed call, effective call; **срочное с.** plug-out connection (*tphny.*); **с. твёрдым припоем** brazed joint; **транзитное с. через два коммутатора** double switch call (*tphny.*); **универсальное шарнирное с.** gymbal; **с. цугом** tandem connection; **шарнирное с.** pin joint; **шаровое шарнирное с.** ball-and-socket joint; **штепсельное с., штыревое с.** plug and socket connector, bullet connection; **шунтовое с.** by-pass (connection), shunt

**соединённый** *adj.* united, combined, connected, joined, coupled, joint, conjugated; poled; banked; **галь**ванически с. d-c coupled; **последовательно с.** cascade-connected

**соединитель** *m.* coupler, connector, connecting device, jointing, sleeve, binder; bond; **линейный с.** final selector, connector, line selector; **многократный координатный с.** cross-bar (selector) switch; **общий линейный с.** rotary hunting connector; **продольный с. кабельной арматуры** continuity cable bond; **рельсовый с.** propulsion bond, rail bond; **стыковой с.** bond, track bond; **стыковой с. контактного рельса** conductor bond

**соединительный** *adj.* connecting, coupling, connector, interconnecting, binding, jointing, joint; matching; connective; conjunctive

**соединить(ся)** *pf. of* **соединять(ся)**

**соединяемый** *adj.* combinable

**соединять, соединить** *v.* to join, to joint, to unite, to connect, to conjugate, to aggregate; to combine, to jumper, to interconnect, to wire up, to bridge, to link up, to couple; to switch on; to clutch, to put in gear; to juxtapose; **взаимно с.** to interlock; **с. встречно** to connect in reverse direction; **с. линии накрест** (*or* **навстречу**) to transpose circuits; **с. навстречу** (*or* **накрест**) to cross-connect; **с. накладками** to overlap, to lash; **с. на прямую** to put through (*tphny.*); **с. перекрёстно** to interchange; **с. распорками** to strut, to stay; **с. сквозно** to jumper

**соединяться, соединиться** *v.* to put through a call; to unite, to combine, to join, to fuse, to coalesce, to congregate

**соединяющий** *adj.* connecting, coupling, joining, linking

**созвездие** *n.* constellation

**созвучие** *n.* assonance, consonance, harmony, accord, accordance

**созвучный** *adj.* consonant (with, to), in harmony (with); sympathetic

**создаваемый** *adj.* created, made, formed, produced, generated

**создавать, создать** *v.* to create, to make, to form, to build, to set up, to stage; to produce, to construct; to generate

**создаваться, создаться** *v.* to be created; to (be) formed

**создание** *n.* production, creation, making, formation, generation; construction, building, erection; **с. фазовой разности (сигналов) по методу тормозящего поля** transit-time excitation of oscillators by means of the retarding-field method

**создать(ся)** *pf. of* **создавать(ся)**

**создающий** *adj.* generating, producing, forming

**сознание** *n.* consciousness, sense; acknowledgement

**сознательно** *adv.* consciously, knowingly

**сознательность** *f.* consciousness, awareness

**сознательный** *adj.* conscious

**соизменимость** *f.* covariance (*math.*)

**соизменимый** *adj.* covariant

**сойтись** *pf. of* **сходиться**

**сокр.** *abbr.* (**сокращение**) abbreviation

**сократимость** *f.* reductibility

**сократимый** *adj.* reductible

**сократительный** *adj.* contracting, contraction

**сократить** *pf. of* **сокращать**

**сокращаемость** *f.* contractibility, contractility

**сокращаемый** *adj.* reducible; contractible, contractile

**сокращать, сократить** *v.* to diminish, to shorten, to curtail; to minimize, to reduce; to cancel; to lay off; to abbreviate; to contract

**сокращающий** *adj.* abbreviating; contracting; reducing

**сокращение** *n.* contraction, constriction; condensation, abbreviation, reduction, decrease; shortcut, shortening; cutting down; cancellation; shrinkage; **с. динамического диапазона** volume contraction; **с. полосы частот** bandwidth compression

**сокращённо** *adv.* briefly, shortly

**сокращённость** *f.* shortness, brevity, compression, abridgement

**сокращённый** *adj.* abbreviated, brief, concise, succinct, abridged; reduced; short-cut; contracted; diminished

**сокрытие** *n.* concealment (system)

**сокрытый** *adj.* concealed, secret

**солариметр** *m.* solarimeter

**«солдат-мотор»** *m.* foot-operated generator

**солдетрон** *m.* soldetron

**солемер** *m.* salinometer, halometer

**соленоид** *m.* solenoid, solenoid coil, magnet coil, air-core coil, actuator

**соленоидальный** *adj.* solenoidal

**соленоидный** *adj.* solenoid

**солёность** *f.* salinity, saltiness

**солёный** *adj.* salt; salty, saline, salted

**солион** *m.* solion (device)

**солистрон** *m.* solistron

**солнечный** *adj.* sun, sunny, solar

**солнце** *n.* sun; sunshine; **«горное с.»** quartz lamp; Finsen arc lamp

**солнцеискатель** *m.* sun-follower, sun-seeking device

**соло** *n. indecl.* solo (*mus.*)

**солодайн** *m.* solodyne

**солодин** *m.* solodyne, unidyne

**соль** *n. indecl.* sol, G (*mus.*)

**соль** *f.* salt

**сольватация** *f.* solvation

**сольмизация** *f.* solmization

**сольный** *adj.* solo (*mus.*)

**сольфеджио** *n. indecl.* solfeggio, solf-fa (*mus.*)

**соляной** *adj.* salt, saline

**соляриметр** *m.* solarimeter

**солярометр** *m.* solarometer

**сомкнутый** *adj.* closed, locked; joined

**сомножество** *n.* corresponding set, coset (*math.*)

**сомножитель** *m.* factor (*math.*); *n*-значный с. *n*-digit factor

**СОН** *abbr.* (**станция орудийной наводки**) gun-laying radar

**сон** *m.* sone

**«сонаграф»** *m.* sonagraph

**«соналатор»** *m.* sonalator

**Сонар** *m.* Sonar

**Сондерс** Saunders

**соника** *f.* sonics

**соникатор** *m.* sonicator

**Сонне** Sonne

**сонолюминесценция** *f.* sonoluminescence

**сонометр** *m.* sonometer, phonometer

**сонохемилюминесценция** *f.* sonochemiluminescence

**соображать, сообразить** *v.* to contrive, to combine; to consider, to take into consideration

**соображение** *n.* consideration, deliberation, calculation, conception, idea, notion

**сообразить** *pf. of* **соображать**

**сообщать, сообщить** *v.* to report, to signal, to notify, to inform, to communicate, to impart, to

transmit; **с. энергию** to energize
**сообщаться, сообщиться** *v.* to be in communication (with), to communicate (with)
**сообщающийся** *adj.* communicating
**сообщение** *n.* report, notice, message, reference, information, communication, announcement; traffic; call; short circuit, cross connection, contact; **автоматическое междугородное с.** automatic long-distance service; **заказное с.** record traffic; **земляное с.** ground fault; **междугородное с.** long-distance communication, trunk traffic, toll traffic; **с. местоположения** position record; **с. начального заряда** priming; **с. обратного хода** reversing; **с. оптимальной частоты** optimum working-frequency prediction; **с. проводов** short, wire fault, cross, line-to-line; **разговорное с.** telephone traffic; **с. с землей** air-ground communication; line-to-ground fault, ground fault; **с. с массой** body contact, ground leakage; **телеграфное с.** teletypewriter service
**сообщённый** *adj.* communicated; imparted, given
**сообщить(ся)** *pf. of* **сообщать(ся)**
**сооружать, соорудить** *v.* to install, to erect, to build, to construct, to establish
**сооружение** *n.* structure, edifice; building, construction, erection, installation; facility, plant; **линейные сооружения** outside plant (*tphny.*); **маломощное транспортное с.** light-duty conveying equipment
**соосность** *f.* coaxial alignment
**соосный** *adj.* coaxial, uniaxial, collinear
**соответственно** *adv.* accordingly, correspondingly, according to, in accordance with, consequently; respectively
**соответственность** *f.* conformance, correspondence, conformity; suitability
**соответственный** *adj.* corresponding, conforming, conformal, related, congruent, homologous; suitable, proper, pertinent
**соответствие** *n.* conformity, congruence, conformance, accordance, compliance, correspondence, correlation; adequacy; expediency, fitness

**соответствовать** *v.* to correspond (to, with), to conform (to), to match, to coincide, to correlate; to be in line (with); to suit, to fit; to respond
**соответствующий** *adj.* corresponding (to), equivalent (to); conformable (to); proper, suitable, appropriate, right, fit; adequate; coincident; specific
**соотносительность** *f.* correlation
**соотносительный** *adj.* correlative
**соотношение** *n.* relation(ship), connection, correspondence, correlation; ratio, rate, proportion; **соотношения обхода** continuation relations; **с. между молекулярным притяжением и взаимодействием диполей** Keesom relationship; **с. неопределённости** uncertainty principles (*math.*); **с. потерь** exchange rate; **с. сторон** (*or* **формат**) aspect ratio (*tv*)
**сопельный** *adj.* nozzle, jet
**сопло** *n.* nipple, nozzle, spout, jet, orifice; muzzle; **с. переменного сечения** variable-area nozzle; **с. постоянного сечения** fixed nozzle; **с. с критическим сечением** critical-flow nozzle
**сопловой** *adj.* nozzle, nipple, jet
**сопоставить** *pf. of* **сопоставлять**
**сопоставление** *n.* comparison, contrast; juxtaposition
**сопоставлять, сопоставить** *v.* to compare (with), to contrast; to juxtapose
**сопоставляющий** *adj.* comparing, contrasting
**сопрано** *n. indecl.* soprano (*mus.*)
**сопредельный** *adj.* adjacent, adjoining, contiguous
**соприкасаться, соприкоснуться** *v.* to be contiguous (to), to border, to be adjacent (to), to adjoin; to come into contact (with), to touch
**соприкасающий(ся)** *adj.* adjoining, abutting; touching, contiguous
**соприкосновение** *n.* contact, touch, touching; contiguity, juxtaposition; osculation (*math.*)
**соприкосновенность** *f.* contiguity
**соприкосновенный** *adj.* contiguous (to)
**соприкоснуться** *pf. of* **соприкасаться**
**сопроводитель** *m.* tracker, tracking device, follower
**сопровождать, сопроводить** *v.* to track, to follow; to escort, to convoy, to accompany; to guide

**сопровождающий** *adj.* tracking, following; accompanying; guiding
**сопровождение** *n.* accompaniment; tracking; convoy, escort; **автоматическое с. по дальности** automatic range control, automatic range tracking; **с. в процессе обзора** track-while-scan; **звуковое с. на разностной несущей** intercarrier sound; **с. « на проходе »** track-while-scan; **с. передаваемого объекта** follow shot (*tv*); **с. по дальности** range tracking **с. по углам** angle tracking; **с. по углу** direct tracking; **с. по углу места** elevation tracking; **полуавтоматическое с. по дальности** range-aided tracking; **с. снаряда в командной системе наведения** joystick control
**сопротивиться** *pf. of* **сопротивляться**
**сопротивление** *n.* resistance, impedance, reactance, opposition, drag; strength; resistor, resistor element; **комплексные сопряжённые сопротивления** conjugate impedances; **с согласованными акустическими сопротивлениями** acoustically matched; **с. автоматического смещения на сетку** un-by-passed cathode resistor; **активное с.** resistance; **акустическое с. при переменном потоке** alternating-flow impedance; **антипаразитное с. в цепи сетки** grid suppressor; **балансное с.** ohmic terminating-resistance; **безразмерное с.** specific impedance; **безразмерное активное акустическое с.** acoustic resistance ratio; **безразмерное реактивное акустическое с.** acoustic reactance ratio; **с. в начале линии** sending-end impedance; **с. в омах** ohmage; **с. в точке возбуждения** (*or* **приложения силы**) driving-point impedance; **с. в цепи накала** filament resistor; **с. в цепи сетки** grid leak; **с. вентильного провода** gate resistance; **внесённое с. (движения)** motional impedance; **внешнее с. усилителя с катодным выходом** external cathode resistance; **с. внешней цепи** terminal resistance; **вносимое (полное) с.** reflected impedance, reflected secondary impedance, insertion impedance, coupled impedance; **внутреннее с.** plate resistance; source resistance, source impedance,

differential resistance, interface resistance, internal resistance; **внутреннее с. фиксирующей цепи** (*or* **схемы**) clamp resistance; **временное с. разрыву** yield point; **временное с. растяжения** tensile strength; **вспомогательное нагрузочное с.** bleeder; **с. входного конца рупора** acoustic throat impedance; **входное с.** entry impedance, input resistance, bleeder resistance; sending-end impedance, line impedance; **входное с. цепи, симметричной относительно « земли »** balanced-to-ground input impedance; **выходное с.** termination, terminating device; **гасящее с.** damping resistance; absorbing resistor, (voltage) dropping resistor, bleeder; **с. гашения** shunt-breaking resistance; **с. двухпроводной линии** loop resistance; **действующее реактивное с.** effective reactance; **с. делителя напряжения** bleeder resistance (*tv*); **дифференциальное с. разрядного промежутка** anode differential resistance, anode resistance; **с. для предупреждения** "preventive" resistance; **с. для устранения** anti-interference resistance; **с. для шунтирования** diverter; **с. для шунтирования амперметра** ammeter shunt; **добавочное с.** preventive lead; instrument multiplier; **добавочное с. вольтметра** (*or* **к вольтметру**) multiplying coil, voltage multiplier; **добавочное с. (к прибору)** range multiplier, instrument multiplier, multiplier resistance, meter multiplier; series resistance; dropping resistor; **дроссельное с.** inductive resistance; **ёмкостное с.** capacitance; condensance, **ёмкостное гасительное с.** capacitor-discharge resistance; **замедляющее с.** timing resistance; **запасное с.** series cut-out; **с. заторможённого преобразователя** blocked impedance (of a transducer); **с. заторможённой системы** clamped impedance; **защитное с. тормоза** protective resistance for rheostatic braking; **с. изгибу при ударной пробе** impact-bending resistance; **с. излучения среды** characteristic acoustic resistance, radiation resistance; **изменяющееся с.** varistor;

импульсное реактивное с. transient reactance, subtransient reactance; **индикаторное с.** current viewing resistor; **инерционное (реактивное) с.** mass reactance (*acoust.*); **комплексное с.** vector impedance; **с. кратер-анод** plate-target resistance; **с. лампы переменному току** anode impedance; **линеаризующее с.** peak resistance; peaking resistor; **с. (линии) в оба конца** go-and-return impedance; **с. линии при холостом ходе** open-circuit impedance of a line; **механическое с. при колебательном смещении** linear mechanical impedance; **механическое с. при поперечных колебаниях** transverse mechanical impedance; **с. на свече** suppressor resistor; **с. навитое на шнур питания** line-cord resistor; **с. нагрузки генератора** oscillator loading; **нагрузочное с.** bleeder, bleeder resistor, terminating resistor; **направленное с.** reactive resistance; **с. ненагружённого преобразователя** short-circuit impedance, free impedance; **непроволочное с.** carbon resistor; metallized resistor; **номинальное с. нагрузки** rated impedance; **объёмное с.** compound resistor, composition resistor; **объёмное углеродистое постоянное с.** carbon composition fixed resistor; **объёмное удельное с. (материала)** volume resistivity (of material); insulativity; **омическое с.** resistor, ohmic conductor, low-frequency resistance, real resistance; **с. освещённого фотоэлемента** light resistance; **отводящее с.** bleeder, drain resistor, drainage resistor; **отнесённое к массе удельное с.** mass resistivity; **отрицательное комплексное с.** expendance; **параллельное полное с.** shunt impedance, leak impedance; **с. переменному току** anode impedance; **с. перехода, переходное с.** contact resistance; **переходное полное с. по продолной (or поперечной) оси** direct-axis transient impedance; **поверхностное углеродистое с.** deposited-carbon resistor; **погонное с.** resistance per unit-length, **погружаемое нагревательное с.** immersion unit; **с. покоя** initial resistance; **полное с.** impedance, total resis-

tance; alternating current resistance; **п. с. обратной последовательности** negative-sequence field impedance; **п. с., определяющее функцию передачи** computing impedance; **п. с. связи** coupled impedance; **п. с. холостого хода** no-load impedance, open-circuit impedance; **п. с. цепи** iterative impedance; **полное динамическое амплитудное с. электрода** electrode impedance; **полное крутильное механическое с.** angular mechanical impedance; **полное синхронное с. по поперечной оси** quadrature-axis synchronous impedance; **полное синхронное с. по продольной оси** direct-axis synchronous impedance; **понижающее с.** voltage dropping resistor; **поперечное с.** leak resistance, inductive resistance; **последовательно включённое с.** dropping resistor; series resistance (of a tube); **последовательное с.** series resistance; series resistor; **постоянно включённое с. нагрузки** bleeder resistor; **с. потоку, полученное при продувании** d-c acoustic resistance; **предельное с. поездного шунта** train shunt resistance; **проволочное (регулировочное) с.** wire-wound rheostat, wire-wound resistor; **с. продольному изгибу** buckling strength, cross breaking strength; **с. разомкнутой схемы** no-load impedance, open-circuit impedance; **разрядное с.** shunt-breaking resistance; **с. растеканию импульсов тока** current-propagation earth resistance; **реактивное с.** reactance; **реактивное с. от жёсткости** stiffness reactance (*acoust.*); **реактивное с. рассеяния** leakage reactance; **реактивное динамическое амплитудное с. электрода** electrode reactance; **с. реакции** reaction impedance of fourpole networks; **регулируемое с.** convergence phase control; **регулируемое угольное с.** carbon regulator, carbon-pile regulator; **реостатное с.** rheostat; **с. ротора** armature resistance; **сверхпереходное полное с. по продольной (or поперечной) оси** direct-axis subtransient impedance; **с. связи** mutual impedance, reflected impedance; **с. сжатию** compressive strength; **сме-**

щающее сеточное с. grid leak;
с. согласования пентода в режиме
A matching impedance of a class-A
pentode; стабилизующее с. (выпрямителя) bleeder of a rectifier; стабильное углеродистое с. cracked-
carbon resistor; теплозависимое с.
thermistor; (трубчатое) эмалированное с. vitreous resistor; с. у
нижней питающей точки base
driving-point resistance; с. у точки
опоры (antenna) base impedance;
удельное магнитное с. reluctivity;
с. упора в цепи двигателя jam
resistor; с. урдокс current-regulator
tube; с. утечки leakage resistance;
bleeder; с. утечки в цепи сетки
incremental grid resistance, a-c resistance; с. холостого хода open-
circuit characteristic, no-load impedance; с. цепи катода cathode
resistor, bias resistor; с. четырёхполюсника при холостом ходе apparent
resistance of four-terminal networks
at no-load conditions, no-load resistance; электрическое с. закреплённой (or заторможённой) системы
clamped electrical impedance; эффективное с. антенны в точке питания antenna-feed impedance
сопротивляемость f. resistibility, resistivity, specific resistance; resistance, strength
сопротивляться, сопротивиться v. to
resist, to oppose; to hold out, to
bear up
сопротивляющийся adj. resisting, resistant
сопрягать, сопрячь v. to gang, to
couple, to join, to conjugate; с.
(контуры) to track (circuits)
сопрягающий adj. tracking, padding
сопряжение n. ganging, coupling, linking, interlinking, union, conjugation,
junction; tracking, padding; с. гетеродина по трём точкам three-point
tracking; с. контуров преобразователя padding of oscillator
сопряжённость f. interlinking; see also
сопряжение
сопряжённый adj. associated, combined, connected, linked, coupled,
paired, conjugate; interconnected,
ganged; adjoint
сопрячь pf. of сопрягать
сопутствующий adj. accompanying,

attendant, concomitant; satellite
соразмерение n. corresponding, matching
соразмерить pf. of соразмерять
соразмерно adv. in proportion (to)
соразмерность f. symmetry, proportionality; matching
соразмерный adj. proportional, proportionate; fit, adequate
соразмерять, соразмерить v. to regulate, to proportion; to relate, to
match, to fit together
сорбционный adj. sorption; sputter
сорбция f. sorption
сороковосьмигранник m. hexakisoctahedron
сороковосьмигранный adj. hexoctahedral
сорт m. quality, sort, grade, kind,
variety, nature; rate
сортамент, сортимент m. assortment;
gage (of wire); set; grading, grades
сортировальный adj. sorting
сортирование n. sorting, classification, assorting, arranging, ranging;
grading
сортированный adj. sorted, graded,
classified; matched; screened
сортировать, рассортировать v. to
classify, to sort, to assort, to grade,
to pick (out), to screen, to size, to
separate
сортировка f. assortment, sorting,
separation, grading, classification,
sizing; matching; batching; sorting
machine, sorter, grader, classifier;
с. чисел по основанию системы
счисления radix sorting
сортировочный adj. sorting, separating, distributing; screening
сортировщик m. sorter, grader
сортирующий adj. sorting, grading,
classifying
сортовой adj. sort, variety; section
(-shaped)
сортрон m. sortron
Сорэ Soret
соседний adj. adjacent, adjoining,
neighboring, next, near by; affine
соскабливание n. scraping off
соскабливать, соскоблить v. to scrape
off
соскакивание n. slipping, jumping off
соскакивать, соскочить v. to slip, to
jump off, to spring off, to come off
соскальзывание n. slip, slide, sliding

**соскальзывать, соскользнуть** *v.* to slide (off, down), to slip (off), to glide, to skid

**соскоблить** *pf. of* **соскабливать**

**соскользнувший** *adj.* slipped (off)

**соскользнуть** *pf. of* **соскальзывать**

**соскочить** *pf. of* **соскакивать**

**сослаться** *pf. of* **ссылаться**

**сосредоточение** *n.* concentration, centering

**сосредоточенно** *adv.* with concentration; intently

**сосредоточенность** *f.* concentration

**сосредоточенный** *adj.* concentrated, centered; focused; lumped

**сосредоточивать, сосредоточить** *v.* to concentrate, to center, to centralize; to fix, to focus; to lump

**состав** *m.* compound, composite; composition, formation, constitution; body; staff; structure; design; set; **амплитудный с. спектра импульса** pulse amplitude spectrum; **с. для заливки муфты** box compound; **консервирующий с.** preservative

**составитель** *m.* compiler; originator (of a radio message); **с. программы** programmer

**составить** *pf. of* **составлять**

**составление** *n.* composition, compilation, compiling, drawing up; construction; preparation, formation, combination; working out; **с. программы вычислений** factorization; **с. уравнения** construction (*math.*)

**составленный** *adj.* composed, made up

**составлять, составить** *v.* to put together, to make up; to compose, to compile, to work out, to prepare, to plan; to design, to set up, to build up, to make, to construct; to comprise, to constitute, to form; **с. схему** to hook up, to connect up

**составляющая** *f. adj. decl.* component, constituent; ingredient, product; **активная с.** resistive component, active component, inphase component, power component; **активная с. тока** active current, wattful current; **горизонтальная с. напряженности земного магнетизма** horizontal intensity of the terrestrial magnetic field; **квадратурная с.** quadrature-phase component; **квадратурная с. поднесущей** QCW

signal, quadrature-continuous-wave signal; **с. ошибки с удвоенной частотой пеленга** quadrantal component; **переменная с.** a-c component, oscillating component; **переходная с. электродвижущей силы по продольной** (*or* **поперечной**) **оси** direct-axis transient electromotive force; **с. по поперечной оси, поперечная с.** quadrature component; **постоянная с.** direct component; d-c component; zero-frequency component; **п. с. анодного тока** feed current, d-c component of plate current; **п. с. сеточного напряжения** direct grid bias; **п. с. тока пучка** d-c beam current; **продольная с.** direct-axis component; **реактивная с.** reactive component, quadrature component, wattless component, idle component, imaginary component; **сверхпереходная с. электродвижущей силы по продольной** (*or* **поперечной**) **оси** direct-axis subtransient electromotive force; **с. электродвижущей силы по продольной** (*or* **поперечной**) **оси** direct-axis component of the electromotive force

**составляющий** *adj.* component, constituent

**составной** *adj.* composite, compound (ed), composed, combined, multipart, aggregate, unitized; complicated; component, constituent, constitutive; complex; plug-in; built-up, joined, jointed, sectional, separable; link, chain

**состаренный** *adj.* aged

**состаривание** *n.* aging

**состарить** *pf. of* **старить**

**состоявшийся** *adj.* formed; completed

**состояние** *n.* state, condition, status, stage, position, situation; behavior; phase; **с двумя состояниями** two-state; **с двумя устойчивыми состояниями** bistable; **исходное с.** reset state; **с. коммутационного аппарата** switching state; **с. пониженного напряжения** operating condition for vacuum tube oscillators; **скрытое с.** latency, abeyance

**состоятельность** *f.* consistency; competence

**состоятельный** *adj.* consistent; responsible

**состоять** *v.* to comprise, to be made up (of), to consist (of)

**состоящий** *adj.* consisting (of), made (of)

**сосуд** *m.* vessel, receiver, container, jar, tank, receptacle

**сосчитанный** *adj.* counted

**сосчитывать, сосчитать** *v.* to add up, to compute, to count, to calculate; to number

**сотая** *f. adj. decl.* hundredth; **выраженный в сотых долях** centesimal

**сотенный** *adj.* hundredth, centesimal

**Соти** Sauty

**сотканный** *adj.* woven

**сотня** *f.* one hundred

**сотовый** *adj.* honeycomb; cellular

**сотообразный** *adj.* honeycombed; cellular

**сотрясатель** *m.* shaker

**сотрясательный** *adj.* shaking

**сотрясать, сотрясти** *v.* to shake, to vibrate

**сотрясающийся** *adj.* shaking, jigging

**сотрясение** *n.* shake, shaking, jarring, vibration; pulsation (of sound); racking, commotion

**сотрясти** *pf. of* **сотрясать**

**сотый** *adj.* hundredth, centesimal

**соударение** *n.* impact, collision; encounter, shock, impingement

**софар** *m.* sofar (sound finding and ranging)

**софит** *m.* studio light board(s); decorative mask, framing mask (*tv*); **ламповый с.** control-light strip

**софитный** *adj.* soffit; festoon

**софокусный** *adj.* confocal

**соффитный** *adj.* soffit

**сохранение** *n.* conservation, preservation, reservation, retention; constancy

**сохранить** *pf. of* **сохранять, хранить**

**сохранно** *adv.* safely, securely

**сохранность** *f.* shelf life; safety; preservation

**сохранный** *adj.* safe, secure

**сохраняемость** *f.* retentivity

**сохранять, сохранить** *v.* to hold, to maintain, to keep; to conserve, to preserve, to retain, to save

**сохраняющий** *adj.* holding, retaining

**сохраняющийся** *adj.* nonvolatile

**сочетание** *n.* combination, joining, union, conjunction; matching; **с. аэродинамических рулей с сервомотором** aerofoil-servomotor system; **с. нескольких видов колебаний** multimode propagation; **разностное с.** subtractive combination (in intermodulation); **сложное с.** multiplexing

**сочетательный** *adj.* combinative, associative

**сочетать** *v.* to combine, to unite, to join, to connect; to match

**сочленённый** *adj.* coupled, connected, linked, interlinked, jointed, articulated

**сочление** *n.* connection, joint, articulation, coupling, link, gang(ing), junction; join; pin joint, concatenation; **угломестное вращательное с.** elevation rotating joint

**сочленять, сочленить** *v.* to aggregate, to join, to joint, to link

**союз** *m.* union, association, alliance; combination, conjunction

**СП** *abbr.* (**светогасильный прибор**) blinker

**спад** *m.*, **спадание** *n.* fall, fall-off, diminution, drop, decrease, abatement, decrement; roll-off; decay, decaying; loss, droop, drooping; sag, sagging; slope, incline; collapse; build(ing)-down; taper

**спадать, спасть** *v.* to fall, to fall off, to diminish, to abate, to drop, to go down, to decrease, to build down, to lower, to recede; to collapse

**спадающий** *adj.* decreasing, drooping, sloping; negative-going

**спаивание** *n.* soldering

**спаивать, спаять** *v.* to solder, to weld, to fuse; to unite

**спай** *m.* joint, seam, solder, soldering; brazing; cohesion; junction; seal; **отлипающий с.** strip, stripped seal; **с. термопары** thermojunction

**спайка** *f.* solder; soldering, brazing; cohesion; joint, seam, jointing, splice, splicing; junction; soldered connection; sealing

**спайщик** *m.* splicer, jointer

**спаренный** *adj.* connected, coupled, paired; twin, duplex, dual; ganged; interlocking

**спаривание** *n.* coupling, pairing, interlinking, twinning, impairment; ganging; **с. строк (чересстрочного растра)** pairing, twinning (*tv*)

**спаривать, спарить** *v.* to couple, to pair, to match

**спаривающий** *adj.* coupling, pairing
**спарить** *pf. of* **спаривать**
**спасательный** *adj.* rescue, (life-)saving, safety
**спасистор** *m.* spacistor
**спасть** *pf. of* **спадать**
**спаянный** *adj.* soldered; united
**спаять** *pf. of* **спаивать**
**СПВРД** *abbr.* (**сверхзвуковой прямоточный воздушно-реактивный двигатель**) supersonic ramjet
**спейсистор** *m.* spacistor, depletion-region device
**спекание** *n.* sticking, burning (of contacts); agglomeration, sintering, caking, agglutination
**спекаться, спечься** *v.* to cake, to clinker; to sinter; to burn, to stick
**спекающийся** *adj.* caking, clinkering, sintering
**спектакль** *m.* show
**спектр** *m.* spectrum; range (of frequencies); **с. комбинационного рассеяния** Raman spectrum; **полосатый с. с группами линий** band spectrum; **частотный с. импульсной посылки** pulse-train spectrum
**спектральный** *adj.* spectrum, spectral
**спектро-** *prefix* spectro-, spectrum
**спектроанализатор** *m.* spectrum analyzer; frequency response analyzer, Fourier analyzer
**спектрогелиограмма** *f.* spectroheliogram
**спектрогелиограф** *m.* spectroheliograph
**спектрогелиоскоп** *m.* spectrohelioscope
**спектрограмма** *f.* spectrogram
**спектрограф** *m.* spectrograph
**спектрографический** *adj.* spectrographic
**спектрометр** *m.* spectrometer
**спектрометрический** *adj.* spectrometric
**спектрометрия** *f.* spectrometry
**спектрорадиометр** *m.* spectroradiometer
**спектрорадиометрический** *adj.* spectroradiometric
**спектроскоп** *m.* spectroscope
**спектроскопический** *adj.* spectroscopic
**спектроскопия** *f.* spectroscopy; **с. на сантиметорых волнах** microwave spectroscopy

**спектрофотометр** *m.* spectrophotometer
**спектрофотометрический** *adj.* spectrophotometric
**спектрофотометрия** *f.* spectrophotometry
**спектрофотоэлектрический** *adj.* spectrophotoelectric
**спектрохимический** *adj.* spectrochemical
**спектрохимия** *f.* spectrochemistry
**спекшийся** *adj.* sintered, caked, baked; molded
**спереди** *adv.* in front, before
**Сперри** Sperry
**спесификация** *see* **спецификация**
**специализированный** *adj.* specialized, special purpose
**специальный** *adj.* special, separate, specific, particular; singular (*math.*)
**спецификатор** *m.* specificator
**спецификация** *f.* specification, statement, description; piece register
**специфический** *adj.* specific
**специфичность** *f.* specificity
**спецлиния** *f.* unlisted line (*tphny.*)
**спецслужба** *f.* special service
**спечённый** *adj.* sintered
**спечься** *pf. of* **спекаться**
**спешный** *adj.* urgent, pressing, hasty
**« спидомакс »** *m.* speedomax
**спидометр** *m.* speedometer, speed counter; **дистанционный с.** odometer
**« спикерфон »** *m.* speakerphone
**спикула** *f.* spicule
**спин** *m.* spin
**спин-обменный** *adj.* spin-exchange
**спиновый** *adj.* spin
**спинор** *m.* spinor
**спинрешёточный** *adj.* spin-lattice
**спин-спиновый** *adj.* spin-spin
**спинтарископ** *m.* spinthariscope, scintillascope
**спин-фононный** *adj.* spin-phonon
**спирализация** *f.* spiralization, spiraling, coiling
**спирализованный** *adj.* coiled
**спираль** *f.* spiral, spire, helix, snail; filament; **с. в волноводе** spiral septum; **встречная с.** double helix; **крутая с.** fast spiral; **простая с.** single-coil filament
**спирально-лучевой** *adj.* spiral-beam
**спирально-сферический** *adj.* helisphere
**спиральный** *adj.* spiral, helical; twist; volute, gyroidal

**спиратрон** *m.* spiratron

**спирометр** *m.* spirometer

**спиротрон** *m.* spirotron

**спирт** *m.* alcohol, spirit

**спиртовой** *adj.* alcohol, alcoholic, spirit

**спиртометр** *m.* alcoholometer

**списать** *pf. of* **списывать**

**список** *m.* list, index, register, record, chart, schedule; directory; **с. абонентов** telephone directory

**списывание** *n.* transcription, copying

**списывать, списать** *v.* to copy, to transcribe

**сплав** *m.* composition, alloy, compound; fusion, sintering; **с. для заливки подшипников** bearing metal

**сплавление** *n.* melting, fusion; alloying; floating; **с. контактов** sticking of contacts

**сплавленный** *adj.* melted, fused; alloyed; drifting, floated

**сплавно-диффузионный** *adj.* post-alloy-diffused

**сплавной** *adj.* alloy-type, alloyed

**сплетать, сплести** *v.* to interlace, to interweave, to weave, to intertwine, to plait, to braid; to joint, to cable joint; to splice

**сплетение** *n.* interweaving, interlacing, meshing; splice

**сплошной** *adj.* continuous, entire, unbroken; solid, massive, compact; uniform; heavy

**сплошность** *f.* (magnetic) continuity

**сплюснутый** *adj.* flattened, stretched out; oblate

**сплюснуть** *pf. of* **сплющивать**

**сплющение** *n.* flattening

**сплющенный** *adj.* flattened (out), stretched out; oblate

**сплющивание** *n.* flattening; telescoping, collapse

**сплющивать, сплюснуть** *or* **сплющить** *v.* to flatten, to compress; to telescope

**спокойно-рабочий** *adj.* make-and-break (*tphny.*)

**спокойный** *adj.* stagnant; calm, quiet, quiescent, resting, at rest; smooth; static

**спокойствие** *n.* quietness, placidity, calm, calmness

**сползание** *n.* slipping off, slipping down, creep, crawl, crawling; drift; **с. тона** wow (*acoust.*)

**сползать, сползти** *v.* to crawl, to creep

**спонжекс** *m.* spongex

**спонтанный** *adj.* spontaneous

**спорадический** *adj.* sporadic

**способ** *m.* way, method, means, manner, process, mode, system, procedure, practice; medium; **буферный с. зарядки** floating method of charging; **с. встречного включения** reverse-rotation testing method, potentiometer method; **с. двойного образца-двойного ярма** double-bar and double-yoke method; **с. двойного угла** two bearings and run between; **с. задания** representation (*math.*); **с. испытания стержневого образца в железном ярме** bar and yoke method; **с. магнитного перешейка** isthmus method (of magnetization); **однопроцессный с.** monobath technique; **с. пробных включений** cut-and-try method; **с. прямого вызова** ringing junction working, signal junction working (*tphny.*); **с. создания растра** technique of display; **степенной с. выражения чисел** exponential notation

**способность** *f.* capability, capacity, aptitude, power, ability, faculty; property; *suffix:* -ability, -ibility; **обладающий способностью самовосстановления** self-repairing; **включающая с.** making-capacity; **задерживающая с.** retentivity, retentiveness; **излучательная с.** emissivity, emittance, radiation capacity; **импульсная пропускная с.** discharge capacity; **исправляющая с.** margin (*tgphy.*); **с. к включению** making capacity; **с. к выпрямлению** detectability, rectifying ability; **с. к модуляции** modulability; **с. к намагничению** magnetizability; **с. к саморегулированию** homeostatic mechanism; **коммутационная с.** switch capacity; **отражательная с., отражающая с.** reflectance, reflectivity, reflective power; **полная излучательная с.** total emissivity (of light); **предельная пропускная с.** channel capacity (information theory); **пропускная с.** current capacity; traffic (carrying) capacity (*tphny.*; carrying capacity; operating capacity; code capacity; communication capacity, informational capacity; «**пропускная с.**» **апертуры** aperture admittance; **разрешающая с.** resolving power, resolution, resolution capability, definition; discrimination, acuity; **р. с. по азимуту и дальности** range and bearing discrimination; **р. с. по горизонтали** resolution in line direction, horizontal resolution; **р. с. по направлению** bearing discrimination; **р. с. по расстоянию** spatial resolution; **р. с. по строкам** resolution in line direction; **р. с. по угловым координатам** angular resolution; **р. с. по форме сигнала** waveform resolution; **р. с. регулятора** control resolution; **р. с. трафарета** mask resolution; **р. с. шкалы** openness of scale; **секундная пропускная с.** amount of information, redundancy; **спектральная излучательная с.** spectral emissivity (of light); **теплотворная с.** calorific value; **фильтрующая с.** filter discrimination; **хроматическая разрешающая с.** chromaticity discrimination

**способный** *adj.* capable, able, apt, fit; **с. к делению** fissionable

**способствование** *n.* assisting, helping, aiding, aid, contribution

**способствовать, поспособствовать** *v.* to further, to promote, to favor, to contribute, to aid, to assist; to be conductive (to); to effect, to cause

**способствующий** *adj.* instrumental

**справа** *adv.* on the right, to the right, from the right side

**справедливо** *adv.* justly, rightly; true; fairly

**справедливость** *f.* validity, truth, correctness, accuracy; justice, fairness

**справедливый** *adj.* valid, true, correct, accurate; just, right

**справиться** *pf. of* **справляться**

**справка** *f.* information, reference

**справляться, справиться** *v.* to manage, to master, to handle, to cope (with), to inquire, to consult, to ask (about)

**справочная** *f. adj. decl.* information office, information desk; **междугородная с.** long-distance information, toll directory desk

**справочник** *m.* reference book, manual; directory

**справочный** *adj.* information, inquiry, reference

**спрессованный** *adj.* pressed

**спрессовать** *pf. of* **прессовать**

**спринклер** *m.* sprinkler

**спринклерный** *adj.* sprinkler

**спроектированный** *adj.* designed, planned

**спроектировать** *pf. of* **проектировать**

**спрос** *m.* requirement, demand, market; inquiry

**спрягать, проспрягать** *v.* to conjugate

**спряжение** *n.* conjugation

**спрямитель** *m.* squarer

**спрямить** *pf. of* **спрямлять**

**спрямление** *n.* squaring; alignment; rectification

**спрямлять, спрямить** *v.* to square; to align; to rectify

**спрямляющий** *adj.* aligning; squaring, squarer

**СПУ** *abbr.* (**самолётное переговорное устройство**) aircraft intercommunication system

**спуск** *m.* lowering; descent, slope, incline; slide, chute; detent; trigger, trigger action, triggering; release; tripping; discharge, drain; tapping, draining, drawing off; **антенный с.** downlead; **с. дивертора** down conductor

**спускать, спустить** *v.* to trigger, to trip, to release; to lower, to let down; to discharge, to drain (off), to run off, to tap; to unwind

**спускаться, спуститься** *v.* to fall; to come down, to descend; to slide, to move down; to be lowered

**спускающийся** *adj.* descending; sloping

**спускной** *adj.* drain, discharge, escape, outlet; rise-and-fall; drop; lowering; release, releasing, trigger

**спусковой** *adj.* trigger

**спустить(ся)** *pf. of* **спускать(ся)**

**спутник** *m.* satellite; **с.-ретранслятор** communication satellite

**спутниковый** *adj.* satellite

**спущенный** *adj.* lowered, let down; drained, run off

**срабатывание** *n.* response; operation, action, actuation, actuating; count (in a radiation counter); firing; wear, abrasion; release, trip, triggering; **с замедленным срабатыванием** slow-operating; **с. без выдержки времени** instantaneous disengagement; **быстрое с.** quick operation; **ложное с.** misoperation; **ложное с. трубки** spurious tube count; **моментное с.** speed releasing; **неправильное с.** malfunction; **паразитное с. трубки** spurious tube count; **преждевременное с.** pretriggering

**срабатывать, сработать** *v.* to respond, to come into action; to wear away, to wear; to operate, to pull up; to fire; to trip

**срабатываться, сработаться** *v.* to wear, to wear away, to deteriorate

**срабатывающий** *adj.* operating

**сработавшийся** *adj.* worn, worn out, used up

**сработанный** *adj.* made; worn out

**сработать(ся)** *pf. of* **срабатывать(ся)**

**сравнение** *n.* comparison; congruence (*math.*); reference; matching; collation

**сравнивание** *n.* leveling; comparison

**сравнивать, сравнить** *v.* to compare, to parallel; to level, to make even

**сравнивающий** *adj.* comparing, comparison; collator

**сравнимость** *f.* comparabiltiy, congruence

**сравнимый** *adj.* comparable, congruous; **с. по модулю N** equal modulo N

**сравнительно** *adv.* comparatively, relatively

**сравнительный** *adj.* comparative, compared, relative

**сравнить** *pf. of* **сравнивать**

**срастание** *n.* adherence, accretion, growth; coalition, coalescence; **двойниковое с.** twinning (of crystals)

**срастаться, срастись** *v.* to concrete, to grow together, to intergrow, to interlock, to coalesce

**срастить(ся)** *pf. of* **сращивать(ся)**, **срастаться**

**сращенный** *adj.* grown together, combined, joined

**сращивание** *n.* joining, jointing; interlinking, fusing; union, binding; splice, joint

**сращиватель** *m.* jointer, splicer

**сращивать, срастить** *v.* to join, to joint, to splice, to combine, to unite

**сращиваться, сраститься** *v.* to interlock, to entangle, to intergrow, to grow together

**сращивающий** *adj.* splicing

**среда** *f.* medium; atmosphere; fluid, agent; surroundings, environment; stratum; **совместно колеблющиеся среды** resonant acoustic medium; **одностороннепропускающая волноводная с.** nonreciprocal waveguide medium

**среди** *prep. genit.* among, amongst, amidst, in the middle; inter-

**срединный** *adj.* middle, medial, mean

**средне-** *prefix* central, middle, medium

**средневзвешенный** *adj.* average-weighted

**средневольтный** *adj.* medium-voltage

**средневременной** *adj.* medium-term

**среднеглубинный** *adj.* shallow-water

**среднее** *n. adj. decl.* mean, average; **взвешенное с.** corrected mean value; **с. по времени** time average; **с. по множеству** assembly average

**среднеквадратический, среднеквадратичный** *adj.* RMS, root-mean-square, mean-root-square

**среднелетальный** *adj.* median-lethal

**среднелинейный** *adj.* midline

**среднепупинизированный** *adj.* medium-loaded

**среднечастотный** *adj.* mid-frequency

**средний** *adj.* average, middle, mean, medial, medium, median, central; moderate, intermediate; center; inner; neutral; **в среднем** at the average, on the average; **выше среднего** above the average

**средоточие** *n.* center, point of concentration, concentration, focus

**средство** *n.* means, way, facility, aid; agent, medium; *pl.* facilities; **радиовспомогательные средства** radio aids; **с. вывода данных** output medium

**срез** *m.* slice, cut; shear, shearing, shearing off; cut-off; droop; crossover; **нижняя частота среза** lower cut-off frequency; **с. X** face perpendicular cut, Curie cut, normal cut, X-cut; **с. Y** face parallel cut, parallel cut

**срезание** *n.* cutting, clipping, shear, shearing; truncation

**срезанный** *adj.* cut, sheared, chopped, clipped; truncated

**срезатель** *m.* clipper

**срéзать** *pf. of* **срезáть, срезывать, резать**

**среза́ть, сре́зать** *v.* to cut (away), to cut off, to shear, to shear off, to trim; to slide

**срезываемый** *adj.* sheared

**срезывание** *n.* shearing, cutting; truncation; beveling

**срезыватель** *m.* clipper

**срезывать** *see* **среза́ть**

**срезывающий** *adj.* shearing; equalizing; attenuation

**СРО** *abbr.* (**самолётный радиолокационный ответчик**) airborne responder

**сровнять** *pf. of* **ровнять**

**сродство** *n.* affinity, relationship

**срок** *m.* date, term, period, fixed time, time, stretch of time; deadline; **с. годности при хранении** shelf life; **с. действительности заказа на переговор** period of validity of a call; **с. износа** life time; **малый с. службы** short-life; **межремонтный с. службы** operating life; **предельный с. службы** age limit; **прослуженный с.** age; **расчётный с. службы** design life; **с. службы** life (of equipment), age, operating life, service life; **с. хранения** shelf life; **эксплуатационный с. службы** service life

**сросток** *m.* joint, splice, junction, junction union, splicing; unsoldered connection; attachment, adhesion; **с. впритык** butt joint; **гильзовый скручиваемый с.** twisted sleeve joint, McIntyre joint; **горячий с.** soldered joint; **ответвительный с.** T-joint, multiple joint; **проходной с.** straight joint, straight splice; **прочный по тяжению с.** tension joint; **разветвительный с.** T-splice, multiple joint; **скрученный с.** spliced joint, twist joint, splayed joint; **тройниковый с.** tee-joint, T-joint, Y-joint, tap joint; branch pipe; **холодный с.** dry joint, solderless joint, solderless splice

**срочно** *adv.* quickly, urgently

**срочность** *f.* urgency

**срочный** *adj.* urgent, pressing

**ср. ск.** *abbr.* (**средняя скорость**) mean velocity, average speed

**сруб** *m.* frame, framework, shell, skeleton, cage

**срыв** *m.* collapse, disruption; breakaway, break-down; shedding; sink (of an oscillator); tearing; **с. колебаний** stopping the oscillations

**срывание** *n.* tearing off (or away); **с. дуги** blowout

**срыватель** *m.* blowout, magnetic blowout

**СС** *abbr.* (**смешанная стойка**) miscellaneous bay (*tphny.*)

**ссаживающий** *adj.* bucking

**ссылаться, сослаться** *v.* to direct, to point at, to indicate, to interpret, to refer (to); to cite, to quote

**ссылка** *f.* reference, citation

**СТ** *abbr.* (**стандарт**) standard

**СтА** *abbr.* (**стартерный аккумулятор**) starter battery

**стабилидин** *m.* stabilidyne

**стабилизатор** *m.* stabilizator, stabilizer, equalizer, balancer; regulator, governor; voltage regulator tube; reactor; regulator circuit; damper; **коронирующий с. напряжения** corona tube regulator; **с. напряжения с лампой накаливания** lamp voltage regulator; **с. с использованием коронирующего разряда** corona tube regulator; **феррорезонансный с.** saturation core regulator; **с. частоты с применением кварца** Weir stabilizer; **электронный с. с обратной связью** degenerative electronic regulator

**стабилизация** *f.* stabilization, settling, stabilizing; regulation, control; equalization, equalizing; fixing; **со сверхвысокой стабилизацией** superregulated; **с. зоны обзора** beam stabilization; **с. от бортовой качки** roll stabilization; **с. по углу места** tilt stabilization; **с. положения** attitude stability; **с. с помощью дифференцирующего звена** derivative equalization, lead equalization; **с. угла наклона** tilt stability

**стабилизированный** *adj.* stabilized, fixed, regulated

**стабилизировать** *v.* to stabilize, to calm, to settle; to fix; to mature; to regulate

**стабилизироваться** *v.* to become stable

**стабилизируемый** *adj.* controllable, controlled; **с. кварцем** crystal-checked

**стабилизирующий** *adj.* holding, stabilizing; equalizing; antihunt

**стабилизованный** *adj.* stabilized, settled; controlled, regulated

**стабилизовать(ся)** *see* **стабилизировать(ся)**

**стабилизуемый** *adj.* stabilized

**стабилизующий** *adj.* stabilizing, steadying

**стабилит** *m.* stabilite

**стабилитрон, стабиловольт** *m.* voltage stabilizing tube, stabilovolt; **с.-делитель напряжения** glow-discharge tube voltage divider

**стабилотрон** *m.* stabilotron tube

**« стабиль »** *m.* metalized mica capacitor

**стабильность** *f.* stability, firmness, rigidity, constancy; stiffness; persistence; **с. положения отметок** positional stability

**стабильный** *adj.* stable, firm, secure; permanent

**стабистор** *m.* stabistor

**ставень** *m.* shutter

**ставить, поставить** *v.* to set, to put, to install, to place; to apply, to put on; to raise; to organize

**ставка** *f.* rate; setting, placing, putting

**стадиальный, стадийный** *adj.* phase, phasic, by stages, stage

**стадия** *f.* stage, phase, step; **по стадиям** by stages, in stages

**стажёр** *m.* engineering graduate, apprentice, probationer

**стакан** *m.* pot, shell, box, bushing, liner; can; glass; **направляющий с. цоколя** base skirt; **развязывающий с.** decoupling choke

**стаканчик** *m.* pot, can, shell; little glass

**стаккато** *adv.* staccato

**стакпол** *m.* stackpole

**сталагмометр** *m.* stalagmometer

**сталагмометрический** *adj.* stalagmometric

**сталбетон** *m.* steel concrete

**сталеалюминиевый** *adj.* steel-aluminum

**сталеалюминий** *m.* steel-aluminum

**сталебетон** *m.* steel concrete

**сталемедный** *adj.* steel-copper

**сталеплавильный** *adj.* steel-melting

**сталепрокатный** *adj.* steel-rolling

**сталкиваться, столкнуться** *v.* to collide, to come into collision, to run into, to encounter; to interfere, to conflict

**сталкивающийся** *adj.* conflicting

**сталь** *f.* steel; **брусковая с.** rod steel; **высококоэрцитивная с.** magnet steel; **с. с ориентированной структурой** anisotropic steel, grain-orien-

ted steel; **с.-серебрянка** silver steel; **фасонная с.** sectional steel

**стальной** *adj.* steel

**стамеска** *f.* chisel, firmer chisel, former

**стан** *m.* mill; height, stature; **заготовочный с.** billet mill; **нотный с.** staff, stave (*mus.*); **обжимный с.** roughing mill; **рельсобалочный с.** structural mill; **с. чернового обжима, черновой с.** breaking down mill, roughing mill; **черновой сутуночный с.** roughing sheet bar mill; **штрипсовый с.** skelp mill

**станд** *m.* stand

**стандарт** *m.* standard, norm, normal; sort; gage; standard specifications; **американский с. высоты строя** American standard pitch; **временный с., предварительный с.** bracket standard

**стандартизация** *f.* standardization, normalization; regularizing, regulation

**стандартизировать** *v.* to standardize, to calibrate, to gage

**стандартизованный** *adj.* standardized, standard

**стандартизовать** *v.* to standardize

**стандартный** *adj.* standard, normal, preferred

**станина** *f.* mount, bed, bed plate, base, pedestal; case, housing, carcas, frame; bench, stand, rack, support; yoke, magnet yoke; column, pillar; **индукторная с.** field structure

**станиолевый** *adj.* (tin-)foil

**станиоль** *m.* (tin)foil

**станок** *m.* machine, tool machine; lathe; bench, stand; processor; **гравитационный с. для нанесения экранов ЭЛТ** gravity pouring machine; **с. для записи на диск** disk recorder; **с. для лобовой обмотки** lapping machine; **с. для накатки** knurling machine; **с. для намотки связанных спиралей** coupled helix machine; **с. для насадки колпачков** capping machine; **с. для подрезки выводов и расплющивания под платой** cut-and-clinch machine; **с. для приварки штенгеля** tubulating machine; **с. для присоединения шейки колбы** bulb-neck splicing machine; **с. для расточки труб** tube seat; **с. для укрепления деталей на ленте** component-belting machine; **с. для установки контактов** terminal machine; **с. для фрезерования пазов** slot cutting machine; **с. для фрезерования торцов рельс** shaping machine for rails; **с. для штенгелёвки ламп** bulb-tubulating machine; **долбёжный с.** vertical planing machine; **карусельный с.** lathe with horizontal face plate; *n*-**коклюшечный оплёточный с.** *n*-carrier braider; **лобовой токарный с.** head turning lathe, facing lathe; **обрезной с. для плоских ножек** flat stem trimmer; **правильный с.** straightening machine; **токарно-карусельный с.** lathe with horizontal face plate; **токарноревольверный с.** turret lathe; **токарный с.** lathe; **токарный с. для обточки длинных валов** shaft turning lathe; **установочный с.** inserter; **цоколёвочный с.** basing machine; **шипорезный с.** dovetail cutting machine; **шлифовальный с.** grinder; **шлифовальный с. с оптическим компаратором** optical contour grinder; **шпоночный с.** grooving machine; key-seating machine

**станционный** *adj.* station, station-type; office; exchange (*tphny.*)

**станция** *f.* station; base; post; exchange (*tphny.*); office, central; set; plant; **аварийная с.** emergency set, emergency station; casual clearing station; **береговая гидроакустическая с.** harbor defense asdic; **биржевая с.** stock-exchange switchboard; **с. быстрой связи** demand working exchange, toll exchange, no-delay telephone exchange; **ведомая с.** drift station, satellite station; slave station; **вспомогательная с.** subexchange, satellite exchange; **вспомогательная руководящая с.** subcontrol station; **входящая (междугородная) с. ТТС**, terminating toll center, terminating office, re-receiving office, called office; **двойная задающая с.** double-master station; **задающая с.** key station, master station; **зарядная с.** battery charger, battery-charging outfit; **индукторная (центральная) с.** magneto central office (*tphny.*); **исходящая (междугородная) с. ОТС**, originating toll center, originating

office, sending office; **контрольно-измерительная с. для замера акустического и гидродинамического полей** acoustic-and-pressure check range; **с. корректирования огня по всплескам** shell-splash set (*rdr.*); **корабельная гидролокационная с.** shipborne sonar; **корреспондирующая с.** distant exchange; **междугородная с. с ИВ** long-distance office with automatic call distribution; **междугородная вызывающая с.** originating trunk exchange; **междугородная узловая с.** toll traffic junction center; **мусоросжигательная с.** destructor station; **с. наводки** direction station; AA director; **наземная дальномерная с.** ground ranging equipment; **с. направления посадки** PAR, precision approach radar, localizer; **неполная с.** discriminating office, switching-selector-repeater office (*dts.*); **низовая с.** lowest-level exchange (*tphny.*); **оконечная междугородная с.** toll center, terminal long-distance office; **с. орудийной наводки** fire control station; **панорамная самолётная радиолокационная с.** airborne ground-mapping radar; **парашютная с.** parachute (radio) set; **переговорная с.** public telephone station; **переговорно-справочная с.** recording board, record office; **предвключённая с.** superimposed station; **приводная с.** field localizer; **пригородная с.** subexchange, dependent exchange, toll exchange; **пригородно-междугородная с.** toll office (*tphny.*); **промежуточная с.** retransmitting station; **промежуточная междугородная с.** through trunk exchange (*tphny.*); **с. радиовещания по проводам** line radio station; **радиолокационная с.** radar; **р. с. дальнего действия** early warning radar station, radar station for long-range searching; **р. с. для вывода самолётов в район аэропорта** approach control radar; **р. с. для облучения Луны** lunar radar, moon radar; **р. с. засечки огневых позиций миномётов** counter-mortar radar; **р. с. кругового обзора** surveillance radar element; **р. с. наведения по** лучу beam-forming radar, beam-rider radar; **р. с. орудийной наводки** gun-laying radar; **р. с. предупреждения о находящихся впереди препятствиях** forward area warning radar; **р. с. с внешней хронизацией** master-oscillator radar set; **р. с. сопровождения** nutating radar; **распорядительная с.** control office (*tphny.*); **ручная с.** manual exchange; **с. с входящими линиями** called exchange; **с. с координатным искателем** crossbar dial central office; **с. с посторонней хронизацией** master-oscillator radar set; **с. с удлинителями** switching pad office (*tphny.*); **своя с.** home station; **с. связи взаимодействия** liaison (radio) set; **старшая с.** rate station; **телефонная с. с местной батареей** local battery exchange, L.B. exchange; **с. точного обнаружения** accurate position finder (*rdr.*); **транзитная с.** through exchange, transit exchange, via office, through switching exchange; **трансляционная с.** repeater station, repeater office (*tphny.*); relay station (*rad.*); **узловая с.** junction center, transit center, multiple center (*tphny.*); **с. управления заходом на посадку** localizer (*av.*)

**старение** *n.* aging, seasoning, preaging; hardening, age hardening; fatigue; deterioration; long-term drift

**стареть, постареть** *v.* to age, to grow old; to season; to become obsolete

**стареющий** *adj.* aging

**старить, состарить** *v.* to make old, to age, to mature; to train

**старт** *m.* start, take-off

**стартер** *m.* starter, self-starter

**стартерный** *adj.* starter

**стартовать** *v.* to start

**стартовый** *adj.* start, starting

**стартстопный** *adj.* start-stop

**старший** 1. *adj.* senior, eldest; major, upper; 2 (*noun*) *m. adj. decl.* chief, head, superior

**старшинство** *n.* seniority; precedence

**старый** *adj.* old

**стат** *m.* stat (unit of radioactivity)

**статив** *m.* bay, rack, frame, stand; surface gage; **с. испытания долговечности** life-test rack; **с. линейных шнуров** line link frame; **с. предох-**

**ранителей для усилительных станций** fuse rack for repeater offices; **пробный с.** test selector rack; **с. разрядников** protector rack; **сборный с.** miscellaneous equipment rack **с. станционных шнуров** office link frame

**статика** *f.* statics

**статикон** *m.* staticon tube

**статистик** *m.* statistician

**статистика** *f.* statistics; record

**статистически** *adv.* statistically

**статистический** *adj.* statistical, statistic

**статический** *adj.* static, statical, steady-state

**статичность** *f.* static character

**статометр** *m.* statometer

**статор** *m.* stator, fixed coil

**статорный** *adj.* stator

**статоскоп** *m.* statoscope

**статутный** *adj.* statute (mile)

**статья** *f.* article; item, clause

**стауроскоп** *m.* stauroscope

**стационарность** *f.* stability

**стационарный** *adj.* stable, stationary, fixed, static; steady, steadystate; equilibrium

**СТВ** *abbr.* (**стойка тонального вызова**) ringer bay

**створ** *m.* range, alignment

**створка** *f.* valve; flap, fold

**створный, створчатый** *adj.* valved, valvate; flap; folding

**стеарин** *m.* stearin

**стеариновый** *adj.* stearin, stearic

**стеатит** *m.* steatite

**стеатитовый** *adj.* steatite, steatitic

**Стейгервальд** Steigerwald

**стекать, стечь** *v.* to leak off, to run off, to flow, to flow off, to drain, to discharge; to trickle; to overflow

**стекаться, стечься** *v.* to run together, to flow together, to converge; to join, to gather, to accumulate, to collect; to concrete

**стекающий** *adj.* draining, discharging, running out

**стекло** *n.* glass; **водомерное с.** water gage, gage glass; **растворимое с.** silicate of potassium; **с. с заданной величиной проводимости** conductive glass; **смотровое с.** sight glass

**стеклование** *n.* vitrification; glass transition

**стекловатый** *adj.* vitreous, glassy

стекловидность *f.* vitreousness, glassiness

стекловидный *adj.* vitreous, vitrified, glassy

стекловолокно *n.* glass fiber; optical fiber

стеклография *f.* screening

стеклодув *m.* glass blower

стеклотекстолит *m.* fiberglass

стеклоткань *f.* glass cloth

стеклянно-жидкостный *adj.* liquid-in-glass

стеклянный *adj.* glass, vitreous; positive (*elec.*)

стеллаж *m.* shelving, shelf, shelves; stand, rack, bearer; **с. для муфт** coupling-box frame

стелларатор *m.* stellarator

стен *m.* sthene

стена *f.* wall, side; **косая с.** splayed wall

стенд *m.* stand, rack, mounting rack; testing unit, testing jig; bench, table; test bed; **с. для испытаний на ударную прочность** shock machine

стендовый *adj.* stand; bench

стенка *f.* wall, partition, side; plate

стенной *adj.* wall, wall-type

стенодный *adj.* stenode

стеноматик *m.* stenomatic

стенометр *m.* stenometer, engymeter

стенотелеграфия *f.* stenotelegraphy

степенный *adj.* exponential, power (*math.*); staid, sedate

степень *f.* stage; degree, grade, extent; rate, rating; step, stage; ratio, power, order (*math.*); digit; depth; **третьей степени** cubic, cubical; **четвёртой степени** biquadratic (*math.*); **с. затягивания** pulling figure; **с. из десяти декад** hundreds digit; **с. изодрома** degree of reset; **с. нагрева** (*or* **накала**) emission factor; **с. надёжности** safety factor; **с. направленности** antenna gain; **с. помола** beating range; **с. преобразования** power of transformation; **с. редукции** reduction gear ratio; **с. связи** coupling value, percentage coupling; **с. соответствия** quality of conformance; **третья с.** cube (*math.*); **с. успокоения** ballistic factor; **с. успокоения измерительного прибора** damping factor of a measuring instrument

стерад, стерадиан *m.* sterad, steradian

стереоадаптер *m.* stereo control unit

стереоакустический *adj.* stereophonic

стереограмма *f.* axonometric chart

стереографический *adj.* stereographic

стереография *f.* stereography

стереодальномер *m.* stereoscopic rangefinder

стереоизомер *m.* stereoisomer

стереоизомерия *f.* stereoisomerism

стереоизомерный *adj.* stereoisomeric

стереокамера *f.* stereo camera

стереометр *m.* stereometer

стереометрический *adj.* stereometric

стереометрия *f.* stereometry, solid geometry

стереомикроскоп *m.* stereomicroscope; **промышленный с.** epitechnoscope

стереопередача *f.* stereo presentation

стереорадиография *f.* stereoradiography

стереорадиоскопия *f.* stereoradioscopy

стереоскоп *m.* stereoscope

стереоскопический *adj.* stereoscopic

стереоснимок *m.* stereoscopic photograph

стереотелевидение *n.* stereoscopic television

стереофильм *m.* three-dimensional film

стереофлюороскопия *f.* stereofluoroscopy

стереофонический *adj.* stereophonic

стереофоничность, стереофония *f.* stereophonism, stereophonics

стереофонный *adj.* stereophonic

стереофотография *f.* stereoscopic photograph (*or* photography)

стереофотометрия *f.* stereophotometry

стереохимический *adj.* stereochemical

стереохимия *f.* stereochemistry

стереть(ся) *pf. of* стирать(ся)

стержень *m.* rod, bar, stick; post, pole; plug; handle, shaft, stem; beam; filament; peg, pin, bolt; leg; shank, stalk, arm; **заделанный с.** clamped bar; **с., заделанный на двух концах** clamped-clamped bar; **заострённый с.** tapered bar; **многосложный с.** laminated bar; **с., стабилизующий частоту** pulling stub (in a magnetron); **стальной с.** steel bolt, steel pin; **сужающийся по экспоненте с.** exponentially tapered rod; **с., шарнирно закреплённый с обоих концов** hinged-hinged bar; **якорный с.** guy rod

**стержневой** *adj.* pivotal; rod, bar; core-type; arm, handle, shaft, stem; filament; interdigital; axial

**стержнеобразный** *adj.* rod-like, bar

**стерилизационный** *adj.* sterilization, sterilizing

**стерилизация** *f.* sterilization

**стерилизировать** *v.* to sterilize

**стерилизованный** *adj.* sterilized

**стерилизовать** *v.* to sterilize

**стерилизующий** *adj.* sterilizing

**стёртый** *adj.* worn out, rubbed off, effaced, obliterated; erased

**стёс** *m.* rake

**стеснённый** *adj.* compressed; restricted, crowded

**стетоскоп** *m.* stethoscope

**стетоскопирование** *n.* auscultation

**стетоскопический** *adj.* stethoscopic

**стехиометрический** *adj.* stoichiometric

**стехиометрия** *f.* stoichiometry

**стечь(ся)** *pf. of* **стекать(ся)**

**Стибиц** Stibitz

**стилистический** *adj.* stylistic

**стилограф** *m.* stylograph

**стилометр** *m.* stylometer

**стиль** *m.* style, manner, fashion

**стильб** *m.* stilb

**Стильсон** Stillson

**стимул** *m.* stimulus, spur (drive)

**стимулирование** *n.* stimulation

**стимулировать** *v.* to stimulate

**стимулирующий** *adj.* stimulating, stimulant

**стимулятор** *m.* stimulator

**стимуляция** *f.* stimulation

**стираемость** *f.* erasibility, erasability, erasing ability

**стираемый** *adj.* erasable

**стираконоый** *adj.* styracon

**стирание** *n.* erasure, erase, erasing, obliteration, obliterating, rubbing off; cancellation, cancel, cancelling; wipe; attrition

**стирать, стереть** *v.* to cancel, to obliterate, to delete, to erase, to rub out; to wipe (off), to clean; to clear

**стираться, стереться** *v.* to wear, to wear away, to rub off

**стирающий** *adj.* erasing; scan-off, play-off

**Стирлинг** Stirling

**стирол** *m.* styrene

**стироловый** *adj.* styrene

**стирон** *m.* styrone

**стирофлекс** *m.* styroflex

**стирофлексовый** *adj.* styroflex

**стирофом** *m.* styrofoam

**СТК** *abbr.* (**стойка тональной коммутации**) voice-frequency patching bay

**сто** *num.* hundred

**Стоби** Stobie

**стоградусный** *adj.* centigrade

**стоечка** *f.* stand, rack; prop, support

**стоечный** *adj.* stand, rack; support

**стоимость** *f.* cost, expense(s), value, expenditure, price; rate, charge

**стоихиометрический** *adj.* stoichiometric

**стойка** *f.* strut, bracing strut, brace, rest, support, prop, holder; stand, pedestal, pedestal base; standard, roof pole, pole, post, rig, pillar, stake; rack, rack frame, counter, bay, frame; stem, shank; cabinet; box; **вводная с. (на крыше)** derrick (on the roof); **с. выделения каналов** derived channel equipment bay (*tphny.*); **с. группового оборудования** group terminal bay; **с. групповых несущих частот** group carrier supply bay; **двусторонняя с.** double-sided rack (*tphny.*); **двутавровая с.** I-pin, I-bracket; **с. дифференциальных систем** voice-frequency terminating equipment bay; **с. для стрелочного фонаря** staff tip adapter; **измерительная с.** repeater test rack; **с. индивидуальных несущих и контрольных частот** channel carrier and pilot supply bay; **с. линейной коммутации** high frequency line and patching bay; **с. линейных фильтров** derived channel equipment bay (*tphny.*); **с. на крыше** roof pole, roof standard; **с. параболоида** paraboloid bracket; **с. переключений** patching bay (*tphny.*); **с. РСЛ** bay trunks (*tphny.*); **с. с гнездовым полем** patching bay (*tphny.*); **с. соединительных линий** bay trunk (*tphny.*); **с. тонального вызова** ringer bay; **с. тональной коммутации** voice frequency patching bay; **с. электрододержателей** electrode mast; **ящичная с.** cabinet rack, console

**стойкий** *adj.* steady, steadfast, firm, resistant, stable, persistent; *suffix:* -proof, -resisting; **с. на свету** light-

proof; **с. против атмосферных условий** weatherproof; **с. против радиолокации** radar-proof; **с. против
сотрясений** shock-proof, shake-proof

**стойкость** *f.* stability, stableness, firmness, sturdiness, steadiness; durability, persistence, endurance, per-
severance

**стоймя** *adv.* erect, upward

**Сток** Stock

**сток** *m.* flow, flowing, drainage, discharge, discharging; drain, channel;
outlet, run-off; **с общим стоком**
common-drain

**стократный** *adj.* centesimal, centuple,
hundredfold

**стокс** *m.* stoke (unit of kinetic vis-
cocity)

**Стокс** Stokes

**стоксов** *poss. adj.* (of) Stokes, Stokes

**стол** *m.* table, desk; board; **с. В В-**
position, B-switchboard, incoming
position, inward position; **с. заказов**
recording board, recording switchboard, record position (*tphny.*); **измерительный и испытательный с.**
circuit chief's desk, wire chief's desk,
test desk (*tphny.*); **коммутационный с.** telephone switchboard; **контрольный с.** observation desk, service observing desk; monitoring position, listening position; **наклонный с.
управления** bench-board; **проверочный с.** listening position (*tphny.*);
**с. прокладки курса** plot table; **с.
распределения заказов по рабочим
местам** ticket distribution desk
(*tphny.*); **расчётный с. переменного
тока** a-c calculating board; **с.
сортировки заказов, сортировочный
с.** ticket distribution desk; routing
desk, net control station (*tphny.*);
**справочный с. междугородной стан-
ции** toll directory desk (*tphny.*);
**с. циркулярного вызова** legging-key
board (*tphny.*)

**столб** *m.* column, pillar, pole, post,
support, strut, rods, mast, peg,
stake; pile (of a battery); **выпрямительный с.** rectifier pile, stack of
rectifier plates; **гидронированный с.**
tarred pole; **натяжной с.** anchoring
pole; **оконечный с. с подкосом**
strutted terminal pole; **первый от
ввода в помещения с.** office pole
(*tphny.*); **разрядный с.** glow column;

**с. с анкерной оттяжкой** stayed pole;
**с. с кронштейнами** bracket pole;
**с. с оттяжкой** guyed pole; **с. с
подкосом, с. с подпоркой** strutted
pole, A-type pole; stayed pole,
braced pole; **скруточный с., с. со
скрещёнными проводами** transposition pole; **с. со шпренгельной оттяжкой** trussed pole

**столбец** *m.* column

**столбик** *m.* stack, pile; post; stub;
column; core; **вентильный с.** metallic rectifier stack; **с. для анкеровки
оттяжки** anchor guy stub; **предельный с.** clearance point, fouling
protection

**столбовидный** *adj.* columnar; pillar-
like

**столбовой** *adj.* pole, pole-type, pillar,
post, column, mast

**столик** *m.* small table; small desk

**столкновение** *n.* collision, impact,
clash, impingement, encounter;
shock; interference; **курс на с.**
collision course

**столкнуться** *pf. of* сталкиваться

**столообразный** *adj.* table-shaped, plateau

**столярный** *adj.* joiner's, carpenter

**Стонели** Stonely

**стоп** *m.* stop; **с.-импульс** stop signal

**стопка** *f.* pile, stack

**стоповый** *adj.* stop

**стопор** *m.* stop, stopper, arrester,
catch, rest, lock, detent, detainer,
fixing device, holding device, locking
device, lock; setting screw; plug,
stopper; **азимутальный с. походного
положения** azimuth stowing lock;
**с. походного положения** stowed lock

**стопорить, застопорить** *v.* to plug, to
stop; to lock, to check

**стопорный** *adj.* stop, stopper, closing,
locking, stopping, blocking; clamping

**стопроцентный** *adj.* hundred per cent

**сторанс** *m.* storance

**сторож** *m.* guard, watchman

**сторожевой** *adj.* watch, watchman's,
guard

**сторожок** *m.* escapement, catch;
tongue, cock

**сторона** *f.* side, flank; place; site;
aspect; party (*tphny.*); end; **на
стороне низкого напряжения** low
tension sided; **с обеих сторон** on
both sides; both party; **с питанием**

с двух сторон dual-feed; вызывающая с. caller (*tphny.*); лицевая с. face (side); с. подачи input; приёмная с. sound-sensitive surface; с. разгрузки delivery end; торцевая с. end-face

стохастический *adj.* stochastic, conjectural

сточный *adj.* discharge, escape, drain, drainage

стояк *m.* lighting riser, power riser, rising main; strut; upright; uprise; с. с громкоговорителями directive speaker-group

стояние *n.* standing

стояночный *adj.* parking

стоять, постоять *v.* to stand, to be standing, to be motionless

стоячий *adj.* standing, stationary; stagnant, still; erect, vertical

стоящий *adj.* costing; standing; с. на готове at stand-by

стр. *abbr.* (страница) page

страбометр *m.* strabometer

« страгглинг » *m.* range straggling

страна *f.* country, land, region; с. света point of the compass

страница *f.* page, leaf

страничный *adj.* page

страта *f.* stria, striation

стратегический *adj.* strategic

стратегия *f.* strategy

стратиметр *m.* stratameter

стратификация *f.* stratification

стратифицированный *adj.* stratified

« стратовидение » *n.* "stratovision"

стратометр *m.* stratometer

« страторадар » *m.* stratoradar

стратостат *m.* stratosphere balloon

стратосфера *f.* stratosphere

стратосферный *adj.* stratospheric, stratosphere

страты *pl.* striae

Страуджер Strowger

страуджеровский *adj.* (of) Strowger, Strowger's

стрейнер *m.* strainer

стрекотание *n.* chirring, chattering, rattling

стрела *f.* arrow, indicator, pointer; shaft, boom, jib, overhang beam, crane arm; camber; cantilever; грузовая с. derrick, davit; монтажная с. провеса (провода) erection dip; с. провеса, с. прогиба dip, sag, deflection, amount of deflection

стрелка *f.* needle, indicator, pointer, arrow, index; hand (of clock); railway point, switch; до отказа против часовой стрелки full counterclockwise; по часовой стрелке clockwise; против часовой стрелки counterclockwise; маневровая с. shunting switch; нецентрализованная с. outlying switch; отводная с. safety switch; путевая с. с вращающимися переводными рельсами contactors switch; путевая с. с неподвижными остряками dumb switch; с. с двумя последовательно расположенными остряками staggered point switch

стрелковый *adj.* arrow, needle

стреловидный *adj.* arrow-shaped, arrow-like, sagittate; swept back, sagittal, sagittary

стрелочник *m.* switcher, switchman

стрелочный *adj.* pointer, needle, arrow, sagittary; indicating; switch

стрельба *f.* shooting, firing, gunfire

стрельчатый *adj.* needle, arrow-shaped, sagittary; lanced, pointed, gabled

стремление *n.* aspiration (for), striving (for), tendency, leaning, inclination, propensity

стремнина *f.* current, race; declivity, steepness, precipice

стремя *f.* stirrup

стремянка *f.* step-ladder, ladder; gangway, footpath

стренга *f.* strand

стренер *m.* strainer, filter

стрикция *f.* striction

стример *m.* streamer

стримерный *adj.* streamer

стриппер *m.* stripper; ingot stripper

строб *m.* gate, strobe; магнитный с. flux gate; с. на нулевой разностной частоте heterostrobe, homostrobe

стробгенератор *m.* gate generator; предварительный с. early-gate generator

стробикон *m.* strobeacon

стробимпульс *m.* strobe pulse, gate pulse, gating pulse; с. засветки индикатора trace kipp; отпирающий с. indicator gate; перемещающийся с. walking strobe

стробирование *n.* gating, strobing, time strobing, time gating; sampling, discrete sampling; с. двойным стробированием double-gate; с. по сетке лампы grid gating

**стробированный** *adj.* gated, strobed
**стробировать** *v.* to strobe, to gate
**стробируемый** *adj.* gated
**стробирующий** *adj.* gating, strobing
**строб-контур** *m.* gate circuit, gating circuit
**стробный** *adj.* gating
**стробо́люкс** *m.* strobolux
**стробоскоп** *m.* stroboscope, optical disk; **с. с десятичным делением** decade stroboscope; **с. с лампой мгновенных вспышек** flashtube stroboscope; **с. с неоновым тиратроном** stroboglow
**стробоскопический** *adj.* stroboscopic
**стробо́телескоп** *m.* strobotelescope
**строботриод** *m.* strobotriode
**строботрон** *m.* strobotron
**строботронный** *adj.* strobotron
**строб-реле** *n. indecl.* relay gate
**строгальный** *adj.* planing
**строгание** *n.* planing, shaping
**строганный** *adj.* planed, shaped
**строгать, выстрогать** *v.* to plane, to shape, to shave; **с. предварительно** to rough-plane
**строгий** *adj.* strict, rigid, severe
**строго** *adv.* strictly, exactly
**строгость** *f.* strictness, severity; rigor (*math.*)
**строевой** *adj.* formation, building, construction
**строение** *n.* formation, construction, building; structure; frame; texture; pattern; **верхнее с.** superstructure; **слоистое с.** lamination; **слоистое с. совокупности двойниковых кристаллов** twin lamination
**строенный** *adj.* triple, triplex, tripartite
**строитель** *m.* builder, constructor, engineer, designer
**строительный** *adj.* building, construction, structural
**строительство** *n.* development, organization; building, construction, erection
**строить, построить** *v.* to build, to construct; to form
**строй** *m.* system, regime, order; formation; pitch (*mus.*); temperament (*mus.*); tuning; **равномерно темперированного строя** equally tempered, of equal temperament; **высокий с.** high pitch; **натуральный с.** just temperament (*mus.*) **чистый с.** just scale, mean tone

**стройка** *f.* development, construction, building
**стройплощадка** *f.* building yard
**строка** *f.* line; row; horizontal line; base line; strip; **с. в графе** measurement; **предыдущая с.** earlier measurement; **тёмная с. развёртки** dimmer sweep trace (of a skiatron)
**стронциевый** *adj.* strontium
**стронций** *m.* strontium, Sr
**Строуд** Stroud
**Строуджер** Strowger
**строуджеровский** *adj.* (of) Strowger, Strowger's
**Строужер** Strowger
**строфотрон** *m.* strophotron
**строчка** *f.* line, row
**строчность** *f.* lininess
**строчный** *adj.* line; horizontal
**струбцина, струбцинка, струбцынга** *f.* clamp, screw-clamp, cramp(-iron), vise
**струве** Struve
**струг** *m.* plane; draw-knife
**стругать** *see* **строгать**
**струевыпрямитель** *m.* straightening vane
**струезащищённый** *adj.* proof against water jets
**струеотклоняющий** *adj.* jet-deflecting
**стружка** *f.* chip, cut, shaving
**струиться** *v.* to run, to stream, to flow; to radiate, to shine
**струйный** *adj.* jet; streamer; current, flow
**структура** *f.* structure, organization, setup; state; pattern, patterning; framework; line; texture; **встречно-штыревая с.** interdigital structure; **грубая с. хода характеристических линий** macrostructure of characteristic, general trend of characteristic; **двухбуквенная с.** diagram *math.* **с. решётки** mask pattern; **стержневая с.** interdigital structure; **трёхбуквенная с.** trigram
**структурирование** *n.* cross-linking
**структурированный** *adj.* cross-linked, structurized
**структурировать** *v.* to cross-link
**структурно-обращённый** *adj.* structurally dual
**структурно-симметричный** *adj.* structurally symmetrical
**структурно-упорядоченный** *adj.* lattice-ordered

структурный *adj.* structural, structure

струна *f.* string, chord; catgut; slide wire; drop; **с. ля а′** string (of a violin); **с. ми е″** string, first string; **с. ре** d' string; **неприжатая с.** open string; **с. с навивкой** overspun string; **с. с ползунком** slide wire; **с. соль** g string

струнный *adj.* string, stringed; filament, thread

струнодержатель *m.* tailpiece (of a violin)

Струхаль Strouhal

струя*f.* jet, spurt, stream, flow, current; spray; spout; ray; **кильватерная с. судна** wake of ship; **сверхзвуковая с.** choked jet

студенистый *adj.* jelly(-like), gelatinous

студийный *adj.* studio

студия *f.* studio, transmitting room; **работаюшая с.** live studio; **с. с заглушённой (задней) стеной** dead-end studio; **с. с незаглушённой (задней) стеной** live-end studio

стук *m.* knock, noise, tap, rap, clatter, chattering, rattling

стукание *n.* knocking

стукать, стукнуть *v.* to beat, to tap, to rap, to bang, to knock, to thump, to hit; to strike

ступенчато *adv.* in steps, in stages, gradually, stepwise

ступенчатость*f.* graduation, gradation

ступенчатый *adj.* graded, graduated, step, step-by-step, gradual, cascade; stepped, step-shaped, step-like; notching; staircase, stairstep; tapering; staggered; multiposition, multistep, multistage

ступень *f.* step, stage, phase, level; degree, grade; digit; notch; amplifier stage; cascade; **ступени на опоре стремянки** pole steps; **ступени сопротивления** working points; degrees of opposition; **ступенями** graded, staggered; by degrees, gradually, step by step; **включать ступенями** to step forward; to cut in gradually; **с. выдержки времени** time differential; **выходная с.** final stage; **с. группового искания** digit; **с. искателей, с. механизмов искания** rank of selectors, rank of switches; **неиспользуемая с.** dead level (*tphny.*); **первая с. усиления** pre-amplifier

stage; **с. переключения** switching cycle, switching step, operating position of a step switch, commutating cycle; wiring phase

ступенька *f.* step; rung, step, spoke

ступица *f.* nave; hub, boss; **с. со спицами для крепления контактных колец** slip ring spider; **якорная с.** armature quill; **с. якоря со спицами** armature spider

стучать, постучать *v.* to knock, to tap, to rap; to chatter, to rattle; to make noise

стык *m.* joint, junction, splice, seam, bond; push, impact; butt joint; **с плотным стыком** gapless; **дроссельный с.** impedance bond; **изолирующий с. путевых рельс** insulated rail joint; **изолирующий с. с двусторонней изоляцией** full-insulated joint; **изолирующий с. с односторонней изоляцией** one-end insulated joint

стыковый *adj.* butt, joint; splicing

Стьюдент Student

стэноматик *m.* stanomatic system

Стюарт Stewart

стягивание *n.* contraction, constriction, tightening; sintering; concentration

стягивать, стянуть *v.* to brace, to shackle; to draw together, to contract, to tighten, to constrict; to close

стягиваться, стянуться *v.* to be drawn together, to tighten, to shrink, to contract; to sinter

стягивающий *adj.* binding, tie; constringent, astringent

стяжка *f.* headband; turn buckle, coach screw; tie piece, tie rod, tightening device, coupler, coupling, draw bar

стяжной *adj.* clamping; coupling, tie

стянутый *adj.* coupled; tightened, drawn together, constricted

стянуть(ся) *pf. of* стягивать(ся)

суб- *prefix* sub-, under-

субгармоника *f.* subharmonic, submultiple, suboctave; mush

субгармонический *adj.* subharmonic

субгруппа *f.* subgroup

субдетерминант *m.* subdeterminant

субдоминанта *f.* subdominant

субзвуковой *adj.* subaudio

субиндекс *m.* subindex

**субкассета** *f.* submatrix; subcassette
**сублимационный** *adj.* sublimation
**сублимация** *f.*, **сублимирование** *n.* sublimation
**сублимированный** *adj.* sublimated
**сублимировать** *v.* to sublimate, to sublime
**сублимирующийся** *adj.* sublimable
**субматрица** *f.* submatrix
**субмикроволна** *f.* submicrowave
**субмикрон** *m.* submicron
**субмиллиметровый** *adj.* submillimeter, submillimetric
**субминиатюризация** *f.* subminiaturization
**субминиатюрный** *adj.* subminiature
**субнаносекундный** *adj.* subnanosecond
**субнесущий** *adj.* subcarrier
**субнормаль** *f.* subnormal
**субнормальный** *adj.* subnormal
**субоктава** *f.* suboctave
**субпанель** *f.* subpanel
**субсеквентный** *adj.* subsequent
**субстандартный** *adj.* substandard
**субстантивный** *adj.* substantive
**субстанциальный** *adj.* substantial
**субституэнд** *m.* substituend
**субтрактивный** *adj.* subtractive
**субшасси** *n. indecl.* subchassis
**субъект** *m.* subject
**субъективность** *f.* subjectivity
**субъективный** *adj.* subjective; visual
**судно** *n.* ship, boat, vessel
**судовождение** *n.* pilotage (*nav.*)
**судовой** *adj.* ship, shipboard, naval, maritime; **судового типа** shipboard-type
**сужающий** *adj.* narrowing
**сужающийся** *adj.* tapered, converging
**суждение** *n.* opinion, proposition, decision; judgment
**сужение** *n.* constriction, contraction, narrowing, restriction, reduction
**суженный** *adj.* narrowed, contracted, constricted, compressed, squeezed
**суживать, сузить** *v.* to narrow, to shrink, to constrict; to taper; to throttle
**суживать(ся), сузить(ся)** *v.* to grow narrow, to get narrow, to shrink, to contract; to taper
**суживающийся** *adj.* tapered, tapering; contracting; **с. по линейному закону** straight-tapered
**Сузерленд** Sutherland
**сузить(ся)** *pf. of* **суживать(ся)**

**СУК** *abbr.* (**солнечный указатель курса**) sun compass
**сукцедент** *m.* succeedent
**сульфат** *m.* sulfate
**сульфатирование** *n.* sulfating
**сульфатировать** *v.* to sulfate
**сульфатирующий** *adj.* sulfating
**сульфатный** *adj.* sulfate, sulfatic
**сульфид** *m.* sulfide
**сульфидно-стронциевый** *adj.* strontium-sulfide
**сульфидный** *adj.* sulfide
**сульфоселенид** *m.* sulfoselenide
**сумеречный** *adj.* crepuscular, twilight, dusk; scotopic
**сумка** *f.* bag, case, sack, pack; **монтерская с.** tool case
**сумма** *f.* sum, amount; **с. по модулю 2** modulo 2 sum
**сумматор** *m.* summator
**суммарный** *adj.* total, overall, gross, combined, cumulative; summary; sum, summed, summarized
**сумматор** *m.* summer, summator, adder, summation device, totalizer; sum circuit, adding circuit; register, accumulator, summing amplifier; **с.-аккумулятор** adder accumulator; **комбинационный с.** coincidence-type adder; **комбинационный с. последовательного типа** serial-type adder; **с. моментов** torque summing member; **с. накапливающего типа, накапливающий с.** counter-type, adder, accumulator, adder-accumulator
**суммирование** *n.* summation, addition, summing up, adding; superimposing; **с. с учётом статистических весов** weighted summation
**суммированный** *adj.* summarized
**суммировать** *v.* to sum up, to summarize, to add; to register
**суммируемый** *adj.* integrable
**суммирующий** *adj.* summing, summation, sum, counting, adding, totalizing; integrating; cumulative; adder
**суммовой** *adj.* summation, sum
**супер** *m.* superheterodyne receiver
**супер-** *prefix* super-
**супервизор** *m.* supervisor
**супергармонический** *adj.* superharmonic
**супергетеродин** *m.* superheterodyne, superheterodyne receiver

**супергетеродинирование** *n.* superheterodyning

**супергетеродинный** *adj.* superheterodyne

**суперградиентный** *adj.* higher-order

**суперизокон** *m.* superisocon

**супериконоскоп** *m.* image iconoscope, supericonoscope

**супериконоскопный** *adj.* supericonoscope

**суперкардиоидный** *adj.* supercardioid

**супермалой** *m.* supermalloy

**супермендур** *m.* supermendur

**супернегадин** *m.* supernegadyne

**суперортикон** *m.* image orthicon, superorthicon (*tv*); **с. с малым расстоянием между сеткой и мишенью** closed-spaced image orthicon

**супероскюляция** *f.* superosculation

**суперпозиция** *f.* superposition, overlaying, overlapping

**суперрегенеративный** *adj.* superregenerative

**суперрегенерация** *f.* superregeneration

**суперрефракция** *f.* superrefraction

**суперсинхродин** *m.* supersynchrodyne

**суперсинхронный** *adj.* supersynchronous

**супертестер** *m.* supertester

**супертурникетный** *adj.* superturnstile

**суперфантомный** *adj.* superphantom, double-phantom

**суперэмитрон** *m.* superemitron

**суппорт** *m.* support, holder, carriage, rest; insulator pin

**суппортный** *adj.* support, holder, carriage, rest

**супрадин** *m.* supradyne

**супрамайка** *f.* supramica

**супрессор** *m.* suppressor

**сургуч** *m.* sealing-wax

**сургучный** *adj.* sealing-wax

**сурдина** *f.* sourdine, mute; muffler, damper, vibration damper, silencer, antihum

**сурдинирующий** *adj.* muting

**сурик** *m.* minium, red lead oxide

**суриковый** *adj.* minium

**суррогат** *m.* substitute, spare, reserve, auxiliary

**суррогатировать** *v.* to substitute

**суррогатный** *adj.* substitute, dummy; random

**сурьма** *f.* antimony, Sb

**сурьмянистосвинцовый** *adj.* antimony-lead

**сурьмянистый** *adj.* antimonous, antimony, antimonial

**сурьмяно-цезиевый** *adj.* antimony-cesium

**сурьмяный** *adj.* antimony, antimonic, antimonial

**Суси** Soucy

**суспензированный** *adj.* suspended; slurry

**суспензия** *f.* suspension; slurry

**сустав** *m.* joint, link, articulation; hinge

**суставной** *adj.* joint, articulation

**суставочный, суставчатый** *adj.* jointed, articulated; hinged; telescopic

**суточный** *adj.* daily, twenty-four-hour, diurnal

**сутунки** *pl.* sheet bars

**суффикс** *m.* suffix

**сухозаряженный** *adj.* drycharged

**сухой** *adj.* dry, arid

**сухопутный** *adj.* land; by land, overland

**сухоразрядный** *adj.* dry flash-over

**сушилка** *f.* dryer, desiccator

**сушильный** *adj.* drying

**сушитель** *m.* dryer, desiccator; **с. газов** acid trap

**сушка** *f.* drying, dehydration, desiccation

**существенно** *adv.* essentially, substantially

**существенность** *f.* importance, significance

**существенный** *adj.* essential, critical, substantial, intrinsic; important, significant

**существо** *n.* nature; being, entity; essence; **по существу** essentially, in essence; in fact

**существование** *n.* existence; occurrence; being; **одновременное с.** compatibility (*math.*); **с. следа** track maintenance

**существовать** *v.* to exist, to be; to be extant

**существующий** *adj.* existing, existent, available

**сущий** *adj.* existing; real, true

**сущность** *f.* entity, substance, nature; essentiality, essence, point; **в сущности** virtually, essentially, in essence

**сфазирование** *n.* phasing-in

**сфазированный** *adj.* phased

**сфалерит** *m.* sphalerit

**сфера** *f.* sphere, realm, range, area, zone, domain, district; scope

**Сфердоп** *m.* Spheredop

**сферический** *adj.* spherical, spheral, globular

**сферичность** *f.* sphericity

**сфероид** *m.* spheroid

**сфероидальный** *adj.* spheroidal, sphere-shaped

**сферометр** *m.* spherometer

**сферометрический** *adj.* spherometric

**сферометрия** *f.* spherometry

**сферофон** *m.* sphaerophone

**сфигмограмма** *f.* sphygmogram

**сфигмограф** *m.* sphygmograph

**сфигмоманометр** *m.* sphygmomanometer

**сфигмометр** *m.* pulsimeter

**сфигмотонометр** *m.* sphygmotonometer

**сфигмофон** *m.* sphygmophone

**сфокусированный** *adj.* focused, narrowed; **с. в точку** point focused

**сформированный** *adj.* shaped, formed, molded

**сформировать** *pf. of* **формировать**

**сформовать** *pf. of* **формовать**

**сформулировать** *pf. of* **формулировать**

**сфотографированный** *adj.* photographed

**сфотографировать** *pf. of* **фотографировать**

**схватить(ся)** *pf. of* **схватывать(ся)**

**схватывание** *n.* pull-in, pulling(-in); setting, hardening

**схватывать, схватить** *v.* to grip, to grasp, to clutch, to seize; to catch; to set, to harden

**схватываться, схватиться** *v.* to set, to bind; to seize

**схватывающий** *adj.* grasping, clutching; catching

**схватывающийся** *adj.* setting, binding

**схема** *f.* scheme, plan, project; chart, sketch, diagram; setup, design; structure; circuit, network, connection; wiring; clamp; layout; system, arrangement; (often omitted in translation: **фокусирующая с.** clamp (*lit.* clamping arrangement), **с. восстановления** restorer (*lit.* restoring arrangement)); **по схеме** often omitted in translation: **генератор по схеме Хартлея** (*or* **по трёхточечной схеме**) Hartley oscillator; **с. анод-**

ного нейтродинирования Hazeltine neutrodyne system; **балансная с. подавления помех** noise-balancing circuit; **биполярная фиксирующая с. на двойном триоде** two-way double-triode clamp; **с. блокировки** single-unit circuit; **с. В** class B amplifier circuit; **с. в канале сигналов цветности** chroma circuit, chrominance circuit, color circuit; **с. введения гасящих импульсов** suppression mixer; **с. введения гасящих импульсов в сигнал зелёного (красного, синего) цветоделительного изображения** green (red, blue) suppression mixer; **вентильная с.** gate; **с. включения** circuit diagram; **с. включения кристаллического триода базой на вход** base input circuit; **внешняя с.** external termination; **с. возвратной работы** pump-back circuit; **с. восстановления постоянной составляющей** clamping circuit, (d-c) restorer; **временная с. соединений студийной аппаратуры** temporary connection of broadcast studio equipment, patch; **с. (временной) развёртки** time base circuit; **с. выделения выбранной цели** lock-following strobe; **с. выделения сигналов цветности** chrominance separation circuit, chrominance take-off circuit; **с. выделения строчных синхронизирующих импульсов** horizontal(-sync) separator; **с. выдержки** delay network; **высокочастотная избирательная с.** high-pass selective circuit; **с. вычитания импульсов** differential amplifier; canceling circuit; **с. Грейнахера** half-wave voltage doubler; **с. Греца** bridge-type rectifier circuit; **групповая с.** group connection; **с. датчика** pickup circuit; **с. двойного действия** reflex circuit arrangement; **с. двойного зигзага** fork connection; **двукратная с.** push-pull circuit; **двухкаскадная с. со сдвинутой настройкой контуров** staggered doublet; **двухполупериодная с.** bridge connection; **с. детектора, реагирующего на полный размах сигнала** peak-to-peak detector circuit; **с. деления частоты импульсов** step divider circuit; **делительная с.** scaling circuit, counting down circuit; analog divider; **с.**

десятичного **сумматора** decimal add circuit; **дифференциальная с.** hybrid coil termination, bifurcation; **с. для проведения совещания по проводам** multiplex connection, conference connection; **с. для смещения точки зажигания дуги** distributor control of grid-controlled rectifiers employing an induction regulator; **с. для сокращения числа воздействующих импульсов** pulse dividing circuit; **с. дополнения до единицы** one's complement circuit; **с. дополнения до нуля** zero's complement circuit; **с. дополнения и прибавления переноса** complement and carry add circuit; **ёмкостная с. создания равносигнальной зоны** capacitance beam switching; **с. ждущей развёртки** triggered time base; **с. задержки типа « умножитель ёмкости »** capacity multiplier; **с. запрещения** AND NOT-gate, EXCEPT-gate, INHIBITORY-gate; **с. измерения напряжения рассогласования** error-indicating circuit, **с. импульсного** (*or* **толчкообразного**) **включения** push-button control of power tools; **инерционная с. синхронизации** flywheel circuit; **с. инерционного звена** time delay circuit; **с. инерционной автоматической подстройки частоты** flywheel frequency control circuit; **с. инерционной синхронизации** flywheel circuit; **искуственная с.** balancing network, phantom connection; **с. кадрового делителя** vertical divider chain; **с. кадровой развёртки** vertical-deflection circuit, vertical-scanning circuit; frame time base (*tv*); **кинематическая с.** pictorial diagram, functional diagram, functional schematic; **с. коммутации, коммутационная с.** circuit diagram, wiring scheme, schematic diagram; **с. компенсации** shading circuit; **с., контролирующая знаки слагаемых** algebraic sign control circuit; **с. коррекции верхних частот параллельным контуром** shunt peaking; **с. к. верхних частот последовательным контуром** series peaking; **с. к. входа** high peaker; **с. к. высоких частот последовательным контуром** series peaking circuit; **с. к. нелинейности** gamma-correcting cir-

cuit; **с. к. по выходу** de-emphasis network; **с. к. последовательным контуром** series corrector; **с. к. с параллельным колебательным контуром** shunt peaking circuit; **с. к. фронта импульсов** pulse corrector; **с. к. частот параллельным контуром** shunt peaking circuit; **« летучая » с. (соединений)** hookup, breadboard hookup, harness; **с. Латура** voltage-doubler rectifier circuit, half-wave circuit, convectional voltage doubler; **с. минимально-фазового фильтра** minimum (net)-phase-shift network; **многоконтурная с. управления** multicircuit control; **многоугольная с.** mesh circuit; **моделирующая с.** analogous circuit; **монтажная с.** wiring diagram, wiring layout; **мостиковая с. звездой** bridged T-network, bridged Y-network; **мостовая с. с лампой тлеющего разряда** (measuring) bridge with neon-bulb indicator; **с. набора на моделирующей установке** analog computer setup; **с. не минимально-фазового фильтра** non-minimum (-net)-phase-shift network; **с. нейтрализации в контуре сетки** Rice circuit, Rice neutralization; **ночная с.** night-service connection; **с. обратной коррекции** de-emphasis circuit; **с. обратной связи** feedback circuit, regenerative circuit; **обращённая с.** dual network, reciprocal network; **общая с.** skeleton diagram; **общая с. включений, общая коммутационная с.** full connection diagram; **однолинейная с. коммутации одной цепи** straight-line diagram; **однопере́кидная с.** single flip-flop (circuit); **однотактная релаксационная с.** one-shot multivibrator; **с. оперативной памяти** topogram; **с., определяющая частоту развёртки** rate-recognition circuit; **опрокидывающая с., опрокидывающаяся с.** flip-flop multivibrator, flip-flop circuit; **параллельная с. коррекции верхних частот** shunt-peaking compensation network; **пассивная пересчётная** (*or* **матричная**) **с.** resistive matrix network; **с. передачи током одного направления** single current system; **перекрёстная с.** lattice network; **пересчётная с.** matrixer, matrix(ing) circuit, scaler, scaling unit, proportional adding circuit, transformation; **пересчётная с. приёмника** receiver matrix; **пересчётная с. с самописцем** scaler-printer; **с. Пикара** phantom circuit, superposed circuit; **пикосрезающая с.** despiking circuit; **с. подвода напряжения к отклоняющим пластинам** deflecting-plate connection; **подробная с. программы** flow chart; **с. подъёма высоких частот** high peaking circuit; **с. подъёма частотной характеристики** peaking network; **с. последовательного действия с линейной зависимостью по модульному значению** linear modular sequential circuit; **с. последовательной коррекции, с. последующей коррекции** series corrector, de-emphasis network; **потенцирующая с.** antilogarithmic circuit; **с. приближенной настройки** compromise network; **с привязки уровня** clamping circuit; **с. приёмника с обратной связью** feedback receiving circuit; **принципальная с.** schematic diagram; **с. проблесковой сигнализации** neon-tube relaxation oscillator; **производственная с.** mnemonic diagram; **с. промежуточного усилителя кабельной линии** repeater circuit; **с. пропускания** OR-gate, OR-circuit; **противоместная с.** antisidetone device, sidetone-reduction wiring (*tphny.*); **противопереходная с.** anti-crossfire network (*tgphy.*); **пятикаскадная с. с взаимно расстроенными контурами** quintuplet, staggered quintuplet; **с. рабочего места телефонистки** operator's speaking circuit; **с. работы контроля клавиатуры** keyboard checking circuit; **с. радиальной развёртки** radial time base; **с. развёртки** time base, time base circuit; **с. развёртывания по дальности** range-sweep circuit; **с. развёртывающего устройства** time base circuit; **развязывающая с.** isolation network; **с. разделения** anti-Rossi circuit; **с. разделения сигналов по времени** time-sharing scheme; **с., различающая полярность импульсов** polarity selector; **с. разновременности** time-anticoincidence circuit; **с. разноимённости**

exclusive OR circuit; **реактивная спусковая с.** one-cycle multivibrator; **с. регенеративного приёмника** regenerator; **рефлексная с.** dual amplification circuit, inverse duplex circuit, reflex circuit; **с. с взаимно расстроенными контурами** staggered circuit; **с. с делением (частоты) на N** scale-of-N circuit; **с. с заземлённым основанием** cathode-tap input circuit; **с. с нулевым выводом** single-way connection; **с. с одним сердечником на разряд** one-core-per-bit circuit; **с. с одним устойчивым положением** monostable circuit; **с. с отрицательной обратной связью** degenerative circuit; **с. с положительной обратной связью** regenerative circuit; **с. с самослышимостью** sidetone wiring (*tphny.*); **с. с фотоэлементом** photoelectric circuit, phototube circuit; **с. связанных цепей** network circuit; **селекторная с.** gate, gating circuit; **сигнальная с. с контрольными жилами кабеля** fault-signaling network; **с. синхронизации с силовой сетью** television mains-hold circuit; **с. сложения младших (or низших) разрядов** low-order add circuit; **с. сложения старших разрядов** high-order add circuit; **с. смягчения действия фиксации уровня** clamp softening circuit; **с. совмещения** keyed clamping circuit (*tv*); **с. совпадений** coincidence amplifier, coincidence circuit, lock-on circuit; gate, AND circuit, parallel gate; **с. совпадений на диодах** diode gate; **с. совпадений с высокой разрешающей способностью** fast coincidence circuit; **с. совпадений с катодным повторителем** cathode gate; **с. «совпадания»** lock-on circuit; **с. соединений** circuit diagram, wiring diagram; **с. сопровождения по дальности** range-tracking circuit; **спусковая с. с двумя устойчивыми состояниями** bistable trigger circuit, Eccles-Jordan multivibrator; **спусковая с. с непосредственной связью** direct-coupled flip-flop; **спусковая с. с отрицательной обратной связью** degenerative trigger circuit; **с. сравнения** subtraction circuit; comparator; **с. стробирования сигнала вспыш-**

ки burst gating circuit; **с. строчной развёртки** horizontal-deflection circuit, horizontal scanning circuit, line scan(ning) circuit; **с. (ступенчатого) деления частоты импульсов** step-divider circuit; **с. суммирования младших разрядов** low-order add circuit; **с. счёта совпадений** coincidence arrangement, coincidence array; **счётная с.** scaler; **с. точной задержки импульсов** phantastron; **трёхкаскадная с. с взаимно расстроенными контурами** triplet, staggered triplet; **ударно-возбуждаемая с.** ringing circuit; **с. удвоения, кратного трём** three multiple doubling circuit; **с. удержания в синхронизме** synchrolock; **с. удлинения импульсов** integrating circuit; **универсальная пересчётная с.** multiscaler; **униполярная фиксирующая с. с диодом и катодом повторителем** one-way diode-cathode-follower clamp; **с. управления в принудительной последовательности** sequence circuit; **с. управления по тангажу** pitch guidance loop; **управляемая с. привязки** clamping circuit, clamper circuit; **с. усиления чередованием каскадов** tuned-aperiodic-tuned circuit; **с. усилителя с катодным выходом** bootstrap circuit; **с. усилителя с отрицательной обратной связью** degenerative-amplifier circuit; **с. установления номера** number checking arrangement (*tphny.*); **с. фазирования цветов** color-indexing circuit; **с. фиксации (or привязки) с двумя триодами** double-diode clamp; **с. фиксации уровня гасящих импульсов** blanking clamper; **с. фиксации (уровня) с диодом и катодным повторителем** diode and cathode-follower clamp; **с. формирования импульсов прямоугольной формы из синусоидального напряжения** sine squaring circuit; **с. централизации Р.В.Х.** circuit; **с. цепей тока** circuit diagram; **с. цепи тока покоя** closed circuit; **цепная с.** ladder network, iterated network, recurrent circuit; **с. частотной коррекции** de-accentuator; preaccentuator; accentuator; **четырёхкаскадная с. с взаимно расстроенными**

**контурами** quadruplet, staggered quadruplet; **с. шнуровых соединений** cording diagram

**схематизировать** *v.* to plan, to schematize

**схематика** *f.* circuitry

**схематически** *adv.* schematically

**схематический** *adj.* schematic, diagrammatic(al), outline

**схематичность** *f.* schematism

**схематичный** *adj.* schematic, outlined

**схемный** *adj.* circuit

**схемотехника** *f.* circuitry, circuit technique(s)

**схлестать** *pf. of* **схлёстывать**

**схлёстывание** *n.* whipping; **с. проводов от раскачки их (ветром)** whipping of wires, swinging cross; **с. проводов при гололёде** sleet jump

**схлёстывать, схлестать** *v.* to whip, to lash

**сход** *m.* drift; tails, tailings; descent, descending

**сходимость** *f.* convergence, converging

**сходиться, сойтись** *v.* to coincide, to agree; to converge, to meet, to join

**сходный** *adj.* similar, analogous, like, allied, cognate; consistent, even; suitable

**сходящийся** *adj.* convergent

**схождение** *n.* convergence, meeting

**схожий** *adj.* alike, like, relative, similar

**схоластический** *adj.* scholastic

**СЦБ** *abbr.* **(сигнализация, централизация и блокировка)** signaling and control techniques for railroads

**сцементировать** *pf. of* **цементировать**

**сцена** *f.* stage, scene; **незаглушённая с.** live stage

**сценарий** *m.* scenario; continuity

**сцениоскоп** *m.* scenioscope

**сценический** *adj.* stage, scenic

**сцепить(ся)** *pf. of* **сцеплять(ся)**

**сцепка** *f.* coupler, clutch; coupling, connecting; **жёсткая автоматическая с.** tight-lock coupler

**сцепление** *n.* cohesion, adhesion, adherence, coherence; coupling, interlinking, interlinkage, linking, linkage; chain, series; clutch, mesh, engagement, grip; connecting, contact, connection; link; docking; concatenation; bond strength; binding; **с зубчатым сцеплением без проскаль-**

зывания positive geared; **с. азиму-тального сельсина** azimuth synchro drive gears; **магнитное с.** magnetic drive; **с. металлической сварки со стеклом** wetting of glass-to-metal; **храповое с.** pawl coupling

**сцеплённый** *adj.* coupled, connected, linked, interlinked; meshed, in gear, engaged; coherent

**сцеплять, сцепить** *v.* to couple, to connect, to link, to interlink, to hook up, to lock; to clutch, to mesh, to put in gear, to engage

**сцепляться, сцепиться** *v.* to interlink, to be coupled; to gear, to mesh, to engage; to cohere, to adhere

**сцепляющий** *adj.* coupling; clutching, engaging

**сцепной** *adj.* coupling

**Сцилард** Szilard

**сцинтиграмма** *f.* scintigram

**сцинтиллирование** *n.* scintillation

**сцинтиллирующий** *adj.* scintillating; scintillation

**сцинтиллограф** *m.* scintillograph

**сцинтиллоскоп** *m.* scintilloscope

**сцинтилляционный** *adj.* scintillation

**сцинтилляция** *f.* scintillation

**СЦМ** *abbr.* (**система центра масс**) center-of-mass system

**счесть** *pf. of* **считать**

**счёт** *m.* score; numeration; count, counting, computation, calculation, reckoning; metering (*tphny.*); account, bill, invoice; tact, time (*mus.*); **за счёт** at the expense (of); by means of; through; due to; **с. в обратную сторону** countdown; **с. в режиме вычитания** counting-down; **с. в течение заданного промежутка времени** preset-time counting; **с. числа повторений** repetition count

**счётно-аналитический** *adj.* computing

**счётно-импульсный** *adj.* pulse-counting

**счётно-решающий** *adj.* computing, digital-computing, analogue

**счётный** *adj.* calculating; counting; computing, countable, enumerable; account

**счетовод** *m.* accountant, ledger clerk, clerk

**счетоводный** *adj.* book-keeping, accounting

**счетоводство** *n.* book-keeping, accounting

**счётчик** *m.* meter, measuring device; register, indicator, recorder; calculator, computer; counter, reckoner; integrating device; scaling unit, scaler; totalizer; hour-meter register mechanism; hour meter; counter tube; **с. амперчасов** current meter; **с. варчасов** reactive energy meter, reactive volt-ampere meter, sine meter, wattless component meter; **с. витков** coil-turn counter; **с. времени** hour meter, chargeable-time indicator; **с. в. движения повозки по инерции** coasting recorder; **с. в. занятости группы** group-occupancy time meter (*tphny.*); **с. в. разговора** chargeable time indicator; **с. вспышек** scintillation counter; **с. вызовов абонента** call meter, message register, call counter, subscriber's meter; **с.-двигатель** motor-meter; **двоично-десятичный с.** decade counter; **двойной с.** two-rate subscriber's meter (*tphny.*); **с. жидкости объёмного типа** positive-displacement meter; **с. занятостей** overflow meter, traffic meter; **с. занятости всех выходов** all-trunk-busy register; **с. излучений с предварительной установкой** predetermined count scaler; **с.-измеритель интервалов времени** counter-timer; **с. импульсов** scaling circuit, scaler, tube scaler; **с.-комутатор** counter-controller; **с. максимального расхода** peak-load hour meter; peak counter; **многокамерный с.** multicellular counter-tube; **многоразрядный двоичный с.** radix 2 counter; **с. моточасов** engine-hour indicator; **нагрузочный с. (на линиях дальней связи)** electromechanical call-counter; position peg-count register; **накапливающий с.** accumulator register; **накапливающий с. для вычислений с учётом порядков** floating accumulator; **недорегулированный на оставание фазы с.** undercompensated meter; **непрерывно подключённый с.** continuous counter; **с. отметок по плотности** density counter; **с. переговоров на последних шнурах** late-choice call meter; **перерегулированный на отставание фазы с.** overcompensated meter; **с. полной энергии** volt-ampere-hour meter, apparent-energy meter; **половинный делительный с. импульсов** scale-of-two counter, scaling couple; **с. порядков** exponent counter; **с. продолжительности разговоров** call control watch, time check, timing register, chargeable-time indicator; **с. рабочего места** position meter, position peg count meter (*tphny*.); **разделённый с.** split-dial feature; **с. разобщений** pulse ionization chamber; **разрядный с.** radiation counter; **ручной с. вызовов** manually operated call meter; **с. с верхними и нижними порогами** up-down counter; **с. с вращающимися деталями** motor-driven hour-meter, motor meter; **с. с горизонтальной счётной характеристикой** flat response meter; **с. с делением на N** scale-of-N counter; **с. с максимальным ваттметром** integrated-demand meter; **с. с периодическим замером мощности** discontinuously integrating meter; **с. с предварительной оплатой** prepayment meter, slot meter; **с. с предварительной установкой, с. с предварительным набором** predetermined counter, preset counter; **с. случаев занятости** congestion meter; **с. совпадений** lock-on counter; **суммирующий с. с часовым механизмом** tachoscope; **сцинтилляционный с.** photomultiplier counter (*phys.*) ;**управляемый с. импульсов** predetermined scaler; **управляемый с. колебаний** gated oscillator counter; **с. числа включений рентгеновской трубки** exposure counter; **с. числа занятий** peg count meter, traffic meter, call meter (*tphny*.); **с. числа состоявшихся разговоров** effective-call meter; **с. числа часов занятости** traffic unit reader; **шаговый с.** step-tube counter; **электрический с.** electric supply meter

**счёты** *pl.* abacus

**счисление** *n.* numeration, numbering; calculation, reckoning; enumeration, computation; **с. координат местоположения** dead reckoning

**счисленный** *adj.* numerated

**счислимый** *adj.* countable, reckoning

**счислитель** *m.* computer

**счислительный** *adj.* computer

**счислять, счислить** *v.* to count, to reckon, to enumerate, to compute, to calculate; to number

**счистить** *pf. of* **счищать**

**считаемый** *adj.* counted

**считание** *n.* counting, calculating

**считать, счесть** *v.* to count, to compute, to reckon, to rate, to calculate; to consider; to register; to number; **с. с регистра** to roll out

**считываемость** *f.* readability

**считывание** *n.* reading, read-out; plotting; computation; sensing, pick-up; playback; read (*info. theo.*); **низкочастотное с.** video pulse reading; **перезарядное с., с. по методу перезарядки (ёмкости)** capacity-discharge reading; **с. показаний** reading; plotting

**считыватель** *m.* reader

**считывать** *v.* to read, to read off, to read out; to take reading; to sense; **с. на ощуп** to sense

**считывающе-печатающий** *adj.* reader-typer

**считывающий** *adj.* scanning, scan-off; reading, reader; viewing

**счищать, счистить** *v.* to bare, to skin, to strip, to clear (away), to take off

**сшабрить** *pf. of* **шабрить**

**съедать, съесть** *v.* to eat, up, to eat away, to corrode; to kill (of voltage)

**съезд** *m.* connecting track; convention, conference

**съём** *m.* pick-up (of a signal); take-off; removal, withdrawal, extraction; **автоматический с. цифровых данных** automatic logging; **полуавтоматический с. данных** semiautomatic data reduction

**съёмка** *f.* survey, surveying; plan; planning; mapping; plotting; pick-up (*tv*); shot, shooting, filming (*tv*); **ближняя с.** close-up view (*tv*); **заключительная с.** final shot (*tv*); **с. картины шумов** noise survey; **с. карты** mapping; **с. крупным планом** close-up shot (*tv*); **с. мелким планом** long shot (*tv*); **с. на плёнку** filming; **непосредственная с.** direct pick-up; **с. общим планом** long shot (*tv*); **с. плана** mapping; **с. под углом** canted shot; **радиолокационная с. плана местности** radar mapping; **с. с движения** follow shot, tracking shot; **с. с натуры** direct pick-up; **с. средним планом** medium shot (*tv*)

**съёмник** *m.* pick-up; stripper, stripping device; extractor, remover

**съёмный** *adj.* movable, withdrawable, removable, detachable, loose, replaceable, demountable, dismountable, renewable, interchangeable; plug-in; hook-on, clip-on, tong-test

**съёмочный** *adj.* surveying; picture-taking; pick-up; camera

**съесть** *pf. of* **съедать**

**сыграть** *pf. of* **играть**

**сыровидный** *adj.* cheese-like, cheesy

**сырой** *adj.* damp, moist, humid; raw, coarse, crude; rough

**сыростестойкий** *adj.* dampproof, mois-tureproof, waterproof

**сырость** *f.* moisture, dampness, humidity, wetness, damp

**сырьё** *n.* raw material; **слюдяное с.** natural mica, block mica

**Сэбин** Sabine

**сэбин** *m.* sabin

**сэбинов** *poss. adj.* Sabine, Sabine's

**Сэйдж** SAGE (system)

**Сэйдик** *m.* Sadic (system)

**сэкономить** *pf. of* **экономить**

**сэмплирование** *n.* sampling

**сэмплированный** *adj.* sampled

**сэмплирующий** *adj.* sampling

**СЭТ** *abbr.* (**самолётный электрический тахометр**) aircraft electric tachometer

**сюита** *f.* suite (*mus.*)

# Т

т *abbr.* 1 (**температура**) temperature; 2 (**точка**) point

**таблетка** *f.* cake, slab, wafer, biscuit; pellet

**таблеточный** *adj.* pellet; pill; tablet, wafer; preforming

**таблица** *f.* table, list, index, chart, sheet, schedule; card; tabular; **т. значений функций** function table (*math.*); **интерферирующая испытательная т.** interfering test pattern; **испытательная т.** test list; test chart; checker board pattern, definition chart, resolution chart; **испытательная т. для оценки числа градаций яркости** brightness test chart; **т. конструктивных размеров** design sheet; **т. нагрузок** stress table; **т. отыскания повреждений** troubleshooting table; **т. последовательности манипуляции** manipulation chart; **т. последовательности переключений** control sequence table; **радиодевиационная т.** calibration table, calibration radio chart; **т. распределения программы по адресам** allocation plan; **т. расцветки** color code; **т. расцветки жил** cable color code; **т. сопряжённости признаков** contingency table; **т. условных сигналов** cue sheet; **эталонная т.** table of corrections

**табличный** *adj.* tabular, table

**табло** *n. indecl.* chart; signal panel; **световое т.** lamp register; **световое т. железнодорожного участка** track-section strain indicator panel

**табулирование** *n.* tabulation

**табулированный** *adj.* tabulated

**табулировать** *v.* to tabulate

**табулирующий** *adj.* tabulating

**тубулограмма** *f.* tabulator form

**табулятор** *m.* tabulator, tabulating machine, accounting machine; **суммирующий т.** rolling total tabulator

**табуляторный, табуляционный** *adj.* tabulating

**тавот** *m.* grease, fat, lubricant, cup grease; **т.-пресс** grease gun

**тавотница** *f.* grease cup

**тавровый** *adj.* T-, tee

**тавтологический** *adj.* tautological

**тавтология** *f.* tautology

**тазиметр** *m.* tasimeter

**тазомер** *m.* pelvimeter

**таймер** *m.* clock, timer

**таймотор** *m.* thymotor

**таймотрол** *m.* thymotrol

**таймтактор** *m.* timing contactor

**тайна** *f.* secrecy, secret

**тайный** *adj.* secret

**тайпотрон** *m.* typotron

**тайристор** *m.* thyristor

**тайристорный** *adj.* thyristor

**тайрит** *m.* thyrite, tyrite

**тайрод** *m.* thyrode

**«тайфон»** *m.* Typhon

**такан** *m.* Tacan

**такелаж** *m.* tackle, ropes, cordage

**такелажный** *adj.* tackle, cordage; guy, strain

**такса** *f.* fee, rate, tariff, toll

**таксировка** *f.* call-fee determination, call-charge determination

**таксометр** *m.* taxameter, taxometer, taximeter

**таксономический** *adj.* taxonomic

**таксономия** *f.* taxonomy

**таксофон** *m.* coin telephone, pay telephone

**такт** *m.* time, beat, bar (*mus.*); measure; time step, time unit; stroke; cycle (of engine rate); **второй т. сложения** second addition time; **т. пеленгации** (periodical) fluctuations of the bearing-indicator mark; **полный т. (вычислений)** complete operation

**тактика** *f.* policy, tactics

**тактильный** *adj.* tactile

**тактирование** *n.* clock timing; **многофазное т.** multiphase clock

**тактирующий** *adj.* clock, clocking, timing; cycling

**тактический** *adj.* tactical

**-тактный** *adj. suffix* -step

**тактовый** *adj.* time, bar, cadence

**тактометр** *m.* tactometer

**тактрон** *m.* tactron

**тали** *pl.* snatch block, block and tackle pulley block, jigger, jam

**таллиевый** *adj.* thallium, thallic

**таллий** *m.* thallium, Tl

**таллий-сульфидный** *adj.* thallofide

**таллофидный** *adj.* thallium, thalofide

**талон** *m.* ticket, coupon, token; **справочный т.** inquiry docket

**талофид** *m.* thalofide

**талофидный** *adj.* thalofide

**тальбот** *m.* talbot (unit of light energy)

**тальк** *m.* talc, talcum

**тамбур** *m.* wire reel, drum barrow; **т. для кабеля** cable drum carriage, drawing drum; **смотровой т.** visor

**тамбурин** *m.* tambourine

**тангенс** *m.* tangent (*math.*); **т.-бусоль** tangent-compass; **т. угла отклонения** ratio of refraction; **т. угла потерь** loss-angle tangent; **т. угла потерь диэлектрика** dielectric power factor

**тангента** *f.* push-to-talk key

**тангенциальный** *adj.* tangent, tangential; centrifugal (force)

**тандем** *m.* tandem; **т.-регистр** tandem sender

**танк** *m.* tank

**танковый** *adj.* tank

**тантал** *m.* tantalum, Ta

**танталатный** *adj.* tantalate

**танталовый** *adj.* tantalum, tantalic

**тарелка** *f.* disk, plate; valve; **торцевая т.** armature end plate

**тарелки** *pl.* cymbals

**тарелкообразный** *adj.* dish-shaped

**тарелочка** *f.* flare, flange

**тарелочный, тарельчатый** *adj.* dish, plate, plate-like, disk

**тарирование** *n.* calibration, calibration test, gaging; taring

**тарированный** *adj.* tared

**тарировать** *v.* to calibrate, to gage; to tare

**тарировочный** *adj.* calibrator

**тариф** *m.* tariff; rate, fee, charge; **дистанционный т.** blanket ratio tariff (*tphny.*); **междугородный т.** long-distance rate, toll rate; **т. по расстоянию** distance rate, distance tariff, blanket ratio; **повременный т.** time rate, flat rate; **подразделён-**

**ный т.** graduated tariff; **простой т.** flat rate (*tphny.*); **твёрдый оптовый т.** contract two-rate tariff

**тарификация** *see* **таксировка**

**тарифный** *adj.* tariff; rate, charge

**тарифопечатающий** *adj.* (call)-charge printer

**тартрат** *m.* tartrate; **т. калия** dipotassium tartrate

**тартриметр** *m.* tartrimeter

**тасиметр** *m.* tasimeter

**таситрон** *m.* tacitron

**тасовка** *f.* randomization

**тастатура** *f.* key pulser (*tphny.*); key assembly, keyboard, strip of keys, key set, digit-key strip; **т. номеров** subscriber dial-pulse repeater

**тастатурный** *adj.* key, keying; pulsing

**Таунс** Townes

**Таунсенд** Townsend

**таунсендовский** *adj.* Townsend

**тахеометр** *m.* tachymeter

**тахеометрия** *f.* tachymetry

**тахиметр** *m.* tachymeter

**тахиметрический** *adj.* tachymetric

**тахистоскоп** *m.* tachistoscope

**тахогенератор** *m.* tachodynamo, tachogenerator, tacho-alternator, speed indicating generator, speed voltage generator; **т. переменного тока** drag-cup alternator

**тахограф** *m.* tachograph

**таходинамо** *n. indecl.* tachogenerator, speed-voltage generator

**тахометр** *m.* tachometer, engine-speed indicator; **т., действующий вихревыми токами** eddy-current revolution counter; **т. обратной связи** cue tachometer; **т. с центробежным маятником** inertia speed counter

**тахометрический** *adj.* tachometer

**тахометрия** *f.* tachometry

**тахоскоп** *m.* tachoscope

**тацитрон** *m.* tacitron

**тащить, потащить** *v.* to draw, to draw along, to drag, to pull, to haul, to tow

**Твайман** Twyman

**ТВД** *abbr.* 1 (**турбина высокого давления**) high-pressure turbine; 2 (**турбовинтовый двигатель**) turbopropeller engine

**твердеть, затвердеть** *v.* to harden, to grow hard, to grow firm, to solidify

**твёрдогазовый** *adj.* solid-gaseous

**твёрдомагнитный** *adj.* permanent-magnet

**твердомер** *m.* durometer, hardometer, hardness tester

**твёрдость** *f.* hardness; rigidity, stiffness, consistency, strength, stability; **т. закалки** temper, degree of hardness; **склерометрическая т.** scratch hardness

**твёрдотелый** *adj.* solid-state, solid

**твёрдотянутый** *adj.* hard-drawn

**твёрдый** *adj.* tough, hard, solid, firm; steady, consistent; rigid, inflexible, stationary; **на твёрдом теле** solid-state

**твики** *pl.* tweeks

**твин** *m.* twin, twin circuit

**твинплекс** *m.* twinplex (*tgphy.*), duplex

**твистор** *m.* twistor

**твисторный** *adj.* twistor

**ТВЧ, т.в.ч.** *abbr.* (**ток высокой частоты**) high-frequency current

**ТВЭ** *abbr.* (**тепловыделяющий элемент**) fuel element

**ТГ** *abbr.* (**телефонный, голый**) telephone-type, bare

**театральный** *adj.* theater

**театранспункт** *m.* theater network televison

**театрофон** *m.* theaterphone, electrophone, program supply system

**Тевенин** Thévenin

**тевотрон** *m.* tevotron

**тезаурус** *m.* thesaurus

**Тейлор** Taylor

**текнетрон** *m.* technetron, tecnetron

**текст** *m.* text

**текстолит** *m.* textolite

**текстолитовый** *adj.* textolite

**текстура** *f.* texture, grain, veining

**текучесть** *f.* fluidity, flow; viscosity

**текучий** *adj.* flowing, fluid, running

**текущий** *adj.* flowing, running, leaking, leaky; current, present

**«телансерфон»** *m.* telanserphone, tel-answerphone

**теле-** *prefix* tele-

**телеавтограмма** *f.* tele-autogram

**телеавтограф** *m.* telautograph, telewriter

**телеавтоматика** *f.* telautomatics, telemechanics, automatic remote control

**телеамперметр** *m.* teleammeter

**телеблокировка** *f.* pilot channel block-ing; **проводная т.** metallic-wire block-ing

**телеваттметр** *m.* telewattmeter

**телевещание** *n.* telecasting, television broadcasting

**телевидение** *n.* television; **т. в инфракрасных лучах** noctovision; **малострочное т.** low-definition television; **т. по выделенной** (*or* **ограниченной**) **сети** closed circuit television; **т. по театральной сети** theater network televison; **т. с высокой чёткостью изображения** high definition television; **т. с малой чёткостью** low definition television; **т. с промежуточной плёнкой** continuous-film tv system; **цветное т. с одновременной передачей цветов** simultaneous color television; **цветное т. с последовательной передачей строк** line sequential television, line sequential color television; **цветное т. с последовательной передачей цветов** sequential color television, field-sequential color television, dot sequential color television; **цветное т. с фиксацией электронного луча** beam-indexing color television

**телевидеть** *v.* to teleview

**телевизионно-звуковой** *adj.* sound-sight

**телевизионный** *adj.* television

**телевизия** *f.* television

**телевизор** *m.* television receiver, television set; **т. с серым фильтром перед экраном** black screen television set

**телевольтметр** *m.* televoltmeter

**телевотер** *m.* tele-voter

**телевыключатель** *m.* teleswitch, remote control switch

**телега** *f.* carriage, wagon, car; truck

**телегеничный** *adj.* telegenic

**телегониометр** *m.* telegoniometer

**телеграмма** *f.* telegram, wire, cable, cablegram; **т. оплачиваемая адресатом** collect message; **т. с обратной проверкой** collated telegram

**телеграф** *m.* telegraph (office); **автоматический т.** для передачи депеш посредством перфорированной ленты perforated-strip-type transmitter; **буквопечатающий т.** printing telegraph, teleprinter exchange; **дистанционный буквопечатающий т.** teletype writer, type printing telegraph; **многократный т. с вилко-образным соединением** forked multiplex telegraph; **общий т. для передачи и приёма** superposed telegraph; **простой т.** closed-circuit telegraph system; **т. с автоматической записью** facsimile recorder; **самопишущий т.** teleprinter; **тональный т.** voice-frequency telegraph; **центральный т.** telegraph exchange

**телеграфирование** n. telegraphy, telegraphing, telegraph operation; **абонентское т.** teletypewriter exchange service; **т. быстродействующими телеграфами** high-speed telegraph system; **встречное четырёхкратное т.** octuplex telegraph; **вч т. на несущей** high-frequency carrier telegraphy; **двустороннее т., двухполюсное т.** double-current working, double-current system (*tgphy*.); **двухчастотное тональное т.** double signal telegraphic keying; **многократное т.** multiplex telegraphy, multiplex working; **многократное встречное т. камертонными аппаратами** syntonic telegraphy; **многократное тональное т.** phonoplex telegraphy; **т. на несущей** carrier telegraphy; **т. на несущей высокой частоты** high frequency carrier telegraphy, carrier current telegraphy; **направленное т.** TT system employing a separate channel for each direction; **т. постоянным током** d-c telegraphy, closed circuit working; **т. с частотной манипуляцией** shift telegraphy, frequency-shift telegraphy; **т. током несущей частоты** carrier telegraphy, carrier-current telegraphy; **тональное т.** audio frequency telegraphy, alternating current telegraphy; **частотное т. с тоном в полосе разговорных частот** inclusion telegraphy

**телеграфировать** v. to telegraph, to wire, to cable

**телеграфист** m., **телеграфистка** f. telegraph operator, telegrapher

**телеграфия** f. telegraphy; **буквопечатающая т.** printing telegraphy; printing telegraphy system, teleprinter, type printing telegraph; **многократная т. по двум каналам** two-channel telegraphy; **т. на звуковой несущей** voice-frequency carrier telegraphy; **т. на несущей**

частоте carrier-current telegraphy; **т. на средних частотах** mid-frequency telegraphy; **т. пишущими телеграфами** writing telegraph system; **т. по суперфантомным цепям** double-phantom telegraphy, superphantom telegraphy; **т. с автографической записью** facsimile, photo telegraphy; **тональная т.** voice frequency carrier telegraphy

**телеграфно-модулированный** *adj.* telegraph-modulated

**телеграфный** *adj.* telegraph, telegraphic, telegraphical

**телеграфон** m. telegraphone

**теледатчик** m. remote pickup

**теледвигатель** m. telemotor

**теледелтос** m. teledeltos

**телediaпроектор** m. slide scanner

**теледьюсер** m. teleducer

**тележечный** *see* **тележный**

**тележка** f. carriage, small cart, chariot, car, truck; dolly; **загрузочная т. с мульдами** charging box car; **кабельная т.** cable-drum carriage, reel carriage; paying-out machine; **т.-кран с длинной стрелой** long-boom dolly; **крановая т.** traveling pulley; **т. оператора** dolly (*tv*); **поворотная т.** swivel truck, bogie truck; **т. с грейфером** bucket trolley; **т. с камерой** dolly (*tv*); **сборочная т.** engine fitting frame

**тележный** *adj.* carriage, cart, car, truck; trolley

**телезритель** m. televiewer

**телеизмерение** n. telemetering, telemetry

**телеизмерительный** *adj.* telemetering

**телеиндикатор** m. tele-indicator

**телеиспытание** n. teletest

**телекино** n. *indecl.* film television, telecinematography, radiomovies

**телекиноаппарат** m. film scanner (*tv*)

**телекиноканал** m. telecine channel

**телекинопередатчик, телекинопроектор** m. telefilm, film scanner, film scanning machine, telecine projector; **т. с линзовым венцом** lens drum scanner

**телекинопроекционный** *adj.* film-scanning

**телекиноустановка** f. telecine unit

**телеконтроль** m. supervisory control, remote control

**телектограф** m. telectrograph

телеметеорограф *m.* telemeteorograph
телеметр *m.* telemeter
телеметрирование *n.* telemetering
телеметрический *adj.* telemetering
телеметрия *f.* telemetry, remote metering; **временная т.** time-division telemetering; **т. морских глубин** bathymetry telemetry; **т. ЧМ-ЧМ** FM-FM telemetering system
телеметрограф *m.* telemetrograph
телемеханика *f.* telemechanics, remote control, tele-automatics, remote-action techniques
телемеханический *adj.* telemechanical
телемотор *m.* telemotor, telecontrolled electric motor
теленабор *m.* telegraphic typesetting
теленаборный *adj.* teletypesetting, teletypesetter
теленадзор *m.* supervisory control
телеобъектив *m.* telephoto lens, tele-objective
телеологический *adj.* teleological
телепередвижка *f.* television car
телеплоттер *m.* teleplotter
телеповторитель *m.* distant repeater
телеподстанция *f.* telesubstation, remote-controlled substation
телеприёмник *m.* long-distance receiver
телепринтер *m.* page teleprinter, teleprinter
Телеран *m.* Teleran
телерегистрация *f.* telerecording
телерегулирование *n.* remote control
телерегулируемый *adj.* remote-controlled
телерецептор *m.* teleceptor, telereceptor
телесамописец *m.* telerecorder, remote recording instrument
телесвязь *f.* telecommunication
телесигнализация *f.* remote signal system, telesignalization
« телесин » *m.* telesyn
телескоп *m.* telescope
телескопировать *v.* to telescope
телескопический *adj.* telescopic
телескопия *f.* telescopy
« телескрайбер » *m.* Telescriber
телесный *adj.* solid (angle); flesh (color)
телестудия *f.* television studio
телесуммирование *n.* remote summation
телесчёт *m.* telecount, telemetering
телесъёмка *f.* telephotography

телетайп *m.* teletype, teletypewriter, type printing telegraph; **т. по телефонной линии** telemixte system; **т. с применением бланков** form teleprinter, page teleprinter
телетайпный *adj.* teletype
телетахометр *m.* tele-tachometer
телетерапия *f.* teletherapy, telecurie therapy
телетермометр *m.* telethermometer
телеуказание *n.* remote indication
телеуказатель *m.* remote indicator, distance gage
телеуправление *n.* remote control, telecontrol, supervisory control; **диспетчерское т.** dispatcher's supervisory control; **т. посредством ведущего луча** beam-(rider) type guidance; **т. при телевидении** television receiver remote control
телеуправляемый *adj.* remote-controlled, telecontrolled, remotely operated
телеучёт *m.* telemetering
телефазометр *m.* telephasemeter
телефон *m.* telephone; **миниатюрные головные телефоны с вкладышами** insert earphones; **безбатарейный т.** sound-powered telephone; **т. внутренней связи** intercom telephone, interphone; **внутританковый т.** tank interphone; **конденсаторный т.** capacitor receiver, electrostatic receiver; **т. МБ** local-battery telephone; **междугородный т.** long-distance telephone, toll telephone; **т. параллельного включения** bridging telephone; **переговорный головной т.** intercommunication headset; **платный т. общественного пользования** public telephone paystation; **т. с двухполюсным магнитом** bipolar receiver; **т. с противоместной схемой** antisidetone telephone set; **т. слухового протеза** hearing aid earphone; **т. ЦБ** common battery telephone, CB telephone; **« чашечный » т.** ear cup, ear-piece, watch receiver; **электродинамический т.** moving-conductor receiver; **электромагнитный т. с якорем** armature magnetic earphone
телефонирование *n.* telephony, telephone operation, telephoning; **т. боковой полосой частот** sideband telephony; **дуплексное т.** two-way

operation, duplexing; **многоканаль-ное т. несущими токами** multichannel carrier system; **т. несущими токами** carrier telephony (system), carrier current telephony, high frequency telephony; **т. разговорными частотами** audio-frequency telephony, voice-frequency telephony

**телефонировать** *v.* to telephone; **т. дуплексом** to duplex; **т. по радио** to radiophone

**телефонист** *m.*, **телефонистка** *f.* telephone operator; **вспомогательная т., входящая т.** inward operator; **т. дальней связи** trunk operator; **исходщая т., т. исходящей связи** outward operator, outgoing operator; **т. коммутатора соединительных линий** trunk-board operator; **междугородная т.** long-distance operator, toll operator; **т. местного коммутатора** PBX operator; **т. распорядительной станции** controlling operator; **т. соединительных линий** trunk operator; **т. стола заказов** recording operator

**телефония** *f.* telephony; **вч т. (с несущей)** carrier (current) telephony; **дуплексная т.** two-way telephone conversation, duplex telephony, phantom telephony; **многократная вч т.** multiplex carrier-current telephony; **т. на боковой полосе** sideband telephony; **т. на звуковых частотах** voice-frequency telephony, a-f telephony; **т. на несущей частоте** carrier telephony; **т. с подавленной несущей в паузах** quiescent carrier telephony; **световая т.** phototelephony

**телефонно-телевизионный** *adj.* videotelephone

**телефонный** *adj.* telephone, telephonic, telephonical

**телефоновидение** *n.* phonovision

**телефонограмма** *f.* telephone message, telephone dispatch, phonogram

**телефонограф** *m.* telephonograph

**телефонометрия** *f.* telephonometry

**телефот** *m.* telephote

**телефото** *n. indecl.* telephoto

**телефотографический** *adj.* telephotographic

**телефотография** *f.* telephotography; telephotograph

**телецентр** *m.* television broadcasting station

**телецентрализация** *f.* remote control interlocking

**теллур** *m.* tellurium, Te

**теллурий** *m.* tellurian; tellurium

**теллуристый, теллурический** *adj.* tellurous, tellurium, telluric

**теллурометр** *m.* tellurometer

**тело** *n.* body; matter, substance, solid; shaft; housing; core; **абсолютно отражающее т.** perfect reflector; **т. накала** luminous element; **т. необтекаемой формы** blunt body; **рабочее т.** actuating medium, working substance; **твёрдое т.** solid; see also **твёрдый; фотометрическое т.** solid of light distribution

**тельфер** *m.* telpher

**тема** *f.* theme, subject

**тембр** *m.* timbre, tone quality, musical quality, tone

**темнеющий** *adj.* growing dark, darkening

**темновой** *adj.* dark

**тёмно-синий** *adj.* mazarine

**тёмный** *adj.* dim, dark; black; obscure

**темп.** *abbr.* (**температура**) temperature

**темп** *m.* tempo, time; rate, frequency

**температура** *f.* temperature; **исходная т.** ambient temperature; **т. крепящего винта** stud temperature; **т. нити накаливания** filament temperature; **т. окружающей среды** ambient temperature; **предельная т. перехода** junction temperature; **т. при длительной непрерывной работе** duration excess temperature; **т. растворения** critical dissolution-temperature; **энергетическая т.** total radiation temperature

**температурно-зависимый** *adj.* temperature-dependent

**температурно-регулировочный** *adj.* temperature-controlled

**температурный** *adj.* temperature; heat, thermal

**температуропроводимость, температуропроводность** *f.* thermal conductivity, thermal diffusivity, temperature exchange

**температурочувствительный** *adj.* temperature-sensitive

**темперация** *f.* temperament (mus)

**темперирование** *n.* tempering

**темперированный** *adj.* tempered

**темперировать** *v.* to temper

**темпернол** *m.* tempernol

**тенденция** *f.* tendency

**тенебресценция** *f.* tenebrescence

**теневой** *adj.* shadow; shaded, dark

**тенеобразный** *adj.* shadow-like

**тенеобразование** *n.* shadiness, obscurity

**тенеобразующий** *adj.* shadow-forming

**тензиметр** *m.* tensimeter

**тензиометр** *m.* tensiometer

**тензодатчик** *m.* strain gage, tensometer; **т. крутящего момента** torsion strain gage; **проволочный д.** bonded strain gage

**тензометр** *m.* strain gage, tensometer, extensometer; **плосконамотанный т.** wrap-around strain gage; **т. с плоской зигзагообразной намоткой** flat-grid strain gage; **т. сопротивления с бумажной подложкой** paper strain gage; **струнный т.** vibrating wire strain gage

**тензометрический** *adj.* strain-gage, tensometric

**тензор** *m.* tensor; **т. напряжения при сжатии** compressive stress tensor; **т. напряженности поля** field tensor; **т. плотности импульса энергии** energy momentum tensor

**тензорезистивный** *adj.* tensoresistive

**тензорный** *adj.* tensor

**тензоэлектрический** *adj.* tensoelectric

**тенистый** *adj.* shaded, shady

**тенор** *m.* tenor

**теноровый** *adj.* tenor

**тенсиометр** *m.* tensiometer

**тень** *f.* shadow, shade, shading

**теодолит** *m.* theodolite

**теорема** *f.* theorem, proposition; **т. для цепей с реактивными сопротивлениями** reactance theorem; **т. импульса вращения** law of moment of momentum; **т. импульсов в гидродинамике** theorem of momentum in fluid dynamics; **т. о выборках** sampling theorem (*math.*); **т. о дискретном представлении** sampling theorem; **т. о дискретном представлении во времени** temporal sampling theorem; **т. о дискретном представлении по частоте** sampling theorem in the frequency domain; **т. о матричном разветвлении, т. о построении схем по заданным матрицам** matrix tree theorem; **т. о среднем значении в дифференциаль-** ном исчислении mean-value theorem of the differential calculus; **т. обратной** (*or* **обратимости**) **энергии** reciprocal-energy theorem; **т. обратной пропорциональности Гельмгольца** Helmholtz reciprocal principle; **основная т. теории функций** fundamental laws of the theory of functions; **т. равномерного распределения** principle of the equipartition of energy; **т. сложения в операторном исчислении** addition of Laplace integrals; **т. сложения скоростей** addition theorem of velocity; **т. умножения в теории вероятностей** multiplication theorem of the probability calculus; **т. четырёхполюсника с реактивным сопротивлением** four-terminal network reactance theorem

**теоремный** *adj.* theoremic

**теоретизировать** *v.* to theorize

**теоретически** *adv.* theoretically

**теоретический** *adj.* theoretical

**теория** *f.* theory; **т. близкодействия** Faraday's theory; **т. вероятности** calculus of probability; **т. выборочного контроля** acceptance sampling; **единая т. поля** unified field theory; **т. многоскоростного распространения волн** multivelocity wave theory; **т. множеств** set theory; **т. «перегиба» симметричных четырёхполюсников** bending theory of symmetrical networks; **т. переноса заряда** theory of charge transport; **т. посторонних примесей** inclusion theory; **т. слуха** auditory theory; **т. трёхцветного зрения** three-sensation theory of color vision, tristimulus theory of vision; **электродинамическая т. абсолютно покоящегося эфира** atomic absolute theory

**Теплер** Toepler

**тепло** *n.* heat, warmth; **т. выделяемое током** Joule's effect, resistance-heating effect

**тепло-** *prefix* thermo-, thermal, heat

**теплоаккумулирующий** *adj.* heat-storing

**теплоаккумулятор** *m.* (water) storage heater

**тепловой** *adj.* heat, thermal, thermic, hot, caloric

**тепловыделение** *n.* heat release, heat liberation

**тепловыделяющий** *adj.* heat-liberating

**теплогашение** *n.* thermoquenching

**теплодиффузия** *f.* thermodynamics of diffusion

**теплоёмкость** *f.* heat capacity, thermal capacity; specific heat; **т. анода** heat storage capacity; **удельная т. при постоянном давлении** specific heat at constant pressure; **удельная т. при постоянном объёме** specific heat at constant volume

**теплоизлучающий** *adj.* heat-radiating

**теплоизлучение** *n.* thermal radiation

**теплоизолирующий, теплоизоляционный** *adj.* heat-insulation, thermal-insulation, heat-insulating

**теплоизолятор** *m.* heat insulator

**теплоизоляция** *f.* thermal insulation

**тепломер** *m.* calorimeter; thermometer

**теплообмен** *m.* heat exchange

**теплообразование** *n.* generation of heat

**теплообразователь** *m.* heat generator, heat producer

**теплообразующий** *adj.* heat-producing, heat-generating

**теплоотбор** *m.* heat take-off

**теплоотвод** *m.* heat-conducting path, heat sink, heat removal, dissipator

**теплоотводящий** *adj.* heat-removing, heat-transmitting

**теплоотдача** *f.* heat transfer; heat emission, heat emissivity

**теплоотдающий** *adj.* heat-liberating, exothermic

**теплопередатчик** *m.* heat transmitter

**теплопередача** *f.* thermal conduction, transmission of heat, heat transfer

**теплопередающий** *adj.* heat-transmitting, heat-transfer

**теплопоглощающий** *adj.* heat-absorbing

**теплопровод** *m.* heat conductor

**теплопроводимость** *see* **теплопроводность**

**теплопроводник** *m.* heat conductor

**теплопроводность** *f.* thermal conduction, thermal conductivity, heat conductivity, heat conduction, heat transfer

**теплопроводный, теплопроводящий** *adj.* thermopositive, heat-conducting, heat-carrying, diathermic, diathermal

**теплопроизводительность** *f.* calorific value, heating power, heating efficiency, heat value

**теплопроизводящий** *adj.* heat-producing

**теплосодержание** *n.* heat content

**теплостойкий** *adj.* fire-resistant, heat-proof, thermostable, heat-fast

**теплостойкость** *f.* temperature constancy, heat endurance, thermal endurance, thermostability, heat resistance

**теплота** *f.* heat, warmth; **т. лучеиспускания** radiant heat; **т. потерь** heat loss, dissipation heat; **скрытая т. превращения** transmutation heat; **удельная т. парообразования** heat equivalent of vaporization of water

**теплотворность** *f.* heat value, heating capacity, calorific value

**теплотворный** *adj.* calorific, heat-producing

**теплоудерживающий** *adj.* heat-conserving

**теплоупорный** *see* **теплостойкий**

**теплоустойчивость** *f.* thermostability, temperature stability

**теплоустойчивый** *adj.* fire-resistance, heatproof, thermostable

**теплочувствительный** *adj.* heat-sensitive

**теплошумовой** *adj.* thermal-noise

**теплоэлектроцентраль** *f.* heating and power plant

**теплоэнергетика** *f.* heat and power engineering

**тёплый** *adj.* warm; thermal

**тера-** *prefix* tera-

**терагерц** *m.* teracycle per second, tera-hertz, fresnel

**тераомметр** *m.* tera ohmmeter

**терапевтический** *adj.* therapeutical

**терапия** *f.* therapy; **лучевая т.** radiation therapy, therapeutic radiology

**тербиевый** *adj.* terbium

**тербий** *m.* terbium, Tb

**терблиг** *m.* therblig

**тереть, потереть** *v.* to rub, to chafe; to grind, to grate

**терм** *m.* term, field; therm (unit of heat); **основной т.** energy level of ground state

**термализация** *f.* thermalization

**термализовать** *v.* to thermalize

**термаллой** *adj.* Thermalloy

**термальный** *adj.* thermal

**терматрон** *m.* thermatron

**терменовокс** *m.* etherophone

**термин** *m.* term, technical term
**терминологический** *adj.* terminological, nomenclature
**терминология** *f.* terminology, nomenclature
**термистор** *m.* thermistor; **т. в виде бусинки** bead thermistor
**термисторный** *adj.* thermistor
**термит** *m.* thermite
**термитный** *adj.* thermite
**термически** *adv.* thermally
**термический** *adj.* thermal, thermic
**термия** *f.* therm (unit of heat)
**термо-** *prefix* thermo-, therm-, heat
**термоакустический** *adj.* thermoacoustic
**термоакцептор** *m.* thermal acceptor
**термоамперметр** *m.* thermoammeter
**термоанализ** *m.* thermal analysis
**термоанемометр** *m.* hot-wire anemometer
**термобаллон** *m.* temperature bulb
**термобарокамера** *f.* space chamber
**термобарометр** *m.* thermobarometer
**термобатарея** *f.* thermobattery, thermopile
**термоваттметр** *m.* thermocouple-type wattmeter
**термовозбуждённый** *adj.* thermally generated
**термовольтметр** *m.* thermocouple-type voltmeter
**термовыключатель** *m.* thermoswitch
**термовысвечивание** *n.* thermal glow
**термогальваномагнитный** *adj.* thermogalvanomagnetic
**термогальванометр** *m.* thermogalvanometer
**термогенерированный** *adj.* thermally generated
**термогидрограф** *m.* thermohydrograph
**термогидрометр** *m.* thermohydrometer
**термоглаз** *m.* heat-radiation head
**термограмма** *f.* thermogram
**термограф** *m.* thermograph
**термодатчик** *m.* temperature-sensitive element; electrical transmitting thermometer
**термодвигатель** *m.* heat engine, thermomotor, thermomagnetic motor
**термодвижущий** *adj.* thermomotive
**термоделение** *n.* thermofission
**термодетектор** *m.* thermodetector, temperature detector
**термодинамика** *f.* thermodynamics

**термодинамический** *adj.* thermodynamic
**термодиффузионный** *adj.* thermal diffusion
**термодиффузия** *f.* thermal diffusion
**термоизмеритель** *m.* thermometer; **т. мощности** thermocouple-type wattmeter
**термоизмерительный** *adj.* thermometer
**термоизоляционный** *adj.* thermal-insulation, heat-insulation
**термоизоляция** *f.* thermal insulation, heat insulation
**термоиндикатор** *m.* heat indicator
**термоион** *m.* thermion
**термоионизация** *f.* thermal ionization
**термоионный** *adj.* thermionic
**термокамера** *f.* thermal camera
**термокатод** *m.* thermionic cathode, hot cathode
**термокатодный** *adj.* hot-cathode
**термоклин** *m.* thermocline
**термокомпенсация** *f.* thermocompensation
**термокомпенсированный** *adj.* temperature-compensated
**термокомпрессия** *f.* thermal compression
**термокрестовина** *f.* thermocouple
**термолиз** *m.* thermolysis
**термолитический** *adj.* thermolytic
**термолюминесцентный** *adj.* thermoluminescent
**термолюминесценция** *f.* thermoluminescence
**термомагнитизм** *m.* thermomagnetism
**термомагнитный** *adj.* thermomagnetic
**термометаморфизм** *m.* thermometamorphism
**термометр** *m.* thermometer, thermodetector; **дистанционный т.** telethermometer; **т., использующий зависимость упругости паров от температуры** vapor pressure thermometer; **манометрический т. с полой пружиной** pressure-spring thermometer **плёночный т. сопротивления** metal-film resistance thermometer **регистрирующий т. с двумя перьями** two-pen recording thermometer; **ртутный т. с отсчётом на расстоянии** mercury distant-reading thermometer; **т. с непосредственным отсчётом** direct-reading thermometer; **т., укрепляемый с**

**помощью магнита** magnetically attached monitoring thermometer

**термометрический** *adj.* thermometer, thermometric

**термометрия** *f.* thermometry

**термометрограф** *m.* thermometrograph

**термомикрофон** *m.* hot-wire microphone, thermal microphone

**термомотор** *see* **термодвигатель**

**термонапряжение** *n.* thermoelectromotive force, thermal voltage

**термонейтральность** *f.* thermoneutrality

**термонол** *m.* thermonol

**термообработка** *f.* heat treatment

**термоотрицательный** *adj.* thermonegative

**термопара** *f.* thermocouple, thermojunction, thermoelement; **т. из неблагородных металлов** base-metal thermocouple; **т. с электрическим нагревом** thermal converter; **уравнительная т.** booster thermocouple

**термопарный** *adj.* thermocouple

**термопатрон** *m.* temperature bulb

**термопластик** *m.* thermoplastic, thermosetting material

**термопластиковый, термопластический, термопластичный** *adj.* thermoplastic

**термопласты** *pl.* thermoplastics

**термоположительный** *adj.* thermopositive

**термопреобразователь** *m.* thermocouple, thermal converter, thermopile

**термореактивный** *adj.* thermoset, thermosetting

**терморегулирование** *n.* temperature regulation

**терморегулятор** *m.* temperature regulator, thermoregulator, heat controller; **т. прямого действия** self-operated thermostatic controller

**терморегуляция** *f.* thermoregulation

**терморезистор** *m.* thermistor

**термореле** *n. indecl.* thermorelay, thermoswitch; **т. с выдержкой времени** thermal time delay relay

**терморелейный** *adj.* thermal-relay

**термосифон** *m.* thermosiphon

**термосифонный** *adj.* thermo-syphon

**термоскоп** *m.* thermoscope

**термосопротивление** *n.* thermal resistance, thermistor

**термоспай** *m.* thermojunction, thermocouple

**термостат** *m.* thermostat; temperature-controlled cabinet

**термостатированный** *adj.* thermostated, temperature-controlled

**термостатический, термостатный** *adj.* thermostatic

**термостимуляция** *f.* thermostimulation

**термостойкость** *f.* heat resistance

**термостолбик** *m.* thermopile

**термосфера** *f.* thermosphere

**термотелефон** *m.* thermal receiver, thermophone, hot-wire telephone

**термотерапия** *f.* thermotherapy

**термоток** *m.* thermoelectric current

**термотрон** *m.* thermotron

**термоупругий** *adj.* thermoelastic

**термоупругость** *f.* thermoelasticity

**термоустойчивость** *f.* thermal stability

**термофизика** *f.* thermophysics

**термофизический** *adj.* thermophysical

**термофон** *m.* thermophone

**термофосфоресценция** *f.* thermophosphorescence

**термофотовольтаический** *adj.* thermophotovoltaic

**термохимический** *adj.* thermochemical

**термохимия** *f.* thermochemistry

**термочувствительный** *adj.* temperature-sensitive, thermo-sensitive, temperature-detecting, temperature-detector

**термошум** *m.* thermal agitation, Johnson noise

**термощуп** *m.* "thermohm", resistance-type temperature detector

**термоэластичный** *adj.* thermoelastic

**термоэлектрический** *adj.* thermoelectric, thermocouple

**термоэлектричество** *n.* thermoelectricity

**термоэлектрод** *m.* thermoelectrode

**термоэлектродвижущий** *adj.* thermoelectromotive

**термоэлектрон** *m.* thermoelectron, thermion

**термоэлектронный** *adj.* thermoelectronic, thermionic

**термоэлемент** *m.* thermoelement, thermocouple, thermopile

**термоядерный** *adj.* thermonuclear

**тернарный** *adj.* ternary

**« терне »** Terne

**тернерный** *adj.* ternary

**терпентинный** *adj.* turpentine

**терратекс** *m.* terratex

**территориально-технический** *adj.* area-technical

**территория** *f.* territory, area, range, zone, district; **т. телефонной сети** telephone exchange area; **т. узловой станции** tandem area

**терция** *f.* third, mediant (*mus.*); **большая т.** major third; **малая т.** minor third

**терять, потерять** *v.* to lose, to drop; to waste; to spill; to give up

**тесемка** *see* **тесьма**

**тесла** *f.* tesla (unit)

**тесный** *adj.* tight, narrow, close

**тессеральный** *adj.* tesseral

**тесситура** *f.* tessitura, texture (*mus.*)

**тест** *m.* test; **т.-таблица** tv test pattern

**тестер** *m.* tester, checker, test probe

**тестовый** *adj.* test

**тестсигнал** *m.* test signal

**тестфильм** *m.* test film; **звуковой т.** audio head aligning tape

**тесьма** *f.* tape, belt, girdle, band, ribbon

**тетрагон** *m.* tetragon

**тетрагональный** *adj.* tetragonal

**тетрада** *f.* tetrad

**тетрадь** *f.* notebook, pad; **т. со схемой линии** pole diagram book (*tphny.*)

**тетракисгексаэдр** *m.* tetrakishexahedron

**тетрахорд** *m.* tetrachord (*mus.*)

**тетраэдр** *m.* tetrahedron

**тетраэдральный** *adj.* tetrahedral

**тетраэдрический** *adj.* tetrahedron, tetrahedral

**тетрод** *m.* tetrode, double grid tube; screened grid tube; **плоский т.** junction tetrode; **плоскостной полупроводниковый т.** junction transistor tetrode; **полупроводниковый усилительный т.** transistor tetrode; **т. с катодной сеткой** space-charge grid tetrode; **т. с отрицательным сопротивлением** negative resistance tetrode, negatron; **т. ярусной конструкции** stacked tetrode

**тефлон** *m.* teflon

**тефлоновый** *adj.* teflon

**тех-; тех.** *abbr.* (**технический**) technical, engineering

**техкадры** *pl.* technical manpower, engineering manpower

**технеций** *m.* technetium, Tc

**техник** *m.* technician, engineer, mechanic; **т.-измеритель** test man, testing technician; **т. по радиоизотопам** isotope technician; **т. по радию** radium technician; **т.-радиограф** radiographer; **т. связи** communication technician, signalman

**техника** *f.* technique, procedure; technology; engineering, technics; equipment, material, hardware, technical devices; **т. атомного ядра** nucleonics; **т. безопасности** safety measures, prevention of accidents; **импульсная т.** pulse work; **импульсная т. измерения дальности** pulse-ranging technique; **т. импульсной манипуляции** gating technique; **инфразвуковая т.** infrasonics; **т. когерентных импульсов** coherence technique; **коммутационная т.** circuitry; **комплексная т.** system engineering; **т. наложения дополнительных данных** interscan technique; **т. настройки высокочастотных блоков** Lowson technique; **т. отражённых сигналов** echo technique; **т. панорамного приёма** radiospectroscopy; **т. передачи на несущих частотах** carrier-current technique; **т. печатных схем** printed circuitry; **т. предварительных импульсов** prepulse techniques; **противошумовая (акустическая) т.** noise-control acoustics; **радиоизмерительная т.** radio direction finding; **радиолокационная т.** radar engineering, radar technology; **т. связи** communication engineering, communication art; **т. селектирования сигнала** gating technique; **телефонная измерительная т.** telephonometry; **ультразвуковая т.** ultrasonics; **фотоэлектрическая т. получения изображений** photoelectric image techniques

**технический** *adj.* technical; engineering

**технолог** *m.* technologist

**технологический** *adj.* technological

**технология** *f.* technology; processing, process engineering; **т. контрольно-измерительных приборов** instrument technology

**техноскоп** *m.* technoscope

**техперсонал** *m.* technical personnel

**течеискание** *n.* leak hunting, leak detection

**течеискатель** *m.* leak detector, leak tester

**течение** *n.* flow, current, stream, run; streaming, flowing; lapse, course (of the time); **дрейфовое т. в ионосфере** ionospheric drift; **потенциальное т.** current of potential

**течь, потечь** *v.* to flow, to run; to leak, to escape

**течь** *f.* leak, leakage; **т. в вакуумной установке** vacuum leak

**ТИ** *abbr.* (**телеизмерение**) telemetering

**тигель** *m.* crucible

**тигельный** *adj.* crucible

**тиккер** *m.* ticker, chopper

**тиндалеметр** *m.* tyndallmeter

**тиноль** *m.* tinol, soldering paste

**тинтометр** *m.* tintometer

**тиокол** *m.* Thiocol

**тиолит** *m.* thiolite

**тип** *m.* type, kind, design, make, pattern; (In gen. case oft. not transl. e.g. **вибратор типа Сен Клер** St. Clair (type) vibrator); **обычные типы колебаний** normal modes; **дифференциального** (*or* **переходного**, *or* **смешанного**) **типа** hybrid-type; **многозвенного типа** ladder-type; **основного типа** dominant-mode; **т. волны** mode, transmission mode; **второй т. тёмного разряда** second Townsend discharge; **т. индикации** presentation type; **т. колебания** mode, transmission mode; **координатный т.** crossbar type (*tphny*.); **основной т. колебаний** dominant mode; **первый т. тёмного разряда** first Townsend discharge; **переходно-временной т. волны** transit-time mode; **т.-штанга** type bar

**типический, типичный** *adj.* typical, type

**типовой** *adj.* type, model, standard, routine, typical

**типовый** *adj.* type

**типографический** *adj.* typographical

**типометр** *m.* typometer

**типотрон** *m.* typotron

**тиратрон** *m.* thyratron, gas-filled triode; **т. дугового разряда** hot-cathode gas-filled tube; **т. с « плюсовой » сеткой** positive-grid thyratron

**тиратронно-двигательный** *adj.* thyratron-motor

**тиратронный** *adj.* thyratron

**тире** *n. indecl.* dash; leg (of filament)

**тиристор** *m.* thyristor

**тирит** *m.* thyrite; tyrite, fergusonite

**тирод** *m.* thyrode

**тиски** *pl.* file bench; vise, jaw vise; **т. для труб** pipe vise; **стуловые т.** bank vise

**титан** *m.* titanium, Ti

**титанат** *m.* titanate

**титанат-бариевый** *adj.* barium-titanate

**титанистый** *adj.* titanium, titanous, titaniferous

**титанит** *m.* titanite

**титановый** *adj.* titanium, titanic, titaniferous

**титансодержащий** *adj.* titaniferous

**титр** *m.* cue; titer

**титратор** *m.* titrator

**титриметр** *m.* titrimeter

**тптрование** *n.* titration, titrimetry

**титрованный** *adj.* titrated

**титровать** *v.* to titrate

**титрометр** *m.* titrimeter

**титрометрия** *f.* titrimetry

**титрометрический** *adj.* titrimetric

**титрующий** *adj.* titrating

**титрующийся** *adj.* titratable

**титульный** *adj.* title; guide

**тихий** *adj.* quiet, noiseless, silent; calm; soft, low

**тихоходный** *adj.* slow-speed, running at low speed

**ТК** *abbr.* 1 (**температурный коэффициент**) temperature coefficient; 2 (**турбокомпрессор**) turbocompressor

**тканеэквивалентный** *adj.* tissue-equivalent

**ткань** *f.* cloth, fabric; tissue; **лакированная хлопчатобумажная т.** cambric; **т. прикрывающая громкоговоритель** grille cloth

**тканьевый** *adj.* cloth; webbing, woven; tissue

**ТКЕ** *abbr.* (**температурный коэффициент ёмкости**) temperature coefficient of capacitance

**ТКИ** *abbr.* (**температурный коэффициент индуктивности**) temperature coefficient of inductance

**ТКЧ** *abbr.* (**температурный коэффициент частоты**) temperature coefficient of frequency

**ТЛГ** *abbr.* (**телеграфия**) telegraphy

**тлеющий** *adj.* glow, glowing, glow-discharge; smoldering

**ТЛФ** *abbr.* (**телефония**) telephone

**ТН** *abbr.* (**точка наводки**) aiming point

**ТНА** *abbr.* (**турбонасосный агрегат**) turbopump unit

**ТНБ** *abbr.* (**технико-нормировочное бюро**) Bureau of Technical Standards

**ТНД** *abbr.* (**турбина низкого давления**) low-pressure turbine

**ТО** *abbr.* 1 (**техническое обслуживание**) technical servicing, maintenance; 2 (**точный отсчёт**) fine reading

**Т-образный** *adj.* T-, T-type, T-shaped, tee-

**товарный** *adj.* freight (train); commercial

**тождественность** *f.* identity

**тождественный** *adj.* identical, homologous

**тождество** *n.* identity (*math.*)

**Този** Tosi

**ток** *m.* current, flow, stream; current (*elec.*); **ползучие токи** sneak currents; **растекающиеся токи** current sheets; **токи утечки** spurious currents, stray currents; **под током, с током** current-carrying; **т. анода в статическом режиме** static operating plate current; **анодный т. покоя** zero signal plate current, steady anode current; **базовый т. открытого триода** turn-on base current; **т. в дуге** electric arc current; **т. в запорном направлении** inverse leakage current; **т. в ответвлении** tee-off current; **т. в приборе** low-voltage current; **т. в провале между двумя пиками** valley current; **т. в проводящем направлении, т. в прямом направлении** forward current; **т. в рабочей точке** quiescent current; **т. в свинцовой оболочке** sheath current; **т. в сети** mains current; **т. в цепи мишени** target current; **т. в числовой линейке** word current; **варьированный т.** bunched current; **т. вентильного провода** gate current; **т. вертикального отклонения** vertical yoke current; **т. включения** in-rush current; **т. возбуждения** field current; driving current, exciting current; **вызывающий судорогу т.** contraction current (*med.*); **вызывной т. наложенный на постоянный**

superposed ringing current; **вызывной т. тональной частоты** voice-frequency signalling current; **выключаемый т.** current-breaking capacity; **выравнивающий т.** circulating current; **высокочастотный т. высокого напряжения** Tesla current (*med.*); **т. горизонтального отклонения** horizontal yoke current; **двусторонний т., двухполюсный т.** double current; **действующий т.** effort current; **т. делителя (напряжения)** bleeder current; **т. динамической устойчивости (трансформатора тока)** instantaneous short-circuit current (of a transformer), mechanical short-time current rating; **т. долины** valley-point current; **допробойный т.** preconduction current; **допускаемый т.** let-go current; **допустимый т.** current-carrying capacity; **допустимый т. по нагреву прибора** rated temperature-rise current of an instrument; **допустимый максимальный т. при выключении** permissive breaking current; **т. дробовых шумов** shot-noise current; **т. зажигания** striking current; **т. замыкания на землю** short circuit current to earth, loss current to earth, earth(ing) current; **запирающий т.** inverse current; **зарядный и разрядный т. пары конденсаторов** static induced current; **затухающий т. проводимости** decaying conduction current; **избыточный т.** spill current; **индуктируемый внешним полем т.** field-generated current; **т. касания** pickup current (of a relay); **т. коллекторного перехода** collector-junction current; **коллекторный т. в рабочей точке** quiescent collector current; **компенсирующий т.** cancellating current; **т. контура** loop current; **контурный т.** mesh current; **т. луча в определённой точке** beam current at a specified point; **максимальный т. катода при коротком замыкании** peak cathode fault current; **многофазный т.** rotary current; **т. на выходе генератора кадровой развёртки** image output, picture output, video time base; **т. на корпус** frame current, fault current, body leakage; **наибольший** выдерживаемый т. предохранителя limiting no-damage current; **наимельший т. плавления предохранителя** minimum fusing-current; **т. накала** heater current; **т. намагничивания** exciting current; **т. нарастания колебаний** preoscillation current; **т. насыщения блокировки** inverse saturation current; **начальный т. эмиссии** initial current flow; **т. начальных скоростей** residual current; **т. несущей частоты** carrier current; **т. неустановившегося режима** transient current, building-up current; **номинально допустимый т.** rated carrying current; **номинальный т. короткого замыкания трансформатора тока** rated short-circuit current of a current transformer; **номинальный первичный т. трансформатора тока** rated primary current of a current transformer; **номинальный пропускной т.** rated current-carrying capacity; **т. нулевого следования** residual current; **т. обоих направлений** bidirectional current; **т. обратного направления** return current, reverse current; **т. обратной связи** feedback current; **обратный т.** inverse current, reverse current, inverse leakage current, reverse leakage current; **обратный сеточный т.** negative grid current, reversed grid current **т. окончания зарядки** finishing rate; **оперативный т.** control current; **т. оседания электронов** landing current; **оставляющий след т.** tracking current; **остаточный т. (фотоэлемента)** transient-decay current (of a photo-electric device); **т. отбойного сигнала** clearing current; **ответвляющийся т.** tee-off current; **т. отклоняющей системы** yoke current; **т. отключения** breaking current; **т. отпускания якоря электромагнита** release current; **параболический т. кадровой частоты** vertical parabolic current; **параболический т. строчной частоты** horizontal parabolic current; **т. первичной цепи** (*or* **обмотки**) primary current; **пережигающий т.** blowing current (of a fuse); **переменный т. частоты высоких тонов** high-pitched alternating current; **т. переходного** разговора unbalance current, cross-talk current; **переходный т. трёхфазного короткого замыкания** transient three-phase short-circuit current; **т. плотного прижатия (якоря реле)** pull-up current (of a relay armature), sealing current; **т. повреждения** fault current; **т. поджигания** ignitor firing current; **т. подпытывания** field current; **т. покоя** quiescent current, spacing current; **полный т. эмиссии** field-free emission current; **полный т. эмиссии лампы** space current; **положительный (обратный) т. анода** anode-ray current; **поражающий сердце т.** fibrillating current; **т. после трансляции** repeat current; **т. последействия** rest current, residual current, aftercurrent; **т., потребляемый делителем напряжения** voltage-divider current; **предельно-допустимый пиковый анодный т.** maximum rated peak anode current; **предельно-допустимый средний анодный т.** maximum-rated mean anode current; **предельно-допустимый эффективный анодный т.** maximum rated effective anode current; **предельный т.** high-peak current; **предельный т. выключения** current-breaking capacity; **предразрядный т.** preconduction current; **т. при втянутом сердечнике** hold-on current; **т. при пиковой нагрузке** peak current; **т. притяжения** pickup current; **пропускной т.** current-carrying capacity; **пространственный т.** cathode current, space current; **т. противоположного направления** countercurrent, reverse(d) current; **прямолинейный т.** rectilinear current; **т. пучка** cathode-ray current, beam current, scanning-beam current; **рабочий т.** marking current (*tgphy*); operating current, open-circuit current, running current; **рабочий т. дающий знак на ленте** (telegraph) signal current; **рабочий т. при нормальной полярности** direct working current; **рабочий т. при обратной полярности** reverse working current; **т. развёртки дальности** range-sweep current; **разговорный переменный т.** voice-frequency current; **разностный т.**

spill current; **разрывной т. (предо-хранителя)** interrupting current (of a fuse contact); **расплавляюший т.** fusing current; **реактивный т.** idle current; **т. с высшими гармониками** harmonic current, complex current; **т. с основной гармоникой** sinusoidal current; **т. с равновеликой площадью** area-balanced current; **т. с управлением от реле** repeater current; **сверхвысокочастотный т.** microwave current; **т. сверхпроводимости** persistent current; **т. свободного колебания** free alternating current; **т. сдвигающей цепи** loop current; **сквозной т.** let-through current; **слабо пульсирующий т.** ripple current; **содействующий т.** additional current of same direction as the existing current; **сопровождающий т.** follow current; **т. срабатывания** minimum working current; **т. строчного отклонения** horizontal yoke current; **темновой т. фотоэлемента** photoelectric dark-current effect; **т. термической устойчивости** short-time current of a switching device; **т. тлеющего разряда** luminous current, glow current; **т. трогания (реле)** pickup current (of a relay); **ударный т.** transient current; **ударный т. короткого замыкания** maximum instantaneous short-circuit current; **ударный т. трёхфазного короткого замыкания** maximum asymmetric three-phase short-circuit current; **уравнительный т.** circulating current, phasing current, restoring current; **устанавливающийся т.** transient current, building-up current; **т. установившегося режима, установившийся т.** steady (-state) current, sustained current; **т. холостого хода, холостой т.** open-circuit current, idle current, quiescent current, running-light current, no-load current; **т. холостой работы трансформатора** exciting current of a transformer; **т. через катод** cathode current; **числовой т.** word current; **т. экранной сетки** screen-grid current; **экранный т. покоя** zero signal screen current; **т. электрода при повреждении** surge electrode current, fault electrode

current; **т. электронного пучка** beam current; **т. электронной эмиссии катода** field-free emission current (of a cathode); **т. эмиссии при отсутствии электрического поля** field-free emission current; **т. эмиттерного перехода** emitter-junction current

**токарный** *adj.* lathe

**токо-** *prefix* current

**токоведущий** *adj.* current-carrying, live

**токовращатель** *m.* d-c ringer, pole changer

**токовый** *adj.* current

**токоизмерительный** *adj.* current-measuring

**токонесущий** *adj.* current-carrying, live

**токоограничивающий** *adj.* current-limiting

**токоограничитель** *m.* current limiter

**токоограничительный** *adj.* current-limiting

**токопитаемый** *adj.* current-fed

**токоподвод** *m.* contact conductor, contact electrode

**токоприёмник** *m.* current collector, trolley; **роликовый т.** trolley

**токоприёмный** *adj.* current-collecting

**токопровод** *m.* current distributor, conductor; conduction

**токопроводящий** *adj.* current-carrying, current-conducting, live, conducting, conductive

**токопрохождение** *n.* current circuit, passage of current

**токораспределение** *n.* current distribution

**токораспределитель** *m.* sequence switch, control switch

**токосниматель** *m.* current collector; **роликовый т.** trolley; **щелевой т.** underground (current) collector

**токоснимательный** *adj.* current-collecting

**токособиратель** *m.* current collector

**токособирательный** *adj.* current-collecting, current-collection, collector

**токостабилизатор** *m.* current regulator tube, ballast lamp, ballast resistor; **накальный т.** heating current regulator tube

**токосъёмник** *m.* slip ring, sliding contact, trolley, sliding shoe, collector-shoe gear; **роликовый т.** trolley-wheel, contact roller

**токосъёмный** *adj.* collector, collector-show, current-collecting

**толерантный** *adj.* tolerance

**толкатель** *m.* pusher, thruster, push rod, poker, follower, tappet; **т. кулачка** tapped actuator, cam follower

**толкать, толкнуть** *v.* to push, to thrust, to slide

**толкач** *m.* tapped, pusher

**толкающий** *adj.* pushing

**толкование** *n.* interpretation

**толковать, истолковать** *v.* to translate, to interpret, to explain

**толстолистовой** *adj.* thick-sheet, thick-leaved

**толстомер** *m.* gap gage

**толстообмазанный** *adj.* heavy-coated

**толстоплёночный** *adj.* thick-film

**толстослойный** *adj.* thick, thick-layered

**толстостенный** *adj.* thick-walled

**толстый** *adj.* thick, heavy

**толуол** *m.* toluol, toluene

**толчкомер** *m.* impact-measuring device

**толчкообразный** *adj.* jerking, jerky; shock (*elec*)

**толчок** *m.* shock, overshot; impulse, thrust, push; surge; excitation; bump; jerk, jar, jolt, kick; burst; **толчки фазы** phase hunting; **т. (напряжения) при размыкании** break (voltage) shock; **т. обратного тока** current kickback; **т. при выключении** transient current, circuit-breaking; **т. тока** overshoot, overshooting

**толща** *f.* thickness, mass; layer; stratum; bulk

**толщина** *f.* thickness, width, depth, size, gage (of wire), caliber; **т. отдельной пластинки** lamination thickness; **т. слоя для поглощения половины излучения** half-value thickness, thickness for half absorption

**толщинный** *adj.* thickness

**толщиномер** *m.* thickness gage, calipers, reflectogage; **предельный т.** thickness-tolerance meter

**толь** *m.* tar-board, tar-paper

**томасовский, томасовый** *adj.* Thomas', (of) Thomas

**томограф** *m.* tomograph, laminagraph

**томография** *f.* tomography, body section roentgenography

**томпак** *m.* tombac, red brass

**Томсон** Thomson

**томсоновский** *adj.* (of) Thomson, Thomson's

**тон** *m.* tone, tint; note; tune; **нижние** (*or* **низкие**) **тона** lower tones, bottom, bass; **высота тона** pitch; **вибрирующий т.** vibrato note; **воющий т.** warble tone; **длительный т.** sustained tone; **клиновой т.** edge tone; **линейно скользящий т.** linear reciprocating sweep; **незатухающий т.** sustained tone; **основной т. аккорда** root (*mus.*); **относительный цветовой т.** relative hue, contrast hue; **разностный т.** difference tone; intermodulation frequency; **т. с вибрацией** vibrato tone; **собственный цветовой т.** intrinsic hue; **созвучный т.** sympathetic tone; **т. сравнения** audiometer note; **т. средней высоты** medium pitch; **т. цвета, цветовой т.** hue, tinge, spectral hue, dominant wavelength (of a colored light, not purple)

**тональномодулированный** *adj.* tone-modulated

**тональность** *f.* tonality, key, keynote; **мажорная т.** major mode

**тональный** *adj.* tonal, tone; audio; tone-modulated; musical; note

**тонарм** *m.* pickup arm, tone arm

**тонгенератор** *m.* tone generator, audio (-frequency) oscillator

**тоника** *f.* keynote, tonic

**тонирование** *n.* toning

**тонический** *adj.* tonic, pitch

**тонкий** *adj.* thin, fine, fine-grained; delicate

**тонковолокнистый** *adj.* fibrous, fine-fibered

**тонковолоченный** *adj.* fine-drawn

**тонкозернистый** *adj.* fine-grained, fine

**тонколистный, тонколистовой** *adj.* fine-sheet, thin-sheet, thin-leaved

**тонконтроль** *m.* tone control; **т. низких частот** bass control; **т. с выделением** (*or* **подчёркиванием**) **низких тонов** bass boosting control, bass boosting circuit

**тонкообмазанный** *adj.* thin-coated, dust-coated

**тонкооттянутый** *adj.* fine-drawn

**тонкоплёночный, тонкоплёнчатый** *adj.* thin-film, thin-filmed

**тонкослоистый, тонкослойный** *adj.* thin-layer; laminated, lamellar

**тонкостенный** *adj.* thin-walled

**тонкость** *f.* thinness, delicacy, fineness

**тонкотянутый** *adj.* thin-drawn

**тонманипулятор** *m.* tone manipulator

**тонмейстер** *m.* sound monitor, monitor, monitor man, (sound) mixer

**тоновый** *adj.* tone, pitch, pitched

**тонометр** *m.* tonometer

**тон-ось** *f.* capstan

**тонотрон** *m.* tonotron

**тонсигнал** *m.* tone signal, audio signal

**тонфильтр** *m.* tone filter

**тончайший** *superl. adj.* ultrathin; very fine

**тончастота** *f.* audio frequency, tone frequency

**топливный** *adj.* fuel

**топливо** *n.* fuel

**топливомер** *m.* fuel gage

**топовый** *adj.* top

**топограмма** *f.* topogram

**топограф** *m.* surveyor; topographer

**топологический** *adj.* topological

**топология** *f.* topology, topological layout

**топор** *m.* axe; **колющий т.** mortise axe

**тор** *m.* torus, torus ring, tore

**Тореус** Thoraeus

**торец** *m.* end, butt, face, end face

**торзиограф** *m.* torsiograph

**торзиометр** *m.* torquemeter, torsion meter

**торзионный** *adj.* torsion

**ториевый** *adj.* thorium

**торий** *m.* thorium, Th

**торийсодержащий** *adj.* thorium-bearing

**торированно-вольфрамовый** *adj.* thorium-tungsten

**торированно-молибденовый** *adj.* thorium- molybdenum

**торированный** *adj.* thoria-coated, thoriated

**торможение** *n.* drag, frictional action, braking; inhibition, retardation, retarding, slowdown, deceleration, drag acceleration; damping; restraining; stopping, jam, jamming; **т. от действия поля** drag of field; **т. постоянным током** injection braking; **т. противовключением** reverse-current braking, plugging; **т. противотоком** regenerative braking, electric braking; dynamic braking; **реостатное т.** dynamic braking; **т. холостого хода** no-load stopping device

**тормо́жённый** *adj.* braked, stopped; inhibited, retarded

**тормоз** *m.* brake, drag; retarder, restrainer; **т., действующий вихревыми токами** eddy-current brake; **т. с приводом от электрогидравлического толкателя** thrustor-operated brake; **скоростной т.** overspeed safety gear; **т. хранения наклона** (*or* **угла места**) elevation stowing lock

**тормозить, затормозить** *v.* to retard, to restrain; to arrest; to brake, to stop; to inhibit; to decelerate

**тормозной** *adj.* retarding, retardation; brake, braking, drag

**тормозящий** *adj.* retarding, decelerating; braking; damping; inhibiting

**тороид** *m.* toroid; torus ring, anchor ring

**тороидально-дисковый** *adj.* pancake-shaped

**тороидальный, тороидный** *adj.* toroidal, toroid, annular, ring-shaped

**торон** *m.* thoron, Tn

**торпеда** *f.* torpode; **акустическая т. с активной системой самонаведения** echo-ranging operated acoustic torpedo

**торпедировать** *v.* to torpedo

**торпедный** *adj.* torpedo

**торр** *m.* torr (unit)

**торсиограф** *m.* torsiograph

**торсиометр** *m.* torsiometer, torque meter, torque pickup

**торсионный** *adj.* torsion

**торус** *m.* torus (*math.*)

**торцевой, торцевый, торцовой** *adj.* end, end-type; front, face

**торшер** *m.* floor standard lamp

**Тоси** Tosi

**тотальный** *adj.* total

**« тофет »** *m.* tophet

**точечно-контактный** *adj.* point-contact

**точечно-плоскостный** *adj.* point-junction

**гочечный** *adj.* point, spot; punctual, lumped; dotted (line); intermittent

**гочило** *n.* oil-stone, grindstone, sharpener

**точить, наточить** *v.* to sharpen, to grind, to point

**точка** *f.* point, dot, period, spot; rest, pause; buggy; **точки квадрантов на круговой диаграмме** quadrantal points; **надёжные точки** fiducial points; **особые точки** singular points; **ретрансляционные точки** interception points; **по точкам** point-by-point; **т. ветвления** transition point; **т. возврата** (*or* **заострения**) **на кривой** cusp (*math.*); **т. второй настройки** repeating point; **выделенная средняя т.** center tap; **т. генерации** (*or* **генерирования**) oscillation point; **граничная т.** threshold; **заземлённая средняя т.** center-point earth; **т. захлопывания** point of closure; **золотая т.** melting point of gold; **т. излома** salient point (*math.*); **изолированная т. кривой** acnode (*math.*); **т. линии развёртки** action spot; **мёртвая т.** dead center; **т. назначения** terminal point; **т. начала сеточного тока** grid-current starting point; **нулевая т. звезды** neutral point, star point; **нулевая т. посредине** center zero; **т. нулевой чувствительности** point of zero response; **особая т.** singular point; **т. останова** break point; **т. отбора давления** pressure tap; **т. ответвления** balance point, tapping point; **т. отвода** distributing point, tapping point, balance point; **т. перевала** passage point, saddle point; **т. перегиба** apex, vertex (*math.*); **т. пересечения двух кривых** cusp (*math.*); **т. пересечения радикальных осей окружностей** radical center (*math.*); **т. перехода** turnover; **т. питания** driving point; **т. питания решётки** array feed point; **т. повторения** rerun point (*math.*); **т. предельного сопротивления** yield point; **т. приложения силы** line of application; **рабочая т. запертого транзистора** (*or* **лампы**) OFF operating point; **рабочая т. лампы** quiescent point (of a tube); **рабочая т. открытого транзистора** (*or* **лампы**) ON operating point; **т. разветвления** branch point, variable connector; **т. разветвления схемы** junction point of a network; **разрезная т.** cut-section; **т. разрыва** breakpoint (*math.*); **т. рассогласования** transition-point; **реперная т.**

reference point, checking point; die-down point (of dial); **т. свиста** oscillation point; **т. сгущения полюсов** accumulation point of poles; **т. скрещивания** transposition point; **т. смыкания** point of closure; **т. сосредоточения массы** discrete mass point (*math.*); **т. сравнения** reference point, datum point; **существенная контрольная т.** strategic check point; **термическая т. возврата** inversion temperature; **узловая т.** nodal point, junction point, branch point

**точно** *adv.* exactly, precisely, punctually, accurately

**точность** *f.* precision, exactness, accuracy, exactitude; closeness, fines; fidelity; validity; delicacy; punctuality; **большой точности** high-performance; **с точностью до N десятичных знаков** accurate to N decimal place; **с удвоенной точностью** double-precise; **т. в измерении углов** angular accuracy; **дальномерная** (*or* **дальностная**, *or* **дистанционная**) **т.** range accuracy; **т. измерения азимута** azimuth accuracy; **т. измерения дальности** distance accuracy; **т. конического сканирования** boresight accuracy; **т. настройки** tuning precision; degree of balance; **т. определения дальности** range accuracy; **т. отсчёта времени** timing precision; **т. отсчёта показаний по шкале** accuracy of reading; **т. посадочной радиолокационной системы** PAP accuracy; **т. при лабораторной проверке** reference-performance accuracy; **т. срабатывания** firing precision

**точный** *adj.* exact, precise, accurate, correct, sharp, distinct, definite, strict, close; fine, sensitive (adjustment); tight; precision

**Т.П., т.пл.** *abbr.* (**точка плавления**) melting point

**Тр** *abbr.* (**трансформатор**) transformer

**т-ра** *abbr.* (**температура**) temperature

**траверза, траверса** *f.* crossarm, pole arm, girder, cross-bar, traverse; interphase connecting pipe; **т. для передвижения щёток** brush-rocker ring, brush-rocker; **кольцевая щёточная т.** brush-holder plate, brush-plate; **несущая т.** supporting rod; **односторонняя т.** side arm;

**отбойная т.** guard arm; **т. щётко-держателя** brush-rocker

**траверс** *see* **траверза**

**травильный** *adj.* etching; pickling

**травитель** *m.* etch

**травить, вытравить** *v.* to pickle, to corrode, to etch, to dip, to scour; to uncoil, to unwind, to pay out cable; **т. кабель** to pay out cable

**травление** *n.* etching, corrosion, pickling, dipping, scouring; **т. на частоту** etching to frequency

**травленный** *adj.* etched, pickled, corroded

**традистор** *m.* tradistor

**траектория** *f.* path, track, trace, trajectory; **криволинейная т.** maneuverable path; **ломано-линейная т.** dog-leg path

**трайджистор** *m.* trigistor

**тракт** *m.* channel, route; circuit, loop; routing; path, track; **волноводный т.** micro-wave plumbing; **входящий т.** incoming train (in a crossbar system); **исходящий т.** outgoing train (*tphny.*); **слуховой т. при радиолокации** audio channel in radio navigation

**трактовка** *f.* treatment

**трактриса** *f.* tractrix

**трал** *m.* sweep, trawl, trawl line

**трама** *f.* tram silk

**трамаг** *m.* tramag

**трамбовать, вытрамбовать** *v.* to ram, to tamp

**трамбовка** *f.* tamping bar, tamper, pile driver, ram; puncture

**трамбовочный, трамбующий** *adj.* tamping, ramming

**транзистанс** *m.* transistance

**транзистор** *m.* transistor, transfer resistor; **на транзисторах, снабжённый транзисторами** transistorized; **т. вплавленного типа** fused-impurity junction transistor; **двухтактный т.** tandem transistor; **диффузионный т.** diffused base transistor; **диффузно-сплавной т.** alloy-diffuse transistor; **каналовый т.** field-effect transistor; **однопереходный т.** unijunction transistor; **плоский** (*or* **плоскостной**) **т.** junction transistor; **плоскостной т., включённый по схеме с общим эмиттером** common-emitter junction transistor; **плоскостной т. с диффузионным пере-**

ходом p-n-p p-n-p diffused-junction transistor; **плоскостной т. с коллекторной ловушкой** hook transistor; **поверхностно-барьерный т.** transistor pellet; **полупроводниковый т. с поверхностным барьером** surface-barrier transistor; **решающий т.** computer transistor; **т. с вплавленным переходом** bonded-barrier transistor; **т. с выращенными переходами** double-doped transistor, grown-junction transistor, rate-grown transistor; **т. с использованием пролётного времени** transit-time oscillator; **т. с наращённым слоем** rate-grown transistor; **т. с обеднённым слоем** depleting-layer transistor; **т. с одним переходом** unijunction transistor; **т. с плавными переходами** graded-junction transistor; **т. с повышенной температурной стабильностью** higher ambient transistor; **т. с проводимостной модуляцией** conductivity modulation transistor; **т. с управлением полем** field effect transistor; **слоистый т.** junction transistor, alloy-junction transistor; **слоистый т., включённый по схеме с общим эмиттером** common-emitter junction transistor; **т. со слоем собственной проводимости** intrinsic-region transistor; **сплавной плоскостной т. лавинного типа** alloyed junction avalanche transistor; **точечный т. с заземлённым эмиттером** ground-base point-contact transistor; **тянутый т.** double-doped transistor, grown-junction transistor, rate-grown transistor; **тянутый т. с диффузионным распределением примеси** grown-diffused transistor; **четырёхслойный т.** hook transistor

**транзисторизация** *f.* transistorization, transistorizing

**транзисторизованный** *adj.* transistorized

**транзисторный** *adj.* transistor, transistorized; **полностью т.** all-transistor, fully transistorized

**транзит** *m.* transit; **автоматический т.** through dialing, automatic through working, dial system tandem operation

**транзитивность** *f.* transitivity, transitive relation

транзитивный *adj.* transitive
транзитный *adj.* through, via, transit
транзитрол *m.* transitrol
транзитрон *m.* transitron
транзитронный *adj.* transitron
«транкор» *m.* trancor
трансатлантический *adj.* transatlantic
трансверсальный *adj.* transversal
трансвертер, трансвертор *m.* transverter
трансдуктор *m.* transducer, transductor, transfer network, sink; **т. вида колебаний** mode transducer; **двухвидовый т.** two-mode transducer
трансдьюсер *see* трансдуктор
транскод *m.* transcode
трансконтинентальный *adj.* transcontinental
транскрибирование *n.* transcription
транскрибировать *v.* to transcribe
транскрипция *f.* transcription
транскристаллический *adj.* transgranular
«транслэй» *m.* transley
транслировать *v.* to transmit, to broadcast, to relay, to reradiate
транслируемость *f.* repeatability
транслирующий *adj.* repeating, relaying, translating
транслитерация *f.* transliteration
транслитерировать *v.* to transliterate
транслятор *m.* translator (of a telegraph system); decoder; converter; translating program; **входящий т.** input translator; **канальный т.** channel converter (of carrier-current system); **т. одностороннего действия** single action translator, one-way translator; **т. покомандного перевода** one-to-one translator; **т. с лампами тлеющего разряда** glow-discharge tube pulse differentiation device; **т. четырёхпроводной системы** four-wire translator
трансляционный *adj.* translation, relaying, rediffusion, broadcasting, relay, retransmitting, transmission
трансляция *f.* rediffusion; relay, relaying, reradiation, rebroadcasting, retransmission, chain broadcasting, simultaneous broadcasting, repeating relay, repeating; repeater; translator; translation; **вилочная т.** forked repeater (*tgphy.*); **восстанавливающая т.** regenerative repeating; **вч т. (с несущей)** carrier repeater

(*tphny.*); **импульсная т.** pulse repeater; **исправляющая т.** regenerative repeating, regenerative repeater; **проволочная т.** electrophone; **т. радиовещания** broadcast relaying; **релейная т.** relaying; **ручная т.** manual relaying, human relaying (*tgphy.*); **т. с одним приёмнопередающим реле для каждого направления** direct-point repeater (*tgphy.*); **телевизионная т.** television relay; **телеграфная т.** telegraph repeater; **т. типа «два-два»** 22-type repeater (*tphny.*); **четырёхпроводная т.** four-wire repeater (*tphny.*)
трансмембранный *adj.* transmembrane
трансмиссионный *adj.* transmitting, transmission
трансмиссия *f.* transmission, transmitting; **промежуточная т.** countershaft
«трансмитайпер» *m.* transmityper
трансмиттер *m.* transmitter, telegraph transmitter, paper tape reader; **ленточный т.** moving tape transmitter
трансмутация *f.* transmutation
трансозонд *m.* transosonde
трансокеанский *adj.* transoceanic
транспарант *m.* transparency
транспозиционный *adj.* transposition
транспозиция *f.* transposition, cross connection, twisting of pairs
трансполярный *adj.* transpolar
транспондер *m.* transponder, transponder-beacon
транспонирование *n.* transposition, frequency conversion; augmentation
транспонированный *adj.* transposed; transposition
транспонировать *v.* to transpose
транспонировка *f.* transposition
транспорт *m.* transport, transportation, transfer, transmission, conveyance; threading, traction; traffic
транспортабельный *adj.* transportable
транспортёр *m.* belt conveyer, transitor; carrier; **ленточный т.** belt-type ticket carrier (*tphny*); conveyer belt
транспортёрный *adj.* conveyer, transporter
транспортир *m.* bow, protractor
транспортировать *v.* to transport, to forward, to relay over, to convey
транспортировка *f.* conveying
транспортируемый *adj.* transportable

**транспортирующий** *adj.* transporting, conveying

**транспортный** *adj.* transport, transporting, conveying

**транспортованный** *adj.* transported, conveyed

**транспортовать** *see* **транспортировать**

**трансурановый** *adj.* transuranic, transuranium

**трансферкар** *m.* transfer car

**трансферный** *adj.* transfer

**трансфинитный** *adj.* transfinite

**трансфлуксор, трансфлюксор** *m.* transfluxor

**трансфокатор** *m.* Zoomar lens

**трансформатор** *m.* transformer, converter; repeating coil, repeater; **т. без железного сердечника** air-core transformer; **безмасляный измерительный т.** dry-type instrument transformer; **т. безопасного напряжения** protective transformer; **т. броневого типа, броневой т.** shell-type transformer; **т. вида колебаний** mode transformer; **воздушный т.** air-core transformer; **встроенный т.** bushing transformer; **втулочный т. тока** bushing-type current transformer; **т. выделения сигнала вспышки** burst-take-off transformer; **выходной т.** output transformer; **выходной дифференциальный т.** repeater-bridging transformer; **т. горшкового типа** insulator-type transformer; **дифференциальный т.** hybrid coil, hybrid transformer, toroidal transformer, toroidal repeating coil; **т. для питания вакуумного насоса** degassing transformer; **т. для питания выпрямителя** rectifier transformer, plate transformer; **т. для питания от сети** mains transformer; **т. для подбора входного сопротивления** lead-in impedance matching transformer; **т. для установки в киосках** vault-type transformer; **т. для установки на столбе** pole-type transformer; **т. для цепей управления** control circuit transformer; **т. затухающих колебаний** magnetic coupling transformer; **измерительный т.** instrument transformer, measuring transformer; **искрогасительный т.** (neutral) compensating transformer; **кабельный т.** slip-over current trans-

former; **кадровый выходной т.** vertical output transformer; **т. кадровой развёртки** frame transformer, frame-scan transformer; **катушечный т.** two-winding-type instrument transformer; **кольцевой (переходный) т.** toroidal repeating coil, ring transformer, hybrid coil; **т. линий дальней связи** trunk-line repeating coil; **масляный т.** oil-filled transformer, oil-immersed transformer; **масляный измерительный т.** oil-filled instrument transformer, oil-cooled instrument transformer; **междуламповый т.** interstage transformer; **механический повышающий т. скорости** mechanical velocity step-up transformer; **многократный т. тока** multi-range current transformer; **мостовой т.** differential transformer; **низковольтный т. для прогрева антенны** aerial heating transformer; **т. низкой частоты** audio-frequency transformer; **т. обратной связи** reaction transformer, flyback transformer; **одновитковый стержневой т.** bar-type-primary instrument transformer; **т. отклоняющей системы** flyback transformer (*tv*); **отправительный т. колебаний** transmitting jigger; **т. переменного полного сопротивления** variable-impedance transformer; **т. перемены фазы на обратную** phase reversal transformer; **переходный т.** intermediate transformer, conversion transformer; line transformer, line repeating coil, insulating transformer (*tphny.*); **переходный т. со средней точкой** phantom transformer (*tphny.*); **переходный линейный т.** line repeater, repeating coil, line transformer; **т. питания** power transformer; **поворотный т.** induction regulator, variable voltage transformer; **т.-повыситель, повысительный т., повышающий т.** step-up transformer, voltage amplifier; **полуволновый т. согласования** half-wave matching stub; **понижающий т., т.-понизитель** step-down transformer; **продуваемый т.** air-blast transformer; **т. проходного типа** through-type transformer; **регулировочный т.** (three-phase) induction regulator; **т.-редуктор** step-down transformer; **резонансный т.** tuned transformer; **рычажный т.** needle-arm transformer; **т. с вводными кабельными муфтами** pothead-type transformer; **т. с выделенной средней точкой обмотки, т. с выводом от средней точки** center-tapped transformer; **т. с коэффициентом 1** one-to-one transformer; **т. с настроенной первичной цепью** tuned-primary transformer; **т. с незамкнутой железной цепью** open-core transformer; **т. с несколькими отпайками, т. с ответвлениями, т. с отводами** tapped transformer, split transformer; **т. с передвижным сердечником** moving coil-type transformer, sliding coil-type transformer; **т. с переключаемыми ответвлениями** tap-changing transformer; **т. с переменной связью** jigger, jigger transformer; **т. с переменным коэффициентом трансформации** (three-phase) induction regulator, variable ratio transformer; **т. с плунжерным сердечником** telescoping coil transformer; **т. с посторонним охлаждением** air-blast transformer; **т. с раздельными цепями** iso-type transformer; **т. с расширительным баком** conservator transformer; **т. с регулируемым коэффициентом трансформации** variac, variable-ratio transformer; **т. с ударным возбуждением** flyback transformer; **т. с фарфоровым блоком для размещения обмоток** slot-type current transformer; **т. связи** coupling transformer, repeating transformer; **секционированный т.** *see* **т. с отводами; т. сельсина** synchro control transformer; **сетевой т.** power transformer; **силовой т.** mains transformer, output transformer, power transformer; **симметрирующий т.** balun, bazooka, balance-to-unbalance transformer, line-balance converter, balancer, balancing transformer; **сменный т. связи** plug-in coupling transformer; **т. со связью больше критической** over-coupled transformer; **т. со средней точкой, т. со средним выводом** center-tap transformer, push-pull transformer; **т. согласования полного сопротивления** impedance-matching trans-

former; **согласовательный** (*or* **согласовывающий**, *or* **согласующий**) **т.** impedance corrector; impedance-matching transformer; **т., согласующий активные сопротивления** resistance-matching transformer; **т.-смеситель** hybrid coil; **стабилизирующий т.** antihunt transformer; **стержневой т.** core-type transformer; **т. строчной развёртки** line transformer, line-scan transformer; **телефонный т. в основной цепи** side circuit repeating coil; **т. тока без первичной обмотки** winding type current transformer; **т. тока горшкового типа** potted current transformer; **фазовращающий т.** goniometer; **фасонный т.** insulator type transformer; **т. частоты** frequency modulator; **шинный т. тока** busbar-type current transformer; **широкополосный т.** video transformer

**трансформаторный** *adj.* transformer, converter

**трансформационный** *adj.* transformational, transformation, conversion

**трансформация** *f.* transformation, conversion, change; **т. проводимости** admittance transformation

**трансформирование** *see* **трансформация**

**трансформированный** *adj.* transformed, changed, converted; modified

**трансформировать** *v.* to transform, to convert, to change, to step down

**трансформирующий** *adj.* transforming, converting

**трансцендента** *f.* transcendence, transcendental function

**трансцендентальный, трансцендентный** *adj.* transcendental

**трансэкваториальный** *adj.* transequatorial

**траншея** *f.* trench, ditch

**трапецевидный** *adj.* tapered, trapeziform

**трапецеидальность** *f.* keystoning

**трапецеидальный** *adj.* trapezoidal, trapezoid, keystone-shaped

**трапециевидный** *adj.* trapeziform, tapered

**трапеция** *f.* trapezium; trapezoid

**трапецойдальность** *f.* keystone effect, keystoning

**трапецойдальный** *adj.* trapezoid(al), keystone-shaped

**трапецоэдр** *m.* trapezohedron

**трасса** *f.* run, way, route, routing; layout; track, path; **т. кабелепроводки** conduit run, duct route

**трассировать** *v.* to mark off, to stake out, to peg out, to take the range; to trace, to locate

**трассировка** *f.* laying out, layout, laying; route of line, direction of line; tracing, location

**трассирующий** *adj.* tracing, marking, marker

**трафарет** *m.* pattern; mask; matrix, stencil; **матричный т.** stencil-cutout matrix

**трафаретный** *adj.* stencil; masking

**трафик** *m.* traffic; **автоматический междугородный т.** automatic long-distance service; **междугородный т.** trunk traffic, long-distance communication; **предпраздничный т.** week-end traffic; **т. пригородной связи** suburban traffic (*tphny.*); **чрезмерный т.** overflow traffic (*tphny.*)

**ТРД** *abbr.* (**турбореактивный двигатель**) turbo-jet engine

**требование** *n.* requirement, demand, requisition, specification; call (*tphny.*); **требования к питанию** power requirements; **технические требования** specifications; **т. на соединение** call order; **т. на соединение между двумя абонентами** person-to-person call; **т. на соединение по авансу** call with indication of charge; **т. на соединение с предварительным приглашением** messenger call (*tphny.*); **т. на соединение с предварительным уведомлением** appointment call, preadvice call; **очередное т. на соединение** call with n prior calls; **предварительное т. на соединение** call booked by prearrangement

**требовать, потребовать** *v.* to demand, to require, to request

**требуемый** *adj.* required, specified, desired, wanted

**требующий** *adj.* requiring, demanding

**тревога** *f.* anxiety, fear; alarm

**тревожный** *adj.* alarm, alarming, emergency, distress

**трезвучие** *n.* triad, common chord

**трезубец** *adj.* trident

**трезубчатый, трезубый** *adj.* trident, tridentate, three-pronged, three-toothed

**трейлер** *m.* trailer

**трель** *f.* warble; trill, shake, quaver, quavering

**трельный** *adj.* trill, quaver, quavering

**тремолирующий** *adj.* trilling; warbling

**тремоло** *n. indecl.* tremolo

**тренажёр** *m.* trainer; simulator; flight simulator

**тренер** *m.* trainer

**трение** *n.* friction, rubbing; **без трения** frictionless; **т. на границе** boundary friction; **начальное т.** friction of rest; **т. покоя** static friction; **т. цапф** journal friction

**тренировать** *v.* to age; to train; to condition

**тренировка** *f.* aging, aging test; seasoning, conditioning; training; **т. на частоту** frequency aging; **повторная т.** re-aging; **прерывистая т.** intermittent life test

**тренировочный** *adj.* aging; conditioning; training

**тренога** *f.* tripod, tripod assembly

**треногий** *adj.* three-legged, tripod

**треножник** *m.* stand, tripod, tripod assembly, trihedral

**треножный** *adj.* tripod, three-legged

**треншальтер** *m.* leakage-protection breaker, leakage-protection cutout, disconnector, isolating switch

**трепещущий** *adj.* fluttering

**треск** *m.* crack, crackling, crackle, clicking noise, noise, bang, rattle, mechanical noise; **т. на линии из-за плохих сростков** line scratches; **пулемётный т.** machine-gun fire (*tphny.*)

**трескотня** *f.* rattle, frying, grinders; **т. щёток** rattling of wipers

**треснувший** *adj.* cracked

**третий** *adj.* third; tertiary

**третичный** *adj.* tertiary, ternary

**третник** *m.* tin solder; **т. в трубках** resin-cored solder

**треть** *f.* third, one third

**третьоктавный** *adj.* third-octave

**треугольник** *m.* triangle, trigon; delta, delta network; delta system; **т. гласных звуков** vowel triangle; **т. с удлинёнными стронами** extended delta; **т. сложения путей волн** wave-path triangle; **т. сопротивлений** vector diagram of complex quantities

треуго́льный *adj.* triangular, triatic, trigonal, trigonous
трёх- *prefix* tri-, three, triple
трёхагрега́тный *adj.* three-unit
трёхадре́сный *adj.* three-address
трёхато́мный *adj.* triatomic; trihydric
трёхбараба́нный *adj.* triple-drum
ртёхбу́квенный *adj.* trigram
ртёхвале́нтность *f.* trivalence
трёхвале́нтный *adj.* tervalent
трёхвариа́нтный *adj.* tervariant
трёхва́ттный *adj.* three-watt
трёхвибра́торный *adj.* tridipole, tripole
трёхгнёздный *adj.* three-cell(ed)
трёхго́рлый *adj.* three-neck(ed)
трёхгра́нник *m.* trihedron
трёхгра́нный *adj.* trihedral; three-surfaced
трёхдете́кторный *adj.* triple-detection
трёхдипо́льный *adj.* tridipole
трёхды́рный, трёхды́рочный *adj.* three-hole
трёхжи́льный *adj.* three-core, triple-core, triple-cored; triplex
трёхзве́нный *adj.* triple-section
трёхзерка́льный *adj.* three-mirror
трёхзна́чный *adj.* three-unit; index-symbol; three-aspect; three-position
трёхзу́бчатый, трёхзу́бый *adj.* trident, three-pronged
трёхимпу́льсный *adj.* triple-pulse
трёхинде́ксный *adj.* three-component
трёхинтегра́торный *adj.* three-integrator
трёхкана́льный *adj.* three-channel, three-terminal
трёхкаска́дный *adj.* three-stage; triple-cascade
трёхке́рновый *adj.* three-legged
трёхколе́нчатый *adj.* three-throw
трёхколо́нный *adj.* three-column, three-legged
трёхкомпане́нтный *adj.* tricolor, tristimulus
трёхкомпоне́нтный *adj.* triple-component, 3-part; ternary; triaxial
трёхконта́ктный *adj.* three-pin, three-point, three-pronged
трёхконту́рный *adj.* three-circuit
трёхкоордина́тный *adj.* three-dimensional
трёхко́рпусный *adj.* three-unit; triple
трёхкра́тный *adj.* triple, threefold, three-stage

трёхкристалли́ческий *adj.* three-crystal
трёхкулачко́вый, трёхкула́чный *adj.* three-jawed, tricam
трёхла́мповый *adj.* three-tube
трёхлине́йный *adj.* trilinear
трёхлистово́й *adj.* three-sheet
трёхлопа́стный *adj.* three-bladed
трёхлучево́й *adj.* three-beam
трёхме́рный *adj.* three-dimensional; triaxial
трёхобмо́точный *adj.* three-winding, triple-wound, three-coil
трёхосево́й *adj.* triaxial
трёхосновно́й, трёхосно́вный *adj.* tribasic
трёхо́сный *adj.* triaxial
трёхпласти́нчатый *adj.* tri-plate
трёхпле́чий *adj.* three-arm(ed)
трёхплоскостно́й *adj.* triplate
трёхпозицио́нный *adj.* three-position
трёхполостно́й *adj.* three-cavity
трёхполю́сник *m.* tri-pole, three-terminal network
трёхполю́сный *adj.* tri-pole, three-polar, triple-pole
трёхпроводно́й *adj.* three-wire, three-conductor, triple
трёхпроже́кторный *adj.* three-beam, three-gun
трёхпу́тный *adj.* three-way
трёхразме́рный *adj.* three-dimensional
трёхрезона́нсный *adj.* triple-tuned
трёхрезона́торный *adj.* three-resonator, three-cavity
трёхря́дный *adj.* three-row, triple; three-range, three-tier
трёхсекцио́нный *adj.* three-unit
трёхсе́точный *adj.* triple-grid
трёхскоростно́й *adj.* three-speed
трёхсло́йный *adj.* three-play, three-layer; triple-covered; tripack
трёхсре́зный *adj.* triple-shear
трёхстаби́льный *adj.* tristable
трёхстержнево́й *adj.* tri-rod; three-legged
трёхсторо́нний *adj.* three-sided, trilateral, three-way
трёхступе́нный, трёхступе́нчатый *adj.* three-stage, three-step(ped); triple-cascade
трёхта́ктный *adj.* three-phase
трёхто́новый *adj.* three-tone
трёхто́чечный *adj.* three-point
« трёхто́чка » *f.* Hartley oscillator
трёхуго́льный *adj.* triangular

трёхуровневый *adj.* three-level

трёхфазный *adj.* three-phase, triphase

трёхфононный *adj.* three-phonon

трёхходовой *adj.* three-way, three-pass, three-throw

трёхцветность *f.* trichromatism

трёхцветный *adj.* three-color(ed), trichromatic, tristimulus; three-aspect

трёхцепный *adj.* three-circuit

трёхчастичный, трёхчастный *adj.* three-piece, three-part

трёхчетвертьволновый *adj.* three-quarter-wave

трёхчлен *m.* trinomial

трёхчленный *adj.* trinomial, trinominal, ternary, three-term

трёхшарнирный *adj.* three-hinged

трёхшлицовой *adj.* three-slot

трёхштырьковый *adj.* three-pin

трёхъярусный *see* трёхэтажный

трёхэлектродный *adj.* three-electrode

трёхэлементный *adj.* three-element

трёхэтажный *adj.* three-tier; three-story, three-storied

трещание *n.* cracking, crackling

трещать, затрещать *v.* to click, to crackle, to crash, to rattle, to buzz; to split, to burst

трещина *f.* crack, flaw, gap, split, break, fracture, cleft, fissure

трещиноватый *adj.* fissured, split, cracked

трещиностойкий *adj.* crackproof, fractureproof

трещотка *f.* rattler

три *num.* three; *prefix* tri-, three

триада *f.* triad, triad of elements; three-digit group

триадный *adj.* triad, ternary

триак *m.* triac

триаморфный *adj.* triamorphous

триангулировать *v.* to triangulate

триангулятор *m.* triangulator

триангуляционный *adj.* triangulable

триангуляция *f.* triangulation

триболюминесценция *f.* triboluminescence, friction luminescence

трибометр *m.* tribometer

трибоэлектрический *adj.* triboelectric

трибоэлектричество *n.* triboelectricity

тривариантный *adj.* trivariant

тригатрон *m.* trigatron, fixed spark-gap

триггер *m.* trigger; flip-flop, flip-flop generator, Eccles-Jordan circuit, binary flip-flop; т. переноса carry flip-flop; потенциально управляемый т. direct-current operated flip-flop; реактивный т. one-cycle multivibrator; т. с автоматическим смещением cathode-biased flip-flop; т. с перекрёстными обратными связями cross-coupled flip-flop; т. со счётным входом complementing flip-flop

триггерный *adj.* trigger; flip-flop

тригистор *m.* trigistor

триглицин *m.* triglycine

тригон *m.* trigon; trigone

тригональный *adj.* trigonal, triangular

тригонометрический *adj.* trigonometric(al)

тригонометрия *f.* trigonometry

триграмма *f.* trigram

тридактор *m.* triductor

тридцатиградусный *adj.* thirty-degree

тридцатидвухричный *adj.* duotricenary

тридцать *f.* thirty; т. вторая thirty-second note

тризистор *m.* trisistor

тризоктаэдр *m.* trisoctahedron

триклинальный, триклинический, триклинный *adj.* triclinic, anorthic, asymmetric

трикон *m.* tricon, tricon radar system, tri-coincidence navigation

трилатерация *f.* trilateration

триммер *m.* trimmer; stabilizer

тримолекулярный *adj.* trimolecular, termolecular

триморфизм *m.* trimorphism

триморфный *adj.* trimorphous

тринидадский *adj.* Trinidad

тринистор *m.* trinistor

трином *m.* trinomial

триноскоп *m.* trinoscope, set of color image tubes

триод *m.* triode; т. в сверхрегенеративной схеме superregenerative detector; генераторный т. transmitting triode; кремниевый однотактный кристаллический т. silicon unijunction transistor; кристаллический т. transistor, triode transistor; кристаллический вычислительный т. computer transistor; плоскостной т. junction-type triode, junction transistor; полупроводниковый т. в вакууме evacuated transistor; прямонакальный т. battery triode;

рентгеновский т. hot-cathode X-ray triode; **решающий т.** computer triode; **т. с большим внутренним сопротивлением** high impedance triode; **т. с низким полным сопротивлением** low impedance triode; **т. с полным внутренним сопротивлением средней величины** medium impedance triode; **т. с тормозящим полем** positive-grid triode; **сдвоенный кристаллический т.** tandem transistor; **сплавной плоскостной т. лавинного типа** alloyed-junction avalanche transistor

**триод-гексод** *m.* triode-hexode; **смесительный т.-г.** triode hexode frequency changer

**триплекс** *m.* triplex

**триплексер** *m.* triplexer

**триплексный** *adj.* triplex

**триплет** *m.* triplet

**триплетный** *adj.* triplet

**триполярный** *adj.* three-polar, three-pole

**тритий** *m.* tritium

**тритон** *m.* tritone (*mus.*); triton

**трифторхлорэтилен** *m.* triflour-chloride-ethylene

**трихлорэтилен** *m.* trichlorine ethylene

**трихроматический** *adj.* tristimulus; trichromatic

**трихромоскоп** *m.* trichromoscope

**трогание** *n.* pickup (of a relay); jogging; touching

**троечный, троичный** *adj.* ternary

**тройка** *f.* three-digit group; triplet, triad; **расстроенная т.** staggered triplet

**тройник** *m.* T-joint, T-piece, tee-joint, Y-joint, tap joint; branch pipe, T-pipe; T-junction box, branch box; triplet; **волноводный т.** hybrid circuit; **т. для Е-Н волн** E-H tee; **шунтированный т.** bridged T-section

**тройниковый** *adj.* three-way, tee, T-piece, T-joint, T-bend

**тройной** *adj.* ternary, triple, three, threefold, triplicate, three-way; trifurcated

**тройня** *f.* triplet

**тролейфон** *m.* trolleyphone

**тролит** *m.* trolite

**тролитул** *m.* trolitul

**тролитуловый** *adj.* trolitul

**троллей** *m.* trolley; **боковой т.** side running trolley

**троллейбус** *m.* trolley bus

**троллейный** *adj.* trolley

**тромбон** *m.* trombone

**тропадин** *m.* tropadyne; **т.-схема** tropadyne circuit, receiver employing a tropadyne mixer

**тропикализация** *f.* tropicalization

**тропикализированный** *adj.* tropicalized

**тропический** *adj.* tropical

**тропопауза** *f.* tropopause

**тропосфера** *f.* troposphere

**тропосферный** *adj.* tropospheric, troposphere

**тропотрон** *m.* tropotron

**трос** *m.* cable, stranded cable, wire cable, rope, line, cord, strand; **т. для воздушных линий** stranded conductor for open wire lines; **т. для подвески** messenger wire; **т. для протягивания кабеля** drawing-in wire; **заземляющий т.** overhead static cable; **натяжной т.** anchoring rope, guy rope; **несущий т.** bearer cable, messenger strand, suspension strand, catenary wire; **оттяжной т.** guy strand, back guy; **поддерживающий т.** *see* **несущий т.**

**тросодержатель** *m.* messenger hanger

**трохоида** *f.* trochoid

**трохоидальный, трохоидный** *adj.* trochoidal

**трохотрон** *m.* trochotron (tube)

**троякий** *adj.* threefold, triple

**ТРС** *abbr.* (**турбореактивный снаряд**) spin-stabilized rocket

**труба** *f.* pipe, tube; trumpet; horn (*mus.*); tunnel, shaft, conduit; tubing; duct; **аэродинамическая т.** wind tunnel; **водоспускная т.** water drain pipe; **защитная т. для кабеля** cable conduit; **измерительная т.** impedance-measuring tube; **т. квадратного сечения** square tube; **коленчатая распределительная т.** dispatch-tube bend; **лабиальная органная т.** closed organ pipe; **т. с абсолютно жёсткими стенками** rigid tube; **т. с акустически жёсткими стенками** sound-hard tube; **т. с акустически мягкими стенками** sound-soft tube; **т. с раструбом** socket tube; **цеметная т. с параллельными каналами для кабелей** cable-duct section (of concrete or break)

**трубка** *f.* tube; receiver, handset, microtelephone; tubing, sleeve, funnel, conduit, duct; pipe, piping; **т. азимута** bearing tube; **т. Бергмана** metal-sheathed insulated tubing; **т. Брауде** image orthicon; **вводная изоляционная т.** inlet funnel, porcelain tube; **т. видения в темноте** night-sight tube, infrared telescope; **т. выпрямителя с подогревным катодом** hot cathode rectifier, high-vacuum rectifier; **высоковакуумная электроннолучевая т.** high-vacuum cathode ray tube; **высоковольтная т. для разложения бегущим лучом** high-voltage scanning tube; **т. высоты (радарной установки)** oscilloscope tube for radar altimeter; **газовая счётная т.** gas-filled radiation counter tube; **газоразрядная коммутирующая т.** grid-switching gas tube; **гибкая изоляционная т.** loom, flexible tubing; **т. дальности** range scope; **дистанционная т.** time fuse; **т. для записи тёмной строкой** skiatron tube; **т. для индикации букв и знаков** character-display tube, character-forming tube; **т. для квантования сигнала во времени** time-sampling tube; **т. для квантования сигнала по уровням и кодирования** quantizing-coding tube; **т. для комнатной проводки** interior tubing, interior conduit; **т. для печатания букв на экране** charactron; **т. для темной записи** dark trace tube; **запоминающая т.** charge-storage tube, information storage tube, memory tube; **защищённая т.** insert tube; **изоляционная т. с оболочкой из освинцованного железа** electric conduit with lead-coated iron sheath; **импульсная разрядная т.** electronic photoflash lamp; **т. индикации азимута** bearing tube; **контрольная электроннолучевая т.** monitoring tube; **люминесцентная т.** cold-cathode-type fluorescent lamp **т. магнитной индукции** magnetic tube of force; **т. мгновенного действия** nonstorage pickup tube; **микротелефонная т.** handset, hand microphone, hand microtelephone; **т. модулятора** grid sleeve; **накопительная т. с регенерацией данных**

по двойной системе regenerative binary storage tube; **т. накопления двоичной информации** binary storage tube; **напорная т.** total head tube; **наставная т.** tonguing; **т. ночного видения** night-sight tube, infrared telescope; **однолучевая «записывающая»** (*or* «запоминающая») **т.** single-beam charge-controlled tube; **однолучевая цветная приёмная т. с затемняющей маской** one-gun shadow-mask color kinescope; **т. осциллографа, служащего для контроля формы сигнала по кадрам** frame (*or* field) monitoring tube; **осциллографическая т.** oscillotron; **переговорная т.** voice tube, interphone; **передающая т. без накопления (заряда)** nonstorage pickup tube (*or* device), nonstorage camera tube; **п. т. с большой контрастностью** (*or* гаммой) high gamma tube; **п. т. с накоплением зарядов** image storing tube, storage camera tube; **п. т. с переносом изображения** image tube; **п. т. с развёрткой пучком быстрых электронов** high-velocity tube, high-velocity pick-up tube; **п. т. эбикон** transmission secondary emission multiplier; **передающая телевизионная т.** camera tube, pick-up tube, picture tube; **передающая телевизионная т. для развёртки бегущим лучом** cathode-ray tube for flying-spot scanner; **т. передающей камеры** camera pick-up tube; **поглощающая т.** lined duct; **приёмная т.** image-viewing tube, reproducing tube; **п. т. для цветного телевидения с сеткой для коммутации цветов** grid-controlled-color kinescope; **п. т. прямого видения с трёхцветным экраном** tricolor direct-view tube; **п. т. с затемняющей маской и трёхцветным экраном** aperture-mask tricolor kinescope; **п. т. с мозаичным экраном** dot-phosphore tube; **п. т. с темновой записью** dark-trace picture tube; **п. т. с теневой маской и трёхцветным экраном** aperture-mask tricolor kinescope; **п. т. с трёхцветным экраном** tricolor picture tube; **приёмная (телевизионная) т.** picture tube, kinescope, display tube, viewing tube, television picture tube; **т. прямого видения** image tube; **т. прямого видения в инфракрасных лучах** night-sight tube, infrared telescope; **т. прямого видения с памятью** direct-viewing memory tube; **т. работающая в режиме второго пересечения** second-crossover tube; **радиолокационная т.** compass tube; **расходомерная т.** Venturi meter; **рентгеновская т. в заземлённом кожухе** shock-proof X-ray tube; **рентгеновская т. для фотографирования вспышкой** flash X-ray tube; **рентгеновская т. с трубчатым анодом** rod-anode X-ray tube; **т. с большим междуэлектродным расстоянием** wide-spaced tube; **т. с большой гаммой** (*or* **контрастностью**) high-gamma tube; **т. с быстро потухающим экраном** fast-screen tube; **т. с двусторонней мишенью** image orthicon; **т. с запасанием заряда** storage tube, memory tube; **т. с записью цветной строкой** color-trace tube; **т. с линейным экраном** strip-phosphor tube; **т. с люминофором белого свечения** white-phosphor tube; **т. с малым междуэлектродным расстоянием** closed-spaced tube; **т. с малым послесвечением** fast-screen tube; **т. с наклонным анодом** angled-anode tube; **т. с накоплением зарядов** memory tube, storage tube, storage camera tube, storage-type camera pick-up tube, memory storage tube; tube; **т. с накоплением модулируемого заряда** charge-controlled storage tube; **т. с « памятью » для перезаписи изображения** signal-converter storage tube; **т. с переносом изображения** electron-image tube; **т. с плавкой вставкой** fuse tube; **т. с плоским дном** flat-ended tube; **т. с подогревным катодом** hot cathode vacuum tube; **т. с подсветкой** intensifier-type tube; **т. с развёрткой пучком быстрых электронов** high-velocity tube; **т. с развёрткой пучком медленных электронов** low-velocity tube; **т. с развёрткой типа A** A-scope; **т. с разделёнными экранами в виде полос** multiband tube; **т. с регули-**

**руемым накоплением заряда** charge-controlled storage tube; **т. с тарелочкой** flare, flange; **т. с тёмной записью** dark-trace tube, dark-trace cathode-ray tube; **т. с цветной записью, т. с цветным (после) свечением** color-trace tube; **т. с экраном белого свечения** white-phosphor tube, white-screen tube; **т. с электростатическим управлением и послеускорением** post-acceleration electrostatic tube; **силовая т., т. силовых линий** field tube; **слуховая т.** Bell receiver, watch receiver (*tphny.*); **т. со световым модулятором** light-valve viewing tube; **соединительная т. манометра** pressure tube; **спектрально-аналитическая рентгеновская т.** X-ray spectrometer tube; **съёмочная т.** camera tube, pickup tube; **телевизионная т.** picture tube; camera tube; **телевизионная передающая т.** camera tube, pickup tube; **телевизионная приёмная т.** picture tube, kinescope; **т. телефона, телефонная т.** telephone receiver, hand receiver; **т. Тимофеева и Шмакова** image iconoscope; **т. тлеющего заряда** neon bulb; **т. точного определения дальности** fine-range scope; **трёхлучевая приёмная т.** three-gun picture tube, trinescope; **трёхлучевая приёмная т. с затемняющей маской для цветного телевидения** three-gun shadow-mask color kinescope; **трёхлучевая приёмная т. с точечным экраном** three-gun masked-dot tube; **трёхцветная приёмная т.** tricolor picture tube, trichromoscope; **электронно-лучевая т.** electron-beam tube, cathode-ray tube; **электронно-лучевая т. с запасанием зарядов** image-storing tube, storage camera tube; **электронно-лучевая т. с накоплением зарядов** iconoscope, beam storage tube; **электронно-лучевая т. с подсветкой** intensifier-type C.R. tube; **электронно-лучевая т. с пучком высокой плотности** high-intensity cathode-ray tube; **электронно-лучевая т. с электростатическим управлением** electrostatic cathode-ray tube

**трубковидный, трубкообразный** *adj.* tubular, tubiform

**трубный** *adj.* pipe, tube; trumpet

**трубодержатель** *m.* tube holder, pipe support

**трубообразный** *adj.* piped, pipe-shaped

**трубопровод,** *m.* **трубопроводка** *f.* conduit, duct; pipe-line: pipe, piping, tubing

**труборез** *m.* pipe cutter, tube cutter

**трубочка** *f.* little tube; sleeve; **волосная т.** capillary tube; **изолирующая т.** insulating sleeve

**трубочный** *adj.* tubed, tube, pipe

**трубчатый** *adj.* tubular, hollow; piping

**трудный** *adj.* hard, difficult

**трудоёмкость** *f.* working hours

**Трутон** Trouton

**трущий, трущийся** *adj.* rubbing, friction; wiping

**тряпичный** *adj.* rag

**трясение** *n.* shaking, trembling

**тряска** *f.* shake, shaking, jolting, jolt, trembling, vibration, jogging, joggling

**тряскостойкий** *adj.* vibration-proof, shake-proof, jar-proof, anti-vibration, shockproof

**трясти, вытрясти, тряхнуть** *v.* to shake, to jolt, to joggle

**« трясучка »** *f.* shaker (unit), tube testing tapper, vibration table

**тряхнуть** *pf. of* **трясти**

**ТС** *abbr.* (**телесигнализация**) telesignalization

**ТСД** *abbr.* (**турбина среднего давления**) medium-pressure turbine

**ТТД** *abbr.* (**тактико-технические данные**) tactical and technical data

**ТТТ** *abbr.* (**тактико-технические требования**) tactical-technical specifications or requirements

**ТТУ** *abbr.* (**территориально-технический участок**) area technical section

**ТУ** *abbr.* 1 (**телеуправление**) telecontrol, remote control; 2 (**технические условия**) technical specifications

**туба** *f.* tuba (*mus.*)

**туго** *adv.* tightly, fast

**тугой** *adj.* tight, stretched, close; stiff

**тугоплавкий** *adj.* tempered, hard; high-melting; infusible

**тугоухость** *f.* relative deafness, dullness of hearing

**ТУиН** *abbr.* (**технические условия и нормы**) technical specifications and norms

**тулий** *m.* thulium, Tm

**туман** *m.* fog, film, mist, haze

**туманность** *f.* haziness, fogginess, mist; nebula

**туманный** *adj.* hazy, foggy, misty; obscure, vague

**туманомер** *m.* fog meter

**тумба** *f.* fender, stub pole: **отбойная т.** pole fender

**тумблер** *m.* toggle switch, tumbler switch; tumbler; **двойной т.** double-toggle actuator

**тунгар** *m.* tungar (rectifier)

**тунгаровый** *adj.* tungar

**туннелирование** *n.* tunneling

**туннель** *m.* tunnel; duct, conduit

**туннельный** *adj.* tunnel; duct, conduit

**туннельтрон** *m.* tunneltron

**тупиковый** *adj.* dead-end(ed); stub, branch; irredundant

**тупой** *adj.* blunt, dull; obtuse; flat

**тупоконечный** *adj.* blunt, blunt-pointed

**тупоугольный** *adj.* obtuse-angled, obtuse

**« турбатор »** *m.* turbator

**турбидиметр** *m.* turbidimeter, opacimeter

**трубидиметрический** *adj.* turbidimetric

**турбидиметрия** *f.* turbidimetry

**турбина** *f.* turbine; **т. с внешним подводом воды** outward-flow turbine; **т. с отъёмом пара и без конденсации** extraction-non-condensing type turbine; **струйная т.** free-jet type turbine; free-deviation turbine; **центростремительная т.** inward-flow central discharge turbine

**турбинный** *adj.* turbine

**турбо-** *prefix* turbo-, turbine

**турбоагрегат** *m.* turbine-driven set, turbo-unit

**турбоальтернатор** *m.* turbo-alternator

**турбобур** *m.* turbo-drill

**турбовозбудитель** *m.* turbo-exciter

**турбовоздуходувка** *f.* turboblower, turbofan

**турбогенератор** *m.* turbogenerator, turbo-alternator; **т. с зубчатой передачей** geared turbogenerator

**турбогенераторный** *adj.* turbogenerator

**турбодинамо** *n. indecl.* turbodynamo

**турбоинвертер** *m.* turbo-inverter

**турбокомпрессор** *m.* turbocompressor, turboblower

**турбомашина** *f.* turbomachine, turbodynamo

**турбомик** *m.* turbomic

**турбонагнетатель** *m.* turbocompressor, turbosupercharger

**турбонасос** *m.* turbo-pump, turbine-driven pump

**турбопреобразователь** *m.* turboconverter

**турбореактивный** *adj.* turbojet

**турборотор** *m.* turborotor

**турботрансформатор** *m.* turbine transformer

**турбоэлектрический** *adj.* turbo-electric

**турбоэлектровоз** *m.* turbo-electric locomotive

**турбулентность** *f.* turbulence

**турбулентный** *adj.* turbulent; eddy

**турбулизация** *f.* turbulization, agitation

**турбулизирование** *n.* turbulization

**турель** *f.* turret; multilens head; **т. объектива** lens turret

**турельный** *adj.* turret, turret-type

**турмалин** *m.* tourmaline

**турмалиновый** *adj.* tourmaline

**турникет** *m.* turnstile

**турникетный** *adj.* turnstile(d); doughnut

**тускло** *adv.* dimly

**тускло-накальный** *adj.* dull-emitting

**тусклость** *f.* dimness, dullness

**тусклый** *adj.* mat, obscure, dim, dull, tarnished

**туфнол** *m.* tufnol

**тучность** *f.* fatness; loop

**туше** *n. indecl.* touch (*mus.*)

**тушение** *n.* extinction, extinguishing, quenching

**тушитель** *m.* extinguisher, quencher

**тушительный** *adj.* extinguishing, quenching

**тушить, затушить, потушить** *v.* to extinguish, to quench, to blow out, to put out; to switch off; to damp

**тфн** *abbr.* (**телефон**) telephone

**ТЦ** *abbr.* (**телецентр**) television center

**тч.** *abbr.* (**точка**) point

**тысячный** *adj.* thousandth

**тычок** *m.* prong, prod, pin, plug,

peg; **испытательный т.** test prod

**Тьюринг** Turing

**т.э.д.с.** *abbr.* (**термоэлектродвижущая сила**) thermo-electromotive force

**Тэйлор** Taylor

**тэта** *f.* theta

**Тэттль** Tuttle

**ТЭЦ** *abbr.* (**теплоэлектроцентраль**) heat and electric power plant

**тюльпанообразный** *adj.* tulip-shaped, funneled, funnel-shaped

**Тюри** Thury

**тяга** *f.* thrust; pull, draft, draught, draw, current, traction; drive; side rod, connecting rod, drawbar, pull rod, pull wire; **воздушная т.** draught, draft; upward pull, head; current of air; **т. защёлки** latch rod; **т. между остряками** front rod; **т. между перьями для привода** head rod; **приводная т.** operating rod, throw rod; **приводная т. семафорного крыла** semaphore-arm operating rod, up-and-down rod; **т. с вилкой** jaw rod; **стрелочная приводная т.** switch operating rod

**тяговый** *adj.* traction, tractive, thrust

**тягомер** *m.* tensometer, draft gage; traction dynamometer; **наклонный т.** inclined draft gage

**тяготение** *n.* gravitation, gravity, attraction, pull; molar attraction; **всемирное т.** general gravitation

**тягучий** *adj.* tenacious, tough; tensile, tractile; malleable; ductile

**тяжёлый** *adj.* weighty, heavy; solid; hard, difficult; dense

**тяжение** *n.* strain, stress, pull

**тяжесть** *f.* gravity, load, weight, heaviness; force

**тянуто-диффузионный** *adj.* grown-diffused, drawn-diffused

**тянутый** *adj.* drawn, pulled, drawn out; grown (-junction)

**тянуть, потянуть** *v.* to stretch, to draw, to draw out; to pull, to drag

**тянуться, потянуться** *v.* to follow (of a contact spring); to stretch, to lengthen, to extend; to drag on, to hold out, to last

**тянучка** *f.* streak; **«белая т.»** pulling on whites, white smear; **«чёрная т.»** black after white, pulling-in on blacks, black smear

# У

**у** *prep. genit.* by, near, at, on; with; of

**Уайт** White

**«уанметр»** *m.* onemeter

**УАТС** *abbr.* (**учрежденская АТС**) agency's dial telephone exchange

**уатт** *m.* watt

**уаттметр** *m.* wattmeter

**убегание** *n.* runaway

**убирающий** *adj.* retractable, retracting; collapsible

**убирающийся** *adj.* collapsible, folding-type, disappearing

**убитрон** *m.* ubitron

**уборщик** *m.* office cleaner; **у. анодов** anode hooker

**убывание** *n.* diminution, subsidance, decrease, falling, sinking, taper

**убывать, убыть** *v.* to diminish, to descend, to decrease, to wane; to sink, to subside, to fade

**убывающий** *adj.* decreasing, decaying, decrescent, diminishing

**убыль** *f.* loss, decay, depreciation; diminution, decrease

**убыток** *m.* loss, damage

**убыть** *pf. of* убывать

**уведомить** *pf. of* уведомлять

**уведомление** *n.* information, notification; **у. о местонахождении** position report

**уведомлять, уведомить** *v.* to communicate, to indicate, to inform, to notify, to advise

**увеличение** *n.* rise, increase, augmentation; extension, expansion; enhancement, amplification, magnification, enlargement, multiplication, increment, growth; **у. в 5 раз** quintupling; **у. динамических оттенков** contrast expansion; **у. индуктивности (линии)** loading; **у. масштаба времени** time magnifying; **у. оптической системы** similarity scale of optical images; **поперечное у.** lateral magnifying power (*tv*); **пропорциональное у.** scaling up; **у. размеров** multiplexing; **у. стрелы провеса** sag magnification; **у. частоты** pull-in;

pulling into tune; **у. числа групп** extra trunking and group switching equipment; **у. числа рядов детерминантов** extension of determinants; **у. эмиссии послесвечения** stimulation (of afterglow)

**увеличенный** *adj.* augmented, increased, magnified, enlarged

**увеличивание** *see* увеличение

**увеличивать, увеличить** *v.* to increase, to enlarge, to magnify, to extend, to augment, to amplify; to boost, to enhance, to intensify; **у. индуктивность цепи** to stiffen (the circuit)

**увеличиваться, увеличиться** *v.* to increase, to rise, to grow, to become larger, to augment

**увеличивающий** *adj.* enlarging, increasing; magnifying

**увеличитель** *m.* enlarger; **у. с плёнки** microfilm projector

**увеличительный** *adj.* magnifying, enlarging, augmentative

**увеличить(ся)** *pf. of* увеличивать(ся)

**увиолевый** *adj.* uviol

**увиоль** *m.* uviol

**увлечение** *n.* pulling; carrying away, entrainment

**увлечь** *pf. of* влечь

**УВС** *abbr.* (**указатель воздушной скорости**) airspeed indicator

**УВЧ** *abbr.* 1 (**ультравысокочастотный**) ultrahigh-frequency; 2 (**усилитель высокой частоты**) high-frequency amplifier

**УГ** *abbr.* (**уголковая горизонтальная**) horizontal V antenna

**угасание** *n.* extinction

**углевод** *m.* carbohydrate

**углеводород** *m.* hydrocarbon

**угледержатель** *m.* carbon holder

**углезернистый** *adj.* granulated-carbon

**углекислый** *adj.* carbonate (of); carbonic acid

**углеочистка** *f.* coal cleaning

**углерод** *m.* carbon, C

**углеродоникелевый** *adj.* nickel-carbon

**углеродистый** *adj.* carbon, carbonic, carbonaceous

**углистый** *adj.* carbonaceous, carbon-like

**угловатость** *f.* angularity

**угловатый** *adj.* angular

**угловой** *adj.* angular; corner; angle, angled

**угломер** *m.* protractor, graphometer, goniometer; **у. с пузырьковым уровнем** bubble quadrant; **у.-транспорт** bevel protractor

**угломестный** *adj.* elevation

**«углоспрямление»** *n.* squaring (of pulses)

**углубить** *pf. of* **углублять**

**углубление** *n.* hollow, depression, deepening, recess, cavity recess, slot; **углубления** pitting; **у. модуляции верхних частот** preaccentuation, preemphasis

**углублённый** *adj.* deep, profound; absorbed (in); deepened, depressed

**углублять, углубить** *v.* to deepen; to recess, to sink, to depress

**угнетённый** *adj.* minority; depressed, oppressed

**угол** *m.* corner; angle; **углы между осями** optic angle, axial angle; **накрест лежащие углы** alternate angles (*math.*); **смежные прилежащие углы** adjacent angles (*math.*); **у. в горизонтальной плоскости** horizontal angle; **у. включения** closing angle; **у. времени пробега** transit angle; **у. вылета ротора** power angle; **у. выхода на траекторию** approach angle; **у. гашения** overlap angle (of rectifier circuits); **головной у. кристалла** cap angle; **у. запаса** margin of commutation; **у. заслонения** angle of cut-off; **у. зрения объектива** camera angle; **у. конического вращения диаграммы** nutation angle; **у. конуса рассеяния** beam angle of scattering; **у. между двумя направлениями** angular separation; **у. между двумя пересекающимися плоскостями** dihedral angle (*math.*); **у. места** elevation tilt, angle of elevation; C-display; **у. наклона иглы** drag angle (of stylus); **у. наклона траектории при полёте** flight path angle; **у. наклона щётки** brush angle; **у. наклонения** (magnetic) dip; **начальный у. траектории**

original inclination; **у. обзора** aspect angle, angle of aspect, visual angle, angle of sight; **у. отклонения луча в приёмной трубке** (*or* **кинескопе**) kinescope deflection angle; **у. отсечки тока** angle of current flow, total conduction angle; **у. отставания по фазе** angle of lag; **у. охвата** angle of wrap; **у. перекоса** angularity; **у. перекрытия** overlap; **у. поворота вала** shaft portion; **у. понижения относительно горизонта** depression angle; **постоянный фазовый у.** epoch angle; **у. потерь** loss angle, dip angle, dielectric loss; **у. при вершине рупора** flaring angle; **у. прицела** angle of sight; **у. пролёта** transit angle; **пространственный многократный у.** polyhedral angle (*math.*); **у. протекания анодного тока** plate-current operating angle; **противолежащий у.** contrary angle, alternate angle: **у. пучка** beam angle; **у. раскрыва рупора** horn flare angle, flaring angle; **у. раскрытия канавки** groove angle; **у. распространения волны** wave angle; **у. рассматривания в горизонтальной плоскости** horizontal viewing angle; **у. рассогласования** displacement angle; **у. рассогласования при коническом сканировании** squint angle; **у. раствора** flare angle, flare (of a horn); field angle; **у. раствора диаграммы направленности антенны** antenna beam angle; **у. раствора дифракционного конуса** angle of incidence; **у. раствора пучка** beam angle, beam width; **у. с оптической осью** angle of emergence; **у. светила** angle of celestial body; **у. сдвига фаз** phase angle; **у. склонения** angle of inclination; **у. скольжения** grazing angle; **у. скрутки** pitch angle; **у. смежности** angle of contingence (*math.*); **телесный у. прихода радиоволн** acceptance angle; **у. у вершины рупора** flaring angle; **у. упреждения** angle of advance; **у. установки** adjusting angle; **фазовый у. диэлектрика** dielectric phase difference; **фазовый у. преобразователя** circuit angle; **фазовый у. пробега** transit phase angle; **у. цветового тона** hue angle

**уголковый** *adj.* corner; angle

**уголок** *m.* corner; small angle; V-antenna; angle iron, elbow

**уголь** *m.* carbon; coal; **угли для токоприёмной дуги** sliding-bow carbon brush; **у. с деполяризатором** bobbin (of a battery); **у. с продольно рифлёной поверхностью** fluted carbon; **у. с фитилём** cored carbon; **у. со сплошным фитилём** solid-cored carbon; **фитильный у.** cored carbon

**угольник** *m.* pipe bend, knee, elbow; corner iron; angled elbow, sharp bend; bevel; square, try-square; **кольцевой у.** spacer ring

**угольноплёночный** *adj.* carbon-film

**ýгольный** *adj.* carbonic, carbon; coal

**уго́льный** *adj.* angle, angular; corner

**угонный** *adj.* runaway

**уд.** *abbr.* (**удельный**) specific

**удаление** *n.* elimination, removal; stripping; rejection, expulsion; escape, departure; extraction; **у. ветвей** pruning, trimming; **у. заусенцев** removal of burrs (*or* ridges); **у. местных помех** clutter rejection

**удалённый** *adj.* remote, distant, outlying; removed, discharged

**удалить** *pf. of* удалять

**удалитель** *m.* eliminator

**удалять, удалить** *v.* to bare, to strip; to drive off; to evacuate; to remove, to take away, to eliminate, to empty, to withdraw; to extract

**удаляющийся** *adj.* outgoing, departing

**удар** *m.* knock, strike, blow, bump, bang, stroke, bolt; impact; shock, kick, impulse; jerk; thrust; impingement

**ударение** *n.* stress, emphasis

**ударить** *pf. of* ударять

**ударно-ступенчатый** *adj.* stress-step

**ударный** *adj.* impact; shock; knock; percussive, percussion; impulse, impulsing, pulsating; **ударное действие** percussion

**удароглушитель** *m.* shock absorber

**ударопрочный** *adj.* shock-resistant

**ударостойкий** *adj.* shock-proof

**ударостойкость** *f.* shock stability, shockproofability

**ударочувствительность** *f.* sensitivity to shock

**ударяемый** *adj.* knocked-on

**ударять, ударить** *v.* to strike, to hit, to knock, to kick; to lash

**ударяющий** *adj.* striking

**удачный** *adj.* successful

**Уда-Яги** Yagi

**уд.в.** *abbr.* (**удельный вес**) specific gravity

**удваивание** *n.* doubling, duplication; splitting

**удваивать, удвоить** *v.* to double, to redouble, to reduplicate, to duplicate; to split

**удваиваться, удвоиться** *v.* to be doubled, to double

**удваивающий** *adj.* doubling

**уд. вес** *see* уд. в.

**удвоение** *n.* doubling, redoubling, duplication, reduplication; splitting; **у. изображения** ghost image (*tv*)

**удвоенный** *adj.* double, doubled, redoubled, reduplicated, twofold

**удвоитель** *m.* doubler; duplicator; **у. выпрямленного напряжения** rectifier doubler

**удвоить(ся)** *pf. of* удваивать(ся)

**удельный** *adj.* specific, unit

**Уден** Oudin

**удержание** *n.* hold, holding, keeping, retention, restraint; deduction; trapping; caging; containment; occlusion; confinement

**удержать(ся)** *pf. of* удерживать(ся); **у. на линии** to hold the line

**удерживаемый** *adj.* confined

**удерживание** *see* удержание

**удерживать, удержать** *v.* to maintain, to hold, to retain, to confine; to suppress; to deduct; to withhold, to keep back, to delay, to detain; **у. постоянным** to keep constant

**удерживаться, удержаться** *v.* to hold on

**удерживающий** *adj.* holding, hold-up; restraining, retaining, retentive, confining

**удешевлённый** *adj.* reduced-rate

**удлинение** *n.* elongation, lengthening, extension, prolongation, stretch, stretching; longitudinal elongation; **относительное у.** constant of elastic strain, reciprocal of Young's modulus; **у. при разрыве** breaking elongation, total extension; **у. телеграфных импульсов** bias telegraph distortion, tailing

**удлинённый** *adj.* oblong; elongated, lengthened, extended, stretched; loaded (of antenna)

**удлинитель** *m.* pad, attenuator, attenuation network; extension rod; artificial line extension, flexible lead connector; lengthener, extender, stretcher; extension, extension piece; **у. без искажения** distortionless pad; **Г-образный у.** L-pad; **у. для пупинизованных линий** building-out section; **у. для регулировки уровня** level-setting attenuator; **у. дополняющий до шага пупинизации** building-out network; **коаксиальный у.** coaxial pad; **магазинный у.** attenuation box; **механический у. линии** line stretcher; **у. с искажением** distortion pad; **сантиметровый у.** microwave attenuator; **цепной у.** ladder attenuator

**удлинительный** *adj.* extension, lengthening; loading

**удлинить** *pf. of* **удлинять**

**удлиняемость** *f.* extensibility

**удлинять, удлинить** *v.* to elongate, to lengthen, to expand, to stretch, to extend, to enlarge; to load

**удлиняющий** *adj.* lengthening, stretching; extensible

**удобный** *adj.* convenient, handy, expedient

**удобообтекаемый** *adj.* streamlined

**удобочитаемость** *f.* legibility, readability

**удовлетворительный** *adj.* fair, satisfactory, adequate; pertinent

**удовлетворять, удовлетворить** *v.* to fit; to satisfy

**удометр** *m.* udometer, rain gage

**удостоверение** *n.* certificate; licence

**удушье** *n.* choking

**Удэн** Oudin

**уединённый** *adj.* lone, isolated, secluded, remote

**ужесточение** *n.* ruggedization

**ужесточенный** *adj.* ruggedized

**УЗ** *abbr.* 1 (**ультразвук**) ultrasonics; 2 (**ультразвуковой**) ultrasonic

**УЗД** *abbr.* (**ультразвуковой дефектоскоп**) ultrasonic flaw detector

**узел** *m.* loop, knot; bundle, pack; junction point, node, nodule, branch point, vertex; group, assembly, block, unit, component, mounting; portion; **вакантные узлы** vacancies; **крупные узлы** subsystem(s); **узлы пучка** cross-over; **автоматический у. связи «кроссбар»** crossbar telephone center; **у. в группе районных сетей** (radial) network-junction exchange; **главный у.** regional center (*tphny.*); **главный телефонный у.** toll traffic junction center; **у. колебаний** vibration node, interference point; **коммутационный у.** switching station; **межрайонный у.** toll center, "interraion" center; **областной у.** sectional center, "oblast" center, principal outlet, primary outlet (*tphny.*); **радиотрансляционный у.** program-line repeater office; **районный у.** "raion" center (*tphny.*); **у. решётки** lattice point; **у. светоделения** dichroic block; **у. сети** network junction-point; **у. скорости, скоростный у.** velocity node; **у. токопроводящих линий** printed-wiring assembly; **центральный у.** national center (*tphny.*); **центральный у. пригородного сообщения, центральный пригородный у.** suburban toll center (*tphny.*)

**узелок** *m.* knot, small knot; **у. ввода** lead wire weld; **сварочный у.** weld knot; **сварочный у. медь-платинит** dumet-to-copper weld knot

**узкий** *adj.* narrow, tight

**узко** *adv.* narrowly, tightly

**узколучевой** *adj.* narrow-beam

**узконаправленный** *adj.* narrow-directed

**узкополосный** *adj.* narrow-band, narrow

**узкостробно-задерживающий** *adj.* narrow-gate delay

**узловатый** *adj.* knotty; nodose, nodulous, nodular, knobby

**узловой** *adj.* nodal, node, knot; main, central; junction; unit

**узлообразование** *n.* toll-switching planning

**узнавание** *n.* recognition

**узнавать, узнать** *v.* to spot; to recognize, to identify

**узор** *m.* pattern, patterning, design, markings, figure, arrangement; **ложные узоры** moire patterns

**узорный** *adj.* pattern; figured

**узорчатый** *adj.* ornamented, figured

**Уиддингтон** Whiddington

**Уилер** Wheeler

**Уильямсон** Williamson

**Уимшерст** Wimshurst

**Уитмор** Whitmore

**Уитстон** Wheatstone

**Уитфильд** Whitfield
**укавист** *m.* ultra-short-wave ham
**указание** *n.* indication; instruction, direction, designation; cue; **у. (об) исправности** normal indication
**указанный** *adj.* indicated
**указатель** *m.* indicator, indicator dial, dial; hairline pointer, marker, pointer, arrow, needle, arm; designator; directive; detector; cursor; directory, index; characteristic (math); **указатели прекращения питания** power-off indicators; **автоматический у. местоположения самолётов** automatic aircraft-position reporter; **у. блок-участка** block indicator, switch indicator; **у. быстроты снижения** rate-of-descent indicator; **у. времени записи** elapsed-time indicator; **вынесенный у. кругового обзора** plan-repeater indicator; **глазовидный у.** eyeball indicator (*tphny.*); **у. горячих букс** hot box detector; **у. дальности по заданной станции** self-distance indicator; **у. действия в реле** operation indicator; flag indicator; **у. действия светофора** light indicator; **у. дистанции и пеленга** range-and-azimuth indicator; **дистанционный у.** remote monitor; **у. для обводки кривой** curve following stylus; **у. заземления, у. замыкания на землю** ground detector, leakage detector, ground indicator; **у. изменения пеленга** lobe-comparison indicator; **у. косинуса фи** phase meter, phase indicator; **у. кругового обзора** plan position indicator; **ленточный у.** ribbon marker; **магнитный у. тока молни** magnetic indicator for lighting currents, surge current indicator; **у. максимального значения** maximum impulse indicator; **у. максимума (нагрузки)** demand attachment, demand indicator, maximum-demand indicator; **у. места муфты** joint marker, sleeve marker; **у. места прокладки кабеля** cable marker; **у. места (спайки)** joint marker, sleeve marker; **у. местоположения самолёта в полёте** air position indicator; **у. минимального значения** level meter of minimum transmission; **у. на вышке** tower indicator; **у. набираемого номера** step indicator; **у.**

**нагрева букс** hot box detector; **у. наличия напряжения** charge indicator; **у. направлений связи** routing plan, routing bulletin; **у. направления вращения поля** phase sequence indicator; **у. направления с цепью задержки** delayed response directional indicator; **у. настройки** "magic eye", indicator tube, (shadow) tuning indicator; **у. неравномерности нагрузки** balance indicator; **у. номера** call indicator; **у. обратного зажигания** arcing indicator; **у. обрыва провода** wire-splice detector; **у. обрыва цепи** open-circuit indicator; **у. огибающей импульса** envelope viewer; **у. отсчёта** reading index; **периодически действующий у. максимума** restricted hour maximum demand indicator; **у. планирования** glidometer; **у. поворота с мигающим светом** flashing direction indicator lamp; **у. поворота семафорного типа** direction indicator arm; **у. поворота фонарного типа** direction indicator lamp; **у.-ползунок** cursor; **у. порядка следования фаз** phase rotation indicator, phase sequence indicator; **у. предшествования** precedence indicator; **у. прекращения питания** power-off indicator; **у. преходящих перенапряжений** transient indicator; **у. программного регулятора** timing index; **у. рассогласования** spot error indicator; **у. с отпадающими клапанами** drop indicator; **самопишущий у.** recorder; **самопишущий у. угловой скорости поворота** rate-of-turn recorder; **у. свободной линии** (automatic) clearing indicator (*tphny.*); **у. сдвига фаз** phase indicator; **у. скорости диска** dial speed indicator; **у. скорости с тахогенератором** magneto speed indicator; **у. скорости сближения** overtaking meter; **у. скорострельности** rate-of-fire indicator; **у. сравнения лепестков** lobe-comparison indicator; **у. степени наполнения бункера** bunker level-indicator, silo level-indicator; **стрелочный у.** switch signal; **у. такс** table of rates, table of trunk rates; **телефонный у.** telephone directory; **точечный у. рассогласования** spot error indicator; **у. узлов колебаний**

nodalizor; **у. уровня контрольного канала** pilot level indicator (*tphny*.); **у. уровня передачи** transmission level measuring set; **фазовый у. изменения пеленга** phase-difference BDI system; **числовой у. векторной группы соединения** numerical index of the vector group

**указательный** *adj.* indicating, indicatory

**указывать, указать** *v.* to indicate, to point at, to denote, to direct, to show, to demonstrate; to signal, to detect

**указывающий** *adj.* pointing, indicating; directing

**УКВ** *abbr.* 1 (**ультракороткие волны**) ultrashort waves; 2 (**ультракоротковолновый**) ultrashort-wave

**укладка** *f.* laying, installation; setting; packing; **волнистая у. (кабеля)** cabling

**укладочный** *adj.* stacking; packing

**укладчик** *m.* stacker, sequencer

**укладывание** *n.* stacking; packing

**укладывать, уложить** *v.* to lay, to lay down; to embed; to pack, to pack up; to stack, to pile

**укладывающийся** *adj.* folding

**уклон** *m.* inclination, bias; drop, slope, pitch, fall, declivity; deviation; fall of potential, potential of gradient; taper, bevel; **у. морского дна** underwater gradient

**уклонение** *n.* deviation, deflection, error, aberration; evasive action

**уклонить** *pf. of* **уклонять**

**уклономер, уклонометр** *m.* inclination meter, inclinometer

**уклонять, уклонить** *v.* to deflect, to deviate

**укомплектовать** *pf. of* **комплектовать, укомплектовывать**

**укомплектовывать, укомплектовать** *v.* to complement

**укорачивание** *n.* shortening, contraction

**укорачивать, укоротить** *v.* to shorten, to diminish, to abridge, to contract, to reduce, to lessen

**укорачивающий** *adj.* shortening

**укоротитель** *m.* shortener; chopper

**укоротить** *pf. of* **укорачивать**

**укорочение** *n.* reduction, contraction, shortening; **у. телеграфных импульсов** bias telegraph distortion

**укороченный** *adj.* chopped, shortened, contracted, abbreviated

**укосина** *f.* cantilever, bracket, hanger, boom

**укрепить** *pf. of* **укреплять**

**укрепление** *n.* fixing, fastening, attachment; consolidation, fortifying, fortification, strengthening; reinforcement; **у. анкерными болтами** anchoring, staying

**укреплённый** *adj.* strengthened; reinforced, stiffened; supported; anchored; embeded, fastened, fixed

**укреплять, укрепить** *v.* to strengthen, to solidify, to brace, to reinforce; to fix, to set, to fasten; to intensify, to amplify, to magnify; to attach; **у. связями** to anchor, to tie, to stay, to guy

**укрепляющий** *adj.* strengthening; bracing; reinforcing

**укрытие** *n.* shelter, housing; cover(ing), sheathing; **трёхмерные укрытия** three-dimensional weavers

**укрытый** *adj.* screened; covered, housed, sheltered

**уксуснокислый** *adj.* acetic acid; acetate (of)

**укулеле** *n. indecl.* ukulele

**улавливание** *n.* catching, collecting, trapping, interception, capture; **у. ионов** ion-trapping effect

**улавливатель** *m.* catcher, catch, interceptor; trap; detector, locator

**улавливать, уловить** *v.* to intercept, to pick up, to collect, to catch, to capture; to discern, to detect, to discover; to locate

**улавливающий** *adj.* catching, intercepting; pickup, catcher

**Уленбек** Uhlenbeck

**улитка** *f.* cochlea; helix

**улиточный** *adj.* helical

**уличный** *adj.* street; outdoor-type

**уловитель** *see* **улавливатель**

**уловить** *pf. of* **улавливать**

**уловление** *see* **улавливание**

**уловленный** *adj.* trapped, caught, collected; detected, located

**уложеный** *adj.* packed; laid

**уложить** *pf. of* **укладывать**

**улучшение** *n.* refinement, improvement, amelioration; **у. структуры металла ультразвуком** ultrasonic grain refinement; **у. сходимости** acceleration by convergence; **у. схо-**

димости путём возведения случайных величин в степень acceleration by powering

**улучшенный** *adj.* refined, improved

**Ульбрихт** Ulbricht

**ульмал** *m.* ulmal

**ульминий** *m.* ulminium

**ультор** *m.* ultor

**ультра-** *prefix* ultra-

**ультраакустика** *f.* ultrasonics, supersonics, magniacoustics

**ультраакустический** *adj.* hypersonic

**ультрааудион** *m.* ultra-audion

**ультравысокий** *adj.* ultrahigh, very high

**ультрагармоники** *pl.* ultraharmonics

**ультрадин** *m.* ultradyne

**ультрадинамический** *adj.* ultradynamic

**ультрадинный, ультрадиновый** *adj.* ultradyne

**ультразвук** *m.* ultra-audible sound, ultrasound, supersound; supersonics, ultrasonics

**ультразвуковой** *adj.* ultrasonic, hypersonic, supersonic, supertonic

**ультразвукоскопия** *f.* ultrasonoscopy

**ультракороткий** *adj.* ultrashort, very short

**ультракоротковолновый** *adj.* ultra-short-wave

**ультралинейный** *adj.* ultra-linear

**ультралюминесценция** *f.* ultraluminescence

**ультрамикровесы** *pl.* ultra-micro-balance

**ультрамикрометр** *m.* ultramicrometer

**ультрамикрон** *m.* ultramicron

**ультрамикроскоп** *m.* ultramicroscope

**ультрамикроскопический** *adj.* ultra-microscopic

**ультрамикроскопия** *f.* ultramicroscopy

**ультрамикроэлектрод** *m.* ultramicro-electrode

**ультраполярный** *adj.* ultrapolar

**ультраскоростной** *adj.* ultrafast

**ультраузкий** *adj.* ultra-narrow, narrow-narrow

**ультраустойчивость** *f.* ultrastability

**ультраустойчивый** *adj.* ultrastable

**«ультрафакс»** *m.* ultrafax

**ультрафильтр** *m.* ultrafilter

**ультрафиолетовый** *adj.* ultraviolet

**ультрафотометр** *m.* ultraphotometer

**ультрацентрифуга** *f.* ultracentrifuge

**ультрон** *m.* ultron

**УМ** *abbr.* (усилитель мощности) power amplifier

**умбра** *f.* umbra

**уменьшаемое** *n. adj. decl.* minuend (*math.*)

**уменьшаемый** *adj.* reducible

**уменьшать, уменьшить** *v.* to decrease, to diminish, to lessen; to minimize, to reduce, to abate, to ease; to narrow; **у. масштаб** to scale down

**уменьшаться, уменьшиться** *v.* to descend, to fall, to drop, to decline, to lessen, to diminish

**уменьшающий** *adj.* reducing, diminishing, decreasing, falling

**уменьшающийся** *adj.* converging

**уменьшение** *n.* diminution, decrease, depletion, lessening, reduction, lowering; bucking-out, abatment, fall-off, drop; decay, decaying; attenuation; compression; **у. глубины модуляции** reduction of detection efficiency; **у. дальности** range attenuation; **у. девиации фазы** phase compression; **у. динамических оттенков** volume contraction; **у. дроссельного действия** unchoking effect; **у. изображения** demagnification; **у. качества передачи** transmission impairment; **у. оптической системы** similarity scale of optical images; **пропорциональное у.** scaling down; **у. размеров** miniaturization; **у. связи** decoupling; **у. чувствительности** desensitization

**уменьшённый** *adj.* diminished, decreased, reduced; undersized

**уменьшитель** *m.* reductor

**уменьшительный** *adj.* diminishing; diminutive

**уменьшить(ся)** *pf. of* уменьшать(ся)

**умеренно** *adv.* moderately

**умеренность** *f.* moderation

**умеренный** *adj.* moderate, medium, mild, temperate

**умерять, умерить** *v.* to damp, to modify, to moderate

**умножать, умножить** *v.* to multiply; to increase, to augment, to magnify, to amplify, to enlarge

**умножающий** *adj.* multiplying

**умножение** *n.* multiplication, multiplying; increase, augmentation, amplification, magnification; shifting and adding; **у. в обратном порядке** pre-multiplication; **у. в обычном по-**

рядке post-multiplication; **у. на обратную величину** reciprocal multiplication; **у. на постоянную величину** multiplication by constant; **у. частоты повторения импульсов** pulse-rate multiplication

**умножитель** *m.* multiplier, variable multiplier; multiplexer; **время-импульсный у.** mark-space multiplier, time division multiplier; **вторично-электронный у.** electron-multiplier tube; **двукратный у. частоты** push-pull multiplier; **у. добротности на полупроводниковых триодах** transistor Q-multiplier; **моделирующий у.** analog multiplier; **у. на основе магнитосопротивления** magnetoresistor multiplier; **у. на четыре** quadrupler; **пятикратный у.** quintupler; **пятикратный у. частоты по двухтактной схеме** push-pull quintupler; **фотоэлектронный у.** photomultiplier tube, electron-multiplier tube; **электронно-лучевой у. со скрещёнными полями** crossed-fields electron-beam multiplier; **электронный у.** secondary-electron amplifier; multiplier tube

**умножительный** *adj.* multiplier

**умножить** *pf. of* **умножать, множить**

**Умов** Umov

**умозаключение** *n.* inference, conclusion; **опосредствованное у.** mediate inference

**умственный** *adj.* mental, intellectual

**умформер** *m.* motor-generator set, converter, umformer; transformer; dynamotor

**умышленно** *adv.* intentionally, on purpose, purposely, deliberately

**умышленный** *adj.* intentional, designed, deliberate

**унарный** *adj.* unary

**унаследованный** *adj.* inherited

**унивариантный** *adj.* univariant

**универсально-применимый** *adj.* all-purpose

**универсальность** *f.* flexibility, versatility

**универсальный** *adj.* universal, general purpose, versatile, multi-purpose; ac/dc, all-mains

**унидинный** *adj.* unidyne

**унимодальный** *adj.* unimodal

**униполярность** *f.* unipolarity

**униполярный** *adj.* unipolar, single-pole, homopolar; acyclic, one-way, unidirectional

**унисон** *m.* unison

**унисонный** *adj.* unison

**унитарный** *adj.* unitary

**унификация** *f.* unitization, unification; normalization

**унифиляр** *m.* declinometer

**унифилярный** *adj.* unifilar

**унифицированный** *adj.* unitized, integrated; modular

**унифицировать** *v.* to unify, to unitize

**уничтожать, уничтожить** *v.* to eliminate, to suppress, to crush, to destroy; to extinguish, to quench; to erase, to obliterate, to demolish; to cancel out

**уничтожаться, уничтожиться** *v.* to cancel out; **взаимно у.** to cancel out

**уничтожающий** *adj.* destructive

**уничтожение** *n.* elimination, destruction, suppression, annihilation, neutralization, cancellation; obliteration, erasure; **встречное у. (волн)** multipath cancellation (of waves); **у. грибка** fungicide; **у. двузначности пеленга** ambiguity resolution; **у. импульсов импульсами** pulse-to-pulse cancellation; **у. незначащих нулей** zero suppression

**уничтоженный** *adj.* nullified, cancelled; destroyed, exterminated

**уничтожитель** *m.* annihilator

**уничтожительный** *adj.* destructive

**уничтожить(ся)** *pf. of* **уничтожать-(ся)**

**унос** *m.* carrying away, entrainment, taking; **у. электролита** drag-out

**УНС** *abbr.* (**указатель наивыгоднейшей скорости**) optimum speed indicator

**унтертон** *m.* subharmonic tone

**унция** *f.* ounce

**УНЧ** *abbr.* (**усилитель низкой частоты**) low-frequency amplifier

**Уодсворт** Wadsworth

**Уокер** Walker

**Уорд** Ward

**Уоткинс** Watkins

**уп** *abbr.* (**управление**) control, steering; management

**УП** *abbr.* 1 (**усилительный пункт**) repeater station; 2 (**указатель поворота**) turn indicator

**упаковать** *pf. of* **упаковывать**

**упаковка** *f.* packaging, packing; pack, package; **у. питания** power pack; **у.**

по типу расположения волокон в древесине cordwood packaging

**упаковочный** *adj.* packing

**упаковывание** *see* упаковка

**упаковывать, упаковать** *v.* to pack (up), to wrap up

**уплотнение** *n.* seal, packing, gasket, gland; sealing, gasketing, tightening; condensation, densification, sharing; condensing, compressing; packaging; multiplex, multiplexing; contraction, squeezing, shrinkage; **быстро вибрирующее у.** oscillating bearing seal; **вращающееся у.** rotating bearing seal; **временное у.** time-division multiplexing; **у., не пропускающее жидкость** liquid tightness; **у. канала временным разделением** time division channeling; **частотное у.** channeling, frequency-division multiplex

**уплотнённость** *f.* multiplexing

**уплотнённый** *adj.* condensed, compressed, thickened; multiplexed; shared; sealed, packed; consolidated

**уплотнитель** *m.* sealant; thickener; packer

**уплотнительный** *adj.* sealing, packing; thickening

**уплотнять, уплотнить** *v.* to seal, to make tight, to tighten, to pack; to consolidate; to squeeze, to compress, to contract; to condense, to thicken

**уплотняющий** *adj.* packing, sealing; blanketing; thickening; impermeable

**уплощение** *n.* flattening

**уплощенный** *adj.* depressed; compressed

**упор** *m.* arrester, stop, rest, latch, detent, dog, guard; rest, support, stay, prop, brace; **автоматический у.** self-adjusting stop; **у. диска** finger stop; **у. заводного диска** finger stop of a dial; **мерный у.** measuring gage; **роликовый у.** roller tappet

**упорка** *see* упор

**упорный** *adj.* stop, thrust

**упорядочение** *n.* ordering, regulation

**упорядоченный** *adj.* ordered, ranked

**упорядочивание** *n.* adjustment

**упорядочивать, упорядочить** *v.* to order, to put in order; to regulate

**употребить** *pf. of* употреблять

**употребление** *n.* usage, use, application, employment

**употреблять, употребить** *v.* to use, to apply, to employ, to make use (of)

**управление** *n.* control; direction, directing; drive, driving, steering, handling, operation, guidance; triggering; management control; government; monitoring; supervision; dispatching; **с дистанционным управлением** remotely controlled; **с программным управлением от перфокарт** card-programmed; **с ручным управлением** manual; **с сеточным управлением** grid-controlled; **с управлением от кристаллических триодов** transistor-driven; **с электронным управлением** electronically steerable; **у. без обратной связи** open-loop control; **у. в истинном масштабе времени** on-line control; **у. выдержкой времени** timing control; **двойное у. для линеаризации характеристики лампы** double control for correction of tube characteristic; **двухступенчатое верньерное у.** dual-ratio control; **дискретное у.** digital process control; **дистанционное рулевое у.** remote-control steering gear; **у. защитным огнём** anti-aircraft fire control; **у. изменением напряжения на якоре** armature control; **коммутационное у.** plugged control; **у. коррекцией низких частот** bass boosting control; **у. крутизной** slope control; **многократное у.** jogging, inching; **у. на конечном участке траектории** terminal guidance; **у. на среднем участке траектории полёта** mid-course guidance; **наземное у. перехватом** ground control of interception, ground-controlled interception; **наземное у. самолётами на трассе** airways ground control; **у. нецентрализованными стрелками и сигналами** outlying switch and signal control; **у. огнём** fire control; **у. от штурвала** steering-column control; **у. орудиями** ordnance control; **у. от поплавка** float(ing) control; **у. по изменению нагрузки** load-responsive control; **у. по разомкнутому контуру** open-loop control; **у. по способу полярных координат** "twist-and-steer" control; **у. по экранирующей сетке** screen control; **у. положением** positioning; **у. положением (ракеты)** altitudinal control; **у. последовательностью операций** operation sequence control; **у. прес-**

сом при помощи перфоленты punch-tape press control; **программное у.** memory stored control; **продольное у.** longitudinal radio guidance; **у. пропусканием** (*or* **отпиранием**) gating; **радиолокационное у.** зенитным огнём с земли ground anti-aircraft control; **у. с зависимой (от тока) выдержкой времени** time-current control; **у. с независимой выдержкой времени** time-element control; **у. с обратной связью** feedback control; **у. с помощью силового привода** power operation; **у. свечением экрана** screen control; **у. сеточным смещением** grid current control; **смешанное у.** electrostatic-magnetic control; **собственное у.** automatic control; **тиратронное у. электродвигателем** thyratron motor control, thymotrol system; **у. у штурвала** steering-column control; **у. углом места** tilt angle control, drift angle control; **у. углом поиска** search angle control; **фотоэлементное у.** phototube control; **центральное у. стрельбой** fire director, fire directing; **центральное техническое у. связи** general office of communication; **у. циклом** loop control; **у. частотой повторения** (*or* **следования импульсов**) repetition-rate control; **шаговое у. с одиночным набором** single-step control; **электронное у. сигналом** electronic gate

**управляемость** *f.* guidance, controllability, control, steering maneuverability

**управляемый** *adj.* controlled, controllable, guided, steerable, operable; operated, slaved, driven; keyed; **у. импульсами** gated; **у. кварцем** crystal-checked; **у. от серводвигателя** pilot-operated; **у. при помощи радиолокационной станции** radar-controlled

**управлять, управить** *v.* (followed by instr. case) to monitor, to control; to manage, to run, to operate, to drive, to regulate; to route (calls); to manipulate, to key; to govern; to guide; **у. пропусканием** to gate; **у. установлением соединения** to advance the call, to route

**управляющий** *adj.* control, controlling, keying, master, operating, steering;

setting; modulating; administrator; correcting

**управщит** *m.* control board

**упразднение** *n.* dismantling, elimination; **у. линии** dismantling the line

**упразднять, упразднить** *v.* to dismantle, to eliminate, to abolish

**упредительный** *adj.* predicted; preventive; prepossessing

**упреждать, упредить** *v.* to lead, to advance; to get out of range; to anticipate, to prevent, to forestall; to precede

**упреждающий** *adj.* anticipatory; predicting

**упреждение** *n.* rate action, lead, advance; derivative action; prediction; anticipation, prevention; **у. прицела** sight offset

**упреждённый** *adj.* predicted, anticipated

**упростить** *pf. of* **упрощать**

**упрощаемый** *adj.* reducible

**упрощать, упростить** *v.* to reduce; to simplify

**упрощение** *n.* reduction; simplification

**упрощённый** *adj.* reduced; simplified, short-cut, abridged

**упругий** *adj.* elastic, resilient, flexible, pliable, springy

**упруго** *adv.* elastically, resiliently

**упругооптический** *adj.* elastooptic

**упругость** *f.* resilience, resiliency, elasticity, springiness, flexibility; tension; extensibility; **акустическая у.** acoustic stiffness

**УПТ** *abbr.* (**усилитель постоянного тока**) direct current amplifier

**УПЧ** *abbr.* (**усилитель промежуточной частоты**) intermediate-frequency amplifier

**упятерённый** *adj.* quintuple

**упятеритель** *m.* quintupler

**УР** *abbr.* (**управляемая ракета**) guided rocket

**уравнение** *n.* equation; equalization, balance, balancing, compensation; leveling; **уравнения преобразования переменных** transformation equations; **уравнения размерности** quantitative equations; **уравнения связи** closing equations (*math.*); **уравнения четырёхполюсника** equations of a network; **автомодельное у.** self-simulating equation; **у. в конечных разностях** difference equation; **у.**

возмущённого движения perturbation equation; **у. возникновения (траектории)** birth equation; **у. выхода** gain equation; **у. дальности действия радиолокационной станции** radar equation; **дифференциальное у. в частных производных** partial differential equation; **у. для распределения в прошлом** retrospective equation; **у. для точки разветвления** nodal equation; **у. исчезновения** death equation; **у. подвижности** Einstein relation; **у. полупроводникового выпрямителя** current equation for semi-conductor rectifier; **у. потока** beam equation; **у. радиолокации** distance equation, range equation, radar equation; **у. радиолокации с учётом влияния земли** surface-propagation radar equation; **у. распределения тока Белова** Below's law of current distribution; **у. регулируемого объекта** plant equation; **у. с буквенными коэффициентами** literal equation; **самосопряженное дифференциальное у.** linear differential equation identical to the adjoining equation; **у. связи между масштабными множителями** closing equation; **у. следящей системы** servo performance equation; **у. согласования** joining equation; **у., содержащее избыточные корни** redundant equation; **у. сопровождения цели** tracking equation; **у. узла** nodal equation, node equation; **у. шумопеленгатора при приёме на слух** direct-listening equation

**уравнённый** *adj.* equated; leveled
**уравнивание** *n.* equalizing, equalization, balance, balancing, compensation; smoothing, filtering; leveling, grading; matching
**уравниватель** *m.* equalizer, balancer, balance gear; regulator; leveler, grader; **у. затухания** simulative network
**уравнивать, уравнять** *v.* to smooth, to level, to even; to adjust, to equal, to equalize, to balance, to compensate, to equilibrate, to equate; to match; to steady
**уравнивающий** *adj.* balancing, adjusting, compensating; leveling
**уравнитель** *m.* equalizer, leveler,

balancer; compensator; **автотрансформаторный у.** compensator balancer; **у. в трёхпроводной системе** three-wire balancer; **у. затухания** simulative network, attenuation compensator; **магнитный у.** magnetic compensator; **у. переменного тока** a-c balancer; **у. тональных частот** tone equalizer
**уравнительный** *adj.* equalizing, balancing, compensating, balancer; leveling, grading; regulating
**уравновесить** *pf. of* **уравновешивать**
**уравновешение** *n.* balancing, equilibration
**уравновешенность** *f.* balanced state, balance, equilibrium, steadiness
**уравновешенный** *adj.* balanced, compensated, counterbalanced, equilibrated, counterpoised, steady, level
**уравновешиваемый** *adj.* balanced
**уравновешивание** *n.* balancing, equilibration, counterpoising, equalization, compensation, neutralizing; **у. пружиной** spring control
**уравновешивать, уравновесить** *v.* to balance, to counterpoise, to counterbalance, to level, to compensate, to equate, to match, to equilibrate, to equalize, to neutralize
**уравновешивающий** *adj.* balancing, equalizing
**уравнять** *pf. of* **уравнивать**
**уралит** *m.* uralite; "Uralit" (wood preservative)
**уран** *m.* uranium, U
**урановый** *adj.* uranium, uranic
**УРВ** *abbr.* (**усилитель-распределитель видеосигнала**) video distribution amplifier
**уреометр** *m.* ureometer, ureameter
**УРИ** *abbr.* (**усилитель-распределитель импульсов**) pulse distribution amplifier
**ур-ние** *abbr.* (**уравнение**) equation
**УРО** *abbr.* (**управляемое реактивное оружие**) guided rocket weapon
**уровень** *m.* level, plan, surface; standard; margin; level gage; base; band; volume; **на одном уровне** even; **на половинном уровне** at half response; **акцепторный у.** shallow trapping level; **у. в определённой полосе частот** band level; **у. в узкой полосе частот** narrow-band level;

верхний у. relatively positive state; входной у. carrier input (of the carrier frequency); высокий у. на входе high input; высокий у. на выходе high output; у. гасящих импульсов pedestal level; у. гашения black-out level, blanketing level; у. грунтовых вод ground water level; допустимый у. ухудшения acceptable malfunction level; у. запирания keying level; заполненный у. filled band; запрещённый у. forbidden band; у. звукового давления в определённой полосе частот band pressure level; контрольный у. reference level; критический у. степени надёжности fiduciary level; маркшейдерский у. leveling instrument; у. несущей, соответствующий (максимальному) белому carrier-reference white level, picture white; у. несущей, соответствующий чёрному carrier-reference black level; нижний у. relatively negative state; низкий у. на входе low input; низкий у. на выходе low output; у. «нумерации» quantization level; общий у. wide-band level; у. ограничения (ограничителя) cut-off level of clipper, clip level; один у. испытания single bias level (math.); опорный у. expected level; у. отпирания keying level; у. ощущения level above threshold, sensation level; у. разговорных токов speech level; у. разделения pedestal level, blanking level; разрешённый у. empty band; у. спектральной плотности spectrum pressure level; у. срабатывания operating level, keying level; у. срабатывания ответчика beacon triggering level; стандартный нулевой у. громкости reference volume; у. угла места elevation level; у. условной отдачи available power efficiency level; черней чёрного у. blacker-than-black level, infra-black level (tv); у. чёрного в изображении picture black; у. чёрного сигнала black level (tv); у. шума в установившемся режиме steady-state noise level; эквивалентный спектральный у. гидроакустического преобразователя transducer equivalent spectrum level; эталонный у. index level (of sound)

уровнеграф m. liquid level indicator, liquid level recorder
уровнемер m. transmission level meter, transmission measuring set; level indicator; liquid level indicator, tank gage; дистанционный у. remote-reading tank gage; регистрирующий у. recording liquid level gage; самопишущий у. recording transmission measuring set; электрический у. electrical-contact liquid-level indicator
урометр m. urinometer, urometer
урон m. loss, damage, harm
УРС abbr. (управляемый реактивный снаряд) guided missile
УРТС abbr. (учрежденская ручная телефонная станция) private manual exchange, PBX, private branch exchange
урчанье n. rumble
УС abbr. 1 (угол сноса) drift angle; 2 (узел связи) communication center; 3 (указатель скорости) speed indicator
усадка f. shrinkage, shrinking, contraction; setting; loss, damage
усадочный adj. shrinkage, contraction
усекать, усечь v. to truncate, to cut off
усечение n. truncation, cutting off; у. по контуру равной интенсивности equi-intensity contour cut
усечённый adj. truncated, cut-off; frustrum; sectional
усечь pf. of усекать
усик m. catwhisker, whisker
усиление n. amplification, intensification, multiplication, magnification; gain, increase, rise, boost, enhancement; repeating (tphny.); compensation; strengthening, reinforcement; у. в глубину longitudinal magnification (tv); вносимое у. insertion gain; допустимое по свисту (or паразитной генерации) у. singing margin; у. за счёт обратной связи regenerative amplification; у. зависимости от реверберации reverberation-controlled gain; у. на сопротивлениях amplification by means of R-C coupling; у. (напряжения) при преобразовании conversion voltage gain; у. низких тонов bass compensation; у., определяемое методом замещения insert voltage gain; относительное у. зеркального канала image ratio; переменное у.

variable mu; **у. по мощности при согласовании на выходе** matched-output power gain; **у. по низкой частоте** audio gain, a.f. amplification; **у. по петле обратной связи** gain around a feedback; **у. по рефлексной схеме** dual amplification, reflex amplification; **у. по току в режиме короткого замыкания** short-circuit current gain; **у. по току в схеме с общей базой** collector-to-emitter current gain; **у. по току в схеме с общим эмиттером, у. по току участка коллектор-основание** collector-to-base current gain; **под-модуляторное у.** subcontrol; **пред-модуляторное у.** submodulator amplification, subcontrol; **у. при наличии оператора** human gain; **у. при холостом ходе** no-load gain; **у. промежуточной частоты** rf amplification, supersonic amplification; **прямое у.** insertion gain; **равномерное у.** flat gain; **регенеративное у.** damping reduction; regenerative amplification; **реостатное у.** see **у. на сопротивлениях; у. с обратной связью** regenerative amplification; **у. с помощью пространственно-моделированного электронного пучка** scalloped-beam amplification; **у. с преобразованием частоты** conversion gain; **у. усилителя** repeater gain

**усиленный** *adj.* strengthened, reinforced, backed up, intensified; increased; amplified, magnified; promoted

**усиливать, усилить** *v.* to magnify, to amplify; to intensify, to increase, to boost; to strengthen, to reinforce, to fortify; to strain, to aggravate

**усиливаться, усилиться** *v.* to fade in; to become stronger, to strengthen; to increase, to intensify

**усиливающий** *adj.* amplifying, magnifying; intensifying; strengthening

**усилие** *n.* force, energy, power, strength; pull; stress, strain, effort, exertion; **касательное у.** shearing strain, tangential force; **осевое у.** longitudinal force; **передаваемое у.** motive power, driving force; **у. при размыкании контактов** throw-off effect; **у. привода** motive force; **разрывное у.** breaking load; **скручивающее у.** torque; **толкающее у.** thrust;

**тяговое у. трения** friction draw-bar

**усилитель** *m.* amplifier; intensifier, enhancer, magnifier, booster; repeater (*tphny.*); accelerator; **амплитронный у. на микроволнах** cross-field microwave amplifier; **у. без дрейфа нуля** drift-free amplifier; **блочный у.** plug-in amplifier; **боковой у.** marginal amplifier; **у. в передающей камере** camera amplifier; **у. в схеме само-писца** data-recording amplifier; **у. в цепи регулировки яркости** intensity-control amplifier; **у. видеоконтрольного устройства** monitor amplifier; **у. воспроизведения (звука)** playback amplifier; **вч у. с несущей** carrier repeater (*tphny.*); **высоко-качественный предварительный у.** preamplifier with presence; **выход-ной у. напряжения** (*or* **тока**) **раз-вёртки** sweep-output amplifier; **вы-ходной у. развёртки** time-base power amplifier; **выходной у. сигналов частоты кадров** vertical output amplifier; **выходной у. строчной частоты** horizontal output amplifier; **гармонический у.** selective amplifier; **у. гасящих импульсов кадровой частоты** frame-blanking amplifier; **у. датчика деформаций** strain gage amplifier; **дающий у.** data repeater; **двухконтурный у.** double-tuned amplifier; **двухконтурный квантово-механический у.** two-cavity maser amplifier; **у. двухстороннего действия** two-wire repeater, two-wire system repeater; **двухтактный у.** balanced valve amplifier; **двухтакт-ный у. с ёмкостным выходом** single-ended push-pull amplifier; **детектор-ный у.** amplifying detector; **у. для выделения опорных импульсов** pedestal-processing amplifier; **у. для параллельного включения** bridging amplifier; **у. для радиовещания по проводам** line radio amplifier, wired-wireless amplifier; **у. для регулиро-вания величины нелинейности** gamma control amplifier (*tv*); **у. для трансляционных линий** program-line repeater; **дроссельный у.** choke-coupled amplifier, impedance-coupled amplifier, inductance-coupled amplifier; **задающий у.** distribution amplifier; **у. звуковой головки** head amplifier; **звукофикационный у.** public-address amplifier; **измерительный у.** phantom repeater; meter matcher; **у. измеряемых величин** measuring amplifier; **импульсный у.** pip amplifier; **инвертирующий у.** complementing amplifier, polarity-inverting amplifier, sign-changing amplifier, sign-reversing amplifier; **искажающий у.** overdriven amplifier; **у. исправляющий импульсы** pulse-regenerative amplifier; **у. кадровой развёртки** vertical-sweep amplifier; **у. кадровых сигналов** frame amplifier (*tv*); **у. компенсации послесвечения люминофора** phosphor amplifier; **у.-компрессор** compandor; **у. конденсаторного микрофона** capacitor-microphone preamplifier; **контрольный у.** monitoring amplifier (*tphny*.); **корректирующий у. с чрезмерным подъёмом усиления на высоких частотах** overpeaked compensating amplifier; **у. коррекции дрейфа нуля** drift-correcting amplifier; **кристаллический у. с сильным полем** junction fieldistor; **магнитный у. исполнительного привода** magnetic servo amplifier; **магнитный у. с регулируемым переходным процессом** transient-controlled magnetic amplifier; **магнитный у. с самовозбуждением** autotransductor; **манипулированный у. «вспышки»** keyed burst-amplifier stage; **манипуляционный у.** pulse power-amplifier, pulse-modulated amplifier; probe amplifier; **у. маркерных сигналов** timing-wave amplifier; **микшерный у., у. микширования наплывом** lap dissolve amplifier; **у. мощности на лампе с бегущей волной** traveling-wave power amplifier; **мощный у. низкой частоты** audio-power amplifier; **мощный у. отклонения** (*or* **развёртки**) **по оси времени** time-base power amplifier; **мощный у. с реостатноёмкостной связью** R-C-coupled power amplifier; **у. на взаимно расстроенных контурах** stagger amplifier; **у. на выходе фотоэлемента** photoelectric cell amplifier; **у. на дросселях** impedance-coupled amplifier, inductor-coupled amplifier; **у. на кристаллических триодах** transistor amplifier; **у. на несколько выходных каналов** trap-valve ampli-

fier; **у. на одно направление** one-way repeater; **у. на печатной схеме** etched amplifier; **у. на полупроводниковых приборах** crystal amplifier; **у. на расстроенных контурах** stagger amplifier; **у. на трансформаторах** transformer-coupled amplifier; **у. напряжения динамической сходимости** (*or* **динамического фокусирования**) dynamic-convergence amplifier; **у. напряжения** (*or* **тока**) **развёртки** sweep amplifier; **у. напряжения** (*or* **тока**) **телевизионной развёртки** television scanning amplifier; **настроенный у.** resonance amplifier; **неинвертирующий у.** non-complementing amplifier; **некорректированный у.** uncompensated amplifier; **нерегулируемый у.** fixed repeater; **у., ограничивающий полосу пропускаемых частот** bandwidth-limiting amplifier; **у.-ограничитель** clipping amplifier; **однотактный у.** single-ended amplifier; **оконечный у. абонентского аппарата** subscriber repeater (*tphny*.); **оконечный линейный у.** terminal repeater, final amplifier; **параметрический у. с повышающим преобразователем** up-converter parametric amplifier; **параметрический у. с продольным пучком** longitudinal-beam amplifier; **перегруженный у.** overdriven amplifier; **переходный у.** bridging amplifier, coupling amplifier; **«печатный» у.** printed-circuit amplifier; **у. пилообразного тока магнитной развёртки** magnetic-sweep amplifier; **плоскорасстроенный у.** flat-staggered amplifier; **у., повышающий крутизну фронтов видеосигнала** sharpener amplifier; **полосный у.** band-pass amplifier; **полупроводниковый параметрический у.** variable-capacitance mavar; **предварительный у. высокой частоты** booster (pre)amplifier; **предварительный у. исполнительного привода** servo pre-amplifier; **у., предотвращающий перегрузку** anti-overloading amplifier; **у. промежуточной частоты** r-f amplifier; **промежуточный у. (линии связи)** line repeater, through repeater; channel amplifier; **промчастотный у.** i-f amplifier; **«прямоугольный» у.** squaring amplifier; **у.**

пульсации ripple amplifier; **у. пускового импульса** trigger amplifier; **у. развёртки** deflection amplifier; **развязывающий у.** isolation amplifier, buffer amplifier; **разделительный у.** bridging amplifier; **разделяющий у.** isolation amplifier; **разрядный у. мощности** digit driver; **у. рассогласования** error amplifier; **у. регулирования величины нелинейности** gamma-control amplifier; **регулировочный у.** amplifier for control purposes; amplifier with variable frequency response; amplifier with automatic gain control, amplifier with variable gain; **резистивный у. (с ёмкостной связью)** resistance-coupled amplifier; **резонансный у.** tuned amplifier; **реостатный у.** resistance amplifier; **решающий у. с параллельной обратной связью** parallel feedback operational amplifier; **решающий у. с положительной обратной связью** regenerative operational amplifier; **у. с автоматической стабилизацией нулевого уровня** self balancing amplifier; **у. с большим смещением** overbiased amplifier; **у. с дроссельно-ёмкостной связью** inductor-capacitor-coupled amplifier; **у. с жатием диапазона громкости** compandor; **у. с катодным выходом** cathode follower, bootstrap amplifier; **у. с комплексной связью** impedance-coupled amplifier; **у. с коррекцией дрейфа нуля** drift-corrected amplifier; **у. с малым временем установления** fast amplifier; **у. с накалённой проволокой** hot-wire magnifier (*tgphy.*); **у. с общим катодом и заземлённым анодом** bootstrap circuit; **у. с однотактным входом и с двухтактным выходом** paraphrase amplifier; **у. с плоской фазовой характеристикой** zero-phase shift amplifier; **у. с подъёмом усиления на высоких частотах** overpeaked compensating amplifier; **у. с потенциальным входом** level amplifier; **у. с регулируемой реактивностью** variable-reactance amplifier; **у. с сжатием динамического диапазона сигнала** compressor amplifier; **у. с управлением по пентодной сетке** suppressor-modulated amplifier; **у. с фиксацией уровня сигнала** clamper

amplifier; **у., связанный с колебательным контуром** tuned amplifier; **у. сигналов вертикального отклонения** vertical amplifier (*tv*); **у. сигналов индикации положения луча** index signal amplifier; **у. сигналов кадров** vertical amplifier; **у. сигналов отметки** notch amplifier; **у. сигналов сведения по строкам** (*or* по горизонтали) line convergence amplifier; **у. силового привода** torque amplifier; **симметричный логарифмический у.** bipolar logarithmic amplifier; **у. системы звукоусиления** (*or* озвучения) public-address amplifier; **скорректированный предварительный у.** self-equalizing amplifier; **у. следящей системы** servo amplifier; **следящий у.** bore-sight servo amplifier; **смесительный у.** adder amplifier; **у. смещения составляющих** mixing amplifier; **у. со взаимно расстроенными контурами** stagger-tuned amplifier; **у. со сжатием диапазона громкости** (volume) compandor, compressor amplifier; **у. со схемой объединения на входе** mixing amplifier; **у. со схемой совпадения** gated amplifier; **у. схемы вычитания** differential amplifier; **у. тока возбуждения** field amplifier; **у. тока отклонения магнитной развёртки** magnetic-sweep amplifier; **у. тока подмагничивания** field amplifier; **у. токов несущей частоты** carrier repeater, carrier amplifier; **трансляционный у.** repeater, intermediate repeater; **трансформаторный у.** transformer-coupled amplifier; **трёхсеточный у. варимю** triple-grid super-control amplifier; **у. узких селекторных импульсов** narrow-gate amplifier; **у. умственных способностей** amplifier of intelligence; **у.-формирователь прямоугольных импульсов** (*or* колебаний) squaring amplifier; **у. хронирующих сигналов** timing-wave amplifier; **у. цепи совмещения** coincidence amplifier; **у. частоты биений** heterodyne amplifier; **у. чересстрочной компенсации** cancellation amplifier; **широкополосный у.** flat-staggered amplifier; **широкополосный многокаскадный у.** stagger-tuned amplifier; **шнуровой у.** cord-circuit repeater; **экономич-**

ный двухтактный у. quiescent push-pull amplifier; электромашинный у. (с поперечным полем) dynamo-electric amplifier, amplidyne, amplidyne generator, rotary amplifier

усилительный *adj.* amplifier, amplifying, magnifier, repeater, amplification, boosting, intensifying

усилить(ся) *pf. of* усиливать(ся)

ускорение *n.* acceleration, speeding up; у. в функции противоэлектродвижущей силы counter electromotive force acceleration; у. за счёт силы тяжести gravitational acceleration; зависящее от тока у. current element acceleration; земное у. acceleration due to gravity; малое у. силы тяжести milligee acceleration; нормальное у. силы тяжести normal gravity; регулируемое по времени у., у. с выдержкой времени timed acceleration; у. с зависимой от тока выдержкой времени time-current acceleration; у. с независимой выдержкой времени time element acceleration, time limit acceleration; у. силы тяжести acceleration due to gravitation, gravitational constant

ускорениемер *m.* accelerometer

ускоренный *adj.* speeded up, fast, accelerated; short-cut

ускоритель *m.* accelerator; smasher; launching vehicle; booster; линеный у. на миллиард вольт billion-volt linear accelerator; последний у. final h-v electrode, ultor

ускорительный *adj.* accelerating; intensifiable

ускорить *pf. of* ускорять

ускоряемый *adj.* accelerative; intensifiable

ускорять, ускорить *v.* to accelerate, to force, to speed, to speed up, to hasten, to quicken; to intensify

ускоряющий *adj.* accelerating

усл. ед. *abbr.* (условная единица) arbitrary unit

условие *n.* condition, requirement, specification(s), stipulation; postulate; if clause; circumstance; environment; линейные условия работы linear conditions; условия совпадения по фазе in-step conditions; у. возникновения генерации start-oscillation condition; «у. половинной громкости» one-half maximum effort; у. получения установившихся колебаний в ламповом генераторе Barkhausen criterion for oscillations; у. срыва колебаний dip condition

условно *adv.* conditionally

условновероятностный *adj.* conditional-probability

условность *f.* conditionality; convention; у. рисунка pictorial liberty

условный *adj.* conditional; relative; conventional; code; arbitrary; nominal; quasi-

усложнение *n.* complication

усложнённый *adj.* complicated

усовершенствование *n.* development, improvement, refinement, perfection; advance, adaptation

усовершенствованный *adj.* improved, perfected, modified, advanced

успешно *adv.* successfully

успешный *adj.* successful, effective

успокаивание *n.* damping; de-excitation; decrement, waning, moderation; quieting

успокаивать, успокоить *v.* to damp; to quiet, to calm, to steady, to soothe

успокаиваться, успокоиться *v.* to calm down, to settle down, to quiet down; to abate, to slacken; to damp

успокаивающий *adj.* damping; quieting; drag

успокаивающийся *adj.* damped

успокоение *n.* amortization, calming, quieting, attenuation, calm; dampening, damping, eddy drag; у. посредством медной оболочки copper (pipe) damping (of a relay)

успокоенный *adj.* damped, dead beat; quieted, quiet

успокоившийся *adj.* abated, moderated

успокоитель *m.* damper; amortisseur, deoscillator, cataract; arrester

успокоить(ся) *pf. of* успокаивать(ся)

усреднение *n.* neutralization; averaging

усреднённый *adj.* averaged, average, mean; neutralized

усреднитель *m.* averager; neutralizer

усреднять, усреднить *v.* to average; to neutralize

усредняющий *adj.* averaging; neutralizing

уставка *f.* setting, placing, putting; у. втяжного сердечника plunger setting (of a relay)

усталостный *adj.* fatigue

**усталость** *f.* fatigue, weariness

**устанавливаемый** *adj.* adjustable, controllable, regulable

**устанавливать, установить** *v.* to mount, to set up, to install, to erect, to fit, to assemble; to make (contact); to regulate, to adjust, to tune; to establish, to determine, to fix, to locate, to define, to ascertain; to state; to set, to place, to put, to arrange; **у. на место** to set, to install, to erect, to position, to build in; **у. на силу тока** to adjust for current; **у. нормально** to set at normal; **у. обратную связь** to provide the feedback, to couple back

**устанавливаться, установиться** *v.* to be settled, to get set, to be fixed, to be determined; to determine

**устанавливающий** *adj.* controlling; adjusting

**устанавливающийся** *adj.* settling

**установившийся** *adj.* steady, steady-state, stable, set, settled, stationary, sustained; balancing

**установить(ся)** *pf. of* **устанавливать (ся)**

**установка** *f.* arrangement, placing, setting; mount, mounting, setting up, assembling, assembly, erecting, erection, installation, installing, fitting; adjustment, adjusting, alignment, regulation; aim, purpose; directions; plant, unit, set, set-up; outfit, equipment, device, apparatus; system; **абонентская у. с двухлинейным коммутатором** subscriber's inter-communication installation; **абонентская у. с добавочными аппаратами** private branch exchange, PBX; **абонентская телефонная у.** telephone connection, subscriber's station, telephone station; **бескоммутаторная телефонная домовая у.** house telephone plant; **бинауральная шумопеленгаторная у.** binaural listening system; **у. в кожухе** enclosing; **ведущая часовая у.** master timekeeper; **групповая у.** party-line station; **у. для внутренней связи** selector installation, intercommunication system; **у. для радиоперекличек** equipment for roundcalls; **у. для управления по радио** radio dispatching system; **добавочная у. для учреждений** preselector attachment

for official teletype connections; **добавочная абонентская у.** extension station; **домовая у.** home wiring; **у. заданного значения** control index setting; **запасная у.** stand-by plant, provisional installation; **запасная генераторная у.** emergency power-system; **заранее сделанная у.** presetting; **зарядная у.** battery charger, battery-charging outfit; **у. зоны регулирования** throttling range adjustment; **у. избирательного вызова** selector installation, inter-communication installation; **у. (изображения) в рамку** framing control *(tv)*; **у. изображения на середину экрана** centering *(tv)*; **у. кадра** framing *(tv)*; **командная у.** signaling device; **командная у. с местной батареей** local-battery order transmission-system; **комплексная контрольно-испытательная у.** integral test system; **у. кренов** pivoting; **у. на общую ось** ganging; **у. на осевых подшипниках** axle mounting, spindle bearing; **у. на расстояние** range adjustment; **у. от руки** manual adjustment; **передающая у. направленного излучения** beam transmitting station; **передвижная у.** mobile unit; **у. по одной прямой** alignment; **у. пожарной сигнализации** fire alarm system; **предварительная у.** preselection; **проблесковая сигнальная у.** blinker light equipment, flashing light equipment; **пусковая у.** launcher; **радиолокационная у.** radio locator; **радиолокационная у. для указания рулёжного поля** airfield surface movement indicator; **у. с ведущим лучом** beam-(rider) type guidance system; **у. с индикатором движущихся целей** MTI system; **у. с последовательным включением** (*or* **расцеплением**) intercommunication system; **у. связи** electric communication equipment, transmitting station; **селекторная у.** selective-ringing system; **сетевая резервная у.** emergency power plant; **у. со скрытой проводкой** underplaster wiring; **у. со штепсельными присоединениями** plug-in units; **собственная генерирующая у.** natural current installation; **согласованная у.** co-alignment; **стрелочная сигнальная у.** semaphore-type traffic signal; **телефонная у. с параллельным включением** PBX system employing parallel connection of the central-office trunk to all extensions; **телефонная учрежденская у.** telephone extension; **у. термопар с параллельным включением** parallel-connected thermocouples; **ультразвуковая у. для очистки** ultrasonic degreaser **учрежденская коммутаторная у.** agency's switchboard; private manual exchange, PBX; **у. фокусного расстояния** focal setting; **фотоэлектрическая у.** light barrier; **централизованная электрочасовая у.** time electrical distribution system; **частная автоматическая у.** private automatic exchange, PAX; dial intercommunication system; **частная коммутаторная у.** private manual exchange, private switchboard; **частная телефонная у. без шнуров** cordless private branch exchange; **у. частного пользования** private exchange, private branch, PBX; **широковещательная у. для вокзалов** railway call system; **у. щёток** adjustment of brushes; **у. электрического нуля в сельсинах** zeroing synchros

**установление** *n.* establishment, institution; determination, determining, fixing; rating; spotting; **у. времени экспозиции** dosage of exposure time; **у. габаритов** sizing; **у. искателя на вызываемый абонент** automatic routing of long-distance calls, automatic routing of toll calls; **у. режима работы лампы** rating of tubes; **у. соединения** setting up a call, completion of a call; **у. точного местонахождения** spotting

**установленный** *adj.* established, fixed, set, constant; mounted; specified, standard; rated; **заранее у.** preset; **недоступно высоко у.** isolated by elevation

**установочный** *adj.* adjustable, regulating, adjusting, tuning, focusing; fixing, setting, locating; installation; feeder

**установщик** *m.* setter, adjuster, fitter, installer, erector

**устареваемость** *f.* obsolescence

**устаревший** *adj.* obsolete

**устарелость** *f.* obsolescence

**устарелый** *adj.* obsolete

**устный** *adj.* oral, verbal

**устойчиво** *adv.* steadily, firmly

**устойчивость** *f.* stability, steadiness, constancy, rigidity; regularity; resistance; **у. к окружающим условиям** environmental tolerance; **у. к переходным явлениям** transient stability; **у. на продольный изгиб** buckling strength; **у. несущей (частоты)** carrier-frequency stability; **у. отметок** positional stability; **у. при продольном изгибе** cross breaking strength, buckling strength; **у. против микрофонного шума** insensitivity to microphonics; **у. против холода** cold constancy; **у. чересстрочной развёртки** interlacing stability (*tv*)

**устойчивый** *adj.* stable, steady, firm, permanent, rigid, solid, strong, resistant; transient-free, settled, quiescent, sustained, non-oscillatory; **у. в одном состоянии** monostable

**устраивать, устроить** *v.* to arrange, to erect, to mount, to make, to construct, to prepare, to install, to set up, to settle, to establish

**устранение** *n.* elimination, removal, clearing, cancellation; correction; suppression, rejection; **у. замыкания на землю** ground-fault neutralizing; **у. короткого замыкания** clearing of a circuit, unshorting; **у. местных помех** clutter rejection; **у. неопределённости при радиопеленговании** sense finding; **у. обратной связи** feedback suppression, regeneration suppression; **у. повторных изображений** echo cancellation; **поляризационное у. помутнения** elimination of polarization-interference; **у. слипания** decoherence

**устранимость** *f.* eliminability

**устранимый** *adj.* eliminable

**устранитель** *m.* suppressor; **у. искажений** correcting device, antidistortion device, equalizer

**устранять, устранить** *v.* to eliminate, to remove; to suppress, to clear; to correct

**устроенный** *adj.* arranged, organized

**устроитель** *m.* organizer

**устроить** *pf. of* **устраивать**

**устройство** *n.* arrangement, installation, equipment; mechanism; system, structure, aid, device, apparatus, machine, gear; oft. replaced in Eng. transl. by name of device, oft. with suffix -er: e.g. **печатающее у.** printer (*lit.* printing device); **запоминающее у.** memory (*lit.* remembering device); outfit, set, unit, instrument, block, element, assembly; station; **внешние устройства ввода** input peripherals; **внешние устройства вывода** output peripherals; **автоматическое у. для периодических испытаний** automatic routine test equipment; **у. автоматической зашифровки речевого сигнала** speech scrambler, speech inverter; **у. автоматической телефонной связи** dial system equipment; **автономное у. (для) обработки данных** off-line processor; **аналоговое счётно-решающее у.** analog computer; **антенное у. из расположенных на одной линии диполей** collinear array, linear array; **антенное у. с излучением вдоль оси** end-fire array; **антенное симметрирующее у.** antenna choke; **асинхронное вычислительное у. с непосредственной связью** asynchronous direct-coupled computer; **ассоциативное запоминающее у.** content-addressable memory; **буферное запоминающее у. на одну строку** row buffer; **быстродействующее запоминающее у. с произвольной выборкой** rapid-random-access memory; **видеоконтрольное у.** monitor, display monitor, viewing monitor, picture monitor; **визуальное у. вывода** display, visual display; **визуальное выходное у. с перекрёстными сетками** cross grid display; **внестанционное коммутационное у.** substation equipment (*tphny.*); **внешнее запоминающее у. (вычислительной машины)** file, computer file, file unit; **внешнее запоминающее у. на дисках с пополняемым (*or* обновляемым) массивом** disk file memory, disk memory, disk storage; **у. воспроизведения звука с оптической фонограммы** optical sound reproducer; **у. вывода на экран буквенно-цифровой информации** alphanumeric display; **у. выдачи (данных)** data logger; **у., вырабатывающее заём** borrow generating device; **выходное**

**у.** terminating device; **вычислительное у. дискретного счёта** digital computer; **вычислительное у. с периодизацией решения** repetitive computer; **вычислительное у. с двумя входами** half subtractor, two-point subtractor; **вычислительное у. с тремя входами** full subtractor; **гироскопическое демпферное у.** gyro antihunt; **главное видеоконтрольное у.** on-the-air monitor, transmission monitor; **глиссадное у.** glide slope facility; **двоичное арифметическое суммирующее у.** binary non-algebraic adder; **дискретное моделирующее у.** sampled-data computer; **дискретное обнаружительное у.** digital detector; **дифференцирующее у.** derivator; **у. для включения контрольного резонатора** echo box actuator; **у. для возбуждения волновода** wave-guide coupler; **у. для выделения звуковой частоты** derived audio equipment; **у. для вычерчивания кривых** plotting board; **у. для групповой двухсторонней связи** conference call installation; **у. для демпфирования** antihunt device; **у. для запоминания переноса** carry storage device; **у. для изменения направления поляризации** polarization rotator; **у. для изменения последовательности команд** sequence alternator; **у. для измерения частоты** frequency-identification unit; **у. для имитаций сигналов от цели** target signal simulator control; **у. для испытания перепайки плавких вставок** resoldering testing device; **у. для контроля размеров изделий** coordinate inspection machine; **у. для коррекции звукозаписи** phonograph equalizer; **у. для линейной связи** linear coupler; **у. для моделирования... ...** simulator; **у. для моделирования запуска снаряда** missile-launching computer; **у. для моделирования структуры поля** analog field mapper; **у. для непрерывного подзаряда аккумуляторов малым током** trickle charger, battery charger; **у. (для) обработки данных в истинном масштабе времени** on-line processor, real-time processor; **у. для определения положения** position-indicating system; **у. для осво-**

бождения соединительных линий releasing selector, arrangement of release; **у. для ответвления** tapper; **у. для отсчёта времени** timing device; **у. для пайки плавких предохранителей обратного тока** soldering device for heat-coil-type fuses; **у. для параллельного включения** device for paralleling of single or multiphase generators; **у. для перехода от коаксиальной линии к волноводу** doorknob transformer; **у. для печати на жёсткую основу** hardcopy printer; **у. для питания системы вращающего анода** motor-control unit; **у. для подсвечивания развёрток** sweep intensifier; **у. для построения характеристик направленности** directivity plotting machine; **у. для разветвления кабельных линий** distribution cable-joint; **у. для регулировки начальной фазы** initiation control device; **у. для сверки двух плёнок** tape comparator; **у. для сжатия диапазона громкости** volume compressor; **у. для сжатия полосы частот при радиолокации** radar-band compressor; **у. для сигнализации о перегорании предохранителей** fuse alarm device; **у. для сложения гармоник** multiharmonigraph; **у. для снятия частотной характеристики** frequency-response display set; **у. для совмещения изображения с картой** chart matching device; **у. для создания активных помех радиолокации** radar jamming device; **у. для телеуправления** telerepeating device; **у. для уменьшения выходного напряжения антенны** strong signal attenuator; **у. для управления циклом** cycler; **у. для фазовой регулировки** power-factor adjustment; **у. для фазовой регулировки счётчика переменного тока** power-factor adjustment for an alternating current meter; **дублирующее печатающее у.** gang printer; **задающее у.** driver (tv); **заземляющее у.** ground plate; **у. записи и выдачи данных о параметрах процесса** data logger; **у. записи на ленту** tape inscriber; **записывающее у. эхолота** sonic recorder; **записывающее счислительное у. для курса** registering

flight log; **запоминающее у.** storage, storage device, storing device, memory, data storage; **з. у. динамического типа с усилением** recirculating amplifier storage device; **з. у. из регистров переадресации** B-store; **з. у. магазинного типа** pushdown storage, stack storage; **з. у. на бездырочных элементах** continuous sheet memory; **з. у. на (магнитных) сердечниках со сложной прошивкой** wired-core (magnetic) storage; **з. у. на электронно-лучевых трубках с отклонением луча** deflection-type cathode-ray tube storage; **з. у. с плоской выборкой слова** two-dimensional word selection memory; **з. у. с поразрядной выборкой** bit-organized memory; **з. у. с прямой выборкой** switch-driven memory, word-organized memory; **з. у., сохраняющее информацию при выключении электропитания** nonvolatile memory; **з. у. типа Z** linear-selection memory; **измерительное у. релейного типа** on-off error detector; **измерительное у. с вращающейся катушкой** moving-coil system, moving-coil drive; **измерительное у. с противодействующей массой** counter-weight-control instrument, gravity-controlled instrument; **импульсное у. с нелинейными катушками** coil pulser; **интерферометрическое у. самонаведения** interferometer homer; **испытательное у. с развёрткой по частоте** wobbulator; **квитирующее у.** revertive communication apparatus; **коаксиальное развязывающее у.** coaxial-line isolator; **кодирующее у. сигнала звукового сопровождения** audio coder; **коммутационное у.** (computer) coupler, switchgear; **коммутирующее у.** decision device; **компандерное у.** audio compressor expander; **компенсирующее у.** bucking-out system, compensator, equalizer, balancer; **комплектное распределительное у.** cubicle switchboard; **у. контроля набора скорости при взлёте** take-off monitor system; **линейное переходное у.** line balance converter; **максимально-вероятностное обнаружительное у.** maximum likelihood detector; **матричное печа-**

тающее у. stylus printer, matrix printer; **у. местного сообщения** short distance trunk equipment; **механическое вычислительное у. гребенчатого типа** lister calculator; **микшерное у.** lap dissolve shutter; **множительное у. непрерывного действия** analog multiplier; **у. моделирования действия помех на радиолокационные системы** radar countermeasure simulator; **моделирующее у.** simulator; analog computer, analog calculator; **моделирующее у. с периодизацией решения** repetitive computer; **моделирующее у. с применением дискретной техники** sampled-data analog computer; **моделирующее у. с проводящей бумагой** resistance paper analogy; **у., моделирующее линию** link simulator; **наблюдающее у.** decision device; **навигационное счётно-решающее у.** bearing-distance computer; **накопительное у. на магнитном барабане** magnetic drum memory; **настроечное у. резонатора сетки** grid-cavity tuner; **настроечное у. с дифференциальным винтом** difference-screw device; **натяжное у. для пружин** spring adjusting tool; **неавтономное печатающее у.** on-line printer; **у. обратного контроля (часов)** testing device for (electric) slave clocks; **у. обхода** alternative trunking, way circuit routing (*tphny*.); **одношлейфное согласующее у.** single-stub tuner; **оконечное видеоконтрольное у.** master monitor, on-the-air monitor, transmission monitor; **оперативное запоминающее у.** immediate-access memory; **опрашивающее у.** scanner; **оптическое вызывное у.** light-signal call system; **основное хронирующее у.** master timer; **открытое распределительное у.** outdoor switchgear; **оттормаживающее у.** brake lifter; **у. падающего клапана** drop indicator equipment; **у. памяти с большим периодом запоминания** long-time memory; **переговорное у.** intercom, duplex installation, talk-back equipment, talk-back circuit; **передающее у. системы цветного телевидения с бегущим лучом** flying-spot color system; **передающее телевизионное**

**у. без накопления заряда** nonstorage pick-up device; **переключающее у.** distribution switchboard; **у. переменной связи** variometer-type coupler, variocoupler; **у. перемножения цифровых и непрерывных величин** digital-analog multiplexer; **пересчётное у.** scaler; **переходное у.** crossbar transition (of a waveguide); adapter; hybrid terminal station (of a four-wire circuit); separating network, dividing network, separating filter; hybrid coil termination, bifurcation; branch connection; **переходное у. УВЧ** bullet transformer; **переходное исказительное у.** distorter; **у. печатания в несколько строк одновременно** surface-at-a-time printer; **печатающее и графопостроительное у.** printer-plotter; **печатающее пересчётное у.** scaler-printer; **плоское синусно-косинусное потенциометрическое решающее у.** flat card resolver; **у., поглощающее толчки при вращении антенны РЛС** radar antenna buffer; **у. подачи магнитной ленты** magnetic-tape file unit; **подающее у.** card-take device; **подсвечивающее у.** sweep magnifier, sweep intensifier; **у. поиска информации** data retrieval system; **последовательное печатающее у.** single-action printer; **построчно печатающее у.** line printer, line-at-a-time printer; **предохранительное автоматически действующее у.** automatic circuit breaker; **приборное штепсельное у.** coupler; **приводное у.** homer, homing device, approach system; **притуляющее у.** despiker; **программное у.** program selector; **противолокационное у.** antiradar device; **пусковое у.** starter, trigger, initiator; **пусковое у. с металлическим реостатом** contactor starter; **радарное у. для измерения высоты угла** radar altimeter, height finder; **«радужное» испытательное у.** chromatic probe (*tv*); **развёртывающее у.** scanner; time base; **развязывающее у.** decoupler, isolator; diplexer (*tv*); **развязывающее у. со смещённым полем** field displacement isolator; **распределительное у.** switching equipment; **распределительное у. закрытого типа** cellular switchboard; **распреде-**

лительное у. энергостанции switch ing station; **расцепляющее у.** tripping device; **регистрирующее у.** logger; **регистровое у.** sending set; **регулировочное у. счётчика при малых нагрузках** low-load (meter) adjustment; **регулировочное у. счётчика при номинальной нагрузке** full-load meter adjustment; **решающее у.** resolver; **ртутное задерживающее у.** mercury tank; **у. с двумя устойчивыми состояниями** bistable device, two-state device; **у. с избранной последовательностью действий** selection sequence calculator; **у. с кольцевым постоянным магнитом** permanent-magnet ring assembly; **у. с коэффициентом нелинейности** unity gamma device; **у. с резонаторной связью** cavity coupling system; **у. с фрикционной передачей** wheel-and-disk device; **у. самонаведения на работающую РЛС** radar seeker; **у. самонаведения снарядов** missile homer; **сверхбыстродействующее запоминающее у.** ultrahigh-access memory; **сверхпроводящее накопительное у. с захватом потока** trapped-flux superconducting memory; **светозащитное у.** framing mask (tv); **у., связывающее клавиатуру с машиной** keyboard coupling device; **сдвигающее (во времени) у.** shifter; **у. сдерживания самохода** anti-creep device; **сигнально-предупредительное у.** alarm device; **симметрирующее у.** bazooka, line-balance converter; balancer; **синхронно-следящее решающее у.** synchronous tracking computer; **у. слабой связи** loose coupler; **следящее у.** tracker, followup system, servosystem; **следящее множительное у.** servomultiplier; **согласующее у.** matching unit, matcher; linkage; **согласующее у. между модулирующим устройством и цифровой вычислительной машиной** analog-digital computer linkage; **согласующее у. УВЧ** bullet transformer; **сопоставляющее (or сравнивающее) у.** collator; **у. стробирования с хранением** holding sampler; **суммирующее у.** adder; **суммирующее-вычитающее у.** adder-subtractor; **счётное у.** calculator; **счётное**

**у. с делением на два** scale-of-two counter; **счётное у. числа знаков** letter-counting device; **счётно-решающее у.** computer, computing machine; **счётно-решающее у. с непрерывной выдачей данных** continuously acting computer; **счётно-решающее у. системы дальности** ranging computer; **счётно-решающее у. шарнирного типа** linkage computer; **у. считывания с магнитного барабана** magnetic-drum playback; **считывающее-перфорирующее у.** reader-punch; **считывающее с бумажной ленты у.** paper-tape reader; **телевизионное передающее у. с линейной характеристикой свет-сигнал** linear image pick-up device; **телеметрическое у. с разделением каналов по фазе** (or амплитуде) ratio-type telemeter; **трёхмерное моделирующее у.** three-dimensional analog computer; **у. умножения с хранимым переносом** stored carry multiplier; **у. управления времени** timing device; **у. управления положением антенны** antenna-positioning system; **ферритовое развязывающее у.** ferrite isolator; **фотоэлектрическое у. для охраны объекта** photoelectric intrusion detector; **храповое у. гнезда** jack closure; **хронирующее у.** timer; **цифровое у. для решения дифференциальных уравнений** digital differential analyzer; **цифровое у. модульной конструкции** modular digital system; **у. череспериодной компенсации** delay line canceller; **четвертьволновое оконечное у.** quarter-wave termination; **читающее с перфоленты у.** perforated tape recorder, punch-card reader; **широкодиапазонное развязывающее у.** broadband isolator; **шумовое сигнализирующее у.** alarm-system actuated by noise; **щёточное у. считывания** brush contact reader; **электродальномерное у.** phase-shift network for DME equipment; **электронное у. для непрерывного измерения и записи величины** electronic pH-reader; **электронное у. для пожарной сигнализации** electronic fire detection device; **электронное моделирующее у.** electronic simulator;

**электронное счётно-решающее у.** electron computer

**уступ** *m.* shoulder; recess; bend; shelf, bank, ledge; **опорный у. для цоколя** base seating shoulder

**уступчивость** *f.* compliance, pliancy, pliability

**уступчивый** *adj.* compliant, pliant, pliable, yielding

**усушка** *f.* shrinkage

**утвердительный** *adj.* affirmative, positive

**утверждение** *n.* affirmation, statement, assertion; strengthening, fixing

**утверждённый** *adj.* affirmed; confirmed; approved

**утекать, утечь** *v.* to leak, to flow away, to run away, to escape; to disperse

**утечка** *f.* leak, leakage, leaking, loss, escape; dissipation, dispersion; creepage; issue; **не дающий утечки** leakproof; **у. в землю** leak to ground, ground leak(age); **у. во входной цепи** input leakage; **у. катод-подогреватель** heater-to-cathode leakage; **у. мишени** target material leakage; **у. на корпус** body leakage, chassis leakage; **у. по высокой частоте** radio-frequency leakage; **регулируемая у.** variable grid leak; **у. с катода на горячую нить** heater-to-cathode leakage; **у. через изоляцию** insulation leakage

**утечь** *pf. of* **утекать**

**утилизация** *f.* utilization

**утилизировать** *v.* to utilize, to recover

**утилизируемый** *adj.* utilizable, available

**утихание** *n.* dying down, abatement, subsidence

**утихать, утихнуть** *v.* to quiet down, to abate, to subside

**«утихающий»** *adj.* muting; abating

**утихнуть** *pf. of* **утихать**

**утихший** *adj.* abated, moderated

**утолщение** *n.* thickening, swelling, expansion, node, bulge, bulging, bulb; lug, camber

**утолщённый** *adj.* fat, thickened; reinforced

**утомительный** *adj.* fatiguing, tiresome

**утомить** *pf. of* **утомлять**

**утомление** *n.* fatigue, weariness; exhaustion

**утомлённый** *adj.* fatigued, tired, weary

**утомляемость** *f.* fatigue; **у. фоточув-**

**ствительной поверхности, у. фотоэлемента** photoelectric fatigue

**утомлять, утомить** *v.* to fatigue, to tire

**утонение** *n.* reduction, tapering, thinning

**утонённый** *adj.* tapered, taper, thinned

**утончающийся** *adj.* tapering

**утончение** *n.* refinement, refining; tapering, thinning

**утончённость** *f.* refinement

**утончённый** *adj.* refined, subtle, fine, specified

**утопление** *n.* embedding, burying; drowning

**утопленный** *adj.* flush(-type), embedded, sunk, recessed, buried, countersunk; built-in

**уточнённый** *adj.* refined, improved

**уточнять, уточнить** *v.* to pin down, to make more exact, to specify

**утраивать, утроить** *v.* to triple, to treble

**утроение** *n.* tripling, trebling

**утроенный** *adj.* triple, threefold

**утроитель** *m.* tripler, trebler

**УТС** *abbr.* 1 (**узловая телефонная станция**) community office; local office; 2 (**учрежденская телефонная станция**) agency's telephone exchange, PBX

**утюг** *m.* (flat) iron

**УУС** *abbr.* (**ультраузкий строб-импульс**) narrow-narrow gate

**УФ** *abbr.* 1 (**ультрафиолетовый**) ultraviolet; 2 (**усилитель фототоков**) photocurrent amplifier

**УФЛ** *abbr.* (**ультрафиолетовая лампа**) ultraviolet lamp

**УФО** *abbr.* (**ультрафиолетовое облучение**) ultraviolet lighting

**ухаживать** *v.* to maintain, to service; to take care of

**ухват** *m.* grip, shank, holder

**ухватный** *adj.* holder, grip

**ухо** *n.* ear; lug, ear, eye, hanger; **по естественному уху** by simple air conduction

**уход** *m.* drift; departure, leaving; servicing, maintenance, upkeep, handling; **длительный у.** long-term drift; **у. за машиной** machine attendance; **у. за сетью в эксплуатации** transmission maintenance work; **медленный у. частоты** long-term frequency drift; **обычный у. в эксплуатации** routine maintenance; **у. фазы** phase pushing figure; **у. час-**

тоты генератора oscillator drift; **у. частоты от номинальной** frequency departure; **у. (частоты) при прогреве** warm-up drift (of frequency)

**ухудшать, ухудшить** *v.* to penalize, to impair, to deteriorate, to make worse, to aggravate

**ухудшение** *n.* impairment, decrease, loss; degradation, deterioration, decline; **у. номиналов от температуры** temperature derating; **у. параметров** derating; **у. параметров при изменении температуры** temperature derating; **у. передачи от искажений** distortion transmission impairment, DTI; **у. передачи от шумов** noise transmission impairment, NTI; **постепенное у. качества детали** long-term drift; **у. свойств** quality impairment

**ухудшить** *pf. of* **ухудшать**

**участковый** *adj.* section, sectional

**участник** *m.* member; participant, participator

**участок** *m.* part, section, region, area, district, tract, sector, zone; piece; spot; distance; locus; **предварительные участки** opposing overlaps; **задний у.** back porch; **линейный у. сеточной характеристики (лампы)** drive range, workable control range, range of uniform control; **мёртвый у. у стрелки** fouling section; **неполно транспонированный у., неуравновешенный у. скрещивания** incomplete transposition section; **у. передержек** shoulder (of an emulsion characteristic); **передний у.** front porch; **пропорциональный у.** proportional region; **у. пупинизации, пупинизированный у.** loading coil section; **у. скрещивания** transposition section; **суженный у.** squeeze section; **усилительный у.** repeater section (*tphny.*); **фазирующий у. строки** phasing line (in facsimile); **чёрный у.** black porch

**учебный** *adj.* training; practice; school

**учение** *n.* teaching, instruction; studying; training; science; **у. о течении fluid mechanics, hydrodynamics**

**учесть** *pf. of* **учитывать**

**учёт** *m.* calculation, estimate; metering, registering, record keeping; counting; accounting; **автоматический у. длительности переговоров** automatic message accounting (*tphny.*); **у. выборочным методом** spot check (*tphny.*); **у. количества разговоров** metering of calls (*tphny.*); **у. по зонам** zone metering (*tphny.*); **у. по зонам и продолжительности** time-zone metering (*tphny.*); **у. по поясам** zone metering (*tphny.*); **у. продолжительности** time metering, timing; **у. разговоров** call metering

**учетверение** *n.* quadrupling, quadruplication

**учетверённый** *adj.* quadruple, quadruplicate

**учетверить** *pf. of* **учетверять**

**учетвертитель** *m.* quadrupler, quadruple

**учетвертительный** *adj.* quadrupler

**учетверять, учетверить** *v.* to quadruple

**учётный** *adj.* accounting; registering

**учитываемый** *adj.* metered; registered

**учитывание** *n.* taking into account

**учитывать, учесть** *v.* to allow, to consider; to take into account, to take into consideration

**учреждение** *n.* office, station; institute; institution, establishment

**учрежденский** *adj.* business; agency

**ушестерённый** *adj.* sextuple

**ушестерять, ушестерить** *v.* to sextuple, to increase sixfold

**уширение** *n.* spread, amplification, broadening, widening, enlargement; **у. кабельного колодца** cable-duct snout

**уширённый** *adj.* widened; enlarged, extended, expanded

**уширитель** *m.* extension

**уширительный** *adj.* enlarging, widening

**уширять, уширить** *v.* to widen, to broaden; to enlarge, to extend, to expand

**ушко** *n.* ear, eye, lug, loop; shackle, catch; sleeve piece, thimble; **припаечное (*or* припаиваемое) у.** soldering tag, soldering lug, soldering terminal; **стрелочное у.** switch adjustment

**ушной** *adj.* ear; aural

**ущерб** *m.* damage, injury, detriment, loss

**Уэкфильд** Wakefield

**уязвимость** *f.* vulnerability; **у. для помех** jamming vulnerability

**уязвимый** *adj.* vulnerable

# Ф

**Ф** *abbr.* 1 (фарада) farad. 2 (**Фаренгейт**) Fahrenheit. 3 (**фот**) phot. 4 (**фронт**) front

**Фабри** Fabry

**фабрика** *f.* factory, plant, works

**фабрикат** *m.* product, article, make, manufacture, manufactured product

**фабрикация** *f.* production, fabrication, output, manufacture

**фабриковать** *v.* to make, to produce, to fabricate, to manufacture

**фабрично-заводский** *adj.* manufacturing, industrial, factory, plant

**фабричный** *adj.* factory, manufacturing

**фагот** *m.* bassoon

**фаза** *f.* phase, stage; period; branch; leg; **в фазе** in phase; **не в фазе** out-of-phase; **с противоположной фазой** antiphase; **ф. во времени** time phase; **вывернутая ф.** reversed phase; **динамическая ф. строчной развёртки** horizontal dynamic phase; **занятая (вч присоединением) ф.** coupling phase (of a high-voltage line); **ф. импульсной модуляции** sampling phase; **начальная ф. генератора** reference oscillator phase; **огибающая ф.** continuous phase; **ф. огибающей** envelope phase; **ф. полёта на среднем участке траектории** midcourse phase; **ф. полного сопротивления** phase of impedance; **ф. сигнала цветности** color phase; **ф. сигналов развёртки** sweep phase; **ф. синхронизации** locking phase; **совпадающая ф.** correct phase; **ф. спуска** terminal phase

**фазирование** *n.* phasing, phasing adjustment, phasing control, phase locking; **ф. по кадрам, ф. полей, ф. по полям** field phasing, vertical phasing; **ф. по строкам, ф. строк** horizontal phasing, line phasing

**фазированный** *adj.* phased

**фазировать** *v.* to phase, to bring in phase, to cohere

**фазировка** *f.* phasing, phasing control, phasing adjustment

**фазировщик** *m.* phaser

**фазируемый** *adj.* phaseable

**фазирующий** *adj.* phasing

**фазис** *see* **фаза**

**фазитрон** *m.* phasitron, phasitron tube

**фазитронный** *adj.* phasitron

**фазный** *adj.* phase, phasic; phase-wound; line-to-neutral

**фазо-, фазово-** *prefix* phase

**фазово-импульсно-модулированный** *adj.* pulsed phase-modulated

**фазовоимпульсный** *adj.* pulse-phase, pulse-position

**фазово-модулированный** *adj.* phase-modulated

**фазово-неустойчивый** *adj.* phase-unstable

**фазовращатель** *m.* phase switcher, phase inverter, phase shifter, phase-shifting circuit, phaser; **тройниковый ф.** magic-tee phase changer; **ферритовый ф. в прямоугольном волноводе** rectangular guide ferrite phase shifter

**фазовращательный** *adj.* phase-shifting

**фазовращающий** *adj.* phase-shifting; phase-shift

**фазовый** *adj.* phase; live

**фазовыравниватель** *m.* phase compensator, phase equalizer, delay equalizer, phase modifier

**фазовыравнивающий** *adj.* phase-equalizing, phase-compensating

**фазограф** *m.* phasograph

**фазозадерживающий** *adj.* phase-delay, phase-lag

**фазоизмерительный** *adj.* phase-measuring

**фазоимпульсный** *adj.* pulse-position

**фазоинверсный** *adj.* phase-inverting

**фазоинвертер** *m.* phase inverter, phase reverser, phase splitter, polarity splitter; **ф. с отрицательной обратной связью** degenerative-amplifier phase inverter

**фазоинвертор** *see* **фазоинвертер**

**фазоиндикатор** *m.* phase indicator, phase monitor, phase meter

**фазоиндикаторный** *adj.* phase-detecting

**фазокомпенсатор** *m.* phase advancer; phase compensator, phase modifier, phase controller; **(коллекторный) ф.** phase adjuster; **синхронный ф.** phase shifting transformer, synchro phase shifter, synchronous phase modifier

**фазоконтрастный** *adj.* phase-contrast

**фазоконтурный** *adj.* phase-contour

**фазокорректор** *m.* phase compensator

**фазолинейный** *adj.* phase-linear

**фазометр** *m.* phase meter, phase indicator, phase measuring instrument; **ф., градуированный в фазовых углах** direct-reading phase analyzer

**фазометрический** *adj.* phase-meter

**фазомодулированный** *adj.* phase-modulated, phase-modulation

**фазомодулятор** *m.* phase modulator

**фазоопрокидыватель** *m.* phase inverter, phase reverser

**фазоопрокидывающий** *adj.* phase-inverter, phase-reversing

**фазопреобразователь** *m.* phase converter

**фазор** *m.* phasor

**фазоразделитель** *m.* phase separator

**фазорасщепитель** *m.* phase splitter; **дифференциальный ф.** incremental phase splitter

**фазорасщепление** *n.* phase splitting

**фазорасщепляющий** *adj.* phase-splitting

**фазорегулирующий** *adj.* phase-regulating, phase-shifting

**фазорегулятор** *m.* phase shifter, phase shifting transformer, synchro phase shifter; **ф. ротора** slide phase valve

**фазосдвигающий** *adj.* phase-shifting, phase-shift

**фазосмещающий** *adj.* phase-shifting

**фазосравнивающий** *adj.* phase-comparing

**фазотрон** *m.* synchrocyclotron, frequency-modulated cyclotron

**фазотропия** *f.* phasotropy

**фазоуказатель** *m.* phase-sequence indicator, phase meter, power-factor meter

**фазочастотный** *adj.* phase-frequency

**фазочувствительный** *adj.* phase-sensitive

**ФАИ** *abbr.* (**Международная авиационная федерация**) International Aeronautical Federation, Fédération Aéronautique Internationale

**фак.** *abbr.* (**факультет**) faculty, department (of a university)

**факел** *m.* flare, torch, flame; **факелы** *pl.* faculae; **посадочный ф.** flare

**факелообразный** *adj.* flame-shaped

**факельный** *adj.* torch, flare, flame

**факсимиле** *n. indecl.* facsimile, replica

**факсимильный** *adj.* facsimile

**факт** *m.* fact; **истинный ф., непреложный ф.** unbiased fact

**фактически** *adv.* in fact, practically, virtually; by facts

**фактический** *adj.* actual; real, factual; virtual, practical, active; present

**фактор** *m.* factor, coefficient, agent, divider, divisor; element factor; **индуктивный внешний ф.** induced environment; **нормализующий ф.** normalized factor; **ф., обуславливающий совместность** compatibility determinant; **ф. эффективности по громкости** loudness efficiency factor

**факториал** *m.* factorial (*math.*); **факториалы** factorial functions

**факториальный** *adj.* factorial, factor

**факторизация** *f.* factorization

**факторный** *adj.* factor

**фактура** *f.* texture, manner of execution; composition; invoice

**фактурный** *adj.* composition; texture; invoice

**факультативный** *adj.* facultative, optional

**факультет** *m.* faculty, department (of a university); factorial (function)

**фальцет** *m.* falsetto, falsetto voice

**фальшивый** *adj.* false; spurious; dummy; pseudo-

**фальшь** *f.* falsity; fault, mistake; mistuning

**ФАН** *abbr.* (**Филиал Академии Наук СССР**) Branch of the Academy of Sciences USSR

**фанатрон** *m.* phanatron

**фанера** *f.* veneer, plywood; **клееная ф., многослойная ф.** plywood

**фанерит** *m.* plywood

**фанерный** *adj.* veneer, plywood

**Фанно** Fanno

**фанотрон** *m.* phanotron

**фантастрон** *m.* phantastron
**фантастронный** *adj.* phantastron
**фантом** *m.* phantom (radiology); model
**фантомный** *adj.* phantom
**фара** *f.* light, headlight, headlamp; landing light; **аэронавигационная ф.** flying lamp; **задняя ф.** reversing light
**фарад** *m.*, **фарада** *f.* farad
**фарадеев** *poss. adj.* Faraday, Faraday's
**Фарадей** Faraday
**фарадей** *m.* faraday (unit of quantity of electricity)
**фарадизатор** *m.* faradizer
**фарадизация** *f.* faradization
**фарадический** *adj.* faradic
**фарадметр** *m.* faradmeter, capacitance meter
**фаралит** *m.* faralit
**фарватер** *m.* channel, fairway, waterway
**фарватерный** *adj.* channel, fairway, waterway
**Фаренгейт** Farenheit
**Фарнсворт, Фарнсуорт** Farnsworth
**фартук** *m.* apron; **ф. из просвинцованной резины** lead-rubber apron
**фартучный** *adj.* apron
**фарфор** *m.* porcelain; **ф. высокотемпературного обжига** electro-technical porcelain; **литой ф.** wet-process porcelain; **пресованный ф.** dry-process porcelain; **формированный** (*or* **формованный**) **ф.** wet-process porcelain; **штампованный ф.** dry-process porcelain
**фарфоровидный** *adj.* porcelain-like, porcelaneous
**фарфоровый** *adj.* porcelain
**фасад** *m.* facade, face; elevation; **боковой ф.** profile, side view; **передний ф.** front elevation, front view
**фасон** *m.* make, style, fashion
**фасонирование** *n.* shaping, fashioning
**фасонировать** *v.* to shape, to fashion
**фасонный** *adj.* shaped, molded, fashioned; shape, form, profile
**фатер-пачинев** *poss. adj.* Vater-Pacini's
**Фаулер** Fowler
**фау-схема** *f.* V-connection
**федерация** *f.* federation
**фединг** *m.* fading, fade-out; **глубокий ф.** blackout fading; **ф. несущей частоты** carrier fading; **продолжительный ф.** blackout (radio propagation)

**фединг-гексод** *m.* fading hexode
**федометр** *m.* fedometer
**фемто-** *prefix* femto-
**фён** *m.* föhn wind, chinook
**фенол** *m.* fenol
**фенолит** *m.* phenolite, phenolic material
**феноловый** *see* **фенольный**
**фенолы** *pl.* phenols
**фенольный** *adj.* phenolic, phenol
**феномен** *m.* phenomen
**феноменальный** *adj.* phenomenal
**феноменологический** *adj.* phenomenological
**феноменология** *f.* phenomenology
**фенопласты** *pl.* phenol base plastics
**Фери** Féry
**Ферма** Fermat
**ферма** *f.* framework, girder, truss, support, beam
**фермер** *m.* farmer
**фермерский** *adj.* farmer
**Ферми** Fermi
**фермиевский** *adj.* Fermi's, Fermi
**фермий** *m.* fermium, Fm
**фермион** *m.* fermion
**фернико** *n. indecl.* fernico, kovar
**ферниковый** *adj.* fernico
**« феррак »** *m.* Ferrac
**феррактор** *m.* ferractor; **мегагерцный ф.** megacycle ferractor
**феррамик** *m.* ferramic
**Ферранти** Ferranti
**Феррарис** Ferraris
**ферри-ионы** *pl.* ferri-ions
**«ферримаг»** *m.* ferrimag
**ферримагнетизм** *m.* ferrimagnetism
**ферримагнитный** *adj.* ferrimagnetic
**ферристор** *m.* ferristor
**феррит** *m.* ferrite; **ферриты для сантиметровых волн** microwave ferrites; **ферриты кубической структуры** ferroxcube; **ф. без подмагничивания** unbiased ferrite; **полупроводящий ф.** ferroxcube; **простой ф.** single component ferrite; **ф. с прямоугольной петлёй гистерезиса** rectangular-loop ferrite, square-loop ferrite; **спекшийся ф.** molded powdered ferrite
**феррит-диодный** *adj.* ferrite-diode
**ферритный, ферритовый** *adj.* ferrite
**феррит-транзисторный** *adj.* ferrite-transistor
**ферро-** *prefix* ferro-, iron
**феррограф** *m.* ferrograph
**ферродинамический** *adj.* ferrodynamic

**феррокарт** *m.* ferrocart, compressed-iron core

**феррокартный, феррокартовый** *adj.* ferrocart

**феррокобальт** *m.* ferrocobalt

**феррокремниевый** *adj.* ferrosilicon

**феррокедур, ферроксдюр** *m.* ferroxdure

**феррокскуб, феррокскюб** *m.* ferroxcube

**феррокспланы** *pl.* ferroxplana; **ф. с ориентированной кристаллической структурой** crystal-oriented ferroxplana

**ферромагнетизм** *m.* ferromagnetism, ferromagnetics

**ферромагнетик** *m.* ferromagnetic (material)

**ферромагнит** *m.* ferromagnet

**ферромагнитность** *f.* ferromagnetism

**ферромагнитный** *adj.* ferromagnetic

**феррометр** *m.* Ferrometer

**феррониккелевый** *adj.* nickel-iron

**феррониккель** *m.* nickel-iron

**ферророезонанс** *m.* ferro-resonance

**ферророезонансный** *adj.* ferroresonant, ferroresonance

**ферросен** *m.* ferrocene

**ферросилициевый** *adj.* iron-silicon

**ферросплав** *m.* ferroalloy

**ферротрон** *m.* ferrotron

**ферроферрит** *m.* ferroferrite

**феррохлорид** *m.* ferric chloride

**феррохром** *m.* chromium-iron

**феррошпинель** *f.* ferrospinel

**ферроэлектрик** *m.* ferroelectric material

**ферроэлектрический** *adj.* ferroelectric

**ферроэлектричество** *n.* ferro-electricity

**Фессенден** Fessenden

**Фетер** Feather

**фетр** *m.* felt

**фетровый** *adj.* felt

**Фехнер** Fechner

**фиберглас** *m.* fiberglass

**фибра** *f.* fiber; **листовая ф.** fiber board; **твёрдая ф.** vulcanized fiber; **тонкая ф.** fish paper

**фибровый** *adj.* fiber, fibrous

**фиброзный** *adj.* fibrous

**фиг.** *abbr.* (**фигура**) figure, pattern, picture, illustration

**фигура** *f.* figure, shape, form, pattern, picture, drawing; diagram; **интерференционные фигуры в сходящемся свете** optical interference figures; **фигуры, образуемые опилками в** электрическом поле dust figures; **ф. в виде зубчатого колеса** gear-wheel pattern; **ф. в виде пунктирной окружности** spot-wheel pattern; **неподвижная ф.** stationary pattern; **образцовая ф.** test pattern

**фигурально** *adv.* figuratively

**фигуральный** *adj.* figurative, symbolic

**фигурный** *adj.* figure; illustration, diagram; shape, form

**фидер** *m.* feeder, feeder line, feeding line, transmission line, main; tie; interconnection; **ф. бегущей волны** nonresonant feeder; **внутренний ф.** feeder in interior wiring; **второй параллельный ф.** duplicate feeder; **главный ф.** mains, main line, trunk main, trunk feeder; **обычный ф. без ответвлений** plain feeder; **однопроводный ф.** voltage feed; **ответвляющийся ф.** open feeder; **отсасывающий ф.** return feeder, outgoing feeder, return feeder cable, single feeder; **отсасывающий рельсовый ф.** negative feeder; **ф. передачи энергии** transmission feeder; **ф. питания направленной антенны** beam antenna feeder; **полый ф.** hollow pipe, waveguide feed; **ф. с автоматическим повторным включением** reclosing feeder; **ф. с ответвлениями** teed feeder; **ф. с параллельными стержнями** pillbox feed; **ф. с полувоздушной изоляцией** semiair-spaced feed; **ф. с проводами большого сечения, сильно нагруженный ф.** heavy feeder; **ф., соединяющий направленную антенну с радиоприёмником** beam receiving antenna feeder; **ф. стоячей волны** resonant feeder; **тупиковый ф.** branch cable, dead-end(ed) feeder, independent feeder; **тупиковый ф. передачи** radial transmission feeder; **ф. цепи передачи энергии** transmission feeder

**фидерный** *adj.* feeder, feeder-line; tie

**физ** *abbr.* (**физика; физический**) physics; physical

**физика** *f.* physics; **доквантовая ф.** prequantum physics; **дорелятивистская ф.** pre-relativity physics; **ф. заряженных частиц высокой энергии** high-energy physics

**физик-радиолог** *m.* radiological physicist, radiation physicist

**физико-** *prefix* physico-, physical

**физиологический** *adj.* physiological

**физиология** *f.* physiology

**физиотерапия** *f.* physiotherapy, physical therapeutics

**физически** *adv.* physically

**физический** *adj.* physical

**Физо** Fizeau

**Фик, Фикк** Fick

**фиксаж** *m.* fixing, fixation; fixative, fixer, fixing bath

**фиксажный** *adj.* fixing (agent, developer)

**фиксатив** *m.* fixative, fixing agent

**фиксатор** *m.* clamp, clamper, catch, detent, lock; steady arm, steady brace; index pin; **ф. верхнего уровня** upper clamp; **ф. на заданную величину** calibration control; **ф. нижнего уровня** lower clamp; **ф. переключателя** catch, lock, detent (of a switch); **ф. пятна** spot fixer

**фиксация** *f.* fixing, fixation, hold, holding, clamp, clamping; registering; **ветровая ф. контактного провода** wind bracing; **ф. во время каждого обратного хода по строкам** line-by-line clamp; **ф. по отрицательному уровню** negative clamping; **ф. решения** clamping; **управляемая ф.** keyed clamp; **ф. уровня** clamp, clamping; **ф. уровня во время каждого обратного хода по строкам** line-by-line clamping; **ф. уровня сигнала с частотой полей** vertical clamping

**фиксирование** *n.* fixing, fixation, clamping

**фиксированный** *adj.* fixed; immobilized

**фиксировать, зафиксировать** *v.* to fix; to lock, to stop, to clamp, to secure, to hold; to rate, to record, to register; to immobilize; **ф. значение** to rate; **ф. уровень** to clamp

**фиксирующий** *adj.* clamp, clamping; interlocking; holding; fixing

**фиктивность** *f.* fictitiousness

**фиктивный** *adj.* fictitious, false, dummy, imitation; theoretical, hypothetical

**Филбрик** Philbrick

**филдистор** *m.* fieldistor, field effect transistor

**филдтрон** *m.* fieldtron

**филд-эффект** *m.* field effect

**филёнка** *f.* jack spacer; panel, end panel, slat

**филёнчатый** *adj.* panel, paneled

**Филипс** Philips

**фильдистор** *m.* fieldistor

**фильм** *m.* film; movies; **свободные фильмы** metachromotype, decalcomania film; **объёмный ф.** three-dimensional film; **художественный ф.** feature film

**фильмит** *m.* filmite

**фильморезный** *adj.* film-cutting

**фильмостат** *m.* film-storage cabinet

**фильмофон** *m.* sound recorder

**фильтр** *m.* filter, sifter, eliminator; weighting network; band-pass filter, wave filter; chocking circuit, reactance circuit; **(объёмные) ёмкостно-резистивные фильтры** distributed components; **антенный ф. телевизионных передатчиков** vestigial sideband antenna filter for television transmitter; **антенный разделительный ф.** diplexer; **анти-интерференционный ф.** beat-interference filter; **ф. в схеме ответвления** ladder-type filter, shunt-arm electric wave filter; **ф. в цепи питания анода** plate supply filter; **ф. верхних частот** high-pass filter, low limiting filter; **ф. вида колебаний** mode filter; **восьмиполюсный ф.** separating filter for light terminals; **вращающийся цветной ф. в виде диска** rotating filter disk; **вращающийся цветовой ф.** color filter disk; **ф. выравнивания данных** data-smoothing network; **вырезающий ф.** band-elimination filter, band-eliminator filter; **вырезывающий ф.** sharp filter; **высокочастотный заградительный** (*or* **режекторный**) **ф.** radio-frequency trap; **ф. гармоник** frequency weighting network, weighting filter; formant filter; harmonic filter, harmonic trap; **грубый сглаживающий ф.** brute-force filter; **двухконтурный ф. с ёмкостной связью** capacity-coupled double-tuned circuit; **двухконтурный ф. УКВ** IF bandpass filters for vhf reception; **ф. для гармоник** smoothing equipment; **ф. для подавления боковой полосы частот** vestigial sideband filter; **ф. для подавления гармоник** harmonic filter; **ф. для подавления низких звуковых**

частот speech filter; **ф. для подавления отражений от неподвижных объектов** fixed target rejection filter; **ф. для подавления телеграфных помех** thump filter; **ф. для устранения фона сети** hum eliminator; **дроссельный ф.** low-pass filter, ultra filter, choke filter, stopper circuit; **ф.-заградитель низких частот** bass-cut filter; **заградительный ф.** surge absorber, rejection trap; **заградительный ф. с сопротивлением** surge absorber, including sometimes a resistor to dissipate part of the surge-energy; **заграждающий ф.** rejector, rejector circuit, eliminator, wave-trap, stopper, band-elimination filter, band-exclusion filter, suppression filter, choke filter, trap, line trap, absorption trap; rejector unit; **заграждающий ф. низких частот** low-pass selective circuit; **заграждающий ф. промежуточной частоты** I.F. trap; **ф. запаздывания** lagging filter; **запирающий ф.** parallel transducer; **защитный ф. для уменьшения внешней засветки** optical light filter; **защитный ф. на трубке для уменьшения внешней засветки** ambient light filter; **ф. из сопротивлений и конденсаторов** R-C filter network; **избирательный ф.** wave filter, notching filter; fixed target rejection filter; selective filter; **кварцевый полосно-пропускающий (** *or* **полосовой) ф.** bandpass crystal filter; **ф. ключевых щелчков** key-click filter; **кольцеобразный ф. типов волн** ring mode filter; **комбинированный ф.** rejector-acceptor circuit; **корректирующий ф.** compensating filter, filter-type equalizer, compensating network, corrective network, correcting filter, trimming filter; **ф.-ловушка** trap, wave trap, absorption trap, band-elimination filter; **многозвенный ф.** iterative filter, multisection filter, ladder-type filter; **многократный волноводный ф.** waveguide multiplexer; **многоячеечный ф.** multisection filter; **мостиковый ф., мостовой ф.** differential filter; **мостовой Т-образный ф.** bridged-T filter; **настраиваемый ф. сантиметровых волн** tunable microwave filter; **ф. нижних частот**

low-pass filter, ultra filter, upper limiting filter; **ф. обратной связи** feedback filter; **ф. ограничения полосы, ф. ограничивающий полосу частот** cut-off filter; **ф. одиночной гармоники** fire line filter; **октавный пропускающий полосовой ф.** octave analyzer; **оптимальный ф., отделяющий полезный сигнал от шума** optimum detecting filter; **ф. ослабляющий шипение иголки** scratch filter; **основной ф.** K-filter, constant-k-filter; **ф. от засвечивания экрана посторонним светом** ambient light filter; **отсасывающий ф.** trap circuit, absorption trap, wave trap, shorting filter, zero-impedance filter; **ф. отстройки** wave selector; **ф. первичных цветов телевизионного приёмника** primary(-color) filter (*tv*); **плавнорегулируемый ф.** continuously variable filter; **П-образный ф.** pi-section filter; **ф.-поглотитель из активированного угля** charcoal bed; **поглощающий ф.** trap; **подключаемый ф.** low-pass connecting filter; **полоснозаграждающий ф. со ступенчатой характеристикой** notch filter; **полосно-пропускающий ф. с максимально плоской характеристикой** maximally flat band-pass filter; **полосовой ф. из объёмных резонаторов** cavity band-pass filter; **полотняный ф.** cloth screen; **ф. постоянного сопротивления** constant-k network, constant-resistance network; **предварительный вч ф.** preselector; **преобразующий ф.** discriminator filter; **ф. пропускания верхних (** *or* **высоких) частот** high-pass filter; **пропускающий ф.** band-pass filter, acceptor circuit, series transducer; **ф., пропускающий постоянный ток и звуковые частоты** direct current and voice pass filter; **ф. против переходных разговоров** crosstalk suppression filter; **ф. против помех от местных предметов** clutter filter; **противоколебательный ф.** antihunt filter; **ф. радиопомех** interference filter; **радиотрансляционный ф.** wired-wireless separating filter; **разделительный ф. для двухканального громкоговорителя** loudspeaker dividing network; **разделительный ф. для синхронизирующего** сигнала sync separator; **режекторный ф. звукового сопровождения** accompanying sound trap; **режекторный ф., подавляющий поднесущую частоту** subcarrier trap; **резистивно-ёмкостный ф.** RC (filter) network, RC (filter) circuit; **резонансный ф. в цепи отрицательной обратной связи** inverse-feedback filter; **резонаторный ф.** cavity filter; **ф. с заданной частотной характеристикой** reactive equalizer; **ф. с контурами высокой добротности** high-Q filter system; **ф. с магнитной перестройкой** ferrite-tunable filter; **ф. с ограниченной полосой пропускания** cut-off filter; **ф. с очень мелкой сеткой** micron filter; **ф. с переменной полосой пропускания** weighting filter; **ф. с П-звеньями** pi-section filter; **ф. с П-образным концом** mid-shunt terminated filter; **ф. с постоянной относительной шириной полосы** constant percentage bandwidth filter; **ф. с прямоугольной характеристикой** sharp-cutoff filter; **ф. с рабочей средой из металлического порошка** sintered filter; **ф. с Т-образным концом** mid-series terminated filter; **ф. с фазовой синхронизацией** tracking filter; **самонастраивающийся ф.** adaptive filter; **сглаживающий ф.** smoother, smoothing filter, smoothing circuit, ripple filter, hum filter, rectifier filter, brute force filter; **сглаживающий ф. сетевого выпрямителя** smoothing elements in power-supply rectifiers; **симметричный ф. нижних частот** balanced low pass filter; **сложный ф.** total filter; **ф. со сверхкритической связью** IF transformer (overcoupled type); **ф. со связью в виде диаграммы** iris-coupled filter; **составной избирательный ф.** composite wave filter; **ступенчатый ф.** notching filter; **телеграфный ф.** anti-induction network; **ф. типов волн, ф. типов колебаний** mode filter; **трубчатый ф.** tubular electrical dust filter; **узкополосовой ф.** fire line filter; **ультракоротковолновый полосовой ф.** IF bandpass filters for vhf reception; **ушной ф.** electric ear; **фазовый ф.** all-pass filter, all-pass network; **фазоопере**

**жающий ф.** phase-advance network; **цветоделительный ф.** dichroic filter; **цепной ф.** iterative filter, iterated network; **частотный (разделительный) ф.** frequency filter; **ф. четырёхполюсника скрещённого типа** lattice-type filter; **шестизвенный ф.** six-stage filter; **щелевой ф. типов колебаний** mode filter slot; **ф. эхосигналов** echo trap, power equalizer

**фильтрация** *f.* filtering, filtration; filter discrimination; smoothing; transmissibility; permeability; **ф. на основе эффекта Допплера** Doppler filtering

**фильтрирующий** *adj.* filtering; smoothing

**фильтровальный** *adj.* filter, filtering

**фильтрование** *n.* filtering, filtration

**фильтрованный** *adj.* filtered

**фильтровать, профильтровать** *v.* to filter, to sift

**фильтровка** *f.* filtering, filtration; sifting

**фильтровочный** *adj.* filter, filtering

**фильтровый** *adj.* filter

**фильтродержатель** *m.* filter ring, filter support

**фильтропоглотитель** *m.* absorption filter

**фильтр-пробка** *f.* band-elimination filter, impedance wave-trap, absorption trap; rejector; **ф. на частоте поднесущей** subcarrier trap

**фильтруемость** *f.* filterability

**фильтруемый** *adj.* filterable

**фильтрующий** *adj.* filtering, filter

**фильтрующийся** *adj.* filtering; filterable

**филярный** *adj.* filar, thread-like

**ФИМ** *abbr.* (**фазово-импульсная модуляция**) pulse-phase modulation, pulse-position modulation

**финзен** *m.* finsen

**финитный** *adj.* finitary (linguistics)

**фирма** *f.* firm, company; **радиопромышленная ф.** radio manufacturing company

**фирменный** *adj.* firm, company

**фирновый** *adj.* firn

**фисгармония** *f.* harmonium

**фитиль** *m.* core, cored-carbon; rosincore; wick

**фитильный** *adj.* core, cored-carbon; rosin-core; wick

**фитинг** *m.* fitting; cross-over bend; **фитинги** fittings

**Фицджеральд** Fitzgerald

**фицджеральдов** *poss. adj.* Fitzgerald's, Fitzgerald

**фишка** *f.* peg, plug, socket, jack plug; **ф. для закрытия многократного гнезда** multiple peg; **запорная ф.** locking-type plug; **изогнутая ф.** angle plug; **ответная ф.** plug receptacle; **переходная ф.** insert; **саморазъёмная ф.** self-release plug; **угoлковая ф.** right-angle connector

**ф-ия** *abbr.* (функция) function

**ф-ка** *abbr.* (фабрика) factory, plant

**ФКП** *abbr.* (фотокинопулемёт) gun camera, camera gun

**ф-ла** *abbr.* (формула) formula

**флаг** *m.* flag; banner; flag burst; **ф. (маятникового сигнада)** banner (*rr. appl.*); **ф.-импульс, ф. -сигнал** flag burst, burst gate, burst gating pulse, burst keying pulse, burst-flag signal; **ф. наверх** flag-up

**флаговый** *adj.* flag; banner

**флажковый** *adj.* vane, flag

**флажок** *m.* flag indicator, operation indicator (of a relay); **ф. (нумератора)** bulleye; **ф. в реле** target

**флажолет** *m.* flageolet (*mus.*)

**флакенол** *m.* flakenol

**фланец** *m.* flange, bush, collar, ring; **ф. ввода** bushing holder; **прямоугольный охватывающий ф.** rectangular cover flange; **соединительный ф.** flange connector, flange coupling, flange sleeve, matching flange connector; **ф. ступицы** nave plate; **цокольный ф.** base fin (*tbs.*)

**фланжировальный** *adj.* flange, flanging

**фланжировать** *v.* to flange

**фланжировочный** *adj.* flange, flanging

**фланцевый** *adj.* flange, flanged

**фланцованный** *adj.* flanged

**флаттер** *m.* flutter effect

**флаштрон** *m.* flashtron

**флегматизатор** *m.* delayer

**флегматизация** *f.* flegmatization

**флегматичный** *adj.* phlegmatic, sluggish

**флейта** *f.* flute; **малая ф.** octave flute; **ф. Папа** Panpipe, syrinx; **ф.-пикколо** octave flute; **прямая ф.** beaked flute; **продольная ф.** fipple flute

**Флек, ван** Van Vleck

**флексия** *f.* desinence, ending, flection

**флексод** *m.* flexode

**флексометр** *m.* flexometer

**флексорайтер** *m.* flexowriter

**флекстрон** *m.* flextron

**флективный** *adj.* inflected, inflectional

**Флеминг** Fleming

**Флетчер** Fletcher

**ФЛИДЕН FLIDEN**

**фликер-эффект** *m.* flicker effect

**флинт** *m.* flint

**флинтглас** *m.* flint glass

**флинтгласный** *adj.* flint-glass

**Флоберт** Flobert

**флогопит** *m.* amber mica, phlogopite; **выветрившийся ф.** amber silver mica

**Флоке** Floquet

**флокены** *pl.* flocculi

**флоккулы** *pl.* flocculi

**флоккуляция** *f.* flocculation

**флот** *m.* navy, fleet

**флотский** *adj.* naval, fleet

**флуксметр** *m.* fluxmeter; **баллистический ф. Грассо** Grassot fluxmeter

**флуктуационно-диссипативный** *adj.* fluctuation-dissipation

**флуктуационный** *adj.* fluctuation

**флуктуация** *f.* fluctuation; jitter; variance, straggling, scintillation; *pl.* noise; **дробные флуктуации** shot noise; **флуктуации от цели на экране радиолокатора** target scintillation; **флуктуации развёртывающего луча** (*or* пучка) scanning-beam noise; **флуктуации цели** target scintillation; **ф. отражённого от цели сигнала** target glint; **ф. скорости** perturbation of velocity

**флуктуировать** *v.* to fluctuate

**флуктуирующий** *adj.* fluctuating

**флуоресцентный** *adj.* fluorescent

**флуоресценция** *f.* fluorescence

**флуоресцирование** *n.* fluorescence

**флуоресцировать** *v.* to fluoresce

**флуоресцирующий** *adj.* fluorescent, phosphorescent

**флуорид** *m.* fluoride

**флуориметр** *m.* fluorimeter

**флуорит** *m.* fluorite

**флуорограф** *m.* photofluorograph, photo-roentgen unit, PR unit

**флуорографический** *adj.* fluorographic

**флуорография** *f.* fluorography, photofluorography

**флуорокарбон** *m.* fluorocarbon

**флуорометр** *m.* fluorometer

**флуороскоп** *m.* fluoroscope, roentgenoscope; **ф. для подгонки обуви** shoe-fitting fluoroscope; **просвечивающий ф.** transmission fluoroscope

**флуороскопический** *adj.* fluoroscope, fluoroscopic

**флуороскопия** *f.* fluoroscopy, photo-fluoroscopy

**флуорофотометр** *m.* fluoro-photometer

**флуороциклобутан** *m.* octafluorocyclo-butane

**флювиограф** *m.* fluviograph

**флюгерный** *adj.* vane, wind vane, weather vane; wind sock

**Флюеллинг** Flewelling

**флюкс** *m.* flux

**флюксметр** *m.* fluxmeter

**флюксор** *m.* fluxor; **ф. с запретом** inhibited fluxor; **ф. с совпадением** coincidence fluxor

**флюксующий** *adj.* flux(ing)

**флюктуационно-диссипативный** *adj.* fluctuation-dissipation

**флюктуационный** *adj.* fluctuation

**флюктуация** *f.* fluctuation

**флюоресценция** *f.* fluorescence

**флюорит** *m.* fluorite

**флюорограф** *see* флуорограф

**флюорографический** *adj.* fluorographic

**флюорография** *see* флуорография

**флюороскоп** *see* флуороскоп

**флюороскопический** *see* флуороскопический

**флюороскопия** *see* флуороскопия

**флюс** *m.* flux, fusing agent

**флюсующий** *adj.* fluxing

**флютбет** *m.* apron, spillway dam; spillway, by-channel

**флянцевый** *adj.* flange, flanged

**фляттер** *m.* flutter index

**ФМ** *abbr.* (**фазовая модуляция**) phase modulation

**фн.** *abbr.* (**фунт**) pound

**Фогель** Vogel

**Фоже** Faugé

**Фойсснер** Feussner

**фокальный** *adj.* focal

**фокометр** *m.* focometer, focimeter

**фокус** *m.* focus, focal point; focal spot (X-rays); **оптический ф.** focus

**фокусирование** *n.* focus, focusing; **ф. в поле ускорения** post-acceleration focusing; **вертикальное динамическое ф.** vertical dynamic convergence; **ионное ф.** gas focusing; **ф. пучка периодическими полями** periodic focusing

**фокусировать** *v.* to focus, to bring in focus, to direct

**фокусировка** *f.* focusing, concentration; **антенная ф. фототелефона** directivity of an optophone reflector; **ф. в углах, ф. в углу растра** corner focus, corner focusing; **плавная ф.** vernier focusing; **предварительная ф.** prefocusing

**фокусировочный** *adj.* focusing

**фокусирующий** *adj.* focus(ing), concentric

**фокусный** *adj.* focus, focal

**фольга** *f.* foil, tinsel; disk; **ф., выбрасываемая для создания радиолокационных помех** radar chaff; **медная ф.** electro-sheet copper; **медная ф., покрытая меднозакисным слоем** coated copper; **металлическая ф.** tinsel

**фольгированный** *adj.* metal-clad

**фольговый** *adj.* foil

**фольгодержатель** *m.* foil holder

**Фольмер** Volmer

**фон** *m.* hum noise, background (noise), ground noise, noise; phone (unit of volume); sensation; backdrop; **ф. выпрямителя** ripple; **ф. на растре (от источника питания)** hum pattern; **ф. на экране телевизионной приёмной трубки** visible noise; **неравномерный ф.** speckled background; **ф., обусловленный космическими лучами** cosmic-ray background; **ф. от подогрева** heater-cathode hum; **ф. от шумов на экране приёмной трубки** noise pattern; **ф. переменного тока** magnetic hum, magnetic ripple, alternating-current hum, hum modulation; **подогревный ф.** heater-cathode hum; **ф. поля сравнения** surround of a comparison field; **ф. помех** noise background, ground noise; **ф. радиоприёмника** set noise; **разрывный ф.** speckled background; **световой ф.** light bias; **точечный шумовой ф.** speckled background

**фонавтограф** *m.* phonoautograph

**фонарик** *m.* lamp, small lantern; **карманный ф.** flash lamp, flashlight

**фонарный** *adj.* lamp

**фонарь** *m.* lantern, light, lamp; connector, connecting piece; **головной ф.** headlamp; **задний габаритный ф.** side light; **карманный ф.** flashlight, battery lamp, pocket lamp; **лобовой ф.** headlamp; **мигающий ф.** blinker; **ф. номерного знака** (rear) number plate light; **опознавательный сигнальный ф.** marker lamp; **передовой ф.** headlight; **переносный ф. с батареей** battery lamp; **проекционный ф.** lantern slide; **рудничный (аккумуляторный) ф.** miner's lamp; **стрелочный ф.** switch lamp; **цветной ф. для опознавания** classification lamp, classification light

**фонация** *f.* phonation, phonics

**фонетика** *f.* phonetics

**фонетический** *adj.* phonetic

**фонетограф** *m.* phonetograph

**фоническая** *f. adj. decl.* mixing booth

**фонический** *adj.* phonic, acoustic

**фоно-** *prefix* phono-

**фоновизия** *f.* phonovision

**фоновызыватель** *m.* phonic ringer

**фоновый** *adj.* phon; background (noise), hum

**фонограмма** *f.* sound record, sound track; phonogram; record tape; **(фотографическая) ф.** sound track; **двухтактная ф. класса A** class A push-pull sound track; **односторонняя ф.** unilateral-area sound track; **оптическая ф. с записью обоих полупериодов** standard track; **противофазная ф.** push-pull recording track; **ф. с двусторонней модуляцией по площади, удвоенная ф.** bilateral area track; **(фотографическая) ф. с переменными плотностью и шириной** matted track

**фонограф** *m.* phonograph

**фонографический** *adj.* phonographic

**фонография** *f.* phonography

**фонографный** *adj.* phonograph

**фонодейк** *m.* phonodeik

**фонокардиограмма** *f.* phonocardiogram (*med.*)

**фонокардиография** *f.* phonocardiography

**фонологический** *adj.* phonologic(al)

**фонометр** *m.* phonometer, acoustimeter; **перемежающийся ф., ф. с перемежающимся тоном сравнения** alternation phonometer

**фонон** *m.* phonon

**фононный** *adj.* phonon

**фонон-электронный** *adj.* phonon-electron

**фоноплекс** *m.* phonoplex

**фоноплёнка** *f.* phono-film

**фонопор** *m.* phonopore

**фоноскоп** *m.* phonoscope

**фонотелеметрия** *f.* phonotelemetry

**фоноэлектрокардиоскоп** *m.* phonoelectrocardioscope

**фонтактоскоп** *m.* fontactoscope

**Фор** Faure

**форвакуум** *m.* initial vacuum, rough exhaust, forvacuum

**форвакуумный** *adj.* forvacuum

**форез** *m.* phoresis

**Форест, де** de Forest

**форкамера** *f.* precombustion chamber, antechamber, prechamber

**форкамерный** *adj.* prechamber, antechamber

**форма** *f.* form, shape, pattern, contour, configuration; state; mode; presentation; make, model; structure; mold, cast; **ф. адреса** address pattern; **ф. волновой кривой, ф. волны** wave shape, wave form; **восковая ф. (для клише)** wax mold; **дискретная ф.** digital form; **ф. для отливки** mold; **ф. излома** structural fracture, grain structure; **ф. кадрового сигнала** frame waveform; **кодированнодесятичная ф.** coded decimal presentation; **ф. колебаний** wave form; **ф. колебаний** $n$-**ной моды** shape of the n-th mode; **ф. коммутирующего сигнала** keying waveform; **ф. кривой без постоянной составляющей** area-balanced waveform; **ф. кривой нагрузки** load pattern; **ф. кривой поля** field form; **ф. кривой с равными площадями по обе стороны от нейтрали** area-balanced waveform; **литейная ф.** mold; **обычная ф.** true form; **ф. огибающей** envelope form; **основная ф. слова** canonical form of word; **ф. переходной характеристики для заднего (для переднего) фронта** trailing (leading) transient; **предварённая ф.** prenex form; **ф. представления порядка со смещением нуля на число N** exponent-plus N form; **ф. пускового импульса** (*or* **сигнала**) trigger waveform; **ф. сигнала опорного генератора** bootstrap waveform; **тупиковая ф.** irredundant form; **ф. управляющего напряжения** (*or* **сигнала**) control waveform, keying waveform; **ф. частотной характеристики в полосе**

пропускания bandpass shape; ф. Эрмита Hermitian form

формализация *f.* formalization

формализировать *v.* to formalize

формализм *m.* formalism

формализованный *adj.* formalized

формальдегид *m.* formaldehyde

формальность *f.* formality

формальный *adj.* formal

форманизотропия *f.* anisotropy of form

форманта *f.* formant; верхняя ф. upper formative sounds; ф. гласного звука речи vowel formant; нижняя ф. sub-formative sound

формантный *adj.* formant

формат *m.* size, form; format, shape; ф. изображения aspect ratio, picture ratio; picture shape, raster shape; ф. кадра aspect ratio, picture ratio

форматный *adj.* format

формация *f.* formation, structure

форменный *adj.* formal, regular, prescribed

формика *f.* formica

формирование *n.* formation, shaping, reshaping; molding; ф. диаграммы излучения beam shaping; предварительное ф. preshaping; ф. прямоугольных импульсов squaring; ф. «рамки изображения» framing (*tv*); ф. сигнала signal normalization, signal standardization; wave shaping; ф. телефонных зон квадратами division of an area by a grid system for simplified calculation of call fees

формированный *adj.* formed, shaped; molded

формирователь *m.* shaper, shaping network; driver amplifier, driver; former; reshaper; shaping unit; ф. возбуждающих импульсов drive-pulse generator; (выходной) ф. управления регистром register driver; ф. записи write driver; ф. импульсов выборки drive-pulse generator; ф. пусковых импульсов trigger shaper; разрядный ф. digit driver; ф. синхронизирующих импульсов clock driver; ф. считывания read driver; ф. тактовых импульсов clock driver; ф. управления регистром register driver

формировать, сформировать *v.* to shape, to form, to model; to reshape; to state; to make; to mold, to cast

формировочный *see* формовочный

формирующий *adj.* shaping, forming; molding

формовальный *see* формовочный

формование *see* формирование, формовка; ф. продавливанием extrusion process

формованный *adj.* formed, shaped; molded, cast

формователь *see* формирователь

формовать, сформовать *v.* to form, to shape, to model; to mold, to cast; to bend the leading wire (*tbs.*)

формовка *f.* forming, bending (of leads); formation, shaping, modeling; molding, casting; ф. растягивания при очень низкой температуре cryogenic stretch forming; ф. точечного транзистора forming (in semiconductors)

формовой *adj.* mold

формовочный *adj.* form, forming; mold, molding; molded

формоизменение *n.* moldability

формообразование *n.* forming

формула *f.* formula; дедуктивно равные формулы interdeducible formulas; формулы прямого и обратного преобразований Фурье reciprocal formulas; формулы расчёта времени реверберации reverberation-time formulas; ф. дальности действия transmission-range formula; исходная ф. assumption formula; ф. Кеплера (для вычисления ёмкости бочек) rough rule for calculation of barrel volume; ф. к.п.д. efficiency formula; математическая ф. formulation, arrangement (of an equation); ф. матричного преобразования matrix formula; общепринятая ф. для ускорения силы тяжести International Gravity Formula; ф. оценки estimator; ф. планиметрирования по трапециям chord-trapezoid formula; пустая ф. null formula; ф. распространения радиоволн radio-transmission formula; структурная ф. constitution formula (atomic structures); configuration formula; ф. трапеций trapezoid(al) rule; элементарная ф. prime formula

формулированный *adj.* formulated; conceived

формулировать, сформулировать *v.* to formulate, to word

**формулировка** *f.* formulation, statement, assertion; **ф. задачи** statement of the problem

**формулятор** *m.* specification form; **расчётный ф.** design sheet

**формфактор** *m.* form factor, crest factor, shape factor

**форометр** *m.* phorometer

**форсаж** *m.* afterburning, afterburner

**форсажный** *adj.* afterburner

**форсирование** *n.* forcing

**форсированный** *adj.* forced

**форсировать** *v.* to force, to speed up, to push, to boost

**форсировка** *f.* forcing; **ф. возбуждения** over excitation

**форсирующий** *adj.* forcing, boosting, pushing

**форстерит** *m.* forsterite

**форсунка** *f.* atomizer; burner, impinging head, injection head, injector nozzle; jet orifice; **ф. для впрыска хладагента-разбавителя** diluent injector

**Фортень** Fortin

**фортепиано** *n. indecl.* piano, pianoforte

**Фортескю** Fortescue

**фортессимо** *adv.* fortissimo, forte forte

**Фортран** Fortran

**форусилитель** *m.* head amplifier, camera amplifier, camera preamplifier

**форшальтер** *m.* toll switching position

**фосвич** *m.* phoswich

**Фостер** Foster

**фосфат** *m.* phosphate; **кислый ф. аммония** ammonium dihydrogen phosphate

**фосфатный** *adj.* phosphate

**фосфóр** *m.* luminophor

**фóсфор** *m.* phosphor, phosphorescent material

**фосфоресцентный** *adj.* phosphorescent

**фосфоресценция** *f.*, **фосфоресцирование** *n.* phosphorescence

**фосфоресцировать** *v.* to phosphoresce

**фосфоресцирующий** *adj.* phosphorescent, luminescent

**фосфористый** *adj.* phosphorus, phosphorous

**фосфорический** *adj.* phosphoric

**фосфорный** *adj.* phosphorus, phosphoric

**фосфороген** *m.* phosphorogen

**фосфороскоп** *m.* phosphoroscope

**фот** *m.* phot

**фот-час** *m.* phot-hour

**ФОТАБ** *abbr.* (**фото-авиационная бомба**) photoflash bomb

**фотикон** *m.* photicon, flashed photicon; **ф. с импульсной подсветкой** flashed photicon

**фотион** *m.* photion

**фото** *abbr.* (**фотографический**) photographic

**фото** *n. indecl.*; *prefix* photograph; photo-, light; photoelectric; photographic

**фотоаппарат** *m.* camera; **ф. для (быстрой) съёмки осциллограмм** oscillo-record camera; **ф. для съёмки показаний приборов** photorecorder; **электростатический печатающий ф.** electrostatic printer

**фотобелок** *m.* master, photomaster

**фотобиение** *n.* photobeat

**фотоваристор** *m.* photovaristor

**фотовизиальный** *adj.* photovisual

**фотовизия** *f.* photovision

**фотовизуальный** *adj.* photovisual

**фотовозбуждение** *n.* photoexcitation

**фотовозбуждённый** *adj.* photoexcited, photogenerated, light-generated

**фотовольтаический** *adj.* photovoltaic

**фотогальванический** *adj.* photovoltaic

**фотогальваномагнитный** *adj.* photoelectromagnetic

**фотогенератор** *m.* photogenerator; **низкочастотный ф.** photoaudio generator

**фотогеология** *f.* photogeology

**фотоголовка** *f.* photoelectric head; **контрольная ф.** scanning head

**фотогониометр** *m.* photogoniometer

**фотогравирование** *n.* photoetch(ing), photolithography; photomasking

**фотогравюра** *f.* photoengraving; photogravure, photoprint

**фотограмма** *f.* photogram

**фотограмметрический** *adj.* photogrammetric

**фотограмметрия** *f.* photogrammetry

**фотограф** *m.* camera man, photographer

**фотографирование** *n.* photographing, photography; **ф. в инфракрасных** (*or* **в тепловых**) **лучах** infrared photography; **ф. с космического летательного аппарата** space photography

**фотографировать, сфотографировать** *v.* to photograph

**фотографически** *adv.* photographically

**фотографический** *adj.* photographic

**фотография** *f.* photography; photograph, picture; **ф. изображения, полученного на люминесцирующем экране** photofluorography; **ф. изображения с экрана радиолокатора** radar photograph; **теневая ф.** shadowgraph; **ф. теневого изображения** shadow photograph

**фотодейтрон** *m.* photodeutron

**фотоделение** *n.* photofission

**фотодетектирование** *n.* photodetection

**фотодетектор** *m.* photodetector; photorectifier

**фотодиод** *m.* photodiode, phototube, photorectifier; **плоскостный ф.** junction photodiode

**фотодиссоциация** *f.* photodissociation

**фотоёмкостный** *adj.* photocapacitor

**фотозапись** *f.* photographic recording, photorecording; **ф. данных** photographic storage; **ф. телевизионных программ** kine-recording, kinescope recording, motion-picture recording

**фотозатвор** *m.* camera shutter

**фотозвукозаписывающий** *adj.* photographic sound recording

**фотозвукорепродуктор** *m.* photographic sound reproducer

**фотоизображение** *n.* facsimile

**фотоимпульс** *m.* photoimpact

**фотоинтерпретация** *f.* photointerpretation

**фотоионизация** *f.* photo-ionization

**фотокамера** *f.* (photo) camera; **ф. для высокоскоростной съёмки, ф. для скоростной съёмки при баллистических испытаниях** ballistic camera; **ф. для съёмки изображений экрана РЛС** radar recording camera; **ф. для съёмки изображений экрана самолётного радиолокатора** airborne radar camera

**фотокардиотахометр** *m.* photocardiotachometer

**фотокатод** *m.* photocathode, photoelectric cathode, photo-sensitive cathode

**фотокатодный** *adj.* photocathode

**фотокинотеодолит** *m.* cinetheodolite

**фотоклистрон** *m.* photoclystron, cavity-type photodiode

**фотокомната** *f.* darkroom

**фотоконтроллер** *m.* photoelectric control, photoelectric controller

**фотокопировальный** *adj.* photocopying

**фотокопия** *f.* photo copy; **позитивная ф.** white print

**фотоксилография** *f.* photoxylography

**фотолаборатория** *f.* photographic laboratory; **ф. войск связи** signal photographic laboratory unit

**фотолампа** *f.* photion; photoflood lamp; photoelectric cell; phototube; **ф. тлеющего разряда** photosensitive discharge tube

**фотолиз** *m.* photolysis

**фотолит** *m.* photolyte

**фотолитический** *adj.* photolytic

**фотолитографический** *adj.* photolithographic

**фотолитография** *f.* photolithography

**фотолюминесцентный** *adj.* photoluminescent

**фотолюминесценция** *f.* photoluminescence

**фотомагнитный** *adj.* photomagnetic

**фотомагнитоэлектрический** *adj.* photomagnetoelectric

**фотомаска** *f.* photographic mask, photomask

**фотомезон** *m.* photomeson

**фотометр** *m.* photometer, photometric receiver; **диффузионный ф.** diffuse-reflection photometer; **ф. для измерения интенсивности дневнего света** hemaraphotometer; **ф. для измерения рассеянного света** light-scattering photometer; **интеграторный ф., интегрирующий ф.** integrating photometer, lumenmeter; **контрастный ф.** equality of brightness photometer, equality of contrast photometer, split-field photometer; **линейный ф.** bench photometer, bar photometer; **мерцающий ф., мигающий ф.** flicker photometer; **объективный ф.** physical photometer; **поглощающий ф.** extinction photometer; **ф. с масляным пятном на экране** grease-spot photometer; **ф. с парафиновыми блоками** Joly photometer; **ф. с равноконтрастными полями** equality-of-contrast photometer; **ф. с равнояркостными полями** equality-of-luminosity photometer; **ф. со смежными полями** equality-of-luminosity photometer, equality-of-contrast photometer; **субъективный ф.** split-field photometer, visual photometer; **трихрома-тический ф.** tristimulus photometer; **ф. Ульбрихта** integrating photometer

**фотометрирование** *n.* photometry, photometric evaluation of light intensity

**фотометрировать** *v.* to measure with a photometer

**фотометрический** *adj.* photometric

**фотометрия** *f.* photometry; **объективная ф.** physical photometry; **субъективная ф.** visual photometry

**фотомеханика** *f.* photomechanics

**фотомеханический** *adj.* photomechanical

**фотомикрометр** *m.* photomicrometer

**фотомодулятор** *m.* photomodulator

**фотомонтаж** *m.* photomontage

**фотон** *m.* photon; **ф. аннигиляционного излучения** annihilation photon

**фотонабор** *m.* phototypesetting

**фотонапряжение** *n.* photoelectric voltage, photovoltage

**фотонейтрон** *m.* photoneutron

**фотонефелометр** *m.* photonephelometer

**фотонный** *adj.* photon

**фотоноситель** *m.* photocarrier

**фотооборудование** *n.* photographic equipment; **разведывательное ф.** photointelligence equipment

**фотообработка** *f.* photoprocessing, photolithography, photoengraving

**фотообразование** *n.* photoproduction

**фотооригинал** *m.* master, photomaster

**фотоосциллограмма** *f.* photooscillogram

**фотоответ** *m.* photoconductive response

**фотоотлипание** *n.* photodetachment

**фотоотрицательный** *adj.* photonegative

**фотоотщепление** *n.* photodetachment

**фотопараметрический** *adj.* photoparametric

**фотопередатчик** *m.* picture transmitter, facsimile transmitter

**фотопечатающий** *adj.* photoprinting

**фотопечать** *f.* photo-offset printing, photoprinting

**фотопластинка** *f.* photographic plate

**фотоплёнка** *f.* photographic film, photographic tape; **высокочувствительная ф.** high-speed film

**фотоположительный** *adj.* photopositive

**фотопочта** *f.* photomail, photomail

system; **воинская ф.** signal photomail organization

**фотоприбор** *m.* photodevice

**фотоприёмник** *m.* recorder (*tv*); photodetector, photoconductive detector

**фотоприлипание** *n.* photoattachment

**фотопроводимость** *f.* photoconductivity, photoconduction, photoconductive effect, photoelectric conductivity, photoconductive properties; **собственная ф.** intrinsic photoconduction

**фотопроводник** *m.* photoconductor

**фотопроводниковый** *adj.* photoconductor

**фотопроводящий** *adj.* photoconductive

**фотопротон** *m.* photoproton

**фоторадиограмма** *f.* photoradiogram

**фоторадиосвязь** *f.* facsimile radio link

**фоторазведка** *f.* photoreconnaissance

**фоторазложение** *n.* photodissociation

**фотораспад** *m.* photodisintegration

**фоторасщепление** *n.* photodisintegration

**фотореакция** *f.* photoreaction

**фоторегистратор** *m.* photorecorder

**фоторегулятор** *m.* photo controller

**фоторезист** *m.* photoresist

**фоторезистивный** *adj.* photoresistor, photoresistive

**фоторезистор** *m.* photoresistor, light-dependent resistor

**фоторекордер** *m.* photographic recorder

**фотореле** *n. indecl.* photorelay, photoswitch, photo-emissive relay, light relay; **ф. времени** photoelectric timer, phototimer; **просмотровое ф.** photoelectric scanner; **резистивное ф.** photoresistive relay, selenium-cell relay

**фоторелейный** *adj.* photorelay; light-barrier

**фоторентгенография** *f.* photoroentgenography

**фоторецептор** *m.* photoreceptor

**фотосамописец** *m.* photographic recorder

**фотосвязь** *f.* facsimile service, facsimile transmission; **штриховая ф.** facsimile service

**фотосинтез** *m.* photosynthesis

**фотосинтетический** *adj.* photosynthetic

**фотосирена** *f.* photoaudio generator

**фотосмесительный** *adj.* photomixer

**фотоснимок** *m.* picture, photograph, photographic print

**фотосопротивление** *n.* photoresistance, photoresistor; photoconductive cell, photoconductor, photoconductive detector; light resistor

**фотосфера** *f.* photosphere

**фотосчётчик** *m.* photoelectric counter

**фотосъёмка** *f.* photographing, taking (of pictures); exposure, take, shot, photograph

**фототелеграмма** *f.* phototelegram, teleautogram, facsimile, radiophotogram

**фототелеграф** *m.* phototelegraph, pantelegraph; wirephoto system, facsimile broadcast station; **ф. Белена** Belin facsimile equipment

**фототелеграфия** *f.* phototelegraphy, facsimile (system), picture telegraphy, wirephoto system, pantelegraphy, telephotography; **ф. изображения** picture transmission; **ф. с использованием радиоканала** radiophototelegraphy

**фототелеграфный** *adj.* facsimile, phototelegraphy

**фототелефония** *f.* phototelephony, picture telegraphy

**фототеодолит** *m.* phototheodolite

**фототермоэлектрический** *adj.* photothermoelectric

**фототипический** *adj.* phototype

**фототипия** *f.* phototype, photoengraving

**фототиристор** *m.* light-activated thyristor

**фототлеющий** *adj.* photoglow

**фототок** *m.* photocurrent, photoelectric current

**фототравление** *n.* photoetching

**фототранзистор** *m.* phototransistor, electrooptical transistor; **слоистый ф.** junction phototransistor

**фототрафарет** *m.* artwork

**фототриод** *m.* phototriode; **полупроводниковый ф.** phototransistor; **слоистый ф.** junction phototransistor

**фототрон** *m.* phototron

**фототропический** *adj.* phototropic

**фототропия** *f.* phototropy

**фотоувеличение** *n.* photo enlarging

**фотоувеличительный** *adj.* photo-enlarger, photo-enlarging

**фотоудар** *m.* photoimpact

**фотоумножитель** *m.* photomultiplier cell, multiplier phototube, electron-multiplier phototube, secondary emission photocell; **ф. с торцевым фотокатодом** head-on photomultiplier tube; **электронный ф.** photomultiplier tube

**фотоумножительный** *adj.* photomultiplier

**фотоупругий** *adj.* photoelastic, piezo-optical

**фотоупругость** *f.* photoelasticity

**фотоусилитель** *m.* photoamplifier

**фотофильтр** *m.* camera filter, taking filter

**фотофлюорография** *f.* photofluorography

**фотофон** *m.* photophone; **ф. с фотоэлементом** pallophotophone

**фотофор** *m.* photophore

**фотофорез** *m.* photophoresis

**фотоформер** *m.* photoformer

**фотохимический** *adj.* photochemical, actinic

**фотохимия** *f.* photochemistry

**фотохром** *m.* photochrome

**фотохроматический** *adj.* photochromatic

**фотохромия** *f.* photochromy

**фотохромный** *adj.* photochromic

**фотоцинкография** *f.* photozincography

**фоточувствительность** *f.* photosensitivity, photoelectric sensitivity

**фоточувствительный** *adj.* photosensitive

**фото-эдс** photo-emf

**фотоэдс-элемент** *m.* photovoltaic cell

**фотоэдс-эффект** *m.* photovoltaic effect

**фотоэлектрический** *adj.* photoelectric, photovoltaic, photovoltage

**фотоэлектричество** *n.* photoelectricity

**фотоэлектродвижущий** *adj.* photoelectromotive

**фотоэлектромагнитный** *adj.* photoelectromagnetic

**фотоэлектрон** *m.* photoelectron

**фотоэлектроника** *f.* photoelectronics

**фотоэлектронный** *adj.* photoelectronic

**фотоэлектропроводящий** *adj.* actinodielectric

**фотоэлемент** *m.* photocell, photoelectric cell, phototube, light cell; photoemissive element, photoelectric tube; electric eye; photodetector; **вентильный ф.** rectifier photocell, photovoltaic cell, barrier-layer photocell, blocking-layer photocell; **ф. в качестве сопротивления** light resistor; **вентильный ф. с тыловым**

**фотоэффектом** back-effect photocell, barrier-layer rear-wall photocell; **вентильный ф. с фронтальным фотоэффектом** front-effect photocell, barrier-layer front-wall photocell; **ф., вырабатывающий бланкирующие** (*or* **гасящие**) **импульсы** blanking gate photocell; **газонаполненный ф.** gas phototube, gas-filled photocell, photoglow tube; **гальванический ф.** primary cell; **ф. для определения положения** aspect photocell; **матричный ф.** phototron; **меднозакисный вентильный ф.** photox cell; **полупроводниковый ф.** photocell, photoelectric cell; **резистивный ф.** photoresistance cell, photoconductive cell; **ф. с внешним фотоэффектом** extrinsic-effect phototube, emission phototube, photoemission cell; **ф. с внутренним фотоэффектом** intrinsic-effect photocell, sandwich photocell, photoconductive cell, photoresistance cell, photovoltaic cell; **ф. с заграждающим слоем, ф. с запирающим слоем** barrier-layer photocell, blocking-layer photocell, photovoltaic cell, barrier-plane photocell; **ф. с запорным слоем** photovoltaic cell, barrage photocell, barrier-film photocell, barrier-layer photocell, blocking-layer photocell, rectifier photocell); **ф. с полным поглощением света** black-body photocell; **ф. с полупрозрачным фотокатодом** front-effect photocell; **ф. с управляемой сеткой** three-electrode cell; **сверхвысокочастотный ф.** microwave phototube, traveling-wave phototube; **ф., чувствительный к инфракрасному излучению** infrared photocell; **электровакуумный ф.** phototube, photovalve

**фотоэлементный** *adj.* phototube, photocell

**фотоэмиссионный** *adj.* photoemissive

**фотоэмиссия** *f.* photoemission, photoemissivity

**фотоэмульсия** *f.* photographic emulsion

**фотоэффект** *m.* photoeffect, photoelectric effect, photoemissive effect; *genit.* **-а** *often* photoelectric: **граница фотоэффекта** photoelectric threshold; **вентильный ф.** photovoltaic effect, barrier-layer effect; **внешний ф.**

photoemission, photoelectric emission, extrinsic photoeffect, photoemissive effect, external photoeffect; **внутренний ф.** photoconductive effect, intrinsic photoeffect, internal photoelectric effect, inner photoeffect; **ф. с запором** photovoltaic effect

**фотоядерный** *adj.* photonuclear

**фотран** *m.* photran

**ФП** *abbr.* (**фильтр-поглотитель**) absorption filter

**ФР** *abbr.* (**фазорегулятор**) phase shifter

**фр.** *abbr.* (**фракция**) fraction

**фрагмент** *m.* fragment

**фрагментарный** *adj.* fragmentary

**фрагментация** *f.* fragmentation

**фраза** *f.* phrase, sentence; **артикуляционная ф.** carrier sentence

**фразеологический** *adj.* phraseological

**фракционирование** *n.* fractionation

**фракционированный** *adj.* fractional, fractionated

**фракционировать** *v.* to fractionate

**фракционировка** *f.* fractionation

**фракционно** *adv.* fractionally; in steps, by degrees

**фракционный** *adj.* fractional; factional

**фракция** *f.* fraction, cut; group, faction

**Фрам** Frahm

**Франке** Franke

**франкий** *m.* frankium, Fr

**франкирование** *n.* prepayment (of postage)

**франкирующий** *adj.* postage-printing

**Франклин** Franklin

**франций** *m.* francium, Fr

**Фраунгофер** Fraunhofer

**фраунгоферов** *poss. adj.*, **фраунгоферовский** *adj.* Fraunhofer's

**фрахт** *m.* cargo, shipment, freight

**фреза** *f.* cutter, milling cutter; **ф. для зубчатых колёс, зуборезная ф.** gear cutter; **комбинированная ф.** gang cutter; **концевая ф.** end bill; **наборная ф.** gang cutter; **радиальная ф.** side milling cutter; **трубчатая ф.** concave cutter; **фасонная ф.** profile cutter; **хвостовая ф.** end bill; **цилиндрическая ф.** plane-milling cutter; **шлицевая ф.** slitting cutter

**фрезер** *m.* cutter, milling cutter; **фасонный ф.** profile cutter

**фрезерный** *adj.* cutting, milling

**фрезерование** *n.* cutting, milling, notching

**фрезерованный** *adj.* milled
**фрезеровать, отфрезеровать** *v.* to mill, to cut
**фрезеровочный** *adj.* milling, cutting
**Фрейштедт** Freystedt
**Френа** Frena
**френелев** *poss. adj.* (of) Fresnel, Fresnel's
**Френель** Fresnel
**френотрон** *m.* frenotron
**Френсис** Francis
**фреон** *m.* freon
**фресканар** *m.* frequent scanning radar
**фригориметр** *m.* frigorimeter
**Фрик** Frick
**фрикативный** *adj.* fricative (sound)
**« фриквента »** *f.* frequenta
**« фриквентит »** *m.* frequentit
**фрикционный** *adj.* friction(al)
**фритта** *f.* frit
**фриттер** *m.* acoustic shock absorber, coherer
**фриттировать, фриттовать** *v.* to frit, to cohere
**фриттовый** *adj.* frit
**фронт** *m.* front; edge; **ф. (импульса)** porch; **бесконечно крутой передний ф.** infinitely sharp leading edge; **ведущий ф.** leading edge; **ф. волны переменной электродвижущей силы** potential front; **задний ф. гасящего (импульса)** back-back porch; **задний ф. (импульса)** trailing edge, back edge, lagging edge; **ф. Найквиста** Nyquist flank; **передний ф.** leading edge, front edge; **пологий ф. волны** sloping wave front; **ф. распространения ударной волны** detonation front; **ф. сигнала** wave front; **ф. скачка уплотнения** pressure front; **ф. ударной волны** shock front, pressure front
**фронтальный** *adj.* front(al), broadside
**фронтовый** *adj.* front; edge
**Фруд** Froude
**Фрум** Froome
**ф-с** *abbr.* (**фот-секунда**) phot-second
**фт.** *abbr.* (**фут**) foot
**фталевый** *adj.* phtalic
**фтор** *m.* fluorine, F
**фторированный** *adj.* fluorinated
**фторокрилаты** *pl.* fluoroacrylates
**фторопласт** *m.* teflon
**фторофосфатный** *adj.* fluorophosphate
**фторохимический** *adj.* fluorochemical

**ФТТ** *abbr.* (**Физика твёрдого тела**) Solid State Physics (Journal)
**Фуко** Foucault
**фулерфон** *m.* fullerphone
**фулкрум** *m.* fulcrum
**фультограф** *m.* Fultograph
**Фультон** Fulton
**фундамент** *m.* foundation, base, bed, groundwork; *pl.* footings; **ф. опоры с подушкой** pad and pedestal type footings
**фундаментально** *adv.* fundamentally
**фундаментальность** *f.* fundamentality, solidity
**фундаментальный** *adj.* fundamental, substantial, solid; foundation; characteristic
**фундаментный** *adj.* base, foundation
**фуникулёр** *m.* funicular (railway)
**функтор** *m.* functor
**функционал** *m.* functional (*math.*)
**функциональный** *adj.* function(al)
**функционирование** *n.* service, functioning, operation
**функционировать** *v.* to function, to operate, to run
**функционирующий** *adj.* functioning, functional
**функция** *f.* function, connective; purpose, service; **функции коры головного мозга** cerebral functions; **многозначные функции** multiform functions; **функции сложения цветов** distribution coefficients; **фундаментальные функции** characteristic functions; **в функции от, как функция от** as a function of; **ф. верности** truth function; **ф. веса** weighting function; **взаимно корреляционная ф.** cross-correlation function; **ф. взятия целой части** entire (function); **ф. включения** step function; **возмущающая ф.** drive; **ф. времени** time variable data, function of time; **вторая ф. распределения вероятностей** second probability distribution; **ф. выгоды** utility function; **ф. выделения** collate function; **вырожденная ф.** confluent function; **дискриминантная ф.** discriminator (of data processing); **добропорядочная ф.** well-behaved function; **ф. единичного вектора** locus function; **зубчатая ф.** spike function; **ф. « и »** AND connective, AND function; **ф. изменения полярности** commutating

function; **изобразимая ф.** reckonable function; **ф. «и-или»** AND-to-OR function; **ф. «или»** OR connective; **интегрируемая ф.** integrand; **исходная ф.** assumed function; antiderivative of a function; **квадратичная ф.** square-law function; **квазипериодическая ф.** quasi transfer function; **классифицирующая ф.** discriminator (of data proc.); **косвенная ф.** sub-function; **ф. логического умножения** collate function; **монотонная ф.** monotonic quantity; **начальная ф.** object function; **невозрастающая ф.** decreasing function; **ф. нескольких переменных** variable function; **неявная ф.** implicit function; **нормальная ф. распределения модуля величины** normal magnitude probability function; **обратная гиперболическая ф.** antihyperbolic function, arc hyperbolic function; **обратная круговая ф.** inverse hyperbolic function; **ф. ориентации в пространстве** spatialization function; **особая ф.** characteristic function, eigenfunction; **ф. от переменных величин** function of variables; **ф. относительной видности** luminosity function; **ф. отрицания** "not" function; **первообразная ф.** antiderivative of a function; **передаточная ф.** transfer function, transmittance; **передаточная ф. замкнутой системы с обратной связью** feedback transfer function, return transfer function; **передаточная ф. по регулирующему воздействию** actuating transfer function; **ф. передачи моделирующего устройства** simulant transfer function; **ф. передачи по току для контура (обратной связи)** loop current transmission; **ф. передачи разомкнутой следящей системы** feedback transfer function; **ф. передачи системы наведения по лучу** beam-rider transfer function; **ф. передачи следящей системы** servo system transfer function; **переходная ф.** transient function; recovery characteristic; **ф. плотности** frequency function (*math.*); **ф. плотности сбоев** failure density function; **подинтегральная (**or **подынтегральная) ф.** integrand; **показательная ф.** чётной степени linear function; **ф.**

**преобразования, преобразующая ф.** transfer function; **производящая ф.** course-of-value function, generating function; **ф. разряда** disruptive function; **ф. распределения по скоростям** speed distribution function, velocity distribution function; **распределительная ф.** Fermi's distribution formula; **результирующая ф. передачи** overall transfer function; **решающая ф., ф. решения** decision function; **симметричная ф. корня** symmetrical radical function; **синхронизирующая ф.** buffer function; **собственная ф.** eigenfunction; **согласующая ф.** buffer function; **степенная ф.** power function; **ступенчатая ф. напряжения** step-function voltage; **ступенчатая пилообразная ф.** ramp function; **угловая ф.** goniometric function; **ф. упругого последействия** retarded elasticity function; **целая ф.** integer function; **частотно-постоянная ф.** part-function; **ф. частотно-фазовой характеристики** frequency response function; **эмпирическая ф. распределения** sum polygon

**фунт** *m.* pound

**фуран** *m.* furan

**фурановый** *adj.* furan

**фургон** *m.* car; truck, wagon, van; **измерительный ф.** cable testing truck, cable testing car

**«фурнатрон»** *m.* furnatron

**фурфуран** *m.* furan

**фурфурол** *m.* furfural

**фурфуроловый** *adj.* furfural

**Фурье** Fourier

**фут** *m.* foot; **«инерционный ф. в квадрате»** slug square feet

**футерованный** *adj.* lined

**футеровать** *v.* to line

**футеровка** *f.* lining

**футеровочный** *adj.* lining

**фут-ламберт** *m.* equivalent foot-candle, foot-lambert

**футляр** *m.* enclosure, box, container; housing; case, slip, jacket, sheath; barrel; **ф. приёмника** container; **ф. телевизионного приёмника** TV cabinet

**футовый** *adj.* foot

**футосвеча, фут-свеча** *f.* foot-candle

**футшток** *m.* tide-gage, depth gage, sounding rod; foot rule

**ф-ч** *abbr.* (**фот-час**) phot-hour

**ФЭ** *abbr.* (**фотоэлемент**) phototube, photocell

**ФЭБ** *abbr.* (**функциональный электронный блок**) functional electronic block

**фэр** *abbr.* (**физический эквивалент рентгена**) rep, physical roentgen equivalent

**фэсом** *m.* fathom

**ФЭУ** *abbr.* (**фотоэлектронный умножитель**) photomultiplier, electron-multiplier phototube, secondary emission photocell

**фюзеляжный** *adj.* fuselage; skin (antenna)

**Фюрстенау** Furstenau

# X

Хаас Haas
Хазелтайн Hazeltine
хайдайн *m.* hidyne, hydyne
Хайдингер Haidinger
хайдродайн *m.* hydraudyne
Хайль Heil
хайперко *n. indecl.* hiperco
хайперник *m.* hipernik
хайперсил *m.* hypersil
хайпертин *m.* hiperthin
хайрад *m.* hyrad
хайтрон *m.* hytron
халкопирит *m.* chalcopyrite
халько- *prefix* chalco-
халькопирит *m.* chalcopyrite
Хан Hahn
Ханкель Hankel
хаотически *adv.* chaotically, at random, in disorder
хаотический *adj.* chaotic, random, disorganized; turbulent
хаотично-неоднородный *adj.* randomly inhomogeneous
хаотичность *f.* randomness
хаоточный *see* хаотический
характер *m.* state, condition, nature, character, type; mode; **х. индикации цели** target information; **х. кривых** rate of curves; **х. работы** functioning, mode of operation
характеризовать *v.* to characterize
характеризующий *adj.* characteristic, essential, specific; characterizing
характериограф *m.* cathode-ray curve tracer
характеристика *f.* characteristic, character, property; pattern; characterization; parameter; rating; curve, characteristic curve; index of logarithm; performance, response; *pl. oft.* operating lines; **характеристики после длительной работы** long-time quality; **реальные характеристики** objective parameters; **стандартные номинальные характеристики** standard rating; **с удлинённой характеристикой** variable-mu; **акустическая х.** acoustic response; **ам-** плитудно-частотная х. amplitude versus frequency response characteristic, amplitude frequency response, gain-frequency characteristic, amplitude-response curve; **анодная х. фотоэлемента** anode characteristic of a phototube; **апертурная х.** aperture response; **в-а х.** E-I characteristic, i-V characteristic; **х. в неустановившемся режиме** transient response; **х. в плоскости магнитного поля** magnetic-plane characteristic; **х. в установившемся режиме** steady-state response; **х. воспроизведения цветов** color-response curve; **входная х.** base characteristic; **высокочастотная х. ионосферы** wide-band ionosphere sounding curves; **градационная х.** grey-tone response, half-tone response; **х. графика направленности** directivity pattern, directional response pattern; **графическая х.** characteristic, characteristic curve; **х. групповой задержки** envelope delay characteristic; **х. дальности** range performance; **динамическая частотная х.** dynamic response curve; **дисперсионная х.** dispersion curve; **х. для качающейся частоты** swept performance; **х. для установившегося режима** steady-state response; **х., зависимая от частоты** (frequency) response; **х. зависимости выходной величины от входной** input-output characteristic; **х. зависимости сдвига фазы от частоты** phase-shift-frequency characteristic, phase-versus-frequency response characteristic; **х. затягивания** pull-in performance; **защитная х.** expulsion-tube characteristic; **х. звукопоглощения различных поверхностей** (sound) damping factor of different walls; **индукционная нагрузочная х.** zero power-factor characteristic (of a synchronous machine); **х. инерционности** lag characteristic, lag curve, persistence

characteristic; **х. контрастности** contrast-response characteristic, contrast-response curve; **контрольная х.** reference performance; **х. коэффициента добротности** figure-of-merit characteristic; **х. крутизны** transconductance characteristic; **х. лампы с пологим шлейфом** tailed characteristic; **линейная фазовая х. системы** linear-phase system function; **логарифмическая амплитудно-частотная х.** "decibel-log-frequency"; **логарифмическая частотная х., выраженная в децибелах** decibel-log frequency characteristic; **механическая х.** speed-torque characteristic; speed-load curve; **х. микрофона, представляя отдачу по давлению в зависимости от частоты** pressure-response frequency characteristic of microphone; **х. момента в функции вылета ротора** power-angle curve; **монотонная х.** monotonic response; **х. на выходе** terminal characteristic; **х. нагрузки** load line; **нагрузочная х.** operating characteristic, load characteristic, full-load saturation curve; **х. направления луча** beam pattern; **х. направленности в электрическом поле** electric-plane characteristic; **х. нарастания амплитуды** transient response; **неколебательная переходная х. регулирования** nonoscillatory control response; **нормальная частотная х.** Gaussian frequency response; **х. обратного тока** graphical representation of multigrid tube characteristic; **х. обратного хода развёртки** retrace characteristic; **х. обратной связи** backward transfer characteristic; **общая х. передачи (телевизионной) системы** overall brightness-transfer characteristic, overall brilliance-transfer characteristic; **х. отдачи** efficiency characteristic; **относительная х.** relative response; **х. отражений от земной поверхности, х. отражений от почвы** ground-echo pattern; **х. отражения** echoing characteristic; **х. отражения цели** coverage diagram; **х. отсечки** shut-down characteristic; **падающая х.** dropping characteristic, falling characteristic, negative characteristic, variable-speed; **падающая**

**вольтамперная х.** negative (resistance) characteristic; **переходная х.** transient response, transfer characteristic, surge characteristic, unit-function response, step response; **переходная х. заднего фронта** trailing transient; **переходная х. переднего фронта** leading transient; **плоская х.** flat response; **плоская частотная х.** flat passband; **х. по высокой частоте** radio-frequency performance; **х. по зеркальному каналу** image response; **х. подавления** rejection performance; **положительная вещественная частотная х.** positive real frequency response; **х. помехоустойчивости** noise-canceling characteristic; **х. послесвечения** light-decay characteristic, persistence characteristic, lag characteristic; **х. потребителя** consumption characteristic; **х. прекращения впуска топлива** shut-down characteristic; **х. преобразования** conversion diagram; **х. при 25% выходной мощности** quarter-power response; **х. при переходном процессе** transient response; **х. пропускания низов** bass response; **пространственная х.** directional characteristic (microphones); **прямая х. нагрузки** load line; **х. работы** performance data, performance curve; **х. работы в переходном режиме** transient properties; **равномерная частотная х.** uniform response; **х. радиолокационной станции** radar performance; **х. радиолокационных изображений** response on radar displays; **х. разряда, разрядная х.** flashover characteristic; **х. распределения энергии Ферми** Fermi characteristic energy; **расчётная х.** estimated performance; **реальная х.** objective parameter; **результирующая х.** combined characteristic; **результирующая х. передачи** overall transfer function; **х. с крутым срезом, х. с крутым фронтом** steep-sided characteristic; **х. с пологим нижним шлейфом** tailed characteristic; **х. свечения** fluorescent characteristic; **х. сигнал-свет** transfer characteristic, current-output-versus-light-input relationship; **х. системы «от света до света»** overall brightness-transfer characteristic;

**скачкообразная x.** square-wave response, transient response; **x., снятая на зажимах** terminal characteristic; **спектральная x. (для видимой части спектра)** screen color characteristic; **спектральная x. передающего устройства** taking characteristic; **спектральная x. телевизионной передающей камеры** pick-up spectral characteristic; **x. стационарного состояния** steady-state response; **степенная x.** power characteristic; **тепловая x.** thermal response; **удлинённая x.** tailed characteristic; **super-control characteristic, variable mu; **x. усиления в области высоких частот** high-frequency response; **x. усиления в области низких частот** low-frequency response; **x. усилителя** amplifier response; **x. установившегося режима** steady-state characteristic; **x. устройства в полосе видеочастот** video characteristic; **x. холостого хода, x. холостой работы** open-circuit characteristic, no-load characteristic; **частотная x. в полосе пропускания** band-pass response; **частотная x. отдачи** efficiency frequency characteristic; **частотная x. по видеосигналу** video response; **частотная x. помещения** room response; **частотная x. при синусоидальном испытательном сигнале** sine-wave response; **частотно-фазовая x.** phase versus frequency response characteristic, phase-shift-frequency characteristic, frequency response curve, frequency plot; **x. чувствительности** response curve; **x. чувствительности измерительного прибора** meter response; **широкополосная частотная x. чувствительности** wide-band frequency response; **x. шумоустойчивости** noise-canceling characteristic; **эквивалентная x.** lumped characteristic, total characteristic; **x. эксплуатационных свойств лампы** performance curve (of tube); **x.: электрон А-электрод В** mutual characteristic, transfer characteristic

**характеристикограф** *m.* oscilloscope used to display operating characteristics of tubes, transformers etc.

**характеристический, характеристичный** *adj.* characteristic(al); performance

**характерно** *adv.* 1. *adv.* characteristically; significantly; 2. *pred. adj.* it is characteristic

**характерный** *adj.* characteristic, representative, specific; typical; distinctive

**характерограф** *m.* cathode-ray curve tracer, automatic recorder for volt-ampere characteristic

**характрон** *m.* charactron

**Харвардское** *n. adj. decl.* Harvard-Groningen Catalogue; Draper Catalogue of the Harvard Observatory

**Харес** Harres

**Харкин** Harkin

**хармодотрон** *m.* harmodotron

**Хартлей, Хартли** Hartley

**хартлей** *m.,* **хартли** *n. indecl.* hartley (unit of information)

**Хартман** Hartman

**Хартри** Hartree

**хартрон** *m.* hartron

**Хатчингс** Hutchings

**Хатчинс** Hutchins

**Хаугвиц** Haugwitz

**Хаукинс** Hawkins

**Хаусдорф** Hausdorff

**x/б** *abbr.* (**хлопчатобумажный**) cotton

**хватать, хватить** *v.* to catch, to catch hold (of), to grasp, to seize, to snatch; to be sufficient, to last; **не хватать** to fall short

**хвост** *m.* tail; shank, shaft; tailing; line, queue; **парафиновые хвосты** wax tailings; **x. импульса** pulse stretching, tail; **ласточкин x.** dovetail; **x. распределения** tail area; **расщепленный x. кометы** bifid; **x. сигнала** wave tail, tailing

**хвостик** *m.* tag, tab (of a cathode); pigtail; tail

**хвостовик** *m.* shaft, stem

**хвостовой** *adj.* tail; rear, posterior

**хевея** *f.* hevea

**Хевисайд** Heaviside

**хевисайдов** *poss. adj.* Heaviside's, (of) Heaviside

**Хегнер** Heegner

**Хейниш** Heinisch

**Хейфорд** Hayford

**Хек** Hoek

**Хелл** Hull

**хемибел** *m.* hemibel

хемилюминесценция *f.* chemilumin-escence
**Хемминг** Hamming
**хемо-** *prefix* chemo-
**хемопауза** *f.* chemopause
**хеморецептор** *m.* chemoreceptor
**хемосорбция** *f.* chemical absorption, chemical adsorption, chemisorption
**хемосфера** *f.* chemosphere
**Хенд** Hund
**Херглотц** Herglotz
**херолд** *m.* Herald system
**Хертер** Hurter
**Хертли** Hartley
**Хефф** Huff
**Хивисайд** Heaviside
**хим.** *abbr.* (**химический; химия**) chemical; chemistry
**химикат** *m.* chemical
**химико-** *prefix* chemico-, chemical
**химикотехнологический** *adj.* chemical-engineering
**химилюминесценция** *f.* chemilumin-escence
**химически** *adv.* chemically
**химический** *adj.* chemical
**химия** *f.* chemistry; **x. веществ с ничтожно малой концентрацией** trace chemistry; **x. высокоактивных веществ** hot chemistry; **космическая x.** space chemistry; **x. поверхностных явлений** surface chemistry
**«химн»** *m.* hymn (alloy)
**химолюминесценция** *f.* chemilumin-escence
**химотронный** *adj.* solion-liquid
**хингидрон** *m.* quinhydrone
**хингидронный, хингидроновый** *adj.* quinhydrone
**Хинчин** Khintchine
**хиперко** *n. indecl.* hiperco
**Хипп** Hipp
**хир.** *abbr.* (**хирургический; хирургия**) surgical; surgery
**Хиран** Hiran
**хирургический** *adj.* surgical
**хирургия** *f.* surgery
**Хисинг** Heising
**хитро** *adv.* cleverly, skillfully
**хитрость** *f.* trick, stratagem; cunning
**хищение** *n.* tampering
**хладагент** *m.* coolant; refrigerant
**хладагент-разбавитель** *m.* diluent
**Хладни** Chladni
**хладниев** *poss. adj.* Chladni's, (of) Chladni

**хладноломкость** *f.* cold shortness
**хладностойкий** *adj.* cold-resistant, anti-freezing
**хладность** *f.* cold brittleness, cold shortness
**хладоагент** *m.* coolant; refrigerant
**хладоломкость** *f.* cold shortness
**хладотехника** *f.* refrigerating technique
**хл.-бум** *abbr.* (**хлопчатобумажный**) cotton
**хлестание** *n.* whipping, lashing, flapping
**хлестать, хлестнуть** *v.* to whip, to lash, to flap
**хлопание, хлопанье** *n.* popping, motor-boating; banging, knocking; flapping, whipping
**хлопать, хлопнуть** *v.* to bang, to knock, to clap; to whip, to flap
**хлопковый** *adj.* cotton
**хлопнуть** *pf. of* хлопать
**хлопок** *m.* cotton
**хлопок** *m.* clap, popping; boom; **сверхзвуковой x.** sonic boom
**хлопчатобумажный** *adj.* cotton
**хлопчатый** *adj.* cotton
**хлопья** *pl.* flakes; flocs
**хлор** *m.* chlorine, Cl, chlorine gas
**хлорат** *m.* chlorate; **x. бария** barium chlorate; **x. калия** potassium per-chlorate
**хлорелла** *f.* chlorella
**хлорид** *m.* chloride
**хлоридный** *adj.* chloride
**хлористоводородный** *adj.* hydro-chloride (of)
**хлористый** *adj.* chlorine, chlorous; chloride (of)
**хлоркаучук** *m.* chlorinated rubber
**хлорнафталин** *m.* chlorine-naphtalene
**хлорнокислый** *adj.* chlorate (of); chloric acid
**хлорометр** *m.* chlorometer
**хлоропрен** *m.* chloroprene
**хлоропреновый** *adj.* chloroprene
**хлорсеребряный** *adj.* silver chloride
**XM** *abbr.* 1 (**хлорат магния**) magnesium chlorate; 2 (**холодная масса**) cold air mass
**ход** *m.* motion, run, running, move, movement; flow, flux, course, path, passage, going; rate, speed, pace, step; operation, functioning; gear, thread; action; stroke; dependence (of a curve), shape, trend; range; **на ходу** on the run, in motion; **анкерный x.** anchor escapement; **бара-**

банный х. scala tympani; **безуспешный х.** unsuccessful feature of behavior; **вековой х. магнитного поля Земли** secular variation; **вестибулярный х.** scala vestibuli; **х. винта** (screw) thread; pitch; **х. витка** lay, twist, twisting; angular momentum, moment of momentum; **годовой х. магнитного поля Земли** annual variation; **х. Грехема** Graham escapement; **задний х.** return movement, back travel, reverse run; retrace, retrace of a sawtooth pulse (*tv*); **х. заслонки** throttling range; **х. иглы изнутри наружу** inside-out stylus motion; **начальный х.** opening move; **недвоичный х.** nonbinary code; **неудачный х.** unsuccessful feature of behavior; **обратный х.** return motion, back stroke, return trace, retrace, reverse run; flyback, flyback retrace, homing action; kickback, back swing; **обратный х. ленты после считывания** tape backspacing; **обратный х. по кадрам** (*or* **по кадру**) frame flyback, vertical flyback; **плавкий х. (искателя)** smooth working (of a selector); **х. плунжера клапана** throttling range; **х. по инерции** idling of a selector; **х. поля** field(-strength) distribution; **поступательный х.** forward run, forward direction; **пробный х.** running-in test; **рабочий х.** working stroke; **х. развёртки** trace of a scan, scan, trace, scan stroke; **х. расчёта** computation procedure; **х. резьбы** (screw) thread; **синхронный х.** synchronism, ganging; **случайный х.** chance move; **совместный х. (пружин реле)** follow (of relay) springs; **суточный х. магнитного поля Земли** diurnal variation; **х. температуры** temperature course; **удачный х.** successful feature of behavior; **улиточный х.** scala media, spiral cochlear canal; **успешный х.** successful feature of behavior; **хитрый х.** syllogism; **холостой х.** idle running, no-load running, light running, no-load idling; lost motion, backlash; open-circuit conditions, open-circuit idling; **чрезмерный х.** overshoot, overstroke; **х. электропечи** furnace process, furnace cycle; **х. якоря** armature stroke

**ходектрон** *m.* hodectron

**ходить** *v.* to go, to run, to work; to pass; **х. в масть** to follow suit

**ходовой** *adj.* going, running; lead. leading; track, path

**ходоскоп** *m.* hodoscope

**ходоуменьшитель** *m.* reducing gear

**ходоуменьшительный** *adj.* speed-reducing

**хозрасчёт** *m.* self-support; cost accounting

**хозяйственно** *adv.* economically

**хозяйственность** *f.* economy

**хозяйственный** *adj.* economical, economic

**хозяйство** *n.* economy; plant; industry; **телефонное х.** telephone property, plant

**Хойль** Hoyle

**Хойт** Hoyt

**хол.** *abbr.* (**холодный**) cold

**Холвек** Holweck

**холинэстераза** *f.* cholinesterase

**Холл** Hall

**холловский, холовский** *adj.* (of) Hall, Hall's

**Холль** Hall

**холод** *m.* cold, coldness, chill

**холод.** *abbr.* (**холодильник**) refrigerator. cooler

**холод(н)еть, похолод(н)еть** *v.* to grow cold, to cool, to chill

**холодильник** *m.* refrigerator, cooler; condenser; **компрессорный х.** compressor-type refrigerator; **х. смешения** mixing condenser, condensing jet

**холодильный** *adj.* refrigerating, cooling, refrigerative, refrigerant; condensing

**холодить, похолодить** *v.* to refrigerate, to cool

**холодно** *adv.* cold, coldly; it is cold

**холоднокатаный** *adj.* cold-rolled

**холодноломкий** *adj.* cold-short (*met.*)

**холоднообработанный** *adj.* cold-wrought, coldworked

**холодность** *f.* coldness

**холоднотянутый** *adj.* cold-drawn, hard-drawn

**холодный** *adj.* cold, cool, chilly, frigid

**холодопроизводительность** *f.* refrigerating capacity

**холодостойкий** *adj.* resistant to cold

**холодостойкость** *f.* cold constancy, resistant to cold; hardness

**холостой** *adj.* idle, free, loose-running; blank, empty, dummy; dead

**холст** *m.* cloth, linen, fabric; canvas; mop

**холстинный, холстяной, холщёвый** *adj.* cloth, linen, fabric; canvas; mop, mopping

**Хольвек** Holweck

**Хольцкнехт** Holzknecht

**Хоман** Hohmann

**хомут** *m.* clamp, clip; collar, bow, ring, hoop; strap; **х. для присоединения к заземлителю** ground clamp; **зажимной х.** binding clip; **закрепляющий х.** fixing strap, mounting bracket; **скрепляющий х.** mounting bracket; **стяжной х.** band

**хомутик** *m.* clamp, clip, clamping; ring, collar; lug; shackle

**хондриосома** *f.* chondriosome

**хондрит** *m.* chondrite

**хорда** *f.* chord; span (of arc)

**хордальный** *adj.* chordal

**хордовый** *adj.* chord, chordal

**хоровой** *adj.* choral, chorus

**хороший** *adj.* good; fine; high

**хорошо** *adv.* well; highly

**хоть, хотя** *conj.* though, although, still, yet; in spite of; **х. бы** even, even though, even if

**Хоули** Hawley

**хранение** *n.* conservation, preservation, care, keeping, holding; storing, storage; accumulation; **х. в массиве** filing; **х. записи** holding; **х. на многопозиционных элементах** multiple-stable-state storage; **х. отметок** track storage; **х. переноса** carry storage

**хранилище** *n.* storage, storehouse; reservoir

**хранимый** *adj.* stored

**хранить, сохранить** *v.* to hold, to keep, to store; to memorize; **х. в массиве (данных)** to file

**хранящийся** *adj.* stored, kept, preserved

**храповик** *m.* ratchet (gear), ratchet wheel, sprocket; **х. и собачка** ratchet and pawl; **х. с собачкой** ratchet gear, cam-and-ratchet (mechanism)

**храповичок** *see* **храповик; вращательный х.** rotary hub, ratchet; **подъёмный х.** vertical hub, ratchet (of a Strowger switch)

**храповой, храповый** *adj.* ratchet

**хребет** *m.* crest, ridge (of mountain); range, chain (of mountains)

**хребтовый** *adj.* ridge, crest; range, chain

**хребтообразный** *adj.* ridged

**хрип** *m.* wheeze; *pl. oft.* mush

**хрипение** *n.* hoarsness; huskiness; rattle, rattling

**хрипеть, прохрипеть** *v.* to wheeze; to rattle, to crackle

**хриплый** *adj.* hoarse, husky

**Христиансен** Christiansen

**Христоффель** Christoffel

**хром** *m.* chromium, Cr; chrome

**хромакодер** *m.* chromacoder

**хромакс** *m.* chromax

**хромалой** *m.* chromaloy

**хромат** *m.* chromate

**хроматермография** *f.* chromathermography

**хроматик** *m.* chromatic tube

**хроматика** *f.* chromatics

**хроматический** *adj.* chromatic, color

**хроматичность** *f.* chromaticity, chrominance

**хроматичный** *adj.* chromatic, color

**хроматограмм** *m.*, **хроматограмма** *f.* chromatogram

**хроматограф** *m.* chromatograph

**хроматографический** *adj.* chromatographic

**хроматография** *f.* chromatography

**хроматометр** *m.* chromatometer

**хроматоскоп** *m.* chromatoscope

**хроматрон** *m.* beam-switching display, focus-mask display, focus-mask tube, chromatron tube; **трёхлучевой х.** three-gun focus-grid tube

**хроматроп** *m.* chromatrope

**хромель** *m.* chromel (alloy)

**хромирование** *n.* chromium-plating; chromizing; **твёрдое х.** hardchrome plating

**хромированный** *adj.* chrome-plated

**хромировать** *v.* to chrome, to chrome-plate

**хромистый** *adj.* chromium, chromous, chrome

**хромо-** *prefix* chromo- (color; chrome; chromium)

**хромовый** *adj.* chromic, chromium, chrome

**хромокобальтовый** *adj.* chrome-cobalt

**хромомарганцевый** *adj.* chromium-manganese

**хромометр** *m.* chromometer, colorimeter

хромомолибденовый *adj.* chrome-molybdenum

хромоникелевый *adj.* chrome-nickel

хромоскоп *m.* chromoscope

хромосфера *f.* chromosphere

хромосферный *adj.* chromospheric

хромофор *m.* chromophore

хромофорный *adj.* chromophore

хромофотометр *m.* chromophotometer

хромоцианид *m.* chromocyanide; **x. калия** potassium chromocyanide

хронаксиметр *m.* chronaximeter

хронаксиметрия *f.* chronaximetry

хронизатор *m.* timer; **главный x.** master timer

хронизация *f.* synchronization

хроника *f.* newsreel; chronicle

хронирование *n.* timing; **x. протягивания плёнки** pulldown timing

хронированный *adj.* timed

хронировать *v.* to time

хронирующий *adj.* timing

хронистор *m.* chronistor

хронически *adv.* chronically

хронический *adj.* chronic, chronical, lingering

хроно- *prefix* chrono- (time)

хроновыключатель *m.* telechron timer

хронограмма *f.* chronogram; **радиальная x.** radial time base

хронограф *m.* chronograph; **магнитный x.** magnetic chronograph; **x. трассы снарядов** shell tracking chronograph

хронокомпаратор *m.* chronocomparator

хронологический *adj.* chronological

хронология *f.* chronology

хронометр *m.* chronometer, timepiece; **x. для суммирования интервалов времени** time totalizer; **фотоэлектрический x.** photo-timer

хронометраж *m.* time metering, timing, time study

хронометрирование *n.* timekeeping, timing

хронометрировать *v.* to time

хронометрический *adj.* chronometric

хронометрия *f.* chronometry, timekeeping

хронореле *n. indecl.* time relay, timing relay

хроносигнал *m.* time signal

хроноскоп *m.* chronoscope

хроносчётчик *m.* time meter, hour meter

хронотрон *m.* chronotron

хронофер *m.* chronopher

хрупкий *adj.* brittle, frail, fragile delicate; inflexible

хрупколомкий *adj.* short-brittle

хрупкость *f.* brittleness, frailness; fragility, frangibility; inflexibility; **водородная x.** hydrogen embrittlement

хрупный *see* хрупкий

хруст *m.* crackle, crackling; scratching; crunch

хрусталик *m.* crystalline lens (*anat.*)

хрусталь *m.* crystal; crystal glass; **английский x.** flint glass

хрустальный *adj.* crystal, crystalline

хрустение *n.* crackling, crunching

хрустеть, хрустнуть *v.* to crackle, to crunch

ХТИ *abbr.* (**Харьковский технологический институт**) Kharkov Technological Institute

художественный *adj.* artistic; high-fidelity

худой *adj.* bad, inferior, ill; thin

худший *compar. and superl. of* плохой *and* худой worse, the worst

хуже *compar. of* худо, худой *and* плохой worse

Хунд Hund

Хупс Hoopes

**x. ч.** *abbr.* (**химически чистый**) chemically pure

Хьюель Hewel

Хьюлетт Hewlett

Хэббль Hubble

Хэвенс Havens

Хэвилэнд, де de Havilland

Хэвисайд Heaviside

Хэй Hay

Хэлл Hull

Хэртли Heurtley

Хэт Huth

ХЭТИ *abbr.* (**Харьковский электротехнический институт**) Kharkov Electrotechnical Institute

Хэуорс Haworth

# Ц

°Ц *abbr.* (градус Цельсия) degree centigrade

ц *abbr.* 1 (центнер) centner, hundredweight; 2 (центральный) central

Ц *abbr.* (центр) center

Цаги, ЦАГИ *abbr.* (Центральный аэрогидродинамический институт) Central Aero-Hydrodynamic Institute

цанги *pl.* tongs, tweezers, forceps; clamp, holder

цанговый *adj.* tongs, tweezers, forceps; clamp, holder; pronged, prong-type

цапон-лак *m.* Zapon enamel

цапоновый *adj.* Zapon

цапфа *f.* plug, pivot, journal; shank; ведущая ц. cross pin, driving pin; ц. кривошипа crank pin; шаровая ц. ball pivot

цапфовый *adj.* plug, pivot, journal, gudgeon; shank

царапание *n.* scratch, scratching, abrasion, mark

царапанный *adj.* scratched, abraded, marked

царапанье *see* царапание

царапать, царапнуть, оцарапать *v.* to abrade, to scratch

царапина *f.* scratch, scratching, abrasion, mark, notch; *pl. oft.* pitting; ц., сделанная напильником cut of a file

царапнуть *pf. of* царапать

царский *adj.* royal, imperial; tsar's

ЦБ *abbr.* (центральная батарея) central battery, common battery, CB; вызывная ЦБ signaling common battery, ringing common battery

ЦБНТИ *abbr.* (Центральное бюро научно-технической информации) Central Bureau of Scientific-Technical Information

ЦБТИ *abbr.* (Центральное бюро технической информации) Central Bureau of Technical Information

ЦВД *abbr.* (цилиндр высокого давления) high-pressure cylinder

цвет *m.* color, tint; цвета мальвы mauve; неприведённые первичные цвета unprimed primaries; основные цвета воспроизводящего устройства display primaries; основные цвета колориметра matching stimuli primaries, instruments stimuli primaries; основные цвета на приёмнике receiver primaries; первичные цвета, принятые Международной комиссией по освещению CIE primaries; первоначальные цвета primary colors; разбавленные основные цвета desaturated primaries; в цветах in full color; белый ц. zero-saturation color; colorless light; голубой ц. cyan color; ц. излучений color of light; ц. изображения pattern color; исходящий ц. reference color; мнимый ц. nonphysical color; ц. наибольшей достижимой чистоты "full color"; ц. накала heat color; ненасыщенный ц. diluted color; нереальный ц. nonphysical color; опознавательный ц. identification color; основной ц. fundamental color, primary color; ц. побежалости oxidation tint, temper color; ц. при последовательном совмещении полукадров field-sequential color; пурпурный ц. magenta color; разбавленный ц. diluted color; реальный ц. physical color; светлый ц. high-luminance color; ц. свечения экрана coloration of screen; синезелёный ц. cyan color; смешанный ц. secondary color; телесный ц. flesh color; тёмносиний ц. cyan color; физически нереальный основной ц. physically unrealizable primary color; фиолетовый ц. violet color, magenta color; яркий ц. high-luminance color; яркий натуральный ц. bright vivid color

цветение *n.* blooming, blossoming, florescence; efflorescence; ц. изображения blooming

цветной *adj.* color, colored, chromatic; non-ferrous

цветнослепой *adj.*; *m. adj. decl.* color-blind (person)

**цветность** *f.* chromaticity, chromaticness, chrominance; **минимальная остаточная ц. на экране ЭЛТ** color background; **ц., соответствующая нулевому значению подне** zero-subcarrier chromaticity

**цветовой** *adj.* color, chromatic; chromaticity

**цветовоспроизводящий** *adj.* color-reproducing, chromatogenic

**цветовыделение** *n.* chromatic selection, component color selection, partial color record

**цветоделение** *n.* color separation

**цветоделённый** *adj.* color-separation

**цветоделительный** *adj.* dichroic

**цветозамещаемый** *adj.* allochromatic

**цветоизбирательный** *adj.* color-selective

**цветоизменение** *n.* change of color, allochroism; **ц. синим светофильтром** blue selection, blue separation

**цветоизменяемость** *f.* change of colors, allochroism

**цветоизменяющий** *adj.* versicolor

**цветомер, цветометр** *m.* colorimeter, tintometer

**цветометрический** *adj.* colorimetric

**цветометрия** *f.* colorimetry

**цветонасыщенность** *f.* color saturation, high light, chroma

**цветоносители** *pl.* chromophore

**цветоощущение** *n.* color sensation

**цветопередача** *f.* color rendering, color rendition

**цветорассеяние** *n.* chromatic dispersion, chromatic aberration

**цветосмеситель** *m.* color-mixer

**цветосмесительный** *adj.* color-mixer

**цветоустойчивость** *f.* color fastness

**ЦВМ** *abbr.* (**цифровая вычислительная машина**) digital computer

**ЦД** *abbr.* (**центр давления**) center of pressure

**ЦДА** *abbr.* (**цифровой дифференциальный анализатор**) digital differential analyzer

**ЦДТС** *abbr.* (**Центральный дом техники связи**) Central House of Communication Engineering

**цевка** *f.* bobbin, reel, spool; spindle

**цевочный** *adj.* bobbin, reel, spool; spindle

**ЦЕГАЗО** *abbr.* (**Центральная государственная авиазенитная оборона**) State Central Antiaircraft Defense

**цезиевочевой** *adj.* cesium-beam

**цезиевосурьмяный** *adj.* cesium-antimony

**цезиевый** *adj.* cesium

**цезий** *m.* cesium, Cs; **хлористый ц.** cesium chloride

**цезированный** *adj.* cesium-coated

**Цейсс** Zeiss

**цейссовский** *adj.* (of) Zeiss

**целевой** *adj.* target, aim, mark; goal, purpose, object; objective

**целенаправленный** *adj.* teleological

**целесообразность** *f.* expediency, expedience, utility; usefulness; efficiency

**целесообразный** *adj.* expedient, suitable; sound

**целеста** *f.* celesta

**целеуказание** *n.* target indication, target designation; acquisition of target; **радиолокационное ц.** radar plot

**целеустановка** *f.* object, objective, aim

**целеустремлённость** *f.* purpose

**целеустремлённый** *adj.* purposeful; goal-seeking

**целиком** *adv.* entirely, completely, wholly, totally

**целить, нацелить, прицелить** *v.* to aim (at), to point (at), to direct (at)

**целлит** *m.* Cellite, secondary cellulose acetate

**целлоидин** *m.* celloidin

**целлоидиновый** *adj.* celloidin

**целлон** *m.* Cellon, hydro-cellulose acetate

**целлоновый** *adj.* Cellon

**целлофан** *m.* cellophane, cellulose hydrate

**целлулоза** *f.* cellulose

**целлулоид** *m.* celluloid

**целлулоидный** *adj.* celluloid

**целлюлоза** *f.* cellulose

**целлюлозность** *f.* cellulosity

**целлюлозный** *adj.* cellulose

**целое** *n. adj. decl.* the whole, the entire, unit; integer

**целостат** *m.* coelostat

**целостность** *f.* wholeness, entirety, integrity, completness

**целостный** *adj.* complete, entire, integral

**целость** *f.* wholeness, entireness, integrity

**целотекс** *m.* Celotex

**целотоновый** *adj.* whole-tone

**целочисленный** *adj.* integer-valued

**целый** *adj.* whole, entire, integral intact, complete, sound, solid

**цель** *f.* aim, mark, goal, target; object; purpose, end; scope; **учебные цели** trial targets; **с целью** with a view (to), in order (to); on purpose; **в целях** with the purpose of, for the purpose of; **ц. измерения** testing purpose; **искусственная ц.** phantom target; **кажущаяся ц.** virtual target; **ложная ц.** false target, decoy device; confusing reflector; **малая ц.** point target; **надводная ц.** surface target; **неравномерно рассеивающая ц.** directional-response target; **околозвуковая ц.** sonic target; **ц., оснащённая средствами радиопомех** countermeasures target; **ц., перемещающаяся с большой скоростью** high-speed target; **прослушиваемая ц.** audible target; **пространственная ц.** three-dimensional target, extended target; **удаляющаяся ц.** outgoing target; **ц. узких строб-импульсов** narrow-gate circuit; **элементарная ц.** simple target

**цельно** *adv.* wholly, entirely, all

**цельно-** *prefix* wholly, entirely, all

**цельно-анкерный** *adj.* whole-anchor

**цельножелезный** *adj.* all-iron

**цельнокатный** *adj.* seamless rolled

**цельнокованый** *adj.* seamless forged

**цельнокрайний** *adj.* entire

**цельнолитой** *adj.* unit-cast, one-piece

**цельнометаллический** *adj.* all-metal

**цельнорезиновый** *adj.* all-rubber

**цельносварной** *adj.* all-welded

**цельностальной** *adj.* all-steel

**цельностеклянный** *adj.* all-glass

**цельность** *f.* wholeness, entirety, totality, integrity

**цельнотянутый** *adj.* seamless (pipe)

**целый** *adj.* whole, entire, integral, total single, one-part; solid, all-in-one-piece

**цельсий** *m.* centigrade; Celsius

**цемент** *m.* cement; **ц. для соединения латуни со стеклом** electric cement; **ц. из смеси гипса с квасцами** Keene's cement; **каменный ц.** oxychloride cement; **магнезиальный ц.** Sorel cement; **ц. с добавлением негашёной извести** electrocement

**цементационный** *adj.* cementing; cementation, carbonization

**цементация** *f.* cementing; surface hardening, case-hardening, carbonization, cementation; **поверхностная ц.** case-hardening

**цементирование** *see* **цементация**

**цементированный** *adj.* cemented; carbonized, casehardened

**цементировать, сцементировать** *v.* to cement; to paste, to gum; to carbonize, to caseharden

**цементирующий** *adj.* cementing

**цементный** *adj.* cement; concrete

**цементовальный** *adj.* cementing, cement

**цементование** *see* **цементация**

**цементовать** *v.* to cement; to gum, to paste

**цементовочный** *adj.* cementation

**цементовый** *adj.* cement; concrete

**цена** *f.* price, worth, value, rate, charge, cost; **ц. деления** multiplier, multiplying factor; scale division; **фабричная ц.** net price

**Ценер** Zener

**ценетический** *adj.* coenetic

**ценность** *f.* value, cost, price; importance

**ценный** *adj.* valuable

**цент** *m.* cent

**центибар** *m.* centibar

**центибел** *m.* centibel

**центиграмм** *m.* centigram

**центнер** *m.* hundredweight

**центр** *m.* center; **со смещённым центром** off-center; **ц. восстановления дырок и элементов** deathnium; **ц. вписанной окружности** incenter; **вращающийся ц.** live center; **вычислительный ц. обработки данных** data processing center; **главный ц. слежения и вычисления данных** primary data-acquisition site; **групповой ц.** joint trunk exchange; **ц. захвата** trapping center, trap; **ц. захвата дырок** hole trap; **зональный ц.** via center; **калибрационный ц. по электронике** Electronic Calibration Center; **ц. линейной воздушной диспетчеризации** airway traffic control center; **ц. массы Земли** geocenter; **ц. накопления и обработки информации** storage computing station; **оперативный ц. ПВО** Air Defense Control Center; **ц. отражения радиолокационной цели** radar center; **передний ц., подвижной ц.**

live center; **ц. приложения силы** center of force; **радиоприёмный ц.** receiving station; **ц. районных сетей** (radial) network center; **распределительный ц. групповых цепей** branch-circuit distribution center; **расчётный ц.** computer center; **ц. симметрии** center; **смещённый ц.** off-center; **телевизионный ц.** television broadcast station, telecast station, television operating center; **теоретический ц. сети** ideal network center; **ц. тяги** center of thrust; **ц. тяжести** center of gravity, centroid; **ц. тяжести сети** ideal network center; **узловой ц.** nodal point; **ц. цели** center target

**централизатор** *m.* interlocker

**централизационный** *adj.* interlocking, interlocker

**централизация** *f.* centralization; interlocking; centralized traffic control; **дальнодействующая ц.** remote-control (interlocking) system; **электрозащелочная ц.** electromechanical interlocking machine

**централизированный** *adj.* centralized; interlocked

**централизировать** *see* **централизовать**

**централизованный** *adj.* centralized; interlocked

**централизовать** *v.* to centralize; to interlock; to fuse, to consolidate

**центральность** *f.* center, equidistance; centrality

**центральный** *adj.* central, center; nodal

**центрирование** *n.* centering, centering control; **двухтактное ц.** push-pull centering; **ц. изображения на экране телевизора** centering of TV picture; **ц. по кадрам** vertical centering; **ц. растра** centering; **ц. растра по кадру** framing; **ц. сетки** grid alignment

**центрированный** *adj.* centered

**центрировать** *v.* to center, to position

**центрирующий** *adj.* centering, positioning

**центрифуга** *f.* centrifuge; **ц. для тренировки космонавтов** human centrifuge

**центрифугирование** *n.* centrifugation, centrifugal action, centrifugal separation

**центрифугированный** *adj.* centrifuged

**центрифугировать, центрифуговать** *v.* to centrifuge

**центробежно** *adv.* centrifugally, by centrifugal means

**центробежный** *adj.* centrifugal

**центровальный** *adj.* centering

**центровать** *v.* to center

**центровка** *f.* centering, centering control, centering adjustment, alignment; **ц. по вертикали** vertical centering control, vertical positioning control

**центровой** *adj.* center, centric, central

**центровочный** *adj.* center, centering

**центровый** *see* **центровой**

**Центрогонти** *abbr.* (Центральный государственный научно-технический институт) State Central Institute of Technology

**центроид** *m.* centroid

**центроида** *f.* poid

**центростремительность** *f.* centripetency, centripence

**центростремительный** *adj.* centripetal; inward radial

**цепеобразный** *adj.* chain-like

**цепкий** *adj.* catching, tenacious, adhesive, sticky

**цепной** *adj.* chain, chaining; linkage; catenary, catenarian; continued, recurrent, iterative; ladder (network)

**цепнореагирующий** *adj.* chain-reacting

**цепочка** *f.* chain, network, iterated series, iterated network; RC circuit, RC filter; string, train; **главная ц.** basic sequence; **ц. делителя напряжения** potentiometer chain; **интегродифференцирующая ц.** lead lag network; **ц. мнимых излучателей** network of image sources; **неразветвлённая ц.** linear chain; **переходная ц.** intermediate filter, RC circuit; **пикинговая ц.** R-C peaking network; **ц. продуктов деления** fission chain; **ц. станций Лоран** loran chain; **ц. станций наведения** home chain; **ц. станций радиолокационной обороны** radar fence; **фазовращающая ц.** phase-shift filter

**цепочный** *see* **цепной**

**Цеппелин** Zeppelin

**цеппелиновый** *adj.* Zeppelin('s)

**цепь** *f.* circuit, network; chain; filter; loop; train; **цепи, включённые после детектора** post detector circuits; **многократные цепи декады** level multiple; **параллельно включённые цепи** multiple circuit(s); **перекрещивающиеся цепи** oblique circuits; **структурно-обращённые цепи** structurally dual networks; **цепи управления электронным сканированием антенны** scan control circuit(s); **цепи частотной коррекции** compensating circuit(s); **ц. аварийного выключения двигателя** destruct circuit; **автоколебательная ц.** astable circuit; **ц. бдительности** acknowledging circuit; **блокируемая ц.** interlocking circuit; **ц. быстрой выборки данных** quick-access loop; **бытовая ц. (во внутренней проводке)** appliance branch circuit; **ц. в виде многоугольника** mesh circuit; **ц. внутренних радиолокационных станций** home chain; **воздушная ц.** open wire circuit; **ц. восстановления постоянной составляющей** direct-current restoring circuit; **ц. вспомогательного зажигания искрового промежутка** trigger circuit; **вторая параллельная ц.** recoil circuit; **ц. выбора кратного** multiple-selection circuit; **ц. выключателя канала цветности** color killer network; **ц. выключающей катушки** trip circuit, tripping circuit; **ц. выравнивания группового времени задержки** group-delay equalizer; **ц. выравнивания отметок** pip-matching circuit; **ц. Галля** sprocket chain; **ц. гарнитуры телефонистки** head circuit (of a telephonist); **ц. гидроакустических станций установленных на дне** bottom-mounted networks; **главная эталонная ц.** telephone transmission reference system; **групповая ц.** branch circuit; **групповая ц. для питания освещения и бытовых приборов** combination lighting and appliance branch circuit; **ц. датчика** sensing circuit; **ц. двойного дифференцирования импульсов** rate-of-rise amplifier; **двухполупериодная ц. с удвоением напряжения** full-wave doubler circuit; **двухпроводная основная (*or* физическая) ц.** two-wire side circuit; **деблокирующая ц.** line freeing circuit; **декадная ц.** cascade of decimal counting units; **ц. деления** multiplier; **ц. деления пополам** halving circuit; **ц. деления частоты** counting-down circuit; **ц. делителя напряжения** bleeder circuit, bleeder network; **дифференцирующая ц.** differentiating circuit, peaker, differentiator, sharpening circuit, phase-lead network; **ц. для отвода статических зарядов** drainage chain; **дроссельная ц.** coil-loaded line; **ёлочная ц.** christmas-tree circuit; **ц. ёмкостной обратной связи** capacitive feedback circuit; **заземляющая ц.** ground circuit; **заказная ц. телефонистки по междугородным переговорам** trunk record circuit; **заказная соединительная ц.** speaker wire, order-wire circuit, service circuit; **ц., залитая специальным компаундом** potted circuit; **ц. замедления** delay train; **(замкнутая) ц. положительной обратной связи** regenerative feedback loop; **замкнутая двухпроводная ц.** closed loop; **замкнутая нервная ц.** neural loop; **ц. записи** writing circuit; **ц. запуска** firing circuit; **ц. звукоснимателя** pickup circuit; **идеальная ц.** all-pass network; **ц. извещения о приближении поезда** train approach circuit; **импульсная рельсовая ц.** half-wave track circuit; **инфразвуковая ц. задержки** subaudio time delay circuit; **ц. искровозбудителя** trigger circuit; **искусственная ц.** phantom circuit, superposed circuit; **искусственная ц. (по средней точке телефонной цепи), искусственная телеграфная ц.** simplexed circuit; **исправная ц.** sound circuit, operable circuit; **ц. кадрового делителя** vertical divider chain; **ц. квантования во времени** sampling circuit; **колебательная ц.** periodic circuit; **ц. компенсации нелинейных искажений типа «гамма»** rooter circuit; **«консольная» ц.** cantilever circuit; **контактная ц.** chain of contacts; **ц. контроля** checkout circuit; **концертная ц.** musical circuit, program circuit; **ц. короткого замыкания** short circuit; **ц. корректирующей задержки** corrective network; **ц. коррекции** stabilizing network; **ц. коррекции искажений** corrective network, equalizing network; **ц. коррекции развёртки (*or* сигналов)** peaking circuit; **ложная ц.** sneak circuit; **ц. манипуляции** keying circuit; **междугородная ц. с дальним набором** dial toll circuit, toll-dialing circuit; **ц. межкаскадной связи** interstage network; **ц. механического решения управлений** mechanizing circuit; **микрофонная ц.** transmitter circuit; **многократная ц.** multiple; **ц. моделирования** computing loop; **модуляционная ц.** tone circuit; **ц. наземных радиолокационных станций системы ВНОС** homer radar chain; **наращённая ц.** built-up circuit, through circuit; **ц. настройки** adjustment network **натяжная ц.** insulator chain, string of insulators; **неправильная ц.** parasitic circuit; **ц. низкочастотной обратной связи** audio-feedback path; **ц. обмотки якоря** path (of an armature winding); **ц. образования точек** dot circuit; **ц. обратной коррекции** deemphasis circuit; **ц. обратной связи** regenerative coupling, regenerative loop, feedback loop, feedback network, feedback circuit, Armstrong circuit, singing path; **ц. обратной связи для приёма по методу биений** ultra-audion circuit; **огневая ц.** powder train, explosive chain; **ц. ограничителя импульсов** clamper circuit, clamping circuit, clipper circuit; **ц. одновременного телефонирования и телеграфирования** composite circuit; **однополупериодная ц. (выпрямления)** half-wave circuit; **ц. одностороннего пропускания** nonreciprocal circuit; **ц. оперативного тока** auxiliary circuit; **освобождающая ц.** clearing circuit; **основная ц.** side circuit, physical circuit; K filter chain; **основная ц. регулирования** basic sequence; **основная стандартная телефонная ц.** telephone transmission reference system; **ц. отбора электрической энергии** take-off circuit; **ц. отключения** deflection circuit; **ц. отметки времени** timing circuit; **ц. отметки начала отсчёта** baseline marker circuit; **ц. отпирания (тиратрона)** trigger circuit; **оттяжная ц.** insulator chain, string of insulators; **педальная ц.** trap circuit; **ц. переноса** carry circuit; **П-образная ц.** pi-network; **ц. повторного включения** reset circuit; **ц. под напряжением** live circuit, alive cir-

cuit; **ц. подачи сигнала готовности к пуску** ready-signal circuit; **поддерживающая ц.** holding wire; **ц. подрыва боевой части** warhead firing circuit; **ц. подслушивания** monitoring circuit, observation circuit; **поперечная ц. (амплидина)** quadrature circuit; **ц. предварительной настройки** presetting circuit; **ц. предионизатора, ц. предыонизатора** keep-alive circuit; **прерывающая ц.** chopping circuit; **ц. прерывистого действия** sampling circuit; **ц. притупления выбросов** despiker circuit; **ц. (проводной) связи** communication circuit; **ц. прямого вызова** ring-down circuit; **ц. прямого действия** forward circuit; **ц. радиолокационных станций** chain home beamed, chain radar system; **разветвлённая ц. (с узлами)** network circuit; **разветвлённая пополам магнитная ц.** double magnetic circuit; **ц. развязки** decoupling circuit; **ц. разговорного тока** telephone circuit; **ц. разделения синхроимпульсов** sync separator circuit; **ц. регулирования по первой производной** rate-of-change circuit; **RC-информирующая ц.** RC-shaping circuit; **ц. с добавочным напряжением** boosted circuit; **ц. с добавочным сопротивлением** economy circuit; **ц. с индуктивностью и ёмкостью** coil-condenser circuit; **простая ц. тока** current path; **ц. с нулевой статической ошибкой** subsidiary loop; **ц. с отводами** tapped circuit; **ц. с повышенной индуктивностью** loaded circuit; **ц. с распределёнными постоянными** distributed circuit; **ц. с сосредоточенными постоянными** lumped circuit, circuit with lumped element; **ц. со встречно действующими электродвижущими силами** subtraction circuit; **ц. со многими устойчивыми состояниями** multistable circuit; **самоудерживающая ц.** stick circuit; **ц. сглаживания пиков** despiker circuit; **селекторная ц.** gate circuit, gating circuit; **ц. сигнализации** warning circuit; **ц. сигналов цветности** chroma circuit; **ц. сигнальной пластины** backplate circuit; **сильноточная ц.** current circuit; **ц. синхронизации цветовых**

**сигналов** color hold circuit; **синхронизационная ц.** drive circuit, sync circuit; **служебная ц.** order circuit order wire, speaker circuit; switching circuit, auxiliary circuit, recording trunk; **ц. смешанного соединения** series-parallel circuit; **ц. смещения** bias circuit; **ц. снятия возбуждения** deenergizing circuit; **собирательная ц.** buffer circuit; **ц., создающая опережение по фазе** phase-advance network; **спрямляющая ц.** squarer circuit, squaring circuit; **ц. срезания пиков** despiker circuit; **ц. станций дальнего обнаружения** long-range detection chain; **стробирующая ц.** gate circuit; **ц. телевизионных ретрансляционных станций** television station link, television chain; **ц. точной калибровки времени** precision timing circuit; **транзитная ц.** built-up circuit, through circuit, via circuit; **трансляционная ц.** radio-telephone circuit; **ц. трансляционного реле** repeat circuit; **ц. удлинения импульсов** integrating network; **уплотнённая ц.** composite circuit; **ц., уплотнённая 12-канальной системой** j-pair, j-circuit; **ц. управления громкостью** volume control circuit; **ц. управления пуском** firing control circuit; **уравновешивающая ц.** compensating chain; **ускоряющая ц.** anticipation network; **фазовращающая ц. с опережением по фазе** phase-lead network; **ферромагнитная ц.** perfect magnetic circuit; **ц. формирования прямоугольных импульсов** squaring circuit; **ц. циркулярной связи** conference circuit; **ц. частотной коррекции** compensating circuit; **ц. четырёхполюсников** recurrent network; **четырёхпроводная основная** (or **физическая**) **ц.** four-wire side line (or circuit); **чистая ц.** silent circuit, quiet circuit, circuit proper; unloaded circuit; **шарнирная ц.** sprocket chain, link chain; **эквивалентная ц. с включёнными по концам конденсаторами** split-condenser circuit; **ц. эталонной передачи** master telephone transmission reference system

**церезин** *m.* ceresin wax
**Церера** *f.* Ceres (*astron.*)
**церонограф** *m.* ceraunograph

**церонометр** *m.* ceraunometer
**церузит, церуссит** *m.* cerussite, lead spar
**цефеида** *f.* cepheid
**цех** *m.* shop, workshop, plant, works; section, department; trade union; **вспомогательный ц.** service department; **кузнечный ц.** forge shop; **опытный ц.** design development shop
**цеховой** *adj.* shop; department; works
**циан-** *prefix* cyan-, cyano-, cyanic
**циан** *m.* cyanogen
**цианид** *m.* cyanide; **общий ц.** total cyanide
**цианидный** *adj.* cyanide
**цианометр** *m.* cyanometer
**цикл** *m.* cycle; period, round; circuit, loop; ring; **ц. Бете-Вайцзеккера** Bethe-Weizsäcker progressive nuclear reaction; **ц. «заряд-разряд»** cycle of operation; **исполнительный ц.** execution cycle; **испытавший ц. а** cycled; **ц. Клаузиуса-Ранкина** Rankin cycle; **командный ц.** instruction cycle; **один ц. развёртки** run-down; **предпусковой ц.** prelaunch cycle; **прерывистый ц. запоминания** break memory cycle; **прерывистый ц. (работы) запоминающего устройства** break storage cycle; **протонный ц.** proton-proton chain; **ц. (работы) запоминающего устройства** storage cycle; **ц. работы механизма** picture cycle; **рабочий ц.** duty cycle; **ц. разряд-заряд** cycle of operation; **ц. с большим заполнением** high-duty cycle; **ц. с малым заполнением** low-duty cycle; **ц. скрещивания** complete transposition section; **сложный ц.** loop-within-loop; **ц. смены дня и ночи** day-night cycle; **ц. установки** insertion cycle; **ц. экспозиции** exposure cycle; **ц. элементарного сигнала** dot cycle
**циклический** *adj.* cycle, cyclic; continuous, repetitive
**цикличность** *f.* cycle of operation
**цикличный** *adj.* cyclical
**цикло-** *prefix* cyclo-
**циклограмма** *f.* cyclogram
**циклограф** *m.* cyclograph
**циклоида** *f.* cycloid
**циклоидальный, циклоидный** *adj.* cycloid, cycloidal

**цикломедианный** *adj.* cyclomedian

**цикломер, циклометр** *m.* cyclometer, cycle counter

**циклометрический** *adj.* cyclometric

**циклометрия** *f.* cyclometry

**циклон** *m.* cyclone

**циклонический, циклонный** *adj.* cyclone, cyclonic

**циклорасщепитель** *m.* cycle splitter

**циклострофический** *adj.* cyclostrophic

**циклотрон** *m.* cyclotron; **малый ц.** omegatron; **сверхмощный ц.** bevatron; **электронный ц.** microtron

**циклотронный** *adj.* cyclotron

**циклофон** *m.* cyclophone

**цикля** *f.* scraper, draw-latch

**цилиндр** *m.* cylinder, roller, roll, drum; (Faraday) cup; **выборочный ц.** access cylinder; **ц. для выборки данных** access cylinder; **контактный ц.** controller cylinder; **ц. контроллера** drum controller, master controller; **параболический ц.** parabolic-cylinder antenna; **полый ц.** slug; **пористый ц.** porous cup; **сеточный ц.** grid conductor; **фокусирующий ц.** concentric cylinder, concentration cylinder, focussing cylinder

**цилиндрический** *adj.* cylindrical, cylindric, cylinder-like

**цилиндровый** *adj.* cylinder

**цилиндро-конический** *adj.* cylindroconical

**цимбалы** *pl.* cembalon, cymbal; dulcimer

**цинк** *m.* zinc, Zn

**цинк-берилий** *m.* zinc-beryllium

**цинквальдит** *m.* zincwaldite

**Цинкен** Zinken

**цинкит** *m.* zincite, spartalite, perikon

**цинкитовый** *adj.* zincite

**цинк-кадмий-сульфид** *m.* zinc-cadmium-sulphide

**цинковальный** *adj.* galvanizing

**цинкование** *n.* galvanizing, hot-dip galvanizing process, zinc plating

**цинкованный** *adj.* galvanized, zinc-plated

**цинковать, оцинковать** *v.* to zinc, to galvanize

**цинково-свинцовый** *adj.* zinc-lead

**цинковый** *adj.* zinc

**цинкосульфидный** *adj.* zinc sulphide

**цинк-сульфид** *m.* zinc sulphide

**цинк-феррит** *m.* zinc ferrite

**Циолковский** Tsiolkovskii

**ЦИП** *abbr.* (**Центральный институт погоды**) Central Weather Institute

**циркон** *m.* zirconium

**цирконат** *m.* zirconate

**циркониевый** *adj.* zirconium

**цирконий** *m.* zirconium, Zr

**цирконистый** *adj.* zirconium

**цирконовый** *adj.* zircon, zirconium, zirconic

**циркулировать** *v.* to circulate, to revolve, to rotate

**циркулирующий** *adj.* circulating, circulation, circuital

**циркуль** *m.* compass, pair of compasses (for drawing); dividers; callipers; **делительный ц., измерительный ц.** callipers, feeler key; **калиберный ц.** callipers; **пропорциональный ц.** sector, reduction compass; **разметочный ц.** callipers, feeler key; **рычажный ц.** beam-compass; **ц. со ставной ножкой** hand compass; **составной ц.** draught-compass

**циркульный** *adj.* circular

**циркуляр** *m.* circular

**циркулярный** *adj.* circular; circulating, circulatory

**циркулятор** *m.* circulator

**циркуляторный** *adj.* circulator

**циркуляционный** *adj.* circulating, circulation

**циркуляция** *f.* circulation; circuit

**циркум** *adv.* circum

**циркумполярный** *adj.* circumpolar

**циртометр** *m.* cyrtometer

**ЦИС** *abbr.* (**Центральный институт связи**) Central Communication Institute

**цис-** *prefix* cis-

**цис-изомер** *m.* cis-isomer

**цис-соединение** *n.* cis-compound; **ц.-соединения** *pl.* cis-constitutions

**цис-транс-изомерия** *f.* cis-trans-isomerism

**циссоида** *f.* Diocles' cissoid

**цистерна** *f.* cistern, tank, reservoir

**цистоскоп** *m.* cystoscope

**цитата** *f.* citation, quotation

**цитирование** *n.* citation, quotation

**цитировать, процитировать** *v.* to cite, to quote

**цитологический** *adj.* cytological

**цитология** *f.* cytology

**цитра** *f.* cittern, cithern

**циферблат** *m.* dial, dial plate, clock

dial, hour plate, face; index plate, index dial, scaleplate

**циферблатный** *adj.* dial, clock dial, hour plate, face

**цифра** *f.* figure, cipher, number, numeral, character; digit, coefficient; **цифры команды** function digits; **цифры одного и того же разряда** digits with like place values; **цифры слагаемого** operand digits, operated digits; **ц. второго слагаемого, ц. второго числа** augend digit; **допустимая ц.** admissive mark; **ц. знака** sign digit; **левая ц.** lowest order digit; **местная кодовая ц.** dial prefix; **ц. младшего разряда** low-order digit; **ц. первого слагаемого** addend digit; **ц., перенесённая из предыдущего разряда** previous-carry digit; **ц. переноса** carry digit, carry number; **правая ц.** high-order digit; **ц. разряда десятков** tens digit; **ц. с удвоенным числом знаков** double-length number, double-precision number; **самая младшая значащая ц.** least significant digit; **ц. (самого) старшего разряда** highest-order digit

**цифратор** *m.* converter; **ц. времени** time-to-number converter; **ц. напряжения** voltage-to-digital converter; **ц. положения** position-to-number converter; **ц. углового положения (вала)** angle-to-digit converter, shaft position(-to-digital) converter; **ц. частоты** frequency-to-number converter

**цифро-аналоговый** *adj.* digital-analog

**цифрователь** *m.* digitizer

**цифровать** *v.* to digitize

**цифровой** *adj.* digital, cipher; numbered, figured; figur, character, numeric

**цифровывать** *v.* to digitize

**цифрозаписывающий** *adj.* digital-recording

**ЦЛА** *abbr.* (**Центральная лаборатория автоматики**) Central Laboratory of Automation

**ЦЛЭМ** *abbr.* (**Центральная лаборатория и экспериментальная мастерская электрических и измерительных приборов**) Central Laboratory and Experimental Workshop of Electrical and Measuring Instruments

**ЦНД** *abbr.* (**цилиндр низкого давления**) low-pressure cylinder

**ЦНИИ** *abbr.* (**Центральный научно-исследовательский институт**) Central Scientific Research Institute

**ЦНИИС** *abbr.* (**Центральный научно-исследовательный институт связи**) Central Scientific Research Institute of Communications

**ЦНИИТ** *abbr.* (**Центральный научно-исследовательский институт техники**) Central Scientific Research Institute of Technology

**ЦНИИТМАШ** *abbr.* (**Центральный научно-исследовательский институт технологии и машиностроения**) Central Scientific Research Institute of Technology and Mechanical Engineering

**ЦНИЛ** *abbr.* (**Центральная научно-исследовательская лаборатория**) Central Scientific Research Laboratory

**ЦНИЛэлектром** *abbr.* (**Центральная научно-исследовательская лаборатория электрической обработки материалов**) Central Scientific Research Laboratory for Electrical Treatment of Materials

**ЦНИМАШ** *abbr.* (**Центральный научно-исследовательский институт машиностроения**) Central Scientific Research Institute of Mechanical Engineering

**ЦНИЭЛ** *abbr.* (**Центральная научно-исследовательская электротехническая лаборатория**) Central Electrical-Engineering Research Laboratory

**ЦНОЛ** *abbr.* (**Центральная научно-опытная лаборатория**) Central Scientific Experimental Laboratory

**ЦНТБ** *abbr.* (**Центральная научно-техническая библиотека**) Central Library of Science and Technology

**ЦНТЛ** *abbr.* (**Центральная научно-техническая лаборатория**) Central Scientific and Technical Laboratory

**Цобель** Zobel

**цоколевка** *f.* basing; **ц. верхнего вывода** top-cap basing; **мастичная ц.** cement basing

**цоколёвочный** *adj.* basing, base

**цоколенабивочный** *adj.* base-filling

**цоколь** *m.* base, foundation, mounting; socle, plinth, block; socket, cap (of bulb); holder; **ц. «бипост»** bipost base; **винтовой ц. Голиафа, винто-**

вой ц. для больших ламп с резьбой диаметром **40** мм Goliath Edison screw cap, G.E.S. cap; **винтовой ц. с резьбой диаметром 10 мм** miniature Edison screw cap; **винтовой ц. с резьбой диаметром 14 мм** small Edison screw cap; **винтовой ц. с резьбой диаметром 27 мм** Edison screw cap; **ц. Голиафа** Goliath Edison screw cap, G.E.S. cap, Mogul base; **ц. диода** cartridge socket for semiconductor diodes; **заделанный ц.** European outside-contact base, European outside-contact socket; **ц. катодной трубки** magnal base; **ламповый ц. «миньон»** mignon socket, midget socket; **ламповый ц. с нарезкой Эдисона** standard (screw) socket; **ц. миньон** miniature Edison screw cap; **наружный контактный ц.** side-contact base; **нормальный ц. Эдисона Е-27** Edison screw cap; **одиннадцатиштырьковый ц.** magnal base; **патронный штепсельный ц.** plug adapter; **переходный ц. с закрылком** skirted adapter; **подходящий ц.** "go" cap; **ц. с витритовой изоляцией** vitrite cap; **ц. с контактом в нижней части патрона** bottom-contact cap; **ц. Свана** bayonet socket, Swan base; **ц. со штырьками** plug-in mounting; **утопленный ц.** European outside-contact base, European outside-contact socket; **фокусирующий ц.** pre-focus cap; **штампованный ц.** wafer base

**цокольный** *adj.* base, foundation; socket, base

**ЦПБ** *abbr.* (**Центральная политехни-**ческая библиотека) Central Polytechnic Library

**ЦПТС** *abbr.* (**центральная пригородная телефонная станция**) suburban central office (*tphny.*)

**ЦПУ** *abbr.* (**центральный пригородный узел**) suburban toll center (*tphny.*)

**ЦРЛ** *abbr.* (**Центральная радиолаборатория**) Central Radio Laboratory

**ЦРУ** *abbr.* (**центральное распределительное устройство**) central distribution system

**ЦРЯ** *abbr.* (**центральный распределительный ящик**) main junction box

**ЦСД** *abbr.* (**цилиндр среднего давления**) medium-pressure cylinder

**Ц-система** *f.* abbr. (**система центра масс**) center-of-mass system

**ЦТ** *abbr.* 1 (**цветное телевидение**) color television; 2 (**центр тяжести**) center of gravity

**ЦТВ** *abbr.* (**цветное телевидение**) color television

**ЦТС** *abbr.* (**центральная телефонная станция**) central office (*tphny.*)

**ЦУ** *abbr.* (**центральное управление**) central administration

**цуг** *m.* train

**ЦЧ** *abbr.* (**цетановое число**) cetane number

**ЦЭС** *abbr.* 1 (**центральная электростанция**) central electric power plant; 2 (**Центральный электротехнический совет**) Central Electro-Technical Council

**ЦЭСС** *abbr.* (**центральная электрическая самолётная станция**) central electric generator unit for aircraft

# Ч

ч. *abbr.* 1 (**час**) hour; 2 (**часть**) part, section, portion; unit; 3 (**число**) number

**Чайльд** Child

**Чальмерс** Chalmers

**чан** *m.* tank, vat, tub; pit trough

**чановый** *adj.* tank, vat, tub; pit trough

**чарджистор** *m.* chargistor

**час** *m.* hour; **часы малой** (*or* **слабой**) **нагрузки, провальные часы** slack hours, nonrush hours; **ч. наибольшей нагрузки** busy hour (*tphny.*); *pl.* rush hours; *see also* **часы** *pl.*

**часовой** 1. *adj.* clock, watch; 2. (*noun*) *m. adj. decl.* watch, guard, sentinel; hour, an hour's, one hour's; **автоматический ч.** automatic time switch; **по ч. стрелке** clockwise; **против ч. стрелки** counterclockwise

**часо-занятие** *n.* traffic unit

**частица** *f.* particle, bit, fraction, corpuscle, grain, speck; molecule; **летучие частицы** flyings; **дочерняя ч.** product particle; **ливневая ч.** shower particle; **небольшая ч.** grain

**частичка** *f.* (small, fine) particle

**частично** *adv.* partially, partly, incompletely; **ч. коаксиальный** partial-to-coaxial

**частично-фокусированный** *adj.* partially focused

**частичный** *adj.* partial, fractional; particle, particulate; molecular; minor

**частное** *n. adj. decl.* quotient (*math.*)

**частность** *f.* particularity; **в частности** in particular

**частный** *adj.* particular, peculiar, exceptional, individual; private; partial

**часто** *adv.* often, frequently, constantly; close; fast; **ч. повторяющийся** frequent

**«частокол»** *m.* railing (*rdr.*); grass (*tv.*)

**частомер** *m.* cymometer, frequency meter

**частость** *f.* density, denseness, thickness; closeness; frequency, rate; quickness; **ч. (события)** relative frequency; **ч. отказов** mission success rate

**частота** *f.* frequency; periodicity; closeness, thickness; wave; **верхние (звуковые) частоты** treble; **частоты вне полосы пропускания** nonpass bands; **нижние частоты диапазона** bass; lower frequencies; **низкие звуковые частоты** bass (*acoust.*); **частоты при которых вектор импеданса находится под углом 45° к диаметру** quadrantal frequencies; **с одинаковой частотой** equifrequent; **алфаграничная ч.** alpha-cut-off frequency; **ч. амплитудно-импульсной модуляции** sampling rate; **ведущая ч.** control frequency, regulation frequency; **верхняя ч.** high-frequency; **верхняя ч. среза** upper cut-off frequency; **верхняя граничная ч.** upper-frequency limit; **ч. (взятия) отсчётов** sampling rate; **ч. включения и выключения** turn-on frequency; **ч. вращения ионов** gyro frequency (in a magnetic field); **вспомогательная ч.** adjacent frequency, side frequency, idler frequency, quenching frequency; **вспомогательная несущая ч.** sub-carrier; **вторая комбинационная ч.** image frequency; **входная ч.** signal frequency; **ч. вынуждающей силы** impressed frequency; **ч. вынужденных колебаний** forcing frequency; foreign frequency; **граничная ч. усиления по току в схеме с общей базой** alpha-cutoff frequency; **ч. десятикилометрового диапазона волн** very low frequency; **ч. десятиметрового диапазона волн** high frequency; **ч. дециметрового диапазона волн** ultrahigh frequency; **ч. длинноволнового диапазона** low frequency; **ч. для (двухсторонней) связи самолёта с землёй** air-ground radio frequency; **ч. для смещения и стирания** bias-erase frequency; **ч. заполнения** pulse-modulated frequency; **звуковая ч. биений** beat-

658

note frequency; **зондирующая ч., ч. зухтона** search frequency; **ч. изгибных колебаний** bending frequency; **изменяющаяся ч.** sliding (test) frequency; **ч. измерений мгновенных значений** sampling rate; **ч. канала** wobble frequency; **ч. квантования** sampling frequency, sampling rate; **ч. километрового диапазона волн** low frequency; **колориметрическая ч. цвета** colorimetric purity; **ч. коммутации (сигналов основных) цветов** color sampling rate; **контрольная ч.** marker frequency, pilot frequency; **ч. коэффициента качества** figure-of-merit frequency; **кратная ч.** multiple frequency; **ч., лежащая в области теплового излучения** infrared frequency; **максимальная ч., воспринимаемая ухом** upper frequency limit; **максимальная воспроизводимая ч.** cutoff frequency; **ч. манипуляции** pulse repetition rate; **ч. маркёра** timing frequency; **ч. метрового диапазона волн** very high frequency; **ч. миллиметрового диапазона волн** extremely-high frequency; **ч. модуляции винта** blade frequency; **ч. модуляции при коническом сканировании** lobing frequency; **навигационная ч.** beacon frequency; **ч. немодулированного сигнала** center frequency, resting frequency; **несинхронизированная ч., ч. несинхронизованного сигнала** free-running frequency; **низшая собственная ч.** fundamental frequency; **низшая употребляемая ч.** LUF; **нормальная ч.** single frequency; **ч., определяющая длину волны** wave frequency; **основная ч.** base frequency, master frequency, fundamental frequency, lowest frequency; fundamental harmonic; **основная хронирующая ч.** master timing frequency; **ч. отклонения электроннолучевой трубки** sweep frequency; **ч. параметрического возбуждения** pump frequency; **ч. побочных колебаний** frequency splitting; **ч. повторения** repetition frequency, recurrent frequency, repetition rate, recurrence rate; **повышающая ч. биений** upsweep frequency; **повышенная ч.** overfrequency, treble; **ч. поглощающего**

**контура** absorbed frequency; **подавляемая ч.** trap frequency; **поднесущая ч., модулируемая сигналом цветности** chrominance subcarrier; **ч. полукадров** field frequency, field repetition rate, frame frequency, vertical frequency; **ч. «полуспада»** half-power frequency; **пониженная ч.** underfrequency; **ч. посылки (импульсов)** sampling rate; **ч. посылок** group rate, repetition rate, train rate, recurrent rate; sampling frequency, sampling rate; **ч. появления данного слова** word frequency; **ч. появления трёхбуквенных структур** trigram frequency; **предельная ч.** limiting frequency, cutoff frequency, critical frequency (of a filter); **предельная ч. усиления по току, предельная рабочая ч. транзистора** alpha-cut-off frequency; **ч. прерываний** chopper frequency; **ч. продольных колебаний** pitch frequency; **ч. проникновения сигналов в ионосферу** penetration frequency; **ч. пульсации (ПуВРД)** frequency of the explosion; **ч. пьезоэлектрического резонатора** crystal frequency; **рабочая ч. РЛС** radar frequency; **ч. рабочей посылки** marking frequency; **ч. равная двойной (или половине) строчной частоте** half-line frequency; **ч. радиовещательного диапазона** broadcast frequency; **разговорная ч.** voice frequency; **ч. разговоров** frequency of calls; **ч. разложения изображения** picture point frequency; **размагничивающая ч.** wiping frequency; **ч. растра** frame-repetition rate; **ч. резонанса токов** antiresonant frequency, antiresonance frequency; **резонансная ч. колебаний в потоке плазмы** Langmuir frequency; **резонансная высокая ч.** tuned radio frequency; **ч. сантиметрового диапазона волн** super high frequency; **ч. свободных колебаний** natural resonant frequency; **ч. связи** working frequency, repetition frequency; **ч. связи между станциями** point-to-point frequency; **ч. серий колебаний искрового разряда** spark frequency; **ч. серий разрядов** group frequency; **симметричная ч.** second-channel frequency; **ч. синхронизации** clock frequency; **ч.**

синхронизирующих импульсов clock rate; **ч. синхронизирующих сигналов в системе ЦТ** burst frequency; **ч. следования** spacing frequency, recurrence rate, repetition rate; **ч. слушания** watch frequency; **ч. смены полей** frame frequency; **собственная ч.** free frequency, allowed frequency, natural frequency, basic frequency; **ч. собственных колебаний электронов в плазме** plasma frequency; **ч. соударений, приходящаяся на единицу объёма** collision frequency per unit volume; **ч. $f_{\alpha}$ спада характеристики** alpha-cut-off frequency; **ч. срабатываний реле** duty classification of a relay; **ч. сравнения** standard note; **средняя ч.** mid-frequency, center frequency, mean frequency, medium frequency; idle frequency, resting frequency; **средняя ч. ЧМ** resting frequency; **ч. срыва колебаний** quench frequency, quenching frequency; **стабилизированная ч.** constant wave; **ч. стокилометрового диапазона волн** extremely-low frequency; **ч. стометрового диапазона волн** medium frequency; **ч. съёмки** frame frequency; **ч. тактирования, тактовая ч.** clock frequency, timing frequency, clock rate, synchronous clock frequency; **угловая ч.** pulsatance, angular frequency, corner frequency, radian frequency; **угловая ч. переменного тока** pulsation; **ч. уравнивания** crossing-over frequency; **условная ч. цвета** excitation purity; **утроенная ч.** treble; **учебная ч.** training frequency, student frequency, practice frequency; **ч. холостого контура** idler frequency; **хронизирующая** (*or* **хронирующая ч.**) timing frequency; **ч. цвета** colorimetric purity; **ч. элементов** abundance of the elements; **ч. элементов мозаики** dot frequency

**частотно-временной** *adj.* time-and-frequency

**частотнозависимый** *adj.* frequency-dependent; frequency-sensitive

**частотно-избирательный** *adj.* frequency-selective

**частотно-импульсно-модулированный** *adj.* pulsed frequency-modulated

**частотно-импульсный** *adj.* pulse-frequency

**частотномодулированный** *adj.* frequency-modulated, FM; warbled

**частотно-преобразовательный** *adj.* frequency conversion

**частотно-селективный** *adj.* frequency-selective

**частотностабилизированный** *adj.* frequency-stable

**частотно-фазовый** *adj.* phase-frequency

**частотный** *adj.* frequency

**частотомер** *m.* frequency meter, frequency teller, periodmeter, cymometer; **ч. задающей частоты** integrating frequency meter, master frequency meter; **ч. н.ч.** a-f meter; **ч. с непосредственным отсчётом** Magmeter; **стрелочный ч.** direct reading frequency meter; **тональный ч.** audio-frequency meter; **ч. Фрама** reed indicator; **язычковый ч. Фрама** Frahm frequency meter

**частотомерный** *adj.* frequency measuring

**частото-преобразовательный** *adj.* frequency-converting

**частый** *adj.* frequent, dense, thick; quick

**часть** *f.* part, portion, fraction, share, element; piece, fragment; branch, department, section; member, link; **вспомогательные части** accessories; **части под напряжением** live parts; **фасонные части** fittings; **анодная ч. разрядного промежутка** anode region; **аппаратная ч.** hardware; **ближняя ч. инфракрасного спектра** near infrared; **вводная ч. программы** program head; **вводная ч. цикла** initializing part of loop; **ведущая ч. прижимного рельса** detector bar; **верхняя ч. колбы** bulb bowl; **вещественная ч. полного сопротивления** real resistance; **внешняя ч. (линии) сети** network spur; **вставная ч.** fitting piece; **вступительная ч. радиотелеграммы** preamble; **выделяющаяся ч.** projection; **выхлопная ч. (турбины)** exhaust casing; **гнездовая ч. двустороннего разъёма** double-sided socket, interconnecting socket; **гнездовая ч. разъёма** socket; **головная ч.** head; **грамматическая ч. семиотики** grammatics; **ч. громоотводного провода на крыше** roof conductor; **действительная ч.** real component of a complex quantity; real component of an AC conductance or resistance; **действительная ч. накапливающего счётчика** real accumulator (computing); **действительная ч. полной проницаемости** effective permeability, real permeability; **ч. декады** sublevel; **добавочная ч.** extension; **дробная ч. числа** mantissa; **дуплексная ч.** counterpart; **задняя ч. корпуса** rear head; **запасная ч.** spare part; **изогнутая ч. вибратора** stub; **изоляционная фасонная ч.** moulded insulation; **катодная ч. разрядного промежутка** cathode region; **кольцевая ч. станции** field ring; **командная ч. цикла** instruction part of cycle; **ч. комплекта** subassembly; **конусная ч. горловины** funnel neck; **логическая ч. вычислительной машины** logic; **материальная ч.** equipment; **механическая ч. музыкального инструмента** action (of a musical instrument); **механическая ч. прибора** measurement mechanism; **младшая ч. произведения** minor product; **многократная фасонная ч.** multiple tile; **ч. набора** subassembly; **находящаяся под напряжением металлическая ч.** live metal; **неподвижная ч. вариометра** stator; **непроводящая ч. периода** idle period; **нижняя ч.** bottom; **ч. области прибора с показаниями достаточной точности** effective range; **оживальная ч.** ogive; **оптическая ч.** opticator; **отрицательная ч. индикаторной диаграммы** negative logic; **передающая ч. телеуправления** transmitter; **передняя ч.** head; **передняя ч. трубки** tube face; **перемыкающая ч.** bridging piece; **переходная ч.** adapter; **повторяющаяся ч. программы** brick; **подвижная ч. с апериодическим успокоением** aperiodic element; **подвижная ч. с периодическим успокоением** damped periodic element; **подводная ч. корабля** bottom (of a boat); **пологая ч. характеристики** plateau; **прагматическая ч. семиотики** pragmatics; **ч. прибора** block; **ч. программы** subroutine, subprogram; **ч. программы вычислительной машины** segment; **рабочая ч.** test section; **рабочая ч. периода**

**(сварки)** arc time; **рабочая ч. шкалы** measurement range, effective range; **разъёмная фасонная ч. трубчатой проводки** split fitting; **растянутая ч. изображения на экране** expanded scope, expanded center; **ч. речи** class; part of speech; **скрученная ч. волновода** waveguide twist; **соединительная ч.** binder, jointing, sleeve; **составная ч.** component; constituent; **составная ч. изображения** term (tv); **сплющенная ч. ножки** pinch of a mount; **старшая ч. произведения** major product; **схемная ч.** hardware, circuitry; **ч. схемы** circuit section; **торцевая ч. обмотки** end winding; **фасонная ч.** molded body; **цветовая ч.** chrominance section; **ч. целого** integrant, component; **ч. цепи из воздушных проводов** aerial circuit; **ч. цепи с отрицательным сопротивлением** third-class conductor; **шарнирная фасонная ч.** split coupling; **штриховая ч. испытательной таблицы (для определения чёткости)** wedge pattern, square-wave test pattern, resolution wedge; **ч. электрической цепи** hot side

**часы** *pl.* clock; watch; timepiece; meter, **автоматические контактные ч.** autotimer; **главные ч.** master clock; **контактные ч.** switch clock; **космические ч.** space clock instrument; **мокрые газовые ч.** water-sealed rotary gas meter; **первичные электрические ч.** master clock; **песочные ч.** hour-glass, sand-glass; **регулируемые вторичные электрические ч.** electric slave-clock; **ч. с запуском от сигнала** triggered time clock; **ч. с переключателем для счётчика** meter changeover clock, time switch; **ч. с самозаводом** self-winding electric clock; **солнечные ч.** sundial; **сухие газовые ч.** bellows-type gas flowmeter; **ч.-указатель** controlled clock, secondary electric clock; **центральные ч.** master clock, clock synchronizer

**«чатертон»** *m.* Chatterton's compound
**чатертоновский** *adj.* Chatterton's
**Чаудхури** Chaudhuri
**чафф** *m.* chaff; window
**чаша** *f.* cup, bowl; basin, pan, dish
**чашевидный, чашеобразный** *adj.* bowl, bowl-shaped; cup, cup-shaped

**чашечка** *f.* small cup; ferrule

**чашечный** *adj.* cup, cup-shaped; ferrule

**чашка** *f.* cup, bowl; housing; dome, gong; vessel; **ч. весов** scale; **выпаривательная ч., выпарная ч.** rectifier bulb; **ч. для капсюля микрофона** handset case (*tphny.*)

**чашкообразный** *adj.* cup, cup-shaped, bowl, bowl-shaped

**чаще** *compar. of* **часто, частый** more often, more frequent(ly); **ч. всего** mostly

**чв-ч** *abbr.* (**человекочас**) man-hour

**Чебышев** Tschebyscheff, Chebyshev

**чебышевский** *adj.* (of) Chebyshev, Chebyshev's, Tschebyscheff's

**чей** *m. poss. pron.* whose

**чека** *f.* cotter, cotter pin, splint pin, key, chock, wedge, pin, pin bolt

**чекан** *m.* calking tool, calker; stamp, die, chisel

**чеканить, отчеканить** *v.* to chisel, to chip, to chase, to engrave, to emboss; to stamp, to hammer

**чеканный** *adj.* calking, stamping; stamped; engraved; calked

**чеканочный** *adj.* calking; stamping

**челеста** *f.* celesta

**челн, челнок** *m.* shuttle, quill; **батарейный ч.** battery chute

**челночный** *adj.* shuttle

**человек** *m.* person, man

**человекочас** *m.* man-hour

**человеческий** *adj.* human

**человечный** *adj.* human, man's

**червяк** *m.* worm, spiral; screw

**червяковый** *adj.* worm

**червячный** *adj.* worm, screw

**чердак** *m.* attic, loft

**чердачный** *adj.* attic, loft

**чередование** *n.* sequence, alternation, alternating, rotation, alternity, interchange, interlace, interlacing, interleave, interleaving; **ч. (периодов) записи и считывания** alternate writing and reading; **последовательное ч.** sequence, switching; **последовательное ч. элементов изображения** dot sequence; **ч. процессов записи и стирания** switching from writing to erasing; **ч. элементов изображения** dot interlace

**чередовать** *v.* to interlace, to interleave, to alternate, to rotate, to take turns, to interchange; to reverse

**чередоваться** *v.* to intersperse, to alternate, to interchange, to interleave, to interlace, to succeed, to follow, to rotate; to reverse

**чередующийся** *adj.* alternating, alternate, interleaved, interlacing; staggered; cycling

**через** *prep. acc.* through, via, per, by; across, over; in, after; within

**черезкожный** *adj.* transcutaneous

**черезстрочный** *adj.* interlaced, interlacing

**Черенков** Cherenkov, Čerenkov

**черенковский** *adj.* Cherenkov's, Čerenkov's

**черенок** *m.* shank, handle, grip; graft, scion; **ч. держателя** support shank; **ч. (держателя) иглы** shank of a needle

**черепаховый** *adj.* tortoise shell

**черепица** *f.* tile, roof tile

**черепичный** *adj.* tile

**чересстрочный** *adj.* interlaced, interlacing

**чересчур** *adv.* excessively, too

**чернее** (*or* **черней**) **чёрного** blacker-than-black

**чернение** *n.* blackening, blacking, carbonization

**чернёный** *adj.* blackened, carbonized

**чернеть, почернеть** *v.* to become black, to blacken, to darken

**чернила** *pl.* ink; **ч. для нанесения данных** coding ink

**чернильница** *f.* inkwell; ink holder, ink trough; **телеграфная ч.** Morse inker

**чернильный** *adj.* ink

**чернить, почернить** *v.* to blacken; to black out

**черно-белый, чёрно-белый** *adj.* black and white, monochrome

**черновой** *adj.* rough, crude, coarse; intermediate; shaping

**чернопишущий** *adj.* black-writing; inkwriter

**чернота** *f.* blackness

**чёрный** *adj.* black; dark; rough, coarse; ferrous (*met.*)

**черпак** *m.* ladle, scoop; bucket, pail

**черпальный** *adj.* ladling, scooping, dredging

**черта** *f.* line, stroke, dash, mark; feature, trait, characteristic; stroke; act; bar; **основные черты** (basic) features; **курсовая ч.** grid wire; **тактовая ч.** bar

**чертёж** *m.* drawing, chart, draft, sketch; design, plan, scheme, layout, blueprint; schedule; **главный монтажный** (*or* **сборный**) **ч.** key-sheet; assembly blueprint; **ч. на кальке** tracing; **рабочий ч.** design drawing

**чертёжный** *adj.* drawing, drafting, graphic

**чертилка** *f.* scratch awl, marking awl, scriber; drop point

**чертить, начертить** *v.* to draw, to design, to sketch, to plot, to trace

**чёрточка** *f.* dot; line; **ч. пунктира** dot

**четверичный** *adj.* quaternary

**четвёрка** *f.* four; four-wire unit, phantom circuit; quad, tetrad; **двойная ч. звездой** pairs cabled in quad-pair formation; **ч. с простой скруткой** spiral quad, spiral four; **скрученная звездой ч.** spiral quad (cables)

**четверной** *adj.* quadruple, fourfold, tetra-; quaternary

**четверо** *num.* four

**четвёрочный** *adj.* quadded

**четвертичный** *adj.* quaternary

**четвёртка** *f.* quarter, one fourth; **центральная ч. кабеля** central quad of cable, four-wire core

**четвертной** *adj.* fourth (*mus.*); quarter, quadruple, tetra-; quadrantal; quaternary, four-component

**четвёртый** *adj.* fourth

**четверть** *f.* quarter, a fourth; quarter note, (one) fourth, crotchet; **ч. волны** quarter-wave length; **ч. длины волны** quarter wavelength; **ч. (окружности) круга** quadrant

**четвертьволновой** *adj.* quarter-wave

**четвертькквадратный** *adj.* quarter-square

**четвертьоборотный** *adj.* quarter-turn

**чёткий** *adj.* clear, legible, sharp; accurate

**чёткость** *f.* clearness, legibility, sharpness, clarity, definition; accuracy; fine detail resolution, resolution; **ч. в углах изображения** corner detail; **ч. изображения** definition, image detail, picture resolution; **ч. разложения** sharpness of definition, fineness of definition; **разрешающая ч. яркостного сигнала** luminance resolution; **ч. разрешающего элемента** spot definition; **ч. шкалы** readability

**чётно-нечётный** *adj.* even-odd

**чётно-симметричный** *adj.* even-symmetric

**чётность** *f.* parity

**чётно-чётный** *adj.* even-even

**чётный** *adj.* even, even-numbered; sharp

**Четтертон** Chatterton

**четыре** *num.* four; **с четырьмя сердечниками** four-limbed

**четырёх-** *prefix* quadri-, tetra-, four

**четырёхадресный** *adj.* four-address

**четырёхатомный** *adj.* tetratomic

**четырёхвалентность** *f.* tetravalence

**четырёхвалентный** *adj.* tetravalent

**четырёхвалковый** *adj.* four-roll(er)

**четырёхгранный** *adj.* tetrahedral

**четырёхжильный** *adj.* four-wire

**четырёхзначный** *adj.* four-digit; four-unit; four-aspect

**четырёхзондовый** *adj.* four-probe

**четырёхкамерный** *adj.* four-cylinder, four-chamber

**четырёхканальный** *adj.* four-channel; four-terminal

**четырёхквадрантный** *adj.* four-quadrant

**четырёхколенчатый** *adj.* four-throw (shaft)

**четырёхколёсный** *adj.* four-wheel(ed)

**четырёхкомпонентный** *adj.* four-component

**четырёхконечный** *adj.* four-point

**четырёхкратный** *adj.* quadruple, quadruplex, fourfold

**четырёхкулачковый** *adj.* four-jaw(ed)

**четырёхкурсовой** *adj.* four-course

**четырёхламповый** *adj.* four-tube

**четырёхлопастный** *adj.* four-blade

**четырёхлучевой** *adj.* four-beam

**четырёхмачтовый** *adj.* four-pole

**четырёхмерный** *adj.* quaternary, four-dimensional

**четырёхокись** *f.* tetroxide

**четырёхосновный** *adj.* tetrabasic

**четырёхосный** *adj.* four-axis, tetra-axial

**четырёхотверстный** *adj.* four-hole

**четырёхплечий** *adj.* four-port

**четырёхпозиционный** *adj.* four-position

**четырёхполюсник** *m.* fourpole, quadripole, quadrupole, quadruple, four-terminal pair network, four-terminal network; O-network, network; transducer, transductor; **Г-образный ч.** L-network; **Г-образный уравнове-**

шенный ч. C-network; **ч. мостового типа, мостовой ч.** lattice network, lattice four-pole; **П-образный ч.** pi-network; **ч., пропускающий все частоты** all-pass lattice; **реактивный ч.** four-terminal network made up by reactive elements; **ч. с линейной фазовой характеристикой** linear phase network; **ч., состоящий из активных сопротивлений** resistance network, resistive network; **спаренный ч.** double-phantom circuit; **стандартный ч. искажения** distortion standard; **ч. типа П** P-network; **Т-образный ч. с продольной перемычкой** bridged T-network; **форсирующий ч.** lead network

**четырёхполюсный** *adj.* quadripole, quadripolar, tetrapolar, four-pole, four-polar, four-terminal, four-contact

**четырёхполярный** *adj.* quadripolar

**четырёхпроводный** *adj.* four-wire; four-way

**четырёхпроцентный** *adj.* four-percent

**четырёхпрядный** *adj.* four-strand

**четырёхразрядный** *adj.* four-digit, four-place

**четырёхрожковый** *adj.* four-pronged

**четырёхрядный** *adj.* four-row, tetra-serial

**четырёхслойный** *adj.* four-layer

**четырёхсплавный** *adj.* four-component (alloy)

**четырёхсторонний** *adj.* quadrilateral, four-sided

**четырёхступенчатый** *adj.* four-stage, four-stepped

**четырёхтактный** *adj.* four-cycle, four-stroke

**четырёхточечный** *adj.* four-point

**четырёхугольник** *m.* quadrangle, tetragon

**четырёхугольный** *adj.* quadrangular, tetragonal, four-angled

**четырёхфазный** *adj.* quadriphase

**четырёхфантомный** *adj.* quadruple-phantom

**четырёхцикловый** *adj.* four-cycle

**четырёхчастичный** *adj.* four-piece

**четырёхчастный** *adj.* four-part

**четырёхчастотный** *adj.* four-frequency

**четырёхчленный** *adj.* four-membered; four-period

**четырёхшкальный** *adj.* four-scale

**четырёхштырьковый** *adj.* four-prong, four-pin

**четырёхщелевой** *adj.* four-slot

**четырёхэлектродный** *adj.* four-electrode, four-element

**четырёхэлементный** *adj.* four-element

**четырёхэтажный** *adj.* four-stacked; four-story

**четырнадцатеричный** *adj.* quarter-denary

**четырнадцатиотрезочный** *adj.* fourteen-bar

**четырнадцатый** *adj.* fourteenth

**четырнадцать** *f.* fourteen

**Чеффи** Chaffee

**чехарда** *f.* leap frog test

**чехол** *m.* cover, case, hood; jacket, can, sheath, container; coat, coating; slip; **ч. от пыли** dust catcher

**чечевица** *f.* lens

**чечевицеобразный** *adj.* lens-shaped, lenticular

**чечевичный** *adj.* lenticular

**чешуйка** *f.* scale, plate, lamella

**чешуйчатый** *adj.* scaly, scaled, flaky, flaked, laminated, lamellar

**чешуя** *f.* scale, plate, lamella

**«чилекс»** *m.* chilex

**ЧИМ** *abbr.* (**частотно-импульсная модуляция**) pulse-frequency modulation

**чинить, починить** *v.* to repair, to mend

**чип** *m.* dice, chip

**чириканье** *n.* warble, warbling, twittering; **ч. при манипуляции** keying chirps

**численно** *adv.* numerically, in number

**численность** *f.* number, numberedness quantity; **численностью** in number; **чётная ч. ядерных частиц** even-numberedness of nuclear particles

**численный** *adj.* numeral, numerical; computational

**числитель** *m.* numerator; **ч. дроби** term of fraction, antecedent, numerator

**числительное** *n. adj. decl.* numeral; **количественное ч.** cardinal (number); **порядковое ч.** ordinal (number)

**числительный** *adj.* numeral

**числить** *v.* to count

**число** *n.* number, cipher, figure, quantity, coefficient; **взаимно простые числа** coprime numbers; **дополняющие числа** complements; **комплексные числа** imaginaries;

числа, над которыми проводится операция operands; **ч. бинарных переключений** binary pattern; **ч. ватт** wattage; **внутреннее квантовое ч.** total angular momentum quantum number; **ч. возбуждённых частиц** excitation number; **ч. возможных считываний** read number; **ч. воспроизводимых градаций яркости** tonal range; **вычитаемое ч.** subtrahend; **ч. двоичных единиц на элемент** bit sample; **ч. допустимых обращений** selection ratio; **ч. дырок** intrinsic number; **замедленное ч. оборотов** crawling speed; **ч. занятий** (telephone) traffic unit; **ч. знаков** digit capacity; **ч. знаков в минуту** lines per minute; **ч. знаков, запасаемое в регистре, ч. знаков, которые регистр может накопить, ч. знаков, накопляемых в регистре** register length; **ч. импульсов фона** background count; **кислотное ч.** acid value; **ч. кодов в секунду** word rate; **ч. колебаний в секунду** number of oscillations per second; **комплексное ч. с модулем, равным единице** complex unit; **кратное ч.** multiple; **кубическое ч.** cube; **ч. M** Mach number; **ч. Маха на конце лопасти** tip Mach number; **механическое передаточное ч.** register ratio; **ч. неразличных логических элементов** metrical information content; **ч. нормального ряда** preferred number; **ч. об/мин** speed of revolution; **ч. оборотов** number of revolutions, speed (r.p.m.), driving speed, rate of revolutions; **ч., образованное по случайному закону** random number; **ч. обращений к накопительной электронно-лучевой трубке, ч. обращений между регенерациями** read-around ratio, read-around number; **однозначное ч., одноразрядное ч.** digit, one-figure number; **ч. осмоления, осмоляное ч.** tar value; **основное ч. оборотов** base speed; **относительное ч.** relative (data processing); **ч. отсчётов** count; counting yield; counting rate; **ч. отсчётов в единицу времени** reading rate; **передаточное ч.** ratio, gear ratio, transmission ratio, reduction rate; **ч. перемен в секунду** cycles per second; **ч. переменной длины** multi-length number;

**ч. переноса** carry quantity; **ч. переноса ионов** migration ratio, transference number, transfer number; **ч. периодов** periodicity; **ч. периодов в секунду** number of cycles; **ч. повторений счётной операции, ч. повторения цикла** cycle criterion; **полное ч. активных проводников якоря** armature factor; **полуцелое ч.** half-integer; **порядковое числительное ч.** designation number; ordinal number; **ч. посылок в секунду** samples per second; **ч. превышающее ёмкость наибольшего регистра** infinity, capacity exceeding number, out-of-range number; **предельное ч.** resolvable number; **приборное ч. M** indicated Mach number; **приемочное ч. изделий** acceptance number; **простое ч.** prime number; **ч. прядок пряжи на дюйм** picks; **ч. разрядов в регистре** register length; **ч. разрядов в слове** digits; **рубиновое ч.** ruby content (protecting colloids); **ч. с большим количеством разрядов** long number; **ч. с двойным количеством разрядов** double-precision number; **ч. с малым количеством разрядов** short number; **ч. с удвоенным количеством знаков** double-length number, double-precision number; **ч. с учётом порядков** floating number; **ч. символов (системы связи)** information supply; **ч. слоёв сосредоточенной обмотки** number of layers (in a concentrated winding); **ч. сложений в минуту** counts per minute; **ч. смен кадров** repetition frequency, field frequency; **ч. смен кадров изображения в секунду** field repetition rate; **среднее ч.** medium; **ч. считываний** read number; **ч. считываний без потери информации** read-around number; **удельное ч. оборотов турбины** type characteristic; **уменшаемое ч.** minued; **условное ч.** call-word, call-number; **ч., участвующее в операции** operand; **ч. Фарадея** faraday; **фокусное ч. объекта** f-number; **ч., характеризующее качество** figure of merit; **характеристическое ч.** eigenvalence; **целое ч.** integer, integral number; **ч. Циолковского** missile mass ratio; **ч. членов** membership (linguistics); **шестнадцатеричное ч.** hexadecimal number; **шумовое ч.**

noise figure; **эквивалентное ч. дво-ичных разрядов** equivalent binary digits

**числовой** *adj.* number, cipher, figure, quantity; numeral, numeric(al)

**чистить, почистить** *v.* to clean, to cleanse; to dredge; to clear; **ч. нажда-ком** to grind with emery

**чистка** *f.* cleaning, cleanup, cleansing; dredging; clearing

**чисто** *adv.* purely, cleanly; **ч. электри-ческий** all-electric

**чистовой** *adj.* finish, finishing, clear; smooth

**чисто-инерциальный** *adj.* pure iner-tial

**чистота** *f.* clearness, clarity, fidelity; cleanliness, purity, fineness; **ч. по-верхности** degree of surface finish

**чистый** *adj.* clean, pure; clear, blank; net; virgin, raw; perfect; finished, smooth

**читаемость** *f.* readability

**читальный** *adj.* reading

**читатель** *m.* reader

**читать, прочитать, прочесть** *v.* to read; to sense

**читающий** *adj.* reading; sensing

**член** *m.* member; term, exponent; article; **ч. «И»** term AND; **остаточ-ный ч. (ряда)** remainder, remainder term; **первый ч. отношения** ante-cedent; **последующий ч.** succeedent; **постоянный ч.** absolute term; **пре-дыдущий ч.** antecedent; **свободный ч. выражения** absolute term in ex-pression; **ч. (уравнения), не содер-жащий переменной величины** ab-solute term; **экспоненциальный ч.** exponent; **экспоненциальный ч. вы-ражения** exponential

**членистый** *adj.* articulate, jointed, segmented

**членораздельность** *f.* articulation; **ч. речи** articulation of sentences, word intelligibility

**членораздельный** *adj.* articulate, dis-tinct

**ЧМ** *abbr.* 1 (**частотная модуляция**) frequency modulation; 2 (**частотно-модулированный**) frequency-modu-lated, f-m

**ЧНН** *abbr.* (**час наибольшей нагрузки**) busy hour

**Чохральский** Czochralski

**ЧПИ** *abbr.* (**частота повторения им-**

**пульсов**) repetition frequency, pulse repetition rate

**чрезвычайно** *adv.* extremely, exceed-ingly, immensely, highly, extra

**чрезвычайный** *adj.* extreme, extra-ordinary, excessive

**чрезмерно** *adv.* excessively

**чрезмерный** *adj.* excessive, extreme, inordinate, immoderate, redundant; over-: **чрезмерная нейтрализация** overneutralization

**ЧРТС** *abbr.* (**частная РТС**) private manual exchange; private switch-board

**ЧТ** *abbr.* (**частотное телеграфирова-ние**) frequency telegraphy

**чтение** *n.* reading; **ч. без разрушения информации** nondestructive read-out; **ч. данных** reading; **ч. записи с магнитной ленты** playback; **пове-рочное** (*or* **проверочное**) **ч.** proof-reading

**чтец** *m.* reader

**чтоб, чтобы** *conj.* in order that, in order to, so that

**чувствительность** *f.* sensitivity, sensi-tiveness, sensibility, susceptibility, susceptiveness; response; excitabi-lity; factor of merit; pick-up; **бал-листическая ч.** Coulomb sensitive-ness; **ч. в диффузном поле** random sensitivity, random-incidence re-sponse; **ч. в режиме излучения** trans-mitting response; **ч. в режиме холостого хода, определённая по полю** open-circuit field response; **вибрационная ч.** pallasthesia; **гру-бая ч.** muting sensitivity; **действи-тельная ч.** real-voice response; **ч. индикатора, ч. индикации** display sensitivity; **интегральная ч.** total sensitivity, photoelectric efficiency, luminous sensitivity; **ч. к обнаруже-нию** detectivity; **ч. к отклонению электроннолучевой трубки в ком-плекте с отклоняющей системой** deflection sensitivity of a magnetic-deflection cathode-ray tube and yoke assembly; **ч. к отклонению элек-троннолучевой трубки с магнитным отклонением** deflection sensitivity of a magnetic-deflection cathode-ray tube; **ч. к отклонению электронно-лучевой трубки с электростатиче-ским отклонением** deflection sensi-tivity of an electrostatic-deflection cathode-ray tube; **катодная световая ч.** cathode luminous sensitivity (of a photo-tube); **малая ч.** sluggishness; **ч. на деление шкалы** unit sensitivity; **обратная ч.** sensibility reciprocal; **относительная ч. в режиме излучения** relative transmitting response; **относительная ч. к ложному** (*or* **паразитному**) **сигналу** spurious response ratio; **ч. по германскому промышленному стандарту DIN** sensitiveness, DIN sensitivity; **ч. по мембране** pressure sensitivity; **ч. по мощности в режиме излучения** transmitting power response; **ч. по напряжению в режиме излучения** transmitting voltage response; **ч. по отклонению** deflection sensitivity; **ч. по отношению к напряжению** volt constant; **ч. по отношению к сопротивлению** megohm constant; **ч. по отношению к току** ampere constant; **ч. по промежуточной частоте** I.F. sensitivity; **ч. по стандарту Американской Ассоциации Стандартов** ASA speed; **ч. по току** current sensitivity, response to current; **ч. по току в режиме излучения** transmitting current response; **ч. по управляющему воздействию** command resolution; **ч. по Шейнеру** sensitivity scale of sensitized surfaces; **предельная ч.** noise figure; **приведённая ч.** figure of merit, factor of merit, normal sensitiveness; **ч. приёмника при «запертом» антенном коммутаторе** lock-on sensitivity; **световая ч. (передающей трубки) при цветовой температуре 2854° К** luminous sensitivity at a color temperature of 2854°K (of a camera tube); **ч. счётчика по мощности** starting power of a meter; **ч.**

**телефона** earphone response; **ч. усилительной схемы** sensitivity of a repeater connection; **ч. (устройства) к отдельным участкам спектра** spectral sensitivity; **фотокатодная световая ч.** photoelectric sensitivity, photoelectric yield

**чувствительный** *adj.* sensitive, sensible, susceptible, responsive; sensing, sensory; painful; quick-response; delicate; critical; **ч. к изменениям фазы** phase-sensitive; **ч. к направлению тока** polar-sensitive; **ч. к облучению** radiosensitive; **ч. к электронной бомбардировке** electron-sensitive

**чувство** *n.* sense, sensation, feeling; **ч. удушья** choking sensation

**чувствование** *n.* sensation, feeling

**чувствовать, почувствовать** *v.* to feel, to experience

**чувствующий** *adj.* sensing, feeling

**чугун** *m.* cast iron

**чугунный** *adj.* cast-iron

**чужеродный** *adj.* alien, foreign

**чужой** *adj.* foreign, strange; extraneous

**чулок** *m.* braiding, sleeving, tubing; mantle; housing; stocking; **ч. для протягивания** cable grip; **кембриковый ч.** cambric tubing, spaghetti; **лакированный ч.** varnished hose; **сквозной ч. для протягивания кабеля** split cable grip

**чулочный** *adj.* stocking; sleeving, tubing

**чурбан** *m.* block, stump; chunk, lump; **анкерный ч.** anchor log

**чуткий** *adj.* delicate, sensitive, responsive

**чуть** *adv.* barely, hardly, scarcely; **ч. не** almost, nearly; **ч. что не** all but

**чучело** *n.* dummy

**чьё** *n.*, **чья** *f.*, **чьи** *pl.* whose

# Ш

ш. *abbr.* 1 (**широкий**) wide, broad;
2 (**широта**) width; latitude
Ш *abbr.* (**школа**) school
**шабер** *m.* shaving knife, scraping knife,
scraper
**шаблон** *m.* master form, pattern, template, model, gage, mold, center;
copy, stencil; bobbin, spool; former;
framework; **ш. для контроля цоколя**
electron tube base gage, tube base
gage; **ш. для намотки секций** winding form; **кабельный ш.** lacing board,
forming board; **ш. колбы** bulb jaw;
**путевой ш.** track gage; **ш. формирования луча** beam-forming matrix
**шаблонный** *adj.* pattern, template;
model, mold; copy; bobbin, spool;
former
**шабот** *m.* stock anvil
**шабрить, сшабрить** *v.* to scrap
**шаброванный** *adj.* scraped
**шаг** *m.* step; pace; unit; tread; pitch,
spacing, measure; **с малым шагом**
closely spaced; **с переменным шагом**
variable-pitch; **безуспешный ш.** unsuccessful feature of behavior; **ш.
винтовой нарезки** screw pitch; **ш.
зацепления, ш. зубцов** tooth pitch;
**ш. импульсов** pulse spacing; **неудачный ш.** *see* **безуспешный ш.;**
**ш. по коллектору** commutator pitch;
**ш. поворота** rotary step; **ш. подачи**
feed pitch; **ш. подъёма** vertical step;
**подъёмный ш.** decade step, decade
selection (*tphny.*); **ш. поиска (гидролокатора)** step angle; **ш. покоя**
neutral position (of a selector);
**полюсный ш. по коллектору** pole
pitch at the commutator; **ш. пупинизации** load-coil spacing; **регулируемый ш.** namotki variable pitch
control; **результирующий ш. обмотки** unit interval in a winding,
winding pitch; **свободный ш.** auxiliary contact of a selector switch;
**ш. сетки** grid pitch, grid-wire
spacing; **ш. скрещивания (проводов)** complete transposition section

(of wires); **ш. скрутки** lay (of a
strand), twist, pitch of strand;
**холостой ш.** *see* **свободный ш.;**
**частичный ш. обмотки** back and
front pitch of a winding; **частичный
ш. с задней стороны** back pitch;
**частичный ш. со стороны коллектора** front pitch
**шаговый** *adj.* step, step-by-step; pitch
**шагомер** *m.* pedometer, distance recorder; pitch counter
**шайба** *f.* washer, disk, bead, grommet;
plate; **базовая ш.** base wafer; **ш.
для впуска проб масла** sump washer;
**ш. для перекрытия** cover disk; **изоляционная ш.** insulating bead, insulating washer, insulating disk,
dielectric bead; **качающаяся ш.**
solenoid-operated pendulum disk;
**перекидная ш.** double coil-type
annunciator drop; **пружинная ш.,
пружинящая ш.** lock washer; **распорная ш.** spacer; **селеновая выпрямительная ш.** selenium cell; **ш.
солнечного фотоэлектрического генератора** silicon disk
**шайбовый** *adj.* washer, disk, bead;
plate
**шальтер** *m.* switch
**шандора** *f.* shutter. (of dam)
**Шапеле** Chapelet
**Шаперон** Chaperon
**шапка** *f.* cap, cover, hood, top; lid; **ш.
цикла** initializing part of cycle
**шапочка** *f.* cap, bonnet; **противомикрофонная ш.** howl arrester
**Шапрон** Chaperon
**шар** *m.* sphere, globe, ball; balloon:
**воздушный ш.** balloon; **ш. кривизны** bevelling sphere (*math.*); **поплавковый ш.** ball float; **светомерный
ш.** photometric integrator, sphere
photometer
**шар-зонд** *m.* sounding balloon, balloon
sonde, registering sonde; **радиолокационный ш.-з.** radar wind
**шарик** *m.* (small) ball; bulb; bead, pellet;
dot; **скользящий ш.** advance ball

шариковый *adj.* ball; bead, pellet

шарикоподшипник *m.* ball bearing

шар-интегратор *m.* integrating-sphere

шарлах *m.* scarlet (dye)

шарлаховый *adj.* scarlet

Шарль Charles

Шарлье Charlier

шарманка *f.* street organ, hand organ

шарнир *m.* hinge, joint, link, articulation; **болтовой ш.** pin joint; **ш. Гука** cardan joint, universal joint; **ш. прижимного рельса** detector bar link; **узловой ш.** hitch joining; **универсальный шаровой ш.** ball and socket joint; **шаровой ш.** spherical calotte

шарнироукреплённый *adj.* hinged

шарнирно-шаровой *adj.* knuckle-and-socket

шарнирный *adj.* joint, hinge, link; articulated, articulate, joined, hinged

шаровидный *adj.* spheric(al), spheroidal, globular; nodular; roundness

шаровый *adj.* ball, sphere, spherical, globular

шарообразный *adj.* spheric(al), spheroidal, globular

Шарп Sharpe

шар-пилот *m.* pilot balloon, sounding balloon, sonde, radar sonde

шасси *n. indecl.* chassis, underframe, carriage, mounting frame, mounting rack; **ш. блока развёрток** deflection chassis; **корытообразное ш.** bathtub unit; **ш. печатной схемы** printed chassis; **ш. приёмника (системы) цветного телевидения** color chassis; **ш. приёмника чёрно-белого телевидения** black-and-white chassis, monochrome chassis; **разъёмное ш.** split frame; **ш. с мотором (в магнитофоне)** tape deck; **ш. с отверстиями** drilled base plate; **ш. с пазами** slotted base plate

шатание *n.* blurring (*tv*); swaying

шатать, шатнуть *v.* to sway, to rock, to shake

шатающийся *adj.* loose

шатнуть *pf. of* шатать

Шауль Chaoul

шахматный *adj.* checkerboard, checkered, staggered, alternate; chess

шахматообразный *adj.* checkered, staggered

шахта *f.* shaft, mine, pit; chamber; manhole; **ш. для пайки** jointing

chamber; **ш. для подъёма кабеля** vertical wall-duct, cable chute

шахтный *adj.* mine, mine-type, pit, shaft, mining

швартов *m.* mooring

швартовать, пришвартовать *v.* to moor

швартовый *adj.* mooring

Шварцшильд Schwarzschild

швеллер *m.* channel, channel iron, channel bar

швеллерный *adj.* channel

ШГ *abbr.* (шумовой генератор) noise generator

шединг, шейдинг *m.* shading

шейдинг-сигнал *m.* shading signal; **кадровый ш.-с.** field shading signal, frame shading signal, vertical shading signal; **строчный ш.-с.** horizontal shading signal, line shading signal

шейка *f.* neck, collar; neck, fingerboard (of a violin), pin, plug, pivot, journal, bolster; **ш. изолятора** insulator groove; **ш. ножки** stem shoulder

шейный *adj.* neck

Шейнер Scheiner

шёлк *m.* silk

шелковидный, шелковистый *adj.* silky, silk-like; tissue

шёлковый *adj.* silk

шёлкография *f.* silk-screen printing, silk-screening

шёлк-сырец *m.* raw silk

шеллак *m.* shellac

шеллаковый *adj.* shellac

шелушение *n.* peeling, husking, shelling

шелушить, нашелушить, ошелушить *v.* to shell, to husk

Шеннон Shannon

шёпот *m.* whisper, unvoiced speech

Шеппард Sheppard

шепчущий *adj.* whispering

Шербиус Scherbius

Шерер Scherer

Шеринг Schering

шероховатость *f.* roughness

шероховатый *adj.* rough, raw, coarse

Шеррер Scherrer

шерстеподобный *adj.* wool-like, wooly

шерсть *f.* wool; **минеральная ш.** rock wool

шерстяной *adj.* wool, woolen, hair

шест *m.* post, rod, (crook) stick; **ш. для протаскивания кабелей** duct

rod; **отходящий ш.** distributing rod, distributing pole
**шестеренка** *f.* pinion, gear, drive gear; **вращательная ш.** rotary hub (of a Strowger switch)
**шестерённый** *adj.* pinion, gear
**шестеричный** *adj.* sixfold, sextuple, senary
**шестерной** *adj.* six, sixfold
**шестерня** *f.* pinion, gear, drive gear, gear wheel, cog wheel; toothed wheel; **ш. без мёртвого хода** antibacklash gear; **ведущая ш.** сельсина selsyn drive gear; **малая ш.** pinion; **ш. на валу якоря** armature pinion; **ш. с торцевыми зубьями** crown wheel
**шести-** *prefix* hex-, hexa-, six
**шестиатомный** *adj.* hexatomic
**шестивалентный** *adj.* hexavalent
**шестивалковый** *adj.* six-roller
**шестигранник** *m.* hexahedron; **ш. на мостике якоря (реле)** armature bushing (of a relay)
**шестигранный** *adj.* hexahedral, cubic
**шестидесятеричный** *adj.* sexagesimal
**шестидесятый** *adj.* sixtieth
**шестикратный** *adj.* sextuple, sixfold
**шестилучевой** *adj.* six-rayed, hexactinal; hexagonal
**шестимачтовый** *adj.* six-mast
**шестиосновный** *adj.* hexabasic
**шестиполюсник** *m.* six-pole (network); **умножающий ш.** multiplying six-pole (network)
**шестиполюсный** *adj.* six-pole
**шестипроводный** *adj.* six-wire
**шестиразмерный** *adj.* six-dimensional; sextuple
**шестисторонний** *adj.* six-sided, hexahedral; hexagonal
**шеститактный** *adj.* six-step
**шестиугольник** *m.* hexagon
**шестиугольный** *adj.* hexagonal
**шестифазный** *adj.* six-phase
**шестицветный** *adj.* six-color
**шестичленный** *adj.* six-membered
**шестнадцатая** *f. decl.* sixteenth note
**шестнадцатеричный, шестнадцатиричный** *adj.* sexadecimal, hexadecimal
**шестнадцатый** *adj.* sixteenth
**шестнадцать** *num.* sixteen
**шестоватый** *adj.* columnar, stalked
**шестовой** *adj.* pole
**шестой** *adj.* sixth
**шесть** *num.* six

**шестьдесят** *num.* sixty
**Шеффер** Sheffer; Scheffer; Schäffer
**шея** *f.* neck; vent
**шильдик** *m.* label
**ШИМ** *abbr.* (широтно-импульсная модуляция) pulse-width modulation, pulse-duration modulation
**шимм** *m.* shim
**шиммирование** *n.* shimming
**шина** *f.* tire; bus, busbar, bar, rail, highway; line, strip, strap; wire; **профильные шины для распределительных устройств** busbars for switch-gear systems; **собирательные шины в коробе** busduct; **ш. бесконечной мощности** infinite bus; **вспомогательная сборная ш.** auxiliary busbar; **входная ш. запоминающего устройства** storage-in bus; **ш. выборки для записи** write select line; **ш. выборки для считывания** read select line; **ш. для крепления кабелей** cable support; **ш. для питания от сети** line-power busbar system; **ш. для скользящего контакта** slide bar; **ш. запрета** inhibit (-ing) line; **ш. одного разряда** bit line, digit line; **ш. передачи чисел** number (transfer) bus; **переключательная ш.** transfer bus; **ш. переноса** carry line; **переходная ш.** transfer bus; **ш. повреждений** fault bus; **разрешающая ш.** enable line; **разрядная ш.** bit line, digit line; **сборная ш.** collecting bar, bus, highway, busbar; **ш. сброса** reset line; **ш. смещения** bias strip, bias wire; **соединительная ш.** junction bar, junction rail; **ш. стирания информации** reset line; **ш. считывания** sense line; **ш. установки нуля** reset line; **числовая ш.** number bus, word line
**Шинеман** Schienemann
**шинный** *adj.* busbar, bar; tire
**шинопровод** *m.* busway, busduct; **ш. в коробе** busduct; **ш. со штепсельными гнёздами** plug-in busduct
**шип** *m.* thorn needle (for playback); dowel (pin), peg, pin, pivot, mandrel, journal; lug, horn, projection; **посадочный ш.** locating pin
**шипение** *n.* hiss(ing), sputtering, frying, mush; scratching; **ш. с потрескиваниями** sputtering, frying
**шипеть, прошипеть** *v.* to hiss

**шипорезный** *adj.* dovetail cutting, dovetailing, tenon-cutting

**шипящий** *adj.* fricative, sibilant, hissing

**шир.** *abbr.* (**ширина**) width

**шире** *comp. of* **широкий, широко** wider, broader

**Шире, Ширекс** Chireix

**ширина** *f.* breadth, width; broadness, wideness; **ш. в свету** internal width; **вертикальная ш.** vertical bandwidth; **геометрическая ш. щели** gap length; **горизонтальная ш. луча** horizontal beamwidth; **действующая ш. щели** effective gap length; **ш. диаграммы по точкам половинной мощности** half-power width; **ш. диапазона звуковых частот** sound band width; **ш. добавочного полюса** interpole arc; **ш. запрещённой зоны** energy gap; **ш. импульса** width of pulse, pulsewidth, pulse length, duration of pulse; **ш. импульса отпирания** gate-width; **ш. импульса посылки** sampling pulse width; **ш. канала** channel width; channel capacity, channel time; **ш. курса** pattern sharpness; **ш. курсового сектора** course width; **ш. лепестка диаграммы направленности** lobe width; **ш. лепестка у половинной мощности** half-power width of a radiation lobe; **ш. луча** beamwidth; **ш. монтированной ножки** mount spread; **ш. отверстия зева** width of jaws; **ш. плеча (катушки)** spread; **ш. плеча фазы на полюс** phasebelt; **ш. полосы** bandwidth; **ш. полосы захвата** sweep width; **ш. полосы пропускания** bandwidth; **ш. полосы цветного сигнала** chromacity bandwidth; **ш. полосы (частот)** (frequency) bandwidth; **ш. полосы частот канала цветности** chrominance channel bandwidth; **ш. поперечного сечения пучка** beam width; **ш. пропускания канала** channel width; **ш. (резонансной) кривой** curve breadth; **ш. селекторного импульса** gate width, trigger-gate width; **ш. таблицы** width of layout; **ш. фазной зоны** belt span; **ш. щётки** brush arc; **энергетическая ш. пучка** energy beam width

**ширительный** *adj.* stretching

**ширить, расширить** *v.* to enlarge, to widen, to expand, to stretch

**ширма** *f.* screen, shield, blind; baffle; mask; **ш. экрана (электронно-лучевой трубки)** barrier (of electron-beam tube)

**широкий** *adj.* broad, wide, extensive, extended

**широковещание** *n.* broadcast(ing)

**широковещательный** *adj.* broadcast (-ing)

**широкодиапазонный** *adj.* wide-range; broadband

**широкоизлучатель** *m.* wide-angle fitting, wide-angle lighting fitting

**широкополосный** *adj.* wide-band, broadband, broad passband, wide-strip; wide-range

**широкораскрывный** *adj.* wide-angle

**широкослойный** *adj.* broad-zoned

**широкостробный** *adj.* wide-gate

**широкоугольный** *adj.* wide-angle

**широкоходовой** *adj.* loose

**широкошляпный** *adj.* broad-headed

**широкоэкранный** *adj.* wide-screen

**широта** *f.* width, breadth; latitude; **ш. распределения** range

**широтно-импульсный** *adj.* pulse-width, pulse-length

**широтно-модулированный** *adj.* width-modulated

**широтный** *adj.* width, breadth; latitude, latitudinal

**шитый** *adj.* sewn

**шифер** *m.* slate

**шиферный** *adj.* slate, slate-like; flaky

**шифр** *m.* code, cipher; cryptography; **шифры движения** traffic figures

**шифратор** *m.* coder, encoder, inscriber; quantizer; **временный ш.** clock-coder; **дифференциальный ш.** incremental coder; **цифровой ш. углового положения** digital angular position encoder

**шифраторный** *adj.* encoder

**шифрация** *f.* coding

**шифровальный** *adj.* code, cipher; cryptographic

**шифровальщик** *m.* code clerk, encoder

**шифрование** *n.* encoding, coding; **ш. ключевыми знаками** vigenere cipher

**шифрованный** *adj.* ciphered, figured, in code, cryptographic, scrambled

**шифровать, зашифровать** *v.* to cipher, to code, to encode

**шифрующий** *adj.* ciphering, scrambling

**шихта** *f.* load, burden, charge

**шихтованный** *adj.* laminated

**шихтовка** *f.* interleaving; **ш. магнитных сердечников** lamination

**шишка** *f.* knob

**шкала** *f.* scale, dial, range, scale plate, index plate; **на всю** (*or* **полную**) **шкалу** full-scale, on a full scale; **со шкалой** graduated; **ш. абсолютных температур** Kelvin scale, absolute scale; **ш. ахроматических тонов** gray tone, shadow tone; **безверньерная ш.** direct-reading scale, direct-drive scale; **ш. без нуля в начале, безнулевая ш.** setup scale, suppressed-zero scale; **верньерная ш.** vernier dial, vernier scale, refined scale, micro-adjustment dial; **гипсометрическая ш. высот** color scale; **ш. двоичных значений** binary scale; **двусторонняя ш.** center zero scale; **ш. делений прицела** sight scale; **ш. задержки по дальности** range-delay dial, range-delay diagram; **круглая ш. с верньером** slow-motion dial, vernier dial, refined scale; **крупная ш.** high range; **логарифмическая ш. времени** logarithmic time-base; **логарифмическая ш. децибел** decibel scale; **меньшая ш.** lower range; **ш. настройки ширины зоны регулирования** throttling range dial; **неподвижная ш. верньера** principal scale; **одноцветная ш. полутонов** monochrome scale; **подвижная ш. времени** triggered time-base; **ш. показания авиагоризонта** horizon dial; **полная ш.** expanded scale; **ш. полутонов** tone control aperture; **ш. приёмника с движком** slide-rule dial; **приставная ш.** reticule; **ш. прицела** tangent scale; **ш. прямого отсчёта** direct-reading scale; **прямолинейная ш.** slide-rule dial; **ш. с боковой подсветкой** floodlight scale; **ш. с движком** slide-rule diagram; **ш. с совмещением указателей** "follow-the-pointer" dial; **ш. с торцевой подсветкой** edge-illuminated scale; **серая ш. полутонов** monochrome scale; **ш. серых тонов** grey scale, monochrome scale, grey tone, shadow tone; **спектральная ш. цветов** chromatic scale; **ш. угла места** elevation scale; **ш. угломера, угломерная ш.** deflection scale; **ш. угломерного инструмента** azimuth scale; **установочная ш.** calibration scale;

**функциональная ш.** monographic scale of functions

**шкальный** *adj.* scale, dial

**шкаф** *m.* cabinet, case, cupboard; switchboard, junction box; rack; **блинкерный ш.** indicator board, drop-indicator panel; **вводной питающий ш.** power entrance cabinet; **микшерно-линейный ш.** mixer line rack; **распределительный ш.** distribution head cubicle, cable distribution head, distribution cabinet, switch cabinet, switchboard panel; **ш. расшивки кабелей** cable rack; **ш. с вызывными ключами** district call box; **ш. синхрокомплекта** synchronizing rack; **участковый кабельный ш.** section cable box

**шкафной** *adj.* cabinet, case, cupboard; switchboard

**шкворень** *m.* (center) pin, stem, bolt, draw bolt

**шкив** *m.* pulley, sheave; **групповой отклоняющий ш.** wheel box; **поворотный ш.** curve sheave

**шкивообразный** *adj.* pulley(-like); disk-like, discoid(al)

**школа** *f.* school; **ш. связи** School of Signals

**шлагбаум** *m.* crossing gate, roadway gate, toll gate, barrier

**шлаковый** *adj.* slag, slaggy, cinder

**шлам** *m.* slime, mud; residue, sediment, silt

**шламовый** *adj.* slime, mud; residue, sediment, silt

**шламообразный** *adj.* slime-like; slurry

**шланг** *m.* hose, jacket; **гибкий металлический ш.** flexible metallic conduit; **изоляционный ш.** insulating tubing

**шланговый** *adj.* hose

**шлейф** *m.* loop, circuit, two-wire circuit, go-and-return circuit; tail (of curve); stub; hair spring; **волноводный настраивающий ш.** waveguide stub tuner; **ш. вызывного тока ЧТС (от городской)** PBX ringing circuit; **короткозамкнутый ш. волноводов** (variable) reactance line for waveguides; **ш. питания ЧТС (от городской)** PBX power circuit; **ш. сигнала** wave tail; **согласующий ш.** matching stub, Q-match; pill transformer

**шлейф-антенна** *f.* folded-dipole antenna; **уголковая ш.-а.** folded-dipole V antenna

**шлейф-вибратор** *m.*, **шлейф-диполь** *m.* folded dipole, bent dipole

**шлейфный** *adj.* loop, circuit

**шлейфовать** *v.* to smooth, to plane

**шлейфовый** *adj.* loop, circuit; loop-type

**шлейф-рупор** *m.* folded horn

**шлейф-система** *f.* loop dialing

**шлейф-трансформатор** *m.* transmission-line conversion transformer

**шлем** *m.* helmet

**Шлемильх** Schloemilch

**шлемовой, шлемовый** *adj.* helmet

**шлепер, шлеппер** *m.* transfer table; dragging device

**Шлефли** Schläfli

**Шлирен** Schlieren

**шлиф** *m.* section, microsection, slide; cross-sectional view; ground-glass joint

**шлифной** *adj.* section, microsection, slide

**шлифовальный** *adj.* grinding, polishing, burnishing, abrasive

**шлифование** *n.* grinding, polishing, abrasion, abrading; **грубое ш.** burnishing

**шлифованный** *adj.* ground, polished

**шлифовать, вышлифовать, отшлифовать** *v.* to grind, to polish, to abrade; **начерно ш.** to fore-grind

**шлифовка** *f.* polishing, grinding, abrasion, stoning; lapping; **ш. кварцевых пластинок** lapping of crystals

**шлифующий** *adj.* polishing, grinding, smoothing, abrasive

**шлихтовальный** *adj.* smoothing, planing, finishing

**шлихтовочный** *adj.* smoothing, planing, finishing

**шлиц** *m.* slit, slot, groove

**шлицевать** *v.* to slit, to slot, to groove

**шлицевой** *adj.* slit, slot, splitting

**шлицованный** *adj.* slit, grooved, splined

**шлюз** *m.* lock, sluice, gate valve; **ш. для газонаполненного кабеля** cable gas-check point; **спускной ш.** flood gate

**шлюзный** *adj.* lock, sluice, gate valve

**шлюпка** *f.* launch, small boat

**шлямбур** *m.* jumper

**шляпка** *f.* cap, head (of nail)

**шляповидный, шляпообразный** *adj.* hat-shaped

**ШМ** *abbr.* (**широтная модуляция**) width modulation

**Шмидт** Schmidt

**шнорхель** *m.* schnorkel

**шнур** *m.* cord, string; flex, flexible cable, twist; lace, lacing, braid; pinch (nucl.); **гибкий ш. со спиральной жилой** spiral-conductor flexible cord; **круглый ш. с мишурными жилами** tinsel cord; **однопроводный ш.** single cord; **переключающий ш.** connecting cord, patch cord, flexible cord; **переносный (соединительный) ш.** jack cord, patch cord; **питающий ш.** power-supply cord, line cord; **розеточный ш.** instrument cord; **ш. с двумя штепселями** double-ended cord, patch cord; **ш. с мишурной жилой** tinsel cord; **ш. со штепселем** cord and plug; **штепсельный ш.** plug-ended cord

**шнурование** *n.* lacing, tying

**шнуровать, зашнуровать, прошнуровать** *v.* to lace, to tie

**шнуровой** *adj.* cord, string, lace

**шнуродержатель** *m.* cord holder

**шнурок** *m.* string, twine, lace; fuse

**шов** *m.* seam, joint, junction; groove, depression; **без шва** seamless; **валиковый ш.** fillet weld; **ш. кладки** wall joint, wall seam; **многорядный заклепочный ш.** multi-series riveting; **многоточечный ш.** multiple projection weld; **ш. на клею** straight glued joint; **несквозной пробочный ш.** plug weld; **нижний ш.** flat weld; **отделанный ш.** finished weld; **палубный ш.** flat weld; **плотнопрочный ш.** composite weld; **ш. с облегчённым валиком** light-fillet weld; **ш. с перекрещивающимися гребнями** ridge projection weld; **сквозной пробочный ш.** rivet weld; **точечный тавровый ш. впритык** tee butt weld; **чешуйчатый ш.** ripple weld; **шахматный заклепочный ш.** zig-zag riveting

**шовный** *adj.* seam, joint, junction; groove, depression

**шок** *m.* shock

**шокировать** *v.* to shock

**Шокли** Shockley

**шоопирование** *n.* Schoope process, metal spraying, metallization process

**шопот** *see* **шёпот**

**Шоран** Shoran

**« Шоринофон »** *m.* Shorinophon (Shorin's medical sound recorder)

**шорох** *m.* noise, rustling, rustle, whirring

**шоссе** *n. indecl.* highway

**шоссейный** *adj.* highway

**Шоттки-эффект** *m.* Schottky effect, shot effect

**шоулак** *m.* showlac

**Шохральский** Czochralski

**шпагат** *m.* binder, twine

**шпаклёвка** *f.* first coat of paint, priming coat; putty knife, spattle; putty, filler; filling up, filling, puttying

**шпала** *f.* crosstie, sleeper

**шпат** *m.* spar; **досчатый ш.** wollastonite; **известковый ш.** Iceland spar; **малиновый ш.** rhodochrosite

**шпатовидный** *adj.* spar, sparry

**шпатовый** *adj.* spar, sparry

**шпахтель** *f.* spattle, putty knife, spatula

**шпенёк** *m.* peg, pin, prong; **фиксирующий ш.** drive pin

**шпиль** *m.* pin, needle, pivot, reel, windlass, capstan; point

**шпилька** *f.* pin, stud, dowel; hairpin; peg, prong; **ш. для настройки диполя** hairpin tuning bar; **направляющая ш.** locating pin; **установочная ш.** dowel pin

**шпиндель** *m.* spindle, shaft, arbor, pivot; axis, axle, center pin; stem (of tube)

**шпиндельный** *adj.* spindle, shaft, arbor, pivot; axis, axle, center pin

**шпинелевый** *adj.* spinel

**шпинель** *f.* spinel

**шпицевание** *n.* pointing (of wires)

**шплинт** *m.* splint pin, cotter pin; pin, key; split

**шпонка** *f.* key, cotter key, spline, dowel, peg, wedge; joint tongue; **врезная ш.** sunk key; **натяжная ш.** gib key; **призматическая ш.** feather key

**шпоночный** *adj.* key, cotter; joint-tongue

**шприц** *m.* injector, syringe, sprayer

**шприцгусс** *m.* die cast

**шприцлак** *m.* spraying varnish

**шприц-пистолет** *m.* spray gun

**шприц-пресс** *m.* insulating machine; **прямоточный ш.-п.** straight-delivery insulating machine; **ш.-п. с боковой подачей** side-delivery insulating machine

**шпуледержатель** *m.* bobbin holder

**шпулька** *f.* spool, bobbin

**шпуля** *f.* spool, bobbin; **ш. для катушки возбуждения** field spool, field bobbin

**ШР** *abbr.* (**штепсельный разъём**) plug-type connector, plug and socket connector

**Шраге** Schrage

**Шредингер** Schrödinger

**шрифт** *m.* print, type, character

**шрифтовой** *adj.* type, character

**шрот-эффект** *m.* schrot-effect, shot effect

**штабик** *m.* molding

**Штайнмец** Steinmetz

**штамп** *m.* stamp, die, punch, puncher; **вырезывающий ш.** blanking die; **вытяжной ш.** extrusion block

**штампование** *n.* stamping, punching; pressing; extrusion

**штампованный** *adj.* stamped, punched; pressed

**штамповать, отштамповать** *v.* to stamp, to punch; to forge; to press; to coin

**штамповка** *f.* stamping, die stamping, punching, swaging; pressing; **ш. металла с электронагревом** electric swaging

**штамповочный** *adj.* stamping, punch (-ing); pressing

**штамповый** *adj.* stamp, punch

**штанга** *f.* rod, bar; pole, beam, arm; stem (of a tube); **защитная ш.** pole fender; **испытательная ш.** buzz stick; **ш. молниеуловителя** discharging rod, elevation rod; **направляющая ш.** mandrel; **оперативная ш.** hook stick; **приводная ш.** operating pipe

**штангенглубиномер** *m.* depth gage; **ш. с нониусом** vernier depth gage

**штанген-глубомер** *see* **штенгенглубиномер**

**штанген-зубомер** *m.* gear-tooth vernier, gear-tooth gage

**штангенрейсмас, штангенрейсмус** *m.* height gage; **ш. с нониусом** vernier height gage

**штангенциркуль** *m.* vernier callipers, slide gage; **простой ш.** slide calliper rule, slide gage; **ш. с нониусом** vernier gage штанговый *adj.* bar, rod; post, pole, beam, arm

**Штарк** Stark

**штарковский** *adj.* Stark

**штат** *m.* staff (of workers); **эксплуатационный ш.** operating force, operating service

**штатив** *m.* clamp stand, stand, support, holder; base, foot, pedestal; tripod; rack frame; **групповой ш.** standard rack; **ш. искателей вызовов** finder rack; **ш. испытательных искателей** test selector rack; **ш. контрольного оборудования** supervisory rack; **низкий ш. для камеры** high hat; **ш. переходных трансформаторов** repeating coil rack; **шарнирный ш.** gate (*tphpy.*)

**штативный** *adj.* stand, holder, support; base, foot, pedestal; tripod; rack

**Штауфер** Stauffer

**штеккер** *m.* plug; telephone plug

**штеккерный** *adj.* plug

**штемпелевание** *n.* stamping

**штемпелевать, заштемпелевать** *v.* to rubber-stamp; to punch, to pressure; to impress, to stamp

**штемпель** *m.* (rubber) stamp, seal; stamp, punch, die; **ш. времени** time check, calculagraph

**штемпельный** *adj.* (rubber) stamp, seal; punch, stamp, die

**штенгель** *m.* stem; exhaust tube; **металлический ш.** metal exhaust tube; **ш. с линзами** button stem; **сквозной ш.** straight-through exhaust tube

**штенгельный** *adj.* stem; exhaust-tube

**штепселевание** *n.* plugging; **ш. занятого гнезда** busy jack plugging; overplugging

**штепселевать** *v.* to plug (in)

**штепсель** *m.* plug, connector, plug adapter; peg, wedge, knob; pin; **ш. бесконечного сопротивления** infinity plug; **ш. в магазине сопротивлений** plug switch; **гаечный ш.** female plug; **двухпроводный ш.** two-pin plug, double-contact plug, double wedge; **дуплексный ш.** counter plug; **избирательный ш.** wander plug; **изолированный ш.** dummy plug; **контрастный ш.** counter plug; **ш. междувагонного электрического соединения** electric coupler plug; **многоштифтовый ш., многоштыревой**

**ш.** multiple plug; **однополюсный ш.** single-pin plug, wander plug; **одноштырьковый ш.** banana plug; **ответвительный ш.** adapter plug; **переключающий ш. для реверса** direction-connection plug; **приборный ш.** coupler plug; **присоединительный ш.** operator's plug; **рельсовый ш.** bond plug, channel pin; **ш. с предохранителем** fused plug; **ш. с фиксированными полюсами** non-interchangeable plug; **соединительный ш.** adapter plug; **справочный ш.** signal plug, indicating plug; **сценический ш.** stage pocket, stage box; **трёхконтактный ш., тройной ш.** three-pin plug; **четырёхконтактный ш., четырёхпроводный ш., четырёхштырьковый ш.** four-pin plug

**штепсельно-зажимный** *adj.* clip-type (prong)

**штепсельно-фланцевой** *adj.* bullet-and-flange

**штепсельный** *adj.* plug

**штифт** *m.* pin, joint pin, peg, prong, prod, dowel, stud; plug; stem; **копирующий ш.** tracer; **мембранный ш.** vibrating pin; **ш. отлипания** core pin, shim; residual stop (of a relay); **пишущий ш. (индикатора)** indicator pencil; **пишущий ш. самописца** pen arm; **ш. против « прилипания »** distance piece (of a relay); **ш. с закруглёнными (*or* скошенными) краями** bevelled stud

**штифтовой** *adj.* pin, peg, stud; bolt; staked-in

**штихель** *m.* engraving tool, recording stylus

**штихмас** *m.* callipers, inside micrometer callipers, pin gage, end gage; **раздвижной ш.** adjustable calliper gage

**шток** *m.* piston rod, coupling rod; stem

**шток-лак** *m.* stick lac

**штольня** *f.* adit, entrance; passage

**штопор** *m.* spin; corkscrew

**штопорный** *adj.* spin; corkscrew

**штора** *f.*, **шторка** *f.* curtain, blind

**шторм** *m.* storm, gale

**штормовой** *adj.* storm, gale

**штрипс** *m.* strip

**штрипсовый** *adj.* strip

**штрих** *m.* prime; mark, line, strip, dash; stroke, touch; hachure; **между-**

кадровый ш. frame line; **нитяной ш.** spider line; **основной ш. (шрифта)** heavy element; **соединительный ш. (шрифта)** light element; **ш. Шеффера** Sheffer function, Sheffer stroke

**штрих-код** *m.* bar code

**штрихованный** *adj.* hatched, shaded, crosshatched

**штриховатый** *adj.* hatched, striated

**штриховать, заштриховать** *v.* to hatch, to shade, to crosshatch

**штриховка** *f.* hatch(ing), hatchure, shading, striation

**штриховой** *adj.* dash(ed) stroke, line; hatch; facsimile (transmission)

**штрихозапись** *f.* facsimile recording

**штрихопередача** *f.* facsimile transmission

**штрихопунктирный** *adj.* dash-and-dot

**штрих-точечный** *adj.* dot-and-dash

**штука** *f.* piece, sample, specimen

**штукатурить, оштукатурить** *v.* to plaster

**штукатурка** *f.* plaster(ing), stucco

**штурвал** *m.* wheel, steering wheel, pilot wheel; **ш. угла места** elevation handwheel

**штурвальчик** *m.* control wheel, handwheel, star knob

**штурман** *m.* navigator; pilot

**штурманский** *adj.* pilot

**штурмовик** *m.* attack aircraft

**штуцер** *m.* connecting pipe, connecting branch, sleeve

**штучный** *adj.* single, individual, piece; thing

**штыковой** *adj.* bayonet; staked-in

**штыревой** *adj.* pin, dowel; spindle, shaft; probe; pivot pin

**штырёк** *m.* pin, peg; **разрезной ш.** split prong

**штырь** *m.* pivot; shaft, spindle; dowel, pin, bolt, stud, peg; spike; probe; post; **ш. в волноводе** line stretcher, probe; **волноводный индукторный ш.** inductive conductor in waveguides; **двойной ш.** double insulator spindle; **ш. для крепления изоляторов** insulator-hook bracket; **поддерживающий ш.** locking pin; **приёмный ш.** pick-up probe; **прямой ш.** straight spindle; **ш. связи** coupling probe; **четвертьволновый ш.** quarterwave antenna

**Штюве** Stüve

**Шулер** Schuler

**шум** *m.* noise, sound, din, rustling, crackle; rumble; **шумы канала с поднимающейся частотной характеристикой** peaked-channel noise; **мелкоструктурные флуктуационные шумы** fine-grain noise; **шумы моря биологического происхождения** noise due to marine life; **шумы непроволочных сопротивлений** carbon noise; **шумы окружающей морской среды** ambient sea noise; **шумы от вибрации диска проигрывателя** turntable rumble; **шумы от завихрений** turbulence noise; **шумы пограничного слоя** boundary layer noise; **шумы при сопровождении реальной цепи** tracking noise; **приведённые ко входу шумы** equivalent input noise; **шумы разной полярности** opposite-polarity noise; **собственные шумы моря** ambient sea noise; **шумы угольных сопротивлений** carbon noise; **ш. в окружающей среде** ambient noise; **ш. всасывания** free-delivery noise; **ш., вызываемый излучением вибрации** structure-generated noise; **геометрический ш.** induced hum impairing geometry of television image; **гидродинамический ш. потока жидкости** flow noise; **гладкий ш.** stochastic noise, white noise; **дифференциальный ш.** incremental noise; **ш. искателя** dialing effect; **капельный ш.** thump; **клиппированный ш.** amplitude-limited noise; **ш. коллектора, коллекторный ш.** brush noise, commutator noise; **ш. крутящего момента** torque noise; **лавинный ш.** avalanche noise; **линейный ш.** circuit noise; **ш. манипуляции** key clicks; **моторный ш.** popping, motorboating; **неразборчивый ш.** babble; **ш. окружающей среды** ambient noise; **ш. от изменения сопротивления контактов** contact-resistance noise; **ш. от конструкции** structure-born noise; **ш. от ленты** tape background noise; **ш. от неустойчивого горения** rumble noise; **ш. от нумерации** quantization noise; **ш. от огибания** translation (playback) noise; **ш. от переприёма тока** interception noise; **ш. от переходных явлений** noise due to transients; **ш. от питания** power-supply noise, hum; **ш. от пульсации пита-ния** ripple noise; **ш. от сети (пита-ния)** *see* **ш. от питания; плотовой ш.** thump; **(подводный) ш. греб-ного винта** underwater propulsion noise; **равномерный по громкости ш.** steady noise; **равноэнергетиче-ский ш.** flat noise, white noise; **речевой ш.** artificial voice; **ш. с равномерным спектром** pure noise; **сетевой ш.** power supply noise; **сеточный ш.** grid noise; **ш. соб-ственной помехи гидро-акустиче-ской станции** sonar self-noise; **соб-ственный ш. аппарата** set noise; **собственный ш. радиоламп** internal noise of tubes, background noise of tubes; **собственный ш. усилителя** amplifier noise; **совместный ш.** thump; **ш. сопротивления** granular resistance noise, noise of resistor; **ш. спутного тока** fluid-coupling noise; **структурный ш.** structure-generated noise; **ш. теплового возбуждения, тепловой ш.** thermal agitation noise, Johnson noise; **термический экви-валентный ш. водной среды** thermal agitation water noise; **ш. токорас-пределения** fluctuation noise, cur-rent distribution noise; **фоновой ш.** background noise, random noise

**шуметь** *pfs.* **зашуметь, нашуметь, по-шуметь, прошуметь** *v.* to make noise, to hiss

**шумность** *f.* noise, noisiness

**шумный** *adj.* noisy, loud, sonorous

**шумовка** *f.* crackle

**шумовой** *adj.* noise

**шумоглушитель, шумозаглушитель** *m.* antihum device, noise silencer, silencer, muffler, sound damper

**шумомер** *m.* audio-noise meter, noise gage, noise meter, sound meter; **аудиометрический ш.** subjective noise meter; **инспекторский ш.** noise-survey meter

**шумопеленгатор** *m.* listening gear, listening sonar, direct listening sonar system, sound locator, hydrophone, hydroacoustic indicator, passive sonar

**шумопеленгаторный** *adj.* hydrophone

**шумопеленгация** *f.* direction finding;

**ш. гидрофона** sonar direction finding

**шумопеленгование** *n.* direct listening, listening; **ш. в ультразвуковом диа-пазоне** ultrasonic listening; **ш. на-правленной системой** directional listening; **ш. ненаправленной систе-мой** nondirectional listening

**шумоподавление** *n.* noise-reduction

**шумостойкий** *adj.* noiseproof

**шумфактор** *m.* noise factor, figure of noise, signal-to-noise merit

**шумящий** *adj.* noisy

**шунт** *m.* shunt, bridge, bypass, branch circuit, derived resistor, shunt resist-ance; **без шунта** unshunted; **шунтом к линии** shunting the line; across the line; **ш. высокого напряжения к конденсатору** static leak; **ш. из магнито-мягкого материала** soft-iron shunt; **ш. измерительного при-бора, измерительный ш.** instru-ment shunt, meter multiplier; **неоно-вый ш. для предохранителей** dead-fuse tattelite; **омический ш.** con-ducting bridge; **ш. с винтовой на-резкой** threaded shunt; **ш. с серие-сной обмоткой** series diverter

**шунтирование** *n.* shunting, bridging, by-passing, derivation; **ш. обмотки возбуждения** field diversion

**шунтированный** *adj.* shunted, bridged, in bridge, by-passed

**шунтировать, зашунтировать** *v.* to shunt, to bridge, to bypass, to span, to connect across, to short

**шунтировка** *see* **шунтирование**

**шунтирующий** *adj.* shunt(ing), bypass, by-passing, bridging

**шунтование** *see* **шунтирование**

**шунтовать** *see* **шунтировать**

**шунтовой** *adj.* shunt, shunt-wound, bypass

**шунтомер** *m.* shunt meter

**шуп** *m.* calliper gage

**шуруп** *m.* wood screw; **ш. с изолирую-щим глазком** insulated screw eye

**шурф** *m.* pit, bore pit; hole, perforation

**шуршание** *n.* whirring, rustle, rustling, crackling

**шуршать, зашуршать** *v.* to crackle, to rustle

**шуршащий** *adj.* crackling, rustling

**Шухарт** Shewhart

# Щ

щадитель *m.* shock absorber

щадительный *adj.* sparing

щадить, пощадить *v.* to spare

щека *f.* jaw, bit; flange, side

щеколда *f.* cam, catch, dog, stop, bolt, latch, locking bar, pawl, trip gear

щекообразный *adj.* jaw-shaped

щел. *abbr.* (щелочной, щелочный) alkaline

щелевидный *adj.* slit-like

щелевой *adj.* slot, slotted, slit, chink

щел.-зем *abbr.* (щёлочноземельный) alkaline-earth

щелина *see* щель

щелистый *adj.* fissured, cracked, chinky; full of cracks, full of flaws

щёлк *m.* snap, click

щёлка *see* щель

щёлкание, щёлканье *n.* clicking, cracking; trilling, snapping

щёлкать, щёлкнуть *v.* to clap, to crack; to snap; to click

щёлок *m.* alkaline solution, lye

щелочение *n.* alkalization; lixiviation

щёлочерастворимый *adj.* alkali-soluble

щёлочестойкий, щёлочеупорный, щёлочеустойчивый *adj.* alkali-resistant

щелочить *v.* to alkalize; to lixiviate

щёлочноземельный *adj.* alkaline-earth

щелочной *adj.* alkaline

щёлочность *f.* alkalinity

щёлочноупорный, щёлочноустойчивый *adj.* alkali-resistant

щёлочь *f.* alkali

щелчковый *adj.* click, snap

щелчок *m.* bang, click; щ. (от плохо сделанной склейки) bloop

щель *f.* chink, cleft, crevice, slit, flaw, crack; aperture, gap, gate, slot; split; joint, seam, groove; nozzle, spout; щ. (волновода) nozzle; с щелью в стене slotted-wall; возбуждаемая зондом щ. probe-fed slot; голосовая щ. glottis; длинная щ. связи long slot coupler; щ. для опускания монет, монетная щ. coin slot; полукруглая щ. crescent slot; щ. связи coupling slit, coupling slot

(of a waveguide); смотровая щ. eye slit; считывающая щ. scanning slit, scanning gate; формообразующая щ. shaped slit (for generating complex waveforms)

щётка *f.* brush; wiper, current-collecting device; commutator brush; shoe; главные щётки, пропускающие энергию energy brushes; щётки под прямым углом cross brushes; щётки, расположенные в шахматном порядке staggered brushes; смещённые на 90° щётки cross brushes; С-щ. private wiper; вспомогательная щ. exploring brush, pilot brush; щ. для кабельных труб, щ. для прочистки каналов duct cleaner, duct cleaning tool; щ. искателя wiper; испытательная щ. private wiper; exploring brush, pilot brush; коллекторная щ. commutator brush; щ. контактного поля bank wiper; набегающая щ. leading brush; наклонённая в сторону вращения коллектора щ. trailing brush; опережающая щ. leading brush; параллельная щ. искателя bridging wiper of a selector; разрезная щ. laminated brush; реактивная щ. leading brush, pilot brush; « С »-щ. pilot brush (of a selector), test wiper; щ. с катком, щ. с роликом rotating brush; сетчатая (медная) щ. gauze brush; скользящая щ. feeder brush; смазочная щ. wiper; собранная щ. brush assembly; трубчатая щ. duct cleaner; «увлекаемая» щ. trailing brush; управляющая щ. private brush; электромеханическая щ. для натирания поля electromechanical brush

щёткодержатель *m.* brush holder, wiper shaft, wiper rod, wiper carriage, selector rod, brush rod, brush gear; щ. в виде плоской пружины spring arm-type brush holder; консольный щ. cantilever brush holder; ножницеобразный щ. scissors-type brush holder; щ. под углом к поверхности коллектора reaction brush holder, trailing brush holder

**щёткоподъёмный** *adj.* brush-lifting

**щёткоустановочный** *adj.* brush-setting

**щёточный** *adj.* brush; wiper(-type)

**щипальный** *adj.* plucking

**щипание** *n.* plucking; cleaving; splitting; pinching, nipping

**щипанный** *ppp.* plucked, pinched; cleaved

**щипаный** *adj.* cleaved; split; plucked, pinched

**щипать** *pfs.* **общипать, ощипать, щипнуть** *v.* to pluck, to pinch, to nip; to cleave

**щипковый** *adj.* plucked

**щипком** *adv.* pizzicato

**щипнуть** *pf. of* **щипать**

**щипок** *m.* pluck, plucking; pinch, nip; grip

**щипцеобразный** *adj.* forceps-like

**щипцовый** *adj.* forceps, tongs

**щипцы** *pl.* tongs, pliers, pincers, nippers; (electrode) holder; **щ. для надевания кабельных подвесок** cable hanger tongs; **щ. для сращивания проводов** jointing clamp

**щипчики** *pl.* large forceps, tweezers, pincers

**щит** *m.* shield, screen, baffle, baffler, shelter, buckler, guard; board, panel, plate; switchboard; gate; slide valve; **аккумуляторный щ., батарейный щ.** battery switchboard; **батарейный коммутационный щ.** battery distribution panel; **воздухоразделительный щ.** air baffler; **вторичный распределительный щ.** subboard, secondary switchboard; **выравнивающий щ.** insulator grading shield; **главный щ.** main switchboard; **главный щ. переключений** main distributing frame; **главный распределительный щ.** main switchboard (-panel); **главный распределительный щ. питания** feeder switchboard; **головной щ.** head gate; **групповой щ.** branch-circuit distribution center; **двойной распределительный щ.** dual switchboard; **деревянный коммутационный щ.** grape-arbor type switchboard; **заградительный щ.** fender; **задний подшипниковый щ.** rear head; **затеняющий щ.** shadow bar; **измерительный щ.** meter board; **испытательный щ.** test board, test

frame; **комбинированный щ. переключений** combined distributing frame; **коммутационный щ.** switchboard; **щ. между источником света и механизмом** cooling plate; **ограждающий щ.** fender; **передвижной акустический щ.** tormentor; **передний подшипниковый щ.** front head; **щ. переключений** distributing frame; **поворотный щ. типа жалюзи** wicket gate; **подшипниковый щ.** bracket; housing, bearing housing; end shield; **предохранительный щ.** fender; **приёмный распределительный щ.** interception frame; **щ. распределения сигналов** signal mixer unit; **распределительный щ.** distribution head; distributing board, distributor, feeder switchboard; **распределительный щ. без токоведущих частей спереди** "dead front" switchboard; **распределительный щ. с выдвижными блоками** draw-out switchboard; **распределительный щ. с коридором обслуживания** corridor switchboard, duplex switchboard; **распределительный щ. с мнемонической схемой** mimic diagram board; **щ. с мимической схемой** diagram board; **щ. с плавкими предохранителями** fuse board; **щ. с проводкой на задней стороне** back-wired panel; **секционный щ.** panel-type board; **статический щ. для выравнивания напряжения в гирлянде** insulator grading shield; **схемный щ.** graphic panel, diagram board; **торцевой щ. для предохранения обмотки** winding shield; **тренировочный щ.** aging rack; **щ. управления** control board, control desk, panel box, control panel, instrument panel

**щитовидный** *adj.* shield-shaped

**щитовой** *adj.* board, panel, plate; screen, shield; switchboard

**щиток** *m.* name plate, plate; screen, shield; small board; panel board, dashboard; hinged valve; **щ. автоматов защиты сети** circuit breaker panel; **выводной щ., выходной щ.** terminal board, **щ. для обозначения** designation card, label; **отражательный щ.** baffle; **щ. переключений** changeover panel; **щ. переключения режимов работы** function-selector panel; **присоединительный щ.** terminal board; **регулирующий щ.** servo flap; **ручной щ.** face shield; **щ. с кнопками** push-button assembly; **щ. с надписью** escutcheon plate; **щ. с предостерегающим знаком, щ. с предостерегающей надписью** caution board, danger board; **щ. с предохранителем на одну линию** one-way board; **щ. с указанием мощности** rating plate; **щ. счётного механизма** register face; **тормозной щ.** dive flap; **щ. управления электросистемой переменного тока** a.c. power panel; **фирменный щ.** nameplate, marking plate

**щитообразный** *adj.* shield-shaped

**щуп** *m.* probe, feeler, sonde; sounding borer, feel gage, pickup, search coil; clearance gage; dip rod; **щ. для нахождения неисправностей** tracing probe; **щ.-индикатор** detecting head; **калиберный щ.** gage feeler; **контактный щ.** feeler; **копирующий щ.** tracer; **щ.-обнаружитель** detecting head; **электромагнитный щ.** magnetic pickup device; **электромагнитный щ. копировального станка** electromagnetic gage head

**щупальный** *adj.* feeling, feeler, probing, sounding

**щупание** *n.* probing, sounding, feeling; touching

**щупать, пощупать** *v.* to probe, to sound, to feel; to touch; to key, to control

# Э

э *abbr.* 1 (эксплуатация) exploitation, operation, running; 2 (электрон) electron; 3 (эрг) erg; 4 (эрстед) oersted; 5 (эскадрилья) squadron

ЭАП *abbr.* 1 (электрический автопилот) electric autopilot; 2 (электроакустический преобразователь) electroacoustic transducer

эбикон *m.* ebicon

эбонит *m.* ebonite, hard rubber, vulcanite

эбонитовый *adj.* ebonite, hard-rubber, vulcanite

эбуллиометр *m.* ebulliometer

эбуллиоскоп *m.* ebullioscope

эбуллиоскопический *adj.* ebullioscope

эбуллиоскопия *f.* ebullioscopy

эв *abbr.* (электрон-вольт) electron-volt

эвакуационный *adj.* evacuation

эвакуация *f.* evacuation, exhaustion

эвакуирование *n.* evacuation

эвакуированный *adj.* evacuated

эвакуировать *v.* to evacuate, to exhaust, to empty

эвапоратор *m.* evaporator

эвапориметр *m.* evaporimeter

эвапорограф *m.* evaporograph

эвапорометр *m.* evaporimeter, atmidometer

эвдиометр *m.* eudiometer

эвдиометрический *adj.* eudiometric

эвдиометрия *f.* eudiometry

эвердур *m.* everdur copper

эвитон *m.* E-viton

Эвклид Euclid

эвкразит *m.* eucrasite

ЭВМ *abbr.* (электронная вычислительная машина) electronic computer

эвокон *m.* evocon

эвольвента *f.* involute, evolute; evolvent

эвольвентный *adj.* involute, evolute; evolvent

эволюта *f.* evolute

эволюционировать *v.* to evolve, to develop; to evolutionize

эволюционный *adj.* evolution(ary)

эволюция *f.* evolution

эвоскоп *m.* evoscope

ЭВП *abbr.* (электровакуумный прибор) electronic apparatus

« эврика » *f.* eureka (alloy)

эвристический *adj.* heuristic

европий *m.* europium, Eu

эвтакситовый *adj.* eutaxic

эвтектика *f.* eutectic

эвтектический *adj.* eutectic

эвтропия *f.* eutropy

ЭД *abbr.* 1 (электродетонатор) electric detonator; 2 (электродиализ) electro-dialysis

эддистат *m.* eddystat

Эджворт Edgeworth

Эдисон Edison

Эдкок Adcock

эдометр *m.* oedometer

эдс *abbr.* (электродвижущая сила) electromotive force, E.M.F.

э. ед. *abbr.* (электронная единица) electronic unit

эжектор *m.* ejector; eductor, lifting out device; jet

эжекторный *adj.* ejector; lifting out device

эжекция *f.* ejection

ЭИН *abbr.* (Энергетический институт Академии наук СССР) Institute of Power Engineering of the Academy of Sciences, USSR

эйгенвеличина *f.*, эйгенверт *m.* proper value, eigenvalue

Эйдкок Adcock

эйдофор *m.* eidophor

Эйкен Eucken

эйкоген *m.* eikonogen

эйконал *m.* eikonal

эйконометр *m.* eikonometer

Эйлер Euler

эйлерианский *adj.* Eulerian, Euler's

эйлеров *poss. adj.* Eulerian, Euler's

Эймбл Amble

Эйнштейн Einstein

эйнштейний *m.* eisteinium, Es

Эйри Airy

Эйхенвальд Eichenwald

экаалюминий *m.* eka-aluminum

**экабор** *m.* ekaboron
**эка-гольмий** *m.* eka-holmium
**эка-йод** *m.* eka-iodine
**экасилиций** *m.* ekasilicon
**экатолит** *m.* ekatolite
**эка-цезий** *m.* eka-cesium
**экаэлемент** *m.* eka-element
**экв.** *abbr.* (эквивалент, эквивалентный) equivalent
**экватор** *m.* equator
**экваториальный** *adj.* equatorial
**экв. ед.** *abbr.* (эквивалентная единица) equivalent unit
**экви-** *prefix* equi-
**эквивалент** *m.* equivalent; artificial device, dummy, mute; **э. антенны, антенный э.** antenna equivalent, dummy antenna, mute antenna, phantom antenna; **э. воздушной линии** open-line balancing network; **э. входной цепи** dummy input circuit; **дифференциальный э.** incremental equivalent; **э. затухания междугородной связи** toll circuit equivalent of attenuation; **э. затухания местного эффекта** sidetone attenuation; **э. затухания полной системы (телефонной) связи** volume equivalent of attenuation; **э. затухания системы, состоящей из аппарата абонента и его связи с междугородной станцией** toll terminal equivalent of attenuation; **э. земли** artificial ground; **имитирующий э., искусственный э.** dummy; **э. линии** imitated line, artificial line, balancing network; **э. нагрузки** dummy load; ohmic terminating-resistance; **э. света, световой э.** lumen equivalent; **тепловой э. работы** thermal equivalent of mechanical energy; thermal value; **транзитный (эффективный) э. затухания** via equivalent of attenuation
**эквивалентность** *f.* equivalence, equivalency; **э. двухполюсников** equivalence of two-terminal networks
**эквивалентный** *adj.* equivalent, equal, identical; reciprocal
**эквиволюминальный** *adj.* equivoluminal, isometric
**эквидистантный** *adj.* equidistant, equispaced
**эквилибр** *m.* equilibrium, balance
**эквилибрировать, эквилибровать** *v.* to equilibrate, to balance

**эквимолекулярный** *adj.* equimolecular
**эквимолярный** *adj.* equimolar
**эквипотенциальный** *adj.* equipotential
**эквифазный** *adj.* equiphase
**эквометр** *m.* equameter
**ЭКГ** *abbr.* (электрокардиограмма) electrocardiogram, EKG
**экзальтация** *f.* exaltation
**экзаменатор** *m.* examinator
**экзаменационный** *adj.* examination
**экзаменация** *f.* examination
**экзаменовать, проэкзаменовать** *v.* to examine
**экземпляр** *m.* copy; specimen; **опытный э.** pattern, specimen, sample
**экзо-** *prefix* exo-
**экзогенный** *adj.* exogenous
**экзоионосферный** *adj.* exoionospheric
**экзосмос** *m.* exosmosis
**экзосмотический** *adj.* exosmotic
**экзосфера** *f.* exosphere
**экзотерический** *adj.* exoteric; external
**экзотермический** *adj.* exothermic, exoenergic, heat-liberating
**экзотермичность** *f.* exothermic nature
**экзотермичный** *see* экзотермический
**экзоэнергетический** *adj.* exoenergetic, exoergic, exoenergic, exothermal
**экипаж** *m.* carriage, car, truck, wagon; crew
**экипированный** *adj.* equipped, furnished
**экипировать** *v.* to equip, to furnish
**экипировка** *f.* equipage, outfit, equipment; **э. реле** relaying
**экипировочный** *adj.* equipment
**Экклес** Eccles
**Эккмиллер** Eckmiller
**эккосорб** *m.* Eccosorb
**эклиметр** *m.* eclimeter
**эклиптика** *f.* ecliptic
**эклиптический** *adj.* ecliptic
**экологический** *adj.* ecological
**экология** *f.* ceology
**экономайзер** *m.* economizer
**эконометр** *m.* econometer
**эконометрика** *f.* econometrics
**экономика** *f.* economics; economy; **э. ядерной энергетики** nuclear-power economics
**экономить, сэкономить** *v.* to economize, to save
**экономический** *adj.* economic(al)
**экономичность** *f.* economy; efficiency; **э. отражателя** reflector savings

**экономичный** *adj.* economical; efficient; antiredundant

**экономия** *f.* economy, saving

**экономный** *adj.* economical

**экосфера** *f.* ecosphere

**экран** *m.* screen, shield, baffle, blind, panel, panel deflector; plate, faceplate; barrier; display; shade; **алюминированный э. с точечным люминофором** aluminized phosphor dot screen; **быстро затухающий э.** fast screen; **э. верхней распорки** top spacer shield; **э. визуального устройства вывода** visual display; **э. вывода сетки** grid-lead screen; **э. для записи тёмной строкой** dark-trace screen; **э. для рентгенопросвечивания** roentgenoscope; **э. для считывания** reading screen; **запоминающий э.** storage surface; **э. из мелкоячеистой сетки** fine(-mesh) screen; **э. из цилиндрических линз** lenticulated screen; **э. индикатора** plan position indicator, indicator screen, scope face; **э. индикатора РЛС слежения** radar tracking screen; **искровой э.** flash-barrier; **кольцевой э.** ring shielding; **э. конечных размеров** finite baffle; **линейчатый э.** strip-type screen; **линзоворастровый э.** lenticulated screen; **медленно затухающий э., медленно потухающий э.** slow screen, long persistance screen; **мозаичный э. с элементами пирамидальной формы** cubical pyramid screen; **мозаичный многоцветный э.** checkerboard color dot screen; **направляющий э.** baffle; **э. нижней распорки** bottom spacer shield; **нормально затухающий э.** normal-speed screen; **ослабляющий э.** reducing screen; **э. отклоняющих катушек** deflection-coil shield; **отражательный э.** baffle plate; **отражающий э.** reflecting screen, reflecting plate, reflecting barrier; **покрытый незернистым люминофором э.** grainless screen; **э., покрытый точечным люминофором** phosphor-dot plate; **полутеневой э.** penumbra zone; **э. приёмной трубки для воспроизведения цветного изображения** color picture screen; **просвечивающий э.** translucent screen, fluoroscope; **э. прямого видения** viewing screen; **э. пульта управления** console scope;

**э., работающий на просвет** rear (-projection) screen, translucent screen; **равномерно белый дифузно-рассеивающий э.** flat-white non-directional screen; **э. с гладким покрытием** grainless screen; **э. с затемняющей сеткой** shadow-mask screen; **э. с малым временем послесвечения** fast screen, short-persistence screen; **э. с подложкой из слюды** mica-supported screen; **э. с послесвечением** afterglow screen, persistence screen, delay screen; **э. с темновой записью** dark-trace screen; **э. с тёмным изображением** dark-trace screen; **э. с торцевой подсветкой** edge-illuminated screen; **э. с цветным свечением** color screen; **светящийся э. с переключаемыми цифрами** number ennunciator; **э. сеточного вывода** grid-lead shield; **смотровой э.** viewing screen; **тепловой э. с плёночным охлаждением** film-cooled heat shield; **торцевой э.** end shield; **точечный люминесцирующий э.** dot fluorescent screen; **э. трубки** tube display, tube face; **трубчатый э.** fluted screen; **управляемый световой э.** intensity-control-type screen; **э., цвет которого зависит от направления возбуждающего пучка** direction-sensitive color screen; **цветной линейчатый э.** color line screen; **цилиндрический э.** cannon tube shield; **чечевицеобразный э.** lenticular screen

**экранирование** *n.* shielding, screening, insulation; shadowing, shading; screening effect; **э. на базе ферритов** ferrite shielding; **э. от электростатических полей** electrostatic shielding

**экранированный** *adj.* shielded, screened, conductively closed

**экранировать** *v.* to shield, to screen; to shade

**экранировка** *f.* shielding, screening

**экранировочный** *adj.* shielding, screening

**экранирующий** *adj.* shielding, screening

**экранный** *adj.* shield, screen

**экс-** *prefix* ex-

**эксенит** *m.* euxenite

**эксикатор** *m.* exsiccator, desiccator; **высоковакуумный э.** high-vacuum jar

эксито́н *m.* exiton
экситонный *adj.* exiton
эксито́н *m.* exitron (tube)
Экснер Exner
экспандер, экспандор *m.* expander
экспансиометр *m.* expansion meter
экспансионный *adj.* expansion
экспансия *f.* expansion, expanse
экспедитор *m.* dispatcher
эксперимент *m.* experiment, test, trail; **э. в подкритическом режиме** subcritical experiment; **э. методом времени пролёта** time-of-flight experiment; **э. методом совпадений** coincidence experiment; **э. по излучению переходных процессов** transient experiment; **э. с монохроматическим излучением** monoenergetic experiment; **э. с разрушением** destructive experiment
экспериментально *adv.* experimentally
экспериментальный *adj.* experimental; experiment; tentative; research
экспериментатор *m.* experimenter
экспериментирование *n.* experimentation
экспериментировать *v.* to experiment
эксперт *m.* expert, appraiser
экспертиза *f.* expert opinion, consultation; estimation, evaluation, appraisal
эксплоатационный, эксплуатационный *adj.* operating, operation(al), exploitation, working; performance, service, field (test)
эксплуатация *f.* exploitation, operation, operating, service, usage, run(ning); **э. без дежурного** unattended operation; **бесперебойная э.** nonstop operation; **э. в аварийном режиме** emergency service; **встречная э. канала** offset operation; **э. (линии) в одном направлении** directional service; **одновременная э. двух аккумуляторных батарей с попеременным зарядом и разрядом** charge-discharge operation; **опытная э.** operation test, running test, reliability field test; **э. при чрезмерных режимах** overrun operation; **э. с автоматическим установлением транзитного соединения** dial system tandem operation; **э. с вызовом по цепи** ringdown operation, "signal" working; **э. с последовательным соединением станций** tandem opera-

tion; **э. с током в состоянии покоя** close-loop operation
эксплуатировать *v.* to exploit, to operate, to use, to run
экспозиметр *m.* exposure meter
экспозиционный *adj.* exposure
экспозиция *f.* exposure; exposition, display
экспонент *m.* exponent, index
экспонента *f.* exponential; **затухающая э.** damped exponential; **э., связанная с запаздывающими нейтронами** delay-controlled exponential
экспонентный *adj.* exponential
экспоненциально *adv.* exponentially
экспоненциально-затухающий *adj.* exponentially damped
экспоненциальный *adj.* exponential; exponent
экспонирование *n.* exposure (of a film)
экспонированный *adj.* exposed
экспонировать *v.* to expose, to show; to irradiate
экспонометр *m.* exposure meter, exponometer; X-ray intensitometer
экспресс *m.* express
экстензометр *m.* extensometer
экстенсивность *f.* extensiveness
экстенсивный *adj.* extensive
экстенсия *f.* extension
экстенсометр *m.* extensometer
экстинкция *f.* extinction
экстра- *prefix* extra
экстрагент *m.* extractant
экстрагирование *n.* extraction, separation
экстрагированный *adj.* extracted
экстрагировать *v.* to extract
экстрагируемость *f.* extractibility
экстрагируемый *adj.* extractable
экстрагирующий *adj.* extracting
экстракод *m.* extracode
экстракт *m.* extract
экстрактивный *adj.* extractive, extractable
экстрактор *m.* extractor; **э. типа градирни** stacked-stage extractor
экстракционный *adj.* extraction
экстракция *f.* extraction; **э. в индикаторных количествах** tracer-stage extraction; **двухжидкостная э.** double-solvent extraction; **э. равными объёмами** equal-volume extraction; **э. растворителем** extraction into solvent, solvent extraction

**экстраординарный** *adj.* extraordinary, uncommon
**экстраполирование** *n.* extrapolation
**экстраполированный** *adj.* extrapolated
**экстраполировать** *v.* to extrapolate
**экстраполяционный** *adj.* extrapolating; extrapolation
**экстраполяция** *f.* extrapolation
**экстраток** *m.* extra-current; **э. замыкания** make induced current; **э. размыкания** break induced current
**экстремальный** *adj.* extreme
**экстремум** *m.* extremum, extreme
**экстренно** *adv.* especially
**экстренный** *adj.* special, urgent; emergency
**экструдинг-процесс** *m.* extrusion process
**экструдирование** *n.* extrusion
**экструдированный** *adj.* extruded
**экструзия** *f.* extrusion
**эксцентрик** *m.* eccentric; cam, cam gear; **э. обратного хода** rear eccentric, reverse eccentric; **передвижной э.** slipping eccentric; **э. переднего хода** forward eccentric; **э. с изменяемым эксцентриситетом** adjustable amplitude eccentric
**эксцентриковый** *adj.* eccentric; cam
**эксцентриситет** *m.* eccentricity; centering error
**эксцентрический** *adj.* eccentric, off-center
**эксцентрично** *adv.* eccentrically
**эксцентричность** *f.* eccentricity, centering error, decentralization
**эксцентричный** *adj.* eccentric, off-center
**эксцентрометр** *m.* eccentricity meter
**эксцесс** *m.* excess; kurtosis; **э. плотности распределения** kurtosis of frequency curve
**эластанс** *m.* elastance
**эластикатор** *m.* elasticator
**эластический** *adj.* elastic(al), flexible, resilient
**эластичность** *f.* elasticity, flexibility, resilience, ductility
**эластичный** *adj.* elastic, flexible, resilient
**эластомер, эластометр** *m.* elastometer
**эл.-графич.** *abbr.* (**электронно-графический**) electron-diffraction
**электрет** *m.* electret
**электретный** *adj.* electret
**электризатор** *m.* electrizer
**электризация** *f.* electrification, elec-

trifying, electrization; excitation; **э. воды при разбрызгивании** waterfall electricity; **э. трением** tribo-electrification
**электризование** *n.* electrification, electrifying
**электризованный** *adj.* electrified
**электризовать, наэлектризовать** *v.* to electrify, to electrize, to charge
**электризуемость** *f.* electrifiableness
**электризуемый** *adj.* electrifiable
**электризующийся** *adj.* electrifiable
**электрик** *m.* electrician
**электрит** *m.* electrit, electret
**электритный** *adj.* electrit, electret
**электрификация** *f.* electrification; **э. магистрали** main-line electrification
**электрифицированный** *adj.* electrified, electrically energized
**электрифицировать** *v.* to electrify
**электрически** *adv.* electrically
**электрически-замкнутый** *adj.* conductively closed
**электрический** *adj.* electrical; **отрицательно э.** electronegative; **положительно э.** electropositive; **чисто э.** all-electric
**электричество** *n.* electricity; **э. вольтовой пары** contact electricity; **связанное э.** disguised electricity, dissimulated electricity, latent electricity; **«стеклянное» э.** vitreous electricity
**электро** *n.* electro, electrotype
**электро-** *prefix* electro-, electric-
**электроаккумуляторный** *adj.* battery-electric
**электроакустика** *f.* electroacoustics
**электроакустический** *adj.* electroacoustic, acousto-electric
**электроанализ** *m.* electro-analysis
**электроанализатор** *m.* electroanalyzer
**электроаналогия** *f.* electrical analogue
**электроаппаратура** *f.* electrical equipment
**электроартериограф** *m.* plethysmograph
**электроаурол** *m.* electric-aurol
**электробаллистика** *f.* electroballistics
**электробаллистический** *adj.* electroballistic, electrical-ballistic
**электробензомер** *m.* electric fuel gage
**электробиология** *f.* electrobiology
**электробиоскопия** *f.* electrobioscopy
**электроблокировка** *f.* electrical locking (device)

электробус *m.* electrobus

электровагонетка *f.* electric truck

электровакуумный *adj.* electric vacuum; electronic-tube

электровалентность *f.* electrovalency, electrovalence

электровалентный *adj.* electrovalent

электровесовой *adj.* electrogravimetric

электроветер *m.* electric breeze, static breeze, aura

электровзвешивание *n.* electric weighing

электровзрыватель *m.* electric fuze, electric exploder; **э. замедленного действия** electric delay fuze

электроводокачка *f.* electric pump

электроводолечебный *adj.* hydroelectric (bath)

электровоз *m.* electric locomotive; **э. без зубчатой передачи** gearless electric locomotive; **взрывобезопасный э.** electric permissible locomotive; **э. для рудничной откатки** electric haulage mine locomotive; **э. зубчатой железной дороги** rack-rail locomotive; **кабельный э.** reel-and-cable locomotive; **маневровый э.** electric switching locomotive; **однофазный э. с преобразовательными устройствами для расщепления фазы** electric split-phase locomotive; **рудничный э. для откачки** electric haulage mine locomotive; **э. с боковым ведущим дышлом** side-rod locomotive; **э. с выпрямительной установкой** rectifier-type motive power locomotive; **э. с зубчатым сцеплением** rack-rail locomotive; **э. с преобразователем** motor-generator locomotive; **сборочный рудничный э.** electric gathering mine locomotive

электровозбудительный *adj.* electromotive

электровоздуходувка *f.* motor blower

электровоспламенение *n.* electrical ignition

электровоспламенитель *m.* electric fuze, electric detonator

электровыделение *n.* electrowinning

электровязкостный *adj.* electroviscous

электрогальванизация *f.* electrogalvanizing

электроген *m.* electrogen

электрогенератор *m.* electric generator

электрогенераторный *adj.* electricity-generating; generator

электрогидравлический *adj.* electrohydraulic

электрогониометр *m.* electrogoniometer

электрогравиметрия *f.* electrogravimetry

электрогравитация *f.* electrogravity, electrogravitics

электрограмма *f.* electrogram

электрограф *m.* electrograph, electric etcher

электрографический *adj.* electrographic

электрография *f.* electrography

электрогрелка *f.* hot pad

электрод *m.* electrode; **плоско-параллельные электроды** parallel-plate electrodes; **электроды, расположенные в одной плоскости** coplanar electrodes; **боковой дежурный э.** side arm keep-alive electrode; **э. в активном состоянии** electrode "in the act"; **вживленный э.** implanted electrode; **воспринимающий э.** collecting electrode; **э., вызывающий отклонение по радиусу** radial-deflecting electrode; **делительный э.** voltage grading electrode; **э. динамического сведения** convergence electrode; **динатронный э.** dynode; **запирающий э.** pick-off electrode; **защитный э.** suppressor electrode; **кистевой э.** spark ball (*med.*) ;**многопрутковый э.** nested electrode; **э., на который подано смещение** biased electrode; **наращиваемый э.** continuous electrode; **ненакалённый э.** cold electrode; **нерасходуемый э.** nonconsumable electrode; **несущий э.** supporting electrode; **обмотанный э., оплётённый э.** covered electrode; **оправочный э. (в машинах для шовной сварки)** contact electrode, contact bar; **осадительный э.** receiving electrode; **отражающий конический э.** reverse-cone electrode; **очищающий э.** sweeping electrode; **передвижной э.** sliding electrode; **переносящий э. (в декатроне)** guide electrode; **«питающий» э. (в чарджисторе)** feeder, feeder electrode; **погружающийся э.** dipper electrode; **поджигательный э.** ignition rod, ignitor; **поджигающий э.** ignitor, starting electrode, trigger electrode, keep-alive electrode; **подовый э.** bottom electrode; **подсвечивающий э.** intensifier electrode, intensifier ring; э. полупроводникового триода emitter electrode; э. последующего ускорения intensifier electrode; э. развёртки записи writing scanning electrode; э. развёртки считывания reading scanning electrode; э. с газообразующим покрытием shielded arc electrode, covered electrode; э. с обмазкой и металлической оболочкой flux-encased electrode; э. с обмазкой флюсами fluxed electrode; э. с флюсовой обмазкой, э. с шлакообразующим покрытием flux-coated electrode, covered electrode; э., создающий послеускорение post-accelerating electrode; э. сравнения reference electrode; струйный ртутный э. streamer mercury electrode; тормозящий э. decelerator, decelerating electrode, retarding electrode; улавливающий э. collecting electrode, interceptor electrode; умножительный э. dynode; управляющий э. «зелёного» прожектора green control grid, green control electrode; управляющий э. приёмной трубки kinescope grid; управляющий э. ЭЛТ Wehnelt cylinder; э., управляющий интенсивностью пучка, э., управляющий силой тока электронного пучка intensity-controlling electrode; ускоряющий э. (секции) переноса изображения image accelerator, image accelerating electrode; фитильный э. cored electrode; фокусирующий (прикатодный) э. focusing cup, focusing electrode; э. формирования пучка, формирующий э. beam-forming electrode; центральный э. (в электронно-лучевой трубке) radial deflection terminal; центрирующий э. focusing electrode; шариковый э. bead electrode

электродальномерный *adj.* electro-distance-measuring (equipment)

электродвигатель *m.* electric motor, electromotor; э. в цилиндрическом корпусе tube-type motor; вентильный э. a.c. electric motor; э. весьма малой мощности subfractional motor; встраиваемый э. drawn-shell motor; встроенный э. motor for independent electric drive; (герметизированный) э. с повышенным давлением внутри корпуса pressur-

ized motor; двухклеточный э. double slot motor, double (squirrel-)cage motor; э. для бытовых приборов appliance motor; э. для наружной установки outdoor motor; э. для растормаживания motor-operated brake-lifter; исполнительный э. actuating motor; коллекторный э. с последовательным возбуждением series commutator motor; короткозамкнутый э. squirrel-cage motor; маломощный (миниатюрный) э. small-type motor, fractional-horsepower motor; э. мощностью выше одной лошадиной силы integral horsepower motor; э. мощностью менее 1/20 л. с. subfractional horsepower motor; э. открытого исполнения open motor; переключаемый э. multi-speed induction motor, series (wound) motor with tapped field winding; прифланцованный э. overhung motor; редукторный э. internally geared motor, motoreducer; э. с встроенной сцепной муфтой clutch(-type) motor; э. с глубокопазным якорем induction motor; э. с двумя концами вала double-shaft motor; э. с двухступенчатым редуктором double-reduction motor; э. с жёсткой характеристикой motor with shunt characteristic; э. с замкнутой системой вентиляции closed-air-circuit motor; э. с контактными кольцами slip-ring motor; э. с короткозамкнутым ротором motor with squirrel-cage rotor; э. с корпусом для встройки drawn-shell motor; э. с мягкой характеристикой motor with series characteristic; э. с охлаждением по замкнутому циклу closed-air-circuit motor; э. (с переключением) на два напряжения dual-voltage motor; э. с переключением полюсов pole-changing motor; э. с повышенным скольжением high-slip motor; э. с последовательным возбуждением series-(wound) motor; э. с поступательно-возвратным движением pulsating motor; э. с регулируемой скоростью вращения adjustable speed motor, variable speed motor; э. с фазным ротором slip-ring motor; э. с шарнирной подставкой для ремённой передачи hinged belted motor; самотормозящийся э. displacement-type-armature motor; **э. со смешанной характеристикой** motor with compound characteristic; **э. со смешанным возбуждением** compound-wound motor; **тяговый э. с опорно-осевыми подшипниками** nose-and-axle-suspended motor; **э., управляющий положением** positioning motor

**электродвигательный** *adj.* electromotive; electric-motor

**электродвижущий** *adj.* electromotive

**электродеионизация** *f.* electrodeionization

**электродекантация** *f.* electrodecantation

**электродермический** *adj.* electrodermal

**электродетонатор** *m.* electric detonator, electric fuze

**электрод-зонд** *m.* sounding electrode

**электродиагноз** *m.* electrodiagnosis

**электродиагностика** *f.* electrodiagnostics, electrodiagnosis

**электродиализ** *m.* electrodialysis

**электродиализатор** *m.* electrodialysis apparatus

**электродинамика** *f.* electrodynamics

**электродинамический** *adj.* electrodynamic

**электродинамометр** *m.* electrodynamometer; **э. для перемножения двух электрических величин** product resolver; **э. с возвратом на нуль** zero-type dynamometer

**электродиффузия** *f.* electrodiffusion

**электродный** *adj.* electrode

**электрододержатель** *m.* electrode holder

**электродренаж** *m.* electrodrainage

**электродуга** *f.* electric arc

**электродуговой** *adj.* electric-arc

**электроёмкость** *f.* capacity

**электроза** *f.* electrose

**электрозаводчик** *m.* electrical manufacturer

**электрозакалка** *f.* electrotempering

**электрозамок** *m.* electric lock

**электрозамыкание** *n.* electric locking; **временное э.** electric time locking; **известительное э.** electric indication locking; **маршрутное э.** electric route locking; **э. нецентрализованной стрелки** outlying switch lock; **э. перемены направления движения** electric traffic locking; **предварительное э.** electric approach locking; **стрелочное э.** electric switch-lever

locking; **участковое э.** electric section locking

**электрозамычка** *f.* electric lock

**электрозапал** *m.* electric primer, electric fuse

**электрозапирание** *n.* electric locking; **э. стрелок** electric switch-lever locking

**электрозапись** *f.* electrical transcription

**электрозащёлка** *f.* electric lock; **известительная э., контрольная э.** electric indication lock; **стрелочная э.** electric switchlever lock

**электрозащита** *f.* cathodic protection

**электрозвуковой** *adj.* electroacoustic

**электрозоль** *m.* electrosol

**электроизгородь** *f.* electric fence

**электроизмеритель** *m.* electric gage; **э. расхода горючего** electric fuel gage

**электроизмерительный** *adj.* electrical measuring, electrically measuring

**электроизолирующий** *adj.* electric insulating

**электроизоляционный** *adj.* electric(al) insulating

**электроиндуктивный** *adj.* electroinductive, inductive

**электроиндукция** *f.* electroinduction, induction

**электроинерция** *f.* electric inertia

**электроинструмент** *m.* electric tool

**электроинтегратор** *m.* differential analyzer

**электроискровый** *adj.* electric-spark, electrospark, electro-arcing

**электрокалориметр** *m.* electrocalorimeter

**электрокалорический** *adj.* electrocaloric

**электрокапиллярность** *f.* electrocapillarity

**электрокапиллярный** *adj.* electrocapillary

**электрокар** *m.* electrocar, electric truck

**электрокардиограмма** *f.* electrocardiogram

**электрокардиограф** *m.* electrocardiograph

**электрокардиографический** *adj.* electrocardiograph(ic)

**электрокардиография** *f.* electrocardiography

**электрокаротаж** *m.* electric logging

**электрокаутер** *m.* electrocautery

**электрокаутеризация** *f.* electrocautery

электрокимограф *m.* electrokymograph
электрокинетика *f.* electrokinetics
электрокинетический *adj.* electrokinetic
электроклапан *m.* electrovalve
электроклёпка *f.* electric riveting
электроклиширование *n.* electrotyping
электрокоагуляция *f.* electrocoagulation
электрокоррозия *f.* electrocorrosion
электрокотёл *m.* electric boiler
электрокран *m.* electric crane; **путевой э.** electric crane truck
электрокультура *f.* electroculture
электролампа *f.* electric lamp
электролечебный *adj.* hydroelectric
электролиз *m.* electrolysis
электролизатор *see* электролизёр
электролизация *f.* electrolyzing
электролизёр *m.* electrolyzer, electrolytic cell, electrolytic furnace; **э. хлора** chlorine cell
электролизирующий *adj.* electrolyzing
электролизный *adj.* electrolysis, electrolytic
электролизованный *adj.* electrolyzed
электролизовать *v.* to electrolyze
электролизуемый *adj.* electrolyzable
электролиния *f.* electric(al) line
электролит *m.* electrolyte, ionogen, electrolytic conductor; **испорченный э., отработанный э.** foul electrolyte; **э., распадающийся на ионы трёх родов** ternary electrolyte; **э. с высоким удельным сопротивлением** high-resistivity electrolyte
электролитически *adv.* electrolitically
электролитический, **электролитный** *adj.* electrolytic
электрология *f.* electrology
электролюминесцентно-фотопроводящий *adj.* electroluminescent-photoconductive
электролюминесцентный *adj.* electroluminescent
электролюминесценция *f.* electroluminescence
электромагнетизм *m.* electromagnetism, electromagnetics
электромагнит *m.* electromagnet; **э. броневого типа, броневой э. горшкого типа** pot electromagnet, pot-type magnet; **э. в стальной оболочке** jacketed electromagnet, ironclad electromagnet; **включающий э.** switching magnet, stepping magnet; **втяж-** 

**ной э.** plunger electromagnet, sucking electromagnet; **втяжной э. со стопорным штифтом** stopped coil electromagnet; **э. для длительной работы** continuous-duty electromagnet; **э. для расторможения постоянным током** d.c. brake-lifting magnet; **исполнительный э.** driving magnet; **кассирующий э.** coin collecting electromagnet; **э. кругового движения (искателя)** rotary magnet (of a switch); **подковообразный э. с катушкой на одном конце** clubfoot electromagnet; **э. привода клапана впуска воздуха в цилиндр** throttle electromagnet; **притягивающий э.** tractive electromagnet; **разобщающий э.** release electromagnet; **э. разового стирания** bulk eraser magnet; **растормаживающий э.** brake-lifting magnet; **э. с большим ходом** long-range electromagnet, long-pull electromagnet; **э. с действием при сбрасывании нагрузки** no-load release electromagnet; **э. с кольцевым якорем** homopolar electromagnet; **э. с откидным якорем** clapper type electromagnet; **э. с перекидывающимся якорем** operating magnet of a stepping relay; **э. с тарелочным якорем** homopolar electromagnet; **э. с тремя сердечниками** E-shaped electromagnet; **самоблокирующийся э.** interlocking electromagnet; **сцепляющий э.** clutch magnet; **тактовый э.** time tapper; **удерживающий э.** locking magnet, hold-up electromagnet, portative electromagnet
электромагнитизация *f.* electromagnetization
электромагнитно-обогащённый *adj.* electromagnetically enriched
электромагнитный *adj.* electromagnetic
электромашинный *adj.* dynamoelectric
электромедицина *f.* electromedicine
электромембранный *adj.* electro-membrane
электромер *m.* electromer
электромерия *f.* electromerism
электромет *m.* electromet
электрометалл *m.* (welding) filler metal
электрометаллургия *f.* electrometallurgy
электрометр *m.* electrometer; **бинантный э.** two-segment electrometer,

Dolezalek electrometer; **вибрационный (язычковый) э.** vibrating reed electrometer; **э. для считывания показаний карманных ионизационных камер** pocket-chamber electrometer; **многокамерный э.** multiple electrometer; **э., нечувствительный к изменению уровня** Lindemann electrometer; **э. с язычковым преобразователем** vibrating reed electrometer
**электрометрический** *adj.* electrometric; electrometer
**электрометрия** *f.* electrometry
**электромеханика** *f.* electromechanics; electrical engineering
**электромеханический** *adj.* electromechanical
**электромиграция** *f.* electromigration
**электромиограф** *m.* electromyograph
**электромиография** *f.* electromyography
**электромобиль** *m.* electric motor car
**электромонтажный** *adj.* electrical installation
**электромонтёр** *m.* electrician
**электромотор** *m.* electric motor
**электромощность** *f.* electric power
**электромузыкальный** *adj.* electrical musical
**электрон** *m.* electron; **э. большой энергии** high-energy electron; **э., возбуждающий лавину** initiating electron; **выбитый э.** knock-on electron; **вылетающий э.** outgoing electron; **вырождающийся э., вырожденный э.** degenerate electron; **испускаемый э., испущенный э.** ejected electron; « **коллективизированный »э.**valence electron; **комптоновский э. отдачи** Compton recoil electron; **наиболее глубокий э.** innermost electron; **э., обладающий большей энергией** high-energy electron; **оболочечный э.** shell electron; **окольный э.** oblique electron; **э. с более высоким квантовым уровнем** promoted electron; **фоновый э.** background electron
**электронагрев** *m.* electrothermics
**электронагреватель** *m.* electric heater, electric radiator; **ребристый э.** electric radiator
**электронагревательный** *adj.* electric-heating
**электронакал** *m.* electrical annealing process

**электронакальный** *adj.* electric-annealing
**электронаркоз** *m.* electronarcosis
**электронасос** *m.* electric pump
**электронатирание** *n.*, **электронатирка** *f.* brush and sponge plating
**электрон-вольт** *m.* electron volt, ev
**электроника** *f.* electronics; **оборонная э.** defense electronics; **э. сверхнизких температур** cryogenic electronics; **э. сложных схем** sophisticated electronics; **э. твёрдого тела, э. твёрдых состояний** solid-state electronics
**электронновибрационный** *adj.* electronic-vibrational
**электронноволновой** *adj.* electron-wave
**электроннодырочный** *adj.* electron-hole, hole-electron
**электроннолучевой** *adj.* electron-beam, electron-ray; cathode-ray
**электронномеханический** *adj.* mechano-electric
**электроннонейтринный** *adj.* electron-neutrino
**электроннооптический** *adj.* electron-optic(al), optical-electronic
**электроннопневматический** *adj.* pneutronic
**электронно-позитронный** *adj.* electron-positron
**электронно-разрядный** *adj.* electron-discharge
**электронно-световой** *adj.* electronic-ray, electron-ray
**электронночувствительный** *adj.* electron-sensitive
**электронноядерный** *adj.* electron-nuclear
**электронный** *adj.* electron, electronic
**электроноген** *m.* electronogen
**электронограмма** *f.* electron diffraction picture
**электронограф** *m.* electron diffraction instrument, electron diffraction camera
**электронографический** *adj.* electron-diffraction
**электронография** *f.* electron diffraction (study)
**электроножорно** *m.* cautery apparatus
**электроно-ионный** *adj.* electron-ion
**электроноподобный** *adj.* electron-like
**электрон-фононный** *adj.* electron-fonon
**электронщик** *m.* electronics man
**электрообогрев** *m.* electrical heating

**электрообогреваемый** *adj.* electrically heated

**электрооборудование** *n.* electrical equipment, electrics

**электроокраска** *f.* electrostatic painting

**электрооптика** *f.* electrooptics

**электрооптический** *adj.* electro-optic (al)

**электрооптоакустический** *adj.* electro-opto-acoustic

**электроорган** *m.* electric organ, Novachrod

**электроосадитель** *m.* electrical precipitator

**электроосаждение** *n.* electrodeposition

**электроосаждённый** *adj.* electrodeposited, plated

**электроосмос** *m.* electro-osmosis

**электроосмотический** *adj.* electro-osmotic

**электроотрицательность** *f.* electronegativity

**электроотрицательный** *adj.* electronegative

**электроочистка** *f.* electrical precipitation (of gases)

**электропайка** *f.* electric welding

**электро-паровой** *adj.* electric-steam

**электропатефон** *m.* electric phonograph

**электропаяльник** *m.* electric soldering iron

**электропередача** *f.* power transmission

**электропечь** *f.* electrical furnace; **э. для быстрого плавления** rapid electromelt furnace; **э. для пайки твёрдым припоем** electric brazing furnace; **отопительная э.** convector; **промышленная э. сопротивления** industrial furnace emplying resistance heating; **э. прямого действия** direct arc furnace

**электропизм** *m.* electropism

**электропиролиз** *m.* electropyrolysis

**электропирометр** *m.* electrical pyrometer

**электропитание** *n.* electric power supply

**электропитающий** *adj.* power-supply (-ing)

**электропишущий** *adj.* electrotype, electrotyping, electrical-type

**электроплавильный** *adj.* electrosmelting

**электроплавка** *f.* electrosmelting

**электроплакирование** *n.* electrocladding

**электроплетизмограмма** *f.* electroplethysmogram

**электроплетизмограф** *m.* electroplethysmograph

**электроплита** *f.* electric stove; **высокочастотная э. на принципе радиолокатора** radar range

**электропневматический** *adj.* electropneumatic

**электропневмоклапан** *m.* electro-pneumatic valve

**электроподушка** *f.* hot pad

**электроподъёмник** *m.* electric elevator

**электропокрытие** *n.* electroplating, electrodeposition

**электрополирование** *n.*, **электрополировка** *f.* electropolishing

**электроположительный** *adj.* electropositive

**электрополотёр** *m.* electromechanical brush, floor polisher

**электрополярный** *adj.* electropolar

**электропреобразовательный** *adj.* electric-conversion

**электроприбор** *m.* electric appliance; **бытовой э.** domestic electric appliance, household electric device

**электропривод** *m.* electrical drive

**электропроводимость** *f.* electrical conductance, conductivity

**электропроводка** *f.* electric wiring, overhead line system; telephone system, communication system

**электропроводность** *f.* electrical conductivity, electric conduction, conductibility

**электропроводящий** *adj.* (electrically) conducting

**электропроигрыватель** *m.* electric record player

**электропроизводительность** *f.* electrical efficiency

**электропромышленник** *m.* electrical manufacturer

**электропромышленность** *f.* electrical industry

**электропсихрометр** *m.* electropsychrometer

**электропунктура** *f.* electrical point spur

**электропушка** *f.* iron-notch gun

**электропылеулавление** *n.* electrical dust precipitation

**электрорадиатор** *m.* electrical radiator

электроразведка *f.* electrical prospecting

электрорассекатель *m.* electrodesiccator, electrodissector

электрорассечение *n.* electrodesiccation, electrodissection

электрорасщепление *n.* electrodisintegration

электрорегулируемый *adj.* electrically controlled

электрорезка *f.* (electric) arc cutting

электроремонтный *adj.* electrical repair

электрореофорез *m.* electrorheophoresis

электроретинограмма *f.* electroretinogram

электросварка *f.* electric welding, arc welding; **контактная э.** electric resistance welding; **э. с импульсной подачей тока** pulsation welding

электросварочный *adj.* electric-welding

электросверло *n.* electric drill

электросвечение *n.* electroluminescence

электросвязь *f.* electrocommunication; **комплексная э.** combined electrocommunication

электросеть *f.* power supply system, main system, distributing network

электросила *f.* electric power

электросиловой *adj.* electric-power (-supply)

электросинтез *m.* electrosynthesis

электроскоп *m.* electroscope; **э. для определения проводимостей веществ** diagometer; **э. с одним золотым листком** Wilson electroscope

электроскопический *adj.* electroscopic

электроснабжение *n.* power supply, current supply; **э. по петлевой системе** looped-in power supply; **э. по системе общих шин** bussed current supply

электросмоз *m.* electro-osmosis

электросодержащий *adj.* electric, containing electricity

электросон *m.* electric sleep

электросопротивление *n.* electric resistance

электросталь *f.* electric steel; **э. на основе жидкого металла** hot metal electric steel

электростанция *f.* power station, power plant, power house; **крупная э., мощная э.** high-power station; **э. на ядерном горючем** nuclear-fuel power station; **районная э.** high-tension

power station; **э. со сборными шипами** radial alarm-system network

электростартер *m.* electric starter

электростатика *f.* electrostatics

электростатический *adj.* electrostatic

электростенолиз *m.* electrostenolysis

электросторож *m.* electric fence

электрострриктивный *adj.* electrostrictive, electrostriction

электрострикционный *adj.* electrostrictive

электрострикция *f.* electrostriction, inverse piezoelectric effect

« электрот » *m.* electret

электротахометр *m.* electric tachometer

электротележка *f.* electric truck

электротеллурограф *m.* electrotellurograph

электротерапия *f.* electrotherapy, electrology

электротермический *adj.* electrothermic, electrothermal

электротермия *f.* electrothermy, electrothermancy, electrothermics

электротермодиффузия *f.* electrothermal diffusion

электротехник *m.* electrician

электротехника *f.* electrical engineering, electrical technology

электротехнический *adj.* electrical, electrotechnical, electrical engineering

электротипия *f.* electrotypy

электротомия *f.* electrodesiccation

электротонус *m.* electrotonus

электроторпеда *f.* electric torpedo

электротяга *f.* electric traction; **э. судов по каналам** canal traction

электроуправление *n.* electrical control

электроуправляемый *adj.* electrically controlled

электроустановка *f.* electrical plant, electrical installation; **домовая э. для тревожной сигнализации** burglary alarm; **собственная э.** natural current installation

электрофабрикант *m.* electrical manufacturer

электрофакс *m.* Electrofax

электроферограмма *f.* electropherogram

электроферография *f.* electropherography

электрофизика *f.* electrophysics

**электрофизиологический** *adj.* electrophysiological

**электрофизиология** *f.* electrophysiology

**электрофильность** *f.* electrophilicity

**электрофильный** *adj.* electrophilic

**электрофильтр** *m.* electric separator, electrostatic precipitator

**электрофлор** *m.* electroflor

**электрофон** *m.* electrophone

**электрофонический** *adj.* electrophonic

**электрофор** *m.* electrophorus

**электрофорез** *m.* electrophoresis

**электрофорезный** *adj.* electrophoretic

**электрофоретический** *adj.* electrophoretic

**электроформовка** *f.* electroforming, electrical forming

**электрофорный** *adj.* electrophorus

**электрохимический** *adj.* electrochemical

**электрохимия** *f.* electrochemistry

**электрохирургический** *adj.* electrosurgical

**электрохирургия** *f.* electrosurgery

**электрохолодильник** *m.* refrigerator

**электроцентраль** *f.* power plant, electric power station

**электрочасовой** *adj.* electric clock

**электрочастица** *f.* electrion

**электрочасы** *pl.* electric clock

**электрошерардизация** *f.* electrosherardizing

**электрошок** *m.* electroconvulsive shock

**электрощуп** *m.* electric feeler

**электроэкстракция** *f.* electroextraction

**электроэндоскопия** *f.* electro-endoscopy

**электроэндосмос** *m.* electroendosmose, electroendosmosis

**электроэнергетика** *f.* electric power engineering

**электроэнергетический** *adj.* electric energy, electric power

**электроэнергия** *f.* electrical energy; **э., вырабатываемая в часы пика графика нагрузки** peak energy; **э., вырабатываемая при базисной работе ГЭС** base energy, off-peak energy

**электроэнцефалограмма** *f.* electroencephalogram

**электроэнцефалограф** *m.* electroencephalograph

**электроэнцефалографический** *adj.* electroencephalographic

**электроэнцефалография** *f.* electroencephalography

**электроэрозионный** *adj.* electroerosion

**электроэрозия** *f.* electroerosion

**электроядерный** *adj.* electronuclear

**электрум** *m.* electrum

**элемент** *m.* element; unit, component, part; cell; member; device; couple; **видимые элементы изображения** resolvable picture elements; **добавочные элементы** additional cells, balancing cells, end cells; regulator cells; **задающие входные элементы** reference input elements; **элементы местоопределения** elements of a fix; **направляющие элементы лентопротяжного механизма** tape guide element; **перемежающиеся по горизонтали элементы изображения** horizontally interlaced dots; **переходные элементы** transmutation elements; **элементы радиоустройств на сантиметровых волнах** microwave assemblies; **различимые (or разрешаемые) элементы изображения** resolvable picture elements; **элементы с пропорциональным суммированием входных сигналов** proportioning elements; **светочувствительные элементы глаза** light-sensitive receptors; **элементы цепи обратной связи** feedback elements; **чередующиеся элементы изображения** alternate picture dots; **элементы штрихового многоцветного экрана** color(-phosphor) stripes; **э. активной зоны** core element; **э. акустического поглощения** acoustic dissipation element; **аналогоцифровой э.** analog-digital element; **э. анодной настройки** plate tuning rod; **э. антенной решётки** elementary antenna; **астатический э.** floating element; **э. в гирлянде** unit insulator; **вакуумный туннельный э.** vacuum tunnel effect device; **вентильный э.** semiconductor device; **э. воздушной деполяризации** air-depolarized cell; **э. волновода** components for waveguides; **воспринимающий э.** detecting element, sensing element, sensor, primary (control) element; **восстанавливающий э.** resetter; **восстанавливающий э. системы** restoring component; **вращающий э.** driving element; **э. выборки слова, э. вы-**

борки числа word selection element; **высокочастотный э. изображения** high-frequency term; **выходной управляющий э.** final control element; **э. гелиобатареи** solar cell; **герметизированный смолой э.** resincast component; **герметизированный тепловыделяющий э.** sealed-in fuel element; **готовый э. машины** building block; **делящийся э.** fissile element; **длительно работающий тепловыделяющий э.** long-exposure fuel element; **э. для эксплуатации с нормально разомкнутой цепью** open-circuit cell; **э. ёмкостной связи** capacity coupler; **э. жёсткости** stiffening element; **жидкостный э.** wet cell; **э. задержки на один разряд** digit delay device; **э. замедлителя** moderator element; **запальный тепловыделяющий э.** source fuel element; **запоминающий э. Гото** Goto-pair memory element; **(запоминающий) э. Кроу** Crowe cell; **излучающий э. трёхсекционной турникетной антенны** superturnstile radiating element; **импедансный э.** ohm unit; **э. информации, (с которым оперирует машина)** information word, machine word; **испольнительный э.** correcting element, effector; **испольнительный э. системы регулирования** final control element; **исходный э.** parent element; **кистевой э.** grip end; **э. класса** member of class; **э. команды** order element, instruction element; **коммутационный э.** circuit element; **э. компенсации** equalizer, equalizing network, distortion-correcting device; **э. конструкции** member, component, structural element; element of symbols in circuit diagrams; **контролируемый э.** monitored element; **кулачковый следящий э.** cam follower; **э. Лекланше** Leclanché cell; **э. линии по типу « Т »** T-section; **логический э. « и »** logical AND component; **э. лопасти** blade section; **люминесцирующий э. мозаичного экрана** color-emitting phosphor dot; **магнитный э. с двумя осями намагничивания** biaxial magnetic element **манометрический э.** pressure element; **материнский э.** parent element; **э. матричной схемы, матричный э.** matrix element; **ма-**

тричный э., отличный от нуля nonvanishing matrix element; **э. Мейдингера** Meidinger cell, balloon cell; **э. мишени** target element; **нагревательный э. с наездниками** clamp-on heater element; **э. наземного оборудования системы** ground element of the system; **накапливающий э., накопительный э.** storage element, memory cell; **накопительный э. сетового типа** honeycomb electrostatic storage element; **наливной э.** wet cell, hydroelectric cell; **э. настройки с помощью феррита** ferrite-tuning element; **э. « не »** NOT element; **э. « не и »** NAND element, NOT AND element; **э. « не или »** NOR element, NOT OR element; **нерабочий э.** unoperated element; **нерегламентированный э.** manual element, unrestricted element; **нерегулярно повторяющийся э.** irregular element, intermittent element; **нечётно-нечётный э.** odd-odd element; **низкочастотный э. изображения** low-frequency term, low-frequency element; **э. обмотки, э. обмотки якоря** armature loop; **обнаружительный э. на инфракрасных лучах** IR detector cell; **э. обшивки с подкрепляющим стрингером** stringer-skin element; **одиночный гальванический э.** monocell, single-cell battery; **окиснортутный э.** zinc mercury cell, oxide of mercury cell; **окисносеребряный э.** silver oxide cell; **э. опорного напряжения** voltage-reference element, reference element; **осколочный э.** fission element; **э. отрицания** NOT element; **э. отрицания эквивалентности** anti-coincidence element, nonequivalent element; **э. памяти** memory cell, storage cell, data location; **э. памяти на сверхпроводниках** persistor cell; **параметрический логический э.** parametron logical element; **пассивный э. антенны** indirectly fed antenna, parasitic antenna, passive antenna; **первичный э.** primary element, detecting element; primary battery, primary cell; **первичный чувствительный э.** primary detector; **э. передаваемого сообщения** piece of information; **э. переменной длительности** variable(-time) element; **э.**

пересчётной схемы matrix element; э. печатной схемы printed component part; плавкий э. fuse-element; плотностный э. gravity cell; э. площади elemental area; подключаемый э. regulating cell; switchgear bay; подключающий э. antenna lead; подстроечный э. для сверхвысокой частоты UHF tuner; полизональный тепловыделяющий э. со спиральными рёбрами polyzonal spiral fuel element; полупроводниковый э. semiconductor device, transistor unit; э. постоянной длительности constant(-time) element; э. постоянной связи fixed coupler; постоянный э. constant voltage cell; преобразующий э. inverting element; примесный э. impurity element, doping element; противовключённый э. regulating cell, stabilizing cell; противовключённый щелочный э. alkaline counter-cell; э., работающий в предельном режиме marginal component; э. развязывающего фильтра в анодной цепи plate-coupling element; разлагающий э. scanning element; э. разложения (изображения) picture element, elemental area; scanning element; разряженный э. exhausted cell; рассеянный э. trace element; э., реагирующий на амплитуду amplitude-sensitive element; реактансный э. ohm unit; ребристый тепловыделяющий э. finned fuel element; регистрирующий э. гальванометра galvanometer recorder; регламентированный э. restricted element; решающий э. decision element, computing element; решающий э. с двумя устойчивыми состояниями bistable element; ртутно-индиевый э. indium-mercuric-oxide cell; э. с воздушной деполяризацией air-depolarizer cell; э. с двумя выводами two-terminal device; э. с двумя устойчивыми состояниями two-state device, bistable element; э. с деполяризатором из полуторахлористого железа с прибавкой брома chloride cell; э. с жидкостью wet cell, hydroelectric cell; э. с «мешочным» деполяризатором agglomerate Leclanché cell; э. с нашатырём salammoniac cell; э. с нечётным A odd-A element;

э. с ограничением bounded element; э. с одной степенью свободы unidirectional element; э. с отрицательным температурным коэффициентом negative-temperature unit; э. с подъёмным электродом plunge cell; э. с потенциальным выходом voltage-output element; э. с продуванием электролита струёй воздуха pneumatic cell; э. с раствором квасцов alum cell; э. с серной кислотой и двухромовокислым калием bichromate cell; э. с токовым выходом current-output device; э. с тремя устойчивыми состояниями three(-stable) state device; э. с цинковой плиткой в ртути и с деполяризатором из хромовой кислоты mercury bichromate cell; э. с чётным A even-A element; сверхбыстродействующий э. ultra-fast element; э. связи coupling element, coupling factor, matching plug; э. связи с волноводной линией line stretcher; сглаживающий э. smoothing filter, steadying circuit; сдвоенный множительный э. paired multiplier; э. системы аварийной защиты safety element; э. системы регулирования regulating element; э. системы управления control element, control component; э. слабой связи loose coupler; э. следящей системы servo element; сменный э. plug-in component; э. со считыванием без разрушения (информации) nondestructive read element; э. согласования adapting component; э. списка цикла for list element; ступенчатый тепловыделяющий э. ladder fuel element; сухой э. без обёртки деполяризатора nonlined construction, unwrapped construction; сухой э. с мешочным деполяризатором bag-type construction; сухой наливной э. inert cell; счётный э. decision element, computing element; считывающий э. sensing element, sensory element; считывающий э. с разрушением destructive read element; съёмный э. plug-in unit; твиновский запоминающий э. Goto-pair memory element; тепловыделяющий э. без оболочки bare fuel element, bare heat-producing element; тепловыделяющий э. без связывающей прослойки unbounded fuel element; **тепловыделяющий э. на ториевой основе** thorium-base fuel element; **тепловыделяющий э. со связывающей прослойкой** bonded fuel element; **тепловыделяющий э. со спиральными рёбрами** spiral-fin fuel element; **тепловыделяющий э. стержневого типа** rod-type fuel element; **термочувствительный э. с сопротивлением** resistance temperature detector; **э. типа арифметической прогрессии** step-until element; **э. типа пересчёта** while element; **токовый э.** current-operated device; **тормозной э. счётчика** meter braking element; **э., требующий профилактической замены** marginal component; **трёхстержневой соединительный э.** tri-rod coupler; **э. трогания (в комбинированном реле)** initiating element; **туннельный э.** tunnel effect device, tunneling device; **туннельный э. на напылённой плёнке** evaporated film tunneling device; **угольный э.** carbon cell, carbon-consuming cell, carbon-combustion cell; **угольный топливый э.** carbon-combustion cell; **управляющий э.** control element, monitoring element, resetter; **фотогальванический э.** photovoltaic cell; **фотогальванический э. с полупрозрачным электродом** backwell photovoltaic cell; **фотоэлектронный э.** photoemissive element; **функциональный э.** functor, functional element; **химический э. с малым порядковым номером** low-Z element; **холостой э.** inert element; **э. централизационного аппарата для запирания защёлки** machine quadrant; **э. цепи с сосредоточенными параметрами** lumped circuit element; **цилиндрический нагревательный э.** cartridge-type heating element; **э. цинк-серебро с деполяризатором из хлористого серебра** chloride-of-silver cell; **чувствительный э.** sensing element, detecting element, primary element, primary sensitive element, sensing head, sensor, feeler, homing eye; **чувствительный э. бокового скольжения** side slip sensor; **чувствительный э. следящей системы** error detector; **чувствительный к**

**земной тяжести э.** gravity-sensitive element; **чувствительный к рассогласованию э.** error sensor; **чувствующий э.** sensor device

**элементарно-симметричный** *adj.* elementary symmetrical

**элементарность** *f.* elementariness

**элементарный** *adj.* elementary, elemental, simple, fundamental; prime

**элементный** *adj.* element, cell; elemental

**элерон** *m.* aileron

**элефантид** *m.* elephantide

**элиминатор** *m.* eliminator

**элиминация** *f.* elimination

**элиминировать** *v.* to eliminate

**элиминируемость** *f.* eliminability

**элиминируемый** *adj.* eliminable

**элинвар** *m.* elinvar

**элконит** *m.* elkonite

**элконитовый** *adj.* elkonite

**эллинг** *m.* car shed; shipyard, dock; hanger

**эллипс** *m.* ellipse; **э. напряжений** stress ellipse

**эллипсис** *see* **эллипс**

**эллипсограф** *m.* ellipsograph

**эллипсоид** *m.* ellipsoid

**эллипсоидальный** *adj.* ellipsoidal, ellipsoid

**эллиптически** *adv.* elliptically

**эллиптический** *adj.* elliptical, elliptic

**эллиптичность** *f.* ellipticity

**эл. магн.** *abbr.* (**электромагнитная единица**) electromagnetic unit

**элонгация** *f.* elongation, stretch

**элотрон** *m.* gamma spectrometer using recoil electrons

**элсвортит** *m.* ellsworthite

**элсикон** *m.* elsicon

**эл.-ст. ед.** *abbr.* (**электростатическая единица**) electrostatic unit

**элт** *abbr.* (**электротехника**) electrical engineering

**ЭЛТ** *abbr.* (**электронно-лучевая трубка**) cathode-ray tube; **ЭЛТ в системе с бегущим лучом** flying-spot cathode-ray tube; **ЭЛТ для квантования сигналов во времени** time-sampling cathode-ray tube; **запасывающая ЭЛТ** scototron; **множительная ЭЛТ с перекрёстными полями** crossed-fields electron-beam multiplier; **ЭЛТ-модулятор света** CRT light tube; **ЭЛТ с магнитным отклонением** EM tube; **ЭЛТ с малым послесвечением**

экрана fast-screen cathode-ray tube; **ЭЛТ с мягкими выводами** wired-in cathode-ray tube; **ЭЛТ с непосредственной записью** direct recording cathode-ray tube; **ЭЛТ с первым анодом, установленным под углом к оси трубки** angled-anode cathode-ray tube; **ЭЛТ с развёртыванием пучком быстрых электронов** high-velocity cathode-ray tube; **ЭЛТ с экраном белого свечения** white-screen cathode-ray tube; **ЭЛТ совмещающая ряд функций приёмника сверхвысокой частоты** Wamoscope tube; **совмещённые ЭЛТ** registered tubes; **цветная ЭЛТ типа « жалюзи »** Venetian blind color tube

**эль-катод** *m.* L-cathode
**эльконит** *m.* elkonite
**эльконитовый** *adj.* elkonite
**Эльмо** Elmo
**элюат** *m.* eluate
**элюирование** *n.* elution
**элюировать** *v.* to elute, to extract
**элюирующий** *adj.* eluting; elutriating
**элютриация** *f.* elutriation
**элюционно-разделительный** *adj.* elution-partition
**элюция** *f.* elution; elutrition
**эмалевый** *adj.* enamel
**эмалепроволока** *f.* enamelled wire
**эмалирование** *n.* enamelling
**эмалированный** *adj.* enamelled
**эмалировать** *v.* to enamel, to glaze
**эмалировка** *f.* enamelling, glazing; **горячая э., э. с обжигом** stove enamelling
**эмалировочный** *adj.* enamelling, glazing
**эмаль** *f.* enamel; **запаечная э.** seal enamel; **э. печной сушки** baking enamel; **светочувствительная э.** photoengraver's enamel; **стекловидная э.** vitreous enamel; **э. холодной сишки** air-drying enamel; **шлифовальная э.** flatting enamel
**эман** *m.* eman
**эманатор** *m.* emanator
**эманационный** *adj.* emanation, emanating
**эманация** *f.* emanation; emission; emanium, Em
**эманирование** *n.* emanating
**эманирующий** *adj.* emanating
**эманометр** *m.* emanometer
**ЭМГ** *abbr.* (**электромиограмма**) electromyogram

**эме** *abbr.* (**электромагнитная единица**) electromagnetic unit
**эмерджентный** *adj.* emergent
**эмиссионный** *adj.* emitting, emissive, emission
**эмиссия** *f.* emission; **вторичная э. экранирующей сетки** screen-grid secondary emission; **э. под действием электрического поля** field emission, autoelectronic emission; **сопряжённая корпускулярная э.** associated corpuscular emission; **фотоэлектронная э.** photoelectric emission, photoemission, photoemissivity
**эмитер** *see* **эмиттер**
**эмитрон** *m.* emitron
**эмиттанс** *m.* emittance
**эмиттер** *m.* emitter; **автоэлектронный э. в форме острия** field emission tip; **вторично-электронный э.** secondary emitter, dynode; **металлический э.** metallic emitting surface; **э. неосновных носителей (в транзисторе)** minority emitter (of a transistor); **э. основных носителей (в транзисторе)** majority emitter (of a transistor)
**эмиттерный** *adj.* emitter
**эмиттировать** *v.* to emit, to give off
**эмиттируемый** *adj.* emitted
**эмиттирующий** *adj.* emitting, emissive
**эмпирически** *adv.* empirically, experimentally
**эмпирический** *adj.* empirical, experimental
**ЭМУ** *abbr.* (**электромашинный усилитель**) dynamoelectric amplifier, amplidyne generator
**эмульгатор** *m.* emulgator; emulsifier
**эмульгация** *f.* emulsification
**эмульгирование** *n.* emulsification
**эмульгировать** *v.* to emulsify, to emulsionize
**эмульгируемый** *adj.* emulsifiable
**эмульгирующий** *adj.* emulsifying
**эмульгирующийся** *adj.* emulsifiable
**эмульсер** *m.* emulsifier
**эмульсин** *m.* emulsin
**эмульсионный** *adj.* emulsive; emulsion
**эмульсирование** *n.* emulsification
**эмульсированный** *adj.* emulsified
**эмульсировать** *v.* to emulsify
**эмульсификатор** *m.* emulsifier
**эмульсификация** *f.* emulsification
**эмульсия** *f.* emulsion; **э. на стеклянной подложке** glass-backed emulsion; **э.,**

**насыщенная** в$^{10}$ B$^{10}$-loaded emulsion; **снимающая э.** base-stripping emulsion; **снятая э.** base-stripped emulsion

**эмульсоид** *m.* emulsoid, emulsion

**эмульсол** *m.* cutting emulsion

**ЭМЭ** *abbr.* (эмпирический матричный элемент) empirical matrix element

**эналит** *m.* enalite

**энантиоморф** *m.* enantiomorph

**энантиотропия** *f.* enantiotropy

**энантиотропный** *adj.* enantiotropic

**энгармонический** *adj.* enharmonic

**Энглер** Engler

**эндекаэдр** *m.* hendecahedron

**эндекаэдрический** *adj.* hendecahedral

**эндемический** *adj.* endemic

**эндо-** *prefix* endo-

**эндовибратор** *m.* endovibrator, resonant cavity; **Г-образный э.** L-shaped cavity; **э. с большим затуханием** lossy cavity

**эндодин** *m.* endodyne, self-heterodyne

**эндолимфа** *f.* endolymph

**эндомицет** *m.* endomycete

**эндорадиозонд** *m.* endoradiosonde

**эндоскоп** *m.* endoscope

**эндосмометр** *m.* endosmometer

**эндосмос** *m.* endosmose, endosmosis

**эндосмотический** *adj.* endosmotic

**эндотермический** *adj.* endothermic, endoenergic, heat-absorbing

**эндотермичность** *f.* endothermic nature

**эндотермный** *adj.* endothermic, heat-absorbing

**эндоэнергетический** *adj.* endoenergic, endothermic

**энергетизм** *m.* energetism

**энергетика** *f.* power engineering; energetics

**энергетический** *adj.* energy, power; energetic, energic

**энергичный** *adj.* energetic, vigorous, high-energy, active

**энергия** *f.* power, energy; **э. активации примеси** impurity activation; **э. боковой полосы частот** sideband energy; **э. в импульсе** pulse energy; peak energy; **э. в лабораторной системе координат** laboratory energy; **э. в системе центра масс** center-of-mass energy; **э. выделенного пучка** beam-out energy; **выделяемая э.** output energy; **э., выделяющаяся в момент деления** prompt energy of fission; **э. гамма-излучения, сопро-** вождающего деление fission gamma energy; **граничная э. бета распада** beta end-point energy; **двигательная э.** motive power; **э. движущегося по орбите объекта** orbital energy; **э. запуска** triggering energy; **затраченная э.** expanded energy; **э., затрачиваемая на преодоление гистерезиса** hysteresis energy; **э. из окружающей среды** environmental energy; **израсходованная э.** expanded energy; **э., используемая для движения** propulsive energy; **э., используемая для создания тяги** thrust-producing energy; **критическая э. деления** activation energy of fission, saddle-point energy, minimax energy; **лучистая э.** radiation energy, radiant energy; **э. магнитной анизотропии кристалла** magnetocrystalline energy; **э. массы покоя** rest-mass energy; **э. мгновенного гамма-излучения** prompt gamma energy; **э. монохроматического излучения** homogeneous radiant energy; **э. морских приливов и отливов** tidal power; **надтепловая э.** epithermal energy; **э. налетающих частиц** incident energy; **э. обмена** interchange energy; **э. однополупериодного выпрямления** half-wave energy; **э. ответного сигнала** re-radiated energy; **э., отвечающая границе поглощения в кадмии** cadmium cut-off energy; **э. падающего потока** incident energy; **э. пары дырка-электрон** hole-electron energy; **э. первичного нейтрона деления** primary fission-neutron energy; **э. первого возбуждения** first excitation energy; **э. поверхностного излучения** energy radiated per unit area; **э. поверхностного натяжения** capillary energy; **поглощённая лучистая э.** absorbed radiation energy; **э. пограничного слоя** boundary layer energy; **подводимая э.** input energy; **потенциальная э. силы тяжести** gravitational energy; **э., потерянная по вторичной цепи** slip energy; **э. при абсолютном нуле** zero-point energy; **просачивающаяся э. пика** spike leakage energy; **э. распада в основном состоянии** ground state disintegration energy; **э. растворения** solvation energy; **э. резонансного поглощения** resonance-absorption energy; **э. рекуперации** restored energy; **э. связи** binding energy, bond(ing) energy; **э. связи на одну частицу** binding energy per particle; **э. соединения** fusion energy; **э., сообщаемая двигательной системой** propulsion-system energy; **э. спин-орбитального взаимодействия** spin-orbit coupling energy; **э. сцепления** cohesive energy, cohesion energy; **тепловая э., образующаяся при входе в плотные слои атмосферы** re-entry thermal energy; **э. тормозного излучения** bremsstrahlung energy; **э. упорядочения** ordering energy; **э. упругих сил** resilient energy; **утрачиваемая э.** depreciation of energy; **электрическая э. на входе** input electrical energy

**энерго-** *prefix* energy; power
**энергобаза** *f.* source of power supply
**энерговыделение** *n.* release of energy, energy liberation
**энергоёмкость** *f.* power capacity, energy capacity, energy content
**эн秘голампа** *f.* power tube
**энерголиния** *f.* power line
**энергомашина** *f.* power engine, motor
**энергонечувствительный** *adj.* energy-insensitive
**энергоотдача** *f.* release of energy, energy liberation
**энергопоглощающий** *adj.* energy-absorbing
**энергопотеря** *f.* loss of energy
**энергореактор** *m.* power reactor
**энергосеть** *f.* power supply system, mains
**энергосистема** *f.* power system
**энергоснабжение** *n.* power supply, energy supply
**энергосодержание** *n.* energy content
**энергостанция** *f.* power-station, power plant
**энергоустановка** *f.* power plant
**энергоцентр** *m.* power center
**энергоцентраль** *f.* central power-station
**энергочувствительный** *adj.* energy-sensitive
**энергоэквивалентный** *adj.* power-equivalent
**эннеод** *m.* enneode
**энтальпия** *f.* enthalpy; **э. торможения** stagnation enthalpy
**энтракометр** *m.* enthrakometer

**энтр. ед.** *abbr.* (**энтропийная единица**) entropy unit
**энтропийный** *adj.* entropy
**энтропия** *f.* entropy
**энцефалограмма** *f.* encephalogram
**энцефалограф** *m.* encephalograph
**энцефалография** *f.* encephalography; **ультраакустическая отражательная э.** echo-encephalography
**ЭО** *abbr.* (**электронный осциллограф**) cathode-ray oscillator
**эолов** *poss. adj.* aeolian
**ЭОП** *abbr.* (**электроннооптический преобразователь**) image converter tube
**эпакта** *f.* epact
**эпатэм** *m.* epatam
**эпи-** *prefix* epi-
**эпигенетический** *adj.* epigenetic
**эпидиаскоп** *m.* epidiascope
**эпиляционный** *adj.* epilation, hair removal
**эпиляция** *f.* epilation, hair removal
**эпископ** *m.* episcope, aphengescope
**эпископический** *adj.* episcopic
**эпитазиметр** *m.* epitasimeter
**эпитаксиально** *adv.* epitaxially
**эпитаксиальный** *adj.* epitaxial
**эпитаксия** *f.* epitaxy
**эпитеория** *f.* epitheory
**эпитермический** *adj.* epithermal
**эпицентр** *m.* epicenter, ground zero
**эпицикл** *m.* epicycle
**эпициклический** *adj.* epicyclic
**эпициклоида** *f.* epicycloid; **удлинённая э.** inflected epicycloid; **укороченная э.** quirked epicycloid
**Эплтон** Appleton
**эпокси-** *prefix* epoxy-
**эпоксид** *m.* epoxyde resines
**эпоксидирование** *n.* epoxidation
**эпоксидированный** *adj.* epoxidized
**эпоксидно-эфирный** *adj.* epoxy-ether
**эпоксидный** *adj.* epoxy
**эпоксиды** *pl.* epoxies
**эпоха** *f.* epoch, age, period, time
**Эппльтон** Appleton
**э.п.р., ЭПР** *abbr.* (**электронный парамагнетический резонанс**) electron paramagnetic resonance
**эпсомит** *m.* epsomite
**Эпштейн** Epstein
**эрбиевый** *adj.* erbium
**эрбий** *m.* erbium, Er
**эрг** *m.* erg
**ЭРГ** *abbr.* (**электроретинограмма**) electroretinogram

**эргметр** *m.* ergmeter
**эрго-** *prefix* ergo-
**эргодический** *adj.* ergodic
**эргодичность** *f.* ergodicity
**эргометр** *m.* ergmeter
**эргон** *m.* ergon
**эргостерин** *m.* ergosterol
**ЭРД** *abbr.* (электрический ракетный двигатель) electrical rocket propulsion
**эректор** *m.* erector
**Эри** Airy
**эриометр** *m.* eriometer
**Эриксон** Ericson
**эрископ** *m.* eriscope
**эритема** *f.* erytheme
**эритемальность** *f.* erythemal factor
**эритемальный** *adj.* erythemal
**эритеметр** *m.* erythemeter
**эритемный** *adj.* erythema; erythematous
**эрланг** *m.* procurement
**Эрмит** Hermite
**эрмитов** *poss. adj.*, **эрмитовский** *adj.* Hermitian
**эрмитовость** *f.* Hermitian (property)
**эрозивный** *adj.* erosive, wearing
**эрозиеустойчивый, эрозионно-устойчивый** *adj.* erosion-resistant
**эрозионный** *adj.* erosion, erosional, erosive
**эрозия** *f.* erosion
**эрратический** *adj.* erratic
**эрс** *abbr.* (эрстед) oersted
**эрстед** *m.* oersted
**эрстедметр** *m.* oerstedmeter
**Эру** Heroult
**эруптивный** *adj.* eruptive, effusive
**Э.С.** *abbr.* (энергия связи) binding energy
**Эсаки** Esaki
**эсе** *abbr.* (электростатическая единица) electrostatic unit
**эскадра** *f.* squadron
**эскадренный** *adj.* squadron
**эскадрилья** *f.* squadron
**эскалатор** *m.* escalator
**эскиз** *m.* sketch, outline, plan, rough drawing
**эскизный** *adj.* sketchy; sketch, outline, plan
**ЭСМ** *abbr.* (электронная счётная машина) electronic computer
**эстезиометр** *m.* esthesiometer
**эстерификация** *f.* esterification
**эстиатрон** *m.* estiatron
**этаж** *m.* floor, story, level

**этажерка** *f.* stack, set of shelves, micromodule stack, stacked wafers
**этажный** *adj.* floor, story, level; tier; graduated, graded, stepped
**эталон** *m.* standard, reference standard; calibration instrument, gage; sample; **э. абсолютно чёрного тела** blackbody standard; **атомный э. частоты** atomic reference oscillator, atomic reference frequency standard; **самолётный э. частоты** airborne-frequency reference; **э. Фабри-Перо** Fabry and Perot interferometer; **э. цветовой несущей** chrominance carrier reference
**эталонирование** *n.* calibration, gaging; standardization, standardizing
**эталонированный** *adj.* standardized; calibrated, gaged
**эталонировать** *v.* to standardize; to calibrate, to gage; to adjust
**эталонный** *adj.* standard; calibrating, calibration; reference; sample
**этап** *m.* step, phase; **начальный э. полёта** launching phase; **э. программы** program step
**этапный** *adj.* step, stage, phase
**Этвеш** Eötvös
**этернит** *m.* eternit
**этикетирование** *n.* labeling
**этикетка** *f.* nameplate, plate, label, tag, tally, escutcheon
**этил** *m.* ethyl
**этилендиаминтартрат** *m.* ethylene diamine tartrate
**этилцеллюлоза** *f.* ethylcellulose
**этоксилиновый** *adj.* ethoxyline
**этриоскоп** *m.* ethrioscope
**Эттингсгаузен** Ettingshausen
**ЭУП** *abbr.* (электрический указатель поворота) electric turn indicator
**эф.** *abbr.* (эфир) ethyl ether
**эфемерида** *f.* ephemeris
**эфемеридный** *adj.* ephemeris, ephemerical
**эфемерный** *adj.* ephemeral, transitory, shortlived
**эфир** *m.* ether; ester; air (as medium for waves); **в эфире** on the air; **э. метакриловой кислоты** methacrylate; **метиловый э. полиметакриловой кислоты** polymethacrylatemethyl-ester; **мировой э.** ether; **сложный э.** ester
**эфирный** *adj.* ether; ester; ethereal
**эфирофон** *m.* etherophon

эфлоресценция *f.* efflorescence

эфлуент *m.* effluent

эфф. *abbr.* (эффективный) effective

эффект *m.* effect, result; capacity; **микрофонные эффекты сетки мишени** target mesh microphonics; **антенной э. пеленгаторной рамки** loop effect; э. **«бегущих световых пятен»** "chaser border" effect; э. **вальцевания** grain effect; **взаимосвязанный** э. coupled effect; э. **вихревого тока** eddy-current effect, Faraday effect; э. **вязкости от электрического заряда частиц в растворе** electroviscous effect; **гальваномагнитный** э. magneto-resistive effect; **глубинный** э. depth magnification; **горьковский** э. Luxemburg effect, Tellegen effect, interaction of radio waves; **грубый маскирующий** э. brute-force blanket effect; э. **деки** sounding-board effect; э. **деления на быстрых нейтронах** fast fission effect; **дробовой** э. shot effect, shrot effect, Schottky effect, fluctuation effect; э. **ёмкости руки** body effect; э. **запасания информации, э. записи информации** memory effect; э. **затенения** blanketing effect; **зенитный** э. zenith-angle effect; э. **изменения знака проводимости** patch effect; э. **Ионсона-Рабека** Johnson-Rahbeck effect, electrostatic adhesion; э. **короткодействия** short-range order effect; э. **косого направления** obliquity effect; **люксембургский** э., **люксембургско-горьковский** э. Luxemburg effect, Tellegen effect; э. **маховика, э. махового колеса, маховой** э. flywheel effect; **местный** э. в **телефоне** sidetone; **микрофонный** э. "pinging" noise, microphomism, microphonics, microphonicity, microphony, microphone effect; **микрофонный** э. **сетки мишени** target mesh microphonics; э. **наложения импульсов** pile-up effect; э. **наложения отражённых ударных волн** Mach's effect; э. **направленности** directional effect, beam-shaping effect; э. **напряжения диссоциации М. Вина** Wien's dissociation voltage effect; э. **обеднения** sweep-out effect; **оболочечный** э. shell effect; **обратный** э. converse effect, feedback; э. **объёмного заряда** charge-volume

effect; э. **опережения** leading effect; э. **органов управления** rate effect of the controls; э. **ослабления яркости** dilution effect; э. **острия** point effect, needle effect; **паразитный супергетеродинный** э. double-super effect; э. **перешейки** isthmus effect; э. **перещёлкивания** crackling effect; **плотностной** э. **при больших энергиях электронов** finite-energy density effect; **плотностной** э. **при очень малых энергиях электронов** zero-energy density effect; **поверхностный** э. surface action, surface charge action, skin effect, Kelvin effect; э. **поглощения** sink effect; э. **подавления** quenching effect; э. **поля рассеяния** stray-field effect; э. **потери части ускоряющего напряжения** "sticking" effect; э. **предварения** precedence effect; **приведённый** э. **массы** reduced mass effect; **приёмный** э. pick-up factor; э. **пространственного заряда** space-charge effect; **псевдоскопический** э. false relief effect; э. **Пуркинье** Purkinje phenomenon; э. **разновремённости** time-diversity effect, diversity effect; **роки-пойнт** э. rocky-point effect, flash arc; э. **свёртывания** coiling effect; **свистящий шумовой** э. whistling effect; э. **сжатия** oblateness effect; э. **смещения** discomposition effect; **совместный** э. co-operative effect; э. **стока** sink effect; **сумеречный** э. night effect, night error, polarization error; **суточный** э. diurnal effect, day-night effect; **термодиэлектрический** э. Costa Riberio effect; **термоэлектронный** э. Richardson's thermionic effect; **тормозной** э. retardation efficiency; **трапецеидальный** э. keystone effect; э. **«трубы Кундта»** Kundt's tube action; э. **увлечения** pulling effect; э. **фонтанирования** fountain effect; **фотодиффузионный** э. Dember effect; э. **Хааса** Haas effect, precedence effect; **хоровой** э. consonant choir effect; э. **шепчущих галерей** whispering gallery effect; э. **Шеффера** Scheffer stroke; э. **экранирования** shadowing principle, screening effect; **энергетический** э. **ядерных процессов** energy shadowing of nuclear processes

**эффективность** *f.* effectiveness, efficiency; activity; relative precision; **актиническая э. экрана ЭЛТ** screen actinic efficiency; **э. анодной цепи** plate-circuit efficiency; **лучеиспускательная э. экрана ЭЛТ** screen radiant efficiency; **э. перезаряда** discharge factor; **э. по громкости** loudness efficiency rating; **э. по площади** area efficiency; **э. подавления, э. помехи** jamming efficiency; **э. постановки помех** jamming effectiveness; **э. профилактики** maintainability; **э. регулирующего стержня** effectiveness of control rod; **световая э. экрана ЭЛТ** screen luminous efficiency; **э. термоэмиссии** thermionic activity; **токовая э. электронной пушки** gun-current efficiency; **э. цепи нагрузки** load-circuit efficiency; **э. эксплуатации** running efficiency

**эффективный** *adj.* effective, useful, efficient; active

**эффектный** *adj.* effective

**эффектор** *m.* effector

**эфферентный** *adj.* efferent

**эффузер** *m.* effuser

**эффузиометр** *m.* effusiometer

**эффузия** *f.* effusion

**эффузор** *m.* effuser

**ЭХВ** *abbr.* (электрохимический взрыватель) electrochemical fuse

**эхо** *n.* echo, reflected sound; **вторичное э.** second trip echo, second-trace echo; **э., вызванное суперрефракцией** spill-over echo; **э. говорящего** talker echo; **двустороннее э.** round trip echo; **э. за счёт сверхрефракции** spill-over echo; **э. от действия туманов** fog echoes; **э. от наземных предметов на индикаторах морских радиолокаторов** land echoes of ship-radar sets; **отражённое э.** secondary echoes, indirect echoes; **пульсирующее э.** fluctuating echo; **э. рассеяния** scatter echo; **сильно-задержанное э.** long-delayed echo; **«трепещущее» э.** flutter echo

**эхобокс** *m.* echo box

**эхоглушитель** *m.* echo suppressor

**эхограмма** *f.* echo-sounder record

**эхо-заградитель** *m.* echo-suppressor; **вентильный э.** valve-type echo-suppressor; **вильчатый э. при переходном устройстве, э. переходного уствойства** terminal echo-suppressor

**эхо-зондирование** *n.* echo sounding

**эхо-изображение** *n.* echo-image, echo-illumination

**эхо-импульс** *m.* echo pulse

**эхо-индикатор** *m.* echo indicator

**эхокамера** *f.* echo chamber, echo unit

**эхо-контроль** *m.* echo checking

**эхоледомер** *m.* ice fathometer, under-ice sonar

**эхолот** *m.* sounding device, echo-sounding device, echo depth finder, acoustic sounder, sonic sound gear; **акустический э.** echo depth sounding sonar; **э. с рекордером, самопишущий э.** echograph; **электронный э.** fathometer

**эхолот-самописец** *m.* recording echo sounder

**эхомессер** *m.* singing point tester

**эхопеленг** *m.* echo bearing

**эхопеленгация** *f.* sounding

**эхо-подавитель** *m.* echo-suppressor

**эхопомеха** *f.* echo trouble

**эхо-резонатор** *m.* echo box

**эхорезонаторный** *adj.* echo-box

**эхо-сигнал** *m.* echo (signal); **э. за счёт заднего лепестка** back echo; **э. морской поверхности** sea return; **э. от суши** ground reflection

**эхо-эффект** *m.* doubling effect

**Эчисон, Эчсон** Acheson

**эш** *abbr.* (эшелон) echelon

**эшелетт** *m.* echelette

**эшелон** *m.* echelon

**эшелонирование** *n.* distribution in depth; echelonment; **э. по высоте** vertical separation; **э. по горизонтали** lateral separation

**эшелонировать** *v.* to distribute in depth; **э. по высоте** to step up; to step down

**эшелонный** *adj.* echelon

**эшинит** *m.* eschynite

**ЭЭГ** *abbr.* (электроэнцефалограмма) encephalogram

# Ю

Ю *abbr.* (юг) south

ю. *abbr.* (юг; южный) south; south, southern

юбка *f.* petticoat, skirt; shell; ю. изолятора petticoat, skirt, cup, shell

юбочный *adj.* petticoat, skirt; shell

Ю.-В. ЮВ *abbr.* (юго-восток; юго-восточный) southeast; southeastern

юг *m.* south

юго-восток *m.* southeast

юго-восточный *adj.* southeastern

юго-запад *m.* southwest

юго-западный *adj.* southwestern

юж., южн. *abbr.* (южный) south, southern

южный *adj.* south; southern, austral

Ю.-З., ю.-з. *abbr.* (юго-запад; юго-западный) southwest; southwestern

Юз Hughes

Юинг Ewing

Юитт Hewitt

Юкава Yukawa

Юлианский Julian

Юм Hume

Юнг Young

Юнгнер Jungner

юнидайн *m.* unidyne

юниметр *m.* unimeter

«юниселектор» *m.* uniselector

Юнкер Junker

ЮП *abbr.* (Южный полюс) south pole, antarctic pole

Юпитер *m.* Jupiter; *gen.* Юпитера *oft.* Jovian

юстирование *n.* registering

юстировать *v.* to adjust, to correct, to align, to set

юстировка *f.* adjustment, adjusting, correction, alignment, rectification, positioning; anodizing to value

юстировочный *adj.* adjusting, setting, aligning

ю. ш. *abbr.* (южная широта) south latitude

# Я

явиться *pf. of* являться

явление *n.* phenomenon, effect; appearance, occurrence, event; **быстропротекающие переходные явления** high-speed transients; **явления в спутной струе** wake phenomenon; **явления, используемые для взведения взрывателя** agents fuze; **явления перенасыщения** refill phenomena; **переходные явления в цепях с ёмкостью** (*or* **с индуктивностью**) single-energy transients; **переходные явления в цепях с индуктивностью и ёмкостью** double-energy transients; **я. анизотропии структуры кристалла** turn-in actions (redirecting process); **я. бахромы** fringing; **я. биений** beating; **я. запасания информации** memory effect; **я. захватывания частоты** entrainment of frequency; **я. изменения знака проводимости** patch effect; **я. истощения носителей** carrier deficit phenomenon; **я. многократного распространения** multipath transmission effect; **я. неустановившегося режима** transient phenomenon; **я. опережения** leading effect; **я. перехода от положительной проводимости к отрицательной** patch effect; **переходное я. с несколькими видами энергии** multiple-energy transient; **я. полярного сияния** auroral display; **пролётное я.** transit time phenomenon; **я. протекающее в сопле** nozzle phenomenon; **я. расширения (импульсов)** stretching effect; **я. рыскания** hunt effect; **я. свечения кристаллов при их трении или разломе** triboluminescence; **я., связанное с циклоном** cyclonic phenomenon; **я. скачка** jump phenomenon; **я. снежной эрозии** nivation phenomenon; **я. теплового прибора** turnover phenomenon; **я. термоэлектронной эмисии** Edison effect; **я. тянучки** streaking effect; **я. чёрного пятна (в передающих трубках)** shadow effect, soot-and-white-wash effect

являться, явиться *v.* to appear, to seem; to be

являющийся *adj.* appearing; being

явно *adv.* clearly, visibly, evidently

явновыраженный *adj.* salient; prominent; noticeable

явнополюсный *adj.* salient-pole

явность *f.* clearness, obviousness, evidence

явный *adj.* clear, obvious, evident, manifest, open, plain, apparent; explicit; salient

явственность *f.* clearness, distinctness

явственный *adj.* clear, distinct

Яги Yagi

ЯД *abbr.* (ядерный двигатель) nuclear engine

яд. ед. *abbr.* (ядерная единица) nuclear unit

ядерно-резонансный *adj.* nuclear-resonance

ядерночистый *adj.* nuclear-pure

ядерно-электрический *adj.* nuclear-electric

ядерный *adj.* nuclear; kernel

ядро *n.* nucleus, kernel; center, core, fulcrum; **я. интегрального уравнения** kernel, kernel function (*math.*); **ионизированное я.** ionization core; **компаунд-я.** intermediate nucleus; **я.-мишень** target nucleus; **я. пламени** flame kernel; **пустотелое я.** hollow core, tubular core

ядровый *adj.* core, heart

ядротехника *f.* nucleonics

яз. *abbr.* (язык) language

язык *m.* language; tongue; clapper; **вспомогательный я.** intermediary language; **выходной я.** object language, target language; **исходный я.** source language; **я. колокола** bell clapper; **компилирующий я.** assembly language; **конфигурационный я.** projective language; **я., ориентированный на использование процедур** procedure-oriented language; **я.-посредник** intermediate language; **я. с гнездовой синтаксической структурой** nested language; **я. с конечным числом состояний** finite state language; **собирающий я.** assembly language; **условный я.** code language; **устный я.** spoken language; **фразеологический я.** phrase structure language

**языковой** *adj.* linguistic

**языковый** *adj.* language; tongue; lingual; bolt (of lock)

**язычковый** *adj.* reed, reeded; tongue

**язычный** *adj.* tongue; lingual

**язычок** *m.* reed; small tongue; lug, catch; tag; **я. для припайки** soldering tag, terminal punching; **контактный я.** contact stud; **пробивающий я.** free-swinging reed; **пружинящий я.** reed; **ударяющий я.** striking reed

**Як-, ЯК-, Як, ЯК, як** *abbr.* (**Яковлев, А. С.**) Yakovlev, A. S. (aircraft models)

**Якоби** Jacobi, Jacobian

**якобиан** *m.* Jacobian

**якобиевский** *adj.* Jacobian, Jacobi's

**якорёк** *m.* reed; anchor; **пружинящий я.** reed

**якорный** *adj.* anchor; armature; reed

**якорь** *m.* anchor; armature; reed; **я. аппарата Морзе** pallet; **воздушный я.** over-road stay; **высоковольтный я.** intensity armature; **гладкий я.** smooth-core armature, surface-wound armature; **дискообразный я.** flat-ring armature; **я. для оттяжки** guy anchor; **качающийся я.** pendulum-type armature (of a relay); armature of a stepping relay; **клеточный я.** squirrel-cage rotor; **короткозамкнутый я.** squirrel-cage rotor; **круглый железный я.** rigid stay; **крыльчатый я.** vane armature; **линейный я.** longitudinal stay; **я. на другой стороне дороги** over-road stay; **натяжной я.** tension guy; **я. Н-образного сечения** H-armature, shuttle armature; **я. оттяжки** guy anchor; **поворотный я.** hinged armature; **я. постоянного магнита** magnet keeper, pole armature, balanced armature; **проволочный я.** wire stay; **я. с вентиляционными каналами** hole armature; **я. с двойной беличьей клеткой** double cage armature; **я. с контактными кольцами** slip ring

rotor; **я. с обмоткой в закрытых каналах** tunnel-wound armature; **я. с простой параллельной обмоткой** multiple-wound armature; **я. с радиальными спицами** spider armature; **свободно подвешенный я.** floating armature; **симметричный я.** balanced armature unit; **стержневой я.** bar-wound armature; **я. телеграфного аппарата** pallet; **я. шагового искателя** armature of a stepping relay; **я. электрической машины** dynamo armature; **явнополюсный я.** radial armature; pole armature, balanced armature

**ЯКР** *abbr.* (**ядерный квадрупольный резонанс**) nuclear quadrupole resonance

**яма** *f.* pit, hole, depression, well; **водоотводная я.** sump (mines); **воздушная я.** air pocket; **я. для столба** pole hole

**ямка** *f.* depression, hole, pit; **центральная я.** fovea

**ямочка** *see* **ямка**

**ЯМР** *abbr.* (**ядерный магнитный резонанс**) nuclear magnetic resonance

**ямчатый** *adj.* pitted

**Янсен** Jansen

**янтарный** *adj.* amber

**янтарь** *m.* amber

**ЯП** *abbr.* (**язык-посредник**) intermediate language

**яп.** *abbr.* (**японский**) Japanese

**ЯПВРД** *abbr.* (**ядерный прямоточный воздушно-реактивный двигатель**) nuclear ramjet engine

**японский** *adj.* Japanese; Japan

**ЯР** *abbr.* (**ядерный резонанс**) nuclear resonance

**ЯРД** *abbr.* (**ядерный реактивный двигатель**) nuclear jet engine

**ярд** *m.* yard

**яркий** *adj.* bright, brilliant, clear, luminous; strong, intense; vivid; rich (color)

**ярко** *adv.* brightly, brilliantly; vividly

**яркогорящий** *adj.* bright-burning

**яркомер** *m.* light-meter, brightness meter, luxometer

**яркостно-модулированный** *adj.* intensity-modulated

**яркостный** *adj.* luminance, brightness, brilliance; intensity

**яркость** *f.* brightness, brilliance, luminance, luminosity; clearness, intensity;

vividness; radiance; actinic quality of light; **я. в условиях сумеречного зрения** dark brightness, dark brilliance; **я. воспроизводимых цветов** vividness of hues; **основная я.** (average) brightness (*tv*); **я. отметки** index intensity; **я. пятна** spot intensity, brightness of cathode-ray spot; **я. самых светлых мест изображения** high-light brightness; **я. света** luminosity; **средняя я. изображения** average video signal intensity; **я. фона** background brightness; **эквивалентная я. поля зрения** equivalent field luminance

**ярлык** *m.* label; tag; ticket; **я. на разговор** docket for a call; **переговорный я.** call ticket, order ticket; **справочный я.** inquiry docket; **учётный я.** dummy ticket (for statistics)

**ярлычок** *see* **ярлык**

**ярмо** *n.* yoke, return pole piece (of relay); magnet frame; carcass, framework; **я. (реле)** heelpiece (of a relay); **я. магнитопровода** magnet frame, magnet yoke; **я. наклона** elevation yoke; **неподвижное я.** fixed yoke; **я. угла места** elevation yoke

**ярус** *m.* story, floor; row, tier, circle, range; stage, layer

**ярусный** *adj.* story, floor; row, tier; stage; graded, gradual, graduated

**ярь** *f.* green color

**ярь-медянка** *f.* verdigris, copper rust

**ясно** *adv.* clearly, brightly; evidently; distinctly; it is clear; it is obvious

**ясность** *f.* clearness, clarity, distinctness; brightness; evidence

**ясный** *adj.* clear, bright; distinct, precise; articulated, intelligible; obvious, apparent, pronounced; transparent

**Яуман** Jaumann

**ячеечный** *adj.* cellular

**ячеистый** *adj.* cellular, porous, alveolar, honeycombed, cavernous

**ячейка** *f.* cell; unit; cabinet, compartment; nucleus; mesh; stage; location; element, component element; **выдвижная я.** drawout unit; **я., вызывающая опережение по фазе** lead network; **я., вызывающая отставание на фазе** lag network; **я. гелиоприёмника** sun battery (cell); **длинная я. запоминающего устройства, длинная я. памяти** long

storage location; **я. для запоминания переноса** carry storage; **запасающая я.** storage unit; **И-я.** AND-circuit, coincidence gate; **ИЛИ-я.** OR-circuit, OR-gate; **И-НЕТ-я.** AND NOT-gate, EXCEPT-gate; **я. команды** location of instruction, location of order; **я. коммутационного устройства** switchboard cell; **кодирующая релейная я.** coding unit; **кондуктометрическая я.** conductivity cell; **логическая я. И** AND-gate; **логическая я. ИЛИ** OR-gate; **я. матрицы** matrix point; **я. многозвенного фильтра** trap, filter; **я. накапливающего счётчика** accumulator stage; **я. накопления адреса** control word; **НЕ-я.** NOT-circuit, NOT-gate; **нечётная я. памяти** odd location; **открытая я. (усилителя)** (repeater) bay; **я. памяти для больших чисел** long storage location; **пентодная триггерная я.** pentode flip-flop; **П-образная я.** P-network; **я. под напряжением** live compartment; **я. распределительного устройства** regulating cell; switchgear bay; **я. с разрушенной информацией** disturbed cell; **я. сетки** mesh; **я. типа «да-нет»** bistable element; **Т-образная я.** bridged-T network; **триггерная я.** flip-flop circuit, trigger circuit, flip-flop stage; **ультразвуковая я. оперативной памяти** supersonic storage; **фазосдвигающая я.** lag-lead network; **я. фильтра** trap, filter; **чётная я. памяти** even location

**ячейковый** *adj.* nuclear; cellular

**ячейкообразный** *adj.* cellular, cellulated, honeycomb

**ячея** *f.* cell (for equipment); bay

**ящ.** *abbr.* (**ящик**) box, case; housing

**ящик** *m.* box, case, chest, kit; cabinet, housing; cage; container; **бронированный я. распределительного устройства** metalclad unit; **я. ввода** service board; **вводный я.** inlet box; **входной я.** intake box; **я. для втаскивания кабеля в канал** draw-in box; **я. для втягивания кабеля** cable hauling box; **добавочный я.** attachment case, duplex box, supplementary box; **я. зависимости** mechanical locking; **кабельный я.** cable junction, cable trough, junction box, feeder box, cross-connection box; **кабельный я. с предохранителями** jointing chamber; **кабельный разветвительный я.** cable distribution box; **муфтовый я.** splicing vault; **ответвительный я.** subcabinet; **я. приёмника** receiver case; **я. присоединения проводов и кабелей** splice box; **проходной я.** pull box, transfer box; **пупиновский я.** loading coil case; **пусковой я., распределительный я.** distribution box, distributor; cable distribution head; control center; **распределительный я. на вводе** splitter-unit; **я. с выдвижными дверцами для монтажа на стане** cut-out box; **я. с лабиринтом** labyrinth baffle; **я. с рубильником с вводом многожильного кабеля и выводом расщеплённых жил** switch splitter; **стандартный я.** container; **я. управления** control pillar; **участковый (кабельный) я.** section box

**ящичный** *adj.* box, box-type, case; cabinet